P9-COP-343

Find out what is happening environmentally around the world.

Go to **INVESTIGATE it!** at **http://www.aw-bc.com/withgott** to access current news stories from *The New York Times* and *ABC News*.

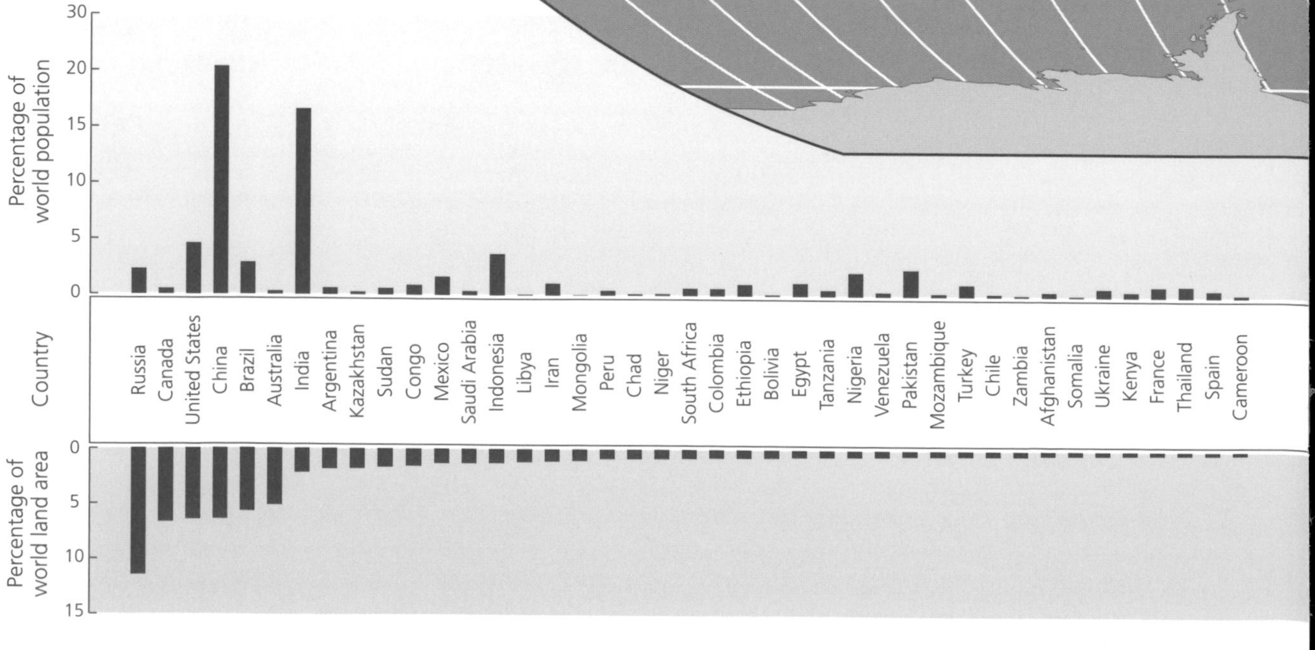

enland
ARCTIC OCEAN
Iceland
Norway
Sweden
Finland
Russia
Kazakhstan
Mongolia
Uzbek.
Kyrg.
Turkmen.
Tajik.
N. Korea
S. Korea
Japan
Iran
Afghan.
China
Kuwait
Pakistan
Nepal
Bhu.
Qatar
PACIFIC OCEAN
West Sahara
Algeria
Libya
Egypt
Saudi Arabia
U.A.E.
India
Bangl.
Mauritania
Niger
Chad
Eritrea
Oman
ARABIAN SEA
Myanmar
Laos
Philippines
Senegal
Gambia
Guinea-Bissau
Mali
Sudan
Yemen
BAY OF BENGAL
Thail.
Vietnam
Guinea
Burk.F.
Benin
Nigeria
Djibouti
Camb.
Sierra Leone
Ivory Coast
Ethiopia
Maldives
Palau
Liberia
Ghana
Togo
Cameroon
Central African Republic
Somalia
Sri Lanka
Malaysia
Equatorial Guinea
Gabon
Congo
Dem. Rep. of the Congo
Uganda
Kenya
Papua New Guinea
Rwanda
Burundi
INDIAN OCEAN
Solomon Islands
Tanzania
Indonesia
Comoros
Malawi
Angola
Zambia
Vanuatu
Mozambique
Madagascar
Zimbabwe
Namibia
Mauritius
Fiji
Botswana
Australia
Swaziland
South Africa
Lesotho
New Zealand
Sweden
Uzbekistan
Morocco
Iraq
Zimbabwe
Japan
Germany
Finland
Malaysia
Vietnam
Poland
Italy
Philippines
Ecuador
New Zealand
United Kingdom
Ghana
Romania
Uganda
Senegal
Syria
Cambodia
Bangladesh
Nicaragua
Honduras
Cuba
Guatemala
Korea, South
Hungary
Portugal
Jordan
Austria
Ireland
Sri Lanka
Latvia
Croatia
Costa Rica
Denmark
Netherlands
Switzerland
Belize
El Salvador
Israel
Kuwait
Maldives

# YOUR ACCESS TO SUCCESS

Students with new copies of Withgott/Brennan, *Environment, Third Edition* have full access to the book's Companion Website—a 24/7 study tool with review exercises, web destinations, and other features designed to help you make the most of your study time.

**Just follow the easy website registration steps listed below...**

## Registration Instructions for *www.aw-bc.com/withgott*

1. Go to *www.aw-bc.com/withgott*
2. Click the cover for Withgott/Brennan, *Environment*, *Third Edition*.
3. Click "Register."
4. Using a coin (not a knife) scratch off the metallic coating below to reveal your Access Code.
5. Complete the online registration form, choosing your own personal Login Name and Password.
6. Enter your pre-assigned Access Code exactly as it appears below.
7. Complete the online registration form by entering your School Location information.
8. After your personal Login Name and Password are confirmed by e-mail, go back to *www.aw-bc.com/withgott*, click your book's cover, enter your new Login Name and Password, and click "Log In".

**Your Access Code is:**

CLE-AD  -C

If there is no metallic coating covering the access code above, the code may no longer be valid and you will need to purchase online access using a major credit card to use the website. To do so, go to *www.aw-bc.com/withgott*, click the cover for Withgott/Brennan, *Environment*, *Third Edition*, then click the "Get Access" button, and follow instructions under "Students".

**Important:** Please read the Subscription and End-User License Agreement located on the "Log In" screen before using the Withgott/Brennan, *Environment*, *Third Edition* Companion Website. By using the website, you indicate that you have read, understood, and accepted the terms of the agreement.

**Minimum system requirements**

**Windows Operating Systems:**

Windows XP Professional sp 2 with Internet Explorer 6 or Firefox 2
Windows Vista Ultimate Edition with Internet Explorer 7

Pentium II 233 MHz processor. 64 MB RAM In addition to the minimum memory required by your OS.

**Macintosh Operating Systems:**

Mac OS 10.3 with Safari 1.2

Mac OS 10.4 with Firefox 2 or Safari 2

In addition to the RAM required by your OS, this product requires 64 MB RAM, with 40MB Free RAM, with Virtual Memory enabled

**Macromedia Shockwave**(TM)

8.50 release 326 plugin

Macromedia Flash Player 6.0.79 & 7.0

Acrobat Reader 6.0.1

800 x 600 pixel screen resolution

**Technical Support** Visit our support site at *http://247.aw.com/*. E-mail support is available 24/7.

JAY WITHGOTT • SCOTT BRENNAN

# ENVIRONMENT

## THE SCIENCE BEHIND THE STORIES

*Third Edition*

San Francisco • Boston • New York
Cape Town • Hong Kong • London • Madrid • Mexico City
Montreal • Munich • Paris • Singapore • Sydney • Tokyo • Toronto

Senior Acquisitions Editor: Chalon Bridges
Publisher: Daniel Kaveney
Editor-in-Chief: Nicole Folchetti
Executive Director of Development: Carol Trueheart
Project Manager: Tim Flem
Development Editor: Ginger Birkeland
Associate Editor: Amanda Brown
Senior Managing Editor: Kathleen Schiaparelli
Marketing Manager: Amy Porubsky
Media Editor: Andrew Sobel
Art Director: Jonathan Boylan
Senior Operations Supervisor: Alan Fischer
Director of Logistics, Operations,
and Vendor Relations: Barbara Kittle
AV Project Manager: Rhonda Aversa
Editorial Assistant: Jessica Neumann
Photo Researchers: Julie Tesser, Rachel Lucas
Composition: GGS Book Services
Production Editor: Holly Henjum, GGS Book Services
Illustrations: Dragonfly Media Group, Imagineering
Interior Design: GGS Book Services
Cover Design: Yvo Riezebos Design

Cover Photograph: Paul Nicklen/National Geographic/Getty

Photo credits continue following the glossary.

Printed using soy-based ink. Paper is recycled and may contain up to 20% post-consumer waste.

**Printed on Recycled Paper**



**Library of Congress Cataloging-in-Publication Data**

Withgott, Jay.
Environment: the science behind the stories.–3rd ed. / Jay Withgott, Scott Brennan.
p. cm.
Includes biographical references and index
ISBN 0-8053-9573-3
1. Environmental sciences. I. Brennan, Scott R. II. Title.

GE105.B74 2008
363.7—dc22

2007033073

ISBN 0–8053–9573–3 [Student Text Component]
978–0–8053–9573–0
ISBN 0–13–241924–6 [Instructor]
978–0–13–241924–6
ISBN 0–13–135705–0 [High School]
978–0–13–135705–1

2 3 4 5 6 7 8 9 10—**QWD**—12 11 10 09
www.aw–bc.com

# Brief Contents

## PART ONE

## FOUNDATIONS OF ENVIRONMENTAL SCIENCE



## PART TWO

## ENVIRONMENTAL ISSUES AND THE SEARCH FOR SOLUTIONS

# Contents

# Preface

We live in extraordinary times. Human impact on our environment has never been so intensive or so far-reaching. The future of our society and the future of Earth's systems depend more critically than ever on the way we interact with the world around us. Fundamental aspects of climate, atmospheric composition, nutrient cycling, and biological diversity are being altered at dizzying speeds. Yet thanks to environmental science, we now understand better than ever how our planet's systems function and how we influence these systems. Environmental science helps us to characterize the problems we create, and it also illuminates the tremendous opportunities we have before us for effecting positive change.

The field of environmental science captures the very essence of this unique moment in history. This interdisciplinary pursuit stands at the vanguard of the current need to synthesize academic disciplines and to incorporate their contributions into a big-picture understanding of the world and our place within it.

We wrote this book because we feel that the vital importance of environmental science in today's world makes it imperative to engage, educate, and inspire a broad audience of today's students—the citizens and leaders of tomorrow. We have therefore tried to implement the very best in modern teaching approaches and to clarify how the scientific process can inform human efforts. We also have aimed to maintain a balanced approach and to encourage critical thinking as we flesh out the social debate over many environmental issues. Finally, in assessing the challenges that our society and our planet face, we focus on providing hope and solutions.

In crafting the third edition of this text, we incorporated the most current information from this fast-moving field and streamlined our presentation to make learning straightforward and appealing. This edition contains more information in fewer pages in a more readable style. To educate more effectively, we added over 100 new figures and revised 200 more. Two central case studies are brand new, as are nine *Science behind the Story* features.

To reflect the rapid change in the scientific understanding and popular awareness of global climate change and our potential responses to it, we have expanded and thoroughly revised our coverage of global climate change, both in the climate change chapter and throughout the book. We also enhanced our coverage of other topics of growing interest, including landscape ecology and ecosystem ecology, petroleum depletion and the issues surrounding "peak oil," alternative fossil fuels, biomass and other renewable energy alternatives, corporate sustainability, environmental justice, and the "triple bottom line" approach to sustainability. We strengthened and extended our applications of the ecological footprint concept and our use of sustainability as an organizing theme. And we remain the only text to treat campus sustainability initiatives in appreciable depth (see Chapter 23), showing students how their peers across the continent are applying principles and lessons from environmental science to forge sustainable solutions on their college and university campuses.

In this edition, we introduce a new pair of features that seek to help students probe the causal reasons behind major issues in environmental science, the consequences these issues have, and the solutions we can bring to bear on them. Five chapters have *Causes and Consequences* concept maps that present a visual display of causal connections, and 11 chapters have interactive *Causes and Consequences* exercises that invite students to pencil in causes, consequences, and solutions they have gleaned from the text. We intend that this pair of features will help students gain a deeper understanding of the causal connections surrounding major issues and will help promote a search for solutions.

**CAUSES AND CONSEQUENCES**

Another change that users of our book will notice is that our *Viewpoints* feature is now in its own stand-alone supplement. To ensure that students are exposed to a diversity of interpretations on key issues, the paired *Viewpoints* essays—one pair per chapter—are authored by invited experts who present divergent points of view on topical questions. The essays provide a taste of informed arguments directly from individuals who are actively involved in work—and debate—on environmental issues. Packaging *Viewpoints* as a supplement allows for the best of both worlds: Instructors who use this feature extensively in their courses gain a format that will allow them to put it to even fuller use, while those

VIEWPOINTS

who use it less gain a leaner main text with enhanced coverage of core issues. The supplement format for *Viewpoints* also permits us to include questions students can use to critically examine and discuss the ideas in the essays, as well as links to websites that expand on contributors' viewpoints. All essays have been revised by their contributors, and many questions and essays are new.

Finally, we have further enhanced opportunities for students to interpret and use graphs—a vital yet often neglected skill. We've beefed up *Appendix A*, our guide to graphing, designed for students to browse at the outset of their course and to use as a reference later. We've also retained our *Interpreting Graphs and Data* feature at the end of each chapter. And our online supplement *GraphIt!* continues to allow students to manipulate data and create graphs on their own.

GRAPHIt!

We have also retained the major features that made the first two editions of our book unique and that are proving so successful in classrooms across North America:

- **Integrated Central Case Studies.** Our teaching experiences, together with feedback from colleagues across the continent, clearly reveal that telling compelling stories about real people and real places is the best way to capture students' interest. Narratives with concrete detail also help teach abstract concepts, because they give students a tangible framework with which to incorporate new ideas. Many textbooks these days serve up case studies in isolated boxes, but we have chosen to integrate each chapter's central case study into the main text, weaving information and elaboration throughout the chapter. In this way, the concrete realities of the people and places of the central case study help to illustrate the topics we cover. We are gratified that students and instructors using our book have consistently applauded this approach, and we hope it can continue to bring about a new level of effectiveness in environmental science education.

- **The Science behind the Story.** Our goal is not simply to present students with facts, but to engage them in the scientific process of testing and discovery. To do this, we discuss the scientific method and the social context of science in our opening chapter, and we describe hundreds of real-life studies throughout the text. We also feature in each chapter *The Science behind the Story* boxes, which elaborate on particular studies, guiding readers through details of the research. In this way we show not merely *what* scientists discovered, but *how* they discovered it. Instructors and students using our book have confirmed that this feature enhances comprehension of chapter material and deepens understanding of the scientific process itself—a key component of effective citizenship in today's science-driven world.

THE SCIENCE BEHIND THE STORY

- **Weighing the Issues.** The multifaceted issues in environmental science often lack black-and-white answers, so students need critical-thinking skills to help navigate the gray areas at the juncture of science, policy, and ethics. We aim to help develop these skills with the *Weighing the Issues* questions dispersed through each chapter. These serve as stopping points for students to absorb and reflect upon what they have read, wrestle with some of the complex dilemmas in environmental science, and engage in spirited classroom discussion.

Weighing THE Issues

- **End-of-chapter features rich in options.** The five features that conclude each chapter are targeted at particular student needs. *Reviewing Objectives* summarizes each chapter's main points and relates them to the learning objectives presented at the chapter's opening, enabling students to confirm that they have understood the most crucial ideas and to review concepts by turning to specified page numbers. *Testing Your Comprehension* questions provide concise study questions targeted to main topics in each chapter, while *Seeking Solutions* questions encourage broader creative thinking aimed at finding solutions. Our new "Think It Through" questions place students in a scenario and empower them to make decisions to resolve problems. *Interpreting Graphs and Data* uses figures from recent scientific studies to help students build quantitative and analytical skills in reading graphs and making sense of data. And *Calculating Ecological Footprints* enables students to calculate the environmental impacts of their own choices and then see how individual impacts scale up to impacts at the societal level.

- **An emphasis on solutions.** The complaint we most frequently hear from students in environmental science courses is that the deluge of environmental problems can seem overwhelming. In the face of so many problems, students often come to feel that there is no hope or that there is little they can personally do to make a difference. We have aimed to counter this impression by drawing out innovative solutions that have worked, are being implemented, or can be tried in the future. For example, Chapter 23 highlights a number of innovative solutions currently under way at campuses around North America. While we do not paint an unrealistically rosy picture of the challenges that lie ahead, we portray dilemmas as opportunities and we try to instill hope and encourage action. Indeed, for every problem that human carelessness has managed to create, human ingenuity can devise one—and likely multiple—solutions. To recognize the efforts of faculty and students toward encouraging sustainable practices on campus and in the community, Pearson Education will award a total of $5,000 in 2008 to the U.S. campus or campuses which best exemplify the principles of sustainability. Visit the book website at www.aw-bc.com/withgott for details on previous Sustainable Solutions Award winners and for entry details.

*Environment: The Science behind the Stories* has grown directly from our professional experiences in teaching, research, and writing. Jay Withgott has synthesized and presented science to a wide readership. His experience in distilling and making accessible the fruits of scientific inquiry has shaped our book's content and the presentation of its material. Scott Brennan has taught environmental science to thousands of undergraduates and has developed an intimate feeling for what works in the classroom. His knowledge and experience have shaped our book's pedagogical approach.

We have been guided in our efforts by extensive input from hundreds of instructors across North America who have served as reviewers for our chapters and as advisors in focus groups arranged by our publisher. The participation of so many learned and thoughtful experts has improved this volume in countless ways.

We sincerely hope that our efforts will come close to being worthy of the immense importance of our subject matter. We invite you, students and instructors alike, to let us know how well we have achieved our goals and where you feel we have fallen short. We are committed to continual improvement, and value your feedback. Please write the authors in care of Chalon Bridges (chalon.bridges@pearson.com), Pearson Publishing, 1301 Sansome Street, San Francisco, California 94111.

At this most historic time to study environmental science, we are honored to serve as your guides in the quest to better understand our world and ourselves.

Jay Withgott and Scott Brennan

# Instructor Supplements

## The Withgott/Brennan Instructor Resource Center on DVD 0-13-601984-6

This powerful media package is organized chapter-by-chapter and includes all teaching resources in one convenient location. You'll find 5-minute *ABC News* Lecture Launcher videos, PowerPoint presentations, Active Lecture questions to facilitate class discussions (for use with or without clickers), and an image library that includes all art and tables from the text.

## Instructor's Guide and Test Bank 0-13-601983-8

This comprehensive resource provides chapter outlines, key terms, a listing of website and media resources, and teaching tips for lecture and classroom activities. A printed version of the Test Bank is conveniently included in the manual, offering hundreds of multiple-choice, short-answer, and essay questions to use on tests and quizzes. In addition, there are unique graphing and scenario-based questions to test students' critical-thinking abilities.

## Computerized Test Bank 0-13-601985-4

Hundreds of multiple-choice, short-answer, essay, graphing, and scenario-based questions on a cross-platform CD-ROM. Categorized by chapter objective for instructor ease in searching for question types.

## Transparency Acetates 0-13-601989-7

Includes 300 full-color acetates of all the art and tables from the text.

## CourseCompass for Environment 0-13-601965-X

This nationally hosted, easy-to-use course management tool allows professors to combine their own material with the material in the book website to create dynamic, online learning environments. Professors can post their syllabus, assign tutorials, customize quizzes, automatically grade them, and track the results instantly in the grade book. Go to www.coursecompass.com

### Blackboard Premium for Environment
### 0-13-601963-3

### Blackboard Open Access
### 0-13-601987-0

### WebCT Premium for Environment
### 0-13-601966-8

### WebCT Open Access
### 0-13-601986-2

## Student Supplements

### Viewpoints
### 0-13-601971-4

*Viewpoints* essays are authored by invited experts who present divergent points of view on topical questions that correspond to each chapter of the book. The essays provide a taste of informed arguments directly from individuals who are actively involved in work—and debate—on environmental issues. To encourage students to think through the issues on their own, each essay is followed by questions that students can use to critically examine and discuss the ideas in the essays, as well as links to websites that support contributors' viewpoints.

### Themes of the Times on the Environment, Volume 1
### 0-8053-4478-0

### Themes of the Times on the Environment, Volume 2
### 0-8053-9608-X

*Themes of the Times on the Environment* is a collection of current *New York Times* articles covering environmental issues in the news today. These articles provide an additional range of case studies beyond those presented in the text.

# Acknowledgments

A textbook is the product of *many* more minds and hearts than one might guess from the names on the cover. The two of us are exceedingly fortunate to be supported and guided by a tremendous publishing team and by a small army of experts in environmental science who have generously shared their time and expertise. Although we alone, as authors, bear responsibility for any inaccuracies, the strengths of this book result from the collective labor and dedication of innumerable people.

We would first like to thank our acquisitions editor, Chalon Bridges. Chalon's commitment and unremitting enthusiasm have inspired our team to relish the challenge of taking a successful and well-received book and making it still better. Her extensive interaction with instructors across North America has helped us define and refine our pedagogy and our innovative features. The approach, design, and essence of this book owe a great deal to Chalon's astute guidance and vision.

A corporate transition during the revision of this text threw some extra challenges our way, but in the end allowed our book to benefit from the expertise of more editors and publishing experts than would otherwise have been the case. In particular, we benefited from two fabulous individuals: senior project editor Mary Ann Murray continued her wonderful work from the second edition, while project manager Tim Flem brought a fresh perspective that enabled numerous innovations. Both worked tirelessly to coordinate the logistics and deadlines for our revision plan, features, reviews, supplements, and contributors. Tim's sharp eyes, sound judgment, and steady hand saw the book through to completion.

Special thanks go to our new publisher, Daniel Kaveney, as well as our former executive editor, Beth Wilbur. We cannot imagine being in better hands. Having benefited from Beth's past guidance, we now are excited about a future steered by Dan's tremendous creative vision and publishing expertise.

Development editor Ginger Birkeland lent us her keen eye as she reviewed every word of our text, improving the presentation of our content in innumerable ways. Our book's enhanced art program benefits from photos researched by Julie Tesser, with help from Rachel Lucas, and its features were strengthened by research from Kristy Manning. Copyeditor Sally Peyrefitte again provided thorough and meticulous examination of our text. We send warm thanks to production editor Holly Henjum and the rest of the staff at GGS Book Services for a tremendous job putting this third edition together.

We also thank development directors Carol Trueheart and Deborah Gale, as well as editorial assistants Jessica Neumann and Anna Amato. We also thank Kathleen Schiaparelli, Andrew Sobel, Jonathan Boylan, and Rhonda Aversa for their helpful contributions to the book and media package. Hilair Chism transformed our concept maps from rough sketches into finished artwork. In addition, we are grateful for lasting contributions to the book's earlier editions by Etienne Benson, Russell Chun, Jonathan Frye, April Lynch, and especially Susan Teahan.

Finally, we would like to acknowledge the people who created our exciting fleet of supplements. Kristy Manning worked hard to create our *Viewpoints* supplement—a formidable logistical and creative task. We also would like to thank the authors of the essays in our *Viewpoints* supplement, each of whom is credited along with his or her essay. Dawn Keller engineered the thorough revision of the *New York Times* case studies in the Investigate It feature on the website, while Chris Brown handled the updates to the *ABC News* video case studies in Investigate It. Our Instructor's Guide was revised by Danielle Wirth, and Thomas Pliske revised our Testbank and PowerPoint slides. Steven Frankel created the Chapter Quizzes on our book's website. Finally and especially, our thanks go to associate editor Amanda Brown for creating and coordinating all of these valuable supplements.

Of course, none of this has any impact on education without the sales and marketing staff to get the book into your hands. Marketing managers Lauren Harp and Amy Porubsky dedicated their talent and enthusiasm to the book's promotion and distribution. And last but surely not least, the many field representatives who help communicate our vision and deliver our product to instructors are absolutely vital, and we deeply appreciate their tireless work and commitment.

In the lists that follow, we acknowledge the many instructors and outside experts who have helped us maximize the quality and accuracy of our presentation through their chapter reviews, feature reviews, class tests, or other services. If the thoughtfulness and thoroughness of these reviewers are any indication, we feel confident that the teaching of environmental science is in excellent hands!

Lastly, Jay gives loving thanks to his wife Susan Masta, who has endured this book's writing and revision with tremendous patience and sacrifice and has provided support and sustenance throughout. Scott would like to thank Angela, Sean, Jonathan, Korby, Karl and Jess, and Jodi and Andy.

We dedicate this book to today's students, who will shape tomorrow's world.

Jay Withgott and Scott Brennan

# Reviewers

**Pre-Revision Reviewers for the Third Edition**

Shamim Ahsan, *Metropolitan State College of Denver*
Kenneth Banks, *University of North Texas*
Barbara Bekken, *Virginia Polytechnic Institute and State University*
David Belt, *Johnson County Community College*
Bonnie L. Brown, *Virginia Commonwealth University*
Michelle Cawthorn, *Georgia Southern University*
Richard Clements, *Chattanooga State Technical Community College*
Randi Darling, *Westfield State College*
Michael L. Draney, *University of Wisconsin–Green Bay*
Robert East, *Washington and Jefferson College*
Corey Etchberger, *Johnson County Community College*
Leslie Fay, *Rock Valley College*
Debra A. Feikert, *Antelope Valley College*
Carl W. Grobe, *Westfield State College*
Robert D. Hollister, *Grand Valley State University*
Robert G. Kremer, *Metropolitan State College of Denver*
Andrew Lapinski, *Reading Area Community College*
Stephen D. Lewis, *California State University, Fresno*
Chun Liang, *Miami University*
Robert L. Mahler, *University of Idaho*
Heidi Marcum, *Baylor University*
Dan McNally, *Bryant University*
Matthew R. Milnes, *University of California, Irvine*
Christopher Pennuto, *Buffalo State College*
Shanna Petermann, *Minnesota State Community and Technical College–Moorhead*
Barbara C. Reynolds, *University of North Carolina–Asheville*
Edward G. Schultz, III, *Valencia Community College*
Roy Sofield, *Chattanooga State Technical Community College*
Jon G. Stanley, *Metropolitan State College of Denver*
Patricia Terry, *University of Wisconsin–Green Bay*
Ray E. Williams, *Rio Hondo College*

**Chapter Reviewers for the Third Edition**

John Campbell, *Northwest College*
Tait Chirenje, *Richard Stockton College*
Roger del Moral, *University of Washington*
JodyLee Estrada Duek, *Pima Community College*
Bonnie Fancher, *Switzerland County High School*
Steven Fields, *Winthrop University*
Linda Fitzhugh, *Gulf Coast Community College*
Navida Gangully, *Oak Ridge High School*
Marcia Gillette, *Indiana University–Kokomo*
Michelle Goldsmith, *Emerson College*
Jack Greene, *Millikan High School*
Amy Gregory, *University of Cincinnati, Clermont College*
Greg Haenel, *Elon University*
David Hong, *Diamond Bar High School*
Daniel Hyke, *Alhambra High School*
Richard Jurin, *University of Northern Colorado*
Brian Kaestner, *Saint Mary's Hall*
Dawn Keller, *Hawkeye Community College*
Sushma Krishnamurthy, *Texas A&M International University*
Jerome Kruegar, *South Dakota State University*
Andrew Lapinski, *Reading Area Community College*
Lissa M. Leege, *Georgia Southern University*
James Lehner, *Taft School*
Kenneth Mantai, *State University of New York at Fredonia*
Christopher Migliaccio, *Miami-Dade College*
Mark Oemke, *Alma College*
Thomas Pliske, *Florida International University*
Shamili A. Sandiford, *College of DuPage*
Jeffrey Schneider, *State University of New York at Oswego*
Pamela Shlachtman, *Miami Palmetto Senior High School*
Ben Smith, *Palos Verdes Peninsula High School*
Patricia Smith, *Valencia Community College*
Debra Socci, *Seminole Community College*
Keith Summerville, *Drake University*
Jamey Thompson, *Hudson Valley Community College*
Todd Tracy, *Northwestern College, Iowa*
Frederick Troeh, *Iowa State University*
Mike Tveten, *Pima Community College*
Michael Vorwerk, *Westfield State College*
Kathryn Weatherhead, *Hilton Head High School*

**Features Reviewers for the Third Edition**

Kelly Cartwright, *College of Lake County*
Steven Frankel, *Northeastern Illinois University*
Richard Jurin, *University of Northern Colorado*
Cindy Klevickis, *James Madison University*
Ned Knight, *Linfield College*
Steven Rudnick, *University of Massachusetts–Boston*
Lorne Wolfe, *Georgia Southern University*

### Reviewers for Previous Editions

David Aborne, *University of Tennessee–Chattanooga;* Jeffrey Albert, *Watson Institute of International Studies;* John V. Aliff, *Georgia Perimeter College;* Mary E. Allen, *Hartwick College;* Dula Amarasiriwardena, *Hampshire College;* Gary I. Anderson, *Santa Rosa Junior College;* Corey Andries, *Albuquerque Technical Vocational Institute;* David M. Armstrong, *University of Colorado–Boulder;* David L. Arnold, *Ball State University;* Joseph Arruda, *Pittsburg State University;* Thomas W. H. Backman, *Linfield College;* Timothy J. Bailey, *Pittsburg State University;* Stokes Baker, *University of Detroit;* Reuben Barret, *Prairie State College;* Morgan Barrows, *Saddleback College;* Henry Bart, *LaSalle University;* James Bartalome, *University of California–Berkeley;* Marilynn Bartels, *Black Hawk College;* David Bass, *University of Central Oklahoma;* Christy Bazan, *Illinois State University;* Christopher Beals, *Volunteer State Community College;* Hans T. Beck, *Northern Illinois University;* Richard Beckwitt, *Framingham State College;* Elizabeth Bell, *Santa Clara University;* Timothy Bell, *Chicago State University;* Terrence Bensel, *Allegheny College;* Gary Beluzo, *Holyoke Community College;* Bob Bennett, *University of Arkansas;* William B. N. Berry, *University of California, Berkeley;* Kristina Beuning, *University of Wisconsin–Eau Claire;* Peter Biesmeyer, *North Country Community College;* Donna Bivans, *Pitt Community College;* Grady Price Blount, *Texas A&M University–Corpus Christi;* Marsha Bollinger, *Winthrop University;* Richard D. Bowden, *Allegheny College;* Frederick J. Brenner, *Grove City College;* Nancy Broshot, *Linfield College;* David Brown, *California State University–Chico;* Evert Brown, *Casper College;* Hugh Brown, *Ball State University;* J. Christopher Brown, *University of Kansas;* Dan Buresh, *Sitting Bull College;* Lee Burras, *Iowa State University;* Christina Buttington, *University of Wisconsin–Milwaukee;* Charles E. Button, *University of Cincinnati, Clermont College;* John S. Campbell, *Northwestern College;* Mike Carney, *Jenks High School;* Kelly S. Cartwright, *College of Lake County;* Jon Cawley, *Roanoke College;* Michelle Cawthorn, *Georgia Southern University;* Linda Chalker-Scott, *University of Washington;* Brad S. Chandler, *Palo Alto College;* Paul Chandler, *Ball State University;* David A. Charlet, *Community College of Southern Nevada;* Sudip Chattopadhyay, *San Francisco State University;* Tait Chirenje, *Richard Stockton College;* Kenneth E. Clifton, *Lewis and Clark College;* Reggie Cobb, *Nash Community College;* John E. Cochran, *Columbia Basin College;* Luke W. Cole, *Center on Race, Poverty, and the Environment;* Thomas L. Crisman, *University of Florida;* Jessica Crowe, *South Georgia College;* Ann Cutter, *Randolph Community College;* Gregory A. Dahlem, *Northern Kentucky University;* Mary E. Davis, *University of Massachusetts, Boston;* Thomas A. Davis, *Loras College;* Lola M. Deets, *Pennsylvania State University–Erie;* Ed DeGrauw, *Portland Community College;* Roger del Moral, *University of Washington;* Craig Diamond, *Florida State University;* Darren Divine, *Community College of Southern Nevada;* Stephanie Dockstader, *Monroe Community College;* Toby Dogwiler, *Winona State University;* Jeffrey Dorale, *University of Iowa;* Tracey Dosch, *Waubonsee Community College;* Iver W. Duedall, *Florida Institute of Technology;* JodyLee Estrada Duek, *Pima Community College;* Jeffrey R. Dunk, *Humboldt State University;* Jean W. Dupon, *Menlo College;* Robert M. East, Jr., *Washington & Jefferson College;* Margaret L. Edwards-Wilson, *Ferris State University;* Dee Eggers, *University of North Carolina–Asheville;* Anne H. Ehrlich, *Stanford University;* Jane Ellis, *Presbyterian College;* Thomas R. Embich, *Harrisburg Area Community College;* Kenneth Engelbrecht, *Metropolitan State College of Denver;* Bill Epperly, *Robert Morris College;* W. F. J. Evans, *Trent University;* Paul Fader, *Freed Hardeman University;* Joseph Fail, *Johnson C. Smith University;* Bonnie Fancher, *Switzerland County High School;* Jiasong Fang, *Iowa State University;* M. Siobhan Fennessy, *Kenyon College;* Francette Fey, *Macomb Community College;* Steve Fields, *Winthrop University;* Brad Fiero, *Pima Community College;* Dane Fisher, *Pfeiffer University;* David G. Fisher, *Maharishi University of Management;* Linda M. Fitzhugh, *Gulf Coast Community College;* Doug Flournoy, *Indian Hills Community College–Ottumwa;* Johanna Foster, *Johnson County Community College;* Chris Fox, *Catonsville Community College;* Nancy Frank, *University of Wisconsin–Milwaukee;* Steven Frankel, *Northeastern Illinois University;* Arthur Fredeen, *University of Northern British Columbia;* Chad Freed, *Widener University;* Robert Frye, *University of Arizona;* Laura Furlong, *Northwestern College;* Sandi B. Gardner, *Triton College;* Kristen S. Genet, *Anoka Ramsey Community College;* Stephen Getchell, *Mohawk Valley Community College;* Marcia Gillette, *Indiana University–Kokomo;* Sue Glenn, *Gloucester County College;* Thad Godish, *Ball State University;* Michele Goldsmith, *Emerson College;* Jeffrey J. Gordon, *Bowling Green State University;* John G. Graveel, *Purdue University;* Cheryl Greengrove, *University of Washington;* Amy R. Gregory, *University of Cincinnati, Clermont College;* Carol Griffin, *Grand Valley State University;* Sherri Gross, *Ithaca College;* David E. Grunklee, *Hawkeye Community College;* Judy Guinan, *Radford University;* Gian Gupta, *University of Maryland, Eastern Shore;* Mark Gustafson, *Texas Lutheran University;* Daniel Guthrie, *Claremont College;* Sue Habeck, *Tacoma Community College;* David Hacker, *New Mexico Highlands University;* Greg Haenel, *Elon University;* Mark Hammer, *Wayne State University;* Grace Hanners, *Huntingtown High School;* Michael Hanson, *Bellevue Community College;* Alton Harestad, *Simon Fraser University;* Barbara Harvey, *Kirkwood Community College;* David Hassenzahl, *University of Nevada Las Vegas;* Jill Haukos, *South Plains College;* Keith Hench, *Kirkwood Community College;* George Hinman, *Washington State University;* Joseph Hobbs, *University of Missouri–Columbia;* Jason Hoeksema, *Cabrillo College;* Curtis Hollabaugh, *University of West Georgia;* David Hong, *Diamond Bar High School;* Catherine Hooey, *Pittsburgh State University;* Kathleen Hornberger, *Widener University;* Debra Howell, *Chabot College;* April Huff, *North Seattle Community College;* Pamela Davey Huggins, *Fairmont State University;* Barbara Hunnicutt, *Seminole Community College;* Jonathan E. Hutchins, *Buena Vista University;* Daniel Hyke, *Alhambra High School;* Juana Ibáñez, *University of New Orleans;* Walter Illman, *University of Iowa;* Daniel Ippolito, *Anderson University;* Bonnie Jacobs, *Southern Methodist University;* Nan Jenks-Jay, *Middlebury College;* Stephen R. Johnson, *William Penn University;* Gina Johnston, *California State University, Chico;* Paul Jurena, *University of Texas–San Antonio;* Richard R. Jurin, *University of Northern Colorado;* Thomas M. Justice, *McLennan Community College;* Stanley S. Kabala, *Duquesne University;* Steve Kahl, *Plymouth State University;* Carol Kearns, *University of Colorado–Boulder;* Richard R. Keenan, *Providence Senior High School;* Dawn G. Keller, *Hawkeye Community College;* John C. Kinworthy, *Concordia University;* Ned J. Knight, *Linfield College;* David Knowles, *East Carolina University;* Penelope M. Koines, *University of Maryland;* Alexander Kolovos, *University of North Carolina–Chapel Hill;* Erica Kosal, *North Carolina Wesleyan College;* Steven Kosztya, *Baldwin Wallace College;* Robert J. Koester, *Ball State University;* Tom Kozel, *Anderson College;* Jim Krest, *University of South Florida–South Florida;* Sushma Krishnamurthy, *Texas A&M International University;* James Kubicki, *Penn State University;* Frank T. Kuserk, *Moravian College;* Diane M. LaCole, *Georgia Perimeter College;* Troy A. Ladine, *East Texas Baptist University;* William R. Lammela, *Nazareth College;* Vic Landrum, *Washburn University;* Tom Langen, *Clarkson University;* Andrew Lapinski, *Reading Area Community College;* Michael T. Lares, *University of Mary;* Kim D. B. Largen, *George Mason University;* John Latto, *University of California–Berkeley;* Lissa Leege, *Georgia Southern University;* John Logue, *University of South Carolina–Sumter;* John F. Looney, Jr., *University of Massachusetts–Boston;* Joseph Luczkovich, *East Carolina University;* Linda Lusby, *Acadia University;* Richard A. Lutz, *Rutgers University;* Jennifer Lyman, *Rocky Mountain College;* Les M. Lynn, *Bergen Community College;* Timothy F. Lyon, *Ball State University;* Sue Ellen Lyons, *Holy Cross School;* Ian R. MacDonald, *Texas A&M University;* James G. March, *Washington and*

*Jefferson College;* Blasé Maffia, *University of Miami;* Keith Malmos, *Valencia Community College;* Kenneth Mantai, *State University of New York–Fredonia;* Anthony J.M. Marcattilio, *St. Cloud State University;* Nancy Markee, *University of Nevada–Reno;* Patrick S. Market, *University of Missouri–Columbia;* Steven R. Martin, *Humboldt State University;* John Mathwig, *College of Lake County;* Allan Matthias, *University of Arizona;* Robert Mauck, *Kenyon College;* Bill Mautz, *University of New Hampshire;* Debbie McClinton, *Brevard Community College;* Paul McDaniel, *University of Idaho;* Jake McDonald, *University of New Mexico;* Gregory McIsaac, *Cornell University;* Dan McNally, *Bryant University;* Richard McNeil, *Cornell University;* Julie Meents, *Columbia College;* Mike L. Meyer, *New Mexico Highlands University;* Steven J. Meyer, *University of Wisconsin–Green Bay;* Patrick Michaels, *Cato Institute;* Christopher Migliaccio, *Miami Dade Community College;* Kiran Misra, *Edinboro University of Pennsylvania;* Mark Mitch, *New England College;* Lori Moore, *Northwest Iowa Community College;* Paul Montagna, *University of Texas–Austin;* Brian W. Moores, *Randolph-Macon College;* James T. Morris, *University of South Carolina;* Sherri Morris, *Bradley University;* Mary Murphy, *Penn State Abington;* William M. Murphy, *California State University–Chico;* Carla S. Murray, *Carl Sandburg College;* Rao Mylavarapu, *University of Florida;* Jane Nadel-Klein, *Trinity College;* Muthena Naseri, *Moorpark College;* Michael J. Neilson, *University of Alabama–Birmingham;* Richard A. Niesenbaum, *Muhlenberg College;* Moti Nissani, *Wayne State University;* Richard B. Norgaard, *University of California–Berkeley;* John Novak, *Colgate University;* Mark P. Oemke, *Alma College;* Niamh O'Leary, *Wells College;* Bruce Olszewski, *San Jose State University;* Brian O'Neill, *Brown University;* Nancy Ostiguy, *Penn State University;* David R. Ownby, *Stephen F. Austin State University;* Eric Pallant, *Allegheny College;* Philip Parker, *University of Wisconsin–Platteville;* Brian Peck, *Simpson College;* Brian D. Peer, *Simpson College;* Clayton Penniman, *Central Connecticut State;* Donald J. Perkey, *University of Alabama–Huntsville;* Raymond Pierotti, *University of Kansas;* Craig D. Phelps, *Rutgers University;* Frank X. Phillips, *McNeese State University;* Elizabeth Pixley, *Monroe Community College;* Thomas E. Pliske, *Florida International University;* Daryl Prigmore, *University of Colorado;* Avram G. Primack, *Miami University of Ohio;* Sarah Quast, *Middlesex Community College;* Loren A. Raymond, *Appalachian State University;* Barbara Reynolds, *University of North Carolina–Asheville;* Thomas J. Rice, *California Polytechnic State University;* Samuel K. Riffell, *Mississippi State University;* Gary Ritchison, *Eastern Kentucky University;* Roger Robbins, *East Carolina University;* Tom Robertson, *Portland Community College, Rock Creek Campus;* Mark Robson, *University of Medicine and Dentistry of New Jersey;* Carlton Lee Rockett, *Bowling Green State University;* Angel M. Rodriguez, *Broward Community College;* Deanne Roquet, *Lake Superior College;* Armin Rosencranz, *Stanford University;* Robert E. Roth, *The Ohio State University;* George E. Rough, *South Puget Sound Community College;* Steven Rudnick, *University of Massachusetts–Boston;* John Rueter, *Portland State University;* Christopher T. Ruhland, *Minnesota State University;* Shamili A. Sandiford; *College of DuPage;* Robert Sanford, *University of Southern Maine;* Ronald Sass, *Rice University;* Carl Schafer, *University of Connecticut;* Jeffery A. Schneider, *State University of New York–Oswego;* Mark Schwartz, *University of California–Davis;* Jennifer Scrafford, *Loyola College;* Richard Seigel, *Towson University;* Julie Seiter, *University of Nevada–Las Vegas;* Wendy E. Sera, *NDAA's National Ocean Service;* Maureen Sevigny, *Oregon Institute of Technology;* Rebecca Sheesley, *University of Wisconsin–Madison;* Brian Shmaefsky, *Kingwood College;* William Shockner, *Community College of Baltimore County;* Christian V. Shorey, *University of Iowa;* Robert Sidorsky, *Northfield Mt. Hermon High School;* Linda Sigismondi, *University of Rio Grande;* Jeffrey Simmons, *West Virginia Wesleyan College;* Cynthia Simon, *University of New England;* Jan Simpkin, *College of Southern Idaho;* Michael Singer, *Wesleyan University;* Diane Sklensky, *Le Moyne College;* Mark Smith, *Chaffey College;* Patricia L. Smith, *Valencia Community College;* Sherilyn Smith, *Le Moyne College;* Debra Socci, *Seminole Community College;* Douglas J. Spieles, *Denison University;* Ravi Srinivas, *University of St. Thomas;* Bruce Stallsmith, *University of Alabama–Huntsville;* Jeff Steinmetz, *Queens University of Charlotte;* Richard J. Strange, *University of Tennessee;* Robert Strikwerda, *Indiana University–Kokomo;* Richard Stringer, *Harrisburg Area Community College;* Andrew Suarez, *University of Illinois;* Keith S. Summerville, *Drake University;* Ronald Sundell, *Northern Michigan University;* Bruce Sundrud, *Harrisburg Area Community College;* Jim Swan, *Albuquerque Technical Vocational Institute;* Mark L. Taper, *Montana State University;* Max R. Terman, *Tabor College;* Julienne Thomas, *Robert Morris College;* Jamey Thompson, *Hudson Valley Community College;* Todd Tracy, *Northwestern College;* Amy Treonis, *Creighton University;* Adrian Treves, *Wildlife Conservation Society;* Frederick R. Troeh, *Iowa State University;* Virginia Turner, *Robert Morris College;* Michael Tveten, *Pima Community College;* Charles Umbanhowar, *St. Olaf College;* G. Peter van Walsum, *Baylor University;* Callie A. Vanderbilt, *San Juan College;* Elichia A. Venso, *Salisbury University;* Rob Viens, *Bellevue Community College;* Michael Vorwerk, *Westfield State College;* Caryl Waggett, *Allegheny College;* Maud M. Walsh, *Louisiana State University;* Daniel W. Ward, *Waubonsee Community College;* Darrell Watson, *The University of Mary Hardin Baylor;* Phillip L. Watson, *Ferris State University;* Lisa Weasel, *Portland State University;* John F. Weishampel, *University of Central Florida;* Barry Welch, *San Antonio College;* James W. C. White, *University of Colorado;* Susan Whitehead, *Becker College;* Richard D. Wilk, *Union College;* Donald L. Williams, *Park University;* Justin Williams, *Sam Houston University;* Ray E. Williams, *Rio Hondo College;* Roberta Williams, *University of Nevada–Las Vegas;* Dwina Willis, *Freed-Hardeman University;* Tom Wilson, *University of Arizona;* James Winebrake, *Rochester Institute of Technology;* Danielle Wirth, *Des Moines Area Community College;* Marjorie Wonham, *University of Alberta;* Wes Wood, *Auburn University;* Jeffrey S. Wooters, *Pensacola Junior College;* Joan G. Wright, *Truckee Meadows Community College;* Michael Wright, *Truckee Meadows Community College;* S. Rebecca Yeomans, *South Georgia College;* Lynne Zeman, *Kirkwood Community College;* Zhihong Zhang, *Chatham College.*

### Class Testers for Previous Editions

David Aborne, *University of Tennessee–Chattanooga;* Reuben Barret, *Prairie State College;* Morgan Barrows, *Saddleback College;* Henry Bart, *LaSalle University;* James Bartalome, *University of California–Berkeley;* Christy Bazan, *Illinois State University;* Richard Beckwitt, *Framingham State College;* Elizabeth Bell, *Santa Clara University;* Peter Biesmeyer, *North Country Community College;* Donna Bivans, *Pitt Community College;* Evert Brown, *Casper College;* Christina Buttington, *University of Wisconsin–Milwaukee;* Tait Chirenje, *Richard Stockton College;* Reggie Cobb, *Nash Community College;* Ann Cutter, *Randolph Community College;* Lola Deets, *Pennsylvania State University–Erie;* Ed DeGrauw, *Portland Community College;* Stephanie Dockstader, *Monroe Community College;* Dee Eggers, *University of North Carolina–Asheville;* Jane Ellis, *Presbyterian College;* Paul Fader, *Freed Hardeman University;* Joseph Fail, *Johnson C. Smith University;* Brad Fiero, *Pima Community College, West Campus;* Dane Fisher, *Pfeiffer University;* Chad Freed, *Widener University;* Sue Glenn, *Gloucester County College;* Sue Habeck, *Tacoma Community College;* Mark Hammer, *Wayne State University;* Michael Hanson, *Bellevue Community College;* David Hassenzahl, *Oakland Community College;* Kathleen Hornberger, *Widener University;* Paul Jurena, *University of Texas–San Antonio;* Dawn Keller, *Hawkeye Community College;* David Knowles, *East Carolina University;* Erica Kosal, *Wesleyan College;* John Logue, *University of Southern Carolina Sumter;* Keith Malmos, *Valencia Community College;* Nancy Markee, *University of Nevada–Reno;* Bill Mautz, *University of New Hampshire;* Julie Meents, *Columbia College;* Stephen Getchell, *Mohawk Valley Community College;* Lori Moore,

*Northwest Iowa Community College;* Elizabeth Pixley, *Monroe Community College;* John Novak, *Colgate University;* Brian Peck, *Simpson College;* Sarah Quast, *Middlesex Community College;* Roger Robbins, *East Carolina University;* Mark Schwartz, *University of California–Davis;* Julie Seiter, *University of Nevada–Las Vegas;* Brian Shmaefsky, *Kingwood College;* Diane Sklensky, *Le Moyne College;* Mark Smith, *Fullerton College;* Patricia Smith, *Valencia Community College East;* Sherilyn Smith, *Le Moyne College;* Jim Swan, *Albuquerque Technical Vocational Institute;* Amy Treonis, *Creighton University;* Darrell Watson, *The University of Mary Hardin Baylor;* Barry Welch, *San Antonio College;* Susan Whitehead, *Becker College;* Roberta Williams, *University of Nevada–Las Vegas;* Justin Williams, *Sam Houston University;* Tom Wilson, *University of Arizona.*

# About the Authors

**Jay H. Withgott** is a science and environmental writer with a background in scientific research and teaching. He holds degrees from Yale University, the University of Arkansas, and the University of Arizona. As a researcher, he has published scientific papers on topics in ecology, evolution, animal behavior, and conservation biology in journals including *Proceedings of the National Academy of Sciences, Proceedings of the Royal Society of London B, Evolution,* and *Animal Behavior.* He has taught university-level laboratory courses in ecology, ornithology, vertebrate diversity, anatomy, and general biology.

As a science writer, Jay has authored articles for a variety of journals and magazines including *Science, New Scientist, BioScience, Smithsonian, Current Biology, Conservation in Practice,* and *Natural History.* He combines his scientific expertise with his past experience as a reporter and editor for daily newspapers to make science accessible and engaging for general audiences.

Jay lives with his wife, biologist Susan Masta, in Portland, Oregon.

**Scott Brennan** has taught environmental science, ecology, resource policy, and journalism at Western Washington University and at Walla Walla Community College. He has also worked as a journalist, photographer, and consultant.

Scott has cultivated his expertise in environmental science and public policy by serving as Executive Vice President and Chief Operating Officer for Alaskans for Responsible Mining, as Executive Conservation Fellow of the National Parks Conservation Association in Washington, D.C., and as a consultant to the U.S. Department of Defense Environmental Security Office at the Pentagon.

When not at work, Scott is likely to be found exploring the Chugach Mountains and the Bristol Bay drainages in southwest Alaska. He lives with his wife, Angela, and their dogs Raven and Hatcher, in south central Alaska's Chester Creek Watershed.

# How are environmental issues currently

**Integrated Central Case Studies** highlight real people in both domestic and international places to bring current environmental issues to life.

Fishermen of the Maldives

CENTRAL CASE

## Rising Seas May Flood the Maldives

**"The impact of global warming and climate change can effectively kill us off, make us refugees. . . ."**
—Ismail Shafeeu, Minister of Environment, Maldives, 2000

**"We simply must do everything we can in our power to slow down global warming before it is too late. The science is clear. The global warming debate is over."**
—Arnold Schwarzenegger, Governor of California, September 2006

A nation of low-lying islands in the Indian Ocean, the Maldives is known for its spectacular tropical setting, colorful coral reefs, and sun-drenched beaches. For visiting tourists it seems to be paradise, and for 370,000 Maldives residents it is home. But residents and tourists alike now fear that the Maldives could soon be submerged by the rising seas that are accompanying global climate change.

Nearly 80% of the Maldives' land area lies less than 1 m (39 in.) above sea level. In a nation of 1200 islands whose highest point is just 2.4 m (8 ft) above sea level, rising seas are a matter of life or death. The world's oceans rose 10–20 cm (4–8 in.) during the 20th century as warming temperatures expanded ocean water and as melting polar ice discharged water into the ocean. According to current projections, sea level will rise another 18–59 cm (7–23 in.) by the year 2100.

Higher seas are expected to flood large areas of the Maldives and cause salt water to contaminate drinking water supplies. Storms intensified by warmer water temperatures will erode beaches, cause flooding, and damage the coral reefs that are so vital to the tourism and fishing industries that drive the nation's economy. Because of such concerns, the Maldives government recently has evacuated residents from several of the lowest-lying islands.

On December 26, 2004, the nation got a taste of what could be in store in the future, when a massive *tsunami*, or tidal wave, devastated coastal areas throughout the Indian Ocean. The tsunami killed 100 Maldives residents and left 20,000 homeless. Schools, boats, tourist resorts, hospitals,

3

# affecting real people around the world?

Integrated Central Case Studies begin each chapter and are further developed throughout the chapter text, making general concepts more understandable and interesting to learn.

### The IPCC summarizes evidence of climate change and predicts future impacts

In recent years, it seems that virtually everyone of us is detecting climatic changes around us. A fisherman in the Maldives notes the seas encroaching on his home island. A rancher in west Texas suffers a multi-year drought. A homeowner in Florida finds it impossible to obtain insurance against the hurricanes and storm surges that increasingly threaten. New Yorkers, Bostonians, Chicagoans, and Los Angelenos marvel over one freakish weather event after another. Are all these impressions part of a real pattern, and is there solid scientific evidence to confirm that climate is indeed already changing?

Over the past century, and particularly in recent years, a wide variety of data sets have shown significant trends in climate conditions. The most thoroughly reviewed and widely accepted synthesis of scientific information concerning climate change is a series of reports issued by the **Intergovernmental Panel on Climate Change** (IPCC). This international panel of scientists and government officials was established in 1988 by the United Nations Environment Programme (UNEP) and the World Meteorological Organization.

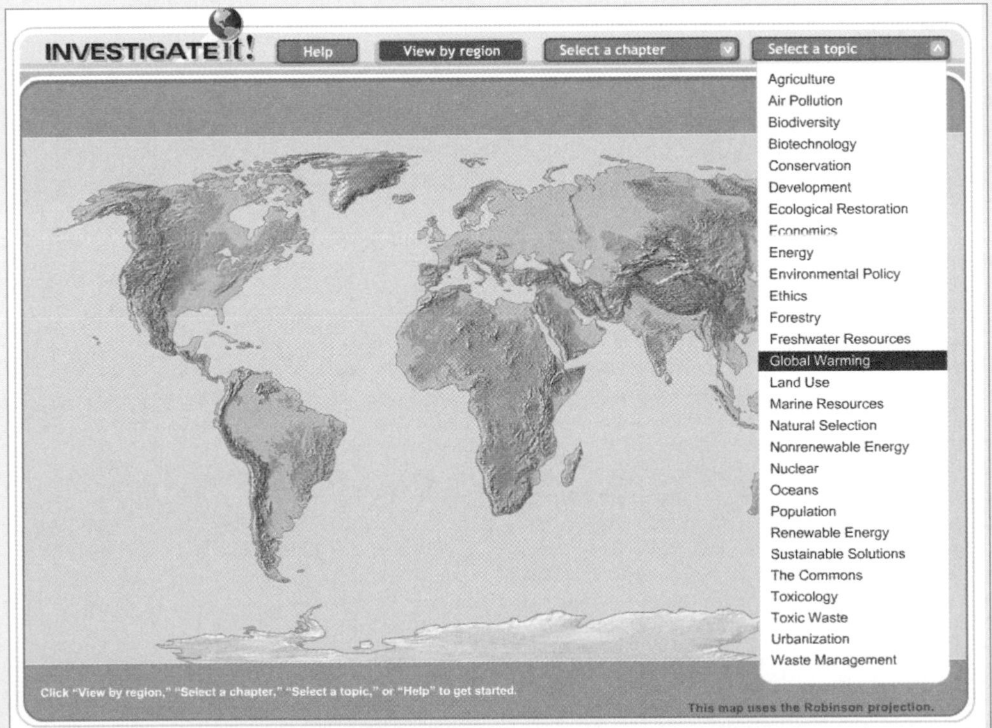


**INVESTIGATE it!** on the Withgott/Brennan Companion Website provides an additional 230 domestic and international case studies beyond those presented in the text. Browse by topic, geographic region, or text chapter to access 200 recent articles from The New York Times and 30 ABCNEWS clips that explore environmental issues in the news today. Every month, two new articles will be posted from The New York Times, giving you access to timely coverage on environmental issues at one convenient site.

# Do you understand the science behind

**The Science behind the Story** highlights how scientists develop hypotheses, test predictions, and analyze and interpret data.

514 PART TWO Environmental Issues and the Search for Solutions

THE SCIENCE BEHIND THE STORY

## Reading History in the World's Longest Ice Core

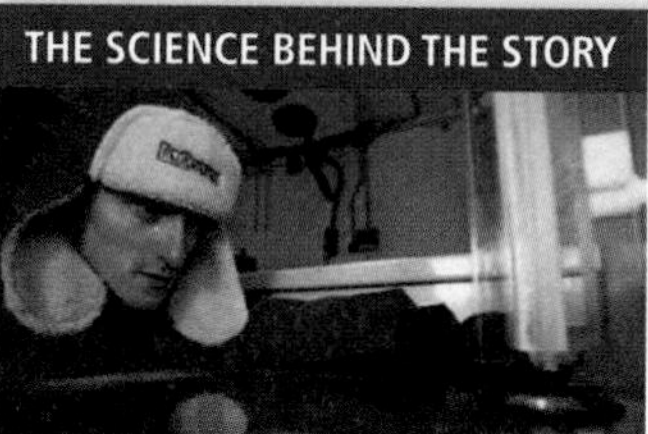

*An EPICA researcher prepares a Dome C ice core sample for analysis*

In the most frigid reaches of our planet, snow falling year after year for millennia compresses into ice and stacks up into immense sheets that scientists can mine for clues to Earth's climate history. The ice sheets of Antarctica and Greenland trap tiny air bubbles, dust particles, and other proxy indicators (•p. 512) of past conditions. By drilling boreholes and extracting ice cores, researchers can tap into these valuable archives.

Recently, researchers drilled and analyzed the deepest core ever. At a remote and pristine site in Antarctica named Dome C, they drilled down 3,270 m (10,728 ft) to bedrock and pulled out more than 800,000 years' worth of ice. The longest previous ice core (from Antarctica's Vostok station) had gone back "only" 420,000 years.

Ice near the top of these cores was laid down most recently, and ice at the bottom is oldest, so by analyzing ice at intervals along the core's length, researchers can generate a timeline of environmental change.

Dome C, a high summit of the Antarctic ice sheet, is one of the coldest spots on the planet, with an annual mean temperature of −54.5° C (−98.1° F). The Dome C ice core was drilled by the European Project for Ice Coring in Antarctica (EPICA), a consortium of researchers from 10 European nations. Antarctic operations are expensive and logistically complicated, ice-drilling requires powerful technology, and the analysis requires a diverse assemblage of experts. When the team published a research paper detailing its data in the journal *Nature* in 2004, the paper had 56 authors.

That landmark paper reported data across 740,000 years. The researchers obtained data on surface air temperature by measuring the ratio of deuterium isotopes (•pp. 91, 93) to normal hydrogen in the ice, because this ratio is temperature-dependent (see top panel of figure). And by examining the density of dust particles, they could tell when arid and/or windy climates sent more dust aloft.

In 2005, two followup papers in the journal *Science* reported analyses of greenhouse gas concentrations from the EPICA ice core. By analyzing air bubbles trapped in the ice, the researchers quantified atmospheric concentrations of carbon dioxide, methane, and nitrous oxide from across 650,000 years.

One finding from the EPICA studies was expected—yet important. The researchers documented that temperature swings in the past were tightly correlated with concentrations of carbon dioxide (see middle panel of figure), as well as methane and nitrous oxide. This bolstered the scientific consensus that greenhouse gas emissions are causing Earth to warm today.

Also clear and expected from the data was that temperature varied with swings in solar radiation due to Milankovitch cycles (• p. 510). The complex interplay of these cycles produced periodic temperature fluctuations on Earth resulting in periods of *glaciation* (when temperate regions of the planet were covered in ice) as well as warm *interglacial* periods. The Dome C ice core spanned eight glacial cycles.

In addition, the EPICA data demonstrate that by increasing greenhouse gas concentrations since the industrial revolution, we have brought them well above the highest levels they reached naturally across 650,000 years. Today's carbon dioxide concentration (383 ppm in 2007) is too recent to show up in the ice core, but is far above previous maximum values (of ~300 ppm) shown in the middle panel of the figure. Present-day concentrations of methane and nitrous oxide are likewise the highest in 650,000 years. These data show that we as a society have brought ourselves deep into uncharted territory.

Other findings from the ice core are not easily explained. Intriguingly, the earlier glacial cycles are of a different character than the more recent cycles (see figure). For the most recent cycles, the Dome C core confirmed what the Vostok data had shown: Glacial periods were long, whereas interglacial periods were brief, with a rapid rise and fall of temperature. Interglacials thus appear on a graph of

# the news stories?

Each The Science behind the Story carefully walks you through the scientific process — not only what scientists know, but how they discovered it.

**522** PART TWO Environmental Issues and the Search for Solutions

THE SCIENCE BEHIND THE STORY

*Outlet glaciers melting into Scoresby Sund, Greenland's largest fjord*

## Timing Greenland's Glaciers as They Race to the Sea

Scientists have known for years that the Arctic is bearing the brunt of global warming and that the massive ice sheet covering Greenland is melting around its edges. But data from 1993 to 2003 showed Greenland's ice loss accounting for only 4–12% of global sea level rise, about 0.21 mm/yr. And as authors of the IPCC's *Fourth Assessment Report* used results from climate models to predict Greenland's future contributions to sea-level rise, the models told them to expect more of the same.

However, some brand-new research hadn't made it into the models. Scientists studying how ice moves were learning that ice sheets can collapse more quickly than expected and that Greenland's ice loss is accelerating. As a result, they said, the IPCC report underestimates the likely speed and extent of future sea level rise.

Greenland's ice sheet is massive, averaging nearly a mile deep and covering as much area as Texas, California, Michigan, and Minnesota combined. If the entire ice sheet were to melt, global sea level would rise by a whopping 7 m (23 ft).

The ice sheet gains mass by accumulating snow during cold weather, which becomes packed into ice over time. It loses mass as surface ice melts in warm weather, generally at the periphery, where ice is thinnest or contacts seawater. If melting and runoff outpace accumulation, then the ice sheet shrinks.

But researchers are now learning that the internal physical dynamics of how ice moves may be more important. These dynamics can speed the flow of immense amounts of ice in *outlet glaciers* downhill toward the coast, where eroding ice sloughs off and melts into the sea.

The first good indication of this process came in 2002 when a team led by Jay Zwally of NASA's Goddard Space Flight Center in Maryland noted that ice in outlet glaciers flows more quickly during warm months, when pools of meltwater form on the surface. In a paper published in the journal *Science* in 2002, Zwally's group proposed that meltwater leaks down through crevasses and vertical tunnels called *moulins* to the bottom of the glacier. There, the water runs downhill in a layer between bedrock and ice, lubricating the bedrock surface and enabling the ice to slide downhill like a car hydroplaning on a wet road. In addition, the meltwater weakens ice on its way down and warms the base of the glacier, melting some of it to create more water (see the figure).

Other scientists proposed an additional mechanism: Warming ocean water melts ice shelves along the coast, depriving outlet glaciers of the buttressing support that holds them in place. Without a floating ice tongue at its terminus, a glacier slides into the ocean more readily.

These physical dynamics represent positive feedback, researchers said; once global warming initiates these processes, they encourage further melting. As such, they said, we should expect that Greenland's melting will accelerate.

Several research groups soon attempted to measure the rates of

# Do you know how to examine the data

**Interpreting Graphs and Data** activities at the end of each chapter give you hands-on experience working with graphs, so you can develop the skills you'll need to understand scientific information when you see it in the news.

## INTERPRETING GRAPHS AND DATA

We burn fossil fuels to generate electricity, to power vehicles for transportation, and as primary energy sources (non-electricity uses, mostly for heating) in homes, businesses, and industry. For each of these uses, the accompanying graph shows trends in the emission of carbon dioxide from fossil fuel combustion in the United States.

1. Calculate the approximate percentage changes in $CO_2$ emissions from transportation; electricity generation; and residential, commercial, and industrial primary energy use (mostly heating) between 1980 and 2006.
2. Between 1980 and 2006, U.S. population increased by 32% and the inflation-adjusted U.S. gross domestic product (GDP) more than doubled. What quantitative conclusions can you draw from these data about $CO_2$ emissions per capita? About $CO_2$ emissions per unit of total economic activity? Create a graph and sketch a trend line of $CO_2$ emissions per capita from 1980 to 2006. Now sketch a trend line of $CO_2$ emissions per unit of total economic activity from 1980 to 2006.
3. Imagine you are put in charge of designing a strategy to reduce U.S. emissions of $CO_2$ from fossil fuel combustion. Based on the data presented here, what approaches would you recommend, and how would you prioritize these? Explain your answers.

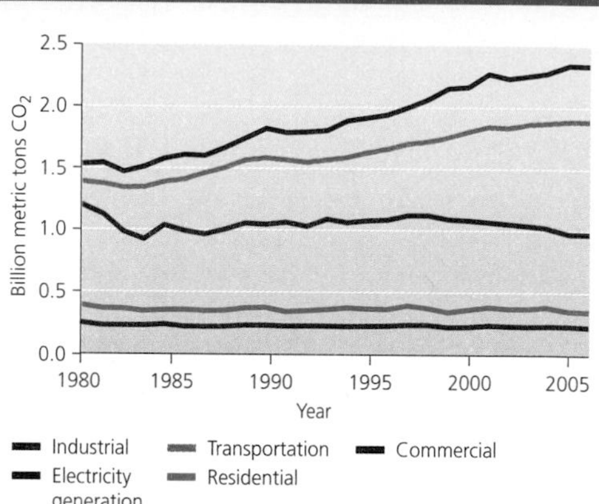


Emissions of $CO_2$ from fossil fuel combustion by end-use sector in the United States, 1980–2006. Data from U.S. Department of Energy, Energy Information Administration. 2007. *Annual energy review 2006.*

**GRAPHIt!** exercises on the Withgott/Brennan Companion Website help you to better understand how to work with and interpret graphs.

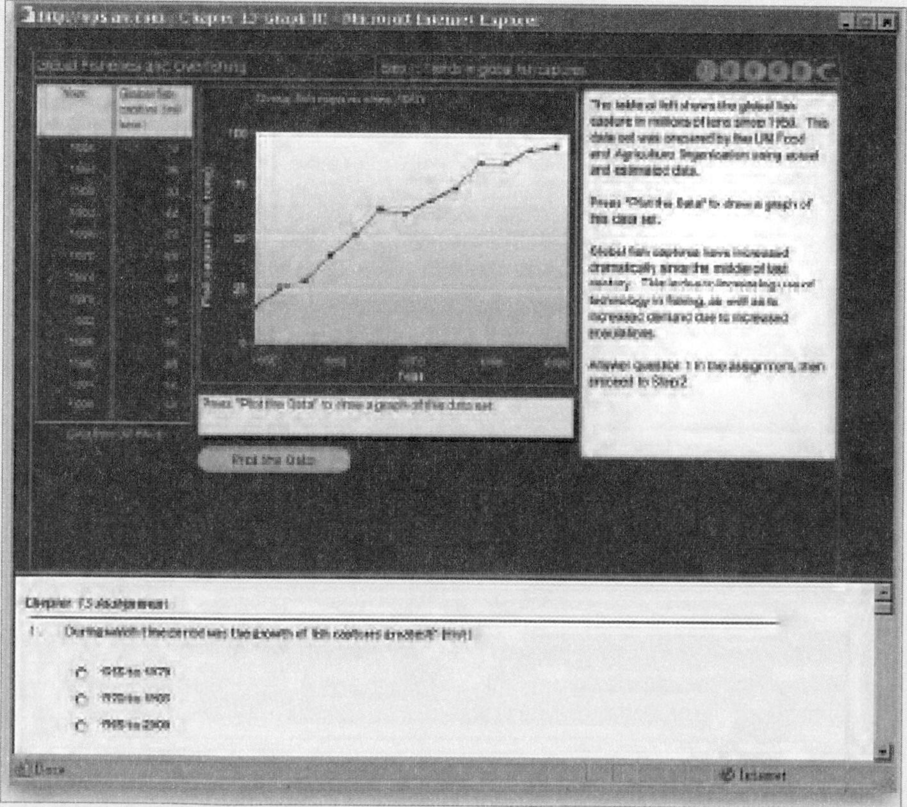

behind environmental issues?

**Environmental Issues** change quickly, so Withgott/Brennan uses the most current data available.

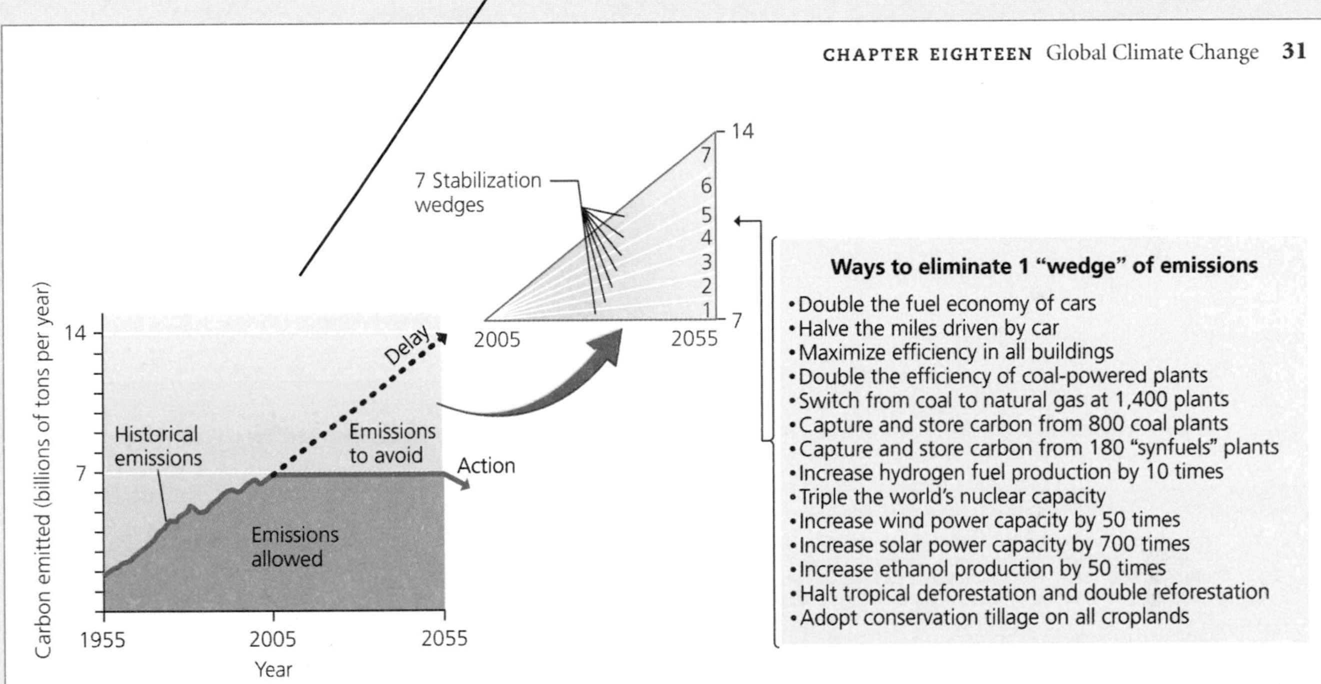


**FIGURE 18.29** Pacala and Socolow began with a standard graph (left) of predicted carbon emissions from $CO_2$ showing the doubling of emissions that scientists expect to occur from 2005 to 2055. They added a flat line to represent the trend if emissions were held constant and separated the graph into emissions allowed (below the line) and emissions to be avoided (the triangular area above the flat line). They then divided this "stabilization triangle" into seven equal-sized portions, which they called "stabilization wedges" (center). Each stabilization wedge represents 1 billion tons of $CO_2$ emissions in 2055 to be avoided. Finally, they identified 15 strategies (box at right), each of which could take care of 1 wedge. If we accomplish just 7 of these 15 strategies, we could halt our growth in emissions for the next half century.
Adapted from Pacala, S. and R. Socolow, 2004. Stabilization wedges: Solving the climate problem for the next 50 years with current technologies. *Science* 305: 968–972; and www.princeton.edu/~cmi/resources/CMI_Resources_new_files/CMI_Stab_Wedges_Movie.swf.

**Environmental Issues** are clearly cited so you can trace the source of the information presented.

# Do you recognize how decisions and

The **Causes and Consequences** figures help you visualize at a glance the connections between the causes and consequences of environmental issues. You are then asked to consider how actions or solutions might affect an issue.

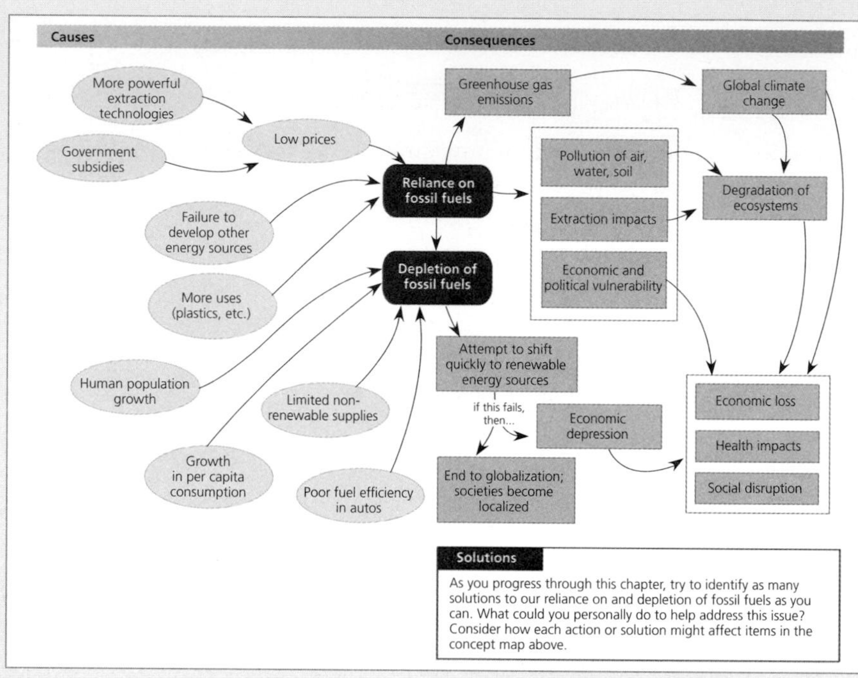


## CAUSES AND CONSEQUENCES

People in different societies consume resources and use land and water at different rates, according to their level of wealth or poverty, the technologies available to them, their society's cultural norms, and other factors. Societies whose people have **large ecological footprints** exert greater per capita impacts on Earth's natural resources and ecological systems. Fortunately, there are many solutions and steps that people can take to address the causes and mitigate the consequences of large ecological footprints.

Write in two causes of large ecological footprints in the spaces provided. Then write in two consequences (impacts on the environment, human health, or quality of life) that result from large ecological footprints. Finally, offer two solutions corresponding to this issue and its consequences. One cause, one consequence, and one solution have been filled in for you, providing examples.

**CAUSES**

*Strong demand for material goods*

____________________

____________________

→ Large Ecological Footprints →

**CONSEQUENCES**

*Depletion of natural resources*

____________________

____________________

**SOLUTIONS**

*Improve efficiency of manufacturing processes*

____________________

____________________

Sometimes solutions can have unintended consequences. Can you think of an undesired consequence that one solution to this issue might have? How might we then deal with *that* consequence?

**Causes and Consequences** exercises help you determine the causes, consequences, and potential solutions to environmental issues.

# actions can impact the environment?

**Weighing THE Issues** | **Environmental Refugees**

Citizens of the Maldives see an omen of their future in the Pacific island nation of Tuvalu, which has been losing 9 cm (3.5 in.) of elevation per decade to rising seas. Appeals from Tuvalu's 11,000 citizens were heard by New Zealand, which began accepting these environmental refugees in 2003. Do you think a national culture can survive if its entire population is relocated? Think of the tens of thousands of refugees from Hurricane Katrina. How are their lives and culture faring in the wake of that tragedy?

Issues in environmental science often lack black-and-white answers, so critical thinking skills help you navigate the gray areas. **Weighing the Issues** questions throughout each chapter encourage you to grapple with questions about science, policy, and ethics.

**CALCULATING ECOLOGICAL FOOTPRINTS**

Global climate change is something to which we all contribute, because fossil fuel combustion plays such a large role in supporting the lifestyles we lead. Conversely, as individuals, each one of us can contribute to mitigating global climate change through personal decisions and actions that affect the way we live our lives. Several online calculators enable you to calculate your own personal *carbon footprint*, the amount of carbon emissions for which you are responsible. Go to one of these, at http://www.carbonfootprint.com, follow the link for the U.S. version, take the quiz, and enter the relevant data in the table.

| | **Carbon footprint (kg per person per year)** |
|---|---|
| World average | |
| Average for industrialized nations | |
| U.S. average | |
| Your footprint | |
| Average needed to halt climate change | |
| Your footprint with three changes | |

**Calculating Ecological Footprints** activities at the end of each chapter let you evaluate the impact of actions, including your own, on a local and global scale.

PART ONE

# FOUNDATIONS OF ENVIRONMENTAL SCIENCE

Researcher studying eucalyptus forest, Australia

CHAPTER

# 1 An Introduction to Environmental Science

Our island, Earth

## Upon completing this chapter, you will be able to:

- Define the term *environment*
- Describe natural resources and explain their importance to human life
- Characterize the interdisciplinary nature of environmental science
- Understand the scientific method and how science operates
- Diagnose and illustrate some of the pressures on the global environment
- Evaluate the concepts of sustainability and sustainable development

# Our Island, Earth

Viewed from space, our home planet resembles a small blue marble suspended against a vast inky-black backdrop. Earth may seem vast to us as we go about our lives on its surface, but the astronaut's perspective reveals that Earth and its natural systems are limited. From this perspective, it becomes clear that as our population, technological powers, and consumption of resources increase, so do our abilities to alter our planet and damage the very systems that keep us alive.

## Our environment is the sum total of our surroundings

A photograph of Earth reveals a great deal, but it does not convey the complexity of our environment. Our **environment** includes all the living and nonliving things around us with which we interact. It includes the continents, oceans, clouds, and ice caps you can see in the photo of Earth from space, as well as the animals, plants, forests, and farms that comprise the landscapes around us. In a more inclusive sense, it also encompasses our built environment—the structures, urban centers, and living spaces that people have created. In its most inclusive sense, our environment also includes the complex webs of social relationships and institutions that shape our daily lives.

People commonly use the term *environment* in the first, most narrow sense—to mean a nonhuman or "natural" world apart from human society. This usage is unfortunate, because it masks the very important fact that humans exist within the environment and are part of nature. As one of many species on Earth, we share with others the same dependence on a healthy, functioning planet. The limitations of language make it all too easy to speak of "people and nature," or "human society and the environment," as though they were separate and did not interact. However, the fundamental insight of environmental science is that we are part of the natural world and that our interactions with its other parts matter a great deal.

## Environmental science explores interactions between humans and the world around us

Appreciating how we interact with our environment is crucial for a well-informed view of our place in the world and for a mature awareness that we are one species among many on a planet full of life. Understanding our relationship with the world around us is also vital because we are altering the very systems we need, in ways we do not yet fully comprehend.

We depend utterly on our environment for air, water, food, shelter, and everything else essential for living. However, our actions modify our environment, whether we intend them to or not. Many of these actions have enriched our lives, bringing us longer life spans, better health, and greater material wealth, mobility, and leisure time. However, these improvements have often degraded the natural systems that sustain us. Impacts such as air and water pollution, soil erosion, and species extinction can compromise human well-being, pose risks to human life, and threaten our ability to build a society that will survive and thrive in the long term. The elements of our environment were functioning long before the human species appeared, and we would be wise to realize that we need to keep these elements in place.

**Environmental science** is the study of how the natural world works, how our environment affects us, and how we affect our environment. We need to understand our interactions with our environment because such knowledge is the essential first step toward devising solutions to our most pressing environmental problems. It can be daunting to reflect on the sheer magnitude of environmental dilemmas that confront us today, but with these problems also come countless opportunities for devising creative solutions.

Environmental scientists study the issues most centrally important to our world and its future. Right now, global conditions are changing more quickly than ever. Right now, through science, we as a civilization are gaining knowledge more rapidly than ever. And right now, the window of opportunity for acting to solve problems is still open. With such bountiful challenges and opportunities, this particular moment in history is indeed an exciting time to be studying environmental science.

## Natural resources are vital to our survival

An island by definition is finite and bounded, and its inhabitants must cope with limitations in the materials they need. On our island, Earth, human beings, like all living things, ultimately face environmental constraints. Specifically, there are limits to many of our **natural resources**, the various substances and energy sources we need to survive. Natural resources that are replenished over short periods are known as **renewable natural resources**. Some renewable resources, such as sunlight, wind, and wave energy, are perpetually available. Others, such as timber, water, and soil, renew themselves over months, years, or decades, if we are careful not to use them up too quickly or destructively. In contrast, resources such as mineral ores and crude oil are in finite supply and are formed much more slowly than we use them. These are known as **nonrenewable natural resources**. Once we deplete them, they are no longer available.

**FIGURE 1.1** Natural resources lie along a continuum from perpetually renewable to nonrenewable. Perpetually renewable resources, such as sunlight and wind energy, will always be there for us. Nonrenewable resources, such as oil and coal, exist in limited amounts that could one day be gone. Resources such as timber, soils, and fresh water can be renewed on intermediate time scales, if we are careful not to deplete them.

We can view the renewability of natural resources as a continuum (**Figure 1.1**). Some renewable resources may turn nonrenewable if we overuse them. For example, overpumping groundwater can deplete underground aquifers and turn lush landscapes into deserts. Populations of animals and plants we harvest from the wild may vanish if we overharvest them. In recent years, our consumption of natural resources has increased greatly, driven by rising affluence and the growth of the largest human population in history.

## Human population growth has shaped our resource use

For nearly all of human history, only a few million people populated Earth at any one time. Although past populations cannot be calculated precisely, **Figure 1.2** gives some idea of just how recently and suddenly our population has grown beyond 6.7 *billion* people.

Two phenomena triggered remarkable increases in population size. The first was our transition from a

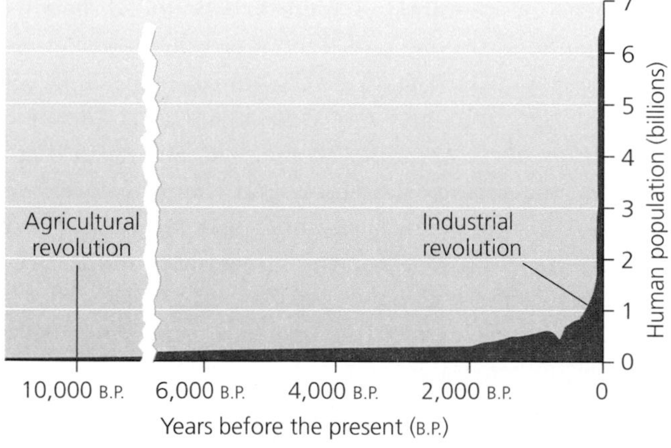



**(a) World population growth**

**(b) Urban society**

**FIGURE 1.2** For almost all of human history, our population was low and relatively stable. It increased significantly (**a**) as a result first of the agricultural revolution and then of the industrial revolution. Our skyrocketing population has given rise to congested urban areas (**b**), such as this city in Java, Indonesia.

hunter-gatherer lifestyle to an agricultural way of life. This change began around 10,000 years ago and is known as the **agricultural revolution**. As people began to grow crops, raise domestic animals, and live sedentary lives in villages, they found it easier to meet their nutritional needs. As a result, they began to live longer and to produce more children who survived to adulthood. The second notable phenomenon, known as the **industrial revolution**, began in the mid-1700s. It entailed a shift from rural life, animal-powered agriculture, and manufacturing by craftsmen, to an urban society powered by **fossil fuels** (nonrenewable energy sources, such as oil, coal, and natural gas (• pp. 544–557). The industrial revolution introduced improvements in sanitation and medical technology, and it enhanced agricultural production with fossil-fuel-powered equipment and synthetic pesticides and fertilizers (• pp. 266–268).

**Thomas Malthus and population growth** At the outset of the industrial revolution in England, population growth was regarded as a good thing. For parents, high birth rates meant more children to support them in old age. For society, it meant a greater pool of labor for factory work.

British economist **Thomas Malthus** (1766–1834) had a different opinion. Malthus claimed that unless population growth were controlled by laws or other social strictures, the number of people would outgrow the available food supply until starvation, war, or disease arose and reduced the population (**Figure 1.3**). Malthus's most influential work, *An Essay on the Principle of Population*, published in 1798, argued that if limits on births (such as abstinence and contraception) were not implemented, deaths would increase through famine, plague, and war.

(a) 18th-century London, England

(b) Thomas Malthus

**FIGURE 1.3** The England of Thomas Malthus's era (1766–1834), shown in this engraving (**a**), favored population growth as society industrialized. Malthus (**b**) argued that population growth could lead to disaster.

Malthus's thinking was shaped by the rapid urbanization and industrialization he witnessed during the early years of the industrial revolution, but debates over his views continue today. As we will see in Chapter 8 and throughout this book, global population growth has indeed contributed to famine, disease, and conflict. However, increasing material prosperity has also helped reduce birth rates—something Malthus did not foresee.

**The Ehrlichs and the "population bomb"** In our day, biologists Paul and Anne Ehrlich of Stanford University have been called "neo-Malthusians" because they too have warned that population growth will have disastrous effects on human welfare. In his 1968 book, *The Population Bomb*, Paul Ehrlich predicted that the rapidly increasing human population would unleash famine and conflict that would consume civilization by the end of the 20th century. Like Malthus, Ehrlich argued that population was growing much faster than our ability to produce and distribute food, and he maintained that population control was the only solution.

Although human population nearly quadrupled in the past 100 years—the fastest it has ever grown (see Figure 1.2a)—Ehrlich's forecasts have not materialized on the scale he predicted. This is due, in part, to agricultural advances made in recent decades (• pp. 264–266). Today, the Ehrlichs and other concerned scientists warn of a postponed, but still impending, global population crisis, as well as impacts from overconsumption of resources.

## Resource consumption exerts social and environmental impacts

Population growth is unquestionably at the root of many environmental problems. However, the growth in consumption is also to blame. The industrial revolution enhanced the material affluence of many of the world's people by considerably increasing our consumption of natural resources and manufactured goods.

**Garrett Hardin and the "tragedy of the commons"** The late Garrett Hardin of the University of California at Santa Barbara disputed the economic theory that unfettered exercise of individual self-interest will serve the public interest (• pp. 38–39). According to Hardin's best-known essay, "The Tragedy of the Commons," published in the journal *Science* in 1968, resources that are open to unregulated exploitation will eventually be depleted.

Hardin based his argument on a scenario described in a pamphlet published in 1833. In a public pasture, or "common," that is open to unregulated grazing, Hardin argued, each person who grazes animals will be motivated to increase the number of his or her animals in the pasture. Ultimately, overgrazing will cause the pasture's food production to collapse. Because no single person owns the pasture, no one has incentive to expend effort taking care of it, and everyone takes what he or she can until the resource is depleted.

Some have argued that private ownership can address this problem. Others point to cases in which people sharing a common resource have voluntarily organized and cooperated in enforcing its responsible use. Still others maintain that the dilemma justifies government regulation of the use of resources held in common by the public, from forests to clean air to clean water.

### Weighing the Issues | The Tragedy of the Commons

Imagine you make your living fishing for lobster. You are free to boat anywhere and set out as many traps as you like. Your harvests have been good, and nothing is stopping you from increasing the number of your traps. However, all the other lobster fishers are thinking the same thing, and the fishing grounds are getting crowded. Catches decline year by year, until one year the fishery crashes, leaving you and all the others with catches too meager to support your families. Some of your fellow fishers call for dividing the waters and selling plot-by-plot access to individuals. Others urge the fishers to team up, set quotas among themselves, and prevent newcomers from entering the market. Still others are imploring the government to get involved and pass laws regulating how much fishers can catch. What do you think is the best way to combat this tragedy of the commons and restore the fishery, and why? Can you think of any other solutions?

FIGURE 1.4 An "ecological footprint" represents the total area of land and water needed to produce the resources a given person or population uses, together with the total amount of land and water needed to dispose of their waste. The footprints of the affluent societies of today's industrialized nations tend to be much larger than the geographic areas these societies take up directly. Adapted from Wackernagel, M., and W. Rees. 1996. *Our ecological footprint: Reducing human impact on the Earth.* Gabriola Island, British Columbia: New Society Publishers.

**Wackernagel and Rees and the ecological footprint**
As global affluence has increased, human society has consumed more and more of the planet's limited resources. We can quantify resource consumption using the concept of the "ecological footprint," developed in the 1990s by environmental scientists Mathis Wackernagel and William Rees. An **ecological footprint** expresses the environmental impact of an individual or population in terms of the cumulative amount of biologically productive land and water required to provide the raw materials the person or population consumes and to dispose of or recycle the waste the person or population produces (**Figure 1.4**). It measures the total area of Earth's surface that a given person or population "uses" once all direct and indirect impacts are totaled up.

For humanity as a whole, Wackernagel and Rees calculated that our species is using 30% more of the planet's resources than are available on a sustainable basis. That is, we are depleting renewable resources 30% faster than they are being replenished—like drawing the principal out of a bank account rather than living off the interest. This excess use has been termed **overshoot**, because we have overshot, or surpassed, Earth's capacity to sustainably support us. Moreover, people from wealthy nations have much larger ecological footprints than do people from poorer nations. Current calculations by Wackernagel and his colleagues indicate that if all the world's people consumed resources at the rate of North Americans, we would need the equivalent of four and a half additional planet Earths.

## Environmental science can help us avoid mistakes made in the past

It remains to be seen whether the direst predictions of Malthus, the Ehrlichs, and others will come to pass for today's global society, but we already have historical evidence that civilizations can crumble when pressures from population and consumption overwhelm resource availability. Easter Island is a classic case (see "The Science behind the Story," • pp. 8–9).

Indeed, many great civilizations have fallen after depleting resources from their environments, and each has left devastated landscapes in its wake. Plato wrote of the deforestation and environmental degradation accompanying ancient Greek cities, for example. Today, archaeologists, historians, and paleoecologists who study past societies and landscapes are accumulating evidence of environmental collapse, from the Greek and Roman Empires to the Maya, the Anasazi, and other civilizations of the New World.

Today's arid deserts of Iraq and other Middle Eastern countries were far more vegetated when the great ancient civilizations thrived there; at that time, these regions were lush enough to support the very origin of agriculture. While deforestation created deserts in temperate regions, in more tropical climates the ancient cities of fallen civilizations became overgrown by jungle. The gigantic stone monuments of the Angkor civilization in Southeast Asia, like those of the Maya in Mexico and Central America, remained unknown to Westerners until the 19th century, and most of these cities remain covered by rainforest. Researchers have learned enough by now, however, that scientist and author Jared Diamond in his 2005 book, *Collapse*, could synthesize this information and formulate sets of reasons why civilizations succeed and persist, or fail and collapse. Success and persistence, it turns out, depend largely on how societies interact with their environments.

Today we are confronted with news and predictions of environmental catastrophes on a regular basis. Studying environmental science will outfit you with the tools that can help you to evaluate information on environmental change and to think critically and creatively about possible actions to take in response.

# The Nature of Environmental Science

Environmental scientists aim to comprehend how Earth's natural systems function, how humans are influenced by those systems, and how we are influencing those systems. Many environmental scientists are motivated by a desire to develop solutions to environmental problems. The solutions themselves (such as new technologies, policy decisions, or resource management strategies) are *applications* of environmental science. The study of such applications and their consequences is, in turn, also part of environmental science.

## Environmental science is an interdisciplinary pursuit

Studying our interactions with our environment is a complex endeavor that requires expertise from many disciplines, including ecology, earth science, chemistry, biology, economics, political science, demography, ethics, and others. Environmental science is thus an **interdisciplinary** field—one that borrows techniques from numerous disciplines and brings research results from these disciplines together into a broad synthesis (**Figure 1.5**).

Traditional established disciplines are valuable because their scholars delve deeply into topics, uncovering new knowledge and developing expertise in particular areas. Interdisciplinary fields are valuable because their practitioners take specialized knowledge from different disciplines, consolidate it, synthesize it, and make sense of it in a broad context to better serve the multifaceted interests of society.

Environmental science is especially broad because it encompasses not only the **natural sciences** (disciplines that study the natural world), but also the **social sciences** (disciplines that study human interactions and institutions). The natural sciences provide us the means to gain accurate information about our world and to interpret it reasonably. Comprehending our interactions with the environment, however, also involves weighing values and understanding human behavior—and this requires the social sciences. Most environmental science programs focus predominantly on the natural sciences as they pertain to environmental issues. Programs that incorporate the social sciences heavily often use the term

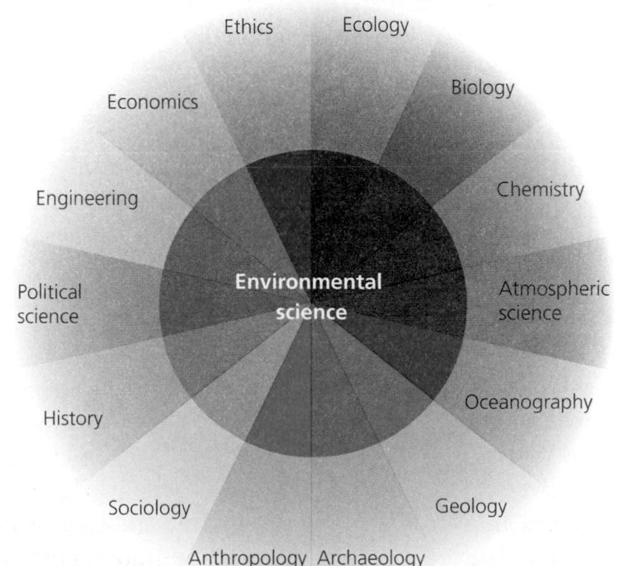


FIGURE 1.5 Environmental science is a highly interdisciplinary pursuit, involving input from many different established fields of study across the natural sciences and social sciences.

THE SCIENCE BEHIND THE STORY

## The Lesson of Easter Island

*The immense moai (statues) of Easter Island*

Easter Island is one of the most remote spots on the globe, located in the Pacific Ocean 3,750 km (2,325 mi) from South America and 2,250 km (1,395 mi) from the nearest inhabited island. When the first European explorers reached the island (today called Rapa Nui) in 1722, they found a barren landscape populated by fewer than 2,000 people, who lived in caves and eked out a marginal existence from a few meager crops. However, explorers also noted that the desolate island featured hundreds of gigantic statues of carved stone, evidence that a sophisticated civilization had once inhabited the island.

Historians and anthropologists long wondered how people without wheels or ropes, on an island without trees, could have moved statues 10 m (33 ft) high weighing 90 metric tons (99 tons) as far as 10 km (6.2 mi) from the quarries where they were chiseled to the coastal sites where they were erected. The explanation, scientists discovered, was that the island did not always lack trees.

Indeed, scientific research tells us that the island had once been lushly forested and had supported a prosperous society of 6,000 to 30,000 people. Tragically, this once-flourishing civilization overused its resources and cut down all its trees, destroying itself in a downward spiral of starvation and conflict. Today Easter Island stands as a parable and a warning for what can happen when a population consumes too much of the limited resources that support it.

To solve the mystery of Easter Island's past, scientists have used various methods. Some, such as British scientist John Flenley, have excavated sediments from the bottom of the island's lakes, drilling cores deep into the mud and examining ancient grains of pollen preserved there.

Because pollen grains vary from one plant species to another, scientists can reconstruct, layer by layer, the history of vegetation in a region through time. By analyzing pollen grains under scanning electron microscopes, Flenley and other researchers found that when Polynesian people arrived (likely between A.D. 300 and 900), the island was covered with a species of palm tree related to the Chilean wine palm, a tall and thick-trunked tree.

Moreover, archaeologists located ancient palm nut casings in caves and crevices, and a geologist found carbon-lined channels in the soil that matched root channels typical of the Chilean wine palm. Scientists deciphering the island people's script on stone tablets discerned characters etched in the form of palm trees.

By studying pollen and the remains of wood from charcoal, scientists such as French archaeologist Catherine Orliac found that at least 21 other species of plants, many of them trees, had also been common but are now completely gone. The island had clearly supported a diverse forest. However, starting around A.D. 750, tree populations declined and ferns and grasses became more common, according to pollen analysis from one lake site. By A.D. 950, the trees were largely gone, and around A.D. 1400 overall pollen levels plummeted, indicating a dearth of vegetation.

The same sequence of events occurred two centuries later at the other two lake sites, which were higher and more remote from village areas. Researchers first hypothesized that the forest loss was due to climate change, but evidence instead supported the hypothesis that the people had gradually denuded their own island.

The trees provided fuelwood, building material for houses and canoes, fruit to eat, fiber for clothing—and, presumably, logs with which to move the stone statues. By

**environmental studies** to describe their academic umbrella. Whichever approach one takes, these fields reflect many diverse perspectives and sources of knowledge.

Just as an interdisciplinary approach to studying issues can help us better understand them, an integrated approach to addressing environmental problems can produce effective and lasting solutions. One example is the dramatic improvement in one aspect of air quality in the United States over the past few decades. Lead had long been added to gasoline to make cars run more smoothly, even though researchers knew that lead emissions from tailpipes could cause health problems, including brain damage and premature death. In 1970 air pollution was severe, and motor vehicles accounted for 78% of U.S. lead emissions. But over the following years, engineers, medical researchers, atmospheric scientists, and politicians all merged their knowledge and skills into a process that eventually resulted in a ban on leaded gasoline. By 1996 all gasoline sold in the United States was unleaded,

The haunting statues of Easter Island were erected by a sophisticated civilization that collapsed after depleting its resource base and devastating its island environment.

hiring groups of men to recreate the feat, several anthropologists have experimentally tested hypotheses about how the islanders moved their monoliths down from the quarries. The methods that have worked involve using numerous tree trunks as rollers or sleds, along with great quantities of rope. The only likely source of rope on the island would have been the fibrous inner bark of the hauhau tree, a species that today is near extinction.

With the trees gone, soil would have eroded away—a phenomenon confirmed by data from the lake bottoms, where large quantities of sediment accumulated. Faster runoff of rainwater would have meant less freshwater available for drinking. Runoff and erosion would have degraded the islanders' agricultural land, lowering yields of bananas, sugar cane, and sweet potatoes. Reduced agricultural production would have led to starvation and population decline.

Archaeological evidence supports a scenario of environmental degradation and civilization decline. Analysis of 6,500 bones by archaeologist David Steadman has shown that at least 6 species of land birds and 25 species of seabirds nested on Easter Island and served as a food source for the islanders. Today, no native land birds and only one seabird species are left.

Remains from charcoal fires aged by radiocarbon dating (• p. 92) show that besides crops and birds, early islanders feasted on the bounty of the sea, including porpoises, fish, sharks, turtles, octopus, and shellfish. But analysis of islanders' diets in the later years indicated that the people consumed little seafood. With the trees gone, the islanders could no longer build the great double canoes their proud Polynesian ancestors had used for centuries to fish and travel among islands.

As resources declined, the islanders' main domesticated food animal, the chicken, became more valuable. Archaeologists found that later islanders kept their chickens in stone fortresses with entrances designed to prevent theft. The once prosperous and peaceful civilization fell into clan warfare, as revealed by unearthed weapons, skeletons, and skulls with head wounds.

Is the story of Easter Island as unique and isolated as the island itself, or does it hold lessons for our world today? Like the Easter Islanders, we are all stranded together on an island with limited resources. Earth may be vastly larger and richer in resources than Easter Island, but Earth's human population is also much greater.

The Easter Islanders must have seen that they were depleting their resources, but it seems that they could not stop. Whether we can learn from the history of Easter Island and act more wisely to conserve the resources on our island, Earth, is entirely up to us.

and the nation's largest source of atmospheric lead emissions had been completely eliminated.

## People vary in their perception of environmental problems

Environmental science arose in the latter half of the 20th century as people sought to better understand environmental problems, their origins, and their solutions. However, the perception of what constitutes a problem may vary from one person or group of people to another, or from one context or situation to another. A person's age, gender, class, race, nationality, employment, and educational background can all affect whether he or she considers a given environmental condition or change to be a "problem."

For instance, today's industrial societies are more likely to view the spraying of the pesticide DDT as a problem than those societies viewed it in the 1950s, because today more is known about the health risks of pesticides (**Figure 1.6**). At the same time, a person living today in a malaria-infested village in Africa or India may welcome the use of DDT if it kills mosquitoes that transmit malaria, because he or she may view malaria as a more immediate health threat. Thus an African and an American who have each knowledgeably assessed the pros and cons may, because of differences in their circumstances, differ in their judgment of DDT's severity as an environmental problem.

Different types of people may also vary in their awareness of problems. For example, in many cultures women are responsible for collecting water and fuelwood. As a result, they are often the first to perceive environmental degradation that affects these resources, whereas men in the same area simply might not "see" the problem. As another example, in most societies information about environmental health risks tends to reach wealthy people more readily than poor people. Thus, who you are, where you live, and what you do can have a huge effect on how you perceive your environment and on how you react to environmental change. In Chapter 2, we will examine the diversity of human values and philosophies and consider their effects on how we define environmental problems.

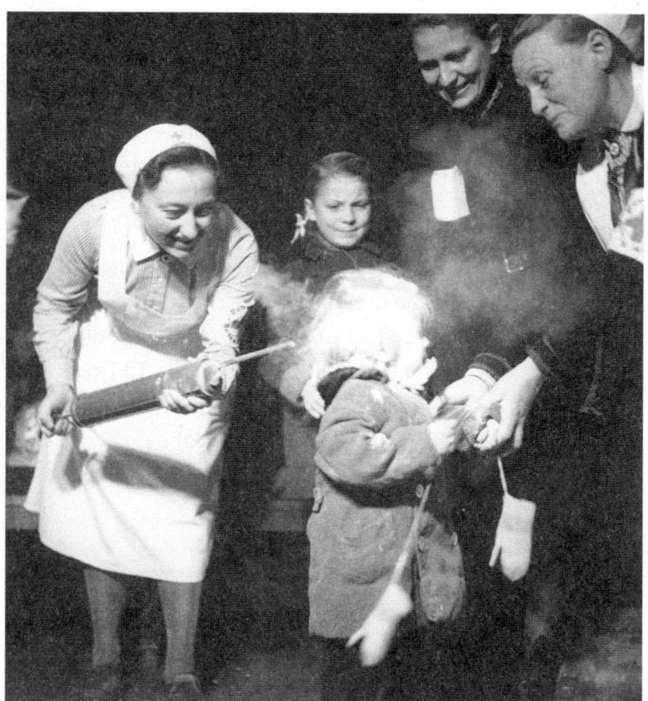

FIGURE 1.6 How a person or a society defines an environmental problem can vary with time and circumstance. In Germany in 1945, health hazards of the pesticide DDT were not yet known, so children were doused with the chemical to treat head lice. Today, knowing of its toxicity to people and wildlife, many developed nations have banned DDT. However, in developing countries where malaria is a threat, DDT is welcomed to combat mosquitoes that transmit the disease.

FIGURE 1.7 Environmental scientists play very different roles from environmental activists such as the protestors shown here. Although many environmental scientists search for workable solutions to environmental problems, they generally try hard to keep their research rigorously objective and to remain free from advocacy.

### Environmental science is not the same as environmentalism

Although many environmental scientists are interested in solving problems, it would be incorrect to confuse environmental science with environmentalism, or environmental activism. They are *not* the same.

Environmental science is the pursuit of knowledge about the workings of the environment and our interactions with it. **Environmentalism** is a social movement dedicated to protecting the natural world—and, by extension, people—from undesirable changes brought about by human actions (**Figure 1.7**). Although environmental scientists may study many of the same issues environmentalists care about, as scientists they try to maintain an objective approach in their work. Attempting to remain free from personal or ideological bias, and open to whatever conclusions the data demand, is a hallmark of the effective scientist.

## The Nature of Science

Modern scientists describe **science** as a systematic process for learning about the world and testing our understanding of it. The term *science* is also commonly used to refer to the accumulated body of knowledge that arises from this dynamic process of observation, testing, and discovery.

Knowledge gained from science can be applied to address societal problems and needs. Among the applications of science are its use in developing technology and in informing policy and management decisions (**Figure 1.8**). Many scientists are motivated by the potential for developing such useful applications, whereas others are motivated simply by a desire to understand how the world works.

**(a) Prescribed burning**

**(b) Methanol-powered fuel-cell car**

**FIGURE 1.8** Scientific knowledge can be applied in policy and management decisions and in technology. Prescribed burning, shown here in the Ouachita National Forest, Arkansas (**a**), is a management practice to restore healthy forests, and it is informed by scientific research into forest ecology. Energy-efficient automobiles, like this methanol-powered fuel-cell car from Daimler–Chrysler (**b**), are technological advances made possible by materials and energy research.

Why does science matter? The late astronomer and author Carl Sagan wrote the following in his 1995 treatise, *The Demon Haunted World: Science as a Candle in the Dark:*

> We've arranged a global civilization in which the most crucial elements—transportation, communications, and all other industries; agriculture, medicine, education, entertainment, protecting the environment; and even the key democratic institution of voting—profoundly depend on science and technology. We have also arranged things so that almost no one understands science and technology. This is a prescription for disaster. We might get away with it for a while, but sooner or later this combustible mixture of ignorance and power is going to blow up in our faces. . . . Science is an attempt, largely successful, to understand the world, to get a grip on things, to get hold of ourselves, to steer a safe course.

Sagan and many other thinkers before and since have argued that science is essential if we hope to sort fact from fiction and develop solutions to the problems—environmental and otherwise—that we face.

## Scientists test ideas by critically examining evidence

The effective scientist thinks critically and does not simply accept the conventional wisdom that he or she hears from others. The scientist becomes excited by novel ideas but is a skeptic and judges ideas by the strength of evidence that supports them. In these ways, scientists are good role models for the rest of us, because every one of us can benefit from learning to think critically in our everyday lives.

Scientists examine ideas about how the world works by designing tests to determine whether these ideas are supported by evidence. Ideas can be refuted by evidence but can never be absolutely proven. If a particular statement or explanation is testable and resists repeated attempts to disprove it, scientists are likely to accept it as a true explanation. Scientific inquiry thus consists of an incremental approach to the truth.

## The scientific method is the key element of science

Scientists generally follow a process called the **scientific method.** A technique for testing ideas with observations, it involves several assumptions and a series of interrelated steps. There is nothing mysterious about the scientific method; it is merely a formalized version of the procedure any of us might naturally take, using common sense, to resolve a question.

Because science is an active, creative, imaginative process, an innovative scientist may find good reason to stray from the traditional scientific method when a particular situation demands it. Moreover, scientists in different fields approach their work differently because they deal with dissimilar types of information. A natural scientist, such as a chemist, will conduct research quite differently from a social scientist, such as a sociologist. Because environmental science includes both natural and social sciences, in our discussion here we use the term *science* in its broad sense, to include both. Despite their many differences, scientists of all persuasions broadly agree on fundamental elements of the process of scientific inquiry.

The scientific method relies on the following assumptions:

- The universe functions in accordance with fixed natural laws that do not change from time to time or from place to place.
- All events arise from some cause or causes and, in turn, cause other events.
- We can use our senses and reasoning abilities to detect and describe natural laws that underlie the cause-and-effect relationships we observe in nature.

As practiced by individual researchers or research teams, the scientific method (**Figure 1.9**) typically consists of the steps outlined below.

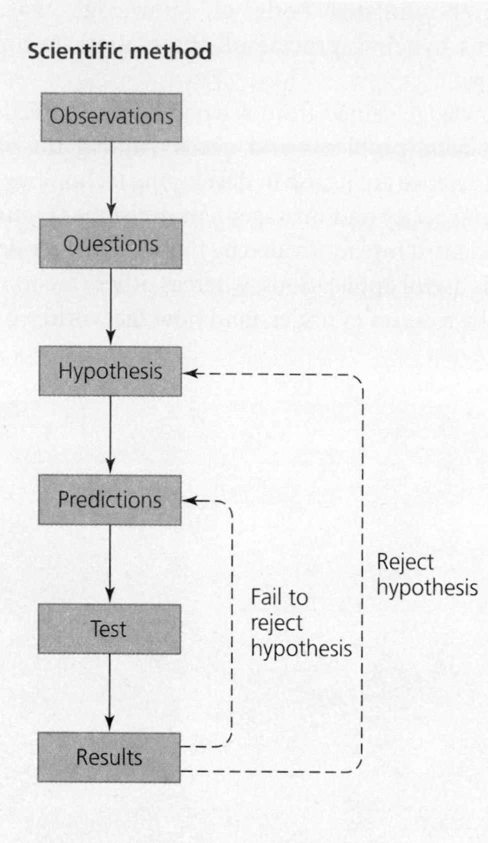


FIGURE 1.9 The scientific method is the observation-based hypothesis-testing approach that scientists use to learn how the world works. This diagram is a simplified generalization that, although useful for instructive purposes, cannot convey the true dynamic and creative nature of science. Moreover, researchers from different disciplines may pursue their work in ways that vary legitimately from this model.

**Make observations** Advances in science typically begin with the observation of some phenomenon that the scientist wishes to explain. Observations set the scientific method in motion and also function throughout the process.

**Ask questions** Curiosity is a fundamental human characteristic; just observe the explorations of a young child in a new environment. Babies want to touch, taste, watch, and listen to anything that catches their attention, and as soon as they can speak, they begin asking questions. Scientists, in this respect, are kids at heart. Why are certain plants or animals less common today than they once were? Why are storms becoming more severe and flooding more frequent? What is causing excessive growth of algae in local ponds? When pesticides poison fish or frogs, does that indicate that people may also be affected? All of these are questions environmental scientists have asked and attempted to answer.

**Develop a hypothesis** Scientists attempt to answer their questions by devising explanations that they can test. A **hypothesis** is a statement that attempts to explain a phenomenon or answer a scientific question. For example, a scientist investigating the question of why algae are growing excessively in local ponds might observe chemical fertilizers being applied on farm fields nearby. The scientist might then state a hypothesis as follows: "Agricultural fertilizers running into ponds cause the amount of algae in the ponds to increase."

**Make predictions** The scientist next uses the hypothesis to generate **predictions**, which are specific statements that can be directly and unequivocally tested. In our algae example, a researcher might predict: "If agricultural fertilizers are added to a pond, the quantity of algae in the pond will increase."

**Test the predictions** Scientists test predictions one at a time by gathering evidence that could potentially refute the prediction and thus refute the hypothesis. The strongest form of evidence comes from experimentation. An **experiment** is an activity designed to test the validity of a prediction or a hypothesis. It involves manipulating **variables**, or conditions that can change.

For example, a scientist could test the prediction linking algal growth to fertilizer by selecting two identical ponds and adding fertilizer to one while leaving the other in its natural state. In this example, fertilizer input is an **independent variable**, a variable the scientist manipulates, whereas the quantity of algae that results is the **dependent variable**, one that depends on the fertilizer input. If the two ponds are identical except for a single independent variable (fertilizer input), then any differences that arise between the ponds can be attributed to that variable. Such an experiment is known as a **controlled experiment** because the scientist controls for the effects of all variables except the one whose effect he or she is testing. In our example, the pond left unfertilized serves as a **control**, an unmanipulated point of comparison for the manipulated **treatment** pond.

Whenever possible, it is best to *replicate* one's experiment; that is, to stage multiple tests of the same comparison of control and treatment. Our scientist could perform a replicated experiment on, say, 10 pairs of ponds, adding fertilizer to one of each pair.

Experiments can establish causal relationships, showing that changes in an independent variable cause changes in a dependent variable. However, experiments are not the only way to test a hypothesis. Sometimes a hypothesis can be convincingly addressed through **correlation**, or statistical relationships among variables. Let's suppose our scientist surveys 50 ponds, 20 of which happen to be fed by fertilizer runoff from nearby farm fields and 30 of which are not. Let's say he or she finds seven times more algal growth in the fertilized ponds than in the unfertilized ponds. The scientist would conclude that algal growth is correlated with fertilizer input; that is, that one tends to increase along with the other. Although this type of evidence is weaker than the causal demonstration that controlled experiments can provide, sometimes it is the best approach or the only feasible one. For example, in studying the effects of global climate change (Chapter 18), we could hardly run an experiment adding carbon dioxide to 10 treatment planets and comparing the result to 10 control planets.

**Analyze and interpret results** Scientists record **data**, or information, from their studies. They particularly value *quantitative* data (information expressed using numbers), because numbers provide precision and are easy to compare. The scientist running the fertilization experiment, for instance, might quantify the area of water surface covered by algae in each pond or might measure the dry weight of algae in a certain volume of water taken from each.

However, even with the precision that numbers provide, a scientist's results may not be clear-cut. Data from treatments and controls may vary only slightly, or different replicates may yield different results. The scientist must therefore analyze the data using statistical tests. With these mathematical methods, scientists can determine objectively and precisely the strength and reliability of patterns they find.

Some research, especially in the social sciences, involves data that is *qualitative*, or not expressible in terms of numbers. Research involving historical texts, personal interviews, surveys, detailed examination of case studies, or descriptive observation of behavior can include qualitative data on which statistical analyses may not be possible. Such studies are still scientific in the broad sense, because their data can be interpreted systematically using other accepted methods of analysis.

If experiments refute a hypothesis, the scientist will reject it and may develop a new hypothesis to replace it. If experiments fail to refute a hypothesis, this outcome lends support to the hypothesis but does not *prove* it is correct. The scientist may choose to generate new predictions to test the hypothesis in a different way and further assess its likelihood of being true. Thus, the scientific method loops back on itself, often giving rise to repeated rounds of hypothesis-revision and new experimentation (see Figure 1.9).

If repeated tests fail to reject a particular hypothesis, evidence in favor of it accumulates, and the researcher may eventually conclude that the idea is well supported. Ideally, the scientist would also want to test different potential explanations. For instance, our scientist might propose an additional hypothesis that algae increase in fertilized ponds because numbers of fish or invertebrate animals that eat algae decrease. It is possible, of course, that both hypotheses could be correct and that each may explain some portion of the initial observation that local ponds were experiencing algal blooms.

## We can test hypotheses in different ways

An experiment in which the researcher actively chooses and manipulates the independent variable is known as a **manipulative experiment** (**Figure 1.10a**). A manipulative experiment provides the strongest type of evidence a scientist can obtain, because it can reveal causal relationships. In practice, however, some modes of scientific inquiry are more amenable to manipulative experimentation than others. Physics and chemistry tend to involve manipulative experiments, but many other fields deal with entities less easily manipulated than physical forces and chemical reagents. This is true of *historical sciences* such as cosmology, which deals with the history of the universe, and paleontology, which explores the history of past life. It is difficult to manipulate experimentally a star thousands of light years away or the fossil tooth from a mastodon. Moreover, many of the most interesting questions in these fields center on the causes and consequences of particular historical events, rather than the behavior of general constants.

Disciplines that do not quite fit the so-called physics model of science sometimes rely on **natural experiments** rather than manipulative ones (**Figure 1.10b**). For instance, an evolutionary biologist might want to test whether animal species isolated on oceanic islands tend

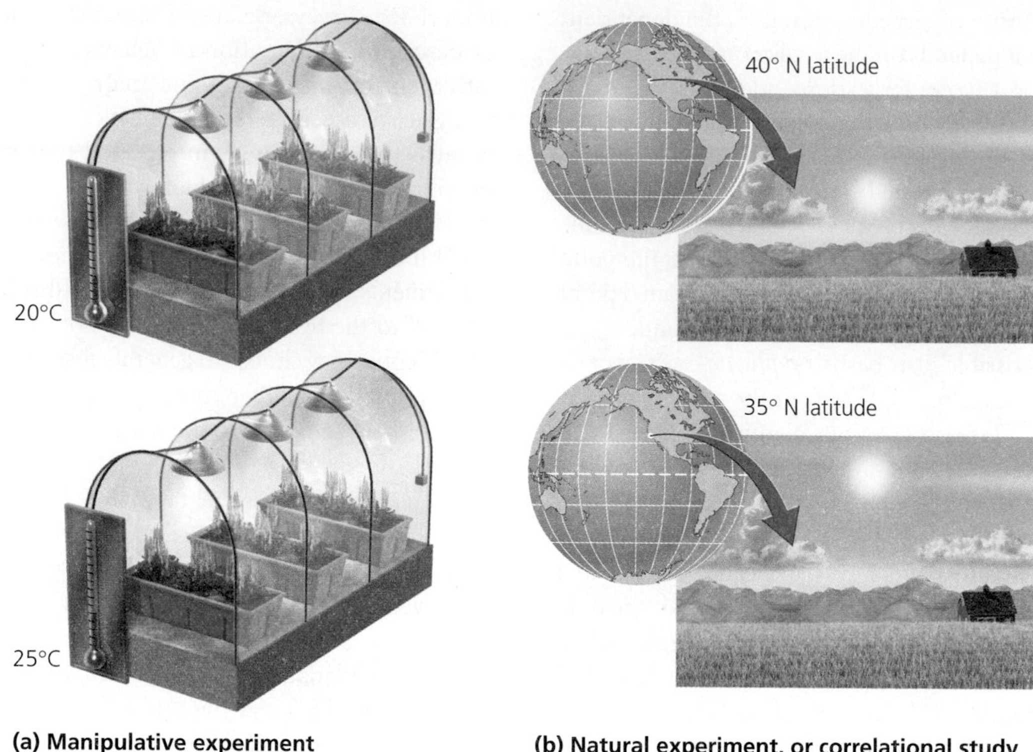


**(a) Manipulative experiment**

**(b) Natural experiment, or correlational study**

FIGURE 1.10 A researcher wishing to test how temperature affects the growth of wheat might run a manipulative experiment (**a**) in which wheat is grown in two identical greenhouses, one kept at 20°C (68°F) and the other kept at 25°C (77°F). Alternatively, the researcher might run a "natural experiment" (**b**) in which he or she compares the growth of wheat in two fields at different latitudes, a cool northerly location and a warm southerly one. Because it would be difficult to hold all variables besides temperature constant in the natural experiment, the researcher might want to collect data on a number of northern and southern fields and correlate data on temperature and wheat growth using statistical methods.

to evolve large body size over time. The biologist cannot run a manipulative experiment by placing animals on islands and continents and waiting long enough for evolution to do its work. However, this is exactly what nature has already done. The biologist might test the idea by comparing pairs of closely related species, in which one of each pair lives on an island and the other on a continental mainland. The experiment has in essence been conducted naturally, and it is up to the scientist to interpret the results.

Both manipulative and natural experiments are used in ecology, one of the disciplines most central to environmental science. The science of **ecology** (Chapters 5–7) deals with the distribution and abundance of organisms (living things), the interactions among them, and the interactions between organisms and their nonliving environments. When possible, ecologists run manipulative experiments. An ecologist wanting to test the importance of a certain insect in pollinating the flowers of a given crop plant might, for example, fit some flowers with a device to keep the insects out while leaving other flowers accessible and later measure the fruit output of each group. In contrast, ecological questions that involve large spatial scales or long time scales may instead require natural experiments.

The social sciences generally involve less experimentation than the natural sciences, depending more on careful observation and interpretation of patterns in data. A sociologist studying how people from different cultures conceive the notion of wilderness, for example, might analyze responses to survey questions, looking for similarities and differences among respondents.

Descriptive observational studies and natural experiments can show correlation between variables, but they cannot demonstrate that one variable *causes* change in another, as manipulative experiments can. Not all variables are controlled for in a natural experiment, so a single result could give rise to several interpretations. However, good correlative studies can make for very convincing science, and they preserve the real-world complexity that manipulative experiments often sacrifice.

Because manipulations are difficult at large scales, some of the biggest questions in environmental science tend to be addressed with correlative data. The large scale and complexity of many questions in environmental science also mean that few studies come up with neat and clean results. As such, scientists are not always able to give policymakers and society black-and-white answers to questions.

## The scientific process does not stop with the scientific method

Scientific work takes place within the context of a community of peers. To have any impact, a researcher's work must be published and made accessible to this community. Thus, the scientific method is embedded within a larger process that takes place at the level of the scientific community as a whole (**Figure 1.11**).

**Peer review** When a researcher's work is done and the results analyzed, he or she writes up the findings and submits them to a journal for publication. Several other scientists specializing in the topic of the paper examine the manuscript, provide comments and criticism (generally anonymously), and judge whether the work merits publication in the journal. This procedure, known as **peer review**, is an essential part of the scientific process.

Peer review is a valuable guard against faulty science contaminating the literature on which all scientists rely. However, because scientists are human and may have their own personal biases and agendas, politics can sometimes creep into the review process. Fortunately, just as individual scientists strive their best to remain objective in conducting their research, the scientific community does its best to ensure fair review of all work. Winston Churchill once called democracy the worst form of government, except for all the others that had been tried. The same might be said about peer review; it is an imperfect system, yet no one has come up with a better one.

**Conference presentations** Scientists frequently present their work at professional conferences, where they interact with colleagues and receive informal comments on their research. Such feedback from colleagues can help improve the quality of a scientist's work before it is submitted for publication.

**Grants and funding** Research scientists spend large portions of their time writing grant applications requesting money to fund their research from private foundations or from government agencies such as the National Science Foundation. Grant applications undergo peer review just as scientific papers do, and competition for funding is often intense.

Scientists' reliance on funding sources can sometimes lead to potential conflicts of interest. A scientist who obtains data showing his or her funding source in an unfavorable light may be reluctant to publish the results for fear of losing funding—or worse yet, may be tempted to doctor the results. This situation can arise, for instance, when an industry funds research to test its products for

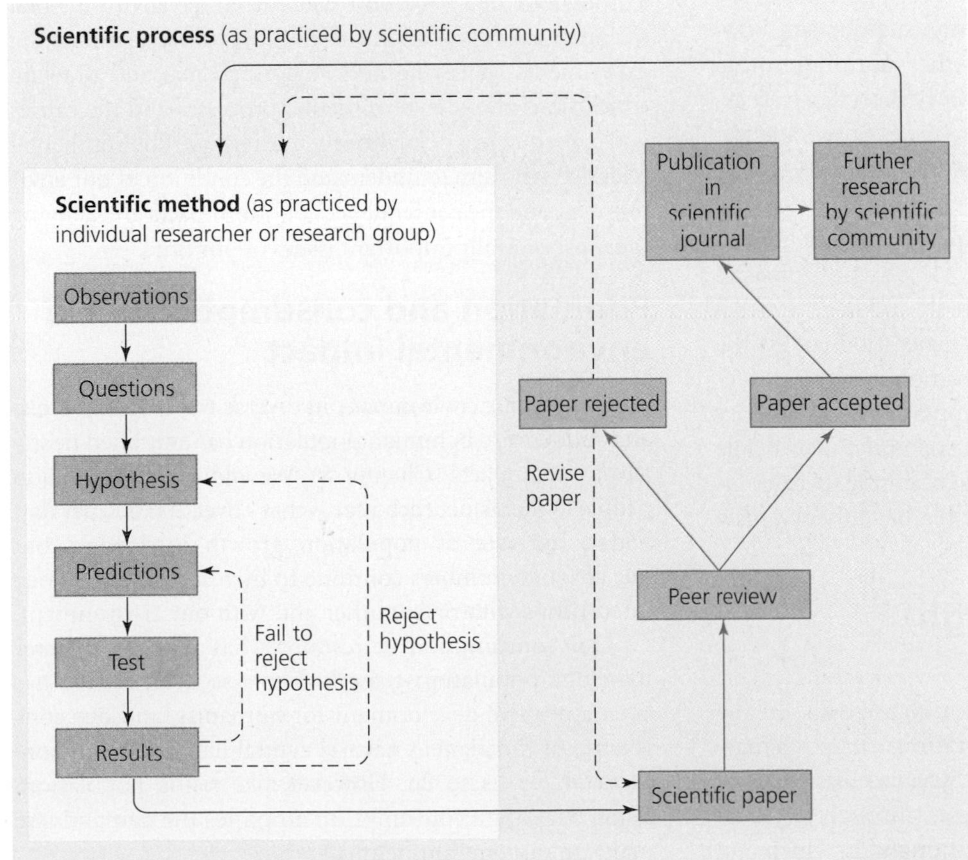


**FIGURE 1.11** The scientific method (inner yellow box) followed by individual researchers or research teams exists within the context of the overall process of science at the level of the scientific community (outer green box). This process includes peer review and publication of research, acquisition of funding, and the elaboration of theory through the cumulative work of many researchers.

safety or environmental impact. Most scientists do not succumb to these temptations, but some funding sources have been known to pressure their scientists for certain results. This is why as a student or informed citizen, when critically assessing a scientific study, you should always try to find out where the researchers obtained their funding.

**Repeatability** Sound science is based on repeatability rather than one-time occurrence. Even when a hypothesis appears to explain observed phenomena, scientists are inherently wary of accepting it. The careful scientist may test a hypothesis repeatedly in various ways before submitting the findings for publication. Following publication, other scientists may attempt to reproduce the results in their own experiments and analyses.

**Theories** If a hypothesis survives repeated testing by numerous research teams and continues to predict experimental outcomes and observations accurately, it may potentially be incorporated into a theory. A **theory** is a widely accepted, well-tested explanation of one or more cause-and-effect relationships that has been extensively validated by a great amount of research. Whereas a hypothesis is a simple explanatory statement that may be refuted by a single experiment, a theory consolidates many related hypotheses that have been tested and have not been refuted.

Note that scientific use of the word *theory* differs from popular usage of the word. In everyday language when we say something is "just a theory," we are suggesting it is a speculative idea without much substance. Scientists, however, mean just the opposite when they use the term; to them, a theory is a conceptual framework that effectively explains a phenomenon and has undergone extensive and rigorous testing, such that confidence in it is extremely strong.

For example, Darwin's theory of evolution by natural selection (• pp. 113–117) has been supported and elaborated by many thousands of studies over 150 years of intensive research. Research has shown repeatedly and in great detail how plants and animals change over generations, or evolve, to express characteristics that best promote survival and reproduction. Because of its strong support and explanatory power, evolutionary theory is the central unifying principle of modern biology. Other prominent scientific theories include atomic theory, cell theory, the big bang theory, plate tectonics, general relativity, and others.

### Science may go through "paradigm shifts"

As the scientific community accumulates data in any given area of research, interpretations may change. Thomas Kuhn's 1962 book *The Structure of Scientific Revolutions* argued that science goes through periodic revolutions: dramatic upheavals in thought, in which one scientific **paradigm**, or dominant view, is abandoned for another.

For example, before the 16th century, scientists believed that Earth was at the center of the universe. Their data on the movements of planets fit that concept quite well, yet the idea eventually was disproved by Nicolaus Copernicus, who showed that placing the sun at the center of the solar system explained the planetary data even better.

Another paradigm shift occurred in the 1960s, when geologists accepted plate tectonics (• pp. 199–201). By this time, evidence for the movement of continents and the action of tectonic plates had accumulated and become overwhelmingly convincing. Such paradigm shifts demonstrate the strength and vitality of science, showing it to be a process that refines and improves itself through time.

Understanding how science works is vital to assessing how scientific ideas and interpretations change through time as new information accrues. This process is especially relevant in environmental science, a young field that is changing rapidly as we learn vast amounts of new information, as human impacts on the planet multiply, and as we gather lessons from the consequences of our actions.

## Sustainability and the Future of Our World

Throughout this book you will encounter environmental scientists asking questions, testing hypotheses, conducting experiments, gathering and analyzing data, and drawing conclusions about environmental processes and the causes and consequences of environmental change. Environmental scientists who aim to understand the condition of our environment and the consequences of our impacts are studying the most centrally important issues of our time.

### Population and consumption drive environmental impact

We modify our environment in diverse ways, but the steep and sudden rise in human population has amplified nearly all of our impacts (Chapter 8). We add about 78 million people to the planet each year—that's over 200,000 per day. Today, the rate of population growth is slowing, but our absolute numbers continue to increase and shape our interactions with one another and with our environment.

Our consumption of resources has risen even faster than our population growth. The rise in affluence has been a positive development for humanity, and our conversion of the planet's natural capital has made life more pleasant for us so far. However, like rising population, rising per capita consumption amplifies the demands we make on our environment.

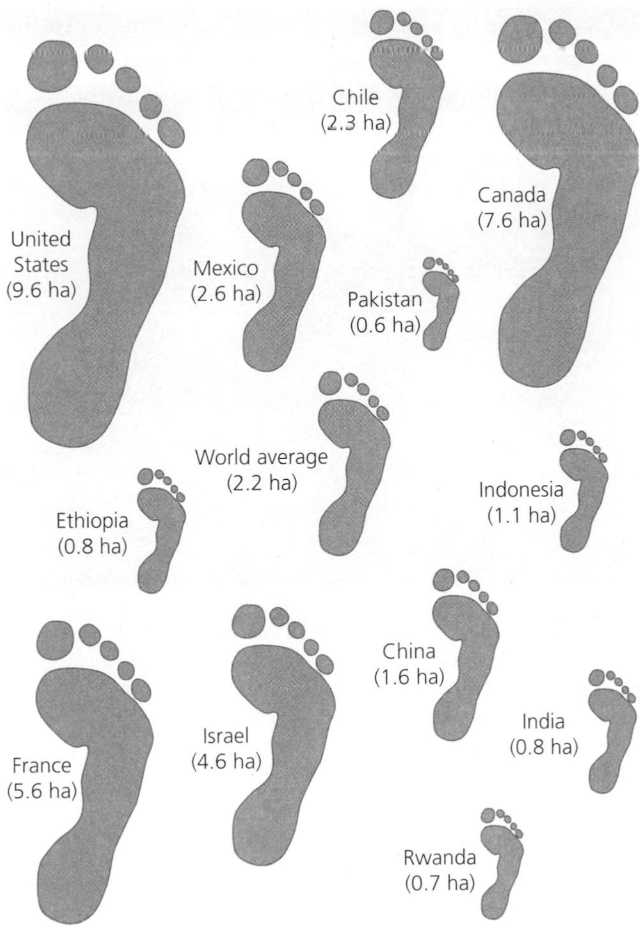


**FIGURE 1.12** The citizens of some nations have much larger ecological footprints than the citizens of others. Shown here are ecological footprints for average citizens of several developed and developing nations, along with the world's average per capita footprint of 2.2 hectares. One hectare (ha) = 2.47 acres. Data are for 2003, from Global Footprint Network, 2006.

Moreover, the world's citizens have not benefited equally from the overall rise in affluence. Today the 20 wealthiest nations boast 40 times the income of the 20 poorest nations—twice the gap that existed four decades ago. The ecological footprint of the average citizen of a developed nation such as the United States is considerably larger than that of the average resident of a developing country (**Figure 1.12**). Within the United States, the richest fifth of people claim nearly half the income, whereas the poorest fifth receive only 5%.

### Weighing the Issues | Ecological Footprints

What do you think accounts for the variation in sizes of ecological footprints among societies? Do you think that nations with larger footprints have an ethical obligation to reduce their environmental impact, so as to leave more resources available for nations with smaller footprints?

## We face challenges in agriculture, pollution, and biodiversity

The dramatic growth in human population and consumption is due in part to our success in expanding and intensifying the production of food (Chapters 9 and 10). Since the agricultural revolution, progressively more powerful technologies have enabled us to grow increasingly more food per unit of land. These advances in agriculture must be counted as one of humanity's great achievements, but they have come at some cost. We have converted nearly half the planet's land surface for agriculture, and our extensive use of chemical fertilizers and pesticides poisons organisms and alters natural systems. Erosion, climate change, and poorly managed irrigation are destroying 5–7 million hectares (ha; 12.5–17.5 million acres) of productive cropland each year.

Meanwhile, pollution from our farms, industries, households, and individual actions dirties our land, water, and air (**Figure 1.13**). Outdoor air pollution, indoor air pollution, and water pollution contribute to the deaths of millions of people each year (Chapters 15–17). Environmental toxicologists are chronicling the impacts on people and wildlife of the many synthetic chemicals and other pollutants we emit into the environment (Chapter 14).

Perhaps our most pressing pollution challenge is to address the looming specter of global climate change (Chapter 18). Scientists have firmly concluded that human activity is altering the composition of the atmosphere and that these changes are affecting Earth's climate. Since the start of the industrial revolution, our combustion of fossil fuels and our deforestation of the landscape have boosted atmospheric carbon dioxide concentrations by 37%, to a level not present in at least 650,000 years. Carbon dioxide and several other gases absorb heat and

**FIGURE 1.13** Outdoor and indoor air pollution contribute to millions of premature deaths each year, and environmental scientists and policymakers are working to reduce this problem in a variety of ways.

warm Earth's surface, which in turn causes glacial melting, sea-level rise, impacts on wildlife and crops, and increased episodes of destructive weather.

The combined impact of human actions such as climate change, overharvesting, pollution, the introduction of nonnative species, and particularly habitat alteration, has driven many species toward extinction (Chapter 11). Today Earth's biological diversity, or **biodiversity** (the cumulative number and diversity of living things), is declining dramatically (**Figure 1.14**). Many biologists say we are already at the outset of a mass extinction event (• pp. 121–123, 301–303) comparable only to five others documented in all of Earth's history. Biologist Edward O. Wilson has warned that the loss of biodiversity is our most serious and threatening environmental dilemma, because extinction is irreversible; once a species has become extinct, it is lost forever.

FIGURE 1.14 Human activities are pushing many organisms, including the panda, toward extinction. Efforts to save endangered species and reduce the loss of biodiversity include many approaches, but all require that adequate areas of appropriate habitat be preserved in the wild.

The most comprehensive scientific assessment of the present condition of the world's ecological systems and their ability to continue supporting our civilization was completed in 2005. In this year, over 2,000 of the world's leading environmental scientists from nearly 100 nations completed the **Millennium Ecosystem Assessment**. The four main findings of this exhaustive project are summarized in Table 1.1. The Millennium Ecosystem Assessment makes clear that our degradation of the world's environmental systems is having negative impacts on all of us but that with care and diligence we can still turn many of these trends around.

## Our energy choices will influence our future immensely

Our reliance on fossil fuels to power our civilization has intensified virtually every impact we have on the environment, from habitat alteration to air pollution to climate change. Fossil fuels have also brought us the material affluence we enjoy. By taking advantage of the richly concentrated energy in coal, oil, and natural gas, we were able to power the machinery of the industrial revolution, produce the chemicals that boosted agricultural yields, run the vehicles and transportation networks of our mobile society, and manufacture and distribute our countless consumer products. It is little exaggeration to say that the lives we live today are a result of the availability of fossil fuels (Chapter 19).

However, in extracting fossil fuels, we are splurging on a one-time bonanza. Scientists calculate that we have depleted half the world's oil supplies and that we are in for a rude awakening very soon, once supply begins to decline while demand continues to rise (• pp. 554–557). We are also approaching peak production of natural gas, and coal is also nonrenewable and in limited supply. How we handle the imminent crises of fossil fuel

**TABLE 1.1 Main Findings of the Millennium Ecosystem Assessment**

- Over the past 50 years, humans have altered ecosystems more rapidly and extensively than in any comparable period of time in human history, largely to meet growing demands for food, fresh water, timber, fiber, and fuel. These changes have caused a substantial and largely irreversible loss in the diversity of life on Earth.
- The changes to ecosystems have contributed to substantial net gains in human well-being and economic development, but these gains have been achieved at growing costs. These costs include the degradation of ecosystems and the services they provide for us, and the exacerbation of poverty for some groups of people.
- This degradation could grow significantly worse during the first half of this century.
- We can reverse the degradation of ecosystems while meeting increasing demands for their services, but doing so will require that we significantly modify many policies, institutions, and practices.

Adapted from *Millennium ecosystem assessment, synthesis report.* 2005.

depletion will largely determine the nature of our lives in the 21st century.

## Fortunately, sustainable solutions abound

We cannot, of course, live without exerting some impact on Earth's systems. Humanity's challenge is to develop solutions that further our quality of life while protecting and restoring the environment that supports us. Fortunately, there are many workable solutions at hand.

In response to agricultural problems, scientists and others have developed and promoted soil conservation, high-efficiency irrigation, and organic agriculture. Technological advances and legislation have greatly reduced the pollution emitted by industry and automobiles in wealthier countries. Although the U.S. government has resisted international efforts to rein in pollutants to halt climate change, American scientists have been at the forefront of climate change science, and other nations are beginning to address the problem, as are some U.S. states and municipalities.

Amid ample reasons for concern about the state of global biodiversity, advances in conservation biology (• pp. 315–323) are enabling scientists and policymakers to work together to protect habitat, slow extinction, and safeguard endangered species. Recycling is helping relieve our waste disposal problems (Chapter 22), and alternative renewable energy sources (Chapters 20 and 21) are being developed to take the place of fossil fuels (**Figure 1.15**). These are but a few of the many solutions we will explore in the course of this book.

**FIGURE 1.15** Our dependence on fossil fuels causes many environmental impacts, and poses risks for disaster once fossil fuel deposits decline. Many renewable energy sources exist, such as solar energy that can be collected with panels like these. Such alternative energy sources could be further developed for sustainable use now and in the future.

Environmental science helps clarify the causes and consequences of human impact on the environment, and it is up to all of us to help develop solutions. To help you think through these connections, we urge you to use the "Causes and Consequences" feature found in most chapters, beginning with the one on page 20.

## Are things getting better or worse?

Despite the myriad challenges we face, many people maintain that the general conditions of human life and the environment are in fact getting better, not worse. A recent proponent of this view, Danish statistician Bjorn Lomborg, wrote the following in his 2001 book *The Skeptical Environmentalist:*

> We are not running out of energy or natural resources. There will be more and more food per head of the world's population. Fewer and fewer people are starving. In 1900 we lived for an average of 30 years; today we live for 67. . . . The air and water around us are becoming less and less polluted. Mankind's lot has actually improved in terms of practically every measurable indicator.

Furthermore, some people maintain that we will find ways to make Earth's natural resources meet all of our needs indefinitely and that human ingenuity will see us through any difficulty. Such views are sometimes characterized as *Cornucopian.* In Greek mythology, *cornucopia*—literally "horn of plenty"—is the name for a magical goat's horn that overflowed with grain, fruit, and flowers. In contrast, people who predict doom and disaster for the world because of our impact upon it have been called *Cassandras,* after the mythical princess of Troy with the gift of prophecy whose dire predictions were not believed.

At least three questions are worth asking each time you are confronted with seemingly conflicting statements from Cassandras and Cornucopians. One question is whether the impacts being debated pertain only to humans or also to other organisms and natural systems. The second question is whether the proponents of each view are thinking in the short term or the long term. The third question is whether they are considering all costs and benefits relevant for the question at hand, or only some. As you proceed through this book and encounter countless contentious issues, consider how one's perception of them may be influenced by these three factors.

## Sustainability is a goal for the future

The primary challenge in our increasingly populated world is how to live within our planet's means, such that Earth and its resources can sustain us—and the rest of

## CAUSES AND CONSEQUENCES

People in different societies consume resources and use land and water at different rates, according to their level of wealth or poverty, the technologies available to them, their society's cultural norms, and other factors. Societies whose people have **large ecological footprints** exert greater per capita impacts on Earth's natural resources and ecological systems. Fortunately, there are many solutions and steps that people can take to address the causes and mitigate the consequences of large ecological footprints.

Write in two causes of large ecological footprints in the spaces provided. Then write in two consequences (impacts on the environment, human health, or quality of life) that result from large ecological footprints. Finally, offer two solutions corresponding to this issue and its consequences. One cause, one consequence, and one solution have been filled in for you, providing examples.

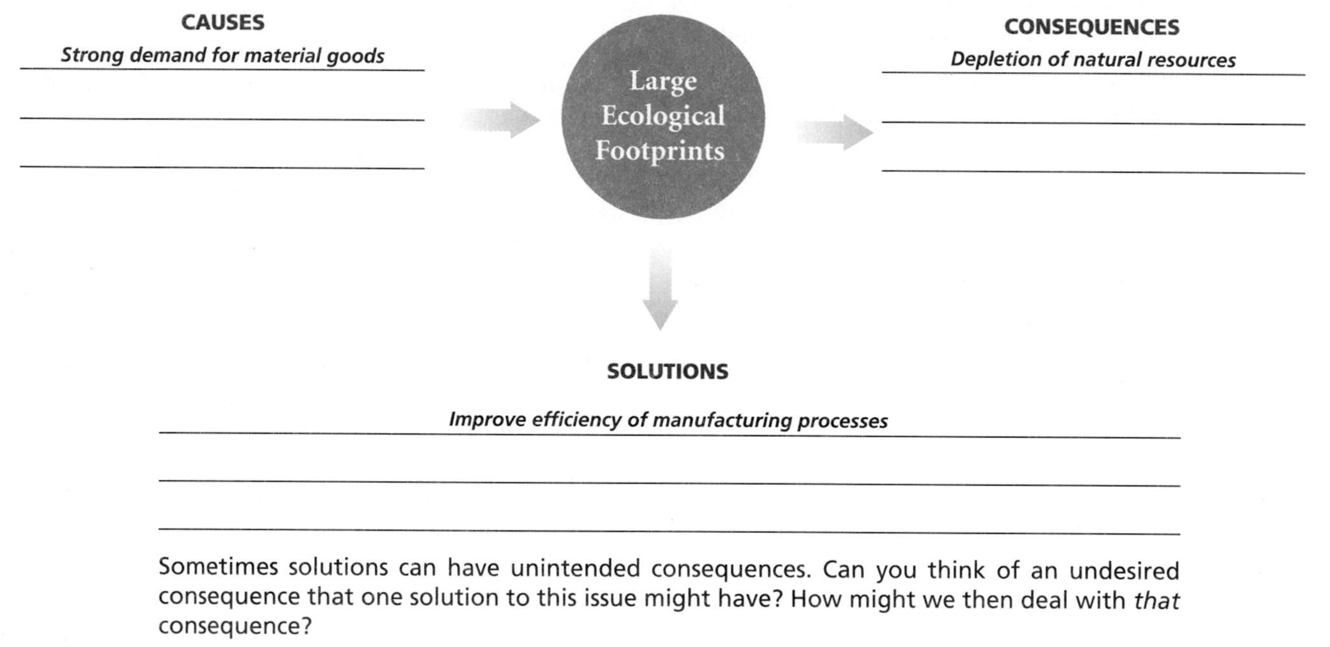


Sometimes solutions can have unintended consequences. Can you think of an undesired consequence that one solution to this issue might have? How might we then deal with *that* consequence?

Earth's life—for the foreseeable future. This is the challenge of **sustainability**, a guiding principle of modern environmental science. Sustainability means leaving our children and grandchildren a world as rich and full as the world we live in now. It means conserving Earth's natural capital, so that after we are gone our descendants will enjoy the use of resources as we have. It means developing solutions that work in the long term. Sustainability requires maintaining fully functioning ecological systems, because we cannot sustain human civilization without sustaining the natural systems that nourish it.

Sustainability is a concept you will encounter throughout this book, infused through issue after issue. Our final chapter (Chapter 23) summarizes and extends our discussion by taking a wide-ranging look at emerging sustainable solutions—on college and university campuses and in the world at large.

## Sustainable development involves environmental protection, economic welfare, and social equity

Environmental protection is often cast as being in opposition to the economic and social needs of human society, but environmental scientists have long recognized that our civilization cannot exist without an intact and functional natural environment. In recent years, people of all persuasions have increasingly realized the connection between environmental quality and human quality of life.

Moreover, we now recognize that often it is society's poorer people who suffer the most from environmental degradation. This realization has led advocates of environmental protection, advocates of economic development, and advocates of social justice to begin working together toward common goals. This cooperative approach has given rise to the modern drive for sustainable development.

Economists employ the term *development* to describe the use of natural resources for economic advancement (as opposed to simple subsistence, or survival). Construction of homes, schools, hospitals, power plants, factories, and transportation networks are all examples of development. **Sustainable development** is the use of renewable and nonrenewable resources in a manner that satisfies our current needs but does not compromise the future availability of resources. The United Nations (• p. 70) defines *sustainable development* as development that "meets the needs of the present without sacrificing the ability of future generations to meet their own needs." This definition is taken from the U.N.-sponsored Brundtland Commission (named after its chair, Norwegian prime minister Gro Harlem Brundtland), which published an influential 1987 report titled *Our Common Future.*

Prior to the Brundtland Report, most people aware of human impact on the environment might have thought "sustainable development" to be an oxymoron—a phrase that contradicts itself. Although development involves making purposeful changes intended to improve the quality of human life, environmental advocates have long pointed out that development often so degrades the natural environment that it threatens the very improvements for human life that were intended. Conversely, many people remain under the impression that protecting the environment is incompatible with serving people's economic needs.

Fortunately, sustainable development efforts by governments, businesses, industries, organizations, and individuals everywhere—from students on campus (• pp. 658–666) to international representatives at the United Nations (**Figure 1.16**)—are beginning to alter these perceptions. These efforts are generating sustainable solutions that meet environmental, economic, and social goals simultaneously, satisfying the so-called **triple bottom line**.

Sustainability and the triple bottom line demand that our current human population limit its environmental impact while also promoting economic well-being and social equity. These aims require us to make an ethical commitment to our fellow citizens and to future generations. They also require that we apply knowledge from the sciences to help us devise ways to limit our impact and maintain the functioning environmental systems on which we depend.

The question "Will we develop in a sustainable way?" may well be the single most important question in the world today. Environmental science holds the key to addressing this question. Because so much remains unstudied and undone, and because it is so central to our modern world, environmental science will remain an exciting frontier for you to explore as a student and as an informed citizen throughout your life.

FIGURE 1.16 South African President Thabo Mbeki hugs a boy who performed in the welcoming ceremony of the U.N.-sponsored World Summit on Sustainable Development, held in Johannesburg, South Africa, in 2002. The conference drew over 10,000 delegates from 200 nations to discuss and set sustainable development goals. Delegates launched hundreds of initiatives and committed millions of dollars at this conference, but most sustainability proponents said these efforts did not go nearly far enough.

## Conclusion

Finding effective ways of living peacefully, healthfully, and sustainably on our diverse and complex planet will require a thorough scientific understanding of both natural and social systems. Environmental science helps us understand our intricate relationship with the environment and informs our attempts to solve and prevent environmental problems.

It is important to keep in mind that identifying a problem is the first step in devising a solution to it. Many of the trends detailed in this book may cause us worry, but others give us reason to hope. One often-heard criticism of environmental science courses and textbooks is that too often they emphasize the negative. In this book we attempt to balance the discussion of environmental problems with a corresponding focus on solutions. Solving environmental problems can move us toward health, longevity, peace, and prosperity. Science in general, and environmental science in particular, can aid us in our efforts to develop balanced and workable solutions to the many environmental dilemmas we face today and to create a better world for ourselves and our children.

## REVIEWING OBJECTIVES

**You should now be able to:**

**Define the term *environment***

- Our environment consists of everything around us, including living and nonliving things. (p. 3)
- Humans are a part of the environment and are not separate from nature. (p. 3)

**Describe natural resources and explain their importance to human life**

- Resources from nature are essential to human life and civilization. (p. 3)
- Some resources are perpetually renewable, others are nonrenewable, and still others are renewable if we are careful not to overexploit them. (pp. 3–4)
- Malthus and the Ehrlichs pointed out risks of human population growth, and Hardin and Wackernagel and Rees pioneered important concepts in resource consumption. (pp. 5–6)

**Characterize the interdisciplinary nature of environmental science**

- Environmental science employs the approaches and insights of numerous disciplines from the natural sciences and the social sciences. (pp. 7–8)

**Understand the scientific method and how science operates**

- Science is a process of using observations to test ideas. (pp. 10–11)
- The scientific method consists of a series of steps, including making observations, formulating questions, stating a hypothesis, generating predictions, testing predictions, and analyzing results obtained from the tests. (pp. 11–13)
- The scientific method is not always followed strictly, and there are different ways to test questions scientifically. (pp. 13–14)
- Scientific research occurs within a larger process that includes peer review of work, journal publication, and interaction with colleagues. (pp. 15–16)
- Scientists use the word *theory* to describe a well-tested explanation validated by a great amount of research. Yet science still goes through paradigm shifts. This openness to change is what gives science its strength. (p. 16)

**Diagnose and illustrate some of the pressures on the global environment**

- Increasing human population and increasing per capita consumption exacerbate human impacts on the environment. (pp. 16–17)
- Human activities such as industrial agriculture and fossil fuel use are having diverse environmental impacts, including resource depletion, air and water pollution, climate change, habitat destruction, and biodiversity loss. (pp. 17–18)

**Evaluate the concepts of sustainability and sustainable development**

- Sustainability means living within the planet's means, such that Earth's resources can sustain us—and other species—for the foreseeable future. (pp. 19–21)
- Sustainable development means pursuing environmental, economic, and social goals in a coordinated way, and it is the most important pursuit in our society today. (pp. 20–21)

## TESTING YOUR COMPREHENSION

1. What do renewable resources and nonrenewable resources have in common? How are they different? Identify two renewable and two nonrenewable resources.
2. How did the agricultural revolution affect human population size? How did the industrial revolution affect human population size? Explain your answers.
3. What is "the tragedy of the commons"? Explain how the concept might apply to an unregulated industry that is a source of water pollution.
4. What is environmental science? Name several disciplines involved in environmental science.
5. What are the two meanings of *science?* Name three applications of science.
6. Describe the scientific method. What is the typical sequence of steps?
7. Explain the difference between a manipulative experiment and a natural experiment.
8. What needs to occur before a researcher's results are published? Why is this process important?
9. Give examples of three major environmental problems in the world today, along with their causes.
10. What is sustainable development, and why is it important? What is meant by the "triple bottom line"?

## SEEKING SOLUTIONS

1. Many resources are renewable if we use them in moderation but can become nonrenewable if we overexploit them. Order the following resources on a continuum of renewability (see Figure 1.1), from most renewable to least renewable: soils, timber, fresh water, food crops, and biodiversity. What factors influenced your decision? For each of these resources, what might constitute overexploitation, and what might constitute sustainable use?
2. Why do you think the Easter Islanders did not or could not stop themselves from stripping their island of all its trees? Do you see similarities between the history of the Easter Islanders and the modern history of our society? Why or why not?
3. What environmental problem do *you* feel most acutely yourself? Do you think there are people in the world who do not view your issue as an environmental problem? Who might they be, and why might they take a different view?
4. If the human population were to stabilize tomorrow and never surpass 7 billion people, would that solve our environmental problems? Which types of problems might be alleviated, and which might continue to become worse?
5. Consider the historic expansion of agriculture and our ability to feed increasing numbers of people, as described in this chapter. Now ask yourself, "Are things getting better or worse?" Ask this question from four points of view: (1) from the human perspective, (2) from the perspective of other organisms, (3) from a short-term perspective, and (4) from a long-term perspective. Do your answers to this question change? If so, how?
6. **THINK IT THROUGH** Imagine you have become head of a major funding agency that disburses funding to researchers pursuing work in environmental science. You must give your staff several priorities to determine what types of scientific research to fund. What environmental problems would you most like to see addressed with research? Describe the research you think would need to be completed so that workable solutions to these problems can be developed. Would more than science be needed to develop sustainable solutions?

## INTERPRETING GRAPHS AND DATA

Environmental scientists study phenomena that range in size from individual molecules (Chapter 4) to the entire Earth (Chapter 7) and that occur over time periods lasting from fractions of a second to billions of years. To simultaneously and meaningfully represent data covering so many orders of magnitude, scientists have devised a variety of mathematical and graphical techniques, such as exponential notation and logarithmic scales. Below are two graphical representations *of the same data*, representing the growth of a hypothetical population from an initial size of 10 individuals at a rate of increase of approximately 2.3% per generation. The graph in part (a) uses a conventional linear scale

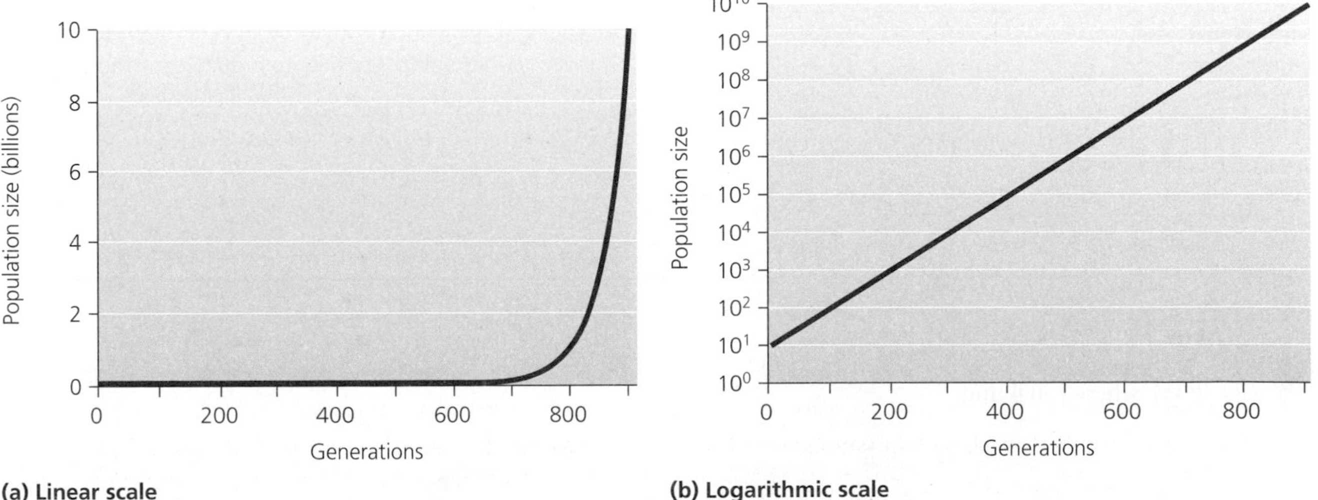


**(a) Linear scale** **(b) Logarithmic scale**

Hypothetical population growth curves, assuming an initial size of 10 and a constant rate of increase of approximately 2.3% per generation.

for the population size; the graph in part (b) uses a logarithmic scale.

1. Using the graph in part (a), what would you say was the population size after 200 generations? After 600? After 800? After 900? How would you answer the same questions using the graph in part (b)?
2. What impression does the graph in part (a) give about population change for the first 600 generations? What impression does the graph in part (b) give? What impression does the graph in part (a) give about population change for the last 100 generations? What impression does the graph in part (b) give?
3. What advantages and disadvantages might there be in using a linear graph like the one in part (a)? What advantages and disadvantages might there be in using a logarithmic graph like the one in part (b)? What types of data would be clarified by each type of graph, and what types would be left unclear?

## CALCULATING ECOLOGICAL FOOTPRINTS

Mathis Wackernagel and his many colleagues at the Global Footprint Network have continued to refine the method of calculating ecological footprints—the amount of biologically productive land and water required to produce the energy and natural resources we consume and to absorb the wastes we generate. According to their 2006 report, there are nearly 1.8 hectares (4.4 acres) available for every person in the world, yet we use on average more than 2.2 ha (5.4 acres) per person, creating a global ecological deficit, or overshoot (• p. 6) of about 25%.

| Country | Ecological footprint (hectares per person) | Proportion relative to world average footprint | Proportion relative to world area available |
|---|---|---|---|
| Bangladesh | 0.5 | | |
| Colombia | 1.3 | | 0.73 (1.3 ÷ 1.78) |
| Mexico | 2.6 | | |
| Sweden | 6.1 | | |
| Thailand | 1.4 | | |
| United States | 9.6 | | |
| World average | 2.23 | 1.0 (2.23 ÷ 2.23) | 1.25 (2.23 ÷ 1.78) |
| Your personal footprint (see Question 3 ) | | | |

Data from *Living planet report 2006*. WWF International, Zoological Society of London, and Global Footprint Network.

Compare the ecological footprints of each of the countries listed in the table. Calculate their proportional relationships to the world population's average ecological footprint and to the land available globally to meet our ecological demands.

1. Why do you think the ecological footprint for people in Bangladesh is so small? Why is it so large for people in the United States?
2. Based on the data in the table, how do you think average per capita income affects ecological footprints?
3. Go to an online footprint calculator such as the one at http://www.myfootprint.org or http://ecofoot.org, and take the test to determine your own personal ecological footprint. Enter the value you obtain in the table, and calculate the other values as you did for each nation. How does your footprint compare to those of people in the United States and in other nations? Name three actions you could take to reduce your footprint. (*Note*: Save this number—you will calculate your footprint again in Chapter 23 at the end of your course!)

## Take It Further

Go to www.aw-bc.com/withgott or the student CD-ROM, where you'll find:

- Suggested answers to end-of-chapter questions
- Quizzes, animations, and flashcards to help you study
- *Research Navigator*™ database of credible and reliable sources to assist you with your research projects
- **GRAPHIt!** Tutorials to help you interpret graphs
- **INVESTIGATEIt!** Current news articles that link the topics you study to case studies from your region to around the world

CHAPTER 2

# Environmental Ethics and Economics: Values and Choices

Kakadu National Park, Australia

## Upon completing this chapter, you will be able to:

- Characterize the influences of culture and worldview on the choices people make
- Outline the nature, evolution, and expansion of environmental ethics in Western cultures
- Describe precepts of classical and neoclassical economic theory and summarize their implications for the environment
- Compare the concepts of economic growth, economic health, and sustainability
- Explain the fundamentals of environmental economics and ecological economics

Protestors against proposed Jabiluka uranium mine

CENTRAL CASE

## The Mirrar Clan Confronts the Jabiluka Uranium Mine

**"For some people, what they are is not finished at the skin, but continues with the reach of the senses out into the land. If the land is summarily disfigured or reorganized, it causes them psychological pain."**
—Barry Lopez, American Nature Writer

**"We say no to uranium mining now and for the future. Our right to say no comes from our ancestors, our heritage, our law and culture, our Native Title."**
—Jacqui Katona, Speaking for The Mirrar

The remote outback of Australia's Northern Territory is a region of subtropical eucalyptus forests, picturesque cliffs, and vast marshes. It is home to native people (Aboriginal people, or Aborigines) who lived here long before the British colonized Australia. Australia established its largest national park here: Kakadu National Park, a World Heritage Site recognized by the United Nations for its irreplaceable natural and cultural resources.

The region's land also holds uranium, the naturally occurring radioactive metal valued for its use in nuclear power plants, nuclear weapons, and various medical and industrial tools. Uranium mining is a key contributor to Australia's economy, accounting for 7% of the nation's economic output.

Many of Australia's uranium deposits occur on lands traditionally claimed by Aboriginal groups. This has given rise to conflict between corporations seeking to develop mining operations and Aboriginal people trying to maintain their traditional cultures, which are deeply tied to the landscape. In the Kakadu region, the Mirrar Clan, an extended family of 30 Aborigines, lives alongside the area's first uranium mine, the Ranger Mine. Although Australian law recognized the Mirrar's traditional claim to the land, the Australian government overruled the clan's objections and approved the Ranger Mine's development in 1978.

In recent years, the Mirrar have fought the proposed development of a second mine, Jabiluka, nearby. Like other Aborigines, the Mirrar hold the landscape to be sacred, and the proposed Jabiluka mine was to be located among several spiritually important sites. The Mirrar also depend on the land's resources for their daily needs, and Jabiluka was to be near their traditional hunting and gathering sites, in the floodplain of a river that provides the clan food and water. The Mirrar viewed Jabiluka as a threat to their health and their environment, given repeated radioactive spills at the Ranger Mine. The Mirrar feared that contaminated water would be released into creeks, that radioactive radon gas would emanate from stored waste, and that dams holding waste could fail catastrophically in an earthquake.

Environmental activists worldwide joined the Mirrar's struggle against Jabiluka; in 1998 nearly 3,000 people traveled to the Kakadu region to protest the mine, and hundreds of protestors were jailed. Two Mirrar leaders, Yvonne Margarula and Jacqui Katona, later received the Goldman Prize, the world's foremost award honoring grassroots environmentalists.

The Mirrar's efforts succeeded at last in 2004, when the mine's owner, Energy Resources of Australia, and its international parent company, Rio Tinto, agreed to give the Mirrar veto power over development at Jabiluka. Under the signed agreement, Jabiluka will not be developed unless the Mirrar agree. The company has now backfilled and cleaned up the site.

Rio Tinto CEO Sir Robert Wilson cited economic factors (declining uranium prices) as well as ethical factors (concerns about developing the mine without Mirrar consent) as reasons for abandoning Jabiluka's development. Today, however, the price of uranium is rising on the world market as many nations look to revive nuclear power. Debates pitting the economic gain from uranium mining versus the ethical preservation of traditional cultures have cropped up across Australia. For instance, at Olympic Dam, a location that holds the world's largest uranium deposit, mining operations are depleting water from a site sacred to the region's Arabunna people, and the mining company has refused to negotiate with the Arabunna.

The Mirrar opposed mining despite the economic benefits the mining company promised them in the form of jobs, income, development, and a higher material standard of living. In formulating their approaches to the mining proposal, the Mirrar and other Australians weighed economic, social, cultural, spiritual, and philosophical questions as well as scientific ones. The story of mining and the Mirrar exemplifies some of the ways in which values, beliefs, and traditions interact with economic interests to influence the choices all of us make about how to live within our environment.

# Culture, Worldview, and the Environment

The Mirrar faced difficult choices. They were offered substantial economic benefits, but they felt that mine development ran counter to their ethical respect for their land. Such trade-offs between economic benefits and ethical concerns arise frequently in environmental issues.

## Ethics and economics involve values

As we saw in Chapter 1, environmental science entails a firm understanding of the natural sciences. To address environmental problems, however, we also need to understand how people perceive their environment, how they value it, and how they relate to it philosophically and pragmatically. Ethics and economics are two very different disciplines. However, each in its own way deals with questions of what we value and how those values influence our decisions and actions. To address any environmental problem, we must aim to understand not only how natural systems work, but also how values shape human behavior. Ethics and economics give us tools we need to pursue the "triple bottom line" (• p. 27) of environmental, economic, and social sustainability.

## Culture and worldview influence our perception of the environment

Every action we take affects our environment. From growing food to building homes to manufacturing products to fueling vehicles, we meet our needs by altering our surroundings. Our decisions about how we manipulate and exploit our environment to meet our needs depend in part on rational assessments of costs and benefits. However, our decisions are also heavily influenced by our culture and our worldview (**Figure 2.1**).

**Culture** can be defined as the ensemble of knowledge, beliefs, values, and learned ways of life shared by a group of people. Culture, together with personal experience, influences each person's perception of the world and his or her place within it, something described as the person's

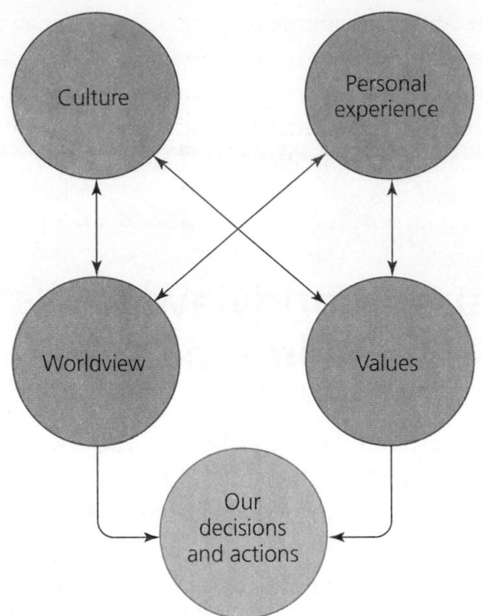


**FIGURE 2.1** Culture and personal experience influence a person's worldview and values, which in turn influence his or her actions and decisions. The disciplines of ethics and economics each examine (in very different ways) factors that influence our values and guide human behavior.

**worldview**. A worldview reflects a person's (or group's) beliefs about the meaning, operation, and essence of the world.

People with different worldviews can study the same situation and review identical data yet draw dramatically different conclusions. For example, many well-meaning people supported the Jabiluka mine while many other well-meaning people opposed it. The officers, employees, and shareholders of the mining company, as well as government officials who supported the mine, view uranium mining as a beneficial source of jobs, income, energy, and economic growth. Mine opponents, in contrast, recognize that uranium mining disturbs the landscape, pollutes air and water, and can expose miners to radiation. Moreover, community disruption, alcoholism, and crime frequently accompany mining booms.

## Many factors shape our worldviews and perception of the environment

The traditional culture and worldview of the Mirrar Clan have played large roles in the group's response to the proposed Jabiluka mine. Australian Aborigines view the landscape around them as the physical embodiment of stories that express the beliefs and values central to their culture. The Australian landscape to them is a sacred text, analogous to the Bible in Christianity, the Koran in Islam, or the Torah in Judaism. Aborigines believe that spirit ancestors possessing human and animal features traveled routes called "dreaming tracks," leaving signs and lessons in the landscape. Modern Aborigines still engage in "walkabouts," long walks that retrace the dreaming tracks. By explaining the origins of specific landscape features, dreaming-track stories assign meaning to notable landmarks and help Aborigines construct detailed mental maps of their surroundings. Passed from one generation to the next, the stories also teach lessons concerning family relations, hunting, food gathering, and conflict resolution. The Mirrar feel that open-pit mining desecrates sacred sites and threatens their culture (**Figure 2.2**).

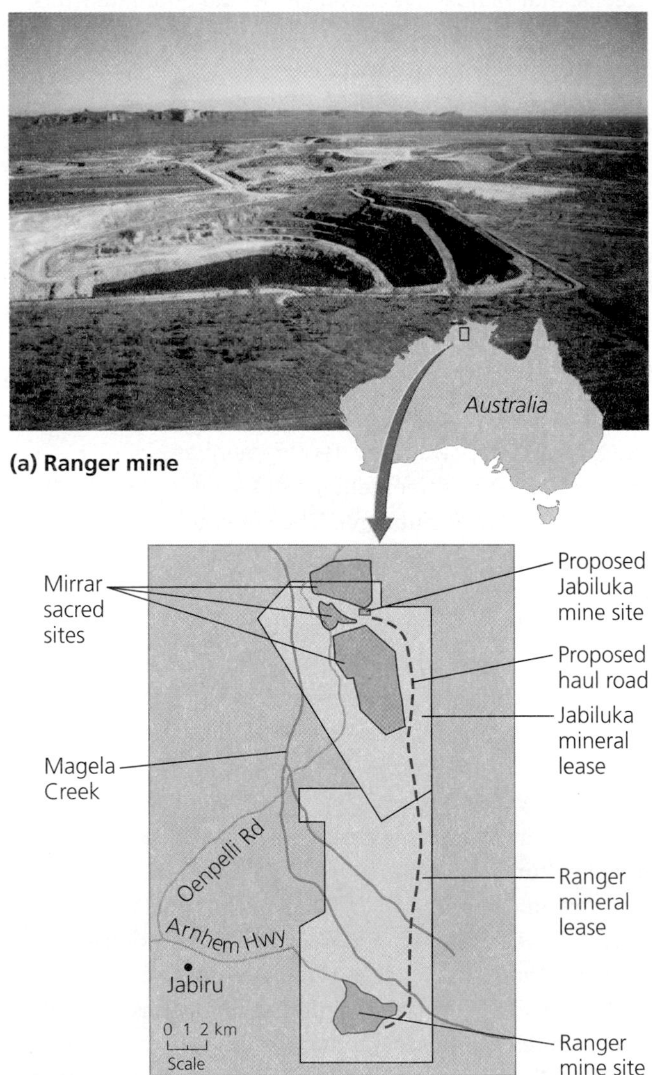


**FIGURE 2.2** The Ranger Mine (**a**), located on traditional Aboriginal land, caused enough environmental impact to spark the Mirrar Clan's opposition to the proposed Jabiluka mine, which would have been developed amid several sites considered sacred (**b**).

### Weighing THE Issues | Uranium Mining in Bethlehem

Suppose a mining company discovered uranium beneath the site in Bethlehem believed to be the birthplace of Jesus—or near the Wailing Wall in Jerusalem, or the mosque at Mecca. What do you think would happen if the company announced plans to develop a mine there, assuring the public that environmental impacts would be minimal and that the mine would create jobs and stimulate economic growth? What aspects of this unlikely situation resemble that of the Mirrar case, and what factors are different? Explain your answers.

Religion is one of many factors that can shape people's worldviews and perception of the environment. A community may also share a particular view of its environment if its members have lived through similar experiences. For example, early European settlers in both Australia and North America viewed their environment as a hostile force because inclement weather and wild animals frequently destroyed crops, killed livestock, and took settlers' lives. Such experiences were shared in stories and songs, and helped shape social attitudes in many frontier communities. The view of nature as a hostile adversary to be overcome has passed from one generation to the next and still influences the way many North Americans and Australians view their surroundings.

A person's political ideology can also shape his or her attitude toward the environment. For instance, one's opinion on the proper role of government may influence whether one wants government to intervene in a market economy to protect environmental quality. Economic factors also sway how people perceive their environment and make decisions. An individual with a strong interest in the outcome of a decision that may result in his or her private gain or loss is said to have a *vested interest.* Mining company executives have a vested interest in a decision to open an area to mining because a new mine can increase profits, to which executive compensation is frequently tied. Likewise, a company's shareholders have vested interests in such a decision because the value of the shares they hold increases with profits.

Throughout this book you will encounter scientific data regarding the environmental impacts of our choices (where to make our homes, how to make a living, what to wear, what to eat, how to travel, how to spend our leisure time, and so on). You will also see that culture, worldviews, and values play critical roles in such choices. Acquiring scientific understanding is only one part of the search for sustainable solutions to environmental problems. We need ethics and economics as well, to help us understand why and how we value those things we value.

# Environmental Ethics

The field of **ethics** is a branch of philosophy that involves the study of good and bad, of right and wrong. The term *ethics* can also refer to the set of moral principles or values held by a person or a society. Ethicists help clarify how people judge right from wrong by elucidating the criteria, standards, or rules that people use in making these judgments. Such criteria are grounded in values—for instance, promoting human welfare, maximizing individual freedom, or minimizing pain and suffering.

People of different cultures or with different worldviews may differ in their values and thus may differ in the specific actions they consider to be right or wrong. This is why some ethicists are **relativists**, who believe that ethics do and should vary with social context. However, different human societies show a remarkable extent of agreement on what moral standards are appropriate. Thus many ethicists are **universalists**, who maintain that there exist objective notions of right and wrong that hold across cultures and situations. For both relativists and universalists, ethics is a *normative* or *prescriptive* pursuit; it tells us how we *ought to* behave.

**Ethical standards** are the criteria that help differentiate right from wrong. One classic ethical standard is *virtue,* which, as the ancient Greek philosopher Aristotle held, involves the personal achievement of moral excellence in character through reasoning and moderation. Another ethical standard is the *categorical imperative* proposed by Immanuel Kant, which roughly approximates Christianity's "golden rule": to treat others as you would prefer to be treated yourself. A third standard is the principle of *utility,* elaborated by British philosophers Jeremy Bentham and John Stuart Mill. The utilitarian principle holds that something is right when it produces the greatest practical benefits for the most people. We all employ such ethical standards as tools for making decisions, consciously or unconsciously, in our everyday lives.

## Environmental ethics pertains to humans and the environment

The application of ethical standards to relationships between humans and nonhuman entities is known as **environmental ethics**. This relatively new branch of ethics arose once people began to perceive environmental changes brought about by industrialization. Human interactions with the environment frequently give rise to ethical questions that can be difficult to resolve. Consider some examples:

- Does the present generation have an obligation to conserve resources for future generations? If so, how much are we obligated to sacrifice?
- Are there situations that justify exposing some communities to a disproportionate share of pollution? If not, what actions are warranted in preventing this problem?
- Are humans justified in driving species to extinction? If destroying a forest would drive extinct an insect species few people have heard of but would create jobs for 10,000 people, would that action be ethically admissible? What if it were an owl species? What if only 100 jobs would be created?

## We have extended ethical consideration to more entities through time

Answers to questions like those above depend partly on what ethical standard(s) a person chooses to use. They also depend on the breadth and inclusiveness of the person's domain of ethical concern. A person who feels responsibility for the welfare of insects would answer the third question very differently from a person whose domain of ethical concern ends with humans. Most of us feel moral obligations to some entities in the world, but by no means to all.

Throughout the history of Western cultures (i.e., European and European-derived societies), people have gradually enlarged the array of entities they feel deserve ethical consideration. The enslavement of human beings was common in many societies until recently, for instance. Women in the United States were not allowed to vote until 1920, and they still face lower pay for equal work. Consider, too, how little ethical consideration citizens of one nation generally extend to those of another on which their government has declared war. Human societies are only now beginning to embrace the principle that all people be granted equal ethical consideration.

Our expanding domain of ethical concern has begun to include nonhuman entities as well. Concern for the welfare of domesticated animals is evident in humane societies and in the lengths many people go to provide for their pets. Animal rights activists voice concern for animals that are hunted, eaten, or used in laboratory testing. Many people now accept that wild animals (at least obviously sentient animals, such as large vertebrates, with which we share similarities) merit ethical consideration. Moreover, today many environmentalists are concerned with the well-being of whole natural communities. Some people go still further, suggesting that all of nature—living and nonliving things alike—should be ethically represented. The historian Roderick Nash illustrated this historical expansion of ethics in his 1989 book, *The Rights of Nature* (**Figure 2.3**).

What is behind this ongoing expansion? Rising economic prosperity in Western cultures, as people gained

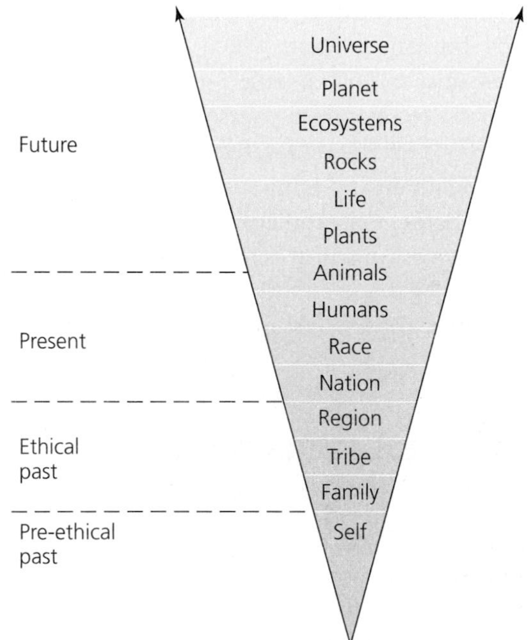


**(a) The evolution of ethics**

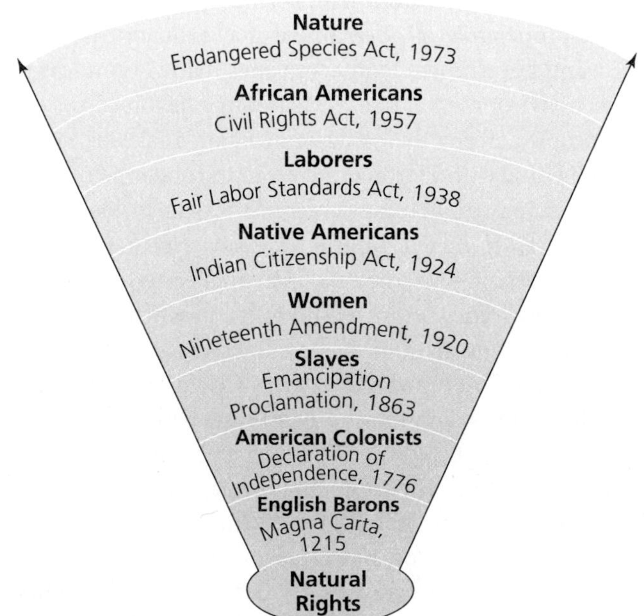


**(b) The expanding concept of rights**

FIGURE 2.3 Through recent history, people in Western (European and European-derived) cultures have broadened the scope of their ethical consideration. We can view a society's domain of ethical concern expanding through time outward from the self (**a**). This historical expansion of ethics is reflected by key legal milestones in the expansion of rights granted by Britain and then the United States (**b**). *Source:* Nash, R. F. 1989. *The rights of nature.* University of Wisconsin Press.

more leisure time and became less anxious about their day-to-day survival, has helped enlarge our ethical domain. Science may also have played a role. Ecology over the past 75 years has made clear that all organisms are interconnected and that what affects plants, animals, and ecosystems can in turn affect people. Evolutionary biology over the past 150 years has shown that humans are merely one species out of millions and have evolved subject to the same pressures as other organisms. Ecology and evolution have demonstrated scientifically that humans do not stand apart from nature, but rather are part of it.

For many non-Western cultures, broad ethical domains are nothing new. Many traditional hunter-gatherer cultures have long granted nonhuman entities ethical standing. The Mirrar, who view their landscape as sacred and alive, are a case in point. Thus, the trends shown in Figure 2.3 do not apply universally to all cultures. However, it is worthwhile for us to examine Western ethical expansion because it is tied to so many of our society's beliefs and actions regarding the environment, and because the global spread of Western influence that has accompanied industrialization has had so much impact on our world. People often simplify the continuum Nash portrayed by dividing it into three ethical perspectives: anthropocentrism, biocentrism, and ecocentrism.

**Anthropocentrism** **Anthropocentrism** describes a human-centered view of our relationship with the environment. An anthropocentrist denies or ignores the notion that nonhuman entities can have rights. An anthropocentrist also measures the costs and benefits of actions solely according to their impact on people (**Figure 2.4**). To evaluate a human action that affects the environment, an anthropocentrist might use criteria such as impacts on human health, economic costs and benefits, and aesthetic concerns. For example, if a proposed mine such as Jabiluka provided a net economic benefit while doing no harm to human health and having little aesthetic impact, the anthropocentrist would conclude it was a worthwhile venture, even if it might drive some native species extinct. If protecting the area would provide spiritual, economic, or other benefits to humans now or in the future, an anthropocentrist might favor its protection. In the anthropocentric perspective, anything not providing benefit to people is considered to be of negligible value.

**Biocentrism** In contrast to anthropocentrism, **biocentrism** ascribes value to certain living things or to the biotic realm in general (see Figure 2.4). In this perspective, nonhuman life has ethical standing, so a biocentrist evaluates actions in terms of their overall impact on living things, human and nonhuman. In the case of a mining

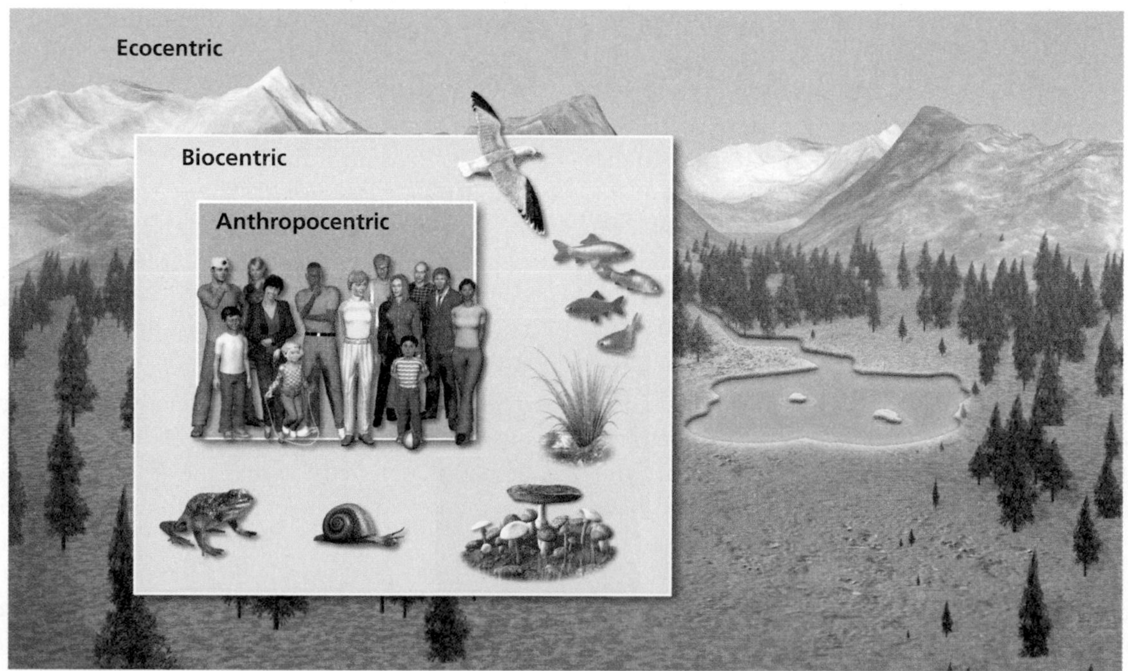


**FIGURE 2.4** We can categorize people's ethical perspectives as anthropocentric, biocentric, or ecocentric. An anthropocentrist extends ethical standing only to humans and judges actions in terms of their effects on humans. A biocentrist values and considers all living things, human and otherwise. An ecocentrist extends ethical consideration to living and nonliving components of the environment. The ecocentrist also takes a holistic view of the connections among these components, valuing the larger functional systems of which they are a part.

proposal, a biocentrist might oppose the mine if it posed a serious threat to the abundance and variety of living things in the area, even if it would create jobs, generate economic growth, and pose no threat to human health. Some biocentrists advocate equal consideration of all living things, whereas others advocate that some types of organisms should receive more than others.

**Ecocentrism** **Ecocentrism** judges actions in terms of their benefit or harm to the integrity of whole ecological systems, which consist of living and nonliving elements and the relationships among them (see Figure 2.4). An ecocentrist would value the well-being of entire species, communities, or ecosystems (we will study these in Chapters 5–7) over the welfare of a given individual. Implicit in this view is that the preservation of larger systems generally protects their components, whereas selective protection of the components may not always safeguard the entire system. Ecocentrism is a more holistic perspective than biocentrism or anthropocentrism. Not only does it encompass a wider variety of entities, but it also stresses preserving the connections that tie entities together into functional systems.

## Environmental ethics has ancient roots

Environmental ethics arose as an academic discipline in the early 1970s, but people have contemplated our ethical relations with nature for thousands of years. Ancient Aboriginal dreaming-track stories treat the environment as a source of sacred teachings, such that boulders, caves, lakes, and other natural features are felt to have moral significance worthy of contemplation and protection. In the Western tradition, the ancient Greek philosopher Plato expressed what he considered humanity's moral obligation to the environment, writing, "The land is our ancestral home and we must cherish it even more than children cherish their mother."

Some ethicists and theologians have pointed to the religious traditions of Christianity, Judaism, and Islam as sources of anthropocentric hostility toward the environment. They point out biblical passages such as, "Be fruitful and multiply, and fill the earth and subdue it; and have dominion over the fish of the sea and over the birds of the air and over every living thing that moves upon the earth." Such wording has justified and encouraged an animosity toward nature that has characterized Western culture over the centuries, some scholars say.

Others interpret sacred texts of these religions to encourage benevolent human stewardship over nature. Consider the directive, "You shall not defile the land in which you live. . . ." The ethic of spiritually based stewardship enjoys strong support in the United States, where a 2003 poll showed that 56% of Americans supported environmental protection because they considered the environment to be "God's creation." Although people have held differing views of their ethical relationship with their environment for millennia, environmental impacts that became apparent during the industrial revolution intensified debate about our species' relationship with its environment.

## The industrial revolution inspired environmental philosophers

As the industrial revolution spread in the 19th century from Great Britain throughout Europe and to North America and elsewhere, it amplified human impacts on the environment. In this period of social and economic transformation, agricultural economies became industrial ones, machines enhanced or replaced human and animal labor, and much of the rural population moved into cities. Consumption of natural resources accelerated, and pollution increased dramatically as coal combustion fueled railroads, steamships, ironworks, and factories.

Many British writers and philosophers of the time criticized the drawbacks of industrialization. Critic **John Ruskin** (1819–1900) called cities "little more than laboratories for the distillation into heaven of venomous smokes and smells." Ruskin also complained that people prized the material benefits that nature could provide but no longer appreciated its spiritual and aesthetic benefits. Motivated by similar concerns, a number of citizens' groups sprang up in 19th-century England that could be considered some of the first environmental organizations (Table 2.1).

In the United States during the 1840s, a philosophical movement called *transcendentalism* flourished, espoused in New England by the American philosophers **Ralph Waldo Emerson** and **Henry David Thoreau** and by poet **Walt Whitman**. The transcendentalists viewed nature as a direct manifestation of the divine, emphasizing the soul's oneness with nature and God. Like Ruskin, the transcendentalists objected to what they saw as their fellow citizens' obsession with material things, and through their writing they promoted their holistic view of nature. The transcendentalist worldview resembled that of the Mirrar in some respects. Both traditions identify a need to experience wild nature, and both view natural entities as symbols or messengers of some deeper truth.

Although Thoreau viewed nature as divine, he also observed the natural world closely and came to understand it in the manner of a scientist; he was in many ways one of

TABLE 2.1 Early Environmental Organizations in 19th-century Great Britain

| Organization | Year established | Purpose |
|---|---|---|
| Scottish Rights of Way Society | 1843 | Protect walking paths in and near cities |
| Commons Preservation Society | 1865 | Preserve forests and other landscapes |
| Society for the Protection of Ancient Buildings | 1877 | Protect the built environment, especially historic buildings |
| Selborne League | 1885 | Protect rare birds, plants, and landscapes |
| Coal Smoke Abatement Society | 1898 | Improve urban air quality |

the first ecologists. His book *Walden,* in which he recorded his observations and thoughts while he lived at Walden Pond away from the bustle of urban Massachusetts, remains a classic of American literature.

## Conservation and preservation arose at the start of the 20th century

One admirer of Emerson and Thoreau was **John Muir** (1838–1914), a Scottish immigrant to the United States who eventually settled in California and made the Yosemite Valley his wilderness home. Although Muir chose to live in isolation in his beloved Sierra Nevada for long stretches of time, he nonetheless became politically active and won fame as a tireless advocate for the preservation of wilderness (**Figure 2.5**).

Muir was motivated by the rapid deforestation and environmental degradation he witnessed throughout North America and by his belief that the natural world should be treated with the same respect that cathedrals receive. Today he is associated with the **preservation** ethic, which holds that we should protect the natural environment in a pristine, unaltered state. Muir argued that nature deserved protection for its own inherent value (an ecocentrist argument), but he also maintained that nature promoted human happiness and fulfillment (an anthropocentrist argument). "Everybody needs beauty as well as bread," he wrote in 1912, "Places to play in and pray in, where nature may heal and give strength to body and soul alike."

Some of the same factors that motivated Muir also inspired the first professionally trained American forester, **Gifford Pinchot** (1865–1946; **Figure 2.6**). Both men opposed the deforestation and unregulated economic development of North American lands that occurred during their lifetimes. However, Pinchot took a more anthropocentric view of how and why we should value nature. He is today the person most closely associated with the **conservation** ethic, which holds that people should put natural resources to use but also that we have a responsibility to manage them wisely. The conservation ethic uses a utilitarian standard, stating that in using resources,

FIGURE 2.5 A pioneering advocate of the preservation ethic, John Muir is also remembered for his efforts to protect the Sierra Nevada from development and for his role in founding the Sierra Club, a leading environmental organization. Here Muir (right) is shown with President Theodore Roosevelt in Yosemite National Park. After his 1903 wilderness camping trip with Muir, the president instructed his interior secretary to increase protected areas in the Sierra Nevada.

FIGURE 2.6 Gifford Pinchot, the first chief of what would become the U.S. Forest Service, was a leading proponent of the conservation ethic. The conservation ethic holds that people should use natural resources, but should strive to ensure the greatest good for the greatest number for the longest time.

we should attempt to provide the greatest good to the greatest number of people for the longest time. Whereas preservation aims to preserve nature for its own sake and for our aesthetic and spiritual benefit, conservation promotes the prudent, efficient, and sustainable extraction and use of natural resources for the benefit of present and future generations.

Pinchot and Muir came to represent different branches of the American environmental movement, and their contrasting ethical approaches often pitted them against one another on policy issues of the day. Nonetheless, they both represented reactions against a prevailing "development ethic," which holds that people are and should be masters of nature, and which promotes economic development without regard to its negative consequences. Pinchot eventually founded what would become the U.S. Forest Service and served as its chief in Theodore Roosevelt's administration. Both Pinchot and Muir left legacies that reverberate today in the different ethical approaches to environmentalism.

**Weighing THE Issues** | **Preservation and Conservation**

With which ethic do you most identify—preservation or conservation? Think of a forest or other important natural resource in your region. Give an example of a situation in which you might adopt a preservation ethic and an example of one in which you might adopt a conservation ethic. Are there conditions under which you'd follow neither, but instead adopt a "development ethic"?

## Aldo Leopold's land ethic arose from the conservation and preservation ethics

As a young forester and wildlife manager, **Aldo Leopold** (1887–1949; **Figure 2.7**) began his career fully in the conservationist camp, having graduated from Yale Forestry School, which Pinchot had helped found just as Roosevelt and Pinchot were advancing conservation on the national stage. As a forest manager in Arizona and New Mexico, Leopold embraced the government policy of shooting predators, such as wolves, to increase populations of deer and other game animals.

At the same time, Leopold followed the development of ecological science. He eventually ceased to view certain species as "good" or "bad" and instead came to see that healthy ecological systems depend on the protection of all their interacting parts. Drawing an analogy to mechanical maintenance, he wrote, "to keep every cog and wheel is the first precaution of intelligent tinkering."

**FIGURE 2.7** Aldo Leopold, a wildlife manager and pioneering environmental philosopher, articulated a new relationship between people and the environment. In his essay "The Land Ethic," he called on people to include the environment in their ethical framework.

It was more than science that pulled Leopold from an anthropocentric perspective toward a more holistic one. One day he shot a wolf, and when he reached the animal, Leopold was transfixed by "a fierce green fire dying in her eyes." The experience remained with him for the rest of his life and helped lead him to a more ecocentric ethical outlook. Years later, as a University of Wisconsin professor, Leopold argued that humans should view themselves and "the land" as members of the same community and that people are obligated to treat the land in an ethical manner. In his 1949 essay "The Land Ethic," he wrote:

> All ethics so far evolved rest upon a single premise: that the individual is a member of a community of interdependent parts. . . . The land ethic simply enlarges the boundaries of the community to include soils, waters, plants, and animals, or collectively: the land. . . . A land ethic changes the role of *Homo sapiens* from conqueror of the land-community to plain member and citizen of it. . . . It implies respect for his fellow-members, and also respect for the community as such.

Leopold intended that the land ethic would help guide decision making. "A thing is right," he wrote, "when it tends to preserve the integrity, stability, and beauty of the biotic community. It is wrong when it tends otherwise." Leopold died before seeing "The Land Ethic" and his best-known book, *A Sand County Almanac,* in print, but today many view him as the most eloquent and important philosopher of environmental ethics.

## Deep ecology extends environmental ethics

One philosophical perspective that goes beyond Leopold's ecocentrism is **deep ecology**, established in the 1970s. Proponents of deep ecology describe the movement as resting on principles of "self-realization" and biocentric equality. They define self-realization as the awareness that humans are inseparable from nature and that the air we breathe, the water we drink, and the foods we consume are both products of the environment and integral parts of us. Biocentric equality is the precept that all living beings have equal value and that, because we are truly inseparable from our environment, we should protect all other living things as we would protect ourselves.

## Ecofeminism critiques male attitudes toward nature and women

As deep ecology and mainstream environmentalism were expanding people's ethical domains during the 1960s and 1970s, major social movements such as the civil rights movement and the feminist movement were in full swing. A number of feminist scholars saw parallels in human behavior toward nature and men's behavior toward women. The degradation of nature and the social oppression of women shared common roots, these scholars asserted.

Ecological feminism, or **ecofeminism**, argues that the patriarchal (male-dominated) structure of society—which traditionally grants more power and prestige to men than to women—is a root cause of both social and environmental problems. Ecofeminists hold that a traditionally female worldview that interprets the world in terms of interrelationships and cooperation is more compatible with nature than a traditionally male worldview that interprets the world in terms of hierarchies and competition. Ecofeminists maintain that a male tendency to try to dominate and conquer what men hate, fear, or do not understand has historically been exercised against both women and the natural environment.

## Environmental justice seeks equal treatment for all races and classes

Our society's domain of ethical concern has been expanding from rich to poor and from majority races and ethnic groups to minority ones. This ethical expansion involves applying a standard of fairness and equality and has given rise to the environmental justice movement. **Environmental justice** involves the fair and equitable treatment of all people with respect to environmental policy and practice, regardless of their income, race, or ethnicity.

The environmental justice movement was fueled by the perception that poor people and minorities tend to be exposed to a greater share of pollution, hazards, and environmental degradation than are richer people and whites. A protest in the early 1980s by African Americans in Warren County, North Carolina, against a toxic waste dump in their community is widely seen as the beginning of the movement (**Figure 2.8**). The state had chosen to site the dump in the county with the highest percentage of African Americans, prompting Warren County residents to suspect "environmental racism." Environmental justice grew to prominence in the early 1990s as more people across North America began fighting environmental hazards in their communities.

In 1983, a U.S. General Accounting Office study found that three of four toxic waste landfills in the southeastern United States were located in communities where the population of racial minorities exceeded that of whites, even though minorities made up only 20% of the region's

FIGURE 2.8 Communities of poor people and people of color have suffered more than their share of environmental problems, a situation that has given rise to the environmental justice movement. The movement gained prominence with this protest of a toxic waste dump in Warren County, North Carolina.

FIGURE 2.9 Hurricane Katrina revealed the need for environmental justice, as the people affected most by the storm and its aftermath were poor and non-white. These girls are playing near where they now live in New Orleans's Lower Ninth Ward. Their mother had moved back to New Orleans after Katrina destroyed her home, but poverty forced her to accept donated gutted housing once the Federal Emergency Management Agency cut off payments. Water in the area is unsafe to drink and many homes are completely destroyed.

population. In 1987, the United Church of Christ Commission for Racial Justice found that the percentage of minorities in areas with toxic waste sites was twice that of areas without toxic waste sites. Researchers studying air pollution, lead poisoning, pesticide exposure, and workplace hazards have found similar patterns.

Despite much progress toward racial equality in Western societies, significant inequities remain. Although our economies have grown, the gaps between rich and poor have widened. And despite stronger enforcement of environmental laws, minorities and the poor still suffer substandard environmental conditions (**Figure 2.9**). Today the environmental justice movement has broadened to encompass equity in transportation options, redevelopment of abandoned urban sites, worker health and safety, and access to parklands.

### Weighing THE Issues | Environmental Justice

Consider the place where you grew up. Where were the factories, waste dumps, and polluting facilities located, and who lived closest to them? Who lives nearest them in the town or city that hosts your campus? Do you think the concerns of environmental justice advocates are justified? If so, what could be done to ensure that poor communities are no more polluted than wealthy ones?

Critics have characterized the attempts of the predominantly white Australian government and uranium mining companies to open mines on traditional lands of Aboriginal people as violations of environmental justice. In Canada, Native Americans of the Dene Nation in the Northwest Territories and Saskatchewan have suffered ill health from working in uranium mines with minimal safeguards.

In the southwestern United States from 1948 through the late 1960s, uranium mines employed many Native Americans, including those of the Navajo nation. Although uranium mining had already been linked to health problems and premature death, the miners had practically no awareness of radiation and its risks. The Navajo language did not even have a word for *radiation,* and for nearly two decades neither the mining industry nor the U.S. government provided the miners information or safeguards. Many Navajo families built homes and bread-baking ovens from the waste rock produced as a by-product of the mining process, not realizing it was radioactive (**Figure 2.10**).

Cases of lung cancer began to appear among Navajo miners in the early 1960s, but scientific studies of radiation's effects on miners at the time excluded Native American workers. The decision to include only white miners in the studies was attributed to the researchers' desire to study a "homogeneous population." A later generation of Americans perceived this as negligence and discrimination, and their desire for justice gave rise to the Radiation Exposure Compensation Act of 1990, a federal law that compensated Navajo miners who suffered health effects from unprotected work in the mines.

FIGURE 2.10 Native Americans employed as uranium miners in Canada and the United States, such as the Navajo miner shown here, have suffered from adverse effects of mining.

### Environmental justice is an international issue

Just as wealthy people often impose their pollution on poorer people, wealthy nations often do the same to poorer nations. One common source of environmental injustice among nations concerns the dumping of hazardous waste (• p. 649). The millions of tons of hazardous waste that we in developed nations produce in our factories, power plants, and incinerators must go somewhere. Proper disposal is expensive, so it is often cheapest for companies to pay cash-strapped nations to take the waste—or to dump it illegally. Often hazardous waste is falsely labeled as harmless or beneficial material, such as fertilizer, fill dirt, or plastics for recycling.

In nations with lax environmental and health regulations, workers and residents are often uninformed of or unprotected against the dangers from this waste. Exposed individuals may fall ill and can die prematurely. Although an international treaty, the Basel Convention, exists to prohibit the international export of waste, trade and illegal dumping has continued. And although 169 nations have ratified the treaty, the United States, the world's largest exporter, has not.

Environmental justice is a key component in pursuing a society that meets the triple bottom line of environmental, economic, and social sustainability (• p. 21). As we explore environmental issues from a scientific standpoint throughout this book, we will also encounter the social aspects of these issues, and the concept of environmental justice will arise again and again.

## Economics: Approaches and Environmental Implications

Questions of environmental justice generally intertwine ethical issues with economic ones, and frequently friction exists between people's ethical and economic impulses. In the case of the Jabiluka mine proposal, people who opposed the mine did so largely on the basis of ethical and environmental justice concerns. Few challenged the mining plan on economic grounds; mine opponents generally recognized uranium as a lucrative resource that generates jobs, income, and electricity. Although some questioned the benefits to local communities, they did not dispute the contribution of uranium exports to the Australian economy. Thus, support for the mine was based primarily on economic factors. Such conflict between ethical and economic motivations is a recurrent theme in environmental issues worldwide.

### Is there a trade-off between economics and the environment?

Although measures to safeguard the environment may frequently mesh well with ethical considerations, we often hear it said that environmental protection works in opposition to economic health. But is this necessarily the case? Growing numbers of economists assert that there need be no such trade-off—that in fact, environmental protection is generally *good* for the economy. As we will see, the view one takes often depends on whether one thinks in the short term or the long term and whether one holds to traditional economic schools of thought or to newer ones that view human economies as coupled to the natural environment.

### Economics studies the allocation of scarce resources

Like ethics, economics examines factors that guide human behavior. **Economics** is the study of how people decide to use scarce resources to provide goods and services in the face of demand for them. By this definition, environmental problems are also economic problems that can intensify as population and per capita resource consumption increase. For example, pollution may be viewed as depletion of the scarce resources of clean air, water, or soil. Indeed, the word *economics* and the word *ecology* come from the same Greek root, *oikos*, meaning "household." In its broadest context, the human "household" is Earth itself. Economists traditionally have studied the household of human society, and ecologists the broader household of all life.

### Several types of economies exist today

An **economy** is a social system that converts resources into **goods**, material commodities manufactured for and bought by individuals and businesses; and **services**, work done for others as a form of business. The oldest type of economy is the **subsistence economy**. People in subsistence economies meet most or all of their daily needs directly from nature and do not purchase or trade for most of life's necessities.

A second type of economy is the **capitalist market economy**. In this system, the interactions among buyers and sellers determine which goods and services are produced, how much are produced, and how these are produced and distributed. Capitalist economies are often contrasted with state socialist economies, or **centrally planned economies**, in which government determines in a top-down manner how to allocate resources. In today's world, capitalism has come to predominate over socialism. However, a pure market economy would operate without government intervention. In reality, all capitalist market economies today, including that of the United States, have borrowed much from state socialism and are in fact hybrid systems (often termed **mixed economies**).

In modern mixed economies, as we will see in Chapter 3, governments typically intervene for several reasons: (1) to eliminate unfair advantages held by single buyers or sellers; (2) to provide social services, such as national defense, medical care, and education; (3) to provide "safety nets" (for the elderly, victims of natural disasters, and so on); (4) to manage the commons (• pp. 5–6); and (5) to mitigate pollution.

## Economies are intricately linked to their environments

All human economies exist within the environment and depend on it in important ways. Economies receive inputs from the environment, process them in complex ways that enable human society to function, then discharge outputs of waste into the environment.

Although these interactions between human economies and the nonhuman environment are readily apparent, traditional economic schools of thought have long overlooked the importance of these connections. Indeed, most conventional economists today still adhere to a worldview that largely ignores the environment (**Figure 2.11a**), and this worldview continues to drive most policy decisions. However, modern economists belonging to the fast-growing fields of environmental economics and ecological economics (• pp. 45–46) explicitly accept that human economies are subsets of the environment and depend crucially on the environment (**Figure 2.11b**).

Economic activity uses resources from the environment. Natural resources (• pp. 3–4) are the various substances and forces we need to survive: the sun's energy, the fresh water we drink, the trees that provide our lumber, the rocks that provide our metals, and the fossil fuels that power our machines and produce our plastics. We can think of natural resources as "goods" produced by nature. Without Earth's natural resources, there would be no human economies and, in fact, no human beings.

Environmental systems also naturally function in a manner that supports economies. Earth's ecological systems purify air and water, cycle nutrients, provide for the pollination of plants by animals, and serve as receptacles and recycling systems for the waste generated by our economic activity. Such essential services, often called **ecosystem services** (Table 2.2), support the life that makes our economic activity possible. Some ecosystem services represent the very nuts-and-bolts of our survival, and others enhance our quality of life.

While the environment enables economic activity by providing ecosystem goods and services, economic activity can affect the environment in return. When we deplete natural resources and produce too much pollution, we can degrade the ability of ecological systems to function. In fact, the Millennium Ecosystem Assessment (• p. 18) concluded in 2005 that 15 of 24 ecosystem services its scientists surveyed globally were being degraded or used unsustainably. The degradation of ecosystem services can in turn negatively affect economies. Currently, ecological degradation is harming poor people more than wealthy people, the Millennium Ecosystem Assessment found. As a result, restoring ecosystem services stands as a prime avenue for alleviating poverty in much of the world.

These interrelationships among economic and environmental conditions have only recently become widely recognized, however. Let's briefly examine how economic thought has evolved, tracing the path that is now beginning to lead economies to become more compatible with natural systems.

## Adam Smith helped found classical economics

Economics shares a common intellectual heritage with ethics, and practitioners of both have long been interested in the relationship between individual action and societal well-being. Some philosophers argued that individuals acting in their own self-interest would harm society. Others believed that such behavior could benefit society, as long as the behavior was constrained by the rule of law and private property rights and operated within fairly competitive markets. The latter view was articulated by Scottish philosopher **Adam Smith** (1723–1790). Known today as the father of **classical economics**, Smith believed that when people are free to pursue their own economic self-interest in a competitive marketplace, the marketplace will behave as if guided by "an invisible hand" that ensures their actions will benefit society as a whole. In his 1776 book *Inquiry into the Nature and Causes of the Wealth of Nations*, Smith wrote:

> It is not from the benevolence of the butcher, the brewer, or the baker that we expect our dinner, but from their regard to their own self-interest. [Each individual] intends only his own security, only his own gain. And he is led in this by an invisible hand to promote an end which was no part of his intention. By pursuing his own interests he frequently promotes that of society more effectually than when he really intends to.

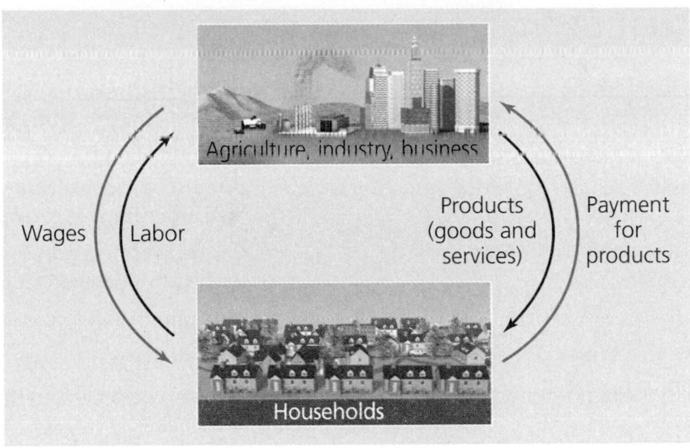


(a) Conventional view of economic activity

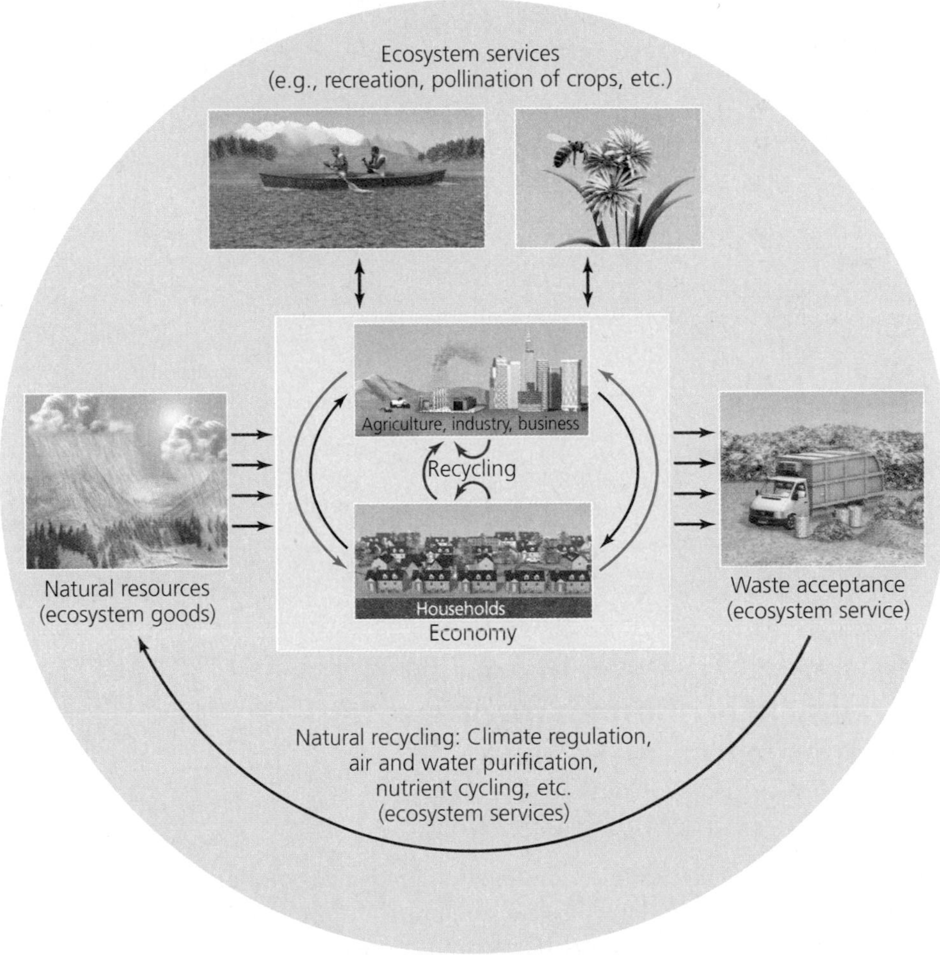


(b) Economic activity as viewed by environmental and ecological economists

**FIGURE 2.11** Standard neoclassical economics focuses on processes of production and consumption between households and businesses (**a**), viewing the environment only as a "factor of production" that helps enable the production of goods. Environmental and ecological economists view the human economy as existing within the natural environment (**b**), receiving resources from it, discharging waste into it, and interacting with it through various ecosystem services.

Smith's philosophy remains a pillar of free-market thought today, and many credit it for the tremendous gains in material prosperity that industrialized nations have experienced in the past few centuries. Others argue that the *laissez-faire* policies spawned by free-market thought worsen inequalities between rich and poor. Still others assert that free-market capitalism is a main cause of environmental degradation. Market capitalism, these critics assert, should be constrained and regulated by democratic government.

TABLE 2.2 Ecosystem Services

| Type of ecosystem service* | Example(s) |
|---|---|
| Regulating atmospheric gases | Maintaining the ozone layer; balancing oxygen, carbon dioxide, and other gases |
| Regulating climate | Controlling global temperature and precipitation through oceanic and atmospheric currents, greenhouse gases, cloud formation, and so on |
| Damping impacts from disturbance | Providing storm protection, flood control, and drought recovery, mainly through vegetation structure |
| Regulating water flow | Providing water for agriculture, industry, transportation |
| Storing and retaining water | Providing water through watersheds, reservoirs, aquifers |
| Controlling erosion and promoting soil retention | Preventing soil loss from wind or runoff; storing silt in lakes and wetlands |
| Forming soil | Weathering rock; accumulating organic material |
| Cycling nutrients | Cycling carbon, nitrogen, phosphorus, sulfur, and other nutrients through ecosystems |
| Treating waste | Removing toxins, recovering nutrients, controlling pollution |
| Pollinating plants | Transporting floral gametes by wind or by pollinating animals, enabling crops and wild plants to reproduce |
| Controlling populations biologically | Controlling prey with predators; controlling hosts with parasites; controlling herbivory on crops with predators and parasites |
| Providing habitat | Providing ecological settings in which creatures can breed, feed, rest, migrate, winter |
| Providing food | Producing fish, game, crops, nuts, and fruits that humans obtain by hunting, gathering, fishing, subsistence farming |
| Supplying raw materials | Producing lumber, fuel, metals, fodder |
| Furnishing genetic resources | Providing unique biological sources for medicine, ornamental species (pets and horticultural plant varieties), and genes for resistance to plant pathogens and crop pests |
| Providing recreational opportunities | Ecotourism, sport fishing, hiking, birding, kayaking, other outdoor recreation |
| Providing cultural or noncommercial uses and goods | Aesthetic, artistic, educational, spiritual, and/or scientific values of ecosystems |

*Ecosystem "goods" are here included in ecosystem services.

Adapted with permission from Costanza, R., et al. 1997. The value of the world's ecosystem services and natural capital. *Nature* 387: 253–260.

## Neoclassical economics incorporates human psychology

Economists subsequently took more quantitative approaches and incorporated human psychology into their work. Modern **neoclassical economics** examines the psychological factors underlying consumer choices, explaining market prices in terms of consumer preferences for units of particular commodities.

In neoclassical economic theory, buyers desire the lowest possible price, whereas sellers desire the highest possible price. This conflict between buyers and sellers results in a compromise price being reached and the "right" quantity of commodities being bought and sold. This is often phrased in terms of *supply,* the amount of a product offered for sale at a given price, and *demand,* the amount of a product people will buy at a given price if free to do so. Theoretically, the market automatically moves toward an equilibrium point, a price at which supply equals demand (**Figure 2.12a**). Similar reasoning can be applied to environmental issues, such that economists can determine "optimal" levels of resource use or pollution control (**Figure 2.12b**).

## Cost-benefit analysis is a widespread tool

Neoclassical economists commonly use a method referred to as **cost-benefit analysis**. In this approach, estimated costs for a proposed action are totaled up and compared to the sum of benefits estimated to result from the action. If benefits exceed costs, the action should be pursued; if costs exceed benefits, it should not. Among

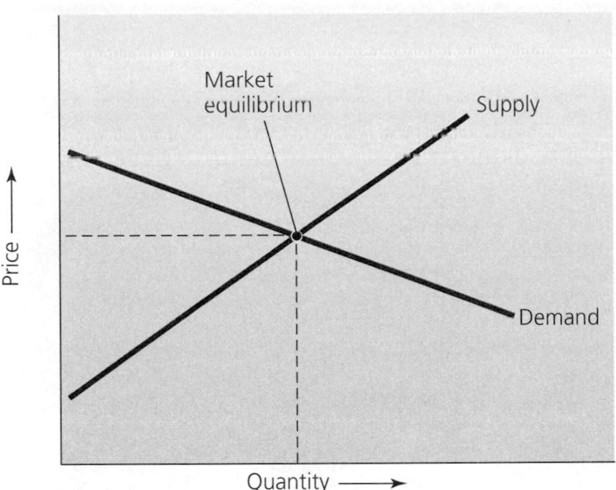


(a) Classic supply–demand curve

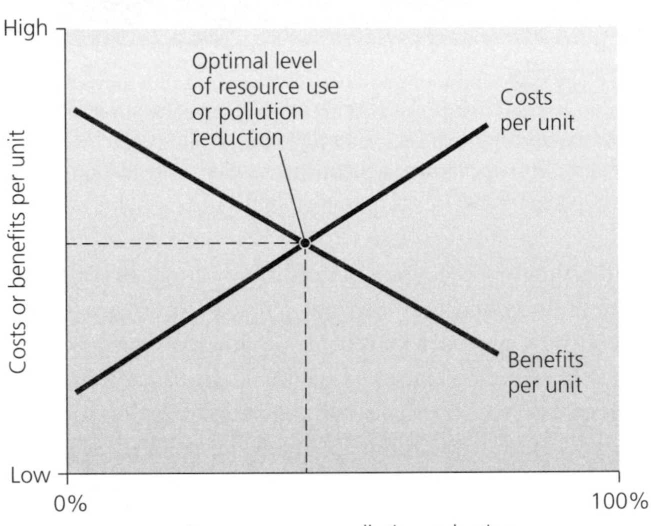


(b) Marginal benefit and cost curves

**FIGURE 2.12** In a basic supply-and-demand graph (**a**), the demand curve indicates the quantity of a given good (or service) that consumers desire at each price, and the supply curve indicates the quantity produced at each price. The market automatically moves toward an equilibrium point at which supply equals demand. We can use a similar graph (**b**) to determine an "optimal" level of resource use or pollution mitigation. In this graph, the cost per unit of resource use or pollution cleanup (blue line) rises as the resource use or pollution cleanup proceeds and it becomes expensive to extract or clean up the remaining amounts. Meanwhile, the benefits per unit of resource use or pollution cleanup (red line) decrease. The point where the lines intersect gives the optimal level.

alternative actions, the one with the greatest excess of benefits over costs should be chosen.

This reasoning seems eminently logical; however, problems often arise because not all costs and benefits can be easily quantified—or even identified or defined. It may be simple to quantify wages paid to uranium miners, the market value of uranium extracted from a mine, or the cost of measures to minimize health risks for miners. But it is difficult to assess the cost of a valued landscape being scarred by mine development, or the cost of radioactive contamination of a stream.

Because some costs and benefits cannot easily be assigned monetary values—and because it is difficult to identify and agree on all costs and benefits—cost-benefit analysis is often controversial. Moreover, because economic benefits are usually more easily quantified than environmental costs, economic benefits tend to be overrepresented in traditional cost-benefit analyses. As a result, environmental advocates often feel that these analyses are biased in favor of economic development and against environmental protection.

## Aspects of neoclassical economics have profound implications for the environment

Today's capitalist market systems operate largely in accord with the precepts of neoclassical economics. These systems have generated unprecedented material wealth, employment, and other desirable outcomes, but they have also contributed to environmental problems. Four fundamental assumptions of neoclassical economics have implications for the environment:

- Resources are infinite or substitutable.
- Costs and benefits are internal.
- Long-term effects should be discounted.
- Growth is good.

**Are resources infinite or substitutable?** Neoclassical economic models generally treat workers and other resources as being either infinite or largely "substitutable and interchangeable." This implies that once we have depleted a resource—natural, human, or otherwise—we should be able to find a replacement for it. Human resources can substitute for financial resources, for instance, or manufactured resources can substitute for natural resources. Theory allows that the substituted resource may be less efficient or more costly, but some degree of substitutability is generally assumed.

Certainly it is true that many resources can be replaced; our societies have made the transition from manual labor to animal labor to steam-driven power to fossil-fuel power, and they may yet make the transition to renewable power sources, such as solar energy. However, Earth's material resources are ultimately limited. Nonrenewable resources such as fossil fuels can be depleted, and many renewable resources can be used up as well if we exploit them faster than they can be replenished. This is what happened to the Easter Islanders (• pp. 8–9) who harvested their forests faster than the forests could regrow.

### Weighing THE Issues | Substitutability and the Environment

Can you think of a natural resource that might be difficult to replace with a substitute? What problems might arise from the assumption that all resources, including clean air and water, are substitutable and interchangeable? Can you think of examples that violate any of the other three assumptions of neoclassical economics? Explain your answers.

**Are costs and benefits internal?** A second assumption of neoclassical economics is that all costs and benefits associated with a particular exchange of goods or services are borne by individuals engaging directly in the transaction. In other words, it is assumed that the costs and benefits of a transaction are "internal" to the transaction, experienced by the buyer and seller alone, and do not affect other members of society.

However, in many situations this is simply not the case. Pollution from a factory, power plant, or mine can harm people living nearby. For instance, the Mirrar have suffered contamination of streams on their land by radioactive material discharged from the Ranger mine. On occasion, uranium concentrations in the water have risen 4,000 times higher than allowed by law. When pollution from an industry or activity affects people not involved in the industry or activity, then they suffer—and often taxpayers end up paying the costs of alleviating it. Market prices do not take the social, environmental, or economic costs of this pollution into account.

Costs or benefits of a transaction that involve people other than the buyer or seller are known as **externalities**. A positive externality is a benefit enjoyed by someone not involved in a transaction, and a negative externality, or **external cost**, is a cost borne by someone not involved in a transaction (**Figure 2.13**). Negative externalities often harm groups of people or society as a whole, while allowing certain individuals private gain. External costs commonly include the following:

- Human health problems
- Property damage
- Declines in desirable elements of the environment, such as fewer fish in a stream
- Aesthetic damage, such as that resulting from air pollution or clear-cutting
- Stress and anxiety experienced by people downstream or downwind from a pollution source
- Declining real estate values resulting from these problems

By ignoring negative externalities, economies create a false idea of the true and complete costs of particular choices

FIGURE 2.13 An Indonesian boy wading in a polluted river suffers external costs: costs that are not borne by the buyer or seller. External costs may include water pollution, health problems, property damage, harm to aquatic life, aesthetic degradation, declining real estate values, and other impacts.

and unjustly subject people to the consequences of transactions in which they did not participate. External costs comprise one reason governments develop environmental legislation and regulations (• p. 60). Unfortunately, external costs are difficult to account for and eliminate. It is tough to assign a monetary value to illness, premature death, or degradation of an aesthetically or spiritually significant site.

**Should long-term effects be discounted?** A third assumption of the neoclassical economic approach is that an event far in the future counts much less than one in the present; in economic terminology, future effects are "discounted." In discounting, short-term costs and benefits are granted more importance than long-term costs and benefits.

This encourages policymakers to play down the long-term consequences of decisions we make today. Because many environmental problems unfold gradually over long time periods, discounting discourages attention to the impacts on future generations of resource depletion, pollution buildup, or other cumulative environmental degradation. As a simple example, applying a 10% annual discount rate to forestry decisions can mean that a stand of ancient trees worth \$500,000 would drop in value by 10% each year and, after 10 years of discounting, would be worth only \$174,339. By this logic, the more quickly the trees are cut, the more they are worth.

The choice of a discount rate involves ethical judgment. This came to light prominently in 2006–2007 as economists worldwide debated the effect of discount rates on estimating the costs of global climate change to our society (see "The Science behind the Story," • pp. 44–45).

**Is growth good?** A fourth neoclassical economic assumption is that economic growth is required to keep employment high and maintain social order. The argument goes something like this: If the poor view the wealthy as the source of their suffering, they may revolt. Promoting economic growth can defuse this situation by creating opportunities for the poor to become wealthier themselves. By making the overall economic pie larger, everyone's slice becomes larger, even if some people still have much smaller slices than others.

In the news media today, increases in an industry's output or percentage growth in a country's economy are always touted as good news, and decreases, stability, or even a minor drop in the *rate* of growth are presented as bad news. For good or ill, economic growth has become the quantitative yardstick by which progress is measured.

## Is the growth paradigm good for us?

The rate of economic growth in recent decades is unprecedented in human history, and the world economy is seven times the size it was half a century ago. All measures of economic activity—trade, rates of production, amount and value of goods manufactured—are higher than they have ever been and are still increasing. This growth has brought many people much greater material wealth (although not equitably, and gaps between rich and poor remain immense).

The modern-day United States may represent the clearest example of the worldview that "more and bigger" is always better. Spurred on by advertising (which has doubled in just the past 20 years) and the increased availability of goods due to technological advances and expanded global trade, American citizens have embarked on a frenzy of consumption unparalleled in history. The dramatic rise in per-person consumption is having numerous consequences (**Figure 2.14**). The many environmental and social impacts of rising per capita consumption of goods and services (and thereby natural resources) will become clear in chapter after chapter throughout this book.

To the extent that rising consumption and economic growth are means to an end—tools with which we can achieve greater human happiness—they can be good things. However, many observers today worry that growth has become an end in itself and is no longer the best tool

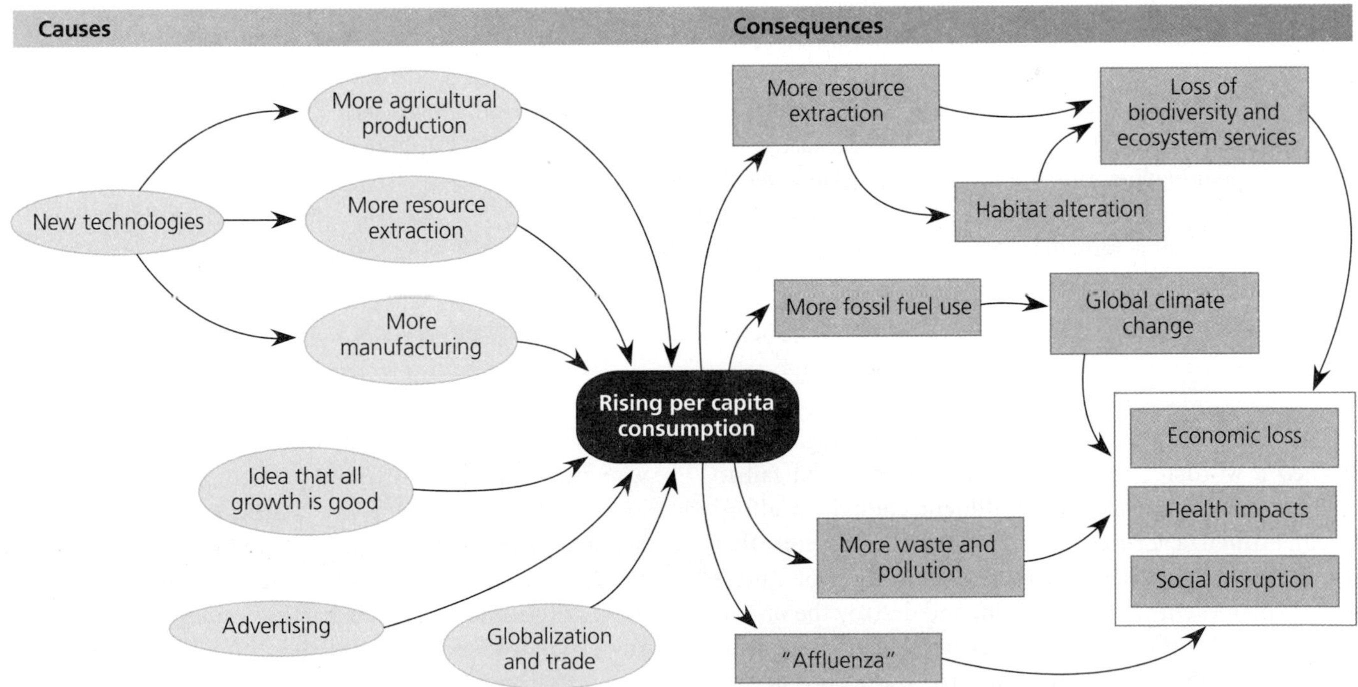


FIGURE 2.14 The rise in per-person consumption of goods and services stems from multiple causes (ovals on left) and results in a diversity of environmental, social, and economic consequences (boxes on right). Arrows in this concept map lead from causes to consequences. Note that items grouped within an outlined box do not necessarily share any special relationship; the outlined box is intended merely to streamline the figure.

**Solutions**

As you progress through this chapter, try to identify as many solutions to rising per capita consumption as you can. What could you personally do to help address this issue? Consider how each action or solution might affect items in the concept map above.

THE SCIENCE BEHIND THE STORY

## Ethics in Economics: Discounting and Global Climate Change

*Sir Nicholas Stern, head of the Government Economic Service, United Kingdom*

To help decide how to respond to global climate change (Chapter 18), the British government commissioned esteemed economist Nicholas Stern to assess the economic costs that a changing climate may impose on society. But once Stern's report was published in November 2006, a debate ensued among economists that had more to do with ethics than with economics. The dispute centered on discounting (• p. 42), as some economists argued that Stern had used an unusually low discount rate—leading to unrealistically high cost estimates.

To produce the *Stern Review on the Economics of Climate Change*, Stern and his research team surveyed the burgeoning literature on the impacts of rising global temperatures, changing regional rainfall patterns, and increasing storminess (see Chapter 18 for our discussion of these issues). They then estimated the economic consequences of these climatic changes and tried to put a price tag on the global cost. The report concluded that without action to forestall it, climate change would cause losses in annual global gross domestic product (GDP) of 5–20% by the year 2200 (see figure).

Stern's team also calculated that by paying just 1% of GDP annually starting now, our society could stabilize atmospheric greenhouse gas concentrations and prevent most of these monetary losses. The bottom line: Spending a relatively small amount of money now will save us much larger expenses in the future.

This conclusion caught the attention of governments worldwide; for the first time, economists were advancing a strong economic argument for tackling climate change immediately.

However, the *Stern Review*'s numbers depend partly on how one chooses to weigh impacts in the future versus impacts happening now. Stern used two discount factors. One accounted for the likelihood that people in the future will be richer than we are today and thus better able to handle economic costs. The other considered whether the future should be discounted simply because it is the future. This latter discount factor (called a "pure time discount factor") is essentially an ethical issue because it places explicit values on the welfare of future versus current generations.

The *Stern Review* used a pure time discount rate of 0.1%. This means that an impact occurring next year is judged 99.9% as important as one occurring this year. It means that the welfare of a person born 10 years from now is 99.0% as valuable as the welfare of one born today. This discount rate treats current and future generations *nearly* equally. Future generations are slightly downweighted only because of the (very small) possibility that our species could go extinct (in which case, there would be no future generations to be concerned about).

with which to pursue happiness. In fact, sociologists have coined a word for the way material goods often fail to bring contentment to people affluent enough to afford them: **affluenza**. Critics of the growth paradigm note that runaway growth resembles the multiplication of cancer cells, which eventually overwhelm and destroy the organism in which they grow. These critics fear that runaway economic growth will likewise destroy the economic system on which we all depend. Resources for growth are ultimately limited, they argue, so nonstop growth is not sustainable and will fail as a long-term strategy.

Defenders of traditional economic approaches reply that Cassandras (• p. 19) have been saying for decades that limited resources would doom growth-oriented economies—yet most of these economies are still expanding dramatically. If resources are dwindling, why are we witnessing the most rapid growth of material wealth in human history? One prime reason is technological innovation. In case after case, improved technology has enabled us to push back the limits on growth. More powerful technology for extracting minerals, fossil fuels, and groundwater has expanded the amounts of these natural resources available to us (**Figure 2.15**). Technological developments such as automated farm machinery, fertilizers, and chemical pesticides have allowed us to grow more food per unit area of land, boosting agricultural output (• pp. 264–268). Faster, more powerful machines in our factories have enabled us to translate our enhanced resource extraction and agricultural production into faster rates of manufacturing.

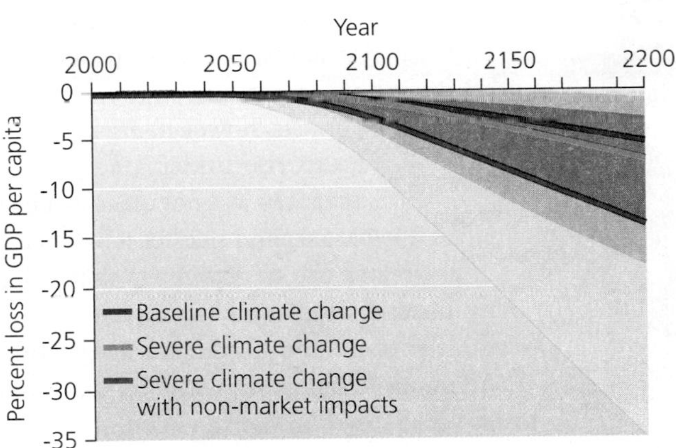


The *Stern Review* estimated that global climate change could cause substantial losses in future wealth. Baseline climate change as predicted by the IPCC Third Assessment Report (• p. 516) could bring losses of 5.3% of per capita GDP annually by the year 2200. Severe climate change, as predicted by research since that report was issued, could bring annual losses of 7.3%, and adding nonmarket values (• pp. 47–48) raises this figure to 13.8%. The gray shaded areas around the three data lines represent ranges of values judged statistically to be 95% likely.

Data from HM Treasury, 2007. *Stern review on the economics of climate change*. London, U.K.

Some economists viewed this discount rate as too low. Prominent Yale economist William Nordhaus critiqued the *Stern Review* in scientific journals and proposed a discount rate starting at 3% and falling to 1% in 300 years. Nordhaus maintained that such numbers were more objective because they are more in line with how people's values are revealed by the prices they pay for things in the free market. He argued that Stern's near-zero discount rate overweights the future, forcing people today to pay far too much to address hypothetical future impacts.

Indeed, when economists use discounting in standard ways to assess capital investments or construction projects (say, development of a railroad, dam, or highway) they typically choose discount rates much closer to those that Nordhaus suggests.

In their response to Nordhaus and other critics, Stern and his team argued that a discount rate of 3% or even 1% is too high. Such rates may be useful for assessing development projects, they maintained, but they are not appropriate for long-term environmental problems that directly affect human well-being.

A 3% rate, Stern pointed out, means that a person born in 1985 is valued only half as much as a person born in 1960. It means a grandchild is judged to be worth far less than a grandparent simply because of the dates they were born. For a variety of reasons, Stern and his supporters argued, the market should not be used to guide ethical decisions.

Stern's group also published sensitivity analyses that analyzed how their conclusions would change if different discount rates were used. These showed that the report's main message—that it's cheaper to prevent climate change now than to pay to deal with it later—was robust to a fair amount of change in the discount rate, up to at least a rate of 1.5%.

At the end of the day, the choice of a discount rate is an ethical decision on which well-intentioned people may differ. As governments, businesses, and individuals begin to invest in addressing climate change, the debate over the *Stern Review* reveals how ethics and economics remain intertwined.

## Economists disagree on whether economic growth is sustainable

Can we conclude, then, that endless improvements in technology are possible and that we will never run into shortages of resources? At one end of the spectrum are Cornucopians (• p. 19), economists, businesspeople, and policymakers who believe technology can solve everything—a philosophy that has greatly influenced economic policy in market economies over the past century.

At the other end of the spectrum, **ecological economists** argue that a couple of centuries is not a very long period of time and that history suggests that civilizations do not, in the long run, overcome their environmental limitations. Ecological economics, which has emerged as a discipline only in the past decade or two, applies principles of ecology and systems science (Chapters 5–7) to the analysis of economic systems. Earth's natural systems generally operate in self-renewing cycles, not in a linear or progressive manner. Ecological economists advocate sustainability in economies and see natural systems as good models. Most ecological economists argue that the growth paradigm will eventually fail and that if nothing is done to rein in population growth and resource consumption, depleted natural systems could plunge our economies into ruin. Many of these economists advocate economies that do not grow and do not shrink, but rather are stable. Such **steady-state economies** are intended to mirror natural ecological systems.

**Environmental economists** tend to agree with ecological economists that economies are unsustainable if population growth is not reduced and resource use is not

FIGURE 2.15 Advances in technology have enabled people to push back the natural limits on growth and continue expanding their economies. Innovations in infrastructure for offshore petroleum extraction, shown here, have allowed us to extract more fossil fuels than ever before and have helped us push our economic productivity far beyond what was possible a century or two ago.

made more efficient. Environmental economists, however, maintain that we can accomplish these changes and attain sustainability within our current economic systems. By retaining the principles of neoclassical economics but modifying them to address environmental challenges, environmental economists argue that we can keep our economies growing and that technology can continue to improve efficiency.

Environmental economists were the first to develop ways to tackle the problems of external costs and discounting. They blazed a trail, and ecological economists then went further, proposing that sustainability requires far-reaching changes leading to a steady-state economy. Thus, whereas environmental economists implemented reform, ecological economists call for revolution.

## A steady-state economy is a revolutionary alternative to growth

The idea of a steady-state economy did not originate with the rise of ecological economics. Back in the 19th century, British economist **John Stuart Mill** (1806–1873) hypothesized that as resources became harder to find and extract, economic growth would slow and eventually stabilize. Individuals and society would subsist sustainably on steady flows of natural resources and on savings accrued during occasional productive but finite periods of growth. Such a model appears to match the economies of Australian Aborigines and many other traditional societies throughout the world before the global expansion of European culture.

Modern proponents of a steady-state global economy, such as American economist Herman Daly, are not so sanguine that a steady state will evolve on its own from a capitalist market system. Instead, most believe we will need to rethink our assumptions and fundamentally change the way we conduct economic transactions.

Critics of a steady-state economy often assume that an end to growth will mean an end to a rising quality of life. Ecological economists, however, argue that quality of life should continue to rise under a steady-state economy. Technological advances will not cease just because growth stabilizes, they argue, and neither will behavioral changes (such as greater use of recycling) that enhance sustainability. Instead, wealth and human happiness can continue to rise after economic growth has leveled off.

Attaining sustainability will certainly require the reforms that environmental economists advocate and may well require the fundamental shifts in thinking, values, and behavior that ecological economists say is necessary. Ecological and environmental economic approaches are still actively being developed, but we will now survey a few strategies for sustainability that have been offered so far.

## We can measure economic progress differently

For decades, economists have assessed the robustness of an economy by calculating its **Gross Domestic Product (GDP)**, the total monetary value of final goods and services it produces each year. However, GDP does not account for nonmarket values (• pp. 47–48), nor is GDP necessarily an expression of *desirable* economic activity. In fact, GDP increases even when economic activity harms society. For example, an oil spill in a U.S. national park could increase the U.S. GDP because oil spills require cleanups, which cost money and, as a result, increase the production of goods and services. A radiation leak at a uranium mine on Aboriginal homelands would likely add to the Australian GDP because of the many monetary transactions required for cleanup and medical care.

Some economists have tried to develop indicators that differentiate between desirable and undesirable economic activity and that can function as better guides to citizens' welfare. One such alternative to the GDP is the **Genuine Progress Indicator (GPI)**, introduced in 1995 by the nonprofit organization Redefining Progress. To calculate GPI, economists begin with conventional economic activity and then add to it positive contributions that are not paid for with money, such as volunteer work and parenting. They then subtract negative impacts, such as crime, pollution, gaps between rich and poor, and other detrimental factors (**Figure 2.16a**). The GPI thereby summarizes more forms of economic activity than does the GDP, and it also differentiates between economic activity that increases societal well-being and economic activity that decreases it. The two indices can differ strikingly; **Figure 2.16b** compares changes in per capita GPI and

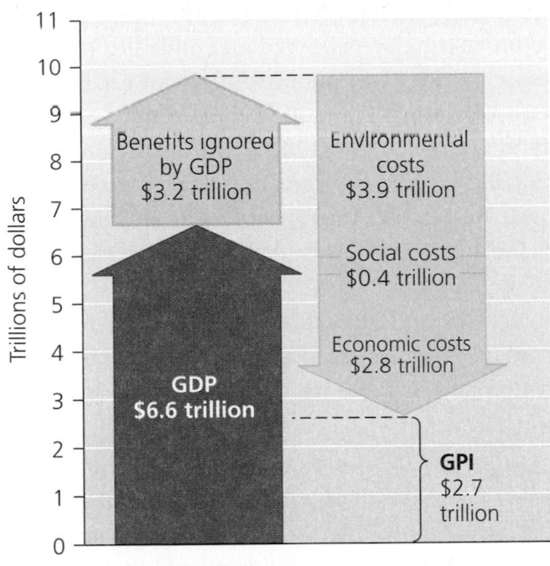


**(a) Components of GDP vs. GPI**

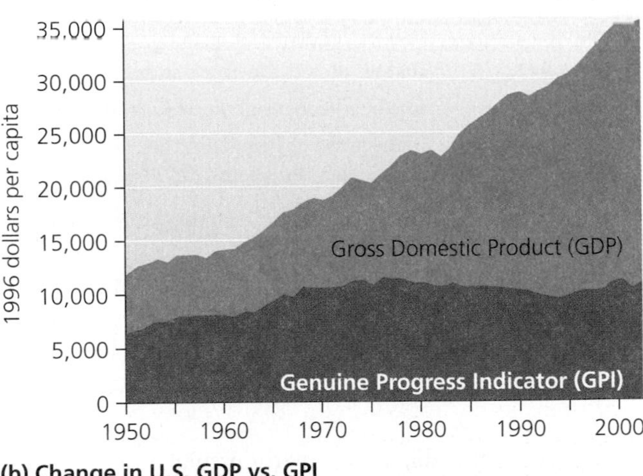


**(b) Change in U.S. GDP vs. GPI**

**FIGURE 2.16** The Genuine Progress Indicator (GPI) adds to the Gross Domestic Product (GDP; red arrow) benefits such as volunteering and parenting (upward pointing gold arrow). The GPI also subtracts external environmental costs such as pollution, social costs such as divorce and crime, and economic costs such as borrowing and the gap between rich and poor (downward pointing gold arrow). Shown in (**a**) are values for GDP and GPI for the United States in 2002. The graph in (**b**) shows that although per capita U.S. GDP has increased dramatically since 1950, per capita U.S. GPI has leveled off since 1970. GPI advocates say this indicates we are spending more money but our lives are not that much better. Data from Venetoulis, J., and C. Cobb. 2004. *The genuine progress indicator, 1950–2002 (2004 update)*. Redefining Progress.

GDP in the United States from 1950 to 2002. On a per-person basis, the nation's GDP has risen greatly, but its GPI has remained largely flat for the past 35 years.

Other "green accounting" indicators include Net Economic Welfare (NEW), which adjusts GDP by adding the value of leisure time and personal transactions while deducting costs of environmental degradation. The Index of Sustainable Economic Welfare (ISEW) is based on income, wealth distribution, natural resource depletion, benefits associated with volunteerism, and environmental degradation. The United Nations' Human Development Index assesses a nation's standard of living, life expectancy, and education. Any of these indices, ecological economists maintain, provide a more accurate portrait of a nation's welfare than the GDP, which policymakers currently use so widely.

Adopting green accounting may prove difficult in practice, however. China introduced a green accounting system in 2006 that subtracted the costs of pollution and resource depletion from its GDP, but it proved so controversial that the government suppressed the 2007 report.

## We can give ecosystem goods and services monetary values

As we have noted, economies receive from the environment vital resources and ecosystem services. However, any survey of environmental problems in the world today—deforestation, biodiversity loss, pollution, collapsed fisheries, climate change, and so on—makes immediately apparent that our society often mistreats the very systems that sustain it. Why is this? From the economist's perspective, people exploit natural resources and systems in large part because the market assigns these entities no quantitative monetary value or, at best, assigns values that underestimate their true worth.

Think for a minute about the nature of some of these services. The aesthetic and recreational pleasure we obtain from natural landscapes, whether wildernesses or city parks, is something of real value. Yet this value is hard to quantify and appears in no traditional measures of economic worth. Or consider Earth's water cycle (• pp. 196–198), by which rain fills our reservoirs with drinking water, rivers give us hydropower and flush away our waste, and water evaporates, purifying itself of contaminants and readying itself to fall again as rain. This natural cycle is absolutely vital to our existence, yet because its value is not quantified, markets impose no financial penalties when we interfere with it.

Ecosystem services are said to have **nonmarket values**, values not usually included in the price of a good or service (Table 2.3 and **Figure 2.17**). Because the market does not assign value to ecosystem services, debates such as that over the Jabiluka mine often involve comparing apples and oranges—in this case, the intangible cultural, ecological, and spiritual arguments of the Mirrar versus the hard numbers of mine proponents.

**TABLE 2.3 Values That Modern Market Economies Generally Do Not Address**

| Nonmarket value | Is the worth we ascribe to things . . . |
|---|---|
| Use value | that we use directly |
| Option value | that we do not use now but might use later |
| Aesthetic value | for their beauty or emotional appeal |
| Cultural value | that sustain or help define our culture |
| Scientific value | that may be the subject of scientific research |
| Educational value | that may teach us about ourselves and the world |
| Existence value | simply because they exist, even though we may never experience them directly (e.g., an endangered species in a far-off place) |

To resolve this dilemma, environmental and ecological economists have sought ways to assign values to ecosystem services. One technique, **contingent valuation**, uses surveys to determine how much people are willing to pay to protect or restore a resource. Such an exercise was conducted with a mining proposal in the Kakadu region in the early 1990s that preceded the Jabiluka proposal. The Kakadu Conservation Zone, a government-owned 50-km$^2$ (19-mi$^2$) plot of land surrounded by Kakadu National Park, was either to be developed for mining or preserved and added to the park. To determine the degree of public support for each option, a government commission sponsored a contingent valuation study to determine how much Australian citizens valued keeping the Kakadu Conservation Zone preserved and undeveloped. Researchers interviewed 2,034 citizens, asking them how much money they would be willing to pay to stop mine development.

The interviewers presented two scenarios: (1) a "major impact" scenario based on predictions of environmentalists who held that mining would cause great harm, and (2) a "minor impact" scenario based on predictions of mining executives who held that development would have few downsides. After presenting both scenarios in detail, complete with photographs, the interviewers asked the respondents how much their households would pay if each scenario, in turn, was correct. Respondents on average said their households would pay \$80 per year to prevent the minor-impact scenario and \$143 per year to prevent the major-impact scenario. Multiplying these figures by the number of households in Australia (5.4 million at the time), the researchers found that preservation was "worth" \$435 million annually to the Australian population under the

(a) Existence values

(b) Use values

(c) Option values

(d) Aesthetic values

(e) Scientific values

(f) Educational values

(g) Cultural values

**FIGURE 2.17** Accounting for nonmarket values such as those shown here may help us make better environmental and economic decisions.

minor-impact scenario, and $777 million under the major-impact scenario. Because both of these numbers exceeded the $102 million in annual economic benefits expected from mine development, the researchers concluded that preserving the land undeveloped was worth more than mining it.

Because contingent valuation relies on survey questions and not actual expenditures, critics complain that in such cases people will volunteer idealistic (inflated) values rather than realistic ones, knowing that they will not actually have to pay the price they name. In part because of such concerns, the Australian government commission decided not to use the Kakadu contingent valuation study's results. (The mine was stopped, but as a result of Aboriginal opposition.)

Whereas contingent valuation measures people's *expressed preferences*, other methods aim to measure people's *revealed preferences*—preferences as revealed by data on actual behavior. For example, researchers may use the amount of money, time, or effort people expend to travel to parks for recreation to measure the value people place on parks. Economists have also analyzed housing prices, comparing homes with similar characteristics but different environmental settings, to infer the dollar value of landscapes, views, and peace and quiet. Yet another approach assigns environmental amenities value by measuring the cost required to restore natural systems that have been damaged or to mitigate harm from pollution.

In 1997 one research team reviewed studies using various valuation methods in an effort to calculate the overall global economic value of all the services that ecosystems provide (see "The Science behind the Story," • pp. 50–51). The researchers came up with a figure of $33 trillion per year ($42 trillion in 2007 dollars)—greater than the combined gross domestic products of all nations in the world. A follow-up study in 2002 concluded that the economic benefits of preserving the world's remaining natural areas outweighed the benefits of exploiting them by a factor of 100 to 1.

## Markets can fail

When they do not reflect the full costs and benefits of actions, markets are said to fail. **Market failure** occurs when markets do not take into account the environment's positive effects on economies (such as ecosystem services) or when they do not reflect the negative effects of economic activity on the environment or on people (external costs).

Traditionally, market failure has been countered by government intervention. Governments can dictate limits on corporate behavior through laws and regulations. They can institute *green taxes*, which penalize environmentally harmful activities. Or, they can design economic incentives that put market mechanisms to work to promote fairness, resource conservation, and economic sustainability. We will examine legislation, regulation, green taxation, subsidies, and market incentives in Chapter 3 as part of our policy discussion.

## Ecolabeling helps address market failure

Legislation, regulation, or taxation by government may not always effectively address market failure, so economists have contrived ways to use aspects of the market to counteract market failure. One technique is to create markets in permits for environmentally harmful activities, and we will examine this approach shortly (• pp. 79–82 and 535).

In a more familiar strategy, manufacturers of certain products designate on their labels how the products were grown, harvested, or manufactured. This method, called **ecolabeling**, serves to tell consumers which brands use environmentally benign processes (**Figure 2.18**).

By preferentially buying ecolabeled products, consumers can provide businesses a powerful incentive to switch to more sustainable processes. One early example of this was labeling cans of tuna as "dolphin-safe," indicating that the methods used to catch the tuna avoid the accidental capture of dolphins. Other examples include labeling recycled paper (• pp. 641–642), organically grown foods (• pp. 285–289), genetically modified foods (• pp. 272–280), and lumber harvested through sustainable forestry (• p. 346).

In a similar vein, individuals who invest their money in the stock market can choose to pursue *socially responsible investing*, which entails investing only in companies that have met certain criteria for environmental or social sustainability.

**FIGURE 2.18** Ecolabeling allows businesses to promote products that have low environmental impact. Organic juices and produce are examples of ecolabeled products that have become widely available in the mass market.

THE SCIENCE BEHIND THE STORY

Dr. Robert Costanza, University of Vermont

## Calculating the Economic Value of Earth's Ecosystems

To Robert Costanza, the problem was like an elephant in the living room, which economists had ignored for decades: Earth's ecosystems provide essential life-support services, including arable soil, waste treatment, clean water, and clean air. However, economists had failed to account for how much those services contribute economically to human welfare.

So Costanza, an environmental economist then at the University of Maryland, joined with 12 colleagues in 1996 at the National Center for Ecological Analysis and Synthesis in Santa Barbara, California, and combed the scientific literature. The team identified over 100 studies that estimated the worth of such ecosystem services as water purification, greenhouse gas regulation, plant pollination, and pollution cleanup.

The studies the team reviewed had estimated values of particular ecosystem services in several ways. Methods such as contingent valuation (•p. 48) had frequently been used because people clearly value such aspects of natural systems as biodiversity and aesthetics, even though we do not generally pay money specifically for them. After poring over studies that examined the value of 17 services provided by oceans, forests, wetlands, and other ecosystems (see figure), Costanza and his colleagues synthesized the results to provide the first comprehensive quantitative estimate of the global value of ecosystem services.

To estimate the worth of services more accurately, Costanza and his team reevaluated the data from the earlier studies using alternative valuation techniques. One method was to calculate the cost of replacing ecosystem services with technology.

For example, marshes protect people from floods and filter out water pollutants. If a marsh were destroyed, the researchers would calculate the value of services it had provided by measuring the cost of the levees and water-purification technology that would be needed to assume those tasks. The researchers then calculated the global monetary value of such wetlands by multiplying those totals by the global area occupied by the ecosystem. By calculating similar totals from coral reefs, deserts, tundra, and other ecosystems, they arrived at a global value for ecosystem services.

By their calculation, the biosphere provides at least \$33 trillion (\$42 trillion in 2007 dollars) worth of ecosystem services each year—more than the gross domestic product (GDP) of all nations combined. Published in the journal *Nature* in 1997, their research paper sparked excitement and controversy.

Some environmental advocates and ethicists argued that it was a bad idea to put a dollar figure on priceless services such as clean air and water. The value of these services cannot be calculated, they held, because we would all perish if they disappeared.

A fast-growing trend, in 2003 U.S. investors sank \$2.16 trillion into socially responsible investing, representing 11% of all investment. As you proceed through this book, you will come across many instances of everyday citizens promoting sustainable solutions through economic means.

### Corporations are responding to sustainability concerns

As more consumers and investors express preferences for sustainable products and services, more industries, businesses, and corporations are finding that they can make money by "greening" their operations.

Some companies have cultivated an eco-conscious image from the start, such as Ben & Jerry's (ice cream), Patagonia (outdoor apparel), and Seventh Generation (household products). The phone company Credo (formerly Working Assets) donates a portion of its proceeds to environmental and other progressive nonprofit groups according to how its customers vote that the funds should be distributed. Today, some newer businesses are trying to go even further than these pioneers. The outdoor apparel startup Nau manufactures stylish items from materials made of corn biomass and recycled bottles, and like Credo, aims to help fund environmental nonprofits. Entrepreneurs are starting thousands of more local sustainably oriented businesses all across the world.

Moreover, in the past few years corporate sustainability has gone mainstream, and some of the world's largest corporations have joined in. They are seeking to

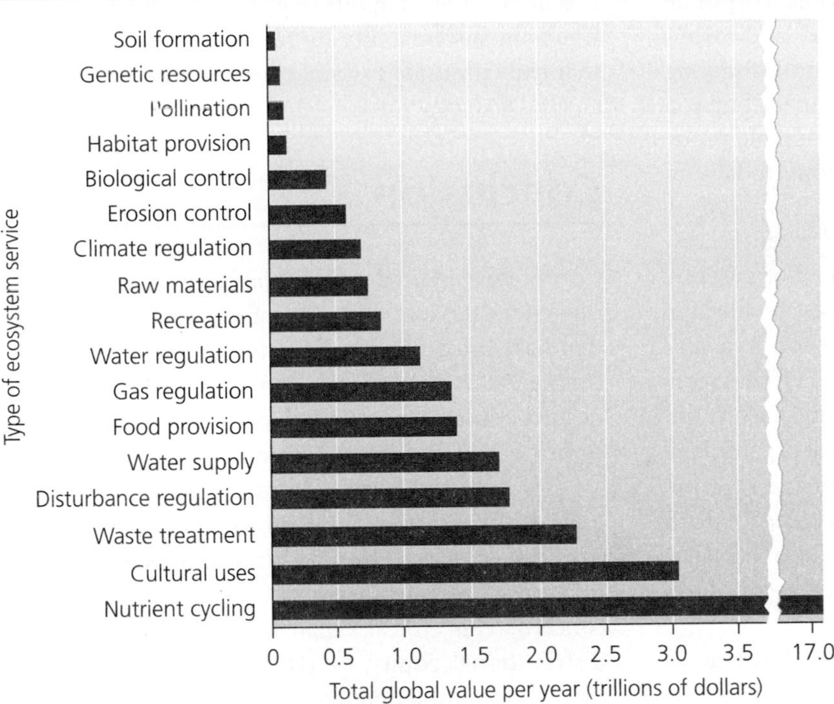


In 1997, Robert Costanza and colleagues estimated the total value of the world's ecosystem services at approximately $33 trillion. Shown are subtotals for each major class of ecosystem service. The $33 trillion figure does not include values from some ecosystems, such as deserts and tundra, for which adequate data were unavailable. Data from Costanza, R., et al. 1997. The value of the world's ecosystem services and natural capital. *Nature* 387: 253–260.

Environmental ethicist Timothy Weiskel of Harvard Divinity School contended that to make a commodity out of biodiversity confused "sacred space with the market place."

Some economists, meanwhile, disparaged the methods by which the researchers calculated values. Replacement costs were not a legitimate way of determining value, they held, because nature's services, like other economic goods and services, are worth only what people will demonstrably pay for them. Critics also argued that combining the value of ecosystem services from various small tracts of land is meaningless because people decide whether to preserve or exploit resources based on particular local considerations, not on generalized global ones.

To address criticisms, Costanza joined Andrew Balmford of Cambridge University and 17 other colleagues to conduct another analysis. They compared the benefits and costs of preserving natural systems intact with those of converting wild lands for agriculture, logging, or fish-farming. Again, they synthesized studies on ecosystem services, this time focusing on just five ecosystems: West African and Malaysian tropical forests, Thai mangrove swamps, Canadian wetlands, and Philippine coral reefs.

In their paper, published in the journal *Science* in 2002, the team reported that a global network of nature reserves covering 15% of Earth's land surface and 30% of the ocean would be worth between $4.4 and $5.2 trillion. This amount is about 100 times the value of the same areas were they to be converted for direct exploitative human use. That 100:1 benefit-cost ratio, they wrote, demonstrates clearly that "conservation in reserves represents a strikingly good bargain."

ride what Daniel Esty and Andrew Winston, authors of the 2006 book *Green to Gold*, call a "green wave" of consumer preference for sustainable products and business practices. These "waveriders" include corporations as varied as Ford Motor Company, Toyota, McDonald's, Starbucks, IKEA, Dow, Dupont, BASF, and British Petroleum.

Carpeting company Interface, Inc., provides one of the best-known cases of a company changing its practices to promote sustainability (• p. 646), but there are many more. Nike, Inc., collects millions of used sneakers each year and recycles their materials to create synthetic surfaces for basketball courts, tennis courts, and running tracks. Nike also uses a high proportion of organic cotton and has developed less-toxic rubber and adhesives. Hewlett-Packard runs ambitious programs to reuse and recycle many of its used products and components, such as toner cartridges, electronics, and plastics. The Gap, Inc., has a sustainably designed headquarters building, has cut energy use in its stores and distribution centers, and runs alternative transportation programs for its employees. In response to media attention and consumer concern, corporations such as Nike, Gap, and others are also working to improve labor conditions in their factories overseas. Today many corporations are finding ways to increase energy efficiency, reduce reliance on toxic substances, increase the use of recycled materials, and minimize greenhouse gas emissions.

Of course, corporations exist to make money for their shareholders, so they cannot be expected to pursue goals that do not turn a profit. Moreover, some corporate greening efforts are more rhetoric than reality, and corporate *greenwashing* may mislead consumers into thinking the companies are acting more sustainably than they are. However, as consumer preferences for sustainable products and practices grow, many corporations today are truly moving toward sustainable models.

Wal-Mart provides the most celebrated recent example of corporate greening efforts. Advocates of sustainability have long criticized the world's largest retailer for its environmental and social impacts, so it came as a surprise when in 2006 and 2007 Wal-Mart CEO Lee Scott began talking up his company's efforts. Under Scott, the company has begun progress in a quest to sell organic and sustainable products, reduce packaging and use recycled materials, enhance fuel efficiency in its truck fleet, reduce energy use in its 6,600 stores, power itself with renewable energy, cut carbon dioxide emissions, and preserve 1 acre of natural land for every acre developed.

Many observers remain skeptical of Wal-Mart's commitment to sustainability. Its environmental stability initiatives will likely save money by reducing waste, but the company has shown little progress toward social sustainability, as its employees often complain of poor treatment. In fact, among many "waverider" companies, not all portions of sustainability's "triple bottom line" (• p. 21) are being met. Yet, even if Wal-Mart achieves a fraction of Scott's stated goals, the social and environmental benefits could be substantial. The corporation's vast global reach and its ability to persuade suppliers to alter their ways in order to retain its business mean that any change Wal-Mart enacts could potentially have far-reaching impacts.

Throughout this book we will encounter further examples of corporate involvement in sustainability efforts. However, corporate actions hinge on consumer behavior. It is up to all of us in our roles as consumers to encourage trends in sustainability by rewarding those corporations that truly promote sustainable solutions.

## Conclusion

Corporate sustainability, ecolabeling, and valuation of ecosystem services are a few of the recent developments that have brought economic approaches to bear on environmental protection and resource conservation. As economics becomes more oriented to sustainability, it renews some of its historic ties to ethics.

Environmental ethics has expanded people's sphere of ethical consideration outward to encompass other societies and cultures, other creatures, and even nonliving entities that were formerly outside the realm of ethical concern. This ethical expansion involves the concept of distributional equity, or equal treatment, which is the aim of environmental justice. One type of distributional equity is equity among generations. The concern of today's generations for the welfare of tomorrow's generations is the force that drives efforts toward sustainability.

Is sustainability a pragmatic pursuit for us? The answer largely depends on whether we believe that economic well-being and environmental well-being are opposed to one another or whether we accept that they can work in tandem. Equating economic well-being with economic growth, as most economists traditionally have, suggests that economic welfare entails a trade-off with environmental quality. However, if economic welfare can be enhanced in the absence of growth, we can envision economic health and environmental quality mutually reinforcing one another.

## REVIEWING OBJECTIVES

**You should now be able to:**

**Characterize the influences of culture and worldview on the choices people make**

- A person's culture strongly influences his or her worldview. Factors such as religion and political ideology are especially influential. (pp. 27–29)

**Outline the nature, evolution, and expansion of environmental ethics in Western cultures**

- Our society's domain of ethical concern has been expanding, such that we have granted more and more entities ethical consideration. (pp. 30–31)
- Anthropocentrism values humans above all else, whereas biocentrism values all life and ecocentrism values ecological systems. (pp. 31–32)
- The preservation ethic (preserving natural systems intact) and the conservation ethic (promoting responsible long-term use of resources) have guided branches of the environmental movement during the past century. (pp. 33–34)
- The environmental justice movement, which seeks equal treatment for people of all races and income levels, is one of several recent outgrowths of environmental ethics. (pp. 35–36)
- Environmental justice is also an international issue between developed and developing nations, and it is vital for social sustainability. (p. 37)

**Describe precepts of classical and neoclassical economics and summarize their implications for the environment**

- Classical economic theory proposes that individuals acting for their own economic good can benefit society as a whole. This view has provided a philosophical basis for free-market capitalism. (pp. 38–39)
- Neoclassical economics focuses on consumer behavior and on supply and demand as forces that drive economic activity. (pp. 40–41)
- Several assumptions of neoclassical economic theory contribute to environmental impact. (pp. 41–43)

**Compare the concepts of economic growth, economic health, and sustainability**

- Conventional economic theory has promoted never-ending economic growth, with little regard to possible environmental impact. (p. 43)
- Economic growth does not necessarily lead to social well-being. (pp. 43–44, 47)
- Some economists believe that developing a steady-state economy will be necessary if we are to achieve sustainability. (pp. 45–46)

**Explain the fundamentals of environmental economics and ecological economics**

- Environmental economists advocate reforming economic practices to promote sustainability. Key approaches are to identify external costs, assign value to nonmonetary items, and attempt to make market prices reflect real costs and benefits. (pp. 45–49)
- Ecological economists support these efforts and others. Many support developing a steady-state economy. (pp. 45–49)
- Consumer choice in the marketplace can help drive businesses and corporations to pursue sustainability goals. (pp. 49–52)

## TESTING YOUR COMPREHENSION

1. What does the study of ethics encompass? Describe three classic ethical standards. What is environmental ethics?
2. Describe the philosophical perspectives of anthropocentrism, biocentrism, and ecocentrism. How would you characterize the perspective of the Mirrar Clan?
3. Differentiate between the preservation ethic and the conservation ethic. Explain the contributions of John Muir and Gifford Pinchot in the history of environmental ethics.
4. Describe Aldo Leopold's "land ethic." How did Leopold define the "community" to which ethical standards should be applied?
5. Define the concept of environmental justice. What inequities relevant to environment justice concerns exist in your city, state, or country?
6. Name four key contributions the environment makes to the economy.
7. Describe Adam Smith's metaphor of the "invisible hand." How did neoclassical economists refine classical economics?
8. Describe four ways in which critics hold that neoclassical economic approaches can negatively affect the environment.
9. Compare and contrast the views of neoclassical economists, environmental economists, and ecological economists.
10. What is contingent valuation, and what is one of its weaknesses? Describe an alternative method that addresses this weakness.

## SEEKING SOLUTIONS

1. Do you feel that an introduction to environmental ethics and worldviews is an important part of a course in environmental science? Should ethics and worldviews be components of other science courses? Explain your answers.
2. Describe your worldview as it pertains to your relationship with the environment. How do you think your culture has influenced your worldview? How do you think your personal experience has influenced it? Do you feel that you fit into any particular category discussed in this chapter? Why or why not?
3. How would you analyze the case of the Mirrar Clan and the proposed Jabiluka uranium mine from each of the following perspectives? In your description, list at least one question that a person of each perspective would likely ask when attempting to decide whether the

mine should be developed. Be as specific as possible, and be sure to identify similarities and differences in approaches:
- Preservationist
- Conservationist
- Deep ecologist
- Environmental justice advocate
- Neoclassical economist
- Ecological economist

4. Do you think that a steady-state economy is a practical alternative to the prevailing paradigm that prioritizes economic growth? Why or why not?
5. Do you think we should attempt to quantify and assign market values to ecosystem services and other entities that have only nonmarket values? Why or why not?
6. **THINK IT THROUGH** A manufacturing facility on a river near your home provides jobs for 200 people in your community and pays $2 million in taxes to the local government each year. Sales taxes from purchases made by plant employees and their families contribute an additional $1 million to local government coffers. However, newspaper reports and recent peer-reviewed studies in well-respected scientific journals reveal that the plant has been discharging large amounts of waste into the river, causing a 25% increase in cancer rates, a 30% reduction in riverfront property values, and a 75% decrease in native fish populations.

   The plant owner says the facility can stay in business only because there are no regulations mandating expensive treatment of waste from the plant. If any such regulations were imposed, he says he would close the plant, lay off its employees, and relocate to a more business-friendly community.

   How would you recommend resolving this situation? What further information would you want to know before making a recommendation? In arriving at your recommendation, how did you weigh the costs and benefits associated with each of the plant's impacts?

## INTERPRETING GRAPHS AND DATA

As described on pages 46–47, economists are beginning to use various indicators of economic well-being as alternatives to the Gross Domestic Product (GDP), the total monetary value of final goods and services produced each year. An alternative measure called the Genuine Progress Indicator (GPI) is calculated as follows:

$$\text{GPI} = (\text{GDP}) + (\text{Benefits ignored by GDP}) - (\text{Environmental costs}) - (\text{Social and economic costs})$$

Benefits include such things as the value of parenting and volunteer work. Environmental costs include the costs of water, air, and noise pollution; loss of wetlands; and depletion of nonrenewable resources, among others. Social and economic costs include investment, lending, and borrowing costs, as well as the costs of crime, family breakdown, underemployment, commuting, pollution abatement, automobile accidents, and loss of leisure time.

1. Describe economic growth as measured by GDP for the United States from 1952 to 2002. Now describe economic growth as measured by GPI over the same time period. To what factors would you attribute the growing difference between these measures?
2. For GPI to grow significantly, one or more things must happen: Either GDP must grow faster; benefits ignored by the GDP must grow faster; or social, economic, and environmental costs must shrink relative to the other terms. Which of these scenarios do you think is most likely? Which would you prefer? How do the data in the graph support your answer?
3. Despite policy enacted in recent decades to regulate air and water pollution, hazardous waste disposal, solid waste management, forestry practices, and species protection, environmental costs continue to increase. Why do you suppose the trend remains in that direction?

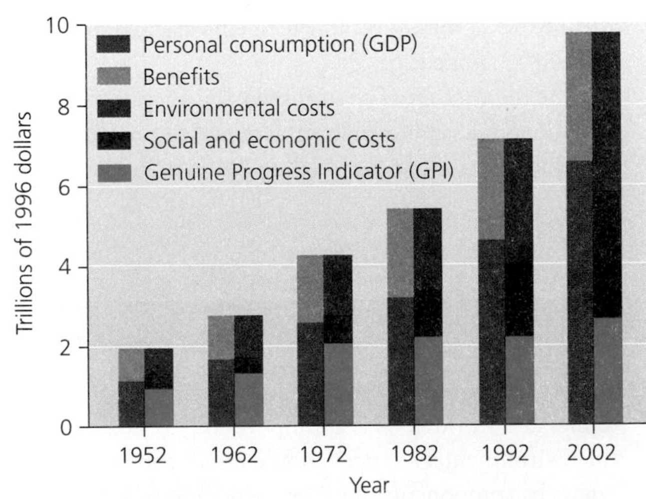


**Components of GPI and GDP for the United States, 1952–2002.** Data from Venetoulis, J., and C. Cobb. 2004. *The genuine progress indicator, 1950–2002 (2004 update)*. Redefining Progress.

## CALCULATING ECOLOGICAL FOOTPRINTS

Although the Gross Domestic Product (GDP) of the United States grew impressively between 1952 and 2002 (see the graph in the "Interpreting Graphs and Data" section), so did the U.S. population. According to the U.S. Census Bureau, the midyear population of the nation was 157,552,740 in 1952, and 288,368,698 in 2002—an 83% increase.

Estimate the values of the components of the Genuine Progress Indicator (GPI) for the United States in 1952 and 2002 from the graph, and enter your estimates into the table to the right. Then, using the population figures, calculate the per capita values for each component in 1952 and 2002.

| Components of GPI | U.S. total in 1952 (trillions of dollars) | Per capita in 1952 (thousands of dollars) | U.S. total in 2002 (trillions of dollars) | Per capita in 2002 (thousands of dollars) |
|---|---|---|---|---|
| GDP | 1.2 | 7.6 | 6.6 | 22.9 |
| Benefits | | | | |
| Environmental costs | | | | |
| Social and economic costs | | | | |
| GPI | | | | |

1. Consider your own life. What would you estimate is the value of the benefits in which you participate? Compare your personal estimate with the national average value in 2002.
2. What would you estimate are the values of the environmental, social, and economic costs for which you are personally responsible? How do they compare with the national average values in 2002?
3. In 2002, social and economic costs were proportionally larger relative to GDP than they were in 1952, and environmental costs were roughly the same in proportion to GDP. How would you account for these trends? What could you do to help improve these trends in your own personal accounting?

## Take It Further

Go to www.aw-bc.com/withgott or the student CD-ROM, where you'll find:

- Suggested answers to end-of-chapter questions
- Quizzes, animations, and flashcards to help you study
- *Research Navigator*™ database of credible and reliable sources to assist you with your research projects
- GRAPHIt! Tutorials to help you interpret graphs
- INVESTIGATEIt! Current news articles that link the topics you study to case studies from your region to around the world

CHAPTER 3

# Environmental Policy: Decision Making and Problem Solving

Border of Tijuana, Mexico (left) and San Diego, California (right)

## Upon completing this chapter, you will be able to:

- Describe environmental policy and assess its societal context
- Identify the institutions important to U.S. environmental policy and recognize major U.S. environmental laws
- List the institutions involved with international environmental policy and describe how nations handle transboundary issues
- Delineate the steps of the environmental policy process and evaluate its effectiveness
- Categorize the different approaches to environmental policy

Wastewater plume in Pacific Ocean near Tijuana

CENTRAL CASE

# San Diego and Tijuana's Sewage Pollution Problems and Policy Solutions

**"Ignorance is an evil weed, which dictators may cultivate among their dupes, but which no democracy can afford among its citizens."**
—William Beveridge, British economist

**"It is the continuing policy of the Federal Government . . . to create and maintain conditions under which man and nature can exist in productive harmony and fulfill the social, economic, and other requirements of present and future generations of Americans."**
—National Environmental Policy Act, 1969

The beaches south of San Diego boast some of the world's best waves for surfing. These days, however, most surfers avoid the temptation. For it is here that the heavily polluted Tijuana River flows across the international border from Mexico and empties into the sea, disgorging millions of gallons of untreated wastewater.

"When it rains, I call it the sewage tsunami," says surfer and environmentalist Serge Dedina. "For 40 square miles, from Imperial Beach to Coronado, there is a brown plume as far as the eye can see."

Such incidents occur when heavy rains overwhelm the ability of sewage treatment plants to process wastewater. San Diego's coastal waters receive stormwater runoff when rains wash pollutants into local rivers. Across the border in the Mexican city of Tijuana, the city's aging, leaky sewer system becomes clogged with debris, causing raw sewage to overflow into the streets and, eventually, into the Tijuana River.

In the summer of 2005, San Diego County officials closed beaches 36 times, issued 446 days' worth of health advisories, and kept three areas off-limits to swimming the entire season. Of these beach closures, 79% were due to pollution from the Tijuana River. The pollution is even worse in the winter, when rains are heavier.

The Tijuana River winds northwestward through the arid landscape of northern Baja California, Mexico, crossing the U.S. border south of San Diego. A river's **watershed** consists of all the land from which water drains into the river, and the Tijuana River's watershed covers 4,500 $km^2$ (1,750 $mi^2$) and is home to 2 million people of two nations. The Tijuana River watershed is a *transboundary* watershed (so named because it crosses a political boundary—in this case, a national border), with approximately 70% of its area in Mexico (**Figure 3.1**). On the Mexican side of the border, the river and the arroyos, or creeks, that flow into it are lined with farms, apartments, shanties, and factories, as well as leaky sewage treatment plants and toxic dump sites. Rains wash pollutants from all these sources through the arroyos into the Tijuana River and eventually onto U.S. and Mexican beaches.

Although pollution has flowed in the Tijuana River for at least 70 years, the problem has grown worse in recent years as the region's population has boomed, outstripping the capacity of sewage treatment facilities. Beach closures and pollution advisories have become commonplace in recent years, and garbage carried by the flows litters the beaches. "Every day I find broken glass, balloons, or can pop-tops. I've even found hypodermic needles. It's really sad," one resident of Imperial Beach told her local newspaper in 2005.

Things are worse on the Mexican side because most Mexican residents of the Tijuana River watershed live in

FIGURE 3.1 The Tijuana River winds northwestward from Mexico into California just south of San Diego, draining 4,500 $km^2$ (1,750 $mi^2$) of land in its watershed (colored green in map). Pollution entering the river affects Mexican residents of the watershed and U.S. citizens on San Diego County beaches. During high-water episodes that overwhelm treatment facilities, millions of gallons of raw sewage have flushed into the river and the Pacific Ocean.

poverty relative to their U.S. neighbors. Close to one-third of Tijuana's homes are not connected to a sewer system, and in poor neighborhoods such as Loma Taurina, river pollution directly affects people's day-to-day lives. The rise of U.S.-owned factories, or *maquiladoras,* on the Mexican side of the border has contributed to the river's pollution, both through direct disposal of industrial waste and by attracting thousands of new workers to the already crowded region.

As impacts have intensified, people in the San Diego and Tijuana areas, from coastal residents to grassroots activists to businesspeople, have pressed policymakers to do something. As we explore environmental policy in this chapter, we will periodically return to the Tijuana River watershed and see how citizens and policymakers together have tried to address these problems.

# Environmental Policy: An Introduction

When a society reaches broad agreement that a problem exists, it may persuade its leaders to try to resolve the problem through the making of policy. **Policy** consists of a formal set of general plans and principles intended to address problems and guide decision making in specific instances. **Public policy** is policy made by governments, including those at the local, state, federal, and international levels. Public policy consists of laws, regulations, orders, incentives, and practices intended to advance societal welfare. **Environmental policy** is policy that pertains to human interactions with the environment. It generally aims to regulate resource use or reduce pollution to promote human welfare and/or protect natural systems.

Forging effective policy requires input from science, ethics, and economics. Science (Chapter 1) provides the information and analysis needed to identify and understand environmental problems and devise potential solutions to them. Ethics and economics (Chapter 2) offer criteria to assess the extent and nature of problems and to help clarify how society might like to address them. Government interacts with individual citizens, organizations, and the private sector in a variety of ways to formulate policy (**Figure 3.2**).

The pollution problems of the Tijuana River watershed illustrate how science, economics, and ethics each inform and motivate the making of policy. Scientific research tells us that raw sewage carries pathogens, organisms that can cause illness in humans and other animals (**Table 3.1**). Science also reveals how untreated sewage can radically alter conditions for aquatic and marine life, lowering concentrations of dissolved oxygen and increasing mortality for many species. Pollution and beach closures

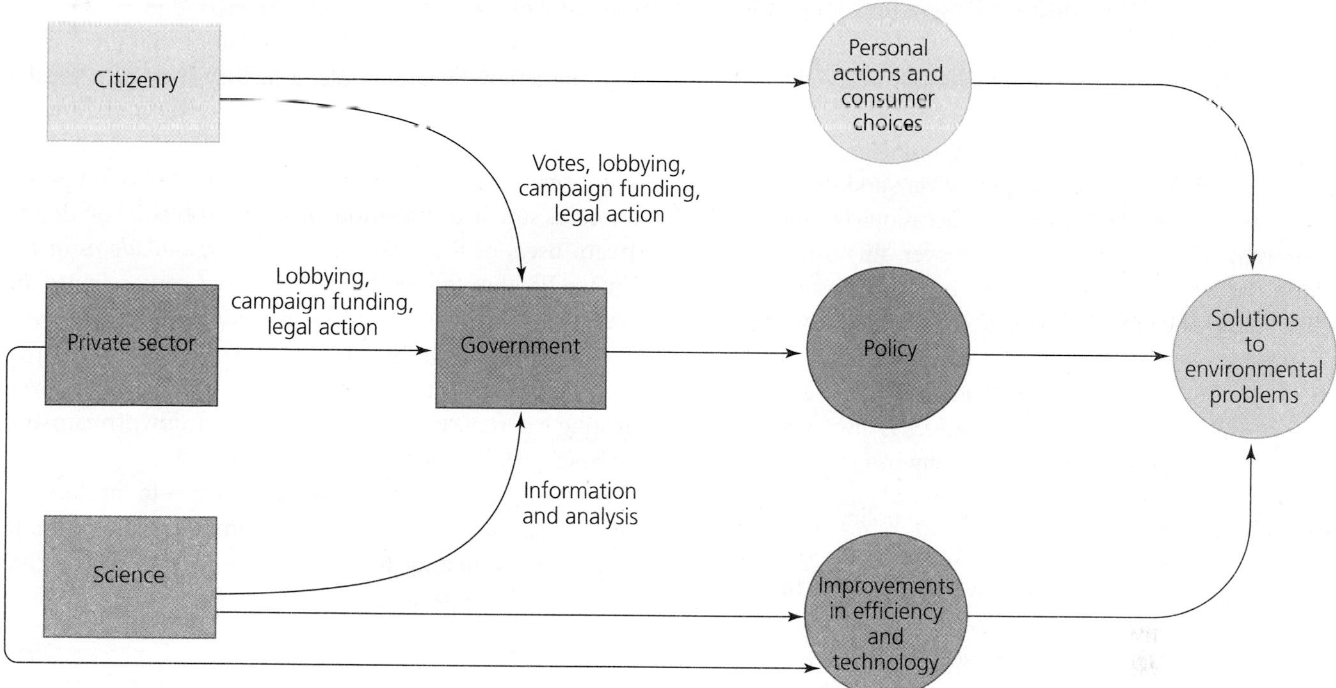


**FIGURE 3.2** Policy plays a central role in how we as a society address environmental problems. Voters, the private sector, and lobbying groups representing various interests influence government representatives. Scientific research also informs government decisions. Governmental representatives and agencies formulate policy that aims to address social problems, including environmental problems. Public policy—along with improvements in technology and efficiency from the private sector and personal actions and consumer purchasing choices exercised by citizens—can produce lasting solutions to environmental problems.

inflict economic loss by reducing recreation, tourism, and other economic activity associated with clean coastal areas. This is a major consideration both in Mexico and in southern California, whose beaches each year host 175 million visitors who spend over $1.5 billion. Ethically, water pollution poses problems because in most cases pollution from upstream users degrades water quality for downstream users not responsible for the pollution. Many phenomena we come to view as environmental problems share this combination of impacts—harming human health, altering ecological systems, inflicting economic damage, and creating inequities among people.

**TABLE 3.1 Pathogens and Health Effects Associated with Untreated Sewage**

| Pathogen | Human health effect |
|---|---|
| *Cryptosporidium* | Diarrhea, vomiting, cramps, dehydration |
| *Giardia lamblia* | Diarrhea, vomiting, cramps, dehydration |
| *Salmonella typhi* | Typhoid fever |
| *Vibrio cholerae* | Cholera |
| Poliovirus | Poliomyelitis |
| *Shigella* spp. | Dysentery |
| *Entamoeba histolytica* | Dysentery |
| Hepatitis A virus | Infectious hepatitis |

## Environmental policy addresses issues of equity and resource use

The capitalist market economic systems of modern constitutional democracies are largely driven by incentives for short-term economic gain rather than long-term social and environmental stability. Market capitalism provides little incentive for businesses or individuals to behave in ways that minimize environmental impact or equalize costs and benefits among parties. Such *market failure* (• p. 49) has traditionally been viewed as justification for government intervention. Environmental policy aims to protect environmental quality and the natural resources people use and also to promote equity in people's use of resources.

**The tragedy of the commons** Policy to protect resources held and used in common by the public is intended to safeguard these resources from depletion or degradation. As Garrett Hardin explained in his essay "The Tragedy of the Commons" (• pp. 5–6), a resource held in common that is accessible to all and is unregulated will eventually become overused and degraded. Therefore, he argued, it is in our best interest to develop guidelines for the use of such resources. In Hardin's illustrative example of a common pasture, guidelines might limit the number of animals each individual can graze or might require pasture users to pay to restore and manage the shared resource. These two concepts, restriction of use and active management, are central to environmental policy today.

The tragedy of the commons does not always play itself out as Hardin predicted. Some traditional societies have devised safeguards against exploitation, and in modern Western societies resource users have occasionally cooperated to prevent overexploitation. Moreover, in many cases Hardin's starting assumptions are not met; resources on public lands may not be equally accessible to everyone, but may instead be more accessible to wealthier or more established resource extraction industries. Nonetheless, the threat of overexploiting public resources is always real and has been a driving force behind much environmental policy.

**Free riders** Another reason to develop policy for publicly held resources is the **free rider** predicament. Let's say a community on a river suffers from water pollution that emanates from 10 different factories. The problem could in theory be solved if every factory voluntarily agreed to reduce its own pollution. However, once they all begin reducing their pollution, it becomes tempting for any one of them to stop doing so. Such a factory, by avoiding the sacrifices others are making, would in essence get a "free ride" on the efforts of others. If enough factories take a free ride, the whole effort will collapse.

Because of the free rider problem, private voluntary efforts are often less effective than efforts mandated by public policy. Public policy can prevent the free rider problem and ensure that all parties sacrifice equitably by enforcing compliance with laws and regulations or by taxing parties to attain funds to pursue societal goals.

**External costs** Environmental policy is also developed to ensure that some parties do not use resources in ways that harm others. One way to promote fairness is by dealing with *external costs* (• p. 42), harmful impacts that result from market transactions but are borne by people not involved in the transactions. For example, a factory may reap greater profits by discharging waste freely into a river and avoiding paying for waste disposal or recycling. Its actions, however, impose external costs (water pollution, decreased fish populations, aesthetic degradation, or other problems) on downstream users of the river. U.S.-owned *maquiladoras* in the Tijuana River watershed dump waste that contaminates the river for downstream users, such as Mexican families who use river water for washing (**Figure 3.3**). Likewise, the detergents these families use for washing further pollute the river, creating external costs for families farther downstream and for beachgoers in Mexico and California.

These goals of environmental policy—to protect resources against the tragedy of the commons and to promote equity by eliminating free riders and addressing external costs—will become apparent in case after case explored in this book.

## Many factors hinder implementation of environmental policy

If the goals of environmental policy are seemingly so noble, why is it that environmental laws are so often challenged, environmental regulations frequently derided, and the ideas of environmental activists repeatedly ignored or rejected by citizens and policymakers?

In the United States, most environmental policy has come in the form of regulations handed down from the federal government or from state or local governments. Businesses and individuals sometimes view these regulations as overly restrictive, bureaucratic, or unresponsive to human needs. For instance, many landowners fear that zoning regulations (• pp. 367–368) or protections for

FIGURE 3.3 River pollution raises many issues that have been viewed as justification for environmental policy. This woman washing clothes in the river may suffer upstream pollution from factories, and her use of detergents may cause further pollution for people living downstream.

endangered species (• pp. 317–320) will impose restrictions on the use of their land. Developers complain of time and money lost to bureaucracy in obtaining permits; reviews by government agencies; surveys for endangered species; and required environmental controls, monitoring, and mitigation. In the eyes of such property owners and businesspeople environmental regulation all too often means inconvenience and economic loss.

Another reason people sometimes do not see a need for environmental policy stems from the very nature of most environmental problems, which often develop gradually. The degradation of ecosystems and public health due to human impact on the environment are long-term processes. Human behavior, however, is generally geared toward addressing short-term needs—even if these conflict with long-term needs—and this tendency is reflected in our social institutions. Businesses usually opt for short-term economic gain over long-term considerations. The news media have a short attention span based on the daily news cycle, whereby new and sudden one-time events are given more coverage than slowly developing long-term trends. Politicians often act out of short-term interest because they depend on reelection every few years. For all these reasons, many environmental policy goals that seem admirable and that attract wide public support in theory may be obstructed in their practical implementation.

# U.S. Environmental Policy

The United States provides a good focus for understanding environmental policy in constitutional democracies worldwide. First, the United States historically pioneered innovative environmental policy. Second, U.S. policies have served as models—both of success and failure—for many other nations and international government bodies. Third, the United States exerts a great deal of influence on the affairs of other nations. Finally, understanding U.S. environmental policy on the federal level enables us to better understand environmental policy at local, state, and international levels.

## U.S. policy arises from the three branches of government

Environmental policy, like all U.S. policy, results from actions of the three branches of government established under the U.S. Constitution (**Figure 3.4**). Statutory law, or **legislation**, is created by the **legislative branch**, or Congress, which consists of the Senate and the House of Representatives. For instance, Congress passed the Tijuana River Valley Estuary and Beach Sewage Cleanup Act in 2000 to fund the treatment of sewage flowing into the United States in the Tijuana River.

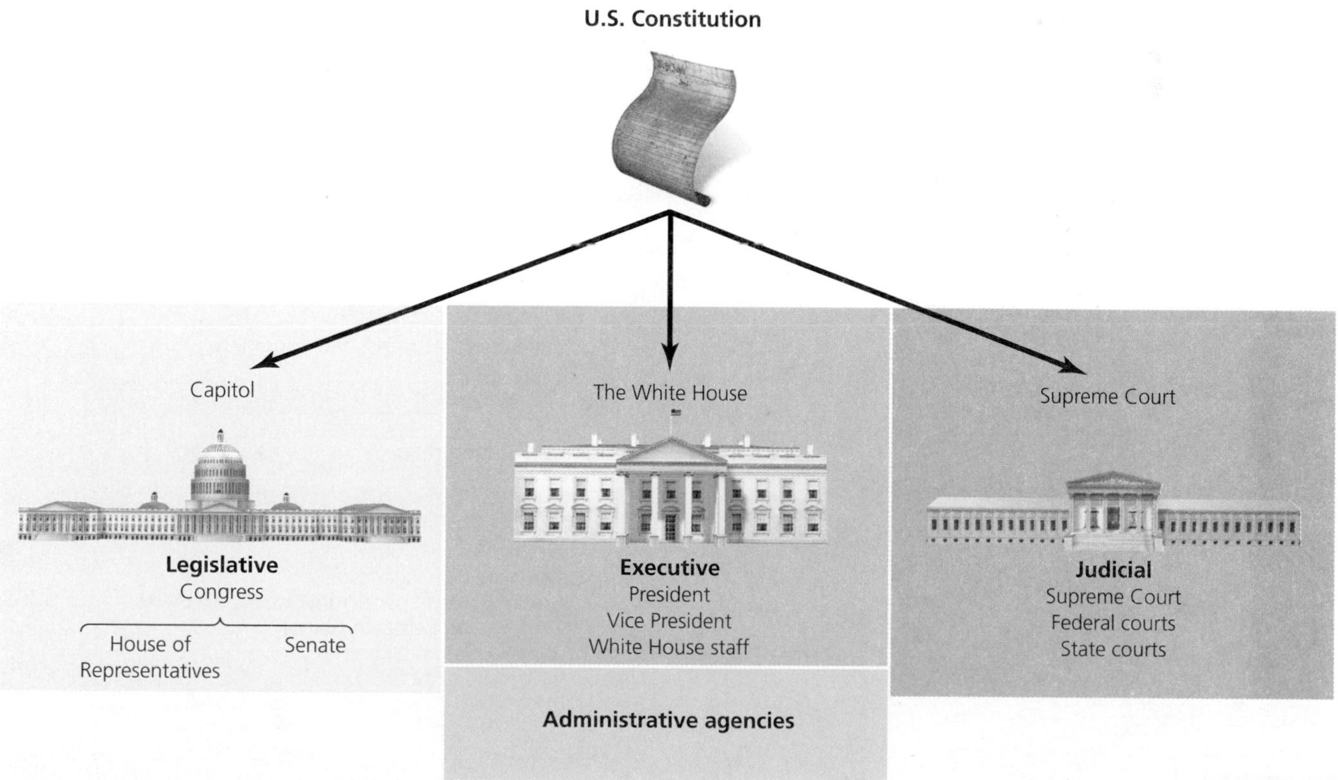


FIGURE 3.4 In the United States, federal powers are shared among the legislative, executive, and judicial branches of government. All three branches (including the administrative agencies of the executive branch) are responsible for aspects of environmental policy.

Legislation is enacted (approved) or vetoed (rejected) by the president, who heads the **executive branch**. The president may also issue *executive orders,* specific legal instructions for government agencies. Once statutory laws are enacted, their implementation and enforcement is assigned to the appropriate administrative agency within the executive branch. Administrative agencies (**Figure 3.5**), which may be established by Congress or by presidential order, are sometimes nicknamed the "fourth branch" of government because they are the source of a great deal of policy, in the form of regulations. **Regulations** are specific rules based on the more broadly written statutory law. Besides issuing regulations, administrative agencies monitor compliance with laws and regulations and enforce them when individuals or corporations violate them.

The **judicial branch**, or judiciary, consisting of the Supreme Court and various lower courts, is charged with interpreting law. This is necessary because social norms, societal conditions, and technologies change over time, and because Congress must write laws broadly to ensure that they apply to varied circumstances throughout the nation. Decisions rendered by the courts make up a body of law known as *case law.* Previous rulings serve as *precedents,* or legal guides, for later cases, steering judicial decisions through time. The judiciary has been an important arena for environmental policy. Grassroots environmental advocates and nongovernmental organizations use lawsuits as tools to help level the playing field with large corporations and government agencies. Conversely, the courts hear complaints from businesses and individuals challenging the constitutional validity of environmental laws they feel to be infringing on their rights.

| Federal Administrative Agencies that Influence Environmental Policy | |
|---|---|
| **Executive Office of the President** | **Department of the Interior** |
| Council on Environmental Quality | Bureau of Indian Affairs<br>Bureau of Land Management (BLM)<br>Bureau of Reclamation<br>Minerals Management Service<br>National Park Service (NPS)<br>Office of Surface Mining<br>U.S. Fish and Wildlife Service (USFWS)<br>U.S. Geological Survey (USGS) |
| **Department of Agriculture (USDA)** | |
| Natural Resources Conservation Service<br>U.S. Forest Service (USFS) | |
| **Department of Commerce** | **Department of Justice** |
| Bureau of the Census<br>National Marine Fisheries Service<br>National Oceanic and Atmospheric Administration (NOAA) | Environmental and Natural Resources Division |
| **Department of Defense** | **Department of Labor** |
| Army Corps of Engineers | Occupational Safety and Health Administration (OSHA) |
| **Department of Energy (DOE)** | **Department of State** |
| Energy Efficiency and Renewable Energy<br>Energy Information Administration (EIA)<br>Federal Energy Regulatory Commission (FERC)<br>National Laboratories and Technology Centers<br>Office of Environmental Management<br>Office of Fossil Energy | Bureau of Oceans and International Environmental and Scientific Affairs<br>Bureau of Population, Refugees, and Migration |
| | **Department of Transportation** |
| | Federal Transit Administration |
| **Department of Health and Human Services** | **Independent Agencies** |
| Agency for Toxic Substances and Disease Registry<br>Centers for Disease Control and Prevention (CDC)<br>Food and Drug Administration (FDA) | Consumer Product Safety Commission<br>Environmental Protection Agency (EPA)<br>National Aeronautics and Space Administration (NASA)<br>National Transportation Research Center<br>Nuclear Regulatory Commission (NRC)<br>Tennessee Valley Authority (TVA)<br>U.S. Agency for International Development (USAID) |

**FIGURE 3.5** The many administrative agencies of the executive branch are the source of most U.S. environmental policy. *Source:* U.S. General Services Administration, Washington, D.C.

## State and local policy also affects environmental issues

The structure of the federal government is mirrored at the state level with governors, legislatures, judiciaries, and agencies. State laws cannot violate principles of the U.S. Constitution, and if state and federal laws conflict, federal laws take precedence.

Many densely populated states, such as California, New York, and Massachusetts, have strong environmental laws and well-funded environmental agencies. Citizens of such states put a premium on safeguarding environmental quality because they have witnessed the extensive environmental impacts their states have already suffered. In contrast, many citizens of less populous states, such as those of the interior West, put less priority on environmental protection and favor unregulated development.

To safeguard public health in locations such as San Diego's beaches, California legislators in 1997 required state environmental health officials to set standards for testing waters for bacterial contamination. Officials issue an advisory, or warning, when bacterial concentrations in nearshore waters exceed health limits established by California law. In 1999, further legislation passed by the California State Assembly mandated another state agency to develop methods for locating sources of contamination. As we proceed through our discussion of federal policy, keep in mind that important environmental policy is also created at the state and local levels.

## Some constitutional amendments bear on environmental law

The U.S. Constitution lays out several principles that have come to be especially relevant to environmental policy. One of these is the clause from the Fourteenth Amendment prohibiting a state from denying "equal protection of its laws" to any person. This amendment provides the constitutional basis for the environmental justice movement (• pp. 35–37).

The Fifth Amendment ensures, in part, that private property shall not "be taken for public use without just compensation." Courts have interpreted this clause, known as the *takings clause*, to ban not only the literal taking of private property but also what is known as regulatory taking. A **regulatory taking** occurs when the government, by means of a law or regulation, deprives a property owner of some or all economic uses of his or her property. Many people cite the takings clause to argue against environmental regulations that restrict land use and development on privately owned land. For example, some would contend that zoning regulations (• pp. 367–368) that prohibit a landowner from opening a hazardous waste dump in a residential neighborhood deprive the landowner of an economically valuable use of the land and, therefore, violate the Fifth Amendment.

In a landmark case in 1992, the Supreme Court ruled that a state land use law intended to "prevent serious public harm" violated the takings clause. The case, *Lucas v. South Carolina Coastal Council*, involved a developer named Lucas who in 1986 purchased beachfront property in South Carolina for $975,000. In 1988, before Lucas began to build, South Carolina's legislature passed a law banning construction on eroding beaches (**Figure 3.6**). A state agency classified the Lucas property as an eroding beach and prohibited residential construction there.

Believing that this constituted a regulatory taking, Lucas asked a state court to overrule the new law and allow him to build homes on his property. The state court agreed and ruled that Lucas was entitled to $1.2 million in compensation. South Carolina appealed the case to the state Supreme Court, which overturned the lower court's decision. Lucas then appealed to the U.S. Supreme Court, which overturned the state Supreme Court's ruling and sided with the lower court, declaring that the state law deprived Lucas of all economically beneficial uses of his land. Today, homes stand on the land, and regulatory taking remains a contentious area of law—a key issue in the sensitive balance between private rights and the public good.

FIGURE 3.6 Beaches subject to erosion and hurricane damage like this one in South Carolina have given rise to environmental policy debates. Should the government be able to prevent development in areas where erosion, storms, and flooding pose risks to property and human life? If so, does such prohibition constitute a taking of private property rights, and should the property owner be compensated? These were the questions addressed in *Lucas v. South Carolina Coastal Council.*

**Weighing THE Issues | Regulatory Takings**

Imagine you have purchased land and plan to make money by clearing its forest and building condominiums on it. Now suppose a local environmental group finds an endangered plant species on the property and petitions the government to prevent development of the land. Use the takings clause to argue why you should be allowed to build or should be financially compensated if you are not allowed to build.

Now imagine that you are a member of the environmental group and also a neighbor of the landowner in question. The landowner's land holds the last stand of forest in this region, which has experienced extensive forest loss over the past decade. You feel your quality of life and that of your neighbors will be compromised if the development goes ahead. Make a case why the landowner should not be allowed to build.

Which argument do you feel has greater merit? Would it make a difference to you if the land in question had been zoned as developable or nondevelopable when the purchase was made? What do you think is the best way to balance private property rights with protection of the public good in cases like this?

## Early U.S. environmental policy addressed public land management

The laws that comprise U.S. environmental policy were created largely in three periods. Laws enacted during the first period, from the 1780s to the late 1800s, dealt primarily with the management of public lands and accompanied the westward expansion of the nation. Environmental laws of this period were intended mainly to promote settlement and the extraction and use of the continent's abundant natural resources.

Among these early laws were the *General Land Ordinances of 1785 and 1787*, which gave the federal government the right to manage unsettled lands and created a grid system for surveying them and readying them for private ownership. Between 1785 and the 1870s, the federal government promoted settlement in the Midwest and West on lands it had appropriated from Native Americans by doling out as many of these lands as possible to its citizens. Western settlement provided these citizens with means to achieve prosperity, while relieving crowding in Eastern cities. It expanded the nation's geographical reach at a time when the young United States was still jostling with European powers for control of the continent. It also wholly displaced the millions of Native Americans who had inhabited these lands. U.S. environmental policy of this era reflected the public perception that Western lands were practically infinite and inexhaustible in natural resources. Following are a few laws typical of this era:

- The Homestead Act of 1862 allowed any citizen to claim, for a $16 fee, 65 ha (160 acres) of public land by living there for 5 years and cultivating the land or building a home (**Figure 3.7a**). A waiting period of only 14 months was available to those who could pay $176 for the land.
- The Mineral Lands Act of 1866 provided land for $5 per acre to promote mining and settlement. It allowed mining to occur subject to local customs, with no government oversight (**Figure 3.7b**).
- The Timber Culture Act of 1873 granted 65 ha (160 acres) to any citizen promising to cultivate trees on one-quarter of that area (**Figure 3.7c**).

Such laws encouraged settlers, entrepreneurs, and land speculators to move west, hastening the closing of the frontier.

## The second wave of U.S. environmental policy addressed impacts of the first

In the late 1800s, as the continent became more populated and its resources were increasingly exploited, public perception and government policy toward natural resources began to shift. Laws of this period aimed to mitigate some of the environmental problems associated with westward expansion. In 1872 Congress designated Yellowstone as the world's first national park. In 1891 Congress, to prevent overharvesting and protect forested watersheds, passed a law authorizing the president to create "forest reserves" that were off-limits to logging. In 1903, President Theodore Roosevelt created the first national wildlife refuge. These acts enabled the creation, over the next few decades, of a national park system, national forest system, and national wildlife refuge system that still stand as global models (• pp. 338–339, 349). These developments reflected a new understanding that the West's resources were exhaustible, and required legal protection.

Land management policies continued through the 20th century, targeting soil conservation in the Dust Bowl years (• pp. 245–246) and extending through the Wilderness Act of 1964 (• p. 350), which sought to preserve pristine lands "untrammeled by man, where man himself is a visitor who does not remain."

**(a) Settlers in Custer County, Nebraska, circa 1860**

**FIGURE 3.7** Settlers (**a**) took advantage of the federal government's early environmental policies, including the Homestead Act of 1862. Early mining activities on public lands (**b**) were largely unregulated. The Mineral Lands Act of 1866 required that mining could occur "subject . . . to the local customs or rules of miners," rather than in a way that would protect the environment. Although the Timber Culture Act of 1873 promoted tree planting on settled agricultural lands, elsewhere the timber industry was allowed to clear-cut the nation's ancient trees (**c**) with little government policy that limited logging or required replanting or conservation.

**(b) Nineteenth-century mining operation, Lynx Creek, Alaska**

**(c) Loggers felling an old-growth tree, Washington**

## The third wave of U.S. environmental policy responded largely to pollution

Further social changes in the mid- to late-20th century gave rise to the third major period of U.S. environmental policy. In a more densely populated country driven by technology, heavy industry, and intensive resource consumption, Americans found themselves better off economically but living amid dirtier air, dirtier water, and more waste and toxic chemicals. During the 1960s and 1970s, several events triggered increased awareness of environmental problems and brought about a shift in public priorities and important changes in public policy.

A landmark event was the 1962 publication of *Silent Spring*, a book by American scientist and writer Rachel

FIGURE 3.8 Scientist, writer, and citizen activist Rachel Carson illuminated the problem of pollution from DDT and other pesticides in her 1962 book, *Silent Spring.*

Carson (**Figure 3.8**). *Silent Spring* awakened the public to the negative ecological and health effects of pesticides and industrial chemicals. (The book's title refers to Carson's warning that pesticides might kill so many birds that few would be left to sing in springtime.)

The Cuyahoga River (**Figure 3.9**) also did its part to raise attention to the hazards of pollution. The Cuyahoga was so polluted with oil and industrial waste that the river actually caught fire near Cleveland, Ohio, more than half a dozen times during the 1950s and 1960s. This spectacle, coupled with an enormous oil spill off the Pacific coast near Santa Barbara, California, in 1969, moved the public to prompt Congress and the president to do more to protect the environment.

Today, largely because of environmental policies enacted since the 1960s, pesticides are more strictly regulated, and the nation's air and water are considerably cleaner. The public enthusiasm for environmental protection that spurred such advances remains strong today. Polls repeatedly show that an overwhelming majority of Americans favor environmental protection. Such support is evident each year in April, when millions of people worldwide celebrate Earth Day in thousands of locally based events featuring speeches, demonstrations, hikes, bird-walks, cleanup parties, and more. Since the first Earth Day, on April 22, 1970, participation in this event has grown and spread to nearly every country in the world (**Figure 3.10**).

## NEPA gives citizens input into environmental policy decisions

Besides Earth Day, two federal actions marked 1970 as the dawn of the modern era of environmental policy in the United States. On January 1, 1970, President Richard Nixon signed the **National Environmental Policy Act (NEPA)** into law (**Figure 3.11**). NEPA created an agency called the Council on Environmental Quality and required that an **environmental impact statement (EIS)** be prepared for any major federal action that might significantly affect environmental quality. An EIS is a report of results from detailed studies that assess the

FIGURE 3.9 In a spectacular display of the need for better control over water pollution, Ohio's Cuyahoga River caught fire several times in the 1950s and 1960s. The Cuyahoga was so polluted with oil and industrial waste that the river would burn for days at a time.

**FIGURE 3.10** April 22, 1970, marked the first Earth Day celebration (**a**). This public outpouring of support for environmental protection sparked the third wave of environmental policy in the United States. Decades later, Earth Day is celebrated by millions of people across the globe (**b**), as shown here in Nepal.

**(a) The first Earth Day, Washington, D.C., 1970**

**(b) Schoolchildren celebrating Earth Day, Katmandu, Nepal, 2002**

potential impacts on the environment that would likely result from development projects undertaken or funded by the federal government.

NEPA's effects have been far-reaching. The EIS process forces government agencies and any businesses that contract with them to evaluate impacts on the environment before proceeding with a new dam, highway, or building project. Although the EIS process generally does not halt such projects, it can serve as an incentive to lessen the environmental damage resulting from a development or activity. NEPA also grants ordinary citizens input in the policy process by requiring that environmental impact statements be made publicly available and that public comment on them be solicited and considered.

## Creation of the EPA marked a shift in environmental policy

Six months after signing NEPA into law, Nixon issued an executive order calling for a new, integrated approach to environmental policy based on the understanding that environmental problems are interrelated. "The Government's environmentally-related activities have grown up piecemeal over the years," the order stated. "The time has come to organize them rationally and systematically." Nixon's order moved elements of agencies regulating water quality, air pollution, solid waste, and other issues into the newly

**FIGURE 3.11** Richard Nixon served as president when modern environmental policy took shape. Here he signs the National Environmental Policy Act (NEPA) into law on January 1, 1970.

created **Environmental Protection Agency (EPA)**. The order charged the EPA with conducting and evaluating research, monitoring environmental quality, setting and enforcing standards for pollution levels, assisting the states in meeting standards and goals, and educating the public.

## Other prominent laws followed

Ongoing public demand for a cleaner environment during this period resulted in a number of key laws that remain linchpins of U.S. environmental policy (**Figure 3.12**). For problems such as the Tijuana River's pollution, a crucial law has been the Clean Water Act of 1977.

Throughout much of U.S. history, water policy was left largely to local and state governments. (An early exception was the federal Rivers and Harbors Act of 1899, which restricted dumping and discharges into navigable waters.) Prior to passage of federal legislation such as the Clean Water Act, pollution problems were subject primarily to *tort law* (law addressing harm caused to one entity by another), and individuals suffering external costs from pollution were limited to seeking redress through lawsuits. Reaction to the ruling in the case *Boomer v. Atlantic Cement Company* in 1970 effectively ended tort law's viability as a tool for preventing pollution. The court ruled that even though residents of Albany, New York, were suffering pollution from a neighborhood cement plant, the pollution could continue because the economic costs of controlling the pollution were greater than the economic costs of the damage the pollution caused.

The flaming waters of the Cuyahoga, however, indicated to many people that tough legislation was needed. Thanks to restrictions on pollutants by the Federal Water Pollution Control Acts of 1965 and 1972, and then the Clean Water Act, U.S. waterways finally began to recover. These laws regulated the discharge of wastes, especially from industry, into rivers and streams. The Clean Water Act also aimed to protect wildlife and establish a system for granting permits for the discharge of pollutants. Today there are thousands of federal, state, and local environmental laws in the United States, and thousands more abroad.

### Weighing THE Issues | Judging Costs in Tort Law

Imagine you are the judge presiding over a lawsuit brought by citizens who live next to an industrial plant. Pollution from the plant regularly covers their homes, yards, and cars with soot, making the people cough and wheeze. In weighing whether to require the plant to clean up its emissions, will you consider in your ruling the financial costs the factory would need to incur? Or will you ignore economic aspects and base your ruling solely on evidence of harm to health? Give reasons for your answer.

## The social context for environmental policy changes over time

Historians have suggested that major advances in environmental policy occurred in the 1960s and 1970s because three factors converged. First, evidence of environmental

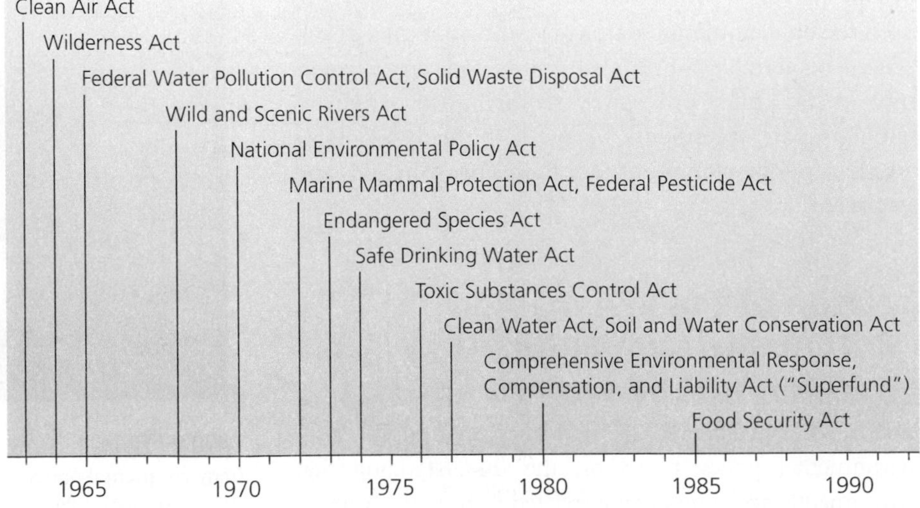


FIGURE 3.12 Most major laws in modern U.S. environmental policy were enacted in the 1960s and 1970s.

problems became widely and readily apparent. Second, people could visualize policies to deal with the problems. Third, the political climate was ripe, with a supportive public and leaders who were willing to act.

By the 1980s, the political climate in the United States had changed. Although public support for the goals of environmental protection remained high, many citizens and policy experts began to feel that the legislative and regulatory means used to achieve environmental policy goals too often imposed economic burdens on businesses and personal burdens on individuals. Since 1980, numerous attempts have been made at the federal level to roll back or weaken environmental laws, culminating in an array of efforts by the George W. Bush administration and by the Republican-controlled Congress in power from 1994 through 2006.

As advocates of environmental protection watched their hard-won gains eroding, many began to question whether new perspectives and strategies were needed. In a provocative 2004 essay titled "The Death of Environmentalism," political consultants Michael Shellenberger and Ted Nordhaus argued that the modern American environmental movement was too wedded to its past and needed to reinvent its approach. Environmental advocates needed to appeal to people's core values and not simply offer technical policy fixes, Shellenberger and Nordhaus maintained. They needed to stop labeling problems as "environmental" and start showing people why these problems are actually human issues that lie at the very heart of our quality of life. They needed to be more responsive to people's needs and to build broad and meaningful alliances with labor unions, advocates for the poor, and other progressive causes in order to establish a political majority that could govern. Most of all, they needed to articulate a positive, inspiring vision for the future.

As the United States retreated from its leadership in environmental policy during the past two decades, other nations increased their political attention to environmental issues. Internationally, environmental advocates were trying to build the kinds of broad alliances and offering the kind of positive vision that Shellenberger and Nordhaus lamented were so lacking in the United States. The 1992 Earth Summit at Río de Janeiro, Brazil, was the largest international diplomatic conference ever held, drawing representatives from 179 nations and unifying these leaders around the idea of sustainable development (• pp. 20–21 and • pp. 666–667).

Indeed, we may now be embarking on a fourth wave of environmental policy, one focused on sustainable development. This new policy approach tries to find ways to safeguard the functionality of natural systems while raising living standards for the world's poorer people (**Figure 3.13**). As the world's nations continue to feel the social, economic, and ecological effects of environmental degradation, environmental policy will without doubt become a more central part of governance and everyday life in all nations in the years ahead.

FIGURE 3.13 Many nations are shifting their policies to support sustainable development efforts, trying to increase standards of living while safeguarding the environment. Here, a woman stirs rice on a solar-powered oven at a restaurant made of recycled drink cans, showcased at the Ubunto Village at the U.N. World Summit on Sustainable Development in Johannesburg, South Africa, in 2002.

# International Environmental Policy

Environmental systems pay no heed to political boundaries, so environmental problems often are not restricted to the confines of particular countries. For instance, most of the world's major rivers cross international borders, so problems like those along the Tijuana River are frequently international in scope. Because U.S. law has no authority in Mexico or any other nation outside the United States, international law is vital to solving transboundary problems.

## Cross-border cooperation can help address environmental problems

Often nations make progress on international issues not through legislation, but through creative bilateral or multilateral agreements hammered out after a lot of hard work and diplomacy. Such was the case with the effort to develop a long-term plan to manage drinking water and wastewater in the Tijuana metropolitan area. This master planning process, funded by the U.S. EPA under Congressional direction, began in January 2002. It involved the Comisión Estatal de Servicios Publicos de Tijuana (CESPT), the city agency that manages water and wastewater in Tijuana; the Mexican National Water Commission; the State Water Commission for Baja California; and the North American Development Bank. The resulting Tijuana Master Plan for Water and Wastewater Infrastructure aims to address a shortage of drinking water and possibilities for its reuse; water infrastructure; and wastewater collection, transport, and treatment.

In 2002 the U.S. Congress also provided funding and authority for the EPA to work with CESPT to upgrade Tijuana's sewer system. The Tijuana Sewer Rehabilitation Project, known as *Tijuana Sana* ("Healthy Tijuana"), was approved in 2001. This cooperative transboundary pollution prevention program will attempt to repair leaky sewer pipes in Tijuana. The $43 million project aims to replace 131 km (81 mi), or 7.5%, of Tijuana's sewer pipes. If it succeeds, the project should reduce or eliminate the most severe sewage spills into the Tijuana River.

## International law includes conventional and customary law

Because solving transboundary dilemmas requires international cooperation, several principles of international law and a number of international organizations have arisen. Whereas U.S. law arises from the Constitution and the Bill of Rights, international environmental law is more nebulous in its origins and authority.

International law known as **customary law** arises from long-standing practices, or customs, held in common by most cultures. International law known as **conventional law** arises from *conventions,* or *treaties,* into which nations enter. One example is the Montreal Protocol, a 1987 accord among more than 160 nations to reduce the emission of airborne chemicals that deplete the ozone layer (• pp. 489–491). Another example is the Kyoto Protocol to reduce fossil fuel emissions that contribute to global climate change (• p. 534).

In 1990, Mexico and the United States signed a treaty and agreed to build an international wastewater treatment plant to handle excess sewage from Tijuana (**Figure 3.14**). In this case the treaty process worked well, but the results fell short of expectations. The facility reached its capacity within 3 years because Tijuana's population grew so quickly, and excess sewage simply flowed downriver. Moreover, a secondary treatment facility was never built, so the water the plant discharged to the ocean often violated water quality standards established by U.S. and Mexican law (see "The Science behind the Story, • pp. 72–73).

FIGURE 3.14 In 1990, the United States and Mexico entered into a treaty and agreed to build the International Wastewater Treatment Plant (IWTP) to handle excess sewage from Tijuana. The IWTP began operating just north of the border in 1997 and treats up to 95 million L (25 million gal) of Mexican wastewater each day, but it reached its capacity within 3 years of opening.

## Several organizations shape international environmental policy

Although there is no real mechanism for enforcing international environmental law, a number of international organizations regularly act to influence the behavior of nations by providing funding, applying peer pressure, and/or directing media attention.

**The United Nations** In 1945, representatives of 50 countries founded the **United Nations (U.N.)**. Headquartered in New York City, this organization's purpose is "to maintain international peace and security; to develop friendly relations among nations; to cooperate in solving international economic, social, cultural and humanitarian problems and in promoting respect for human rights and fundamental freedoms; and to be a centre for harmonizing the actions of nations in attaining these ends."

The United Nations has taken an active role in shaping international environmental policy (**Figure 3.15**). Of several agencies within it that influence environmental policy, most notable is the *United Nations Environment Programme (UNEP),* created in 1972, which helps nations understand and solve environmental problems. Based in Nairobi, Kenya, its mission is sustainability, enabling countries and their citizens "to improve their quality of life without compromising that of future generations." UNEP's extensive research and outreach activities provide a wealth of information useful to policymakers and scientists throughout the world.

**The European Union** The **European Union (EU)** seeks to promote Europe's unity and its economic and social progress

FIGURE 3.15 The United Nations is active in international environmental policymaking. For instance, it sponsored the 2002 Earth Summit in Johannesburg, South Africa.

(including environmental protection) and to "assert Europe's role in the world." The EU can sign binding treaties on behalf of its 27 member nations and can enact regulations that have the same authority as national laws in each member nation. It can also issue *directives,* which are more advisory in nature. The EU's European Environment Agency works to address waste management, noise pollution, water pollution, air pollution, habitat degradation, and natural hazards. The EU also seeks to remove trade barriers among member nations. It has classified some nations' environmental regulations as barriers to trade because some northern European nations have traditionally had more stringent environmental laws that prevent the import and sale of environmentally harmful products from other member nations.

**The World Trade Organization** Based in Geneva, Switzerland, the **World Trade Organization (WTO)** was established in 1995, having grown from a 50-year-old international trade agreement. The WTO represents multinational corporations and promotes free trade by reducing obstacles to international commerce and enforcing fairness among nations in trading practices. Whereas the United Nations and the European Union have limited influence over nations' internal affairs, the WTO has real authority to impose financial penalties on nations that do not comply with its directives. These penalties can on occasion play major roles in shaping environmental policy.

Like the EU, the WTO has interpreted some national environmental laws as unfair barriers to trade. For instance, in 1995 the U.S. EPA issued regulations requiring cleaner-burning gasoline in U.S. cities, following Congress's amendments of the Clean Air Act. Brazil and Venezuela filed a complaint with the WTO, saying the new rules unfairly discriminated against the petroleum they exported to the United States, which did not burn as cleanly. The WTO agreed, ruling that even though the South American gasoline posed a threat to human health in the United States, the EPA rules represented an illegal trade barrier. The ruling forced the United States to alter its approach to regulating gasoline. Not surprisingly, critics have frequently charged that the WTO aggravates environmental problems.

### Weighing THE Issues | Trade Barriers and Environmental Protection

If Nation A has stricter laws for environmental protection than Nation B, and if these laws restrict the ability of Nation B to export its goods to Nation A, then by the policy of the WTO and the EU, Nation A's environmental protection laws could be overruled in the name of free trade. Do you think this is right? What if Nation A is a wealthy industrialized country and Nation B is a poor developing country that needs every economic boost it can get?

**Nongovernmental organizations** A number of nongovernmental organizations (NGOs) have become international in scope and exert influence over international environmental policy (**Figure 3.16**). The nature of these advocacy groups is diverse. Some, such as the Nature Conservancy, focus on accomplishing conservation objectives on the ground (in its case, purchasing and

FIGURE 3.16 Pursuing different visions of what makes for international environmental progress, nongovernmental organizations, such as the environmental advocacy group Greenpeace, sometimes clash with international institutions, such as the World Bank and the World Trade Organization.

THE SCIENCE BEHIND THE STORY

## Assessing the Environmental Impact of Treating Transboundary Sewage

*California State Assemblywoman Lori Saldaña*

In 1990 the United States and Mexico agreed to construct a wastewater treatment plant in the U.S. portion of the Tijuana River Valley. Before the South Bay International Wastewater Treatment Plant (IWTP) could be built, however, the U.S. government was legally required by the National Environmental Policy Act (NEPA) to assess its environmental impact.

The IWTP's planners also had other laws to take into account. The Clean Water Act requires that harmful toxins and bacteria be removed from wastewater discharged into U.S. rivers, lakes, and oceans, and the Endangered Species Act protects species such as the Pacific pocket mouse, which inhabits the Tijuana River Valley.

International treaties also dictated how the IWTP should be built. For example, an agreement between the United States and Mexico required that the plant be funded by the EPA, be built on U.S. territory, and treat at least 1,095 L/sec (25 million gal/day). Moreover, state and local laws regulated the impact of the plant on nearby communities and on the quality of California's coastal waters. Before construction could begin, all these constraints had to be addressed by the EPA and the U.S. section of the International Boundary and Water Commission (IBWC), the agency that would own and operate the plant.

In 1991 a draft EIS for the plant was released for public comment. The draft provided a preliminary assessment of the plant's impact on biological and cultural resources, public health and safety, scenic and recreational values, water quality, and other environmental factors. Three years later, after extensive research and public discussion, a final EIS was released. In it, the EPA and the IBWC endorsed the plant as the "preferred alternative," and the EPA confirmed the endorsement with an official Record of Decision.

In its Record of Decision, the EPA chose to open the plant in phases. In the first phase, large solids and some suspended particles would be filtered out of wastewater using an advanced primary treatment process. However, other pollutants—including toxic metals and bacteria—would remain untreated until facilities were constructed for a secondary treatment process known as activated sludge, in which microorganisms convert highly toxic waste into less toxic substances (see • pp. 435–438 for information on wastewater treatment). As long as those facilities remained uncompleted, the plant would be releasing polluted water into the ocean, in violation of the Clean Water Act.

In response, two environmental groups sued the EPA and the IBWC, arguing that the EIS had failed to consider a type of secondary treatment known as a completely mixed aerated pond system. The suit was eventually settled out of court, but it helped spur the EPA and IBWC to conduct a supplemental EIS in which they considered seven different secondary treatment alternatives. Based on this analysis, the EPA eventually decided to endorse the pond system.

In the supplemental EIS, the EPA also reaffirmed its decision to release treated wastewater through the South Bay Ocean Outfall, a 5.8-km (3.6-mi) underwater tunnel with outlets 29 m (95 ft) below the ocean surface (see figure). The decision was based on a computer model of ocean currents and pollution levels off the California coast, which indicated that wastewater released through the outfall—once given secondary treatment—

managing land and habitat for rare species) without becoming politically involved. Other groups, including Conservation International, the World Wide Fund for Nature, Greenpeace, Population Connection, and many others, attempt to shape policy directly or indirectly through research, education, lobbying, or protest. NGOs apply more funding and expertise to environmental problems—and conduct more research intended to solve them—than do many national governments.

**The World Bank** Established in 1944 and based in Washington, D.C., the **World Bank** is one of the globe's largest sources of funding for economic development. This institution can shape environmental policy through its funding of dams, irrigation infrastructure, and other major development projects. In fiscal year 2005, the World Bank provided over $22.3 billion in loans for projects designed to benefit the poorest people in the poorest countries around the world.

The IWTP gives sewage primary treatment but still needs to fund facilities to provide secondary treatment. Currently, treated waste is discharged into the Pacific Ocean at the South Bay Ocean Outfall, shown here during its construction.

would be sufficiently diluted to meet federal and state pollution standards.

However, tests in 1997 and 1998 indicated that wastewater treated to the advanced primary level was still acutely toxic to fish and other marine life. In 1999, local activist (and now State Assemblywoman) Lori Saldaña and oceanographer Tim Baumgartner studied water quality above the outfall. They found a noticeable decrease in water quality compared to nearby areas. Saldaña and Baumgartner posted their results on the Internet and urged policymakers to speed construction of the secondary treatment facilities.

Of even greater concern, the IWTP was unable to cope with the increasing amounts of sewage that rapidly growing Tijuana was generating. "The day the plant became operational, it was already obsolete," noted Marco Gonzalez, a lawyer for a surfer's group that advocates for clean water.

In 1999, Gonzalez's group and the state sued the IWBC for violating the Clean Water Act. The settlement led to extensive testing and a mandate to construct a secondary treatment facility in Mexico.

Today, all parties are debating a proposal from a private consortium of investors, called the Bajagua Project. These investors propose to build a plant in Mexico that would provide secondary treatment for the IWTP's wastewater and would also treat an additional 34 million gallons of sewage per day. The investors will finance the plant, and if the effluent meets Clean Water Act standards, the U.S. Congress will be asked to cover the costs of construction and operation.

Bajagua Project proponents say that with this plan, the private sector can solve a pollution problem that has stymied two national governments for years— and can do it in a cost-efficient manner. The IWBC backs the project and has signed an agreement with the consortium, but local people are split. Some, like Lori Saldaña, voice distrust of the venture and point out that the company won a sole-source contract without competitive bidding, thanks to political lobbying that extended all the way up to Vice President Dick Cheney. Others, like Marco Gonzalez, want to give the private effort a chance to improve water quality and believe there are enough safeguards to prevent private abuse of public dollars.

Debates over the merits and dangers of such partnerships between the public and private sectors are becoming more common during a period in which the U.S. government has increasingly promoted the role of private enterprise in addressing public problems. If given final approval, the Bajagua Project could begin operation in September 2008.

Despite its admirable mission, the World Bank has frequently been criticized for funding unsustainable projects that cause more environmental problems than they solve. Providing for the needs of growing human populations in poor nations while minimizing damage to the environmental systems on which people depend can be a tough balancing act. Environmental scientists today agree that the concept of sustainable development must be the guiding principle for such efforts.

## International institutions and dynamics wield influence in a globalizing world

As globalization proceeds, our world is becoming ever more interconnected. As a result, human societies and Earth's ecological systems are being altered at unprecedented rates. Trade and technology have expanded the global reach of all societies, especially those such as the

United States, which consume resources from across the world. Highly consumptive nations that import goods and resources from far and wide exert extensive impacts on the planet's environmental systems. Multinational corporations operate outside the reach of national laws and rarely have incentive to conserve resources or conduct their business sustainably in the nations where they operate. For all these reasons, in today's globalizing world the organizations and institutions that influence international policy are becoming increasingly vital.

# The Environmental Policy Process

Anyone can become involved in helping ideas become public policy. In constitutional democracies such as the United States, it is true that each and every person has a political voice and can make a difference. Unfortunately, it is also true that money wields influence and that some people and organizations are far more politically connected and influential than others. We will explore some of the ways people make themselves influential as we examine the main steps of the policymaking process. Our discussion pertains both to citizens at the grassroots level and to large organizations and corporations.

## The environmental policy process begins when a problem is identified

The first step in the environmental policy process is to identify an environmental problem (**Figure 3.17**). Identifying a problem requires curiosity, observation, record keeping, and an awareness of our relationship with the environment. For example, assessing the contamination of San Diego- and Tijuana-area beaches required understanding the ecological and health impacts of untreated wastewater. It also required being able to detect contamination on beaches and understanding water flow dynamics among the beaches, the Pacific Ocean, and the Tijuana River watershed.

## Identifying causes of the problem is the second step in the policy process

Once an individual or group has defined a particular environmental problem, the next step is to discover specific causes of the problem. A person seeking causes for pollution in the Tijuana River watershed might notice that transboundary pollution began to worsen in the 1960s, when U.S.-based companies started opening *maquiladoras* on the

1 Identify problem

2 Identify specific causes of the problem

3 Envision solution and set goals

4 Get organized

5 Cultivate access and influence

6 Manage development of policy

FIGURE 3.17 Understanding the steps of the policy process is an essential element of solving environmental problems.

Mexican side of the border. Advocates of the *maquiladora* system argue that these factories provide much-needed jobs south of the border while keeping companies' costs low by paying Mexican workers far less than U.S. workers. Critics argue that the factories are waste-generating, water-guzzling polluters whose transboundary nature makes them particularly difficult to regulate.

Identifying problems and their causes requires that scientific research play a major role in the policy process. Much of this work takes place in the arena of *risk assessment* (• p. 403), in which scientists evaluate the extent and nature of problems and judge the risks that they pose to public health or environmental quality.

## The third step is envisioning a solution

The better one can identify specific causes of a problem, the more effectively one will be able to envision solutions to it and argue for implementing those solutions. Science plays a role here too, through the process of *risk management* (• pp. 403–404), in which scientists develop strategies to minimize risk. Frequently, however, solutions involve primarily social or political action. In San Diego, citizen activists wanted Tijuana to enforce its own pollution laws more effectively—something that, once visualized, began to happen when San Diego city employees started training and working with their Mexican counterparts to keep hazardous wastes out of the sewage treatment system.

## Getting organized is the fourth step

When it comes to gaining the ear of elected officials and influencing policy, organizations are generally more effective than lone individuals. The sole critic or crusader is easily dismissed as a crackpot or troublemaker, but a group of hundreds or thousands of individuals is not as easily dismissed. Furthermore, organizations are more effective at raising funds, which by U.S. law they may contribute to political campaigns.

As effective as large organizations can be, small coalitions and even individual citizens who are motivated, informed, and organized can solve environmental problems. As anthropologist Margaret Mead once remarked, "Never doubt that a small group of thoughtful, committed citizens can change the world—indeed it is the only thing that ever has."

San Diego–area resident Lori Saldaña provides an example. Concerned about the Tijuana River's pollution, Saldaña reviewed plans for the international wastewater treatment plant that the U.S. government proposed to build (see "The Science behind the Story," • pp. 72–73). She concluded that it would merely shift pollution from the river to the ocean, where sewage would be released 5.6 km (3.5 mi) offshore. Working with her local Sierra Club chapter, Saldaña protested the plant's design and participated in a lawsuit that forced the government to conduct further studies and the EPA to agree to a design change. For her efforts, Saldaña received awards and was appointed to a commission on border environmental issues by President Bill Clinton. After a decade of activism, Saldaña ran for the California State Assembly in 2004 and won. She is now the representative from California's 76th district.

## Gaining access to political powerbrokers is the fifth step

The fifth step in the policy process entails gaining access to policymakers who have the clout to help enact the desired changes. People gain access and influence through lobbying, campaign contributions, and the revolving door.

**Lobbying** Anyone who spends time or money trying to change an elected official's mind is engaged in **lobbying.** Although anyone can lobby, it is much more difficult for an ordinary citizen than for the full-time professional lobbyists employed by the many businesses and organizations seeking a voice in politics. Environmental advocacy organizations are not the most influential of lobbying groups. According to a 2002 report by *Fortune* magazine, the highest-ranking environmental organization, the Sierra Club, ranked 52nd among such groups, far back from such heavyweights as the National Rifle Association (NRA) and the American Association of Retired Persons (AARP). Indeed, the American Petroleum Institute spends on lobbying an amount that is nearly as great as the entire budgets of the top five U.S. environmental advocacy groups combined.

**Campaign contributions** Supporting a candidate's reelection efforts with money is another way to make one's voice heard. Because environmental policy often regulates the activities of corporations and industries, they have a strong interest in shaping it. Although corporations and industries may not legally make direct campaign contributions, they are allowed to establish *political action committees (PACs)* for that purpose. PACs raise money and distribute it to political campaigns, helping likeminded candidates win elections, in hope of gaining access to those individuals once they are elected.

**The revolving door** Some individuals employed by government-regulated industries gain political influence when they take jobs with the very government agencies responsible for regulating their industry. Businesses also often hire former government bureaucrats who regulated their industries. This movement of individuals between the private sector and government agencies is known as the **revolving door**.

Proponents of the revolving door system assert that corporate executives who take government jobs regulating their own industry bring with them an intimate knowledge of that industry. This inside experience makes them highly qualified and likely to benefit society with especially well-informed policy, proponents maintain. Critics of the system contend that taking a job regulating your former employer is a clear conflict of interest that undermines the effectiveness of the regulatory process. Regulators from the private sector will be biased toward assisting their industry, critics say, and may fail to enforce regulations, thus acting against the interests of the taxpayers who pay their salaries.

All three of these methods—lobbying, campaign contributions, and the revolving door—were employed by the private-sector consortium that now seeks to contract with the U.S. government to build a wastewater treatment plant to supplement the IWTP (see "The Science behind the Story," pp. 72–73). Backers of this effort, the Bajagua Project, lobbied government officials, gave campaign contributions to California congressmen, and employed former government officials, all of whom they hoped would be supportive of their cause. Following years of such efforts, in February 2006 the International Boundary and Water Commission agreed to support the Bajagua Project.

## Shepherding a solution into law is the sixth step in the policy process

Whether you're a corporate lobbyist or a grassroots activist, once your organization has the access and clout to influence policymakers, the better-known parts of the policy process come into play. Having gained access to elected officials and convinced them to hear your requests, you may be asked to prepare a bill, or draft law, that embodies the solutions you seek. Anyone can draft a bill. The hard part is finding members of the House and Senate willing to introduce the bill and shepherd it from subcommittee through full committee and on to passage by the full Congress. Lobbying and media attention intensify as the bill progresses through this process (**Figure 3.18**). If it passes through all of these steps, the bill may become law, but it can die in countless fashions along the way.

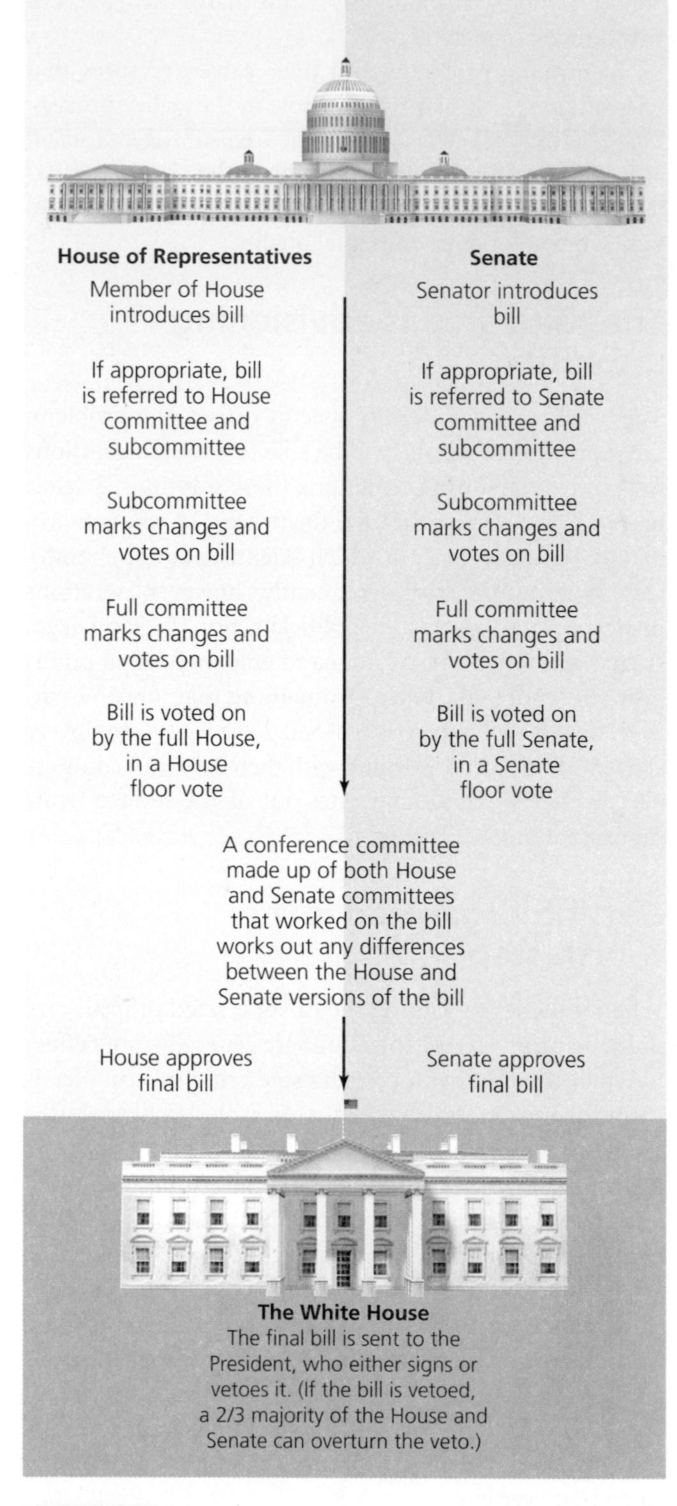


**FIGURE 3.18** Before a bill becomes U.S. law, it must clear a number of hurdles in both legislative bodies. If the bill passes the House and Senate, a conference committee must work out any differences between the House and Senate versions before the bill is sent to the president. The president may then sign or veto the bill.

The policy process does not end with the enactment of legislation. Following a law's enactment, administrative agencies implement regulations. Policymakers also evaluate the policy's successes and failures and may revise the policy as necessary. Moreover, the judicial branch interprets law in response to suits in the courts, and much environmental policy has lived and died by judicial interpretation. The full policy process is long and often cumbersome, but it has yielded effective results in constitutional democracies in the United States and many other nations.

## Science plays a role in policy

Ethical values, economic interests, and political ideology influence most policy decisions. However, environmental policy decisions that are effective are generally those informed by scientific research. For instance, when deciding whether and how to regulate a substance that may pose a public health risk, regulatory agencies such as the EPA comb the scientific literature and may commission new studies, seeking to gain as full an understanding of the health risks as science can reasonably provide. When trying to win votes for a bill to reduce pollution, a legislator may use data from scientific studies to quantify the cost of the pollution or the predicted benefits of its reduction. The more information a policymaker can glean from science, the better policy he or she will be able to create. In today's world, a nation's strength depends on its commitment to science, and this is the reason that governments devote a portion of our taxes to fund scientific research.

## Science can be "politicized"

Unfortunately, sometimes policymakers choose to ignore science and instead allow political ideology alone to determine policy. This complaint has been lodged against the George W. Bush administration by an unprecedented number of scientists. In 2004, the Union of Concerned Scientists released a statement titled, "Restoring Scientific Integrity in Policy Making," which faulted the Bush Administration for manipulating scientific information for political ends; censoring, suppressing, and editing reports from government scientists; placing people who are unqualified or who have clear conflicts of interest in positions of power; ignoring scientific advice; and misleading the public by misrepresenting scientific knowledge. By mid-2007, more than 12,000 scientists had signed on to this letter.

A few examples that have come to light include the following:

- Bush appointee Philip Cooney, a lawyer with no scientific background and a former oil industry lobbyist, edited several key U.S. government reports on climate change to portray scientific findings as far more uncertain than they actually were and to downplay the threats most scientists see in climate change. Cooney resigned when this was revealed, and the Exxon-Mobil corporation hired him just days later.
- The National Aeronautics and Space Administration (NASA) ordered one of its employees, world-renowned climate scientist James Hansen, to refrain from speaking out after he publicly recommended that the United States reduce its greenhouse gas emissions to combat climate change. He refused to back down, and after the matter garnered media attention, NASA changed its policies, allowing its scientists more freedom to speak out. However, NASA also quietly deleted a portion of its own mission statement that Hansen had cited in his defense, which read that NASA sought "to understand and protect our home planet."
- Interior Department deputy assistant secretary Julie MacDonald, a Bush appointee with no background in wildlife biology, edited scientific reports assessing sage grouse and a number of other species being considered for protection under the Endangered Species Act, disparaging her scientists' conclusions without offering evidence of her own. Following MacDonald's edits, species that Fish and Wildlife Service scientists recommended for protection were not granted protection.

Such high-profile cases appear to be the tip of an iceberg. In the past several years, scientists at government agencies—particularly those working on politically sensitive issues such as climate change or endangered species protection—say they have repeatedly found their work suppressed or discredited, and their jobs threatened. Many have chosen self-censorship. When taxpayer-funded science is suppressed or distorted for political ends, by either the right or the left, we all lose.

It is important to note that these abuses of power came to light only because brave government employees risked their careers to alert the public, and because journalists worked hard to uncover and publicize these issues. Moreover, problems do not generally get addressed unless they elicit considerable public scrutiny. We cannot simply take for granted that science will play a role in policy. As scientifically literate citizens of a democracy, we all need to keep guard to make sure our government representatives are making proper use of the tremendous scientific assets we have at our disposal.

# Approaches to Environmental Policy

A great deal of environmental policy has functioned by setting rules or limits and threatening punishment for violating these rules or limits, in what is often called a **command-and-control** approach. The command-and-control approach has resulted in some major successes. Without doubt, our environment would be in far worse shape today were it not for this type of government regulatory intervention.

However, many people have grown disenchanted with the top-down, sometimes heavy-handed nature of the command-and-control approach. Criticism of environmental legislation and regulation, and the means by which they often have been implemented, has spurred political scientists, economists, and others to invent alternative approaches for attaining environmental policy goals. The most widely developed alternatives involve the creative use of economic incentives to encourage desired outcomes, discourage undesired outcomes, and set market dynamics in motion to achieve goals in an economically efficient manner.

## Command-and-control policy has improved our lives

Most of the major environmental laws of recent decades, and most regulations enforced by agencies today, use the command-and-control approach. This simple and direct approach to policymaking has brought citizens of the United States and many other nations cleaner air, cleaner water, safer workplaces, healthier neighborhoods, and many other improvements in quality of life. The relatively safe, healthy, comfortable lives most of us enjoy today owe much to the environmental policy of the past few decades.

Command-and-control policy has also largely worked as intended. In 2003, the White House Office of Management and Budget undertook an extensive analysis of U.S. regulatory policy to determine whether the economic benefits of regulations really did exceed the economic costs. The analysis revealed that the benefits outweighed the costs by even more than most people would have guessed (see "The Science behind the Story," • pp. 80–81).

## The command-and-control approach has its critics

Despite the successes of command-and-control policy, it is not without its drawbacks. Although policy responds to market failure (• p. 49), it is clear that government intervention sometimes fails every bit as badly as markets can fail. Sometimes government actions are well intentioned but not well enough informed, so they can lead to unforeseen consequences. Policy can also fail if a government does not live up to its responsibilities to protect its citizens or treat them equitably. This may occur when leaders allow themselves to be unduly influenced by *interest groups,* small groups of people seeking private gain that work against the larger public interest. In addition, the command-and-control approach can fail when it generates opposition among citizens to government policy. If citizens view laws and regulations primarily as restrictions on their freedom, those policies will not last long in a constitutional democracy.

The most common critique of command-and-control policy is that it achieves its goals in a more costly and less efficient manner than the free market can. By mandating particular solutions to problems, command-and-control policy fails to take advantage of the fact that private entities competing in the free market can often produce better solutions at lower cost. Many minds that are economically motivated to compete in the market are more likely to innovate and find optimal solutions than a smaller number of policymakers with no such economic incentive.

As a result, policymakers now often try to combine the advantages of government and the private sector. The challenge of crafting economic policy tools is to channel the innovation and economic efficiency of the free market in directions that benefit the public.

## Subsidies are a widespread economic policy tool

One set of economic policy tools aims to encourage industries or activities that are deemed desirable. Governments may give *tax breaks* to certain types of businesses or individuals, for instance. Relieving the tax burden lowers costs for the business or individual, thus assisting the desirable industry or activity.

A similar economic policy tool is the **subsidy**, a government giveaway of cash or publicly owned resources that is intended to encourage a particular activity. National governments commonly provide subsidies to industries they judge to benefit the nation in some way. Subsidies can be used to promote environmentally sustainable activities, but all too often they have been used to prop up unsustainable ones. Subsidies judged to be harmful to the environment and to the economy total roughly $1.45 *trillion* yearly across the globe, according to environmental scientists Norman Myers and Jennifer Kent—an amount larger than the economies of all but five nations.

The average U.S. taxpayer pays $2,000 per year in environmentally harmful subsidies, plus $2,000 more through increased prices for goods and through degradation of ecosystem services, Myers and Kent estimate. U.S. subsidies for industries and activities promoting automobile transportation alone amount to $1,700 per taxpayer per year, and the nation's heavily subsidized gasoline is generally cheaper than bottled water. The most recent Green Scissors Report estimates that in 2003, $58 billion of U.S. taxpayers' money was budgeted for 68 environmentally harmful subsidies. The Green Scissors Report is a project of 22 nongovernmental organizations such as Friends of the Earth, Taxpayers for Common Sense, and the U.S. Public Interest Research Group. Among the subsidies highlighted in recent reports are the following:

- *General Mining Law of 1872.* Each year, mining companies extract $500 million to $1 billion in minerals from U.S. public lands without paying a penny in royalties to the taxpayers who own these lands. Since this law was enacted, the U.S. government has given away over $245 billion of mineral resources, and mining activities have polluted more than 40% of the watersheds in the West. The 135-year-old act still allows mining companies to buy public lands for $5 or less per acre.

- *Coal subsidies.* Since 1984, Congress has made $1.8 billion available to the coal industry through the Clean Coal Technology Program, paying for industry research that may lead to technologies to reduce air pollution from coal combustion. An additional $800 million per year goes to the coal industry for further research and development. These amounts dwarf subsidies granted to less-polluting renewable energy sources.
- *Forest Service road-building subsidies.* The U.S. Forest Service manages taxpayer-owned forests for various uses, including timber harvesting. Although many people assume that the corporations that harvest trees from public forests cover the costs of roads required to get the logs out, this is not the case. From 1992 to 1997, the Forest Service spent more than $387 million in tax dollars for road construction for timber companies (**Figure 3.19**). Despite legislative reform, subsidies are currently estimated at $170 million over 5 years.

Advocates of sustainable resource use have long urged governments to subsidize environmentally sustainable activities instead. To some extent, this is being done. For instance, subsidies for renewable energy sources totaled nearly $1.1 billion in the United States in 1999, according to the U.S. Department of Energy. But this amount falls short of the $2.8 billion that went to nonrenewable energy sources in that same year.

FIGURE 3.19 When companies cut timber in U.S. national forests, roads must be built and maintained to enable access. The costs of these roads are paid by taxpayers.

## Green taxes discourage undesirable activities

Another economic policy tool—taxation—can be used to discourage undesirable activities. Taxing undesirable activities helps to "internalize" external costs by making them part of the overall cost of doing business. Taxes on environmentally harmful activities and products are called **green taxes**. By taxing activities and products that cause undesirable environmental impacts, a tax becomes a tool for policy as well as simply a way to fund government.

Green taxes have yet to gain widespread support in the United States, although similar "sin taxes" on cigarettes and alcohol are tools of U.S. social policy. Taxes on pollution have been widely instituted in Europe, where many nations have adopted the *polluter pays principle.* This principle specifies that the price of a good or service should include all its costs, including costs of environmental degradation that would otherwise be passed on as external costs. This is the main reason that gasoline prices are so cheap in the United States relative to those in European nations.

Under green taxation, a factory that pollutes a waterway would pay taxes based on the amount of pollution it discharges. The idea is to give companies a financial incentive to reduce pollution, while allowing the polluter the freedom to decide how best to minimize its expenses. One polluter might choose to invest in technologies to reduce its pollution if doing so is less costly than paying the taxes. Another polluter might find abating its pollution more costly and could choose to pay the taxes instead funds the government might then apply toward mitigating pollution in some other way.

Green taxation provides incentive for industry to lower emissions not merely to a level specified in a regulation, but to still-lower levels. However, green taxes do have disadvantages. One is that businesses will most likely pass on their tax expenses to consumers, and these increased costs may affect low-income consumers disproportionately more than high-income ones.

## Markets in permits can save money and produce results

A more innovative market-based approach is **permit trading**. In a permit trading system, the government creates a market in permits for an environmentally harmful activity, and companies, utilities, or industries are allowed to buy, sell, or trade rights to conduct the activity. For instance, to

THE SCIENCE BEHIND THE STORY

## Comparing Costs and Benefits of Environmental Regulations

Truck with cleaner diesel exhaust thanks to federal regulations

Federal regulations that aim to protect environmental quality often result in clearer air or cleaner water. They also usually have financial implications for affected parties. Are such regulations worth the costs to industry, businesses, and consumers? In 2003, a federal study determined that the answer to that question is a definitive yes.

The study, conducted by the White House Office of Management and Budget (OMB), weighed the costs of a range of recently created federal regulations against the economic benefits resulting from the regulations. The OMB examined 107 federal regulations enacted in the United States from 1992 to 2002, many of which dealt with environmental issues.

The researchers found that the regulations they reviewed carried a sizable annual cost, estimated at \$36–42 billion. However, their estimated value in terms of public good totaled \$146–230 billion. Thus, the economic benefits of these regulations far exceeded their costs. Moreover, of all the regulations, those dealing with environmental protection were found to be especially cost-effective.

Most regulatory benefits were reflected in health and social gains resulting from clean air. Over the studied 10-year period, the benefits of clean-air regulations were found to be 6–10 times greater than the costs of complying with the regulations. Reductions in hospitalization and emergency room visits, premature deaths, and lost workdays together accounted for savings of \$118 billion to \$177 billion.

The authors of the OMB review relied on cost-benefit analyses (• pp. 40–41) previously made by experts in the EPA and other agencies and scrutinized them for appropriateness and accuracy before using them. To see how the EPA performed its economic analysis and how the OMB double-checked the EPA's findings, we can examine one EPA regulation included in the OMB report—an air pollution regulation called the Heavy Duty Engine/Diesel Fuel Rule, which today is bringing much cleaner diesel fuel to the nation's highways.

The diesel regulation aimed to address diesel engines that emit fine soot, sulfur compounds, and nitrogen oxides ($NO_X$). In the 1990s, the EPA regulation was intended to make diesel trucks and buses run more cleanly by lowering the fuel's sulfur content. However, achieving cleaner diesel fuel would mean substantial—and potentially costly—changes to the way that fuel is made and handled.

To determine potential costs, EPA analysts examined how various parties involved with diesel fuel use—from fuel processors to engine manufacturers to vehicle operators—would need to change their processes or equipment to clean up the fuel. For example, a heavy-duty engine redesigned to run more efficiently on cleaner fuel would cost approximately \$4,600 more to operate, the analysts calculated. Total research and development costs for emission control were predicted to exceed \$600 million. The EPA asked for input from affected parties, scientists, and the public, finally determining that removing sulfur from diesel fuel would be neither cheap nor

decrease emissions of air pollutants, a government might grant emissions permits and set up an *emissions trading* system. The government first determines the overall amount of pollution it will accept and then issues permits to polluters that allow them each to emit a certain fraction of that amount. Polluters may buy, sell, and trade these permits with other polluters. Each year, the government may reduce the amount of overall emissions allowed.

In such a *cap-and-trade* system, a polluting party that is able to reduce its pollution receives credit for the amount it did not emit and can sell this credit to other parties. Suppose, for example, you are a plant owner with permits to release 10 units of pollution, but you find that you can become more efficient and release only 5 units of pollution instead. You then have a surplus of permits, which might be very valuable to some other plant owner who is having trouble reducing pollution or who wants to expand production. In such a case, you can sell your extra permits. Doing so generates income for you and meets the needs of the other plant, while preventing any increase in the total amount of pollution. Moreover, environmental organizations can buy up surplus permits and "retire" them, thus reducing the overall amount of pollution.

Marketable permits provide companies an economic incentive to find ways to reduce emissions. If successful, permit trading can end up costing both industry and government much less than a conventional regulatory system.

An emissions trading system has been in place in the United States, established by the 1990 amendments to the Clean Air Act (• p. 482). This legislation introduced a cap-and-trade program to reduce emissions of sulfur dioxide

Estimates of the Total Annual Benefits and Costs of Major Federal Rules, October 1, 1992 to September 30, 2002

| Agency | Benefits (millions of 2001 dollars) | Costs (millions of 2001 dollars) |
|---|---|---|
| Agriculture | 3,094–6,176 | 1,643–1,672 |
| Education | 655–813 | 361–610 |
| Energy | 4,700–4,768 | 2,472 |
| Health & Human Services | 9,129–11,710 | 3,165–3,334 |
| Housing & Urban Development | 551–625 | 348 |
| Labor | 1,804–4,185 | 1,056 |
| Transportation | 6,144–9,456 | 4,220–6,718 |
| Environmental Protection Agency | 120,753–193,163 | 23,359–26,604 |
| **Total** | **146,812–230,896** | **36,625–42,813** |

Source: U.S. Office of Management and Budget. 2003. *Informing regulatory decisions: 2003 report to Congress on the costs and benefits of federal regulations and unfunded mandates on state, local, and tribal entities.* Washington, D.C.: U.S. OMB.

quick, with total annualized costs of about $4.2 billion by 2030.

EPA analysts then considered the benefits of reduced diesel pollution. Using air quality data from more than 940 smog monitors around the country, the EPA estimated diesel-related air pollution levels with and without the proposed diesel regulation. The analysts examined public health statistics, especially data related to respiratory illnesses such as asthma and chronic bronchitis. They then estimated how many cases of disease or premature death might be prevented if people breathed fewer diesel-related pollutants. They determined the economic impacts of such health problems by tallying up medical, workforce, and social costs. Each premature death was assigned an impact of $6 million, and each case of chronic bronchitis was pegged at $331,000. Many of the estimates were regionalized to account for different economic conditions in various parts of the country.

In the end, the EPA analysis determined that the benefits of cleaner diesel fuel far exceeded the costs. If sulfur content of diesel were cut from 500 to 15 parts per million, then 2.6 million tons of smog-causing $NO_x$ emissions would be eliminated each year. An estimated 8,300 premature deaths, 5,500 cases of chronic bronchitis, over 360,000 asthma attacks, and about 1.5 million lost workdays would be prevented each year. By 2030, the analysis determined, benefits would reach about $70 billion.

Following this analysis, the Heavy Duty Engine/ Diesel Fuel Rule took effect in early 2001, just before President Bill Clinton left office. The George W. Bush administration stalled it for three years, then let it proceed. The program is being phased in gradually through 2010, and already trucks are using much cleaner diesel fuel on the nation's highways, for only 3–5 cents per gallon more.

The OMB, in compiling its own review of federal rules, scrutinized the EPA study and concluded that most of the estimates did indeed make economic sense. The OMB analysts made some minor changes when calculating their own figures. For example, although the EPA said every 1-ton reduction in $NO_x$ emissions represented a benefit of $10,200, the OMB used its own updated figures and determined $NO_x$ reductions to be worth $5,500 per ton. Overall, however, the OMB confirmed the validity of the EPA estimates and used most of the EPA's economic data in the OMB report.

The OMB report, in totaling and highlighting the value of many recent environmental regulations, is now being used to evaluate and support efforts at environmental regulation around the country.

(**Figure 3.20**), a major contributor to acidic deposition (• pp. 491–495). Starting in 1995, permits were issued to power plants, and fewer emissions were allowed year by year.

By most measures, the program has been a success. Sulfur dioxide emissions from sources involved in the program declined by 35% between 1990 and 2005 (**Figure 3.21**). As a result, sulfate deposition has been reduced nationally, and air quality and visibility have improved. The 35% reduction in pollution was greater than the amount actually required by the legislation, offering evidence that cap-and-trade systems can cut pollution more effectively than command-and-control regulations. Moreover, the cuts were attained at much less cost than was predicted, and with no apparent effect on electricity supply or economic growth. Savings from the permit trading system have been estimated at billions of dollars per year, and the EPA calculates that by 2010, benefits of the program will outweigh costs by about 40 to 1. Other similar programs have also shown success, including one in the Los Angeles Basin to reduce smog, and one among northeastern states aimed at nitrogen oxides.

Cap-and-trade programs are no panacea, however. Although they can reduce pollution overall, they do allow hotspots of pollution to occur around plants that buy permits to pollute more. Moreover, large firms can hoard permits, deterring smaller new firms from entering the market, and thereby suppressing competition. Nevertheless, permit trading has shown promise for safeguarding environmental quality while granting industries the flexibility to lessen their impacts in ways that are economically palatable.

FIGURE 3.20 Emissions trading has helped decrease the sulfur dioxide pollution that contributes to acid rain in the United States. Currently, nations and industries are developing markets in carbon emissions trading, following ratification of the Kyoto Protocol.

Presently, a market in carbon emissions is operating among European nations as a result of the Kyoto Protocol to address climate change (• pp. 534–535). Under the Kyoto Protocol, nations have targets for reducing their carbon emissions from power plants, automobiles, and other sources that are driving climate change.

Each nation participating in the European Union Emission Trading Scheme (• p. 535) takes the emissions permits it is allowed and allocates them to its industries according to their emissions at the start of the program. The industries then can trade permits freely, establishing a market whereby the price of a carbon emissions permit fluctuates according to supply and demand.

Unfortunately, European nations allocated too many permits in the program's first phase, destroying industries' financial incentive to cut emissions and causing the permits to become nearly worthless. These nations plan to correct their over-allocation in 2008 when the program enters its next phase.

The United States has not ratified the Kyoto Protocol, but more and more U.S. industries are taking part in carbon trading through the Chicago Climate Exchange and the Regional Greenhouse Gas Initiative (• p. 535), and other such markets are on the horizon.

### Weighing THE Issues | Emissions Trading

Some environmental activists oppose emissions trading because they view it as giving polluters "a license to pollute." How do you feel about emissions trading as a means of reducing air pollution? Would you favor command-and-control regulation or market-based permit trading? What advantages and disadvantages do you see in each?

## Market incentives are being tried widely on the local level

You may well have already taken part in transactions involving financial incentives as policy tools. Many municipalities charge residents for waste disposal according to the amount of waste they generate. Other cities place taxes or disposal fees on items that require costly safe disposal, such as tires and motor oil. Still others give rebates to residents who buy water-efficient toilets and appliances, because the rebates can cost the city less than upgrading its sewage treatment system. Likewise, power companies sometimes offer discounts to customers who buy high-efficiency lightbulbs and appliances, because doing so is cheaper for the utilities than expanding the generating capacity of their plants.

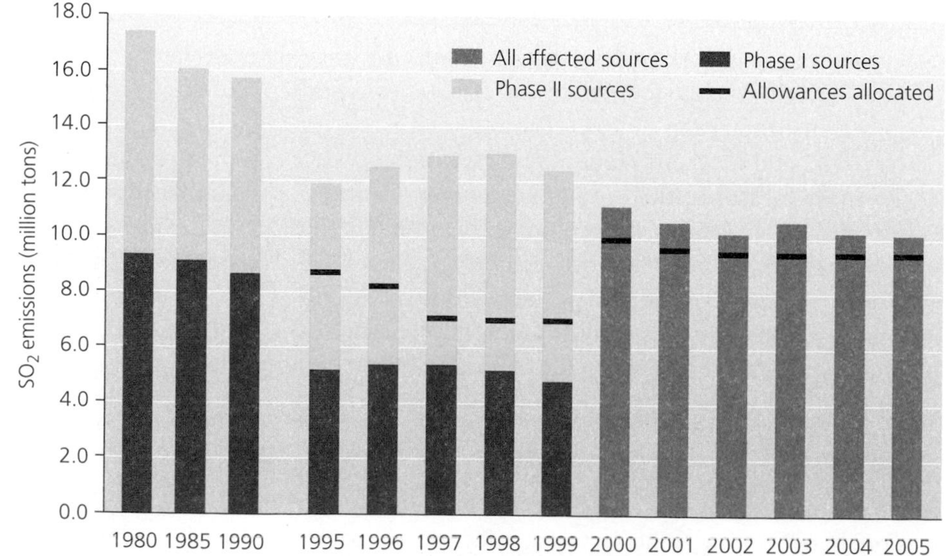


FIGURE 3.21 Emissions of sulfur dioxide from sources participating in the emissions trading program mandated by the 1990 Clean Air Act amendments have fallen 35% since 1990. Emissions dropped steeply in 1995, the first year of the program, as a result of reductions among large major sources included in phase I of the program (red portions of bars). Emissions fell again in 2000, when phase II of the program began, which required reductions from smaller sources (yellow portions of bars). As of 2005, emissions from both types of sources (orange bars) slightly exceeded the amounts allocated in permits (black lines through bars), but they were greatly below emissions levels that were predicted had there not been a permit trading program. Data from U.S. Environmental Protection Agency.

At all levels, from the local to the international, market-based incentives can reduce environmental impact while minimizing overall costs to industry and easing concerns about the intrusiveness of government regulation. Command-and-control policy is straightforward to implement, easy to monitor, and frequently works. Market-based approaches can be more complicated, but if they work, they can lessen environmental impact at a lower overall cost.

## Conclusion

Environmental policy is a problem-solving tool that makes use of science, ethics, and economics and that requires an astute understanding of the political process. Conventional command-and-control approaches of legislation and regulation are the most common approaches to policymaking, but various innovative economic policy tools are also being developed. As we have seen in the case of the Tijuana River, environmental issues often overlap political boundaries and require international cooperation. Through the hard work of concerned citizens interacting with their government representatives, the political process eventually produced promising solutions in the Tijuana River Valley Estuary and Beach Sewage Cleanup Act (renewed in 2004) and in binational agreements and management plans.

We will draw on the fundamentals of environmental policy introduced in this chapter throughout the remainder of this book. By understanding these fundamentals, you will be well equipped to develop your own creative solutions to many of the challenging problems we will encounter.

## REVIEWING OBJECTIVES

**You should now be able to:**

**Describe environmental policy and assess its societal context**

- Policy is a tool for decision making and problem solving that makes use of information from science and values from ethics and economics. (pp. 58–59)
- Environmental policy is designed to protect natural resources and environmental amenities from degradation or depletion and to promote equitable treatment of people. (pp. 59–60)

**Identify the institutions important to U.S. environmental policy and recognize major U.S. environmental laws**

- The legislative, executive, and judicial branches, together with administrative agencies, all play roles in U.S. environmental policy. (pp. 61–62)
- U.S. environmental policy came in three waves. The first encouraged frontier expansion and resource extraction. The second aimed to mitigate impacts of the first. The third targeted pollution and gave us many of today's major environmental laws. (pp. 64–68)
- Some major U.S. laws include the National Environmental Policy Act, the Clean Air Act, and the Clean Water Act. (pp. 66–68)
- Currently, a fourth wave of environmental policy, occurring internationally, centers around sustainable development. (p. 69)

**List the institutions involved with international environmental policy and describe how nations handle transboundary issues**

- Many environmental problems cross political boundaries and thus must be addressed internationally. (pp. 69–70)
- International policy includes customary law (law by shared traditional custom) and conventional law (law by treaty). (p. 70)
- Institutions such as the United Nations, European Union, World Bank, World Trade Organization, and nongovernmental organizations all play roles in international policy. (pp. 70–73)

**Delineate the steps of the environmental policy process and evaluate its effectiveness**

- The policy process entails several steps: (1) identifying the problem, (2) identifying causes of the problem, (3) envisioning solutions, (4) getting organized, (5) gaining access to power, and (6) guiding a solution into law. (pp. 74–76)
- In a democracy, anyone can use the policy process, although corporations and organizations with money and resources tend to have the most clout. (pp. 75–76)
- Science plays a role in policymaking, although some policymakers may ignore or distort it for political ends. (p. 77)

**Categorize the different approaches to environmental policy**

- Legislation from Congress and regulations from administrative agencies make up most federal policy. These top-down approaches are referred to as command-and-control. (pp. 77–78)
- Shortcomings of the command-and-control approach have led many economists to advocate economic policy tools. (p. 78)
- Market-based approaches include subsidies, green taxation, and permit trading. (pp. 78–83)

## TESTING YOUR COMPREHENSION

1. Describe two common justifications for environmental policy, and discuss three problems that environmental policy commonly seeks to address.
2. Outline the primary responsibilities of the legislative, executive, and judicial branches of the U.S. government. What is the "fourth branch" of the U.S. government?
3. What is meant by a *regulatory taking*?
4. Summarize the differences between the first, second, and third waves of environmental policy in U.S. history. What approach appears to be the fourth wave?
5. What did the National Environmental Policy Act accomplish? Briefly describe the origin and mission of the U.S. Environmental Protection Agency.
6. What is the difference between customary law and conventional law? What special difficulties do transboundary environmental problems present?
7. Why are environmental regulations sometimes considered to be unfair barriers to trade?
8. Describe the environmental policy process, from identification of a problem through enactment of a federal law.
9. What kinds of things can an individual citizen do to become influential in the policymaking process?
10. Differentiate among a green tax, a subsidy, a tax break, and a marketable emissions permit.

## SEEKING SOLUTIONS

1. Many free-market advocates maintain that environmental laws and regulations are an unnecessary government intrusion into private affairs. As you may recall from Chapter 2 (• pp. 38–39), Adam Smith argued that individuals can benefit society by pursuing their own self-interest. Do you agree? Can you describe a situation in which an individual acting in his or her self-interest could harm society by causing an environmental problem? Can you describe how environmental policy might rectify the situation? What are some advantages and disadvantages of instituting environmental laws and regulations, versus allowing unfettered exchange of materials and services?
2. Reflect on the causes for the transitions in U.S. history from one type of environmental policy to another. Now peer into the future, and think about how life might be different in 25, 50, or 100 years. What would you speculate about the environmental policy of the future? What issues might it address? Do you predict we will have more or less environmental policy?
3. Compare the roles of the United Nations, the European Union, the World Bank, the World Trade Organization, and nongovernmental organizations. If you could gain the support of just one of these institutions for a policy you favored, which would you choose? Why?
4. Think of one environmental problem that you would like to see solved. From what you've learned about the policymaking process, describe how you think you could best shepherd your ideas through the process to address this problem.
5. Compare the main approaches to environmental policy—command-and-control, tort law, and economic or market-based approaches. Can you describe an advantage and a disadvantage of each? Do you think any one approach is most effective? Could we do with just one approach, or does it help to have more than one?
6. **THINK IT THROUGH** You have just been named head of the U.S. Environmental Protection Agency. New legislation from Congress mandates reductions in water pollution from untreated municipal wastewater, chemical discharges from factories, and oil spillage from commercial and recreational boats. The new law mandates a 25% reduction in these pollution sources over 10 years, but it does not specify how these reductions are to be accomplished and instead leaves the details up to the EPA. What policy approaches would you choose to pursue to carry out the mandates of the legislation? Give reasons for your choices.

## INTERPRETING GRAPHS AND DATA

The Clean Air Act legislation of 1970, 1977, and 1990 was designed to improve air quality in the United States by monitoring and reducing the emissions of air pollutants judged to pose threats to human health, such as carbon monoxide, nitrogen dioxide, sulfur dioxide, ozone, particulate matter, and lead (• pp. 482–484). The main source of lead emissions in 1970 was the exhaust of vehicles burning gasoline to which tetra-ethyl lead had been added to improve combustion. By 1985, leaded gasoline was phased out of use, although airplanes and racecars were exempted.

The 1990 amendments addressed the growing problem of urban smog by requiring the use of reformulated gas (RFG) in cities with the worst smog problems. One of the RFG requirements specifies 2% oxygen content in fuel, which has been met by adding either ethanol (• pp. 590–591) or methyl tert butyl ether (MTBE). Although it burns cleanly, MTBE is water-soluble and may cause cancer, so groundwater contamination from fuel spills is a concern. Twenty-five states have now passed legislation banning or restricting the use of MTBE. The following graph shows trends in U.S. lead emissions and MTBE consumption since 1970.

1. Did policy resulting from Clean Air Act legislation succeed in reducing the public health risk from exposure to lead? Use data provided in the graph to support your answer.
2. In the 1990s, use of MTBE in RFG increased rapidly on the east and west coasts of the United States. Ethanol blends were used mostly in the corn-growing states of the central part of the country; ethanol is made from corn, so it provides a market for in-state farmers while low transportation costs make it cheaper for in-state consumers. As MTBE use is being phased out in many states, use of ethanol is increasing. Name three ways in which these trends illustrate how state government actions may differ from federal government actions.
3. Do you think it is better policy to regulate air and water quality separately (e.g., under the Clean Air Act and Clean Water Act) or together under a single, comprehensive act to prevent and control pollution (as some European nations do)? How do the data in the graph support your reasoning?

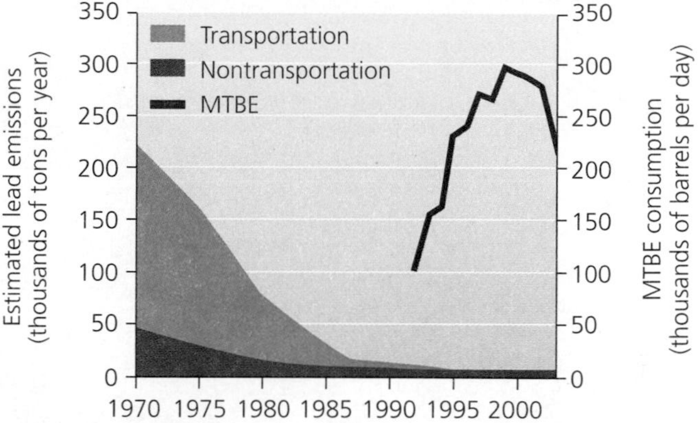


**Estimated lead emissions from transportation and nontransportation sources, and consumption of MTBE in the United States.** Data from U.S. Department of Transportation, Bureau of Transportation Statistics.

## CALCULATING ECOLOGICAL FOOTPRINTS

In 2000, the population of the United States was approximately 285 million. According to the U.S. Geological Survey, the country withdrew about 408 billion gallons of fresh water per day for a variety of uses, and average per capita use was 1,430 gallons per day. Seventy-nine percent of 2000 withdrawals were from surface water, and 21 percent of withdrawals were from groundwater. Thermoelectric power plants accounted for 48% of total withdrawals, irrigation 34%, public water supplies 11%, and industrial, livestock, and mining uses 8% (percentages add up to 101% because of rounding).

| Daily Water Use in 2000 | | | | |
|---|---|---|---|---|
| | **Electricity** | **Irrigation** | **Public water supply** | **Industrial / livestock / mining** |
| **You** | 686 gal | | | |
| **Your class** | | | | |
| **Your hometown** | | | | |
| **United States** | | | | |

Data from U.S. Geological Survey, *Summary of water use in the United States, 2000.* http://ga.usgs.gov/edu/wateruse2000.html

1. How many gallons of the country's daily water use come from surface water?
2. What types of policies would you recommend to reduce overall water use most aggressively? Provide one example of a command-and-control policy and one example of an economic policy tool.
3. Describe how these recommended policies would affect withdrawals in one of the four water use categories listed above.

## Take It Further

Go to www.aw-bc.com/withgott or the student CD-ROM, where you'll find:

- Suggested answers to end-of-chapter questions
- Quizzes, animations, and flashcards to help you study
- *Research Navigator*™ database of credible and reliable sources to assist you with your research projects
- **GRAPHIt!** Tutorials to help you interpret graphs
- **INVESTIGATEIt!** Current news articles that link the topics that you study to case studies from your region to around the world

CHAPTER

# 4 From Chemistry to Energy to Life

*Exxon Valdez* oil tanker in Prince William Sound, Alaska

## Upon completing this chapter you will be able to:

- Explain the fundamentals of environmental chemistry and apply them to real-world situations
- Describe the molecular building blocks of living organisms
- Differentiate among the types of energy and recite the basics of energy flow
- Distinguish photosynthesis, respiration, and chemosynthesis, and summarize their importance to living things
- Itemize and evaluate the major hypotheses for the origin of life on Earth
- Outline our knowledge regarding early life and give supporting evidence for each major concept

Workers spraying fertilizer on an oil-coated Alaska beach

CENTRAL CASE

# Bioremediation of the *Exxon Valdez* Oil Spill

**"There is a dramatic difference. . . . It really cleaned the oil off the rock. It looked like someone brought in new rock."**
—EPA PROGRAM MANAGER CHUCK COSTA, DESCRIBING EXPERIMENTAL BIOREMEDIATION RESULTS IN 1989

**"The rush to bioremediation in Alaska was a function of the size of the problem and limited availability of options. . . . It did not turn out to be the silver bullet that many hoped it would be."**
—ALASKA DEPARTMENT OF ENVIRONMENTAL CONSERVATION REPORT, 1993

On March 24, 1989, the tanker *Exxon Valdez* struck a reef in Alaska's Prince William Sound and spilled 42 million L (11 million gal) of crude oil, which eventually coated 2,100 km (1,300 mi) of Alaskan coastline. The largest oil spill in U.S. history, it killed an estimated 100,000–400,000 seabirds, 2,600–5,500 sea otters, 200–300 harbor seals, and countless fish. The oil smothered intertidal plants and animals and defiled the area's relatively pristine environment. The local economy took a nosedive as hundreds of fishermen were thrown out of work and tourism plummeted.

The massive spill was met with a massive response. Thousands of workers employed by Exxon (now ExxonMobil) and by government agencies, together with local volunteers, launched a cleanup effort of unprecedented scope. The cleanup crews corralled the oil with booms, skimmed it from the water, soaked it up with absorbent materials, and dispersed it with chemicals. They pressure-washed the beaches, removed contaminated sand with backhoes and tractors, and even tried burning the oil.

Scientists also used the opportunity to test a new cleanup strategy that enlisted nature to help take care of the mess. They stimulated naturally occurring bacteria to biodegrade, or break down, the oil. About 5% of the single-celled microbes present on Alaskan beaches feed on chemical compounds called *hydrocarbons* that are produced by the region's conifer trees. Hydrocarbons from conifers are chemically similar to the hydrocarbons that make up crude oil, so scientists predicted that the microbes might also be able to degrade oil. Scientists from the EPA and Exxon decided to put the bacteria to work in a process called **bioremediation**, the attempt to clean up pollution by enhancing natural processes of biodegradation by living organisms.

Although the bacteria were presented with an abundant new food source, they were not immediately able to consume the oil washing up on the beaches. The oil contained plenty of carbon, but not enough nitrogen and phosphorus. To remedy this imbalance of nutrients, scientists applied a fertilizing mixture containing nitrogen and phosphorus to several beaches, leaving other areas of the shore as untreated controls. The fertilizing treatment seemed to work; bacterial numbers increased, and oil residues decreased visibly. Encouraged, scientists put the program into full swing, and by the end of the year workers had treated more than 113 km (70 mi) of contaminated beach. They expanded the applications over the next 2 years.

Because there were many complicating factors, experts have interpreted the results of the study differently, and they still debate how much the treatments increased chemical breakdown of the oil. Some say degradation was sped up fivefold, whereas others think it made no difference. However, the well-publicized *Valdez* operation served as a model effort, and today bioremediation is actively researched and increasingly applied in many situations. Bioremediation has many practical limitations, but, when feasible, it can accomplish much good with a minimum of expense and environmental disturbance.

## Chemistry and the Environment

The *Exxon Valdez* oil spill ignited a wide array of ecological, economic, political, and ethical concerns. Today many wildlife populations have recovered, but some have not, and pockets of oil remain throughout the region. Lawsuits against the company (some still pending) testify to the concerns of fishermen whose livelihoods were wrecked and of cleanup workers who say their health was affected. The U.S. Congress in the year following the spill

FIGURE 4.1 The *Exxon Valdez* spill coated hundreds of thousands of seabirds with oil, impairing their ability to insulate themselves with their feathers and bringing on fatal hypothermia. Rescue workers labored tirelessly to clean oil from those birds they could capture and treat.

passed the Oil Pollution Control Act, which required the Coast Guard and the U.S. Environmental Protection Agency (EPA) to strengthen regulations on tankers and their operators.

At the root of all these diverse impacts is the chemical makeup of the oil. It is the chemistry of crude oil that causes it to gum up birds' feathers and mammals' fur, impairing their insulating abilities and bringing on hypothermia (**Figure 4.1**). It is oil's chemistry that causes it to float on water and accumulate on beaches. It is certain hydrocarbons from oil that, mixed in the water column or volatile in the air, are harmful to wildlife and carcinogenic to humans. Yet the chemistry of crude oil also provides the energy that powers our remarkable civilization and modern way of life—a way of life that allows us the luxury to study, reflect on, and act to address these very issues.

Examine any environmental issue, and you will likely discover chemistry playing a central role. Chemistry is crucial to understanding how gases such as carbon dioxide and methane contribute to global climate change, how pollutants such as sulfur dioxide and nitric oxide cause acid rain, and how pesticides and other artificial compounds we release into the environment affect the health of wildlife and people. Chemistry is central, too, in understanding water pollution and wastewater treatment, hazardous waste and its cleanup and disposal, atmospheric ozone depletion, and most energy issues.

Chemistry is also central to developing solutions to environmental problems. Bioremediation is one clear illustration of this, and organisms are now used to clean up pollution in a variety of situations. Hydrocarbon-consuming bacteria and fungi are used to clean up soil beneath leaky gasoline tanks that threaten drinking water supplies. Other kinds of microbes are used to degrade

**TABLE 4.1 Earth's Most Abundant Chemical Elements, by Mass**

| Earth's crust | Oceans | Air | Organisms |
|---|---|---|---|
| Oxygen (O), 49.5% | Oxygen (O), 88.3% | Nitrogen (N), 78.1% | Oxygen (O), 65.0% |
| Silicon (Si), 25.7% | Hydrogen (H), 11.0% | Oxygen (O), 21.0% | Carbon (C), 18.5% |
| Aluminum (Al), 7.4% | Chlorine (Cl), 1.9% | Argon (Ar), 0.9% | Hydrogen (H), 9.5% |
| Iron (Fe), 4.7% | Sodium (Na), 1.1% | Other, <0.1% | Nitrogen (N), 3.3% |
| Calcium (Ca), 3.6% | Magnesium (Mg), 0.1% | | Calcium (Ca), 1.5% |
| Sodium (Na), 2.8% | Sulfur (S), 0.1% | | Phosphorus (P), 1.0% |
| Potassium (K), 2.6% | Calcium (Ca), <0.1% | | Potassium (K), 0.4% |
| Magnesium (Mg), 2.1% | Potassium (K), <0.1% | | Sulfur (S), 0.3% |
| Other, 1.6% | Bromine (Br), <0.1% | | Other, 0.5% |

pesticide residues in soil. Plants as diverse as wheat, tobacco, water hyacinth, and cattails have been employed to clean up toxic waste sites by letting them draw up heavy metals through their roots (see "The Science behind the Story," • pp. 90–91).

Sometimes suitable plants or bacterial cultures are introduced to a site. Sometimes naturally existing ones are fertilized, as at Prince William Sound. And sometimes the best way to mitigate pollution is simply to monitor naturally occurring organisms as they do their work, without disrupting the system. Bioremediation can be far less expensive, less environmentally intrusive, and more effective than conventional methods for cleaning up pollution. However, bioremediation does not always work, and it can sometimes be very slow, leave a job uncompleted, or introduce new problems. Scientists today are seeking ways to take bioremediation to the next level, by genetically engineering microbes and plants to become more efficient at the specific metabolic tasks we ask of them. Environmental chemists are excited about the countless future applications of chemistry that may help us address environmental problems.

## Atoms and elements are chemical building blocks

To appreciate the chemistry involved in environmental science, we must begin with a grasp of the fundamentals. All material in the universe that has mass and occupies space is termed **matter**. Matter may be transformed from one type of substance into others, but it cannot be created or destroyed. This principle is referred to as the **law of conservation of matter**. In environmental science, this principle helps us understand that the amount of matter stays constant as it is recycled in nutrient cycles and ecosystems (• pp. 180, 186–198). It also makes clear that we cannot simply wish away the matter (such as waste and pollution) that we want to get rid of. Every drop of oil spilled in a pristine bay will end up somewhere, whether it sinks into the sediment, coats a bird's feathers, or is consumed by bacteria. Every piece of garbage or billow of smokestack pollution or canister of nuclear waste that we dispose of will not simply disappear; instead, we will need to take responsible initiative to mitigate its impacts.

The **carbon**, **nitrogen**, and **phosphorus** that played such key roles in the bioremediation of the oil spill in Prince William Sound are each elements. An **element** is a fundamental type of matter, a chemical substance with a given set of properties, which cannot be broken down into substances with other properties. Chemists currently recognize 92 elements occurring in nature, as well as more than 20 others that scientists have artificially created. Besides carbon and nitrogen, elements especially abundant in living organisms include **hydrogen** and **oxygen** (Table 4.1). Each element is assigned an abbreviation, or chemical symbol. The *periodic table of the elements* (see Appendix C) summarizes information on the elements in a comprehensive and elegant way.

Elements are composed of **atoms**, the smallest components that maintain the chemical properties of the element. Every atom has a nucleus of **protons** (positively charged particles) and **neutrons** (particles lacking electric charge). The atoms of each element have a defined number of protons, called the *atomic number.* (Elemental carbon, for instance, has six protons in its nucleus; thus, its atomic number is 6.) An atom's nucleus is surrounded by negatively charged particles known as **electrons**, which balance the positive charge of the protons (**Figure 4.2**).

**Isotopes** Although all atoms of a given element contain the same number of protons, they do not necessarily contain the same number of neutrons. Atoms with differing numbers of neutrons are referred to as **isotopes** (**Figure 4.3a**). Isotopes are denoted by their elemental

THE SCIENCE BEHIND THE STORY

## Letting Plants Do the Dirty Work

*Cattails in a wetland*

When soil is contaminated with heavy metals from mining, manufacturing, or other causes, the standard solution has been to dig up tons of soil and pile it into a hazardous waste dump. Bulldozing so much dirt, however, is expensive, destructive, and can release toxic chemicals to the air. As an alternative, scientists are developing methods of **phytoremediation,** using plants (*phyto* means "plant") to detoxify contaminated soils.

Most heavy metals are toxic to plants, just as they are to wildlife and people. But some plants can take up toxic elements and store them without harm. To determine which plants are most effective, scientists run studies in the field and lab.

A group of five researchers from four nations, led by Israel's Hassan Azaizeh, compared the performance of the common reed (*Phragmites australis*) and the cattail (*Typha latifolia*) in removing the toxic metal selenium from water. Environmental engineers seek to use marsh plants such as these in artificially constructed wetlands (• pp. 436–437) to treat wastewater.

Azaizeh's research team used outdoor beds 7.0 m long for its experiments. One bed was filled with reeds, one with cattails, and the third was a control without plants. For four months, the team pumped water spiked with selenium through the soil of each bed from one end to the other. The researchers took periodic samples of water, soil, and plant tissues from six sampling points spaced along the bed and chemically analyzed them.

The group found that most selenium passed through the control bed, giving water at its outlet high concentrations of selenium. In contrast, the cattails removed about half the selenium, so only half appeared in the outlet water. In the reed bed, virtually no selenium was in the outlet water; the reeds had removed nearly all of it (see first figure). The researchers concluded that the reed was most effective at removing selenium.

They also determined that cattails immobilize selenium in their fine roots and nearby organic matter, whereas reeds transport much of the

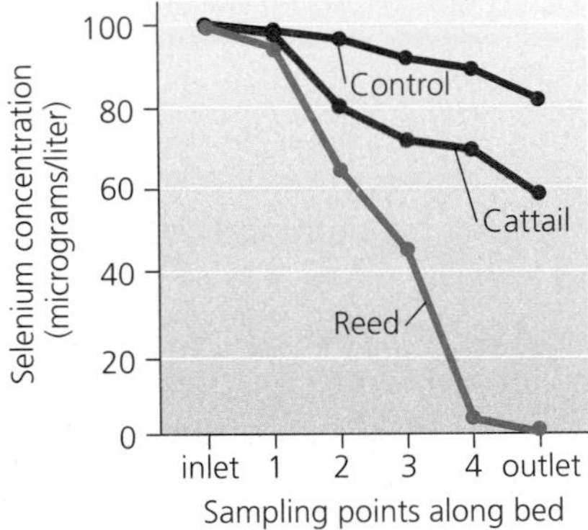


**Water from the outlet (right) of a bed contains less selenium than water from the inlet (left). Cattails removed more selenium than the control bed, but reeds removed the most.** Data are for 1 week after start of experiment, from Azaizeh, H., et al. 2006. Phytoremediation of selenium using subsurface-flow constructed wetland. *International Journal of Phytoremediation* 8:187–198.

symbol preceded by the *mass number,* or combined number of protons and neutrons in the atom. For example, $^{14}C$ (carbon-14) is an isotope of carbon with 8 neutrons (and 6 protons) in the nucleus rather than the normal 6 neutrons (and 6 protons) of $^{12}C$ (carbon-12).

Because they differ slightly in mass, isotopes of an element differ slightly in their behavior. This fact has turned out to be very useful for researchers. Scientists have been able to use isotopes to study a number of phenomena that help illuminate the history of Earth's physical environment.

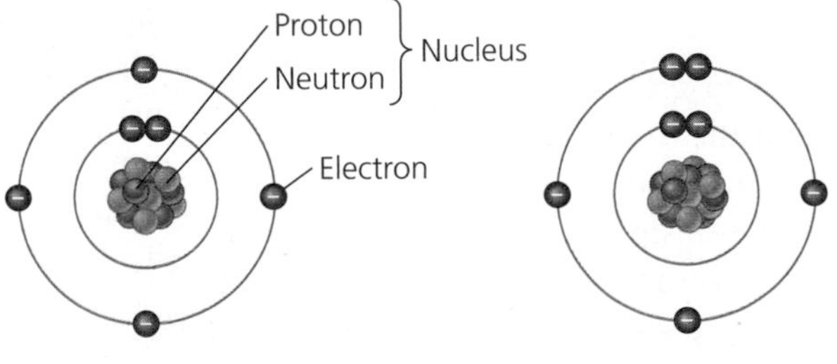


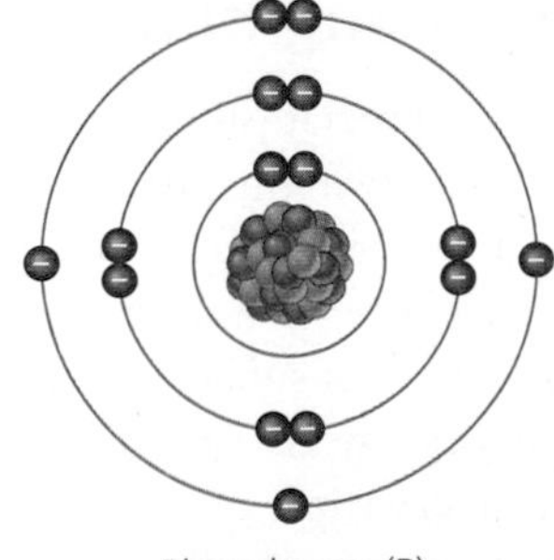


**FIGURE 4.2** In an atom, protons and neutrons are held in the nucleus, and electrons move around the nucleus. Each chemical element has a different number of protons, neutrons, and electrons. Carbon possesses 6 of each, nitrogen 7, and phosphorus 15. These schematic diagrams are meant to clearly show and compare numbers of electrons for these three elements. In reality, however, electrons do *not* orbit the nucleus in rings as shown; they move through space in more complex ways.

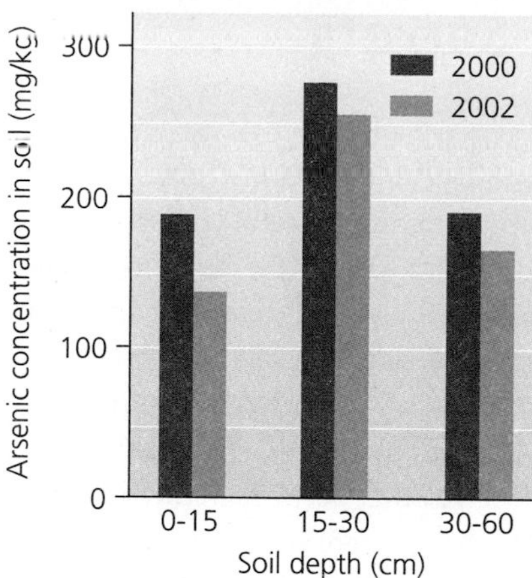


Chinese brake fern removes arsenic from soil. Bars show arsenic concentrations in 2 years at three soil depths. Data from Kertulis-Tartar, G. M., et al. 2006. Phytoremediation of an arsenic-contaminated site using *Pteris vittata* L.: A two-year study. *International Journal of Phytoremediation* 8:311–322.

metal up into their stems and leaves. Plants that extract metals from soil or water and store them aboveground (called *phytoextraction*) can be harvested, gradually reducing contamination with each harvest. Harvested plants are smelted to recover metals or are disposed of as hazardous waste (• pp. 646–653).

The Azaizeh team published its study in 2006 in the *International Journal of Phytoremediation*. Another study in the same journal and year examined phytoextraction by the Chinese brake fern (*Pteris vittata*), a plant known to accumulate the highly toxic element arsenic.

Gina Kertulis-Tartar of Dalton State College in Georgia and three colleagues ventured to a site in Florida where chromated copper arsenate (CCA), a wood preservative, had contaminated the soil with arsenic. Kertulis-Tartar's team planted 324 Chinese brake ferns at the site. Over 2 years, they harvested fern fronds several times. Fronds were dried in an oven, weighed, ground up, and chemically analyzed. The researchers also took periodic soil samples.

The team harvested 7,725 g of dry fern biomass, containing 26.4 g of arsenic—a concentration 1,000—10,000 times greater than that found naturally in Florida soil. During the study period, soil concentrations of arsenic decreased by 26% in the top 15 cm of soil, 43% at depths of 15–30 cm, and 12% at depths of 30–60 cm (see second figure).

The researchers also found that living fronds contained more arsenic than senescing (dying) fronds because as fronds age, they leach arsenic. Thus, if ferns are used for phytoextraction, their fronds should be harvested before they senesce.

Discoveries from studies such as these are being applied widely. At military bases in Iowa, Tennessee, and Nebraska, the U.S. Army Corps of Engineers is using vegetation in artificial wetlands to minimize contamination of groundwater by ammunition. A Virginia company, Edenspace, uses plants to extract lead from residential sites, arsenic from military and energy facilities, zinc and cadmium at Superfund sites (• pp. 652–653), and tungsten from abandoned mines.

Phytoremediation does have limitations. Metals need to be water-soluble. Cleanup is limited to depths reached by plants' roots. Many harvests may be required (8 years were estimated to be necessary in the Chinese brake fern study). Some plants (including Chinese brake fern and common reed) can escape and turn invasive. And plants that accumulate metals may harm insects that eat the plants and animals that eat the insects. Despite such obstacles, phytoremediation is expanding and is helping us mitigate many types of pollution.

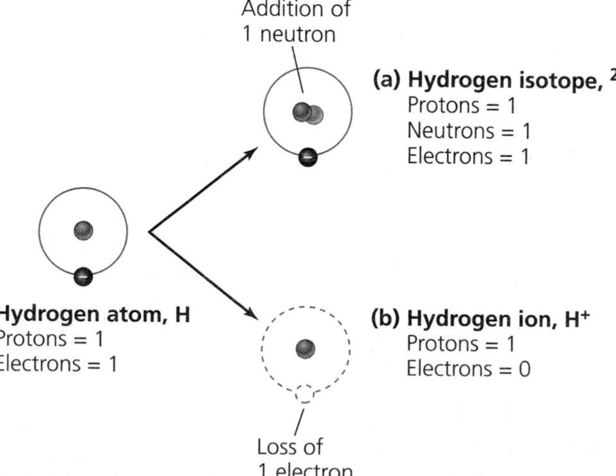


**FIGURE 4.3** The isotope of hydrogen, hydrogen-2 ($^2H$), or deuterium **(a)** contains a neutron as well as a proton, and thus it has greater mass than a typical hydrogen atom. Shown in **(b)** is the hydrogen ion, $H^+$. By losing its electron, it gains a positive charge.

Researchers also have used them to study the flow of nutrients within and among organisms and the movement of organisms from one geographic location to another (see "The Science behind the Story," • pp. 92–93).

Some isotopes are **radioactive** and "decay," changing their chemical identity as they shed subatomic particles and emit high-energy radiation. Such *radioisotopes* decay into lighter and lighter radioisotopes, until they become *stable isotopes,* isotopes that are not radioactive. Each radioisotope decays at a rate determined by that isotope's **half-life,** the amount of time it takes for one-half the atoms to give off radiation and decay. Different radioisotopes have very different half-lives, ranging from fractions of a second to billions of years. The radioisotope uranium-235 ($^{235}U$) is our society's source of energy for commercial nuclear power (• pp. 575–588). It decays into a series of daughter isotopes, eventually forming lead-207 ($^{207}Pb$), and has a half-life of about 700 million years.

THE SCIENCE BEHIND THE STORY

*Dr. Keith Hobson, Environment Canada and the University of Saskatchewan*

## How Isotopes Reveal Secrets of Earth and Life

Isotopes have become one of the most powerful instruments in the environmental scientist's toolkit. These alternate versions of chemical elements enable scientists interested in the past to date ancient materials, reconstruct the climate of past ages, and study the lifestyles of prehistoric people. They allow researchers focused on the here-and-now to work out photosynthetic pathways, measure animals' diets and health, and trace nutrient flows through organisms and ecosystems.

Researchers studying the past often use *radiocarbon dating.* Carbon's most abundant isotope is $^{12}C$, but $^{13}C$ and $^{14}C$ also occur. Carbon-14 is radioactive and occurs in organisms at the same low concentration that it occurs in the atmosphere. Once an organism dies, no new $^{14}C$ is incorporated into its tissues, and the radioactive decay process (• p. 91) gradually reduces its store of $^{14}C$, converting these atoms to $^{14}N$ (nitrogen-14). The decay is slow and steady enough to act as a kind of clock, so that scientists can date ancient organic materials by measuring the percentage of carbon that is $^{14}C$ and matching this value against the clocklike progression of decay.

In this way, archaeologists and paleontologists have dated prehistoric human remains; charcoal, grain, and shells found at ancient campfires; and bones and frozen tissues of recently extinct animals, such as mammoths. Scientists can also estimate the age of a fossil by radiocarbon dating the rock, peat, or sediment that surrounds it. Researchers have dated the most recent ice age with $^{14}C$ analysis of trees overrun by glacial ice sheets. Ice drilled from glaciers today can be aged by measuring the $^{14}C$ in air bubbles trapped during its formation.

Because the half-life (• p. 91) of $^{14}C$ is 5,730 years, radiocarbon dating is not useful for items over 50,000 years old; too little $^{14}C$ remains to permit accurate analysis. For dating older items, scientists use other isotopes. Uranium-238 (with a half-life of 4.5 billion years) has been used to age very early fossils. For dating geological formations, potassium-argon dating is useful (potassium-40 decays to argon-40). Oxygen-18 has been widely used to infer changes in climate and sea level.

Researchers interested in present-day ecology can use stable isotopes (• p. 91) that occur in nature in constant ratios. For instance, nitrogen occurs as 99.63% nitrogen-14 and 0.37% nitrogen-15. Ratios of isotopes are called *isotopic signatures,* and by analyzing these signatures scientists can gain valuable information. For example, organisms tend to retain $^{15}N$ in their tissues but readily excrete $^{14}N$. As a result, animals higher in a food chain (• p. 150) show isotopic signatures biased toward $^{15}N$, as do animals that are starving. Keith Hobson, an ecologist with Environment Canada and the University of Saskatchewan, has used nitrogen signatures to analyze the diets of seabirds and marine mammals, to show that geese fast while nesting, and to trace artificial contaminants in food chains.

Hobson and other scientists have also used stable carbon isotopes for ecological studies. Plants produce food through three photosynthetic pathways, and the isotopic signature

**Ions** Atoms may also gain or lose electrons to become **ions**, electrically charged atoms or combinations of atoms (**Figure 4.3b**). Ions are denoted by their elemental symbol followed by their ionic charge. For instance, a common ion used by mussels and clams to form shells is $Ca^{2+}$, a calcium atom that has lost two electrons and so has a charge of positive 2.

### Atoms bond to form molecules and compounds

Atoms can bond together and form **molecules**, combinations of two or more atoms. Common molecules containing only a single element include those of oxygen gas ($O_2$) and nitrogen gas ($N_2$), both of which are abundant in air. A molecule composed of atoms of two or more different elements is called a **compound**. Water is a compound; composed of two hydrogen atoms bonded to one oxygen atom, it is denoted by the chemical formula $H_2O$. Another compound is **carbon dioxide**, consisting of one carbon atom bonded to two oxygen atoms; its chemical formula is $CO_2$.

Atoms bond together because of an attraction for one another's electrons. Because the strength of this attraction varies among elements, atoms may be held together in different ways, according to whether and how they share or transfer electrons. When atoms in a molecule share electrons, they generate a *covalent bond.* For instance, two atoms of hydrogen bond to form hydrogen gas, $H_2$, by

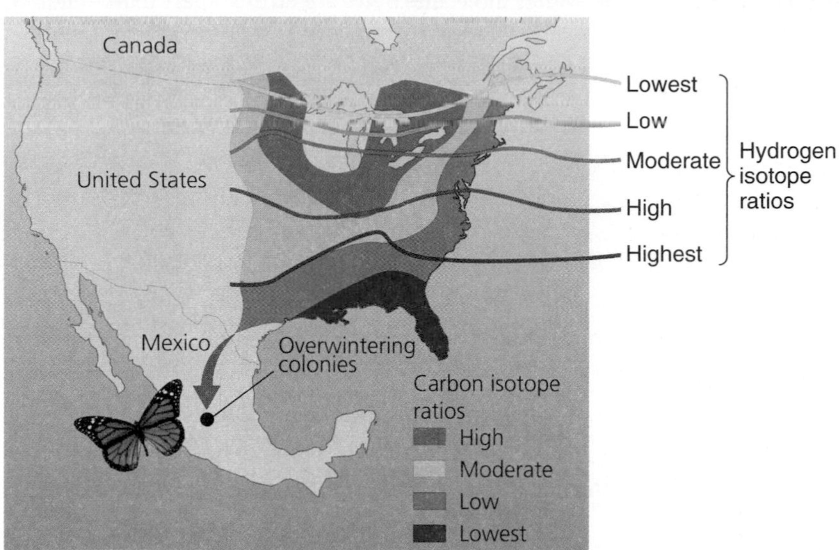


Plants in different geographic areas show different isotopic ratios for elements such as carbon and hydrogen. Caterpillars of monarch butterflies incorporate into their tissues carbon and hydrogen in the isotopic ratios present in the plants they eat. When these caterpillars metamorphose into butterflies and migrate, they carry these isotopic signals with them, providing scientists clues to their origin. Shown is a map of isotopic ratios across eastern North America produced from measurements of monarchs in summer. The four colored bands show ratios of $^{13}C$ to $^{12}C$. The five gray lines show ratios of $^{2}H$ (heavy hydrogen or deuterium) to $^{1}H$. By measuring isotope ratios in monarchs wintering in Mexico and matching them against this map, researchers were able to pinpoint the geographic origin of many of the butterflies.
*Source*: Wassenaar, L. I., and K. A. Hobson, 1998. Natal origins of migratory monarch butterflies at wintering colonies in Mexico: New isotopic evidence. *Proceedings of the National Academy of Sciences USA* 95:15436–15439.

of carbon in plants varies according to the pathway used. Grasses show higher ratios of $^{13}C$ to $^{12}C$ than oak trees do, for instance, whereas cacti have intermediate ratios. When animals eat plants, they incorporate the plants' isotopic signatures into their own tissues, and this signal passes up the food chain. As a result, carbon isotope studies can tell ecologists what an animal has been eating. Archaeologists have used isotopic signatures in human bone to determine when ancient people switched from a hunter-gatherer diet to an agricultural one.

Isotopic data can even tell a scientist where an animal has been. For example, nectar-feeding bats have been shown to move seasonally between regions dominated by cacti and regions dominated by trees. Movements have also been inferred for seals and salmon, migrating warblers, and elephants hunted for ivory.

Researchers are now using isotopes to track movements of birds and other animals that migrate thousands of miles. This is possible because the isotopic signature of hydrogen in rainfall varies systematically across large areas. This signature gets passed from rainwater to plants, and from plants to animals, leaving a fingerprint of geographic origin in an animal's tissues. Hobson and colleagues in 1998 used isotopic data from hydrogen and carbon to pinpoint the geographic origins of monarch butterflies that had migrated to communal roosts in Mexico, providing important information for their conservation (see the figure).

Other chemical elements show similar standing patterns of natural variation, researchers say, so there remains much more we can learn from the use of these subtle chemical clues.

sharing electrons equally. Atoms in a covalent bond can also share electrons unequally, with one atom exerting a greater pull. Such is the case with water, in which oxygen attracts electrons more strongly than hydrogen, forming what are termed *polar* covalent bonds. In contrast, if the strength of attraction is unequal enough, an electron may be transferred from one atom to another. Such a transfer creates oppositely charged ions that are said to form *ionic bonds.* These associations are not considered molecules, but instead are called *ionic compounds,* or *salts.* Table salt (NaCl) contains ionic bonds between positively charged sodium ions ($Na^+$), each of which donated an electron, and negatively charged chloride ions ($Cl^-$), each of which received an electron.

Elements, molecules, and compounds can also come together in mixtures without chemically bonding. Homogenous mixtures of substances are called *solutions,* a term most often applied to liquids, but also applicable to some gases and solids. Air in the atmosphere is a solution formed of constituents such as nitrogen, oxygen, water, carbon dioxide, **methane** ($CH_4$), and **ozone** ($O_3$). Human blood, ocean water, plant sap, and metal alloys such as brass are all solutions. Crude oil at high pressure may carry natural gas in solution and often contains other substances distributed unevenly. It is a heavy liquid mixture of many kinds of molecules consisting primarily of carbon and hydrogen atoms. Its physical properties vary with its temperature, pressure, and composition.

## The chemical structure of the water molecule facilitates life

Water dominates Earth's surface, covering over 70% of the globe, and its abundance is a primary reason Earth is hospitable to life. Scientists think life originated in water and stayed there for 3 billion years before moving onto land. Today every land-dwelling creature remains critically tied to water for its existence.

The water molecule's amazing capacity to support life results from its unique chemical properties. A water molecule's single oxygen atom attracts electrons more strongly than its two hydrogen atoms, resulting in a polar molecule with partial negative charge at the oxygen end and partial positive charge at the hydrogen end. Because of this configuration, water molecules can adhere to one another in a special type of interaction called a *hydrogen bond,* in which the oxygen atom of one water molecule is weakly attracted to one or two hydrogen atoms of another (**Figure 4.4**). These loose connections among molecules give water several properties important in supporting life and stabilizing Earth's climate:

- Water exhibits strong cohesion. (Think of how water droplets on a surface join when you touch them to one another.) This cohesion facilitates the transport of chemicals, such as nutrients and waste, in plants and animals.
- Heating weakens hydrogen bonds but does not initially speed molecular motion. As a result, water can absorb a large amount of heat with only small changes in its temperature. This capacity to resist temperature change helps stabilize systems, be they organisms, ponds, lakes, or climate systems.

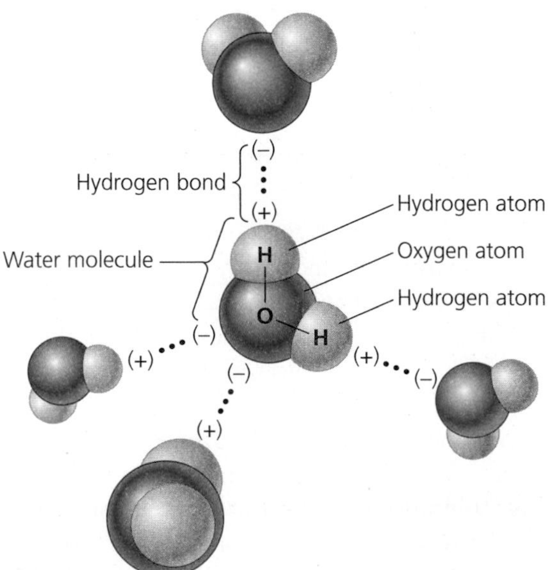


FIGURE 4.4 Water is a unique compound that has several properties crucial for life. Hydrogen bonds give water cohesion by enabling water molecules to adhere loosely to one another.

- Water molecules in ice are farther apart than in liquid water (**Figure 4.5a**), so ice is less dense than liquid water—the reverse pattern of most other compounds, which become denser as they freeze. This is why ice floats on liquid water. Floating ice insulates water bodies, preventing them from freezing solid in winter.
- Water molecules bond well with other polar molecules, because the positive end of one molecule bonds readily to the negative end of another. As a result, water can hold in solution, or dissolve, many other molecules, including chemicals vital for life (**Figure 4.5b**). It follows that most biologically important solutions involve water.

### Weighing the Issues | Water's Properties for Life

Water has several special properties that accommodate life. Can you generate examples of specific ways in which each of these properties might help a particular organism, such as a fish in a pond? Can you think of ways in which any of water's properties might sometimes bring harm to the fish?

## Hydrogen ions determine acidity

In any aqueous solution, a small number of water molecules dissociate, each forming a hydrogen ion ($H^+$) and a hydroxide ion ($OH^-$). The product of hydrogen and hydroxide ion concentrations is always $10^{-14}$. As the concentration of one increases, the concentration of the other decreases, and the product of their concentrations remains constant. Pure water contains equal numbers of these ions, each at a concentration of $10^{-7}$, and we say that this water is neutral. Most aqueous solutions, however, contain different concentrations of these two ions. Solutions in which the $H^+$ concentration is greater than the $OH^-$ concentration are **acidic**, whereas solutions in which the $OH^-$ concentration exceeds the $H^+$ concentration are **basic.**

The **pH** scale (**Figure 4.6**) was devised to quantify the acidity or basicity of solutions. It runs from 0 to 14, because these numbers reflect the negative logarithm of the hydrogen ion concentration. Thus pure water has a pH of 7, because its hydrogen ion concentration is $10^{-7}$. Seawater has a greater concentration of hydroxide ions, close to $10^{-6}$. This means that its hydrogen ion concentration is about $10^{-8}$, and thus its pH is close to 8. Solutions with pH less than 7 are acidic, those with pH greater than 7 are basic, and those with pH of 7 are neutral. Because the pH scale is logarithmic, each step on the scale represents a tenfold difference in hydrogen ion concentration. Thus, a substance with pH of 6 contains 10 times as many hydrogen ions as a substance with pH of 7, and a substance with pH of 5 contains 100 times as many

(a) Why ice floats on water

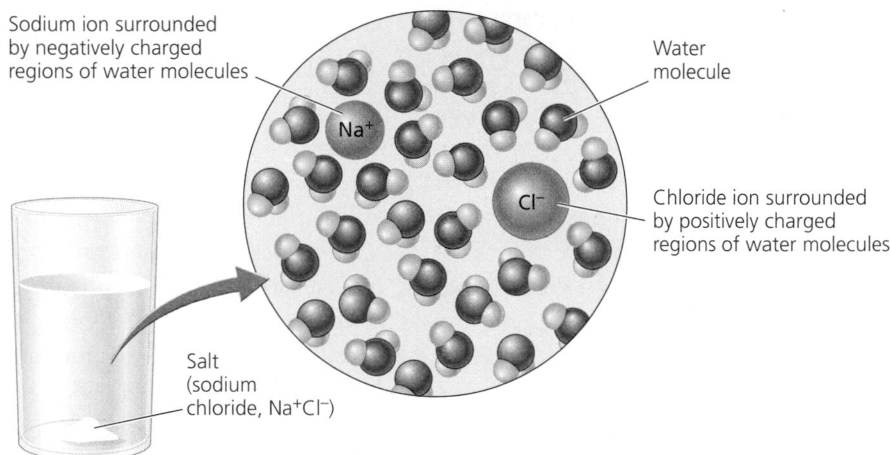


(b) Water as a solvent; how water dissolves salt

**FIGURE 4.5** Ice (**a**) floats in liquid water because ice is less dense. In ice, molecules are connected by stable hydrogen bonds, forming a spacious crystal lattice. In liquid water, hydrogen bonds frequently break and re-form, and the molecules are closer together and less organized. Water is often called the "universal solvent" because it can dissolve so many chemicals, especially polar and ionic compounds (**b**). Seawater holds sodium and chloride ions, among others, in solution.

hydrogen ions as one with pH of 7. Figure 4.6 shows pH for a number of common substances. Industrial air pollution has intensified the acidity of precipitation (• pp. 491–495), and rain in parts of the northeastern and midwestern United States now frequently dips to pH of 4 or lower.

## Matter is composed of organic and inorganic compounds

Beyond their need for water, living things also depend on organic compounds, which they create and of which they are created. **Organic compounds** consist of carbon atoms (and generally hydrogen atoms) joined by covalent bonds, and they may include other elements, such as nitrogen, oxygen, sulfur, and phosphorus. Carbon's unusual ability to build elaborate molecules has resulted in millions of different organic compounds that show various degrees of complexity. Because of the diversity of organic compounds and their importance in living organisms, chemists differentiate organic compounds from inorganic compounds, which lack carbon–carbon bonds.

Crude oil and petroleum products are made up of organic compounds called hydrocarbons. **Hydrocarbons** consist solely of atoms of carbon and hydrogen. The simplest hydrocarbon is methane ($CH_4$), the key component of natural gas; it has one carbon atom bonded to four hydrogen atoms (**Figure 4.7a**). Adding another carbon atom and two more hydrogen atoms gives us ethane ($C_2H_6$), the next-simplest hydrocarbon (**Figure 4.7b**). The smallest (and therefore lightest-weight) hydrocarbons (those consisting of four or fewer carbon atoms) exist in a gaseous state at normal temperatures and pressures. Larger (therefore heavier) hydrocarbons are liquids, and those containing over 20 carbon atoms are normally solids. Some hydrocarbons from petroleum are known to pose health hazards to wildlife and people. For example, *polycyclic aromatic hydrocarbons,* or PAHs (**Figure 4.7c**), which are volatile molecules with a structure of multiple carbon

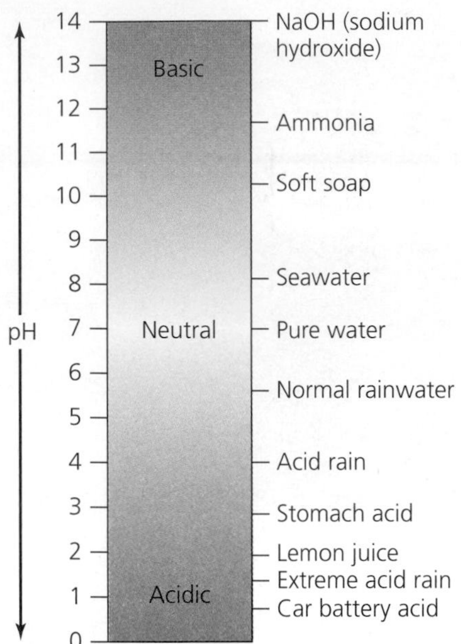


**FIGURE 4.6** The pH scale measures how acidic or basic a solution is. The pH of pure water is 7, the midpoint of the scale. Acidic solutions have higher hydrogen ion concentrations and lower pH, whereas basic solutions have lower hydrogen ion concentrations and higher pH.

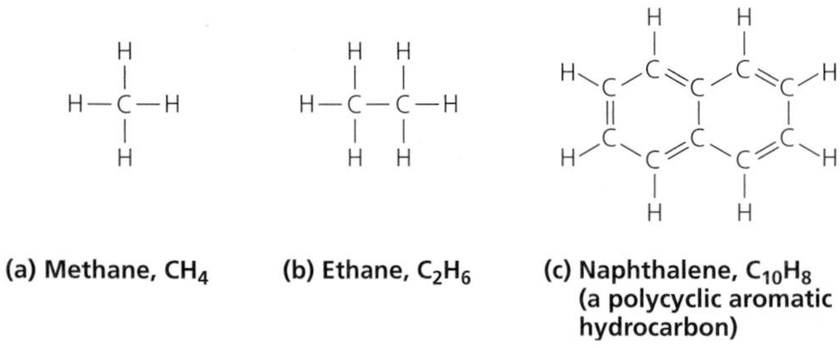

**(a) Methane, $CH_4$** **(b) Ethane, $C_2H_6$** **(c) Naphthalene, $C_{10}H_8$ (a polycyclic aromatic hydrocarbon)**

**FIGURE 4.7** Hydrocarbons are a major class of organic compound, and mixtures of them make up fossil fuels such as crude oil. The simplest hydrocarbon is methane (**a**). Many hydrocarbons consist of linear chains of carbon atoms with hydrogen atoms attached; the shortest of these is ethane (**b**). Volatile hydrocarbons with multiple rings, such as naphthalene (**c**), are called polycyclic aromatic hydrocarbons (PAHs).

rings, can evaporate from spilled oil and gasoline and can mix with water. The eggs and young of fish and other aquatic creatures are often most at risk. PAHs also occur in particulate form in various combustion products, including cigarette smoke, wood smoke, and charred meat.

Bacteria used in the bioremediation of petroleum spills do not actually consume the entire hydrocarbon molecules they attack. Rather, the bacteria, facilitated by oxygen, break down complex hydrocarbon structures into simpler ones, or into their simplest components, hydrogen and carbon. Often this involves a series of chemical reactions. For example, bacterial degradation of the PAH naphthalene shown in Figure 4.7c involves 11 steps, during which atoms are added, removed, and rearranged, eventually producing the simpler products pyruvate and acetaldehyde (**Figure 4.8**).

## Macromolecules are building blocks of life

Just as the carbon atoms in hydrocarbons may be strung together in chains, other organic compounds sometimes combine to form long chains of repeated molecules. Some of these chains, called **polymers**, play key roles as building blocks of life. Three types of polymers are essential to life: proteins, nucleic acids, and carbohydrates. Lipids are not considered polymers but are also essential. These four types of molecules are referred to as **macromolecules** because of their large size.

**Proteins** **Proteins** serve many functions. Some help produce tissues and provide structural support for the organism. For example, animals use proteins to generate skin, hair, muscles, and tendons. Some proteins help store energy, and others transport substances. Some function as components of the immune system, defending the organism against foreign attackers. Still others act as hormones, molecules that serve as chemical messengers within an organism. Finally, proteins can serve as enzymes, molecules that catalyze, or promote, certain chemical reactions. For example, bacteria used for bioremediation use specialized

11 steps (10 intermediate compounds)

Naphthalene ($C_{10}H_8$)

Pyruvate ($C_3H_3O_3^-$)

Acetaldehyde ($C_2H_4O$)

**FIGURE 4.8** Bacterial breakdown of naphthalene involves 11 steps and results in the simpler organic compounds pyruvate and acetaldehyde. The many chemicals involved in this complicated process are omitted from this simplified diagram.

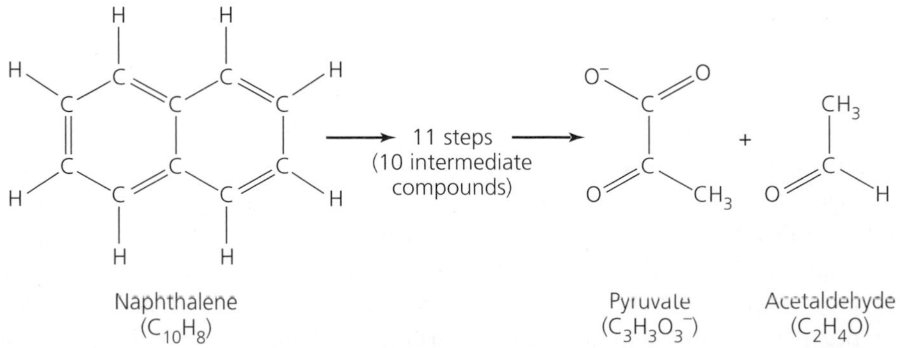

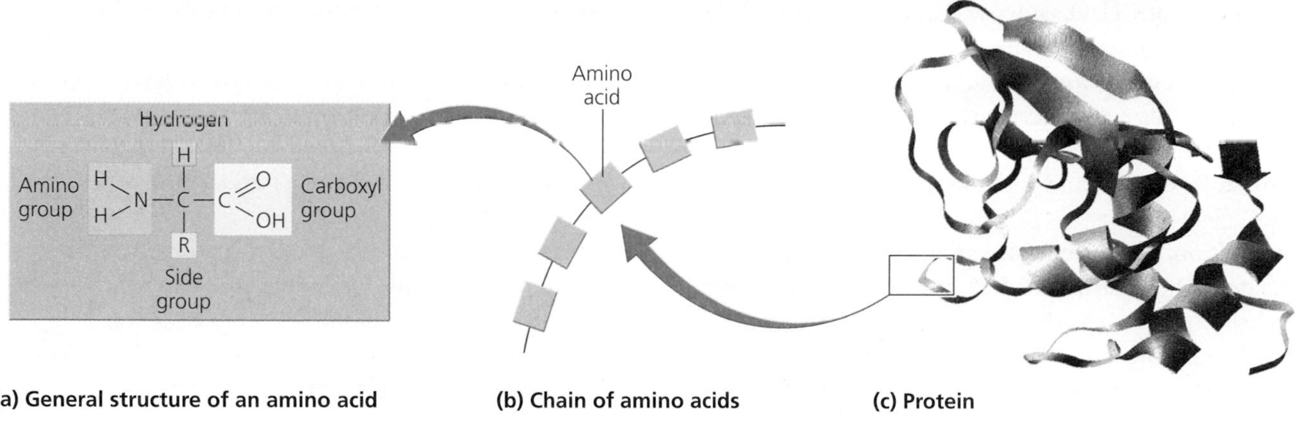


**FIGURE 4.9** Proteins are made up of long chains of amino acids. Amino acids (**a**) contain a central carbon linked to a hydrogen atom, an acidic carboxyl group (−COOH), a basic amine group ($-NH_2$), and an organic side group unique to each type of amino acid. Chains of amino acids (**b**) comprise proteins, which fold into complex convoluted shapes (**c**) that help determine their functions.

enzymes to break down hydrocarbons, just as we use enzymes to digest our food.

Proteins consist of long chains of organic molecules called amino acids (**Figure 4.9a**). Organisms combine up to 20 different types of amino acids into long chains (**Figure 4.9b**) to build proteins. A protein's identity is determined by its sequence of amino acids and by the shape the protein molecule assumes as it folds. Protein molecules typically have highly convoluted shapes (**Figure 4.9c**). A protein's folding pattern affects its function, because the position of each chemical group helps determine how it interacts with cell surfaces and with other molecules.

**Nucleic acids** Protein production is directed by **nucleic acids**. The two nucleic acids—**deoxyribonucleic acid (DNA)** and **ribonucleic acid (RNA)**—carry the hereditary information for organisms and are responsible for passing traits from parents to offspring. Nucleic acids are composed of long chains of nucleotides, each of which contains a sugar molecule, a phosphate group, and a nitrogenous base (**Figure 4.10a**). The double strands of DNA can be pictured as rungs of a ladder twisted into a spiral, giving the molecule a shape called a double helix (**Figure 4.10b**). RNA is similar, but it is generally single-stranded and uses ribose (instead of deoxyribose) as its sugar group. Hereditary information encoded in the nucleotide sequence of DNA is rewritten to a molecule of RNA. RNA then directs the order in which amino acids assemble to build proteins, which go on to influence the structure and maintenance of the organism. Genetic information in DNA is passed from one generation to another as the strands replicate during cell division and egg or sperm formation. Regions of DNA coding for particular proteins that perform particular functions are called **genes**. In most organisms, the *genome*—the set of all an organism's genes—is divided into chromosomes. Different types of organisms have different numbers of genes and chromosomes. Most bacteria have a single circular chromosome, for instance, whereas humans have 46 linear ones.

**Carbohydrates** A third type of biologically vital polymer, **carbohydrates** consist of atoms of carbon, hydrogen, and oxygen. Simple carbohydrates, called sugars or monosaccharides, have structures, or skeletons, that are three to seven carbon atoms long, and formulas that are

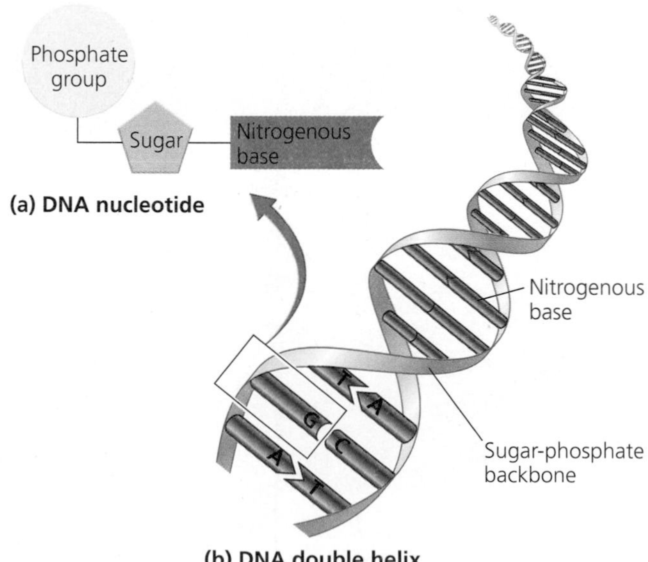


**FIGURE 4.10** Nucleic acids encode genetic information in the sequence of nucleotides (**a**), small molecules that pair together like rungs of a ladder. DNA includes four types of nucleotides, each with a different nitrogenous base: adenine (A), guanine (G), cytosine (C), and thymine (T). Adenine (A) pairs with thymine (T), and cytosine (C) pairs with guanine (G). In RNA, thymine is replaced by uracil (U). DNA twists into the shape of a double helix (**b**).

some multiple of $CH_2O$. Glucose ($C_6H_{12}O_6$) is one of the most common and important sugars, providing energy that fuels plant and animal cells. Glucose also serves as a building block for complex carbohydrates, or polysaccharides. Plants use starch, a glucose-based polysaccharide, to store energy, and animals eat plants to acquire starch.

In addition, both plants and animals use complex carbohydrates to build structure. Insects and crustaceans form hard shells from the carbohydrate chitin. Cellulose, the most abundant organic compound on Earth, is a complex carbohydrate found in the cell walls of leaves, bark, stems, and roots.

**Lipids** **Lipids** are a chemically diverse group of compounds, classified together because they do not dissolve in water:

- *Fats and oils* store energy, especially for mobile animals. Their hydrocarbon structures somewhat resemble gasoline, a similarity echoed in their function: to amass energy and release it when burned.
- *Phospholipids* are similar to fats but consist of one water-repellant side and one water-attracting side. This allows them, when arranged in a double layer, to make up the primary component of cell membranes.
- *Waxes* are digestible by some but not all organisms, and they can play structural roles (for instance, beeswax in bees' hives).
- *Steroids* are used in animal cell membranes and in production of hormones, including estrogen and androgen, vital to sexual maturation.

## We create synthetic polymers

The polymers in nature that are so vital to our survival have inspired chemists. These scientists have taken the polymer concept and run with it, creating innumerable types of synthetic (human-made) polymers, which we call **plastics**. Polyethylene, polypropylene, polyurethane, and polystyrene are just a few of the many synthetic polymers in our manufactured products today (we often know brand names better, such as Nylon, Teflon, and Kevlar). Plastics, many of them derived from hydrocarbons in petroleum, are all around us

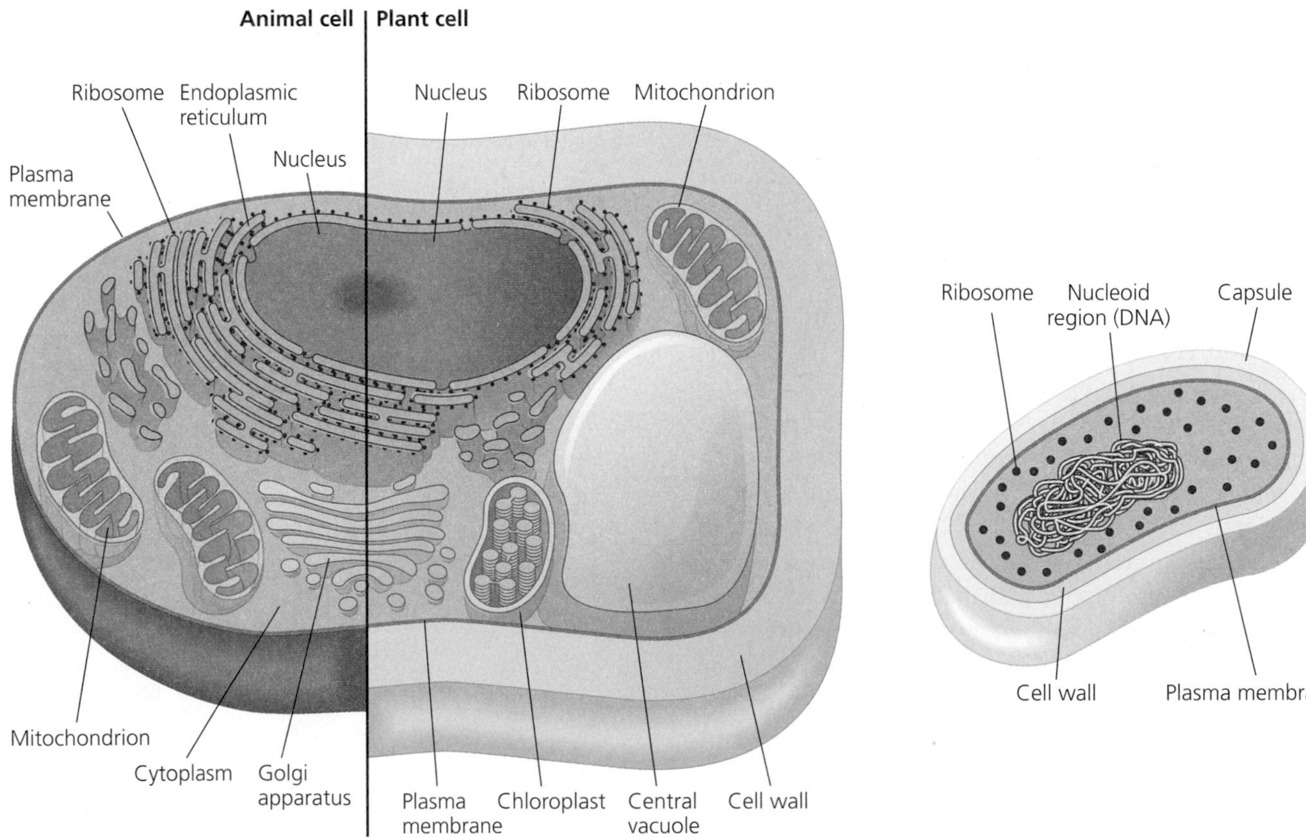


FIGURE 4.11 Cells are the smallest unit of life that can function independently. Eukaryotic cells (**a**) contain organelles, such as mitochondria and chloroplasts, as well as a membrane-bound nucleus containing DNA. Plant cells (right half of left diagram) have rigid cell walls of cellulose, whereas animal cells (left half of left diagram) have more flexible cell membranes. Prokaryotic cells (**b**) are simpler, lacking membrane-bound organelles and an enclosed nucleus.

in our everyday lives, from furniture to food containers to fiber optics to fleece jackets (see Figure 19.10, • p. 556).

We value synthetic polymers because they resist chemical breakdown. Plastics are so useful to us, ironically, for the same reason that an oil spill is so persistent and difficult to remediate. Although long-lasting plastic products make our lives easier, the waste and pollution they create when we discard them is long-lasting as well. In future chapters we will see how pollutants that resist breakdown can cause problems for wildlife and human health (• pp. 394–397), for water quality (• pp. 429–435), for marine animals (• pp. 455–456), and for waste management (• pp. 647–653). Fortunately, chemists, policymakers, and citizens are finding more ways to design less-polluting substances and to recycle materials more effectively.

## Organisms use cells to compartmentalize macromolecules

Natural polymers and macromolecules help to build **cells,** the most basic unit of life's organization. All living things are composed of cells, and organisms range in complexity from single-celled bacteria to plants and animals that contain millions of cells. Cells vary greatly in size, shape, and function.

Biologists classify organisms into two groups based on the structure of their cells. **Eukaryotes** include plants, animals, fungi, and protists. The cells of eukaryotes (**Figure 4.11a**) consist of an outer membrane of lipids and an inner fluid-filled chamber containing *organelles,* internal structures that perform specific functions. These organelles include (among others) ribosomes, which synthesize proteins, and mitochondria, which help extract energy from sugars and fats. Eukaryotic cells also contain a membrane-enclosed nucleus that houses DNA. Eukaryotes generally have many cells.

**Prokaryotes** are generally single-celled, and prokaryotic cells lack membrane-bound organelles and a nucleus (**Figure 4.11b**). All bacteria are prokaryotes, as are the lesser-known microorganisms called archaea. Bacteria are diverse and are ubiquitous in the environment, and of course they do far more than attack oil spills. Many types of bacteria perform functions vital to human life—for instance, aiding in digestion and preventing the buildup of harmful wastes.

In eukaryotes, cells specialize in different roles and are organized into collections of cells performing the same function, called *tissues.* Tissues make up *organs,* and organisms are composed of *organ systems.* We have now completed a (very quick!) review of the hierarchy in which matter is organized in living things on Earth (**Figure 4.12**). Over the next three chapters, we will explore the levels of this hierarchy above the organismal level, as we study the science of ecology. But first we will examine energy, something that underlies every process in environmental science.

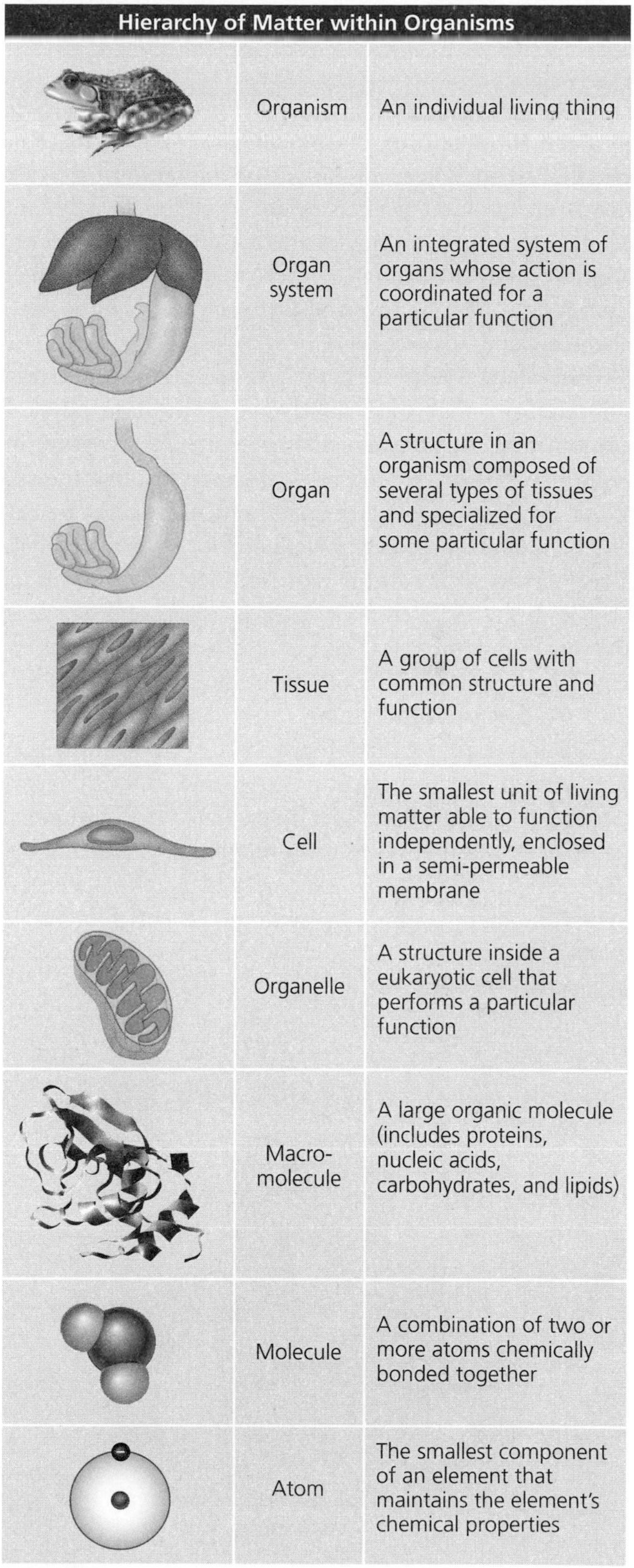

**Hierarchy of Matter within Organisms**

| |  |  |
|---|---|---|
| | Organism | An individual living thing |
| | Organ system | An integrated system of organs whose action is coordinated for a particular function |
| | Organ | A structure in an organism composed of several types of tissues and specialized for some particular function |
| | Tissue | A group of cells with common structure and function |
| | Cell | The smallest unit of living matter able to function independently, enclosed in a semi-permeable membrane |
| | Organelle | A structure inside a eukaryotic cell that performs a particular function |
| | Macro-molecule | A large organic molecule (includes proteins, nucleic acids, carbohydrates, and lipids) |
| | Molecule | A combination of two or more atoms chemically bonded together |
| | Atom | The smallest component of an element that maintains the element's chemical properties |

FIGURE 4.12 Within an organism, matter is organized in a hierarchy of levels, from atoms through cells through organ systems.

## Energy Fundamentals

Creating and maintaining organized complexity, whether of a cell, an organism, or an ecological system, requires energy. Energy is needed to organize matter into complex forms such as biological polymers, to build and maintain cellular structure, to power interactions among species, and to power the geological forces that shape our planet. Energy is somehow involved in nearly every biological, chemical, and physical event.

But what, exactly, is energy? An intangible phenomenon, **energy** is that which can change the position, physical composition, or temperature of matter. A sparrow in flight expends energy to propel its body through the air. When the sparrow lays an egg, its body uses energy to create the calcium-based eggshell and color it with pigment. The sparrow sitting on its nest transfers energy from its body to heat the developing chicks inside its eggs. Some of the most dramatic releases of energy in nature do not involve living things; think of volcanoes erupting or tornadoes sweeping across the plains.

Scientists differentiate between two types of energy: **potential energy**, energy of position; and **kinetic energy**, energy of motion. Consider river water held behind a dam. By preventing water from moving downstream, the dam causes the water to accumulate potential energy. When the dam gates are opened, the potential energy is converted to kinetic energy, in the form of water's motion as it rushes downstream.

Such energy transfers take place at the atomic level every time a chemical bond is broken or formed. **Chemical energy** is potential energy held in the bonds between atoms. Bonds differ in their amounts of chemical energy, depending on the atoms they hold together. Converting a molecule with high-energy bonds (such as the carbon–carbon bonds of petroleum products) into molecules with lower-energy bonds (such as the bonds in water or carbon dioxide) releases energy by changing potential energy into kinetic energy and produces motion, action, or heat. Just as automobile engines split the hydrocarbons of gasoline to release chemical energy and generate movement, our bodies split glucose molecules in our food for the same purpose (**Figure 4.13**).

### Energy is always conserved . . .

Although energy can change from one form to another, it cannot be created or destroyed. Just as matter is conserved (• p. 89), the total energy in the universe remains constant and thus is said to be conserved. Scientists have dubbed this principle the **first law of thermodynamics.** The potential energy of the water behind a dam will equal the kinetic energy of its eventual movement down the riverbed. Similarly, burning converts the potential energy in a log of firewood to an equal amount of energy produced as heat and light. We obtain energy from the food we eat, which we expend in exercise, put toward the body's maintenance, or store as fat. We do not somehow create additional energy or end up with less than the food gives us. Any individual system can temporarily increase or decrease in energy, but the total amount in the universe remains constant.

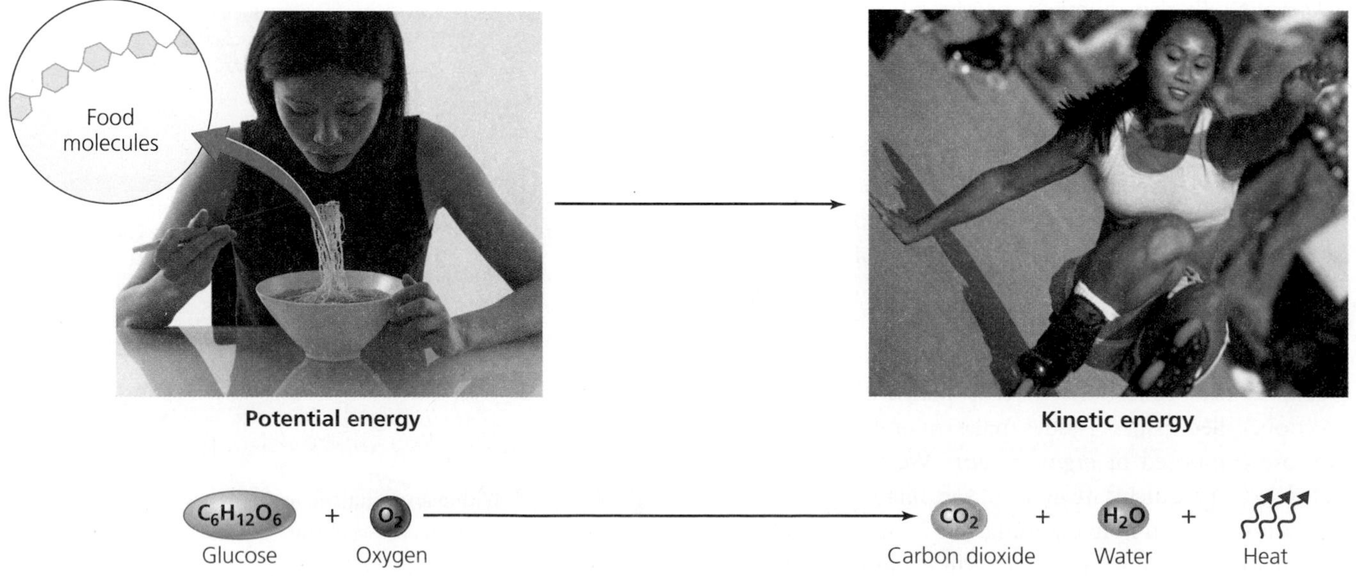


**FIGURE 4.13** Energy is released when potential energy is converted to kinetic energy. Potential energy stored in sugars, such as glucose, in the food we eat (**a**), combined with oxygen, becomes kinetic energy when we exercise (**b**), releasing carbon dioxide, water, and heat as by-products.

## . . . But energy changes in quality

Although the overall amount of energy is conserved in any process of energy transfer, the **second law of thermodynamics** states that the nature of energy will change from a more-ordered state to a less-ordered state, if no force counteracts this tendency. That is, systems tend to move toward increasing disorder, or *entropy*. For instance, after death every organism undergoes decomposition and loses its structure. A log of firewood—the highly organized and structurally complex product of many years of slow tree growth—transforms in the campfire to a residue of carbon ash, smoke, and gases such as carbon dioxide and water vapor, as well as the light and the heat of the flame (**Figure 4.14**). With the help of oxygen, the complex biological polymers making up the wood are converted into a disorganized assortment of rudimentary molecules and heat and light energy.

The nature of an energy source helps determine how easily people can harness it. Sources such as petroleum products and high-voltage electricity contain concentrated energy that we can readily release. It is relatively easy for us to gain large amounts of energy efficiently from these high-quality sources. In contrast, sunlight and the heat stored in ocean water are considered low-quality energy sources. Each day the world's oceans absorb heat energy from the sun equivalent to that of 250 billion barrels of oil—more than 3,000 times as much as our global society uses in a year. But because this energy is spread out across such vast spaces, it is diffuse and difficult to harness.

In every transfer of energy we attempt, some portion that we are unable to harness escapes. We can express the degree to which we successfully capture energy in terms of the *energy conversion efficiency*, the ratio of useful output of energy to the amount that needs to be input. When we burn gasoline in an automobile engine, only about 16% of the energy released is used to power the automobile, and the rest of the energy is converted to heat (see Figure 18.27, • p. 531). Incandescent light bulbs are even less efficient; only 5% of their energy is converted to the light that we use them for, while the rest escapes as heat. Viewed in this context, the 15% efficiency of much current solar energy technology does not look bad at all.

Although the second law of thermodynamics specifies that systems tend to move toward disorder, the order of an object or system can be increased through the input of energy from outside the system. This is precisely what living organisms do. Organisms maintain their structure and function by consuming energy. They represent a constant struggle to maintain order and combat the natural tendency toward disorder.

## Light energy from the sun powers most living systems

The energy that powers Earth's ecological systems comes primarily from the sun. The sun releases radiation from large portions of the electromagnetic spectrum, although our atmosphere filters much of this out, and we see only some of this radiation as visible light (**Figure 4.15**). Most of the sun's energy is reflected, or else absorbed and re-emitted, by the atmosphere, land, or water (see Figure 18.1, • p. 507). Solar energy drives our weather and climate patterns, including winds and ocean currents. A small amount (less than 1% of the total) powers plant growth, and a still smaller amount flows from plants into the organisms that eat them and the organisms that decompose dead organic matter. A minuscule percentage of this energy is eventually deposited below ground in the chemical bonds in fossil fuels.

Some organisms use the sun's radiation directly to produce their own food. Such organisms, called **autotrophs** or **primary producers**, include green plants, algae, and cyanobacteria. (Cyanobacteria are a type of bacteria named for their cyan, or blue-green, color.) Autotrophs turn light energy from the sun into chemical energy in a

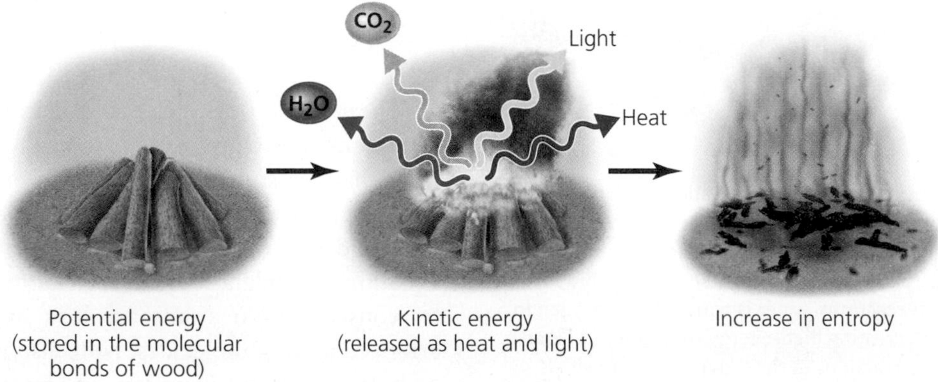


**FIGURE 4.14** The burning of firewood demonstrates energy transfer leading from a more-ordered to a less-ordered state. This increase in entropy reflects the second law of thermodynamics.

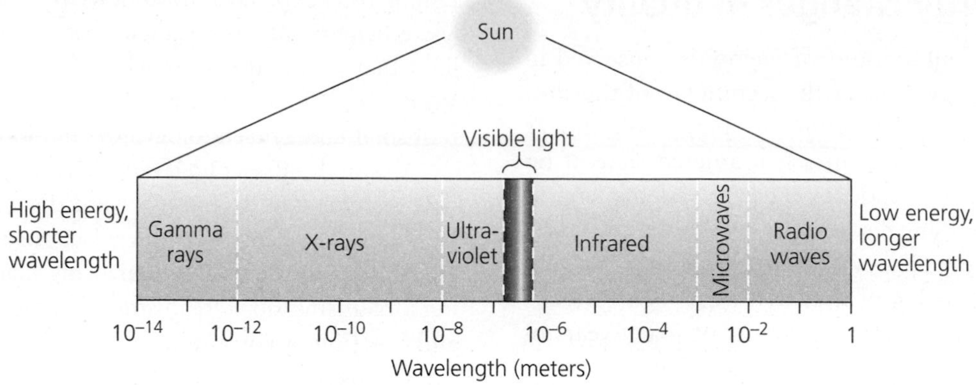


**FIGURE 4.15** The sun emits radiation from many portions of the electromagnetic spectrum, and visible light makes up only a small proportion of this energy. Some radiation that reaches our planet is reflected back, some is absorbed by air, land, and water, and a small amount powers photosynthesis.

process called **photosynthesis** (**Figure 4.16**). In photosynthesis, sunlight powers a series of chemical reactions that convert carbon dioxide and water into sugars, transforming low-quality energy from the sun into high-quality energy the organism can use. It is an example of moving toward a state of lower entropy, and as such it requires a substantial input of outside energy.

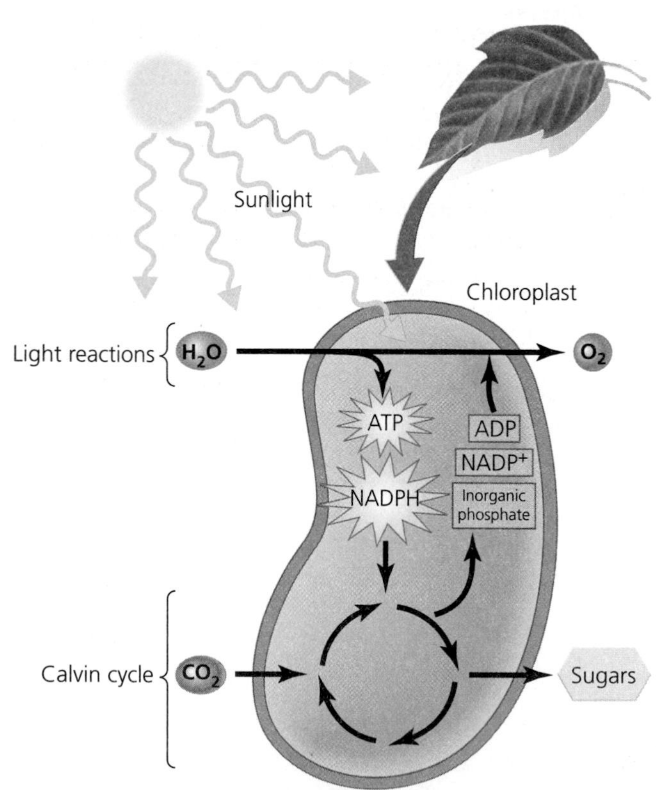


**FIGURE 4.16** In photosynthesis, autotrophs such as plants, algae, and cyanobacteria use sunlight to convert carbon dioxide and water into sugars and oxygen. This schematic diagram summarizes the complex sets of chemical reactions that take place within chloroplasts. In the light reactions, water is converted to oxygen in the presence of sunlight, creating high-energy molecules (ATP and NADPH) that help drive reactions in the Calvin cycle, in which carbon dioxide is used to produce sugars. Molecules of ADP, $NADP^+$, and inorganic phosphate created in the Calvin cycle in turn help power the light reactions, creating an endless loop.

## Photosynthesis produces food for plants and animals

Photosynthesis occurs within cell organelles called *chloroplasts,* where the light-absorbing pigment *chlorophyll* (which is what makes plants green) uses solar energy to initiate a series of chemical reactions called *light reactions.* During these reactions, water molecules are split, and they react to form hydrogen ions ($H^+$) and molecular oxygen ($O_2$), thus creating the oxygen that we breathe. The light reactions also produce small, high-energy molecules that are used to fuel reactions in the *Calvin cycle.* In these reactions, carbon atoms from carbon dioxide are linked together to manufacture sugars. Photosynthesis is a complex process, but the overall reaction can be summarized with the following equation:

$$6CO_2 + 12H_2O + \text{the sun's energy} \longrightarrow C_6H_{12}O_6\ (\text{sugar}) + 6O_2 + 6H_2O$$

The numbers preceding each molecular formula indicate how many of each molecule are involved in the reaction. Note that the sum of the numbers on each side of the equation for each element are equal; that is, there are 6 C, 24 H, and 24 O on each side. This illustrates how chemical equations are balanced, with each atom recycled and matter conserved. No atoms are lost; they are simply rearranged among molecules. Note also that water appears on both sides of the equation. The reason is that for every 12 water molecules that are input and dissociated in the process, 6 water molecules are newly created. We can streamline the photosynthesis equation by showing only the net loss of 6 water molecules:

$$6CO_2 + 6H_2O + \text{the sun's energy} \longrightarrow C_6H_{12}O_6 \text{ (sugar)} + 6O_2$$

Thus in photosynthesis, water, carbon dioxide, and light energy from the sun are transformed to produce sugar (glucose) and oxygen. To accomplish this, green plants draw up water from the ground through their roots, suck in carbon dioxide from the air through their leaves, and harness sunlight. With these ingredients, they create sugars for their growth and maintenance, and they release oxygen as a by-product.

Animals, in turn, depend on the sugars and oxygen from photosynthesis. Animals survive by eating plants, or by eating animals that have eaten plants, and by taking in oxygen. In fact, it is thought that animals appeared on Earth's surface only after the planet's atmosphere had been supplied with oxygen by the earliest autotrophs, the cyanobacteria.

FIGURE 4.17 This geyser in the Black Rock Desert of Nevada propels scalding water into the air, powered by geothermal energy from deep below ground. The bright colors of the rocks are from colonies of bacteria that thrive in the hot, mineral-laden water.

## Cellular respiration releases chemical energy

Organisms make use of the chemical energy created by photosynthesis in a process called **cellular respiration**. To release the chemical energy of glucose, cells employ oxygen to convert glucose back into its original starting materials, water and carbon dioxide. The energy released during this process is used to form chemical bonds or to perform other tasks within cells. The net equation for cellular respiration is the exact opposite of that for photosynthesis:

$$C_6H_{12}O_6 \text{ (sugar)} + 6O_2 \longrightarrow 6CO_2 + 6H_2O + \text{energy}$$

However, the energy gained per glucose molecule in respiration is only two-thirds of the energy input per glucose molecule in photosynthesis—a prime example of the second law of thermodynamics. Respiration occurs in the autotrophs that create glucose and also in **heterotrophs**, organisms that gain their energy by feeding on other organisms. Heterotrophs include most animals, as well as the fungi and microbes that decompose organic matter. In most ecological systems, plants, algae, or cyanobacteria form the base of a food chain through which energy passes to heterotrophs (• pp. 148–150).

## Geothermal energy also powers Earth's systems

Although the sun is Earth's primary power source, it is not the only one. A minor additional source is the gravitational pull of the moon, which causes ocean tides. This low-quality energy is weak and diffuse in comparison to solar energy. A more significant additional energy source is radiation emanating from inside Earth, powered by radioactivity (• p. 91). Radiation from radioisotopes deep inside Earth heats the inner portion of our planet, and this heat gradually makes its way to the surface. There it drives plate tectonics (• pp. 199–201), heats magma that erupts from volcanoes, and warms groundwater, which in some locations shoots out of the ground in the form of geysers (**Figure 4.17**). Called **geothermal energy**, this heating from deep within the planet is now being harnessed for commercial power (• pp. 617–620).

Long before humans came along, however, geothermal energy was powering other biological communities. On the floor of the ocean, jets of geothermally heated water—essentially underwater geysers—gush into the icy-cold depths. In one of the more amazing scientific discoveries of recent decades, scientists realized that these *hydrothermal vents* can host entire communities of organisms that thrive in the extreme high-temperature, high-pressure conditions. Gigantic clams, immense tubeworms, and odd mussels, shrimps, crabs, and fish all flourish in the seemingly hostile environment near scalding water that shoots out of tall chimneys of encrusted minerals (**Figure 4.18**).

These locations are so deep underwater that they completely lack sunlight, so their communities cannot be fueled through photosynthesis. Instead, bacteria in deep-sea vents use the chemical-bond energy of hydrogen sulfide ($H_2S$) to transform inorganic carbon into organic carbon compounds in a process called **chemosynthesis**:

$$6CO_2 + 6H_2O + 3H_2S \longrightarrow C_6H_{12}O_6 \text{ (sugar)} + 3H_2SO_4$$

FIGURE 4.18 Hydrothermal vents on the ocean floor (**a**) send spouts of hot, mineral-rich water into the cold blackness of the deep sea. Specialized biological communities thrive in these unusual conditions. Odd creatures such as giant tubeworms (**b**) survive thanks to bacteria that produce food from hydrogen sulfide.

Chemosynthesis occurs in various ways, but note how this particular reaction for chemosynthesis closely resembles the photosynthesis reaction. These two processes use different energy sources, but each uses water and carbon dioxide to produce sugar and a by-product, and each produces potential energy that is later released during respiration. Energy from chemosynthesis passes through the deep-sea-vent animal community as heterotrophs such as clams, mussels, and shrimp gain nutrition from chemoautotrophic bacteria. Hydrothermal vent communities excited scientists not only because they were novel and unexpected, but also because some researchers think they may help us understand how life itself came to be.

# The Origin of Life

How life originated is one of the most centrally important—and intensely debated—questions in modern science. In searching for the answer, scientists have learned a great deal about the history of life on our planet.

## Early Earth was a very different place

Earth formed about 4.5 billion years ago in the same way as the other planets of our solar system; dispersed bits of material whirling through space around our sun were drawn together by gravity, coalescing into a series of spheres. For several hundred million years after the planets formed, enough stray material remained in the solar system that Earth and the other young planets were regularly bombarded by large chunks of debris. The greatest impacts were likely so explosive that they vaporized our planet's newly formed oceans. Moreover, volcanic and tectonic activity was severe, and ultraviolet radiation from the sun was intense, making early Earth a hostile place (**Figure 4.19**). Any life that got under way during this "bombardment stage" might easily have been killed off. Only after most debris was cleared from the solar system was life able to gain a foothold and keep it.

Earth's early atmosphere largely lacked oxygen until photosynthesizing microbes started producing it. Whereas today's atmosphere is dominated by nitrogen and oxygen (see Table 4.1), Earth's early atmosphere likely contained large amounts of carbon dioxide, carbon monoxide (CO), hydrogen, ammonia ($NH_3$), methane, and water vapor.

## Several hypotheses have been proposed to explain life's origin

Most scientists interested in life's origin think that life must have begun when inorganic chemicals linked themselves into small molecules and formed organic compounds. Some of these compounds gained the ability to replicate, or reproduce themselves, whereas others found ways to group together into proto-cells. Much debate and ongoing research, however, focuses on the nature and location of the first chemical reactions and the energy source(s) that powered them.

**Primordial soup: The heterotrophic hypothesis** The hypothesis traditionally favored is that life originated from a "primordial soup" of simple inorganic chemicals dissolved in the ocean's surface waters or tidal shallows. Scientists since the 1930s have suggested how amino acids might have formed under these conditions and how more complex organic compounds could have followed, including ribonucleic acids that could replicate themselves. This hypothesis is

FIGURE 4.19 Early Earth was a very different place from our planet today. Microbial life originated amid sulfur-spewing volcanoes, intense ultraviolet radiation, frequent extraterrestrial impacts, and an atmosphere containing ammonia.

termed *heterotrophic* because it proposes that the first life forms used organic compounds as an energy source.

Lab experiments have provided evidence that such a process can work. In 1953, biochemists Stanley Miller and Harold Urey passed electricity through a mixture of water vapor, hydrogen, ammonia, and methane (believed at that time to represent the early atmosphere). They were able to produce many organic compounds, including amino acids. Subsequent experiments confirmed these findings, but scientists since then have modified their ideas about early atmospheric conditions, so these experiments seem less likely to represent what actually happened.

**"Seeds" from space: The panspermia hypothesis** Another hypothesis proposes that microbes from elsewhere in the solar system traveled on meteorites that crashed to Earth, seeding our planet with life. Scientists had long rejected this idea, believing that even if organisms or biological compounds exist in space, the searing temperatures that comets and meteors attain as they enter our atmosphere should destroy them before they reach the surface.

However, the Murchison meteorite, which fell in Australia in 1969, was found to contain many amino acids, suggesting that amino acids within rock can survive impact. Since then, experiments simulating impact conditions have shown that organic compounds and some bacteria can withstand a surprising amount of abuse. Planetary scientists calculate that large asteroid impacts on one planet (such as Mars) can throw up so much material that some eventually may make its way to other planets (such as Earth). And recent astrobiology research suggests that comets have brought large amounts of water, and possibly organic compounds, to Earth. As a result of such findings, long-distance travel of microbes through space and into our atmosphere now seems more plausible than previously thought.

**Life from the depths: The chemoautotrophic hypothesis** In the 1970s and 1980s, some scientists proposed that life originated at deep-sea hydrothermal vents, where sulfur was abundant. In this scenario, the first organisms were chemoautotrophs, creating their own food from hydrogen sulfide.

Genetic analysis of the relationships of present-day organisms (• pp. 106–107) suggests that some of the most ancient ancestors of today's life forms lived in extremely hot, wet environments. The scorching heat of hydrothermal vents could have acted to speed up chemical reactions that link atoms together into long molecules, a necessary early step in life's formation. Scientists have shown experimentally that it is possible to form amino acids and begin a chain of steps that might potentially lead to the formation of life under high-temperature, high-pressure conditions similar to those of hydrothermal vents.

**Weighing THE Issues | Hypotheses on Life's Origin**

Which lines of evidence in the debate over the origin of life strike you as the most convincing, and why? Which strike you as the least convincing, and why? Can you think of any further scientific research that could be done to address the question of how life originated?

## The fossil record has taught us much about life's history

Whether the first life arose from deep-sea vents, tidal pools, or comet craters, we know that life diversified into countless forms over Earth's long history, leading eventually to the planet full of life that we know today. The earliest evidence of life on Earth comes from rocks about 3.5 billion years old. Although these earliest traces are controversial, there is ample evidence that single-celled bacteria were present on Earth well over 3 billion years ago. Remains of these microscopic life forms (or their chemical by-products) have been preserved just as have later, much larger creatures such as dinosaurs—by fossilization.

As organisms die, some are buried by sediment. Under the right conditions, the hard parts of their bodies— such as bones, shells, and teeth—may be preserved as the sediments are compressed into rock (• pp. 198–199). Minerals replace the organic material, leaving behind a **fossil**, an imprint in stone of the dead organism (**Figure 4.20**).

FIGURE 4.20 The fossil record has helped reveal the history of life on Earth. The numerous fossils of trilobites suggest that these animals, now extinct, were abundant in the oceans from roughly 540 million to 250 million years ago.

Throughout the world, geological processes over millions of years have buried sedimentary rock layers and later brought them to the surface, revealing assemblages of fossils of plants and animals from different time periods. The cumulative body of fossils worldwide is known as the **fossil record.** Paleontologists study the fossil record to infer the history of past life on Earth.

The fossil record clearly shows that:

- Earlier types of organisms changed, or evolved, into later ones.
- The species living today are a tiny fraction of all species that ever lived; the vast majority are long extinct.
- The number of species existing at any one time has generally increased through history.
- Many organisms present early in history were smaller and simpler than modern organisms.
- There have been several episodes of *mass extinction,* or simultaneous loss of great numbers of species (• pp. 121–123, 301–303).

The fossil record also tells us that for most of life's history, microbes, like the bacteria that consume hydrocarbons or the cyanobacteria that produce oxygen, were the only life on Earth. It was not until about 600 million years ago that large and complex organisms such as animals, land plants, and fungi appeared.

The crude oil with which we began this chapter is itself a kind of fossil. Matter from dead organisms can be preserved as it sinks to the seafloor and is buried in the

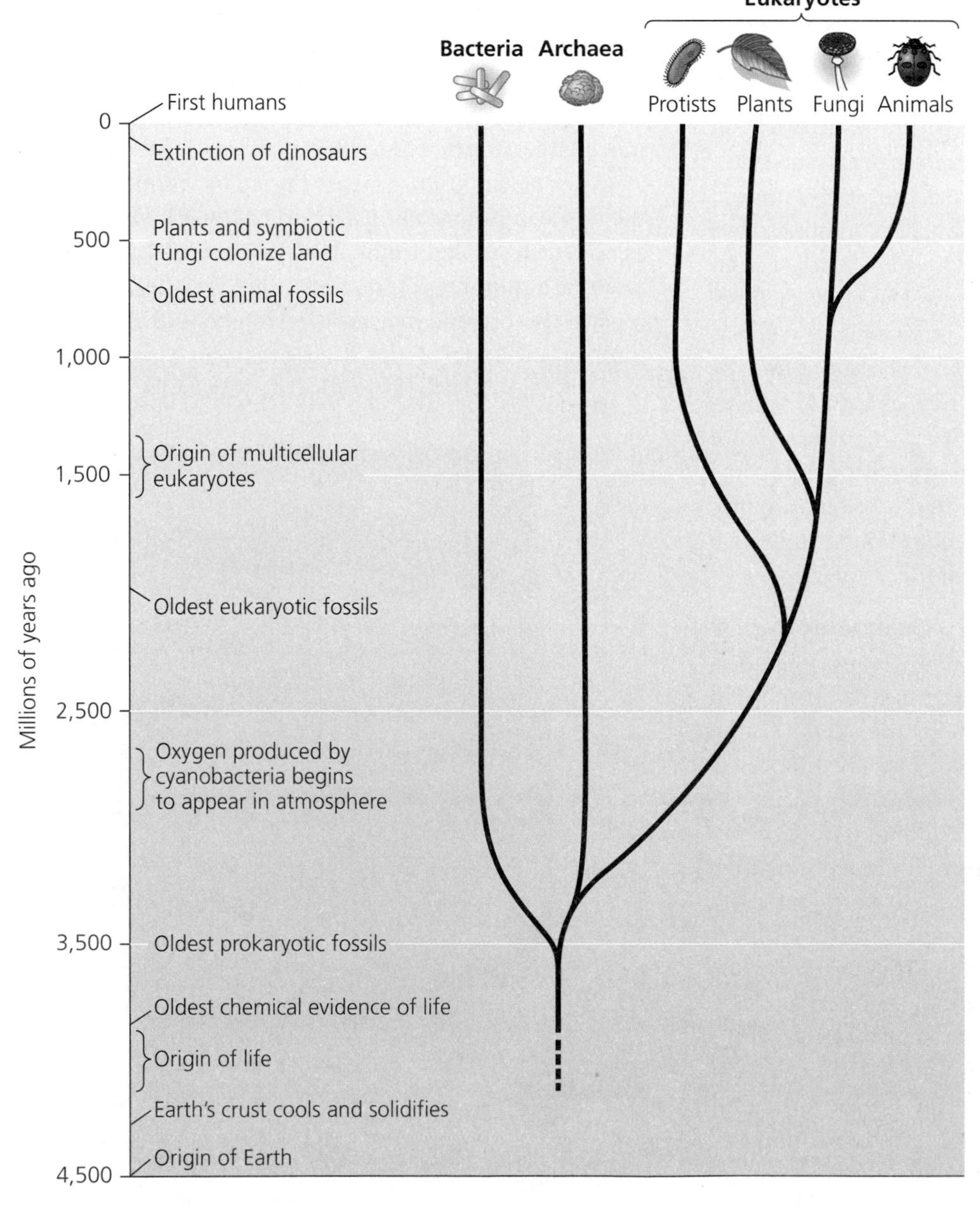


**FIGURE 4.21** The fossil record and the analysis of present-day organisms and their genes allow scientists to reconstruct evolutionary relationships among organisms and to build a "tree of life." As you progress upward from the trunk of the tree to the tips of its branches, you move forward in time. Each fork denotes the divergence of major groups of organisms (and in this greatly simplified diagram, each group includes many thousands of species). The tree of life consists of three main groups—the bacteria, the recently discovered archaea, and the diverse eukaryotes. Protists and ancestral eukaryotes are poorly known, and future study may produce discoveries that further refine our understanding of life's history.

absence of oxygen; eventually it becomes compressed and turns into the amorphous mixes of hydrocarbons we call fossil fuels (• p. 544). Coal, oil, and natural gas are the fossil fuels we use to power our civilization. When we drive a car, ignite a stove, or flick on a light switch, we are using energy from life buried millions of years ago.

### Present-day organisms and their genes help us decipher life's history

Besides fossils, biologists also use present-day organisms to infer how evolution proceeded in the past. By comparing the genes and/or the external characteristics of organisms, scientists can create branching trees, similar to family genealogies, that show the relationships among organisms and thus their history of divergence through time. As you follow such a tree from its trunk to the tips of its branches, you proceed forward through time, tracing the history of life.

A major advance was made in recent years as scientists discovered an entire new domain of life, the archaea, single-celled prokaryotes that are genetically very different from bacteria. Today most biologists view the tree of life as a three-pronged edifice consisting of the bacteria, the archaea, and the eukaryotes (**Figure 4.21**).

We will examine the flowering of the diversity of life on our planet in Chapter 5 and further in Chapter 11. The associations among organisms, and the associations between organisms and their environments, form the basis for ecology, a discipline of primary importance to environmental science.

## Conclusion

Life has flourished on Earth for over 3 billion years, stemming from an origin that scientists are eagerly attempting to understand. Deciphering how life originated depends in part on understanding energy, energy flow, and chemistry. Knowledge in these areas also enhances our understanding of how present-day organisms interact, how they relate to their nonliving environment, and how environmental systems function. Energy and chemistry are in some way tied to nearly every significant process in environmental science.

Chemistry can also be a tool for finding solutions to environmental problems. Cleaning up chemical pollution through bioremediation with microbes or plants is just one example. Chemistry can be a powerful ally, whether one wants to analyze agricultural practices, manage water resources, reform energy policy, conduct toxicological studies, or find ways to mitigate global climate change.

## REVIEWING OBJECTIVES

**You should now be able to:**

**Explain the fundamentals of environmental chemistry and apply them to real-world situations**

- Understanding chemistry provides a powerful tool for developing solutions to many environmental problems. (pp. 88–89)
- Atoms form molecules, and changes at the atomic level can result in alternate forms of elements, such as ions and isotopes. (pp. 89–92)
- Characteristics of the water molecule help facilitate life. (pp. 94–95)
- Living things depend on organic compounds, which are carbon-based. (pp. 95–96)

**Describe the molecular building blocks of living organisms**

- Macromolecules, including proteins, nucleic acids, carbohydrates, and lipids, are key building blocks of life. (pp. 96–98)
- Organisms use cells to compartmentalize macromolecules. (pp. 98–99)

**Differentiate among the types of energy and recite the basics of energy flow**

- Energy can change between potential and kinetic energy. Chemical energy is potential energy in the bonds between atoms. (p. 100)
- The total amount of energy in the universe is conserved; it cannot be created or lost. (p. 100)
- Systems tend to increase in entropy, or disorder, unless energy is added to build or maintain order and complexity. (p. 101)
- Earth's systems are powered by radiation from the sun and by geothermal heating from the planet's core. (pp. 101–104)

**Distinguish photosynthesis, respiration, and chemosynthesis, and summarize their importance to living things**

- In photosynthesis, autotrophs use carbon dioxide, water, and solar energy to produce the sugars they need, as well as oxygen. (pp. 102–103)

- In respiration, organisms extract energy from sugars by converting them in the presence of oxygen into carbon dioxide and water. (p. 103)
- In chemosynthesis, specialized autotrophs use carbon dioxide, water, and chemical energy from minerals to produce sugars. (pp. 103–104)

**Itemize and evaluate the major hypotheses for the origin of life on Earth**

- The heterotrophic hypothesis proposes that life arose from chemical reactions in surface or shallow waters of the ocean. (pp. 104–105)
- The panspermia hypothesis proposes that life or substances needed for life's origin on Earth arrived from space. (p. 105)
- The chemoautotrophic hypothesis proposes that life arose from chemical reactions at deep-sea hydrothermal vents. (p. 105)

**Outline our knowledge regarding early life and give supporting evidence for each major concept**

- The fossil record has revealed many patterns in the history of life, including that species evolve, most species are extinct, and species numbers on Earth have increased. (pp. 105–106)
- By comparing modern-day organisms, scientists can infer genetic relationships among them and understand their evolutionary history. (pp. 106–107)

## TESTING YOUR COMPREHENSION

1. What are the basic building blocks of matter? Provide examples using chemicals common in living organisms.
2. Name four ways in which the chemical nature of the water molecule facilitates life.
3. What are the three classes of biological polymer, and what are their functions?
4. Describe the two major forms of energy, and give examples of each.
5. State the first law of thermodynamics, and describe some of its implications.
6. What is the second law of thermodynamics, and how might it affect our interactions with the environment?
7. Describe the two major sources of energy that power Earth's environmental systems.
8. What substances are produced by photosynthesis? By cellular respiration? By chemosynthesis?
9. Compare and contrast three competing hypotheses for the origin of life.
10. Name three things scientists have learned from the fossil record.

## SEEKING SOLUTIONS

1. Under what types of conditions might bioremediation be a successful strategy, and when might it not be?
2. Can you think of an example of an environmental problem not mentioned in this chapter that a good knowledge of chemistry could help us solve?
3. Describe an example of energy transformation from one form to another that is not mentioned in this chapter.
4. Give three examples of ways in which the input of energy can resist the tendency toward disorder that the second law of thermodynamics describes.
5. Referring to the chemical reactions for photosynthesis and respiration, provide an argument for why increasing amounts of carbon dioxide in the atmosphere due to global climate change (Chapter 18) might potentially increase amounts of oxygen in the atmosphere. Give an argument for why it might potentially decrease amounts of atmospheric oxygen. What would you need to know to determine which of these two outcomes might occur?
6. **THINK IT THROUGH** Your state's Department of Environmental Protection has put you in charge of cleaning up an old industrial site so that it meets safety standards and does not contaminate drinking water supplies. Your staff's initial inspection of surface soil at the site shows that it is contaminated with oil and with lead, a toxic heavy metal. Your job allows you to engage experts in bioremediation and phytoremediation at your state's universities, as well as environmental engineers in the private sector. You have a budget of several million dollars and 5 years to get the job done. What steps will you take to get the site cleaned up? Describe scientific research you would commission, economic questions you would ask, and engineering options you might consider.

## INTERPRETING GRAPHS AND DATA

In phytoremediation, plants are used to clean up soil or water contaminated by heavy metals such as lead (Pb), arsenic (As), zinc (Zn), and cadmium (Cd). For plants to absorb these metals from soil, the metals must be dissolved in soil water. For any given instance, all metal can be accounted for as either remaining bound to soil particles, being dissolved in soil water, or being stored in the plant.

In a study on the effectiveness of alpine penny-cress (*Thlaspi caerulescens*) for phytoremediation, Enzo Lombi and his colleagues grew crops of this small perennial plant for approximately one year in pots of soil from contaminated sites. They then measured the amount of zinc and cadmium in the soil and in the plants when they were harvested.

1. What were the zinc and cadmium concentrations in the soil prior to phytoremediation? What were the zinc and cadmium concentrations in the soil after one year of phytoremediation?
2. How much zinc and cadmium were removed from the soil? If the plants continue to remove zinc and cadmium from the soil at the rates shown, how long would it take to remove all the zinc and cadmium?

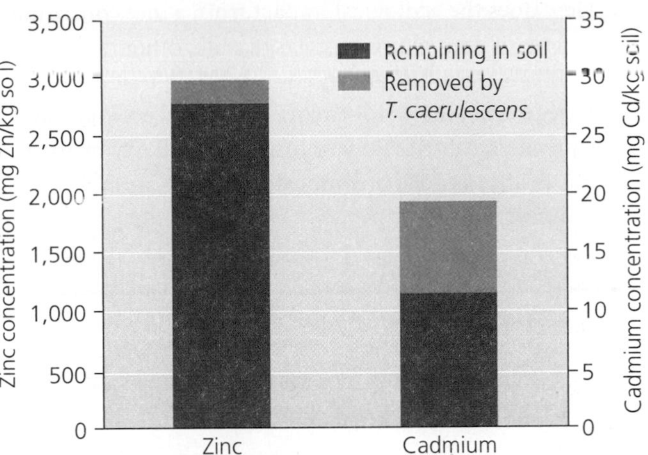


**Removal of zinc and cadmium from contaminated soil by alpine penny-cress, *Thlaspi caerulescens*.**
Data from Lombi, E., et al. 2001. Phytoremediation of heavy metal-contaminated soils: Natural hyperaccumulation versus chemically enhanced phytoextraction. *Journal of Environmental Quality* 30:1919–1926.

3. Alpine penny-cress produces natural compounds that increase the solubility of metals in soil water. If these dissolved metals are not taken up by the plants, what may be an unintended consequence of having increased their solubility?

## CALCULATING ECOLOGICAL FOOTPRINTS

In ecological systems, a rough rule of thumb is that when energy is transferred from plants to plant-eaters or from prey to predator, the efficiency is only about 10% (• pp. 149–150). Much of this inefficiency is a consequence of the second law of thermodynamics. Another way to think of this is that eating 1 calorie of meat from an animal is the ecological equivalent of eating 10 calories of plant material.

Humans are considered omnivores because we can eat both plants and animals. The choices we make about what to eat have significant ecological consequences. With this in mind, calculate the ecological energy requirements for four different diets, each of which provides a total of 2,000 dietary calories per day.

| Diet | Source of calories | Number of calories consumed | Ecologically equivalent calories | Total ecologically equivalent calories |
|---|---|---|---|---|
| **100% plant** | Plant | | | |
| **0% animal** | Animal | | | |
| **90% plant** | Plant | 1,800 | 1,800 | 3,800 |
| **10% animal** | Animal | 200 | 2,000 | |
| **50% plant** | Plant | | | |
| **50% animal** | Animal | | | |
| **0% plant** | Plant | | | |
| **100% animal** | Animal | | | |

1. How many ecologically equivalent calories would it take to support you for a year, for each of the four diets listed?
2. How does the ecological impact from a diet consisting strictly of animal products (e.g., milk, other dairy products, eggs, and meat) compare with that of a strictly vegetarian diet? How many additional ecologically equivalent calories do you consume each day by including as little as 10% of your calories from animal sources?
3. What percentages of the calories in your own diet do you think come from plant versus animal sources? Estimate the ecological impact of your diet, relative to a strictly vegetarian one.
4. Describe some challenges of providing food for the growing human population, especially as people in many poorer nations develop a taste for an American-style diet rich in animal protein and fat.

## Take It Further

Go to www.aw-bc.com/withgott or the student CD-ROM, where you'll find:

- Suggested answers to end-of-chapter questions
- Quizzes, animations, and flashcards to help you study
- *Research Navigator*™ database of credible and reliable sources to assist you with your research projects
- GRAPHit! Tutorials to help you interpret graphs
- INVESTIGATEit! Current news articles that link the topics that you study to case studies from your region to around the world

CHAPTER

# 5 Evolution, Biodiversity, and Population Ecology

Monteverde cloud forest, Costa Rica

## Upon completing this chapter, you will be able to:

- ▶ Explain the process of natural selection and cite evidence for this process
- ▶ Describe the ways in which evolution influences biodiversity
- ▶ Discuss reasons for species extinction and mass extinction events
- ▶ List the levels of ecological organization
- ▶ Outline the characteristics of populations that help predict population growth
- ▶ Assess logistic growth, carrying capacity, limiting factors, and other fundamental concepts in population ecology
- ▶ Identify efforts and challenges involved in the conservation of biodiversity

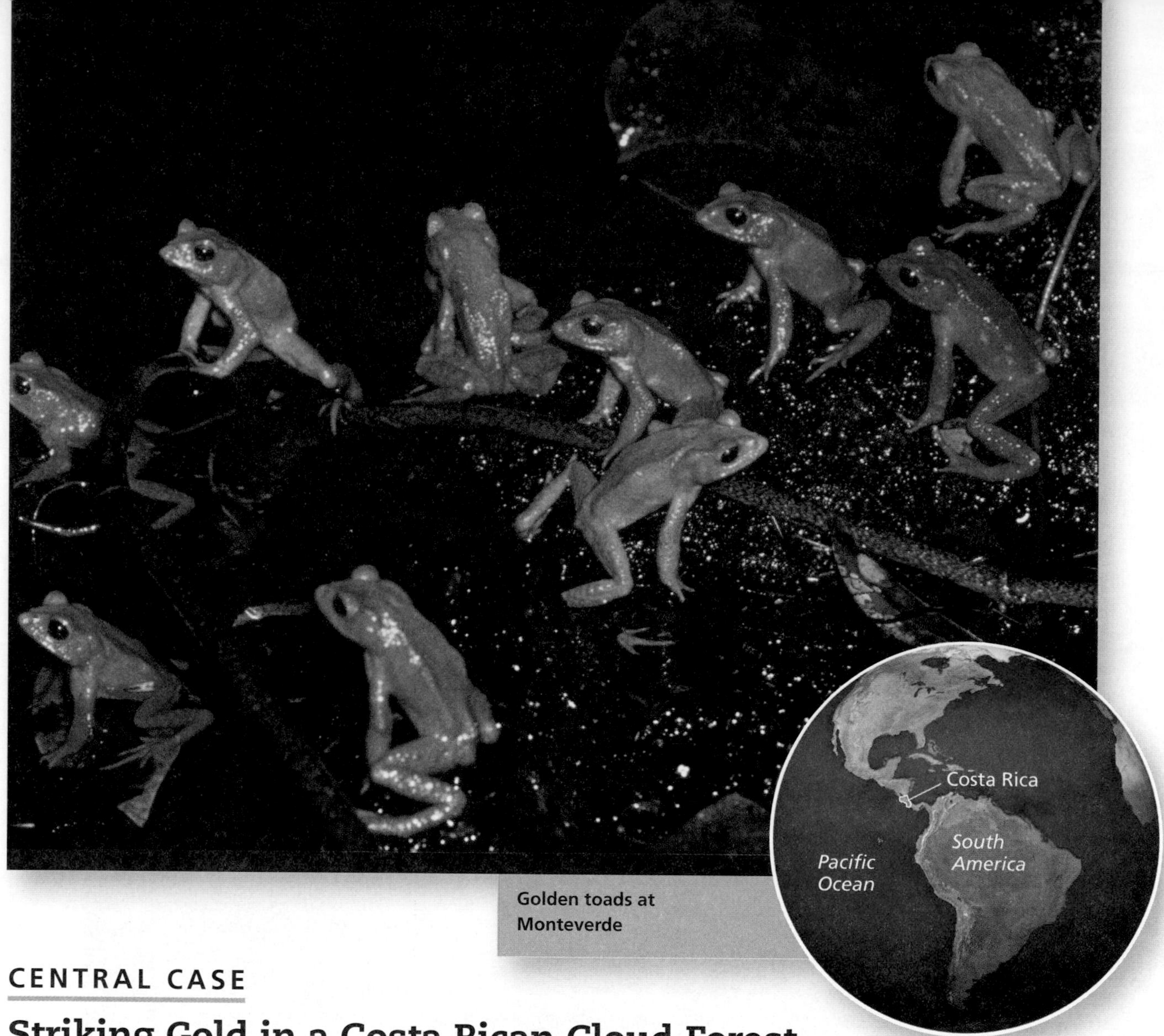


Golden toads at Monteverde

CENTRAL CASE

# Striking Gold in a Costa Rican Cloud Forest

**"I must confess that my initial response when I saw them was one of disbelief and suspicion that someone had dipped the examples in enamel paint."**
—DR. JAY M. SAVAGE, DESCRIBING THE GOLDEN TOAD IN 1966

**"What a terrible feeling to realize that within my own lifetime, a species of such unusual beauty, one that I had discovered, should disappear from our planet."**
—DR. JAY M. SAVAGE, DESCRIBING THE GOLDEN TOAD IN 1998

During a 1963 visit to Central America, biologist Jay M. Savage heard rumors of a previously undocumented toad living in Costa Rica's mountainous Monteverde region. The elusive amphibian, according to local residents, was best known for its color: a brilliant golden yellow-orange. Savage was told the toad was hard to find because it appeared only during the early part of the region's rainy season.

*Monteverde* means "green mountain" in Spanish, and the name couldn't be more appropriate. The village of Monteverde sits beneath the verdant slopes of the Cordillera de Tilarán, mountains that receive over 400 cm (157 in.) of annual rainfall. Some of the lush forests above Monteverde, which begin around 1,500 m (4,920 ft, just under a mile high), are known as *cloud forests* because much of their moisture arrives with low-moving clouds that blow inland from the Caribbean Sea. Monteverde's cloud forest was not fully explored at the time of Savage's first visit, and researchers who had been there described the area as pristine, with a rich bounty of ferns, liverworts, mosses, clinging vines, orchids, and other organisms that thrive in cool, misty environments. Savage knew that such conditions create ideal habitat for many toads and other amphibians.

In May of 1964, Savage organized an expedition into the muddy mountains above Monteverde to try to document the existence of the previously unknown toad species in its natural habitat. Late on the afternoon of May 14, he and his colleagues found what they were looking for. Approaching the mountain's crest, they spotted bright orange patches on the forest's black floor. In one area that was only 5 m (16.4 ft) in diameter, they counted 200 golden toads. Savage gave the creature the scientific name *Bufo periglenes* (literally, "the brilliant toad").

The discovery received international attention, making a celebrity of the tiny toad and helping to make a travel destination of its mountain home, which was soon protected within the Monteverde Cloud Forest Preserve.

At the time, no one knew that the Monteverde ecosystem was about to be transformed. No one foresaw that the oceans and atmosphere would begin warming because of global climate change (Chapter 18) and cause Monteverde's moisture-bearing clouds to rise, drying the forest. No one could guess that this newly discovered species of toad would become extinct in less than 25 years.

## Evolution as the Wellspring of Earth's Biodiversity

The golden toad was new to science, and countless species still await discovery, but scientists understand quite well how the world became populated with the remarkable diversity of organisms we see today. We know that the process of biological evolution has brought us from a stark planet inhabited solely by microbes to a lush world of 1.5 million (and likely millions more) species (**Figure 5.1**).

The term *evolution* in the broad sense means change over time, but scientists most often use the term to refer specifically to biological evolution. Biological **evolution** consists of genetic change in populations of organisms across generations. These changes in genes (• p. 97) often lead to modifications in the appearance, functioning, or behavior of organisms from generation to generation through time. Biological evolution results from random genetic changes, and it may proceed randomly or may be directed by natural selection. **Natural selection** is the process by which traits that enhance survival and reproduction are passed on more frequently to future generations than those that do not, thus altering the genetic makeup of populations through time.

Evolution by natural selection is one of the best-supported and most illuminating concepts in all of science, yet it has remained socially controversial among some nonscientists who fear it threatens their religious beliefs. Although scientists sometimes disagree about the specific mechanisms thought to drive evolution in particular cases, or about the time scales over which it takes place, this routine scientific debate should not be equated with the socially driven opposition of some nonscientists. From a scientific standpoint, evolutionary theory is indispensable, because it is the foundation of modern biology.

Understanding evolution is also vital for a full appreciation of environmental science. Perceiving how organisms adapt to their environments and change over time is crucial for understanding the history of life. It is also needed for ecology, a central component of environmental science. Evolutionary processes are relevant to many aspects of environmental science, including pesticide resistance, agriculture, medicine, and environmental health.

**FIGURE 5.1** Much of our planet's biological diversity resides in tropical rainforests. Monteverde's cloud-forest community includes organisms such as this (**a**) resplendent quetzal *(Pharomachrus mocinno)*, (**b**) puffball mushroom *(Calostoma cinnabarina)*, (**c**) harlequin frog *(Atelopus varius)*, and (**d**) scutellerid bug *(Pachycoris torridus)*

## Natural selection shapes organisms and diversity

In 1858, **Charles Darwin** and **Alfred Russell Wallace** each independently proposed the concept of natural selection as a mechanism for evolution and as a way to explain the great variety of living things. Darwin and Wallace were each exceptionally keen naturalists from England who had studied plants and animals in such exotic locales as the Galapagos Islands and the Malay Archipelago.

Natural selection is a simple concept that offers an astonishingly powerful explanation for patterns evident in nature. The idea of natural selection follows logically from a few straightforward premises that are readily apparent to anyone who observes the life around us (**Table 5.1**). One is that organisms face a constant struggle to survive and reproduce. Another is that organisms tend to produce more offspring than can survive. A third is that individuals of a species vary in their characteristics. Although not known in Darwin and Wallace's time, we now know that variation is due to differences in genes, the environments within which genes are expressed, and the interactions between genes and environment. As a result of this variation, some individuals within a species will happen to be better suited to their environment than others and thus will be able to reproduce more.

Many characteristics are passed from parent to offspring through the genes, and a parent that produces many offspring will pass on more genes to the next generation than a parent that produces few or no offspring. In the next generation, therefore, the genes of better-adapted individuals will be more prevalent than those of less well-adapted individuals. From one generation to another through time, species will evolve to possess characteristics that lead to better and better reproductive success in a given environment. A trait that promotes success is called an **adaptive trait,** or an **adaptation.**

**TABLE 5.1 The Logic of Natural Selection**

- Organisms struggle to survive and reproduce
- Organisms tend to produce more offspring than can survive
- Individuals vary in their characteristics
- Many characteristics are inherited by offspring from parents

**Therefore,**

- Some individuals will be better suited to their environment than other individuals
- Some individuals will produce more offspring or offspring of higher quality than others, thus transmitting more genes to future generations
- Future generations will contain more genes, and thus more characteristics, of the better-reproducing individuals. As a result, characteristics evolve across generations through time

## Natural selection acts on genetic variation

For an organism to pass a trait along to future generations, genes in the organism's DNA (• p. 97) must code for the trait. Accidental alterations that arise during DNA replication give rise to genetic variation among individuals. In an organism's lifetime, its DNA will be copied millions of times by millions of cells. In all this copying and recopying, sometimes a mistake is made. Accidental changes in DNA, called **mutations**, can range in magnitude from the addition, deletion, or substitution of single nucleotides (• p. 97) to the insertion or deletion of large sections of DNA. If a mutation occurs in a sperm or egg cell, it may be passed on to the next generation. Most mutations have little effect, but some can be deadly, whereas others can be beneficial. Those that are not lethal provide the genetic variation on which natural selection acts.

Sexual reproduction also generates variation. When organisms reproduce through sex, they mix, or recombine, their genetic material, so that a portion of each parent's genome (• p. 97) is included in the genome of the offspring. This process of *recombination* produces novel combinations of genes, generating variation among individuals.

Natural selection can act on genetic variation and alter organismal characteristics through time in three main ways (**Figure 5.2**). Selection that drives a feature in one direction rather than another—for example, toward larger or smaller, faster or slower—is called *directional selection.* In contrast, *stabilizing selection* produces intermediate traits, in essence preserving the status quo. Under *disruptive selection,* traits diverge from their starting condition in two or more directions.

An organism's environment determines what pressures natural selection will exert on the organism. However, environments change, and organisms may move to new places and encounter new conditions. In either case, a trait that is adaptive in one location or season may not be so in another. Golden toads that had adapted to the moist conditions of Monteverde's cloud forest would not have survived in drier forests—and apparently did not persist after Monteverde's climate began to become drier starting 25 years ago (•pp. 132, 134–135). Differences in environmental conditions in time and space make adaptation a moving target for organisms.

In all these ways, variable genes and variable environments interact as organisms engage in a perpetual

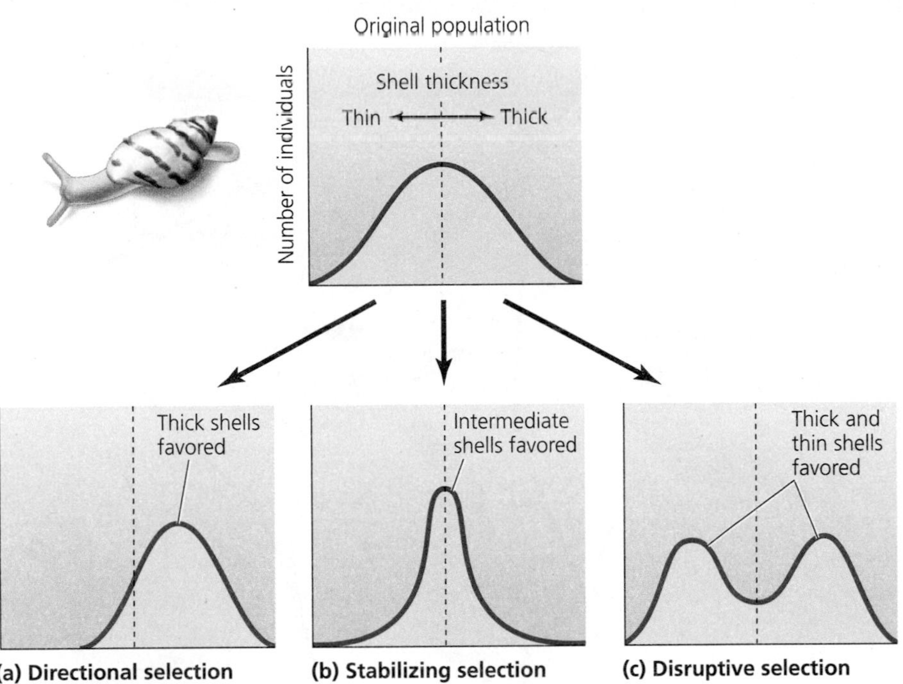


FIGURE 5.2 Selection can act in three ways. Consider snails living in a tropical cloud forest, and assume that we begin with a population of snails with shells of different thicknesses (top graph). Because shells protect snails against predators, snails with thick shells may be favored over those with thin shells, through *directional selection* (**a**). Alternatively, suppose that a shell that is too thin breaks easily, whereas a shell that is too thick wastes resources that are better used for feeding or reproduction. In such a case, *stabilizing selection* (**b**) could favor snails with shells that are neither too thick nor too thin. Under *disruptive selection* (**c**), extreme traits are favored. For example, if thin-shelled snails are so resource-efficient that they outreproduce intermediate-shelled snails, and thick-shelled snails are so well protected from predators that they also outreproduce intermediate-shelled snails, then each "extreme" strategy works more effectively than a compromise between the two.

process of adapting to the changing conditions around them. During this process, natural selection does not simply weed out unfit individuals. It also helps to elaborate and diversify traits that in the long term may help lead to the formation of new species and whole new types of organisms. In this way, natural selection has helped bring about the wondrous flowering of life on our planet.

## Evidence of natural selection is all around us

The results of natural selection are all around us, visible in every adaptation of every organism (**Figure 5.3**). In addition, countless lab experiments (mostly with fast-reproducing organisms, such as bacteria and fruit flies) have demonstrated rapid evolution of traits. The evidence for selection that may be most familiar to us is that which Darwin himself cited prominently in his work 150 years ago: our breeding of domesticated animals. In our dogs, our cats, and our livestock, we have conducted selection under our own direction. We have chosen animals with traits we like and bred them together, while not breeding those with variants we do not like. Through such *selective breeding*, we have been able to exaggerate particular traits we prefer.

Consider the great diversity of dog breeds (**Figure 5.4a**), all of which comprise variations on a single species. From Great Dane to Chihuahua, they can interbreed freely and produce viable offspring, yet breeders maintain striking differences among them by allowing only like individuals to breed with like. This process of selection conducted under human direction is termed **artificial selection**.

Artificial selection has also given us the many crop plants we depend on for food, all of which were domesticated from wild ancestors and carefully bred over years, centuries, or millennia (**Figure 5.4b**). Through selective breeding, we have created corn with larger, sweeter kernels; wheat and rice with larger and more numerous grains; and apples, pears, and oranges with better taste. We have diversified single types into many, for instance, breeding variants of the plant *Brassica oleracea* to create broccoli, cauliflower, cabbage, and brussels sprouts. Our entire agricultural system is based on artificial selection.

## Evolution generates biological diversity

When Charles Darwin wrote about the wonders of a world full of diverse animals and plants, he conjured up the vision of a "tangled bank" of vegetation harboring all kinds of creatures. Such a vision fits well with the arching vines, dripping leaves, and mossy slopes of the tropical cloud forest of Monteverde. Indeed, tropical forests worldwide teem with life and harbor immense biological diversity (see Figure 5.1).

**Biological diversity**, or **biodiversity** for short, refers to the sum total of all organisms in an area, taking into account the diversity of species, their genes, their

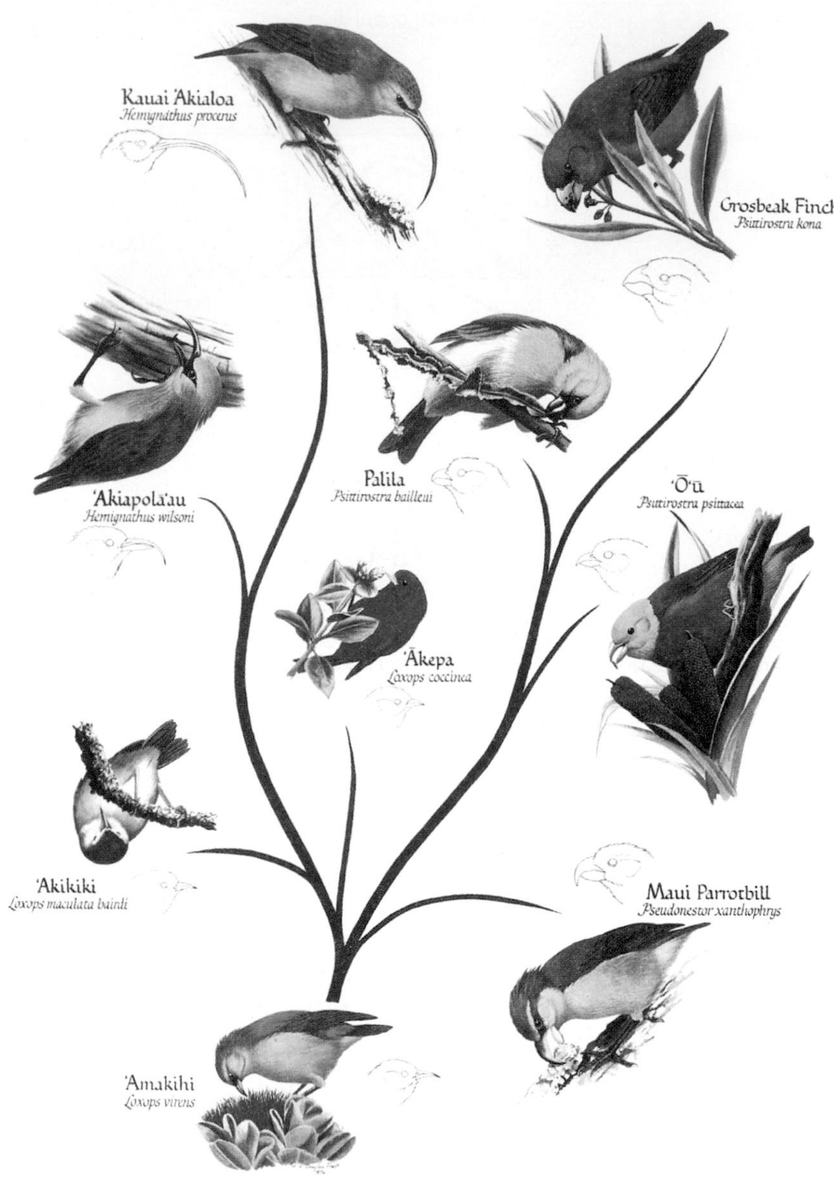


**FIGURE 5.3** Natural selection has produced tremendous diversity among organisms in the wild. In the group of birds known as Hawaiian honeycreepers, closely related species have adapted to different food resources, habitats, or ways of life, as indicated by the diversity in their plumage colors and the shapes of their bills. Such a burst of species formation due to natural selection is known as an *adaptive radiation.*

populations, and their communities. A **species** is a particular type of organism or, more precisely, a population or group of populations whose members share certain characteristics and can freely breed with one another and produce fertile offspring. A **population** is a group of individuals of a particular species that live in the same area. We have already discussed genes (• p. 97), and we will introduce communities shortly (• p. 123; and Chapter 6).

Scientists have described between 1.5 million and 1.8 million species, but many more remain undiscovered or unnamed. Estimates for the total number of species in the world range up to 100 million, with many of them thought to occur in tropical forests. In this light, the discovery of a new toad species in Costa Rica in 1964 seems far less surprising. Although Costa Rica covers a tiny fraction (0.01%) of Earth's surface area, it is home to 5–6% of all species known to scientists. And of the 500,000 species scientists estimate exist in the country, only 87,000 (17.4%) have been inventoried and described.

Tropical rainforests such as Costa Rica's, however, are by no means the only places rich in biodiversity. Step outside anywhere on Earth, even in a major city, and you will find numerous species within easy reach. They may not always be large and conspicuous like Yellowstone's bears or Africa's elephants, but they will be there. Plants poke up from cracks in asphalt in every city in the world, and even Antarctic ice harbors microbes. In a handful of backyard soil there may exist an entire miniature world of life, including several insect species, several types of mites, a

**(a) Ancestral wolf and derived dog breeds**

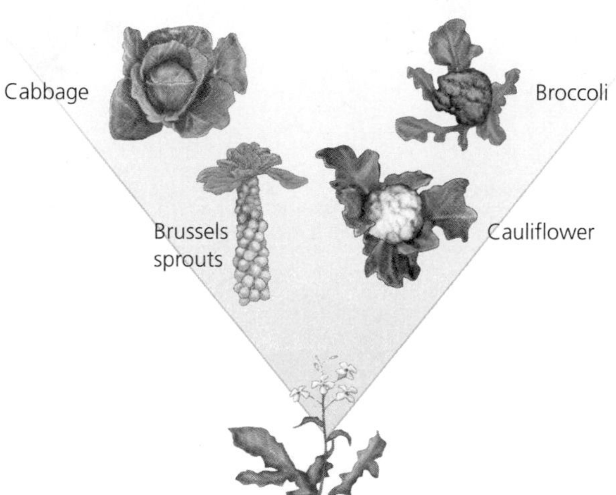


**(b) Ancestral *Brassica oleracea* and derived crops**

**FIGURE 5.4** Selection imposed by humans (selective breeding, or artificial selection) has resulted in the numerous breeds of dogs (**a**). By starting with the gray wolf *(Canis lupus)* as the ancestral wild species, and by breeding like with like and selecting for the traits we prefer, we have evolved breeds as different as Great Danes and Chihuahuas. By this same process we have created the immense variety of crop plants we depend on for sustenance (**b**). Cabbage, brussels sprouts, broccoli, and cauliflower were all evolved from a single ancestral species, *Brassica oleracea.*

millipede or two, many nematode worms, a few plant seeds, countless fungi, and millions upon millions of bacteria. We will examine Earth's biodiversity in detail in Chapter 11.

## Speciation produces new types of organisms

How did Earth come to have so many species? Whether there are 1.5 million or 100 million, such large numbers require scientific explanation. The process by which new species are generated is termed **speciation**. Speciation can occur in a number of ways, but most biologists consider the main mode of species formation to be *allopatric speciation*, species formation due to the physical separation of populations over some geographic distance. To understand allopatric speciation, begin by picturing a population of organisms. Individuals within the population possess many similarities that unify them as a species because they are able to reproduce with one another and share genetic information. However, if the population is broken up into two or more populations that become isolated from one another, individuals from one population cannot reproduce with individuals from the others.

When a mutation arises in the DNA of an organism in one of these isolated populations, it cannot spread to the other populations. Over time, each population will independently accumulate its own set of mutations. Eventually, the populations may diverge, or grow different enough, that their members can no longer mate with one another. Once this has happened, there is no going back; the two populations cannot interbreed, and they have embarked on their own independent evolutionary trajectories as separate species (**Figure 5.5**). The populations will continue diverging in their characteristics as chance mutations accumulate that confer traits causing the populations to become different in random ways. If environmental conditions happen to be different for the two populations, then natural selection may accelerate the divergence. Through the speciation process, single species can generate multiple species, each of which can in turn generate more.

## Populations can be separated in many ways

The long-term geographic isolation of populations that can lead to allopatric speciation can occur in various ways (**Table 5.2**). Glacial ice sheets may move across continents during ice ages and split populations in two. Major rivers may change course and do the same. Mountain ranges may rise and divide regions and their organisms. Drying climate may partially evaporate lakes, subdividing them into multiple smaller bodies of water. Warming or cooling temperatures may cause whole plant communities to move northward or southward, or upslope or downslope, creating new patterns of plant and animal distribution. Regardless of the mechanism of separation, for speciation to occur, populations must remain isolated for a long time, generally thousands of generations.

If the geological or climatic process that has isolated populations reverses itself—if the glacier recedes, or the

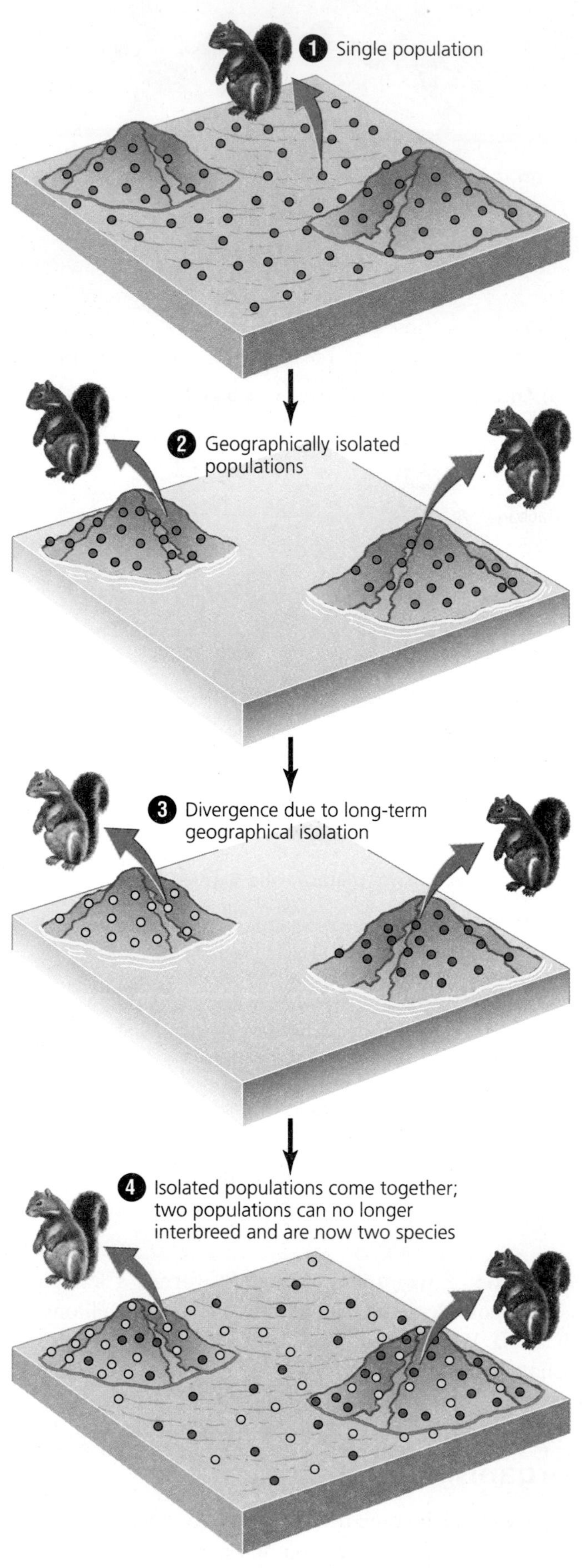


FIGURE 5.5 Allopatric speciation has generated much of Earth's diversity. In this process, some geographical barrier splits a population. In this diagram, two mountaintops (1) are turned into islands by rising sea level (2), isolating populations of squirrels. Each isolated population accumulates its own independent set of genetic changes over time, until individuals become genetically distinct and unable to breed with individuals from the other population (3). The two populations now represent separate species and will remain so even if the geographical barrier is removed and the new species intermix (4).

river returns to its old course, or warm temperatures turn cool again—then the populations can come back together. This is the moment of truth for speciation. If the populations have not diverged enough, their members will begin interbreeding and reestablish gene flow, mixing the mutations that each population accrued while isolated. However, if the populations have diverged sufficiently, they will not interbreed, and two species will have been formed, each fated to continue on its own evolutionary path.

Although allopatric speciation has long been considered the main mode of species formation, speciation appears to occur in other ways as well. *Sympatric speciation* occurs when species form from populations that become reproductively isolated within the same geographic area. For example, populations of some insects may become isolated if they feed and mate exclusively on different types of plants. Or they may mate during different seasons, isolating themselves in time rather than space. In some plants, speciation apparently has occurred as a result of hybridization between species. In others, it seems to have resulted from mutations that changed the numbers of chromosomes. Garnering solid evidence for mechanisms of speciation is difficult, so biologists still actively debate the relative prevalence of each of these modes of speciation.

**TABLE 5.2 Mechanisms of Population Isolation That Can Give Rise to Allopatric Speciation**

- Glacial ice sheets advance
- Mountain chains are uplifted
- Major rivers change course
- Sea level rises, creating islands (see Figure 5.5)
- Climate warms, pushing vegetation up mountain slopes and fragmenting it
- Climate dries, dividing large single lakes into multiple smaller lakes
- Ocean current patterns shift
- Islands are formed in the sea by volcanism

## Life's diversification results from numerous speciation events

Innumerable speciation events have generated complex patterns of diversity at levels above the species level. Evolutionary biologists study such patterns, examining how groups of organisms arose and how they evolved the characteristics they show. For instance, how did we end up with plants as different as mosses, palm trees, daisies, and redwoods? Why do fish swim, snakes slither, and sparrows sing? How and why did the ability to fly evolve independently in birds, bats, and insects? To address such questions, we need to know how the major groups diverged, and this pattern ultimately results from the history of individual speciation events.

We saw in Chapter 4 how the history of divergence can be represented in a treelike diagram (Figure 4.21, • p. 106). Such branching diagrams, called cladograms, or **phylogenetic trees**, illustrate scientists' hypotheses as to how divergence took place (**Figure 5.6**). Phylogenetic trees can show relationships among species, among major groups of species, among populations within a species, or even among individuals. In addition, by mapping traits onto a tree according to which organisms possess them, one can trace how the traits themselves may have evolved. For instance, the tree of life shows that birds, bats, and insects are distantly related, with many other flightless groups between them. So, it is far simpler to conclude that the three groups evolved flight independently than to conclude that the many flightless groups all lost an ancestral ability to fly. Because phylogenetic trees help biologists make such inferences about so many traits, they have become one of the modern biologist's most powerful tools.

Life's history, as revealed by phylogenetic trees and by the fossil record (• pp. 105–107), is complex indeed, but a few big-picture trends are apparent. As we mentioned in Chapter 4, life in its 3.5 billion years has evolved complex structures from simple ones, and large sizes from small ones. However, these are only generalizations. Many organisms have evolved to become simpler or smaller when natural selection favored it. Many very complex life forms have disappeared (**Figure 5.7**), and it is easy to argue that Earth still belongs to the bacteria and other microbes, some of them little changed over eons.

Even fans of microbes, however, must marvel at some of the exquisite adaptations that animals, plants, and fungi have evolved: The heart that beats so reliably for an animal's entire lifetime that we take it for granted. The complex organ system to which the heart belongs. The stunning plumage of a peacock in full display. The ability of each and every plant on the planet to lift water and nutrients from the soil, gather light from the sun, and turn it into food. The staggering diversity of beetles and other insects. The human brain and its ability to reason. All these and more have resulted from the process of evolution as it has generated new species and whole new branches on the tree of life.

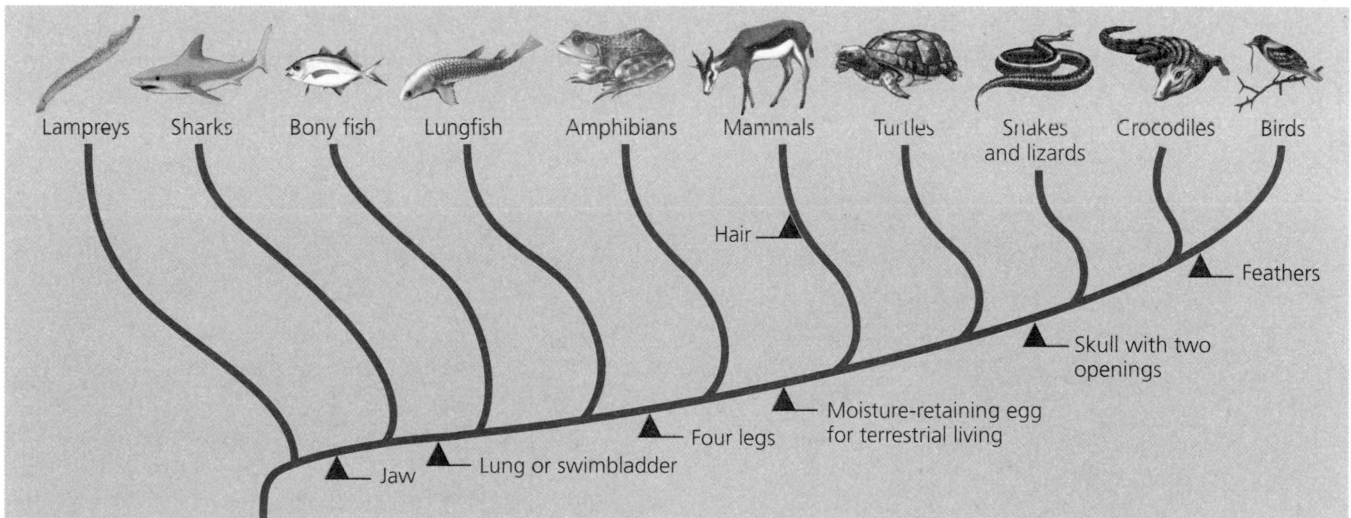


FIGURE 5.6 Phylogenetic trees show the history of life's divergence. Similar to family genealogies, these diagrams illustrate relationships among groups of organisms, as inferred from the study of similarities and differences among present-day creatures. The tree shown here is a greatly simplified representation of relationships among groups of vertebrates—one small portion of the huge and complex "tree of life." (See Figure 4.21, • p. 106, for a tree showing life's major groups.) Each branch results from a speciation event, and time proceeds upward from bottom to top. By mapping traits onto phylogenetic trees, biologists can study how traits have evolved over time. In this diagram, major traits are mapped using arrows to indicate when they originated. For instance, all vertebrates "above" the point at which jaws are indicated have jaws, whereas lampreys diverged before jaws originated and thus lack them.

FIGURE 5.7 Life has not always progressed from simple to complex during evolution. Many complex organisms have gone extinct, taking their designs and innovations with them. For example, the strange creatures portrayed in this painting were found fossilized in the Burgess Shale of the Canadian Rockies in British Columbia. They lived in marine environments 530 million years ago and vanished without leaving descendants.

## Speciation and extinction together determine Earth's biodiversity

Although speciation generates Earth's biodiversity, it is only one part of the equation—for, as you will recall from Chapter 4 (• p. 106), the vast majority of species that once lived are now gone. The disappearance of a species from Earth is called **extinction**. From studying the fossil record, paleontologists calculate that the average time a species spends on Earth is 1–10 million years. The number of species in existence at any one time is equal to the number added through speciation minus the number removed by extinction.

Extinction is a natural process, but human impact can profoundly affect the rate at which it occurs (**Figure 5.8**). The apparent extinction of the golden toad made headlines worldwide, but unfortunately it was not such an unusual occurrence. As we will see in Chapter 11, the biological diversity that makes Earth such a unique planet is being lost at an astounding rate. This loss affects people directly, because other organisms provide us with life's necessities—food, fiber, medicine, and ecosystem services (• pp. 38, 40). Species extinction brought about by human impact may well be the single biggest environmental problem we face, because the loss of a species is irreversible.

## Some species are more vulnerable to extinction than others

In general, extinction occurs when environmental conditions change rapidly or severely enough that a species cannot adapt genetically to the change; natural selection simply does not have enough time to work. All manner of environmental events can cause extinction—climate change, the rise and fall of sea level, the arrival of new harmful species, severe weather events such as extended droughts, and more. In general, small populations and species narrowly specialized on some particular resource or way

FIGURE 5.8 Until 10,000 years ago, the North American continent teemed with a variety of large mammals, including mammoths, camels, giant ground sloths, lions, saber-toothed cats, and various types of horses, antelope, bears, and others. Nearly all of this megafauna went extinct suddenly about the time that humans first arrived on the continent. Similar extinctions occurred in other areas simultaneously with human arrival, suggesting to many scientists that overhunting or other human impacts were responsible. See also Figure 11.9, • p. 303.

of life are most vulnerable to extinction from environmental change.

The golden toad was a prime example of a vulnerable species. It was **endemic** to the Monteverde cloud forest, meaning that it occurred nowhere else on the planet. Endemic species face relatively high risks of extinction because all their members belong to a single, sometimes small, population. At the time of its discovery, the golden toad was known from only a 4-$km^2$ (988-acre) area of Monteverde. It also required very specific conditions to breed successfully. During the spring at Monteverde, water collects in shallow pools within the network of roots that span the cloud forest's floor. The golden toad gathered to breed in these root-bound reservoirs, and it was here that Jay Savage and his companions collected their specimens in 1964. Monteverde provided ideal habitat for the golden toad, but the minuscule extent of that habitat meant that any environmental stresses that deprived the toad of the resources it needed to survive might doom the entire world population of the species.

In the United States, a number of amphibians are limited to very small ranges and thus are vulnerable to extinction. The Yosemite toad is restricted to a small region of the Sierra Nevada in California, the Houston toad occupies just a few areas of Texas woodland, and the Florida bog frog lives in a tiny region of Florida wetland. Fully 40 salamander species in the United States are restricted to areas the size of a typical county, and some of these live atop single mountains (**Figure 5.9**).

## Earth has seen several episodes of mass extinction

Most extinction occurs gradually, one species at a time. The rate at which this type of extinction occurs is referred to as the *background extinction rate.* However, Earth has seen five events of staggering proportions that killed off massive numbers of species at once. These episodes, called **mass extinction events**, have occurred at widely spaced intervals in Earth history and have wiped out half to 95% of our planet's species each time.

The best-known mass extinction occurred 65 million years ago and brought an end to the dinosaurs (although birds are modern representatives of dinosaurs). Evidence suggests that the impact of a gigantic asteroid caused this event, called the Cretaceous-Tertiary, or K-T, event (see "The Science behind the Story," • pp. 122–123).

The K-T event, as massive as it was, was moderate compared to the mass extinction at the end of the Permian period 250 million years ago (see Appendix D for geologic periods). Paleontologists estimate that 75–95% of all species on Earth may have perished during this event. Precisely what caused the event scientists don't know. The hypothesis with the most support so far is that massive volcanism threw into the atmosphere a global blanket of soot and sulfur, smothering the planet, reducing sunlight, and inducing severe climate change.

## The sixth mass extinction is upon us

Many biologists have concluded that Earth is currently entering its sixth mass extinction event—and that we are the cause. The changes to Earth's natural systems set in motion by human population growth, development, and resource depletion have driven many species extinct and are threatening countless more. The alteration and outright destruction of natural habitats, the hunting and harvesting of species, and the introduction of species from one place to another where they can harm native species—these processes and many more have combined to threaten Earth's biodiversity (• pp. 304–310).

When we look around us, it may not appear as though a human version of an asteroid impact is taking place, but we cannot judge such things on a human timescale. On

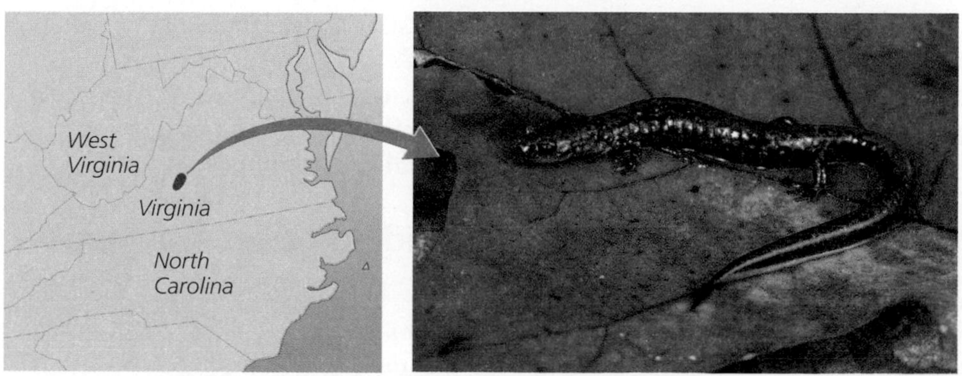


**FIGURE 5.9** Forty salamander species in the United States are restricted in range to areas the size of a typical county. Small range sizes leave these creatures vulnerable to extinction if severe local environmental changes occur. The Peaks of Otter salamander (*Plethodon hubrichti*) pictured here lives on only a few peaks in Virginia's Blue Ridge Mountains.

THE SCIENCE BEHIND THE STORY

## The K-T Mass Extinction

*Dr. Luis Alvarez (L) and Dr. Walter Alvarez (R) with a sample of the iridium layer*

When he first started working at Bottaccione Gorge in Italy, American geologist Walter Alvarez had no idea he would soon help discover what killed off the dinosaurs.

Scientists already knew that 65 million years ago, at the transition from the Cretaceous to the Tertiary periods (see Appendix D for geologic timescale), about 70% of the species then living, including the dinosaurs, disappeared. But they didn't know why.

Alvarez was developing a new method to determine the age of sedimentary rocks, and he'd chosen Bottaccione Gorge because it formed an ideal geological archive. Its 400-m (1,300-ft) walls are stacked like layer cake with beds of rose-colored limestone that formed between 100 million and 50 million years ago from dust that had settled to the bottom of an ancient sea.

While analyzing these layers, Alvarez noticed a band of reddish clay 1 cm (0.4 in.) thick sandwiched between two layers of limestone. The older layer just below it was packed with fossils of globotruncana, a sand-sized animal that lived in the late Cretaceous period. The newer layer just above it contained just a few scattered fossils of a cousin of globotruncana, typical of sedimentary rock formed in the early Tertiary period. The intermediate clay layer, which had formed just as the dinosaurs went extinct, had no fossils at all.

Walter and his father, physicist Luis Alvarez, analyzed the Cretaceous-Tertiary (K-T) clay layer to determine how long it had taken to form, using the rare metal iridium as a kind of clock. Almost all the iridium on Earth's surface comes from dust from meteorites (shooting stars) that burn up in the atmosphere. Because the same amount of meteorite dust rains down each year, the Alvarezes measured iridium levels in the clay layer to determine how many years' worth of iridium had accumulated.

Iridium levels in the limestone were typical for sedimentary rocks, about 0.3 parts per billion. In the clay layer, however, the Alvarezes found levels 30 times higher.

To make sure the finding was not unique to Bottaccione Gorge, they checked the K-T clay layer at a Danish sea cliff. It had 160 times more iridium than surrounding rock. The Alvarezes hypothesized that the excess iridium had come from a massive asteroid that smashed into Earth, causing a global environmental catastrophe that had caused a mass extinction.

To convince themselves and the scientific community that an asteroid impact caused the K-T event, the Alvarezes had to rule out other possible explanations. For example, the extra iridium could have come from seawater. Calculations, however, proved that seawater did not contain enough iridium to account for the high levels in the clay layers.

the geological timescale, extinction over 100 years or over 10,000 years appears every bit as instantaneous as extinction over a few days.

**Weighing THE Issues** **Should We Care about Extinction?**

Although many scientists say biodiversity loss is our biggest environmental problem, some critics say we should not be concerned about biodiversity loss or a new mass extinction. What do you think can account for such a disparity of viewpoints? Thinking back to our discussion of ethics and economics in Chapter 2, can you elaborate reasons why we should (or should not) be concerned about biodiversity loss and species extinction?

## Levels of Ecological Organization

The extinction of species, their generation through speciation, and other evolutionary mechanisms and patterns have substantial influence on ecology. It's often said that ecology provides the stage on which the play of evolution unfolds. The two, it's clear, are tightly intertwined in many ways. As we discussed in Chapter 1 (• p. 14), ecology is the study of interactions among organisms and between organisms and their environments.

### Ecology is studied at several levels

Life occurs in a hierarchy of levels, from the atoms, molecules, and cells we reviewed in Chapter 4 (see Figure 4.12, • p. 99) up through the **biosphere**, which is the

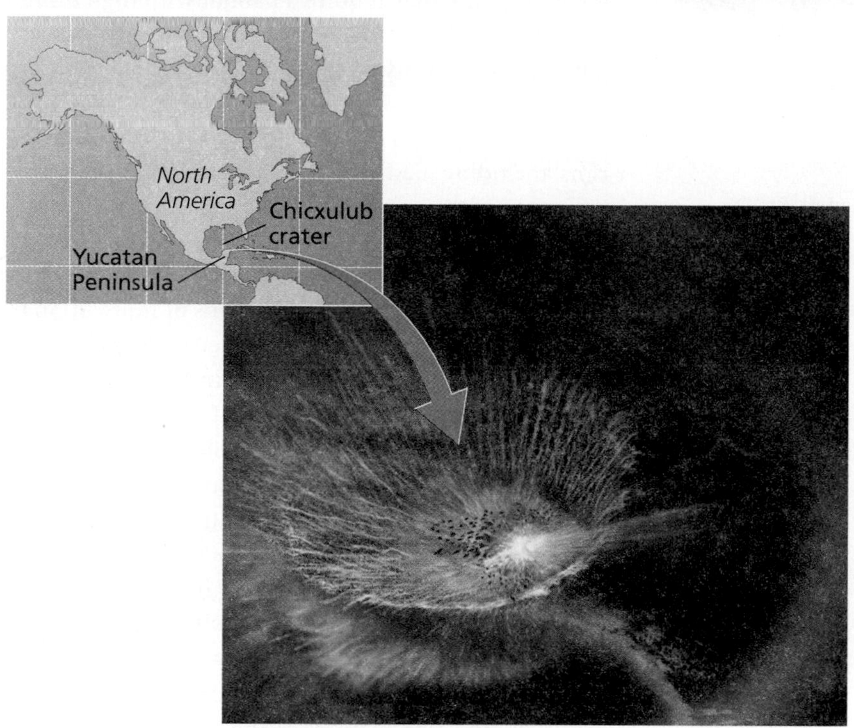


A colossal asteroid impact 65 million years ago is thought to have caused the Cretaceous-Tertiary mass extinction.

An asteroid 10 km (6.2 mi) wide strikes Earth, on average, every 100 million years. Such an impact would unleash an explosion 1,000 times more forceful than the 1883 eruption of the Indonesian volcano Krakatau, which scattered so much dust around the world that global temperatures cooled and sunsets were intense for 2 years afterward.

An asteroid impact 65 million years ago, they suggested, kicked up enough soot to blot out the sun for several years. This would have inhibited photosynthesis, causing plants to die, animals to starve, and food webs to collapse. Only a few smaller animals survived, feeding on rotting vegetation. When sunlight finally returned, plants sprouted from dormant seeds, and a long recovery began.

Published in the journal *Science* in 1980, the Alvarezes' explanation was immediately challenged by other geologists, who claimed that spectacular volcanic eruptions more likely explained the high iridium levels in Bottaccione Gorge.

But throughout the 1980s, scientists kept finding evidence supporting the asteroid-impact hypothesis. Iridium-enriched clay turned up at K-T layers around the world, as did bits of minerals called shocked quartz and stishovite, which form only under the extreme pressure of thermonuclear explosions and asteroid impacts.

The Alvarezes' hypothesis became widely accepted after 1991, once scientists pinpointed a 65-million-year-old crater of the predicted size in ocean sediments off the coast of Mexico. Today, scientists broadly agree that an asteroid impact 65 million years ago caused our planet's most recent mass extinction.

cumulative total of living things on Earth and the areas they inhabit. Ecologists study relationships on the higher levels of this hierarchy (**Figure 5.10**), namely on the organismal, population, community, and ecosystem levels. **Communities** are made up of multiple interacting species that live in the same area. A population of golden toads, a population of resplendent quetzals, populations of ferns and mosses, together with all of the other interacting plant, animal, fungal, and microbial populations in the Monteverde cloud forest, would be considered a community. **Ecosystems** encompass communities and the abiotic (nonliving) material and forces with which their members interact. Monteverde's cloud-forest ecosystem consists of the community plus the air, water, soil, nutrients, and energy the community's organisms use.

At the organismal level, the science of ecology describes relationships between organisms and their physical environments. It helps us understand, for example, what aspects of the golden toad's environment were important to it, and why. **Population ecology** investigates the quantitative dynamics of how individuals within a species interact with one another. It helps us understand why populations of some species (such as the golden toad) decline while populations of others (such as ourselves) increase. **Community ecology** focuses on interactions among species, from one-to-one interactions to complex interrelationships involving entire communities. In the case of Monteverde, it allows us to study how the golden toad and many other species of its cloud-forest community interact. Finally, **ecosystem ecology** reveals patterns, such as energy and nutrient flow, by studying living and nonliving components of systems in conjunction. As we will see, changing climate has had a strong influence on the organisms of Monteverde's cloud-forest ecosystem.

As improving technologies allow scientists to learn more about the complex operations of natural systems on a

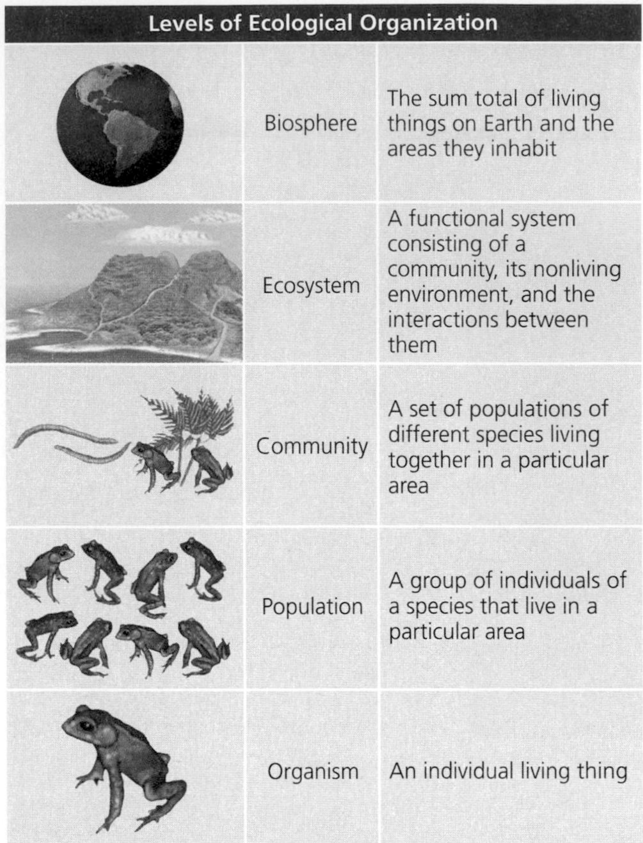


FIGURE 5.10 Life exists in a hierarchy of levels. Ecology includes the study of the organismal, population, community, and ecosystem levels and, increasingly, the level of the biosphere. Levels below the organismal level were illustrated in Figure 4.12, • p. 99.

global scale, ecologists are increasingly expanding their horizons beyond ecosystems to the biosphere as a whole. In this chapter we explore ecology up through the population level. In Chapter 6 we examine the community level, and in Chapter 7 we explore the ecosystem and biosphere levels.

## Habitat, niche, and specialization are important in organismal ecology

On the organismal level, each organism relates to its environment in ways that tend to maximize its survival and reproduction. One key relationship involves the specific environment in which an organism lives—its **habitat**. A species' habitat consists of the living and nonliving elements around it—for instance, rock, soil, leaf litter, humidity, plant life, and more. The golden toad lived in a habitat of cloud forest—more specifically, on the moist forest floor, using seasonal pools for breeding and burrows for shelter.

The plants known as epiphytes use other plants as habitat; they grow on trees for physical support, obtaining water from the air and nutrients from organic debris that collects among their leaves. Epiphytes thrive in cloud forests because they require a habitat with high humidity, and Monteverde hosts more than 330 species of epiphytes, mostly ferns, orchids, and bromeliads (pineapple relatives). By collecting pools of rainwater and pockets of leaf litter, epiphytes create habitat for many other organisms, including many invertebrates and even frogs that lay their eggs in the rainwater pools.

Habitats are scale-dependent. A tiny soil mite may perceive its habitat as a mere square meter of soil. A vulture, in contrast, may view its habitat in terms of miles upon miles of hills and valleys that it easily traverses by air.

Each organism thrives in certain habitats and not in others, leading to nonrandom patterns of **habitat use**. Mobile organisms actively select habitats in which to live from among the range of options they encounter, a process called *habitat selection*. In the case of plants and sessile animals, whose progeny disperse passively, patterns of habitat use result from success in some habitats and failure in others. The criteria by which organisms favor some habitats over others can vary greatly. The soil mite may judge available habitats in terms of the chemistry, moisture, and compactness of the soil and the percentage and type of organic matter. The vulture may ignore not only soil but also topography and vegetation, focusing solely on the abundance of dead animals in the area that it scavenges for food. Every species judges habitats differently because every species has different needs.

Habitat selection is important in environmental science because the availability and quality of habitat are crucial to an organism's well-being. Indeed, because habitats provide everything an organism needs, including nutrition, shelter, breeding sites, and mates, the organism's very survival depends on the availability of suitable habitats. Often this need engenders conflict with people who want to alter or develop a habitat for their own purposes.

Another way in which an organism relates to its environment is through its niche. A species' **niche** reflects its use of resources and its functional role in a community. This includes its habitat use, its consumption of certain foods, its role in the flow of energy and matter, and its interactions with other organisms. The niche is a multidimensional concept, a kind of summary of everything an organism does. We will examine the niche concept further in Chapter 6 (• p. 143).

Organisms vary in the breadth of their niche. Species with narrow breadth, and thus very specific requirements, are said to be **specialists**. Those with broad tolerances, able to use a wide array of habitats or resources, are **generalists**. For example, in a study of eight Costa Rican bird species that feed from epiphytes, ornithologist T. Scott Sillett found that four were generalists. The other four were specialists on the insect resources the epiphytes provided and spent more than 75% of their foraging efforts feeding from ephiphytes.

Specialist and generalist strategies each have advantages and disadvantages. Specialists can be successful over evolutionary time by being extremely good at the things they do, but they are vulnerable when conditions change and threaten the habitat or resource on which they have specialized. Generalists succeed by being able to live in many different places and weather variable conditions, but they may not thrive in any one situation as much as a specialist does. An organism's habitat, niche, and degree of specialization each reflect the adaptations of the species and are products of natural selection.

# Population Ecology

Individuals of the same species inhabiting a particular area make up a population. Species may consist of multiple populations that are geographically isolated from one another. This is the case with a species characteristic of Monteverde—the resplendent quetzal (*Pharomachrus mocinno*), considered one of the world's most spectacular birds (see Figure 5.1a). Although it ranges from southernmost Mexico to Panama, the resplendent quetzal lives only in high-elevation tropical forest and is absent from low-elevation areas. Moreover, human development has destroyed much of its forest habitat. Thus, the species today exists in many separate populations scattered across Central America.

In contrast, humans have spread into nearly every corner of the planet. As a result, it is difficult to define a distinct human population on anything less than the global scale. Some would maintain that in the ecological sense of the word, all 6.7 billion of us comprise one population.

## Populations exhibit characteristics that help predict their dynamics

Whether one is considering humans or quetzals or golden toads, all populations show characteristics that help population ecologists predict the future dynamics of that population. Attributes such as density, distribution, sex ratio, age structure, and birth and death rates all help the ecologist understand how a population may grow or decline. The ability to predict growth or decline is useful in monitoring and managing threatened and endangered species (• pp. 317–321). It is also vital in applying to human populations (Chapter 8). Understanding human population dynamics, their causes, and their consequences is one of the central elements of environmental science and one of the prime challenges for our society today.

**Population size** Expressed as the number of individual organisms present at a given time, **population size** may increase, decrease, undergo cyclical change, or remain the same over time. Extinctions are generally preceded by population declines. As late as 1987, scientists documented a golden toad population at Monteverde in excess of 1,500 individuals, but in 1988 scientists sighted only 10 toads, and in 1989 they found only a single individual. By 1990, the species had disappeared.

The passenger pigeon *(Ectopistes migratorius)*, also now extinct, illustrates the extremes of population size (**Figure 5.11**). It was once the most abundant bird in

**(a) Passenger pigeon**

**(b) 19th-century lithograph of pigeon hunting in Iowa**

**FIGURE 5.11** The passenger pigeon (**a**) was once North America's most numerous bird, and its flocks literally darkened the skies when millions of birds passed overhead (**b**). However, human cutting of forests and hunting drove the species to extinction within a few decades.

North America; flocks of passenger pigeons literally darkened the skies. In the early 1800s, ornithologist Alexander Wilson watched a flock of 2 billion birds 390 km (240 mi) long that took 5 hours to fly over and sounded like a tornado. Passenger pigeons nested in gigantic colonies in the forests of the upper Midwest and southern Canada. Once people began cutting the forests, however, the birds made easy targets for market hunters, who gunned down thousands at a time and shipped them to market by the wagonload. By the end of the 19th century, the passenger pigeon population had declined to such a low number that they could not form the large colonies they apparently needed to breed effectively. In 1914, the last passenger pigeon on Earth died in the Cincinnati Zoo, bringing the continent's most numerous bird species to extinction within just a few decades.

**Population density** The flocks and breeding colonies of passenger pigeons showed high population density, another attribute that ecologists assess to better understand populations. **Population density** describes the number of individuals within a population per unit area. For instance, the 1,500 golden toads counted in 1987 within 4 $km^2$ (988 acres) indicated a density of 375 toads/$km^2$. In general, larger organisms have lower population densities because they require more resources—and thus more area—to survive.

High population density can make it easier for organisms to group together and find mates, but it can also lead to conflict in the form of competition if space, food, or mates are in limited supply. Overcrowded organisms may also become more vulnerable to the predators that feed on them, and close contact among individuals can increase the transmission of infectious disease. For these reasons, organisms sometimes leave an area when densities become too high. In contrast, at low population densities, organisms benefit from more space and resources but may find it harder to locate mates and companions.

Overcrowding at high population densities is thought to have doomed Monteverde's harlequin frog (*Atelopus varius*; see Figure 5.1c), an amphibian that disappeared at the same time as the golden toad. The harlequin frog is a habitat specialist, favoring "splash zones," areas alongside rivers and streams that receive spray from waterfalls and rapids. As Monteverde's climate grew warmer and drier in the 1980s and 1990s, water flow decreased and many streams dried up. Splash zones grew smaller and fewer, and harlequin frogs were forced to cluster together in what remained of the splash-zone habitat. Researchers J. Alan Pounds and Martha Crump recorded frog population densities up to 4.4 times higher than normal, with more than 2 frogs per meter (3.3 ft) of stream. Such overcrowding likely made the frogs vulnerable to disease transmission, predator attack, and assault from parasitic flies. From their field research, during which Pounds and Crump witnessed 40 frogs dead or dying, the researchers concluded that these factors led to the harlequin frog's disappearance from Monteverde.

A new population of harlequin frogs was found in 2003 on a private reserve elsewhere in Costa Rica, so there is still hope that the species may survive in the country. The frog was rediscovered by University of Delaware student Justin Yeager, who was doing field research during his study abroad trip that summer. However, the global population of the harlequin frog has declined in size by 80% in recent years, and it survives in appreciable numbers only in neighboring Panama.

**Population distribution** It was not simply the harlequin frog's density, but also its distribution in space that led to its demise at Monteverde. **Population distribution**, or **population dispersion**, describes the spatial arrangement of organisms within an area. Ecologists define three distribution types: random, uniform, and clumped (**Figure 5.12**). In a *random distribution*, individuals are located haphazardly in space in no particular pattern. This type of distribution can occur when the resources an organism needs are found throughout an area and other organisms do not strongly influence where members of a population settle.

A *uniform distribution* is one in which individuals are evenly spaced. This can occur when individuals hold territories or otherwise compete for space. In a desert where there is little water, each plant needs a certain amount of space for its roots to gather adequate moisture. As a result, each plant may be equidistant from others.

In a *clumped distribution*, the pattern most common in nature, organisms arrange themselves according to the availability of the resources they need to survive. Many desert plants grow in patches around isolated springs or along arroyos that flow with water after rainstorms. During their mating season, golden toads were found clumped at seasonal breeding pools. Humans, too, exhibit clumped distribution; people frequently aggregate together in urban centers. Clumped distributions often indicate habitat selection.

Distributions can depend on the scale at which one measures them. At very large scales, all organisms show clumped or patchy distributions, because some parts of the total area they inhabit are bound to be more hospitable than others.

**Sex ratios** For organisms that reproduce sexually and have distinct male and female individuals, the sex ratio of a population can help determine whether it will increase or decrease in size over time. A population's **sex ratio** is its proportion of males to females. In monogamous species

(a) Random

(b) Uniform

(c) Clumped

FIGURE 5.12 Individuals in a population can spatially distribute themselves over a landscape in three fundamental ways. In a random distribution (**a**), organisms are dispersed at random through the environment. In a uniform distribution (**b**), individuals are spaced evenly, at equal distances from one another. Territoriality can result in such a pattern. In a clumped distribution (**c**), individuals occur in patches, concentrated more heavily in some areas than in others. Habitat selection or flocking to avoid predators can result in such a pattern.

(in which each sex takes a single mate), a 50/50 sex ratio maximizes population growth, whereas an unbalanced ratio leaves many individuals of one sex without mates.

**Age structure** Populations most often consist of individuals of different ages. **Age distribution**, or **age structure**, describes the relative numbers of organisms of each age within a population. A population made up mostly of individuals past reproductive age will tend to decline over time. In contrast, a population with many individuals of age or pre-productive age is likely to increase. A population with an even age distribution will likely remain stable as births keep pace with deaths.

*Age structure diagrams,* often called *age pyramids,* are visual tools scientists use to show the age structure of populations (**Figure 5.13**). The width of each horizontal bar represents the relative size of each age class. A pyramid with a wide base has a relatively large age class that has not yet reached its reproductive stage, indicating a population capable of rapid growth. In this respect, a wide base of an age pyramid is like an oversized engine on a rocket—the bigger the booster, the faster the increase. We will examine age pyramids further in Chapter 8 (• pp. 216–217) in reference to human populations.

**Birth and death rates** All the preceding factors can influence the rates at which individuals within a population are born and die. A convenient way to express birth and death rates is to measure the number of births and deaths per 1,000 individuals for a given time period. Such a rate is termed a *crude birth rate* or *crude death rate.*

Just as individuals of different ages have different abilities to reproduce, individuals of different ages show different probabilities of dying. For instance, people are more likely to die at old ages than young ages; if you were to follow 1,000 10-year-olds and 1,000 80-year-olds for a year, you would find that at year's end more 80-year-olds had died than 10-year-olds. However, this pattern does not hold for all organisms. Amphibians such as the golden toad produce large numbers of young, which suffer high death rates. For a toad, death is less likely (and survival more likely) at an older age than at a very young age.

To show how the likelihood of death can vary with age, ecologists use graphs called **survivorship curves** (**Figure 5.14**). There are three fundamental types of survivorship curves. Humans, with higher death rates at older ages, show a *type I* survivorship curve. Toads, with highest death rates at young ages, show a *type III* survivorship curve. A *type II* survivorship curve is intermediate and indicates equal rates of death at all ages. Many birds are thought to show type II curves.

## Populations may grow, shrink, or remain stable

Now that we have outlined some key attributes of populations, we are ready to take a quantitative view of population change by examining some simple mathematical

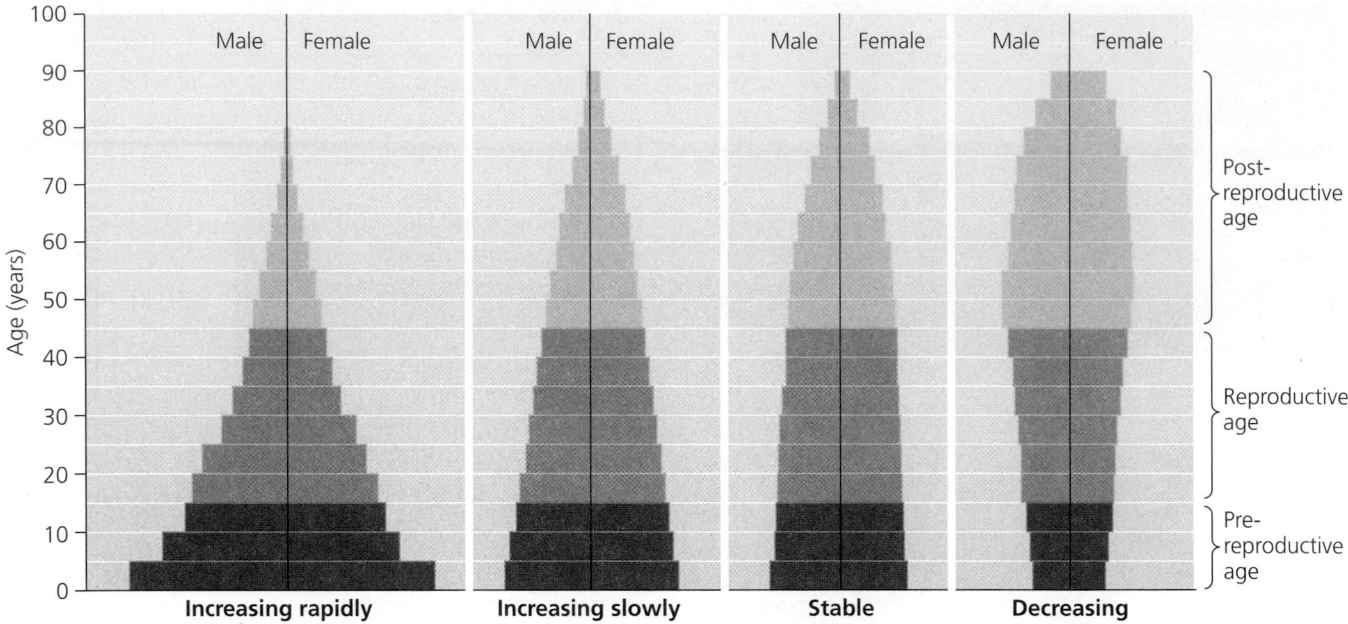


FIGURE 5.13 Age structure diagrams show relative frequencies of individuals of different age classes in a population. In this example for humans, populations weighted toward young age classes (at left) grow most quickly, whereas those weighted toward old age classes (at right) decline.

concepts used by population ecologists and *demographers* (scientists who study human populations). Population growth, or decline, is determined by four factors:

- Births within the population (*natality*)
- Deaths within the population (*mortality*)
- **Immigration** (arrival of individuals from outside the population)
- **Emigration** (departure of individuals from the population)

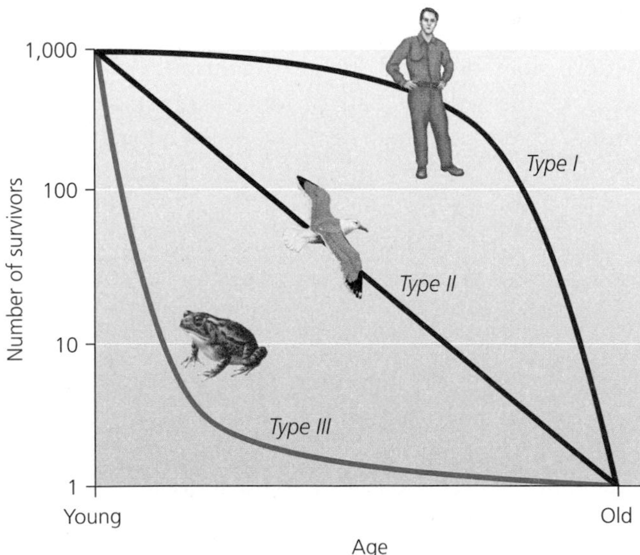


FIGURE 5.14 In a type I survivorship curve, survival rates are high when organisms are young and decrease sharply when organisms are old. In a type II survivorship curve, survival rates are equivalent regardless of an organism's age. In a type III survivorship curve, most mortality takes place at young ages, and survival rates are greater at older ages. Examples include humans (type I), birds (type II), and amphibians (type III).

Births and immigration add individuals to a population, whereas deaths and emigration remove individuals. To measure a population's **growth rate**, we calculate the crude birth rate plus the immigration rate, minus the crude death rate plus the emigration rate, each expressed as the number per 1,000 individuals per year:

$$(\text{Crude birth rate} + \text{immigration rate}) - (\text{Crude death rate} + \text{emigration rate}) = \text{Growth rate}$$

The resulting number tells us the net change in a population's size per 1,000 individuals. For example, a population with a crude birth rate of 18 per 1,000, a crude death rate of 10 per 1,000, an immigration rate of 5 per 1,000, and an emigration rate of 7 per 1,000 would have a growth rate of 6 per 1,000:

$$(18/1{,}000 + 5/1{,}000) - (10/1{,}000 + 7/1{,}000) = 6/1{,}000$$

Thus, a population of 1,000 in one year will reach 1,006 in the next. If the population is 1,000,000, it will reach 1,006,000 the next year. These population increases are often expressed as percentages, which we can calculate using the following formula:

$$\text{Growth rate} \times 100\%$$

Thus, a growth rate of 6/1,000 would be expressed as:

$$6/1{,}000 \times 100\% = 0.6\%$$

By measuring population growth in terms of percentages, scientists can compare increases and decreases in species that have far different population sizes. They can also project changes that will occur in the population over longer periods, much like you might calculate the amount of interest your savings account will earn over time.

## Unregulated populations increase by exponential growth

When a population, or anything else, increases by a fixed percentage each year, it is said to undergo **exponential growth**. A savings account is a familiar frame of reference for describing exponential growth. If at the time of your birth your parents had invested $1,000 in a savings account earning 5% interest compounded each year, you would have only $1,629 by age 10, and $2,653 by age 20, but you would have over $30,000 when you turn 70. If you could wait just 10 years more, that figure would rise to nearly $50,000 (Table 5.3). Only $629 was added during your first decade, but approximately $19,000 was added during the decade between ages 70 and 80. The reason is that a fixed percentage of a small number makes for a small increase, but that same percentage of a large number produces a large increase. Thus, as savings accounts (or populations) become larger, each incremental increase likewise gets larger. Such acceleration is a characteristic of exponential growth.

We can visualize changes in population size by using population growth curves. The J-shaped curve in **Figure 5.15** shows exponential increase. As Thomas Malthus (• p. 5) realized, populations of all organisms increase exponentially unless they meet constraints. Each organism reproduces by a certain amount, and as populations get larger, more individuals reproduce by that amount. If there are

**TABLE 5.3 Exponential Growth in a Savings Account with 5% Annual Compound Interest**

| Age (in years) | Principal |
|---|---|
| 0 (birth) | $ 1,000 |
| 10 | $ 1,629 |
| 20 | $ 2,653 |
| 30 | $ 4,322 |
| 40 | $ 7,040 |
| 50 | $11,467 |
| 60 | $18,679 |
| 70 | $30,426 |
| 80 | $49,561 |

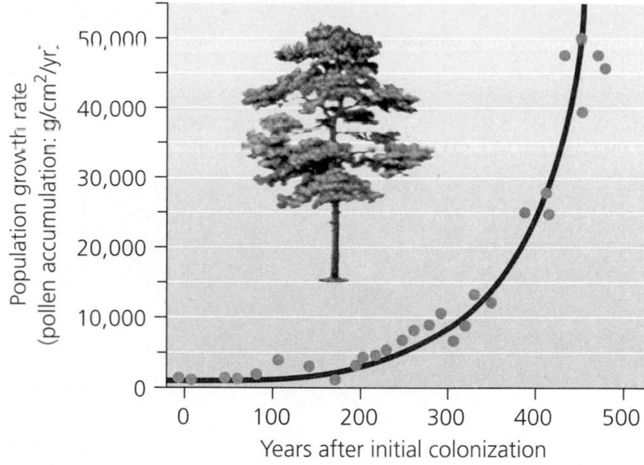


**FIGURE 5.15** Although no species can maintain exponential growth indefinitely, some may grow exponentially for a time when colonizing an unoccupied environment or exploiting an unused resource. Scientists have used pollen records to determine that the Scots pine (*Pinus sylvestris*) increased exponentially after the retreat of glaciers following the last ice age around 9,500 years ago. Go to GRAPHIT! at www.aw-bc.com/withgott or on the student CD-ROM. Data from Bennett, K. D. 1983. Postglacial population expansion of forest trees in Norfolk, U.K. *Nature* 303:164–167.

no external limits, ecologists theoretically expect exponential growth.

Normally, exponential growth occurs in nature only when a population is small and environmental conditions are ideal for the organism in question. Most often, these conditions occur when organisms are introduced to a new environment. Mold growing on a piece of bread or fruit, or bacteria colonizing a recently dead animal, are cases in point. But species of any size may show exponential growth under the right conditions. A population of the Scots pine, *Pinus sylvestris*, grew exponentially when it began colonizing the British Isles after the end of the last ice age (see Figure 5.15). Receding glaciers had left conditions ideal for its exponential expansion.

## Limiting factors restrain population growth

Exponential growth rarely lasts long. If even a single species in Earth's history had increased exponentially for very many generations, it would have blanketed the planet's surface, and nothing else could have survived. Instead, every population eventually is constrained by **limiting factors**, physical, chemical, and biological characteristics of the environment that restrain population growth. The interaction of these factors determines the **carrying capacity**, the maximum population size of a species that a given environment can sustain.

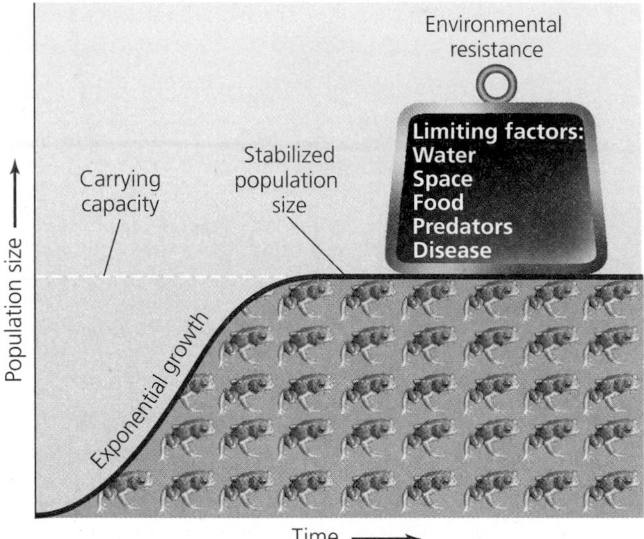


FIGURE 5.16 The logistic growth curve shows how population size may increase rapidly at first, then grow more slowly, and finally stabilize at a carrying capacity. Carrying capacity is determined both by the *biotic potential* of the organism and by various external limiting factors, collectively termed *environmental resistance.*

Ecologists use the curve in **Figure 5.16** to show how an initial exponential increase is slowed and finally brought to a standstill by limiting factors. Called the **logistic growth curve**, it rises sharply at first but then begins to level off as the effects of limiting factors become stronger. Eventually the force of these factors—which taken together are termed *environmental resistance*—stabilizes the population size at its carrying capacity.

The logistic curve is a simplified model, and real populations can behave differently. Some may cycle indefinitely above and below the carrying capacity. Some may show cycles that become less extreme and approach the carrying capacity. Others may overshoot the carrying capacity and then crash, fated either for extinction or recovery (**Figure 5.17**).

Many factors contribute to environmental resistance and influence a population's growth rate and carrying capacity. Space is one factor that limits the number of individuals a given environment can support; if there is no physical room for additional individuals, they are unlikely to survive. Other limiting factors for animals in a terrestrial environment include the availability of food, water, mates, shelter, and suitable breeding sites; temperature extremes; prevalence of disease; and abundance of predators. Plants are often limited by amounts of sunlight and moisture and the type of soil chemistry, in addition to disease and attack from plant-eating animals. In aquatic systems, limiting factors include salinity, sunlight, temperature, dissolved oxygen, fertilizers, and pollutants. To determine limiting factors, ecologists may conduct experiments in which they increase or decrease a hypothesized limiting factor and observe its effects on population size.

## Carrying capacities can change

Because limiting factors can be numerous, and because environments are complex and ever-changing, carrying capacity can vary. Moreover, organisms may sometimes be capable of altering their environments so as to reduce environmental resistance and raise their carrying capacity.

Our own species has proved particularly effective at this. When our ancestors began to build shelters and use fire for heating and cooking, they reduced the environmental resistance of areas with cold climates and were able to expand into new territory. As limiting factors are overcome (through development of new technologies or through natural environmental change), the carrying capacity for a species may increase. We humans have managed so far to increase the planet's carrying capacity for ourselves, but we have done so by appropriating immense proportions of the planet's resources, and in the process we have reduced the carrying capacities for many other organisms.

### Weighing the Issues | Carrying Capacity and Human Population Growth

As we have seen (• p. 4), the global human population has risen from fewer than 1 billion 200 years ago to 6.7 billion today, and we have far exceeded our historic carrying capacity. What accounts for the increase in Earth's carrying capacity for us? Do you think there are limiting factors for the human population? What might they be? Do you think we can keep raising our carrying capacity in the future? Might Earth's carrying capacity for us decrease?

## The influence of some factors depends on population density

Just as carrying capacity is not a fixed entity, the influence of limiting factors can vary with changing conditions. In particular, the density of a population can increase or decrease the impact of certain factors. Recall that high population density can help organisms find mates but can also increase competition and the risk of predation and disease. Such factors are said to be **density-dependent** factors, because their influence waxes and wanes with population density. The logistic growth curve in Figure 5.16 represents the effects of density dependence. The

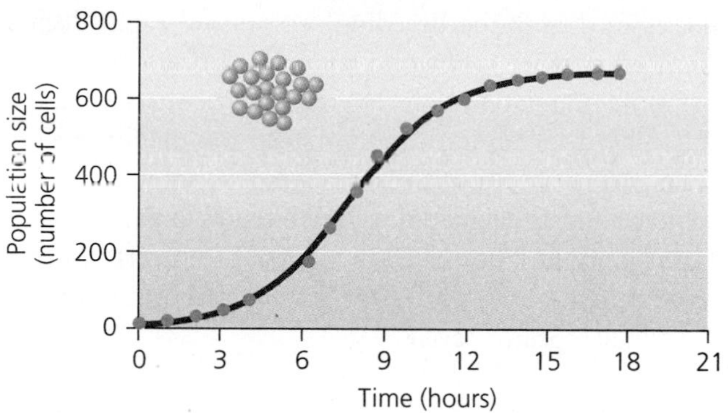

**(a) Yeast cells, *Saccharomyces cerevisiae***

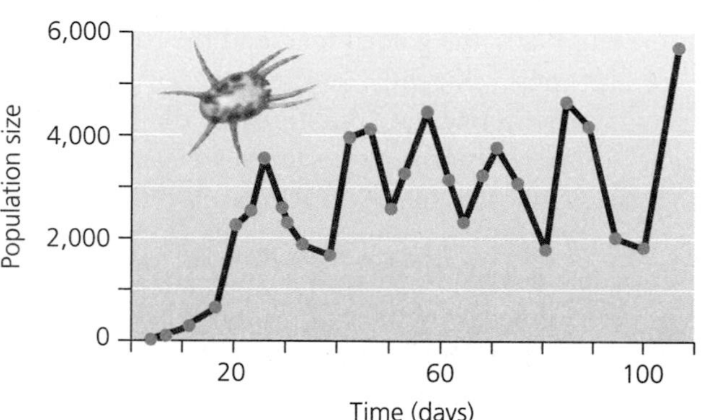


**(b) Mite, *Eotetranychus sexmaculatus***

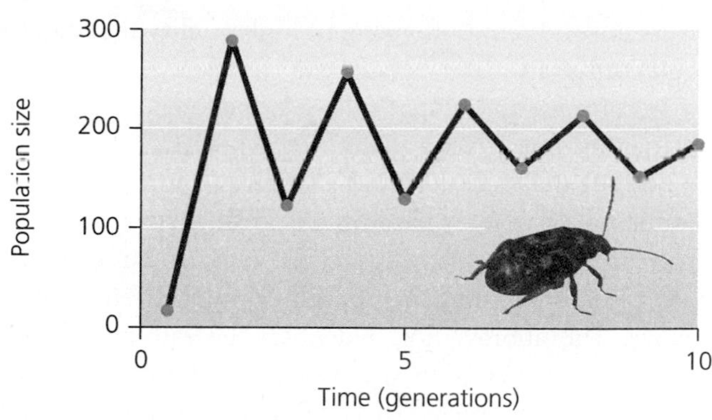


**(c) Stored-product beetle, *Callosobruchus maculatus***

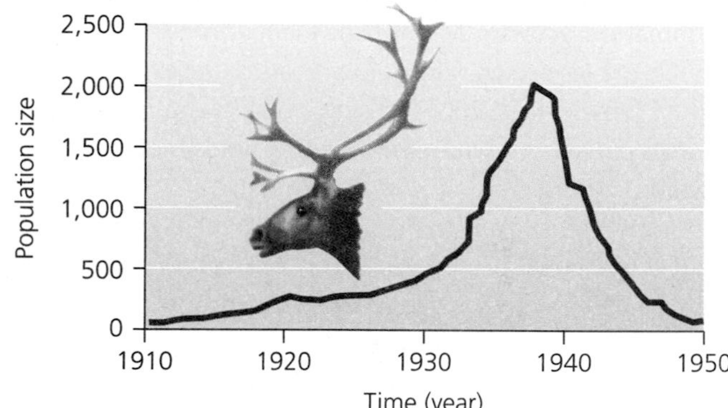


**(d) St. Paul reindeer, *Rangifer tarandus***

FIGURE 5.17 Population growth in nature often departs from the stereotypical logistic growth curve, and it can do so in several fundamental ways. Yeast cells from an early lab experiment show logistic growth **(a)** that closely matches the theoretical model. Some organisms, like the mite shown in **(b)**, show cycles in which population fluctuates indefinitely above and below the carrying capacity. Population oscillations can also dampen, lessening in intensity and eventually stabilizing at carrying capacity **(c)**, as in a lab experiment with the stored-product beetle. Populations that rise too fast and deplete resources may crash just as suddenly **(d)**, like the population of reindeer introduced to the Bering Sea island of St. Paul. Data from Pearl, R. 1927. The growth of populations. *Quarterly Review of Biology* 2: 532–548, (a); Huffaker, C. B. 1958. Experimental studies on predation: Dispersion factors and predator-prey oscillations, *Hilgardia* 27: 343–383, (b); Utida, S. 1967. Damped oscillation of population density at equilibrium, *Researches on Population Ecology* 9: 1–9, (c); Scheffer, V. C. 1951. Rise and fall of a reindeer herd, *Scientific Monthly* 73: 356–362, (d).

more population size rises, the more environmental resistance kicks in.

**Density-independent** factors are limiting factors whose influence is not affected by population density. Temperature extremes and catastrophic events such as floods, fires, and landslides are examples of density-independent factors, because they can eliminate large numbers of individuals without regard to their density.

## Biotic potential and reproductive strategies vary from species to species

Limiting factors from an organism's environment provide only half the story of population regulation. The other half comes from the attributes of the organism itself. For example, organisms differ in their *biotic potential,* or ability to produce offspring. A fish with a short gestation period that lays thousands of eggs at a time has high biotic potential, whereas a whale with a long gestation period that gives birth to a single calf at a time has low biotic potential. The interaction between an organism's biotic potential and the environmental resistance to its population growth helps determine the fate of its population.

Giraffes, elephants, humans, and other large animals with low biotic potential produce relatively few offspring and take a long time to gestate and raise each one. Species that take this approach to reproduction compensate by devoting large amounts of energy and resources to caring for and protecting the relatively few offspring they produce during their lifetimes. Such species are said to be **K-selected.** K-selected species are so named because their populations tend to stabilize over time at or near their carrying capacity, and $K$ is an abbreviation for carrying capacity. Because their populations stay close to carrying capacity, these organisms must compete to hold their own in a crowded world. In these species, natural selection favors individuals that invest in producing offspring of high quality that can be good competitors.

In contrast, species that are **r-selected** focus on quantity, not quality. Species considered to be r-selected have high biotic potential and devote their energy and resources to producing as many offspring as possible in a relatively short time. Their offspring do not require parental care after birth, so r-strategists simply leave their survival to chance. The abbreviation *r* denotes the rate at which a population increases in the absence of limiting factors. Populations of r-selected species fluctuate greatly, such that they are often well below carrying capacity. This is why natural selection in these species favors traits that lead to rapid population growth. Many fish, plants, frogs, insects, and others are r-selected. The golden toad is one example. Each adult female laid 200–400 eggs, and its tadpoles spent 5 weeks unsupervised in the breeding pools metamorphosing into adults.

Table 5.4 summarizes stereotypical traits of r-selected and K-selected species. However, it is important to note that these are two extremes on a continuum and that most species fall somewhere between these endpoints. Moreover, some organisms show combinations of traits that do not clearly correspond to a place on the continuum. A redwood tree (*Sequoia sempervirens*), for instance, is large and long-lived, yet it produces many small seeds and offers no parental care.

Populations of K-selected species are generally regulated by density-dependent factors such as disease, predation, and food limitation. In contrast, density-independent factors tend to regulate populations of r-selected species, whose success or failure is often determined by large-scale environmental change. Many r-selected species frequently experience large swings in population size, such as rapid increases during the breeding season and rapid declines soon after, when unfit and unlucky young are removed from the population. For this reason, scientists often have difficulty determining whether steep population declines are a part of natural cycles or a sign of serious trouble. For years, scientists debated the golden toad's apparent extinction. Now that it has failed to reappear for nearly 20 years, most agree that the toad's population crash was not part of a normal, repeating cycle.

**TABLE 5.4 Traits of r-selected and K-selected species**

| r-selected species | K-selected species |
|---|---|
| Small size | Large size |
| Fast development | Slow development |
| Short-lived | Long-lived |
| Reproduction early in life | Reproduction later in life |
| Many small offspring | Few large offspring |
| Fast population growth rate | Slow population growth rate |
| No parental care | Parental care |
| Weak competitive ability | Strong competitive ability |
| Variable population size, often well below carrying capacity | Constant population size, close to carrying capacity |
| Variable and unpredictable mortality | More constant and predictable mortality |

### Changes in populations influence the composition of communities

In the late 1980s, the golden toad and the harlequin frog were the most diligently studied species affected by changing environmental conditions in the Costa Rican cloud forest. However, once scientists began looking at populations of other species at Monteverde, they began to notice more troubling changes. By the early 1990s, not only had golden toads, harlequin frogs, and other organisms been pushed from their cloud-forest habitat into apparent extinction, but many species from lower, drier habitats had also begun to appear at Monteverde. These immigrants included species tolerant of drier conditions, such as blue-crowned motmots (*Momotus momota*) and brown jays (*Cyanocorax morio*). By the year 2000, 15 dry-forest species had moved into the cloud forest and begun to breed.

Meanwhile, population sizes of several cloud-forest bird species had declined. After 1987, 20 of 50 frog species vanished from one part of Monteverde, and ecologists later reported more disappearances, including those of two lizards native to the cloud forest. Scientists hypothesized that the climatic trends that researchers were documenting (see "The Science behind the Story," • pp. 134–135) were causing population fluctuations and disease outbreaks and were unleashing changes in the composition of the community.

## The Conservation of Biodiversity

Changes in populations and communities have been taking place naturally as long as life has existed, but today human development, resource extraction, and population pressure are speeding the rate of change and altering the types of change. The ways we modify our environment cannot be fully understood in a scientific vacuum, however. The actions that threaten biodiversity have complex social, economic, and political roots, and environmental scientists appreciate that we must understand these aspects if we are to develop solutions.

Fortunately, millions of people around the world are already taking action to safeguard the biodiversity and ecological and evolutionary processes that make Earth such a unique place (Chapter 11). Costa Ricans have been confronting the challenges to their nation's biodiversity, and their actions so far show what even a small country of modest means can do.

## Social and economic factors affect species and communities

Many of the threats to Costa Rica's species and ecological communities result from past economic and social forces whose influences are still evident. European immigrants and their descendants viewed Costa Rica's lush forests as an obstacle to agricultural development, and timber companies saw them simply as a source of wood products. Costa Rica's leading agricultural products have long included beef and bananas, whose production and cultivation require extensive environmental modification.

Between 1945 and 1995, the country's population grew from 860,000 to 3.34 million, and the percentage of land devoted to pasture increased from 12% to 33%. With much of the formerly forested land converted to agriculture, the proportion of the country covered by forest decreased from 80% to 25%. In 1991, Costa Rica was losing its forests faster than any other country in the world—nearly 140 ha (350 acres) per day. As a result, populations of most species were declining, and some were becoming endangered (**Figure 5.18**). As had occurred in the history of the United States, few people foresaw the need to conserve biological resources until it became clear that they were being rapidly lost.

## Costa Rica took steps to protect its environment

During the 1950s, a group of Quakers, Christian pacifists who opposed the U.S. military draft, emigrated from Alabama to Costa Rica and founded the village of Monteverde. The Quakers relied on milk and cheese for much of their economic activity, but they also set aside one-third of their land for conservation. The Quakers' efforts, along with contributions from international conservation organizations, provided the beginnings of what is today the Monteverde Cloud Forest Reserve. This privately managed 10,500-ha (26,000-acre) reserve was established in 1972 to protect the forest and its populations of 2,500 plant species, 400 bird species, 500 butterfly species, 100 mammal species, and 120 reptile and amphibian species, including the golden toad.

**(a) Golden-cheeked warbler**

**(b) Green sea turtle**

**(c) Red-backed squirrel monkey**

**FIGURE 5.18** Costa Rica is home to a number of species classified as globally threatened or endangered. The golden-cheeked warbler, *Dendroica chrysoparia* (**a**), winters in Central America but breeds in the Texas hill country, where its habitat is being lost to housing development. The green sea turtle, *Chelonia mydas* (**b**), occurs throughout the world's oceans and lays eggs on beaches in Costa Rica and elsewhere, but it has undergone steep population declines. The red-backed squirrel monkey, *Saimiri oerstedii* (**c**), is endemic to a tiny area in Costa Rica; its small geographic range makes it vulnerable to forest loss.

THE SCIENCE BEHIND THE STORY

# Climate Change and Its Effects on Monteverde

*Dr. J. Alan Pounds (L) with Dr. Luis Coloma, looking for harlequin frogs*

Soon after the golden toad's disappearance, scientists began to investigate the potential role of global climate change (Chapter 18) in driving cloud-forest species toward extinction. They noted that the period from July 1986 to June 1987 was the driest on record at Monteverde, with unusually high temperatures and record-low stream flows. These conditions had caused the golden toad's breeding pools to dry up in the spring of 1987, likely killing nearly all of the eggs and tadpoles in the pools.

By reviewing reams of weather data, scientists found that the number of dry days and dry periods each winter in the Monteverde region had increased between 1973 and 1998. Because amphibians breathe and absorb moisture through their skin, they are susceptible to dry conditions. Based on these facts, herpetologists J. Alan Pounds and Martha Crump in 1994 hypothesized that hot, dry conditions were to blame for high adult mortality and breeding problems among golden toads and other amphibians.

Throughout this period, scientists worldwide were realizing that the atmosphere and oceans were warming because of human release of carbon dioxide and other greenhouse gases (Chapter 18). With this in mind, Pounds and others reviewed the scientific literature on ocean and atmospheric science to analyze the effects on Monteverde's local climate of warming patterns in the ocean regions around Costa Rica.

Warmer oceans, the researchers found, caused clouds to pass over at higher elevations, where they were no longer in contact with the trees. Once the cloud forest's moisture supply was pushed upward, out of reach of the mountaintops, the forest began to dry out.

In a 1999 paper in the journal *Nature*, Pounds and two colleagues reported that broad-scale climate modification was causing local changes at the species, population, and community levels. Higher clouds and decreasing moisture in the forest could explain not only the disappearance of the golden toad and harlequin frog, but also the concurrent population crashes and subsequent disappearance of 20 other species of frogs and toads from the Monteverde

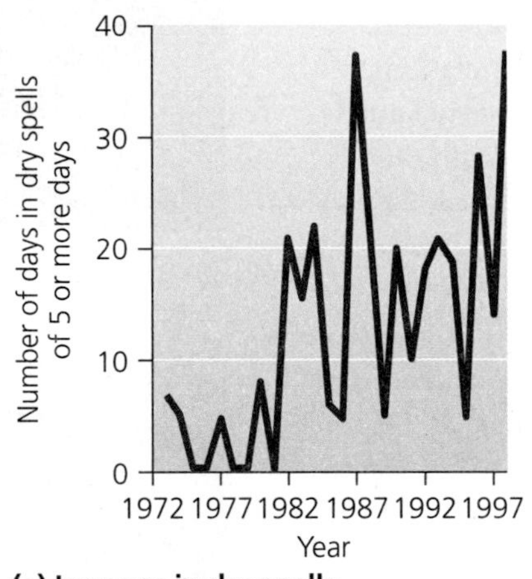


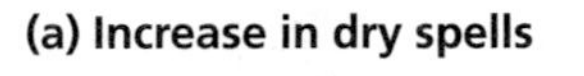
**(a) Increase in dry spells**

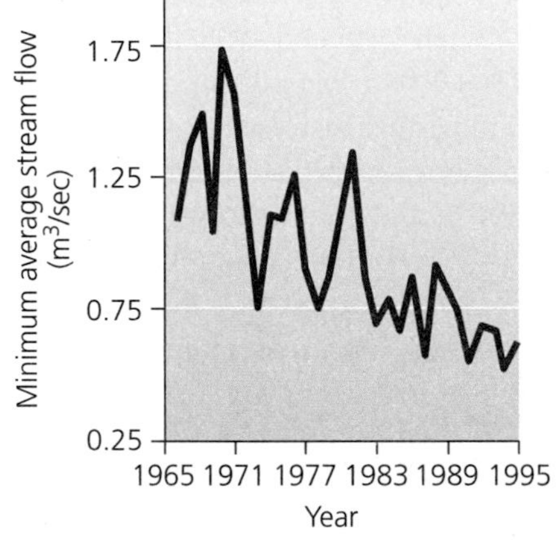

**(b) Decrease in stream flow**

Warming and drying trends in Monteverde's climate may have contributed to the region's amphibian declines. Evidence gathered over three decades shows **(a)** an increase in the annual number of dry days and **(b)** a decrease in the amount of annual stream flow. Data from Pounds, J. A., et al. 1999. Biological response to climate change on a tropical mountain. *Nature* 398:611–615.

In 1970, the Costa Rican government and international representatives came together to create the country's first national parks and protected areas. The first parks centered on areas of spectacular scenery, such as Poas Volcano National Park. Santa Rosa National Park encompassed valuable tropical dry forest, Tortuguero National Park contained essential nesting beaches for the green turtle (see Figure 5.18b), and Cahuita National Park was meant to protect a prominent coral reef system. Initially the government gave the parks little real support. According to Costa Rican conservationist Mario Boza, in their early years the parks were granted only five guards, one vehicle, and no funding.

Today government support is greater. Fully 12% of the nation's area is contained in national parks, and a further 16% is devoted to other types of wildlife and conservation reserves. Costa Ricans, along with international biologists, are working to protect endangered species and recover their populations. Costa Rica and its citizens are now reaping the benefits of their conservation efforts—not region. Amphibians that survived underwent population crashes in each of the region's 3 driest years.

Moreover, whole communities were being altered. As the montane forests dried out, dry-tolerant species of birds and reptiles shifted upslope, and moisture-dependent species were stranded at the mountaintops by a rising tide of aridity. Although organisms may in general move in response to changing environmental conditions, if a species has nowhere to go, then extinction may result.

Pounds and his colleagues have recently fleshed out the story still further. Although clouds had moved higher in the sky, the extra moisture evaporating from warming oceans was increasing cloud cover overall, blocking sunlight during the day and trapping heat at night. As a result, at Monteverde and other tropical locations, daytime and nighttime temperatures were becoming more similar.

Such conditions are optimal for chytrid fungi, pathogens that can lethally infect amphibians. The chytrid *Batrachochytrium dendrobatidis* is thought to have contributed to the likely extinction of 30 of the world's 113 species of harlequin frogs.

Pounds's team presented these observations in *Nature* in 2006, arguing that at Monteverde and elsewhere, climate change is promoting disease epidemics that are driving extinct many of the world's amphibians.

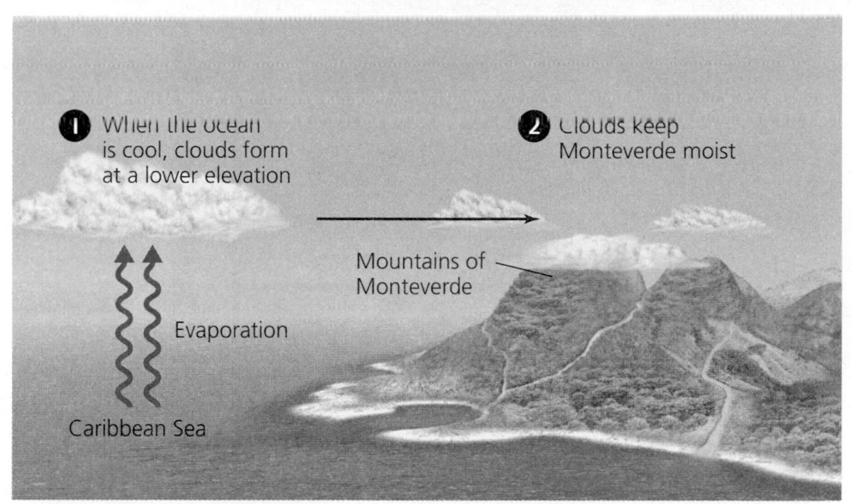


**(a) Cool ocean conditions**

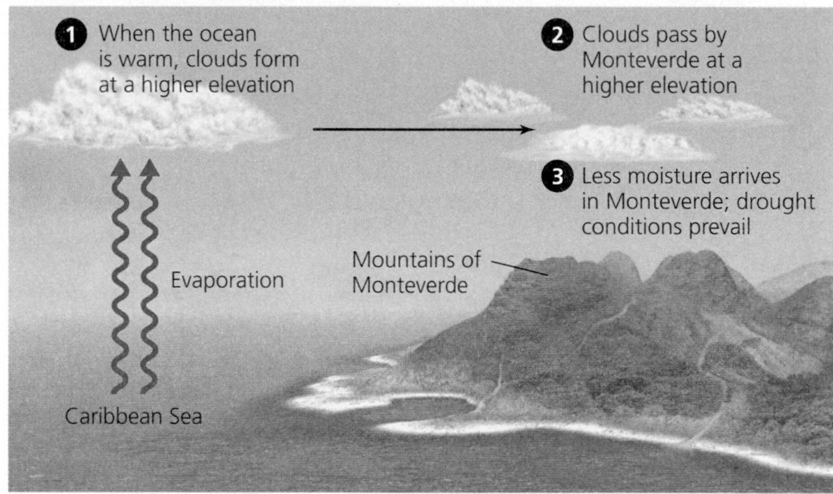


**(b) Warm ocean conditions**

Monteverde's cloud forest gets its name and life-giving moisture from clouds that sweep inland from the oceans. When ocean temperatures are cool **(a)**, the clouds keep Monteverde moist. Warmer ocean conditions **(b)** resulting from global climate change cause clouds to form at higher elevations and pass over the mountains, drying the cloud forest.

only ecological benefits, but also economic ones. Because of its parks and its reputation for conservation, tourists from around the world now visit Costa Rica, a phenomenon called **ecotourism** (**Figure 5.19**). The ecotourism industry draws more than 1 million visitors to Costa Rica each year, provides thousands of jobs to Costa Ricans, and is a major contributor to the country's economy. Today's Costa Rican economy is fueled in large part by commerce and tourism, whose contributions (40%) outweigh those of industry (22%) and agriculture (13%) combined.

### Weighing THE Issues | How Best to Conserve Biodiversity?

Most people view national parks and ecotourism as excellent ways to help keep ecological systems intact. Yet the golden toad went extinct despite living within a reserve established to protect it, and climate change does not pay attention to park boundaries. What lessons can we take from this about the conservation of biodiversity? Are parks and preserves sufficient? What other approaches should we pursue to save declining species?

FIGURE 5.19 Costa Rica has protected a wide array of its diverse natural areas. This protection has stimulated the nation's economy through ecotourism. Here, visitors experience a walkway through the forest canopy in one of the nation's parks.

It remains to be seen how effectively ecotourism can help preserve natural systems in Costa Rica in the long term. As forests outside the parks disappear, the parks are beginning to suffer from illegal hunting and timber extraction. Boza and other conservationists say the parks are still underprotected and underfunded. Ecotourism will likely need to generate still more money to preserve habitat, protect endangered species, and restore altered communities to their former condition. Restoration is being carried out in Costa Rica's Guanacaste Province, for instance, where scientists are restoring dry tropical forest from grazed pasture. Restoration of ecological communities is one phenomenon we will examine in our next chapter, as we move from populations to communities.

## Conclusion

The golden toad and other organisms of the Monteverde cloud forest have helped illuminate the fundamentals of evolution and population ecology that are integral to environmental science. The evolutionary processes of natural selection, speciation, and extinction help determine Earth's biodiversity. Understanding how ecological processes work at the population level is crucial to protecting biodiversity threatened by the mass extinction event that many biologists maintain is already underway.

## REVIEWING OBJECTIVES

**You should now be able to:**

**Explain the process of natural selection and cite evidence for this process**

- Because organisms produce excess young, individuals vary in their traits, and many traits are inherited, some individuals will prove better at surviving and reproducing. Their genes will be passed on and become more prominent in future generations. (pp. 113–114)
- Mutations and recombination provide the genetic variation for natural selection. (pp. 114–115)
- We have produced our pets, farm animals, and crop plants through artificial selection. (pp. 115, 117)

**Describe the ways in which evolution results in biodiversity**

- Natural selection can act as a diversifying force as organisms adapt to their environments in myriad ways. (pp. 114–117)
- Speciation (by geographic isolation and other means) produces new species. (pp. 117–118)
- Once they have diverged, lineages continue diverging, a process represented in a phylogenetic tree. (p. 119)

**Discuss reasons for species extinction and mass extinction events**

- Extinction often occurs when species that are highly specialized or that have small populations encounter rapid environmental change. (pp. 120–121)
- Earth's life has experienced five known episodes of mass extinction, which were due to asteroid impact and possibly volcanism and other factors. (pp. 121–123)
- Today, human impact may be initiating a sixth mass extinction. (pp. 121–122)

**List the levels of ecological organization**

- Ecologists study phenomena on the organismal, population, community, and ecosystem levels—and, increasingly, at the level of the biosphere. (pp. 122–124)

**Outline the characteristics of populations that help predict population growth**

- Populations are characterized by population size, population density, population distribution, sex ratio, and age structure. (pp. 125–127)
- Immigration and emigration, as well as birth and death rates, determine how a population will grow or decline. (pp. 128–129)

**Assess logistic growth, carrying capacity, limiting factors, and other fundamental concepts of population ecology**

- Populations unrestrained by limiting factors will undergo exponential growth until they meet environmental resistance. (pp. 132–130)
- Logistic growth describes the effects of density dependence; exponential growth slows as population size increases, and population size levels off at a carrying capacity. (pp. 130–131)
- K-selection and r-selection describe theoretical extremes in how organisms can allocate growth and reproduction. (pp. 131–132)

**Identify efforts and challenges involved in the conservation of biodiversity**

- Social and economic factors influence our impacts on natural systems. (pp. 132–136)
- Extensive efforts to protect and restore species and habitats will be needed to prevent further erosion of biodiversity. (pp. 133–136)

## TESTING YOUR COMPREHENSION

1. Explain the premises and logic that support the concept of natural selection.
2. Name two examples of evidence for natural selection.
3. How does allopatric speciation occur?
4. Name three organisms that have gone extinct, and give a probable reason for each extinction.
5. What is the difference between a species and a population? Between a population and a community?
6. Contrast the concepts of habitat and niche.
7. List and describe each of the five major population characteristics discussed in this chapter. Explain how each shapes population dynamics.
8. Can any species undergo exponential growth forever? Explain your answer.
9. Describe how limiting factors relate to carrying capacity.
10. Explain the difference between K-selected species and r-selected species. Can you think of examples of each that were not mentioned in the chapter?

## SEEKING SOLUTIONS

1. In what ways has artificial selection changed people's quality of life? Give examples. Can you imagine a way in which artificial selection could be used to improve our quality of life further? Can you imagine a way it could be used to reduce our environmental impact?
2. What types of species are most vulnerable to extinction, and what kinds of factors threaten them? Can you think of any species in your region that are threatened with extinction today? What reasons lie behind their endangerment?
3. Do you think the human species can continue raising its global carrying capacity? How so, or why not? Do you think we *should* try to keep raising our carrying capacity? Why or why not?
4. Describe the evidence suggesting that changes in temperature and precipitation led to the extinction of the golden toad and to population crashes for other amphibians at Monteverde. What do you think could be done to help make future such declines less likely?
5. What are the advantages of ecotourism for a country like Costa Rica? Can you think of any disadvantages? What would you recommend that Costa Rica do to prevent the loss of its biodiversity?
6. **THINK IT THROUGH** You are a population ecologist studying animals in a national park, and your government is asking for advice on how to focus its limited conservation funds. How would you rate the following three species, from most vulnerable (and thus most in need of conservation attention) to least vulnerable? Give reasons for your choices.
   - A bird with an even sex ratio that is a habitat generalist
   - A salamander endemic to the park that lives in high-elevation forest
   - A fish that specializes on a few types of invertebrate prey and has a high population size

## INTERPRETING GRAPHS AND DATA

Amphibians are sensitive biological indicators of climate change because their reproduction and survival are so closely tied to water. One way in which drier conditions may affect amphibians is by reducing the depth of the pools of water in which their eggs develop. Shallower pools offer less protection from UV-B (ultraviolet-B) radiation, which some scientists maintain may kill embryos directly or make them more susceptible to disease.

Herpetologist Joseph Kiesecker and colleagues conducted a field study of the relationships among water depth, UV-B radiation, and survivorship of western toad (*Bufo boreas*) embryos in the Pacific Northwest. In manipulative experiments, the researchers placed toad embryos in mesh enclosures at three different depths of water. The researchers placed protective filters that blocked all UV-B radiation over some of these embryos, while leaving other embryos unprotected without the filters. Some of the study's results are presented in the accompanying graph.

1. If the UV-B radiation at the surface has an intensity of 0.27 watts/m$^2$, approximately what is its intensity at depths of 10 cm, 50 cm, and 100 cm?
2. Approximately how much did survival rates at the 10-cm depth differ between the protected and unprotected treatments? Why do you think survival rates differed significantly at the 10-cm depth but not at the other depths?
3. What do you think would be the effect of drier-than-average years on the western toad population, if the average depth of pools available for toad spawning dropped? How do the data in the graph address your hypothesis? Do they support cause-and-effect relationships among water depth, UV-B exposure, disease, and toad mortality?

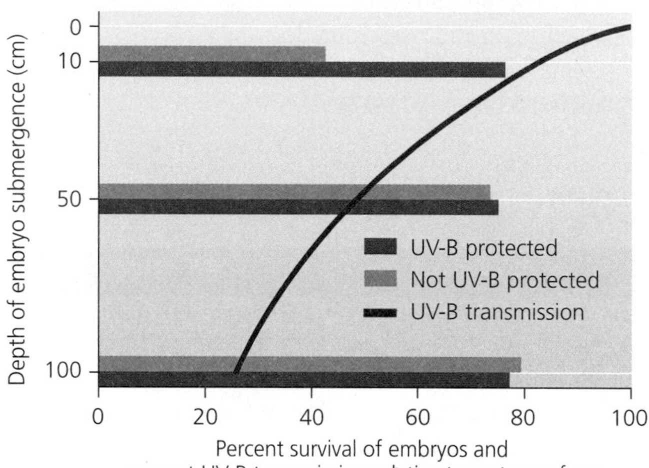


Embryo survivorship in western toads (*Bufo boreas*) at different water depths and UV-B light intensities. Red bars indicate embryos protected under a filter that blocked UV-B light; orange bars indicate unprotected embryos. The blue line indicates the amount of UV-B light reaching different depths in the water column, expressed as a percentage of the UV-B radiation at the water surface. Data from Kiesecker, J. M., et al. 2001. Complex causes of amphibian population declines. *Nature* 410: 681–684.

## CALCULATING ECOLOGICAL FOOTPRINTS

In 2004, coffee consumption in the United States topped 2.7 billion pounds (out of 14.8 billion pounds produced globally). Next to petroleum, coffee is the most valuable commodity on the world market, and the United States is its leading importer. Most coffee is produced in large tropical plantations, where coffee is the only tree species and is grown in full sun. However, approximately 2% of coffee is produced in small groves where coffee trees and other species are intermingled. These *shade-grown* coffee forests maintain greater habitat diversity for tropical rainforest wildlife. Given the information above, estimate the coffee consumption rates in the table to the right.

| | Population | Pounds of coffee per day | Pounds of coffee per year |
|---|---|---|---|
| You (or the average American) | 1 | 0.025 | 9 |
| Your class | | | |
| Your hometown | | | |
| Your state | | | |
| United States | | | |

Data from O'Brien, T. G., and M. F. Kinnaird. 2003. Caffeine and conservation. *Science* 300:587; and International Coffee Organization.

1. What percentage of global coffee production is consumed in the United States? If only shade-grown coffee were consumed in the United States, how much would shade-grown production need to increase to meet that demand?
2. How much extra would you be willing to pay for a pound of shade-grown coffee, if you knew that your money would help to prevent habitat loss or extinction

for animals such as Sumatran tigers, rhinoceroses, and the many songbirds that migrate between Latin America and North America each year?

3. If everyone in the United States were willing to pay as much extra per pound for shade-grown coffee as you are, how much additional money would that provide for conservation of biodiversity in the tropics each year?

## Take It Further

Go to www.aw-bc.com/withgott or the student CD-ROM, where you'll find:

- Suggested answers to end-of-chapter questions
- Quizzes, animations, and flashcards to help you study
- *Research Navigator*™ database of credible and reliable sources to assist you with your research projects
- GRAPHit! Tutorials to help you interpret graphs
- INVESTIGATEit! Current news articles that link the topics that you study to case studies from your region to around the world

CHAPTER

# 6 Species Interactions and Community Ecology

Marsh community along the shore of Lake Ontario

## Upon completing this chapter, you will be able to:

- Compare and contrast the major types of species interactions
- Characterize feeding relationships and energy flow, using them to construct trophic levels and food webs
- Distinguish characteristics of a keystone species
- Characterize the process of succession and the debate over the nature of communities
- Perceive and predict the potential impacts of invasive species in communities
- Explain the goals and methods of ecological restoration
- Describe and illustrate the terrestrial biomes of the world

Aggregation of zebra mussels

## CENTRAL CASE

# Black and White, and Spread All Over: Zebra Mussels Invade the Great Lakes

**"We are seeing changes in the Great Lakes that are more rapid and more destructive than any time in the history of the Great Lakes."**
—Andy Buchsbaum, National Wildlife Federation

**"When you tear away the bottom of the food chain, everything that is above it is going to be disrupted."**
—Tom Nalepa, National Oceanic and Atmospheric Administration

As if the Great Lakes hadn't been through enough already, the last thing they needed was the zebra mussel. The pollution-fouled waters of Lake Erie and the other Great Lakes shared by Canada and the United States had become gradually cleaner in the years following the Clean Water Act of 1970. As government regulation brought industrial discharges under control, people once again began to use the lakes for recreation, and populations of fish rebounded.

Then the zebra mussel arrived. Black-and-white-striped shellfish the size of a dime, zebra mussels attach to hard surfaces and open their paired shells, feeding on algae by filtering water through their gills. This mollusc, given the scientific name *Dreissena polymorpha,* is native to the Caspian Sea, Black Sea, and Azov Sea in western Asia and eastern Europe. It made its North American debut in 1988, when it was discovered in Canadian waters at Lake St. Clair, which connects Lake Erie with Lake Huron. Evidently ships arriving from Europe had discharged ballast water containing the mussels or their larvae into the Great Lakes.

Within just two years of their discovery in Lake St. Clair, zebra mussels had reached all five of the Great Lakes. The next year, these invaders entered New York's Hudson River to the east, and the Illinois River at Chicago

(a) Clogging a pipe

(b) Suffocating native clams

FIGURE 6.1 Zebra mussels clog water intake pipes (**a**) of power plants and industrial facilities. They also starve and suffocate native clams (**b**) by adhering to their shells and sealing them shut.

to the west. From the Illinois River and its canals, they soon reached the Mississippi River, giving them access to a vast watershed covering 40% of the United States. By 1994 they had colonized waters in 19 U.S. states and two Canadian provinces.

How could a mussel spread so quickly? The zebra mussel's larval stage is well adapted for long-distance dispersal. Its tiny larvae drift freely for several weeks, traveling as far as the currents take them. Adults that attach themselves to boats and ships may be transported from one place to another, even to small isolated lakes and ponds well away from major rivers. Moreover, in North America the mussels encountered none of the particular species of predators, competitors, and parasites that had evolved to limit their population growth in the Old World.

Zebra mussels are best known for clogging up water intake pipes at factories, power plants, municipal water supplies, and wastewater treatment facilities (**Figure 6.1a**). At one Michigan power plant, workers counted 700,000 mussels per square meter of pipe surface. Great densities of these organisms can damage boat engines, degrade docks, foul fishing gear, and sink buoys that ships use for navigation. Through such impacts, it is estimated that zebra mussels cost the U.S. economy hundreds of millions of dollars each year.

Zebra mussels also have severe impacts on the ecological systems they invade. They eat primarily **phytoplankton**: microscopic photosynthetic algae, protists, and cyanobacteria that drift in open water. Because each mussel filters a liter or more of water every day, they consume so much phytoplankton that they can deplete populations. Phytoplankton form the foundation of the Great Lakes food web, so their depletion is bad news for **zooplankton**, the tiny aquatic animals that eat phytoplankton—and for the fish that eat both. Water bodies with zebra mussels have fewer zooplankton and open-water fish than water bodies without them, researchers are finding. Zebra mussels also interfere with native molluscs, suffocating them by attaching to their shells (**Figure 6.1b**).

However, zebra mussels also benefit some bottom-feeding invertebrates and fish. By filtering algae and organic matter from open water and depositing nutrients in their feces, they shift the community's nutrient balance to the bottom and benefit the species that feed there. Once they have cleared the water, sunlight penetrates more deeply, spurring the growth of underwater plants and algae.

In the past several years, scientists have noticed a surprising new twist: One invader is being displaced by another. The quagga mussel (*Dreissena bugensis*), a close relative of the zebra mussel from the Dneiper River in Ukraine, is spreading through the Great Lakes and beyond. This species, named after an extinct zebra-like animal, appears to be replacing the zebra mussel in many locations. What consequences this shift may have for ecological communities scientists are only beginning to understand.

## Species Interactions

By interacting with many species in a variety of ways, zebra mussels have set in motion an array of changes in the ecological communities they have invaded. Interactions among species are the threads in the fabric of communities. Ecologists have organized species interactions into several fundamental categories. Most prominent are competition, predation, parasitism, herbivory, and mutualism. **Table 6.1** summarizes the positive and negative impacts of each type of interaction for each participant.

**TABLE 6.1 Effects of Species Interactions on Their Participants**

| Type of interaction | Effect on species 1 | Effect on species 2 |
|---|---|---|
| Mutualism | + | + |
| Commensalism | + | 0 |
| Predation, parasitism, herbivory | + | – |
| Neutralism | 0 | 0 |
| Amensalism | – | 0 |
| Competition | – | – |

"+" denotes a positive effect; "–" denotes a negative effect; "0" denotes no effect.

## Competition can occur when resources are limited

When multiple organisms seek the same limited resource, their relationship is said to be one of **competition**. Competing organisms do not usually fight with one another directly and physically. Competition is generally more subtle, involving the consequences of one organism's ability to match or outdo others in procuring resources. The resources for which organisms compete can include just about anything an organism might need to survive, including food, water, space, shelter, mates, sunlight, and more. Competitive interactions can take place among members of the same species *(intraspecific competition)* or among members of two or more different species *(interspecific competition)*.

We have already discussed intraspecific competition in Chapter 5, without naming it as such. Recall that density dependence (• pp. 130–131) limits the growth of a population; individuals of the same species compete with one another for limited resources, such that competition is more intense when there are more individuals per unit area (denser populations). Thus, intraspecific competition is really a population-level phenomenon.

In contrast, interspecific competition can have substantial effects on the composition of communities. Interspecific competition can give rise to different types of outcomes. If one species is a very effective competitor, it may exclude another species from resource use entirely. This outcome, called *competitive exclusion*, occurred in Lake St. Clair and western Lake Erie as the zebra mussel outcompeted a native mussel species.

Alternatively, if neither competing species fully excludes the other, the species may continue to live side by side. This result, called *species coexistence*, may produce a stable point of equilibrium, in which the population size of each remains fairly constant through time.

Coexisting species that use the same resources tend to adjust to their competitors to minimize competition with them. Individuals can do this by changing their behavior so as to use only a portion of the total array of resources they are capable of using. In such cases, individuals do not fulfill their entire *niche*, or ecological role (• p. 124). The full niche of a species is called its **fundamental niche** (**Figure 6.2a**). An individual that plays only part of its role because of competition or other species interactions is said to display a **realized niche** (**Figure 6.2b**), the portion of its fundamental niche that is actually filled, or realized. The quagga mussel can occupy a wider range of water conditions and substrates than the zebra mussel, so it is thought to have a larger fundamental niche. It also appears to be reducing the zebra mussel's realized niche, as it displaces the zebra mussel in many areas.

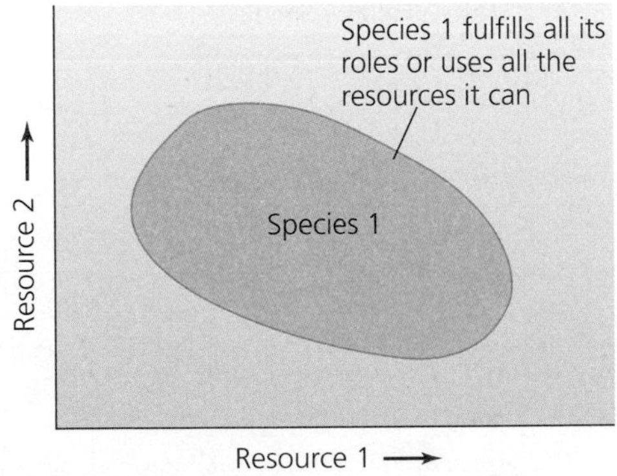


**(a) Fundamental niche**

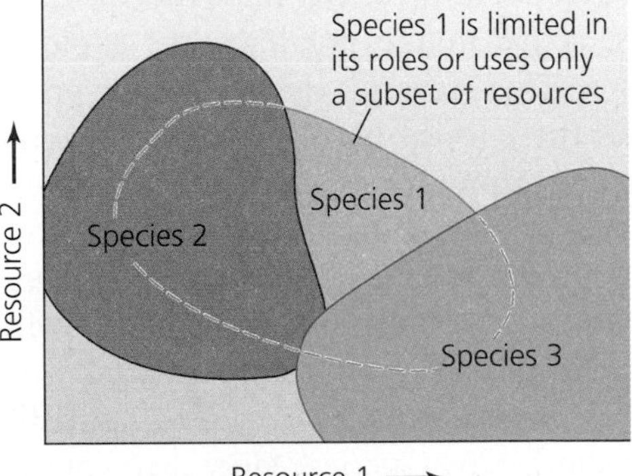


**(b) Realized niche**

**FIGURE 6.2** An organism facing competition may be forced to play a more limited ecological role or use fewer resources than it would in the absence of its competitor. With no competitors, an organism can exploit its full fundamental niche (**a**). But when competitors restrict what an organism can do or what resources it can use, the organism is limited to a realized niche (**b**), which covers only a subset of its fundamental niche. In considering niches, ecologists have traditionally focused on competition, but they now recognize that other species interactions also are influential.

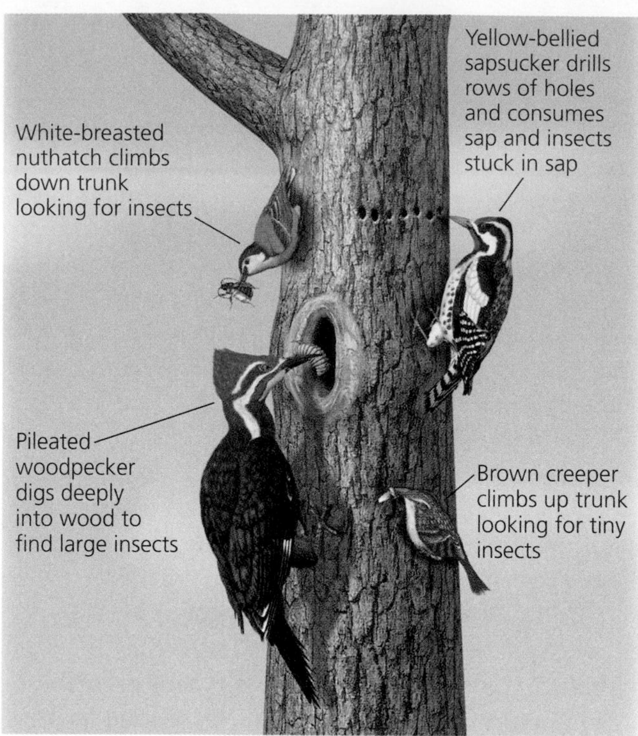


FIGURE 6.3 When species compete, they tend to partition resources, each specializing on a slightly different resource or way of attaining a shared resource. Many types of birds—including the woodpeckers, creeper, and nuthatch shown here—feed on insects from tree trunks, but they use different portions of the trunk, seeking different foods in different ways.

Species make similar adjustments over evolutionary time. They adapt to competition by evolving to use slightly different resources or to use their shared resources in different ways. If two bird species eat the same type of seeds, one might come to specialize on larger seeds and the other to specialize on smaller seeds. Or one bird might become more active in the morning and the other more active in the evening, thus avoiding direct interference. This process is called **resource partitioning** because the species partition, or divide, the resource they use in common by specializing in different ways (**Figure 6.3**).

Resource partitioning can lead to *character displacement*, which occurs when competing species evolve physical characteristics that reflect their reliance on the portion of the resource they use. By becoming more different from one another, two species reduce their competition. Through natural selection (• pp. 113–116), birds that specialize on larger seeds may evolve larger bills that enable them to make best use of the resource, whereas birds specializing on smaller seeds may evolve smaller bills. This is precisely what extensive research has revealed about the finches from the Galapagos Islands that were first described by Charles Darwin.

## Several types of interactions are exploitative

In competitive interactions, each participant has a negative effect on other participants, because each takes resources the others could have used. This is reflected in the two minus signs shown for competition in Table 6.1. In other types of interactions, some participants benefit while others are harmed (note the +/− interactions in Table 6.1). We can think of interactions in which one member exploits another for its own gain as exploitative interactions, or *exploitation.* Such interactions include predation, parasitism, herbivory, and related concepts, as outlined below.

## Predators kill and consume prey

Every living thing needs to procure food, and for most animals, that means eating other living organisms. **Predation** is the process by which individuals of one species, a *predator*, hunt, capture, kill, and consume individuals of another species, the *prey* (**Figure 6.4**). Along with competition, predation has traditionally been viewed as one of the primary organizing forces in community ecology. Interactions between predators and prey structure the food webs that we will examine shortly, and they influence community composition by helping determine the relative abundance of predators and prey.

Zebra mussel predation on phytoplankton has reduced phytoplankton populations by up to 90%, according to many studies in the Great Lakes and Hudson River. Zebra mussels also consume the smaller types of zooplankton. This predation, combined with the competition mentioned above, has caused zooplankton population sizes and biomass to decline by up to 70% in Lake Erie

FIGURE 6.4 Predator-prey interactions have ecological and evolutionary consequences for both prey and predator. Here, a fire-bellied snake *(Liophis epinephalus)* devours a frog in the Monteverde cloud forest we studied in Chapter 5.

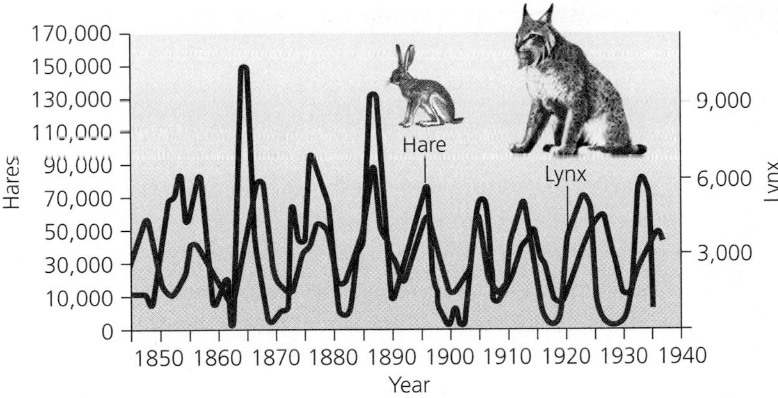


FIGURE 6.5 Predator-prey systems sometimes show paired cycles, in which increases and decreases in the population of one organism apparently drive increases and decreases in the population of the other. Although such cycles are predicted by theory and are seen in lab experiments, they are difficult to document conclusively in natural systems. Data from MacLulich, D.A. 1937. Fluctuation in the numbers of varying hare *(Lepus americanus). University of Toronto Studies in Biology Series 43,* Toronto: University of Toronto Press.

and the Hudson River since zebra mussels arrived. Meanwhile, the mussels do not eat some cyanobacteria (• p. 101), so concentrations of these cyanobacteria rise in lakes with zebra mussels. Most predators are also prey, however, and zebra mussels have become a food source for a number of North American diving ducks, muskrats, crayfish, flounder, sturgeon, and eels, as well as several types of fish with grinding teeth, such as carp and freshwater drum.

Predation can sometimes drive population dynamics by causing cycles in population sizes. An increase in the population size of prey creates more food for predators, which may survive and reproduce more effectively as a result. As the predator population rises, additional predation drives down the population of prey. Fewer prey in turn causes some predators to starve, so that the predator population declines. This allows the prey population to begin rising again, starting the cycle anew. Most natural systems involve so many factors that such cycles don't last long, but in some cases we see extended cycles (**Figure 6.5**).

Predation also has evolutionary ramifications. Individual predators that are more adept at capturing prey will likely live longer, healthier lives and be better able to provide for their offspring than will less adept individuals. Thus, natural selection on individuals within a predator species leads to the evolution of adaptations that make them better hunters. Prey face an even stronger selective pressure—the risk of immediate death. For this reason, predation pressure has caused organisms to evolve an elaborate array of defenses against being eaten (**Figure 6.6**).

## Parasites exploit living hosts

Organisms can exploit other organisms without killing them. **Parasitism** is a relationship in which one organism, the *parasite*, depends on another, the *host*, for nourishment or some other benefit while simultaneously doing the host harm. Unlike predation, parasitism usually does not result in an organism's immediate death, although it sometimes contributes to the host's eventual death.

**(a) Cryptic coloration**

**(b) Warning coloration**

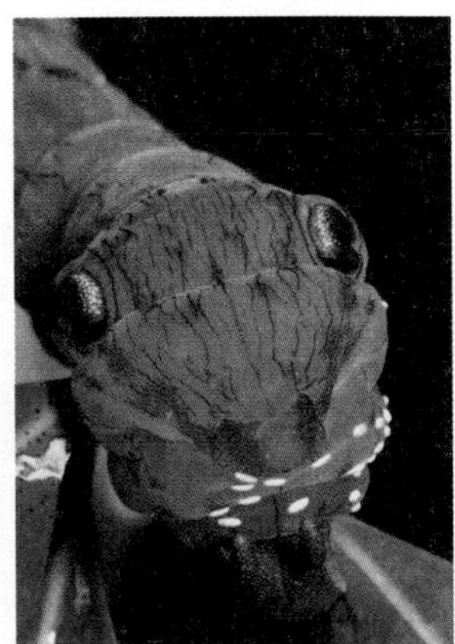

**(c) Mimicry**

FIGURE 6.6 Natural selection to avoid predation has resulted in many fabulous adaptations. Some prey hide from predators by *crypsis*, or camouflage, such as this gecko on tree bark (**a**). Other prey are brightly colored to warn predators that they are toxic or distasteful, such as this monarch butterfly (**b**). Still others fool predators with mimicry. Some, such as walking sticks imitating twigs, mimic for crypsis, whereas others mimic toxic, distasteful, or dangerous organisms. This caterpillar (**c**), when disturbed, swells and curves its tail end and shows false eyespots to look like a snake's head.

FIGURE 6.7 With its suction-like mouth and rasping tongue, the parasitic sea lamprey *(Petromyzon marinus)* attaches itself to fish and sucks their blood for days or weeks, sometimes killing the fish. Sea lampreys wreaked havoc on Great Lakes fisheries after entering the lakes through human-built canals.

Many parasites live in close contact with their hosts. These parasites include disease pathogens, such as the protists that cause malaria and dysentery, as well as animals, such as tapeworms, that live in the digestive tracts of their hosts. Other parasites live on the exterior of their hosts, such as the ticks that attach themselves to their hosts' skin, and the sea lamprey *(Petromyzon marinus)*, another invader of the Great Lakes (**Figure 6.7**). Sea lampreys are tube-shaped vertebrates that grasp the bodies of fish with a suction-cup mouth and a rasping tongue, sucking blood from the fish for days or weeks. Sea lampreys invaded the Great Lakes from the Atlantic Ocean after people dug canals to connect the lakes for shipping, and the lampreys soon devastated economically important fisheries of chubs, lake herring, whitefish, and lake trout. Since the 1950s, U.S. and Canadian fisheries managers have reduced lamprey populations by applying chemicals that selectively kill lamprey larvae.

Other types of parasites are free-living and come into contact with their hosts only infrequently. For example, the cuckoos of Eurasia and the cowbirds of the Americas parasitize other birds by laying eggs in their nests and letting the host bird raise the parasite's young.

Many insects parasitize other insects, often killing them in the process, and are called *parasitoids.* Various species of parasitoid wasps lay eggs on caterpillars. When the eggs hatch, the wasp larvae burrow into the caterpillar's tissues and slowly consume them. The wasp larvae metamorphose into adults and fly from the body of the dying caterpillar.

Just as predators and prey evolve in response to one another, so do parasites and hosts, in a process termed *coevolution.* Hosts and parasites can become locked in a duel of escalating adaptations, a situation sometimes referred to as an *evolutionary arms race.* Like rival nations racing to stay ahead of one another in military technology, host and parasite may repeatedly evolve new responses to the other's latest advance. In the long run, though, it may not be in a parasite's best interest to become too harmful to its host. Instead, a parasite might leave more offspring in the next generation—and thus be favored by natural selection—if it allows its host to live a longer time, or even to thrive.

## Herbivores exploit plants

One of the most common types of exploitation is **herbivory**, which occurs when animals feed on the tissues of plants. Insects that feed on plants are the most widespread type of herbivore; just about every plant in the world is attacked by some type of insect (**Figure 6.8**). In most cases,

FIGURE 6.8 Herbivory is a common way to make a living. The world holds many thousands, and perhaps millions, of species of plant-eating insects, such as this larva (caterpillar) of the death's head hawk moth *(Acherontia atropos)* from western Europe.

herbivory does not kill a plant outright but may affect its growth and reproduction.

Like animal prey, plants have evolved a wide array of defenses against the animals that feed on them. Many plants produce chemicals that are toxic or distasteful to herbivores. Others arm themselves with thorns, spines, or irritating hairs. In response, herbivores may evolve ways to overcome these defenses, and the plant and the animal may embark on an evolutionary arms race.

Some plants go a step further and recruit certain animals as allies to assist in their defense. Many such plants encourage ants to take up residence by providing thorns or swelled stems for the ants to nest in or nectar-bearing structures for the ants to feed from. These ants protect the plant in return by attacking other insects that land or crawl on it. Still other plants respond to herbivory by releasing volatile chemicals when they are bitten or pierced. The airborne chemicals attract predatory insects that may attack the herbivore. Such cooperative strategies, trading food for defense, are examples of our next type of species interaction, mutualism.

## Mutualists help one another

**Mutualism** is a relationship in which two or more species benefit from interaction with one another. Generally each partner provides some resource or service that the other needs.

Many mutualistic relationships—like many parasitic relationships—occur between organisms that live in close physical contact. (Indeed, biologists hypothesize that many mutualistic associations evolved from parasitic ones.) Such physically close association is called **symbiosis**. Thousands of terrestrial plant species depend on mutualisms with fungi; plant roots and some fungi together form symbiotic associations called mycorrhizae. In these relationships, the plant provides energy and protection to the fungus, while the fungus assists the plant in absorbing nutrients from the soil. In the ocean, coral polyps, the tiny animals that build coral reefs, share beneficial arrangements with algae known as zooxanthellae (• p. 450). The coral provide housing and nutrients for the algae in exchange for a steady supply of food—90% of their nutritional requirements.

You, too, are part of a symbiotic association. Your digestive tract is filled with microbes that help you digest food—microbes you are providing a place to live. Indeed, we may owe our very existence to symbiotic mutualisms. It is now widely accepted that the eukaryotic cell (• pp. 98–99) originated after certain prokaryotic cells engulfed other prokaryotic cells and established mutualistic symbioses. Scientists have inferred that some of the engulfed cells eventually evolved into cell organelles.

Not all mutualists live in close proximity. One of the most important mutualisms in environmental science involves free-living organisms that may encounter each other only once in their lifetimes. This is *pollination* (**Figure 6.9**), an interaction of key significance to agriculture and our food supply (• pp. 271–272). Bees, birds, bats, and other creatures transfer pollen (male sex cells) from one flower to another, fertilizing eggs that become embryos within seeds. Most pollinating animals visit flowers for their nectar, a reward the plant uses to entice them. The pollinators receive food, and the plants are pollinated and reproduce. Various types of bees pollinate 73% of our crops, one expert has estimated—from soybeans to potatoes to tomatoes to beans to cabbage to oranges.

FIGURE 6.9 In mutualism, organisms of different species benefit one another. An important mutualistic interaction for environmental science is pollination. This hummingbird visits flowers to gather nectar and in the process transfers pollen between flowers, helping the plant reproduce. Pollination is of key importance to agriculture, ensuring the reproduction of many crop plants.

### Some interactions have no effect on some participants

Two other types of species interaction get far less attention. **Amensalism** is a relationship in which one organism is harmed and the other is unaffected. In **commensalism**, one species benefits and the other is unaffected. Amensalism has been difficult to confirm because it is hard to prove that the organism doing the harm is not in fact besting a competitor for a resource. Some experts suggest that *allelopathy*, a phenomenon whereby certain plants release poisonous chemicals that harm others nearby, is an example of amensalism. However, allelopathy can also be viewed as one plant investing in chemicals to outcompete others for space.

An example of commensalism occurs when the conditions created by one plant happen to make it easier for another plant to establish and grow. For instance, palo verde trees in the Sonoran Desert create shade and leaf litter that allow the soil beneath them to hold moisture longer, creating an area that is cooler and moister than the surrounding sunbaked ground. Young plants find it easier to germinate and grow in these conditions, so seedling cacti and other desert plants generally grow up directly beneath "nurse" trees such as palo verde. This phenomenon, called *facilitation*, influences the structure and composition of communities and how they change through time.

## Ecological Communities

As we saw in Chapter 5 (Figure 5.10, • p. 124), a **community** is an assemblage of species living in the same place at the same time. The members of a community interact with one another in the ways described above, and the direct interactions among species often have indirect effects that ripple outward to affect other community members. The strength of interactions also varies, and together species interactions determine the structure, function, and species composition of communities. *Community ecologists* are interested in which species coexist, how they relate to one another, how communities change through time, and why these patterns exist.

### Energy passes among trophic levels

The interactions among members of a community are many and varied, but some of the most important involve who eats whom. As organisms feed on one another, matter and energy move through the community, from one **trophic level**, or rank in the feeding hierarchy, to another (**Figure 6.10**).

**Producers** Producers, or autotrophs ("self-feeders"), comprise the first trophic level, as we saw in Chapter 4 (• pp. 101–102). Terrestrial green plants, cyanobacteria, and algae capture solar energy and use photosynthesis to produce sugars. The chemosynthetic bacteria of hot springs and deep-sea hydrothermal vents use geothermal energy in a similar way to produce food.

**Consumers** Organisms that consume producers are known as *primary consumers* and comprise the second trophic level. Grazing animals, such as deer and grasshoppers, are primary consumers. The third trophic level consists of *secondary consumers*, which prey on primary consumers. Wolves that prey on deer are considered secondary consumers, as are rodents and birds that prey on grasshoppers. Predators that feed at even higher trophic levels are known as *tertiary consumers*. Examples of tertiary consumers include hawks and owls that eat rodents that have eaten grasshoppers. Note that most primary consumers are *herbivores* because they consume plants, whereas secondary and tertiary consumers are *carnivores* because they eat animals. Animals that eat both plant and animal food are referred to as *omnivores*.

**Detritivores and decomposers** *Detritivores* and *decomposers* consume nonliving organic matter. Detritivores, such as millipedes and soil insects, scavenge the waste products or the dead bodies of other community members. Decomposers, such as fungi and bacteria, break down leaf litter and other nonliving matter further into simpler constituents that can then be taken up and used by plants. These organisms enhance the topmost soil layers (• pp. 239–240) and play essential roles as the community's recyclers, making nutrients from organic matter available for reuse by living members of the community.

In Great Lakes communities, phytoplankton are the main producers, floating freely and photosynthesizing with sunlight that penetrates the upper layer of the water. Zooplankton are primary consumers, feeding on the phytoplankton. Phytoplankton-eating fish are primary consumers, and zooplankton-eating fish are secondary consumers. At higher trophic levels are tertiary consumers such as larger fish and birds that feed on zooplankton-eating fish. Zebra mussels and quagga mussels, by eating both phytoplankton and zooplankton, function on multiple trophic levels. When any of these organisms dies and sinks to the bottom, detritivores scavenge its tissues, and microbial decomposers recycle its nutrients.

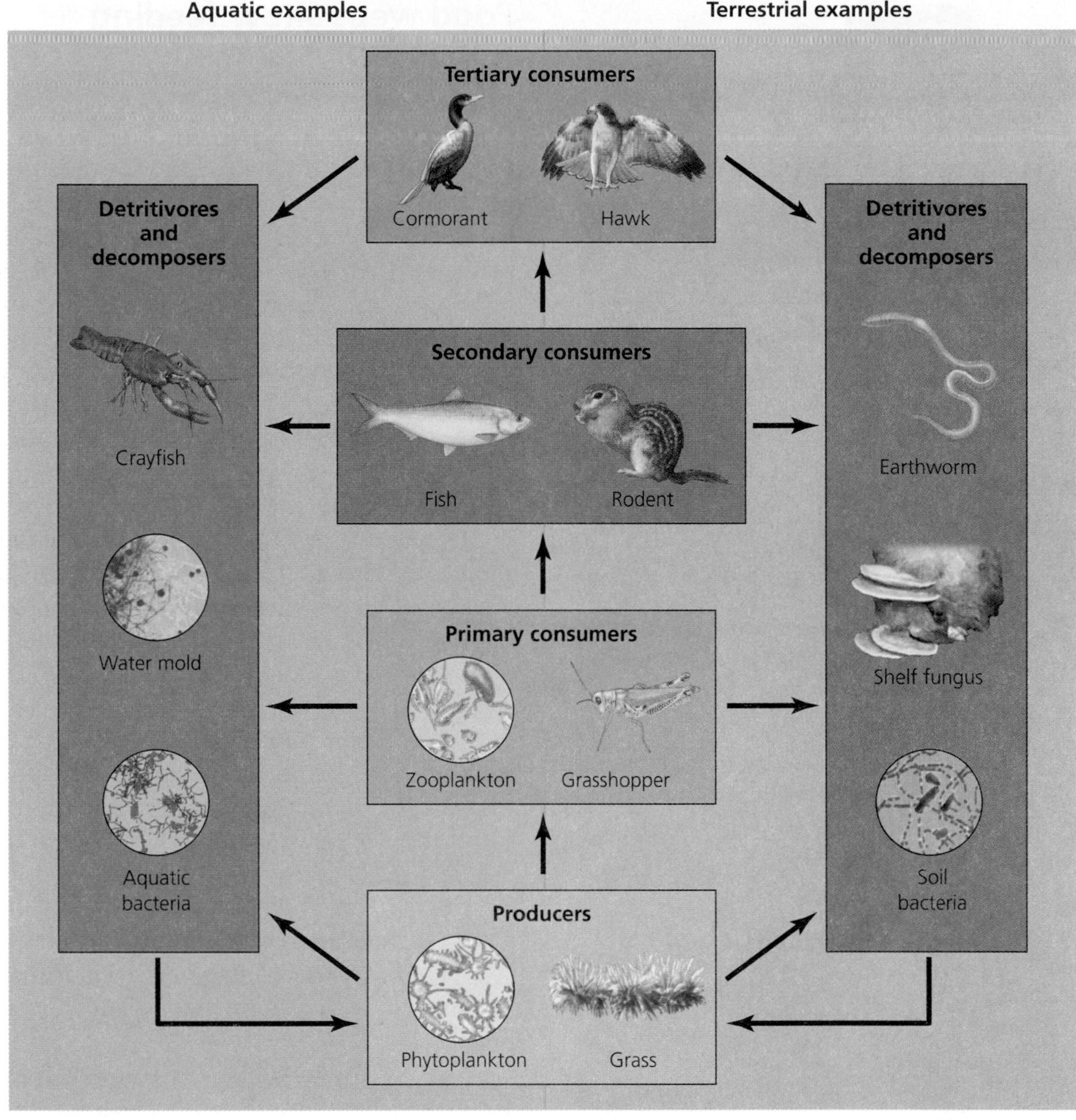


**FIGURE 6.10** Ecologists organize species hierarchically by their feeding rank, or trophic level. The diagram shows aquatic (left) and terrestrial (right) examples at each level. Arrows indicate the direction of energy flow. Producers produce food by photosynthesis, primary consumers (herbivores) feed on producers, secondary consumers eat primary consumers, and tertiary consumers eat secondary consumers. Communities can have more or fewer trophic levels than in this example. Detritivores and decomposers feed on nonliving organic matter and the remains of dead organisms from all trophic levels, and they "close the loop" by returning nutrients to the soil or the water column for use by producers.

## Energy, biomass, and numbers decrease at higher trophic levels

At each trophic level, most of the energy that organisms use is lost as waste heat through respiration. Only a small amount of the energy is transferred to the next trophic level through predation, herbivory, or parasitism. The first trophic level (producers) contains a large amount of energy, but the second (primary consumers) contains less energy—only that amount gained from consuming producers. The third trophic level (secondary consumers) contains still less energy, and higher trophic levels (tertiary consumers) contain the least. A general rule of thumb is that each trophic level contains just 10% of the energy of the trophic level below it, although the actual proportion can vary greatly.

This pattern can be visualized as a pyramid (**Figure 6.11**). The pattern also tends to hold for the numbers of

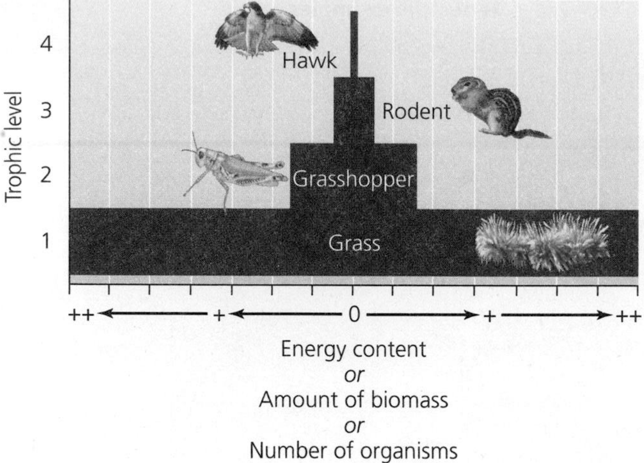


FIGURE 6.11 A *pyramid of energy* (or numbers, or biomass) illustrates a rough rule of thumb for the way ecological communities are structured. Organisms at lower trophic levels generally exist in far greater numbers, with greater energy content and greater biomass, than organisms at higher trophic levels. The reason is that when one organism consumes another, most energy gets used up in respiration rather than in building new tissue. The example shown here is generalized for illustrative purposes; the actual shape of any given pyramid may vary greatly.

organisms at each trophic level; generally, fewer organisms exist at higher trophic levels than at lower ones. A grasshopper eats many plants in its lifetime, a rodent eats many grasshoppers, and a hawk eats many rodents. Thus, for every hawk in a community there must be many rodents, still more grasshoppers, and an immense number of plants. Because the difference in numbers of organisms among trophic levels tends to be large, the same pyramid-like relationship also often holds true for biomass.

This pyramid-like pattern illustrates why eating at lower trophic levels—being a vegetarian rather than a meat-eater, for instance—decreases a person's ecological footprint. Each amount of meat or other animal product we eat requires the input of a considerably greater amount of plant material (see Figure 10.19, • p. 283). Thus, when we eat meat, we are taking in the end product of far more energy consumption, per calorie of energy that we gain, than when we eat plant products.

### Weighing THE Issues | The Footprints of Our Diets

What proportion of your diet would you estimate consists of meat, milk, eggs, or other animal products? Would you be willing to decrease this proportion in order to reduce your ecological footprint? Name some other ways in which you could reduce your footprint through your food choices.

## Food webs show feeding relationships and energy flow

As energy is transferred from species on lower trophic levels to species on higher trophic levels, it is said to pass up a *food chain.* Plant, grasshopper, rodent, and hawk make up a food chain, a linear series of feeding relationships. Thinking in terms of food chains is conceptually useful, but in reality ecological systems are far more complex than simple linear chains. A more accurate representation of the feeding relationships in a community is a **food web**, a visual map of feeding relationships and energy flow, showing the many paths by which energy passes among organisms as they consume one another.

**Figure 6.12** shows a food web from a temperate deciduous forest of eastern North America. Like virtually all diagrams of ecological systems, it is greatly simplified, leaving out the vast majority of species and interactions that occur. Note, however, that even within this simplified diagram we can pick out a number of different food chains involving different sets of species.

A Great Lakes food web would involve the phytoplankton and cyanobacteria that photosynthesize near the water's surface, the zooplankton that eat them, the fish that eat all these, the larger fish that eat the smaller fish, and the lampreys that parasitize the fish. It would include a number of native mussels and clams and, since 1988, the zebra mussels and quagga mussels that are crowding them out. It would include diving ducks that used to feed on native bivalves and now prey on the mussels.

This food web would also show that an array of bottom-dwelling invertebrates feed from the refuse of the exotic mussels. The mussels' waste products promote bacterial growth and disease pathogens that harm native bivalves, but they also provide nutrients that nourish crayfish and many smaller *benthic* (bottom-dwelling) invertebrate animals. Finally, the food web would include underwater plants and macroscopic algae, whose growth is promoted by the non-native mussels. The mussels clarify the water by filtering out phytoplankton, and sunlight penetrates more deeply into the water column, spurring photosynthesis. Thus, zebra and quagga mussels alter this food web by shifting productivity from open-water regions to benthic and *littoral* (nearshore) regions. In so doing, the mussels affect fish indirectly, helping benthic and littoral fishes and making life harder for open-water fishes (see "The Science behind the Story," • pp. 152–153). You may wish to jump ahead and examine Figure 6.16a (• p. 159) for an illustration of some of these effects.

**FIGURE 6.12** Food webs are conceptual representations of feeding relationships in a community. This food web pertains to eastern North America's temperate deciduous forest and includes organisms on several trophic levels. In a food web diagram, arrows are drawn from one organism to another to indicate the direction of energy flow as a result of predation, parasitism, or herbivory. For example, an arrow leads from the grass to the cottontail rabbit to indicate that cottontails consume grasses. The arrow from the cottontail to the tick indicates that parasitic ticks derive nourishment from cottontails. Communities include so many species and are complex enough, however, that most food web diagrams are bound to be gross simplifications.

## Some organisms play bigger roles in communities than others

"Some animals are more equal than others," George Orwell wrote in his 1945 book *Animal Farm.* Although Orwell was making wry sociopolitical commentary, his remark hints at a truth in ecology. In communities, ecologists have found, some species exert greater influence than do others. A species that has strong or wide-reaching impact far out of proportion to its abundance is often called a **keystone species.** A keystone is the wedge-shaped stone at the top of an arch that is vital for holding the structure together; remove the keystone, and the arch will collapse (**Figure 6.13a**). In an ecological community, removal of a keystone species will have substantial ripple effects and will alter a large portion of the food web.

Often, large-bodied secondary or tertiary consumers near the tops of food chains are considered keystone

THE SCIENCE BEHIND THE STORY

# Inferring Zebra Mussels' Impacts on Fish Communities

*Dr. David Strayer, Institute of Ecosystem Studies, sampling aquatic invertebrates*

Food webs are complicated systems, and disentangling them to infer the effects of any one species is fraught with difficulty. When zebra mussels appeared in the Great Lakes, people feared for sport fisheries and estimated that fish population declines could cost billions of dollars. The mussels would deplete the phytoplankton and zooplankton that fish depended on, people reasoned, and many fewer fish would survive. Yet even after 15 years, there was no solid evidence of widespread harm to fish populations.

So, aquatic biologist David Strayer of the Institute of Ecosystem Studies in Millbrook, New York, joined Kathryn Hattala and Andrew Kahnle of New York State's Department of Environmental Conservation (DEC). They mined datasets on fish populations in the Hudson River, which zebra mussels had invaded in 1991.

Strayer and others had already been studying effects of zebra mussels on aspects of the community for years. Their data showed that since the species' introduction to the Hudson:

- Biomass of phytoplankton had fallen by 80%.
- Biomass of small zooplankton had fallen by 76%.
- Biomass of large zooplankton had fallen by 52%.

Zebra mussels increased filter-feeding in the community 30-fold, thereby depleting the phytoplankton and small zooplankton and leaving all sizes of zooplankton with less phytoplankton to eat. Overall, the zooplankton and invertebrate animals of the open water that are eaten by open-water fish declined by 70%.

However, Strayer's work had also found that *benthic,* or bottom-dwelling, invertebrates in shallow water (especially in the nearshore, or *littoral,* zone) had increased by 10% and likely much more, because the mussels' shells provide habitat structure and their feces provide nutrients.

These contrasting trends in the benthic shallows and the open deep water led Strayer's team to hypothesize that zebra mussels would harm open-water fish that ate plankton but would help littoral-feeding fish. They predicted that after zebra mussels were introduced, larvae and juveniles of six common open-water fish species would decline in number,

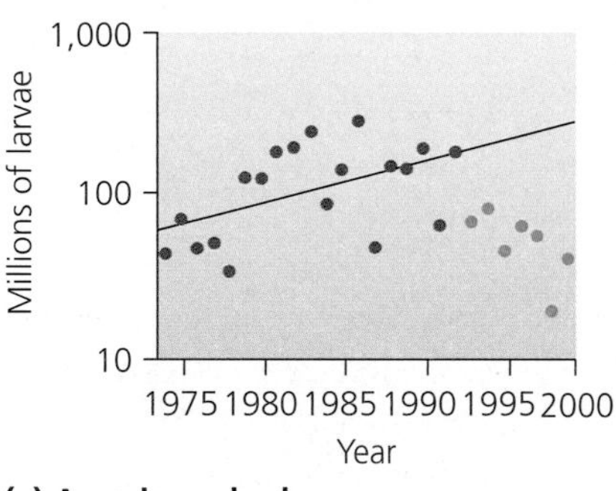


**(a) American shad**

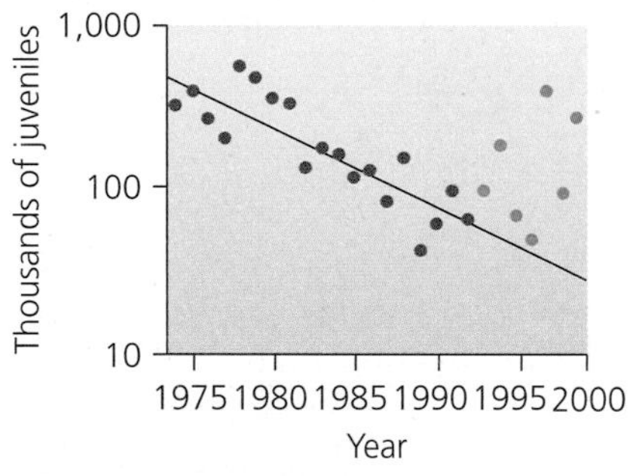


**(b) Tessellated darter**

Larvae of American shad **(a)**, an open-water fish, had been increasing in abundance before zebra mussels were introduced (red points and trend line). After zebra mussels were introduced, shad larvae decreased in abundance (orange points). Juveniles of the tessellated darter **(b)**, a littoral zone fish, had been decreasing in abundance before zebra mussels were introduced (red points and trend line). After zebra mussels were introduced, they increased in abundance (orange points).
*Source:* Strayer, D., et al. 2004. Effects of an invasive bivalve (*Dreissena polymorpha*) on fish in the Hudson River estuary. *Canadian Journal of Fisheries and Aquatic Sciences* 61: 924–941.

species. Top predators control populations of herbivores, which, in the absence of the predators, could multiply and greatly modify the plant community (**Figure 6.13b**). Thus, predators at high trophic levels can indirectly promote populations of organisms at low trophic levels by keeping species at intermediate trophic levels in check, a phenomenon ecologists refer to as a **trophic cascade**. In the United States, government bounties promoted the hunting of wolves and mountain lions, which were largely exterminated by the middle of the 20th century. In the absence of these predators, unnaturally dense deer populations have overgrazed forest-floor vegetation and eliminated whole cohorts of tree seedlings, causing major changes in forest structure.

The removal of top predators in the United States was an uncontrolled large-scale experiment with unintended

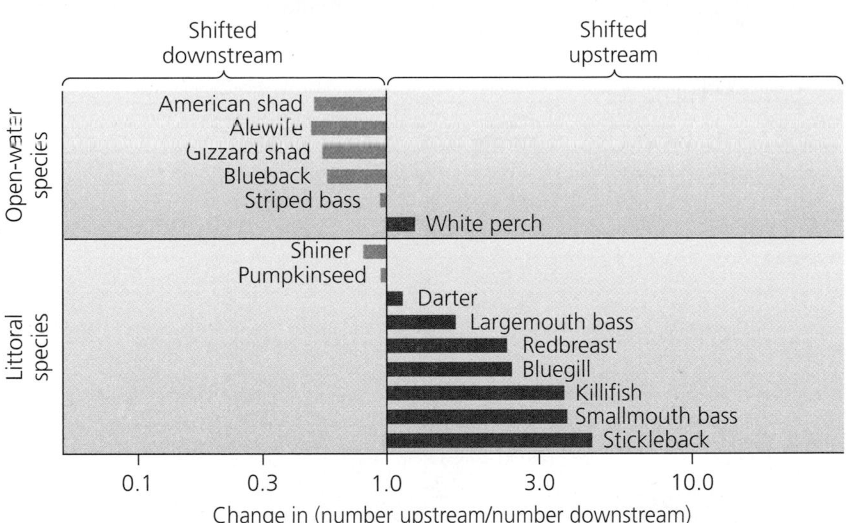


**Young of open-water fish, such as American shad, blueback herring, and alewife, tended to shift downstream toward areas with fewer zebra mussels in the years following zebra mussel arrival. Young of littoral fish, such as killifish, bluegill, and largemouth bass, tended to shift upstream toward areas with more zebra mussels.** *Source:* Strayer, D., et al. 2004. Effects of an invasive bivalve (*Dreissena polymorpha*) on fish in the Hudson River estuary. *Canadian Journal of Fisheries and Aquatic Sciences* 61: 924–941.

decline in growth rate, and shift downriver toward saltier water, where mussels are absent. Conversely, they predicted that larvae and juveniles of 10 littoral fish species would increase in number, increase in growth rate, and shift upriver to regions of greatest zebra mussel density.

To test their predictions, the researchers analyzed data from three types of fish surveys carried out over 26 years spanning periods before and after the zebra mussel's arrival. One data set came from surveys conducted from 1985 to 1999 by the DEC. The other two came from surveys conducted from 1974 to 1999 by biologists hired by electric utilities, which in New York are required to monitor fish populations in return for using the Hudson's water for cooling at their power plants.

The researchers compared values for abundance, growth, and distribution for young fish before 1991 with those after 1991. The results supported their predictions. Larvae and juveniles of open-water fish, such as American shad, blueback herring, and alewife, tended to decline in abundance in the years after zebra mussels were introduced; see the first figure, part (a). Those of littoral fish, such as tessellated darter, bluegill, and largemouth bass, tended to increase; see the first figure, part (b).

Growth rates showed the same trend: Open-water fish grew more slowly after zebra mussels were introduced, whereas littoral fish grew more quickly. In terms of distribution in the 248-km (154-mi) stretch of river studied, open-water fish shifted downstream toward areas with fewer zebra mussels, whereas littoral fish shifted upstream toward areas with more zebra mussels (second figure). The results were published in 2004 in the *Canadian Journal of Fisheries and Aquatic Sciences.*

Overall, the data supported the hypothesis that the fish community would respond to changes in its food resources caused by zebra mussels. The results are correlative, and correlation does not prove causation. However, previous attempts to address these questions experimentally were so limited in time and scale that they could not reflect true effects in natural systems.

Research such as this helps illuminate the often obscure connections and impacts that particular species interactions have on communities as a whole. In this case, the research may also help fisheries managers predict changes in commercially and recreationally important fish populations. With this knowledge, biologists may be able to manage fisheries more effectively in the Hudson and other areas invaded by zebra mussels.

consequences. But ecologists have verified the keystone species concept in controlled experiments. Classic research by marine biologist Robert Paine established that the predatory starfish *Pisaster ochraceus* has great influence on the community composition of intertidal organisms (• pp. 451–452) on the Pacific coast of North America. When *Pisaster* is present in this community, species diversity is high, with several types of barnacles, mussels, and algae. When *Pisaster* is removed, the mussels it preys on become numerous and displace other species, suppressing species diversity.

More recent work off the U.S. Atlantic coast published in 2007 suggests that the reduction of shark populations by commercial fishing has allowed populations of certain skates and rays to increase, which has depressed numbers of bay scallops and other bivalves they eat.

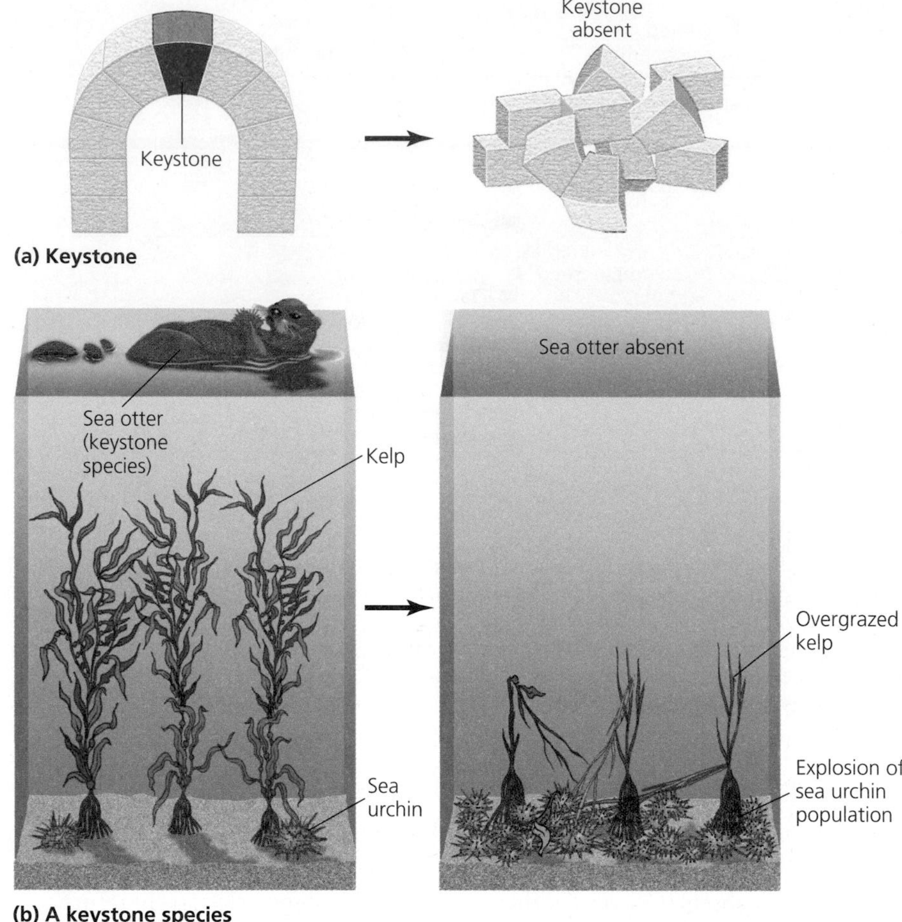


FIGURE 6.13 A keystone is the wedge-shaped stone at the top of an arch that holds its structure together (**a**). A keystone species, such as the sea otter, is one that exerts great influence on a community's composition and structure (**b**). Sea otters consume sea urchins that eat kelp in marine nearshore environments of the Pacific. When otters are present, they keep urchin numbers down, which allows lush underwater forests of kelp to grow and provide habitat for many other species. When otters are absent, urchin populations increase and the kelp is devoured, destroying habitat and depressing species diversity. See "The Science behind the Story," • pp. 156–157.

Animals at high trophic levels, such as wolves, starfish, sharks, and sea otters (see Figure 6.13), are most often seen as keystone species that can induce trophic cascades. However, other types of organisms also can exert strong community-wide effects. "Ecosystem engineers" physically modify the environment shared by community members. Beavers build dams and turn streams into ponds, flooding acres of dry land and turning them to swamp. Prairie dogs dig burrows that aerate the soil and serve as homes for other animals. Less conspicuous organisms and those toward the bottoms of food chains may have even greater impact. Remove the fungi that decompose dead matter, or the insects that control plant growth, or the phytoplankton that are the base of the marine food chain, and a community may change very rapidly indeed. Because there are usually more species at lower trophic levels, however, it is less likely that any one of them alone might have wide influence; if one species is removed, other species that remain may be able to perform many of its functions.

Identifying keystone species is no simple task, and there is no cut-and-dried definition of the term to help us. Community dynamics are complex, species interactions differ in their strength, and the strength of species interactions can vary through time and space. "The Science behind the Story" (• pp. 156–157) gives an idea of the surprises that are sometimes in store for ecologists studying these interactions.

## Communities respond to disturbance in different ways

The removal of a keystone species and the spread of an invasive species are just two of many types of disturbance that can modify the composition, structure, or function of an ecological community. Over time, a given community may experience natural disturbances ranging from gradual phenomena such as climate change to sudden events such as hurricanes, floods, or avalanches. Today, human impacts are major sources of disturbance for ecological communities worldwide.

Communities are dynamic systems and may respond to disturbance in several ways. A community that resists change and remains stable despite disturbance is said to show **resistance** to the disturbance. Alternatively, a community may show **resilience**, meaning that it changes in response to disturbance but later returns to its original state. Or, a community may be modified by disturbance permanently and may never return to its original state.

## Succession follows severe disturbance

If a disturbance is severe enough to eliminate all or most of the species in a community, the affected site will undergo a somewhat predictable series of changes that ecologists call **succession.** In the traditional view of this process, ecologists described two types of succession. **Primary succession** follows a disturbance so severe that no vegetation or soil life remains from the community that occupied the site. In primary succession, a biotic community is built essentially from scratch. In contrast, **secondary succession** begins when a disturbance dramatically alters an existing community but does not destroy all living things or all organic matter in the soil. In secondary succession, vestiges of the previous community remain, and these building blocks help shape the process.

Primary succession takes place after a bare expanse of rock, sand, or sediment becomes newly exposed to the atmosphere. This can occur when glaciers retreat, lakes dry up, or volcanic lava or ash spreads across the landscape (**Figure 6.14**). Species that arrive first and colonize the new substrate are referred to as **pioneer species.** Pioneer species are well adapted for colonization, having traits such as spores or seeds that can travel long distances. The pioneers best suited to colonizing bare rock are the mutualistic aggregates of fungi and algae known as *lichens.* Lichens succeed because their algal component provides food and energy via photosynthesis while the fungal component takes a firm hold on rock and captures the moisture that both organisms need to survive. As lichens grow, they secrete acids that break down the rock surface. The resulting waste material forms the beginnings of soil, and once soil begins to form, small plants, insects, and worms find the rocky outcrops more hospitable. As new organisms arrive, they provide more nutrients and habitat for future arrivals. As time passes, larger plants establish themselves, the amount of vegetation increases, and species diversity rises.

FIGURE 6.14 As this Alaskan glacier retreats, small plants (foreground) begin the process of primary succession.

Secondary succession begins when a fire, a hurricane, logging, or farming removes much of the biotic community. Consider an abandoned agricultural field in eastern North America (**Figure 6.15**). In the first few years after farming ends, the site will be colonized by pioneer species of grasses, herbs, and forbs that were already in the vicinity and that disperse effectively. As time passes, shrubs and fast-growing trees such as aspens rise from the field. Pine trees subsequently rise above the

FIGURE 6.15 Secondary succession occurs after a disturbance, such as fire, landslides, or farming, removes most vegetation from an area. Here is shown a typical series of changes in a plant community of eastern North America following the abandonment of a farmed field.

THE SCIENCE BEHIND THE STORY

Dr. James Estes, University of California at Santa Cruz, at sea in Alaska

## Otters, Urchins, Kelp, and a Whale of a Chain Reaction

Ecologists required years of careful study to comprehend the relationship among sea otters, sea urchins, and kelp forests diagrammed in Figure 6.13. And once they thought they understood it, more surprises were in store.

Sea otters live in coastal waters of the Pacific Ocean. These mammals float on their backs amid the waves, feasting on sea urchins that they pry from the bottom and bring to the surface. Once abundant, sea otters were hunted nearly to extinction for their fur. Protection by international treaty in 1911 allowed their numbers to regrow. Otters returned to high densities in some regions (such as off Alaska and the Aleutian Islands), but they failed to return in others.

Biologists noted that regions with otters hosted dense "forests" of kelp. Kelp is a brown alga (seaweed) that anchors to the seafloor and reaches toward sunlit surface waters, growing up to 60 m (200 ft) high. Kelp forests (• p. 450) provide complex physical structure in which diverse communities of fish and invertebrates find shelter and food.

In regions without sea otters, scientists found kelp forests absent. Urchins in such areas become so numerous that they may eat every last bit of kelp, creating empty seafloors called "urchin barrens" that are relatively devoid of life.

Ecologists observed that once areas with urchin barrens were recolonized by sea otters, urchin numbers declined and kelp forests returned. Through comparative research of this kind, ecologists determined that otters were largely responsible for the presence of the kelp forest community, simply by keeping urchin numbers in check through predation. This research—mostly by James Estes of the University of California at Santa Cruz and his colleagues—established sea otters as a prime example of a keystone species.

But the story did not end there. In the 1990s, otter populations dropped precipitously in Alaska and the Aleutians. No one knew why. So, Estes and his co-workers placed radio tags on Aleutian otters and studied them at sea. They hypothesized that fertility rates had dropped, but the radio-tracking observations showed that females were raising pups without problem. They then tested a second hypothesis: that otters were simply moving to other locations. But the radio tracking showed no unusual dispersal, and the population declines were taking place evenly over large areas. They were left with one viable hypothesis: increased mortality.

Biologists were finding no evidence of disease or starvation, however. Then one day in 1991, Estes's team witnessed something never seen before. They watched as a sea otter was killed and eaten by an orca, or killer whale. These striking black-and-white predators grow up to 10 m (32 ft) long, hunt in groups, and had always attacked larger prey. A sea otter to them is a mere snack. Yet over the following years, Estes's team saw nine more cases of orca predation on otters. Could killer whales be killing off the otters?

The researchers calculated the hours spent studying otters at sea before and after 1991 to assess the likelihood that they had simply missed seeing predation by orcas earlier. Their statistical analysis convinced them the rate of attacks had indeed increased. Then they compared a bay where otters were vulnerable to orcas with a lagoon where they were protected. Otter numbers in the lagoon remained stable over 4 years, whereas those in the bay dropped by 76%. Radio tracking showed no movement between these locations.

Finally, using data on otter birth rates, death rates, and population age structure (• pp. 127–128), they estimated that to account for the otter decline, 6,788 orca attacks per year would have had to occur in their study area. This produced an *expected* rate of observed attacks that matched their *actual* number of observed attacks well. These lines of evidence led the researchers to propose that predation by orcas was eliminating otters.

aspens and shrubs, forming a pine-dominated forest. This pine forest develops an understory of hardwood trees, because pine seedlings do not grow well under a canopy, whereas some hardwood seedlings do. Eventually the hardwoods outgrow the pines, creating a hardwood forest.

Processes of succession occur in many diverse ecological systems, from ponds to rocky intertidal areas

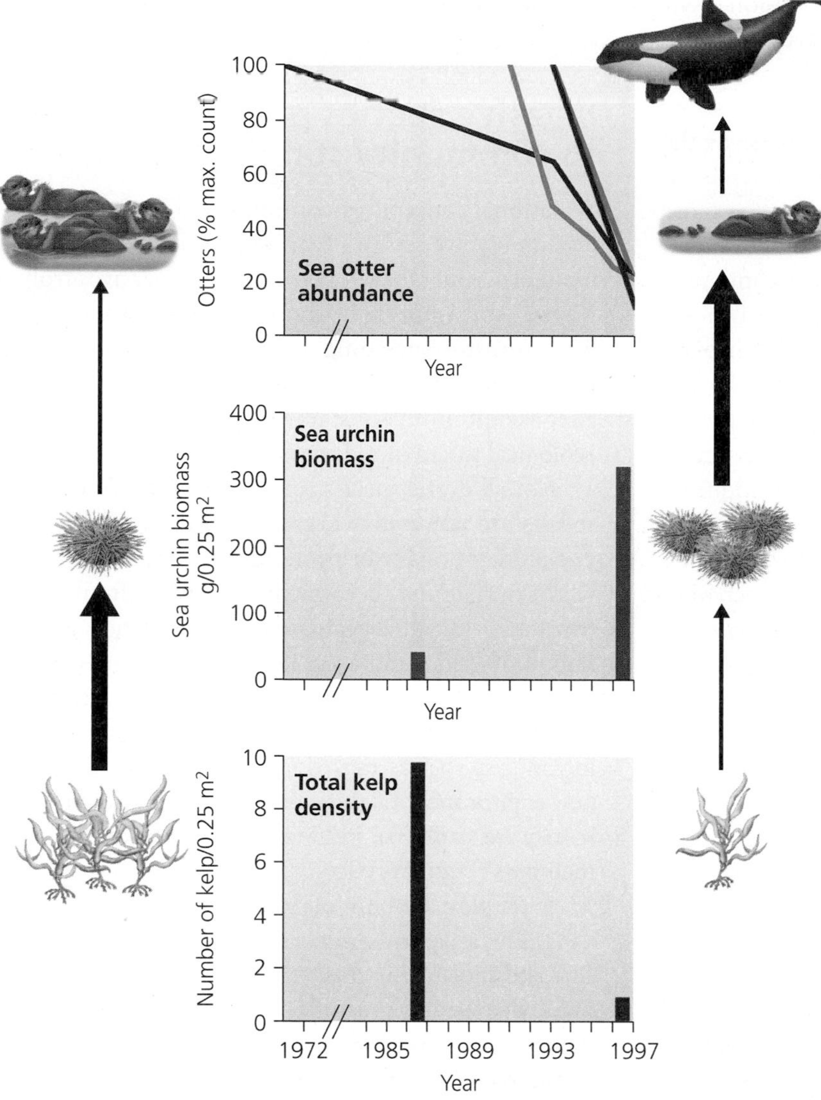


Before the 1990s (left side of figure), otters kept urchin numbers in check, allowing kelp forests to grow. By the late 1990s (right side of figure), orcas (killer whales), deprived of their usual food sources, were eating otters. This set off a chain reaction across trophic levels: Otters decreased, sea urchins increased, and kelp decreased. The lines in the top graph indicate trends in otter populations from four Aleutian islands. Width of arrows indicates strength of interaction. *Source:* Estes, J., et al. 1998. Killer whale predation on sea otters linking oceanic and nearshore ecosystems. *Science* 282: 473–476.

As otters declined in the Aleutians, urchins increased, and kelp density fell dramatically (see the figure). This *trophic cascade* supported the idea that otters were keystone species, but now it seemed that one keystone species was being controlled by another.

Why had orcas suddenly started eating otters? A possible answer came in 2003, after Alan Springer, Estes, and others determined that sea otters were only the latest in a series of population crashes in the northern Pacific. Harbor seals had declined by more than 90% in the late 1970s and 1980s. Fur seals had fallen by 60% since the 1970s. Sea lions crashed by 80% in the 1980s. And preceding these declines were the collapses of the great whales—gray whales, blue whales, humpback whales, and others.

Historically, orcas ate great whales, which they would kill in groups, like a wolf pack taking down an elk. But industrial whaling by ships from Japan, Russia, and other nations since the 1950s caused populations of great whales in the northern Pacific to plummet by 99% between 1965 and 1973. Once human hunting decimated the great whales, the orcas had to turn elsewhere. Springer, Estes, and their colleagues argued that they had shifted their diet to smaller, less-favored seals and sea lions. When their predation had depleted those populations, the chain reaction continued, and they turned to smaller prey still—sea otters.

Not all scientists accept that industrial whaling began this trophic cascade, and the debate among experts has been vigorous. The ongoing discussion highlights just how difficult it can be to study and interpret the species interactions that affect ecological communities.

to the carcasses of animals. For instance, some ponds may undergo succession as algae, microbes, plants, and zooplankton grow, reproduce, and die, gradually filling the water body with organic matter. The pond may acquire further organic matter and sediments from streams and surface runoff. Eventually, the water body fills in, becoming a bog (• p. 415) or even a terrestrial system.

In the traditional view of succession that we have described, the transitions between stages of succession eventually lead to a *climax community*, which remains in place with little modification until some disturbance restarts succession. Early ecologists felt that each region had its own characteristic climax community, determined by the region's climate.

Today, ecologists recognize that succession is far more variable and less predictable than originally thought. In some cases, conditions at one stage may promote progression to another stage through facilitation (• p. 148), but in other cases organisms at one stage may, through competition, inhibit a community's progression to another stage. The trajectory of succession can vary greatly according to chance factors, such as which particular species happen to gain an early foothold. In addition, climax communities are not predetermined solely by climate, but rather may vary with soil conditions and other factors from one time or place to another. Moreover, once a climax community is disturbed and succession is set in motion, there is no guarantee that the community will ever return to that climax state. Many communities disturbed by human impact have not returned to their former conditions.

## How cohesive are communities?

Ecologists who have studied how communities change in response to disturbance have conceptualized communities in different ways. Early in the 20th century, botanist Frederick Clements promoted the view that communities are cohesive entities whose members remain associated over time and space. Communities, he argued, are discrete units with integrated parts, much like organisms. Clements's view implied that the many varied members of a community share similar limiting factors and evolutionary histories.

Henry Gleason disagreed. Gleason, also a botanist, maintained that each species responds independently to its own limiting factors and that species can join or leave communities without greatly altering the composition of the communities. Communities, Gleason argued, are not cohesive units, but temporary associations of individual species that can reassemble into different combinations.

Today ecologists side largely with Gleason, although most see validity in aspects of both men's ideas. Indeed, many ecologists still find it useful to refer to communities by names that highlight certain key plants (such as "oak-hickory forest," "tallgrass prairie," and "pine-bluestem community"). Ecologists find labeling communities as though they were cohesive units to be a pragmatic tool, even though they know the associations could change radically decades or centuries hence.

## Invasive species pose new threats to community stability

Traditional concepts of community cohesion and of succession involve sets of organisms understood to be native to an area. But what if a new organism arrives from elsewhere? And what if this non-native, or *exotic*, organism turns *invasive*, spreading widely and becoming dominant in a community? Such **invasive species** can potentially alter a community substantially and are one of the central ecological forces in today's world.

Not all exotic species turn invasive, but most invasive species are non-native organisms that people have *introduced*, intentionally or by accident, from elsewhere in the world. Species may become invasive when limiting factors (• pp. 129–130) that regulate their population growth are removed. Plants and animals brought to a new area may leave behind the predators, parasites, and competitors that had evolved to exploit them in their native land. The introduced species may do well if few organisms in the new environment fill these roles. As the species proliferates, it may exert diverse influences on its fellow community members (**Figure 6.16**).

Examples abound of introduced species that have turned invasive and have exerted major ecological effects. The chestnut blight, an Asian fungus, killed nearly every mature American chestnut *(Castanea dentata)*, the dominant tree species of many eastern North America forests, in the quarter-century preceding 1930. Asian trees had evolved defenses against the fungus over long millennia of coevolution, but the American chestnut had not. A different fungus caused Dutch elm disease, destroying most of the American elms *(Ulmus americana)* that once gracefully lined the streets of many U.S. cities. Fish introduced into streams for sport compete with and exclude native fish. Hundreds of island-dwelling animals and plants worldwide have been driven extinct by the goats, pigs, and rats introduced by human colonists. We will examine more examples in our discussion of biodiversity in Chapter 11 (• pp. 305–307).

Ecologists generally view the impacts of invasive species—and introduced species in general—as overwhelmingly negative. Yet many people enjoy the beauty of introduced ornamental plants in their gardens. And some introduced species that have turned invasive provide benefits to our economy, such as the European honeybee, which pollinates many of our crops (• p. 272). Whatever view one takes, the impacts of invasive species on native species and ecological communities are significant, and

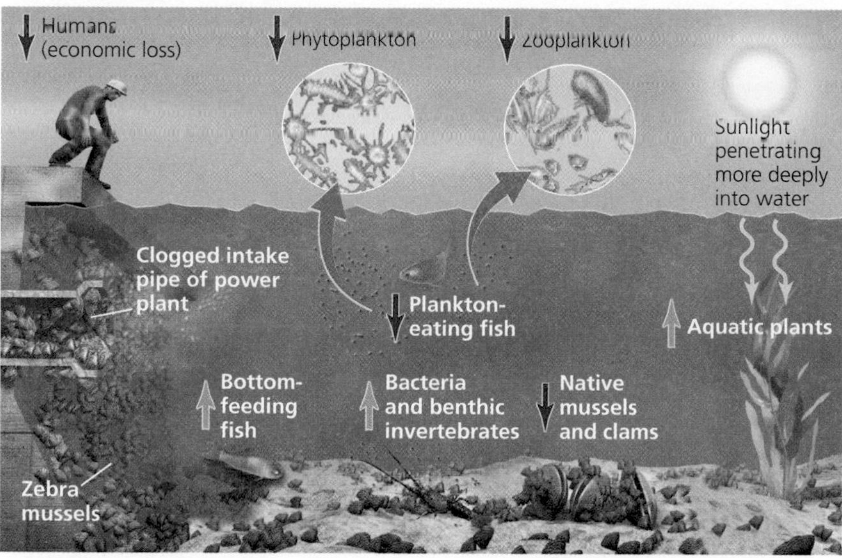


(a) Impacts of zebra mussels on members of a Great Lakes nearshore community

FIGURE 6.16 The zebra mussel is a prime example of a biological invader that has modified an ecological community. By filtering phytoplankton and small zooplankton from open water, it exerts impacts on other species, both negative (red downward arrows) and positive (green upward arrows) (**a**). The map in (**b**) shows the range of the zebra mussel in North America as of 2005. In less than two decades it spread from the Great Lakes east to Vermont and Connecticut; west to Nebraska and Kansas; and south to Louisiana and Mississippi (shaded states and red dots), and it has been transported by people to other states as well (blue dots). The map in (**c**) shows data from 2007 on the known range of the quagga mussel, which is rapidly following the zebra mussel's spread. *Source* (b, c): U.S. Geological Survey.

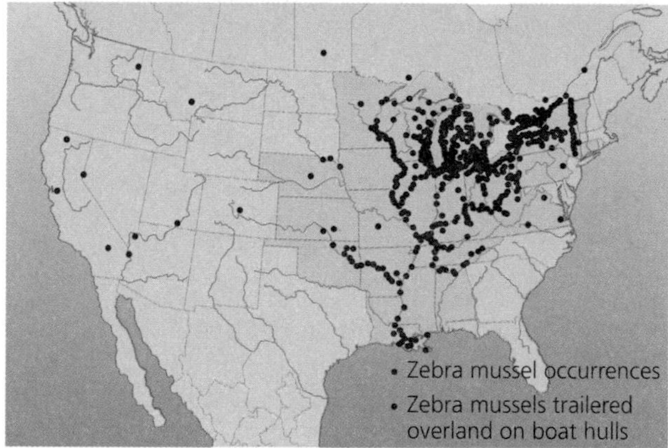


(b) Occurrence of zebra mussels in North America, 2005

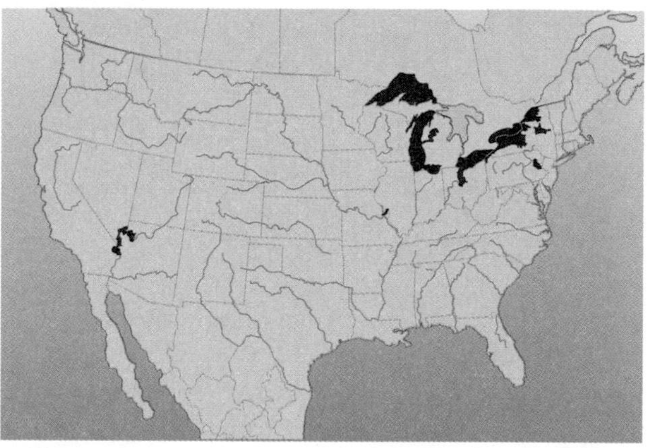

(c) Occurrence of quagga mussels in North America, 2007

they are growing with the increasing mobility of people and the globalization of our society.

Our global trade helped spread zebra and quagga mussels, which were unintentionally transported in the ballast water of cargo ships. To maintain stability at sea, ships take water into their hulls as they begin their voyage and discharge that water at their destination. Decades of unregulated exchange of ballast water have ferried hundreds of species across the oceans.

In North America, zebra and quagga mussels—and the media attention they continue to generate—helped put invasive species on the map as a major environmental and economic problem. Scientific research proliferated, and many ecologists came to view invasive species as the second-greatest threat to species and natural systems, behind only habitat destruction. In 1990 the U.S. Congress passed the Nonindigenous Aquatic Nuisance Prevention and Control Act, which became the National Invasive Species Act of 1996. Among other things, this law directed the Coast Guard to ensure that ships dump their freshwater ballast at sea and exchange it with saltwater before entering the Great Lakes.

Since then, funding has become widely available for the control and eradication of invasive species. Managers have tried a wide variety of techniques to control the spread of zebra mussels—removing them manually, applying toxic chemicals, drying them out, depriving them of oxygen, introducing predators and diseases, and stressing them with heat, sound, electricity, carbon dioxide, and ultraviolet light. However, most of these are localized and short-term fixes that are not capable of making a dent in the huge populations at large in the environment. In case after case, managers are finding that controlling and eradicating invasive species are so difficult and expensive that preventive measures (such as ballast water regulations) represent a much better investment.

## CAUSES AND CONSEQUENCES

As globalization proceeds, as we become more mobile, and as our societies become more interconnected, people are transporting plants and animals among regions, nations, and continents. Of the many non-native species introduced, some become invasive and spread, displacing native flora and fauna. **Invasive species** are exerting ever-greater impacts on our natural communities and societies. Thankfully, we can take steps to prevent introductions, control the spread of invasives, and even eradicate certain exotic species.

Write in two causes of invasive species in the spaces provided. Then write in two consequences (impacts on the environment, human health, or quality of life) that result from invasive species. Finally, offer two solutions to this issue and its consequences. One cause, one consequence, and one solution have been filled in for you, providing examples.

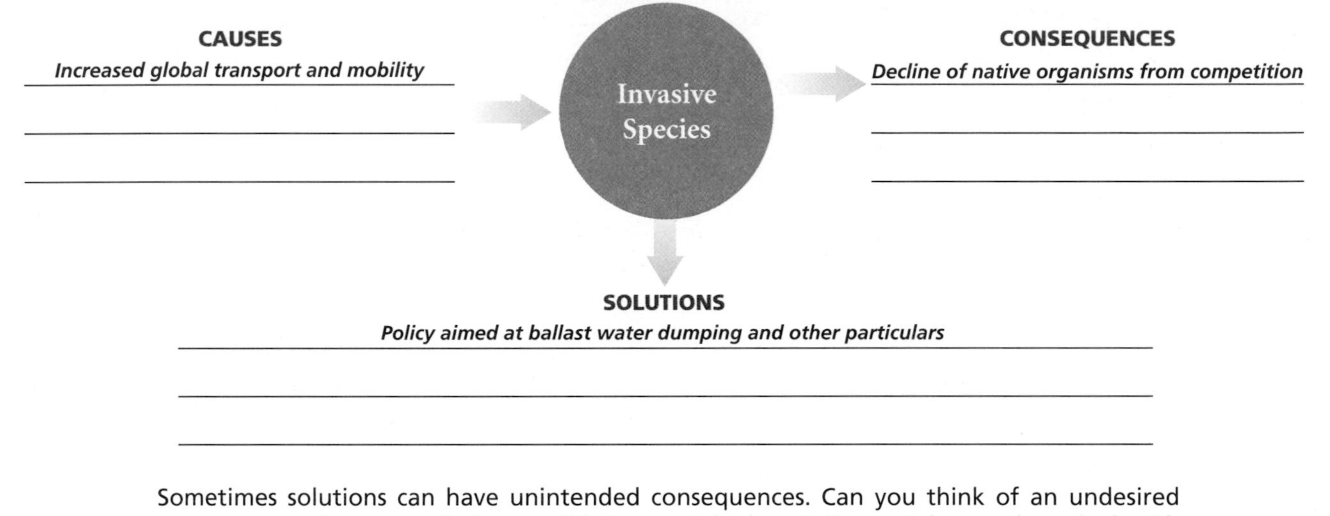


Sometimes solutions can have unintended consequences. Can you think of an undesired consequence that one solution to this issue might have? How might we then deal with *that* consequence?

### Weighing the Issues | Are Invasive Species All Bad?

Some ethicists have questioned the notion that all invasive species should automatically be considered bad. If we introduce a non-native species to a community and it greatly modifies the community, do you think that is a bad thing? What if it drives another species extinct? What if the invasive species arrived on its own, rather than through human intervention? What ethical standard(s) (• p. 29) would you apply to determine whether we should battle or accept an invasive species?

## We can try to restore an altered community to its former condition

Invasive species are adding to the tremendous transformations that humans have already forced on natural landscapes and communities through habitat alteration, deforestation, hunting of keystone species, pollution, and other activities. With so much of Earth's landscape altered by human impact, it is impossible to find areas that are truly pristine. This realization has given rise to the conservation effort known as **ecological restoration.** The practice of ecological restoration is informed by the science of **restoration ecology**. Restoration ecologists research the historical conditions of ecological communities as they existed before our industrialized civilization altered them. They then try to devise ways to restore some of these areas to an earlier condition, often to a natural "presettlement" condition.

For instance, in the United States nearly every last scrap of tallgrass prairie that once covered the eastern Great Plains and parts of the Midwest was converted to agriculture in the 19th century. Now a number of efforts are underway to restore small patches of prairie by planting native prairie vegetation, weeding out invaders and competitors, and introducing controlled fire to mimic the fires that historically maintained this community.

FIGURE 6.17 Dr. Robert Betz pioneered the practice of prairie restoration at the Fermi National Accelerator Laboratory (Fermilab) in Illinois. Dr. Betz died in 2007, but work such as his has inspired thousands of people to help restore disturbed ecological communities to their natural conditions.

The region outside Chicago, Illinois, boasts several of the largest prairie restoration projects so far, including a 184-ha (455-acre) area inside the massive ring of the Fermilab nuclear accelerator in Batavia (**Figure 6.17**).

The world's largest restoration project is the ongoing effort to restore the Florida Everglades, a 7,500-$km^2$ (4,700-$mi^2$) ecosystem of marshes and seasonally flooded grasslands. In recent decades, flood control practices and water diversions for irrigation and development have dried out this naturally wet region. Populations of wading birds have dropped by 90–95%, and economically important fisheries have suffered greatly as a result. The 30-year, $7.8-billion restoration project intends to restore water by undoing damming and diversions of 1,600 km (1,000 mi) of canals, 1,150 km (720 mi) of levees, and 200 water control structures. Because the Everglades provides drinking water for millions of Florida citizens, as well as considerable tourism revenue, restoring its ecosystem services (• pp. 38–40, 50–51) should prove economically beneficial as well as ecologically valuable.

One of the most intriguing new restoration efforts is the drive to restore the Mesopotamian marshes between the Tigris and Euphrates Rivers in Iraq, formerly one of the world's greatest wetlands. The government of Saddam Hussein diverted water from the marshes in an effort to debilitate the minority peoples who had lived here for thousands of years. Following the U.S. occupation of Iraq, ecologists from many nations joined together to try to restore the marshes and give their human residents a place to live again in their traditional lifestyle. Whether the funds, resources, and access are granted to allow this project to succeed remains to be seen.

As our population grows and development spreads, ecological restoration is becoming an increasingly vital conservation strategy. However, restoration is difficult, time-consuming, and expensive, and it does not always succeed in creating systems that function as effectively as natural systems. It is therefore best, whenever possible, to protect natural systems from degradation in the first place, so that restoration does not become necessary.

### Weighing THE Issues | Restoring "Natural" Communities

Practitioners of ecological restoration in North America aim to restore communities to their natural state. But what is meant by "natural"? Does it mean the state of the community before industrialization? Before Europeans came to the New World? Before any people laid eyes on the community? Let's say Native Americans altered a forest community 8,000 years ago by burning the underbrush regularly to improve hunting, and continued doing so until Europeans arrived 400 years ago and cut down the forest for farming. Today the area's inhabitants want to restore the land to its "natural" forested state. Should restorationists try to recreate the forest of the Native Americans or the forest that existed before Native Americans arrived? What values do you think underlie the desire for restoration?

# Earth's Biomes

Across the world, each portion of each continent has different sets of species, leading to endless variety in community composition. However, communities in far-flung places often share strong similarities in their structure and function. This allows us to classify communities into broad types. A **biome** is a major regional complex of similar communities—a large ecological unit recognized primarily by its dominant plant type and vegetation structure. The world contains a number of biomes, each covering large contiguous geographic areas (**Figure 6.18**).

## Biomes are groupings of communities that cover large geographic areas

Which biome covers any particular portion of the planet depends on a variety of abiotic factors, including temperature, precipitation, atmospheric circulation,

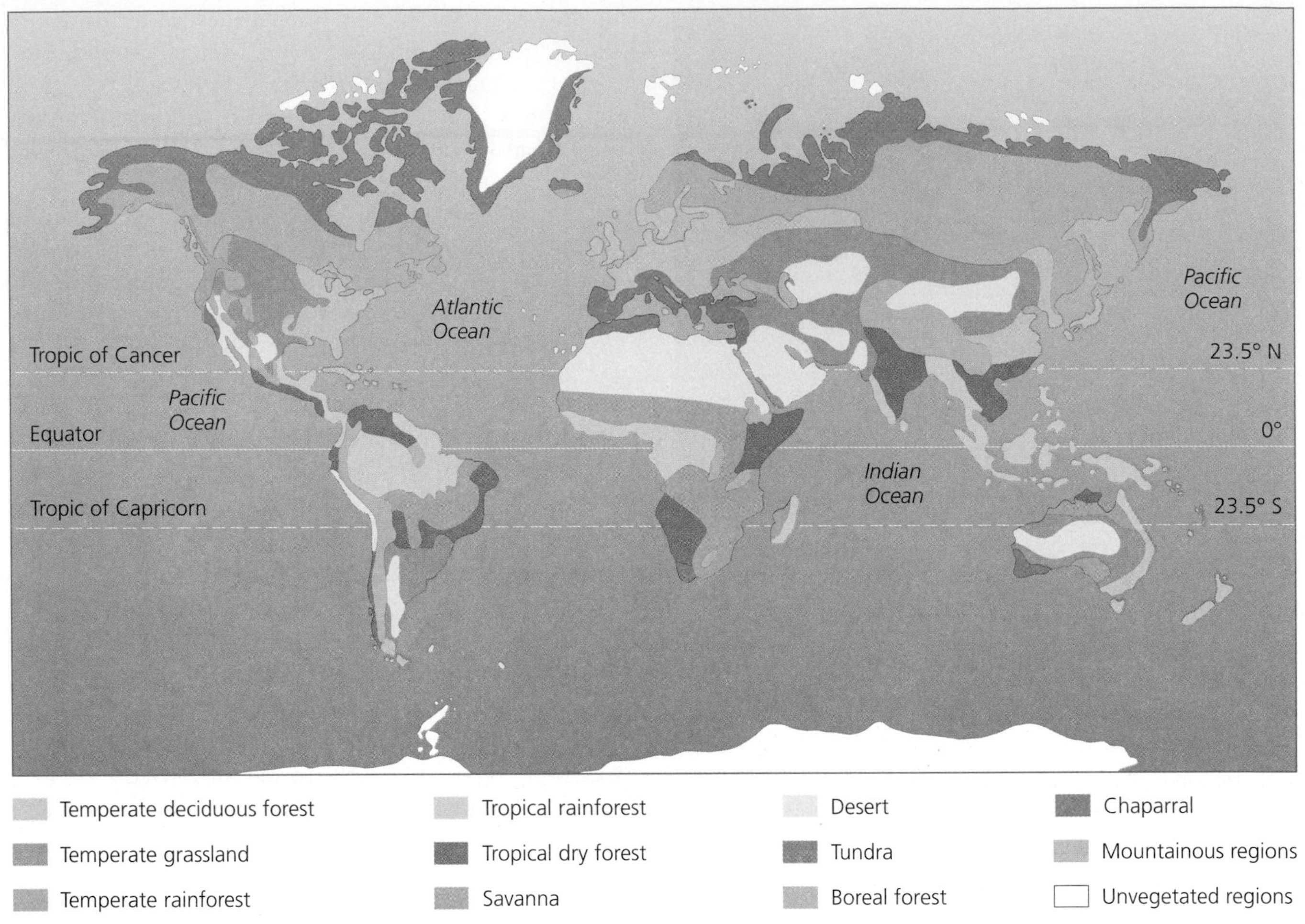


**FIGURE 6.18** Biomes are distributed around the world according to temperature, precipitation, atmospheric and oceanic circulation patterns, and other factors.

and soil characteristics. Among these factors, temperature and precipitation exert the greatest influence (**Figure 6.19**). Because biome type is largely a function of climate, and because average monthly temperature and precipitation are among the best indicators of an area's climate, scientists often use climate diagrams, or **climatographs**, to depict such information. Global climate patterns cause biomes to occur in large patches in different parts of the world. For instance, temperate deciduous forest occurs in eastern North America, north-central Europe, and eastern China. Note in Figure 6.18 that patches representing the same biome tend to occur at similar latitudes. This is due to the north-south gradient in temperature and to atmospheric circulation patterns (• pp. 479–480).

Each biome encompasses a variety of communities that share similarities. For example, the eastern United States supports part of the temperate deciduous forest biome. From New Hampshire to the Great Lakes to eastern Texas, precipitation and temperature are similar enough that most of the region's natural plant cover consists of broad-leafed trees that lose their leaves in winter. Within this region, however, there exist many different types of temperate deciduous forest, such as oak-hickory, beech-maple, and pine-oak forests, each sufficiently different to be designated a separate community.

## Aquatic systems also show biome-like patterns

In our discussion of biomes, we will focus exclusively on terrestrial systems because the biome concept, as traditionally developed and applied, has been limited to terrestrial systems. Areas equivalent to biomes also exist in the oceans, but their geographic shapes would look very different from those of terrestrial biomes if plotted on a world map. One might consider the thin strips along the world's coastlines to represent one aquatic system, the

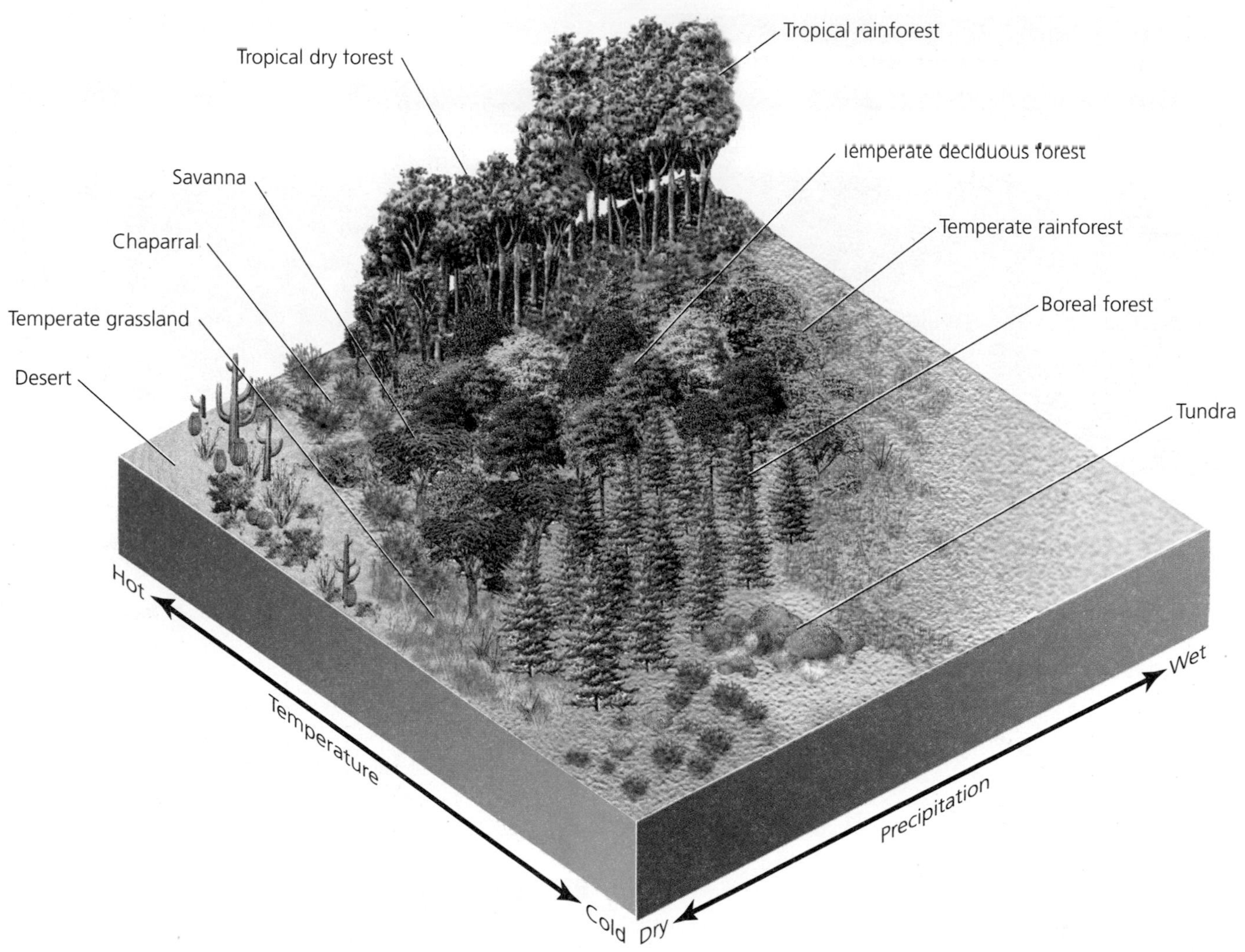


**FIGURE 6.19** As precipitation increases, vegetation generally becomes taller and more luxuriant. As temperature increases, types of plant communities change. Together, temperature and precipitation are the main factors determining which biome occurs in a given area. For instance, deserts occur in dry regions, tropical rainforests occur in warm wet regions, and tundra occurs in the coldest regions.

continental shelves another, and the open ocean, the deep sea, coral reefs, and kelp forests as still other distinct sets of communities. There are also many coastal systems that straddle the line between terrestrial and aquatic, such as salt marshes, rocky intertidal communities, mangrove forests, and estuaries. And of course there are freshwater systems such as those of the Great Lakes.

Unlike terrestrial biomes, aquatic systems are shaped not by air temperature and precipitation, but by factors such as water temperature, salinity, dissolved nutrients, wave action, currents, depth, and type of substrate (e.g., sandy, muddy, or rocky bottom). Marine communities are also more clearly delineated by their animal life than by their plant life. We will examine freshwater and marine systems in the greater detail they deserve in Chapters 15 and 16.

## We can divide the world into roughly ten terrestrial biomes

**Temperate deciduous forest** The **temperate deciduous forest** (Figure 6.20) that dominates the landscape around the central and southern Great Lakes is characterized by broad-leafed trees that are *deciduous*, meaning that they lose their leaves each fall and remain dormant during winter, when hard freezes would endanger leaves. These midlatitude forests occur in much of Europe and eastern China as well as in eastern North America—all areas where precipitation is spread relatively evenly throughout the year. Although soils of the temperate deciduous forest are relatively fertile, the biome generally consists of far fewer tree species than are found in

**(a) Temperate deciduous forest**

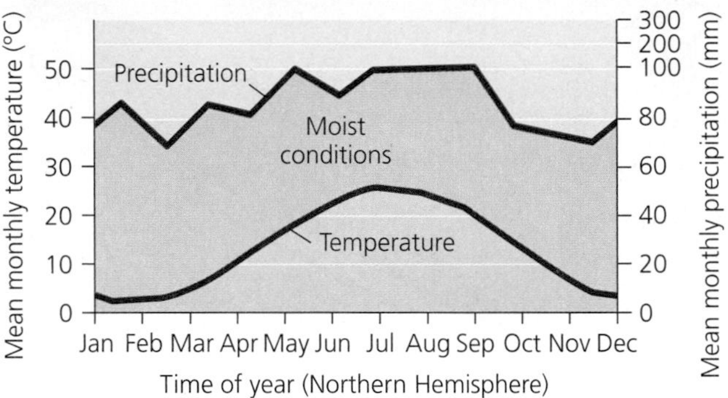


**(b) Washington, D.C., USA**

FIGURE 6.20 Temperate deciduous forests experience relatively stable seasonal precipitation but more varied seasonal temperatures. Scientists use climate diagrams to illustrate an area's average monthly precipitation and temperature. Typically in these diagrams, the *x* axis marks months of the year (beginning in January for regions in the Northern Hemisphere and in July for regions in the Southern Hemisphere). Paired *y* axes denote average monthly temperature and average monthly precipitation. The twin curves plotted on a climate diagram indicate trends in precipitation (blue) and in temperature (red) from month to month. When the precipitation curve lies above the temperature curve, as is the case throughout the year in the temperate deciduous forest biome around Washington, D.C., the region experiences relatively "moist" conditions, which we indicate with green coloration. Climatograph adapted from Breckle, S.W., 2002. *Walter's vegetation of the Earth: The ecological systems of the geo-biosphere*, 4th ed. Berlin: Springer-Verlag.

tropical rainforests. Oaks, beeches, and maples are a few of the most abundant types of trees in these forests. A sampling of typical animals of the temperate deciduous forest of eastern North America is shown in Figure 6.12 (• p. 151).

**(a) Temperate grassland**

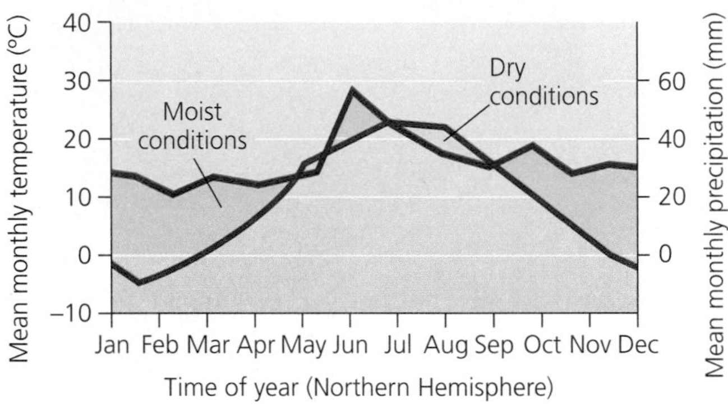


**(b) Odessa, Ukraine**

FIGURE 6.21 Temperate grasslands experience temperature variations throughout the year and too little precipitation for many trees to grow. Constructed for Odessa, Ukraine, this climatograph indicates both "moist" (green) and "dry" (yellow) climate conditions. When the temperature curve is above the precipitation curve, as is the case in May and mid-June through September, the climate conditions are "dry." Climatograph adapted from Breckle, S.W., 2002.

**Temperate grassland** Traveling westward from the Great Lakes, we find **temperate grasslands** (**Figure 6.21**). Temperate grasslands occur where temperature differences between winter and summer become more extreme and rainfall diminishes. The limited amount of precipitation in the Great Plains region west of the Mississippi River can support grasses more easily than trees. Also known as *steppe* or *prairie*, temperate grasslands were once widespread throughout parts of North and South America and much of central Asia. Today people have converted most of the world's grasslands for agriculture. Characteristic vertebrate animals of the native North American grasslands include American bison *(Bison bison)*, prairie dogs, pronghorn antelope *(Antilocapra americana)*, and ground-nesting birds such as meadowlarks.

**(a) Temperate rainforest**

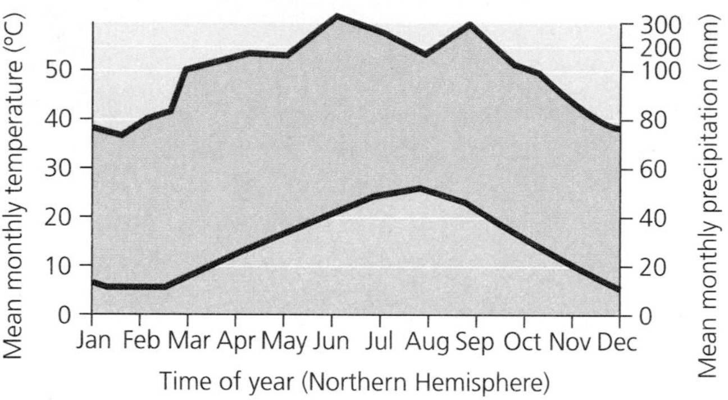


**(b) Nagasaki, Japan**

**FIGURE 6.22** Temperate rainforests receive a great deal of precipitation and feature moist, mossy interiors. Climatograph adapted from Breckle, S.W., 2002.

**(a) Tropical rainforest**

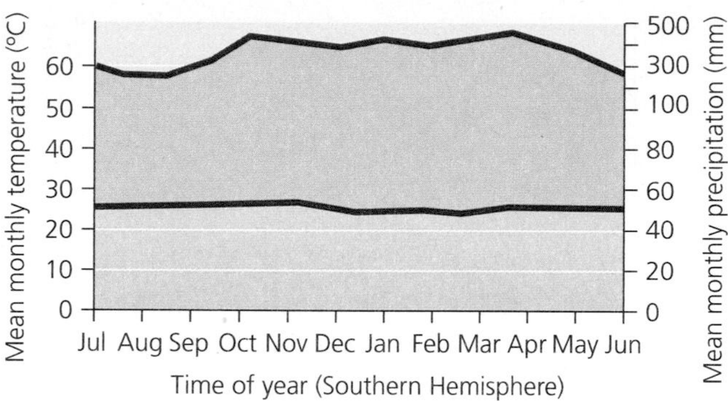


**(b) Bogor, Java, Indonesia**

**FIGURE 6.23** Tropical rainforests, famed for their biodiversity, grow under constant, warm temperatures and a great deal of rain. Climatograph adapted from Breckle, S.W., 2002.

**Temperate rainforest** Further west in North America, the topography becomes varied, and biome types intermix. The coastal Pacific Northwest region, with its heavy rainfall, features **temperate rainforest** (Figure 6.22). Coniferous trees, such as cedars, spruces, hemlocks, and Douglas fir *(Pseudotsuga menziesii)* grow very tall in the temperate rainforest, and the forest interior is shaded and damp. In the Pacific Northwest, moisture-loving animals such as the bright yellow banana slug *(Ariolimax columbianus)* are common, and old-growth stands hold the endangered spotted owl *(Strix occidentalis)*. The soils of temperate rainforests are usually quite fertile but are susceptible to landslides and erosion if forests are cleared. Temperate rainforests can produce large volumes of commercially important forest products such as lumber and paper. Controversy has surrounded these forests in the Pacific Northwest, where overharvesting has driven some species toward extinction and pushed many forest-dependent human communities toward economic stagnation.

**Tropical rainforest** In tropical regions we see the same pattern found in temperate regions: Areas of high rainfall grow rainforests, areas of intermediate rainfall host dry or deciduous forests, and areas of lower rainfall are dominated by grasses. However, tropical biomes differ from their temperate counterparts in other ways because they are closer to the equator and therefore warmer on average year-round. For one thing, they hold far greater biodiversity.

**Tropical rainforest** (Figure 6.23) is found in Central America, South America, southeast Asia, west Africa, and other tropical regions and is characterized by year-round rain and uniformly warm temperatures. Tropical rainforests have dark, damp interiors, lush vegetation, and highly diverse biotic communities, with greater numbers of species of insects, birds, amphibians, and various other animals than any other biome. These forests are not dominated by single species of trees, as are forests closer to the poles, but instead consist of very high

(a) Tropical dry forest

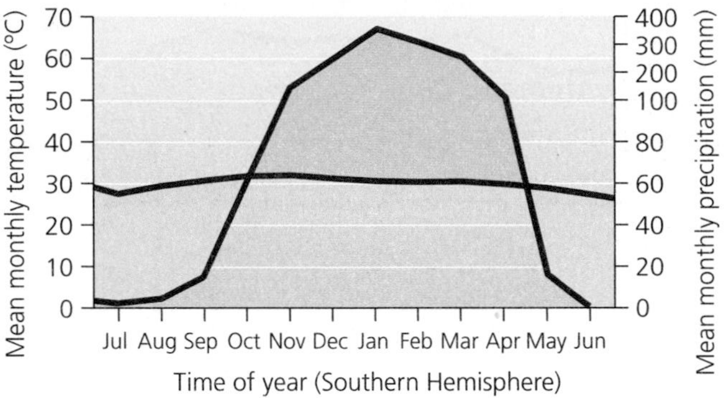


(b) Darwin, Australia

FIGURE 6.24 Tropical dry forests experience significant seasonal variations in precipitation and relatively stable warm temperatures. Climatograph adapted from Breckle, S.W., 2002.

(a) Savanna

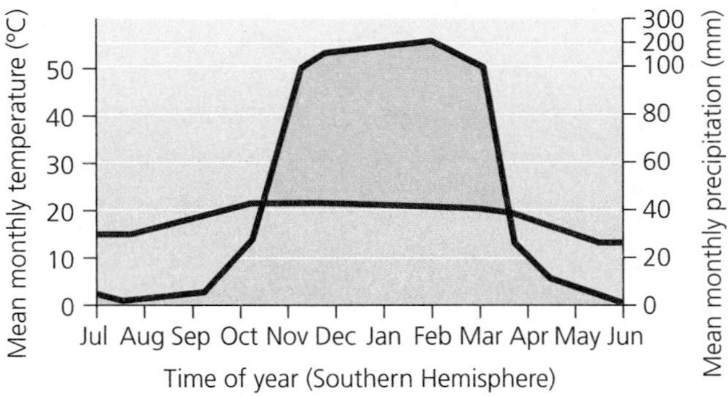


(b) Harare, Zimbabwe

FIGURE 6.25 Savannas are grasslands with clusters of trees. They experience slight seasonal variation in temperature but significant variation in rainfall. Climatograph adapted from Breckle, S.W., 2002.

numbers of tree species intermixed, each at a low density. Any given tree may be draped with vines, enveloped by strangler figs, and loaded with epiphytes (orchids and other plants that grow in trees), such that trees occasionally collapse under the weight of all the life they support.

Despite this profusion of life, tropical rainforests have very poor, acidic soils that are low in organic matter. Nearly all nutrients present in this biome are contained in the trees, vines, and other plants—not in the soil. An unfortunate consequence is that once tropical rainforests are cleared, the nutrient-poor soil can support agriculture for only a short time. As a result, farmed areas are abandoned quickly, and the soil and forest vegetation recover slowly.

**Tropical dry forest** Tropical areas that are warm year-round but where rainfall is lower overall and highly seasonal give rise to **tropical dry forest**, or *tropical deciduous forest* (**Figure 6.24**), a biome widespread in India, Africa, South America, and northern Australia. Wet and dry seasons each span about half a year in tropical dry forest. Rains during the wet season can be extremely heavy and, coupled with erosion-prone soils, can lead to severe soil loss where people have cleared forest. Across the globe, we have converted a great deal of tropical dry forest to agriculture. Clearing for farming or ranching is made easier by the fact that vegetation heights are much lower and canopies less dense than in tropical rainforest. Organisms that inhabit tropical dry forest have adapted to seasonal fluctuations in precipitation and temperature. For instance, plants are deciduous and often leaf out and grow profusely with the rains, then drop their leaves during the driest times of year.

**Savanna** Drier tropical regions give rise to **savanna** (**Figure 6.25**), tropical grassland interspersed with clusters of acacias or other trees. The savanna biome is found today across stretches of Africa (the ancestral home of our species), South America, Australia, India, and other dry tropical regions. Precipitation in savannas usually arrives during distinct rainy seasons and concentrates grazing animals near widely spaced water holes. Common herbivores

**(a) Desert**

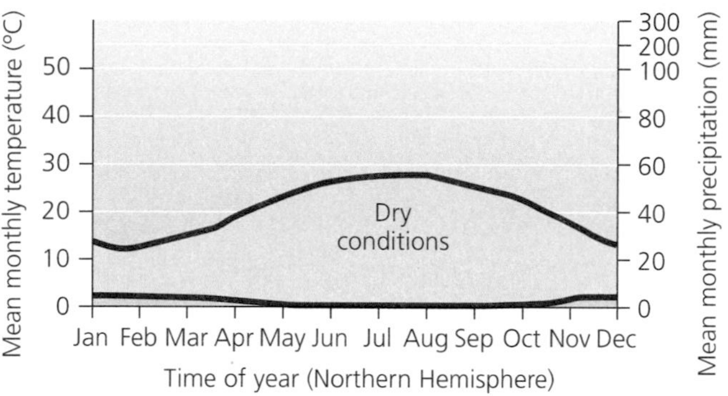


**(b) Cairo, Egypt**

FIGURE 6.26 Deserts are dry year-round, but they are not always hot. Precipitation can arrive in intense, widely spaced storm events. The temperature curve is consistently above the precipitation curve in this climatograph of Cairo, Egypt, indicating that the region experiences "dry" conditions all year. Climatograph adapted from Breckle, S.W., 2002.

**(a) Tundra**

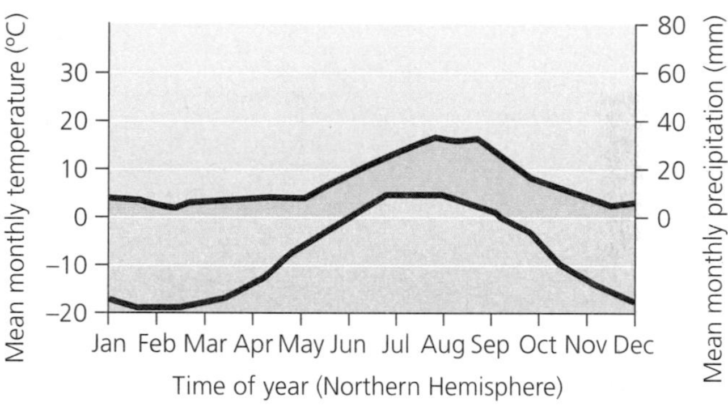


**(b) Vaigach, Russia**

FIGURE 6.27 Tundra is a cold, dry biome found near the poles and atop high mountains at lower latitudes. Climatograph adapted from Breckle, S.W., 2002.

on the African savanna include zebras, gazelles, and giraffes, and the predators of these grazers include lions, hyenas, and other highly mobile carnivores.

**Desert** Where rainfall is very sparse, **desert** (**Figure 6.26**) forms. This is the driest biome on Earth; most deserts receive well under 25 cm (9.8 in.) of precipitation per year, much of it during isolated storms months or years apart. Depending on rainfall, deserts vary greatly in the amount of vegetation they support. Some, like the Sahara and Namib deserts of Africa, are mostly bare sand dunes; others, like the Sonoran Desert of Arizona and northwest Mexico, are quite heavily vegetated.

Deserts are not always hot; the high desert of the western United States is one example. Because deserts have low humidity and relatively little vegetation to insulate them from temperature extremes, sunlight readily heats them in the daytime, but daytime heat is quickly lost at night. As a result, temperatures vary widely from day to night and across seasons of the year. Desert soils can often be quite saline and are sometimes known as lithosols, or stone soils, for their high mineral and low organic-matter content.

Desert animals and plants show many adaptations to deal with a harsh climate. Most reptiles and mammals, such as rattlesnakes and kangaroo mice, are active in the cool of night, and many Australian desert birds are nomadic, wandering long distances to find areas of recent rainfall and plant growth. Many desert plants have thick, leathery leaves to reduce water loss, or green trunks so that the plant can photosynthesize without leaves, which would lose water. The spines of cacti and many other desert plants guard those plants from being eaten by herbivores desperate for the precious water they hold.

**Tundra** Nearly as dry as desert, **tundra** (**Figure 6.27**) occurs at very high latitudes along the northern edges of Russia, Canada, and Scandinavia. Extremely cold winters with little daylight and moderately cool summers with lengthy days characterize this landscape of lichens and low, scrubby vegetation without trees. The great seasonal variation in temperature and day length results from this biome's

(a) Boreal forest

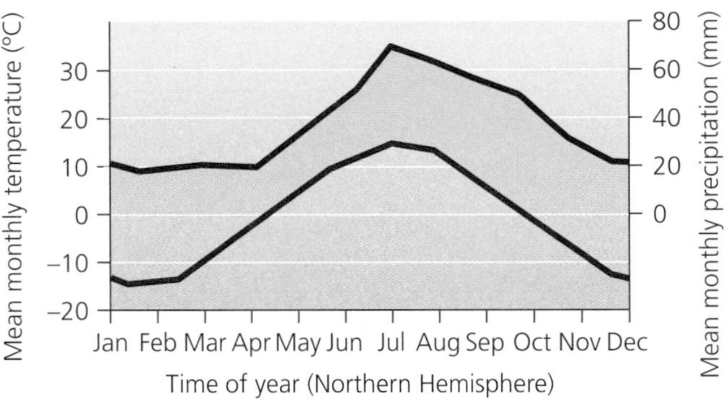


(b) Archangelsk, Russia

**FIGURE 6.28** Boreal forest is characterized by long, cold winters, relatively cool summers, and moderate precipitation. Climatograph adapted from Breckle, S.W., 2002.

(a) Chaparral

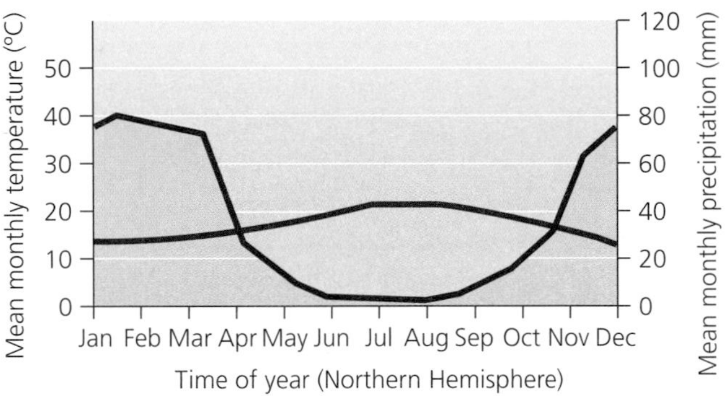


(b) Los Angeles, California, USA

**FIGURE 6.29** Chaparral is a highly seasonal biome dominated by shrubs, influenced by marine weather, and dependent on fire. Climatograph adapted from Breckle, S.W., 2002.

position close to the poles, which are angled toward the sun in the summer and away from the sun in the winter.

Because of the cold climate, underground soil remains more or less permanently frozen, and is called *permafrost.* During the long, cold winters, the surface soils freeze as well; then, when the weather warms, they melt and produce seasonal accumulations of surface water that make ideal habitat for mosquitoes and other biting insects. The swarms of insects benefit bird species that migrate long distances to breed during the brief but productive summer. Caribou also migrate to the tundra to breed and then leave for the winter. Only a few animals, such as polar bears *(Ursus maritimus)* and musk oxen *(Ovibos moschatus)*, can survive year-round in this extreme climate. Tundra also occurs as *alpine tundra* at the tops of high mountains in temperate and tropical regions.

**Boreal forest** The northern coniferous forest, or **boreal forest**, often called *taiga* (**Figure 6.28**), stretches in a broad band across much of Canada, Alaska, Russia, and Scandinavia. It consists of a few species of evergreen trees, such as black spruce *(Picea mariana)*, that dominate large stretches of forests interspersed with occasional bogs and lakes. The boreal forest's uniformity over huge areas reflects the climate common to this latitudinal band of the globe: These forests develop in cooler, drier regions than do temperate forests, and they experience long, cold winters and short, cool summers. Soils are typically nutrient-poor and somewhat acidic. As a result of the strong seasonal variation in day length, temperature, and precipitation, many organisms compress a year's worth of feeding, breeding, and rearing of young into a few warm, wet months. Year-round residents of boreal forest include mammals such as moose *(Alces alces)*, wolves *(Canis lupus)*, bears, lynx *(Felis lynx)*, and many burrowing rodents. This biome also hosts many insect-eating birds that migrate from the tropics to breed during the brief, intensely productive, summer season.

**Chaparral** In contrast to the boreal forest's broad, continuous distribution, **chaparral** (**Figure 6.29**) is limited to fairly small patches widely flung around the globe. Chaparral consists mostly of evergreen shrubs and is densely thicketed. This biome is also highly seasonal, with mild, wet winters and warm, dry summers. This type of climate is induced by oceanic influences and is often termed "Mediterranean." In addition to ringing the Mediterranean Sea, chaparral occurs along the coasts of California, Chile, and southern Australia. Chaparral communities experience frequent fire, and their plant species are adapted to resist fire or even to depend on it for germination of their seeds.

## Altitude creates patterns analogous to latitude

As any hiker or skier knows, climbing in elevation causes a much more rapid change in climate than moving the same distance toward the poles. Vegetative communities change along mountain slopes in correspondence with this small-scale climate variation (**Figure 6.30**). It is often said that hiking up a mountain in the southwestern United States is like walking from Mexico to Canada. A hiker ascending one of southern Arizona's higher mountains

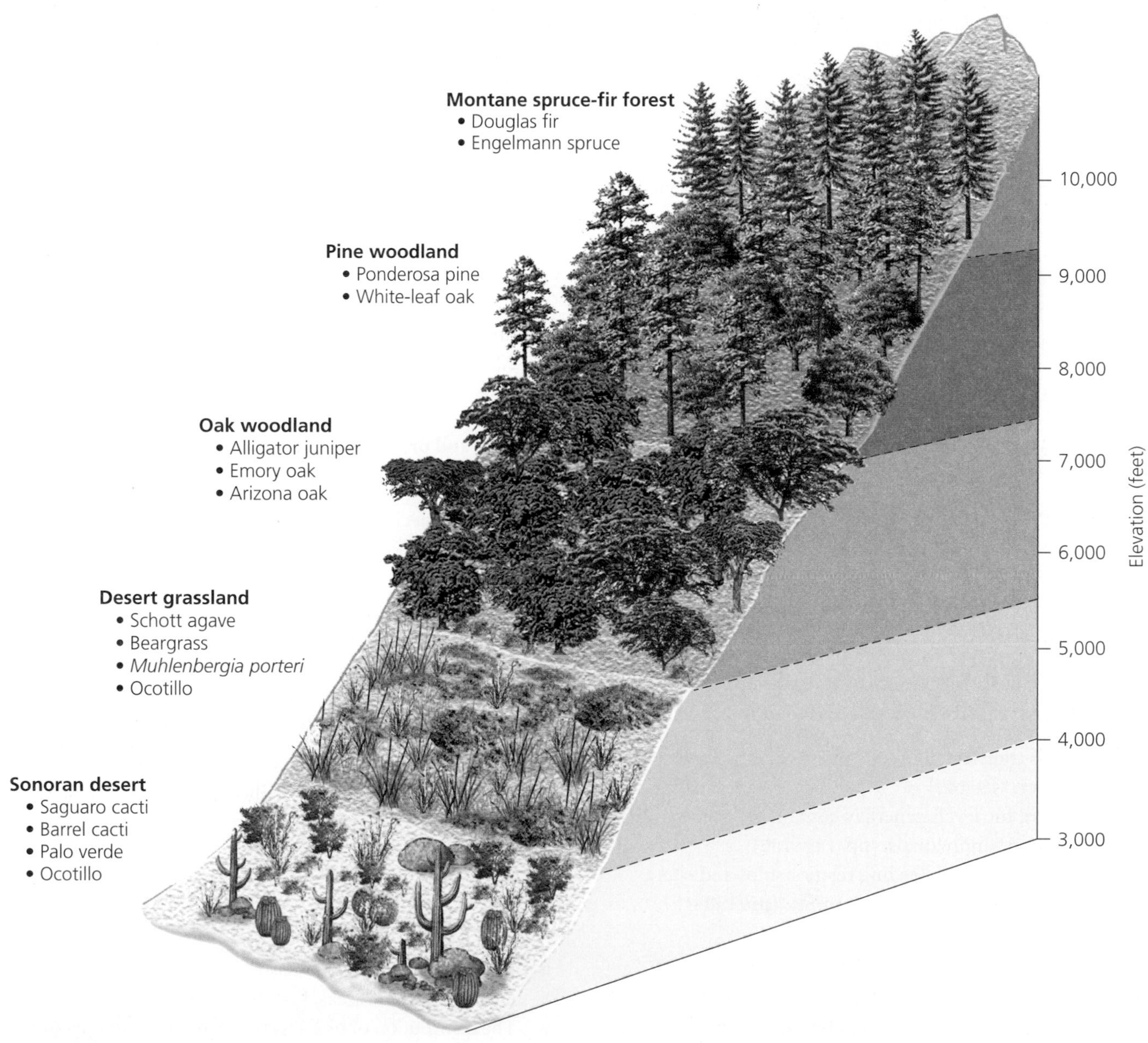


FIGURE 6.30 As altitude increases, vegetation changes in ways similar to the ways it changes as one moves toward the poles. Climbing a mountain in southern Arizona, as pictured here, can be likened to traveling from Mexico to Canada, taking the hiker through the local equivalent of several biomes.

would begin in Sonoran Desert or desert grassland and proceed through oak woodland, pine forest, and finally spruce-fir forest—the equivalent of passing through several biomes. A hiker scaling one of the great peaks of the Andes in Ecuador could begin in tropical rainforest and end amid glaciers in alpine tundra.

## Conclusion

The natural world is so complex that we can visualize it in many ways and at various scales. Dividing the world's communities into major types, or biomes, is informative at the broadest geographic scales. Understanding how communities function at more local scales requires understanding how species interact with one another. Species interactions such as predation, parasitism, competition, and mutualism give rise to effects both weak and strong, direct and indirect. Feeding relationships can be represented by the concepts of trophic levels and food webs, and particularly influential species are sometimes called keystone species. Increasingly, humans are altering communities, in part by introducing non-native species that sometimes turn invasive. But more and more, through ecological restoration, we are also attempting to undo the changes we have caused.

## REVIEWING OBJECTIVES

**You should now be able to:**

**Compare and contrast the major types of species interactions**

- Competition results when individuals or species vie for limited resources. It can occur within or among species and can result in coexistence or exclusion. It also can lead to realized niches, resource partitioning, and character displacement. (pp. 143–144)
- In predation, one species kills and consumes another. Predation is the basis of food webs and can influence population dynamics and community composition. (pp. 144–145)
- In parasitism, one species derives benefit by harming (but usually not killing) another. (pp. 145–146)
- Herbivory is an exploitative interaction whereby an animal feeds on a plant. (pp. 146–147)
- In mutualism, species benefit from one another. Some mutualisms are symbiotic, whereas other mutualists are free-living. (p. 147)

**Characterize feeding relationships and energy flow, using them to construct trophic levels and food webs**

- Energy is transferred in food chains among trophic levels. (pp. 148–149)
- Lower trophic levels generally contain more energy, biomass, and individuals. (pp. 149–150)
- Food webs illustrate feeding relationships and energy flow among species in a community. (pp. 150–151)

**Distinguish characteristics of a keystone species**

- Keystone species have impacts on communities that are far out of proportion to their abundance. (pp. 151–154)
- Top predators are frequently considered keystone species, but other types of organisms also exert strong effects on communities. (pp. 151–154)

**Characterize the process of succession and the debate over the nature of communities**

- Succession is a stereotypical pattern of change within a community through time. (p. 155)
- Primary succession begins with an area devoid of life. Secondary succession begins with an area that has been severely disturbed. (pp. 155–158)
- Clements held that communities are discrete, cohesive units. His view has largely been replaced by that of Gleason, who held that species may be added to and deleted from communities through time. (p. 158)

**Perceive and predict the potential impacts of invasive species in communities**

- Invasive species such as the zebra mussel have altered the composition, structure, and function of communities. (pp. 158–160)
- Humans are the cause of most modern species invasions, but we can also respond to invasions with prevention and control measures. (pp. 158–160)

**Explain the goals and methods of ecological restoration**

- Ecological restoration aims to restore communities to a more "natural" state, variously defined as before human or industrial interference. (pp. 160–161)
- Restoration efforts are informed by the growing science of restoration ecology. (pp. 160–161)

**Describe and illustrate the terrestrial biomes of the world**

- Biomes represent major classes of communities spanning large geographic areas. (pp. 161–169)
- The distribution of biomes is determined by temperature, precipitation, and other factors. (pp. 161–162)
- Aquatic systems can be classified in similar ways, determined by different factors. (pp. 162–163)

## TESTING YOUR COMPREHENSION

1. How does competition lead to a realized niche? How does it promote resource partitioning?
2. Contrast the several types of exploitative species interactions. How do predation, parasitism, and herbivory differ?
3. Give examples of symbiotic and nonsymbiotic mutualisms. Describe at least one way in which mutualisms affect your daily life.
4. Compare and contrast trophic levels, food chains, and food webs. How are these concepts related and how do they differ?
5. What is meant by a keystone species, and what types of organisms are most often considered keystone species?
6. Explain primary succession. How does it differ from secondary succession? Give an example of each.
7. Explain and contrast Clements's and Gleason's views of ecological communities.
8. Name five changes to Great Lakes communities that have occurred since the invasion of the zebra mussel.
9. What factors most strongly influence the type of biome that forms in a particular place on land? What factors determine the type of aquatic system that may form in a given location?
10. Draw climate diagrams for a tropical rainforest and for a desert. Label all parts of the diagram, and describe all of the types of information an ecologist could glean from such a diagram.

## SEEKING SOLUTIONS

1. Imagine that you spot two species of birds feeding side by side, eating seeds from the same plant, and that you begin to wonder whether competition is at work. Describe how you might design scientific research to address this question. What observations would you try to make at the outset? Would you try to manipulate the system to test your hypothesis that the two birds are competing? If so, how?
2. Spend some time outside on your campus or in your yard or in the nearest park or natural area. Find at least 10 species of organisms, and observe them long enough to watch them feed or to make an educated guess about what they feed on. Now, using Figure 6.12 as a model, draw a simple food web involving all the organisms you observed.
3. Can you think of one organism not mentioned in this chapter as a keystone species that you believe may be a keystone species? For what reasons do you suspect this? How could you experimentally test whether an organism is a keystone species?
4. Why do scientists consider invasive species to be a problem? What makes a species "invasive," and what ecological effects can invasive species have?
5. From year to year, biomes are stable entities, and our map of world biomes appears to be a permanent record of patterns across the planet. But are the locations and identities of biomes permanent, or could they change over time? Provide reasons for your answers.
6. **THINK IT THROUGH** A federal agency has put you in charge of devising responses to the zebra mussel invasion. Based on what you know from this chapter, how would you seek to control this species' spread and reduce its impacts? What strategies would you consider pursuing immediately, and for which strategies would you commission further scientific research? For each of your ideas, name one benefit or advantage, and identify one obstacle it might face in being implemented. What additional steps might you suggest to deal with the unfolding quagga mussel invasion?

## INTERPRETING GRAPHS AND DATA

Each spring, citizen volunteers join representatives of the United States Geological Survey, the California Department of Fish and Game, and the Monterey Bay Aquarium to survey populations of the sea otter (*Enhydra lutris*) along 600 km (375 mi) of California's coastline, from Half Moon Bay to Santa Barbara. The information gathered from the survey is used by federal and state wildlife agencies in making decisions about how to manage the sea otter—a keystone species that in turn influences the coast's ecological communities.

1. In percentage terms, how does the most recent population estimate of sea otters (from 2007) compare with the survey's earliest estimate (from 1983)?
2. Referring back to "The Science behind the Story" (• pp. 156–157), describe how you think recent population trends for sea otters are likely to affect the composition and structure of the nearshore community in regard to kelp, sea urchins, and other organisms.

3. Given your hypotheses in Question 2, sketch three of your own graphs for the period 1983–2007 showing population trends for (a) sea urchins, (b) kelp, and (c) fish that depend on dense kelp forest habitat. Explain how your graphs illustrate predictions from your hypotheses.

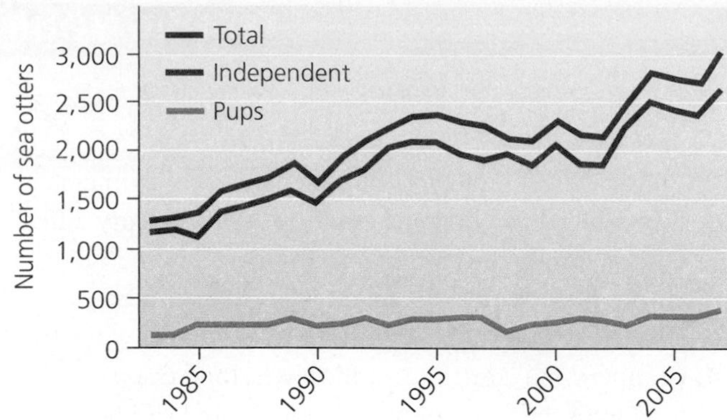


**Numbers of sea otters counted during spring surveys, 1983–2007. Shown are numbers of independent adults and subadults, dependent pups, and total individuals.** Data from USGS Western Ecological Research Center.

## CALCULATING ECOLOGICAL FOOTPRINTS

In 2003, scientist David Pimentel reviewed estimates for the economic and ecological costs of certain introduced and invasive species. He noted that over 50,000 species have been introduced in the United States (see Table 1) and estimated that damages and costs associated with controlling aquatic invasive species in the United States total $9 billion annually (see Table 2). Calculate values missing from these tables.

**Table 1. Introduced Species**

| Taxon | Percentage of total introduced | Number of species introduced |
|---|---|---|
| Plants | 50 | 25,000 |
| Mammals | 0.04 | |
| Birds | 0.19 | |
| Molluscs | 0.18 | |
| Arthropods | 9 | |
| Microbes | 40 | |

1. Of the over 50,000 species introduced into the United States, approximately half are plants. Describe how non-native plants may sometimes turn invasive and pose threats to native communities.

**Table 2. Aquatic Invaders**

| Invader | Percentage of total costs | Economic cost |
|---|---|---|
| Fish species | 60 | $5.4 billion |
| Zebra/quagga mussels | 11 | |
| Asiatic clam | 11 | |
| West Nile virus | 11 | |
| Aquatic plant species | 5 | |
| Shipworm | 2 | |
| Green crab | 1 | |

*Source:* Pimentel, D., 2003. Economic and ecological costs associated with aquatic invasive species. *Proceedings of the Aquatic Invaders of the Delaware Estuary Symposium*, Malvern, Pennsylvania, May 20, 2003.

2. Compare the costs from aquatic invaders shown in Table 2 with the percentages shown for all non-native species in Table 1. Speculate on the costs you might expect from terrestrial introduced species, and explain your answers.
3. How might your own behavior influence the influx and ecological impacts of non-native species like those listed above? What could you do to minimize the impacts of invasive species?

## Take It Further

Go to www.aw-bc.com/withgott or the student CD-ROM, where you'll find:

- Suggested answers to end-of-chapter questions
- Quizzes, animations, and flashcards to help you study
- *Research Navigator*™ database of credible and reliable sources to assist you with your research projects
- **GRAPHIt!** Tutorials to help you interpret graphs
- **INVESTIGATEIt!** Current news articles that link the topics that you study to case studies from your region to around the world

CHAPTER

# 7 Environmental Systems and Ecosystem Ecology

The Mississippi River as it enters the Gulf of Mexico

## Upon completing this chapter, you will be able to:

- Describe the nature of environmental systems
- Define ecosystems and evaluate how living and non-living entities interact in ecosystem-level ecology
- Outline the fundamentals of landscape ecology
- Compare and contrast how carbon, phosphorus, nitrogen, and water cycle through the environment
- Explain how plate tectonics and the rock cycle shape the landscape around us and the earth beneath our feet

Fisherman in the Gulf of Mexico's "dead zone"

CENTRAL CASE

# The Gulf of Mexico's "Dead Zone"

**"In nature there is no 'above' or 'below,' and there are no hierarchies. There are only networks nesting within other networks."**
—Fritjof Capra, Theoretical Physicist

**"Let's say you put Saran Wrap over south Louisiana and suck the oxygen out. Where would all the people go?"**
—Nancy Rabalais, Biologist for the Louisiana Universities Marine Consortium

Louisiana fishermen have long hauled in more seafood than those of any other U.S. state except Alaska. Each year they have plied the rich waters of the northern Gulf of Mexico and have sent nearly 600 million kg (1.3 billion lb) of shrimp, fish, and shellfish to our dinner tables.

Then in 2005, Hurricane Katrina and Hurricane Rita pummeled the Gulf Coast and left Louisiana's fisheries in ruin. Boats, docks, marinas, and fueling stations were destroyed, and thousands were suddenly out of work. Today, fishermen are struggling to reestablish their livelihoods as the industry continues to rebuild itself.

But for years before the hurricanes hit, fishing had become increasingly difficult. In the words of longtime Louisiana fisherman Johnny Glover, it was "getting harder and harder to make a living." The reason? Each year billions of organisms were suffocating in the Gulf's "dead zone," a region of water so depleted of oxygen that marine organisms are killed or driven away.

The low concentrations of dissolved oxygen in the bottom waters of this region represent a condition called **hypoxia** (see Figure 7.5, • p. 179, and "The Science behind the Story," • pp. 194–195). Aquatic animals obtain oxygen by respiring through their gills, and, like us, these animals will asphyxiate if deprived of oxygen. Fully oxygenated water contains up to 10 parts per million (ppm) of oxygen, but when concentrations drop below 2 ppm, creatures that can leave an affected area will do so. Below 1.5 ppm, most marine organisms die. In the Gulf's hypoxic zone, oxygen concentrations frequently drop well below these levels.

The dead zone appears each spring and grows through the summer and fall, starting near the mouths of the Mississippi and Atchafalaya rivers off the Louisiana coast. In 2002 the dead zone reached a record 22,000 km$^2$ (8,500 mi$^2$)—an area larger than New Jersey. Shrimp boats came up with nets nearly empty. One shrimper derided his meager catch as "cat food." Others, ironically, said they hoped a hurricane would strike and stir some oxygen into the Gulf's stagnant waters.

What's starving these waters of oxygen? Scientists studying the dead zone have identified modern Midwestern farm practices and other human impacts hundreds of kilometers away. The Gulf, they say, is being excessively enriched by nitrogen and phosphorus flushed down the Mississippi River. This nutrient pollution comes from fertilizers used on farms far upriver, as well as from other sources, including urban runoff, industrial discharges, fossil fuel emissions, and municipal sewage outflow. The excess nutrients spur plankton blooms in the Gulf, nourishing bacteria that then deplete the dissolved oxygen in bottom waters.

The U.S. government has acted on these findings, proposing that farmers in states such as Ohio, Iowa, and Illinois cut down on fertilizer use. Farmers' advocates protest that farmers are being singled out while urban pollution sources are ignored. Meanwhile, scientists have documented coastal dead zones in 200 other areas throughout the world, from Chesapeake Bay to Oregon to Denmark to the Black Sea. The story of how researchers determine the causes of these dead zones involves understanding environmental systems and the often complex behavior they exhibit.

# Earth's Environmental Systems

Our planet's environment consists of complex networks of interlinked systems. In the realm of community ecology (Chapter 6), these systems include the ecological webs of relationships among species. At the ecosystem level, they include the interaction of living species with the nonliving entities around them. Earth's systems also include cycles that shape the landscapes around us and guide the flow of key chemical elements and compounds that support life and regulate climate. We depend on these systems for our very survival.

Assessing questions holistically by taking a "systems approach" is helpful in environmental science, in which so many issues are multifaceted and complex. Such an approach poses a challenge, however, because systems often show behavior that is difficult to understand and predict. The scientific method operates best when researchers isolate and manipulate small parts of complex systems, focusing in depth on manageable components. However, environmental scientists are rising to the challenge of studying systems broadly and are beginning to find solutions to problems such as the Gulf of Mexico's hypoxic zone.

## Systems show several defining properties

A **system** is a network of relationships among parts, elements, or components that interact with and influence one another through the exchange of energy, matter, or information. Systems receive inputs of energy, matter, or information, process these inputs, and produce outputs. Energy inputs to Earth's environmental systems include solar radiation as well as heat released by geothermal activity, organismal metabolism, and human activities such as fossil fuel combustion. Information inputs can come in the form of sensory cues from visual, olfactory (chemical), magnetic, or thermal signals. Inputs of matter occur when chemicals or physical material moves among systems, such as when seeds are dispersed long distances, migratory animals deposit waste far from where they consumed food, or plants convert carbon in the air to living tissue by photosynthesis.

As a system, the Gulf of Mexico receives inputs of freshwater, sediments, nutrients, and pollutants from the Mississippi and other rivers. Shrimpers and fishermen harvest some of the Gulf system's output: matter and energy in the form of shrimp and of fish such as menhaden. This output subsequently becomes input to the human economic system and to the digestive systems of the many people who consume the seafood.

Sometimes a system's output can serve as input to that same system, a circular process described as a **feedback loop**. Feedback loops are of two types, negative and positive. In a **negative feedback loop** (**Figure 7.1a**), output that results from a system moving in one direction acts as input that moves the system in the other direction. Input and output essentially neutralize one another's effects, stabilizing the system.

A thermostat, for instance, stabilizes a room's temperature by turning the furnace on when the room gets cold and shutting it off when the room gets hot. Similarly, negative feedback regulates our body temperature. If we get too hot, our sweat glands pump out moisture that evaporates to cool us down, or we may move from sun to shade. If we get too cold, we shiver, creating heat, or we move into the sun or put on more clothing. Another example of negative feedback is a predator-prey system in which predator and prey populations rise and fall in response to one another (see Figure 6.5, • p. 145). Most systems in nature

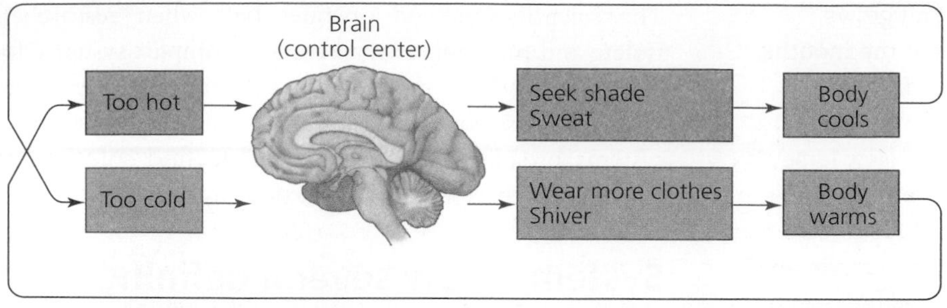


**(a) Negative feedback**

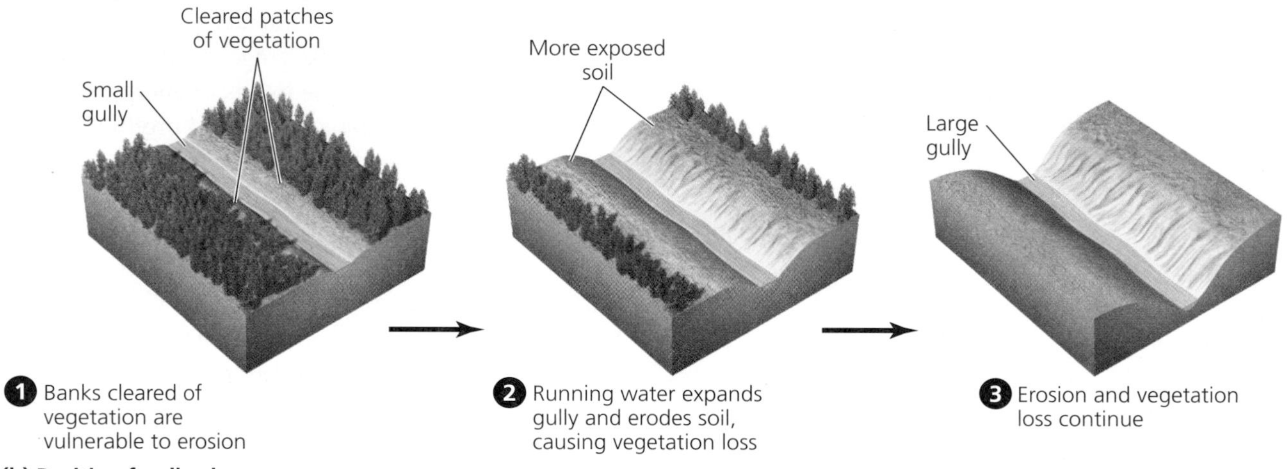


**(b) Positive feedback**

**FIGURE 7.1** Negative feedback loops (**a**) exert a stabilizing influence on systems and are common in nature. The human body's response to heat and cold involves a negative feedback loop. Positive feedback loops (**b**) have a destabilizing effect on systems and push them toward extremes. For example, the clearing of vegetated land can lead to a runaway process of soil erosion: water flowing through an eroded gully may expand the gully and lead to further erosion. Rare in nature, positive feedback loops are common in natural systems altered by human impact.

involve negative feedback loops. Negative feedback loops enhance stability, and in the long run, only those systems that are stable will persist.

**Positive feedback loops** have the opposite effect. Rather than stabilizing a system, they drive it further toward one extreme or another. Exponential growth in human population (• pp. 208–209) provides an example. The more people are born, the more there are to give birth to further people; increased output leads to increased input, leading to further increased output. Another example is the spread of cancer; as cells multiply out of control, the process is self-accelerating.

Positive feedback can also occur with the process of erosion, the removal of soil by water or wind (• pp. 242–243). Once vegetation has been cleared to expose soil, erosion may become progressively more severe if the forces of water or wind surpass the rate of vegetative regrowth (**Figure 7.1b**). Positive feedback can alter a system substantially. Positive feedback loops are rare in nature, but they are common in natural systems altered by human impact.

The inputs and outputs of complex natural systems usually occur simultaneously, keeping the system constantly active. Earth's climate system, for instance, does not ever stop. When processes within a system move in opposing directions at equivalent rates so that their effects balance out, the process is said to be in **dynamic equilibrium.**

Processes in dynamic equilibrium can contribute to **homeostasis**, the tendency of a system to maintain constant or stable internal conditions. When homeostasis exists, organisms and other systems can keep their internal conditions within a range that allows them to function. Homeostatic systems are often thought of as being in a steady state; however, the steady state itself may change slowly over time while the system maintains its ability to stabilize conditions internally. For instance, organisms grow and mature. Similarly, Earth has experienced a gradual increase in atmospheric oxygen over its history (• p. 104), yet life has adapted, and Earth remains, by most definitions, a homeostatic system.

It is difficult to understand systems fully by focusing on their individual components because systems can show **emergent properties,** characteristics not evident in the components alone. Stating that systems possess emergent properties is a lot like saying, "The whole is more than the sum of its parts." For example, if you were to reduce a tree to its component parts (leaves, branches, trunk, bark, roots, fruit, and so on) you would not be able to predict the whole tree's emergent properties, which include the role the tree plays as habitat for birds, insects, parasitic vines, and other organisms (**Figure 7.2**). You could analyze the tree's chloroplasts (photosynthetic cell organelles), diagram its branch structure, and evaluate its fruit's nutritional content, but you would still be unable to understand the tree as habitat, as part of a forest landscape, or as a reservoir for carbon storage.

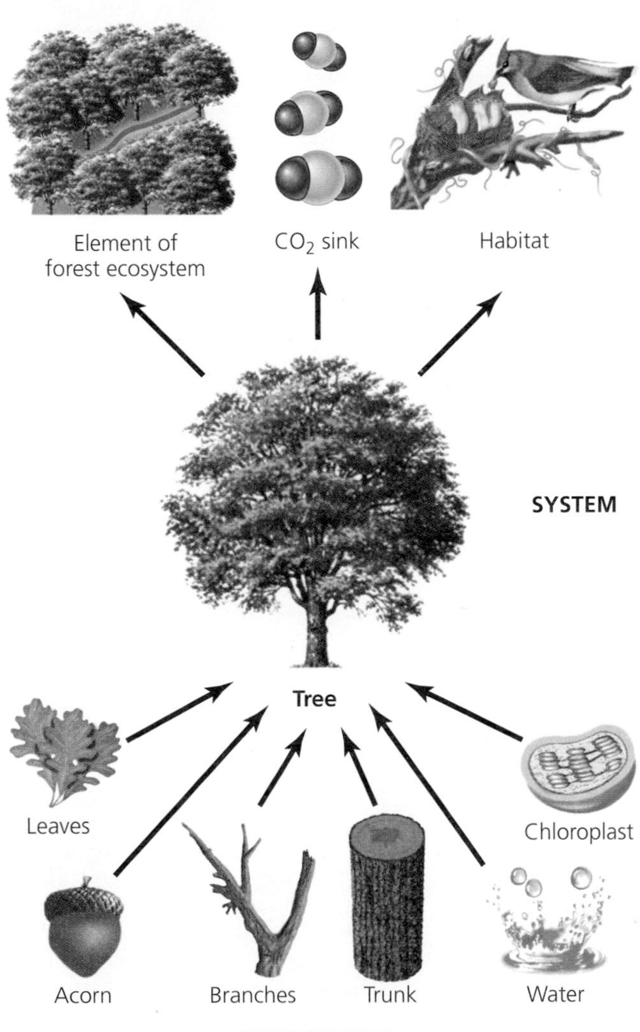


FIGURE 7.2 A system's emergent properties are not evident when we break the system down into its component parts. For example, a tree serves as wildlife habitat and plays roles in forest ecology and global climate regulation, but you would not know that from considering the tree only as a collection of leaves, branches, and chloroplasts. If we try to understand systems solely by breaking them into component parts, we will miss much of what makes them important.

Systems seldom have well-defined boundaries, so deciding where one system ends and another begins can be difficult. Consider a desktop computer system. It is certainly a network of parts that interact and exchange energy and information, but what are its boundaries? Is the system what arrives in a packing crate and sits on top of your desk? Or does it include the network you connect it to at school, home, or work? What about the energy grid you plug it into, with its distant power plants and transmission lines? And what of the Internet? Browsing the Web, you are drawing in digitized text, light, and sound from around the world.

No matter how we attempt to isolate or define a system, we soon see that it has many connections to systems larger and smaller than itself. Systems may exchange energy, matter, and information with other systems, and they may contain or be contained within other systems—so where we draw boundaries may depend on the spatial or temporal scale at which we choose to focus.

## Understanding the dead zone requires considering multiple systems

The Gulf of Mexico and the Mississippi River are systems that interact with one another. On a map, the Mississippi River appears as a branched and braided network of water channels. But where are this system's boundaries? You might argue that the Mississippi consists primarily of water, originates in Minnesota, and ends in the Gulf of Mexico near New Orleans. But what about the rivers that feed it and the farms, cities, and forests that line its banks (**Figure 7.3**)? Major rivers such as the Missouri, Arkansas, and Ohio flow into the Mississippi. Hundreds of smaller tributaries drain vast expanses of farmland, woodland, fields, cities, towns, and industrial areas before their water joins the Mississippi's. These waterways carry with them millions of tons of sediment, hundreds of species of plants and animals, and numerous pollutants.

For an environmental scientist interested in runoff and the flow of water, sediment, or pollutants, it may make the best sense to view the Mississippi River's *watershed* (• p. 58) as a system. As with the Tijuana River (see Chapter 3), one must consider the entire area of land a river drains to comprehend and solve problems of river pollution. However, for a scientist interested in the Gulf of Mexico's dead zone, it may make the best sense to view the Mississippi River watershed together with the Gulf as the system of interest, because their interaction is central to the problem. In environmental science, one's delineation of a system can and should depend on the questions one is addressing.

The reason for the Gulf of Mexico's dangerously low levels of oxygen, scientists have concluded, is abnormally

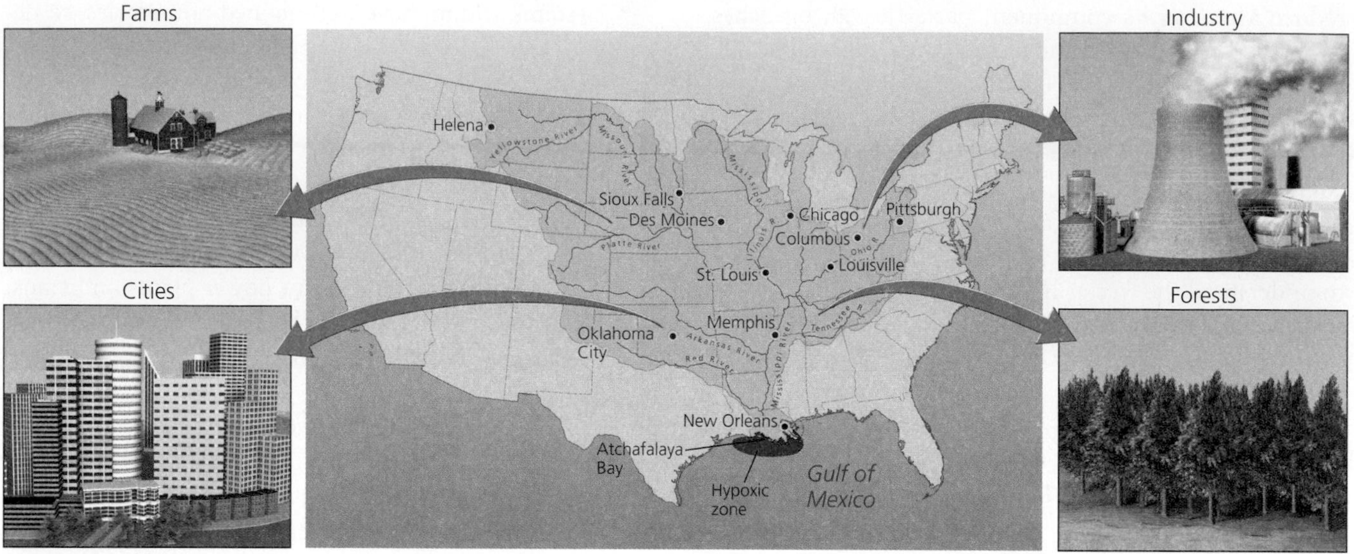


FIGURE 7.3 The Mississippi River system is the largest in North America. The river's watershed encompasses 3.2 million $km^2$ (1.2 million $mi^2$), or 41% of the area of the lower 48 U.S. states. The river carries water, sediment, and pollutants from a variety of sources downriver to the Gulf of Mexico, where nutrient pollution has given rise to a hypoxic zone.

high levels of nutrients such as nitrogen and phosphorus. The excess nutrients originate with sources in the Mississippi River watershed, particularly nitrogen- and phosphorus-rich fertilizers applied to crops. Inorganic nitrogen fertilizers used on Midwestern farms account for roughly 30% of total nitrogen contributions to the river, an amount equal to that from natural soil decomposition. The remainder comes from various sources, including animal manure, nitrogen-fixing crops, sewage treatment facilities, street runoff, and industrial and automobile emissions. Altogether, agricultural sources are thought to contribute 74% of the nitrate and 65% of the total nitrogen carried in the river. Much of the nitrate originates in the upper portions of the watershed from farms growing corn and soybeans in Iowa, Illinois, Indiana, Minnesota, and Ohio.

Nitrogen fertilizer input to farmland in the Mississippi River watershed has increased dramatically since 1950 (**Figure 7.4**). Since 1980, the Mississippi River and the Atchafalaya River (which drains a third of the Mississippi's water through a second delta) have pumped over 1 million metric tons of nitrate into the Gulf of Mexico each year, three times as much as during the 1960s.

The enhanced nitrogen input to the Gulf boosts the growth of *phytoplankton* (• p. 142)—photosynthetic algae, protists, and cyanobacteria that drift near the surface. Phytoplankton ordinarily are limited in their growth by scarcity of nutrients such as nitrogen and phosphorus. As phytoplankton flourish at the surface and as zooplankton (• p. 142) consume them, more dead phytoplankton and waste products of phytoplankton and zooplankton drift to the bottom, providing food for organisms (mainly bacteria) that decompose them. The result is a population explosion of bacteria. These decomposers consume enough oxygen to cause oxygen concentrations in bottom waters to plummet, suffocating shrimp and fish that live at the bottom and creating the dead zone.

The fresh water from the river remains naturally stratified in a layer at the surface that mixes only very slowly with the denser salty ocean water, so that oxygenated surface

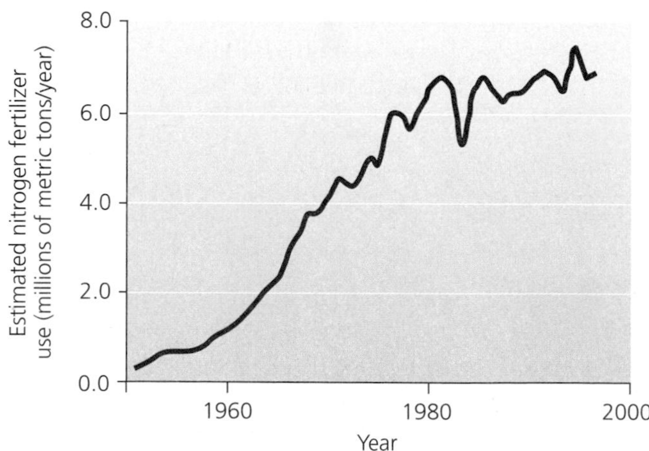


FIGURE 7.4 Use of nitrogen-based fertilizer in the Mississippi-Atchafalaya River basin skyrocketed after 1950. About 15% of nitrogen from fertilizer applied to farms reaches rivers that flow to the Gulf. Data from National Science and Technology Council: Committee on Environment and Natural Resources. May 2000. Hypoxia: An integrated assessment in the northern Gulf of Mexico.

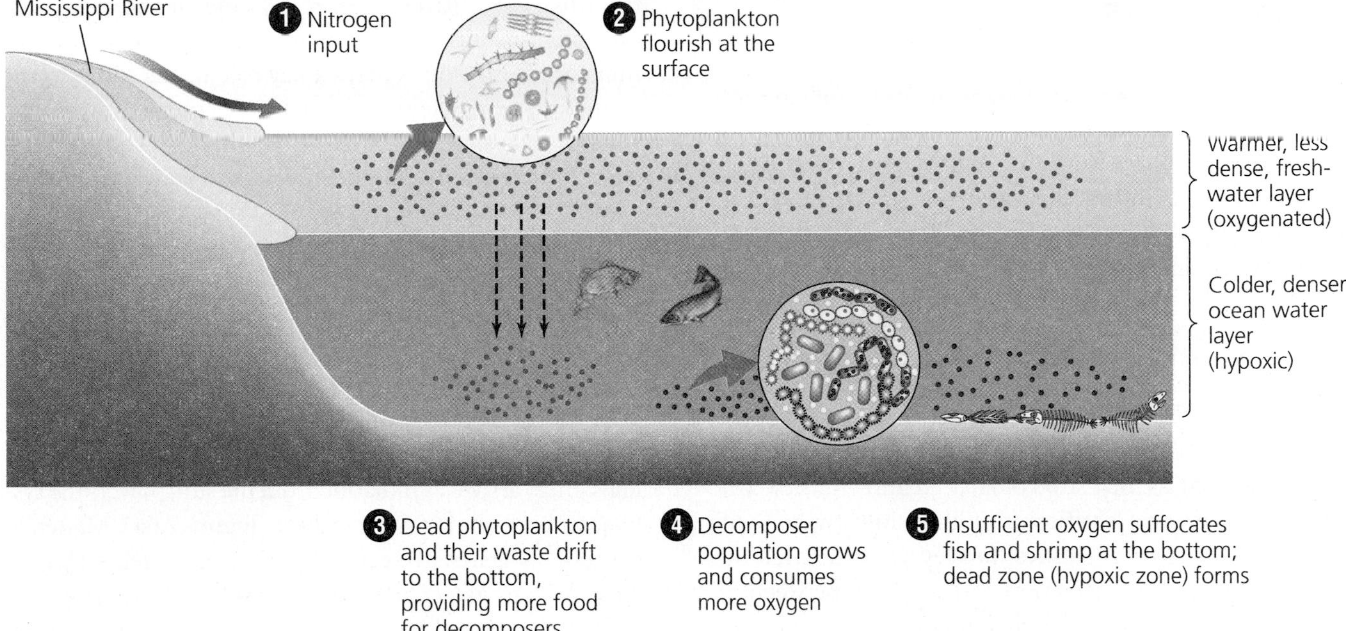


FIGURE 7.5 Excess nitrogen causes eutrophication in coastal marine systems such as the Gulf of Mexico. Coupled with stratification (layering) of water, eutrophication can severely deplete dissolved oxygen. Nitrogen from river water (1) boosts growth of phytoplankton (2), which die and are decomposed at the bottom by bacteria (3). Stability of the surface layer prevents deeper water from absorbing oxygen to replace oxygen consumed by decomposers (4), and the oxygen depletion suffocates or drives away bottom-dwelling marine life (5). This process gives rise to hypoxic zones like that of the Gulf of Mexico.

water does not make its way down to the bottom-dwelling life that needs it. This process of nutrient overenrichment, blooms of algae, increased production of organic matter, and subsequent ecosystem degradation is known as **eutrophication** (**Figure 7.5**). As we will see in Chapters 15 and 16, eutrophication can take place in both freshwater and saltwater systems.

Moderate amounts of additional nutrients may increase the productivity of fisheries, but at higher concentrations this fertilizing effect is offset by hypoxia, and fishery yields decline.

## Environmental systems may be perceived in various ways

There are many ways to delineate natural systems, and your choice will depend on the particular issues in which you are interested. Categorizing environmental systems can help make Earth's dazzling complexity comprehensible to the human brain and accessible to problem solving.

For instance, scientists sometimes divide Earth's components into structural spheres. The **lithosphere** is the rock and sediment beneath our feet, in the planet's uppermost layers. The **atmosphere** is composed of the air surrounding our planet. The **hydrosphere** encompasses all water—salt or fresh, liquid, ice, or vapor—in surface bodies, underground, and in the atmosphere. The **biosphere** consists of the total of all the planet's living organisms and the abiotic (nonliving) portions of the environment with which they interact. We will examine the workings of various environmental systems within these structural spheres in more detail in later chapters.

Although these categories can be useful, their boundaries overlap, so the systems interact. Picture a robin plucking an earthworm from the ground after a rain. You are witnessing an organism (the robin) consuming another organism (the earthworm) by removing it from part of the lithosphere (the soil) that the earthworm had been modifying—all this made possible because rain (from the hydrosphere) recently wet the ground. The robin might then fly through the air (the atmosphere) to a tree (an organism), in the process respiring (combining oxygen from the atmosphere with glucose from the organism, and adding water to the hydrosphere and carbon dioxide and heat to the atmosphere). Finally, the bird might defecate, adding nutrients from the organism to the lithosphere below.

The study of such interactions among living and nonliving things is a key part of ecology at the ecosystem level. As scientists become more inclined to approach systems holistically, ecology at the ecosystem level and beyond has become increasingly important.

# Ecosystems

An **ecosystem** consists of all organisms and nonliving entities that occur and interact in a particular area at the same time. The ecosystem concept builds on the idea of the biological community (Chapter 6), but ecosystems include abiotic components as well as biotic ones. In ecosystems, energy flows and matter cycles among these components.

## Ecosystems are systems of interacting living and nonliving entities

The idea of ecosystems originated early last century with scientists such as British ecologist Arthur Tansley, who saw that biological entities are tightly intertwined with chemical and physical entities. Tansley and others felt that there was so much interaction and feedback between organisms and their abiotic environments that it made most sense to view living and nonliving elements together. For instance, the flow of water, sediment, and nutrients from the Mississippi River plays a key role in the nearshore ecosystem of the northern Gulf of Mexico. The input of moisture from clouds plays a key role in the Monteverde cloud forest ecosystem we visited in Chapter 5. And abiotic factors such as temperature, precipitation, latitude, and elevation have substantial influence over which biomes exist in a given locality (• pp. 161–163).

Ecologists soon began analyzing ecosystems as an engineer might analyze the operation of a machine. In this view, ecosystems are systems that receive inputs of energy, process and transform that energy while cycling matter internally, and produce a variety of outputs (such as heat, water flow, and animal waste products) that can move into other ecosystems. Energy flows in one direction through ecosystems; most arrives as radiation from the sun, powers the system, and exits in the form of heat (**Figure 7.6a**). Matter, in contrast, is generally recycled within ecosystems (**Figure 7.6b**). We saw in Chapter 6 (• pp. 148–151) how energy and matter are passed among organisms (producers, consumers, and decomposers) through food web relationships. Matter is recycled because when organisms die and decay

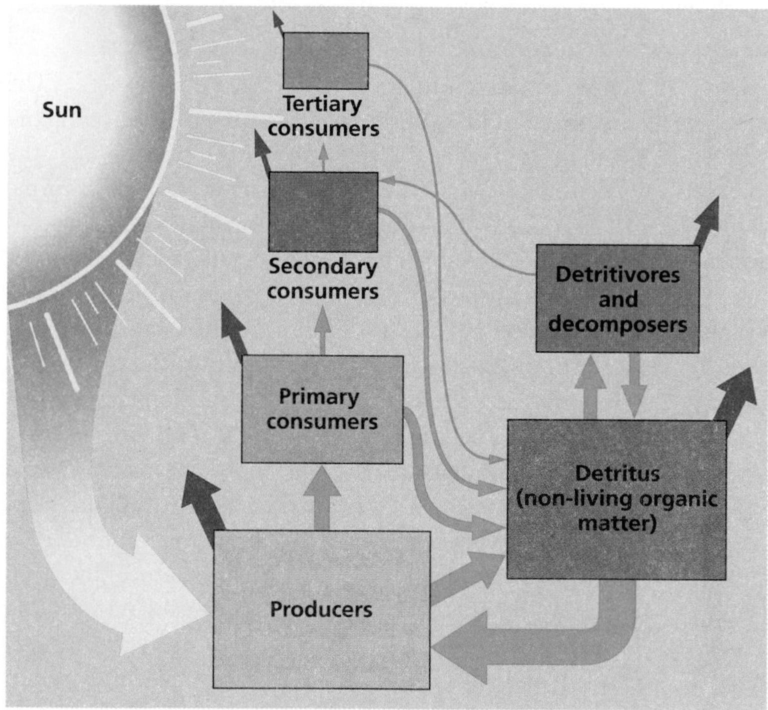


**(a) Energy flowing through an ecosystem**

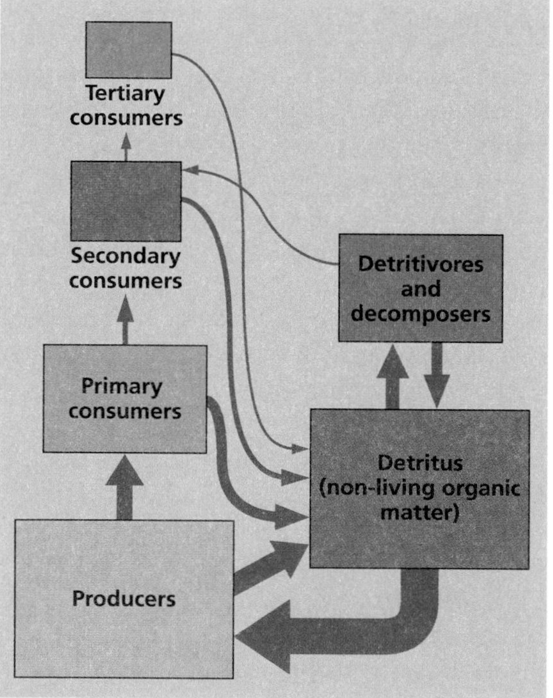


**(b) Matter cycling within an ecosystem**

FIGURE 7.6 Energy enters, flows through, and exits an ecosystem. In (**a**), light energy from the sun (yellow arrow) drives photosynthesis in producers, which begins the transfer of chemical energy (green arrows) among trophic levels and detritus. Energy exits the system through respiration in the form of heat (red arrows). In contrast, matter cycles within an ecosystem. In (**b**), blue arrows show the movement of nutrients among trophic levels and detritus. In both diagrams, box sizes represent relative magnitudes of energy or matter content, and arrow widths represent relative magnitudes of energy or matter transfer. Such magnitudes may vary tremendously from one ecosystem to another. For simplicity, various abiotic components (such as water, air, and inorganic soil content) of ecosystems are omitted from these schematic diagrams.

their nutrients remain in the system. In contrast, most energy that organisms take in is later lost through respiration.

## Energy is converted to biomass

Energy flow in most ecosystems begins with radiation from the sun. In Chapter 4 (• p. 101–103) we explored how autotrophs, such as green plants and phytoplankton, use photosynthesis to capture the sun's energy and produce food. We then saw in Chapter 6 (• pp. 148–150) how organisms at higher trophic levels consume organisms at lower trophic levels, transferring energy.

As autotrophs convert solar energy to the energy of chemical bonds in sugars, they perform *primary production.* Specifically, the assimilation of energy by autotrophs is termed **gross primary production.** Autotrophs use a portion of this production to power their own metabolism by respiration (• p. 103). The energy that remains after respiration and is used to generate biomass ecologists call **net primary production.** Thus, net primary production equals gross primary production minus respiration. Net primary production can be measured by the energy or the organic matter stored by plants after they have metabolized enough for their own maintenance.

Another way to think of net primary production is that it represents the energy or biomass available for consumption by heterotrophs. Plant matter not eaten by herbivores becomes fodder for detritivores and decomposers once the plant dies or drops its leaves. Heterotrophs use the energy they gain from plants for their own metabolism, growth, and reproduction. The total biomass that heterotrophs generate by consuming autotrophs is termed *secondary production.*

Ecosystems vary in the rate at which plants convert energy to biomass. The rate at which production occurs is termed **productivity,** and ecosystems whose plants convert solar energy to biomass rapidly are said to have high **net primary productivity.**

Freshwater wetlands, tropical forests, coral reefs, and algal beds tend to have the highest net primary productivities, whereas deserts, tundra, and open ocean tend to have the lowest (**Figure 7.7a**). Variation among ecosystems and among biomes in net primary productivity results in geographic patterns of variation across the globe (**Figure 7.7b**). In terrestrial ecosystems, net primary productivity tends to increase with temperature and precipitation. In aquatic ecosystems, net primary productivity tends to rise with light and the availability of nutrients.

## Nutrients can limit productivity

**Nutrients** are elements and compounds that organisms consume and require for survival. Organisms need several dozen naturally occurring chemical elements to survive. Elements and compounds required in relatively large amounts are called *macronutrients* and include nitrogen, carbon, and phosphorus. Nutrients needed in small amounts are called *micronutrients.*

Nutrients stimulate production by plants, and lack of nutrients can limit production. The availability of nitrogen or phosphorus frequently is a limiting factor (• pp. 129–130) for plant or algal growth. When these nutrients are added to a system, producers show the greatest response to whichever nutrient has been in shortest supply. Phosphorus tends to be limiting in freshwater systems, and nitrogen in marine systems. Thus the Gulf of Mexico's hypoxic zone is thought to result primarily from excess nitrogen, whereas ponds and lakes in the Mississippi River Valley tend to suffer eutrophication when they contain too much phosphorus.

Canadian ecologist David Schindler and others demonstrated the effects of phosphorus on freshwater systems in the 1970s by experimentally manipulating entire lakes. In one experiment, the researchers divided a 16-ha (40-acre) lake in Ontario in half with a plastic barrier. To one half the researchers added carbon, nitrate, and phosphate; to the other they added only carbon and nitrate. Soon after the experiment began, they witnessed a dramatic increase in algae in the half of the lake that received phosphate, whereas the other half hosted algal levels typical for lakes in the region (**Figure 7.8**). This difference held until shortly after they stopped fertilizing 7 years later, when algae decreased to normal levels in the half that had previously received phosphate. Such experiments showed clearly that phosphorus addition can markedly increase primary productivity in lakes.

Similar experiments in coastal ocean waters show nitrogen to be the more important limiting factor for primary productivity. In experiments in the 1980s and 1990s, Swedish ecologist Edna Granéli took samples of ocean water from the Baltic Sea and added phosphate, nitrate, or nothing. Chlorophyll and phytoplankton increased greatly in the flasks with nitrate, whereas those with phosphate did not differ from the controls. Experiments in Long Island Sound by other researchers show similar results. For open ocean waters far from shore, research indicates that iron is a highly effective nutrient.

Because nutrients run off from land in the Baltic, Long Island Sound, the Gulf of Mexico, and worldwide, primary productivity in the oceans tends to be greatest in nearshore waters, which receive nutrient runoff from land, and lowest in open ocean areas far from land (see Figure 7.7b). Satellite imaging technology that reveals phytoplankton densities has given scientists an improved view of productivity at regional and global scales, which has helped them track blooms of algae that contribute to coastal hypoxic zones (**Figure 7.9**).

The number of known dead zones is increasing globally, with about 200 documented so far. Most are located off the coasts of Europe and the eastern United

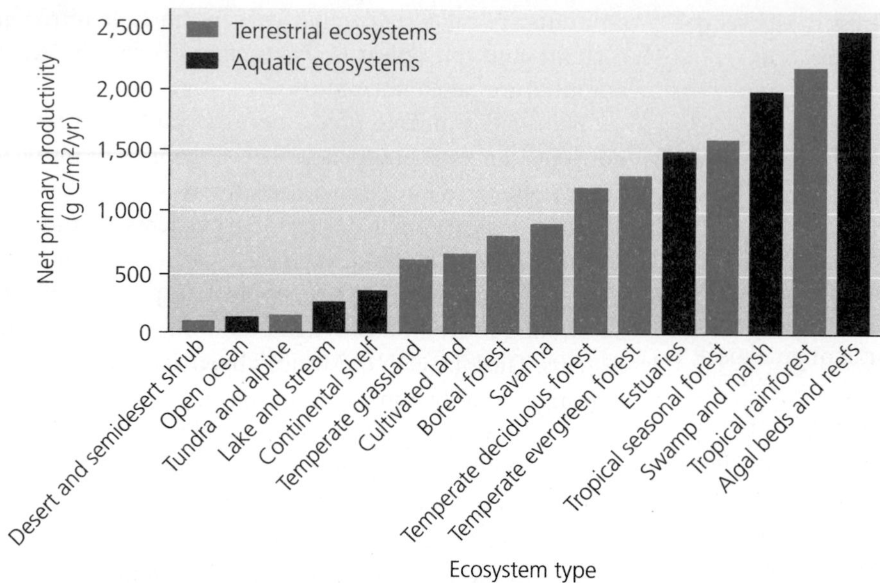


**(a) Net primary productivity for major ecosystem types**

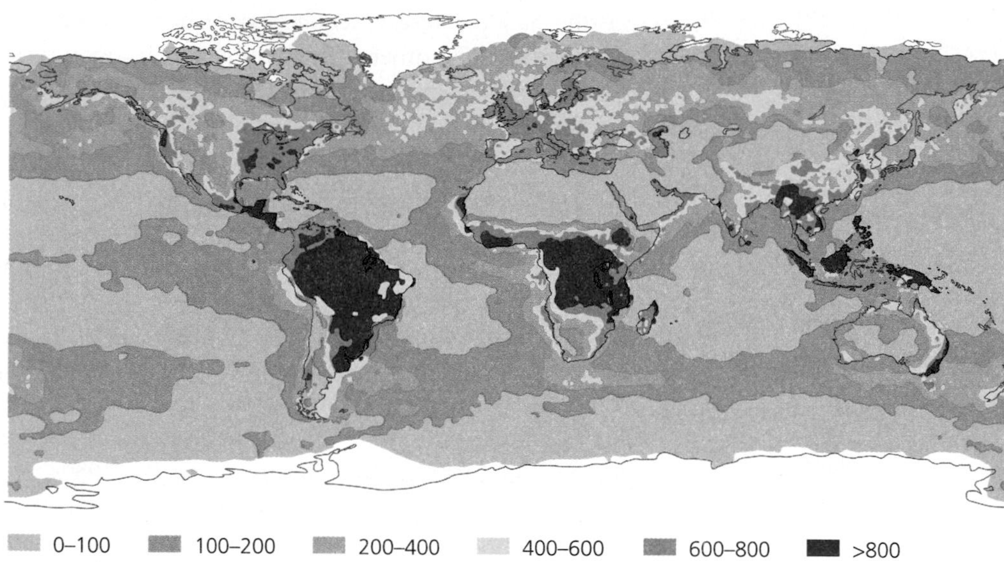


**(b) Global map of net primary productivity**

FIGURE 7.7 **(a)** Freshwater wetlands, tropical forests, coral reefs, and algal beds show high net primary productivities on average, whereas deserts, tundra, and the open ocean show low values. **(b)** On land, net primary production varies geographically with temperature and precipitation, which also influence the locations of biomes (• pp. 161–163). In the world's oceans, net primary production is highest around the margins of continents, where nutrients (of both natural and human origin) run off from land. Data in (a) from Whittaker, R. H. 1975. *Communities and ecosystems*, 2nd ed. New York: MacMillan. Map in (b) from satellite data presented by Field, C. B., et al. 1998. Primary production of the biosphere: Integrating terrestrial and oceanic components. *Science* 281: 237–240.

States. Specific causes vary from place to place, but most result from rising nutrient pollution from farms, cities, and industry. In North America, Chesapeake Bay may be the most severely affected area besides the Gulf of Mexico. Decades of pollution and human impact here have devastated fisheries and greatly altered the bay's ecology.

Good news of a sort comes from the Black Sea, which borders Ukraine, Russia, Turkey, and eastern Europe. This immense inland sea had long suffered one of the world's worst hypoxic zones, which destroyed many of its fisheries and seagrass beds. Then in the 1990s, after the Soviet Union collapsed, industrial agriculture in the region declined drastically. With fewer fertilizers running off into it, the Black Sea began to recover, and today fish and mussel beds are reviving. However, agricultural collapse is not a strategy anyone would choose to alleviate hypoxia. Rather, scientists are proposing a variety of innovative and economically acceptable ways to reduce nutrient runoff.

## Ecosystems integrate spatially

We can conceptualize ecosystems at different scales. An ecosystem can be as small as an ephemeral puddle of water where brine shrimp and tadpoles feed on algae and detritus with mad abandon as the pool dries up. Or

FIGURE 7.8 A portion of this lake in Ontario was experimentally treated with the addition of phosphate. This treated portion experienced an immediate, dramatic, and prolonged algal bloom, visible in the opaque water in the topmost part of this photo. Photo by David W. Schindler.

an ecosystem might be as large as a bay, lake, or forest. For some purposes, scientists even view the entire biosphere as a single all-encompassing ecosystem. The term is most often used, however, to refer to systems of moderate geographic extent that are somewhat self-contained. For example, the salt marshes that line the outer delta of the Mississippi where its waters mix with those of the Gulf of Mexico may be classified as an ecosystem.

Adjacent ecosystems may share components and interact extensively. For instance, a pond ecosystem is very different from a forest ecosystem that surrounds it, but salamanders that develop in the pond live their adult lives under logs on the forest floor until returning to the pond to breed. Rainwater that nourishes forest plants may eventually make its way to the pond, carrying with it nutrients from the forest's leaf litter. Likewise, coastal dunes, the ocean, and a lagoon or salt marsh all may interact, as do forests and prairie where they converge. Areas where ecosystems meet may consist of transitional zones called **ecotones,** in which elements of each ecosystem mix.

### Weighing THE Issues | Ecosystems Where You Live

Think about the area where you live. How would you describe that area's ecosystems? How do these systems interact with one another? If one ecosystem were greatly disturbed (say, if a wetland or forest were replaced by a shopping mall), what impacts might that have on nearby natural systems?

## Landscape ecologists study geographic areas with multiple ecosystems

Because components of different ecosystems may intermix, ecologists often find it useful to view these systems on a larger geographic scale that encompasses multiple ecosystems. For instance, if you are studying large mammals such as black bears, which move seasonally from mountains to valleys or between mountain ranges, you had better consider the overall landscape that includes all these areas. If you study fish such as salmon, which move between marine and fresh-water ecosystems, you need to know how these systems interact.

FIGURE 7.9 Satellite images like this one have helped scientists track runoff and phytoplankton blooms. Shown is the Louisiana coast on December 30, 1997. Greenish brown at the top is land, and deep blue at the bottom right is water of the Gulf of Mexico. Along the coast, two light brown plumes of sediment can be seen expanding into the Gulf's waters from the mouths of the Mississippi and Atchafalaya rivers. The shades of turquoise represent phytoplankton blooms that result from nutrient inputs and that foster the dead zone. Image from the SeaWiFS Project, NASA/Goddard Space Flight Center, and ORBIMAGE.

In such a broad-scale approach, called **landscape ecology**, scientists study how landscape structure affects the abundance, distribution, and interaction of organisms. A landscape-level approach is also proving useful for scientists, citizens, planners, and policymakers to plan for sustainable regional development (see the "Science behind the Story," • pp. 186–187).

For a landscape ecologist, a landscape is made up of a spatial array of **patches**. Depending on the researcher's perspective, patches may be ecosystems, or communities, or areas of habitat for a particular organism. Patches are spread spatially over a landscape in a **mosaic**. This metaphor reflects how natural systems often are arrayed across landscapes in complex patterns, like an intricate work of art. Thus, a forest ecologist may refer to a mosaic of forested patches remaining in an agricultural landscape, whereas a butterfly biologist might speak of a mosaic of patches of grassland habitat for a particular species of butterfly.

In referring to a *landscape*, ecologists generally imply a spatial scale larger than an ecosystem but smaller than a biome (• pp. 161–169). However, one can view a landscape at different scales. **Figure 7.10** illustrates a landscape consisting of four ecosystem types, with ecotones along their borders indicated by thick red lines. At this scale, we perceive a mosaic consisting of four patches and a river. However, the inset shows a magnified view of an ecotone. At this finer resolution, we see that the ecotone consists of patches of forest and grassland in a complex arrangement. The scale at which an ecologist focuses will depend on the questions he or she is interested in, or on the organisms he or she is studying.

Every organism has specific habitat needs, so when its habitat is distributed in patches across a landscape, individuals may need to expend energy and risk predation traveling from one to another. If the patches are distant enough, the organism's population may become divided into subpopulations, each occupying a different patch in

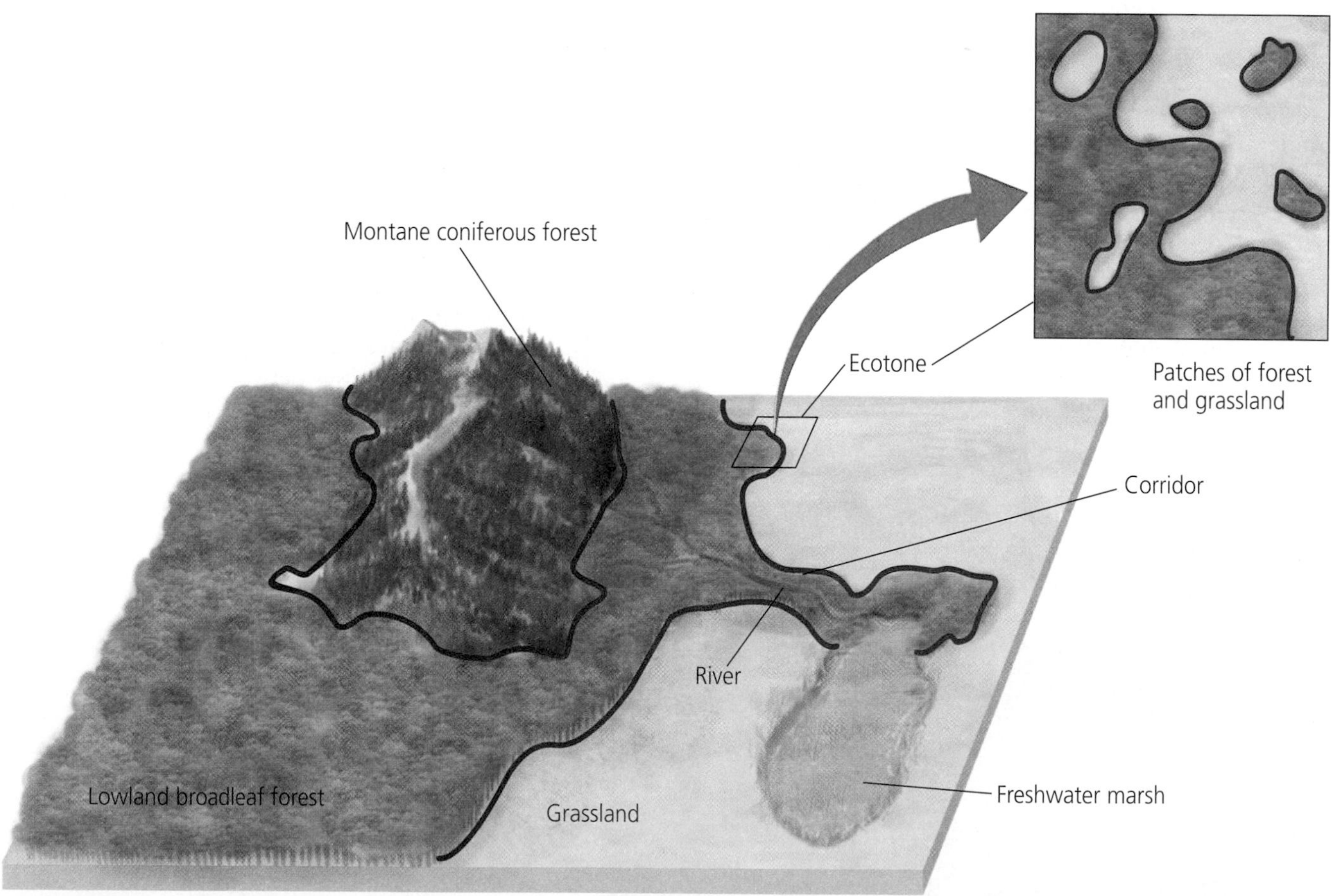


**FIGURE 7.10** Landscape ecology deals with spatial patterns above the ecosystem level. This generalized diagram of a landscape shows a mosaic of patches of five ecosystem types (three terrestrial types, a marsh, and a river). Thick red lines indicate ecotones. A stretch of lowland broadleaf forest running along the river serves as a corridor connecting the large region of forest on the left to the smaller patch of forest alongside the marsh. The inset shows a magnified view of the forest-grassland ecotone and how it consists of patches on a smaller scale.

the mosaic. Such a network of subpopulations, most of whose members stay within their respective patches but some of whom move among patches or mate with members of other patches, is called a **metapopulation.** When patches are still more isolated from one another, individuals may not be able to travel between them at all. In such a case, smaller subpopulations may be at risk of extinction.

Because of this extinction risk, metapopulations and landscape ecology are of great interest to **conservation biologists** (• p. 315), who study the loss, protection, and restoration of biodiversity. Of particular concern is the fragmentation of habitat into small and isolated patches (• pp. 316–317, 352)—something that often results from human impact.

Establishing corridors of habitat (see Figure 7.10) to link patches is one approach that conservation biologists pursue as they attempt to maintain biodiversity in the face of human impact. We will return to these issues when we study conservation biology and habitat fragmentation in more detail in Chapters 11 and 12 (• pp. 315–323, 352–353).

## Remote sensing helps us apply landscape ecology

More and more scientists are taking a landscape perspective these days, and remote sensing technologies are improving our ability to do so. Satellites orbiting Earth are sending us more and better data than ever before on what the surface of our planet looks like (see Figure 7.9). By helping us monitor our planet from above, satellite imagery has become a vital tool in modern environmental science.

A common tool for research in landscape ecology is the **geographic information system (GIS).** A GIS consists of computer software that takes multiple types of data (for instance, on geology, hydrology, vegetation, animal species, and human development) and combines them on a common set of geographic coordinates. The idea is to create a complete picture of a landscape and to analyze how elements of the different data sets are arrayed spatially and how they may be correlated.

**Figure 7.11** illustrates in a simplified way how different datasets of a GIS are combined, layer upon layer, to form a composite map. GIS has become a valuable tool used by geographers, landscape ecologists, resource managers, and conservation biologists. GIS technology also brings insights that affect planning and land-use decisions.

Some conservation groups, such as the Nature Conservancy, now apply the landscape ecology approach widely in their land acquisition and management strategies. Principles of landscape ecology, and tools such as GIS, are increasingly used in regional planning processes, such as the Sonoran Desert Conservation Plan (see the "Science behind the Story," • pp. 186–187).

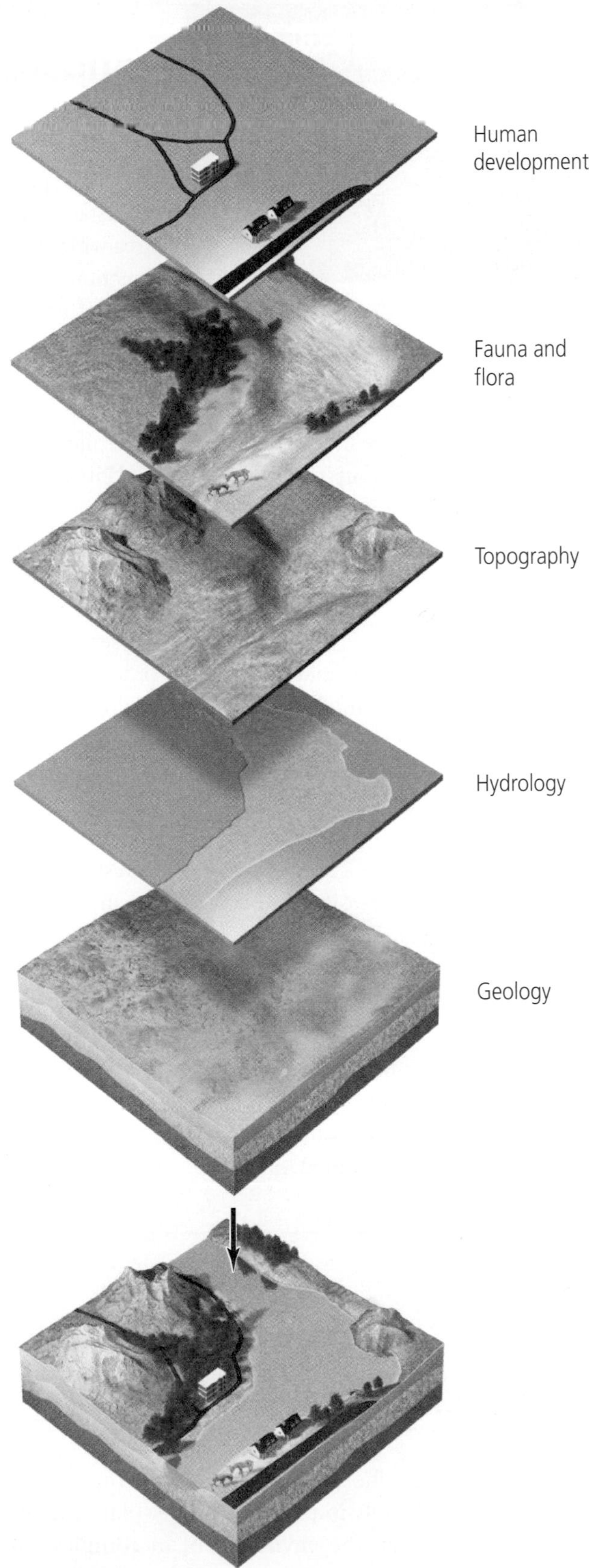


**FIGURE 7.11** Geographic information systems (GIS) allow us to layer different types of data on natural landscape features and human land uses so as to produce maps integrating this information. GIS can be used to explore correlations among these data sets and to help in regional planning.

THE SCIENCE BEHIND THE STORY

## Landscape-level Planning in the Sonoran Desert

*Scenic Sonoran Desert outside Tucson*

People who live in southeastern Arizona know they reside in a special place—where mountains rise like "sky islands" up from the stark beauty of the Sonoran Desert, and where cottonwood-lined creeks snake through golden grasslands.

But as more people flock to this alluring region, Tucson and its surrounding communities are booming in population as development sprawls across the landscape, gobbling up 26 km$^2$ (10 mi$^2$) of scenic desert each year. Water tables are falling, rivers are running dry, pollution is increasing, and it takes longer and longer to get across town.

As residents of Tucson and the rest of Pima County voiced concern over losing their special environment and quality of life, the Sonoran Desert Conservation Plan (SDCP) was born. Adopted in 2001, the SDCP puts forth a vision of how to sustain a thriving metropolitan area within a healthy, intact, and functioning natural environment. In particular, it spells out how land and water in different areas of Pima County should be used and managed.

Policymakers, businesspeople, and citizens of all stripes participated in the regional planning (• p. 367) process that culminated in the SDCP, but scientists played a major role—particularly geographers, conservation biologists, and others practicing aspects of landscape ecology.

Contributing were over 150 scientists from colleges and universities such as the University of Arizona, government agencies such as the U.S. Forest Service, and nongovernmental organizations such as the Nature Conservancy. Together they worked to determine what lands would be best to preserve and what areas could most easily be developed without losing precious natural or cultural heritage.

Scientists identified high-priority sites for conservation by drawing together data on biodiversity patterns, important riparian areas, habitat corridors used by wildlife, cultural and archaeological sites, and locations with endangered species. They analyzed published reports, natural history databases, information from experts convened at workshops, and data from satellite imagery. Experts nominated areas for conservation based on how well each area fulfilled the SDCP's goals, and scientists and planners then refined the areas' boundaries.

Researchers used geographic information systems (GIS; • p. 185) to combine and map data on topography, hydrography, vegetation cover, biodiversity, archaeology, and land use. With all these data sets layered together, scientists, planners, and policymakers could envision which regions were the richest in natural and cultural value. These lands were given higher priority for conservation.

Taking this landscape-level view also enabled researchers to study spatial patterns such as the configuration of patches in the landscape mosaic and the connectivity among patches. Conservation biologists argued for

# Biogeochemical Cycles

Just as nitrogen and phosphorus from Minnesota soybean fields end up in Gulf shrimp on our dinner plates, all nutrients move through the environment in complex and fascinating ways. Whereas energy enters an ecosystem from the sun, flows from one organism to another, and is dissipated to the atmosphere as heat, the physical matter of an ecosystem is circulated over and over again.

## Nutrients circulate through ecosystems in biogeochemical cycles

Nutrients move through ecosystems in **nutrient cycles** or **biogeochemical cycles.** In these cycles they travel through the atmosphere, hydrosphere, and lithosphere, and from one organism to another, in dynamic equilibrium. A carbon atom in your fingernail today might have helped comprise the muscle of a cow a year ago, may have

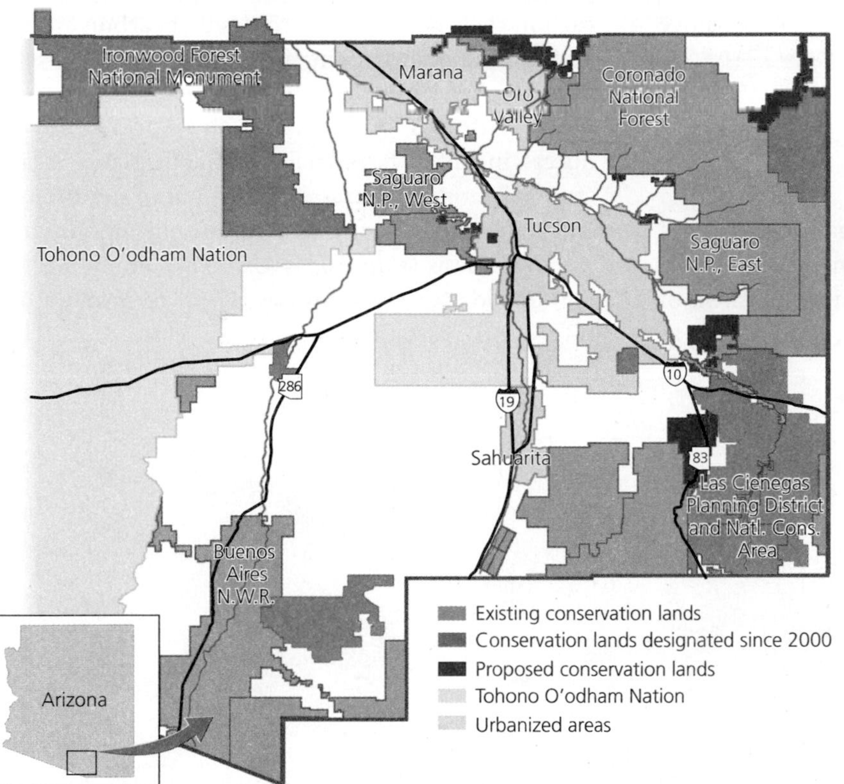


As a result of the Sonoran Desert Conservation Plan process, many lands in eastern Pima County, Arizona, have been designated for conservation since 2000 (orange), and more are currently proposed for public acquisition (red).

preserving segments of land that could serve as corridors for the movement of wildlife such as mountain lions, bobcats, coatimundis, and migratory birds.

The Sonoran Desert Conservation Plan has produced results. As the process was underway in 2000, Pima County helped federal agencies establish the 52,000-ha (129,000-acre) Ironwood National Monument and the 58,000 ha (143,000 acre) Las Cienegas National Conservation Area and Planning District. Since that time, the county has purchased land from a number of ranchers, so that the land can be managed for the public rather than subdivided and developed without regulation.

Then in 2004, Pima County voters passed a $174 million open space bond measure, providing funding for the county to acquire more land for conservation under the SDCP. So far, the county has purchased over 8,500 additional hectares (21,000 acres) with this funding and is considering further parcels for acquisition (see figure).

The SDCP has also spurred other efforts. For instance, the Regional Transportation Authority approved $45 million to retrofit existing roads and design new roads with structures to allow wildlife to cross under them safely without risk of being hit by vehicles.

Through the hard work and dedication of researchers, citizens, and policymakers, the science of landscape ecology has helped to inform the planning process in Pima County—a process that has won praise and is being studied by other regions across North America. As a result, residents of the growing Tucson area will have a healthier environment in which to live, far into the future.

resided in a blade of grass a month before that, and may have been part of a dinosaur's tooth 100 million years ago. After we die, the nutrients in our bodies will spread widely through the environment, eventually being incorporated by an untold number of organisms far into the future.

Nutrients move from one *pool*, or *reservoir*, to another, remaining for varying amounts of time *(residence time)* in each. The dinosaur, the grass, the cow, and you are each reservoirs for carbon atoms. The movement of nutrients among pools is termed *flux*, and the rates of flux between any given pools can change over time. Human activity has influenced certain flux rates. We have increased the flux of nitrogen from the atmosphere to pools on the Earth's surface and have shifted the flux of carbon in the opposite direction. Reservoirs that release more nutrients than they accept are called *sources*, and reservoirs accepting more nutrients than they are releasing are called *sinks*. As we discuss biogeochemical cycles, think about how they involve negative feedback loops that promote dynamic equilibrium, and also consider how some human actions can generate positive feedback loops.

## The carbon cycle circulates a vital organic nutrient

As the definitive component of organic molecules (• pp. 95–96), carbon is an ingredient in carbohydrates, fats, and proteins and in the bones, cartilage, and shells of all living things. From fossil fuels to DNA, from plastics to pharmaceuticals, carbon (C) atoms are everywhere. The **carbon cycle** describes the routes that carbon atoms take through the environment (**Figure 7.12**).

**Photosynthesis, respiration, and food webs** Producers, including terrestrial and aquatic plants, algae, and cyanobacteria, pull carbon dioxide out of the atmosphere and out of surface water to use in photosynthesis. Photosynthesis (• pp. 102–103) breaks the bonds in carbon dioxide ($CO_2$) and water ($H_2O$) to produce oxygen ($O_2$) and carbohydrates (e.g., glucose, $C_6H_{12}O_6$).

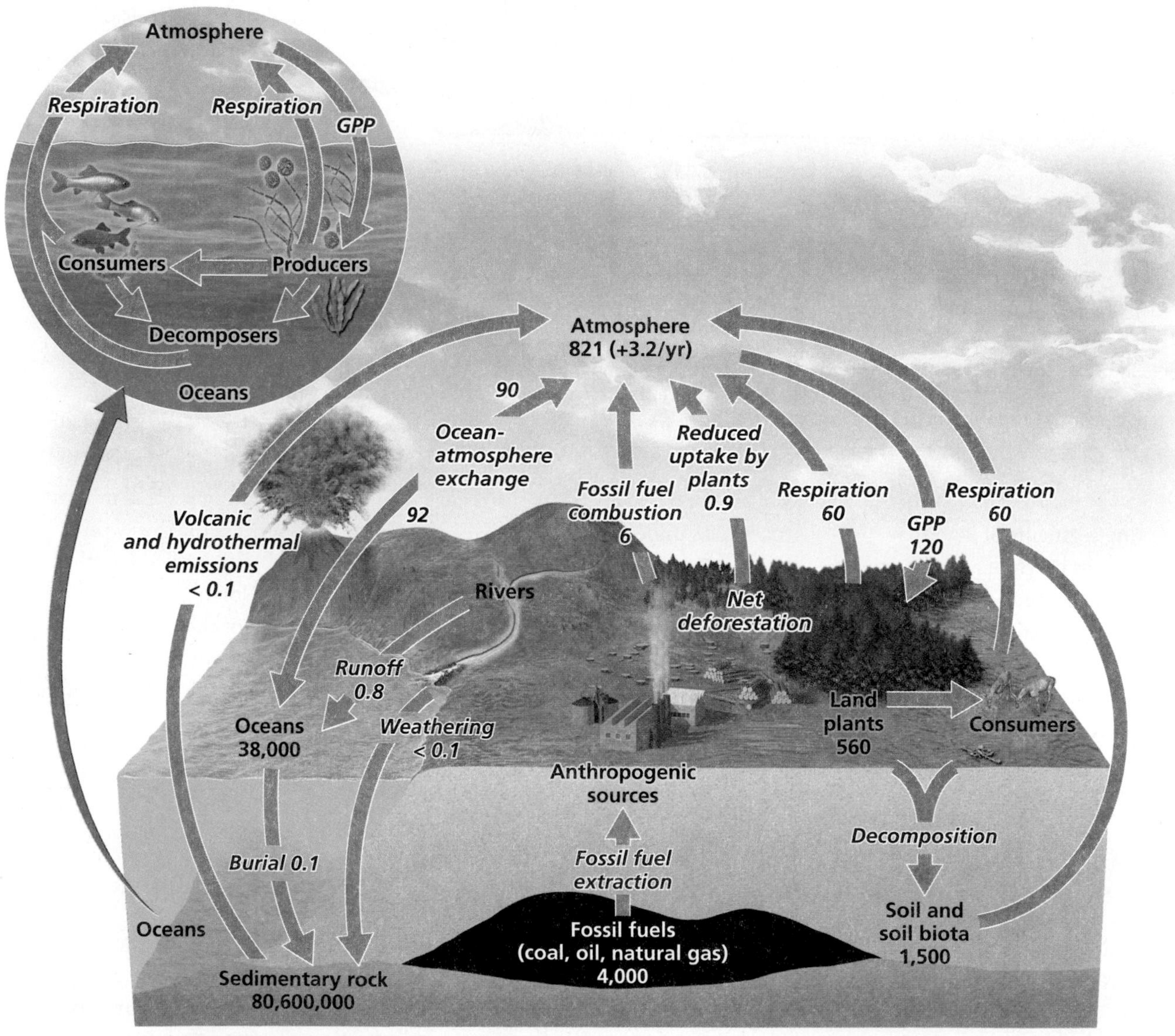


FIGURE 7.12 The carbon cycle summarizes the many routes that carbon atoms take as they move through the environment. Gray arrows represent fluxes among reservoirs, or pools, for carbon. In the carbon cycle, plants use carbon dioxide from the atmosphere for photosynthesis (gross primary production, or "GPP" in the figure). Carbon dioxide is returned to the atmosphere through respiration by plants, their consumers, and decomposers. The oceans sequester carbon in their water and in deep sediments. The vast majority of the planet's carbon is stored in sedimentary rock. In the figure, pool names are printed in black type, and numbers in black type represent pool sizes expressed in petagrams (units of $10^{15}$ g) of C. Processes, printed in italic red type, give rise to fluxes, printed in italic red type and expressed in petagrams of C per year. Data from Schlesinger, W. H. 1997. *Biogeochemistry: An analysis of global change*, 2nd ed. London: Academic Press.

Autotrophs use some of the carbohydrates to fuel their own respiration, thereby releasing some of the carbon back into the atmosphere and oceans as $CO_2$. When producers are eaten by primary consumers, which in turn are eaten by secondary and tertiary consumers, more carbohydrates are broken down in respiration, producing carbon dioxide and water. The same process occurs when decomposers consume waste and dead organic matter. Respiration from all these organisms releases carbon back into the atmosphere and oceans.

All organisms use carbon for structural growth, so a portion of the carbon an organism takes in becomes incorporated into its tissues. The abundance of plants and the fact that they take in so much carbon dioxide for photosynthesis makes plants a major reservoir for carbon. Because $CO_2$ is a greenhouse gas of primary concern (• pp. 508–509), much research on global climate change is directed toward measuring the amount of $CO_2$ that plants tie up. Scientists are working toward understanding exactly how much this portion of the carbon cycle influences Earth's climate.

**Sediment storage of carbon** As organisms die, their remains may settle in sediments in ocean basins or in freshwater wetlands. As layers of sediment accumulate, older layers are buried more deeply, experiencing high pressure over long periods of time. These conditions can convert soft tissues into fossil fuels—coal, oil, and natural gas—and shells and skeletons into sedimentary rock, such as limestone. Sedimentary rock comprises the largest single reservoir in the carbon cycle. Although any given carbon atom spends a relatively short time in the atmosphere, carbon trapped in sedimentary rock may reside there for hundreds of millions of years.

Carbon trapped in sediments and fossil fuel deposits may eventually be released into the oceans or atmosphere by geological processes such as uplift, erosion, and volcanic eruptions. It also reenters the atmosphere when we extract and burn fossil fuels.

**The oceans** The world's oceans are the second-largest reservoir in the carbon cycle. They absorb carbon-containing compounds from the atmosphere, from terrestrial runoff, from undersea volcanoes, and from the waste products and detritus of marine organisms. Some carbon atoms absorbed by the oceans—in the form of carbon dioxide, carbonate ions ($CO_3^{2-}$), and bicarbonate ions ($HCO_3^-$)—combine with calcium ions ($Ca^{2+}$) to form calcium carbonate ($CaCO_3$), an essential ingredient in the skeletons and shells of microscopic marine organisms. As these organisms die, their calcium carbonate shells sink to the ocean floor and begin to form sedimentary rock. The rates at which the oceans absorb and release carbon depend on many factors, including temperature and the numbers of marine organisms converting $CO_2$ into carbohydrates and carbonates.

## We are shifting carbon from the lithosphere to the atmosphere

By mining fossil fuel deposits, we are essentially removing carbon from an underground reservoir with a residence time of millions of years. By combusting fossil fuels in our automobiles, homes, and industries, we release carbon dioxide and greatly increase the flux of carbon from the ground to the air. Since the mid-18th century, our fossil fuel combustion has added about 250 billion metric tons (276 billion tons) of carbon to the atmosphere. Meanwhile, the movement of $CO_2$ from the atmosphere back to the hydrosphere, lithosphere, and biosphere has not kept pace.

In addition, cutting down forests and burning fields removes carbon from the pool of vegetation and releases it to the air. And if less vegetation is left on the surface, there are fewer plants to draw $CO_2$ back out of the atmosphere.

As a result, scientists estimate that today's atmospheric carbon dioxide reservoir is the largest that Earth has experienced in the past 650,000 years, and perhaps in the past 20 million years. The ongoing flux of carbon out of the fossil fuel reservoir and into the atmosphere is a driving force behind global climate change (Chapter 18).

Our understanding of the carbon cycle is not yet complete. Scientists have long been baffled by the so-called missing carbon sink. Of the carbon dioxide we emit by fossil fuel combustion and deforestation, scientists have measured how much goes into the atmosphere and oceans, but there remain roughly 1–2 billion metric tons unaccounted for. Many researchers think this must be taken up by plants or soils of the northern temperate and boreal forests (• pp. 163–164, 168). But they'd like to know for sure—because if certain forests are acting as a major sink for carbon (and thus restraining global climate change), we'd like to be able to keep it that way. For if forests that today are sinks were to turn into sources and begin releasing the "missing" carbon, climate change could accelerate drastically.

## The phosphorus cycle involves mainly lithosphere and ocean

The element phosphorus (P) is a key component of cell membranes and of several molecules vital for life, including DNA, RNA, ATP, and ADP (• pp. 97, 102). Although phosphorus is indispensable for life, the amount of phosphorus in organisms is dwarfed by the vast amounts in rocks, soil, sediments, and the oceans.

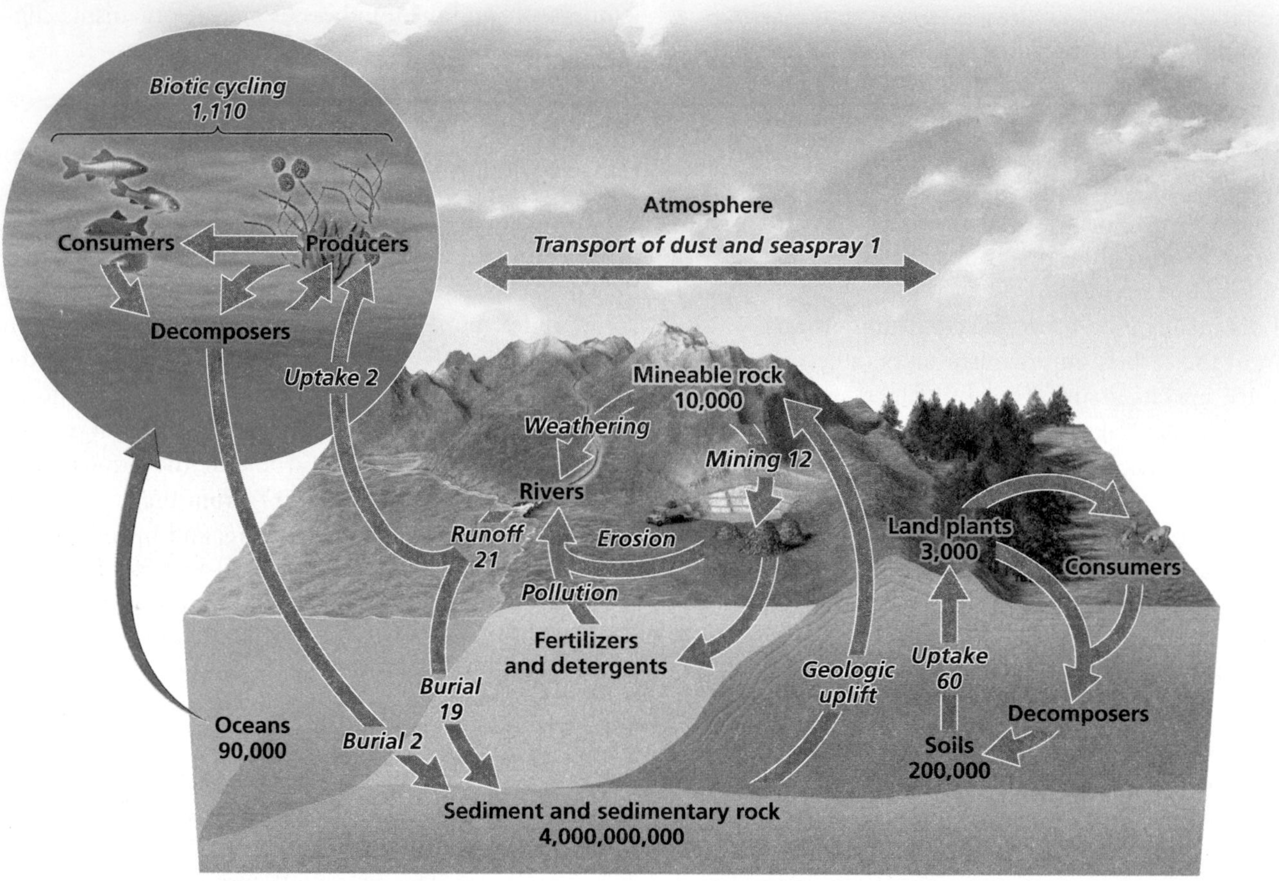


FIGURE 7.13 The phosphorus cycle summarizes the many routes that phosphorus atoms take as they move through the environment. Gray arrows represent fluxes among reservoirs, or pools, for phosphorus. Most phosphorus resides underground in rock and sediment, but the phosphorus cycle moves this element through the soil, the oceans, and freshwater and terrestrial ecosystems. Rocks containing phosphorus are uplifted geologically and weathered away in this slow process, and small amounts of phosphorus cycle through food webs, where this nutrient is often a limiting factor for plant growth. In the figure, pool names are printed in black type, and numbers in black type represent pool sizes expressed in teragrams (units of $10^{12}$ g) of P. Processes, printed in italic red type, give rise to fluxes, printed in italic red type and expressed in teragrams of P per year. Data from Schlesinger, W. H. 1997. *Biogeochemistry: An analysis of global change*, 2nd ed. London: Academic Press.

Unlike the carbon and nitrogen cycles, the **phosphorus cycle** (**Figure 7.13**) has no appreciable atmospheric component besides the transport of tiny amounts of wind-blown dust and seaspray.

**Geology and phosphorus availability** The vast majority of Earth's phosphorus is contained within rocks and is released only by weathering (• pp. 237–238), which releases phosphate ions ($PO_4^{3-}$) into water. Phosphates dissolved in lakes or in the oceans precipitate into solid form, settle to the bottom, and reenter the lithosphere's phosphorus reservoir in sediments. Because most phosphorus is bound up in rock and only slowly released, environmental concentrations of phosphorus available to organisms tend to be very low. This relative rarity explains why phosphorus is frequently a limiting factor for plant growth and why an artificial influx of phosphorus can produce immediate and dramatic effects.

**Food webs** Plants can take up phosphorus through their roots only when phosphate is dissolved in water. Primary consumers acquire phosphorus from water and plants and pass it on to secondary and tertiary consumers. Consumers also pass phosphorus to the soil through the excretion of waste. Decomposers break down phosphorus-rich organisms and their wastes and, in so doing, return phosphorus to the soil.

## We affect the phosphorus cycle

Humans influence the phosphorus cycle in several ways. We mine rocks containing phosphorus to extract this nutrient for the inorganic fertilizers we use on crops and lawns. Our wastewater discharge also tends to be rich in phosphates. Phosphates that run off into waterways can boost algal growth and cause eutrophication, leading to murkier waters

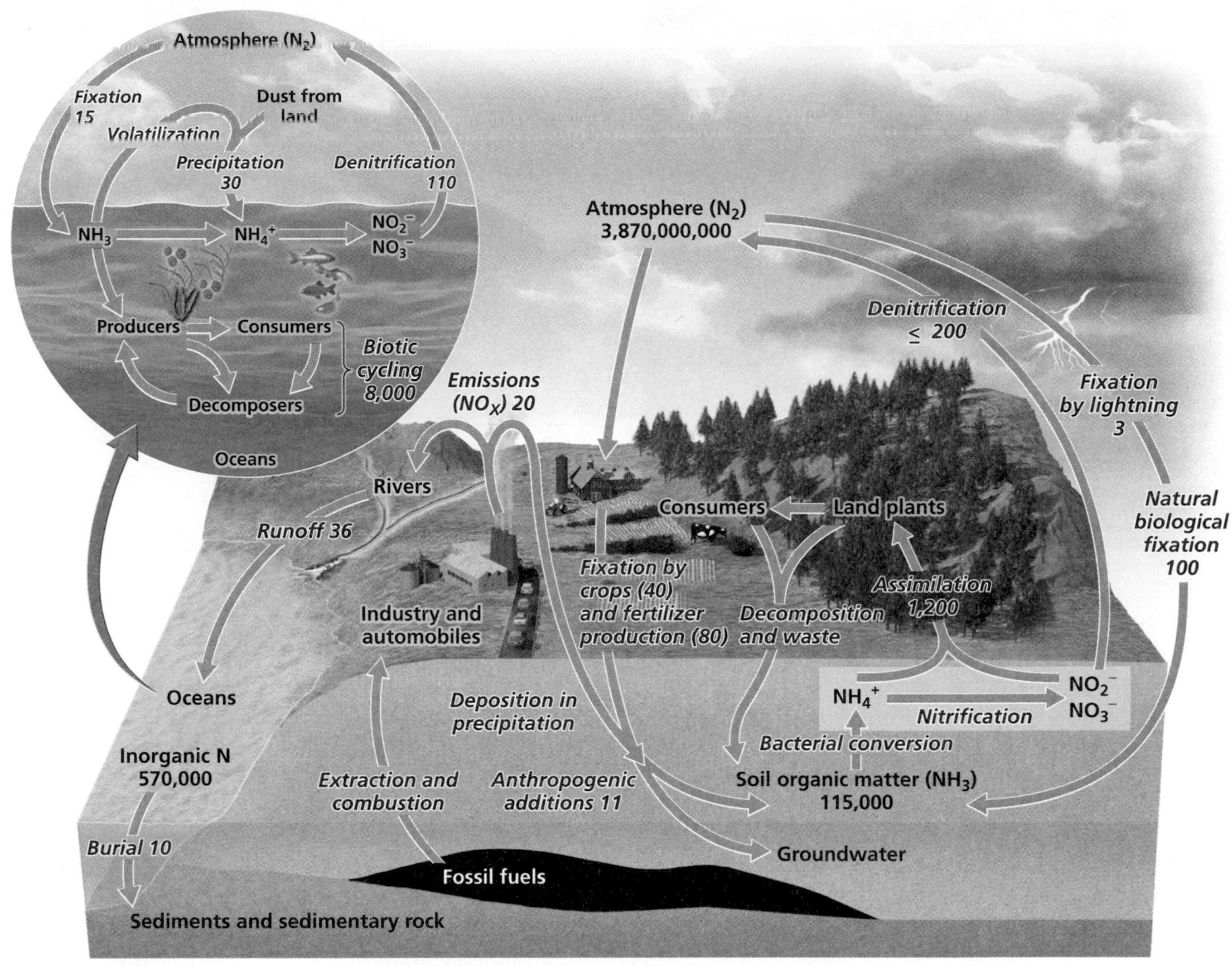


FIGURE 7.14 The nitrogen cycle summarizes the many routes that nitrogen atoms take as they move through the environment. Gray arrows represent fluxes among reservoirs, or pools, for nitrogen. In the nitrogen cycle, specialized bacteria play key roles in "fixing" atmospheric nitrogen and converting it to chemical forms that plants can use. Other types of bacteria convert nitrogen compounds back to the atmospheric gas $N_2$. In the oceans, inorganic nitrogen is buried in sediments, whereas nitrogen compounds are cycled through food webs as they are on land. In the figure, pool names are printed in black type, and numbers in black type represent pool sizes expressed in teragrams (units of $10^{12}$ g) of N. Processes, printed in italic red type, give rise to fluxes, printed in italic red type and expressed in teragrams of N per year. Data from Schlesinger, W. H. 1997. *Biogeochemistry: An analysis of global change*, 2nd ed. London: Academic Press.

and altering the structure and function of aquatic ecosystems. Phosphates are also present in detergents, so one way each of us can reduce phosphorus input into the environment is to purchase phosphate-free detergents.

## The nitrogen cycle involves specialized bacteria

Nitrogen (N) makes up 78% of our atmosphere by mass, and is the sixth most abundant element on Earth. It is an essential ingredient in the proteins, DNA, and RNA that build our bodies. Like phosphorus, nitrogen is an essential nutrient for plant growth. Thus the **nitrogen cycle** (**Figure 7.14**) is of vital importance to us and to all other organisms. Despite its abundance in the air, nitrogen gas ($N_2$) is chemically inert and cannot cycle out of the atmosphere and into living organisms without assistance from lightning, highly specialized bacteria, or human intervention. For this reason, the element is relatively scarce in the lithosphere and hydrosphere and in organisms. However, once nitrogen undergoes the right kind of chemical change, it becomes biologically active and available to the organisms that need it, and it can act as a potent fertilizer. Its scarcity makes biologically active nitrogen a limiting factor for plant growth.

**Nitrogen fixation** To become biologically available, inert nitrogen gas ($N_2$) must be "fixed," or combined with hydrogen in nature to form ammonia ($NH_3$), whose water-soluble ions of ammonium ($NH_4^+$) can be

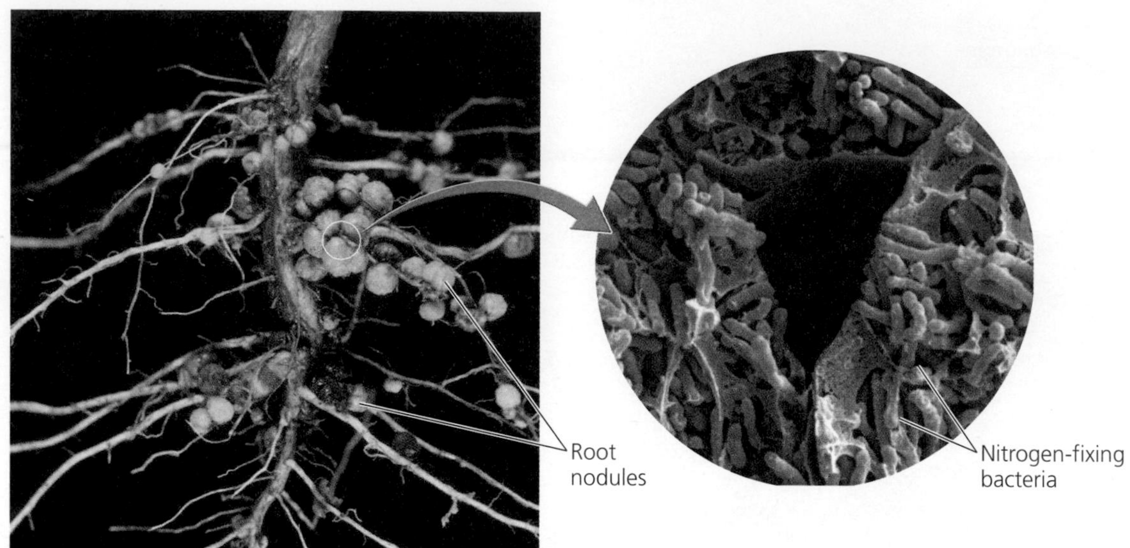


FIGURE 7.15 Specialized bacteria live in nodules on the roots of this legume plant. In the process of nitrogen fixation, the bacteria convert nitrogen to a form that the plant can take up into its roots.

taken up by plants. **Nitrogen fixation** can be accomplished in two ways: by the intense energy of lightning strikes, or when air in the top layer of soil comes in contact with particular types of **nitrogen-fixing bacteria.** These bacteria live in a mutualistic relationship (• p. 147) with many types of plants, including soybeans and other legumes, providing them nutrients by converting nitrogen to a usable form. As we will see in Chapter 9, farmers have long nourished their soils by planting crops that host nitrogen-fixing bacteria among their roots (**Figure 7.15**).

**Nitrification and denitrification** Other types of specialized bacteria then perform a process known as **nitrification.** In this process, ammonium ions are first converted into nitrite ions ($NO_2^-$), then into nitrate ions ($NO_3^-$). Plants can take up these ions, which also become available after atmospheric deposition on soils or in water or after application of nitrate-based fertilizer.

Animals obtain the nitrogen they need by consuming plants or other animals. Recall from Chapter 4 ("The Science behind the Story," • pp. 92–93) how scientists use stable isotopes of nitrogen to study the trophic level and nutritional condition of animals. Decomposers obtain nitrogen from dead and decaying plant and animal matter and from animal urine and feces. Once decomposers process the nitrogen-rich compounds they take in, they release ammonium ions, making these available to nitrifying bacteria to convert again to nitrates and nitrites.

The next step in the nitrogen cycle occurs when **denitrifying bacteria** convert nitrates in soil or water to gaseous nitrogen via a multistep process. Denitrification thereby completes the cycle by releasing nitrogen back into the atmosphere as a gas.

## We have greatly influenced the nitrogen cycle

The impacts of excess nitrogen from agriculture and other human activities in the Mississippi River watershed have become painfully evident to shrimpers and scientists with an interest in the Gulf of Mexico (see "The Science behind the Story" on • pp. 194–195). But hypoxia in the Gulf and other coastal locations around the world is hardly the only problem resulting from human manipulation of the nitrogen cycle.

Historically, nitrogen fixation was a *bottleneck,* a step that limited the flux of nitrogen out of the atmosphere. This changed when the research of two German chemists enabled us to fix nitrogen on an industrial scale. Fritz Haber worked in the German army's chemical weapons program during World War I. Shortly before the war, Haber found a way to combine nitrogen and hydrogen gases to synthesize ammonia, a key ingredient in modern explosives and agricultural fertilizers. Several years later, Carl Bosch built on Haber's work and devised methods to produce ammonia on an industrial scale. The work of these two scientists enabled people to overcome the limits on productivity long imposed by nitrogen scarcity in nature. The widespread application of their findings has enhanced agriculture and thereby contributed to the enormous increase in human population over the past 90 years. Farmers, golf course managers, and homeowners have all taken advantage of the fertilizers made possible by the **Haber-Bosch process.** These developments have led to a dramatic alteration of the nitrogen cycle. Today, using the Haber-Bosch process, our species is fixing at least as much nitrogen artificially as is being fixed naturally. We have effectively doubled the natural rate of nitrogen fixation on Earth (**Figure 7.16**).

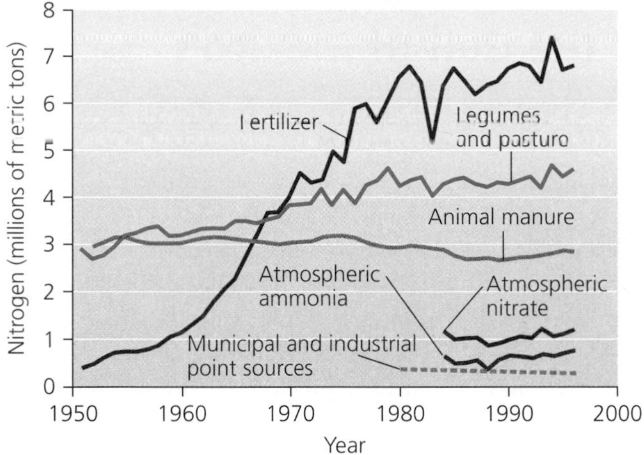


**FIGURE 7.16** In the past half century, human inputs of nitrogen into the environment have greatly increased, and today fully half of the nitrogen entering the environment is of human origin. These data for the Mississippi River basin show that agricultural fertilizer has for the past 40 years been the leading source of all nitrogen inputs, natural and artificial. Data from National Science and Technology Council: Committee on Environment and Natural Resources. May 2000. *Hypoxia: An integrated assessment in the northern Gulf of Mexico.*

By fixing atmospheric nitrogen, we increase its flux out of the atmosphere and into other reservoirs. In addition, we have affected fluxes in other parts of the cycle. When we burn forests and fields, we force nitrogen out of soils and vegetation and into the atmosphere. When we burn fossil fuels, we increase the rate at which nitric oxide (NO) enters the atmosphere and reacts to form nitrogen dioxide ($NO_2$). This compound is a precursor to nitric acid ($HNO_3$), a key component of acid precipitation (• pp. 491–495). We introduce another nitrogen-containing gas, nitrous oxide ($N_2O$), by allowing anaerobic bacteria to break down the tremendous volume of animal waste produced in agricultural feedlots (• pp. 281–282). We have also accelerated the introduction of nitrogen-rich compounds into terrestrial and aquatic systems by destroying wetlands and cultivating more legume crops that host nitrogen-fixing bacteria in their roots.

These activities increase amounts of nitrogen available to aquatic plants and algae, boosting their growth. Algal populations soon outstrip the availability of other required nutrients and begin to die and decompose. As in the Gulf of Mexico, this large-scale decomposition can rob other aquatic organisms of oxygen, leading to shellfish die-offs and other significant impacts on ecosystems. Increased nitrate pollution, such as that found in groundwater in the Mississippi River basin, can also lead to human health effects.

In 1997, a team of scientists led by Peter Vitousek of Stanford University summarized the dramatic changes humans have caused in the global nitrogen cycle. According to the Vitousek team's report, we have

- Doubled the rate at which fixed nitrogen enters terrestrial ecosystems (and the rate is still increasing).
- Increased atmospheric concentrations of the greenhouse gas $N_2O$ and of other oxides of nitrogen that produce smog.
- Depleted essential nutrients, such as calcium and potassium, from soils, because fertilizer helps flush them out.
- Acidified surface water and soils.
- Greatly increased transfer of nitrogen from rivers to oceans.
- Encouraged plant growth, causing more carbon to be stored within terrestrial ecosystems.
- Reduced biological diversity, especially plants adapted to low nitrogen concentrations.
- Changed the composition and function of estuaries and coastal ecosystems.
- Harmed many coastal marine fisheries.

## Weighing THE Issues | Nitrogen Pollution and Its Financial Impacts

Most nitrate that enters the Gulf of Mexico originates from farms and other sources in the upper Midwest, yet many of its negative impacts are borne by downstream users, such as Gulf Coast fishermen. Who do you believe should be responsible for addressing this problem? Should environmental policies on this issue be developed and enforced by state governments, the federal government, both, or neither? Explain the reasons for your answer.

In 1998, the U.S. Congress passed the Harmful Algal Bloom and Hypoxia Research and Control Act. This law called for an "integrated assessment" of hypoxia in the northern Gulf to address the extent, nature, and causes of the dead zone, as well as its ecological and economic impacts. The assessment report published 2 years later outlined potential solutions and their estimated social and economic costs. The report proposed that the federal government work with Gulf Coast and Midwestern communities to

- Reduce nitrogen fertilizer use on Midwestern farms.
- Change the timing of fertilizer application to minimize rainy-season runoff.
- Use alternative crops.
- Manage nitrogen-rich livestock manure more effectively.
- Restore nitrogen-absorbing wetlands in the Mississippi River basin.
- Use artificial wetlands to filter farm runoff.
- Improve technologies in sewage treatment plants.
- Restore frequently flooded lands to reduce runoff.
- Restore wetland ecosystems near the Mississippi River's mouth to enhance nitrogen-absorbing ability.
- Evaluate how these approaches work.

THE SCIENCE BEHIND THE STORY

# Hypoxia and the Gulf of Mexico's "Dead Zone"

*Dr. Nancy Rabalais, LUMCON*

She was prone to seasickness, but Nancy Rabalais cared too much about the Gulf of Mexico to let that stop her. Leaning over the side of an open boat idling miles from shore, she hauled a water sample aboard—and helped launch the long effort to breathe life back into the Gulf's "dead zone."

Since that expedition in 1985, Rabalais, her colleague and husband Eugene Turner, and fellow scientists at the Louisiana Universities Marine Consortium (LUMCON) and Louisiana State University have made great progress in unraveling the mysteries of the region's hypoxia—and in getting it on the political radar screen.

Rabalais and other researchers started tracking oxygen levels at nine sites in the Gulf every month and continued those measurements for 5 years. At dozens of other spots near the shore and in deep water, they took less frequent oxygen readings. For some of this work, the researchers have relied on mobile oxygen probes. Sensors, as they are lowered into the water, measure oxygen levels and send continuous readings back to a shipboard computer. Further data have come from fixed, submerged oxygen meters that continuously measure dissolved oxygen and store the data.

The team also collected hundreds of coastal and Gulf water samples, using lab tests to measure levels of nitrogen, salt, bacteria, and phytoplankton. LUMCON scientists logged hundreds of miles in their boats, regularly monitoring more than 70 sites in the Gulf. They also donned scuba gear to view firsthand the condition of shrimp, fish, and other sea life. Such a range of long-term data allowed the scientists to build a "map" of the dead zone, tracking its location and effects.

In 1991, Rabalais made that map public, earning immediate headlines. That year, her group mapped the size of the zone at more than 10,000 $km^2$ (about 4,000 $mi^2$). Bottom-dwelling shrimp were stretching out of their burrows, straining for oxygen. Many fish had fled. The bottom waters, infused with sulfur from bacterial decomposition, smelled of rotten eggs.

The group's years of continuous tracking also explained the dead zone's predictable emergence. As rivers rose each spring (and as fertilizers were applied in the Midwestern farm states), oxygen would start to disappear in the northern Gulf. The hypoxia would last through the summer or fall, until seasonal storms mixed oxygen into hypoxic areas. Over time, monitoring linked the dead zone's size to the volume of river flow and its nutrient load; the 1993 flooding of the Mississippi created a zone much larger than the year before. Conversely, a drought in 2000 brought lower river flows, lower nutrient loads, and a smaller dead zone (see figure).

The source of the problem, Rabalais said, lay back on land. The Mississippi and Atchafalaya rivers draining into the Gulf were polluted from agricultural runoff, and the nutrient pollution from fertilizers spurred algal blooms whose decomposition snuffed out oxygen in wide stretches of ocean water.

Many Midwestern farming advocates and some scientists, such as Derek Winstanley, chief of the Illinois State Water Survey, challenged the findings. They argued that the Mississippi naturally carries high loads of nitrogen from the rich prairie soil and that the Rabalais team had not ruled out upwelling in the Gulf as a source of nutrients.

But sediment analyses showed that Mississippi River mud was much lower in nitrates early in the century, and Rabalais and Turner found that silica residue from phytoplankton blooms increased in Gulf sediments between 1970 and 1989, paralleling rising nitrogen levels. In 2000, the federal integrative assessment involving dozens of scientists laid the blame for the dead zone on nutrients from fertilizers and other sources.

Then in 2004, while representatives of farmers and fishermen bickered over political fixes, Environmental Protection Agency (EPA) water quality scientist Howard Marshall suggested that to alleviate the dead zone, we'd be best off reducing phosphorus pollution from industry and sewage treatment. His reasoning: phytoplankton need both nitrogen and phosphorus, but there is now so

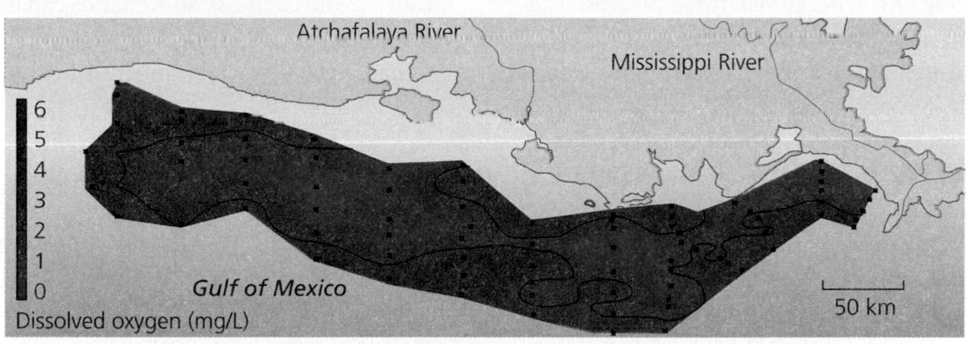


**(a) Dissolved oxygen at bottom, July 2006**

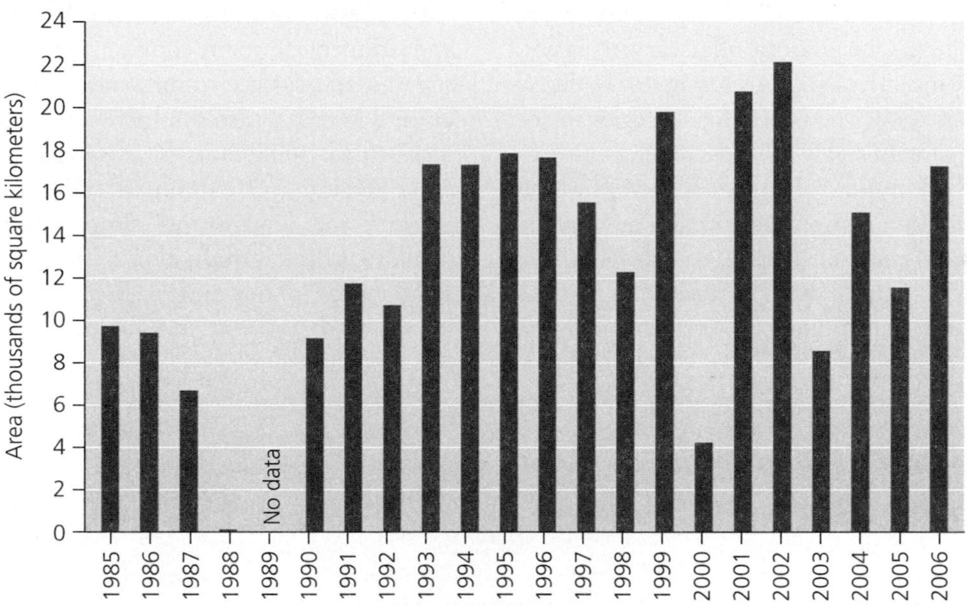


**(b) Area of hypoxic zone in the northern Gulf of Mexico**

The map in **(a)** shows dissolved oxygen concentrations in bottom waters of the Gulf of Mexico off the Louisiana coast from the July 2006 survey. Areas in red indicate the lowest oxygen levels, and areas in blue indicate the highest levels. Regions considered hypoxic (< 2 mg/L) are encircled with a black line. Black dots indicate sampling points. The size of the Gulf's hypoxic zone varies **(b)** as a result of several factors. Floods increase its size by bringing additional runoff (as with the Mississippi River floods of 1993), whereas tropical storms decrease its size by mixing oxygen-rich water into the dead zone (as in 2003). Between 1985 and 2006, the hypoxic zone averaged 13,000 km$^2$ (5,000 mi$^2$) in size. Data from Nancy Rabalais, LUMCON.

much nitrogen in the Gulf that phosphorus has become the limiting factor on phytoplankton growth.

Many other scientists now propose that nitrogen and phosphorus should be managed jointly. Further research indicates that the federally mandated 30% reduction in nitrogen in the river will not be enough. Some scientists also suggest that large-scale restoration of wetlands along the river and at the river's delta would best filter pollutants before they reached the Gulf. All this research is guiding a federal plan to reduce farm runoff, clean up the Mississippi, restore wetlands, and shrink the Gulf's dead zone—and it has also led to a better understanding of hypoxic zones around the world.

A combined state and federal task force proposed reducing nitrogen flowing down the Mississippi River by 30% by 2015, although some scientists later estimated it will require cuts of 50% to decrease the size of the dead zone. Farmers' advocates argued that farmers were being unfairly singled out, because many other artificial and natural sources contribute to the problem. They also argued that severe restrictions on fertilizer use would hurt farmers economically and decrease crop yields.

In 2003, Congress reauthorized the Harmful Algal Bloom and Hypoxia Research Control Act for 5 more years, providing funds for scientists, farmers, and policymakers to search for innovative solutions to alleviate pollution while not hurting agriculture. One proposal offers farmers insurance and economic incentives for not using excess fertilizer. Another program is testing new farming strategies to see whether any can maintain yields while decreasing fertilizer use. A third approach involves planting cover crops in the off-season to reduce runoff from bare fields. Yet another encourages farmers to maintain artificial wetlands on their lands that serve as natural buffers against pollution. Wetland plants host denitrifying bacteria that convert nitrates to nitrogen gas, so wetlands can effectively clean up a large amount of nitrogen pollution. Many Midwestern farmers are taking part in these strategies.

## The hydrologic cycle influences all other cycles

Water is so integral to life that we frequently take it for granted. The essential medium for all manner of biochemical reactions (• pp. 94–95), water plays key roles in nearly every environmental system, including each of the nutrient cycles we have just discussed. Water carries nutrients and sediments from the continents to the oceans via rivers, streams, and surface runoff, and it distributes sediments onward in ocean currents. Increasingly, water also distributes artificial pollutants. The water cycle, or **hydrologic cycle** (**Figure 7.17**), summarizes how water—in liquid, gaseous, and solid forms—flows through our environment. Our brief introduction to the hydrologic cycle here sets the stage for our more in-depth discussion of freshwater and marine systems in Chapters 15 and 16.

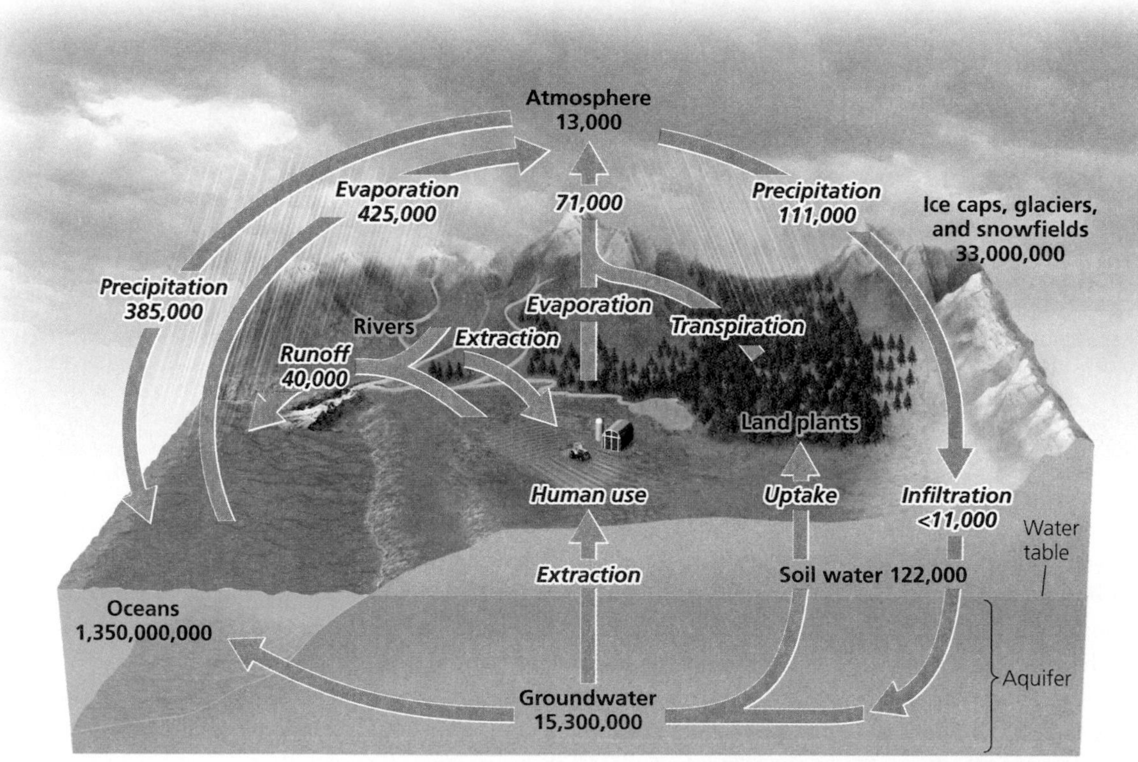


FIGURE 7.17 The hydrologic cycle summarizes the many routes that water molecules take as they move through the environment. Gray arrows represent fluxes among reservoirs, or pools, for water. The hydrologic cycle is a system unto itself but also plays key roles in other biogeochemical cycles. Oceans hold 97% of our planet's water, whereas most freshwater resides in groundwater and icecaps. Water vapor in the atmosphere condenses and falls to the surface as precipitation then evaporates from land and transpires from plants to return to the atmosphere. Water flows downhill into rivers, eventually reaching the oceans. In the figure, pool names are printed in black type, and numbers in black type represent pool sizes expressed in units of cubic kilometers ($km^3$). Processes, printed in italic red type, give rise to fluxes, printed in italic red type and expressed in $km^3$ per year. Data from Schlesinger, W. H. 1997. *Biogeochemistry: An analysis of global change,* 2nd ed. London: Academic Press.

## CAUSES AND CONSEQUENCES

The hypoxic zone in the Gulf of Mexico is not unique; at least 200 **dead zones** exist at estuaries and along seacoasts worldwide. These low-oxygen regions share many of the same causes, and they all have ecological and economic repercussions for ecosystems and people. Fortunately, we can pursue strategies to lessen the number, severity, and impacts of dead zones.

Write in two causes of dead zones in the spaces provided. Then write in two consequences (impacts on the environment, human health, or quality of life) that result from dead zones. Finally, offer two solutions to this issue and its consequences. One cause, one consequence, and one solution have been filled in for you, providing examples.

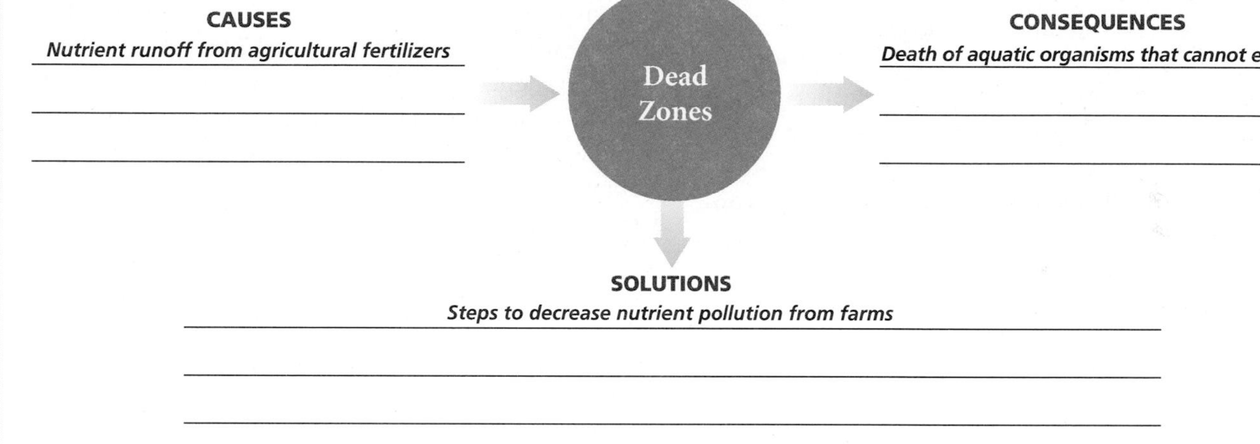


Sometimes solutions can have unintended consequences. Can you think of an undesired consequence that one solution to this issue might have? How might we then deal with *that* consequence?

The oceans are the main reservoir in the hydrologic cycle, holding 97% of all water on Earth. The fresh water we depend on for our survival accounts for less than 3%, and two-thirds of this small amount is tied up in glaciers, snowfields, and ice caps (• p. 413). Thus, considerably less than 1% of the planet's water is in a form that we can readily use—groundwater, surface fresh water, and rain from atmospheric water vapor.

**Evaporation and transpiration** Water moves from oceans, lakes, ponds, rivers, and moist soil into the atmosphere by **evaporation,** the conversion of a liquid to gaseous form. Warm temperatures and strong winds speed rates of evaporation. A greater degree of exposure has the same effect; an area logged of its forest or converted to agriculture or residential use will lose water more readily than a comparable area that remains vegetated. Water also enters the atmosphere by **transpiration,** the release of water vapor by plants through their leaves. Transpiration and evaporation act as natural processes of distillation, effectively creating pure water by filtering out minerals carried in solution.

**Precipitation, runoff, and surface water** Water returns from the atmosphere to Earth's surface as **precipitation** when water vapor condenses and falls as rain or snow. Precipitation may be taken up by plants and used by animals, but much of it flows as **runoff** into streams, rivers, lakes, ponds, and oceans. Amounts of precipitation vary greatly from region to region globally, helping give rise to the variety of biomes.

**Groundwater** Some precipitation and surface water soaks down through soil and rock to recharge underground reservoirs known as **aquifers.** Aquifers are spongelike regions of rock and soil that hold **groundwater,** water found underground beneath layers of soil. The upper limit of groundwater held in an aquifer is referred to as the **water table.** Aquifers may hold groundwater for long periods of time, so the water may be quite ancient. In some cases groundwater can take hundreds or even thousands of years to recharge fully after being depleted. Groundwater becomes exposed to the air where the water table reaches the surface, and the exposed water can run off toward the ocean or evaporate into the atmosphere.

### Our impacts on the hydrologic cycle are extensive

Human activity affects every aspect of the water cycle. By damming rivers to create reservoirs, we increase evaporation and, in some cases, infiltration of surface water into aquifers. By altering Earth's surface and its vegetation, we increase surface runoff and erosion. By spreading water on agricultural fields, we can deplete rivers, lakes, and streams and can increase evaporation. By removing forests and other vegetation, we reduce transpiration and may lower water tables. And by emitting into the atmosphere pollutants that dissolve in water droplets, we change the chemical nature of precipitation, in effect sabotaging the natural distillation process that evaporation and transpiration provide. Perhaps most threatening to our future, we are overdrawing groundwater to the surface for drinking, irrigation, and industrial use and have thereby begun to deplete groundwater resources. Water shortages have already given rise to numerous conflicts worldwide, from the Middle East to the American West (• pp. 412–413, 426).

**Weighing THE Issues | Your Water**

Are you aware of any evidence of water shortages or conflict over water use in your region? What is the quality of your water, and what pollution threats does it face? Given your knowledge of the hydrologic cycle, what solutions would you propose for water problems in your region?

## Geological Systems: How Earth Works

Biogeochemical cycles are not the planet's only cyclical environmental systems. Physical processes of geology determine Earth's landscape and form the foundation for the biotic patterns that overlay the landscape.

### The rock cycle is a fundamental environmental system

We tend to think of rock as pretty solid stuff. Yet in the long run, over geological time, rocks do change. Rocks and the minerals that comprise them are heated, melted, cooled, broken down, and reassembled in a very slow process called the **rock cycle** (**Figure 7.18**). The type of rock in a given region helps determine soil chemistry and thereby influences the biotic components of the region's ecosystems. Understanding the rock cycle enables us to appreciate more clearly the formation and conservation of soils, mineral resources, fossil fuels, and other natural resources.

**Igneous rock** All rocks can melt. At high enough temperatures, rock will enter a molten, liquid state called **magma.** If magma is released from the lithosphere (as in a volcanic eruption), it may flow or spatter across Earth's surface as **lava.** Rock that forms when magma cools is called **igneous** (from the Latin *ignis,* meaning "fire") **rock** (**Figure 7.18a**).

Igneous rock comes in several different types, because magma can solidify in different ways. Magma that cools slowly while it is well below Earth's surface is known as *intrusive* igneous rock. Half Dome and many other famous rock formations at Yosemite National Park in California were formed in this way and later exposed above the surface. Granite is the best-known type of intrusive rock. A slow cooling process allows minerals of different types to segregate from one another and aggregate with minerals of their own type, forming the crystals that give granite its multicolored, coarse-grained appearance. In contrast, when magma is ejected from a volcano, it cools quickly, so minerals have little time to differentiate into clusters. This kind of igneous rock is called *extrusive* igneous rock, and its most common representative is basalt.

**Sedimentary rock** All rock weathers away with time. The relentless forces of wind, water, freezing, and thawing eat away at rocks, stripping off one tiny grain (or large chunk) after another. Particles of rock blown by wind or washed away by water finally come to rest downhill, downstream, or downwind from their sources, forming **sediments.** These eroded remains of rocks usually are deposited very slowly, but floods can accelerate the process. Floods that sweep down the Mississippi River deposit sediments and nutrients across the floodplain, where they enrich soils, and at the river's mouth, building up its delta. Sediment layers accumulate over time, causing the weight and pressure of overlying layers to increase.

**Sedimentary rock** (**Figure 7.18b**) is formed when dissolved minerals seep through sediment layers and act as a kind of glue, crystallizing and binding sediment particles together. The formation of rock through these processes of compaction, binding, and crystallization is termed **lithification.** Similar processes of physical compaction and chemical transformation create the fossils of organisms (• p. 105) and the fossil fuels we use for energy (• pp. 544–557).

Like igneous rock, the several types of sedimentary rock are classified by the way they form and the size of particles they contain. Rock such as limestone and rock salt forms by

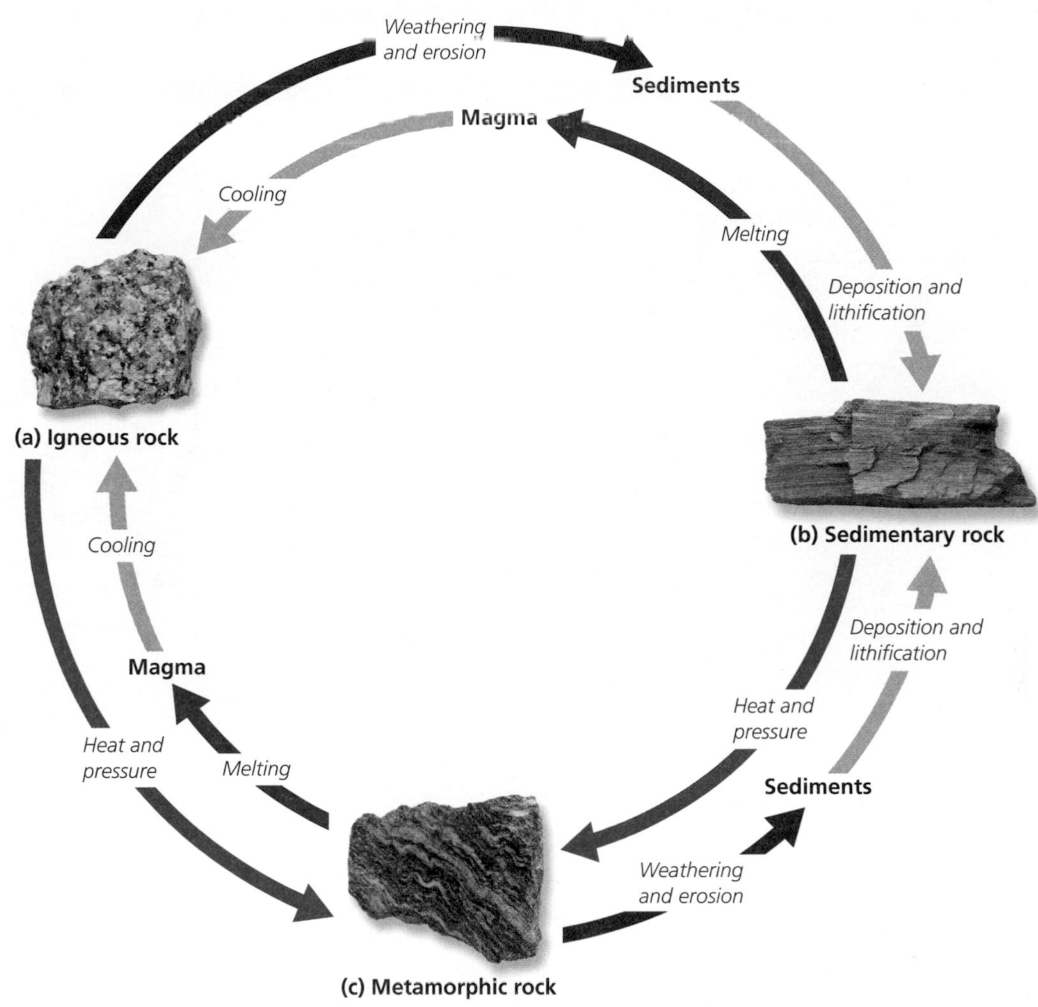


FIGURE 7.18 In the rock cycle, igneous rock (**a**) is formed when rock melts to form magma and the magma then cools. Sedimentary rock (**b**) is formed when rock is weathered and eroded and the resulting sediments are compressed to form new rock. Metamorphic rock (**c**) is formed when rock is subjected to intense heat and pressure underground. Through these processes (shown by differently colored arrows), each type of rock can be converted into either of the other two types.

chemical means when rocks dissolve and their components crystallize to form new rock. A second type of sedimentary rock forms when layers of sediment compress and physically bond to one another. Examples include conglomerate, made up of large particles that give it the appearance of nougat; sandstone, made of cemented sand particles; and shale, composed of still smaller mud particles.

**Metamorphic rock** Geological forces may bend, uplift, compress, or stretch rock. When great heat or pressure is exerted on rock, the rock may change its form, becoming **metamorphic** (from the Greek for "changed form" or "changed shape") **rock** (**Figure 7.18c**). The forces that metamorphose rock occur at temperatures lower than the rock's melting point but high enough to reshape crystals within the rock and change its appearance and physical properties. Common types of metamorphic rock include marble, formed when limestone is heated and pressurized, strengthening its structure; and slate, formed when shale is heated and metamorphosed.

## Plate tectonics shapes Earth's geography

The rock cycle takes place within the broader context of **plate tectonics,** a process that underlies earthquakes and volcanoes and that determines the geography of the Earth's surface. Earth's surface consists of a lightweight thin **crust** of rock floating atop a malleable **mantle,** which in turn surrounds a molten heavy **core** made mostly of iron. Earth's internal heat drives convection currents that flow in loops in the mantle, pushing the mantle's soft rock cyclically upward (as it warms) and downward (as it cools), like a gigantic conveyor belt. As the mantle material moves, it drags large plates of crust along its surface edge.

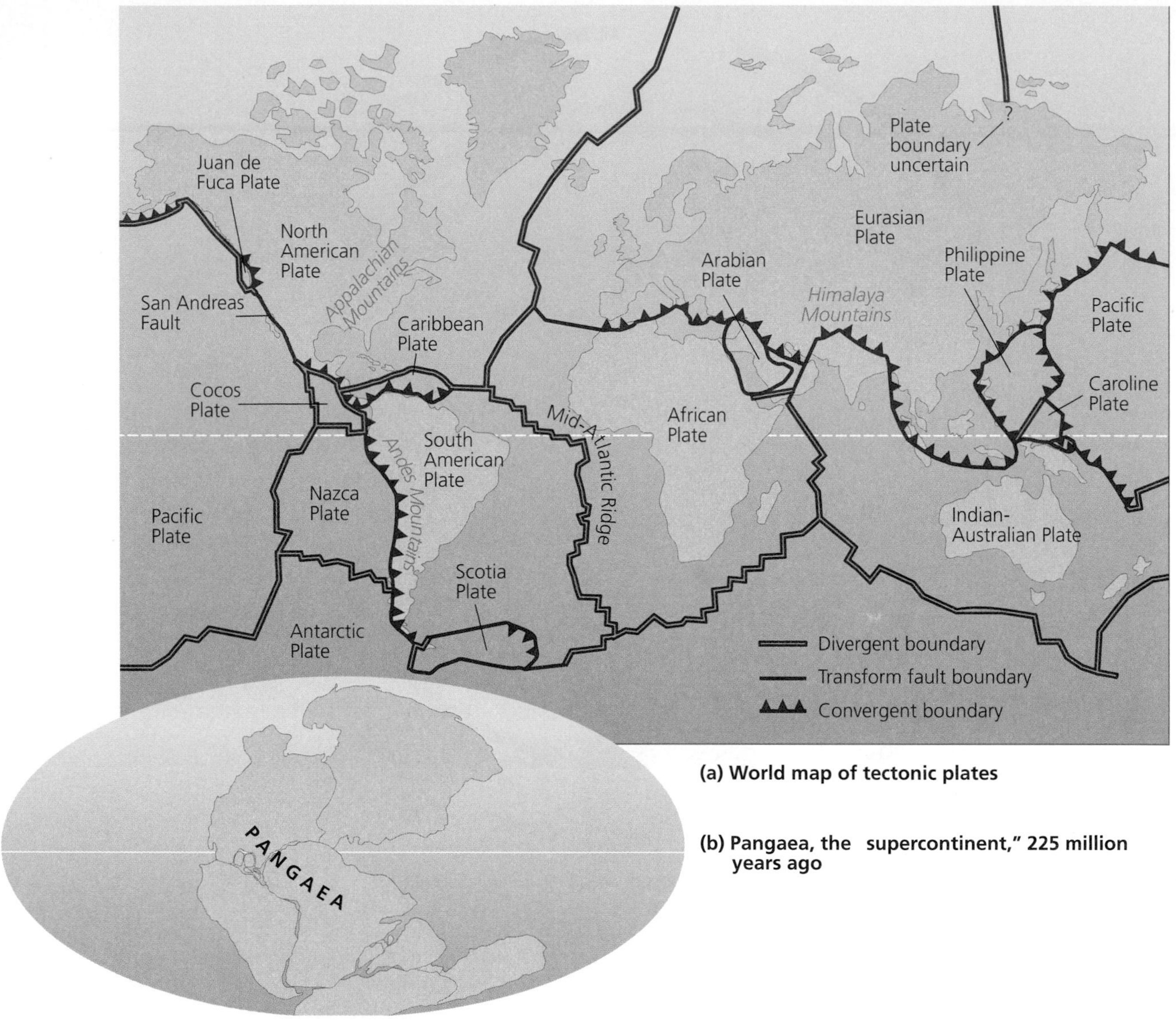


(a) World map of tectonic plates

(b) Pangaea, the supercontinent," 225 million years ago

FIGURE 7.19 Earth's crust consists of roughly 15 major plates (**a**) that move very slowly by the process of plate tectonics. Today's continents were joined together in the landmass Pangaea (**b**) about 225 million years ago.

Earth's surface consists of about 15 major tectonic plates, most including some combination of ocean and continent (**Figure 7.19a**). Imagine peeling an orange and putting the pieces of peel back onto the fruit; the ragged pieces of peel are like the plates of crust riding atop Earth's surface. These plates move at rates of roughly 2–15 cm (1–6 in.) per year. This movement has influenced Earth's climate and life's evolution throughout our planet's history as the continents combined, separated, and recombined in various configurations. By studying ancient rock formations throughout the world, geologists have determined that at least twice, all landmasses were joined together in a supercontinent scientists have dubbed *Pangaea* (**Figure 7.19b**).

At **divergent plate boundaries,** magma surging upward to the surface divides plates and pushes them apart, creating new crust as it cools and spreads (**Figure 7.20a**). A prime example is the Mid-Atlantic Ridge, part of a 74,000-km (46,000-mi) system of magmatic extrusion cutting across the seafloor. Plates expanding outward from divergent plate boundaries at midocean ridges bump against other plates, creating different types of plate boundaries.

When two plates meet, they may slip and grind alongside one another, forming a **transform plate boundary** (**Figure 7.20b**) and creating friction that spawns earthquakes along slipstrike faults. The Pacific Plate and the North American Plate rub against each other along California's San Andreas Fault. Southern California is slowly inching its way toward northern California along this fault, and Los Angeles will eventually reach San Francisco.

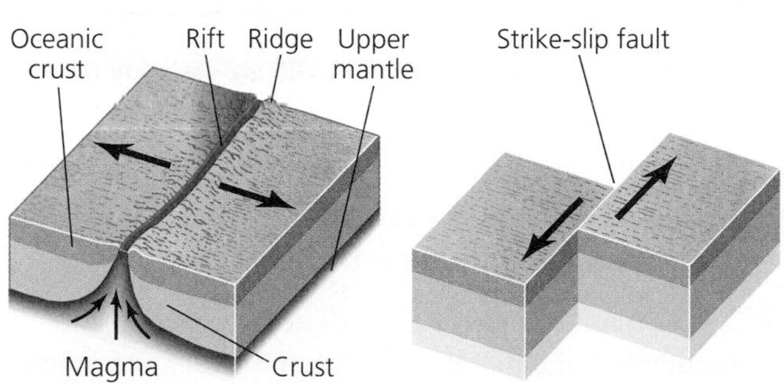


(a) Divergent plate boundary (b) Transform plate boundary

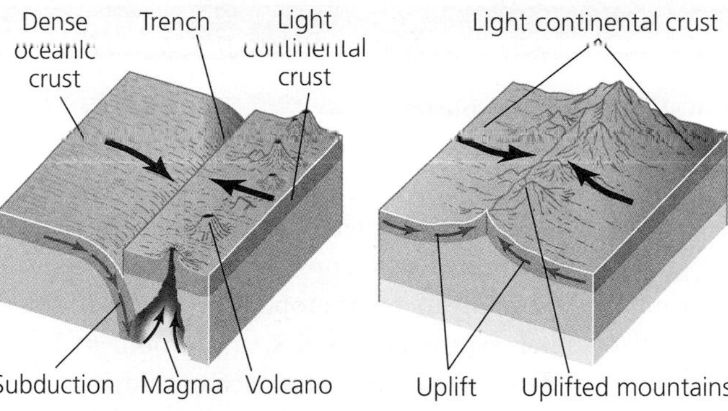


(c) Convergent plate boundary

FIGURE 7.20 Different types of boundaries between tectonic plates result in different geologic processes. At a divergent plate boundary, such as a midocean ridge on the seafloor (**a**), magma extrudes from beneath the crust, and the two plates move gradually away from the boundary in the manner of conveyor belts. At a transform plate boundary (**b**), two plates slide alongside one another, creating friction that leads to earthquakes. Where plates collide at a convergent plate boundary (**c**), one plate may be subducted beneath another, leading to volcanism, or both plates may be uplifted, causing mountain ranges to form.

When plates collide at **convergent plate boundaries,** either of two consequences may result (**Figure 7.20c**). First, one plate of crust may slide beneath another in a process called **subduction.** The subducted crust is heated as it dives into the mantle, and it may send up magma that erupts through the surface in volcanoes. Mount Saint Helens in Washington, which erupted violently in 1980 and renewed its activity in 2004, is fueled by magma from subduction. When denser ocean crust slides beneath lighter continental crust, volcanic mountain ranges are formed that parallel coastlines. Examples include the Cascades (which include Mount Saint Helens) and South America's Andes Mountains (where the Nazca Plate slides below the South American Plate). When one plate of oceanic crust is subducted beneath another, the resulting volcanism may form arcs of islands, such as Japan and the Aleutians. In addition, deep trenches may be created, such as the Mariana Trench, the planet's deepest abyss.

Alternatively, two colliding plates of continental crust may slowly lift material from both plates. The Himalayas, the world's highest mountains, are the result of the Indian-Australian Plate's collision with the Eurasian Plate 40–50 million years ago, and these mountains are still being uplifted today. The Appalachian Mountains of the eastern United States, once the world's highest mountains themselves, resulted from a much earlier collision with the edge of what is today Africa.

Amazingly, this environmental system of such fundamental importance was completely unknown to us just half a century ago. Our civilization was sending humans to the moon by the time our geologists were explaining the movement of land under our very feet. It is humbling to reflect on this; what other fundamental systems might we not yet appreciate or understand while our technology—and our ability to affect Earth's processes—continues racing ahead?

## Conclusion

Earth hosts many interacting systems, and the way one perceives them depends on the questions in which one is interested. Physical systems and processes such as the hydrologic cycle, the rock cycle, and plate tectonics lay the groundwork for the ways in which life spreads itself across the planet. Life interacts with its abiotic environment in ecosystems, systems through which energy flows and materials are recycled. Understanding the biogeochemical cycles that describe the movement of nutrients within and among ecosystems is crucial, because human activities are causing significant changes in the ways those cycles function.

Thinking in terms of systems is important in understanding how Earth works, so that we may learn how to avoid disrupting its processes and how to mitigate any disruptions we cause. By studying the environment from a systems perspective and by integrating scientific findings with the policy process, people who care about the Mississippi River and the Gulf of Mexico are working today to address the dead zone. Their model is one that we can apply to many other issues in environmental science.

We might also consider adopting other models more generally. Unperturbed ecosystems use renewable solar energy, recycle nutrients, exhibit dynamic equilibrium, and involve negative feedback loops. The environmental systems we see on Earth today are those that have survived the test of time. Our industrialized civilization is young in comparison. Might we not be able to take a few lessons about sustainability from a careful look at the natural systems of our planet?

## REVIEWING OBJECTIVES

**You should now be able to:**

**Describe the nature of environmental systems**

- Systems are networks of interacting components that generally involve feedback loops, show dynamic equilibrium, and result in emergent properties. (pp. 175–177)
- Earth's natural systems are complex, so environmental scientists often take a holistic approach to studying environmental systems. (pp. 175, 177–179)
- Because environmental systems interact and overlap, one's delineation of systems depends on the questions in which one is interested. (pp. 177–179)

**Define ecosystems and evaluate how living and nonliving entities interact in ecosystem-level ecology**

- Ecosystems consist of all organisms and nonliving entities that occur and interact in a particular area at the same time. (p. 180)
- Energy flows in one direction through ecosystems, whereas matter is recycled. (p. 180)
- Energy is converted to biomass, and ecosystems vary in their productivity. (pp. 180–182)
- Input of nutrients can boost productivity, but an excess of nutrients can alter ecosystems in ways that cause severe ecological and economic consequences. (pp. 181–183)

**Outline the fundamentals of landscape ecology**

- Landscape ecology studies how landscape structure influences organisms. (p. 183)
- Landscapes consist of patches spatially arrayed in a mosaic. Organisms dependent on certain types of patches may occur in metapopulations. (pp. 183–184)
- Remote sensing technology and GIS are assisting the use of landscape ecology in conservation and regional planning. (pp. 185–187)

**Compare and contrast how carbon, phosphorus, nitrogen, and water cycle through the environment**

- Most carbon is contained in sedimentary rock. Substantial amounts also occur in the oceans and in soil. Carbon flux between organisms and the atmosphere occurs via photosynthesis and respiration. (pp. 188–189)
- Phosphorus is most abundant in sedimentary rock, with substantial amounts in soil and the oceans. Phosphorus has no appreciable atmospheric pool. It is a key nutrient for plant growth. (pp. 189–191)
- Nitrogen is a vital nutrient for plant growth. Most nitrogen is in the atmosphere, so it must be "fixed" by specialized bacteria or lightning before plants can use it. (pp. 191–193)
- Water moves widely through the environment in the hydrologic cycle. (pp. 196–198)
- Humans are causing substantial impacts to Earth's biogeochemical cycles. These impacts include shifting carbon from fossil fuel reservoirs into the atmosphere, shifting nitrogen from the atmosphere to the planet's surface, and depleting groundwater supplies, among many others. (pp. 189, 190, 192–195, 197–198)

**Explain how plate tectonics and the rock cycle shape the landscape around us and the earth beneath our feet**

- Matter is cycled within the lithosphere, and rocks transform from one type to another. (pp. 198–199)
- Plate tectonics is a fundamental system that produces earthquakes and volcanoes and guides Earth's physical geography. (pp. 199–201)

## TESTING YOUR COMPREHENSION

1. Which type of feedback loop is most common in nature, and which more commonly results from human action? How might the emergence of a positive feedback loop affect a system in homeostasis?
2. Describe how hypoxic conditions can develop in coastal marine ecosystems such as the northern Gulf of Mexico.
3. What is the difference between an ecosystem and a community?
4. Describe the typical movement of energy through an ecosystem. Describe the typical movement of matter through an ecosystem.
5. What role do each of the following play in the carbon cycle?
   - Cars
   - Photosynthesis
   - The oceans
   - Earth's crust
6. Contrast the function performed by nitrogen-fixing bacteria with that performed by denitrifying bacteria.
7. How has human activity altered the carbon cycle? The phosphorus cycle? The nitrogen cycle? To what environmental problems have these changes given rise?

8. What is the difference between evaporation and transpiration? Give examples of how the hydrologic cycle interacts with the carbon, phosphorus, and nitrogen cycles.
9. Name the three main types of rocks, and describe how each type may be converted to the others via the rock cycle.
10. How does plate tectonics account for mountains? For volcanoes? For earthquakes? Why do you think it took so long for scientists to discover such a fundamental environmental system as plate tectonics?

## SEEKING SOLUTIONS

1. As global warming (Chapter 18) melts icecaps in the Arctic, it exposes darker-colored surfaces—and dark surfaces absorb more sunlight and heat than light surfaces. Would you expect this to result in a feedback process? If so, which type—negative or positive? Explain your answer.
2. Consider the ecosystem(s) that surround(s) your campus. How do some of the principles from our discussion on ecosystems apply to the ecosystem(s) around your campus?
3. For a conservation biologist interested in sustaining populations of each organism below, why would it be helpful to take a landscape ecology perspective? Explain your answer in each case.
   - A forest-breeding warbler that suffers poor nesting success in small fragmented forest patches
   - A bighorn sheep that must move seasonally between mountains and lowlands
   - A toad that lives in upland areas but travels cross-country to breed in localized pools each spring
4. A simple change in the flux rate between just two reservoirs in a single nutrient cycle can potentially have major consequences for ecosystems and, indeed, for the globe. Explain how this can be, using one example from the carbon cycle and one example from the nitrogen cycle.
5. How do you think we might solve the problem of eutrophication in the Gulf of Mexico? Assess several possible solutions, your reasons for believing they might work, and the likely hurdles we might face. Explain who should be responsible for implementing solutions, and why.
6. **THINK IT THROUGH** Imagine you are a shrimper on the Louisiana coast and that your income is decreasing because the dead zone is making it harder to catch shrimp. One day your senator comes to town, and you have a one-minute audience with her. What steps would you urge her to take in Washington, D.C., to try to help alleviate the dead zone and bring back the shrimp fishery?

   Now imagine that you are an Iowa farmer who has learned that the federal government is insisting that you use 30% less fertilizer on your crops each year. You know that in good growing years you could do without that fertilizer, and you'd be glad not to have to pay for it. But in bad growing years, you need the fertilizer to ensure a harvest so that you can continue making a living. And you must apply the fertilizer each spring before you know whether it will be a good or bad year. What would you tell your senator when she comes to town?

## INTERPRETING GRAPHS AND DATA

Scientists are debating what effects global climate change (Chapter 18) may have on nutrient cycles. As soil becomes warmer, especially at far northern latitudes, nutrients in the soil should become more available to plants, stimulating plant growth. One hypothesis is that more carbon will end up stored in the soil as a result, because plants will pull carbon from the atmosphere and transfer it to the soil reservoir as they shed leaves or die.* Under this hypothesis, increased flux of carbon from the atmosphere to the soil would act as negative feedback counteracting climate warming, because less carbon in the atmosphere would lead to less warming.

To test whether the carbon flux actually changes in this way when nutrients are made more available in a tundra ecosystem, researchers are conducting a long-term study in Alaska.† For 20 years, they have added fertilizer to treatment plots while leaving control plots unfertilized. Recently, they estimated amounts of carbon by measuring biomass aboveground and belowground in both sets of plots. Aboveground biomass consists of living plant material, whereas belowground biomass consists mostly of nonliving organic material stored in the soil and not yet decomposed. Some of the research team's results are presented in the graph below.

*Hobbie, S. E., et al. 2002. A synthesis: The role of nutrients as constraints on carbon balances in boreal and arctic regions. *Plant and Soil* 242: 163–170.

†Mack, M. C., et al. 2004. Ecosystem carbon storage in arctic tundra reduced by long-term nitrogen fertilization. *Nature* 431: 440–443.

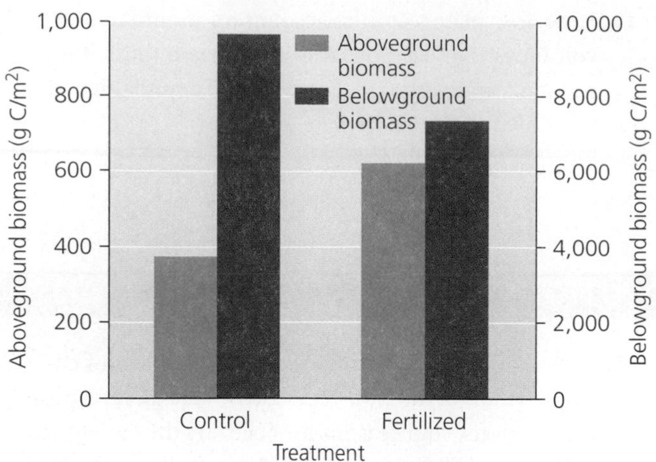


Effects of 20 years of fertilization (10 g N and 5 g P per $m^2$ per year) on C pools in tundra near Toolik Lake, Alaska. Differences are statistically significant for the aboveground, belowground, and total carbon pools.

1. Calculate the sizes of the aboveground, belowground, and total carbon pools (in g $C/m^2$) for the Control and Fertilized treatment groups.
2. What do the aboveground data indicate about the effect of fertilizer on plant growth? What do the belowground data indicate about the effect of fertilizer on organic material stored in the soil?
3. Do the data support the hypothesis that the net effect of increased nutrient availability will be to remove carbon from the atmosphere and store it in the soil? Using Figure 7.12 as a reference, can you suggest a different hypothesis that might explain the data better? Based on this data, would you predict that the warming of tundra soil will decrease the atmospheric concentration of $CO_2$ and act as negative feedback to climate change, or increase it and act as positive feedback?

## CALCULATING ECOLOGICAL FOOTPRINTS

In the United States, a common dream is to own your own home, surrounded by a weed-free, green lawn. Nationwide there are about 20 million acres of lawn grass, making it the nation's largest single crop!

Assuming that all of the populations indicated in the following table have lawns and will fertilize them at a typical fertilizer application rate of 45 lb of nitrogen per acre, calculate the total amount of nitrogen that will be applied to their lawns. When estimating the number of lawns, assume that the typical household includes three people.

| Fertilizer application | Number of lawns | Pounds of nitrogen |
|---|---|---|
| To your 1/3-acre lawn | 1 | 15 |
| To the lawns of your classmates | | |
| To all the lawns in your hometown | | |
| To all the lawns in your state | | |
| To all the lawns in the United States | 60,000,000 | |

1. Where does all of this nitrogen come from? Where does it go?
2. What other environmental impacts are caused by fertilizer production, transport, and application?
3. What steps could you take to reduce nitrogen pollution?

## Take It Further

Go to www.aw-bc.com/withgott or the student CD-ROM, where you'll find:

- Suggested answers to end-of-chapter questions
- Quizzes, animations, and flashcards to help you study
- *Research Navigator*™ database of credible and reliable sources to assist you with your research projects
- GRAPHIt! Tutorials to help you interpret graphs
- INVESTIGATEIt! Current news articles that link the topics that you study to case studies from your region to around the world

PART TWO

# Environmental Issues and the Search for Solutions

Canal Street, New Orleans, after Hurricane Katrina

CHAPTER

# 8 Human Population

Crowded street in Guangzhou, one of China's largest cities

## Upon completing this chapter, you will be able to:

- Assess the scope of human population growth
- Evaluate how human population, affluence, and technology affect the environment
- Explain and apply the fundamentals of demography
- Outline and assess the concept of demographic transition
- Describe how wealth and poverty, the status of women, and family planning programs affect population growth
- Characterize the dimensions of the HIV/AIDS epidemic

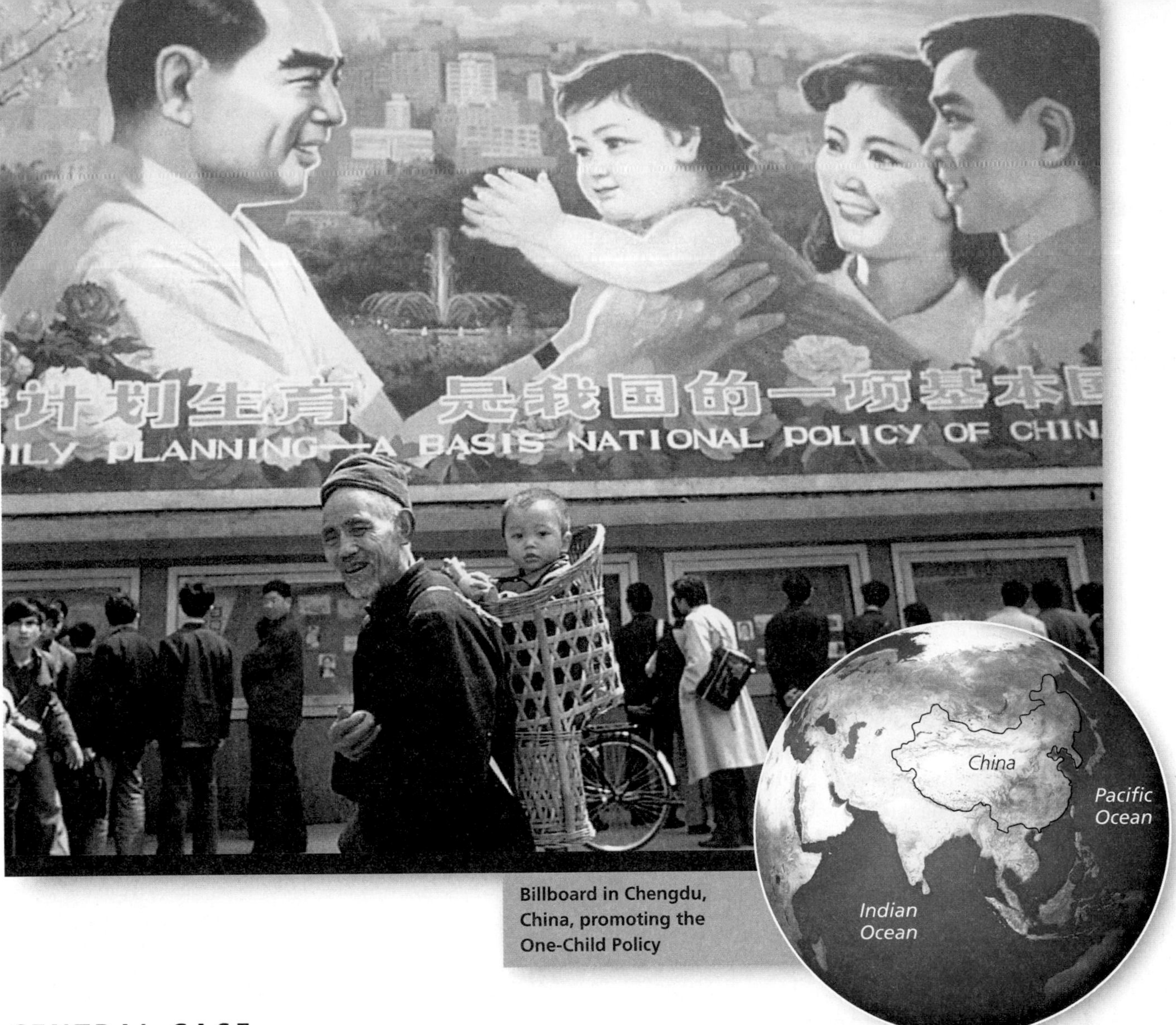


Billboard in Chengdu, China, promoting the One-Child Policy

CENTRAL CASE

# China's One-Child Policy

**"Population growth is analogous to a plague of locusts. What we have on this earth today is a plague of people."**

—Ted Turner, media magnate and supporter of the United Nations Population Fund

**"There is no population problem."**

—Sheldon Richman, senior editor, Cato Institute

The People's Republic of China is the world's most populous nation, home to one-fifth of the 6.7 billion people living on Earth at the start of 2008. When Mao Zedong founded the country's current regime 59 years earlier, roughly 540 million people lived in a mostly rural, war-torn, impoverished nation. Mao believed population growth was desirable, and under his leadership China grew and changed. By 1970, improvements in food production, food distribution, and public health allowed China's population to swell to approximately 790 million people. At that time, the average Chinese woman gave birth to 5.8 children in her lifetime.

Unfortunately, the country's burgeoning population and its industrial and agricultural development were eroding the nation's soils, depleting its water, leveling its forests, and polluting its air. Chinese leaders realized that the nation might not be able to feed its people if their numbers grew much larger. They saw that continued population growth could exhaust resources and threaten the stability and economic progress of Chinese society. The government decided to institute a population-control program that precluded large numbers of Chinese couples from having more than one child.

The program began with education and outreach efforts encouraging people to marry later and have fewer children. Along with these efforts, the Chinese government increased the accessibility of contraceptives and abortion. By 1975, China's annual population growth rate had dropped from 2.8% to 1.8%. To further decrease birth rates, in 1979 the government took the more drastic step of instituting a system of rewards and punishments to enforce a one-child limit. One-child families received better access to schools, medical care, housing, and government jobs, and mothers with only one child were given longer maternity leaves. Families with more than one child, meanwhile, were subjected to social scorn and ridicule, employment discrimination, and monetary fines. In some cases, the fines exceeded half the offending couple's annual income.

Beginning in 1984, the one-child policy was loosened, strengthened, and then loosened again as government leaders sought to find ways to maximize population control while minimizing public opposition. Today the one-child program applies mostly to urban couples, whereas many rural farmers and ethnic minorities are exempted.

In enforcing its policies, China has been conducting one of the largest and most controversial social experiments in history. In purely quantitative terms, the experiment has been a major success; the nation's growth rate is now down to 0.6%, making it easier for the country to deal with its many social, economic, and environmental challenges. However, China's population control policies have also produced unintended consequences, such as widespread killing of female infants, an unbalanced sex ratio, and a black-market trade in teenaged girls. Moreover, the policies have elicited intense criticism from people who oppose government intrusion into personal reproductive choices.

As other nations become more and more crowded, might their governments also feel forced to turn to drastic policies that restrict individual freedoms? In this chapter, we examine human population dynamics worldwide, consider their causes, and assess their consequences for the environment and our society.

# Human Population Growth: Approaching 7 Billion

While China works to slow its population growth, populations continue to rise in most nations of the world. Most of this growth is occurring in poverty-stricken developing nations that are ill-equipped to handle it. India (**Figure 8.1**) currently is on course to surpass China as the world's most populous nation. Although the *rate* of global growth is slowing, we are still increasing in absolute numbers, and today 6.7 billion of us inhabit the planet.

Just how much is 6.7 billion? We often have trouble conceptualizing huge numbers. Keep in mind that one billion is 1,000 times greater than one million. If you were to count once each second without ever sleeping, it would take over 30 years to reach a billion. To put a billion miles on your car, you would need to drive from New York to Los Angeles more than 350,000 times.

FIGURE 8.1 Population growth in developing countries is leading our global population toward 7 billion and beyond. India is on course to surpass China soon as the world's most populous nation. In this photo, Indian women wait in line for immunizations for their babies.

## The human population is growing nearly as fast as ever

As we saw in Chapter 1 (• pp. 4–5), the human population is growing at a tremendous rate. World population has doubled just since 1966 and now grows by roughly 80 million people annually (2.5 people every *second*). This is the equivalent of adding all the people of California, Texas, and New York to the world each year. It took until after 1800, virtually all of human history, for our population to reach 1 billion. Yet we reached 2 billion by 1930, and 3 billion in just 30 more years, in 1960. Our population added its next billion in just 15 years (1975), its next billion in a mere 12 years (1987), and its most recent billion in another 12 years (**Figure 8.2**). Think about when you were born and the number of people added to the planet just since that time. No previous generations have ever lived amid so many other people.

What accounts for such unprecedented growth? We saw in Chapter 5 (• p. 129) how exponential growth—the increase in a quantity by a fixed percentage per unit time—accelerates the absolute increase of population size over time, just as compound interest accrues in a savings account. The reason, you will recall, is that a given percentage of a large number is a greater quantity than the same percentage of a small number. Thus, even if the

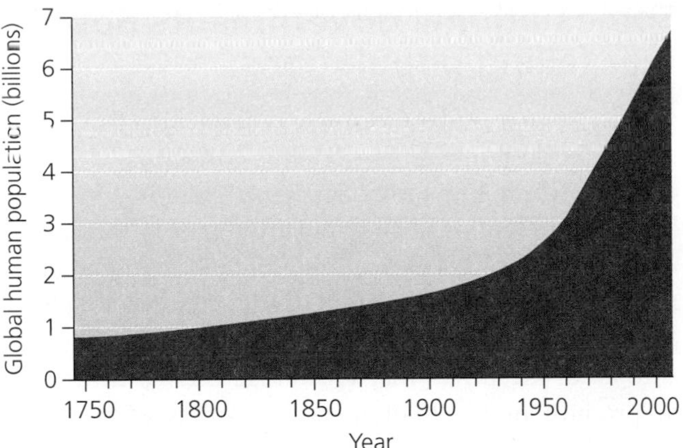


FIGURE 8.2 The global human population has grown exponentially, rising from less than 1 billion in 1800 to over 6.7 billion today. Data from U.S. Bureau of the Census.

growth rate remains steady, population size will increase by greater increments with each successive generation.

In fact, our growth rate has not remained steady. Instead, for much of the 20th century, the growth rate of the human population actually rose from year to year. It peaked at 2.1% during the 1960s and has declined to 1.2% since then. Although 1.2% may sound small, exponential growth endows small numbers with large consequences. A hypothetical population starting with one man and one woman that grows at 1.2% gives rise to a population of 2,939 after 40 generations and 112,695 after 60 generations. In today's world, rates of annual growth vary greatly from region to region (**Figure 8.3**).

At a 2.1% annual growth rate, a population doubles in size in only 33 years. We can roughly estimate doubling times with a handy rule of thumb. Just take the number 70, and divide it by the annual percentage growth rate: $70 \div 2.1 = 33.3$. Had China not instituted its one-child policy—that is, had its growth rate remained unchecked at 2.8%—it would have taken only 25 years to double in size. Had population growth continued at this rate, China's population would have surpassed 2 billion people in 2004.

## Is population growth really a "problem"?

Our ongoing population growth has resulted largely from technological innovations, improved sanitation, better medical care, increased agricultural output, and other factors that have led to a decline in death rates, particularly a drop in rates of infant mortality. Birth rates have not declined as much, so births have outpaced deaths for many years now. Thus, the so-called population problem actually arises from a very good thing—our ability to keep more of our fellow human beings alive longer.

Indeed, just as the mainstream view in the day of Thomas Malthus (• p. 5) held that population increase was a good thing, some people today argue that population growth poses few problems. Under the Cornucopian view that many economists hold, resource depletion due to population increase is not a problem if new resources can be found to replace depleted ones (• pp. 41–42). In contrast, environmental scientists recognize that few resources are actually created by people and that not all resources

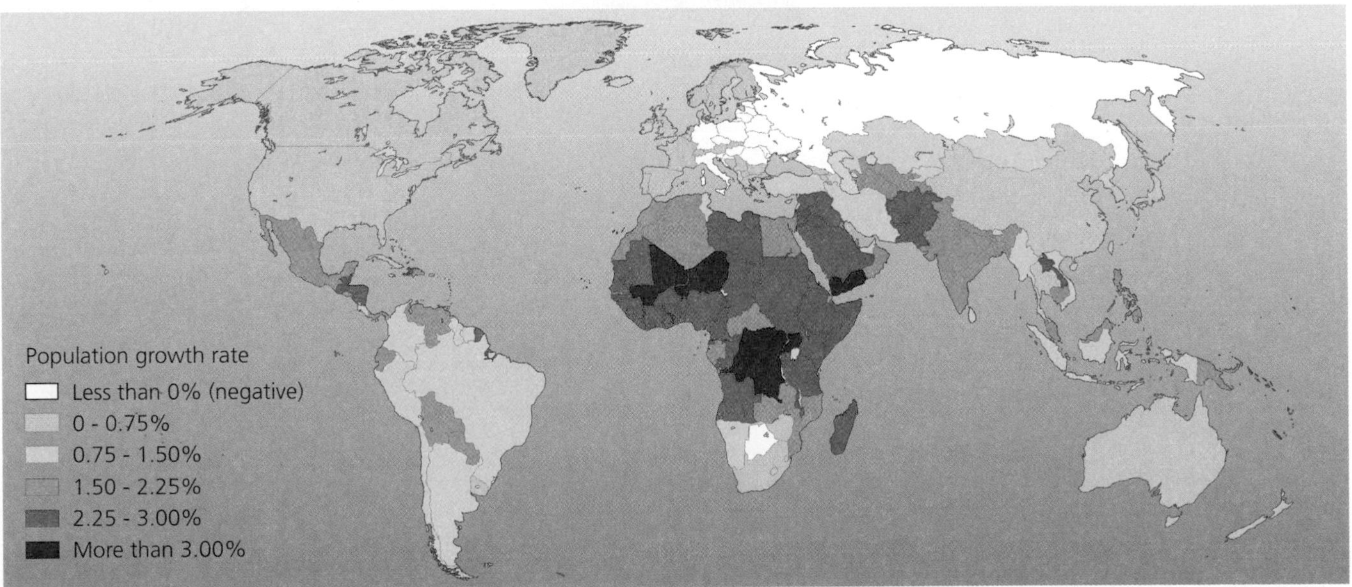


FIGURE 8.3 Population growth rates vary greatly from place to place. Population is growing fastest in poorer nations of the tropics and subtropics but is now beginning to decrease in some northern industrialized nations. Shown are natural rates of population change (• p. 219) as of 2006. Data from Population Reference Bureau. 2006. *2006 World population data sheet.*

can be replaced once they are depleted. For example, once species have gone extinct, we cannot replicate their exact function in ecosystems, or know what benefits we might have obtained from them. Another irreplaceable resource is land, that is, space in which to live; we cannot expand Earth like a balloon to increase its surface area.

Even if resource substitution could hypothetically enable population growth to continue indefinitely, could we maintain the *quality* of life that we would desire for ourselves and our descendants? Surely some of today's resources are bound to be easier or cheaper to use, and less environmentally destructive to harvest or mine, than any resources that can replace them. Unless the availability and quality of all resources keeps pace with population growth, the average person in the future will have less space in which to live, less food to eat, and less material wealth than the average person does today. Thus population increases are indeed a problem if they deplete resources, stress social systems, or degrade the natural environment, such that our quality of life declines (**Figure 8.4**).

## Some national governments now fear falling populations

Despite these considerations—and despite the fact that in today's world population growth is correlated with poverty, not wealth—many policymakers find it difficult to let go of the notion that population growth increases a nation's economic, political, or military strength. Many national governments, even those that view global population increase as a problem, still offer financial and social incentives that encourage their own citizens to produce more children. Governments of countries currently experiencing population declines (such as many in Europe) feel especially uneasy. According to the Population Reference Bureau, two out of every three European national governments now take the view that their birth rates are too low, and none state that theirs is too high. However, outside Europe, 49% of national governments feel their birth rates are too high, and only 12% feel they are too low.

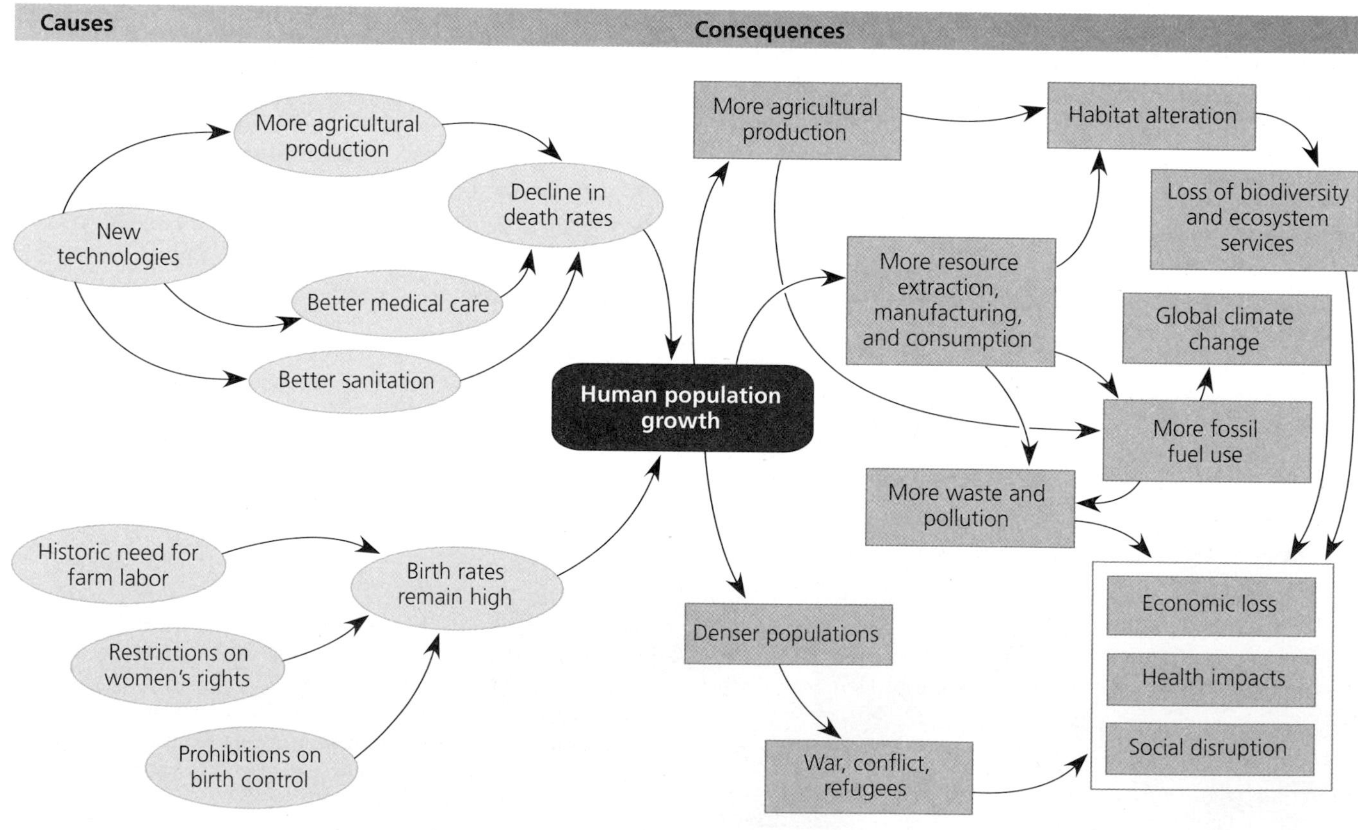


**FIGURE 8.4** Human population growth has diverse causes and consequences, as we shall see throughout this chapter. The consequences of rapid population growth are generally negative for both society and the environment. Arrows in this concept map lead from causes to consequences. Note that items grouped within outlined boxes do not necessarily share any special relationship; the outlined boxes are intended merely to streamline the figure.

**Solutions**

As you progress through this chapter, try to identify as many solutions to human population growth as you can. What could you personally do to help address this issue? Consider how each action or solution might affect items in the concept map above.

## Population is one of several factors that affect the environment

The extent to which population increase can be considered a problem involves more than just numbers of people. One widely used formula gives us a handy way to think about factors that affect environmental quality. Nicknamed the **IPAT model**, it is a variation of a formula proposed in 1974 by Paul Ehrlich (• p. 5) and John Holdren, a professor of environmental policy at Harvard University. The IPAT model represents how our total impact (I) on the environment results from the interaction among population (P), affluence (A), and technology (T):

$$I = P \times A \times T$$

Increased population intensifies impact on the environment as more individuals take up space, use resources, and generate waste. Increased affluence magnifies environmental impact through the greater per capita resource consumption that generally has accompanied enhanced wealth. Technology that enhances our abilities to exploit minerals, fossil fuels, old-growth forests, or fisheries generally increases impact, but technology to reduce smokestack emissions, harness renewable energy, or improve manufacturing efficiency can decrease impact.

We might also add a sensitivity factor (S) to the equation to denote how sensitive a given environment is to human pressures:

$$I = P \times A \times T \times S$$

For instance, the arid lands of western China are more sensitive to human disturbance than the moist regions of southeastern China. Plants grow more slowly in the arid west, making the land more vulnerable to deforestation and soil degradation. Thus, adding an additional person to western China should have more environmental impact than adding one to southeastern China.

We could refine the IPAT equation further by adding terms for the effects of social institutions such as education, laws and their enforcement, stable and cohesive societies, and ethical standards that promote environmental well-being. Such factors all affect how population, affluence, and technology translate into environmental impact.

Impact can be thought of in various ways, but we can generally boil it down either to pollution or resource consumption. The depletion of resources by larger and hungrier populations has been a focus of scientists and philosophers since Malthus's time. Today, researchers calculate that humanity is appropriating for its own use up to one-third of Earth's terrestrial net primary production (see "The Science behind the Story," • pp. 212–213).

Recall how the inhabitants of Easter Island (• pp. 8–9) brought down their own civilization by depleting their most important limited resource, trees. History offers other cases in which resource depletion helped end civilizations, from the Mayans to the Mesopotamians. Some environmental scientists have predicted similar problems for our global society in the near future if we do not embark on a path toward sustainability (**Figure 8.5**).

However, as we noted in Chapter 1, Malthus and present-day "neo-Malthusians" have not yet seen their direst predictions come true. One reason is that we have developed

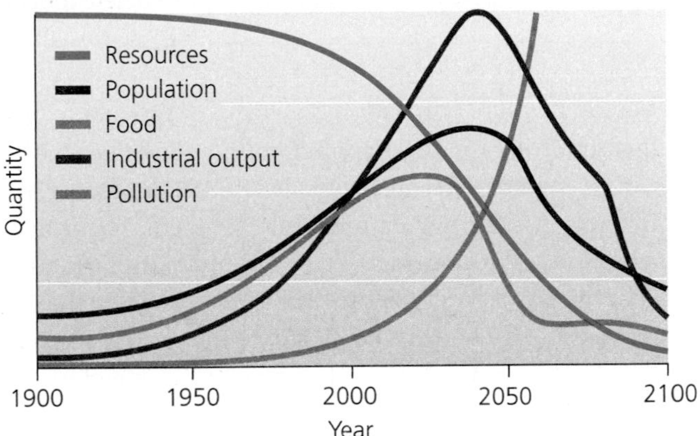


**(a) Projection based on status quo policies**

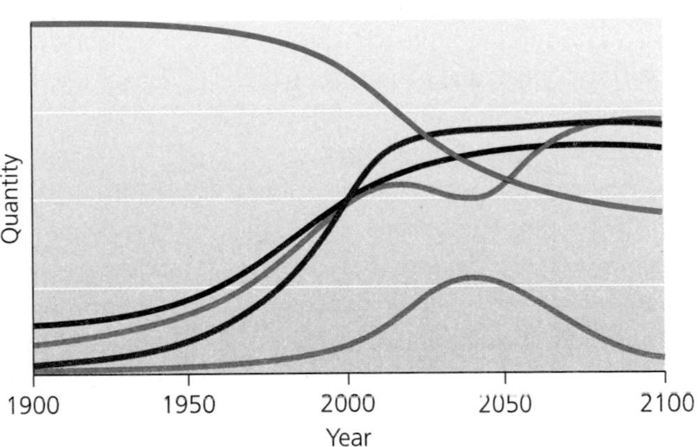


**(b) Projection based on policies for sustainability**

FIGURE 8.5 Environmental scientists Donella Meadows, Jorgen Randers, and Dennis Meadows used computer simulations to project future trends in human population, resource availability, food production, industrial output, and pollution. Their projections are based on past data and current scientific understanding of the environment's biophysical limits. Shown in **(a)** is their projection for a world in which "society proceeds in a traditional manner without any major deviation from the policies pursued during most of the twentieth century" and allowing for twice as many resources to be discovered and extracted as are now known to be accessible. In this projection, population and production increase until declining nonrenewable resources make further growth impossible, causing population and production to decline rather suddenly. Shown in **(b)** is a scenario with policies aimed at sustainability. In this projection, population levels off below 8 billion, production and resource availability stabilize at medium-high levels, and pollution declines to low levels. Data from Meadows, D., et al. 2004. *Limits to growth: The 30-year update.* White River Junction, VT: Chelsea Green Publishing.

THE SCIENCE BEHIND THE STORY

## Measuring the Environmental Impact of Human Population

Dr. Marc Imhoff of NASA.

We can measure population growth quantitatively and precisely, but how can we quantify the environmental impacts that our population (and affluence and technology) exert?

One way is to ask: Of all the biomass that Earth's plants produce, what percentage goes to feed, clothe, shelter, or otherwise serve human beings? That is, what proportion of the planet's plant production are we using up for ourselves?

This was the approach taken by six earth scientists and conservation biologists led by Marc Imhoff of the NASA Goddard Space Flight Center in Maryland, who teamed up to examine our consumption of *net primary production* (NPP; • pp. 181–182), the net amount of energy stored in plant matter as a result of photosynthesis.

Human overuse of NPP diminishes resources for other species; alters habitats, communities, and ecosystems; and can threaten our future ability to derive ecosystem services from our environment. No previous studies had examined how our consumption of NPP might vary from place to place across the globe.

Imhoff's team first determined how much terrestrial NPP people use to derive food and fiber products. To do this, they gathered nation-by-nation data from the U.N. Food and Agriculture Organization on food and fiber consumed. They then used estimates of harvesting and processing efficiencies to calculate how much plant matter is actually required to produce the amounts consumed in the end. Once they came up with totals for each nation, they took a global map of population density like the one shown in Figure 8.9 (• p. 215) and apportioned NPP consumption spatially within each nation in accordance with population densities, crediting high-density areas with greater consumption than low-density areas.

In this global map of NPP consumption, densely populated regions such as India, eastern China, and Indonesia showed the greatest use of NPP. Appropriation of NPP was also heavy in affluent regions of high resource consumption, such as Europe, Japan, and the United States.

To understand environmental impact, however, it's also necessary to know the *proportion* of NPP that people consume in each region. Thus, the team next mapped geographic patterns of terrestrial NPP using data on climate and vegetation from satellite photos. They then overlaid this data on their map of NPP consumption and created a global map of the proportion of NPP consumed by people (see the figure).

The researchers estimated that humans use 20.3% of the planet's NPP overall, with low and high estimates of 14% and 26%. However, their methods left out certain items included in two previous studies, such as NPP lost to land clearing and agriculture, and underground plant biomass on grazed lands. If these are added, they noted, their mean estimate rises to 36.6%.

Their map shows that in regions of the world that are sparsely inhabited (such as the taiga, Arctic tundra, Sahara Desert, Australian desert, and Amazon basin), humans consume almost none of the NPP. In many heavily populated areas such as India, eastern China, Japan, and Europe, people use more than 100% of the NPP produced in their area; that is, they are importing food, fiber, and energy from other places. In many large urban centers, people use over 30,000% of the local NPP; that is, they need to import 300 times as much production as their own region can produce.

technology—the T in the IPAT equation—time and again to alleviate our strain on resources and allow us to further expand our population. For instance, we have employed technological advances to increase global agricultural production faster than our population has risen (• p. 265).

Modern-day China shows how all elements of the IPAT formula can combine to cause tremendous environmental impact in very little time. The world's fastest-growing economy over the past two decades, China is "demonstrating what happens when large numbers of poor people rapidly become more affluent," in the words of Earth Policy Institute president Lester Brown. While millions of Chinese are increasing their material wealth and their consumption of resources, the country is battling unprecedented environmental challenges brought about by its pell-mell economic development. Intensive agriculture has expanded westward out of the country's historic moist rice-growing areas, causing farmland to erode and literally blow away, much like the Dust Bowl tragedy that befell the U.S. agricultural heartland in the 1930s (• pp. 245–246). China

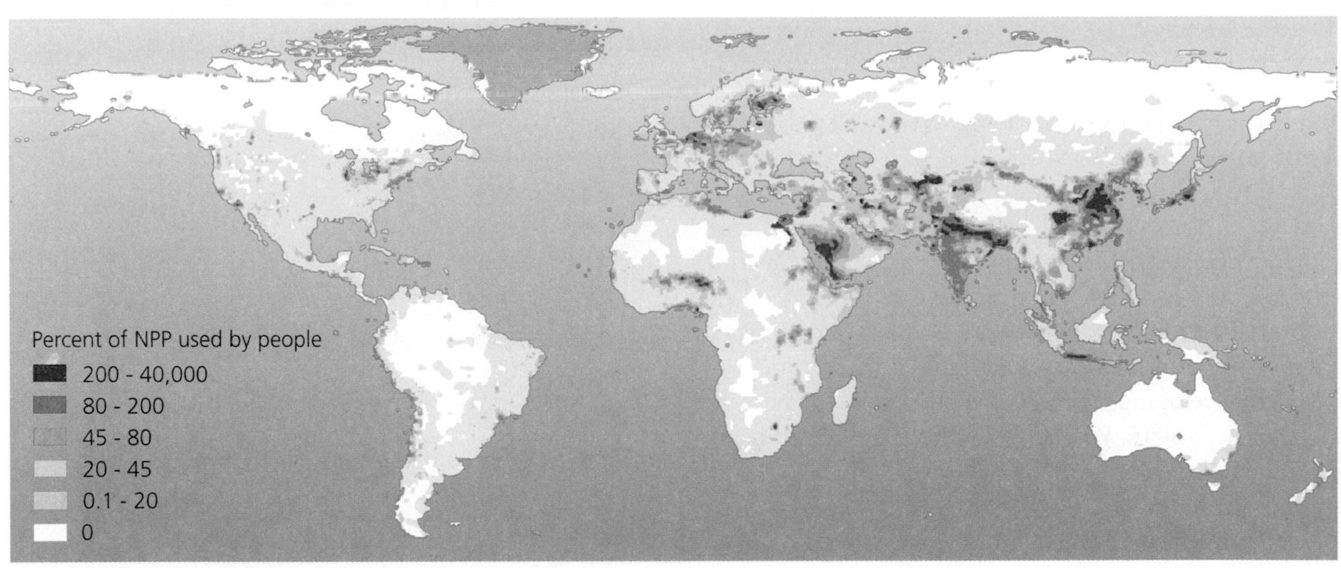


The proportion of the net primary production (NPP) of each region that people appropriate for human consumption varies by region. Regions using high percentages must import their production from elsewhere. *Source:* Imhoff, Marc L., et al. 2004. Global patterns in human consumption of net primary production. *Nature* 429:870–873.

This is why Imhoff's team calls its map "a spatially explicit balance sheet of 'supply' and 'demand.'" Through regional and global trade, we move the products we gain from the planet's NPP far and wide, distributing them unevenly. As a result, the environmental impacts of our consumption are often felt far from where we consume products.

Although the map identifies areas that require net imports of NPP, it does not attempt to build in data on global trade. This important next step should reveal that much of the NPP consumed in populated areas is being taken from sparsely populated areas with high NPP. For instance, North Americans and Europeans import timber logged from the Amazon basin, as well as soybeans and beef grown in areas where Amazonian forest has been cleared. Thus, some of the consumption the map attributes to North America and Europe actually originates in and affects South America.

All elements of the IPAT equation (• p. 211) are visible in the Imhoff team's map. The influence of population (P) is clear; for instance, although people in east and south-central Asia have the lowest per capita consumption of any region, dense populations here result in an average 72% use of regional NPP. Affluence (A) exerts impacts because industrialized nations consume nearly twice as much per capita NPP as developing nations. And technology (T) affects the efficiency with which NPP is appropriated—something the researchers measured in preparing their data. For instance, in developing countries it takes 2 tons of tree biomass to produce 1 ton of milled lumber, but in industrialized countries, better technology has reduced that to only 1.3 tons.

Maps like the one produced in this project can help guide efforts to build a sustainable society. By showing areas of high and low consumption and impact, such maps can help us to make better decisions and minimize our impacts on ecosystems and ecosystem services.

has overpumped many of its aquifers and has drawn so much water for irrigation from the Yellow River that the once-mighty waterway now dries up in many stretches. Although China is reducing its air pollution from industry and charcoal-burning homes, the country faces new urban pollution and congestion threats from rapidly increasing numbers of automobiles. As the world's developing countries try to attain the level of material prosperity that industrialized nations enjoy, China is a window on what much of the rest of the world could soon become.

### Weighing the Issues | China's Reproductive Policy

Consider the benefits as well as the problems associated with a reproductive policy such as China's. Do you think a government should be able to enforce strict penalties for citizens who fail to abide by such a policy? If you disagree with China's policy, what alternatives can you suggest for dealing with the resource demands of a quickly growing population?

# Demography

As we have seen, it is a fallacy to think of people as being somehow outside nature. Humans exist within their environment as one species out of many. As such, all the principles of population ecology we outlined in Chapter 5 that apply to toads, frogs, and passenger pigeons apply to humans as well. Environmental factors set limits on our population growth, and the environment has a carrying capacity (• pp. 129–130) for our species, just as it does for every other.

We happen to be a particularly successful organism, however—one that has repeatedly raised its carrying capacity by developing technology to overcome the natural limits on its growth. We did so with the agricultural and the industrial revolutions (• pp. 4–5) and likely before that with our invention of tools (**Figure 8.6**).

Environmental scientists who have tried to pin a number to the human carrying capacity have come up with wildly differing estimates. The most rigorous estimates range from 1–2 billion people living prosperously in a healthy environment to 33 billion living in extreme poverty in a degraded world of intensive cultivation without natural areas. As our population climbs toward 7 billion and beyond, we may yet continue to find ways to raise our carrying capacity. Given our knowledge of population ecology, however, we have no reason to presume that human numbers can go on growing indefinitely. Indeed, as we have seen (see Figure 5.17d, • p. 131), populations that exceed their carrying capacity can crash.

## Demography is the study of human population

The application of principles from population ecology to the study of statistical change in human populations is the focus of **demography**. Data gathered by demographers help us understand how differences in population characteristics and related phenomena (for instance, decisions about reproduction) affect human communities and their environments. Demographers study population size, density, distribution, age structure, sex ratio, and rates of birth, death, immigration, and emigration of humans, just as population ecologists study these characteristics in other organisms. Each of these characteristics is useful for predicting population dynamics and environmental impacts.

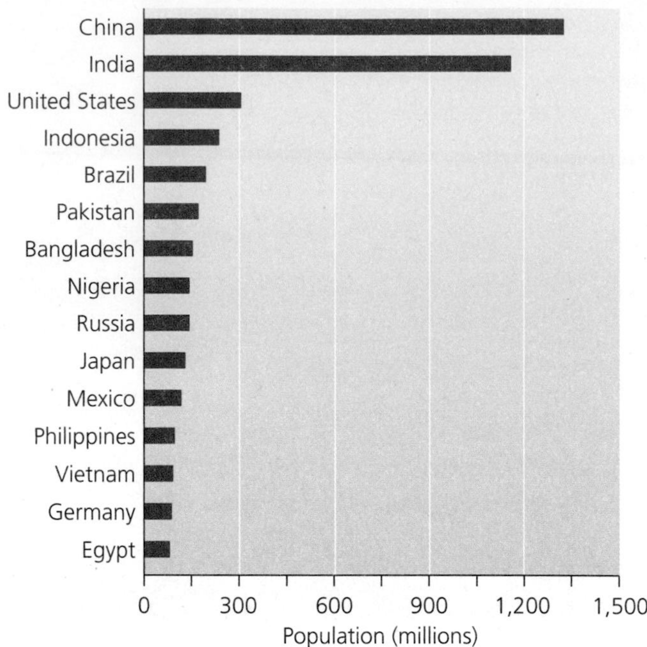


FIGURE 8.7 The world's nations range in human population up to China's 1.3 billion. Shown here are estimates for mid-2008 populations of the world's most populous 15 countries. Data from Population Reference Bureau. 2007. *2007 world population data sheet.*

**Population size** Our global human population of roughly 6.7 billion is spread among 200 nations with populations ranging from several thousand for some Pacific island nations up to China's 1.3 billion, India's 1.15 billion, and the 300 million of the United States (**Figure 8.7**). The size that

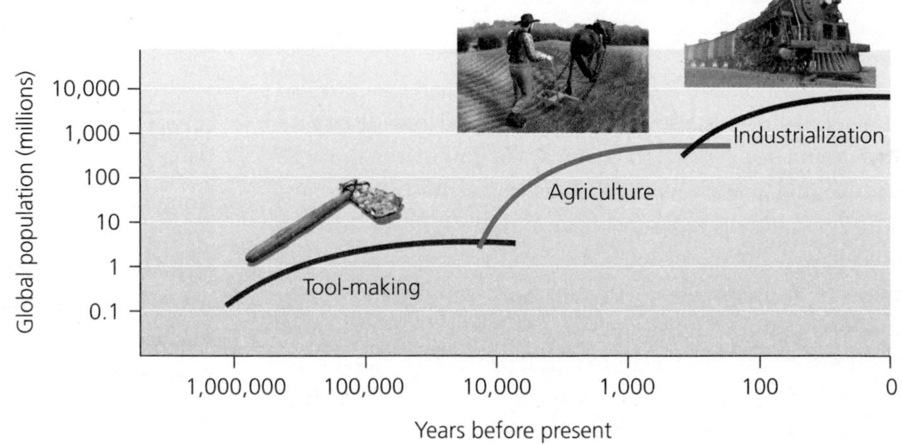


FIGURE 8.6 Tool making, the advent of agriculture, and industrialization each allowed our species to raise its global carrying capacity. The logarithmic scale of the axes makes it easier to visualize this pattern. Data from Goudie, A. 2000. *The human impact.* Cambridge, MA: MIT Press.

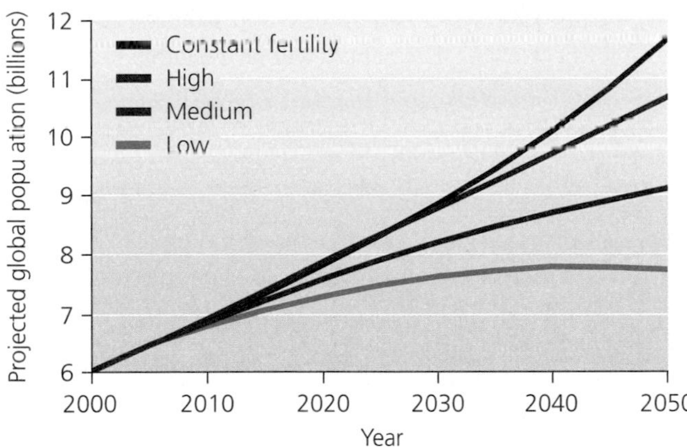


**FIGURE 8.8** The United Nations predicts trajectories of world population growth, presenting several estimates based on different assumptions of fertility rates. In this 2006 projection, population is estimated to reach 11.9 billion in the year 2050 if fertility rates remain constant at 2006 levels (top line in graph). However, U.N. demographers expect fertility rates to continue falling, so they arrived at a best guess (*medium* scenario) of 9.2 billion for the human population in 2050. In the *high* scenario, if women on average have 0.5 child more than in the medium scenario, population will reach 10.8 billion in 2050. In the *low* scenario, if women have 0.5 child less than in the medium scenario, the world will contain 7.8 billion people in 2050. Data from United Nations Population Division. 2006. *World population prospects: The 2006 revision.*

our global population will eventually reach remains to be seen (**Figure 8.8**). However, population size alone—the absolute number of individuals—doesn't tell the whole story. Rather, a population's environmental impact depends on its density, distribution, and composition (as well as on affluence, technology, and other factors outlined earlier).

**Population density and distribution** People are distributed very unevenly over the planet. In ecological terms, our distribution is clumped (• pp. 126–127) at all spatial scales. At the global scale (**Figure 8.9**), population density is highest in regions with temperate, subtropical, and tropical climates, such as China, Europe, Mexico, southern Africa, and India. Population density is lowest in regions with extreme-climate biomes, such as desert, deep rainforest, and tundra. Dense along seacoasts and rivers, human population is less dense at locations far from water. At regional scales, we cluster together in cities and suburbs and are spread more sparsely across rural areas. At local scales, we aggregate in certain neighborhoods and in individual households.

This uneven distribution means that certain areas bear far more environmental impact than others. Just as the Yellow River has experienced intense pressure from millions of Chinese farmers, the world's other major rivers, from the Nile to the Danube to the Ganges to the Mississippi, have all received more than their share of human impact. Urbanization (Chapter 13) entails the packaging and transport of goods, intensive fossil fuel consumption, and hotspots of pollution. However, people's concentration in cities also relieves pressure on ecosystems in less-populated areas by releasing them from some human development (• pp. 375–376).

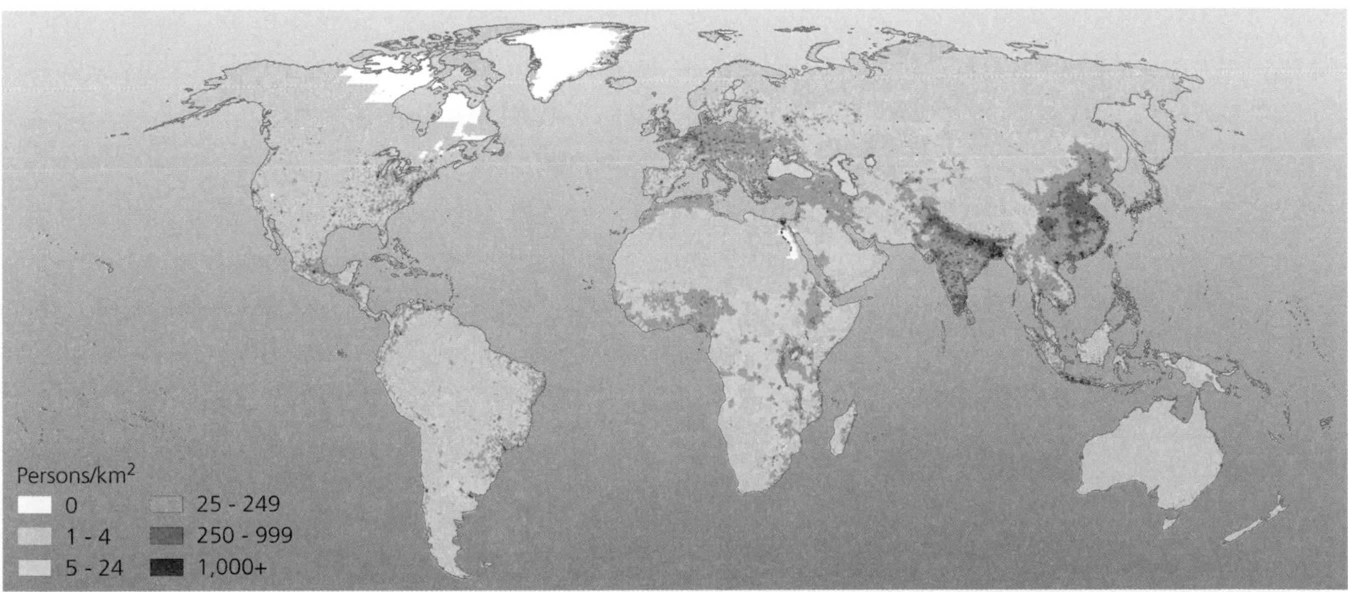


**FIGURE 8.9** Human population density varies tremendously from one region to another. Arctic and desert regions have the lowest population densities, whereas areas of India, Bangladesh, and eastern China have the densest populations. Data are for 2000, from Center for International Earth Science Information Network (CIESIN), Columbia University; and Centro Internacional de Agricultura Tropical (CIAT), 2005.

At the same time, areas with low population density are often vulnerable to environmental impacts, because the reason they have low populations in the first place is that they are sensitive and cannot support many people (a high S value in our revised IPAT model). Deserts, for instance, are easily affected by development that commandeers a substantial share of available water. Grasslands can be turned to deserts if they are farmed too intensively, as has happened across vast stretches of the Sahel region bordering Africa's Sahara Desert, in the Middle East, and in parts of China and the United States.

**Age structure** Data on the age structure or age distribution of human populations are especially valuable to demographers trying to predict future dynamics of populations. As we saw in Chapter 5 (• pp. 127–128), large proportions of individuals in young age groups portend a great deal of reproduction and, thus, rapid population growth. Examine age pyramids for the nations of Canada and Madagascar (**Figure 8.10**). Not surprisingly, it is Madagascar that has the greater population growth rate. In fact, its annual growth rate, 2.7%, is 9 times that of Canada's 0.3%.

By causing dramatic reductions in the number of children born since 1970, China virtually guaranteed that its population age structure would change. Indeed, in 1995 the median age in China was 27; by 2030 it will be 39. In 1997 there were 125 children under age 5 for every 100 people 65 or older in China, but by 2030 there will be only 32. The number of people older than 65 will rise from 100 million in 2005 to 236 million in 2030 (**Figure 8.11**). This dramatic shift in age structure will challenge China's economy, health care systems, families, and military forces because fewer working-age people will be available to support social programs that assist the increasing number of older people.

This pattern of population aging is occurring in many countries, including the United States (**Figure 8.12**). Older populations will present new challenges for many nations, as increasing numbers of older people require the care and financial assistance of relatively fewer working-age citizens.

However, a shift in age structure toward an older population also reduces the proportion of dependent children. Fewer young adults may mean a decrease in crime rates. Moreover, older people are often productive members of society, contributing volunteer activities and services to their children and grandchildren. In many ways both good and bad, graying populations will affect societies in China, the United States, and elsewhere throughout the lifetimes of all of us alive today.

**Sex ratios** The ratio of males to females also can affect population dynamics. Imagine two islands, one populated by 99 men and 1 woman and the other by 50 men and 50 women. Where would we be likely to see the greatest population increase over time? Of course, the island with an equal number of men and women would have a greater number of potential mothers and thus a greater potential for population growth.

The naturally occurring sex ratio in human populations at birth features a slight preponderance of males; for every 100 female infants born, about 106 male infants are born. This phenomenon is an evolutionary adaptation to

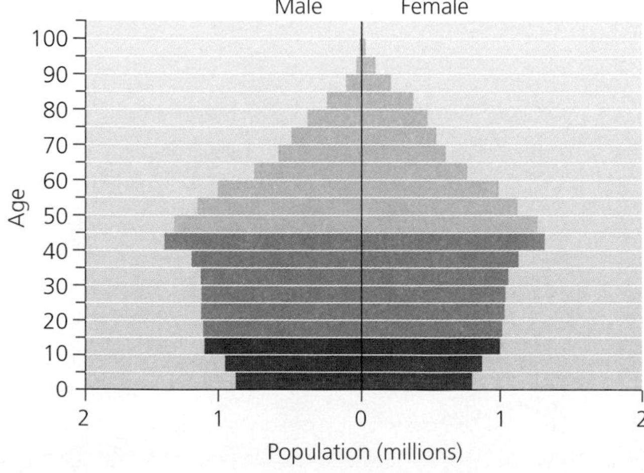


**(a) Age pyramid of Canada in 2005**

Male Female

Age 100 90 80 70 60 50 40 30 20 10 0

2 1 0 1 2

Population (millions)

**(b) Age pyramid of Madagascar in 2005**

**FIGURE 8.10** Canada (**a**) shows a balanced age structure, with relatively even numbers of individuals in various age classes. Madagascar (**b**) shows an age distribution heavily weighted toward young people. Madagascar's population growth rate is nine times that of Canada's. Go to GRAPHIt! at www.aw-bc.com/withgott or the student CD-ROM.
Data from U.N. Population Division.

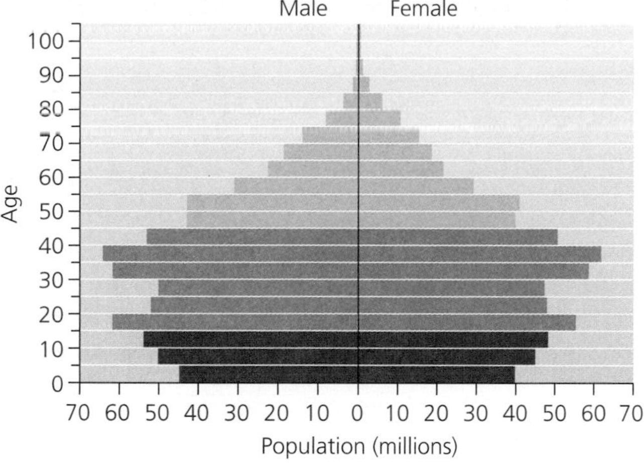


**(a) Age pyramid of China in 2005**

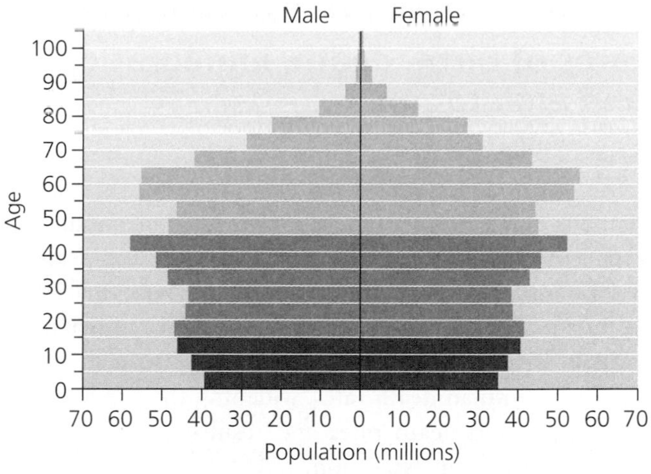


**(b) Projected age pyramid of China in 2030**

**FIGURE 8.11** As China's population ages, older people will outnumber the young. Age pyramids show the predicted graying of the Chinese population between 2005 (**a**) and 2030 (**b**). Today's children may, as working-age adults (**c**), face pressures to support greater numbers of older citizens than has any previous generation. Data from U.N. Population Division.

**(c) Young female factory workers in Hong Kong**

the fact that males are slightly more prone to death during any given year of life. It tends to ensure that the ratio of men to women will be approximately equal at the time people reach reproductive age. Thus, a slightly uneven sex ratio at birth may be beneficial. However, a greatly distorted ratio can lead to problems.

In recent years, demographers have witnessed an unsettling trend in China: The ratio of newborn boys to girls has become strongly skewed. In the 2000 census, 120 boys were reported born for every 100 girls. Some provinces reported sex ratios as high as 138 boys for every 100 girls. The leading hypothesis for these unusual sex ratios is that

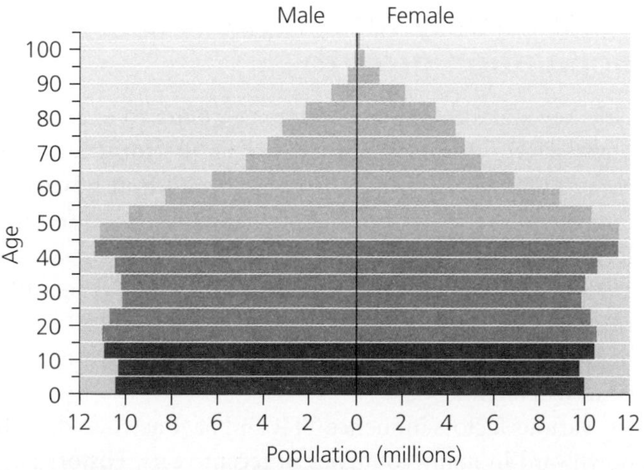


**FIGURE 8.12** The "baby boom" is visible in the 2005 age pyramid for the United States, in the age brackets between 40 and 50. The nation is experiencing an aging population as baby-boomers grow older. Data from U.N. Population Division.

many parents, having learned the sex of their fetuses by ultrasound, are selectively aborting female fetuses.

Traditionally, Chinese culture has valued sons because they carry on the family name, assist with farm labor in rural areas, and care for aging parents. Daughters, in contrast, will most likely marry and leave their parents, as the culture dictates. As a result, they cannot provide the same benefits to their parents as will sons. Sociologists hold that this cultural gender preference, combined with the government's one-child policy, has led some couples to selectively abort female fetuses or to abandon or kill female infants. Moreover, the Chinese government reinforced this gender discrimination when in 1984 it exempted rural peasants from the one-child policy if their first child was a girl, but not if the first child was a boy.

China's skewed sex ratio may further lower population growth rates. However, it has proved tragic for the "missing girls." It is also beginning to have the undesirable social consequence of leaving many Chinese men single. This, in turn, has resulted in a grim new phenomenon. In parts of rural China, teenaged girls are being kidnapped and sold to families in other parts of the country as brides for single men.

## Population growth depends on rates of birth, death, immigration, and emigration

Rates of birth, death, immigration, and emigration determine whether a human population grows, shrinks, or remains stable. The formula for measuring population growth that we used in Chapter 5 (• p. 128) also pertains to humans: birth and immigration add individuals to a population, whereas death and emigration remove individuals. Technological advances have led to a dramatic decline in human death rates, widening the gap between birth rates and death rates and resulting in the global human population expansion.

In today's ever-more-crowded world, immigration and emigration are playing increasingly large roles. Refugees, people forced to flee their home country or region, have become more numerous in recent decades as a result of war, civil strife, and environmental degradation. The United Nations puts the number of refugees who flee to escape poor environmental conditions at 25 million per year and possibly many more. Often the movement of refugees causes environmental problems in the receiving region as these desperate victims try to eke out an existence with no livelihood and with no cultural or economic attachment to the land or incentive to conserve its resources. The millions who fled Rwanda following the genocide there in the 1990s, for example, inadvertently destroyed large areas of forest while trying to obtain fuelwood, food, and shelter to stay alive once they reached the Democratic Republic of Congo (**Figure 8.13**).

FIGURE 8.13 The flight of refugees from Rwanda into the Democratic Republic of Congo in 1994 following the Rwandan genocide caused unimaginable hardship for the refugees and tremendous stress on the environment into which they moved.

TABLE 8.1 Trends in China's Population Growth

| Measure | 1950 | 1970 | 1990 | 2007 |
|---|---|---|---|---|
| Total fertility rate | 5.8 | 5.8 | 2.2 | 1.6 |
| Rate of natural population increase (% per year) | 1.9 | 2.6 | 1.4 | 0.5 |
| Doubling time (years) | 37 | 27 | 49 | 140 |
| Population (billions) | 0.56 | 0.83 | 1.15 | 1.32 |

Data from China Population Information and Research Center; and Population Reference Bureau. 2007. *2007 World population data sheet.*

For most of the past 2,000 years, China's population has been relatively stable. The first significant increases resulted from enhanced agricultural production and a powerful government during the Qing, or Manchu, Dynasty in the 1800s. Population growth began to outstrip food supplies by the mid-1850s, and quality of life for the average Chinese peasant began to decline. Over the next 100 years, China's population grew slowly, at about 0.3% per year, amid food shortages and political instability. As we have seen, population growth rates rose again following Mao's establishment of the People's Republic in 1949 and have declined since the establishment of the one-child policy (**Table 8.1**).

In recent decades, falling growth rates in many countries have led to an overall decline in the global growth rate (**Figure 8.14**). These declines have come about, in part, from steep drops in birth rates. Note that falling rates of growth do not necessarily mean that population sizes overall are decreasing, but only that rates of increase are slowing.

## Total fertility rate influences population growth

One key statistic demographers calculate to examine a population's potential for growth is the **total fertility rate (TFR)**, the average number of children born per female member of a population during her lifetime. **Replacement fertility** is the TFR that keeps the size of a population stable. For humans, replacement fertility roughly equals a TFR of 2.1. When the TFR drops below 2.1, population size, in the absence of immigration, will shrink.

Various factors influence TFR and have acted to drive it downward in many countries in recent years. Historically, people tended to conceive many children, which helped ensure that at least some would survive, but today's improved medical care has reduced infant mortality rates. Increasing urbanization has also driven TFR down; whereas rural families need children to contribute to farm labor, in urban areas children are usually excluded from the labor market, are required to go to school, and impose economic costs on their families. Moreover, if a government provides some form of social security, as most do these days, parents need fewer children to support them in their old age when they can no longer work. Finally, with greater education and changing roles in society, women tend to shift into the labor force, putting less emphasis on child rearing.

All these factors have come together in Europe, where TFR has dropped from 2.6 to 1.5 in the past half century. Every European nation now has a fertility rate below the replacement level, and populations are declining in 14 of 44 European nations. In 2007, Europe's overall annual **natural rate of population change** (change due to birth and death rates alone, excluding migration) was −0.1%. Worldwide by 2007, 71 countries had fallen below the replacement fertility of 2.1. These countries make up roughly 45% of the world's population and include China (with a TFR of 1.6). **Table 8.2** shows TFRs of major continental regions.

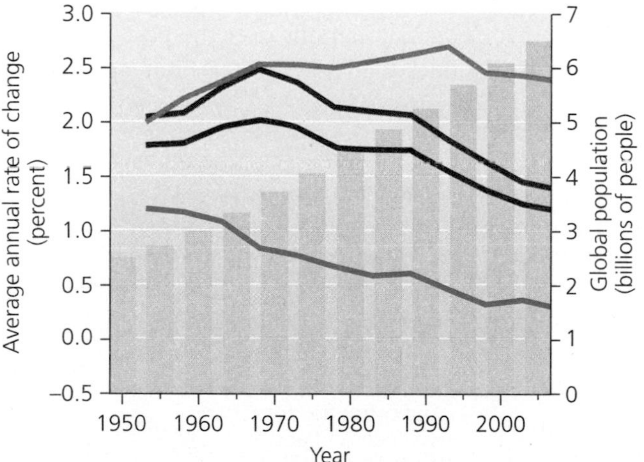


**FIGURE 8.14** The annual growth rate of the global human population (dark blue line) peaked in the late 1960s and has declined since then. Growth rates of industrialized nations (green line) have fallen since 1950, while those of developing nations (red line) have fallen since the global peak in the late 1960s. For the world's least-developed nations (orange line), growth rates began to fall in the 1990s. Although growth *rates* are declining, global population size (gray bars) is still growing about the same amount each year, because smaller percentage increases of ever-larger numbers produce roughly equivalent additional amounts. Data from U.N. Population Division. 2006. *World population prospects: The 2006 revision.*

**TABLE 8.2 Total Fertility Rates for Major Continental Regions**

| Region | Total fertility rate (TFR) |
|---|---|
| Africa | 5.0 |
| Latin America and the Caribbean | 2.5 |
| Asia | 2.4 |
| Oceania | 2.1 |
| North America | 2.0 |
| Europe | 1.5 |

Data from Population Reference Bureau. 2007. *2007 World population data sheet.*

### Weighing THE Issues | Consequences of Low Fertility?

In the United States, Canada, and almost every European nation, the total fertility rate has now dipped below the replacement fertility rate. What economic or social consequences do you think might result from below-replacement fertility rates? Would you rather live in a society with a high TFR or a low TFR? Why?

## Some nations have experienced the demographic transition

Many nations with lowered birth rates and TFRs are experiencing a common set of interrelated changes. In countries with good sanitation, effective health care, and reliable food supplies, more people than ever before are living long lives. As a result, over the past 50 years the life expectancy for the average person has increased from 46 to 68 years as the global crude death rate has dropped from 20 deaths per 1,000 people to 9 deaths per 1,000 people. Strictly speaking, **life expectancy** is the average number of years that an individual in a particular age group is likely to continue to live, but often people use this term to refer to the average number of years a person can expect to live from birth. Much of the increase in life expectancy is due to reduced rates of infant mortality. Societies going through these changes are generally those that have undergone urbanization and industrialization and have generated personal wealth for their citizens.

To make sense of these trends, demographers developed a concept called the **demographic transition**. This is a model of economic and cultural change first proposed in the 1940s and 1950s by demographer Frank Notestein to explain the declining death rates and birth rates that have occurred in Western nations as they became industrialized. Notestein believed nations move from a stable pre-industrial state of high birth and death rates to a stable post-industrial state of low birth and death rates. Industrialization, he proposed, causes these rates to fall naturally by first decreasing mortality and then lessening the need for large families. Parents thereafter choose to invest in quality of life rather than quantity of children. Because death rates fall before birth rates fall, a period of net population growth results. Thus, under the demographic transition model, population growth is seen as a temporary phenomenon that occurs as societies move from one condition to another.

**The pre-industrial stage** Notestein's demographic model describing the population impacts of industrialization proceeds in several stages (**Figure 8.15**). The first is the **pre-industrial stage,** characterized by conditions that have defined most of human history. In pre-industrial societies, both death rates and birth rates are high. Death rates are high because disease is widespread, medical care rudimentary, and food supplies unreliable and difficult to obtain. Birth rates are high because people must compensate for infant mortality by having several children. In this stage, children are valuable as workers who can help meet a family's basic needs. Populations within the pre-industrial stage are not likely to experience much growth, which is why the human population was relatively stable until the industrial revolution.

**Industrialization and falling death rates** Industrialization initiates the second stage of the demographic transition, known as the **transitional stage**. This transition from the pre-industrial stage to the industrial stage is generally characterized by declining death rates due to increased food production and improved medical care. Birth rates in the transitional stage remain high, however, because people have not yet grown used to the new economic and social conditions. As a result, population growth surges.

**The industrial stage and falling birth rates** The third stage in the demographic transition is the **industrial stage.** Industrialization increases opportunities for employment outside the home, particularly for women. Children become less valuable, in economic terms, because they do not help meet family food needs as they did in the pre-industrial stage. If couples are aware of this, and if they have access to birth control, they may choose to have fewer children. Birth rates fall, closing the gap with death rates and reducing population growth.

**The post-industrial stage** In the final stage, the **post-industrial stage**, both birth and death rates have fallen to low and stable levels. Population sizes stabilize or decline slightly. The society enjoys the fruits of industrialization without the threat of runaway population growth.

## Is the demographic transition a universal process?

The demographic transition has occurred in many European countries, the United States, Canada, Japan, and several other developed nations over the past 200 to 300 years. It is a model that may or may not apply to all

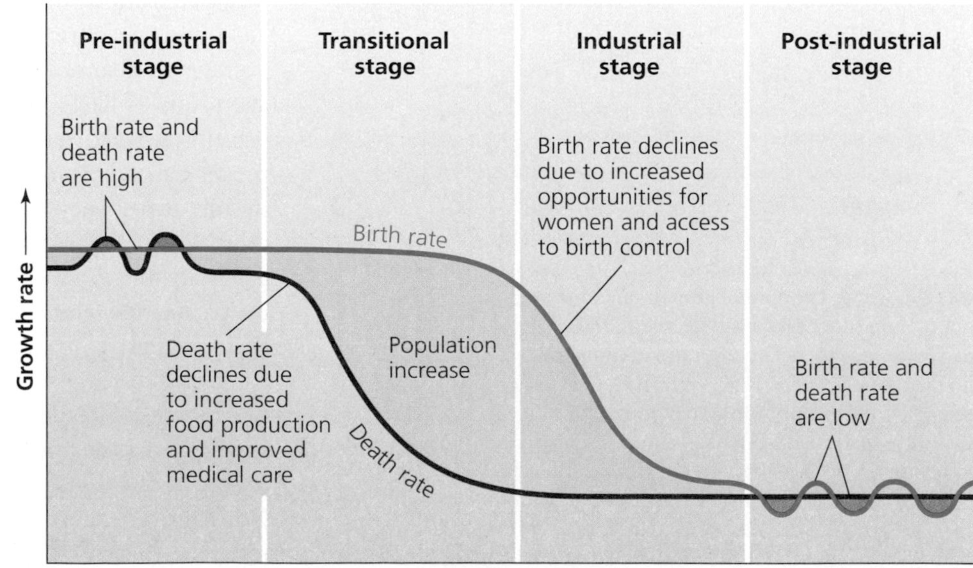


FIGURE 8.15 The demographic transition is an idealized process that has taken some populations from a pre-industrial state of high birth rates and high death rates to a post-industrial state of low birth rates and low death rates. In this diagram, the wide green area between the two curves illustrates the gap between birth and death rates that causes rapid population growth during the middle portion of this process. Adapted from Kent, M. M. and K. A. Crews. 1990. *World population: Fundamentals of growth.* Population Reference Bureau.

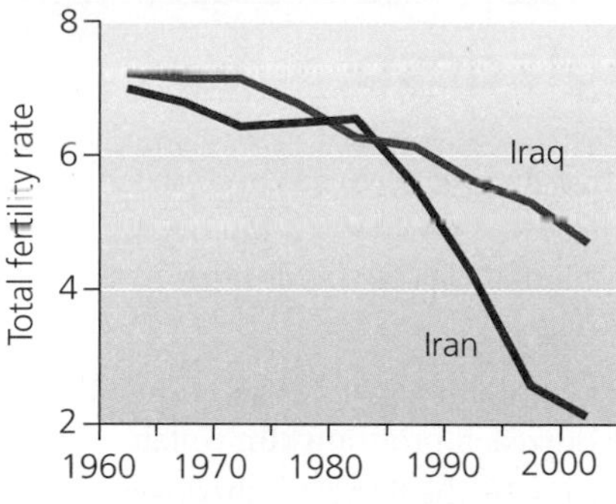


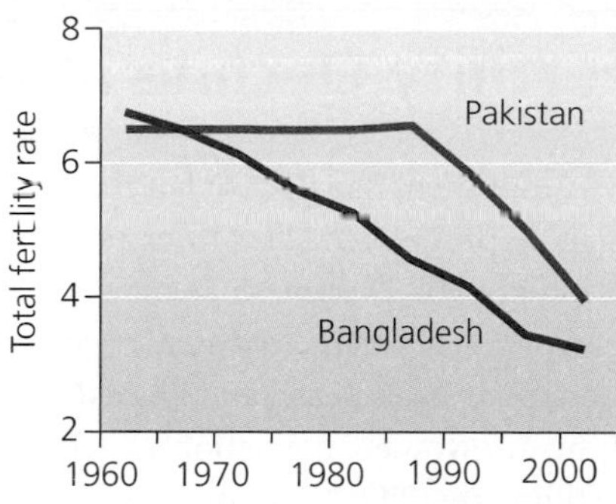


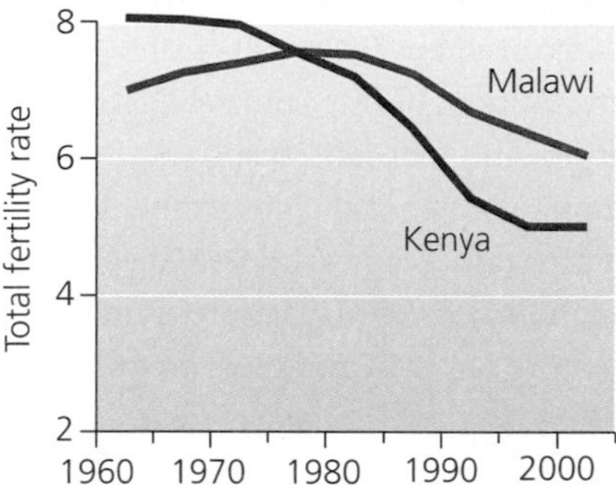


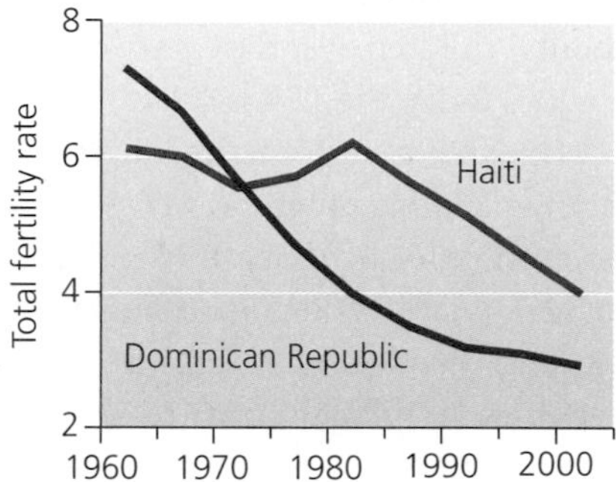


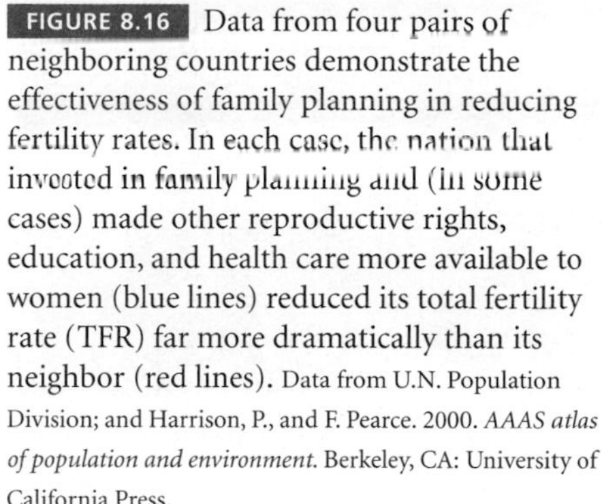
**FIGURE 8.16** Data from four pairs of neighboring countries demonstrate the effectiveness of family planning in reducing fertility rates. In each case, the nation that invested in family planning and (in some cases) made other reproductive rights, education, and health care more available to women (blue lines) reduced its total fertility rate (TFR) far more dramatically than its neighbor (red lines). Data from U.N. Population Division; and Harrison, P., and F. Pearce. 2000. *AAAS atlas of population and environment.* Berkeley, CA: University of California Press.

developing nations as they industrialize now and in the future. On the one hand, note in Figure 8.14 (•p. 219) how growth rates fell first for industrialized nations, then for less-developed nations, and finally for least-developed nations, suggesting that it may merely be a matter of time before all nations experience the transition. On the other hand, some social scientists point out that population dynamics may differ for developing nations that try to adopt the Western world's industrial model rather than devising their own. And some demographers assert that the transition will fail in cultures that place greater value on childbirth or grant women fewer freedoms.

Moreover, natural scientists estimate that for people of all nations to attain the material standard of living that North Americans now enjoy, we would need the natural resources of four and a half more planet Earths. Whether developing nations, which include the vast majority of the planet's people, pass through the demographic transition is one of the most important and far-reaching questions for the future of our civilization and Earth's environment.

# Population and Society

Demographic transition theory links the statistical study of human populations with societal factors that influence, and are influenced by, population dynamics. Let's examine a few of these societal factors more closely.

## Women's empowerment greatly affects population growth rates

Fertility rates have dropped most noticeably in countries where women have gained improved access to contraceptives and family-planning programs (**Figure 8.16**; and see "The Science behind the Story," • pp. 222–223), and where they have gained better educational opportunities (**Figure 8.17**).

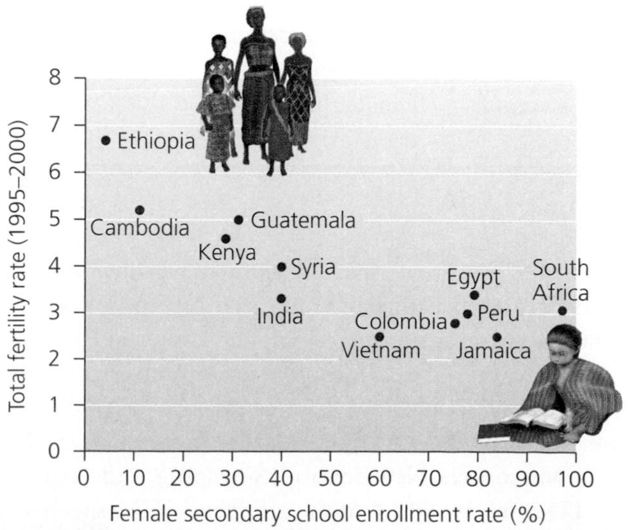


**FIGURE 8.17** Increasing female literacy is strongly associated with reduced birth rates in many nations. Data from McDonald, M., and D. Nierenberg. 2003. Linking population, women, and biodiversity. *State of the world 2003.* Washington, D.C.: Worldwatch Institute.

THE SCIENCE BEHIND THE STORY

# Fertility Decline in Bangladesh

*A woman in Matlab, Bangladesh.*

Research in developing countries indicates that poverty and overpopulation can create a vicious cycle, in which poverty encourages high fertility and high fertility obstructs economic development. Are there policy steps that such countries can take to bring down fertility rates? Scientific analysis of family-planning programs in the South Asian nation of Bangladesh suggests that the answer is yes.

Bangladesh is one of the poorest, most densely populated countries on the planet. Its 145 million people live in an area about the size of Wisconsin, and 45% of them live below the poverty line. With few natural resources and 1,000 people per km$^2$ (over 2,500 per mi$^2$—more than twice the population density of New Jersey), limiting population growth is critically important. Back in 1976, Bangladeshi president Ziaur Rahman declared, "If we cannot do something about population, nothing else that we accomplish will matter much."

Since then, Bangladesh has made striking progress in controlling population growth. Despite stagnant economic development, low literacy rates, poor health care, and limited rights for women, the nation's total fertility rate (TFR) has dropped markedly. In the 1970s, the average woman in Bangladesh gave birth to about 6 children over the course of her life. Today, the TFR is 3.2.

Researchers hypothesized that family-planning programs were responsible for Bangladesh's rapid reduction in TFR. Because conducting an experiment to test such a hypothesis is difficult, some researchers took advantage of a natural experiment (• pp. 13–14). By comparing Bangladesh to countries that are socioeconomically similar but have had less success in lowering TFR, such as Pakistan (see Figure 8.16, • p. 221), researchers concluded that Bangladesh succeeded because of aggressive, well-funded outreach efforts that were sensitive to the values of its traditional society.

However, no two countries are identical, so it is difficult to draw firm conclusions from such broad-scale comparisons. This is why the Matlab Family Planning and Health Services Project, in the isolated rural area of Matlab, Bangladesh, has become one of the best-known experiments in family planning in developing countries.

The Matlab Project was an intensive outreach program run collaboratively by the Bangladeshi government and international aid organizations. Each household in the project area received biweekly visits from local women offering counseling, education, and free contraceptives. Compared to a similar government-run program in a nearby area, the Matlab Project featured more training, more services, and more frequent

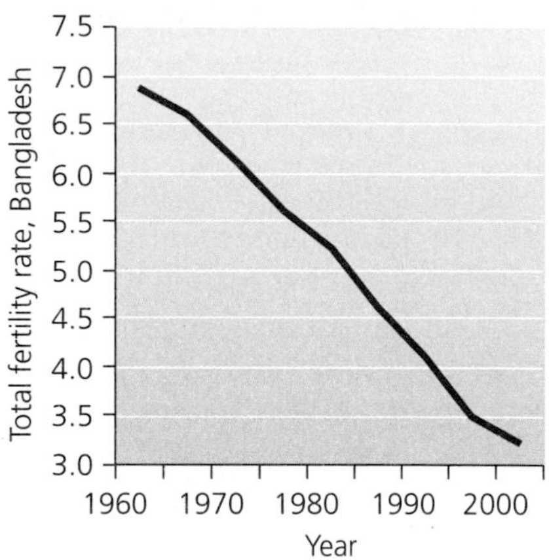


**Total fertility rate has declined markedly in Bangladesh in the past 40 years, in part because of family-planning programs.** Data from U.N. Population Division.

In 2007, 54% of married women worldwide (aged 15–49) reported using some modern method of contraception to plan or prevent pregnancy. China, at 86%, had the highest rate of contraceptive use of any nation. Five European nations showed rates of contraceptive use above 70%, as did Australia, Brazil, Canada, Costa Rica, Cuba, Micronesia, New Zealand, Singapore, and Thailand (the U.S. rate was 68%). At the other end of the spectrum, 20 African nations had rates below 10%. These low rates of contraceptive use contribute to high fertility rates in sub-Saharan Africa, where the region's TFR is 5.5 children per woman. By comparison, in Asia, where the TFR in 1950 was 5.9, today it is 2.4—in part a legacy of the population control policies of China and some other Asian countries.

These data clearly indicate that in societies where women have little power, substantial numbers of pregnancies are unintended. Unfortunately, many women still lack the information and personal freedom of choice to allow them to make their own decisions about when to have children and how many to have. Today, many social scientists and policymakers recognize that for population growth to slow and stabilize, women need to achieve equal power with men in

In the Matlab Project, Bangladeshi households received visits from local women offering counseling, education, and free contraceptives.

visits. In both areas, a highly organized health surveillance system gave researchers detailed information about births, deaths, and health-related behaviors such as contraceptive use. The result was an experiment comparing the Matlab Project with the government-run project.

When Matlab Project director James Phillips and his colleagues reviewed a decade's worth of data in 1988, they found that fertility rates had declined in both areas. The decline appeared to be due almost entirely to a rise in contraceptive use, because other factors—such as the average age of marriage—remained the same. Phillips and his colleagues also found the declines to be significantly greater in the Matlab area than in the government-run area. These findings suggested that high-intensity outreach efforts can affect TFR even in the absence of significant improvements in women's status, education, or economic development.

But why exactly was the outreach program successful? One hypothesis was that visits from health care workers helped convince local women that small families are desirable. However, in 1999, Mary Arends-Kuenning, a graduate student in economics at the University of Michigan, and her colleagues reported that there was no relationship between women's perception of the ideal family size and the number of visits made by outreach workers, either in Matlab or nearby comparison areas. Ideal family size declined equally in all areas.

Instead of creating new demand for birth control, the Matlab Project appears to have helped women convert an already-existing desire for fewer children into behaviors, such as contraceptive use, that reduce fertility.

Bangladesh's ability to rein in fertility rates despite unfavorable social and economic conditions bodes well for impoverished nations facing explosive population growth. However, significant challenges remain. If rates fail to decline further, the country's population could double to 290 million—nearly the size of today's U.S. population—in 35–40 years. Scientific research has helped illuminate the impact of family-planning programs on fertility, but further reductions may require fundamental social, political, and economic changes that are difficult to implement in traditional, resource-strapped countries such as Bangladesh.

societies worldwide. Studies show that in societies in which women are freer to decide whether and when to have children, fertility rates have fallen, and the resulting children are better cared for, healthier, and better educated.

Unfortunately, we are still a long way from achieving gender equality. Over two-thirds of the world's people who cannot read, and 60% of those living in poverty, are women. Violence against women remains shockingly common. In many societies, men restrict women's decision-making abilities, including decisions as to how many children they will bear.

The gap between the power held by men and by women is just as obvious at the highest levels of government. Worldwide, only 17% of elected government officials in national legislatures are women (**Figure 8.18**). The United States lags behind not only Europe but also many developing nations in the proportion of female representatives in its government. As more women win positions of power, perhaps gender equality will become a more tangible reality. Such equality would have environmental consequences, for when women have economic and political power and access to education, they gain the option, and often the motivation, to limit the number of children they bear.

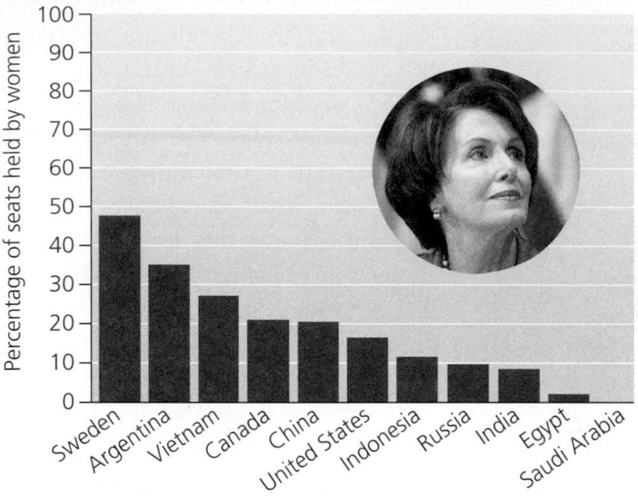


FIGURE 8.18 Although women make up half of the world's population, they are vastly underrepresented in positions of political power. Measured by the percentages of seats held by women in national legislatures, women in some nations fare better than those in others. Although the United States finally gained a female Speaker of the House (Nancy Pelosi, inset), the United States lags behind the world average in the proportion of women in its legislative branch. Data from Inter-Parliamentary Union, Women in national parliaments (July 2007).

## Population policies and family-planning programs are working around the globe

Data show that funding and policies that encourage family planning can effectively lower population growth rates in all types of nations, even those that are least industrialized. No nation has pursued a sustained population control program as intrusive as China's, but other rapidly growing nations have implemented less-restrictive programs.

India was the first nation to implement population control policies. However, when some policymakers introduced forced sterilization in the 1970s, the resulting outcry brought down the government. Since then, India's efforts have been more modest and far less coercive, focusing on family planning and reproductive health care. However, today a number of Indian states also run programs of incentives and disincentives promoting a "two-child norm," and current debate centers on whether this is a just and effective approach. Regardless, unless India strengthens its efforts to slow population growth, it seems set to overtake China and become the world's most populous nation about the year 2030.

The government of Thailand relies on an education-based approach to family planning that has reduced birth rates and slowed population growth. In the 1960s, Thailand's growth rate was 2.3%, but today it stands at 0.7%. This decline was achieved without a one-child policy. It has resulted, in large part, from government-sponsored programs devoted to family-planning education and increased availability of contraceptives. Brazil, Mexico, Iran, Cuba, and many other developing countries have instituted active programs to reduce their population growth. These programs entail setting targets and providing incentives, education, contraception, and reproductive health care.

These programs are working. The data shown in Figure 8.17 are not the only cases in which family-planning programs have helped lower fertility rates. One study in 2000 examined four different pairs of nations located in the same parts of the world, with one country in each pair having a stronger program: Thailand and the Philippines, Pakistan and Bangladesh, Tunisia and Algeria, and Zimbabwe and Zambia. The demographers concluded that in all four cases, the country with the stronger program (Thailand, Bangladesh, Tunisia, and Zimbabwe) initiated or accelerated a decline in fertility with its policies. In the case of Thailand and the Philippines, the researchers also concluded that the Catholic Church's strong presence in the Philippines held back the success of family planning there.

In 1994, the United Nations hosted a milestone conference on population and development in Cairo, Egypt, at which 179 nations endorsed a platform calling on all governments to offer universal access to reproductive health care within 20 years. The conference marked a turn away from older notions of command-and-control population policy geared toward pushing contraception and lowering population to preset targets. Instead, it urged governments to offer better education and health care and to address social needs that bear indirectly on population (such as alleviating poverty, disease, and sexism).

Despite the successes of family planning internationally, recent Republican administrations in the United States have declined to fund family-planning efforts by the United Nations. Canceling this funding was one of George W. Bush's first acts on becoming U.S. president in 2001. Moreover, worldwide, funding for family planning programs fell from $723 million to $442 million in the decade following the Cairo conference.

### Weighing THE Issues | U.S. Involvement in International Family Planning

From 1998 to 2001, the United States provided $46.5 million to the United Nations Population Fund (UNFPA), which advises governments on family planning, as well as sustainable development, poverty reduction, reproductive health, and HIV/AIDS prevention in many nations, including China. Starting in 2001, the Bush administration withheld funds, saying that U.S. law prohibits funding any organization that "supports or participates in the management of a program of coercive abortion or involuntary sterilization," and claiming that the Chinese government has been implicated in both. Many nations criticized the U.S. decision, and the European Union offered UNFPA additional funding to offset the loss of U.S. contributions. What do you think of the U.S. decision? Should the United States fund family-planning efforts in other nations? What conditions, if any, should it place on the use of such funds?

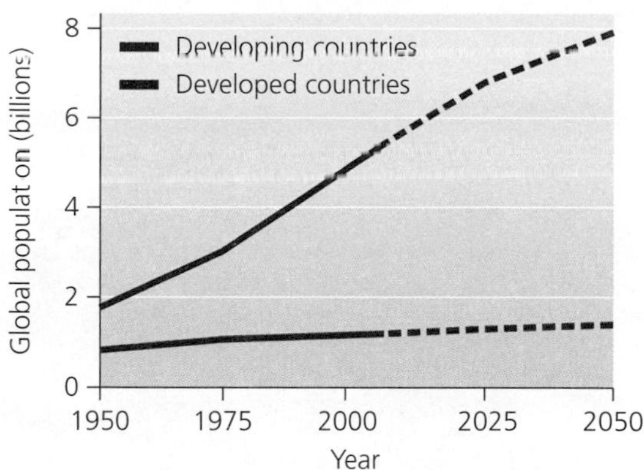


FIGURE 8.19 Over 99% of the next 1 billion people added to Earth's human population will reside in the less developed, poorer parts of the world. The dashed line indicates projected future trends. Data from U.N. Population Division.

## Poverty is strongly correlated with population growth

The alleviation of poverty was a prime target of the Cairo conference because poorer societies tend to show higher population growth rates than do wealthier societies. This pattern is consistent with demographic transition theory. Note in Table 8.3 how poorer nations tend to have higher fertility and growth rates, along with higher birth and infant mortality rates and lower rates of contraceptive use.

Trends such as these influence the distribution of people on the planet. In 1960, 70% of all people lived in developing nations. By 2007, 82% of the world's population was living in these countries. Moreover, fully 99% of the next billion people to be added to the global population will be born in these poor, less developed regions (**Figure 8.19**).

This is unfortunate from a social standpoint, because these people will be added to the countries that are least able to provide for them. It is also unfortunate from an environmental standpoint, because poverty often results in environmental degradation. People who depend on agriculture in an area of poor farmland, for instance, may need to try to farm even if doing so degrades the soil and is not sustainable. This is largely why Africa's once-productive

**TABLE 8.3 Per Capita Wealth, with Rates of Fertility, Population Growth, and Contraceptive Use, for Selected Nations**

| Nation | Per Capita GNI PPP (U.S. $)* | Population increase (% per year) | Children born per woman (TFR) | Population density (per km$^2$) | Infant mortality (per 1,000) | Percentage of couples using birth control |
|---|---|---|---|---|---|---|
| Tanzania | 740 | 2.6 | 5.4 | 41 | 78 | 20 |
| Niger | 830 | 3.4 | 7.1 | 11 | 126 | 5 |
| Ethiopia | 1,190 | 2.5 | 5.4 | 70 | 77 | 14 |
| Haiti | 1,490 | 1.8 | 4.0 | 323 | 57 | 25 |
| Pakistan | 2,500 | 2.3 | 4.1 | 213 | 78 | 22 |
| India | 3,800 | 1.6 | 2.9 | 344 | 58 | 49 |
| Syria | 3,920 | 2.5 | 3.5 | 108 | 19 | 35 |
| China | 7,730 | 0.5 | 1.6 | 138 | 27 | 86 |
| Brazil | 8,800 | 1.4 | 2.3 | 22 | 27 | 70 |
| Romania | 9,820 | −0.2 | 1.3 | 90 | 14 | 34 |
| Mexico | 11,330 | 1.7 | 2.4 | 54 | 21 | 59 |
| Spain | 28,420 | 0.3 | 1.4 | 90 | 4 | 53 |
| Japan | 33,730 | 0.0 | 1.3 | 338 | 3 | 48 |
| Canada | 34,610 | 0.3 | 1.5 | 3 | 5 | 73 |
| United Kingdom | 35,690 | 0.3 | 1.8 | 251 | 5 | 79 |
| United States | 44,260 | 0.6 | 2.1 | 31 | 7 | 68 |

*GNI PPP is "gross national income in purchasing power parity," a measure that standardizes income and makes it comparable among nations by converting income to "international" dollars using a conversion factor. International dollars indicate the amount of goods and services one could buy in the United States with a given amount of money. Data from Population Reference Bureau. 2007. *2007 World population data sheet.*

FIGURE 8.20 In the semi-arid Sahel region of Africa, where population is increasing beyond the land's ability to handle it, drought and dependence on grazing agriculture have led to environmental degradation.

Sahel region, like many areas of western China, is turning to desert (**Figure 8.20**). Poverty also drives people to hunt large mammals in Africa's forests, including the great apes that are now disappearing as settlers and miners kill them for their "bush meat."

## Consumption from affluence creates environmental impact

Poverty can lead people into environmentally destructive behavior, but wealth can produce even more severe and far-reaching environmental impacts. The affluence of a society such as the United States, Japan, or the Netherlands is built on levels of resource consumption unprecedented in human history. Much of this chapter has dealt with numbers of people rather than on the amount of resources each member of the population consumes or the amount of waste each member produces. The environmental impact of human activities, however, depends not only on the number of people involved but also on the way those people live. Recall the A for affluence in the IPAT equation. Affluence and consumption are spread unevenly across the world, and affluent societies generally consume resources from other localities as well as from their own—as we saw in "The Science behind the Story" (• pp. 212–213).

In Chapter 1 (• pp. 6, 17), we introduced the concept of the *ecological footprint,* the cumulative amount of Earth's surface area required to provide the raw materials a person or population consumes and to dispose of or recycle the waste produced. Individuals from affluent societies leave considerably larger per capita ecological footprints (see Figure 1.12, • p. 17). In this sense, the addition of one American to the world has as much environmental impact as the addition of six Chinese or of 12 Indians or Ethiopians. This fact should remind us that the "population problem" does not lie entirely with the developing world.

Indeed, just as population is rising, so is consumption, and some environmental scientists have calculated that we are already living beyond the planet's means to support us sustainably. One recent analysis concludes that humanity's global ecological footprint surpassed Earth's capacity to support us in 1987 and that our species is now living more than 25% beyond its means (**Figure 8.21**). The rising consumption that is accompanying the rapid industrialization of China, India, and other populous nations makes it all the more urgent for us to find a path to global sustainability.

## The wealth gap and population growth contribute to conflict

The stark contrast between affluent and poor societies in today's world is the cause of social as well as environmental stress. Over half the world's people live below the internationally defined poverty line of U.S. $2 per day. The richest one-fifth of the world's people possesses over 80 times the income of the poorest one-fifth (**Figure 8.22**). The richest one-fifth also uses 86% of the world's resources. That leaves only 14% of global resources—energy, food, water, and other essentials—for the remaining four-fifths of the world's people to share.

As the gap between rich and poor grows wider and as the sheer numbers of those living in poverty continue to increase, it seems reasonable to predict increasing tensions between "haves" and "have-nots." This is why the inequitable distribution of wealth is one of the key factors the U.S. Departments of Defense and State take into

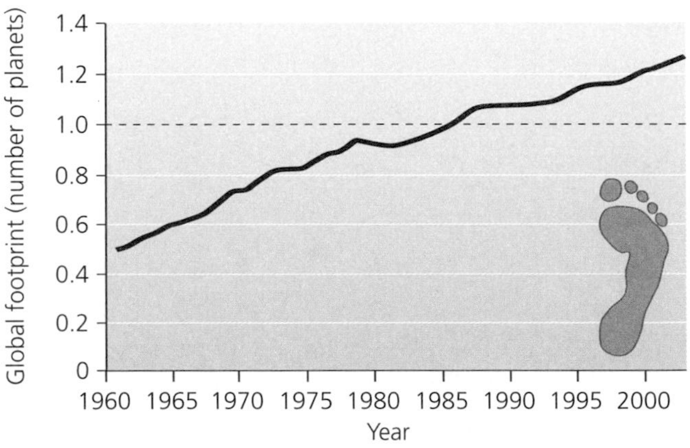


FIGURE 8.21 The global ecological footprint of the human population is over 2.5 times larger than in 1961 and now exceeds what Earth can bear in the long run, scientists have calculated. The estimate shown here indicates that we have already overshot our carrying capacity by at least 25%; that is, we are using renewable natural resources 25% faster than they are being replenished. Data from WWF—World Wide Fund for Nature, 2006. *Living planet report.* Gland, Switzerland: WWF.

(a) A family living in the United States

(b) A family living in India

FIGURE 8.22 A typical U.S. family (**a**) may own a large house with numerous material possessions. A typical family in a developing nation such as India (**b**) may live in a small, sparsely furnished dwelling with few material possessions and little money or time for luxuries.

account when assessing the potential for armed conflict, whether it be conventional warfare or terrorism.

## HIV/AIDS is exerting major impacts on African populations

The rising material wealth and falling fertility rates of many industrialized nations today is slowing population growth in accordance with the demographic transition model. Nations in which the epidemic of the human immunodeficiency virus (HIV) and acquired immunodeficiency syndrome (AIDS) has taken hold, however, are not following Notestein's script. Instead, in these countries mortality is increasing, presenting a scenario more akin to Malthus's fears (**Figure 8.23**). African nations are being hit hardest. Of the 40 million people around the world infected with HIV/AIDS as of 2005, 27 million live in the nations of sub-Saharan Africa. The low rate of contraceptive use, which contributes to this region's high fertility rate, also fuels the expansion of AIDS. One in every 16 people aged 15 to 49 in sub-Saharan Africa is infected with HIV, and for southern African nations, the figure is 1 in 5.

The AIDS epidemic is having the greatest impact on human populations of any disease since the Black Death killed roughly one-third of the population in 14th-century Europe and since smallpox brought by Europeans to the New World wiped out perhaps millions of native people. As AIDS takes roughly 6,000 lives in Africa every day, the epidemic is unleashing a variety of demographic changes. Infant mortality in sub-Saharan Africa has risen to 9 deaths out of 100 live births—15 times the rate in the developed world. The high numbers of infant deaths and premature deaths of young adults has caused life expectancy in parts of southern Africa to fall from a high of close to 59 years in the early 1990s back down to less than 40 years, where it stood in the early 1950s. AIDS is also leaving behind many millions of orphans.

Africa is not the only region with reason to worry. HIV is well-established in the Caribbean and in Southeast Asia, and it is spreading quickly in eastern Europe and central Asia. In China, where injected drug use is fueling its spread, the World Health Organization estimated that 10 million people could become infected between 2005 and 2010.

## Severe demographic changes have social, political, and economic repercussions

Everywhere in sub-Saharan Africa, AIDS is undermining the ability of developing countries to make the transition to modern technologies because it is removing

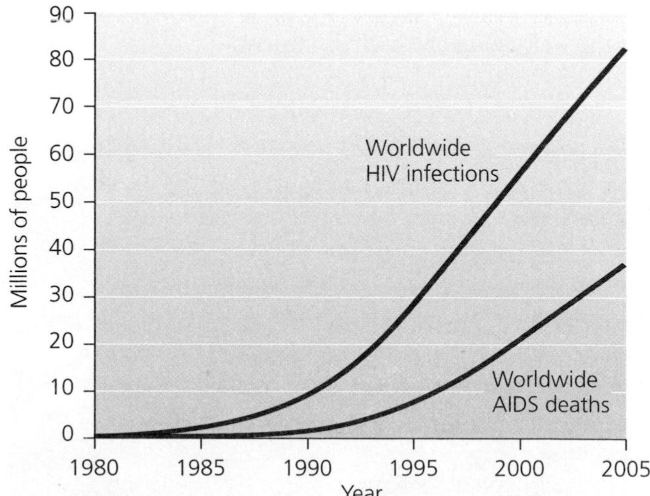


FIGURE 8.23 AIDS cases are increasing rapidly in much of the world. As of 2005, total cumulative HIV infections since 1980 were estimated at nearly 83 million, and 37 million people are estimated to have died from the disease so far. Data from UNAIDS; and *Vital signs 2006–2007*. Washington, D.C.: Worldwatch Institute.

many of the youngest and most productive members of society. For example, in 1999 Zambia lost 600 teachers to AIDS, and only 300 new teachers graduated to replace them. In Rwanda, more than one in three college-educated residents of the city of Kigali are infected with the virus. South Africa loses an estimated $7 billion per year to declines in its labor force as AIDS patients fill the nation's hospitals (**Figure 8.24**). The loss of productive household members to AIDS causes families and communities to break down as income and food production decline while medical expenses and debt skyrocket.

These problems are hitting many countries at a time when their governments are already experiencing what has been called *demographic fatigue.* Demographically fatigued governments face overwhelming challenges related to population growth, including educating and finding jobs for their swelling ranks of young people. With the added stress of HIV/AIDS, these governments face so many demands that they are stretched beyond their capabilities to address problems. As a result, the problems grow worse, and citizens lose faith in their governments' abilities to help them.

If nations in sub-Saharan Africa—and other regions where the disease is spreading fast, such as India and southeast Asia—do not take aggressive steps soon, and if the rest of the world does not step in to help, these countries could fail to advance through the demographic transition. Instead, their rising death rates could push birth rates back up, potentially causing these countries to fall back to the pre-industrial stage of the demographic transition model. Such an outcome would lead to greater population growth while economic and social conditions worsen. It would be a profoundly negative outcome, both for human welfare and for the well-being of the environment.

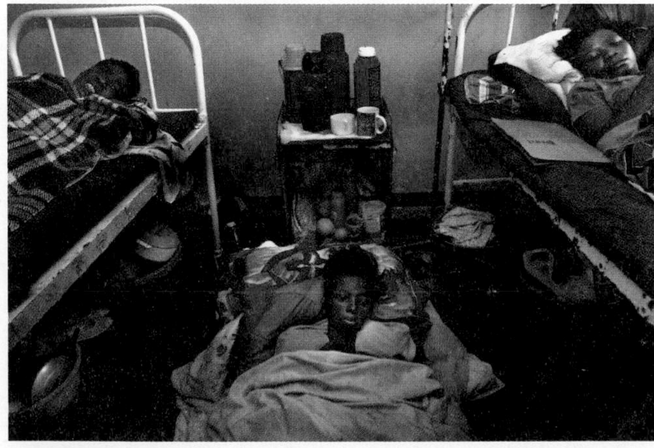

FIGURE 8.24 AIDS patients occupy 60% of South Africa's hospital beds. By 2010, AIDS in Africa may orphan an estimated 40 million children.

If one of humanity's goals is to generate a high standard of living and quality of life for all the world's people, then developing nations must find ways to reduce their population growth. However, those of us living in the industrialized world must also be willing to reduce our consumption. Earth does not hold enough resources to sustain all 6.7 billion of us at the current North American standard of living, nor can we go out and find extra planets; so, we must make the best of the one place that supports us all.

## Conclusion

Today's human population is larger than at any time in the past. Our growing population, as well as our growing consumption, affects the environment and our ability to meet the needs of all the world's people. Approximately 90% of children born today are likely to live their lives in conditions far less healthy and prosperous than most of us in the industrialized world are accustomed to.

However, there are at least two major reasons to be encouraged. First, although global population is still rising, the *rate* of growth has decreased nearly everywhere, and some countries are even seeing population declines. Most developed nations have passed through the demographic transition, showing that it is possible to lower death rates while stabilizing population and creating more prosperous societies. A second reason to feel encouraged is the progress in expanding rights for women worldwide. Although there is still a long way to go, women are receiving better education, more economic independence, and more ability to control their reproductive decisions. Aside from the clear ethical progress these developments entail, they are helping slow population growth.

Human population cannot continue to rise forever. The question is how it will stop rising: through the gentle and benign process of the demographic transition, through restrictive governmental intervention such as China's one-child policy, or through the miserable Malthusian checks of disease and social conflict caused by overcrowding and competition for scarce resources. Moreover, sustainability demands a further challenge—that we stabilize our population size in time to avoid destroying the natural systems that support our economies and societies. We are indeed a special species. We are the only one to achieve such dominance as to alter so much of Earth's landscape and even its climate system. We are also the only species with the intelligence needed to turn around an increase in our own numbers before we destroy the very systems on which we depend.

## REVIEWING OBJECTIVES

**You should now be able to:**

**Assess the scope of human population growth**

- Our global population of 6.7 billion people adds about 80 million people per year (2.5 people every second). (p. 208)
- Our growth rate peaked at 2.1% in the 1960s and now stands at 1.2%. Growth rates vary among regions of the world. (p. 209)
- Rising population can deplete resources, intensify pollution, stress social systems, or degrade ecosystems, such that the natural environment or our quality of life decline. (pp. 209–210)

**Evaluate how human population, affluence, and technology affect the environment**

- The IPAT model summarizes how environmental impact (I) results from interactions among population size (P), affluence (A), and technology (T). (p. 211)
- Rising population and rising affluence (leading to greater consumption) each increase environmental impact. Technological advances have frequently worsened environmental degradation, but they can also help mitigate our impact. (pp. 211–213)

**Explain and apply the fundamentals of demography**

- Demography applies principles of population ecology to the statistical study of human populations. (p. 214)
- Demographers study size, density, distribution, age structure, and sex ratios of populations, as well as rates of birth, death, immigration, and emigration. (pp. 214–218)
- Total fertility rate (TFR) contributes greatly to change in a population's size. (pp. 218–219)

**Outline and assess the concept of demographic transition**

- The demographic transition model explains why population growth has slowed in industrialized nations. Industrialization and urbanization have reduced the economic need for children, while education and the empowerment of women have decreased unwanted pregnancies. Parents in developed nations choose to invest in quality of life rather than quantity of children. (pp. 219–220)
- The demographic transition may or may not proceed to completion in all of today's developing nations. Whether it does is of immense importance for the quest for sustainability. (pp. 220–221)

**Describe how wealth and poverty, the status of women, and family planning programs affect population growth**

- When women are empowered and achieve equality with men, fertility rates fall, and children tend to be better cared for, healthier, and better educated. (pp. 221–223)
- Family-planning programs and reproductive education have successfully reduced population growth in many nations. (pp. 221–224)
- Poorer societies tend to have higher population growth rates than do wealthier societies. (p. 225)
- The high consumption rates of affluent societies may make their ecological impact greater than that of poorer nations with larger populations. (pp. 226–227)

**Characterize the dimensions of the HIV/AIDS epidemic**

- About 40 million people worldwide are infected with HIV/AIDS, of which 27 million live in sub-Saharan Africa. (p. 227)
- Epidemics that claim large numbers of young and productive members of society influence population dynamics and can have severe social and political ramifications. (pp. 227–228)

## TESTING YOUR COMPREHENSION

1. What is the approximate current human global population? How many people are being added to the population each day?
2. Why has the human population continued to grow despite environmental limitations? Do you think this growth is sustainable? Why or why not?
3. Contrast the views of environmental scientists with those of Cornucopian economists and policymakers regarding whether population growth is a problem. Name several reasons why population growth is commonly viewed as a problem.
4. Explain the IPAT model. How can technology either increase or decrease environmental impact? Provide at least two examples.
5. What characteristics and measures do demographers use to study human populations? How does each of these help determine the impact of human populations on the environment?
6. What is the total fertility rate (TFR)? Can you explain why the replacement fertility for humans is approximately 2.1? How is Europe's TFR affecting its natural rate of population change?

7. Why have fertility rates fallen in many countries?
8. How does the demographic transition model explain the increase in population growth rates in recent centuries? How does it explain the decrease in population growth rates in recent decades?
9. Why are the empowerment of women and the pursuit of gender equality viewed as being important to controlling population growth? Describe the aim of family-planning programs.
10. Why do poorer societies have higher population growth rates than wealthier societies? How does poverty affect the environment? How does affluence affect the environment?

## SEEKING SOLUTIONS

1. China's reduction in birth rates is leading to significant change in the nation's age structure. Review Figure 8.11, which portrays the projected change. You can see that the population is growing older, based on the top-heavy age pyramid for the year 2030. What sorts of effects might this ultimately have on Chinese society? What steps could be taken in response?
2. The World Bank estimates that half the world's people survive on less than the equivalent of U.S. $2 per day. How would you expect this situation to affect the political stability of the world? Explain your answer.
3. Apply the IPAT model to the example of China provided in the chapter. How do population, affluence, technology, and ecological sensitivity affect China's environment? Now consider your own country or your own state. How do population, affluence, technology, and ecological sensitivity affect your environment? How can we regulate the relationship between population and its effects on the environment?
4. Do you think that all of today's developing nations will complete the demographic transition and come to enjoy a permanent state of low birth and death rates? Why or why not? What steps might we as a global society take to help ensure that they do? Now think about developed nations such as the United States and Canada. Do you think these nations will continue to lower and stabilize their birth and death rates in a state of prosperity? What factors might affect whether they do so?
5. **THINK IT THROUGH** Imagine that India's prime minister puts you in charge of that nation's population policy. India has a population growth rate of 1.7% per year, a TFR of 2.9, a 46% rate of contraceptive use, and a population that is 72% rural. What policy steps would you recommend, and why?
6. **THINK IT THROUGH** Now imagine that you have been tapped to design population policy for Germany. Germany is losing population at an annual rate of 0.2%, has a TFR of 1.3, a 72% rate of contraceptive use, and a population that is 88% urban. What policy steps would you recommend, and why?

## INTERPRETING GRAPHS AND DATA

In 2006, the population of the United States reached 300 million. As the nation's population has increased, its demographics have shifted as well, including the numbers of people per household. The bar graph depicts the percentage of U.S. households of different sizes in 1970 and in 2000.

1. Did household sizes generally increase, decrease, or stay the same from 1970 to 2000?
2. How many times more common were one-person households than households of five people or more in 2000?
3. What percentage of households had fewer than three people in 1970 and in 2000?
4. Describe the impacts you think these changes in household size may have had, in terms of (**a**) land use and development, and (**b**) consumption of energy and natural resources. Do you feel that household size is an important element to consider when determining the environmental impacts of population growth?

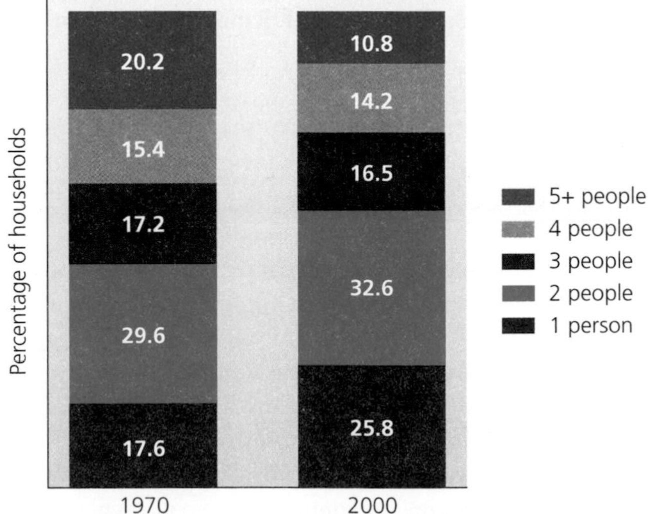


Percentage of U.S. households with one, two, three, four, and five or more people, in 1970 and in 2000. Data from U.S. Census Bureau.

## CALCULATING ECOLOGICAL FOOTPRINTS

A nation's population size and the affluence of its citizens each influence its resource consumption and environmental impact. In 2006, the world's population passed 6.5 billion. Average per capita income was $9,190 per year, and the average ecological footprint was 2.2 hectares (ha) per person. The sampling of data in the table will allow you to explore patterns in how population, affluence, and environmental impact are related.

| Nation | Population (millions of people) | Affluence (per capita income, in GNI PPP) | Personal impact (per capita footprint, in ha/person) | Total impact (national footprint, in millions of ha) |
|---|---|---|---|---|
| **Brazil** | 186.8 | $8,230 | 2.1 | 391.9 |
| **Ethiopia** | 74.8 | $1,000 | 0.8 | |
| **Japan** | 127.8 | $31,410 | 4.4 | |
| **Mexico** | 108.3 | $10,030 | 2.6 | |
| **Russia** | 142.3 | $10,640 | 4.4 | |
| **United States** | 299.1 | $41,950 | 9.6 | 2871.4 |

*Data sources:* Population Reference Bureau. 2006. *World population data sheet 2006*; and WWF—World Wide Fund for Nature, 2006. *Living planet report.* Gland, Switzerland: WWF.

1. Calculate the total impact (national ecological footprint) for each country.
2. Draw a graph illustrating per capita impact (on the *y* axis) vs. affluence (on the *x* axis). What do the results show? Explain why the data look the way they do.
3. Draw a graph illustrating total impact in relation to population. What do the results suggest to you?
4. Draw a graph illustrating total impact in relation to affluence. What do the results suggest to you?
5. You have just used three of the four variables in the IPAT equation. Can you give one example of how the T (technology) variable might increase the total impact of the United States, and one example of how it might decrease the U.S. impact?

## Take It Further

Go to www.aw-bc.com/withgott or the student CD-ROM, where you'll find:

- Suggested answers to end-of-chapter questions
- Quizzes, animations, and flashcards to help you study
- *Research Navigator*™ database of credible and reliable sources to assist you with your research projects
- **GRAPHit!** Tutorials to help you interpret graphs
- **INVESTIGATEit!** Current news articles that link the topics that you study to case studies from your region to around the world

CHAPTER

# 9 Soil and Agriculture

Wheat fields in Paraná, Brazil

## Upon completing this chapter, you will be able to:

- ▶ Explain the importance of soils to agriculture, and describe the impacts of agriculture on soils
- ▶ Outline major historical developments in agriculture
- ▶ Delineate the fundamentals of soil science, including soil formation and soil properties
- ▶ State the causes and predict the consequences of soil erosion and soil degradation
- ▶ Recite the history and explain the principles of soil conservation

No-till farm in Ceara, Brazil

CENTRAL CASE

# No-Till Agriculture in Southern Brazil

**"The nation that destroys its soil destroys itself."**
—U.S. PRESIDENT FRANKLIN D. ROOSEVELT

**"There are two spiritual dangers in not owning a farm. One is the danger of supposing that breakfast comes from the grocery, and the other that heat comes from the furnace."**
—CONSERVATIONIST AND PHILOSOPHER ALDO LEOPOLD

In southernmost Brazil, hundreds of thousands of people make their living farming. The warm climate and rich soils of this region's rolling highlands and coastal plains have historically made for bountiful harvests. However, repeated cycles of plowing and planting over many decades diminished the productivity of the soil. More and more topsoil—the valuable surface layer of soil richest in organic matter and nutrients—was eroded away by water and wind. Meanwhile, the synthetic fertilizers used to restore nutrients polluted area waterways. Crop yields fell, and by 1990 farmers were looking for help.

As a result, many of southern Brazil's farmers abandoned the conventional practice of tilling the soil after harvests. In its place, they turned to *zero-tillage*, or *no-till*, farming. Turning the earth by tilling (plowing, disking, harrowing, or chiseling) aerates the soil and works weeds and old crop residue into the soil to nourish it. However, tilling also leaves the surface bare, allowing wind and water to erode precious topsoil. Tilling historically boosted the productivity of agriculture in Europe, but in subtropical regions such as southern Brazil, heavy rainfall promotes erosion, causing tilled soils to lose organic matter and nutrients, and hot weather can overheat tilled soil.

Working with agricultural scientists and government extension agents, southern Brazil's farmers began leaving crop residues on their fields after harvesting, and planting "cover crops" to keep soil protected during periods when they weren't raising a commercial crop. When they went to plant the next crop, they merely cut a thin, shallow groove into the soil surface, dropped in seeds, and covered them. They did not invert the soil, but rather kept it covered with plants or their residues at all times, reducing erosion by 90%.

With less soil eroding away and more organic material being added to it, the soil held more water and encouraged better plant growth. In the state of Santa Catarina, maize yields per hectare increased by 47% between 1991 and 1999, wheat yields rose by 82%, and soybean yields by 83%. In the states of Paraná and Rio Grande do Sul, maize yields were up 67% over 10 years, and soybean yields were up 68%.

No-till farming methods also reduced farmers' costs, because farmers now used less labor and less fuel. No-till agriculture spread quickly in the region as farmers saw their neighbors' successes and traded information through "Friends of the Land" clubs organized on local, municipal, regional, and statewide levels. In Paraná and Rio Grande do Sul, the area being farmed with no-till methods shot up from 700,000 ha (1.7 million acres) in 1990 to 10.5 million ha (25.9 million acres) in 1999, when it involved 200,000 farmers. In Santa Catarina, where farms are generally smaller, over 100,000 farmers now apply no-till methods to 880,000 ha (2.2 million acres) of farmland. No-till farming is now spreading to other parts of Latin America.

By enhancing soil conditions and reducing erosion, no-till techniques have benefited southern Brazil's society and environment as well; its air, waterways, and ecosystems are less polluted. Similar effects are being felt in parts of the United States and elsewhere in the world where no-till and reduced-tillage methods are increasingly being applied.

Reduced tillage is not a panacea for all areas of the world. But in regions suitable for reduced tillage, proponents say these approaches can help make agriculture sustainable. We will need sustainable agriculture if we are to feed the world's human population while protecting the natural environment, including the soils that vitally support our production of food.

# Soil: The Foundation for Agriculture

As the human population has grown, so have the amounts of land and resources we devote to agriculture, which currently covers 38% of Earth's land surface. We can define **agriculture** as the practice of raising crops and livestock for human use and consumption. We obtain most of our food and fiber from **cropland**, land used to raise plants for human use, and **rangeland** or pasture, land used for grazing livestock.

Healthy soil is vital for agriculture, for forestry (Chapter 12), and for the functioning of Earth's natural systems. **Soil** is not merely lifeless dirt; it is a complex plant-supporting system consisting of disintegrated rock, organic matter, water, gases, nutrients, and microorganisms. Soil is a renewable resource; once depleted of its productivity, it may renew itself over time. However, this renewal generally occurs very slowly. If we abuse soil through careless or uninformed practices, we can greatly reduce its ability to support life for long time periods.

## As population and consumption increase, soils are being degraded

If we are to feed the world's rising human population, we will need to change our diet patterns or increase agricultural production—and do so sustainably, without degrading the environment and its ability to support agriculture. However, we cannot simply keep expanding agriculture into new areas, because land suitable and available for farming is running out. Instead, we must find ways to improve the efficiency of food production in areas that are already in agricultural use.

Today many lands unsuitable for farming are being farmed, causing considerable environmental damage. Mismanaged agriculture can turn grasslands into deserts; remove ecologically precious forests; diminish biodiversity; encourage invasive species; pollute soil, air, and water with toxic chemicals; and allow fertile soil to be blown and washed away.

As our planet gains 80 million people each year, we lose 5–7 million ha (12–17 million acres) of productive cropland annually. Especially in drier regions, degrading soil conditions (**Figure 9.1**) have increased the difficulty of raising crops and grazing livestock. Soil degradation around the globe has resulted roughly equally from forest removal, cropland agriculture, and overgrazing of livestock (**Figure 9.2**).

Over the past 50 years, scientists estimate that soil degradation has reduced potential rates of global grain production by 13% on cropland and 4% on rangeland. By the middle of this century, there will likely be 3 billion more mouths to feed. For these reasons, it is imperative that we learn to practice agriculture in sustainable ways that maintain the integrity of soil.

## Agriculture arose starting 10,000 years ago

During most of the human species' 160,000-year existence, we were hunter-gatherers, depending on wild plants and animals for our food and fiber. Then about 10,000 years ago, as the climate warmed following a

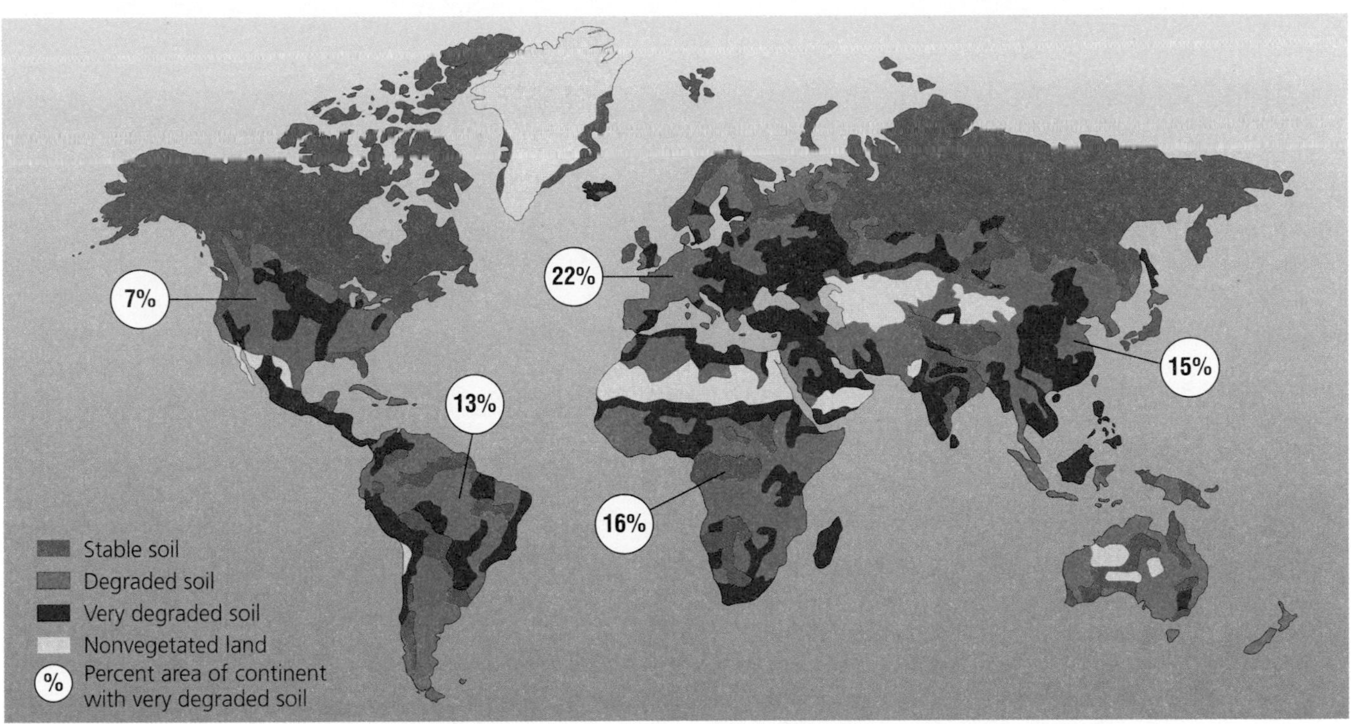


FIGURE 9.1 Soils are becoming degraded in many areas worldwide. Europe currently has a higher proportion of degraded land than other continents because of its long history of intensive agriculture, but degradation is rising quickly in developing countries in Africa and Asia. Go to GRAPHIt! at www.aw-bc.com/withgott or on the student CD–ROM.
Data from United Nations Environment Programme (UNEP), 2002. *Global environmental outlook 3.* London: UNEP and Earthscan Publications.

period of glaciation, people in some cultures began to raise plants from seed and to domesticate animals.

Agriculture most likely began as hunter-gatherers brought back to their encampments wild fruits, grains, and nuts. Some of these foods fell to the ground, were thrown away, or were eaten but survived passage through the digestive system. The plants that grew from these seeds likely produced fruits larger and tastier than those in the wild, because they sprang from seeds of fruits that people had selected because they were especially large and delicious. As these plants bred with others nearby that shared their characteristics, they gave rise to subsequent generations of plants with large and flavorful fruits.

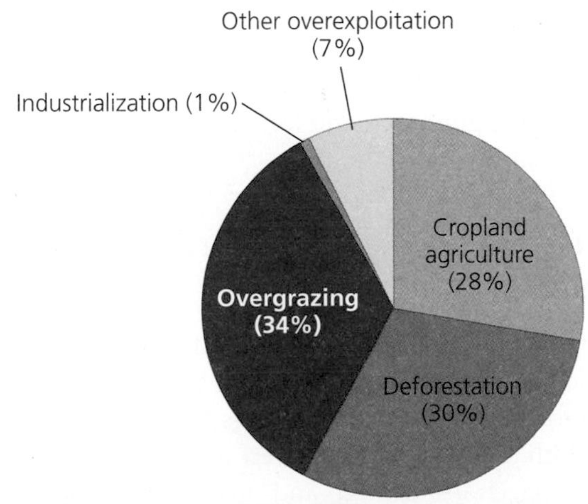


FIGURE 9.2 Most of the world's soil degradation results from cropland agriculture, overgrazing by livestock, and deforestation.
Data from Wali, M. K, et al. 1999. Assessing terrestrial ecosystem sustainability: Usefulness of regional carbon and nitrogen models. *Nature and Resources* 35: 21–33.

Eventually, people realized they could guide this selective process through conscious effort, and our ancestors began intentionally planting seeds from plants whose produce was most desirable. This practice of selective breeding (• pp. 115–117) continues to the present day and has produced the many hundreds of crops we enjoy, all of which are artificially selected versions of wild plants. People followed the same process of artificial selection with animals, creating livestock from wild species.

Evidence from archaeology and paleoecology suggests that agriculture was invented independently by different cultures in at least 5 areas of the world and possibly 10 or more (**Figure 9.3**). The earliest widely accepted evidence for plant and animal domestication is from the "Fertile Crescent" region of the Middle East about 10,500 years ago. Using methods such as radiocarbon dating (• p. 92) of crop remains, scientists have determined that wheat and barley originated here, as did rye, peas, lentils, onions, garlic, carrots, grapes, and other food plants

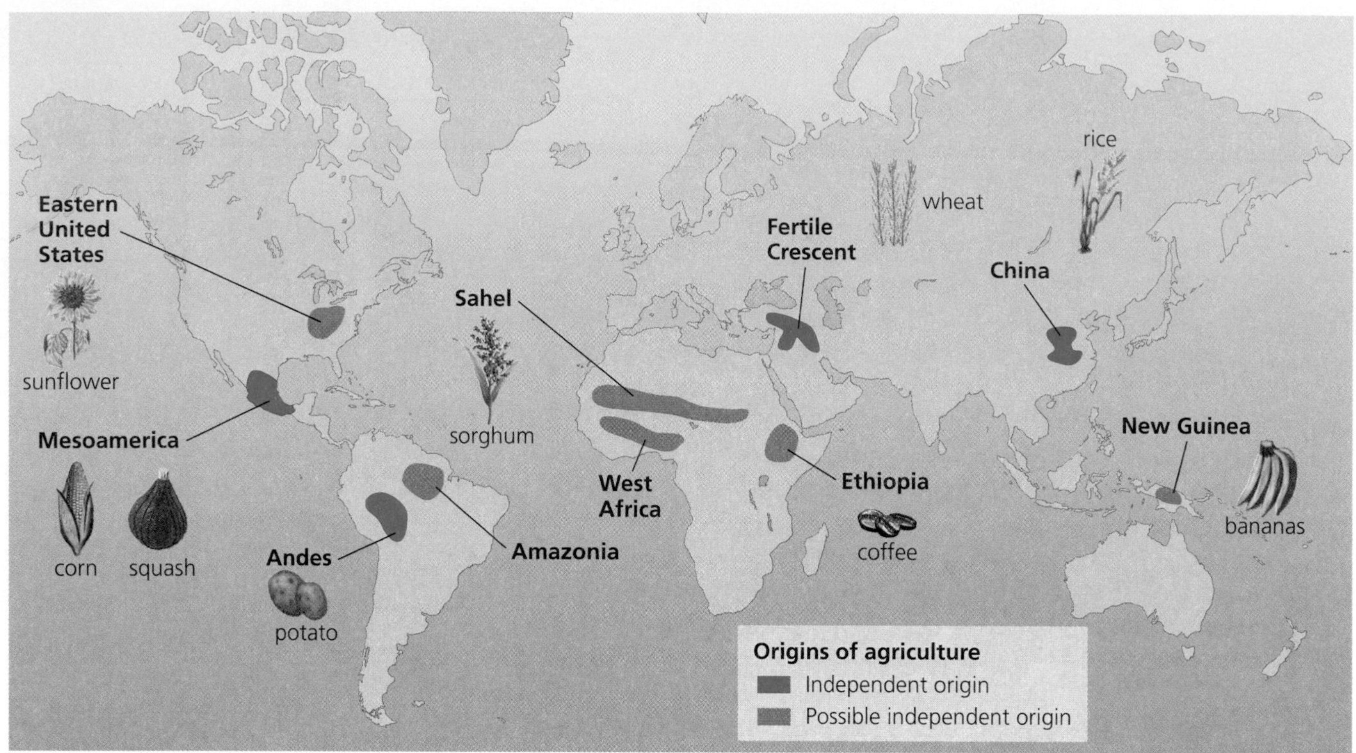


FIGURE 9.3 Agriculture originated independently in multiple locations throughout the world as different cultures domesticated certain plants and animals from wild species living in their environments. This depiction summarizes conclusions from diverse sources of research on early agriculture. Areas where people are thought to have invented agriculture independently are colored green. (China may represent two independent origins.) In areas colored blue, people either invented agriculture independently or obtained the idea from cultures of other regions. A few of the many crop plants domesticated in each region are shown. Data from syntheses in Diamond, J. 1997. *Guns, germs, and steel.* New York: W. W. Norton; and Goudie, A. 2000. *The human impact,* 5th ed. Cambridge, MA: MIT Press.

familiar to us today. The people of the Fertile Crescent also domesticated goats and sheep. In China, domestication began 9,500 years ago, leading to the rice, millet, and pigs we know today. Agriculture in Africa (coffee, yams, sorghum, and more) and the Americas (corn, beans, squash, potatoes, llamas, and more) developed in several regions 4,500–7,000 years ago.

Once our ancestors learned to cultivate crops and raise animals, they began to settle in more permanent camps and villages, often near water sources. Agriculture and a sedentary lifestyle likely reinforced one another in a positive feedback cycle (• pp. 175–176). The need to harvest crops kept people sedentary, and once they were sedentary, it made sense to plant more crops. Population increased as a result of these developments and further promoted them. Moreover, the ability to grow excess farm produce enabled some people to live off the food that others produced, leading to the development of professional specialties, commerce, technology, densely populated urban centers, social stratification, and politically powerful elites. For better or worse, the advent of agriculture eventually brought us the civilization we know today.

For thousands of years, the work of cultivating, harvesting, storing, and distributing crops was performed by human and animal muscle power, along with hand tools and simple machines (**Figure 9.4**). This biologically powered agriculture is known as **traditional agriculture**. In the oldest form of traditional agriculture, known as *subsistence agriculture,* farming families produce only enough food for themselves. In contrast, *intensive traditional agriculture* aims to produce excess food to sell at market. To do so, this type of agriculture sometimes uses teams of laboring animals and employs significant quantities of irrigation water and fertilizer, but it stops short of using fossil fuels.

## Industrialized agriculture is newer still

The industrial revolution introduced large-scale mechanization and fossil fuel combustion to agriculture just as it did to industry, enabling farmers to replace horses and oxen with faster and more powerful means of cultivating, harvesting, transporting, and processing crops. Such **industrialized agriculture** boosted yields by intensifying irrigation and introducing synthetic fertilizers, while the advent of chemical pesticides reduced competition from weeds and herbivory by crop pests. To be efficient, however,

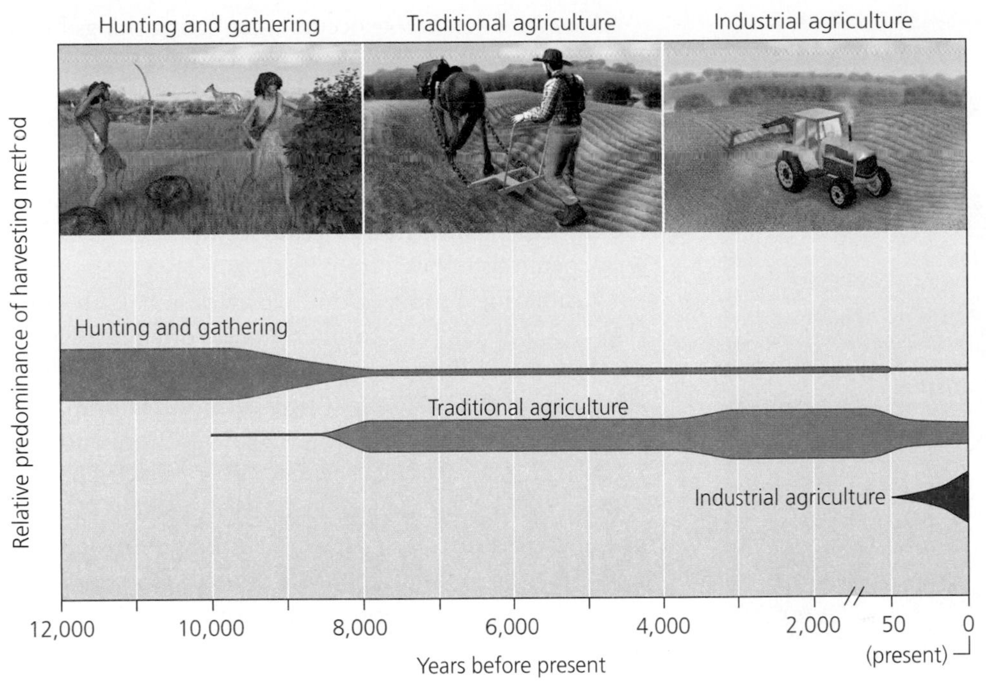


**FIGURE 9.4** Hunting and gathering comprised the predominant human lifestyle until the onset of agriculture and sedentary living centered around farms, villages, and cities, beginning nearly 10,000 years ago. Over the millennia, societies practicing traditional agriculture gradually replaced hunter-gatherer cultures. Only within the past century has industrialized agriculture spread to replace much of traditional agriculture.

industrialized agriculture demands that vast areas be planted with single types of crops. The uniform planting of a single crop, termed **monoculture**, is distinct from the *polyculture* approach of much traditional agriculture, such as Native American farming systems that mixed maize, beans, squash, and peppers in the same fields. Today, industrialized agriculture occupies over 25% of the world's cropland.

Industrialized agriculture spread from developed nations to developing nations with the advent of the **green revolution**, a phenomenon we will explore in Chapter 10 (• pp. 266–268). Beginning around 1950, the green revolution introduced new technology, crop varieties, and farming practices to the developing world. These advances dramatically increased yields per acre of cropland and helped millions avoid starvation. But despite its successes, the green revolution is exacting a high price. The intensive cultivation of farmland is creating new problems and worsening old ones. Many of these problems pertain to the integrity of soil, the very foundation of our terrestrial food supply.

# Soil as a System

We generally overlook the startling complexity of soil. Derived from rock, soil also contains a large biotic component, supports plant growth, and is molded by life (**Figure 9.5**).

By volume, soil consists very roughly of half mineral matter and up to 5% organic matter. The rest consists of pore space taken up by air or water. The organic matter in soil includes living and dead microorganisms as well as decaying material derived from plants and animals. Most of us tend to think of soil as inert and lifeless, but a single teaspoon of soil can contain 100 million bacteria, 500,000 fungi, 100,000 algae, and 50,000 protists. Soil also provides habitat for earthworms, insects, mites, millipedes, centipedes, nematodes, sow bugs, and other invertebrates, as well as burrowing mammals, amphibians, and reptiles. The composition and quality of a region's soil can have as much influence on the region's ecosystems as do the climate, latitude, and elevation. In fact, because soil is composed of living and nonliving components that interact in complex ways, soil itself meets the definition of an ecosystem (• p. 180).

## Soil formation is slow and complex

The formation of soil plays a key role in terrestrial primary succession (• p. 155), which begins when the lithosphere's parent material is exposed to the effects of the atmosphere, hydrosphere, and biosphere. **Parent material** is the base geological material in a particular location. It can include lava or volcanic ash; rock or sediment deposited by glaciers; wind-blown dunes; sediments deposited by rivers, in lakes, or in the ocean; or **bedrock**, the continuous mass of solid rock that makes up Earth's crust.

The processes most responsible for soil formation are weathering, erosion, and the deposition and decomposition of organic matter. **Weathering** describes the physical, chemical, and biological processes that break down rocks and minerals, turning large particles into smaller particles (**Figure 9.6**).

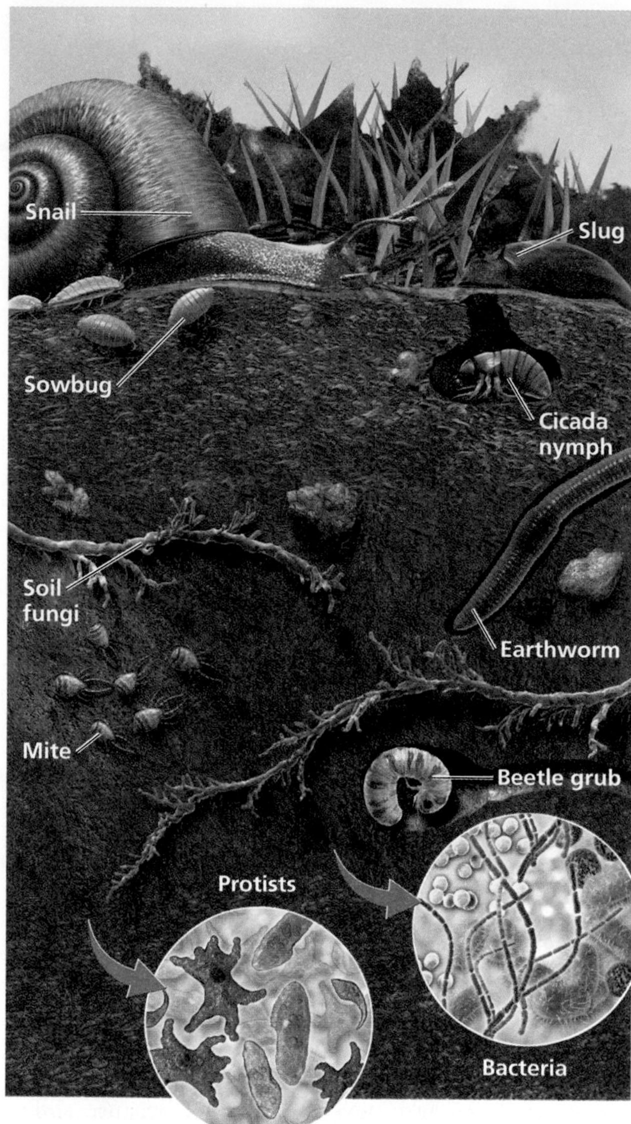


FIGURE 9.5 Soil is a complex mixture of organic and inorganic components and is full of living organisms whose actions help keep it fertile. In fact, entire ecosystems exist in soil. Most soil organisms, from bacteria to fungi to insects to earthworms, decompose organic matter. Many, such as earthworms, also help to aerate the soil.

*Physical* or *mechanical weathering* breaks rocks down without triggering a chemical change in the parent material. Wind and rain are two main forces of physical weathering. Daily and seasonal temperature variation aids their action by causing the thermal expansion and contraction of parent material. Areas with extreme temperature fluctuations experience rapid rates of physical weathering. Water freezing and expanding in cracks in rock also causes physical weathering.

*Chemical weathering* results when water or other substances chemically interact with parent material. Warm, wet conditions usually accelerate chemical weathering.

*Biological weathering* occurs when living things break down parent material by physical or chemical means. For instance, lichens initiate primary succession by producing acid, which chemically weathers rock. A tree may accelerate weathering by the physical action of its roots as they grow and rub against rock. It may also accelerate weathering chemically through the decomposition of its leaves or with chemicals released from its roots.

Weathering produces fine particles and is the first step in soil formation. Another process often involved is **erosion**, the dislodging and movement of soil from one area to another. Erosion may help form soil in one locality by depositing material it has depleted from another. Erosion occurs particularly when soil is denuded of vegetation, leaving the surface exposed to water and wind that may wash or blow it away. Although erosion can sometimes help build new soil in the long term, on the timescale of human lifetimes and for the natural systems on which we depend, erosion is generally perceived as a destructive process that reduces the amount of life that a given area of land can support.

Biological activity contributes to soil formation through the deposition, decomposition, and accumulation of organic matter. As plants, animals, and microbes die or deposit waste, this material is incorporated into the substrate, mixing with minerals. The deciduous trees of temperate forests, for example, drop their leaves each fall, making leaf litter available to the detritivores and decomposers (• p. 148) that break it down and incorporate its nutrients into the soil. In decomposition, complex organic molecules are broken down into simpler ones, including those that plants can take up through their roots. Partial decomposition of organic matter creates *humus*, a dark, spongy, crumbly mass of material made up of complex

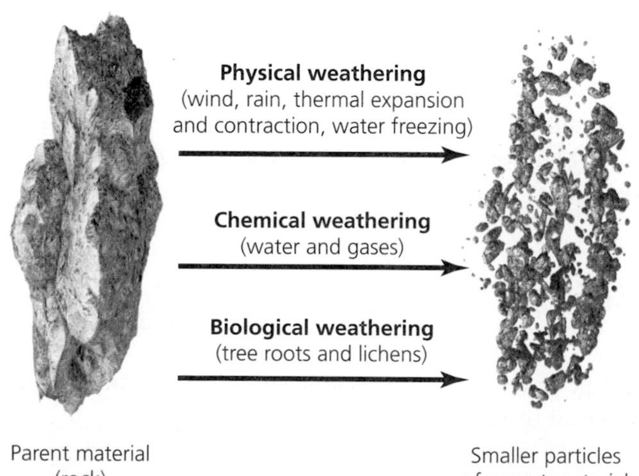


FIGURE 9.6 The weathering of parent material is the first step in soil formation. Rock is broken down into finer particles by physical, chemical, or biological means.

TABLE 9.1 Five Factors That Influence Soil Formation

| Factor | Effects |
|---|---|
| Climate | Soil forms faster in warm, wet climates. Heat speeds chemical reactions and accelerates weathering, decomposition, and biological growth. Moisture is required for many biological processes and can speed weathering. |
| Organisms | Earthworms and other burrowing animals mix and aerate soil, add organic matter, and facilitate microbial decomposition. Plants add organic matter and affect a soil's composition and structure. |
| Topographical relief | Hills and valleys affect exposure to sun, wind, and water, and they influence how soil moves. Steeper slopes promote runoff and erosion and hinder leaching, accumulation of organic matter, and differentiation of soil layers. |
| Parent material | Chemical and physical attributes of the parent material influence properties of the resulting soil. |
| Time | Soil formation takes decades, centuries, or millennia. The four factors above change over time, so the soil we see today may be the result of multiple sets of factors. |

Adapted from Jenny, H. 1941. *Factors of soil formation: A system of quantitative pedology.* New York: McGraw-Hill, Inc. Reprinted 1994 by Dover Publications, Mineola, New York.

organic compounds. Soils with high humus content hold moisture well and are productive for plant life.

Weathering, erosion, the accumulation and transformation of organic matter, and other processes that contribute to soil formation are all influenced by outside factors. Soil scientists cite five primary factors that influence the formation of soil (Table 9.1).

### Weighing THE Issues | Earth's Soil Resources

It can take hundreds to thousands of years to produce just 1 inch of topsoil. Do you think such a resource should be considered "renewable"? How would you define a "renewable resource"? How might your definition change according to whether you take an anthropocentric, biocentric, or ecocentric ethical perspective (• pp. 31–32)? How do you think soil's long renewal time should influence its management? What types of practices encourage the formation of new topsoil?

## A soil profile consists of layers known as horizons

Once weathering produces small particles from parent material, then wind, water, and organisms begin to move and sort them. Eventually, distinct layers develop. Each layer of soil is termed a **horizon**, and the cross-section as a whole, from surface to bedrock, is known as a **soil profile**.

The simplest way to categorize soil horizons is to recognize A, B, and C horizons corresponding to topsoil, subsoil, and parent material. However, soil scientists often find it useful to subdivide the layers more finely. Six major horizons in a typical soil profile are the O, A, E, B, C, and R horizons (Figure 9.7). Soils from different locations

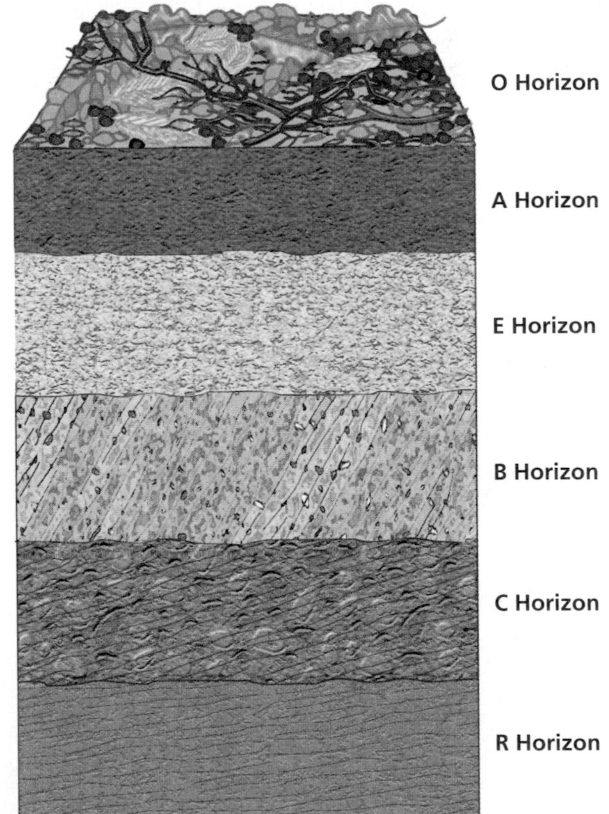


FIGURE 9.7 Mature soil consists of layers, or horizons, that have different compositions and characteristics. The uppermost layer (the **O horizon**, or litter layer) consists mostly of organic matter deposited by organisms. Below it lies the **A horizon**, or topsoil, consisting of some organic material mixed with mineral components. Minerals and organic matter tend to leach out of the **E horizon** down into the **B horizon**, or subsoil, where they accumulate. The **C horizon** consists largely of weathered parent material unaltered or only slightly altered by the processes of soil formation. The C horizon may overlie an **R horizon** of pure parent material.

vary, and few soil profiles contain all six of these horizons, but any given soil contains at least some of them.

A crucial horizon for agriculture and ecosystems is the A horizon, or **topsoil**. Topsoil consists mostly of inorganic mineral components such as weathered substrate, with organic matter and humus from above mixed in. Topsoil is the portion of the soil that is most nutritive for plants, and it takes its loose texture, dark coloration, and strong water-holding capacity from its humus content. The O and A horizons are home to most of the countless organisms that give life to soil. Topsoil is vital for agriculture, but agriculture practiced unsustainably over time will deplete organic matter, reducing the soil's fertility and ability to hold water.

Generally, the degree of weathering and the concentration of organic matter decrease as one moves downward in a soil profile. Minerals are generally transported downward as a result of **leaching**, the process whereby solid particles suspended or dissolved in liquid are transported to another location. Soil that undergoes leaching is a bit like coffee grounds in a drip filter. When it rains, water infiltrates the soil, dissolves some of its components, and carries them downward into the lower horizons. Minerals commonly leached from the E horizon include iron, aluminum, and silicate clay. In some soils, minerals may be leached so rapidly that plants are deprived of nutrients. Minerals that leach rapidly from soils may be carried into groundwater and can pose human health threats when the water is extracted.

## Soil can be characterized by color, texture, structure, and pH

The six horizons presented in Figure 9.7 depict an idealized soil, but soils display great variety. U.S. soil scientists classify soils into 12 major groups, based largely on the processes that form them. Within these 12 "orders," there are dozens of "suborders," hundreds of "great groups," and thousands of soils belonging to lower categories, all arranged in a hierarchical system. Scientists classify soils using properties such as color, texture, structure, and pH.

**Soil color** The color of soil can indicate its composition and sometimes its fertility. Black or dark brown soils are usually rich in organic matter, whereas a pale gray to white color often indicates leaching or low organic content. Long before modern analytical tests of soil content were developed, the color of topsoil provided farmers and ranchers with information about a region's potential to support crops and provide forage for livestock.

**Soil texture** Soil texture is determined by the size of particles and is the basis on which the United States Department of Agriculture (USDA) assigns soils to one of three general categories (**Figure 9.8**). **Clay** consists of

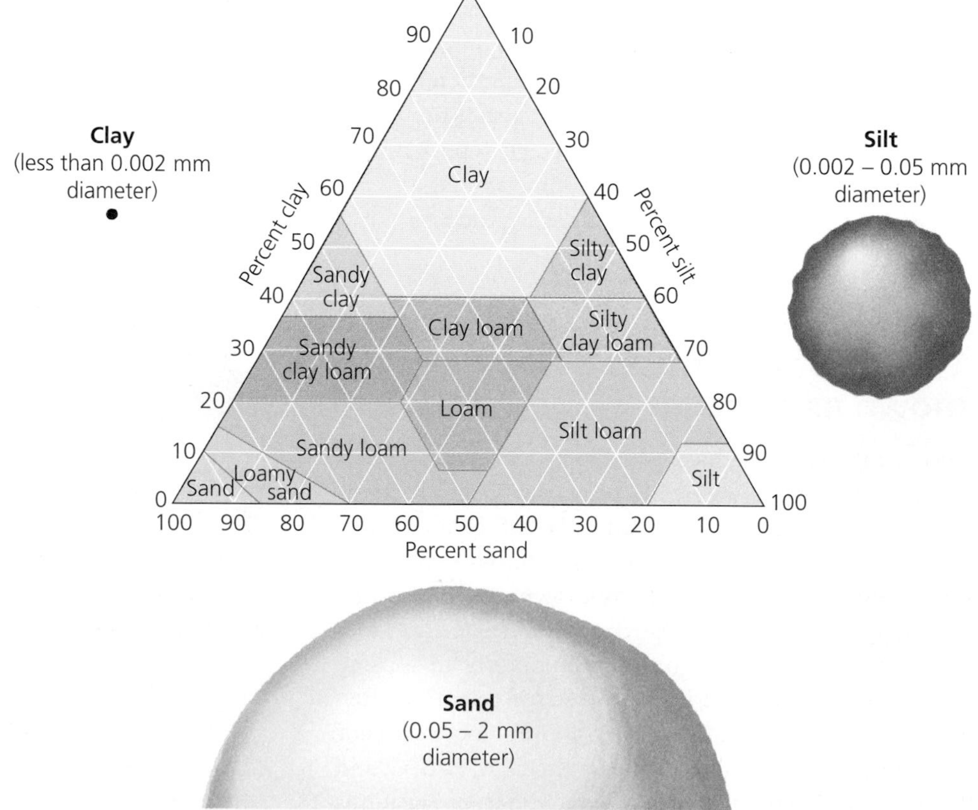


**FIGURE 9.8** The texture of soil depends on its mix of particle sizes. Using the triangular diagram shown, scientists classify soil texture according to the relative proportions of sand, silt, and clay. After measuring the percentage of each type of particle size in a soil sample, a scientist can trace the appropriate white lines extending inward from each side of the triangle to determine what type of soil texture that particular combination of values creates. Loam is generally the best for plant growth, although some types of plants grow better in other textures of soil.

particles less than 0.002 mm in diameter, **silt** of particles 0.002–0.05 mm, and **sand** of particles 0.05–2 mm. Sand grains, as any beachgoer knows, are large enough to see individually and do not adhere to one another. Clay particles, in contrast, readily adhere to one another and give clay a sticky feeling when moist. Soil with a relatively even mixture of the three particle sizes is known as **loam**.

For the farmer, soil texture influences a soil's "workability," its relative ease or difficulty of cultivation. Soil texture also influences *soil porosity*, a measure of the size of spaces between particles. In general, the finer the particles, the smaller the spaces between them and the harder it is for water and air to travel through the soil, slowing infiltration and reducing the amount of oxygen available to soil biota. Conversely, soils with large particles allow water to pass through (and beyond the reach of plants' roots) too quickly. Thus, crops planted in sandy soils require frequent irrigation. For this reason, silty soils with medium-sized pores, or loamy soils with mixtures of pore sizes, are generally best for plant growth and crop agriculture.

**Soil structure** Soil structure is a measure of the "clumpiness" of soil. Some degree of structure encourages soil productivity, and biological activity helps promote this structure. However, soil clumps that are too large can discourage plant roots from establishing if soil particles are compacted too tightly together. Repeated tilling can compact soil and make it less able to absorb water. When farmers repeatedly till the same field at the same depth, they may end up forming *plowpan*, a hard layer that resists the infiltration of water and the penetration of roots.

**Soil pH** The degree of acidity or alkalinity (• pp. 94–96) influences a soil's ability to support plant growth. Plants can die in soils that are too acidic or alkaline, but moderate variation influences the availability of nutrients for plants' roots. During leaching, for instance, acids from organic matter may remove some nutrients from the sites of exchange between plant roots and soil particles, and water carries these nutrients deeper.

## Cation exchange is vital for plant growth

The characteristics of soil affect its ability to provide plants with nutrients. Plants gain many nutrients through a process called *cation exchange*. Soil particle surfaces that are negatively charged hold cations, or positively charged ions (• pp. 91–92), such as those of calcium, magnesium, and potassium. In cation exchange, plant roots donate hydrogen ions to the soil in exchange for these nutrient ions, which the soil particles then replenish by exchange with soil water. *Cation exchange capacity* expresses a soil's ability to hold cations (preventing them from leaching and thus making them available to plants) and is a useful measure of soil fertility. Soils with fine texture (e.g., clay) and soils rich in organic matter have the greatest *cation exchange capacity*. As soil pH becomes lower (more acidic), cation exchange capacity diminishes, nutrients leach away, and soil instead may supply plants with harmful aluminum ions. This is one way in which acid precipitation (• pp. 491–495) can harm soils and plant communities.

## Regional differences in soil traits can affect agriculture

Soil characteristics and soil profiles vary from place to place. One example that bears on agriculture is the difference between soils of tropical rainforests and those of temperate grasslands. Although rainforest ecosystems have high primary productivity (• pp. 181–182), most of their nutrients are tied up in plant tissues and not in the soil. The soil of Amazonian rainforest in northern Brazil is in fact much less productive than the soil of grassland in Kansas.

To understand how this can be, consider the main differences between the two regions: temperature and rainfall. The enormous amount of rain that falls in the Amazon readily leaches minerals and nutrients out of the topsoil and E horizon. Those not captured by plants are taken quickly down to the water table, out of reach of most plants' roots. High temperatures speed the decomposition of leaf litter and the uptake of nutrients by plants, so amounts of humus remain small, and the topsoil layer remains thin.

Thus when forest is cleared for farming, cultivation quickly depletes the soil's fertility. This is why the traditional form of agriculture in tropical forested areas is *swidden* agriculture, in which the farmer cultivates a plot for one to a few years and then moves on to clear another plot, leaving the first to grow back to forest (**Figure 9.9**). This method is effective at low population densities, but with today's high human populations, soils may not be allowed enough time to regenerate. As a result, intensive agriculture has ruined the soils and forests of many tropical areas.

In temperate grassland areas such as the Kansas prairie, in contrast, rainfall is low enough that leaching is reduced and nutrients remain high in the soil profile, within reach of plants' roots. Plants take up nutrients and then return them to the topsoil when they die; this cycle maintains the soil's fertility. The thick, rich topsoil of temperate grasslands can be farmed repeatedly with minimal loss of fertility if proper farming techniques are used. However, growing and harvesting crops without returning adequate organic matter to the soil gradually depletes organic material, and leaving soil exposed to the elements increases erosion of topsoil. It is such consequences that

FIGURE 9.9 In tropical forested areas, the traditional form of farming is *swidden* agriculture, as seen here in Surinam. In this practice, forest is cut, the plot is farmed for one to a few years, and the farmer then moves on to clear another plot, leaving the first to regrow into forest. This frequent movement is necessary because tropical soils are nutrient-poor, with nearly all nutrients held in the vegetation. Burning the cut vegetation adds nutrients to the soil, which is why this practice is often called *slash-and-burn* agriculture. At low population densities, this form of farming had little large-scale impact on forests, but at today's high population densities, it is a leading cause of deforestation.

farmers in southern Brazil, the U.S. Midwest, and other locations have sought to forestall through the use of reduced tillage.

# Soil Degradation and Soil Conservation

Scientists' studies of soil and the practical experience of farmers have shown that the most desirable soil for agriculture is a loamy mixture with a pH close to neutral that is workable and capable of holding nutrients. Many soils deviate naturally from this ideal, but increasingly, human impact is setting limits on productivity by degrading many once-excellent soils. Common problems affecting soil productivity include erosion, desertification, salinization, waterlogging, nutrient depletion, structural breakdown, and pollution.

## Erosion can degrade ecosystems and agriculture

Erosion, as we have noted, is the removal of material from one place and its transport toward another by the action of wind or water. *Deposition* is the arrival of eroded material at its new location. Erosion and deposition are natural processes that in the long run help create soil. Flowing water can deposit eroded sediment in river valleys and deltas, producing rich and productive soils. This is why floodplains are excellent for farming and why flood-control measures can decrease long-term farming productivity.

However, erosion often becomes a problem locally for ecosystems and agriculture because it nearly always occurs much more quickly than soil is formed. Furthermore, erosion tends to remove topsoil, the most valuable soil layer for living things. People have increased the vulnerability of fertile lands to erosion through three widespread practices:

- Overcultivating fields through poor planning or excessive tilling
- Overgrazing rangelands with more livestock than the land can support
- Clearing forests on steep slopes or with large clear-cuts (• pp. 340–341)

Erosion can be gradual and hard to detect. For example, an erosion rate of 12 tons/ha (5 tons/acre) removes only a penny's thickness of soil. In many parts of the world, scientists, farmers, and extension agents are measuring erosion rates in hopes of identifying areas in danger of serious degradation before they become too badly damaged (see "The Science behind the Story," • pp. 340–341).

## Soil erodes by several mechanisms

Grasslands, forests, and other plant communities protect soil from wind and water erosion. Vegetation breaks the wind and slows water flow, and plant roots hold soil in place and take up water. Removing plant cover nearly always accelerates erosion. Several types of erosion can occur, including wind erosion and four principal kinds of water erosion (**Figure 9.10**). Research indicates that rill erosion has the greatest potential to move topsoil, followed by sheet erosion and splash erosion, respectively. All types of water erosion—particularly gully erosion—are more likely to occur where slopes are steeper. In general, steeper slopes, greater precipitation intensities, and sparser vegetative cover all lead to greater water erosion.

One study conducted in the early 1990s determined that at erosion rates typical for the United States, U.S. croplands lose about 2.5 cm (1 in.) of topsoil every 15–30 years, reducing corn yields by 4.7–8.7% and wheat yields by 2.2–9.5%. According to U.S. government figures, erosion rates in the United States declined from 9.1 tons/ha (3.7 tons/acre) in 1982 to 5.9 tons/ha (2.4 tons/acre) in

(a) Splash erosion

(b) Sheet erosion

(c) Rill erosion

(d) Gully erosion

FIGURE 9.10 *Splash erosion* (**a**) occurs as raindrops dislodge soil particles that fill in gaps between remaining clumps, decreasing the soil's ability to absorb water. In *sheet erosion* (**b**), water flows in thin sheets over broad surfaces, washing topsoil away in uniform layers. In *rill erosion* (**c**), water runs along small furrows, deepening and widening them into channels called rills. Rills can merge to form larger channels and eventually gullies. *Gully erosion* (**d**) cuts deeply into soil, leaving large gullies that expand as erosion proceeds.

2001, thanks to soil conservation measures discussed below. Yet in spite of these measures, U.S. farmlands still lose 6 tons of soil for every ton of grain harvested.

## Soil erosion is a global problem

In today's world, humans are the primary cause of erosion, and we have accelerated it to unnaturally high rates. In a 2004 study, geologist Bruce Wilkinson analyzed prehistoric erosion rates from the geologic record and compared these with modern rates. He concluded that human activities move over 10 times more soil than all other natural processes on the surface of the planet combined.

More than 19 billion ha (47 billion acres) of the world's croplands suffer from erosion and other forms of soil degradation resulting from human activities. Between 1957 and 1990, China lost as much arable farmland as exists in Denmark, France, Germany, and the Netherlands combined. In Kazakhstan, industrial cropland agriculture was imposed on land better suited for grazing, and wind erosion then degraded tens of millions of hectares. For Africa, soil degradation over the next 40 years could reduce crop yields by half. Couple these declines in soil quality and crop yields with the rapid population growth occurring in many of these areas, and we begin to see why some observers describe the future of agriculture as a crisis situation.

THE SCIENCE BEHIND THE STORY

## Measuring Erosion

*Dr. Jerry Ritchie, measuring soil movement with Carole Ritchie*

Can a hedge of grass help stop soil erosion? Grass hedges are widely used, especially in the tropics, to trap eroding soil by slowing runoff from rain. But Jerry Ritchie, a researcher with the U.S. Department of Agriculture, wanted to measure how well they actually work. Going beyond the visible signs of soil loss, he used techniques ranging from simple measuring pins to complex radiation detectors to calculate just how much soil has eroded.

In the 1990s, Ritchie and his team of researchers began by measuring erosion around hedges planted near a set of gullies in Maryland. They relied on cheap, simple tools known as erosion pins (see the figure), which were developed in the 1960s and 1970s by scientists working for the U.N. Food and Agriculture Organization. Erosion pins are spikes that can be made from almost anything, including bamboo stakes or pieces of plastic pipe. The pins, each cut to a uniform length, are driven into the soil until their tops are level with the ground's surface. Over time, if soil in the area is eroding, the soil surface will recede, and the erosion pins will be increasingly exposed. By using many pins over a wide area and averaging their readings, scientists can determine an overall erosion rate for the area.

Researchers often dig expansive holes, called catchpits, nearby. These pits are lined with plastic and serve to collect eroding soil. Researchers measure the volume of soil that accumulates in the catchpits and compare that data with the extent of exposure on erosion pins. In the Maryland experiment, Ritchie used such techniques and found that 1–2 cm (0.4–0.8 in.) of soil was accumulating upslope from the hedges per year, indicating that the grass was indeed trapping soil.

Erosion pins and catchpits work well in one spot but are impractical over a large region. To get evidence of erosion on a wide scale, scientists have turned to measuring a modern-day leftover rarely considered an environmental benefit—nuclear fallout from atomic weapons testing. In 1945, the United States exploded the world's first nuclear bomb, and since then more than 2,000 nuclear devices have been tested by the United States, the former Soviet Union, and other nations. Nuclear weapons testing has spread radioactive material through the atmosphere worldwide, and fallout has covered Earth's surface with minuscule but measurable amounts of nuclear debris.

## Desertification reduces productivity of arid lands

Soil degradation is especially severe in arid environments, where **desertification** is a concern. This term describes a loss of more than 10% productivity due to erosion, soil compaction, forest removal, overgrazing, drought, salinization, climate change, depletion of water sources, and other factors. Severe desertification can expand desert areas and create new ones in once-fertile regions. This process has occurred in areas of the Middle East that have been inhabited, farmed, and grazed for long periods of time—including the Fertile Crescent region, where agriculture first originated (• pp. 235–236). These arid lands—in present-day Iraq, Syria, Turkey, Lebanon, and Israel—are not so fertile anymore.

Arid and semiarid lands are prone to desertification because their precipitation is too meager to meet the demand for water from growing human populations. About 40% of Earth's land surface can be classified as *drylands*, arid areas that are particularly subject to degradation. Declines of soil quality in these areas have endangered the food supply or well-being of more than 1 billion people worldwide. Of the affected lands, most degradation results from wind and water erosion (**Figure 9.11**).

By some estimates, desertification affects one-third of the planet's land area, costing people in over 100 countries tens of billions of dollars in income each year. China alone loses $6.5 billion annually from desertification. In its western reaches, desert areas are expanding and joining one another because of overgrazing from over 400 million goats, sheep, and cattle. In the Sistan Basin along the border of Iran and Afghanistan, an oasis that supported a million livestock recently turned barren in just 5 years, and windblown sand buried more than 100 villages. The African nation of Nigeria loses an amount of land equal to half the state of Delaware each year to the expanding Sahara Desert. In Kenya, overgrazing and deforestation fueled by rapid population growth has left 80% of its land vulnerable to desertification. In a positive feedback cycle, soil degradation forces ranchers to crowd onto more marginal land and farmers to reduce fallow periods, both of which further worsen soil degradation.

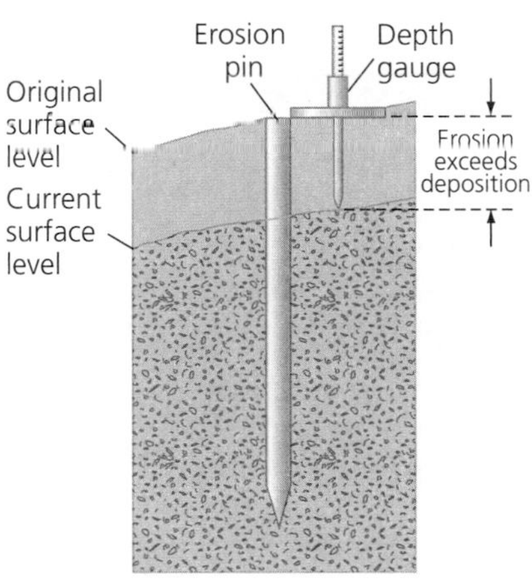


As soil erodes around an erosion pin, more of the pin is exposed, enabling the soil scientist to measure the amount of soil loss.

Fallout includes cesium-137, a radioactive isotope of the element cesium. As discussed in Chapter 4 (• pp. 92–93), environmental scientists often track and measure isotopes of chemical elements. Cesium-137, a product of nuclear fuel and weapons reactions, has a half-life of 30 years—a duration that enables soil scientists to use the isotope as a universal environmental tracer for erosion and sediment deposits. Soil tends to absorb cesium-137 quickly and evenly, so if soil in an area hasn't moved or been heavily disturbed, testing will show fairly uniform levels of the isotope. But if some areas show low concentrations of cesium-137 and others show high concentrations, then erosion may be at work. Ritchie decided to sample for cesium-137 in the area around the hedges and compare the results against his physical soil measurements.

In studies involving cesium-137, soil samples are tested using a gamma spectrometer. This device measures gamma rays, which serve as unique signatures of energy emitted by chemical elements in soil or rock. As they are emitted, gamma rays show up as sharp emission lines on a spectrum. The energy represented in these emissions reflects which elements are present, and the intensity of the lines reveals the concentration of each element. Thus it is possible to calculate the amount of an isotope such as cesium-137 in a test sample. In erosion tests, each sample from the study area is measured for cesium-137, and those levels are compared to baseline levels for the region. By pinpointing places in the study area with lower or higher levels of accumulated cesium, scientists can detect where soil has moved and how much has shifted.

In Maryland, the radioactive testing helped Ritchie determine that hedges may offer only partial help against erosion. Although his team's physical measurements of soil accumulation showed soil being deposited near the hedges, the cesium-137 tests revealed that the area around the hedges had nonetheless undergone a net loss of soil over four decades. Grass hedges can help, Ritchie wrote when releasing his findings for the Federal Agricultural Research Service in 2000, but they "should not be seen as a panacea."

A 2007 United Nations report estimated that desertification, worsened by climate change, could displace 50 million people in 10 years. The report suggested that industrialized nations fund reforestation projects in dryland areas of the developing world to slow desertification while gaining carbon credits in emissions trading programs (• pp. 79–82, 535).

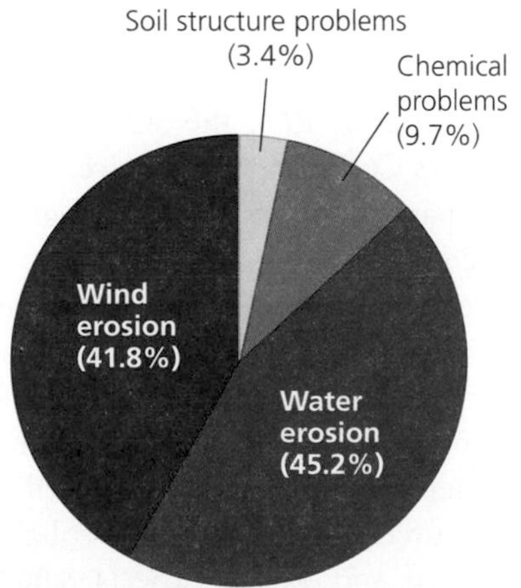


**FIGURE 9.11** Soil degradation on drylands is due primarily to erosion by wind and water. Data from U.N. Environment Programme. 2002. *Tackling land degradation and desertification.* Washington and Rome: Global Environment Facility and International Fund for Agricultural Development.

As a result of desertification, in recent years gigantic dust storms from denuded land in China have blown across the Pacific Ocean to North America, and dust storms from Africa's Sahara Desert have blown across the Atlantic Ocean to the Caribbean Sea (see Figure 17.10a, • p. 481). Such massive dust storms occurred in the United States during the Dust Bowl days of the early 20th century, when desertification shook American agriculture and society to their very roots.

## The Dust Bowl was a monumental event in the United States

Prior to large-scale cultivation of North America's Great Plains, native prairie grasses of this temperate grassland region held soils in place. In the late 19th and early 20th centuries, many homesteading settlers arrived in Oklahoma, Texas, Kansas, New Mexico, and Colorado with hopes of making a living there as farmers. Between 1879 and 1929, cultivated area in the region soared from around 5 million ha (12 million acres) to 40 million ha (100 million acres). Farmers grew abundant wheat, and ranchers grazed many thousands of cattle, sometimes expanding onto unsuitable land. Both types of agriculture contributed to erosion by removing native grasses and breaking down soil structure.

In the early 1930s, a drought exacerbated the ongoing human impacts, and the region's strong winds began to erode millions of tons of topsoil (**Figure 9.12**). Dust storms traveled up to 2,000 km (1,250 mi), blackening rain and snow as far away as New York and Vermont. Some areas lost as much as 10 cm (4 in.) of topsoil in a few years. The most-affected region in the southern Great Plains became known as the **Dust Bowl**, a term now also used for the historical event itself. The "black blizzards" of the Dust Bowl forced thousands of farmers off their land, and many who remained had to rely on government assistance programs to survive.

**(a) Kansas dust storm, 1930s**

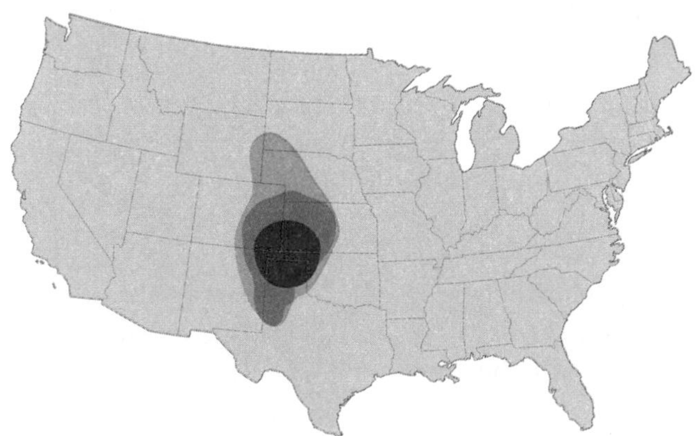

**(b) Dust Bowl region**

FIGURE 9.12 Drought combined with poor agricultural practices brought devastation and despair to millions of U.S. farmers in the 1930s in the Dust Bowl region of the southern Great Plains. The photo (**a**) shows towering clouds of dust approaching houses near Dodge City, Kansas, in a 1930s dust storm. The map (**b**) shows the Dust Bowl region, with darker colors indicating the areas most affected.

## The Soil Conservation Service pioneered measures to slow soil degradation

In response to the devastation in the Dust Bowl, the U.S. government, along with state and local governments, increased support for research into soil conservation measures. The U.S. Congress passed the Soil Conservation Act of 1935, establishing the Soil Conservation Service (SCS). This new agency worked closely with farmers to develop conservation plans for individual farms, following several aims and principles:

- Assess the land's resources, problems, and opportunities for conservation.
- Draw on science to prepare an integrated plan for each property.
- Work closely with land users to ensure that conservation plans harmonize with the users' objectives.
- Implement conservation measures on individual properties to contribute to the overall quality of life in the watershed or region.

The early teams that the SCS formed to combat erosion typically included soil scientists, forestry experts, engineers, economists, and biologists. These teams were among the earliest examples of interdisciplinary approaches to environmental problem solving. The first director of the SCS, Hugh Hammond Bennett, was an innovator and evangelist for soil conservation. Under his leadership, the agency promoted soil-conservation practices through county-based **conservation districts**. These districts operate with federal direction, authorization, and funding, but they are organized by the states. The districts implement soil conservation programs locally and aim to empower local residents to plan and set priorities in their home areas. In 1994 the SCS was renamed the **Natural Resources Conservation Service**, and its responsibilities were expanded to include water quality protection and pollution control.

The SCS served as a model for similar efforts elsewhere in the world (**Figure 9.13**). Southern Brazil's no-till movement came about through local grass-roots organization by farmers, with the help of agronomists and government extension agents who provided them information and resources. In this model of collaboration between local farmers and trained experts, 8,000 Friends of the Land clubs now exist in Paraná and Rio Grande do Sul, and 7,700 in Santa Catarina. Many of these groups are delineated by the boundaries of the more than 3,000 small-scale watersheds *(microbacias)* in which they farm.

FIGURE 9.13 Government agricultural extension agents assist farmers by providing information on new research and techniques that can help them farm productively while minimizing damage to the land. Such specialists have helped U.S. farmers since the Dust Bowl and now assist farmers worldwide. Here, an extension agent from Colombia's Instituto Colombiano Agropecuario inspects yuca plants grown by farmer Pedro Gomez on a farm in Valle del Cauca.

No-till agriculture as practiced in southern Brazil is one of many approaches to soil conservation. Hugh Hammond Bennett advocated a complex approach, combining techiques such as crop rotation, contour farming, strip-cropping, terracing, grazing management, and reforestation, as well as wildlife management. Such measures are now widely applied in many places around the world.

## Farmers can protect soil against degradation in various ways

A number of farming techniques can reduce the impacts of conventional cultivation on soils (**Figure 9.14**). Some of these have been promoted by the SCS since the Dust Bowl. Some, like no-till farming in Brazil, are finding popularity more recently. Others have been practiced by certain cultures for centuries.

**Crop rotation** In **crop rotation**, farmers alternate the type of crop grown in a given field from one season or year to the next (**Figure 9.14a**). Rotating crops can return nutrients to the soil, break cycles of disease associated with continuous cropping, and minimize the erosion that can come from letting fields lie fallow. Many U.S. farmers rotate their fields between wheat or corn and soybeans from one year to the next. Soybeans are legumes, plants that have specialized bacteria on their roots that fix nitrogen (• pp. 191–192), revitalizing soil that the previous crop had partially depleted of nutrients. Crop rotation also reduces insect pests; if an insect is adapted to feed and lay eggs on one crop, planting a different type of crop will leave its offspring with nothing to eat.

In a practice similar to crop rotation, farmers like those in southern Brazil plant "cover crops" designed to prevent erosion and forestall nitrogen loss from leaching during times of year when the main crops are not growing. Santa Catarina's extension agents have worked with farmers to test over 60 species as cover crops, including legumes, oats, and turnips.

**Contour farming** Water running down a hillside with little vegetative cover can easily carry soil away, so sloped agricultural land is especially vulnerable to erosion. Several methods have been developed for farming on slopes. **Contour farming** (**Figure 9.14b**) consists of plowing furrows sideways across a hillside, perpendicular to its slope, to help prevent formation of rills and gullies. The technique is so named because the furrows follow the natural contours of the land. In contour farming, the downhill side of each furrow acts as a small dam that slows runoff and catches soil before it is carried away. Contour farming is most effective on gradually sloping land with crops that grow well in rows. Extension agents from Santa Catarina in Brazil have helped farmers who still plow to switch to contour plowing and to plant barriers of grass along their contours.

**Terracing** On extremely steep terrain, terracing (**Figure 9.14c**) is the most effective method for preventing erosion. Terraces are level platforms, sometimes with raised edges, that are cut into steep hillsides to contain water from irrigation and precipitation. **Terracing** transforms slopes into series of steps like a staircase, enabling farmers to cultivate hilly land without losing huge amounts of soil to water erosion. Farmers have used terracing for centuries in ruggedly mountainous regions, such as the foothills of the Himalayas and the Andes. Terracing is labor-intensive to establish but in the long term is likely the only sustainable way to farm in mountainous terrain.

FIGURE 9.14 The world's farmers have adopted various strategies to conserve soil. Rotating crops such as soybeans and corn (**a**) helps restore soil nutrients and reduce impacts of crop pests. Contour farming (**b**) reduces erosion on hillsides. Terracing (**c**) minimizes erosion in steep mountainous areas. Intercropping (**d**) can reduce soil loss while maintaining soil fertility. Shelterbelts (**e**) protect against wind erosion. In (**f**), corn grows up from amid the remnants of a "cover crop" used in no-till agriculture.

**Intercropping** Farmers may also minimize erosion by **intercropping**, planting different types of crops in alternating bands or other spatially mixed arrangements (**Figure 9.14d**). Intercropping helps slow erosion by providing more ground cover than does a single crop. Like crop rotation, intercropping reduces vulnerability to insects and disease and, when a nitrogen-fixing legume is used, replenishes the soil. Some southern Brazilian farmers intercrop food crops with cover crops. The cover crops are physically mixed with primary crops, which include maize, soybeans, wheat, onions, cassava, grapes, tomatoes, tobacco, and orchard fruit.

**Shelterbelts** A widespread technique to reduce erosion from wind is to establish **shelterbelts** or *windbreaks* (**Figure 9.14e**). These are rows of trees or other tall, perennial plants that are planted along the edges of fields to slow the wind. On the Great Plains, fast-growing species such as poplars are often used.

Shelterbelts can be combined with intercropping in a practice known as *alley cropping*, one type of *agroforestry*. In this approach, fields planted in rows of mixed crops are surrounded by or interspersed with rows of trees that provide fruit, wood, or protection from wind. Such methods have been used in India, Africa, and in Brazil, where coffee growers near a national conservation area have established farming systems combining farming and forestry.

**Reduced tillage** To plant using the no-till method (**Figure 9.14f**), a tractor pulls a "no-till drill" (**Figure 9.15**) that cuts furrows through the O horizon of dead weeds and crop residue and the upper levels of the A horizon. The device drops seeds into the furrow and closes the furrow over the seeds. Often a localized dose of fertilizer is added to the soil along with the seed. *Reduced-tillage* agriculture disturbs the soil surface more than no-tillage does, but less than conventional cultivation does. By increasing organic matter and soil

FIGURE 9.15 Farmers practice no-till agriculture with a no-till drill like the one shown here. The drill cuts a furrow through the soil surface, drops in a seed, and closes the furrow over the seed.

TABLE 9.2 No-Till Farming in Brazil

**Benefits of no-till farming**

- Conserves biodiversity in soil and in ecosystems
- Produces sustainable, high crop yields
- Heightens environmental awareness among farmers
- Provides shelter and winter food for animals
- Reduces irrigation demands by 10–20%
- Crop residues act as a sink for carbon (1 metric ton/ha)
- Reduces fossil fuel use by 40–70%
- Enhances food security by increasing drought resistance
- Reduces erosion by 90%

**Other benefits arising from the reduction in erosion**

- Reduces silt deposition in reservoirs
- Reduces water pollution from chemicals
- Increases groundwater recharge and lessens flooding
- Increases sustained crop yields and lowers food prices
- Lowers costs of treating drinking water
- Reduces costs of maintaining dirt roads
- Eliminates dust storms in towns and cities
- Increases efficiency in use of fertilizer and machinery

Modified from Shaxson, T. F. 1999. The roots of sustainability: Concepts and practice: Zero tillage in Brazil. *ABLH Newsletter ENABLE; World Association for Soil and Water Conservation (WASWC) Newsletter.*

biota while reducing erosion, no-till and reduced tillage farming can restore and improve soil quality. Based on the Brazilian experience, proponents of no-till farming have claimed that the practice offers a number of benefits (**Table 9.2**).

As the appeal of no-till farming spread in Brazil, it also spread in neighboring Argentina and Paraguay. In Argentina, the area under no-till farming rose 70-fold during the 1990s and now covers over 30% of the nation's arable land. Results there parallel those in Brazil: increased crop yields, reduced erosion, enhanced soils, and a healthier environment. Maize yields grew by 37% and soybean yields by 11%, while costs to farmers fell by 40–57%. Erosion, pesticide use, and water pollution declined. As in Brazil, the techniques spread largely because of the actions of farmers and their national no-till farmers' organization.

In the United States, no-till farming is no less prevalent. Indeed, no-till and reduced tillage methods were pioneered in the United States and the United Kingdom. Today nearly half of U.S. acreage is farmed using no-till and reduced-tillage methods. No-till agriculture in the United States is most prevalent in the

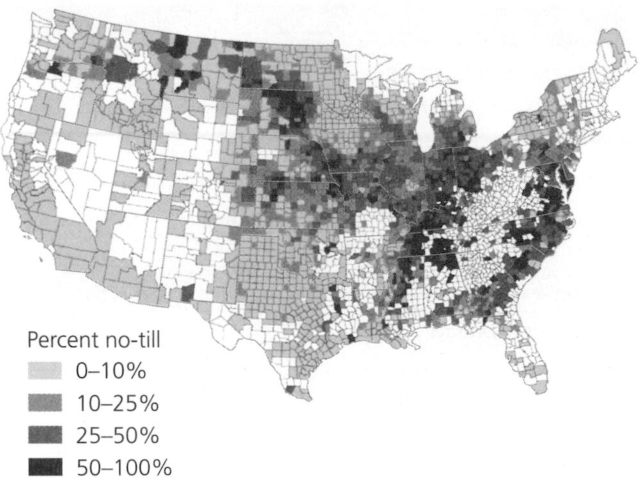


FIGURE 9.16 No-till farming has become widespread across the United States. As shown by the red and orange colors on this county-by-county map, no-till methods are practiced most in the Midwest and along the central and southern Atlantic seaboard.

Midwest and along the central and southern Atlantic seaboard (**Figure 9.16**).

Critics of no-till and reduced-tillage farming in the United States have noted that these techniques often require substantial use of chemical herbicides (because weeds are not physically removed from fields) and synthetic fertilizer (because other plants take up a significant portion of the soil's nutrients). In many industrialized countries, this has indeed been the case. Proponents, however, point out that southern Brazil's farmers have departed somewhat from the industrialized model by relying more heavily on *green manures* (dead plants as fertilizer) and by rotating fields with cover crops, including nitrogen-fixing legumes. The manures and legumes nourish the soil, and cover crops reduce weeds by taking up space the weeds might occupy.

Critics maintain, however, that green manures are generally not practical for large-scale intensive agriculture. Certainly, reduced tillage methods work well in some areas but not in others, and they work better with some crops than with others. Farmers will do best by educating themselves on the options and doing what is best for their particular crops on their own land.

**Weighing THE Issues** | **How Would You Farm?**

You are a farmer owning land on both sides of a steep ridge. You want to plant a sun-loving crop on the sunny, but very windy, south slope of the ridge and a crop that needs a great deal of irrigation on the north slope. What type of farming techniques might maximize conservation of your soil? What other factors might you want to know about before you decide to commit to one or more methods?

## Erosion-control practices protect and restore plant cover

Farming methods to control erosion make use of the general principle that maximizing vegetative cover will protect soils, and this principle has been applied widely beyond farming. It is common throughout the developed world to stabilize eroding banks along creeks and roadsides by planting plants to anchor the soil. In areas with severe and widespread erosion, some nations have planted vast plantations of fast-growing trees. China has embarked on the world's largest tree-planting program to slow its soil loss (**Figure 9.17**). Although such "reforestation" efforts do help slow erosion, they do not at the same time produce ecologically functioning forests, because tree species are selected only for their fast growth and are planted in monocultures.

## Irrigation boosts productivity but can cause long-term soil problems

Erosion is not the only threat to the health and integrity of soils. Soil can also be degraded when we apply water to crops. The artificial provision of water to support agriculture is known as **irrigation**. Some crops, such as rice and cotton, require large amounts of water, whereas others, such as beans and wheat, require relatively little. Other factors influencing the amount of water

FIGURE 9.17 Vast swaths of countryside in northern and western China have been planted with fast-growing poplar trees. These "reforestation" efforts do not create ecologically functional forests—the plantations are too biologically simple—but they do reduce soil erosion.

## CAUSES AND CONSEQUENCES

Soil erodes naturally, but human activities can greatly speed the rate of **soil erosion**, leading to diverse ecological effects and to potentially severe impacts on agricultural production. Fortunately, farmers, scientists, and policymakers have devised a number of ways to conserve soil and prevent or mitigate erosion.

Write in two causes of soil erosion in the spaces provided. Then write in two consequences (impacts on the environment, human health, or quality of life) that result from soil erosion. Finally, offer two solutions to this issue and its consequences. One cause, one consequence, and one solution have been filled in for you, providing examples.

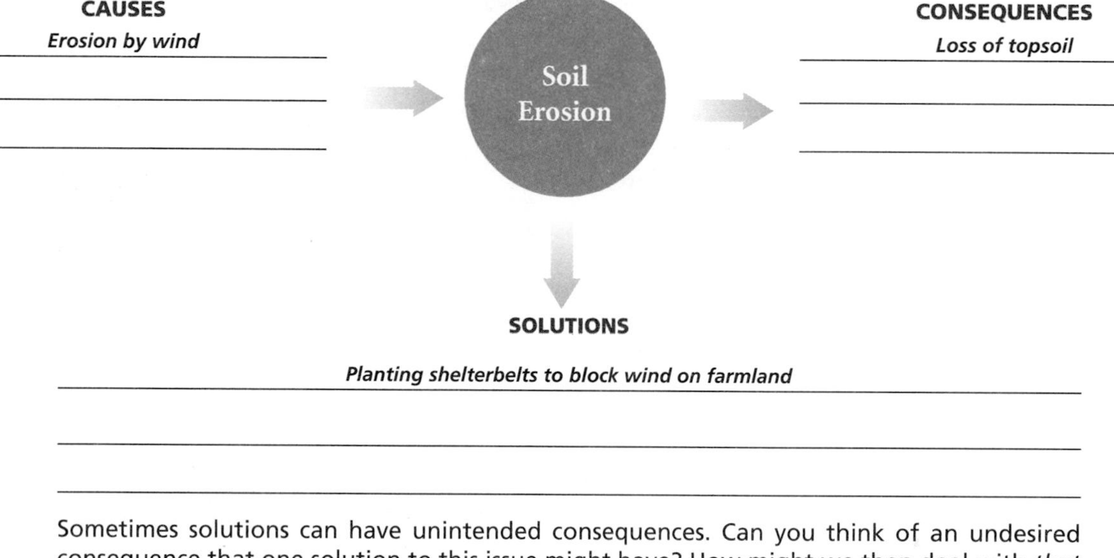


Sometimes solutions can have unintended consequences. Can you think of an undesired consequence that one solution to this issue might have? How might we then deal with *that* consequence?

required for growth include the rate of evaporation, as determined by climate, and the soil's ability to hold water and make it available to plant roots. If the climate is too dry or too much water evaporates or runs off before it can be absorbed into the soil, crops may require irrigation. By irrigating crops, people have managed to turn previously dry and unproductive regions into fertile farmland.

Seventy percent of all fresh water that people withdraw is used for irrigation. Irrigated acreage has increased dramatically worldwide, reaching 277 million ha (684 million acres) in 2003, greater than the entire area of Mexico and Central America. We will examine irrigation further in Chapter 15 (• pp. 423–424).

If some water is good for plants and soil, it might seem that more must be better. But we can supply too much water. Overirrigated soils saturated with water may experience **waterlogging** when the water table rises to the point that water bathes plant roots, depriving them of access to gases and essentially suffocating them. If it lasts long enough, waterlogging can kill plants.

A more frequent problem is **salinization**, the buildup of salts in surface soil layers. In dryland areas where precipitation is minimal and evaporation rates are high, water evaporating from the soil's A horizon may pull water from lower horizons upward by capillary action. As this water rises through the soil, it carries dissolved salts. When the water evaporates at the surface, those salts precipitate, often turning the soil surface whitish with encrusted salt.

Irrigation in arid areas generally hastens salinization because it provides repeated doses of moderate amounts of water, which dissolve salts in the soil and gradually raise them to the surface. Moreover, because irrigation water often contains some dissolved salt in the first place, irrigation introduces new salt to the soil. Salinization now inhibits production on one-fifth of all irrigated cropland globally, costing more than $11 billion each year.

## Salinization is easier to prevent than to correct

The remedies for mitigating salinization once it has occurred are more expensive and difficult to implement than the techniques for preventing it in the first place. The best way to prevent salinization is to avoid planting crops that require a great deal of water in areas that are prone to the problem. A second way is to irrigate with water low in salt content. A third way is to irrigate efficiently, supplying no more water than the crop requires. This minimizes the amount of water that evaporates and hence the amount of salt that accumulates in the topsoil.

Currently, irrigation efficiency worldwide is low; plants end up using only 43% of the water that we apply. Drip irrigation systems (**Figure 9.18**) that target water directly to plants are one solution. These systems allow more control over where water is aimed and waste far less water. Once considered expensive to install, they are becoming cheaper, such that more farmers in developing countries will be able to afford them.

If salinization has occurred, one potential way to mitigate it is to stop irrigating and wait for rain to flush salts from the soil. However, this solution is unrealistic because salinization generally becomes a problem in dryland areas where precipitation is never adequate to flush soils. A better option may be to plant salt-tolerant plants, such as barley, that can be used as food or pasture. A third option is to bring in large quantities of less-saline water with which to flush the soil. However, using too much water may cause waterlogging. As with many environmental problems, it is easier to prevent salinization than to correct it after the fact.

### Weighing THE Issues | Measuring and Regulating Soil Quality

The U.S. Environmental Protection Agency (EPA) has adopted measures of air quality and water quality and has set legal standards for allowable levels of various pollutants in air and water. Could such standards be developed for soil quality? If so, what properties should be measured to inform such standards? Should such standards be developed? Why or why not?

## Fertilizers boost crop yields but can be overapplied

Salinization is not the only way we may chemically damage soil. Overapplying fertilizers can do so as well. Plants require nitrogen, phosphorus, and potassium to

**(a) Conventional irrigation**

**(b) Center-pivot irrigation, aerial view**

**(c) Drip irrigation**

FIGURE 9.18 Currently, plants take up less than half the water we apply in irrigation. Conventional methods (**a**) lose a great deal of water to evaporation. Center-pivot irrigation (**b**) is commonly used in large-scale industrial agriculture. Huge pipes on rollers pivot around a central point and spray water on crops, creating gigantic green circles on the landscape. In more efficient irrigation approaches, water is precisely targeted to plants. In drip irrigation systems, such as this one watering grape vines in California (**c**), hoses are arranged so that water drips directly onto the plants.

(a)

(b)

FIGURE 9.19 Farmers apply fertilizers to soils to add nitrogen, phosphorus, and other nutrients that encourage plant growth. In (**a**), a British farmer applies synthetically manufactured granules of inorganic fertilizer. In (**b**), an American farmer plows rye into the soil. Rye is often grown as a source of green manure for this purpose.

grow, as well as smaller amounts of over a dozen other nutrients. Plants remove these nutrients from soil as they grow, and leaching likewise removes nutrients. If agricultural soils come to contain too few nutrients, crop yields decline. Therefore, a great deal of effort has aimed to enhance nutrient-limited soils by adding **fertilizer**, any of various substances that contain essential nutrients (**Figure 9.19**).

There are two main types of fertilizers. **Inorganic fertilizers** are mined or synthetically manufactured mineral supplements. **Organic fertilizers** consist of the remains or wastes of organisms and include animal manure; crop residues; fresh vegetation (*green manure*); and *compost,* a mixture produced when decomposers break down organic matter, including food and crop waste, in a controlled environment. Organic fertilizers can provide some benefits that inorganic fertilizers cannot. The proper use of compost improves soil structure, nutrient retention, and water-retaining capacity, helping to prevent erosion. As a form of recycling, composting reduces the amount of waste consigned to landfills and incinerators (• p. 640).

However, organic fertilizers are no panacea. For instance, manure, when applied in amounts needed to supply sufficient nitrogen for a crop, may introduce excess phosphorus that can run off into waterways. Inorganic fertilizers are generally more susceptible than are organic fertilizers to leaching and runoff, and they are somewhat more likely to cause unintended off-site impacts. Inorganic fertilizer use worldwide has grown tremendously in recent decades (**Figure 9.20**). Its use has greatly boosted our global food production, but its overapplication is causing increasingly severe pollution problems.

Applying fertilizer to croplands can have impacts far beyond the boundaries of the fields (**Figure 9.21**). We saw in Chapter 7 how nitrogen and phosphorus runoff from farms and other sources in the Mississippi River basin each year spurs phytoplankton blooms in the Gulf of Mexico and creates an oxygen-depleted "dead zone" that kills fish and shrimp. Such eutrophication

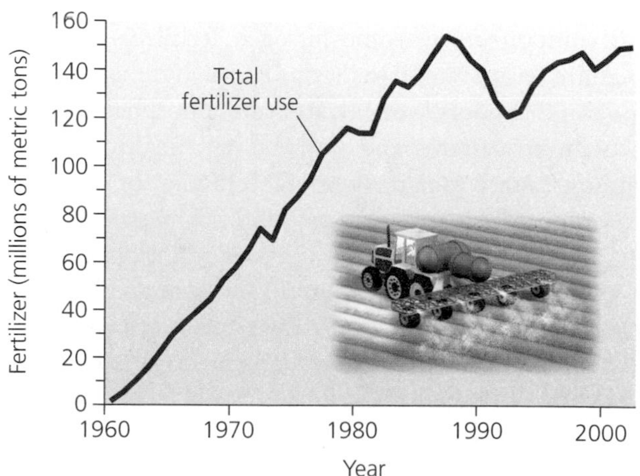


FIGURE 9.20 Use of synthetic fertilizers has risen sharply over the past half century and now stands at over 147 million metric tons annually. (The temporary drop during the early 1990s was due to economic decline in countries of the former Soviet Union following that nation's dissolution.) Data from Food and Agriculture Organization of the United Nations (FAO).

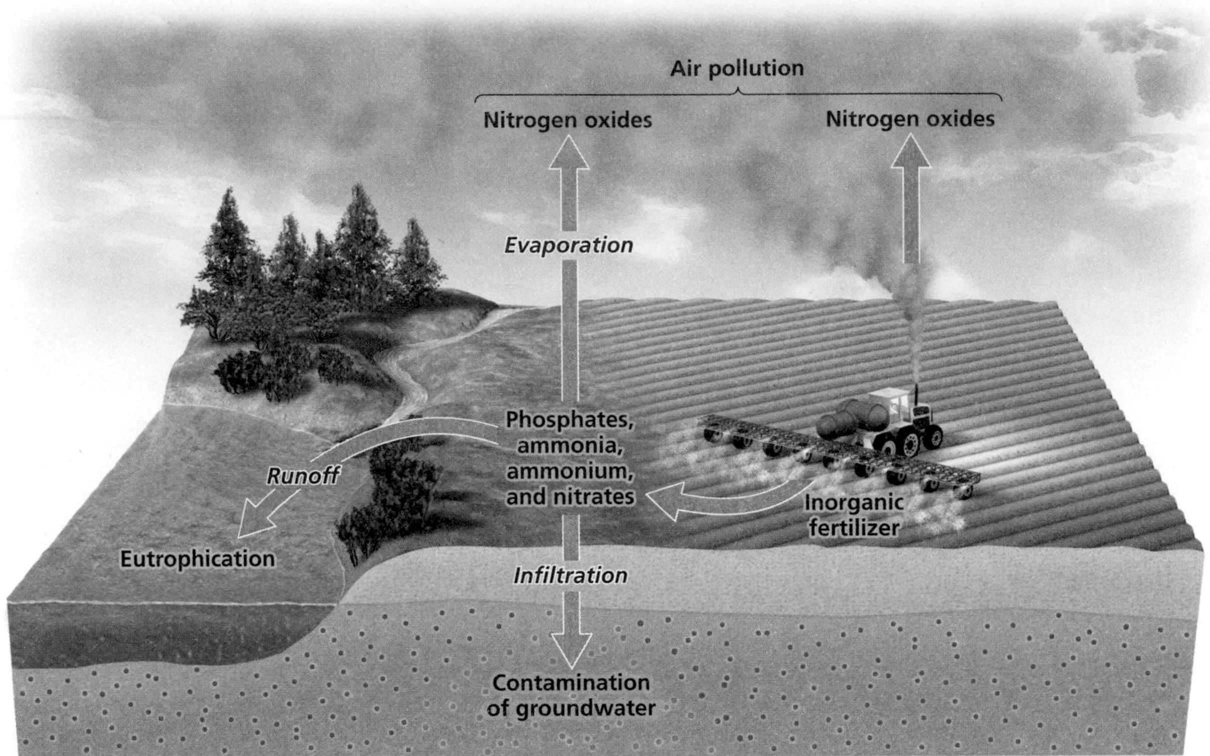


FIGURE 9.21 The overapplication of inorganic (or organic) fertilizers can have effects beyond the farm field, because nutrients not taken up by plants end up elsewhere. Nitrates can leach into groundwater, where they can pose a threat to human health in drinking water. Phosphates and some nitrogen compounds can run off into surface waterways and alter the ecology of streams, rivers, ponds, and lakes through eutrophication. Some compounds, such as nitrogen oxides, can even enter and pollute the air. Human inputs of nitrogen have greatly modified the nitrogen cycle (• pp. 191–193) and now account for one-half the total nitrogen flux on Earth.

(• pp. 179, 429–430) occurs at countless river mouths, lakes, and ponds throughout the world. Moreover, nitrates readily leach through soil and contaminate groundwater, and components of some nitrogen fertilizers can even volatilize (evaporate) into the air. Through these processes, unnatural amounts of nitrates and phosphates spread through ecosystems and pose human health risks, including cancer and methemoglobinemia, or blue-baby syndrome, which can asphyxiate and kill infants. The U.S. EPA has determined that nitrate concentrations in excess of 10mg/L for adults and 5 mg/L for infants in drinking water are unsafe, yet many sources around the world exceed even the looser standard of 50 mg/L set by the World Health Organization.

## Grazing practices and policies can contribute to soil degradation

We have focused in this chapter largely on the cultivation of crops as a source of impacts on soils and ecosystems, but raising livestock also has impacts. When sheep, goats, cattle, or other livestock graze on the open range, they feed primarily on grasses. As long as livestock populations do not exceed a range's carrying capacity (• pp. 129–130) and do not consume grasses faster than grasses can be replaced, grazing may be sustainable. However, when too many animals eat too much of the plant cover, impeding plant regrowth and preventing the replacement of biomass, the result is **overgrazing**.

Rangeland scientists have shown that overgrazing causes a number of impacts, some of which give rise to positive feedback cycles that exacerbate damage to soils, natural communities, and the land's productivity for grazing (**Figure 9.22**). When livestock remove too much plant cover, more soil is exposed and made vulnerable to erosion. Soil erosion makes it difficult for vegetation to regrow, perpetuating the lack of cover and giving rise to more erosion. Moreover, non-native weedy plants may invade denuded soils (**Figure 9.23**). These invasive plants are usually less palatable to livestock and can outcompete native vegetation in the new, modified environment, further decreasing native plant cover.

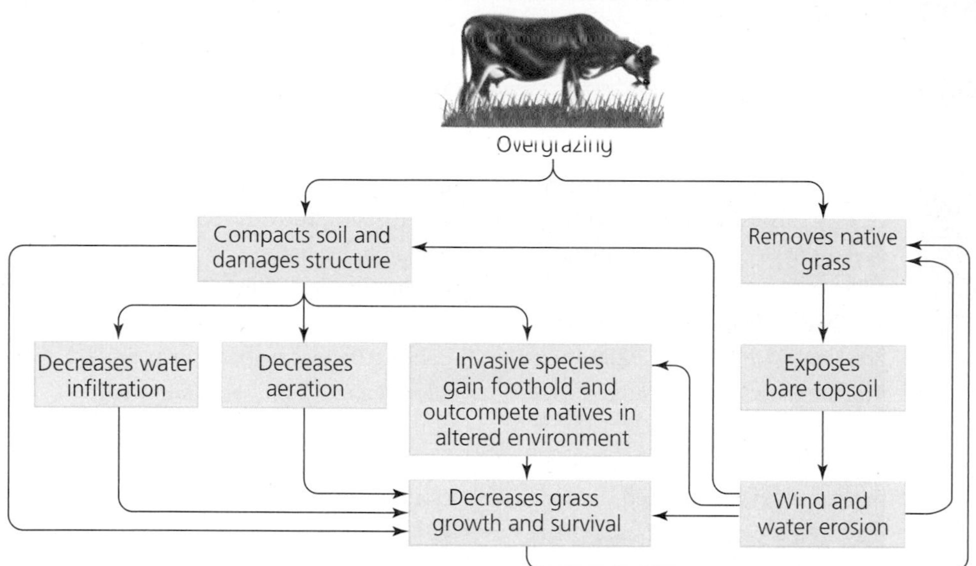


FIGURE 9.22 When grazing by livestock exceeds the carrying capacity of rangelands and their soil, overgrazing can set in motion a series of consequences and positive feedback loops that degrade soils and grassland ecosystems.

Overgrazing can also compact soils and alter their structure. Soil compaction makes it harder for water to infiltrate, harder for soils to be aerated, harder for plants' roots to expand, and more difficult for roots to conduct cellular respiration (• p. 103). All of these effects further decrease the growth and survival of native plants.

As a cause of soil degradation worldwide, overgrazing is equal to cropland agriculture, and it is a greater cause of desertification. Humans keep a total of 3.4 billion cattle, sheep, and goats. Rangeland classified as degraded now adds up to 680 million ha (1.7 billion acres), five times the area of U.S. cropland, although some estimates put the number as high as 2.4 billion ha (5.9 billion acres), fully 70% of the world's rangeland area. Rangeland degradation is estimated to cost $23.3 billion per year. Grazing exceeds the sustainable supply of grass in India by 30% and in parts of China by up to 50%. To relieve pressure on rangelands, both nations are now beginning to feed crop residues to livestock.

Range managers in the United States do their best to assess the carrying capacity of rangelands and inform livestock owners of these limits, so that herds are rotated from site to site as needed to conserve grass cover and soil integrity. Managers also can establish and enforce limits on grazing on publicly owned land when necessary. U.S. ranchers have traditionally had little incentive to conserve rangelands because most grazing has taken place on public lands leased from the government, not on lands privately owned by ranchers. In addition, the U.S. government has heavily subsidized grazing. As a result, extensive overgrazing has caused many environmental problems in the American West.

Today increasing numbers of ranchers are working cooperatively with government agencies, environmental scientists, and even environmental advocates to find ways to ranch more sustainably and safeguard the health of the land (see "The Science behind the Story," • pp. 256–257).

FIGURE 9.23 The effects of overgrazing can be striking, as shown in this photo along a fence line separating a grazed plot (right) from an ungrazed plot (left) in the Konza Prairie Reserve in Kansas. The overgrazed plot has been invaded by weedy plants that outcompete native grasses in degraded soils.

## Forestry, too, has impacts on soil

Farming and grazing are agricultural practices that help feed human populations, that depend on healthy soils, and that affect the conditions of those soils. Forestry,

THE SCIENCE BEHIND THE STORY

## Restoring the Malpai Borderlands

*A prescribed burn in the Malpai Borderlands*

In the high desert of southern Arizona and New Mexico, scientists and ranchers trying to heal the scars left by decades of cattle ranching and overgrazing found they had to contend with a creature even more damaging than a hungry steer: Smokey Bear.

Wildfires might seem a natural enemy of grasslands, but researchers in the Malpai Borderlands realized that people's efforts to suppress fire had done far more harm. Before large numbers of Anglo settlers and ranchers arrived more than a century ago, this semiarid Western landscape thrived under an ecological cycle common to many grasslands. Trees such as mesquite grew near creeks. Hardy grasses such as black grama covered the drier plains. Periodic wildfires, usually sparked by lightning, burned back shrubs and trees and kept grasslands open. Native grazing animals such as deer, rabbits, and bighorn sheep ate the grasses but rarely ate enough to deplete the range. Fed by seasonal rains, new grasses sprouted without being overeaten or crowded out by larger plants.

By the 1990s, however, those grasslands were increasingly scarce. Ranchers had brought large cattle herds to the area in the late 1800s. The cows chewed through vast expanses of grass, trampled the soil, and scattered mesquite seed into areas where grasses had dominated. Ranchers fought wildfires to keep their herds safe, and the federal government joined in the firefighting efforts.

Gradually, the Malpai's ranching families found themselves struggling to make a living. Decades of photos taken by ranchers and by botanist Raymond Turner showed how soil had eroded, and how trees and brush had overgrown the grass. The ranchers knew their cattle were part of the problem, but they also suspected that firefighting efforts were to blame.

In 1993, a group of ranchers launched an innovative plan. They formed the Malpai Borderlands Group, designating about 325,000 ha (800,000 acres) of land for protection and study. Ranchers joined government agencies, environmentalists, and scientists to study the region's ecology, bring back grasses, and restore native animal species. They sought to return periodic fire to the landscape by conducting prescribed burns and by allowing natural fires to run their course (see the figure).

The group's research efforts have centered on the Gray Ranch, a 121,000-ha (300,000-acre) parcel in the heart of the borderlands. At McKinney Flats within the Gray Ranch, scientists led by researcher Charles Curtin divided pastureland on the ranch into four study areas of about 890 ha (2,200 acres) each. Each area is further divided into four "treatments," or areas with varying land management techniques:

- In Treatment 1, land is burned and grazed.
- In Treatment 2, land is burned but not grazed.
- In Treatment 3, land is grazed but not burned.
- In Treatment 4, land is neither grazed nor burned.

Treatments 1 and 3, which allow grazing, also feature small fenced-off

the cultivation of trees, is a similar practice that we will discuss in Chapter 12. Forestry can have substantial impacts on soils, just as farming and ranching can. As with farming and ranching techniques, forestry practices have been modified over the years to try to minimize damage to soils. Some methods, such as clear-cutting—the removal of all trees from an area at once—can lead to severe erosion, particularly on steep slopes (**Figure 9.24**). Alternative timbering methods that remove fewer trees over longer periods of time are more successful in minimizing erosion.

## A number of U.S. and international programs promote soil conservation

In recent years, the U.S. Congress has enacted a number of laws promoting soil conservation. The Food Security Act of 1985 required farmers to adopt soil conservation plans and practices as a prerequisite for receiving price supports and other government benefits.

The Conservation Reserve Program, also enacted in 1985, pays farmers to stop cultivating highly erodible cropland and instead place it in conservation reserves

To restore native grasslands, the Malpai Borderlands Group reinstated fire as a natural landscape process, conducting controlled burns. Monitoring indicates that restoring fire has improved ecological conditions in the region.

areas that prevent animals from eating grass. These "exclosures" allow scientists to make precise side-by-side comparisons of how grazing affects grasses. Scientists measure rainfall in each area and monitor soils for degradation and erosion. Teams of wildlife and vegetation specialists monitor each treatment for the distribution and abundance of birds, insects, animals, and vegetation.

By comparing areas where fire is suppressed to those where fire can burn, researchers have documented how the suppression of fire leads to more brush and trees and less grass. When fire burns an area, woody plants such as mesquite are damaged and their seed production is disrupted. The flames are often followed by the return of grass. Such benefits follow both natural fires and carefully monitored, deliberately set controlled burns.

The researchers added a new dimension to the ongoing study by introducing black-tailed prairie dogs to certain pastures. Ranchers have traditionally sought to rid their land of prairie dogs because of concerns that they devour grass that cattle could consume. But the Malpai research is showing that prairie dogs and cattle may instead benefit one another in various ways.

All told, the landscape-level (• pp. 183–184) McKinney Flats project may be the largest replicated terrestrial ecological experiment in North America. It has helped spur a number of other research initiatives, and in 2007 the Malpai Borderlands Group held a scientific conference featuring research from the region.

The scientists have also helped ranchers develop a cycle of "grassbanking," in which cattle are allowed to graze on shared plots of land while other areas recover. And ranchers must work with the weather. Scientists have found that controlled burns or grassbanking should track with cycles of rain and drought to bring back grass.

Because of such research, controlled burns are now a regular part of the Malpai landscape. Ranchers have burned more than 100,000 ha (250,000 acres) since 1994, and natural fires are often allowed to run their course with little or no intervention. Damaged areas have been reseeded with native grasses.

Scientists increasingly believe that ranching, if managed properly, can help bring damaged grazing areas back to life. "We cannot assume rangelands will recover on their own," Curtin wrote in a recent study on the Malpai Borderlands. "Conservation of grazed lands requires restoring and sustaining natural processes."

planted with grasses and trees. The USDA estimates that for an annual cost of $1 billion, this program saves 700 million metric tons (771 million tons) of topsoil each year. Besides reducing erosion, the Conservation Reserve Program has generated income for farmers and has provided habitat for native wildlife.

In 1996, Congress extended the program by passing the Federal Agricultural Improvement and Reform Act. Also known as the "Freedom to Farm Act," this law aimed to reduce subsidies and government influence over many farm products. It also created the Environmental Quality Incentive Program and the Natural Resources Conservation Foundation to promote and pay for the adoption of conservation practices in agriculture. In 1988, the USDA initiated the Low-Input Sustainable Agriculture Program to provide funding for individual farmers to develop and practice sustainable agriculture.

Internationally, the United Nations promotes soil conservation and sustainable agriculture through a variety of programs of its Food and Agriculture Organization (FAO). The FAO's Farmer-Centered Agricultural Resource Management Program (FAR) supports innovative

FIGURE 9.24 Deforestation, discussed further in Chapter 12, can cause severe erosion, particularly when trees are clear-cut for timber on steep slopes.

approaches to resource management and sustainable agriculture in China, Thailand, Vietnam, Indonesia, Sri Lanka, Nepal, the Philippines, and India. This program studies agricultural success stories and tries to help farmers elsewhere duplicate successful efforts. Rather than following a top-down, government-controlled approach, the FAR program calls on the creativity of local communities to educate and encourage farmers to conserve soils and secure their food supply.

## Conclusion

Our species has enjoyed a 10,000-year history with agriculture, yet despite all we have learned about soil degradation and conservation, challenges remain. Many of the policies enacted and the practices developed to combat soil degradation in the United States and worldwide have been quite successful, particularly in reducing topsoil erosion. It is clear, however, that even the best-conceived soil conservation programs require research, education, funding, and commitment from both farmers and governments if they are to fulfill their potential. In light of continued population growth, we will likely need better technology and wider adoption of soil conservation techniques to avoid an eventual food crisis. Increasingly, it seems relevant to consider whether the universal adoption of Aldo Leopold's land ethic (• pp. 34–35) will also be required if we are to feed the 9 billion people expected to crowd our planet by mid-century.

## REVIEWING OBJECTIVES

**You should now be able to:**

**Explain the importance of soils to agriculture, and describe the impacts of agriculture on soils**

- Successful agriculture requires healthy soil. (p. 234)
- As the human population grows and consumption increases, pressures from agriculture are degrading Earth's soil, and we are losing 5–7 million ha (12–17 million acres) of productive cropland annually. (p. 234)

**Outline major historical developments in agriculture**

- Beginning about 10,000 years ago, people began breeding crop plants and domesticating animals. (pp. 234–236)
- Domestication takes place through the process of selective breeding, or artificial selection. (pp. 235–236)
- Agriculture originated multiple times independently in different cultures across the world. (pp. 235–236)
- Industrial agriculture is replacing traditional agriculture, which largely replaced hunting and gathering. (pp. 236–237)

**Delineate the fundamentals of soil science, including soil formation and the properties of soil**

- Soil includes diverse biotic communities that decompose organic matter. (pp. 237–238)
- Climate, organisms, relief, parent material, and time are factors influencing soil formation. (p. 239)
- Soil profiles consist of distinct horizons with characteristic properties. (pp. 239–240)

- Soil can be characterized by color, texture, structure, and pH. (pp. 240–241)
- Soil properties affect plant growth and agriculture. (pp. 241–242)

**State the causes and predict the consequences of soil erosion and soil degradation**

- Some agricultural practices have resulted in high rates of erosion, lowering crop yields. (pp. 242–243)
- Desertification affects a large portion of the world's soils, especially those in arid regions. (pp. 244–245)
- Overirrigation can cause salinization and waterlogging, which lower crop yields and are difficult to mitigate. (pp. 251–252)
- Overapplying fertilizers can cause pollution problems that affect ecosystems and human health. (pp. 253–254)
- Overgrazing can cause soil degradation on grasslands and diverse impacts to native ecosystems. (pp. 254–257)
- Careless forestry practices, such as deforesting steep slopes, are a major cause of erosion. (pp. 256, 258)

**Recite the history and explain the principles of soil conservation**

- The Dust Bowl in the United States and similar events elsewhere have inspired scientists and farmers to develop better ways to conserve topsoil. (pp. 245–247)
- Farming techniques such as crop rotation, contour farming, intercropping, terracing, shelterbelts, and reduced tillage enable farmers to reduce soil erosion and boost crop yields. (pp. 247–250)
- In the United States and across the world, governments are devising innovative policies and programs to deal with soil degradation. (pp. 256–258)

## TESTING YOUR COMPREHENSION

1. How did the practices of selective breeding and human agriculture begin roughly 10,000 years ago? Summarize the influence of agriculture on the development and organization of human communities.
2. Describe the methods used in traditional agriculture, and contrast subsistence agriculture with intensive traditional agriculture. How does industrialized agriculture differ from traditional agriculture?
3. What processes are most responsible for the formation of soil? Describe the three types of weathering that may contribute to soil formation.
4. Name the five primary factors thought to influence soil formation, and describe one effect of each.
5. How are soil horizons created? What is the general pattern of distribution of organic matter in a typical soil profile?
6. Why is erosion generally considered a destructive process? Name three human activities that can promote soil erosion. Describe four kinds of soil erosion by water. What factors affect the intensity of water erosion?
7. List innovations in soil conservation introduced by Hugh Hammond Bennett, first director of the SCS. What other farming techniques can help reduce the risk of erosion due to conventional cultivation methods?
8. How does terracing effectively turn steep and mountainous areas into arable land? Explain the method of no-till farming. Why does this method reduce soil erosion?
9. How do fertilizers boost crop growth? How can large amounts of fertilizer added to soil also end up in water supplies and the atmosphere?
10. What policies can be linked to the practice of overgrazing? Describe the effects of overgrazing on soil. What conditions characterize sustainable grazing practices?

## SEEKING SOLUTIONS

1. How do you think a farmer can best help to conserve soil? How do you think a scientist can best help to conserve soil? How do you think a national government can best help to conserve soil?
2. How and why might actual soils differ from the idealized six-horizon soil profile presented in the chapter? How might departures from the idealized profile indicate the impact of human activities? Provide at least three examples.
3. Discuss how the methods of no-till or reduced tillage farming, as described in this chapter, can enhance soil quality. What drawbacks or negative effects might no-till or reduced-tillage practices have on soil quality, and how might these be prevented?

4. Discuss how methods of locally based sustainable agriculture described in this chapter are promoting the science of soil conservation. In reference to Aldo Leopold's land ethic (• pp. 34–35), how are humans and the soil members of the same community?
5. **THINK IT THROUGH** You are a land manager with the U.S. Bureau of Land Management (BLM, • p. 347) and have just been put in charge of 500,000 acres of public lands that have been degraded by decades of overgrazing. Soil is eroding, creating large gullies. Invasive weeds are replacing native grasses. Shrubs are encroaching on grassland areas because fire was suppressed. Environmentalists want an end to ranching on the land. Ranchers want continued grazing, but they are concerned about the land's condition and are willing to entertain new ideas. What steps would you take to assess the land's condition and begin restoring its soil and vegetation? Would you allow grazing, and if so, would you set limits on it?
6. **THINK IT THROUGH** You are the head of an international granting agency that assists farmers with soil conservation and sustainable agriculture. You have $10 million to disburse. Your agency's staff has decided that the funding should go to (1) farmers in an arid area of Africa prone to salinization, (2) farmers in a fast-growing area of Indonesia where swidden agriculture is practiced, (3) farmers in southern Brazil practicing no-till agriculture, and (4) farmers in a dryland area of Mongolia undergoing desertification. What types of projects would you recommend funding in each of these areas, how would you apportion your funding among them, and why?

## INTERPRETING GRAPHS AND DATA

Kishor Atreya and his colleagues at Kathmandu University in Nepal conducted a field experiment to test the effects of reduced tillage versus conventional tillage on erosion and nutrient loss in the Himalayan Mountains in central Nepal. The region in which they worked has extremely steep terrain (with an average slope of 18%), and receives over 138 cm (55 in.) of rain per year, with 90% of it falling between May and September. Atreya's team measured the amounts of soil, organic carbon, and nitrogen lost from the research plots (which were not terraced) over the course of a year. Some of their results are presented in the graph.

1. Under the conditions of the study reported above, how much soil, organic carbon, and nitrogen would be saved annually in fields with reduced tillage relative to fields with conventional tillage? Express your answers both in absolute units and as percentages.
2. Given that annual crop yields in each of the study plots were approximately 4 metric tons/ha, what is the ratio of soil lost to crop yield under conventional tillage? Under reduced tillage?
3. Is reduced tillage a sustainable management practice for Nepalese farmers? If so, what data from the study above would you cite in support of your answer? If not, or if you cannot say, then what concerns raised by the data above would need to be addressed, or what additional data would be needed, to answer the question?

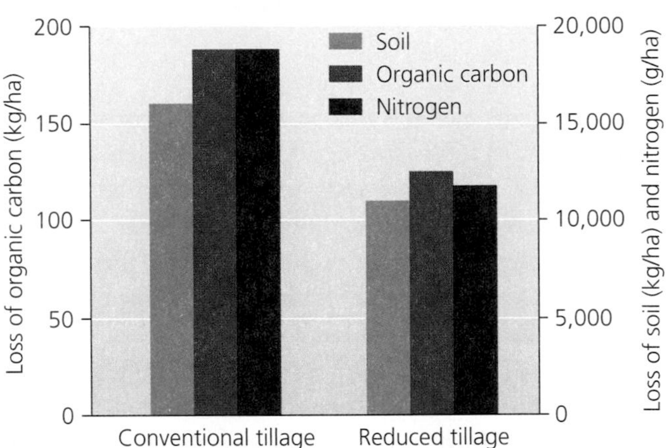


Annual soil and nutrient losses in plots under conventional and reduced tillage systems. All reduced tillage values are significantly different from their conventional tillage counterparts.

Data from Atreya, K., et al. 2006. Applications of reduced tillage in hills in central Nepal. *Soil & Tillage Research* 88:16–29.

## CALCULATING ECOLOGICAL FOOTPRINTS

In the United States, approximately 6 pounds of topsoil are lost for every 1 pound of grain harvested. Erosion rates vary greatly with soil type, topography, tillage method, and crop type. For simplicity let us assume that the 6:1 ratio applies to all plant crops and that a typical diet includes 1 pound of plant material or its derived products (sugar, for example) per day. In the first two columns of the table, calculate the annual topsoil losses associated with growing this food for you and for other groups, assuming the same diet.

| | Plant products consumed (lb) | Soil loss at 6:1 ratio (lb) | Soil loss at 4:1 ratio (lb) | Reduced soil loss at 4:1 relative to 6:1 ratio (lb) |
|---|---|---|---|---|
| You | 365 | 2,190 | 1,460 | 730 |
| Your class | | | | |
| Your state | | | | |
| United States | | | | |

1. Improved soil conservation measures reduced erosion by one-third from 1982 to 1997. If additional measures were again able to reduce the current rate of soil loss by a third, the ratio of soil lost to grain harvested would fall from 6:1 to 4:1. Calculate the soil losses associated with food production at a 4:1 ratio, and record your answers in the third column of the table.
2. Calculate the amount of topsoil hypothetically saved by the additional conservation measures in question 1, and record your answers in the fourth column of the table.
3. Define a "sustainable" rate of soil loss. Describe how you might determine whether a given farm was practicing sustainable use of soil.

## Take It Further

Go to www.aw-bc.com/withgott or the student CD-ROM, where you'll find:

- Suggested answers to end-of-chapter questions
- Quizzes, animations, and flashcards to help you study
- *Research Navigator*™ database of credible and reliable sources to assist you with your research projects
- GRAPHit! Tutorials to help you interpret graphs
- INVESTIGATEit! Current news articles that link the topics that you study to case studies from your region to around the world

CHAPTER 10

# Agriculture, Biotechnology, and the Future of Food

Organic vegetable farm in Whatcom County, Washington

## Upon completing this chapter, you will be able to:

- Explain the challenge of feeding a growing human population
- Identify the goals, methods, and environmental impacts of the "green revolution"
- Categorize the strategies of pest management
- Discuss the importance of pollination
- Describe the science behind genetically modified food
- Evaluate the debate over genetically modified food
- Ascertain approaches for preserving crop diversity
- Assess feedlot agriculture for livestock and poultry
- Weigh approaches in aquaculture
- Evaluate sustainable agriculture

Farmer in maize field in Oaxaca, Mexico

CENTRAL CASE

## Possible Transgenic Maize in Oaxaca, Mexico

**"Worrying about starving future generations won't feed them. Food biotechnology will. . . . At Monsanto, we now believe food biotechnology is a better way forward."**

—Advertising Campaign of the Monsanto Company

**"Industrial agriculture has not produced more food. It has destroyed diverse sources of food, and it has stolen food from other species . . . using huge quantities of fossil fuels and water and toxic chemicals in the process."**

—Vandana Shiva, Director of the Research Foundation for Science, Technology, and Natural Resource Policy, India

Corn is a staple grain of the world's food supply. We can trace its ancestry back roughly 5,500 years, when people in the highland valleys of what is now the state of Oaxaca in southern Mexico first domesticated that region's wild maize plants. The corn we eat today arose from some of the many varieties that evolved from the early selective crop breeding conducted by the people of this region.

Today Oaxaca remains a world center of biodiversity for maize, with many native varieties (called *cultivars* or *landraces*) growing in the rich, well-watered soil. Preserving such varieties of crops in their ancestral homelands is important for securing the future of our food supply, scientists maintain, because these varieties serve as reservoirs of genetic diversity—reservoirs we may need to draw on to sustain or advance our agriculture.

Thus, when Mexican government scientists conducting routine genetic tests of Oaxacan farmers' maize announced in 2001 that they had turned up DNA that matched genes from genetically modified (GM) corn, it caused global consternation. GM corn was widely grown in the United States, but Mexico had banned its cultivation in 1998. Corn is one of many crops that scientists have genetically engineered to express desirable traits such as large size, fast growth, and resistance to insect pests. To genetically engineer crops, scientists extract genes from the DNA of one organism and transfer them into the DNA of another. The aim is to improve crop performance and feed the world's hungry, but many people worry that the **transgenes** from these **transgenic** plants could cause harm. One concern is that transgenic crops might crossbreed with native crops and thereby "contaminate" the genetic makeup of native crops.

Two researchers at the University of California at Berkeley, Ignacio Chapela and his postdoctoral associate David Quist, shared these concerns, and they ventured to Oaxaca to test samples of native maize. Their analyses seemed to confirm the government scientists' findings, revealing what they argued were traces of DNA from genetically engineered corn in the genes of native maize plants. They also suggested that the invading genes had split up and spread throughout the maize genome. Quist and Chapela published their findings in the scientific journal *Nature* in November 2001.

Activists opposed to GM food trumpeted the news and urged a ban on imports of transgenic crops from producer countries such as the United States into developing nations. The agrobiotech industry defended the safety of its crops and questioned the validity of the research—as did many of Quist and Chapela's peers. Responding to criticisms from researchers, *Nature* took the unprecedented step of stating that Quist and Chapela's paper should never have been published—a move that ignited a firestorm of controversy (see "The Science behind the Story," • pp. 278–279).

Subsequent research by Mexican government scientists reported that in 15 localities, 3–60% of maize contained transgenes and that 37% of maize grains distributed by the government to farmers were transgenic. These results were never published in a peer-reviewed scientific journal. Nonetheless, a commission of experts convened under the North American Free Trade Agreement (NAFTA) to study the issue reasoned that corn shipments from the United States contain a mix of GM and non-GM grain, that small farmers plant some of this seed, and that transgenes entering Mexico could spread by wind pollination and by interbreeding with native cultivars of maize.

Meanwhile, a team of Mexican and American scientists was conducting the most extensive study yet, expecting to find transgenes to be widespread in the Oaxacan maize they surveyed in 2003 and 2004. In August 2005, they published their results. To their surprise, they found no transgenes at all, despite analyzing 154,000 maize seeds (see "The Science behind the Story," • pp. 278–279). Had Quist and Chapela's transgenes somehow disappeared since 2001? Were they so rare or localized that this large survey had missed them? Had Quist and Chapela's research simply been in error? Scientists are still debating the issue, but most agree on one thing: If transgenes are not yet present in Mexican maize, they probably soon will be, for there is little to stop them.

A harder question to answer is how the genetic modification of crops may affect people and the environment (potentially in both positive and negative ways). In this chapter, we take a wide-ranging view of the ways people have devised to increase our agricultural output, the environmental effects of these efforts, and the food of the future.

# The Race to Feed the World

Although human population growth has slowed, we can still expect our numbers to swell to 9 billion by the middle of this century. For every two people living today, there will be three in 2050. Feeding 50% more mouths half a century from now while protecting the integrity of soil, water, and ecosystems will require sustainable agriculture. This could involve approaches as diverse as organic farming and the genetically modified crops that are eliciting so much controversy in Oaxaca and elsewhere.

## We are producing more food per person

Over the past half century, our ability to produce food has grown even faster than global population (**Figure 10.1**). However, largely because of political obstacles and inefficiencies in distribution, today 850 million people in developing countries do not have enough to eat. Every 5 seconds, somewhere in the world, a child starves to death. Agricultural scientists and policymakers pursue a goal of **food security,** the guarantee of an adequate and reliable food supply available to all people at all times. Making a food supply sustainable depends on maintaining healthy soil, water, and biodiversity. As we saw in Chapter 9, careless agricultural practices can have devastating effects on the environment and on the ability of soils to continue supporting crops and livestock.

Starting in the 1960s, a number of scientists, including Paul Ehrlich (• p. 5), predicted widespread starvation and a catastrophic failure of agricultural systems, arguing that the human population could not continue to grow without outstripping its food supply. However, the human population has continued to increase well past

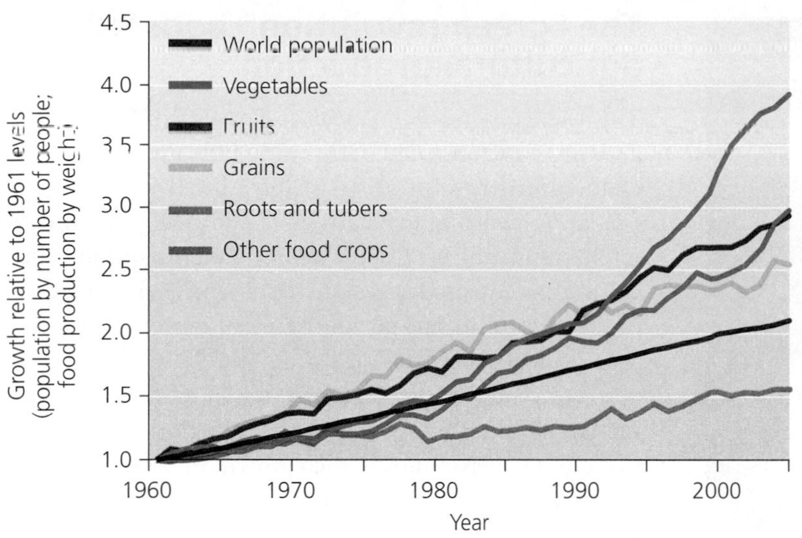


FIGURE 10.1 Global agricultural food production rose by over two-and-a-half times in the past four decades, growing at a faster rate than world population. Production of all types of foods, particularly vegetables, increased from 1961 to 2005. Trend lines show cumulative increases relative to 1961 levels. Data from U.N. Food and Agriculture Organization.

their predictions. Although it is tragic that 850 million people are hungry today, this number is smaller than the 960 million who lacked reliable and sufficient food in 1970. In percentage terms, we have reduced hunger by half, from 26% of the population in 1970 to 13% today.

We have achieved these advances in part by increasing our ability to produce food (see Figure 10.1). We have increased food production by devoting more fossil fuel energy to agriculture; by planting and harvesting more frequently; by increasing the use of irrigation, fertilizer, and pesticides; by increasing the amount of cultivated land; and by developing (through crossbreeding and genetic engineering) more productive crop and livestock varieties.

However, with grain crops, the world's staple foods, we are producing slightly less food per person each year. Since 1985, world grain production per person has fallen by 9% (**Figure 10.2**). Moreover, the world's soils are in decline, and nearly all the planet's arable land has already been claimed. Simply because agricultural production has outpaced population growth so far, there is no guarantee that it will continue to do so.

## We face undernourishment, overnutrition, and malnutrition

Although many people lack access to adequate food, others are affluent enough to consume more than is healthy. Most people who are **undernourished,** receiving less than 90% of their daily caloric needs, live in the developing world. Meanwhile, in the developed world, many people suffer from **overnutrition,** receiving too many calories each day. In the United States, where food is available in abundance, junk food is cheap, and people tend to lead sedentary lives with little exercise, more than three out of five adults are technically overweight, and over one out of four is obese. Worldwide, obesity now affects over 300 million people.

For most people who are undernourished, the reasons are economic. One-fifth of the world's people live on less than \$1 per day, and over half live on less than \$2 per day, the World Bank estimates. Hunger is a problem even in the United States, where the U.S. Department of Agriculture (USDA) has classified 31 million Americans

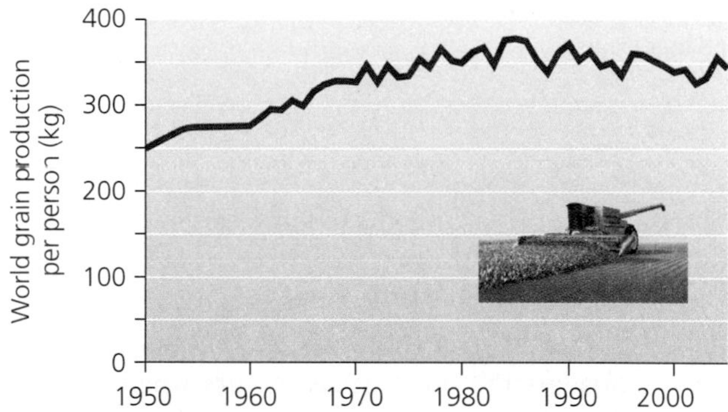


FIGURE 10.2 World per capita production of grain rose until 1985. Since then it has declined slightly, because grain production has grown more slowly than global population. Some have pointed to this trend as a sign that our agriculture is suffering from the deteriorating soil and environmental conditions it has caused. Others propose that the decline is due to two factors: (1) populations in developed nations, where people may be near their limit of grain consumption already, have stabilized; and (2) population growth is greatest in developing nations, where people consume far less than in developed nations. As a result, more and more of the world's people live in societies in which people consume less, thus dragging down the global average demand for grain production. Data from U.N. Food and Agriculture Organization and U.S. Department of Agriculture.

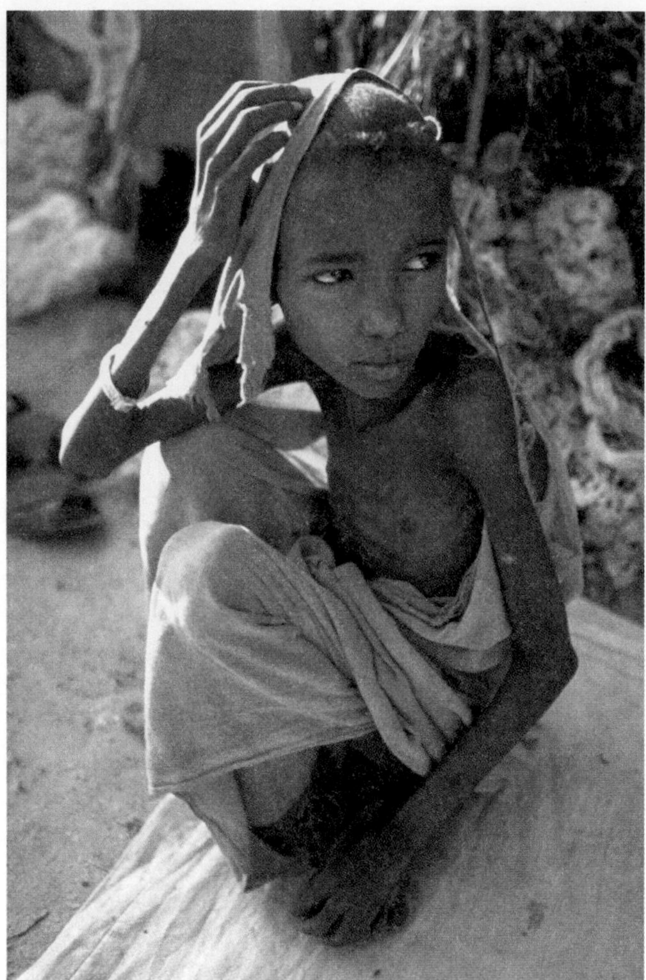

FIGURE 10.3 Millions of children, including this child in Somalia, suffer from forms of malnutrition, such as kwashiorkor and marasmus.

as "food insecure," lacking the income required to procure sufficient food at all times.

Just as the *quantity* of food a person eats is important for health, so is the *quality* of food. **Malnutrition,** a shortage of nutrients the body needs, occurs when a person fails to obtain a complete complement of vitamins and minerals. Malnutrition can afflict both undernourished and overnourished individuals, and it can lead to disease (**Figure 10.3**). When people eat a high-starch diet but not enough protein or essential amino acids (• p. 97), then *kwashiorkor* results. Children who have recently stopped breast-feeding are most at risk for developing kwashiorkor, which causes bloating of the abdomen, deterioration and discoloration of hair, mental disability, immune suppression, developmental delays, anemia, and reduced growth. Protein deficiency together with a lack of calories can lead to *marasmus*, which causes wasting or shriveling among millions of children in the developing world.

## The "green revolution" boosted agricultural production

The desire for greater quantity and quality of food for our growing population led in the mid- and late-20th century to the **green revolution** (first introduced in Chapter 9, • p. 237). Realizing that farmers could not go on indefinitely cultivating more land to increase crop output, agricultural scientists devised methods and technologies to increase crop output per unit area of existing cultivated land. Industrialized nations dramatically increased their per-area yields; the average hectare of U.S. cornfield during the 20th century, for instance, upped its corn output fivefold. Many people saw such growth in production and efficiency as key to ending starvation in developing nations.

The transfer of technology to the developing world that marked the green revolution began in the 1940s, when U.S. agricultural scientist Norman Borlaug introduced Mexico's farmers to a specially bred type of wheat (**Figure 10.4**). This strain of wheat produced large seed heads, was short in stature to resist wind, was resistant to diseases, and produced high yields. Within two decades of planting and harvesting this specially bred crop, Mexico tripled its wheat production and began exporting wheat. The stunning success of this program inspired others. Borlaug—who won the Nobel Peace Prize for his work—took his wheat to India and Pakistan and helped transform agriculture there. Soon many developing countries were increasing their crop yields using selectively bred strains of wheat, rice, corn, and other crops from

FIGURE 10.4 Norman Borlaug holds examples of the wheat variety he bred that helped launch the green revolution. The high-yielding, disease-resistant wheat helped increase agricultural productivity in many developing countries.

industrialized nations. Some varieties yielded three or four times as much per acre as did their predecessors.

## The green revolution has brought the environment both benefit and harm

Along with the new grains, developing nations imported the methods of industrialized agriculture. They began applying large amounts of synthetic fertilizers and chemical pesticides on their fields, irrigating crops with generous amounts of water, and using heavy equipment powered by fossil fuels. From 1900 to 2000, humans expanded the world's total cultivated area by 33% while increasing energy inputs into agriculture by 80 *times*.

This high-input agriculture succeeded dramatically in allowing farmers to harvest more corn, wheat, rice, and soybeans from each hectare of land. Intensive agriculture saved millions in India from starvation in the 1970s and eventually turned that nation into a net exporter of grain (**Figure 10.5**).

These developments had mixed effects on the environment. On the positive side, the intensified use of already-cultivated land reduced pressures to convert additional natural lands for new cultivation. Between 1961 and 2003, food production rose 150% and population rose 100%, while area converted for agriculture increased only 10% (see "Interpreting Graphs and Data," • pp. 291–292). For this reason, the green revolution prevented some degree of deforestation and habitat conversion just when many countries were experiencing their fastest population growth rates. In this sense, the green revolution was a boon for the preservation of biodiversity and natural ecosystems.

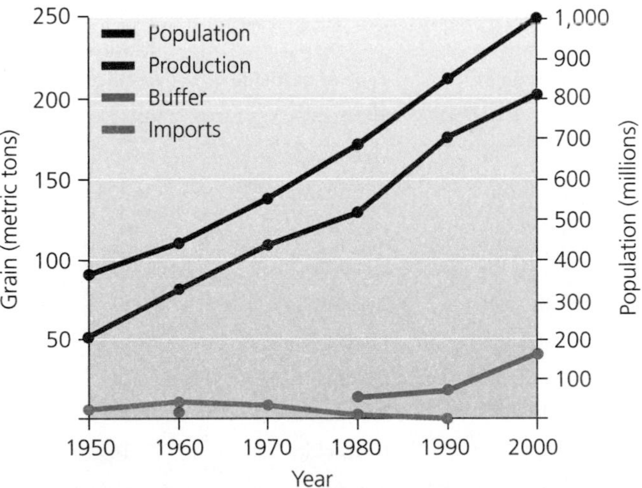


FIGURE 10.5 Thanks to green revolution technology, India's grain production has kept pace with its rapid population growth. This has enabled India to stop importing grain and to maintain extra grain in reserve as a buffer against food shortages.

On the negative side, the intensive application of water, fossil fuels, inorganic fertilizers, and synthetic pesticides worsened pollution, erosion, salinization, and desertification (Chapter 9).

### Weighing THE Issues — The Green Revolution and Population

In the 1960s, India's population was skyrocketing, and its traditional agriculture was not producing enough food to support the growth. By adopting green revolution agriculture, India sidestepped mass starvation. In the years since intensifying its agriculture, India has added several hundred million more people and continues to suffer widespread poverty and hunger. Norman Borlaug called his green revolution methods "a temporary success in man's war against hunger and deprivation," something to give us breathing room in which to deal with what he called the "Population Monster."

Do you think we can call the green revolution a success? Do you think the green revolution has solved problems, delayed our resolution of problems, or created new ones? How sustainable are green revolution approaches? Consider our discussion of demographic transition theory from Chapter 8 (• pp. 219–221). Have we been dealing with the "Population Monster" during the breathing room that the green revolution has bought for us?

---

The planting of crops in *monocultures*, large expanses of single crop types (• p. 237; **Figure 10.6**), has made planting and harvesting more efficient and has thereby increased output. However, monocultural planting has reduced biodiversity over huge areas, because many fewer wild organisms are able to live in monocultures than in native habitats or in traditional small-scale polycultures. Moreover, when all plants in a field are genetically similar, as in monocultures, all will be equally susceptible to viral diseases, fungal pathogens, or insect pests that can spread quickly from plant to plant (see Figure 10.6). For this reason, monocultures bring significant risks of catastrophic failure.

Monocultures have also contributed to a narrowing of the human diet. Globally, 90% of the food we consume now comes from just 15 crop species and eight livestock species—a drastic reduction in diversity from earlier times. The nutritional risks of such dietary restriction have been alleviated by the fact that expanded global trade has provided many people access to a wider diversity of foods from around the world. However, this effect

FIGURE 10.6 Most agricultural production in industrialized countries comes from monocultures—large stands of single types of crop plant, such as this wheat field in Washington. Clustering crop types in uniform fields on large scales greatly improves the efficiency of planting and harvesting, but it also decreases biodiversity and makes crops susceptible to outbreaks of pests that specialize on particular crops. Armyworms (inset) are agricultural pests whose outbreaks can substantially reduce crop yields. The caterpillars of these moths defoliate a wide variety of crops, including wheat, corn, cotton, alfalfa, and beets.

has benefited wealthy people far more than poor people. One reason farmers and scientists were so concerned about transgenic contamination of Oaxaca's native maize is that Oaxacan maize varieties serve as a valuable source of genetic variation in a world where so much variation is being lost to monocultural practices.

## Pests and Pollinators

Throughout the history of agriculture, the insects, fungi, viruses, rats, and weeds that eat or compete with our crop plants have taken advantage of the ways we cluster food plants into agricultural fields. As just one example, various species of moth caterpillars known as armyworms (see Figure 10.6) lower yields of everything from beets to sorghum to millet to canola to pasture grasses. Pests and weeds have always posed problems for traditional agriculture, and they pose an even greater threat to monocultures, where a pest adapted to specialize on the crop can move easily from one individual plant to many others of the same type.

What people term a *pest* is any organism that damages crops that are valuable to us. What we term a *weed* is any plant that competes with our crops. These are subjective categories that we define entirely by our own economic interests. There is nothing inherently malevolent in the behavior of a pest or a weed. These organisms are simply trying to survive and reproduce. From the viewpoint of an insect that happens to be adapted to feed on corn, grapes, or apples, a grain field, vineyard, or orchard represents an endless buffet.

### We have developed thousands of chemical pesticides

To prevent pest outbreaks and to limit competition with weeds, people have developed thousands of artificial chemicals to kill insects *(insecticides)*, plants *(herbicides)*, and fungi *(fungicides)*. Such poisons that target pest organisms are collectively termed **pesticides.** Table 10.1 shows the 10 most widely used pesticides in U.S. agriculture. All told, roughly 400 million kg (900 million lb) of

**TABLE 10.1 Most Commonly Used Pesticides* in Agriculture in the United States**

| Active ingredient† | Type of pesticide | Millions of kg applied per year |
|---|---|---|
| Glyphosate | Herbicide | 39–41 |
| Atrazine | Herbicide | 34–36 |
| Metam sodium | Fumigant | 26–28 |
| Acetochlor | Herbicide | 14–16 |
| 2,4-D | Herbicide | 13–15 |
| Malathion | Insecticide | 9–11 |
| Methyl bromide | Fumigant | 9–11 |
| Dichloropropene | Fumigant | 9–11 |
| Metolachlor-s | Herbicide | 9–11 |
| Metolachlor | Herbicide | 7–10 |

*Includes only "conventional pesticides" used in agriculture, and not other types such as disinfectants and wood preservatives.

†Includes only active ingredients, not additional ingredients such as oil, sulfur, and sulfuric acid.

Data from Kiely, T., et al. 2004. *Pesticides industry sales and usage: 2000 and 2001 market estimates.* Washington, DC: U.S. Environmental Protection Agency.

active ingredients from conventional pesticides are applied in the United States each year. Three-quarters of this total is applied on agricultural land. Since 1960, pesticide use has risen fourfold worldwide. Usage in industrialized nations has leveled off in the past two decades, but it continues to rise in the developing world. Today more than $32 billion is expended annually on pesticides, with one-third of that total spent in the United States. We will address the health consequences of synthetic pesticides for people and other organisms in Chapter 14.

## Pests evolve resistance to pesticides

Despite the toxicity of these chemicals (Chapter 14), their usefulness tends to decline with time as pests evolve resistance to them. Recall from our discussion of natural selection (• pp. 113–116) that organisms within populations vary in their traits. Because most insects and microbes occur in huge numbers, it is likely that a small fraction of individuals may by chance have genes that confer some degree of immunity to a given pesticide. Even if a pesticide application kills 99.99% of the insects in a field, 1 in 10,000 survives. If an insect survives by being genetically resistant to a pesticide, and if it mates with other resistant individuals of the same species, the insect population may grow. This new population will consist of individuals that are genetically resistant to the pesticide. As a result, pesticide applications will cease to be effective (**Figure 10.7**).

In many cases, industrial chemists are caught up in an "evolutionary arms race" with the pests they battle, racing to increase or retarget the toxicity of their chemicals while the armies of pests evolve ever-stronger resistance to their efforts. The number of species known to have evolved resistance to pesticides has grown over the decades. As of 2007, there were more than 2,700 known cases of resistance by 550 species to over 300 pesticides. Insects such as the green peach aphid, Colorado potato beetle, and diamondback moth have evolved resistance to multiple insecticides. Resistant pests can take a significant economic toll on crops. As just one example, gummy stem blight destroyed two-thirds of Texas's melon crop in 1997 after the blight evolved resistance to the pesticide Benlate.

1 Pests attack crops

2 Pesticide is applied

3 Most pests are killed. A few with innate resistance survive

4 Survivors breed and produce a pesticide-resistant population

5 Pesticide is applied again

6 Pesticide has little effect. New, more toxic, pesticides are developed

FIGURE 10.7 Through the process of natural selection, crop pests may evolve resistance to the poisons we apply to kill them. This simplified diagram shows that when a pesticide is applied to an outbreak of insect pests, it may kill virtually all individuals except those few with an innate immunity to the poison. Those surviving individuals may found a population with genes for resistance to the poison. Future applications of the pesticide may then be ineffective, forcing us to develop a more potent poison or an alternative means of pest control.

## Biological control pits one organism against another

Because of pesticide resistance and the health risks from some synthetic chemicals, agricultural scientists increasingly battle pests and weeds with organisms that eat or infect them. This strategy, called **biological control,** or **biocontrol** for short, operates on the principle that "the enemy of one's enemy is one's friend." For example, parasitoid wasps (• p. 146) are natural enemies of many caterpillars. These wasps lay eggs on a caterpillar, and the larvae that hatch from the eggs feed on the caterpillar, eventually killing it. Parasitoid wasps have been used as biocontrol agents in many situations. Some such efforts have succeeded at pest control and have led to steep reductions in chemical pesticide use.

One classic case of successful biological control is the introduction of the cactus moth, *Cactoblastis cactorum,* from Argentina to Australia in the 1920s to control invasive prickly pear cactus that was overrunning rangeland. Within just a few years, the moth managed to free millions of hectares of rangeland from the cactus (**Figure 10.8**).

A widespread modern biocontrol effort has been the use of ***Bacillus thuringiensis*** **(Bt),** a naturally occurring soil bacterium that produces a protein that kills many caterpillars and the larvae of some flies and beetles. Farmers have used the natural pesticidal activity of this bacterium to their advantage by spraying spores of this bacterium on their crops. If used correctly, Bt can protect crops from pest-related losses. (In addition, as we will see shortly (• p. 274), scientists have managed to isolate the bacterium's gene responsible for its toxic effects and engineer it into crop plants so that they can protect themselves against insect attack.)

**(a) Before cactus moth introduction**

**(b) After cactus moth introduction**

FIGURE 10.8 In a classic case of biocontrol, larvae of the cactus moth, *Cactoblastis cactorum,* were used to clear non-native prickly pear cactus from millions of hectares of rangeland in Queensland, Australia. These photos from the 1920s show an Australian ranch before (**a**) and after (**b**) introduction of the moth.

## Biological control agents themselves may become pests

In most cases, biological control involves introducing an animal or microbe from a foreign ecosystem, often from another continent. Such relocation helps ensure that the target pest has not already evolved ways to deal with the biocontrol agent. However, it also means that no one can know for certain in advance what effects the agent might have. In some cases, biocontrol agents have become invasive and harmed nontarget organisms. Following the cactus moth's success in Australia, for example, it was introduced in other countries to control prickly pear. Moths introduced to Caribbean islands spread to Florida on their own and are now eating their way through rare native cacti in Florida and spreading to other states. If these moths reach Mexico and the southwestern United States, they could decimate many native and economically important species of prickly pear there.

One recent study in Hawaii revealed the extent to which some biocontrol agents miss their targets. Wasps and flies have been introduced to control agricultural pests in Hawaii at least 122 times over the past century, and biologists Laurie Henneman and Jane Memmott suspected that some of these might be adversely affecting native Hawaiian caterpillars that were not pests. They sampled parasitoid wasp larvae from 2,000 caterpillars of various species in a remote mountain swamp far from farmland. In this area, designated as a wilderness preserve, they found that fully 83% of the parasitoids were biocontrol agents that had been intended to combat lowland agricultural pests.

If biological control works as planned, it can be a permanent solution that requires no further maintenance and is environmentally benign. However, if the agent has nontarget effects, the harm done may also be permanent, because removing the agent from the system once it is established is far more difficult than simply halting a chemical pesticide application. Moreover, like all invasive species, invasive biocontrol agents can have wide-ranging ecological and economic impacts (• pp. 158–160, 305–307). One noted skeptic of biocontrol, ecologist Daniel Simberloff (• p. 318), has remarked that biocontrol "should be used with our eyes wide open—and as a last resort." However, two British scientists reviewing cases as of the year 2000 concluded that only a small percentage of efforts have resulted in demonstrable nontarget effects, and perhaps fewer than 10% of these effects were substantial.

Because of concerns about unintended impacts, researchers now study biocontrol proposals carefully before putting them into action, and government regulators must approve these efforts. However, there will never be a sure-fire way of knowing in advance whether a given biocontrol program will work as planned.

## Integrated pest management combines biocontrol and chemical methods

As it became clear that both chemical and biocontrol approaches have their drawbacks, agricultural scientists and farmers began developing more sophisticated strategies, trying to combine the best attributes of each approach. In **integrated pest management (IPM),** numerous techniques are integrated to achieve long-term suppression of pests, including biocontrol, use of chemicals when needed, close monitoring of populations, habitat alteration, crop rotation, transgenic crops, alternative tillage methods, and mechanical pest removal.

In recent decades, IPM has become popular in many parts of the world. Indonesia stands as an exemplary case (**Figure 10.9**). The nation had subsidized pesticide use heavily for years, but its scientists came to understand that pesticides were actually making pest problems worse. They were killing the natural enemies of the brown planthopper, which began to devastate rice fields as its populations exploded. Concluding that pesticide subsidies were costing money, causing pollution, and apparently decreasing yields, the Indonesian government in 1986 banned the importation of 57 pesticides, slashed pesticide subsidies, and encouraged IPM. Within 4 years, pesticide production fell to below half its 1986 level, imports fell to one-third, and subsidies were phased out (saving $179 million annually). Rice yields rose 13%.

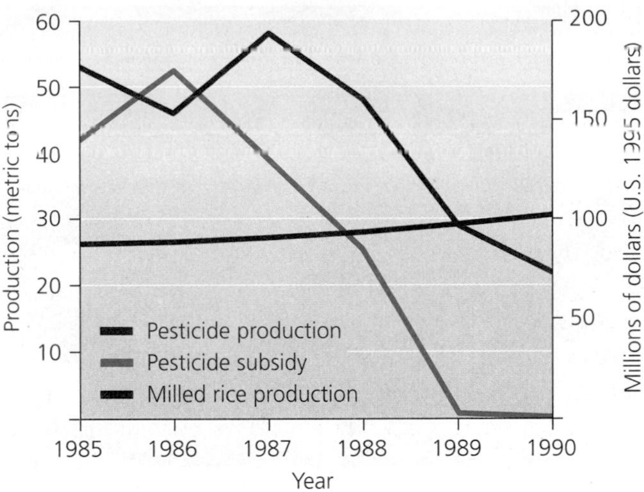


FIGURE 10.9 The Indonesian government threw its weight behind integrated pest management starting in 1986. Within just a few years, pesticide production and pesticide imports were down drastically, pesticide subsidies were phased out, and yields of rice increased slightly.

## We depend on insects to pollinate crops

Managing insect pests is such a major issue in agriculture that it is easy to fall into a habit of thinking of all insects as somehow bad or threatening. But in fact, most insects are harmless to agriculture, and some are absolutely essential. The insects that pollinate agricultural crops are among the most vital, yet least understood and least appreciated, factors in cropland agriculture. Pollinators are the unsung heroes of agriculture.

**Pollination** is the process by which male sex cells of a plant (pollen) fertilize female sex cells of a plant; it is the botanical version of sexual intercourse. Without pollination, no plants could reproduce sexually, and no plant species would persist for long.

Plants such as ferns, conifer trees, and grasses achieve pollination by the wind. Millions of minuscule pollen grains are blown long distances, and by chance a small number land on the female parts of other plants of their species. The many kinds of plants that sport showy flowers, however, typically are pollinated by animals, such as hummingbirds, bats, and insects (**Figure 10.10**; see also Figure 6.9, p. 147). Flowers are, in fact, evolutionary adaptations that function to attract pollinators. The sugary nectar and protein-rich pollen in flowers serve as rewards to lure these sexual intermediaries, and the sweet smells and bright colors of flowers are signals to advertise these rewards.

Our staple grain crops are derived from grasses and are wind-pollinated, but many other crops depend on

FIGURE 10.10 Many agricultural crops depend on insects to pollinate them. Our food supply, therefore, depends partly on conservation of these vital animals. Flowers such as these apple blossoms use colors and sweet smells to advertise nectar and pollen to pollinators such as the European honeybee.

insects for pollination. The most complete survey to date, by tropical bee biologist Dave Roubik, documented 800 species of cultivated plants that rely on bees and other insects for pollination. An estimated 73% of cultivars are pollinated, at least in part, by bees; 19% by flies; 5% by wasps; 5% by beetles; and 4% by moths and butterflies. In addition, bats pollinate 6.5% and birds 4%. Overall, native species of bees in the United States alone are estimated to provide $3 billion of pollination services each year to crop agriculture.

Populations of native pollinators have declined precipitously, however. As one example of many, the U.S. Great Basin states are a world center for the production of alfalfa seed, and alfalfa flowers are pollinated mostly by native alkali bees that live in the soil as larvae. In the 1940s to 1960s, farmers began plowing the land and increasing pesticide use in an effort to boost yields. These measures killed vast numbers of the soil-dwelling bees, and alfalfa seed production plummeted.

## Conservation of pollinators is vital

Preserving the biodiversity of native pollinators is especially important today because the domesticated workhorse of pollination, the honeybee *(Apis mellifera)*, is also declining. North American farmers regularly hire beekeepers to bring colonies of this introduced Old-World honeybee to their fields when it is time to pollinate crops (**Figure 10.11**). Honeybees pollinate over 100 crops that comprise one-third of the U.S. diet, thereby contributing many billions of dollars in services.

In recent years, two accidentally introduced parasitic mites have swept through honeybee populations, decimating hives and pushing many beekeepers toward financial ruin. On top of this, starting in 2006, entire hives inexplicably began dying off. Scientists are racing to discover the reasons for what is being called "colony collapse disorder."

Farmers and homeowners alike can help maintain populations of pollinating insects by reducing or eliminating pesticide use. All insect pollinators, including honeybees, are vulnerable to the vast arsenal of insecticides that modern industrial agriculture applies to crops and that many homeowners apply to lawns and gardens. Some insecticides are designed to specifically target certain types of insects, but most are not. Without full and detailed information on the effects of pesticides, farmers and homeowners trying to control the "bad" bugs that threaten the plants they value all too often kill the "good" insects as well.

Homeowners, even in the middle of a large city, can encourage populations of pollinating insects by planting gardens of flowering plants that nourish pollinating insects and by providing nesting sites for bees. These can be simple contraptions made of wood that have holes and plastic straws. And by allowing noncrop flowering plants (such as clover) to grow around the edges of their fields, farmers can maintain a diverse community of insects—some of which will pollinate their crops.

FIGURE 10.11 Honeybees are widely used to pollinate crop plants, and beekeepers transport hives of bees to crops when it is time for flowers to be pollinated. Recently honeybees have suffered devastating epidemics of parasitism, making it increasingly important to conserve native species of pollinators.

## Genetically Modified Food

The green revolution enabled us to feed a greater number and proportion of the world's people, but relentless population growth demands still more innovation. A new set of potential solutions began to arise in the 1980s and 1990s as advances in genetics enabled scientists to directly alter the genes of organisms, including crop plants and livestock. The genetic modification of organisms that provide us food holds promise to increase nutrition and the efficiency of agriculture while lessening impacts on the planet's environmental systems. However, genetic modification may also pose risks that are not yet well understood. This possibility has given rise to protest around the globe from consumer advocates, small farmers, environmental activists, and opponents of big business.

### Genetic modification of organisms depends on recombinant DNA

The genetic modification of crops and livestock is one type of genetic engineering. **Genetic engineering** is any process whereby scientists directly manipulate an organism's genetic material in the laboratory by adding, deleting, or changing segments of its DNA (• p. 97). **Genetically modified (GM) organisms** are organisms that have been genetically engineered using a technique called recombinant DNA technology. **Recombinant DNA** is DNA that has been patched together from the DNA of multiple organisms. In this process, scientists break up DNA from multiple organisms and then splice segments together, placing genes that produce certain proteins and code for certain desirable traits (such as rapid growth, disease and pest resistance, or higher nutritional content) into the genomes of organisms lacking those traits.

Recombinant DNA technology was developed by scientists studying the bacterium *Escherichia coli.* As shown in **Figure 10.12**, scientists first isolate *plasmids*, which are small, circular DNA molecules, from a bacterial culture. At the same time, DNA containing a gene of interest is removed from the cells of another organism. Scientists insert the gene of interest into the plasmid to form recombinant DNA. This recombinant DNA enters new bacteria, which then reproduce, generating many copies of the desired gene.

When scientists use recombinant DNA technology to develop new varieties of crops, they can often introduce the recombinant DNA directly into a plant cell and regenerate an entire plant from that single cell. Some plants, including many grains, are not receptive to plasmids, in which case scientists may use a "gene gun" to shoot DNA directly into plant cells. An organism that contains DNA from another species is called a *transgenic* organism, and the genes that have moved between them are called *transgenes.* The creation of transgenic organisms is one type of **biotechnology,** the material application of biological science to create products derived from organisms. Biotechnology has helped us develop medicines, clean up pollution, understand the causes of cancer and other diseases, dissolve blood clots after heart attacks, and make better beer and cheese. **Figure 10.13** details several of the most notable developments in GM foods. These examples and the stories behind them illustrate both the promises and pitfalls of food biotechnology.

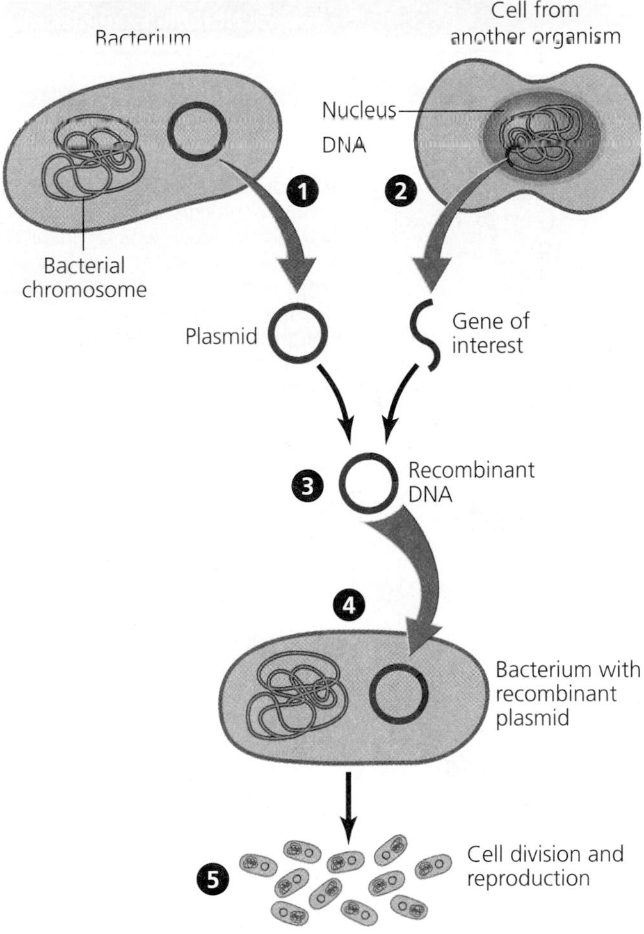


FIGURE 10.12 To create recombinant DNA, a gene of interest is excised from the DNA of one type of organism and is inserted into a stretch of bacterial DNA called a *plasmid.* The plasmid is then introduced into bacterial cells that reproduce and generate more copies, or it may be introduced directly into cells of the organism to be modified. If all goes as planned, the new gene will be expressed in the GM organism as a desirable trait, such as rapid growth or high nutritional content in a food crop.

| Several Notable Examples of Genetically Modified Food Technology | |
|---|---|
| Food | Development |
| Golden rice | Millions of people in the developing world get too little vitamin A in their diets, causing diarrhea, blindness, immune suppression, and even death. The problem is worst with children in east Asia, where the staple grain, white rice, contains no vitamin A. Researchers took genes from plants that produce vitamin A and spliced the genes into rice DNA to create more-nutritious "golden rice" (the vitamin precursor gives it a golden color). Critics charged that biotech companies hyped their product, which contains only small amounts of the nutrient and may not be the best way to combat vitamin A deficiency. India's foremost critic of GM food, Vandana Shiva, charged that "vitamin A rice is a hoax . . . a very effective strategy for corporate takeover of rice production, using the public sector as a Trojan horse." Backers of the technology counter that the nutritive value can be further improved and could enhance the health of millions of people. |
| Flavr Savr tomato | By reversing the function of a normal tomato gene, the Calgene Corporation created the Flavr Savr tomato, which Calgene maintained would ripen longer on the vine, taste better, stay firm during shipping, and last longer in the produce department. The U.S. Food and Drug Administration approved the Flavr Savr tomato for sale in the United States in 1994. Calgene stopped selling the Flavr Savr in 1996, however, for several reasons, including problems with the technique and public safety concerns. |
| Ice-minus strawberries | University of California–Berkeley researcher Steven Lindow removed a gene that facilitated the formation of ice crystals from the DNA of a particular bacterium, *Pseudomonas syringae*. The modified, frost-resistant bacteria could then serve as a kind of antifreeze when sprayed on the surface of frost-sensitive crops such as strawberries. The multiplying bacteria would coat the berries, protecting them from frost damage. However, early news coverage of this technique showed scientists spraying plants while wearing face masks and protective clothing, an image that caused public alarm. |
| Bt crops | By equipping plants with the ability to produce their own pesticides, scientists hoped to boost crop yields by reducing losses to insects. By the late 1980s, scientists working with *Bacillus thuringiensis* (Bt) had pinpointed the genes responsible for producing that bacterium's toxic effects on insects, and had managed to insert the genes into the DNA of crops. The USDA and EPA approved Bt versions of 18 crops for field testing, from apples to broccoli to cranberries. Corn and cotton are the most widely planted Bt crops today. Proponents say Bt crops reduce the need for chemical pesticides. However, critics worry that the continuous presence of Bt in the environment will induce insects to evolve resistance to the toxins and that Bt crops might cause allergic reactions in humans. Another concern is that the crops may harm nontarget species. A 1999 study reported that pollen from Bt corn can kill the larvae of monarch butterflies, a nontarget species, when corn pollen drifts onto milkweed plants monarchs eat. Another study that year showed that the Bt toxin could leach from corn roots and poison the soil. |

**FIGURE 10.13** The early development of genetically modified foods has been marked by a number of cases in which these products ran into trouble in the marketplace or were opposed by activists. A selection of these cases serves to illustrate some of the issues that proponents and opponents of GM foods have been debating.

## Genetic engineering is like, and unlike, traditional agricultural breeding

The genetic alteration of plants and animals by humans is nothing new; through artificial selection (• pp. 115–117), we have influenced the genetic makeup of our livestock and crop plants for thousands of years. As we saw in Chapter 9 (• pp. 235–236), our ancestors altered the gene pools of our domesticated plants and animals through selective breeding by preferentially mating individuals with favored traits so that offspring would inherit those traits. Early farmers selected plants and animals that grew faster, were more resistant to disease and drought, and produced large amounts of fruit, grain, or meat.

Proponents of GM crops often stress this continuity with our past and say there is little reason to expect that today's GM food will be any less safe than selectively bred food. Dan Glickman, head of the USDA from 1995 to 2001, remarked:

> Biotechnology's been around almost since the beginning of time. It's cavemen saving seeds of a high-yielding plant. It's Gregor Mendel, the father of genetics, cross-pollinating his garden peas. It's a diabetic's insulin, and the enzymes in your yogurt. . . . Without exception, the biotech products on our shelves have proven safe.

However, as biotech critics are quick to point out, the techniques geneticists use to create GM organisms differ from traditional selective breeding in several ways. For

| Several Notable Examples of Genetically Modified Food Technology | |
|---|---|
| Food | Development |
| StarLink corn | StarLink corn, a variety of Bt corn, had been approved and used in the United States for animal feed but not for human consumption. In 2000, StarLink corn DNA was discovered in taco shells and other corn products, causing fears that the corn might cause allergic reactions. No such health effects have been confirmed, but the corn's French manufacturer, Aventis CropScience, chose to withdraw the product from the market. Although StarLink corn was grown on only a tiny portion of U.S. farmland, its transgene apparently spread widely to other corn through cross-pollination. This episode cost U.S. taxpayers, because the U.S. government spent $20 million to purchase contaminated corn and remove it from the food supply. |
| Sunflowers and superweeds | Sunflowers have also been engineered to express the Bt toxin. Research on Bt sunflowers suggests that their transgenes might spread to other plants and turn them into vigorous weeds that compete with the crop. This is most likely to happen with crops like squash, canola, and sunflowers that can breed with their wild relatives. In 2002, Ohio State University researcher Allison Snow and colleagues bred wild sunflowers with Bt sunflowers and found that hybrids with the Bt gene produced more seeds and suffered less herbivory than hybrids without it. They concluded that if Bt sunflowers were planted commercially, the Bt gene would spread into wild sunflowers, potentially turning them into superweeds. Researcher Norman Ellstrand of the University of California–Riverside, meanwhile, had found that transgenes from radishes were transferred to wild relatives 1 km (0.6 mi) away and that hybrids produced more seeds, so the gene could be expected to spread in wild populations. He found the same results with sorghum and its weedy relative, johnsongrass. Such results suggest that transgenic crops can potentially create superweeds that can compete with crops and harm nontarget organisms. |
| Roundup Ready crops | The Monsanto Company manufactures a widely used herbicide called Roundup. Roundup kills weeds, but kills crops too, so farmers must apply it carefully. Thus, Monsanto engineered Roundup Ready crops, including soybeans, corn, cotton, and canola, that are immune to the effects of its herbicide. With these variants, farmers can spray Roundup on their fields without killing their crops, in theory making the farmer's life easier. Of course, this also creates an incentive for farmers to use Monsanto's Roundup herbicide rather than a competing brand. Unfortunately, Roundup is not completely benign; its active ingredient, glyphosate, is the third-leading cause of illness for California farm workers. It also harms nitrogen-fixing bacteria and desirable fungi in soils that are essential for crop production. Biotech proponents have argued that GM crops are good for the environment because they reduce pesticide use. This may often be the case, but some studies have shown that farmers apply more herbicide when they use Roundup Ready crops. |
| Terminator seeds | In the late 1990s the USDA worked with Delta and Pine Land Company to engineer a line of crop plants that can kill their own seeds. This so-called "terminator" technology would ensure that farmers buy seeds from seed companies every year rather than planting seeds saved from the previous year's harvest. Because GM crops require a great deal of research and development, seed companies reason that they need to charge farmers annually for seeds in order to recoup their investment. Critics worried that pollen from terminator plants might fertilize normal plants, damaging the crops of farmers who save seeds from year to year. Some nations, like India and Zimbabwe, banned terminator seeds. These countries saw the efforts of biotech seed companies to sell them terminator seeds as a ploy to make poor farmers dependent on multinational corporations for seeds. In the face of this opposition, in 1999 agrobiotech companies Monsanto and AstraZeneca announced that they would not bring their terminator technologies to market. |

FIGURE 10.13 *Continued*

one, selective breeding most often mixes genes from individuals of the same or similar species, whereas with recombinant DNA technology, scientists routinely mix genes of organisms as different as viruses and crops, or spiders and goats. For another, selective breeding deals with whole organisms living in the field, whereas genetic engineering involves lab experiments dealing with genetic material apart from the organism. And whereas traditional breeding selects from among combinations of genes that come together on their own, genetic engineering creates the novel combinations directly. Thus, traditional breeding changes organisms through the process of selection (• pp. 113–114), whereas genetic engineering is more akin to the process of mutation (• p. 114).

## Biotechnology is transforming the products around us

In just three decades, GM foods have gone from science fiction to big business. As recombinant DNA technology first developed in the 1970s, scientists debated among themselves whether the new methods were safe. They collectively regulated and monitored their own research until most scientists were satisfied that reassembling genes in bacteria did not create dangerous superbacteria. Once the scientific community declared itself confident in the 1980s that the technique was safe, industry leaped at the chance to develop hundreds of applications, from improved medicines (such as hepatitis B vaccine and insulin for diabetes) to designer plants and animals.

TABLE 10.2 Percentage of Major Crops in the United States That Are Genetically Modified

| Crop | Percent herbicide-resistant | Percent insect-resistant |
|---|---|---|
| Canola | 95 | 0 |
| Soybeans | 93 | 0 |
| Cotton | 61 | 52 |
| Corn | 50 | 40 |

Data is for 2006, from the International Service for the Acquisition of Agri-Biotech Applications (ISAAA). 2007.

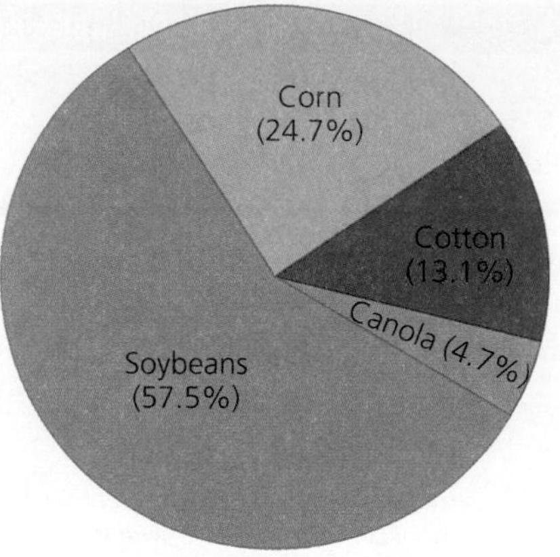


**(a) GM crops by type**

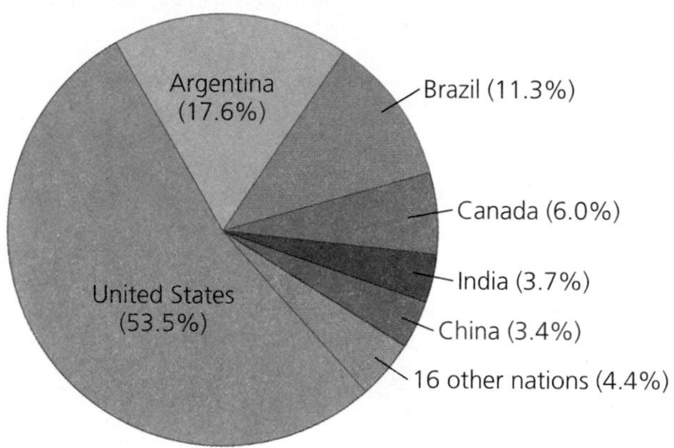


**(b) GM crops by nation**

FIGURE 10.14 Of the world's genetically modified crops (**a**), soybeans constitute the majority so far. Of the world's nations (**b**), the United States devotes the most land area to GM crops. Data is for 2006, from the International Service for the Acquisition of Agri-Biotech Applications (ISAAA). 2007.

Most GM crops today are engineered to resist herbicides, so that farmers can apply herbicides to kill weeds without having to worry about killing their crops. Other crops are engineered to resist insect attack. Some are modified for both types of resistance. Resistance to herbicides and pests makes it efficient, and in some cases more economical, for large-scale commercial farmers to do their jobs. As a result, sales of GM seeds to these farmers in the United States and other countries have risen quickly.

Today most of the U.S. harvests of soybeans, corn, cotton, and canola consist of genetically modified strains (**Table 10.2**). Worldwide, three of every five soybean plants is now transgenic, as is one of every four cotton plants, one of every five canola plants, and one of every six corn plants. Globally in 2006, it was estimated that over 10 million farmers grew GM crops on 102 million ha (252 million acres) of farmland—an area the size of Kansas, Nebraska, Oklahoma, and California combined.

Soybeans account for most of the world's GM crops (**Figure 10.14a**). Of the 22 nations growing GM crops in 2006, six (the United States, Argentina, Brazil, Canada, India, and China) accounted for 96% of production, with the United States alone growing over half of the global total (**Figure 10.14b**). The global area planted in GM crops has jumped by more than 10% annually every year since 1996, and over half the world's people now live in nations in which GM crops are grown. The market value of GM crops in 2006 was estimated at $6.15 billion, and over the previous decade their adoption is estimated to have increased farmers' incomes by $27 billion.

## What are the impacts of GM crops?

As GM crops were adopted, as research proceeded, and as biotech business expanded, many citizens, scientists, and policymakers became concerned. Some feared the new foods might be dangerous for people to eat. Others were concerned that transgenes might escape and pollute ecosystems and damage nontarget organisms. Still others worried that pests would evolve resistance to the supercrops and become "superpests" or that transgenes would be transferred from crops to other plants and turn them into "superweeds." Some, like Quist and Chapela, worried that transgenes might ruin the integrity of native ancestral races of crops (see "The Science behind the Story, • pp. 278–279).

Plenty of evidence exists that conventional crops can interbreed with their wild relatives (rice can breed with wild rice, for instance), so there seems little reason to believe that transgenic crops would not interbreed with closely related wild plants. In the first confirmed case, GM oilseed rape was found hybridizing with wild mustard. In another case, creeping bentgrass engineered for use on golf courses—a GM plant not yet approved by the USDA—was found to have pollinated wild grass up to 21 km (13 mi) away from its test growing site in Oregon. Most scientists think transgenes will inevitably make their way from GM crops into wild plants, but the consequences of this are open to debate.

Supporters of GM crops maintain that no ill health effects on people have been demonstrated and that transgenic crops are, in fact, beneficial for the environment. Adoption of insect-resistant Bt crops appears to reduce the use of chemical insecticides, because farmers use fewer chemicals if their crops do not need them.

Some researchers maintain that herbicide use also declines in the wake of GM crops, and that herbicide-tolerant crops enable no-till farming (• pp. 233–234, 248–250), which can bring various environmental benefits. Other researchers conclude that GM crops result in more herbicide use because farmers can afford to use more herbicide if their crops can withstand it and because weeds may evolve resistance to herbicides, making heavier applications necessary. It is a difficult question because there are many ways to analyze the data and because many factors influence the choices farmers make. For instance, a steep fall in the price of Monsanto's Roundup herbicide may account for much of the increase in its application to Roundup Ready soybeans.

Researchers and industry supporters who claim environmental benefits from GM crops also say these crops reduce carbon emissions, for two reasons: (1) if fewer pesticide applications are made, then tractors use less fuel; and (2) if herbicide-resistant crops encourage the adoption of no-till farming, then more carbon is sequestered in the soil. One GM crop research agency estimates that in 2005, GM crops reduced carbon emissions equivalent to taking 3.6 million cars off the road.

Because GM technology is new and changing, and because its large-scale introduction into our environment is newer still, there remains a great deal that scientists don't know about how transgenic crops behave in the field. Certainly, millions of Americans eat GM foods every day without outwardly obvious signs of harm, and evidence for negative ecological effects is limited so far. However, it is still too early to dismiss concerns about environmental impacts without further scientific research. Therefore, critics argue that we should proceed with caution, adopting the **precautionary principle**, the idea that one should not undertake a new action until the ramifications of that action are well understood.

The best efforts so far to test the ramifications of GM crops produced results in 2003 through 2005. The British government, in considering whether to allow the planting of GM crops, commissioned three large-scale studies. The first study, on economics, found that GM crops could produce long-term financial benefits for Britain, although short-term benefits would be minor. The second study addressed health risks and found little to no evidence of harm to human health, but it also noted that effects on wildlife and ecosystems should be tested before crops are approved. The third study (involving 19 researchers, 200 sites, and $8 million in funding) tested effects on bird and invertebrate populations from four GM crops modified for herbicide resistance. Results showed that fields of GM beets and GM spring oilseed rape supported less biodiversity than fields of their non-GM counterparts. Fields of GM maize supported more, however, and fields of winter oilseed rape showed mixed results. Policymakers had hoped that the biodiversity study would end the debate, but the science showed that the impacts of GM crops are complex.

## Debate over GM foods involves more than science

Far more than science is involved in the debate over GM foods. Ethical issues play a large role. For many people, the idea of "tinkering" with the food supply seems dangerous or morally wrong. Even though our agricultural produce is the highly artificial product of thousands of years of selective breeding, people tend to think of food as natural. Furthermore, because every person relies on food for survival and cannot choose *not* to eat, the genetic modification of dietary staples such as corn, wheat, and rice essentially forces people to consume GM products or to go to special effort to avoid them.

The perceived lack of control over one's own food has driven widespread concern that the global food supply is being dominated by a few large agrobiotech corporations that develop GM technologies, among them Monsanto, Syngenta, Bayer CropScience, Dow, DuPont, and BASF. Many activists say these multinational corporations threaten the independence and well-being of the small farmer. This perceived loss of democratic local control is a driving force in the opposition to GM foods, especially in Europe and the developing world. Critics of biotechnology also voice concern that much of the research into the safety of GM organisms is funded, overseen, or conducted by the corporations that stand to profit if their transgenic crops are approved for human consumption, animal feed, or ingredients in other products.

So far, GM crops have not lived up to their promise of feeding the world's hungry. Nearly all commercially available GM crops have been engineered to express either pesticidal properties (e.g., Bt crops) or herbicide tolerance. Often, a company's GM crops are tolerant to herbicides that the same company manufactures and profits from (e.g., Monsanto's Roundup Ready crops). Crops with traits that might benefit poor small-scale farmers of developing countries (such as increased nutrition, drought tolerance, and salinity tolerance) have not been widely commercialized, perhaps because corporations have less economic incentive to do so. Whereas the green revolution was a largely public venture, the "gene revolution" promised by GM crops is largely driven by market considerations of companies selling proprietary products.

In North America, farmers have found themselves battling the aggressive tactics of the St. Louis–based Monsanto

THE SCIENCE BEHIND THE STORY

# Transgenic Contamination of Native Maize?

*Dr. Ignacio Chapela (left) and Dr. David Quist (right), University of California, Berkeley*

David Quist and Ignacio Chapela's *Nature* paper reporting DNA from genetically engineered corn in native Mexican maize (• p. 264) was controversial from the moment it was published. The idea that a transgene had entered native maize in remote areas of Oaxaca was provocative enough. But their claim that the transgene also had been jumping throughout the genome came under particular fire.

Soon after publication, several geneticists pointed out flaws in the methods Quist and Chapela used to determine the location of the transgene. Interpreting results of an experimental technique known as inverse polymerase chain reaction (i-PCR), Quist and Chapela had reported that the transgene was surrounded by essentially random sequences of DNA. Critics argued that i-PCR was unreliable and that Quist and Chapela had used insufficient controls. Their results, the critics suggested, could have arisen from similarities between the transgene and stretches of maize DNA.

A number of geneticists also attacked Quist and Chapela's more fundamental claim that transgenes had been integrated into the genome of native maize. Some argued that the low levels of transgenic DNA detected indicated, at most, first-generation hybrids between local cultivars and transgenic varieties. Other critics, such as Paul Christou, editor of the journal *Transgenic Research,* noted that PCR was a highly sensitive technique and that all of Quist and Chapela's findings could be unreliable if laboratory practices had been careless.

On April 11, 2002, *Nature* published two letters from scientists critical of the study, along with an editorial note concluding that "the evidence available is not sufficient to justify the publication of the original paper."

Quist and Chapela responded by acknowledging that some of their initial findings, particularly those based on the i-PCR technique, were probably invalid. But they also presented new analyses to support their fundamental claim and pointed to a Mexican government study that also found high rates of transgenic contamination. However, this did little to convince skeptics, especially because the Mexican government's results remained unpublished.

The correspondence published in *Nature* was part of a broader debate that took place on the editorial pages of scientific journals, in newspapers and magazines, over the Internet, and within the halls of academe. The debate sometimes turned personal. GM supporters pointed out that Chapela and Quist had long opposed transgenic crops and biotechnology corporations. Chapela had spoken out against his university's plan to enter into a $25 million partnership with the biotechnology firm Novartis (now Syngenta), sparking a debate that divided the Berkeley faculty.

---

company. This first came to light in Canada, where Monsanto was engaged in a high-publicity struggle with Saskatchewan farmer Percy Schmeiser (**Figure 10.15**). Schmeiser maintained that pollen from Monsanto's Roundup Ready canola (see Figure 10.13) used by his neighbors blew onto his land and pollinated his non-GM canola. When he harvested his seed and replanted it the next year, he thus unwittingly grew some plants containing Monsanto's patented herbicide-resistance gene. Schmeiser never purchased Monsanto's patented seed and said he did not want the crossbreeding. Monsanto investigators took seed samples from his plants and charged him with violating Canada's law that makes it illegal for farmers to reuse patented seed or grow the seed without a contract with the company. Monsanto sued Schmeiser, and the court sided with Monsanto, ordering the farmer to pay the corporation roughly $238,000. Schmeiser appealed the case to Canada's Supreme Court, which in 2004 ruled by a 5–4 vote that Schmeiser had violated Monsanto's patent (although it spared the 74-year-old farmer from having to make payments to the company). Schmeiser received wide public support, a government committee called for revising the patent law, and the National Farmers Union of Canada called for a moratorium on GM food.

As of January 2005, Monsanto had launched 90 such lawsuits against 147 farmers and 39 farm companies in 25 U.S. states, winning recorded judgments averaging $412,000, according to a report by the Center for Food Safety. Monsanto says it is merely demanding that farmers heed the U.S. and Canadian patent laws. North Dakota farmer Tom Wiley sees it differently, saying, "Farmers are being sued for having GMOs on their property that they did not buy, do not want, will not use, and cannot sell."

Given such developments, the future of GM foods seems likely to hinge on social, economic, legal, and political factors as well as scientific ones. European consumers have

A farmer harvests maize above the Oaxacan village of Los Naranjos.

GM opponents countered that Quist and Chapela's most vocal critics received funding from biotechnology corporations. Even *Nature,* whose impartiality is critical to its reputation as a first-tier science journal, was engaged in commercial partnerships with biotechnology corporations.

Chapela was denied tenure by Berkeley's administration, even though his department and dean had recommended him. Alleging unfair treatment, Chapela waged a 3-year fight to win tenure and gained a following that included 230 academics who signed a letter of support. In May 2005, a new chancellor awarded him tenure.

Then in August 2005, a research team led by Sol Ortiz-García of Mexico's Instituto Nacional de Ecologia and Allison Snow of Ohio State University published results from an extensive survey of maize from across Oaxaca. They collected and analyzed 154,000 seeds from 870 maize plants in 125 fields at 18 localities in 2003 and 2004. Out of all these samples, they detected no evidence for transgenes from GM corn. Their statistical analysis indicated with 95% certainty that if transgenes are present at all, they likely occur in fewer than 1 out of 10,000 seeds.

García and Snow's team suggested several explanations for how their results could differ so greatly from the earlier studies:

- Farmers may have stopped planting GM seed once scientists and extension agents began warning them about it in 2001.
- Transgenes found in 2000–2001 may have later become rare by chance or after hybridization with local landraces.
- Transgenic maize may be ill-suited to the local environment and may have been selected against.

For many people, the new data eased concerns about transgenic "contamination" of local landraces. However, most researchers agree that given global trade and the rapidly increasing use of GM corn worldwide, gene flow between transgenic corn and Mexico's native landraces is bound to happen someday, if it has not already.

expressed widespread unease about possible risks of GM technologies. In contrast, U.S. consumers have largely accepted the GM crops approved by U.S. agencies, generally without even realizing that the majority of their food contains GM products.

Opposition in nations of the European Union resulted in a de facto moratorium on GM foods from 1998 to 2003, blocking the import of hundreds of millions of dollars in U.S. agricultural products. This prompted the United States to bring a successful case before the World Trade Organization (• p. 71), complaining that Europe's resistance was hindering free trade. Europeans now demand that GM foods be labeled and criticize the United States for not joining 100 other nations in signing the Cartagena Protocol on Biosafety, a treaty that lays out guidelines for open information about exported crops.

Transnational spats between Europe and the United States will surely affect the future direction of agriculture, but the world's developing nations could exert the most influence in the end. Recent decisions by the governments of India and Brazil to approve GM crops (following long and divisive debate) are already adding greatly to the world's transgenic agriculture. Moreover, China is aggressively expanding its use of transgenic crops.

A counterexample is Zambia, one of several African nations that refused U.S. food aid meant to relieve starvation during a drought in late 2002. The governments of these nations worried that their farmers would plant some of the GM corn seed meant to be eaten and that GM corn would thereby establish itself in their countries. They viewed this as undesirable because African economies depend on exporting food to Europe, which has put severe restrictions on GM food.

In the end, Zambia's neighbors accepted the grain after it had been milled (so none could be planted), but Zambia held out. Citing health and environmental risks,

FIGURE 10.15 Saskatchewan farmer Percy Schmeiser was accused by the Monsanto Company of planting its patented Roundup Ready canola in his field without a contract with the company. Schmeiser said his non-GM plants were contaminated with Monsanto's transgenes from neighboring farms. Monsanto won the David-and-Goliath case in Canada's Supreme Court, but Schmeiser became a hero to small farmers and anti-GM food activists worldwide.

uncertain science, and the precautionary principle, the Zambian government declined the aid, despite the fact that 2–3 million of its people were at risk of starvation. Intense debate followed within the country and around the world. Eventually the United Nations delivered non-GM grain, and in April 2003 the Zambian government announced a plan to coordinate a comprehensive long-term policy on GM foods.

The Zambian experience demonstrates some of the ethical, economic, and political dilemmas modern nations face. The corporate manufacturers of GM crops naturally aim to maximize their profits, but they also aim to develop products that can boost yields, increase food security, and reduce hunger. Although industry, activists, policymakers, and scientists all agree that hunger and malnutrition are problems and that agriculture should be made environmentally safer, they often disagree about the solutions to these dilemmas and the risks that each proposed solution presents.

# Preserving Crop Diversity

As the excitement over Quist and Chapela's findings in Oaxaca demonstrate, one concern many people harbor about transgenic crops is that transgenes might move, by pollination, into local native races of crop plants.

## Crop diversity provides insurance against failure

Preserving the integrity of native variants gives us a bulwark against commercial crop failure. Monocultures essentially place all our eggs in one basket, such that any single catastrophic cause could potentially wipe out entire crops. Although modern industrial agriculture relies on a small number of plant types, its foundation lies in places like Oaxaca that remain important repositories of crop biodiversity. The wild relatives of crop plants and their domesticated varieties, or cultivars, contain genes that, through conventional crossbreeding or genetic engineering, might confer resistance to disease, pests, inbreeding, or other unforeseen challenges.

Because accidental interbreeding can decrease the diversity of local variants, many scientists argue that we need to protect areas like Oaxaca. For this reason, the Mexican government helped create the Sierra de Manantlan Biosphere Reserve around an area harboring the localized plant thought to be the direct ancestor of maize. For this reason, too, it imposed a national moratorium in 1998 on the planting of transgenic corn (although that ban was lifted in 2005).

We have lost a great deal of genetic diversity in our crop plants already. The number of wheat varieties in China is estimated to have dropped from 10,000 in 1949 to 1,000 by the 1970s. Mexico's famed maize varieties now number only 30% of what existed in the 1930s. In the United States, many fruit and vegetable crops have decreased in diversity by 90% in less than a century. A primary cause of this loss of diversity is that market forces have discouraged diversity in the appearance of fruits and vegetables. Commercial food processors prefer items to be similar in size and shape, for convenience. Consumers, for their part, have shown preferences for uniform, standardized food products over the years. Now that local and organic agriculture (• pp. 285–289) is growing in affluent societies, however, consumer preferences for diversity are increasing.

## Seed banks are living museums

Protecting areas with high crop diversity is one way to preserve genetic assets for our agricultural systems. Another is to collect and store seeds from crop varieties and periodically plant and harvest them to maintain a diversity of cultivars. This is the work of **seed banks,** institutions that preserve seed types as a kind of living museum of genetic diversity (**Figure 10.16**). These facilities keep seed samples in cold, dry conditions to encourage long-term viability, and they grow them periodically to renew the stocks. Most funding for these facilities is not adequate, however, so it is questionable how many of the seeds will actually be preserved.

(a) Traditional food plants of the Desert Southwest

(b) Pollination by hand

FIGURE 10.16 Seed banks preserve genetic diversity of traditional crop plants. Native Seeds/SEARCH of Tucson, Arizona, preserves seeds of food plants important in traditional diets of Native Americans of Arizona, New Mexico, and northwestern Mexico. Beans, chiles, squashes, gourds, maize, cotton, and lentils are all in its collections, as well as lesser-known plants such as amaranth, lemon basil, and devil's claw (**a**). Traditional foods such as mesquite flour, prickly pear pads, chia seeds, tepary beans, and cholla cactus buds help fight diabetes, which Native Americans frequently suffer after adopting a Western diet. At the farm where seeds are grown, care is taken to pollinate varieties by hand (**b**) to protect their genetic distinctiveness.

The Royal Botanic Garden's Millennium Seed Bank in Britain holds over 1 billion seeds and aims to bank seed from 10% of the world's plants by 2010. In Arctic Norway, construction has begun on a "doomsday vault" seed bank intended to hold seeds from around the world as a safeguard against global agricultural calamity. Other major efforts include the U.S. National Seed Storage Laboratory at Colorado State University, Seed Savers Exchange in Iowa, and the Wheat and Maize Improvement Center (CIMMYT) in Mexico.

# Feedlot Agriculture: Livestock and Poultry

Food from cropland agriculture makes up a large portion of the human diet, but most of us also eat animal products. People don't *need* to eat meat or other animal products to live full, active, healthy lives, but for many it is difficult to obtain a balanced diet without incorporating animal products. Choosing to eat animal products has significant environmental, social, agricultural, and economic impacts.

## Consumption of animal products is growing

As wealth and global commerce increase, so does our consumption of meat, milk, eggs, and other animal products (**Figure 10.17**). The world population of domesticated animals raised for food rose from 7.3 billion animals to 20.6 billion animals between 1961 and 2000. Most of these animals are chickens. Global meat production has increased fivefold since 1950, and per capita meat consumption has nearly doubled.

## High consumption has led to feedlot agriculture

In traditional agriculture, livestock were kept by farming families near their homes or were grazed on open grasslands by nomadic herders or sedentary ranchers. These traditions have survived, but the advent of industrial agriculture has brought a new method. **Feedlots,** also known as *factory farms* or *concentrated animal feeding operations (CAFOs)*, are essentially huge warehouses or pens designed to deliver energy-rich food to animals living at extremely

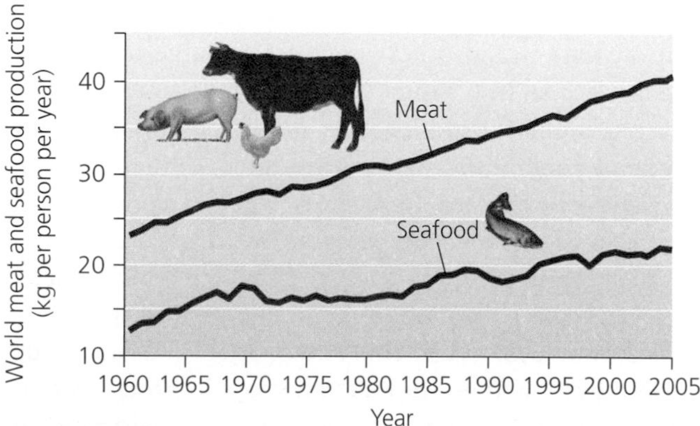


FIGURE 10.17 Per capita consumption of meat from farm animals has risen steadily worldwide, as has per capita consumption of seafood (marine and freshwater, harvested and farmed). Data from U.N. Food and Agriculture Organization.

FIGURE 10.18 These chickens at a Pennsylvania factory farm are housed several to a cage and have been "debeaked," the tips of their beaks cut off to prevent them from pecking one another. The hens cannot leave the cages and essentially spend their lives eating, defecating, and laying eggs, which roll down slanted floors to collection trays. The largest U.S. chicken farms house hundreds of thousands of individuals.

high densities (**Figure 10.18**). Today over half of the world's pork and poultry come from feedlots, as does much of its beef.

Feedlot operations allow for greater production of food and are probably necessary for a country with a level of meat consumption like that of the United States. Feedlots have one overarching benefit for environmental quality: Taking cattle, sheep, goats, and other livestock off the land and concentrating them in feedlots reduces the impact they would otherwise exert on large portions of the landscape. In Chapter 9 (• pp. 254–257) we saw how overgrazing has degraded soils and vegetation over hundreds of millions of hectares of land. Animals that are densely concentrated in feedlots will not contribute to soil degradation through overgrazing.

However, environmental advocates have criticized feedlots for their contributions to water and air pollution. Waste from feedlots can emit strong odors and can pollute surface water and groundwater, because livestock produce prodigious amounts of feces and urine. One dairy cow can produce about 20,400 kg (44,975 lb) of waste in a single year. Greeley, Colorado, is home to North America's largest meatpacking plant and two adjacent feedlots. Each feedlot has room for 100,000 cattle that are fed surplus grain and injected with anabolic steroids to stimulate growth. During its stay at the feedlot, a typical steer will eat 1,360 kg (3,000 lb) of grain, gain 180 kg (400 lb) in body weight, and generate 23 kg (50 lb) of manure each day. The amount of manure that 200,000 such animals generate exceeds the amount of waste produced by all the human residents of Atlanta, St. Louis, Boston, and Denver combined.

Poor waste containment practices at feedlots in North Carolina, Maryland, and other states have been linked to outbreaks of disease, including virulent strains of *Pfiesteria*, a microbe that poisons fish. The crowded and dirty conditions under which animals are often kept necessitates heavy use of antibiotics to control disease. These chemicals can be transferred up the food chain, and their overuse can cause microbes to evolve resistance to them.

Feedlot impacts can be minimized when properly managed, and both the EPA and the states regulate U.S. feedlots. Most feedlot manure is applied to farm fields as fertilizer, reducing the need for chemical fertilizers. Manure in liquid form can be injected into the ground where plants need it, and farmers can conduct tests to determine amounts that are appropriate to apply.

### Weighing THE Issues | Feedlots and Animal Rights

Animal rights activists decry factory farming because they say it mistreats animals. Chickens, pigs, and cattle are kept crowded together in small pens their entire lives, fattened up, and slaughtered. Chickens are often "debeaked." Do you think animal rights concerns should be given weight as we determine how best to raise our food? Do you think these are as important as the environmental issues? Are conditions at feedlots a good reason for being vegetarian?

## Our food choices are also energy choices

What we choose to eat has ramifications for how we use energy and the land that supports agriculture. Recall our discussions of thermodynamics, trophic levels, and pyramids of energy (• pp. 148–150). Every time energy moves from one trophic level to the next, as much as 90% is lost. For example, if we feed grain to a cow and then eat beef from the cow, we lose most of the grain's energy to the cow's metabolism. Energy is used up as the cow converts the grain to tissue as it grows, and as the cow uses its tissues and respires on a daily basis to maintain itself. For this reason, eating meat is far less energy-efficient than relying on a vegetarian diet. The lower in the food chain from which we take our food sources, the greater the proportion of the sun's energy we put to use as food, and the more people Earth can support.

Some animals convert grain feed into milk, eggs, or meat more efficiently than others (**Figure 10.19**). Scientists have calculated relative energy-conversion efficiencies for different types of animals. Such energy efficiencies have ramifications for land use because land and water are required to raise food for the animals, and some animals require more than others. **Figure 10.20** shows the area of land and weight of water required to produce 1 kg (2.2 lb) of food protein for milk, eggs, chicken, pork, and beef. Producing eggs and chicken meat requires the least space and water, whereas producing beef requires the most. Such differences make clear that when we choose what to eat, we are also indirectly choosing how to make use of resources such as land and water.

In 1900 we fed about 10% of our global grain production to animals. In 1950 this number reached 20%, and by the beginning of the 21st century we were feeding 45% of our global grain production to animals. Although much of the grain fed to animals is not of a quality suitable for human consumption, the resources required to grow it could have instead been applied toward growing food for people. One partial solution is to feed livestock crop residues, plant matter such as stems and stalks that we would not consume anyway, and this is increasingly being done.

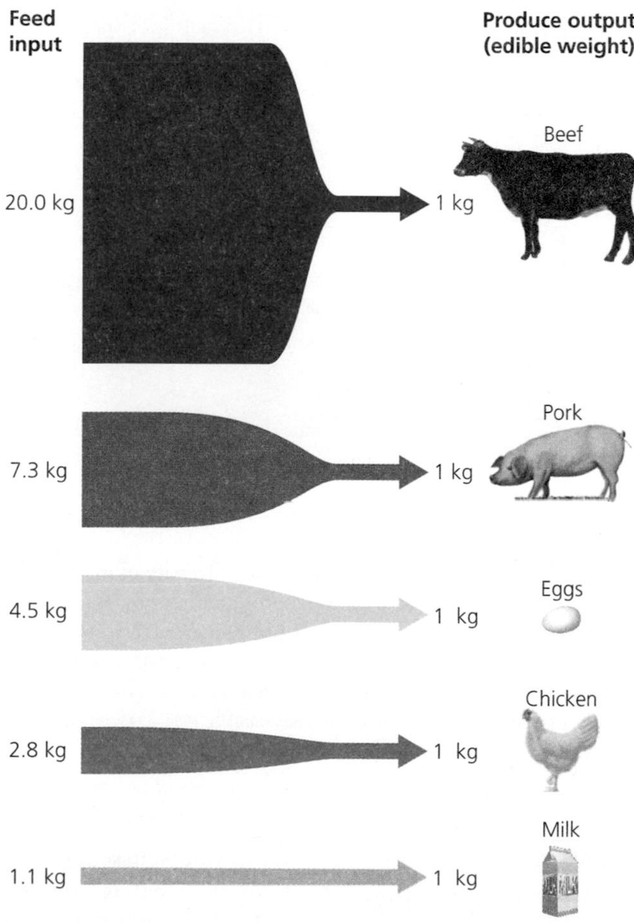


FIGURE 10.19 Different animal food products require different amounts of input of animal feed. Chickens must be fed 2.8 kg of feed for each 1 kg of resulting chicken meat, for instance, whereas 20 kg of feed must be provided to cattle to produce 1 kg of beef. Go to GRAPHIt! at www.aw-bc.com/withgott or on the student CD-ROM. Data from Smil, V. 2001. *Feeding the world: A challenge for the twenty-first century.* Cambridge, MA: MIT Press.

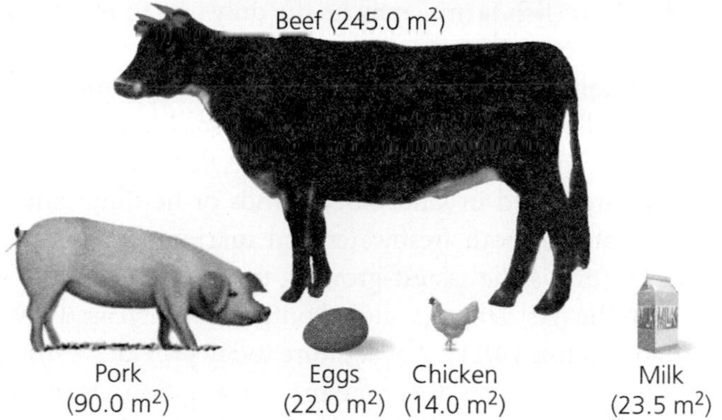


**(a) Land required to produce 1 kg of protein**

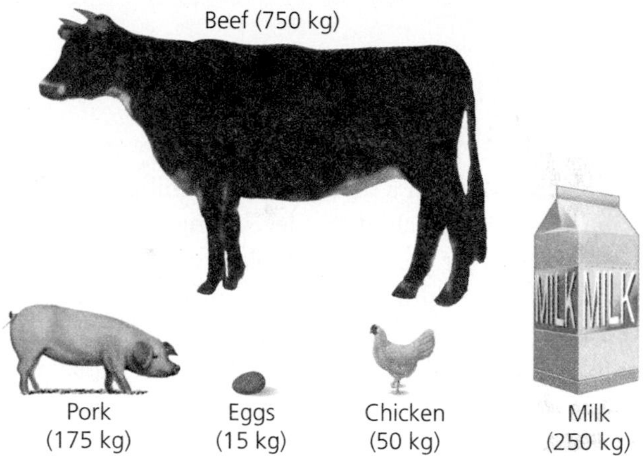


**(b) Water required to produce 1 kg of protein**

FIGURE 10.20 Producing different types of animal products requires different amounts of land and water. Raising cattle for beef requires by far the most land and water of all animal products. Go to GRAPHIt! at www.aw-bc.com/withgott or on the student CD-ROM. Data from Smil, V. 2001. *Feeding the world: A challenge for the twenty-first century,* Cambridge, MA: MIT Press.

## Aquaculture

Besides plants grown as crops and animals raised on rangelands and in feedlots, we rely on aquatic organisms for food. Wild fish populations are plummeting throughout the world's oceans as increased demand and new technologies have led us to overharvest most marine fisheries (• pp. 457–464). This means that raising fish and shellfish on "fish farms" may be the only way to meet our growing demand for these foods.

We call the raising of aquatic organisms for food in controlled environments **aquaculture.** Many aquatic species are grown in open water in large, floating net-pens. Others are raised in land-based ponds or holding tanks. People pursue both freshwater and marine aquaculture. Aquaculture is the fastest-growing type of food production; in the past 20 years, global output has increased sevenfold (**Figure 10.21**). Aquaculture today provides a third of the world's fish for human consumption, is most widespread in Asia, and involves over 220 species. Some, such as carp, are grown for local consumption, whereas others, such as salmon and shrimp, are exported to affluent countries.

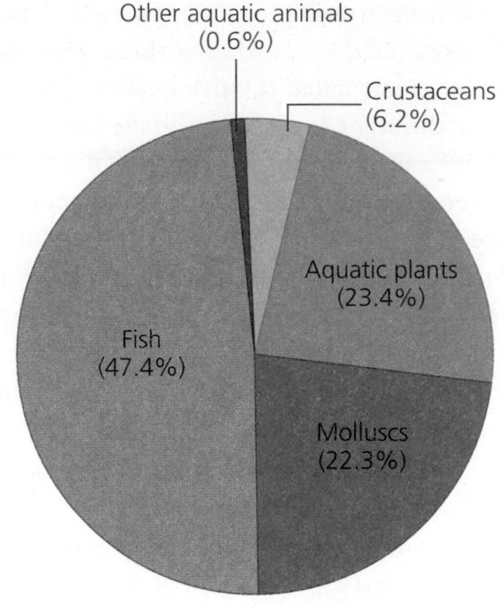


**(a) World aquaculture production by groups**

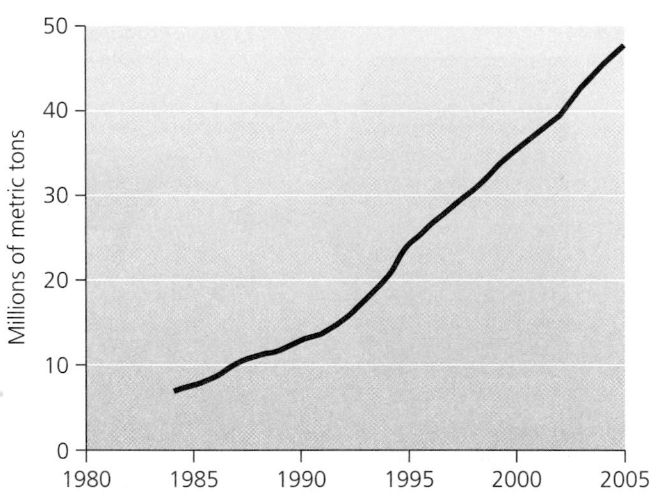


**(b) World aquaculture production**

**FIGURE 10.21** Aquaculture involves a wide diversity of marine and freshwater organisms **(a).** Global production of meat (all species other than plants) from aquaculture has risen steeply in the past two decades **(b).** Data in (a) from U.N. Food and Agriculture Organization. 2007. *The state of world fisheries and aquaculture, 2006;* and (b) U.N. Food and Agriculture Organization. *Aquaculture production statistics, 1984–1993* and *Fishery statistics: Aquaculture production.*

## Aquaculture brings a number of benefits

When conducted on a small scale by families or villages, as in China and much of the developing world, aquaculture helps ensure people a reliable protein source. Such small-scale aquaculture can be sustainable, and it is compatible with other activities. For instance, uneaten fish scraps make excellent fertilizers for crops. Aquaculture on larger scales can help improve a region's or nation's food security by increasing overall amounts of fish available.

Aquaculture on any scale also helps reduce fishing pressure on overharvested and declining wild stocks. Reducing fishing pressure also lessens the *by-catch* (• p. 459; the unintended catch of nontarget organisms) that results from commercial fishing. Furthermore, aquaculture relies far less on fossil fuels than do fishing vessels and provides a safer work environment than does commercial fishing. Fish farming can also be remarkably energy-efficient, producing as much as 10 times more fish per unit area than is harvested from waters of the continental shelf and up to 1,000 times as much as is harvested from the open ocean.

## Aquaculture has negative impacts

Along with its benefits, aquaculture has disadvantages. Dense concentrations of farmed animals can increase the incidence of disease, which reduces food security, necessitates antibiotic treatment, and results in additional expense. A virus outbreak wiped out half a billion dollars in shrimp in Ecuador in 1999, for instance. Aquaculture can also produce prodigious amounts of waste, from the farmed organisms and from the feed that goes uneaten and decomposes in the water. Farmed fish often are fed grain, and as we have discussed, growing grain to feed animals that we then eat is energy-inefficient. In other cases, farmed fish are fed fish meal made from wild ocean fish such as herring and anchovies, whose harvest may place additional stress on wild fish populations.

If farmed aquatic organisms escape into ecosystems where they are not native (as several carp species have

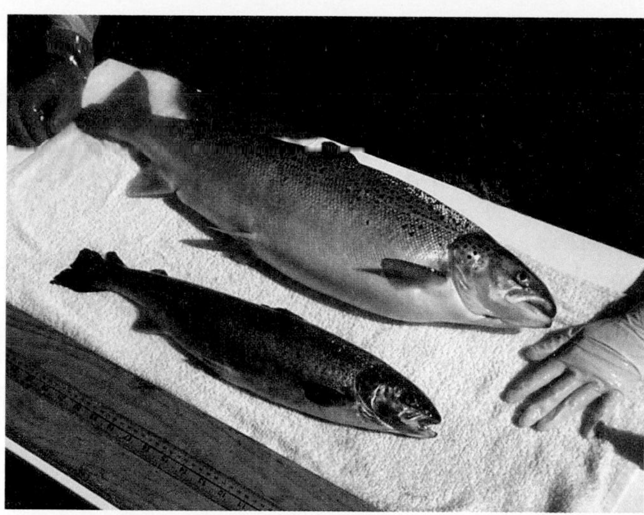

FIGURE 10.22 Efforts to genetically modify important food fish have resulted in the creation of transgenic salmon (top), which can be considerably larger than wild salmon of the same species.

done in U.S. waters), they may spread disease to native stocks or may outcompete native organisms for food or habitat. These possibilities raise special concern when the farmed animals are genetically modified. Genetic engineering of Atlantic and Pacific salmon has produced transgenic fish that grow to 5–50 times the normal size for their species (**Figure 10.22**). GM fish such as these may outcompete their wild cousins while also spreading disease to them. They may also interbreed with native and hatchery-raised fish and weaken already troubled stocks. Researchers have concluded that under certain circumstances, escaped transgenic salmon may increase the extinction risk for native populations of their species, in part because the larger male fish (such as those carrying a gene for rapid and excessive growth) have better odds of mating successfully.

# Sustainable Agriculture

Industrialized agriculture involves many adverse environmental impacts, from the degradation of soils (Chapter 9) to reliance on fossil fuels (Chapter 19) to problems arising from pesticide use, genetic modification, and intensive feedlot and aquaculture operations. Although aspects of intensive commercial agriculture help alleviate certain environmental pressures, they often exacerbate others. Industrial agriculture in some form seems necessary to feed our planet's 6.7 billion people, but many feel we will be better off in the long run by practicing less-intensive methods of raising animals and crops.

Farmers and researchers have made great advances toward sustainable agriculture in recent years. **Sustainable agriculture** is agriculture that does not deplete soils faster than they form. It is farming and ranching that does not reduce the amount of healthy soil, clean water, and genetic diversity essential to long-term crop and livestock production. Simply put, it is agriculture that can be practiced in the same way far into the future. For example, the no-till agriculture practiced in southern Brazil that we examined in Chapter 9 appears to fit the notion of sustainable agriculture, as does the traditional Chinese practice of carp aquaculture in small ponds.

Sustainable agriculture is closely related to *low-input agriculture*, agriculture that uses smaller amounts of pesticides, fertilizers, growth hormones, water, and fossil fuel energy than are currently used in industrial agriculture. Food-growing practices that use no synthetic fertilizers, insecticides, fungicides, or herbicides—but instead rely on biological approaches such as composting (• p. 640) and biocontrol—are termed **organic agriculture.**

## Organic agriculture is on the increase

Citizens, government officials, farmers, and agricultural industry representatives have debated the meaning of the word *organic* for many years. In 1990, Congress passed the Organic Food Production Act, establishing national standards for organic products and facilitating the sale of organic food. As required by this law, the USDA in 2000 issued criteria by which crops and livestock could be officially certified as organic (**Table 10.3**). California legislation passed the same year established stricter state guidelines for labeling foods organic, and Washington and Texas followed with still stricter guidelines. Today 17 U.S. states have laws spelling out standards for organic products.

### Weighing THE Issues | Do You Want Your Food Labeled?

The USDA issues labels to certify that produce claiming to be organic has met the government's organic standards. Critics of genetically modified food want GM products to be labeled as well. Given that 70% of processed food now contains GM ingredients, labeling would add costs. But the European Union currently labels such foods. Do you want your food to be labeled? Would you choose among foods based on whether they are organic or genetically modified? Do you feel your food choices have environmental impacts? Is purchasing power an effective way to make your views heard?

Long viewed as a small niche market, the market for organic foods is on the increase. Although it accounts for just over 1% of food expenditures in the United States and

**TABLE 10.3 USDA Criteria for Certifying Crops and Livestock as Organic**

**For crops to be considered organic . . .**

- The land where they are grown must be free of prohibited substances for at least 3 years.
- They must not be genetically engineered.
- They must not be treated with ionizing radiation (a means of eliminating bacteria in packaged food).
- The use of sewage sludge is prohibited.
- They must be produced without fertilizer containing synthetic ingredients.
- Fertility and crop nutrients must be achieved through crop rotations, cover crops, animal and crop wastes, or synthetic materials approved by the National Organic Standards Board.
- Use of most conventional pesticides is prohibited.
- Use of organic seeds and other planting stock is preferred.
- Crop pests, weeds, and diseases should be controlled through physical, mechanical, and biological management practices or with synthetic substances approved by the National Organic Standards Board.

**For livestock to be considered organic . . .**

- Mammals must be raised under organic management from the last third of gestation; poultry, no later than the second day of life.
- Producers must feed livestock 100% organic agricultural feed; however, vitamin and mineral supplements are allowed.
- Producers of existing dairy herds must provide 80% organically produced feed for 9 months, followed by 3 months of 100% organically produced feed.
- Use of hormones or antibiotics is prohibited, although vaccines are permitted.
- Animals must have access to the outdoors.

Data from the National Organic Program. 2002. *Organic production and handling standards.* Washington, DC: U.S. Department of Agriculture.

Canada, sales of organic products increased 20% annually from 1989 to 2005, when global sales of organic products reached $35 billion. In 2003, 3–5% (close to $13 billion) of Europe's food market was organic. Although 3–5% may not seem like much, organic agriculture expanded by a factor of 40 between 1985 and 2003 in Europe, representing an annual growth rate of nearly 30%.

Production is increasing along with demand (**Figure 10.23**). Although organic agriculture takes up less than 1% of cultivated land worldwide (24 million ha [59 million acres] in 2004), this area is rapidly expanding. In the United States and Canada, the amount of land used in organic agriculture has recently increased 10–35% each year. As of 2005 in the United States, 690,000 ha (1.7 million acres) of cropland and 930,000 ha (2.3 million acres) of pasture and rangeland were in organic production. Numbers of organic livestock rose 17-fold from 1992 to 2005, and numbers of organic poultry rose 224-fold (although these animals still comprise less than 1% of the national total). Today farmers in all 50 U.S. states and more than 130 nations practice organic farming commercially to some extent.

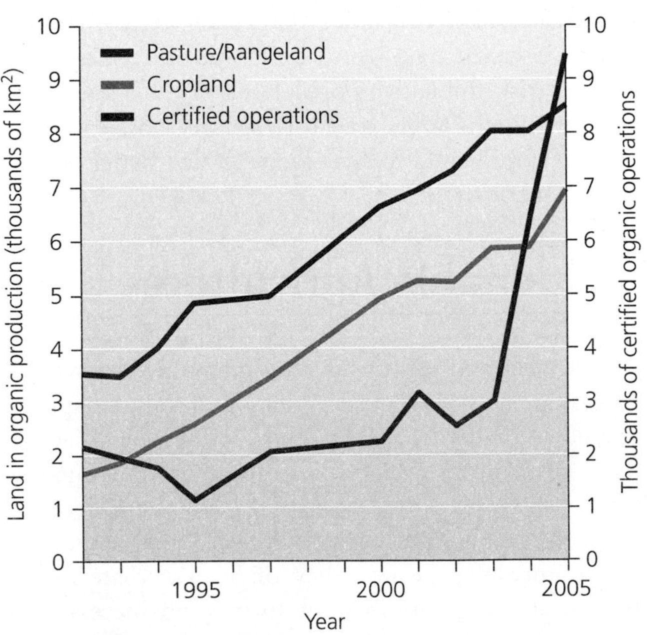


FIGURE 10.23 All aspects of organic agriculture have grown rapidly in the United States since the mid-1990s. Acreage devoted to organic crops and livestock has each quadrupled, and the number of certified operations has doubled.

For farmers, organic farming can bring the benefits of lower input costs, enhanced income from higher-value produce, and reduced chemical pollution and soil degradation (see "The Science behind the Story," • p. 288). Although in many cases more insects attack organic crops because of the lack of chemical pesticides, biocontrol methods (• pp. 270–271) can often keep pests in check. Moreover, the lack of synthetic chemicals maintains soil quality and encourages helpful pollinating insects. Obstacles to organic farming for farmers include risks and costs of shifting to new farming methods and less market infrastructure for selling their products.

Among consumers, many favor organic products because of concern that consuming produce grown with the use of pesticides may pose health risks. Consumers also buy organic produce out of a desire to improve environmental quality. Many shoppers, however, will not buy organic produce because it usually is more expensive and often looks less uniform and aesthetically appealing in the supermarket aisle compared with the standard produce of high-input agriculture. Overall, enough consumers are willing to pay more for organic meat, fruit, and vegetables that businesses are making such foods more widely available. In early 2000, one of Britain's largest supermarket brands announced that it would sell only organic food—and that the new organic products would cost their customers no more than had nonorganic products. Many textile makers (among them The Gap, Levi's, and Patagonia) are increasing their use of organic cotton.

Government initiatives have also spurred the growth of organic farming. Several million hectares of land have converted to organic farming in Europe since the European Union adopted a policy in 1993 to support farmers financially during conversion. The United States offers no such support, which may be why the U.S. organic market lags behind that of Europe. Such support is important, because conversion often means a temporary loss in income for farmers. More and more studies, however, suggest that reduced inputs and higher market prices can, in the long run, make organic farming more profitable for the farmer than conventional methods.

## Organic agriculture can succeed in cities

One surprising place that organic agriculture is making inroads is within our cities. As we will see in Chapter 13 (• pp. 373–374), many urban areas now offer community gardens in which residents can grow small plots of fruits and vegetables.

Organic urban agriculture is deeply entrenched in the nation of Cuba, though for very different reasons. Long a close ally of the former Soviet Union, Cuba lost 75% of its total imports, 53% of its oil imports, and 80% of its fertilizer and pesticide imports in 1989 as the USSR was breaking up. Faced with such losses and with a U.S. trade embargo, Cuba's farmers had little choice but to "go organic." Because far less oil was available to fuel Cuba's transportation system, farmers began growing food closer to cities and even within them. By 1998 the Cuban government's Urban Agriculture Department had helped develop more than 8,000 gardens in the capital city of Havana (**Figure 10.24**). Over 30,000 people worked in these gardens, which covered 30% of the city's available land. Cuba also began using oxen instead of tractors and relying on biocontrol and integrated pest management.

Cuba's agriculture requires more human labor than do the industrialized farms of developed nations, and Cuba's policies are guided by top-down control in a rigid state socialist system. Nevertheless, Cuba's low-input farming has produced some positive achievements, such as complete

**FIGURE 10.24** Organic gardening takes place within the city limits of Havana, Cuba, out of necessity. With little money to pay for the large amounts of fertilizers and pesticides required for industrialized agriculture, Cubans get much of their food from local urban agriculture without these inputs.

THE SCIENCE BEHIND THE STORY

# Organic Farming

*Dr. Mark A. Liebig, USDA Agricultural Research Service*

Fields of wheat and potatoes, some grown organically and some cultivated with the synthetic chemicals favored by industrialized agriculture, stand side by side on an experimental farm in Switzerland. Although conventionally farmed fields receive up to 50% more fertilizer, they produce only 20% more food than organically farmed fields. How are organic fields able to produce decent yields without synthetic agricultural chemicals? The answer, scientists have found, lies in the soil.

Researchers have long documented that organic farming is better for the environment, but farmers have long found that synthetic chemicals help them fight pests and increase crop yields. Although organic farming might put fewer synthetic chemicals into the air and water, some have contended that it will never contribute much to the world's food supply.

To address concerns about crop yields, Swiss researchers at the Research Institute of Organic Agriculture have compared organic and conventional fields since 1978, using a series of growing areas that feature four different farming systems.

One group of plots mirrors conventional farms, in which large amounts of chemical pesticides, herbicides, and fertilizer are applied to soil and plants. Another set of fields is treated with a mixed approach of conventional and organic practices, including chemical additives, synthetic sprays, and livestock manure as fertilizer. Organic plots use only manure, mechanical weeding machines, and plant extracts to control pests. A fourth group of plots follows similar organic practices but also uses extra natural boosts, such as adding herbal extracts to compost. The two organic plots receive about 35–50% less fertilizer than the conventional fields and 97% fewer pesticides.

Over more than 20 years of monitoring, the organic fields yielded 80% of what the conventional fields produced, researchers reported in the journal *Science* in 2002. Organic crops of winter wheat yielded about 90% of the conventional wheat crop yield. Organic potato crops averaged about 68% of the conventional potato yields. The comparatively low potato yield was due to nutrient deficiency and a fungus-caused potato blight.

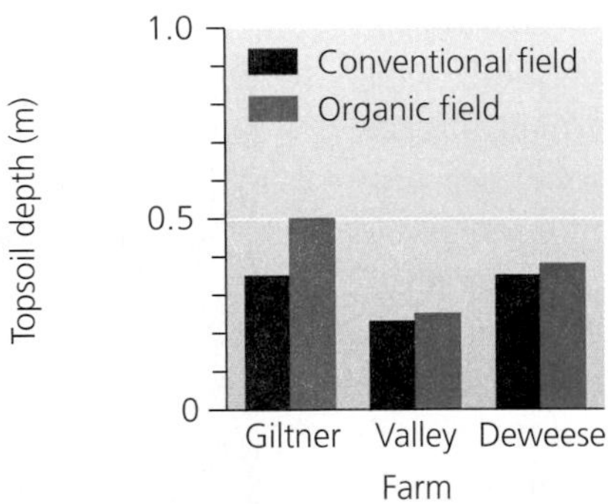


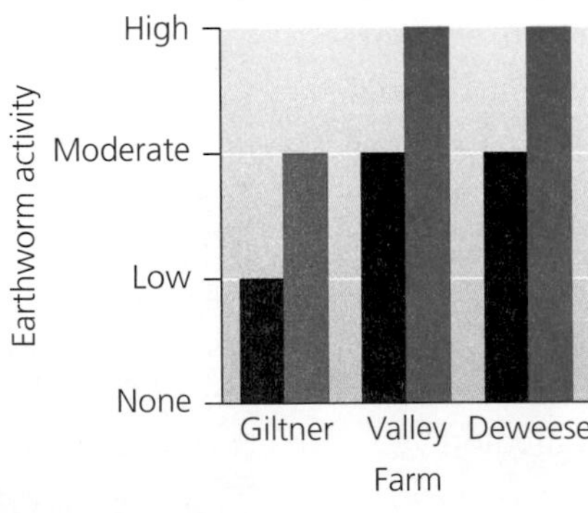


**Researchers in Nebraska and North Dakota demonstrated that organic farming at three sites increased topsoil depth moderately and increased activity of earthworms dramatically.** Data from Liebig, M. A., and J. W. Doran. 1999. Impact of organic production practices on soil quality indicators. *Journal of Environmental Quality* 28: 1601–1609.

Scientists have hypothesized that organic farms keep their yields high because organic agricultural practices better conserve soil quality, keeping soil fertile over the long term. In one study that appears to back this hypothesis, U.S. researchers compared five pairs of organic and conventional farms in the mid-1990s in North Dakota and Nebraska (see the figure). After extracting soil samples from each of the farms, the researchers analyzed the soil for water-holding ability, microbial biomass, and nutrients such as carbon and nitrogen. Organic farming, they found, produced soils that contained more naturally occurring nutrients, held greater quantities of water, and had higher concentrations of microbial life than conventionally farmed soil. Organic farms also had deeper nutrient-rich topsoil and greater earthworm activity—all signs of soil healthy enough to produce impressive crops without help from synthetic chemicals.

Scientists at the Swiss research project have found similar signs of soil fertility on their organic research plots. Increasingly, researchers are concluding that organically managed soil supports a more diverse range of microbial and plant life, which translates into increased biodiversity, self-sustaining fields, and strong crop yields. Such findings may be pivotal as large growers increasingly debate whether to turn to organic farming.

FIGURE 10.25 Farmers' markets, such as this one in San Francisco, have become more widespread as consumers have rediscovered the benefits of buying fresh, locally grown produce.

control of the sweet-potato borer (a significant pest insect) and record yields for 10 crops in 1996–1997. Although Cuba's move toward organic agriculture was involuntary, its achievements illustrate how other nations might, by choice, begin to farm in ways that rely less on enormous inputs of fossil fuels and synthetic chemicals.

## Locally supported agriculture is growing

In developed nations, increasing numbers of farmers and consumers are supporting local small-scale agriculture. Farmers' markets (**Figure 10.25**) are becoming more numerous throughout North America as consumers rediscover the joys of fresh, locally grown produce. The average food product sold in U.S. supermarkets travels at least 2,300 km (1,400 mi) between the farm and the shelf, and supermarket produce is often chemically treated to preserve freshness and color. At farmers' markets, consumers can buy fresh produce in season from local farmers and often have a wide choice of organic items and unique local varieties.

Some consumers are even partnering with local farmers in a phenomenon called *community-supported agriculture.* In this practice, consumers pay farmers in advance for a share of their yield, usually in the form of weekly deliveries of produce. Consumers get fresh seasonal produce, while farmers get a guaranteed income stream up front to invest in their crops—an alternative to taking out loans and being at the mercy of the weather. As of 2005, about 1,700 U.S. farms were supplying 340,000 families per week in community-supported agriculture programs.

# Conclusion

Many of the intensive commercial agricultural practices we have discussed exert substantial negative environmental impacts. At the same time, it is important to realize that many aspects of industrialized agriculture have helped relieve certain pressures on land or resources. Whether Earth's natural systems would be under more pressure from 6.7 billion people practicing traditional agriculture or from 6.7 billion people living under our modern industrialized model is a very complicated question.

What is certain is that if our planet is to support 9 billion people by mid-century without further degradation of the soil, water, pollinators, and other resources and ecosystem services that support our food production, we must find ways to shift to sustainable agriculture. Approaches such as biological pest control, organic agriculture, pollinator conservation, preservation of native crop diversity, sustainable aquaculture, and likely some degree of careful and responsible genetic modification of food may all be parts of the game plan we will need to set in motion to work together toward a sustainable future.

## REVIEWING OBJECTIVES

**You should now be able to:**

**Explain the challenge of feeding a growing human population**

- Our food production has outpaced our population growth, yet 850 million people still go hungry. (pp. 264–266)

**Identify the goals, methods, and environmental impacts of the "green revolution"**

- The green revolution aimed to increase agricultural productivity per unit area of land in developing nations without further degrading natural lands. (p. 266)
- Scientists used selective breeding to develop crop strains that grew quickly, were more nutritious, or were resistant to disease or drought. (pp. 266–267)
- The expanded use of fossil fuels and chemical fertilizers and pesticides has increased pollution. However, the increased efficiency of production has reduced the amount of natural land converted for farming. (pp. 267–268)

**Categorize the strategies of pest management**

- Most "pests" and "weeds" are killed with synthetic chemicals that can pollute the environment and pose health hazards. (pp. 268–269)
- Pests can evolve resistance to chemical pesticides, forcing us to design more toxic poisons. (p. 269)
- We employ natural enemies of pests against them in the practice of biological control. (pp. 270–271)
- Integrated pest management combines various techniques and minimizes the use of synthetic chemicals. (p. 271)

**Discuss the importance of pollination**

- Insects and other organisms are essential for the reproduction of many crop plants. (pp. 271–272)
- Conservation of pollinating insects is vitally important to our food security. (p. 272)

**Describe the science behind genetically modified food**

- Genetic modification uses the technology of recombinant DNA. Genes for desirable traits are moved from one type of organism to another. (p. 273)
- Modification through genetic engineering is both like and unlike traditional selective breeding. (pp. 274–275)
- GM crops may have ecological impacts, including the spread of transgenes, an increase or decrease in chemical pollution, and indirect impacts on biodiversity. More research is needed to determine how widespread or severe these impacts may be. (pp. 276–277)

**Evaluate the debate over genetically modified food**

- Anxiety over (yet-unproven) health impacts inspires opposition to GM foods. (pp. 274–275, 277)
- Many people have ethical qualms about altering food through genetic engineering. (p. 277)
- Opponents of GM foods view multinational biotechnology corporations as a threat to the independence of small farmers. (pp. 277–280)

**Ascertain approaches for preserving crop diversity**

- Protecting diversity of native crop varieties provides insurance against failure of major crops. (p. 280)
- Seed banks preserve rare and local varieties of seed, acting as storehouses for genetic diversity. (pp. 280–281)

**Assess feedlot agriculture for livestock and poultry**

- Increased consumption of animal products has driven the development of high-density feedlots. (pp. 281–282)
- Feedlots create tremendous amounts of waste and other environmental impacts, but they also relieve pressure on lands that could otherwise be overgrazed. (p. 282)
- Eating animal products consumes more resources than eating plant products. (pp. 282–283)

**Weigh approaches in aquaculture**

- Aquaculture provides economic benefits and food security, relieves pressures on wild fish stocks, and can be sustainable. (pp. 283–284)
- Aquaculture also creates pollution, habitat loss, and other environmental impacts. (pp. 284–285)

**Evaluate sustainable agriculture**

- Organic agriculture exerts fewer environmental impacts than industrial agriculture. It is a small part of the market but is growing rapidly. (pp. 285–289)
- Locally supported agriculture, as shown by farmers' markets and community-supported agriculture, is also growing. (p. 289)

## TESTING YOUR COMPREHENSION

1. What kinds of techniques have people employed to increase agricultural food production? How did agricultural scientist Norman Borlaug help inaugurate the green revolution?
2. Explain how pesticide resistance occurs.
3. Explain the concept of biocontrol. List several components of a system of integrated pest management (IPM).

4. About how many and what types of cultivated plants are known to rely on insects for pollination? Why is it important to preserve the biodiversity of native pollinators?
5. What is recombinant DNA? How is a transgenic organism created? How is genetic engineering different from traditional agricultural breeding? How is it similar?
6. Describe several reasons why many people support the development of genetically modified organisms, and name several uses of such organisms that have been developed so far.
7. Describe the scientific concerns of opponents of GM crops. Describe some of their other concerns.
8. Name several positive and negative environmental effects of feedlot operations. Why is beef an inefficient food from the perspective of energy consumption?
9. What are some economic benefits of aquaculture? What are some negative environmental impacts?
10. What are the objectives of sustainable agriculture? What factors are causing organic agriculture to expand?

## SEEKING SOLUTIONS

1. Assess several ways in which high-input agriculture can be beneficial for the environment and several ways in which it can be detrimental to the environment. Now suggest several ways in which we might modify industrial agriculture to lessen its environmental impact.
2. What factors make for an effective biological control strategy of pest management? What risks are involved in biocontrol? If you had to decide whether to use biocontrol against a particular pest, what questions would you want to have answered before you decide?
3. People who view GM foods as solutions to world hunger and pesticide overuse often want to speed their development and approval. Others adhere to the precautionary principle and want extensive testing for health and environmental safety. How much caution do you think is warranted before a new GM crop is introduced?
4. Cuba adopted low-input organic agriculture out of necessity. If the country were to become economically prosperous once more, do you think Cubans would maintain this form of agriculture, or do you think they would turn to intensive, high-input farming instead? What path do you think they should pursue, and why?
5. **THINK IT THROUGH** You are with the USDA, and it is your job to make the regulatory decision whether to allow the planting of a new genetically modified strain of cabbage that produces its own pesticide and has twice the vitamin content of regular cabbage. What questions would you ask of scientists before deciding whether to approve the new crop? What scientific data would you want to see, and how much would be enough? Would you also consult nonscientists or take ethical, economic, and social factors into consideration?
6. **THINK IT THROUGH** You are a farmer in the U.S. Midwest, own 1,000 acres, and intend to farm corn for a living. Would you choose to grow genetically modified corn? How would you manage pests and weeds? Would you grow corn for people, livestock, or both? Think of the various ways corn is grown, purchased, and valued in different places—such as the United States, Europe, Oaxaca, and Zambia—as you formulate your answer.

## INTERPRETING GRAPHS AND DATA

In the year 2000, over 80 million metric tons of nitrogen fertilizer were used in producing food for the world's 6 billion people. Food production, use of nitrogen fertilizers, and world population all had grown over the preceding 40 years, but at somewhat different rates. During this time, food production grew slightly faster than population while relatively little additional land was converted to agricultural use. Fertilizer use grew most rapidly.

1. Express the year 2003 values of the four graphed indices as percentages of the value of each index in 1961.
2. Calculate the ratio of the food production index to the nitrogen fertilizer use index in 1961 and in 2003. What does comparing these two ratios tell you about how the efficiency of nitrogen use in agriculture has changed? Is this an example of the principle of diminishing returns?

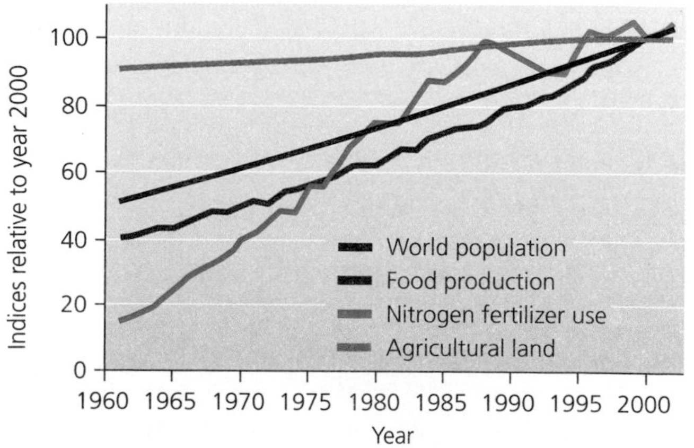


Global food production, nitrogen fertilizer use, human population, and land in agriculture, 1961–2003, relative to 2000 levels (2000 = 100). Data from U.N. Food and Agriculture Organization.

3. As world population has grown, so has the demand for food, yet we have devoted little additional land to food production. Calculate the ratio of the agricultural land index to the population index for 1961 and for 2003. What does comparing the ratios from these two years tell you about how the per capita demand on agricultural land has changed over time? To what factors would you attribute this change?

## CALCULATING ECOLOGICAL FOOTPRINTS

As food production became more industrialized during the 20th century, several trends emerged. One trend, documented in this chapter, was a loss in the number of varieties of crops grown. A second trend was the increasing amount of energy expended to store food and ship it to market. In the United States today, food travels an average of 1,400 miles from the field to your table. The price you pay for the food covers the cost of this long-distance transportation, which in 2004 was approximately one dollar per ton per mile. Assuming that the average person eats 2 pounds of food per day, calculate the food transportation costs for each category in the table below.

| Consumer | Daily cost | Annual cost |
|---|---|---|
| You | $1.40 | $511 |
| Your class | | |
| Your town | | |
| Your state | | |
| United States | | |

1. What specific challenges to environmental sustainability are imposed by a food production and distribution system that relies on long-range transportation to bring food to market?
2. A study by Pirog and Benjamin* noted that locally produced food traveled only 50 miles or so to market, thus saving 96% of the transportation costs. Locally grown foods may be fresher and cause less environmental impact as they are brought to market, but what are the disadvantages to you as a consumer in relying on local food production? Do you think the advantages outweigh those disadvantages?
3. What has happened to gasoline prices recently? How would future increases in the price of gas affect your answers to the preceding questions?

*Pirog, R., and A. Benjamin. 2003. *Checking the food odometer: Comparing food miles for local versus conventional produce sales to Iowa institutions.* Ames, IA: Leopold Center for Sustainable Agriculture, Iowa State University.

## Take It Further

Go to www.aw-bc.com/withgott or the student CD-ROM, where you'll find:

- Suggested answers to end-of-chapter questions
- Quizzes, animations, and flashcards to help you study
- *Research Navigator*™ database of credible and reliable sources to assist you with your research projects
- **GRAPHIt!** Tutorials to help you interpret graphs
- **INVESTIGATEIt!** Current news articles that link the topics you study to case studies from your region to around the world

CHAPTER

# 11 Biodiversity and Conservation Biology

The Sikhote-Alin Mountains meet the Pacific Ocean

## Upon completing this chapter, you will be able to:

- Characterize the scope of biodiversity on Earth
- Contrast background extinction rates with periods of mass extinction
- Evaluate the primary causes of biodiversity loss
- Specify the benefits of biodiversity
- Assess the science and practice of conservation biology
- Compare and contrast traditional and innovative biodiversity conservation efforts

A Siberian tiger in the Sikhote-Alin Mountains

## CENTRAL CASE

# Saving the Siberian Tiger

**"Future generations would be truly saddened that this century had so little foresight, so little compassion, such lack of generosity of spirit for the future that it would eliminate one of the most dramatic and beautiful animals this world has ever seen."**
—George Schaller, Wildlife Biologist, on the Tiger

**"If you kill a tiger, you can buy a motorbike."**
—Anonymous Poacher, on Selling Tiger Parts

Historically, tigers roamed widely across Asia from Turkey to northeast Russia to Indonesia. Within the past 200 years, however, people have driven the majestic striped cats from most of their historic range. Today, tigers are exceedingly rare and are creeping toward extinction.

Of the tigers that still survive, those of the subspecies known as the Siberian tiger are the largest cats in the world. Males can reach 3.7 m (12 ft) in length and weigh up to 360 kg (800 lb). Also named Amur tigers for the watershed they occupied along the Amur River, which divides Siberian Russia from Manchurian China, these cats now find their last refuge in the forests of the remote Sikhote-Alin Mountains of the Russian Far East.

For thousands of years the Siberian tiger coexisted with the region's native people and held a prominent place in their lore. These people referred to the tiger as "Old Man" or "Grandfather" and viewed it as a guardian of the mountains and forests. They rarely killed a tiger unless it had preyed on a person.

The Russians who moved into the region and exerted control in the early 20th century had no such cultural traditions. They hunted tigers for sport and hides, and some Russians reported killing as many as 10 tigers in a single hunt. Later, poachers began killing tigers to sell their body parts to China and other Asian countries, where they are used in traditional medicine and as aphrodisiacs. Meanwhile, road building, logging, and agriculture began to fragment tiger habitat and provide easy access for poachers. The tiger population dipped to perhaps 20–30 animals.

International conservation groups began to get involved, working with Russian biologists to try to save the dwindling tiger population. One such group was the Hornocker Wildlife Institute, now part of the Wildlife Conservation Society (WCS). In 1992 the group helped launch the Siberian Tiger Project, devoted to studying the tiger and its habitat. The team put together a plan to protect the tiger, began educating people regarding the tiger's importance and value, and worked closely with those who live in proximity to the big cats.

Today, WCS biologists track tigers with radio-collars, monitor their movements and health, determine causes of death when they die, and provide funding for local wildlife officials to deter and capture poachers.

Thanks to such efforts, today Siberian tigers in the wild number roughly 330–370, and about 1,500 more survive in zoos and captive breeding programs around the world. Poaching is still rampant, and the outlook for the species' survival looks challenging. But many people are trying to save these endangered animals. It is one of numerous efforts around the world today to stem the loss of our planet's priceless biological diversity.

## Our Planet of Life

Growing human population and resource consumption are putting ever-greater pressure on the flora and fauna of the planet, from tigers to tiger beetles. We are diminishing Earth's diversity of life, the very quality that makes our planet so special. In Chapter 5 (• pp. 115–116) we introduced the concept of **biological diversity,** or **biodiversity,** as the sum total of all organisms in an area, taking into account the diversity of species, their genes, their populations, and their communities. In this chapter we will probe further, examining current biodiversity trends and their relevance to our lives. We will then explore science-based solutions to biodiversity loss.

### Biodiversity encompasses several levels

Biodiversity is a concept as multifaceted as life itself, and different biologists employ different working definitions according to their own aims, interests, and values. Nonetheless, there is broad agreement that the concept applies across several major levels in the organization of life (**Figure 11.1**). The level that is easiest to visualize and most commonly used is species diversity.

**FIGURE 11.1** The concept of biodiversity encompasses several levels in the hierarchy of life. Species diversity (middle frame of the figure) refers to the number or variety of species. Genetic diversity (bottom frame) refers to variation in DNA composition among individuals within a species. Ecosystem diversity (top frame) and related concepts refer to variety at levels above the species level, such as ecosystems, communities, habitats, or landscapes.

**Species diversity** As you will recall (• p. 116), a *species* is a distinct type of organism, a set of individuals that uniquely share certain characteristics and can breed with one another and produce fertile offspring. Biologists may use differing criteria to delineate species boundaries; some emphasize characteristics shared because of common ancestry, whereas others emphasize ability to interbreed. In practice, however, scientists broadly agree on species identities. We can express **species diversity** in terms of the number or variety of species in the world or in a particular region. One component of species diversity is *species richness,* the number of species. Another is *evenness* or *relative abundance,* the extent to which numbers of individuals of different species are equal or skewed.

As we have seen (• pp. 117–119), speciation generates new species, adding to species richness, whereas extinction decreases species richness. Although immigration, emigration, and local extinction may increase or decrease species richness locally, only speciation and extinction change it globally.

*Taxonomists,* the scientists who classify species, use an organism's physical appearance and genetic makeup to determine its species. Taxonomists also group species by their similarity into a hierarchy of categories meant to reflect evolutionary relationships. Related species are grouped together into *genera* (singular, *genus*); related genera are grouped into families; and so on (**Figure 11.2**). Every species is given a two-part Latin or Latinized scientific name denoting its genus and species. The tiger, *Panthera tigris,* differs from the world's other species of large cats, such as the jaguar *(Panthera onca),* the leopard *(Panthera pardus),* and the African lion *(Panthera leo).* These four species are closely related in evolutionary terms, as indicated by the genus name they share, *Panthera.* They are more distantly related to cats in other genera such as the cheetah *(Acinonyx jubatus)* and the bobcat *(Felis rufus),* although all cats are classified together in the family Felidae.

Biodiversity exists below the species level in the form of *subspecies,* populations of a species that occur in different geographic areas and differ from one another in some characteristics. Subspecies are formed by the same processes that drive speciation but result when divergence stops short of creating separate species. Scientists denote subspecies with a third part of the scientific name. The Siberian tiger, *Panthera tigris altaica,* is one of five subspecies of tiger still surviving (**Figure 11.3**). Tiger subspecies differ in color, coat thickness, stripe patterns, and size. For example, *Panthera tigris altaica* is 5–10 cm (2–4 in.) taller at the shoulder than the Bengal tiger *(Panthera tigris tigris)* of India and Nepal, and it has a thicker coat and larger paws.

**Genetic diversity** Scientists designate subspecies when they recognize substantial, genetically based differences

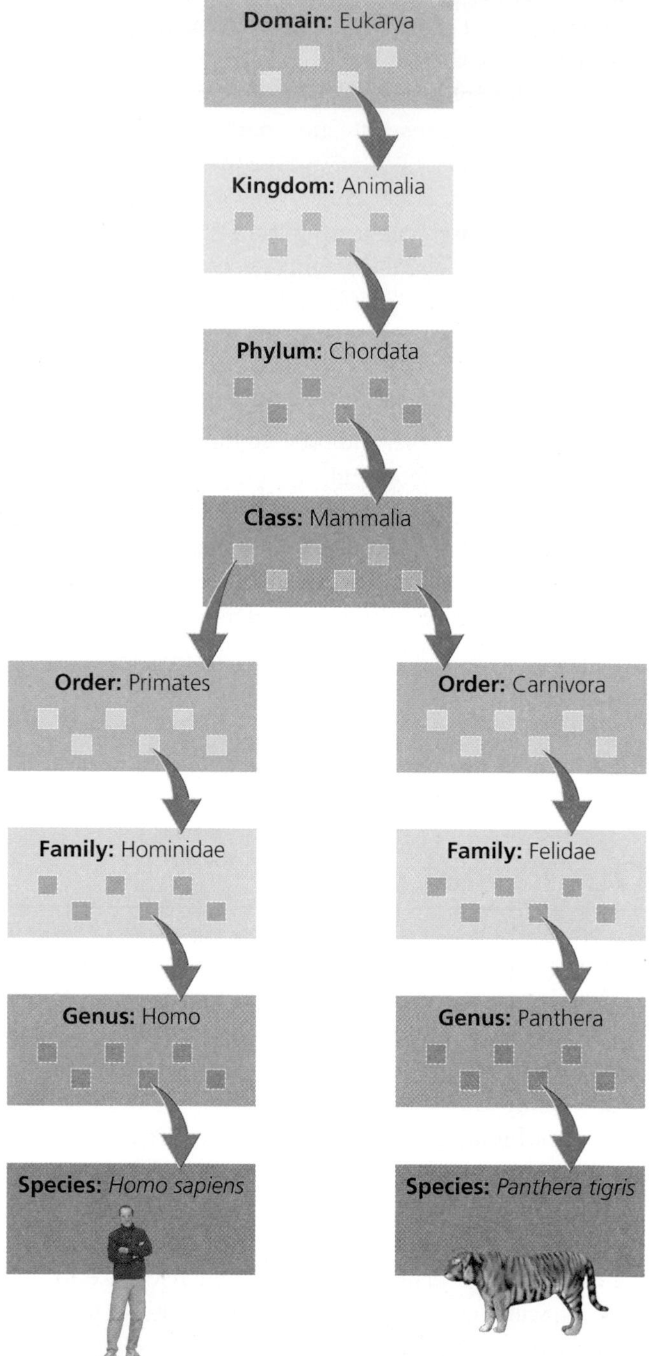


FIGURE 11.2 Taxonomists classify organisms using a hierarchical system meant to reflect evolutionary relationships. Species that are similar in their appearance, behavior, and genetics (because of recent common ancestry) are placed in the same genus. Organisms of similar genera are placed within the same family. Families are placed within orders, orders within classes, classes within phyla, phyla within kingdoms, and kingdoms within domains. For instance, humans (*Homo sapiens,* a species in the genus *Homo*) and tigers (*Panthera tigris,* a species in the genus *Panthera*) are both within the class Mammalia. However, the differences between our two species, which have evolved over millions of years, are great enough that we are placed in different orders and families.

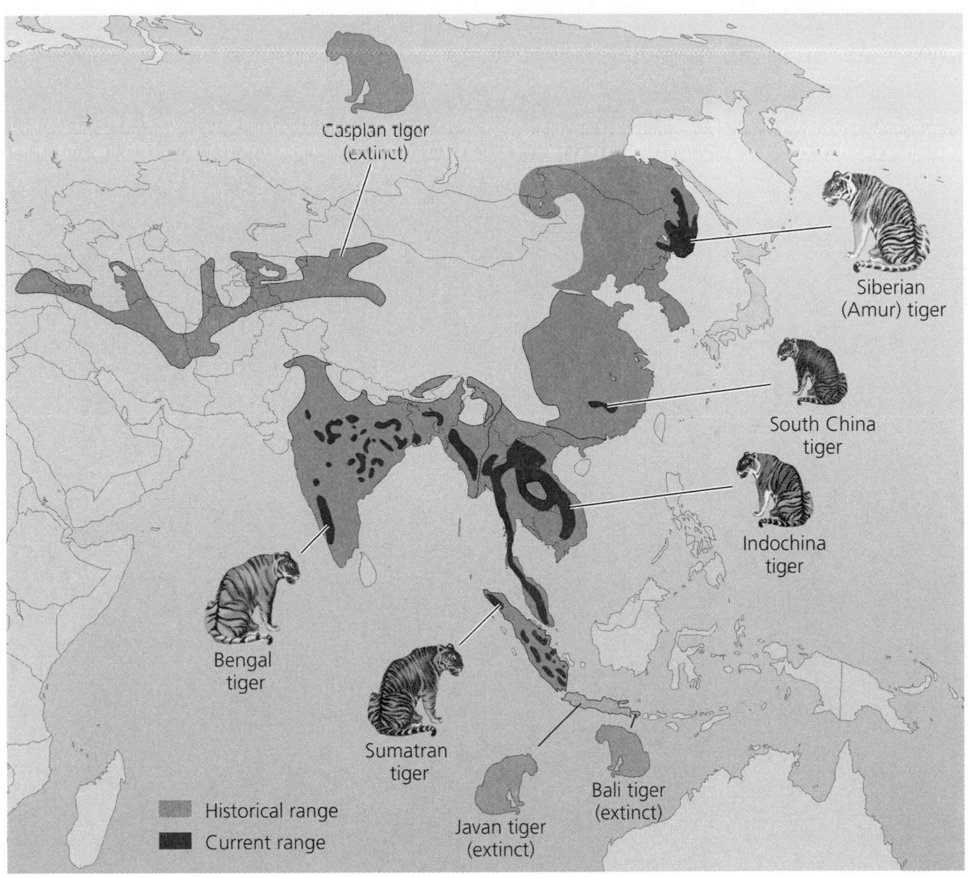


FIGURE 11.3 Three of the eight subspecies of tiger—the Bali, Javan, and Caspian tigers—went extinct during the 20th century. Today only the Siberian (Amur), Bengal, Indochina, Sumatran, and South China tigers persist, and fewer than 30 individuals of the South China tiger survive. Deforestation, hunting, and other pressures from people have caused tigers of all subspecies to disappear from most of the geographic range they historically occupied. This map contrasts the ranges of the eight subspecies in the years 1800 (orange) and 2000 (red). Data from the Tiger Information Center.

among individuals from different populations of a species. However, all species consist of individuals that vary genetically from one another to some degree, and this genetic diversity is an important component of biodiversity. **Genetic diversity** encompasses the differences in DNA composition among individuals within species and populations.

Genetic diversity provides the raw material for adaptation to local conditions. A diversity of genes for coat thickness in tigers allowed natural selection to favor genes for thin coats of fur in Bengal tigers living in warm regions, and genes for thick coats of fur for Siberian tigers living in cold regions. In the long term, populations with more genetic diversity may stand better chances of persisting, because their variation better enables them to cope with environmental change.

Populations with little genetic diversity are vulnerable to environmental change for which they are not genetically prepared. Populations with depressed genetic diversity may also be more vulnerable to disease and may suffer *inbreeding depression,* which occurs when genetically similar parents mate and produce weak or defective offspring. Scientists have sounded warnings over low genetic diversity in species that have dropped to low population sizes in the past, including cheetahs, bison, and elephant seals, but the full consequences of reduced diversity in these species remain to be seen. Diminishing genetic diversity in our crop plants also is a prime concern to humanity (• pp. 280–281).

**Ecosystem diversity** Biodiversity also encompasses levels above the species level. *Ecosystem diversity* refers to the number and variety of ecosystems, but biologists may also refer to the diversity of biotic community types or habitats within some specified area. If the area is large, scientists may also consider the geographic arrangement of habitats, communities, or ecosystems at the landscape level, including the sizes, shapes, and interconnectedness of patches of these entities. Under any of these concepts, a seashore of rocky and sandy beaches, forested cliffs, offshore coral reefs, and ocean waters would hold far more biodiversity than the same acreage of a monocultural cornfield. A mountain slope whose vegetation changes from desert to hardwood forest to coniferous forest to alpine meadow would hold more biodiversity than an area the same size consisting of only desert, forest, or meadow.

## Some groups hold more species than others

Species are not evenly distributed among taxonomic groups. In terms of number of species, insects show a staggering predominance over all other forms of life

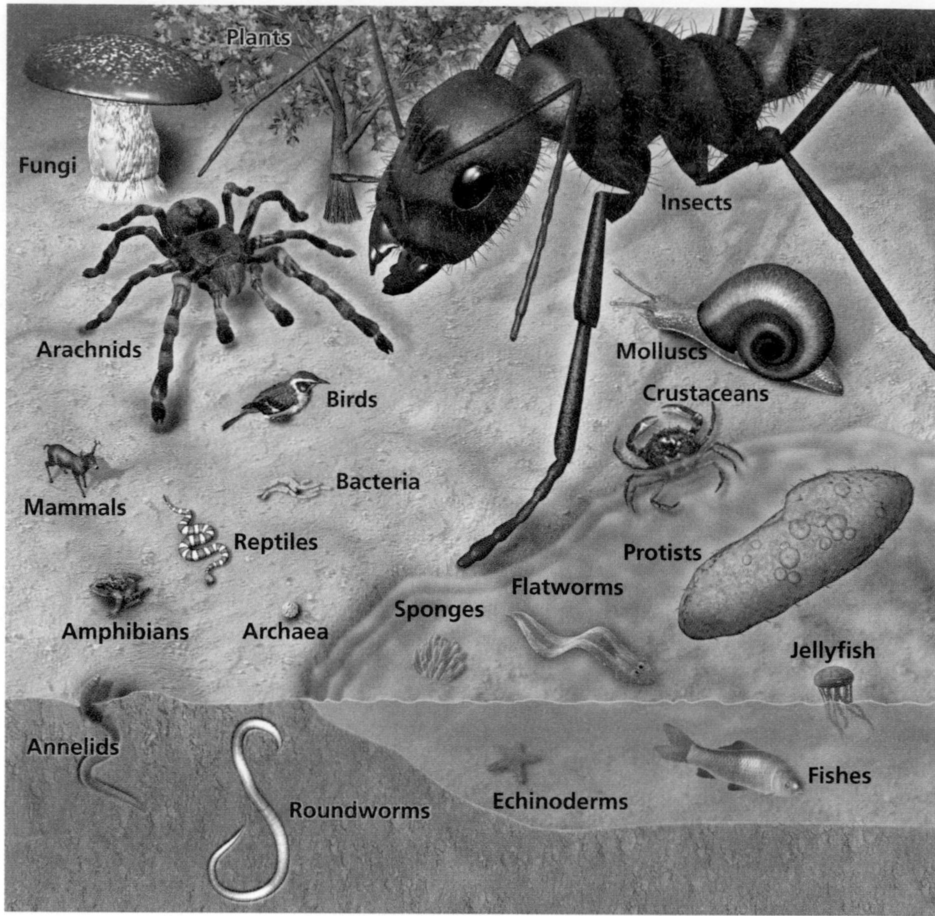


FIGURE 11.4 This illustration shows organisms scaled in size to the number of species known from each major taxonomic group. This gives a visual sense of the disparity in species richness among groups. However, because most species are not yet discovered or described, some groups (such as bacteria, archaea, insects, nematodes, protists, fungi, and others) may contain far more species than we now know of. Data from Groombridge, B., and M. D. Jenkins. 2002. *Global biodiversity: Earth's living resources in the 21st century.* UNEP-World Conservation Monitoring Centre. Cambridge, U.K.: Hoechst Foundation.

(**Figure 11.4** and **Figure 11.5**). Within insects, about 40% are beetles. Beetles outnumber all non-insect animals and all plants. No wonder the 20th-century British biologist J. B. S. Haldane famously quipped that God must have had "an inordinate fondness for beetles."

Some groups have given rise to many species in a relatively short period of time through the process of adaptive radiation (see Figure 5.3, • p. 116). Others have diversified because of a tendency to become separated by barriers that promote allopatric speciation (• pp. 117–118). Still other groups have accumulated species through time because of low rates of extinction.

## Measuring biodiversity is not easy

Coming up with precise quantitative measurements to express a region's biodiversity is difficult. Scientists often express biodiversity in terms of its most easily measured component, species diversity, and, in particular, species richness. Species richness is a good gauge for overall biodiversity, but we still are profoundly ignorant of the number of species that exist worldwide. So far, scientists have identified and described 1.7–2 million species of plants, animals, and microorganisms. However, estimates for the total number that actually exist range from 3 million to 100 million, with our best educated guesses ranging from 5 million to 30 million.

Our knowledge of species numbers is incomplete for several reasons. First, some areas of Earth remain little explored. We have barely sampled the ocean depths, hydrothermal vents (• pp. 103–104), or the tree canopies and soils of tropical forests. Second, many species are tiny and easily overlooked. These inconspicuous organisms include bacteria, nematodes (roundworms), fungi, protists, and soil-dwelling arthropods. Third, many organisms are so difficult to identify that ones thought to be identical sometimes turn out, once biologists look more closely, to be multiple species. This is frequently the case with microbes, fungi, and small insects, but also sometimes with organisms as large as birds, trees, and whales.

Smithsonian Institution entomologist Terry Erwin pioneered one method of estimating species numbers. In 1982, Erwin's crews fogged rainforest trees in Central America with clouds of insecticide and then collected insects, spiders, and other arthropods as they died and fell from the treetops. Erwin concluded that 163 beetle species specialized on the tree species *Luehea seemannii.*

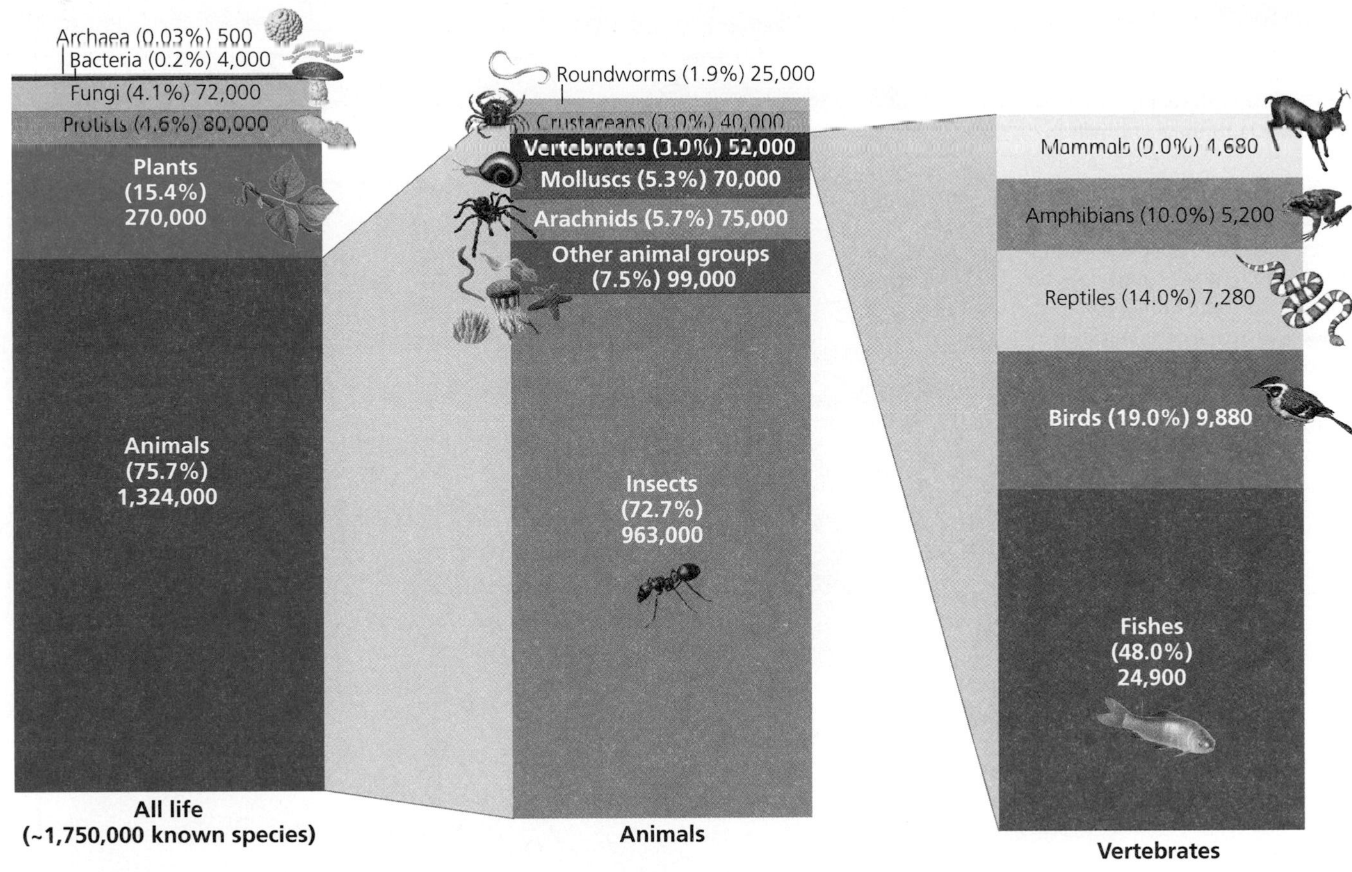


FIGURE 11.5 In the left portion of the figure, we see that three-quarters of known species are animals. The central portion subdivides animals, revealing that nearly three-quarters of animals are insects and that vertebrates comprise only 3.9% of animals. Among vertebrates (right portion of figure), nearly half are fishes, and mammals comprise only 9%. As noted, most species are not yet discovered or described, so some groups may contain far more species than we now know of. Data from Groombridge, B., and M. D. Jenkins. 2002. *Global biodiversity: Earth's living resources in the 21st century.* UNEP-World Conservation Monitoring Centre. Cambridge, U.K.: Hoechst Foundation.

If this were typical, he figured, then the world's 50,000 tropical tree species would hold 8,150,000 beetle species and—because beetles represent 40% of all arthropods—20 million arthropod species. If canopies hold two-thirds of all arthropods, then arthropod species in tropical forests alone would number 30 million. Many assumptions were involved in this calculation, and several follow-up studies have revised Erwin's estimate downward, but it remains one of the better efforts at estimating species numbers.

## Biodiversity is unevenly distributed

Numbers of species tell only part of the story of Earth's biodiversity. Living things are distributed across our planet unevenly, and scientists have long sought to explain the distributional patterns they see.

One of our planet's most striking patterns of diversity is the fact that species richness generally increases as one approaches the equator (**Figure 11.6**). This pattern of variation with latitude, called the *latitudinal gradient,* is one of the most obvious patterns in ecology, and one of the most difficult for scientists to explain.

Hypotheses abound for the cause of the latitudinal gradient in species richness, but it seems likely that plant productivity and climate stability play key roles in the phenomenon (**Figure 11.7**). Greater amounts of solar energy, heat, and humidity at tropical latitudes lead to more plant growth, making areas nearer the equator more productive and able to support larger numbers of animals. The relatively stable climates of equatorial regions—their similar temperatures and rainfall from day to day and season to season—help ensure that single species won't dominate ecosystems, but that instead numerous species can coexist. Whereas varying environmental conditions favor generalists—species that can deal with a wide range of circumstances but that do no single thing very well—stable conditions favor organisms with specialized niches that do particular things

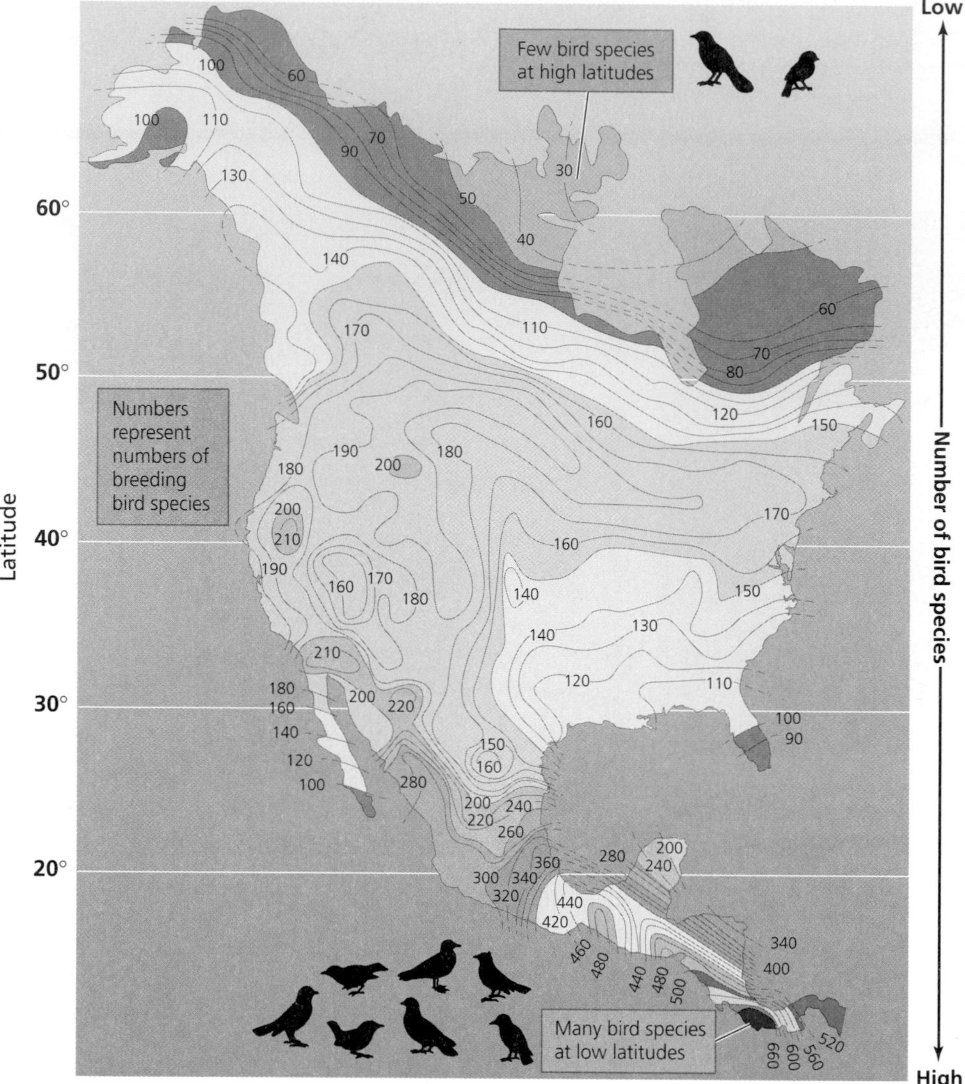


FIGURE 11.6 For many types of organisms, number of species per unit area tends to increase as one moves toward the equator. This trend, the latitudinal gradient in species richness, is one of the most readily apparent—yet least understood—patterns in ecology. One example is bird species in North and Central America: In any one spot in arctic Canada or Alaska, 30 to 100 breeding species can be counted; in areas of Costa Rica and Panama, the number rises to over 600. Adapted from Cook, R. E. 1969. Variation in species density in North American birds. *Systematic Zoology* 18: 63–84.

very well. An additional historical factor exists: Polar and temperate regions may be relatively lacking in species because glaciation events repeatedly forced organisms out of these regions and toward more tropical latitudes.

The latitudinal gradient influences the species diversity of Earth's biomes (• pp. 161–169). Tropical dry forests and rainforests tend to support far more species than tundra and boreal forests, for instance. Tropical biomes typically show more evenness as well, whereas in high-latitude biomes with low species richness, particular species may greatly outnumber others. For example, Canadian boreal forest is dominated by immense expanses of black spruce, whereas Panamanian tropical forest contains hundreds of tree species, no one of which greatly outnumbers the others.

At smaller scales, diversity patterns vary with habitat type. Generally, habitats that are structurally diverse allow for more ecological niches (• pp. 124, 143) and support greater species richness and evenness. For instance, forests generally support greater diversity than grasslands.

For any given geographic area, species diversity tends to increase with diversity of habitats, because each habitat supports a somewhat different community of organisms. Thus, ecotones (• pp. 183–185), where habitats intermix, often have high biodiversity. Because human disturbance (such as clearing plots of forest) can sometimes increase habitat diversity (which ecologists call *habitat heterogeneity*), species diversity may be higher in disturbed areas. However, this is true only at local scales. At larger scales, human disturbance decreases diversity because species that rely on large unbroken expanses of single habitat will disappear.

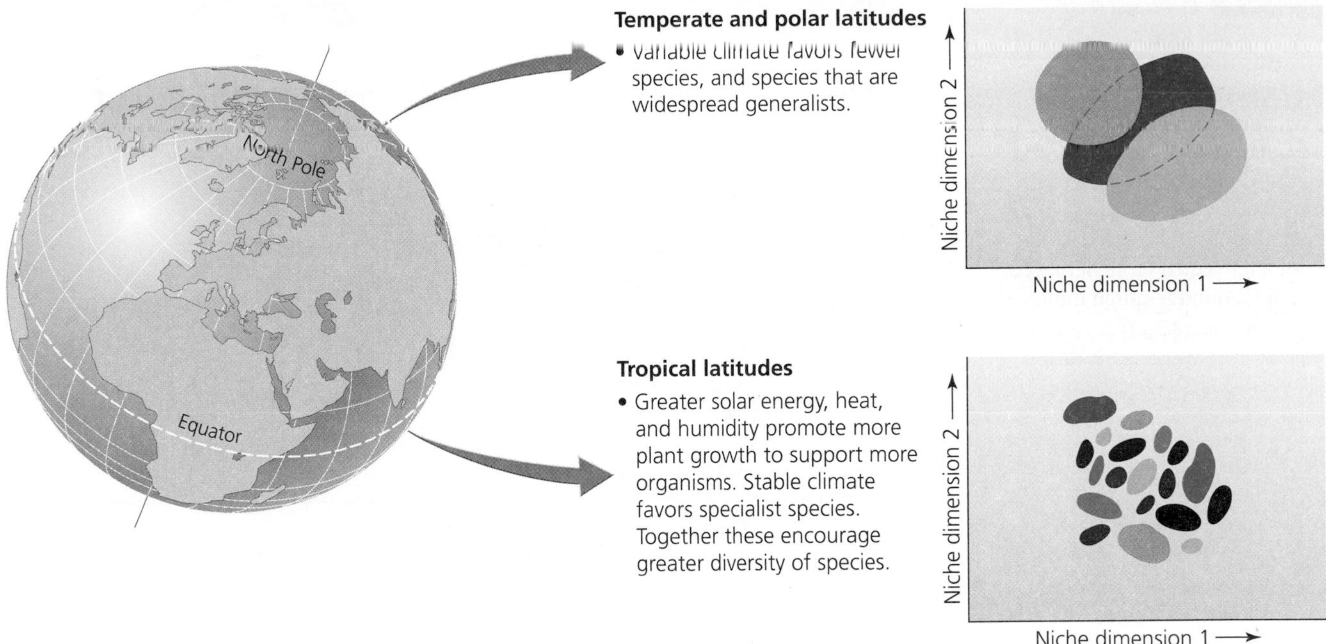


FIGURE 11.7 Ecologists have offered many hypotheses for the latitudinal gradient in species richness, and one set of ideas is summarized here. The variable climates (across days, seasons, and years) of polar and temperate latitudes favor organisms that can survive a wide range of conditions. Such generalist species have expansive niches; they can do many things well enough to survive, and they spread over large areas. In tropical latitudes, the abundant solar energy, heat, and humidity induce greater plant growth, which supports more organisms. The stable climates of equatorial regions favor specialist species, which have restricted niches but do certain things very well. Together these factors are thought to promote greater species richness in the tropics.

Understanding such patterns of biodiversity is vital for landscape ecology (• pp. 183–185), regional planning (• pp. 186–187 and 367), and forest management (• pp. 331–346). We will discuss further geographic patterns later in this chapter, when we explore solutions to the ongoing loss of biodiversity that our planet is currently experiencing.

# Biodiversity Loss and Species Extinction

Biodiversity at all levels is being lost to human impact, most irretrievably in the extinction of species. Once vanished, a species can never return. **Extinction** (• p. 120) occurs when the last member of a species dies and the species ceases to exist, as apparently was the case with Monteverde's golden toad. The disappearance of a particular population from a given area, but not the entire species globally, is referred to as **extirpation.** The tiger has been extirpated from most of its historic range, but it is not yet extinct. Extirpation is an erosive process that can, over time, lead to extinction.

## Extinction is a natural process

Extirpation and extinction occur naturally. If organisms did not naturally go extinct, we would be up to our ears in dinosaurs, trilobites, ammonites, and the millions of other types of creatures that vanished from Earth long before humans appeared. Paleontologists estimate that roughly 99% of all species that ever lived are now extinct, leaving the remaining 1% as the wealth of species on our planet today.

Most extinctions preceding the appearance of humans occurred one by one for independent reasons, at a pace that paleontologists refer to as the **background rate of extinction.** For example, the fossil record indicates that for mammals and marine animals, one species out of 1,000 would typically become extinct every 1,000 to 10,000 years. This translates to an annual rate of one extinction per 1 to 10 million species.

## Earth has experienced five mass extinction episodes

Extinction rates rose far above this background rate during several mass extinction events in Earth's history. In the past 440 million years, our planet experienced five major

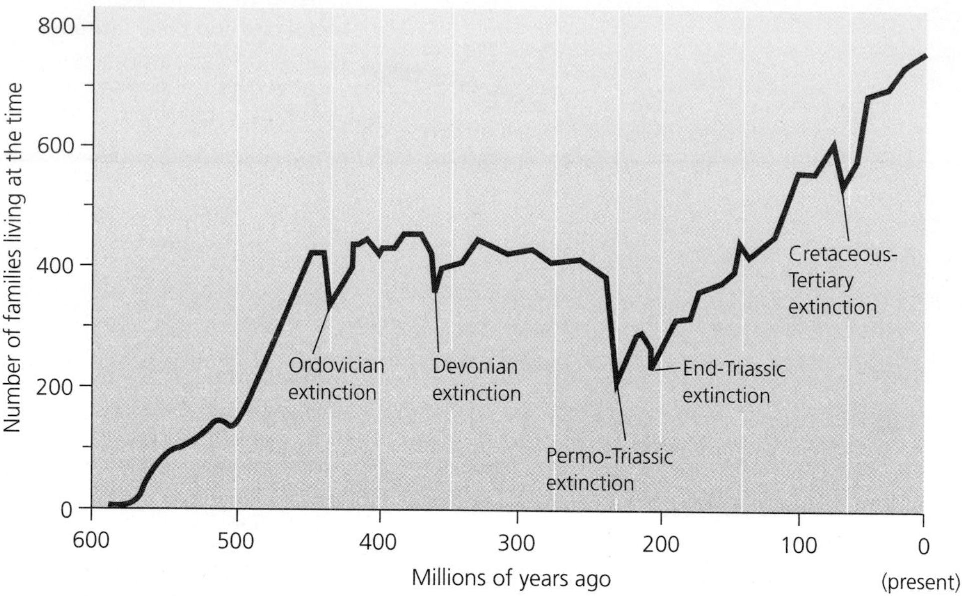


**FIGURE 11.8** The fossil record shows evidence for five episodes of mass extinction during the past half-billion years of Earth history. At the end of the Ordovician, Devonian, Permian, Triassic, and Cretaceous periods, 50–95% of the world's species appear to have gone extinct. Each time, biodiversity later rebounded to equal or higher levels, but each rebound required millions of years. Data from Raup, D. M., and J. J. Sepkoski. 1982. Mass extinctions in the marine fossil record. *Science* 215: 1501–1503.

episodes of **mass extinction** (• pp. 121–123; **Figure 11.8**). Each of these events eliminated more than one-fifth of life's families and at least half its species (**Table 11.1**). The most severe episode occurred at the end of the Permian period, 248 million years ago, when close to 54% of all families, 90% of all species, and 95% of marine species went extinct. The best-known episode occurred at the end of the Cretaceous period, 65 million years ago, when an apparent asteroid impact brought an end to the dinosaurs and many other groups (• pp. 122–123). Evidence exists for further mass extinctions in the Cambrian period and earlier, more than half a billion years ago.

If current trends continue, the modern era, known as the Quaternary period, may see the extinction of more than half of all species. Although similar in scale to previous mass extinctions, today's ongoing mass extinction is different in two primary respects. First, humans are causing it. Second, humans will suffer as a result of it.

## Humans are setting the sixth mass extinction in motion

We have recorded hundreds of instances of human-induced species extinction over the past few centuries. Sailors documented the extinction of the dodo on the Indian Ocean island of Mauritius in the 17th century, and today we have only a few of the dodo's body parts in museums. Among North American birds in the past two centuries alone, we have driven into extinction the Carolina parakeet, great auk, Labrador duck, and passenger pigeon (• pp. 125–126), and probably the Bachman's warbler and

**TABLE 11.1 Mass Extinctions**

| Event | Date (millions of years ago [mya]) | Cause | Types of life most affected | Percentage of life depleted |
|---|---|---|---|---|
| Ordovician | 440 mya | Unknown | Marine organisms; terrestrial record is unknown | >20% of families |
| Devonian | 370 mya | Unknown | Marine organisms; terrestrial record is unknown | >20% of families |
| Permo-Triassic | 250 mya | Possibly volcanism | Marine organisms; terrestrial record is less known | >50% of families; 80–95% of species |
| End-Triassic | 202 mya | Unknown | Marine organisms; terrestrial record is less known | 20% of families; 50% of genera |
| Cretaceous-Tertiary | 65 mya | Asteroid impact | Marine and terrestrial organisms, including dinosaurs | 15% of families; >50% of species |
| Current | Beginning 0.01 mya | Human impacts | Large animals, specialized organisms, island organisms, organisms harvested by people | Ongoing |

Eskimo curlew. Several more species, including the whooping crane, Kirtland's warbler, and California condor (• p. 320), teeter on the brink of extinction. Recent possible sightings of the ivory-billed woodpecker in the wooded swamps of Arkansas, Florida, and Louisiana provide some hope that this charismatic species may still persist.

However, species extinctions caused by humans precede written history. Indeed, people may have been hunting species to extinction for thousands of years. Archaeological evidence shows that in case after case, a wave of extinctions followed close on the heels of human arrival on islands and continents (**Figure 11.9**). After Polynesians reached Hawaii, half its birds went extinct. Birds, mammals, and reptiles vanished following human arrival on many other oceanic islands, including large island masses such as New Zealand and Madagascar. The pattern appears to hold for at least two continents, as well. Dozens of species of large vertebrates died off in Australia after Aborigines arrived roughly 50,000 years ago, and North America lost 33 genera of large mammals after people arrived on the continent at least 10,000 years ago.

## Current extinction rates are much higher than normal

Today, species loss is accelerating as our population growth and resource consumption put increasing strain on habitats and wildlife. In 2005, scientists with the Millennium Ecosystem Assessment (• p. 18) calculated that the current global extinction rate is 100 to 1,000 times greater than the background rate. Moreover, they projected that the rate would increase tenfold or more in future decades.

To keep track of the current status of endangered species, the World Conservation Union maintains the **Red List,** an updated list of species facing high risks of extinction. The 2006 Red List reported that 23% (1,093) of mammal species and 12% (1,206) of bird species are threatened with extinction. Among other major groups (for which assessments are not fully complete), estimates of the percentage of species threatened ranged from 31% to 86%. Since 1970, at least 58 fish species, 9 bird species, and 1 mammal species have gone extinct, and in the United States alone over the past

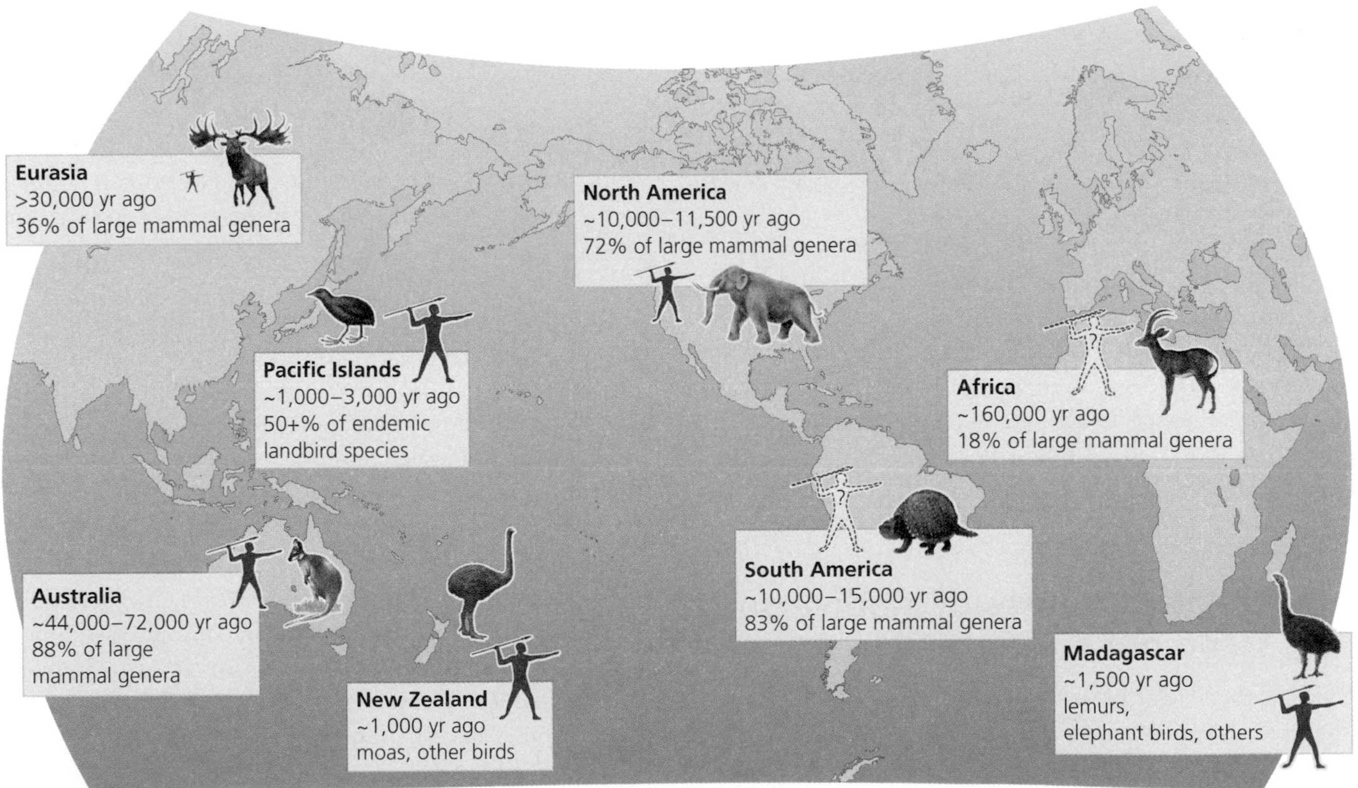


**FIGURE 11.9** This map shows for each region the time of human arrival and the extent of the recent extinction wave. Illustrated are representative extinct megafauna from each region. The human hunter icons are sized according to the degree of evidence that human hunting was a cause of extinctions; larger icons indicate more certainty that humans (as opposed to climate change or other factors) were the cause. Data for South America and Africa are so far too sparse to be conclusive, and future archaeological and paleontological research could well alter these interpretations. Adapted from Barnosky, A. D., et al. 2004. Assessing the causes of late Pleistocene extinctions on the continents. *Science* 306: 70–75; and Wilson, E. O. 1992. *The diversity of life*. Cambridge, MA: Belknap Press.

500 years, 236 animals and 17 plants are confirmed to have gone extinct. For all of these figures, the *actual* numbers of species extinct and threatened, like the actual number of total species in the world, are doubtless greater than the *known* numbers.

Among the 1,093 mammals facing possible extinction on the Red List is the tiger, which despite—or perhaps because of—its tremendous size and reputation as a fierce predator, is one of the most endangered large animals on the planet. In 1950, eight tiger subspecies existed (see Figure 11.3). Today, three are extinct. The Bali tiger, *Panthera tigris balica,* went extinct in the 1940s; the Caspian tiger, *Panthera tigris virgata,* during the 1970s; and the Javan tiger, *Panthera tigris sondaica,* during the 1980s.

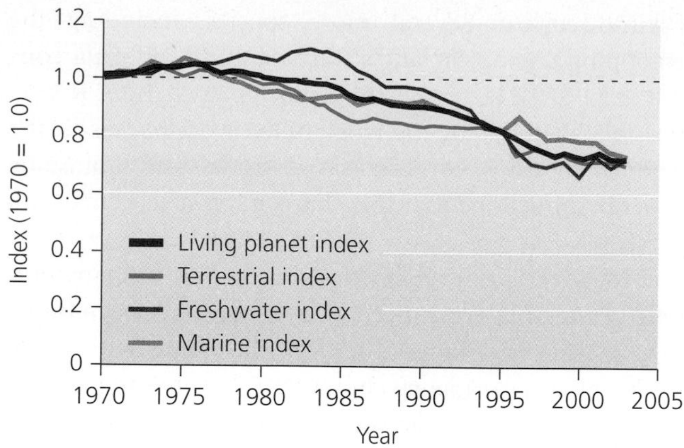


**FIGURE 11.10** The Living Planet Index serves as an indicator of the state of global biodiversity. Index values summarize population trends for 1,313 species. Between 1970 and 2003, the Living Planet Index fell by roughly 30%. The index for terrestrial species fell by 31%; for freshwater species, 28%; and for marine species, 27%. Data from World Wide Fund for Nature and U.N. Environment Programme. 2006. *The Living Planet Report, 2006.* Gland, Switzerland: WWF.

## Biodiversity loss involves more than extinction

Extinction is only part of the story of biodiversity loss. The larger part of the story is the decline in population sizes of many organisms. Declines in numbers are accompanied by shrinkage of species' geographic ranges. Thus, many species today are less numerous and occupy less area than they once did. Tigers numbered well over 100,000 worldwide in the 19th century but number only about 5,000 today. Such declines mean that genetic diversity and ecosystem diversity, as well as species diversity, are being lost.

To measure and quantify this degradation, scientists at the World Wildlife Fund and the United Nations Environment Programme (UNEP) developed a metric called the *Living Planet Index.* This index summarizes trends in the populations of 695 terrestrial species, 344 freshwater species, and 274 marine species that are well enough monitored to provide reliable data. Between 1970 and 2003, the Living Planet Index fell by roughly 30% (**Figure 11.10**).

## Several major causes of biodiversity loss stand out

Reasons for the decline of any given species are often multifaceted, complex, and difficult to determine. The current precipitous decline in populations of amphibians throughout the world provides an example. Frogs, toads, and salamanders worldwide are decreasing drastically in abundance. As we saw in Chapter 5 with the golden toad, some have already gone extinct—and scientists are struggling to explain why. Recent studies have implicated a wide array of factors, and most scientists now suspect that such factors may be interacting synergistically (see "The Science behind the Story," • pp. 308–309).

Overall, scientists have identified four primary causes of population decline and species extinction: habitat alteration, invasive species, pollution, and overharvesting. Global climate change (Chapter 18) now is becoming the fifth. Each of these factors is intensified by human population growth and by our increase in per capita consumption of resources.

**Habitat alteration** Nearly every human activity can alter the habitat of the organisms around us. Farming replaces diverse natural communities with simplified ones of only one or a few plant species. Grazing modifies the structure and species composition of grasslands. Either type of agriculture can lead to desertification. Clearing forests removes the food, shelter, and other resources that forest-dwelling organisms need to survive. Hydroelectric dams turn rivers into reservoirs upstream and thereby affect water conditions and floodplain communities downstream. Urbanization and suburban sprawl supplant diverse natural communities with simplified human-made ones, driving many species from their homes (**Figure 11.11**).

Because organisms are adapted to the habitats in which they live, any major change in their habitat is likely to render it less suitable for them. Of course, human-induced habitat change may benefit some species. Animals such as house sparrows, pigeons, gray squirrels, and cockroaches do very well in urban and suburban environments and benefit from our modification of natural habitats. However, the species that benefit are relatively few; for every species that gains, more lose. Furthermore, the species that do well in our midst tend to be weedy, cosmopolitan generalists that are in little danger of disappearing any time soon.

Habitat alteration is by far the greatest cause of biodiversity loss today. It is the primary source of population declines for 83% of threatened mammals and 85% of threatened birds, according to UNEP data. As just one example of thousands, the prairies native to North America's Great Plains are today almost entirely converted to agriculture. Less than 1% of original prairie habitat remains. As a result, grassland bird populations have declined by an estimated 82–99%. Many grassland species have been extirpated from large areas, and the two species of prairie chickens still persisting in pockets of the Great Plains could soon go extinct.

Habitat destruction has occurred widely in nearly every biome. Over half of temperate forests, grasslands, and shrublands had been converted by the year 1950 (mostly for agriculture). Across Asia, scientists estimate that 40% of the tiger's remaining habitat has disappeared just in the last decade. Today habitat is being lost most rapidly in tropical rainforests, tropical dry forests, and savannas (**Figure 11.12**).

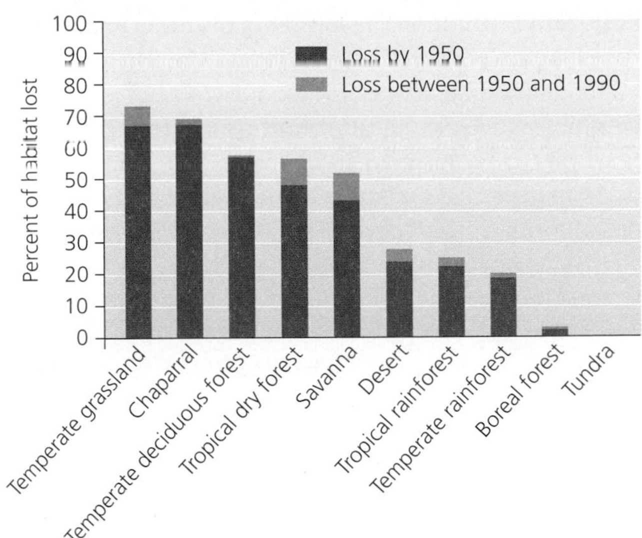


**FIGURE 11.12** Human impact has caused varying extents of habitat loss across the world's biomes. Shown for each biome are percentages of original area that were converted for human use before 1950 (red portion of bars) and during 1950–1990 (orange portion of bars). Biomes such as temperate grassland and chaparral lost over 70% of their area by 1950, whereas biomes such as tundra and boreal forest lost virtually none. In recent decades, tropical dry forest and savanna lost the greatest fraction. Data from Millennium Ecosystem Assessment. 2005. *Ecosystems and human well-being: Biodiversity synthesis.* World Resources Institute, Washington, D.C. Categories used in this figure are modified from the source so as to match the biome categorization used in this book (• pp. 161–169).

**Invasive species** Our introduction of non-native species to new environments, where some may become invasive (**Figure 11.13**), has also pushed native species toward extinction. Some introductions are accidental. Examples include aquatic organisms (such as zebra mussels; Chapter 6) that are transported in the ballast water of ships, animals that escape from the pet trade, and the weed seeds that cling to our socks as we travel from place to place. Other introductions are intentional. People transport food crops, farm animals, and other organisms as they colonize new places, generally unaware of the ecological consequences that could result.

**FIGURE 11.11** Development of land for new housing is one of many ways in which habitat is altered or destroyed.

Species native to islands are especially vulnerable to disruption from introduced species because the native species exist in isolation with relatively few parasites, predators, and competitors. As a result, they have not evolved the defenses necessary to resist invaders that are better adapted to these pressures.

Most organisms introduced to new areas perish, but the few types that survive may do very well, especially if they find themselves without the predators and parasites that attacked them back home or without the competitors that had limited their access to resources. Once released from the limiting factors of predation, parasitism, and competition, an introduced species may increase rapidly, spread, and displace native species. Invasive species cause billions of dollars in economic damage each year.

**Pollution** Pollution can harm organisms in many ways. Air pollution (Chapter 17) can degrade forest ecosystems. Water pollution (Chapter 15) can adversely affect fish and amphibians. Agricultural runoff (including fertilizers, pesticides, and sediments; Chapters 7, 9, and 10) can harm many terrestrial and aquatic species. Heavy metals, polychlorinated biphenyls (PCBs), endocrine-disrupting compounds, and various other toxic chemicals can poison people and wildlife (Chapter 14), and the effects of oil and chemical spills on wildlife are dramatic and well known.

Although pollution is a substantial threat, it tends to be less significant than public perception holds it to be. The damage to wildlife and ecosystems caused by

**Invasive Species**

| Species | Native to... | Invasive in... | Effects |
|---|---|---|---|
| Mosquito fish *(Gambusia affinis)*  | North America | Africa, Asia, Europe, and Australia | Introduced to control mosquito populations, the mosquito fish outcompetes native fish, eats their eggs, and does no better than native species in controlling mosquitoes. |
| Zebra mussels *(Dreissenna polymorpha)*  | Caspian Sea | Freshwater ecosystems including the Great Lakes of Canada and the United States | Zebra mussels (Chapter 6) most likely made their way from their home by traveling in ballast water taken on by cargo ships. They compete with native species and clog water treatment facilities and power plant cooling systems. |
| Kudzu *(Pueraria montana)*  | Japan | Southeastern United States | Kudzu is a vine that can grow 30 m (100 ft) in a single season. The U.S. Soil Conservation Service introduced kudzu in the 1930s to help control erosion. Adaptable and extraordinarily fast-growing, kudzu has taken over thousands of hectares of forests, fields, and roadsides in the southeastern United States. |
| Asian long-horned beetles *(Anoplophora glabripennis)*  | Asia | United States | Having first arrived in the United States in imported lumber in the 1990s, these beetles burrow into hardwood trees and interfere with the trees' ability to absorb and process water and nutrients. They may wipe out the majority of hardwood trees in an area. Several U.S. cities, including Chicago in 1999 and Seattle in 2002, have cleared thousands of trees after detecting these invaders. |
| Rosy wolfsnail *(Euglandina rosea)* 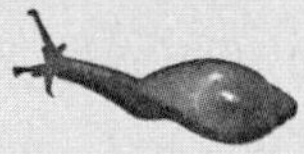 | Southeastern United States and Latin America | Hawaii | In the 1950s, well-meaning scientists introduced the rosy wolfsnail to Hawaii to prey upon and reduce the population of another invasive species, the giant African land snail (*Achatina fulica*), which had been introduced early in the 20th century as an ornamental garden animal. Within a few decades, however, the carnivorous rosy wolfsnail had instead driven more than half of Hawaii's native species of banded tree snails to extinction. |
| Cane toad *(Bufo marinus)*  | Southern United States to tropical South America | Northern Australia and other locations | Since being introduced 70 years ago to control insects in sugarcane fields, the cane toad has wreaked havoc across northern Australia (and other locations). The skin of this tropical American toad can kill its predators, and the cane toad outcompetes native amphibians. |
| Bullfrog *(Rana catesbiana)*  | Eastern North America | Western North America | The bullfrog is contributing to amphibian and reptile declines in western North America. Bullfrog tadpoles grow large and can outcompete and prey on other tadpoles, but need to grow a long time in permanent water to do so. Historically most water bodies in the arid West dried up part of the year, making it impossible for bullfrogs to live there, but artificial impoundments—dams, farm ponds, canals—gave the bullfrogs bases from which they could spread. |

**FIGURE 11.13** Invasive species are species that thrive in areas where they are introduced, outcompeting, preying on, or otherwise harming native species. Of the many thousands of invasive species, this chart shows a few of the best known.

pollution can be severe, but it tends to be less than the damage caused by habitat alteration or invasive species.

**Overharvesting** For most species, a high intensity of hunting or harvesting by humans will not *in itself* pose a threat of extinction, but for some species it can. The Siberian tiger is one such species. Large in size, few in number, long-lived, and raising few young in its lifetime—a classic K-selected species (• pp. 131–132)—the Siberian tiger is just the type of animal to be vulnerable to population reduction by hunting. The advent of Russian hunting nearly drove the animal extinct, whereas decreased hunting during and after World War II allowed a population increase. By the

| Invasive Species | | | |
|---|---|---|---|
| **Species** | **Native to...** | **Invasive in...** | **Effects** |
| Gypsy moth *(Lymantria dispar)* | Eurasia | Northeastern United States | In the 1860s, a scientist introduced the gypsy moth to Massachusetts in the mistaken belief that it might be bred with others to produce a commercial-quality silk. The gypsy moth failed to start a silk industry, and instead spread through the northeastern United States and beyond, where its outbreaks defoliate trees over large regions every few years. |
| European starling *(Sturnus vulgaris)* | Europe | North America | The bird was first introduced to New York City in the late 19th century by Shakespeare devotees intent on bringing every bird mentioned in Shakespeare's plays to the new continent. It only took 75 years for the birds to spread to the Pacific coast, Alaska, and Mexico, becoming one of the most abundant birds on the continent. Starlings are thought to outcompete native birds for nest sites. |
| Indian mongoose *(Herpestes auropunctatus)* | Southeast Asia | Hawaii | Rats that had invaded the Hawaiian islands from ships in the 17th century were damaging sugarcane fields, so in 1883 the Indian mongoose was introduced to control rat populations. Unfortunately, the rats were active at night and the mongooses fed during the day, so the plan didn't work. Instead mongooses began preying on native species like ground-nesting seabirds and the now-endangered Nene or Hawaiian goose (*Branta sandvicensis*). |
| A green alga *(Caulerpa taxifolia)* | Tropical oceans and seas | Mediterranean Sea | Dubbed the "killer algae," *Caulerpa taxifolia* has spread along the coasts of several Mediterranean countries since it apparently escaped from Monaco's aquarium in 1984. Creeping underwater over the sand and mud like a green shag carpet, it crowds out other plants, is inedible to most animals, and tangles boat propellers. It has been the focus of intense eradication efforts since arriving recently in Australia and California. |
| Cheatgrass *(Bromus tectorum)* | Eurasia | Western United States | In just 30 years after its introduction to Washington state in the 1890s, cheatgrass has spread across much of the western United States. Its secret: fire. Its thick patches that choke out other plants and use up the soil's nitrogen burn readily. Fire kills many of the native plants, but not cheatgrass, which grows back even stronger amid the lack of competition. |
| Brown tree snake *(Boiga irregularis)* | Southeast Asia | Guam | Nearly all native forest bird species on the South Pacific island of Guam have disappeared. The culprit is the brown tree snake. The snakes were likely brought to the island inadvertantly as stowaways in cargo bays of military planes in World War II. Guam's birds had not evolved with tree snakes, and so had no defenses against the snake's nighttime predation. The snakes also cause numerous power outages each year on Guam and have spread to other islands where they are repeating their ecological devastation. The arrival of this snake is the greatest fear of conservation biologists in Hawaii. |

THE SCIENCE BEHIND THE STORY

## Amphibian Diversity and Amphibian Declines

*Dr. Madhava Meegaskumbura searching for frogs at night in Sri Lanka*

Amphibians illustrate the two most salient aspects of Earth's biodiversity today. Scientists are discovering more and more species while more and more populations and species are vanishing.

New species of most classes of vertebrates are discovered at a rate of only one or a few per year, but the number of known amphibian species (which include frogs, salamanders, and others)—about 6,150 as of mid-2007—has jumped by nearly 35% just since 1985.

At the same time, however, at least 2,500 of these species are in decline. Researchers feel that they may be naming some species just before they go extinct and losing others before they are even discovered. Nearly 170 species of frogs, toads, and salamanders studied just years or decades ago, including the golden toad (Chapter 5), are now thought to be gone.

These losses are especially worrying because amphibians are widely regarded as "biological indicators" that can tell us whether an ecosystem is in good shape. Amphibians rely on both aquatic and terrestrial environments and may breathe and absorb water through their skin, so they are sensitive to pollution and other environmental stresses. Studying the reasons for amphibian declines can tell us much about the state of our environment.

From the American tropics to Africa to southern Asia, scientific scrutiny and improved technology have revealed amphibian "hot spots." In the 1990s, an international team of scientists set out to determine whether Sri Lanka, a large tropical island off the coast of India, held more than the 40 frog species that were already known. Researcher Madhava Meegaskumbura and his team combed through trees, rivers, ponds, and leaf litter for 8 years, collecting more than 1,400 frogs at 300 study sites. The scientists analyzed the frogs' physical appearance, habitat use, and vocalizations. They also examined the frogs' genes by obtaining sequences of nucleotides in several regions of their DNA (• p. 97). They then compared these genetic, physical, and behavioral characteristics to those of known species of frogs.

The team found that the DNA from many of their frogs didn't match that of known species. And they found that many of their frogs looked different, sounded different, or behaved differently from known species. Clearly, they had discovered new species unknown to science. Some of these novel species live on rocks and sport leg fringes and markings that help disguise them as clumps of moss. Others are tree frogs that lay their eggs in baskets they construct. In all, more than 100 new species of amphibians were discovered—all on an island only slightly larger than the state of West Virginia! When reported in the journal *Science* in 2002, the study caught the attention of biologists worldwide.

Such promising discoveries, however, come against a backdrop of distressing declines. Observed numbers

mid-1980s, the Siberian tiger population was likely up to 250 individuals.

The political freedom that came with the Soviet Union's breakup in 1989, however, brought with it a freedom to harvest Siberia's natural resources, the tiger included, without regulations or rules. This coincided with an economic expansion in many Asian countries, where tiger penises are traditionally used to try to boost human sexual performance and where tiger bones, claws, whiskers, and other body parts are used to treat a wide variety of maladies (**Figure 11.14**). Thus, the early 1990s brought a boom in poaching (poachers killed at least 180 Siberian tigers between 1991 and 1996). Sale of the parts from one tiger fetches roughly $15,000 in today's black market—a powerful economic incentive for poachers in poor regions. The modern rise in poaching has been accompanied by a dramatic increase in logging of the Korean pine forests on which the tigers and their prey depend.

Over the past century, hunting has led to steep declines in the populations of many other K-selected animals. The Atlantic gray whale has gone extinct, and several other whales remain threatened or endangered. Gorillas and other primates that are killed for their meat may face extinction soon. Thousands of sharks are killed each year simply for their fins, which are used in soup. Today the oceans contain only 10% of the large animals they once did (• pp. 460, 462).

The red-eyed tree frog (*Agalychnis callidryas*) resides in the Neotropics, where new amphibian species are being discovered but where others are mysteriously declining.

of amphibians are down around the globe. Scientists are racing to pin down the causes and have found evidence for causes as varied as habitat destruction, chemical pollution, disease, invasive species, and climate change. Most worrisome is that many populations are vanishing even when no direct damage is apparent. In some cases, researchers surmise that a combination of factors may be at work.

In one study, researchers Rick Relyea and Nathan Mills presented young frogs with two common dangers—pesticides and predators—to see how the mix affected their survival. The team collected 10 pairs of gray treefrogs from a wildlife area in Missouri and placed their eggs in clean water. When tadpoles emerged from the eggs, groups of 10 were each put in different tubs of water. Some tubs contained pure water, others contained varying levels of the pesticide carbaryl, and others contained the harmless solvent acetone as a control. To some of each of these three types of tubs, the researchers added a hungry predator—a young salamander. The salamander was caged and couldn't reach the tadpoles, but the tadpoles were aware of its presence. In a series of experiments, over up to 16 days, researchers watched to see how many tadpoles survived the different combinations of stress factors.

Their results, published in the *Proceedings of the National Academy of Sciences* in 2001, revealed that tadpoles that withstood one type of stress might not survive two. As expected, all tadpoles in clean water with no predators survived, and all tadpoles exposed to high concentrations of carbaryl died within several days, regardless of predator presence. But when carbaryl levels were lower, the presence of the salamander made a noticeable difference. In one trial, about 75% of tadpoles survived the pesticide if no predator was present, but in the presence of the salamander, survival rates dropped to about 25%. Related experiments showed that predator presence alone did not account for this difference. Thus, when both stresses were present (a condition likely in the tadpoles' natural habitat), death rates increased by two to four times.

One year later, a study published in the same journal by herpetologist Joseph Kiesecker found similar results with pathogens and pesticides. His field and lab experiments revealed that wood frogs were more vulnerable to parasitic infections that cause limb deformities when they were exposed to water containing pesticides.

As scientists learn more about how such factors combine to threaten amphibians, they are gaining a clearer picture of how the fate of these creatures may foreshadow the future for other organisms. "Amphibians have been around for 300 million years. They're tough, and yet they're checking out all around us," says David Wake, a biologist at the University of California at Berkeley, who was among the first to note the creatures' decline. "We really do see amphibians as biodiversity bellwethers."

**FIGURE 11.14** Body parts from tigers are used as traditional medicines and aphrodisiacs in some Asian cultures. Poachers have illegally killed countless tigers through the years to satisfy the surging market demand for these items. Here a street vendor in northern China displays tiger body parts for sale.

**Climate change** The preceding four types of human impacts affect biodiversity in discrete places and times. In contrast, our manipulation of Earth's climate system (Chapter 18) is beginning to have global impacts on biodiversity. As our emissions of greenhouse gases from fossil fuel combustion cause temperatures to warm worldwide, we are modifying global weather patterns and increasing the frequency of extreme weather events.

Extreme weather events such as droughts increase stress on populations, and warming temperatures are forcing organisms to shift their geographic ranges toward the poles and higher in altitude. Some species will be able to adapt, but others will not. Like the cloud-forest fauna at Monteverde (Chapter 5), mountaintop organisms cannot move further upslope to escape warming temperatures, so they will likely perish. Trees may not be able to move poleward fast enough. Animals and plants may find themselves among different communities of prey, predators, and parasites to which they are not adapted.

In the Arctic, where warming has been greatest, the polar bear (**Figure 11.15**) is being considered for inclusion on the U.S. endangered species list (• p. 317) because thawing ice hinders its ability to hunt seals (• pp. 518–519). All in all, scientists now predict that a 1.5–2.5°C global temperature increase could put 20–30% of the world's plants and animals at increased risk of extinction.

All five of these primary causes of population decline are intensified by human population growth and rising per capita consumption. More people and more consumption mean more habitat alteration, more invasive species, more pollution, more overharvesting, and more climate change. Growth in population and growth in consumption are the ultimate reasons behind the proximate threats to biodiversity. Just as we now have a solid scientific understanding of the causes of biodiversity loss, we are also coming to appreciate its consequences (**Figure 11.16**) as we begin to erode the many benefits that biodiversity brings us.

FIGURE 11.15 The polar bear (*Ursus maritimus*) is thought to be threatened by climate change as Arctic warming melts the sea ice from which it hunts seals, forcing the bears to swim farther for food. In response to a lawsuit from environmental groups, the U.S. Fish and Wildlife Service began considering in 2006 whether to add the polar bear to the endangered species list.

## Benefits of Biodiversity

Scientists worldwide are presenting us with data that confirm what any naturalist who has watched the habitat change in his or her hometown already knows: From amphibians to tigers, biodiversity is being lost rapidly and visibly within our lifetimes.

This suggests the question, "Does it matter?" There are many ways to answer this question, but we can begin by considering the ways that biodiversity benefits people. Scientists have offered a number of tangible, pragmatic reasons for preserving biodiversity, showing how biodiversity directly or indirectly supports human society. In addition, many people feel that organisms have an intrinsic right to exist and that ethical and aesthetic dimensions to biodiversity preservation cannot be ignored.

### Biodiversity provides ecosystem services free of charge

Contrary to popular opinion, some things in life can indeed be free—as long as we choose to protect the ecological systems that provide them. Intact forests provide clean air and buffer hydrologic systems against flooding and drought. Native crop varieties provide insurance against disease and drought. Abundant wildlife can attract tourists and boost the economies of developing nations. Intact ecosystems provide these and other valuable processes, known as *ecosystem services* (• pp. 30, 40, 50–51), for all of us, free of charge.

Maintaining these ecosystem services is one clear benefit of protecting biodiversity. According to UNEP, biodiversity

- Provides food, fuel, and fiber
- Provides shelter and building materials
- Purifies air and water
- Detoxifies and decomposes wastes
- Stabilizes and moderates Earth's climate
- Moderates floods, droughts, wind, and temperature extremes
- Generates and renews soil fertility and cycles nutrients
- Pollinates plants, including many crops
- Controls pests and diseases

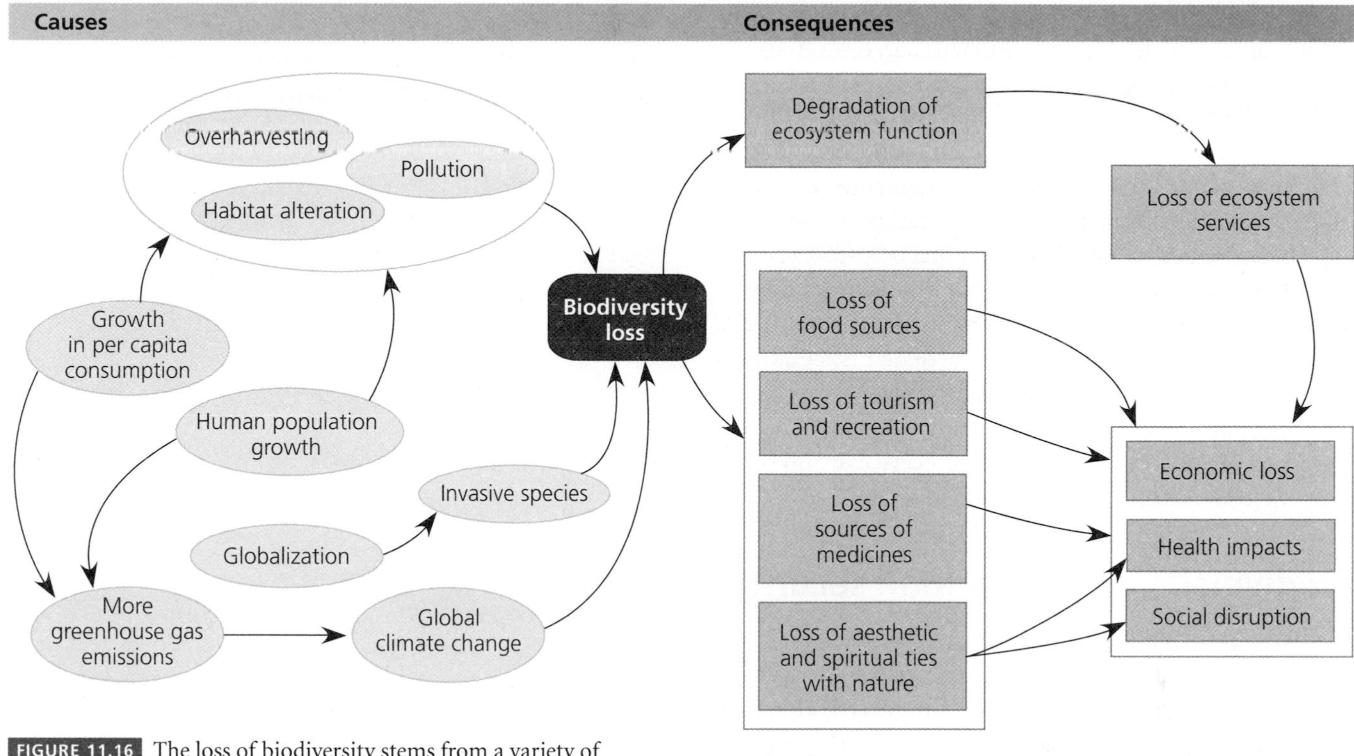


**FIGURE 11.16** The loss of biodiversity stems from a variety of causes (ovals on left) and results in a number of consequences (boxes on right) for ecological systems and human well-being. Arrows in this concept map lead from causes to consequences. Note that items grouped within outlined boxes do not necessarily share any special relationship; the outlined boxes are intended merely to streamline the figure.

**Solutions**

As you progress through this chapter, try to identify as many solutions to biodiversity loss as you can. What could you personally do to help address this issue? Consider how each action or solution might affect items in the concept map above.

- Maintains genetic resources as inputs to crop varieties, livestock breeds, and medicines
- Provides cultural and aesthetic benefits
- Gives us the means to adapt to change

Organisms and ecosystems support a vast number of vital processes that humans could not replicate or would need to pay for if nature did not provide them. As we have seen (• pp. 50–51), the annual value of just 17 of these ecosystem services may be in the neighborhood of $16–54 trillion per year.

## Biodiversity helps maintain ecosystem function

Functioning ecosystems are vital, but does biodiversity really help them maintain their function? Ecologists are finding that the answer appears to be yes. Research demonstrates that high levels of biodiversity tend to increase the *stability* of communities and ecosystems. Research has also found that high biodiversity tends to increase the *resilience* (• p. 153) of ecological systems—their ability to weather disturbance, bounce back from stresses, or adapt to change. Most of this research has dealt with species diversity, but new work is finding similar effects for genetic diversity. Thus, a decrease in biodiversity could diminish a natural system's ability to function and to provide services to our society.

What about the extinction of selected species, however? Skeptics have asked whether the loss of a few endangered species will really make much difference in an ecosystem's ability to function. Ecological research suggests that the answer to this question depends on which species are removed. Removing a species that can be functionally replaced by others may make little difference. Recall, however, our discussion of keystone species (• pp. 151–154). Like the keystone that holds together an arch, removal of a keystone species results in significant changes in an ecological system. If a keystone species is extirpated or driven extinct, other species may decline or disappear as a result.

Top predators such as tigers are often considered keystone species. A single top predator may prey on many other carnivores, each of which may prey on many herbivores, each of which may consume many plants. Thus the removal of a single individual at the top of a food chain can have impacts that multiply as they cascade down the food chain. Moreover, top predators such as tigers, wolves, and grizzly bears are among the species most vulnerable to human impact. Large animals are frequently hunted and also need large areas of habitat, making them susceptible to habitat loss and fragmentation. Top predators are also vulnerable to the buildup of toxic pollutants in their tissues through the process of biomagnification (• pp. 396–397).

The influence of "ecosystem engineers" (• p. 154) such as ants and earthworms can be every bit as far-reaching as those of keystone species. Ecosystems are complex, and it is difficult to predict which particular species may be important. Thus, many people prefer to apply the precautionary principle in the spirit of Aldo Leopold (• pp. 34–35), who advised, "To keep every cog and wheel is the first precaution of intelligent tinkering."

## Biodiversity enhances food security

Biodiversity benefits our agriculture. As our discussion of native landraces of corn in Oaxaca, Mexico, in Chapter 10 showed, genetic diversity within crop species and their ancestors is enormously valuable. In 1995, Turkey's wheat crops received at least \$50 billion worth of disease resistance from wild wheat strains. California's barley crops annually receive \$160 million in disease resistance benefits from Ethiopian strains of barley. During the 1970s a researcher discovered a maize species in Mexico known as *Zea diploperennis.* This maize is highly resistant to disease, and it is a perennial, meaning it will grow back year after year without being replanted. At the time of its discovery, its entire range was limited to a 10-ha (25-acre) plot of land in the mountains of the Mexican state of Jalisco.

In addition, new potential food crops are waiting to be used (**Figure 11.17**). The babassu palm *(Orbignya phalerata)* of the Amazon produces more vegetable oil than any other plant. The serendipity berry *(Dioscoreophyllum cumminsii)* produces a sweetener that is 3,000 times sweeter than table sugar. Several species of salt-tolerant grasses and trees are so hardy that farmers can irrigate them with saltwater. These same plants also produce animal feed, a substitute for conventional vegetable oil, and other economically important products.

**Food Security and Biodiversity: Potential new food sources**

| Species | Native to... | Potential uses and benefits |
|---|---|---|
| Amaranths (three species of *Amaranthus*) | Tropical and Andean America | Grain and leafy vegetable; livestock feed; rapid growth, drought resistant |
| Buriti palm *(Mauritia flexuosa)* | Amazon lowlands | "Tree of life" to Amerindians; vitamin-rich fruit; pith as source for bread; palm heart from shoots |
| Maca *(Lepidium meyenii)* | Andes Mountains | Cold-resistant root vegetable resembling radish, with distinctive flavor; near extinction |
| Tree tomato *(Cyphomandra betacea)* | South America | Elongated fruit with sweet taste |
| Babirusa *(Babyrousa babyrussa)* | Indonesia: Moluccas and Sulawesi | A deep-forest pig; thrives on vegetation high in cellulose and hence less dependent on grain |
| Capybara *(Hydrochoeris hydrochoeris)* | South America | World's largest rodent; meat esteemed; easily ranched in open habitats near water |
| Vicuna *(Lama vicugna)* | Central Andes | Threatened species related to llama; valuable source of meat, fur, and hides; can be profitably ranched |
| Chachalacas (*Ortalis*, many species) | South and Central America | Birds, potentially tropical chickens; thrive in dense populations; adaptable to human habitations; fast-growing |
| Sand grouse (*Pterocles*, many species) | Deserts of Africa and Asia | Pigeon-like birds adapted to harshest deserts; domestication a possibility |

**FIGURE 11.17** By protecting biodiversity, we can enhance food security. The wild species shown here are a tiny fraction of the many plants and animals that could someday supplement our food supply. Adapted from Wilson, E. O. 1992. *The diversity of life.* Cambridge, MA: Belknap Press.

## Organisms provide drugs and medicines

People have made medicines from plants for centuries, and many of today's widely used drugs were discovered by studying chemical compounds present in wild plants, animals, and microbes (**Figure 11.18**). Each year pharmaceutical products owing their origin to wild species generate up to \$150 billion in sales.

It can truly be said that every species that goes extinct represents one lost opportunity to find a cure for cancer or AIDS. The rosy periwinkle *(Catharanthus roseus)* produces compounds that treat Hodgkin's disease and a particularly deadly form of leukemia. Had this native plant of Madagascar become extinct prior to its discovery by medical researchers, two deadly diseases would have claimed far more victims than they have to date.

In Australia, where the government has placed high priority on research into products from rare and endangered species, a rare species of cork, *Duboisia leichhardtii,* now provides hyoscine, a compound that physicians use to treat cancer, stomach disorders, and motion sickness. Another Australian plant, *Tylophora,* provides a drug that treats lymphoid leukemia. Researchers are now exploring the potential of the compound prostaglandin E2 in treating gastric ulcers. This compound was first discovered in two frog species unique to the rainforest of Queensland, Australia. Scientists believe that both species are now extinct.

### Weighing the Issues | Bioprospecting in Costa Rica

Bioprospectors working for pharmaceutical companies scour biodiversity-rich countries, searching for organisms that can provide new drugs, foods, medicines, or other valuable products. Many have been criticized for "biopiracy"—harvesting indigenous species to create commercial products without compensating the country of origin. To make sure it would not lose the benefits of its own biodiversity, the nation of Costa Rica reached an agreement with the Merck pharmaceutical company in 1991. The nonprofit National Biodiversity Institute of Costa Rica (INBio) allowed Merck to evaluate a number of Costa Rica's species for their commercial potential in return for $1.1 million, a small royalty rate on any products developed, and training for Costa Rican scientists.

Do you think that both sides win in this agreement? What if Merck discovers a compound that could be turned into a billion-dollar drug? Does this provide a good model for other countries? For other companies?

**Medicines and Biodiversity: Natural sources of pharmaceuticals**

| Plant | Drug | Medical application |
|---|---|---|
| Pineapple (*Ananas comosus*)  | Bromelain | Controls tissue inflammation |
| Autumn crocus (*Colchicum autumnale*)  | Colchicine | Anticancer agent |
| Yellow cinchona (*Cinchona ledgeriana*)  | Quinine | Antimalarial |
| Common thyme (*Thymus vulgaris*)  | Thymol | Cures fungal infection |
| Pacific yew (*Taxus brevifolia*)  | Taxol | Anticancer (especially ovarian cancer) |
| Velvet bean (*Mucuna deeringiana*)  | L-Dopa | Parkinson's disease suppressant |
| Common foxglove (*Digitalis purpurea*)  | Digitoxin | Cardiac stimulant |

**FIGURE 11.18** By protecting biodiversity, we can enhance our ability to treat illness. Shown here are just a few of the plants that have so far been found to provide chemical compounds of medical benefit. Adapted from Wilson, E. O. 1992. *The diversity of life.* Cambridge, MA: Belknap Press.

## Biodiversity generates economic benefits through tourism and recreation

Besides providing for our food and health, biodiversity can represent a direct source of income through tourism, particularly for developing countries in the tropics that boast impressive species diversity. Many people like to travel to experience protected natural areas, and in so doing they create economic opportunity for residents living near those areas. Visitors spend money at local businesses, hire local people as guides, and support parks that employ local residents. Ecotourism (• pp. 135–136) thus can bring jobs and income to areas that otherwise might be poverty-stricken.

Ecotourism has become a vital source of income for nations such as Costa Rica, with its rainforests; Australia, with its Great Barrier Reef; Belize, with its reefs, caves, and rainforests; and Kenya and Tanzania, with their savanna wildlife. The United States, too, benefits from ecotourism; its national parks draw millions of visitors domestically and from around the world. Ecotourism serves as a powerful financial incentive for nations, states, and local communities to preserve natural areas and reduce impacts on the landscape and on native species.

As ecotourism increases in popularity, however, critics have warned that too many visitors to natural areas can degrade the outdoor experience and disturb wildlife. Anyone who has been to Yosemite, the Grand Canyon, or the Great Smokies on a crowded summer weekend can attest to this. Ecotourism's effects on species living in parks and reserves are much debated, and likely they vary enormously from one case to the next. As ecotourism continues to increase, so will debate over its costs and benefits for local communities and for biodiversity.

FIGURE 11.19 Edward O. Wilson is the world's most recognized authority on biodiversity and its conservation and has inspired many people who study our planet's life. A Harvard professor and world-renowned expert on ants, Wilson has written over 20 books and has won two Pulitzer prizes. His books *The Diversity of Life* and *The Future of Life* address the value of biodiversity and its outlook for the future.

## People value and seek out connections with nature

Not all of the benefits of biodiversity to humans can be expressed in the hard numbers of economics or the day-to-day practicalities of food and medicine. Some scientists and philosophers argue that there is a deeper importance to biodiversity. E. O. Wilson (**Figure 11.19**) has described a phenomenon he calls **biophilia,** "the connections that human beings subconsciously seek with the rest of life." Wilson and others have cited as evidence of biophilia our affinity for parks and wildlife, our keeping of pets, the high value of real estate with a view of natural landscapes, and our interest—despite being far removed from a hunter-gatherer lifestyle—in hiking, bird-watching, fishing, hunting, backpacking, and similar outdoor pursuits.

In a 2005 book, writer Richard Louv adds that as today's children are increasingly deprived of outdoor experiences and direct contact with wild organisms, they suffer what he calls "nature-deficit disorder." Although it is not a medical condition, this alienation from biodiversity and the natural environment, Louv argues, may damage childhood development and lie behind many of the emotional and physical problems young people in developed nations face today.

### Weighing the Issues | Biophilia and Nature-Deficit Disorder

What do you think of the concepts of biophilia and "nature-deficit disorder"? Have you ever felt a connection to other living things that you couldn't explain in scientific or economic terms? Do you think that an affinity for other living things is innately human?

## Do we have ethical obligations toward other species?

If Wilson, Louv, and others are right, then biophilia not only affects ecotourism and real estate prices, but also influences our ethics. When Maurice Hornocker and his associates first established the Siberian Tiger Project, he wrote: "Saving the most magnificent of all the cat species and one of the most endangered should be a global responsibility. . . . If they aren't worthy of saving, then what are we all about? What is worth saving?"

We humans are part of nature, and like any other animal we need to use resources and consume other organisms to survive. In that sense, there is nothing immoral about our doing so. However, we have conscious reasoning ability and are able to control our actions and make conscious decisions. Our ethical sense has developed from this intelligence and ability to choose. As our society's sphere of ethical consideration has widened over time, and as more of us take up biocentric or ecocentric worldviews (• pp. 31–32), more people have come to feel that other organisms have intrinsic value and an inherent right to exist.

Despite our ethical convictions and despite biodiversity's many benefits, the future of biodiversity remains far from secure. Even our protected areas and national parks are not large enough or well enough protected to ensure that biodiversity is fully safeguarded within their borders. The search for solutions to today's biodiversity crisis is an exciting and active one, and scientists are playing a leading role in developing innovative approaches to maintaining the diversity of life on Earth.

## Conservation Biology: The Search for Solutions

Today, more and more scientists and citizens perceive a need to do something to stem the loss of biodiversity. In his 1994 autobiography, *Naturalist*, E. O. Wilson wrote:

> When the [20th] century began, people still thought of the planet as infinite in its bounty. The highest mountains were still unclimbed, the ocean depths never visited, and vast wildernesses stretched across the equatorial continents. . . . In one lifetime exploding human populations have reduced wildernesses to threatened nature reserves. Ecosystems and species are vanishing at the fastest rate in 65 million years. Troubled by what we have wrought, we have begun to turn in our role from local conqueror to global steward.

### Conservation biology arose in response to biodiversity loss

The urge to act as responsible stewards of natural systems, and to use science as a tool in that endeavor, helped spark the rise of conservation biology. **Conservation biology** is a scientific discipline devoted to understanding the factors, forces, and processes that influence the loss, protection, and restoration of biological diversity. It arose as biologists became increasingly alarmed at the degradation of the natural systems they had spent their lives studying.

Conservation biologists choose questions and pursue research with the aim of developing solutions to such problems as habitat degradation and species loss (**Figure 11.20**).

FIGURE 11.20 Conservation biologists integrate lab and field research to develop solutions to biodiversity loss. Here, a conservation biologist checks on a jabiru stork nest in the Pantanal region of Brazil.

Conservation biology is thus an applied and goal-oriented science, with implicit values and ethical standards. This perceived element of advocacy sparked some criticism of conservation biology in its early years. However, as scientists have come to recognize the scope of human impact on the planet, more of them have directed their work to address environmental problems. Today conservation biology is a thriving pursuit that is central to environmental science and to achieving a sustainable society.

### Conservation biologists work at multiple levels

Conservation biologists integrate an understanding of evolution and extinction with ecology and the dynamic nature of environmental systems. They use field data, lab data, theory, and experiments to study the impacts of humans on other organisms. They also attempt to design, test, and implement ways to mitigate human impact.

These researchers address the challenges facing biological diversity at all levels, from genetic diversity to species diversity to ecosystem diversity. At the genetic level, *conservation geneticists* study genetic attributes of organisms, generally to infer the status of their populations. If two populations of a species are found to be genetically distinct enough to be considered subspecies, they may have different ecological needs and may require different types of management.

In addition, as a population dwindles, genetic variation is lost from the gene pool. Conservation geneticists ask how small a population can become and how much genetic variation it can lose before running into problems such as inbreeding depression. By determining a *minimum viable population size* for a given population, conservation geneticists and population biologists provide wildlife managers with an indication of how vital it may be to increase the population.

Problems for populations and subspecies spell problems for species, because declines and local extirpation generally precede range-wide endangerment and extinction. As we will soon see, it is at the species level that much of the funding and resources for conservation biology exist.

Many efforts also revolve around habitats, communities, ecosystems, and landscapes—and these efforts are often informed by studies of genes, populations, and species. As we saw in our discussion of landscape ecology (• pp. 183–185), organisms are sometimes distributed across a landscape as a *metapopulation*, or a network of subpopulations. Because small and isolated subpopulations are most vulnerable to extirpation, conservation biologists pay special attention to them. By examining how organisms disperse from one habitat patch to another, and how their genes flow among subpopulations, conservation biologists try to learn how likely a population is to persist or succumb in the face of habitat change or other threats.

## Island biogeography theory is a key component of conservation biology

Safeguarding habitat for species and conserving communities and ecosystems requires thinking and working at the landscape level. One key conceptual tool for doing so is the **equilibrium theory of island biogeography.** This theory, introduced by E. O. Wilson and ecologist Robert MacArthur in 1963, explains how species come to be distributed among oceanic islands. Since then, researchers have also applied it to "habitat islands"—patches of one habitat type isolated within "seas" of others. The Sikhote-Alin Mountains, last refuge of the Siberian tiger, are a habitat island, isolated from other mountains by deforested regions, a seacoast, and populated lowlands.

Island biogeography theory explains how the number of species on an island results from an equilibrium balance between the number added by immigration and the number lost through extirpation. It predicts an island's species richness based on the island's size and its distance from the mainland:

- The farther an island is located from a continent, the fewer species tend to find and colonize it. Thus, remote islands host fewer species because of lower immigration rates (**Figure 11.21a**). This is called the *distance effect.*
- Large islands have higher immigration rates because they present fatter targets for dispersing organisms to encounter (**Figure 11.21b**).
- Large islands have lower extinction rates because more space allows for larger populations, which are less vulnerable to dropping to zero by chance (**Figure 11.21c**).

Together, these latter two trends give large islands more species at equilibrium than small islands—a phenomenon called the *area effect.* Large islands also tend to contain more species because they generally possess more habitats than smaller islands, providing suitable environments for a wider variety of arriving species. Very roughly, the number of species on an island is expected to double as island size increases tenfold. This effect can be illustrated with *species-area curves* (**Figure 11.22**).

These theoretical patterns have been widely supported by empirical data (see "The Science behind the Story," • pp. 318–319). The patterns hold up for terrestrial habitat islands, such as forests fragmented by logging and road building (**Figure 11.23**). Small islands of forest lose their diversity fastest, starting with large species that were few in number to begin with. In a landscape of fragmented habitat, species requiring the habitat will gradually disappear, winking out from one fragment after another over time.

Fragmentation of forests and other habitats constitutes one of the prime threats to biodiversity. In response to habitat fragmentation, conservation biologists have designed landscape-level strategies to try to optimize the

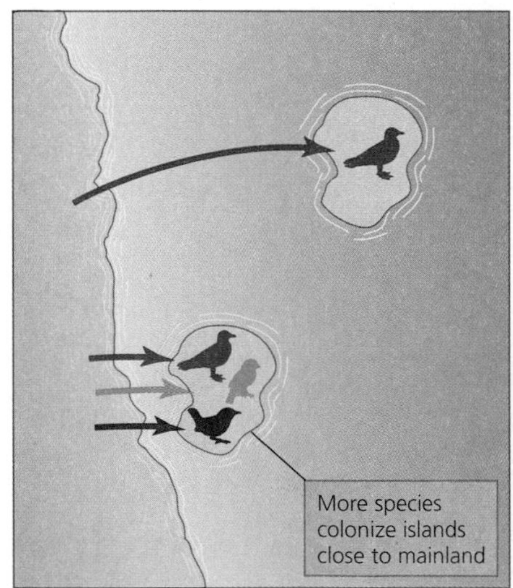


**(a) Distance effect**



**(b) Target size**

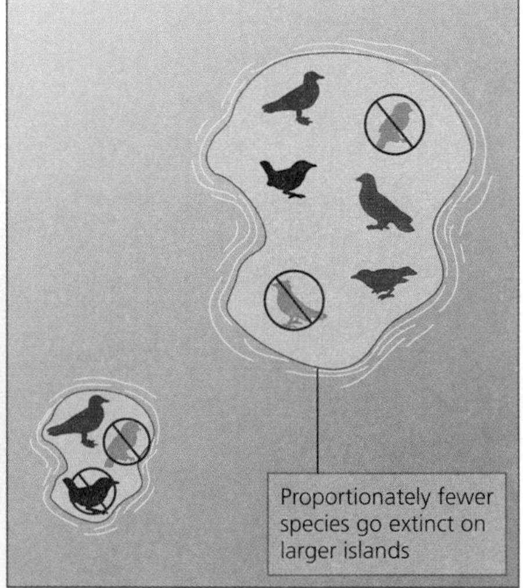


**(c) Differential extinction**

**FIGURE 11.21** Island biogeography theory explains species richness on islands as a function of immigration and extinction rates interacting with island size and distance from the mainland. Islands located close to a continent receive more immigrants than islands that are distant (**a**), so that near islands end up with more species. Large islands present fatter targets for dispersing organisms to encounter (**b**), so that more species immigrate to large islands than to small islands. Large islands also experience lower extinction rates (**c**), because their larger area allows for larger populations.

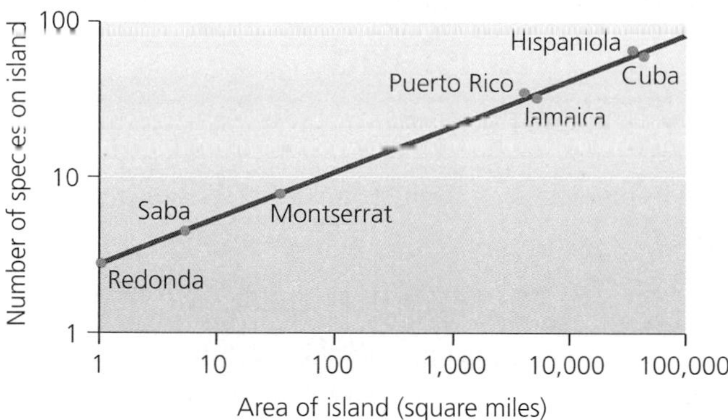


FIGURE 11.22 The larger the island, the greater the number of species—a prediction borne out by data from around the world. By plotting the number of amphibians and reptile species on Caribbean islands as a function of the areas of these islands, this species-area curve shows that species richness increases with area. The increase is not linear, but logarithmic; note the scales of the axes. Go to GRAPHit! at www.aw-bc.com/withgott or on the student CD-ROM. Data from MacArthur, R. H., and E. O. Wilson. 1967. *The theory of island biogeography.* Princeton University Press.

arrangement of areas to be preserved. We will examine a few of these strategies in our discussion of parks and preserves in Chapter 12 (• pp. 352–353).

**Weighing THE Issues** | **Fragmentation and Biodiversity**

Suppose a critic of conservation tells you that human development increases biodiversity, pointing out that when a forest is fragmented, new habitats, such as grassy lots and gardens, may be introduced to an area and allow additional species to live there. How would you respond?

## Should endangered species be the focus of conservation efforts?

The primary legislation for protecting biodiversity in the United States is the **Endangered Species Act (ESA).** Passed in 1973, the ESA forbids the government and private citizens from taking actions (such as developing land) that destroy endangered species or their habitats. The ESA also forbids trade in products made from endangered species. The aim is to prevent extinctions, stabilize declining populations, and enable populations to recover. As of mid-2007, there were 1,312 species in the United States listed as "endangered" or as "threatened," the latter status considered one notch less severe than endangered.

The ESA has had a number of notable successes. Following the ban on the pesticide DDT and years of intensive effort by wildlife managers, the peregrine falcon, brown pelican, bald eagle, and other birds have recovered and are no longer listed as endangered (• p. 396). Intensive management programs with other species, such as the red-cockaded woodpecker, have held formerly declining populations steady in the face of continued pressure on habitat. In fact, roughly 40% of declining populations are now stable.

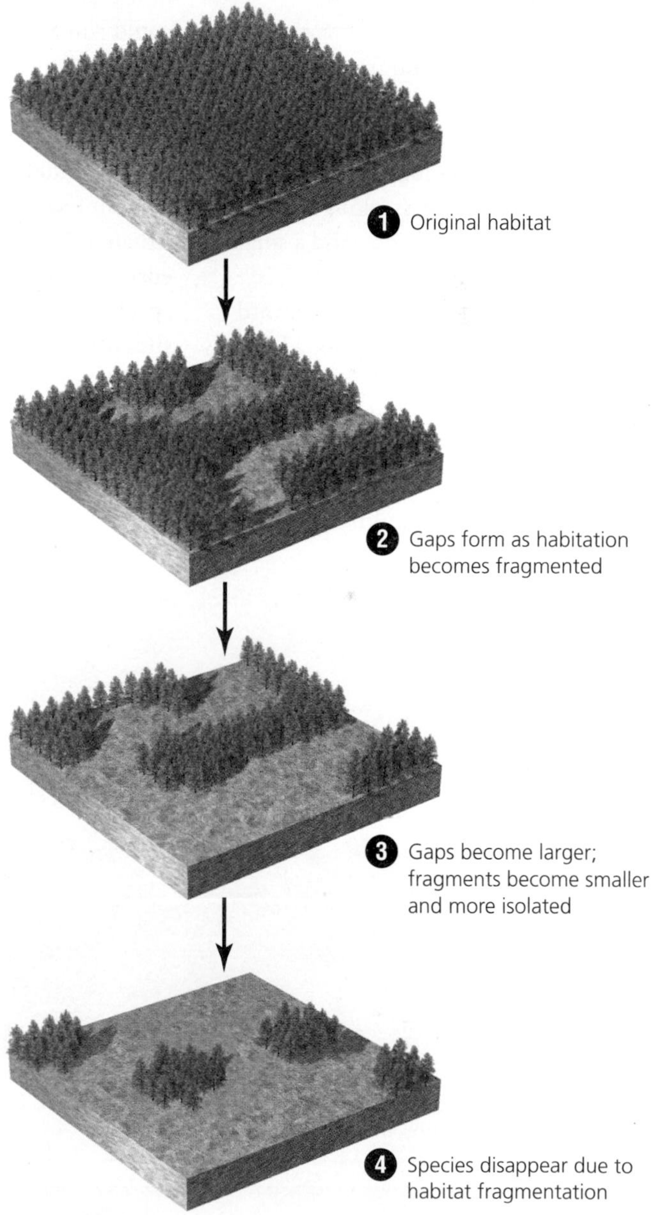


FIGURE 11.23 Forest clearing, farming, road building, and other types of human land use and development can fragment natural habitats. Habitat fragmentation usually begins when gaps are created within a natural habitat. As development proceeds, these gaps expand, join together, and eventually dominate the landscape, stranding islands of habitat in their midst. As habitat becomes fragmented, fewer populations can persist, and numbers of species in the fragments decrease with time.

THE SCIENCE BEHIND THE STORY

# Testing and Applying Island Biogeography Theory

*Dr. Daniel Simberloff, today a distinguished professor at the University of Tennessee*

The researchers who first experimentally tested the equilibrium theory of island biogeography and applied it to conservation biology leaned on their own resourcefulness, along with some plastic, some pesticides, and a small station wagon.

Robert MacArthur and Edward O. Wilson originally developed island biogeography theory by using observational data from oceanic islands, correlating numbers of species found on islands with island size and distance between landmasses. Yet as of 1966, no one had tested its precepts in the field with a manipulative experiment. Wilson decided to remedy that.

Wilson began looking for islands in the United States where he could run an experimental test: to remove all animal life from islands and then observe and measure recolonization. To be suitable, the islands would need to be small, contain few forms of life, and be situated close to the mainland to ensure an influx of immigrating species. Wilson found his research sites off the tip of Florida: six small mangrove islands 11–18 m (36–59 ft) in diameter, home only to trees and a few dozen species of insects, spiders, centipedes, and other arthropods.

Daniel Simberloff and E. O. Wilson used mangrove islands in the Florida Keys to test island biogeography theory.

Daniel Simberloff, Wilson's graduate student at the time, painstakingly counted each island's arthropods, breaking up bark and poking under branches to find every mite, midge, and millipede. Then with the help of professional exterminators, Wilson and Simberloff wrapped each island in a plastic tarpaulin and gassed the interior with the pesticide methyl bromide. After removing the tarpaulins, Simberloff checked to make sure no creatures were left alive. The researchers then waited to see how life on the islands would return.

Over the next 2 years, Simberloff scrambled up trees and turned over leaves, looking for newly arrived organisms. His monitoring showed that life recovered on most islands within a year, regaining about the same number of species and total number of arthropods the islands had sheltered originally. Larger islands once again became home to a greater number of species than smaller islands. Outlying islands recovered more slowly and reached lower species diversities than did islands near the mainland. These results provided the first evidence from a manipulative experiment for the predictions of island biogeography theory.

Published in the journal *Ecology* in 1969 and 1970, Simberloff and Wilson's research gave new empirical

This success comes despite the fact that the U.S. Fish and Wildlife Service and the National Marine Fisheries Service, the agencies responsible for upholding the ESA, are perennially underfunded for the job. In recent years, the Fish and Wildlife Service has been caught between political influence (e.g., • p. 77) from a presidential administration opposed to endangered species conservation and lawsuits from environmental groups demanding

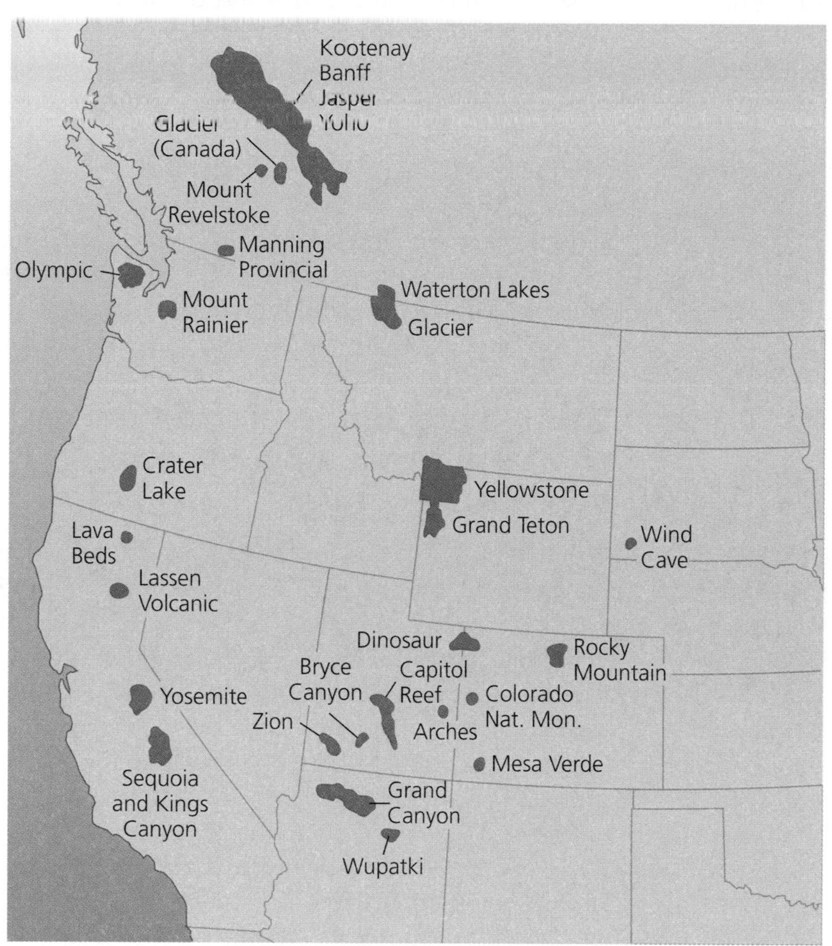


William Newmark visualized national parks of western North America as islands of natural habitat in a sea of development. His data showed that mammal species were disappearing from these terrestrial "islands," in accordance with island biogeography theory. Adapted from Quammen, D. 1997. *The song of the dodo.* New York: Simon & Schuster.

rigor to a set of ideas that was increasingly helping scientists understand geographic patterns of biodiversity.

Their research also fueled a question of pressing concern for conservation biology: Could island biogeography theory also be applied to isolated "islands" of habitat on continents? At the University of Michigan, biology graduate student William Newmark set out to address this question.

Newmark had learned that many North American national parks kept records that documented sightings of wildlife over the course of the parks' existence. He surmised that by examining these historical records, he could infer which species had vanished from parks, which were new arrivals, and roughly when these changes occurred. The parks, increasingly surrounded by development, were islands of natural habitat isolated by farms, roads, towns, and cities, so Newmark hypothesized that island biogeography theory would apply.

In 1983, Newmark drove his Toyota station wagon to 24 parks in the western United States and Canada. At each park, he studied the wildlife sighting records, focusing on larger mammals such as bear, lynx, and river otter (but not species, such as wolves, that had been deliberately eradicated by hunting).

Newmark found a few species missing from many parks, and they added up to a troubling total. Forty-two species had disappeared in all, and not as a result of direct human action. The red fox and river otter had vanished from Sequoia and Kings Canyon National Parks, for example, and the white-tailed jackrabbit and spotted skunk no longer lived in Bryce Canyon National Park. As theory predicted, the smallest parks showed the greatest number of losses, and the largest parks retained a greater number of species.

These species disappeared because the parks, Newmark concluded, were too small to sustain their populations in the long term. Moreover, the parks had become island habitats too isolated to be recolonized by new arrivals. Newmark's findings, published in the journal *Nature* in 1987, placed island biogeography theory squarely on the mainland. Today, this theory helps inform national park policy and biodiversity conservation plans around the world.

enforcement of current law. Moreover, efforts to reauthorize the ESA faced stiff opposition in Republican Congresses from the 1990s until 2006. Efforts to weaken the ESA by stripping it of its ability to safeguard habitat were narrowly averted in 2006 after 5,700 scientists sent Congress a letter of protest.

Polls repeatedly show that most Americans support the idea of protecting endangered species. Yet some oppose and resent provisions of the ESA. Many opponents feel that the ESA places more value on the life of an endangered organism than it does on the livelihood of a person. This was a common perception in the Pacific Northwest in the 1990s, when protection for the northern spotted owl (see Figure 23.8, • p. 667) slowed logging in old-growth rainforest and many loggers began to fear for their jobs.

Resentment toward the ESA also comes from landowners worried that federal officials will restrict the use of private land if threatened or endangered species are found on it. This has led in many cases to a practice described as "shoot, shovel, and shut up," among landowners who want to conceal the presence of such species on their land.

ESA supporters maintain that such fears are overblown, pointing out that the ESA has stopped few development projects. Moreover, a number of provisions of the ESA and its amendments promote cooperation with landowners. *Habitat conservation plans* and *safe harbor agreements* allow landowners to harm species in some ways if they voluntarily improve habitat for the species in others.

Debate over the U.S. law has influenced other nations' approaches to species protection. When Canada enacted its long-awaited endangered species law in 2002, the *Species at Risk Act (SARA),* the Canadian government was careful to stress cooperation with landowners and provincial governments, rather than presenting the law as a decree from the national government. Canada's environment minister, David Anderson, wanted none of the hostility the U.S. act had unleashed. Environmentalists and many scientists, however, protested that SARA was too weak and failed to protect habitat adequately.

Today a number of nations have laws protecting species, although they are not always well enforced. In Russia, the government issued Decree 795 in 1995, creating a Siberian tiger conservation program and declaring the tiger one of the nation's most important natural and national treasures. However, funding from the state for tiger conservation is so meager that the Wildlife Conservation Society feels it necessary to help pay for Russians to enforce their own anti-poaching laws.

## Captive breeding, reintroduction, and cloning are single-species approaches

In the effort to save threatened and endangered species, zoos and botanical gardens have become centers for **captive breeding,** in which individuals are bred and raised in controlled conditions with the intent of reintroducing them into the wild. One example is the program to save the California condor, North America's largest bird (**Figure 11.24**). Condors were persecuted in the early 20th century, collided with electrical wires, and succumbed to lead poisoning from scavenging carcasses of animals killed with lead shot. By 1982, only 22 condors remained, and biologists decided to take all the birds into captivity, in hopes of boosting their numbers and then releasing them. The ongoing program is succeeding. So far, over 100 of the 250 birds raised in captivity have been released into the wild at sites in California and Arizona, where a few pairs have begun nesting.

FIGURE 11.24 In efforts to save the California condor *(Gymnogyps californianus)* from extinction, biologists have raised hundreds of chicks in captivity with the help of hand puppets designed to look and feel like the heads of adult condors. Using these puppets, biologists feed the growing chicks in an enclosure and shield them from all contact with humans, so that when the chick is grown it does not feel an attachment to people.

Other reintroduction programs have been more controversial. The successful program to reintroduce wolves to Yellowstone National Park has proven popular with the American public but met stiff resistance from ranchers, who fear the wolves will attack their livestock. In Arizona and New Mexico, a wolf reintroduction program is making slow headway amid occasional shootings of wolves.

For the Siberian tiger, China is considering a similar reintroduction program. The Chinese government says it is preparing 600 captive Siberian tigers for release into its forests in the far northeastern portion of the country, but critics note that the forests are so fragmented that efforts would be better focused on improving habitat first.

The newest idea for saving species from extinction is to create more individuals by cloning them. In this technique, DNA from an endangered species is inserted into a cultured egg without a nucleus, and the egg is implanted into a closely related species that can act as a surrogate mother. So far two Eurasian mammals have been cloned in this way. With future genetic technology, some scientists even talk of recreating extinct species from DNA recovered from preserved body parts. However, even if cloning can succeed from a technical standpoint, most biologists agree that such efforts are not an adequate response to biodiversity loss. Without ample habitat and protection in the wild, having cloned animals in a zoo does little good.

## Some species act as "umbrellas" for protecting habitat and communities

Protecting habitat and conserving communities, ecosystems, and landscapes are the goals of many conservation biologists. Often, they use particular species essentially as tools to conserve communities and ecosystems. This is because the ESA provides legal justification and resources for species conservation, but no such law exists for communities or ecosystems. Large species that roam great distances, such as the Siberian tiger, require large areas of habitat. Meeting the habitat needs of these so-called *umbrella species* automatically helps meet those of thousands of less charismatic animals, plants, and fungi that would never elicit as much public interest.

Environmental advocacy organizations have found that using large and charismatic vertebrates as spearheads for biodiversity conservation is an effective strategy. This approach of promoting particular *flagship species* is evident in the longtime symbol of the World Wide Fund for Nature (World Wildlife Fund in North America), the panda. The panda is a large endangered animal requiring sizeable stands of undisturbed bamboo forest. Its lovable appearance has made it a favorite with the public—and an effective tool for soliciting funding for conservation efforts that protect far more than just the panda.

At the same time, many conservation organizations today are moving beyond the single-species approach. The Nature Conservancy, for instance, has in recent years focused more on whole communities and landscapes. The most ambitious effort may be the Wildlands Project, a group proposing to restore huge amounts of North America's land to its presettlement state.

### Weighing THE Issues | Single-Species Conservation?

What would you say are some advantages of focusing on conserving single species, versus trying to conserve broader communities, ecosystems, or landscapes? What might be some of the disadvantages? Which do you think is the better approach, or should we use both?

## International conservation efforts include widely signed treaties

At the international level, the United Nations has facilitated several treaties to protect biodiversity. The 1973 **Convention on International Trade in Endangered Species of Wild Fauna and Flora (CITES)** protects endangered species by banning the international transport of their body parts. When nations enforce it, CITES can protect tigers and other rare species whose body parts are traded internationally.

In 1992, leaders of many nations agreed to the **Convention on Biological Diversity.** This treaty embodies three goals: to conserve biodiversity, to use biodiversity in a sustainable manner, and to ensure the fair distribution of biodiversity's benefits. The Convention aims to help

- Provide incentives for biodiversity conservation
- Manage access to and use of genetic resources
- Transfer technology, including biotechnology
- Promote scientific cooperation
- Assess the effects of human actions on biodiversity
- Promote biodiversity education and awareness
- Provide funding for critical activities
- Encourage every nation to report regularly on their biodiversity conservation efforts

The treaty's many accomplishments so far include ensuring that Ugandan people share in the economic benefits of wildlife preserves, increasing global markets for "shade-grown" coffee and other crops grown without removing forests, and replacing pesticide-intensive farming practices with sustainable ones in some rice-producing Asian nations. As of 2007, 188 nations had become parties to the Convention on Biological Diversity. Those choosing *not* to do so include Iraq, Somalia, the Vatican, and the

United States. This decision is just one example of why the U.S. government is no longer widely regarded as a leader in biodiversity conservation efforts.

## Biodiversity hotspots pinpoint areas of high diversity

One international approach oriented around geographic regions, rather than single species, is the effort to map **biodiversity hotspots.** The concept of biodiversity hotspots was introduced in 1988 by British ecologist Norman Myers as a way to prioritize regions that are most important globally for biodiversity conservation. A hotspot supports an especially great number of species that are **endemic** (• p. 121) to the area, that is, found nowhere else in the world (**Figure 11.25**). To qualify as a hotspot, a location must harbor at least 1,500 endemic plant species, or 0.5% of the world total. In addition, a hotspot must have already lost 70% of its habitat as a result of human impact and be in danger of losing more.

The nonprofit group Conservation International maintains a list of 34 biodiversity hotspots (**Figure 11.26**). The ecosystems of these areas together once covered 15.7% of the planet's land surface but today, because of habitat loss, cover only 2.3%. This small amount of land is the exclusive home for 50% of the world's plant species and 42% of all terrestrial vertebrate species. The hotspot concept gives incentive to focus on these areas of endemism, where the greatest number of unique species can be protected with the least amount of effort.

FIGURE 11.25 The golden lion tamarin *(Leontopithecus rosalia)*, a species endemic to Brazil's Atlantic rainforest, is one of the world's most endangered primates. Captive breeding programs have produced roughly 500 individuals in zoos, but the tamarin's habitat is fast disappearing.

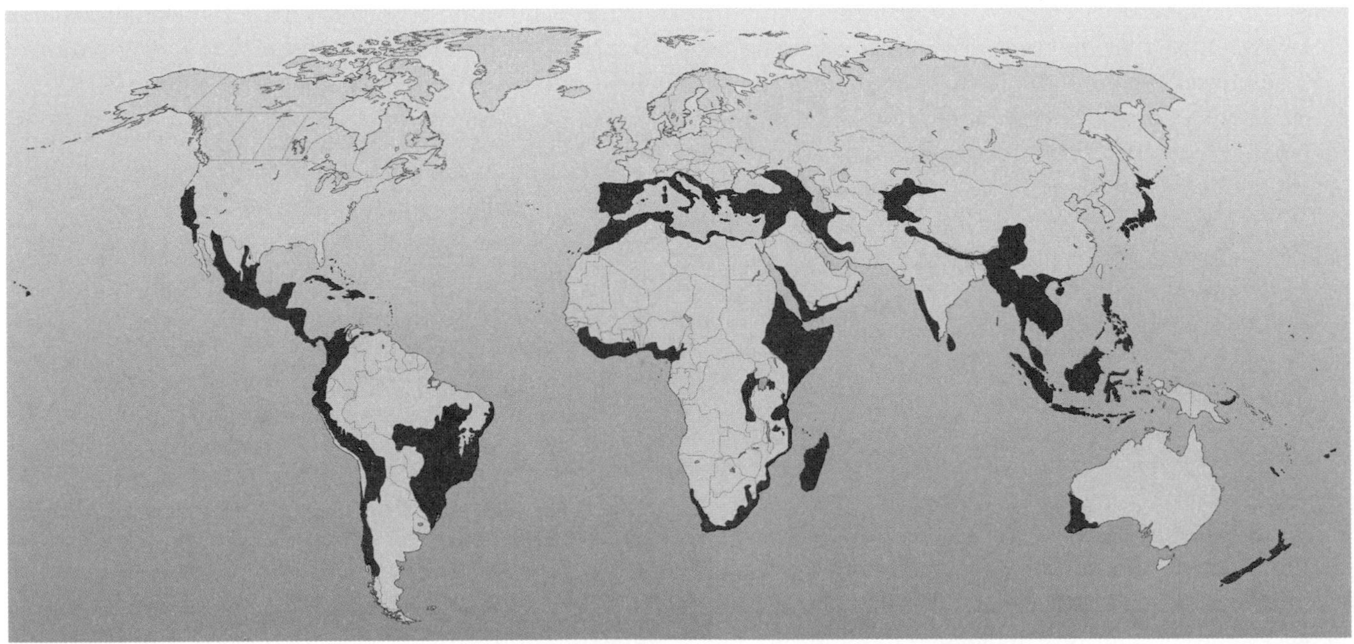

FIGURE 11.26 Some areas of the world possess exceptionally high numbers of species found nowhere else. Many conservation biologists have supported prioritizing habitat preservation in these areas, dubbed *biodiversity hotspots.* Shown in red are the 34 biodiversity hotspots mapped by Conservation International. Data from Conservation International, 2005.

## Community-based conservation is increasingly popular

Taking a global perspective and prioritizing optimal locations to set aside as parks and reserves makes good sense, but setting aside land for preservation affects the people that live in and near these areas. In past decades, many conservationists from developed nations, in their zeal to preserve ecosystems in other nations, too often neglected the needs of people in the areas they wanted to protect. Many developing nations came to view this international environmentalism as a kind of neocolonialism.

Today this has largely changed, and many conservation biologists actively engage local people in efforts to protect land and wildlife in their own backyards, in an approach sometimes called **community-based conservation.** Setting aside land for preservation deprives local people of access to natural resources, but it can also guarantee that these resources will not be used up or sold to foreign corporations and can instead be sustainably managed. Moreover, parks and reserves draw ecotourism that can support local economies.

In the small Central American nation of Belize, conservation biologist Robert Horwich and his Wisconsin-based group Community Conservation, Inc., have helped start a number of community-based conservation projects. The Community Baboon Sanctuary consists of tracts of riparian forest that farmers have agreed to leave intact, to serve as homes and traveling corridors for the black howler monkey, a centerpiece of ecotourism. The reserve uses the local nickname for the monkey, signaling respect for residents, and today a local women's cooperative is running the project. Residents receive income for guiding and housing visiting researchers and tourists in a new museum built with Community Conservation funds.

Some other projects have not turned out as well, however. Nearby on the Belizean coast, efforts to create a locally run reserve for the manatee have struggled because of shortfalls in funding. Community-based conservation has not always been successful, but in a world of increasing human population, locally based management that meets people's needs sustainably will be essential.

## Innovative economic strategies are being employed

As conservation moves from single-species approaches to the hotspot approach to community-based conservation, innovative economic strategies are also being attempted. One strategy is the *debt-for-nature swap.* In such a swap, a conservation organization raises money and offers to pay off a portion of a developing country's international debt in exchange for a promise by the country to set aside reserves, fund environmental education, and better manage protected areas.

A newer strategy that Conservation International has pioneered is the *conservation concession.* Nations often sell concessions to foreign multinational corporations, allowing them to extract resources from the nation's land. A nation can, for instance, earn money by selling to an international logging company the right to log its forests. Conservation International has stepped in and paid nations for concessions for conservation rather than resource extraction. The nation gets the money *and* keeps its natural resources intact. The South American country of Surinam, which still has extensive areas of pristine rainforest, entered into such an agreement and has virtually halted logging while pulling in $15 million. It remains to be seen how large a role such strategies will play in the future protection of biodiversity.

# Conclusion

The erosion of biological diversity on our planet threatens to result in a mass extinction event equivalent to the mass extinctions of the geological past. Human-induced habitat alteration, invasive species, pollution, overharvesting of biotic resources, and climate change are the primary causes of biodiversity loss. This loss matters, because human society could not function without biodiversity's pragmatic benefits. As a result, conservation biologists are rising to the challenge of conducting science aimed at saving endangered species, preserving their habitats, restoring populations, and keeping natural ecosystems intact. The innovative strategies of these scientists hold promise to slow the erosion of biodiversity that threatens life on Earth.

## REVIEWING OBJECTIVES

**You should now be able to:**

**Characterize the scope of biodiversity on Earth**

- Biodiversity can be thought of at three levels commonly called species diversity, genetic diversity, and ecosystem diversity. (pp. 295–297)
- Roughly 1.7–2 million species have been described so far, but scientists agree that the world holds millions more. (pp. 298–299)
- Some taxonomic groups (such as insects) hold far more diversity than others. (pp. 298–299)

- Global estimates of biodiversity may be based on extrapolations from local areas and certain taxonomic groups. (pp. 298–299)
- Diversity is unevenly spread across different habitats, biomes, and regions of the world. (pp. 299–301)

**Contrast background extinction rates with periods of mass extinction**

- Species have gone extinct at a background rate of roughly one species per 1 to 10 million species each year. Most species that have ever lived are now extinct. (p. 301)
- Earth has experienced five mass extinction events in the past 440 million years. (pp. 301–302)
- Human impact is presently initiating a sixth mass extinction. (pp. 302–304)

**Evaluate the primary causes of biodiversity loss**

- Habitat alteration is the main cause of current biodiversity loss. (pp. 304–305)
- Invasive species, pollution, and overharvesting are also important causes. (pp. 305–309)
- Climate change threatens to become a major cause very soon. (p. 310)

**Specify the benefits of biodiversity**

- Biodiversity is vital for functioning ecosystems and the services they provide us. (pp. 310–312)
- Wild species are sources of food, medicine, and economic development. (pp. 312–313)
- Many people feel that we have a psychological need to connect with the natural world and an ethical duty to preserve nature. (p. 314)

**Assess the science and practice of conservation biology**

- Conservation biology studies biodiversity loss and seeks ways to protect and restore biodiversity. (p. 315)
- Conservation biologists integrate research at the genetic, population, species, ecosystem, and landscape levels. (p. 315)
- Island biogeography theory explains how size and distance influence species richness on islands. It also applies to terrestrial islands of habitat in fragmented landscapes. (pp. 316–319)

**Compare and contrast traditional and innovative biodiversity conservation efforts**

- Most conservation efforts and laws so far have focused on threatened and endangered species. The U.S. Endangered Species Act has been effective but controversial. (pp. 317–320)
- Single-species recovery efforts include captive breeding and reintroduction programs. (pp. 320–321)
- Charismatic and well-known species are often used as tools to conserve habitats and ecosystems. Increasingly, landscape-level conservation is being pursued. (p. 321)
- International conservation approaches include treaties, biodiversity hotspots, community-based conservation, debt-for-nature swaps, and conservation concessions. (pp. 321–323)

## TESTING YOUR COMPREHENSION

1. What is biodiversity? List and describe three levels of biodiversity.
2. What are the five primary causes of biodiversity loss? Can you give a specific example of each?
3. List five invasive species, and describe their impacts.
4. Define the term *ecosystem services.* Give three examples of ecosystem services that humans would have a hard time replacing if their natural sources were eliminated.
5. What is the relationship between biodiversity and food security? Between biodiversity and pharmaceuticals? Give three examples of potential benefits of biodiversity conservation for food security and medicine.
6. Describe four reasons why people suggest biodiversity conservation is important.
7. What is the difference between an umbrella species and a keystone species? Could one species be both an umbrella species and a keystone species?
8. Explain the theory of island biogeography. Use the example of the Siberian tiger to describe how this theory can be applied to fragmented terrestrial landscapes.
9. Name two successful accomplishments of the U.S. Endangered Species Act. Now name two reasons some people have criticized it.
10. What is a biodiversity hotspot? Describe community-based conservation.

## SEEKING SOLUTIONS

1. Many arguments have been advanced for the importance of preserving biodiversity. Which argument do you think is most compelling, and why? Which argument do you think is least compelling, and why?
2. Some people declare that we shouldn't worry about endangered species because extinction has always occurred. How would you respond to this view?
3. Environmental advocates from developed nations who want to preserve biodiversity globally have long pushed to set aside land in biodiversity-rich regions of developing nations. Leaders of developing nations have responded by accusing these advocates of neocolonialism. "Your nations attained their prosperity and power by overexploiting their environments decades or centuries ago," these leaders ask, "so why should we now sacrifice our development by setting aside our land and resources?" What would you say to these leaders? What would you say to the environmental advocates? Do you see ways that both preservation and development goals might be reached?
4. Compare the biodiversity hotspot approach to the approach of community-based conservation. What are the advantages and disadvantages of each? Can we—and should we—follow both approaches?
5. **THINK IT THROUGH** You are an influential legislator in a country that has no endangered species act, and you want to introduce legislation to protect your country's vanishing biodiversity. Consider the U.S. Endangered Species Act and the Canadian Species At Risk Act, as well as international efforts such as CITES and the Convention on Biological Diversity. What strategies would you write into your legislation? How would your law be similar to and different from the U.S., Canadian, and international efforts?
6. **THINK IT THROUGH** As a citizen and resident of your community, and a parent of two young children, you attend a town meeting called to discuss the proposed development of a shopping mall and condominium complex. The development would eliminate a 100-acre stand of forest, the last sizeable forest stand in the town. The developers say the forest loss will not matter because plenty of 1-acre stands still exist scattered throughout town. One of the town's decision-makers recognizes you as someone who hikes, bird-watches, and fishes with your children in the forest, and she asks you to comment about the development's possible impacts on the community's biodiversity. What will you choose to tell your fellow citizens and the town's decision-makers at this meeting?

## INTERPRETING GRAPHS AND DATA

Habitat alteration is the primary cause of present-day biodiversity loss, and of all human activities, the one that has altered the most habitat is agriculture. Between 1850 and 2000, 95% of the grasslands of the Midwestern United States were converted to agricultural use, and industrialized farming practices replaced diverse natural communities with greatly simplified ones. The vast monocultures of industrialized agriculture produce bountiful harvests, but the intensive use of water, chemical fertilizers, and pesticides bring substantial costs in lost ecosystem services.

Recently, researchers surveyed the scientific literature, reviewing all research that compared the effects on biodiversity of organic farming (• pp. 285–289) versus conventional industrialized farming practices. The scientists predicted organic farming would benefit most organisms because of reduced chemical pollution. The graph shows the overall number of studies that showed positive effects (such as increases in abundance or species richness) and negative effects of organic farming on biodiversity.

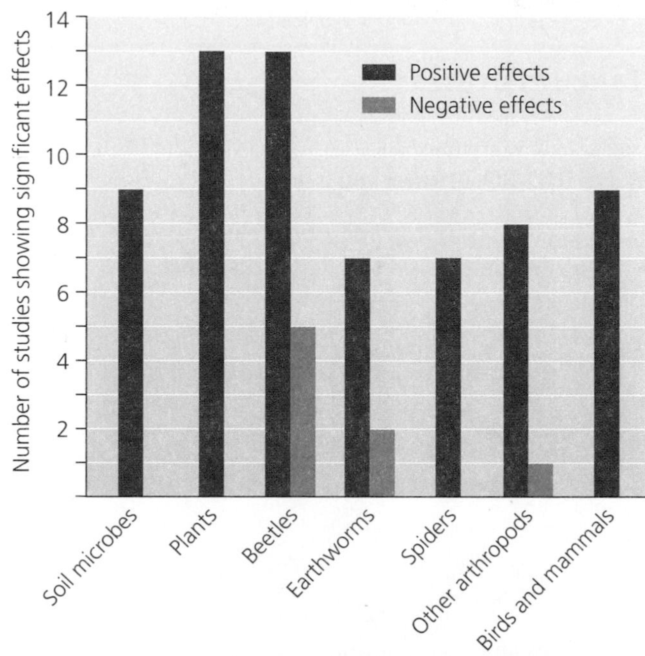


Numbers of scientific studies reporting positive and negative effects on biodiversity of organic agriculture (red bars) versus conventional farming practices (orange bars). Data from Hole, D., et al. 2005. Does organic farming benefit biodiversity? *Biological Conservation* 122: 113–130.

1. Overall, how many studies showed a positive effect of organic farming on biodiversity? How many studies reported a negative effect?

2. For which group or groups of organisms is evidence of positive effects the strongest? Reference the numbers to support your choice(s).

3. Recall the ecosystem services provided by biodiversity (• pp. 310–311). What services do the group(s) you chose in Question 2 provide?

## CALCULATING ECOLOGICAL FOOTPRINTS

Of the five major causes of biodiversity loss discussed in this chapter, habitat alteration arguably has the greatest impact. In their 1996 book introducing the ecological footprint concept, authors Mathis Wackernagel and William Rees present a consumption/land-use matrix for an average North American. Each cell in the matrix lists the number of hectares of land of that type required to provide for the different categories of a person's consumption (food, housing, transportation, consumer goods, and services). Of the 4.27 hectares required to support this average person, 0.59 hectares are forest, with most (0.40 hectares) being used to meet the housing demand. Using this information, calculate the missing values in the table.

|  | Hectares of forest used for housing | Total forest hectares used |
|---|---|---|
| You | 0.40 | 0.59 |
| Your class |  |  |
| Your state |  |  |
| United States |  |  |

Data from Wackernagel, M., and W. Rees. 1996. *Our ecological footprint: Reducing human impact on the earth.* British Columbia, Canada: New Society Publishers.

1. Approximately two-thirds of the forests' productivity is consumed for housing. To what use(s) would you speculate that most of the other third is put?
2. If the harvesting of forest products exceeds the sustainable harvest rate, what will be the likely consequence for the forest? For communities surrounding the forest?
3. What impacts would you expect on biodiversity in each of the following cases, and why?
   a. The cutting of small plots of forest within a large forest
   b. The clear-cutting (• pp. 340–341) of an entire forest
   c. The clear-cutting of an entire forest followed by planting of a monocultural plantation of young trees

Does the spatial scale at which you judge the effects on biodiversity affect your answer? In what ways?

## Take It Further

Go to www.aw-bc.com/withgott or the student CD-ROM, where you'll find:

- Suggested answers to end-of-chapter questions
- Quizzes, animations, and flashcards to help you study
- *Research Navigator*™ database of credible and reliable sources to assist you with your research projects
- GRAPHit Tutorials to help you interpret graphs
- INVESTIGATEit! Current news articles that link the topics that you study to case studies from your region to around the world

CHAPTER

# 12 Resource Management, Forestry, Land Use, and Protected Areas

Clear-cuts on timber industry land adjacent to Gifford Pinchot National Forest, Washington

## Upon completing this chapter, you will be able to:

- Identify the principles, goals, and approaches of resource management
- Summarize the ecological roles and economic contributions of forests, and outline the history and scale of forest loss
- Explain the fundamentals of forest management and describe the major methods of harvesting timber
- Analyze the scale and impacts of agricultural land use
- Identify major federal land management agencies and the lands they manage
- Recognize types of parks and reserves, and evaluate issues involved in their design

Anti-logging protest at Clayoquot Sound

CENTRAL CASE

# Battling Over the Last Big Trees at Clayoquot Sound

**"What we have is nothing less than an ecological Holocaust occurring right now in British Columbia."**
—Mark Wareing, Western Canada Wilderness Committee, 1990

**"Clear-cutting . . . may be either desirable or undesirable, acceptable or unacceptable, according to the type of forest and the management objectives."**
—Dr. Hamish Kimmins, University of British Columbia, 1992

It was the largest act of civil disobedience in Canadian history, and it played out along a seacoast of stunning majestic beauty, at the foot of some of the world's biggest trees. At Clayoquot Sound on the western coast of Vancouver Island, British Columbia, protestors blocked logging trucks, preventing them from entering stands of ancient temperate rainforest. The activists chanted slogans, sang songs, and chained themselves to trees.

Loggers complained that the protestors were keeping them from doing their jobs and making a living. In the end, 850 of the 12,000 protestors were arrested in full view of the global media, and this remote, mist-enshrouded land of cedars and hemlocks became ground zero in the debate over how we manage forests.

That was in 1993, and the activists were opposing **clear-cutting**, the logging practice that removes all trees from an area. Most of Canada's old-growth temperate rainforest had already been cut, and the forests of Clayoquot Sound were among the largest undisturbed stands of temperate rainforest left on the planet.

Timber from old-growth forests had powered British Columbia's economy since the province's early days. Historically, one in five jobs in British Columbia depended on its $13 billion timber industry, and many small towns would have gone under without it. By 1993, however, the timber industry was cutting thousands of jobs a year because of mechanization, and the looming depletion of old growth threatened to slow the industry.

Meanwhile, the environmental group Greenpeace was convincing overseas customers to boycott products made from British Columbia trees clear-cut by multinational timber company MacMillan Bloedel. Soon British Columbia's premier found himself trying to persuade European nations not to boycott his province's main export.

In 1995, the provincial government called for an end to clear-cutting at Clayoquot Sound, after its appointed scientific panel of experts submitted a new forestry plan for the region. The plan recommended reducing harvests, retaining 15–70% of old-growth trees in each stand, decreasing the logging road network, designating forest reserve areas, and managing riparian zones. Yet 2 years later, the provincial government reversed many of these regulations on logging, and a new premier pronounced forest activists "enemies of British Columbia."

The antagonists called a truce and struck a deal; wilderness advocates and MacMillan Bloedel agreed to log old growth in limited areas, using environmentally friendly practices. In 1998, Native people of the region formed a timber company, Iisaak, in agreement with MacMillan Bloedel's successor, Weyerhaeuser, and began logging in a more environmentally sensitive manner.

Meanwhile, leaving most of the trees standing had accomplished just what forest advocates had predicted: People from all over the world—1 million each year—were now visiting Clayoquot Sound for its natural beauty and kayaking and whale-watching in its waters. Ecotourism (along with fishing and aquaculture) surpassed logging as the driver of local economies. The United Nations designated the site as an international biosphere reserve, encouraging land protection and sustainable development. The trees appeared to be worth more left standing than cut down.

Tensions continue today, however. Another timber company, Interfor, is harvesting areas near park and biosphere reserve boundaries. Local forest advocates worry that the provincial government's new Working Forest Policy will increase logging in the region, and the town of Tofino petitioned the province to exempt Clayoquot Sound's forests from the policy. After Iisaak found it difficult to make money doing sustainable forestry, it entered into an agreement with Ecotrust Canada in 2006. Today logging is being done more sustainably, and at a profit, under this arrangement. Meanwhile, the provincial government is considering new forestry plans that would shift logging out of old-growth forests and into younger forests that were already logged in the past. As long as our demand for lumber, paper, and forest products keeps increasing, pressures will keep building on the remaining forests on Vancouver Island and around the world.

# Resource Management

Debates over forest resources reflect broader questions about how to manage natural resources in general. We need to manage the resources we take from the natural world because many of them are limited. We have seen (• pp. 3–4) how resources such as fossil fuels are nonrenewable on human time scales, whereas resources such as the sun's energy are perpetually renewable. Between these extremes lie resources that are renewable if they are not exploited too rapidly or carelessly—resources such as soils, fresh water, wildlife and fisheries, rangeland, and timber.

**Resource management** is the practice of harvesting potentially renewable resources in ways that do not deplete them. Resource managers are guided in their decision making by available research in the natural sciences, but their decisions are also often influenced by political, economic, and social factors. A key question in managing resources is whether to focus narrowly on the resource of interest or to look more broadly at the environmental system of which the resource is a part. Taking a broader view can often help avoid damaging the system and can thereby help sustain the availability of the resource in the long term.

## Several natural resources are vital to us

Besides timber, several other types of natural resources are vital to our civilization. These include soils, fresh water, wildlife and fisheries, rangeland, and minerals (**Figure 12.1**). All natural resources also serve functions in the ecosystems of which they are a part.

FIGURE 12.1 Our civilization depends on a number of natural resources, including timber, soils, fresh water, wildlife and fisheries, rangeland, and minerals. Minerals are a nonrenewable resource that is mined, but the others are renewable through responsible resource management.

**Soils** Soil resources, particularly topsoil, are of direct importance to us because they support the plants we grow for food and fiber and thus play a central role in agriculture. As we saw in Chapter 9 (• pp. 247–250), certain farm practices, such as terracing and use of windbreaks, can help guard against loss of topsoil to erosion, and other farm practices, such as planting nitrogen-fixing crops, can help maintain soil quality. Healthy soils also support forests and other natural communities, serving as a site for decomposition and a reservoir for nutrients.

**Fresh water** Each of us depends directly on fresh water, so ensuring a dependable supply of drinking water is a life-or-death issue. Fresh water also is necessary for agriculture; indeed, we use most fresh water not for drinking but for irrigating crops. In addition, waterways and wetlands are crucial for wildlife and properly functioning ecosystems, so people who manage water resources try to maintain supplies for all these reasons. Water managers also try to protect the quality of these supplies by guarding against pollution. We will examine fresh water resources in depth in Chapter 15.

**Wildlife and fisheries** People have long hunted animals for food. Managers regulate the hunting of game animals, such as deer and quail, for food and sport in an effort to maintain populations of these animals at desired levels. As populations of many other terrestrial species decline from habitat loss and other causes (• pp. 304–310), management for nongame species is becoming increasingly important. Marine animals are also harvested, and despite extensive management of fisheries, many stocks of fish and shellfish have declined throughout the world's oceans from overharvesting, as we will see in Chapter 16. Besides their use as food, living organisms provide people with many other benefits, from materials to medicines to aesthetic appreciation (• pp. 310–314). Organisms are also, of course, vital components of the ecosystems that sustain our world.

**Rangeland** Most of our food from animals today comes not from wild animals but from domesticated ones that are farmed. Most cattle in North America are raised in crowded feedlots, but they have traditionally been raised by grazing on open grassland. As we saw in Chapter 9 (• pp. 254–257), grazing can be sustainable, but overgrazing damages soils, waterways, and vegetative communities. Range managers are responsible for regulating ranching on public lands, and they advise ranchers on sustainable grazing practices.

**Minerals** Our civilization depends on numerous minerals, and until we achieve a closed-loop economy (• pp. 645–646, 672–673), we will rely on the mining of minerals. Iron is mined and processed to make steel. Copper is used in pipes, electrical wires, and a variety of other applications. Aluminum is extracted via bauxite ore and used in packaging and other end products. Lead is used in batteries, to shield medical patients from radiation, and in many other ways. Zinc, tungsten, phosphate, uranium, gold, silver—the list goes on and on.

Although we rely on these mineral resources, we do not manage their extraction in the way we manage renewable natural resources. Like fossil fuels, minerals are nonrenewable resources that are mined rather than harvested. Therefore, the mining industry has no built-in incentive to conserve. Instead, it benefits by extracting as much as it can as fast as it can and then, once extraction becomes too inefficient to be profitable, moving on to new sites.

The process of mining also has historically degraded surrounding environments, sometimes on massive scales. Mining may directly remove vegetation from large areas, cause erosion, and produce acidic runoff that poisons area waterways. In addition, the smelting of metals following mining can create severe air pollution. We touch on some of the environmental and social costs of coal mining in Chapter 19 (• pp. 560–561), but these issues exist with all types of mining.

Improved technology and more environmentally sensitive extraction procedures can help minimize environmental impacts. Such advances come as a result of public pressure, government legislation, or economic savings. In contrast, industries extracting potentially renewable resources have an added incentive to reduce ecological impacts: doing so may help sustain their ability to harvest resources—and this is where resource management comes in.

## Managers have tried to achieve maximum sustainable yield

Traditionally, one guiding principle in resource management has been **maximum sustainable yield**. The aim is to achieve the maximum amount of resource extraction without depleting the resource from one harvest to the next. Recall the logistic growth curve (Figure 5.16, • p.129–130), which shows that a population grows most quickly when it is at an intermediate size. A fisheries manager aiming for maximum sustainable yield will prefer to keep fish populations at intermediate levels so that they rebound quickly after each harvest. Doing so should result in the greatest amount of fish harvested over time, while the population sustains itself (**Figure 12.2**).

This management approach, however, keeps the fish population at only about half its carrying capacity (•pp. 129–130)—well below the level it would attain in the absence of fishing. Reducing one population in this way will likely have effects on other species and on the food

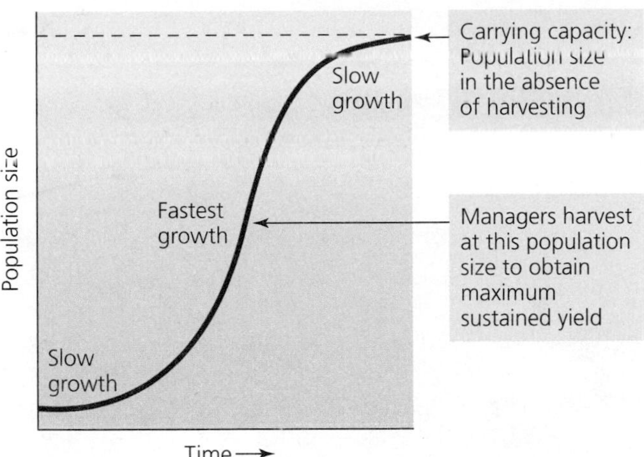


FIGURE 12.2 Using the concept of maximum sustainable yield, resource managers attempt to maximize the amount of resource harvested while keeping the harvest sustainable in perpetuity. In the case of a wildlife population or fisheries stock that grows according to a logistic growth curve, managers aim to keep the population at an intermediate level, well below the carrying capacity, because populations grow fastest at intermediate sizes.

web dynamics of the community (• pp. 150–151). From an ecological point of view, management for maximum sustainable yield may thereby set in motion complex and significant ecological changes. Thus, whereas the maximum sustainable yield approach may at first sound ideal, in practice it can alter the ecosystems from which the resource is derived. If those ecosystems cease to function effectively as a result, then in time this may decrease the availability of the resource.

In forest management, maximum sustainable yield argues for cutting trees shortly after they have gone through their fastest stage of growth, and trees often grow most quickly at intermediate ages. Thus, trees may be cut long before they have grown as large as they would in the absence of harvesting. Although this practice may maximize timber production over time, it can cause drastic changes in the ecology of a forest by eliminating habitat for species that depend on mature trees.

## Today many managers pursue ecosystem-based management

Because of these dilemmas, increasing numbers of managers today espouse ecosystem-based management. **Ecosystem-based management** attempts to manage the harvesting of resources in ways that minimize impact on the ecosystems and ecological processes that provide the resource. The plan proposed in 1995 by the Scientific Panel for Sustainable Forest Practices on Clayoquot Sound and approved by British Columbia's government was essentially a plan for ecosystem-based management. By carefully managing ecologically important areas such as riparian corridors, by considering patterns at the landscape level (• pp. 183–185), and by affording protection to some forested areas, the plan aimed to allow continued timber harvesting at reduced levels while preserving the functional integrity of the ecosystem.

Although ecosystem-based management has gained a great deal of support in recent years, it is challenging for managers to determine how best to implement this type of management. Ecosystems are complex, and our understanding of how they operate is limited. Thus, ecosystem-based management has often come to mean different things to different people.

## Adaptive management evolves and improves

Some management actions will succeed, and some will fail. A wise manager will try new approaches if old ones are not effective. **Adaptive management** involves systematically testing different management approaches and aiming to improve methods as time goes on. It entails monitoring the results of one's practices and continually adjusting them as needed, based on what is learned. This approach is intended as a true fusion of science and management, because hypotheses about how best to manage resources are explicitly tested. Adaptive management can be time-consuming and complicated, however. It has posed a challenge for many managers, because those who adopt new approaches must often overcome inertia and resistance to change from proponents of established practices.

Adaptive management was featured in the *Northwest Forest Plan*, a 1994 plan crafted by the Clinton administration to resolve disputes between loggers and preservationists over the last remaining old-growth temperate rainforests in the continental United States. In western Oregon, western Washington, and northwestern California, this plan sought to allow limited logging to continue with adequate protections for species and ecosystems and to let science guide management.

# Forest Management

Forests cover roughly 30% of Earth's land surface (**Figure 12.3**). They provide habitat for countless organisms; help maintain soil, air, and water quality; and play key roles in our planet's biogeochemical cycles (• pp. 186–198). Forests have also long provided humanity with wood for fuel, construction, paper production, and more. Foresters, those professionals who manage forests through the practice of **forestry**, must balance the central

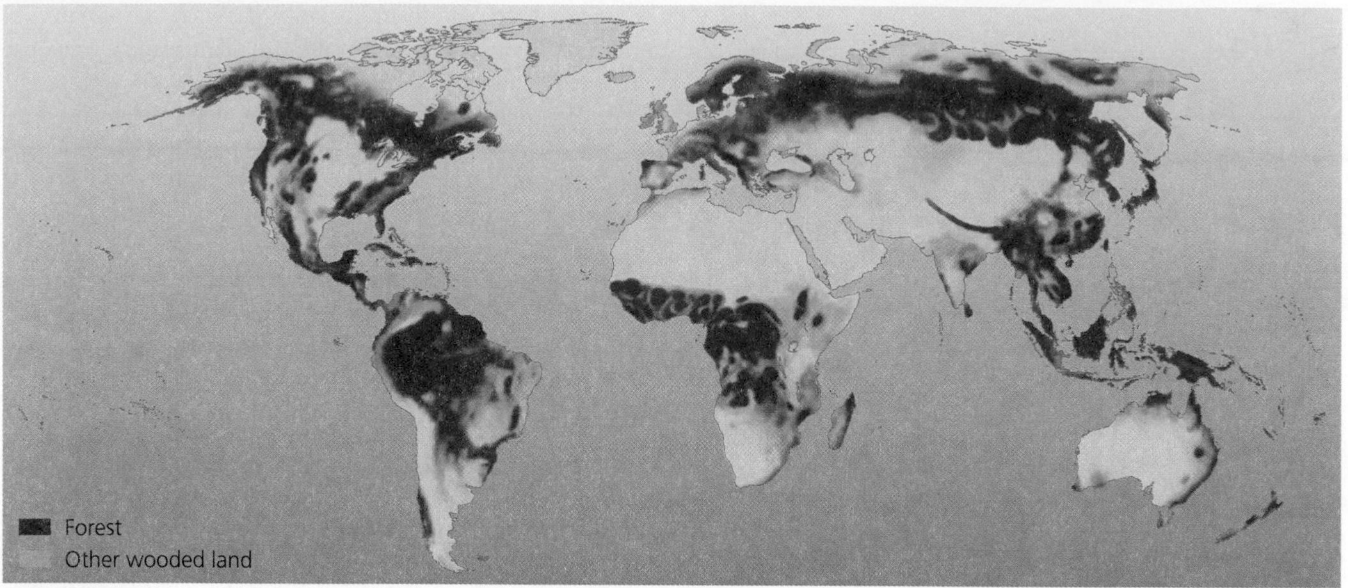


FIGURE 12.3 About 30% of Earth's land surface is covered by forest. Most of this consists of the boreal forests of the north and the tropical forests of South America and Africa. Further land can be classified as "wooded land," which supports trees at sparser densities. Data from U.N. Food and Agriculture Organization (FAO). 2005. *Global forest resources assessment.*

importance of forests as ecosystems with civilization's demand for wood products.

Most of the world's forests occur as boreal forest, a biome (• pp. 161–169) that stretches across much of Canada, Russia, and Scandinavia, or as tropical rainforest, a biome that occurs in South and Central America, equatorial Africa, and Indonesia and southeast Asia. Temperate forests cover less area globally, in part because people have already cleared so many of them.

## Forests are ecologically valuable

Because of their structural complexity and their ability to provide many niches for organisms, forests comprise some of the richest ecosystems for biodiversity (**Figure 12.4**). Trees furnish food and shelter for an immense diversity of vertebrate and invertebrate animals. Countless insects, birds, mammals, and other organisms subsist on the leaves, fruits, and seeds that trees produce.

Some animals are adapted for living in dense treetop canopies. Here beetles, caterpillars, and other leaf-eating insects abound, providing food for birds such as tanagers and warblers, while arboreal mammals from squirrels to sloths to monkeys consume fruit and leaves. Other animals specialize on the subcanopies of trees, and still others utilize the bark, branches, and trunks. Cavities in trunks provide nest and shelter sites for a wide variety of vertebrates. Dead and dying trees are valuable for many species; these snags are decayed by insects that are eaten by woodpeckers and other animals.

Understory shrubs and groundcover plants give a forest structural complexity and provide habitat for still more organisms. Moreover, the leaves, stems, and roots of forest plants are colonized by an extensive array of fungi and microbes, in both parasitic and mutualistic relationships (• pp. 145–147). And much of a forest's diversity resides in the forest floor, where the soil is generally nourished by leaf litter. As we saw in Chapter 9, myriad soil organisms help decompose plant material and cycle nutrients.

In general, forests with a greater diversity of plants host a greater diversity of organisms overall. And in general, fully mature forests, such as the undisturbed old-growth forests remaining at Clayoquot Sound, contain more biodiversity than younger forests, because older forests contain more structural diversity and thus more microhabitats and resources for more species.

Forests also provide all manner of vital ecosystem services (• pp. 38, 40). Forest vegetation stabilizes soil and prevents erosion. Plants help regulate the hydrologic cycle, slowing runoff, lessening flooding, and purifying water as they take it in from the soil and release it to the atmosphere. Forest plants also store carbon, release oxygen, and moderate climate. By performing such ecological functions, forests are indispensable for our survival.

## Forest products are economically valued

Forests also provide people with economically valuable wood products. For millennia, wood from forests has fueled our fires, keeping people warm and well fed. It has housed people, keeping us sheltered. It built the ships that carried people and cultures from one continent to

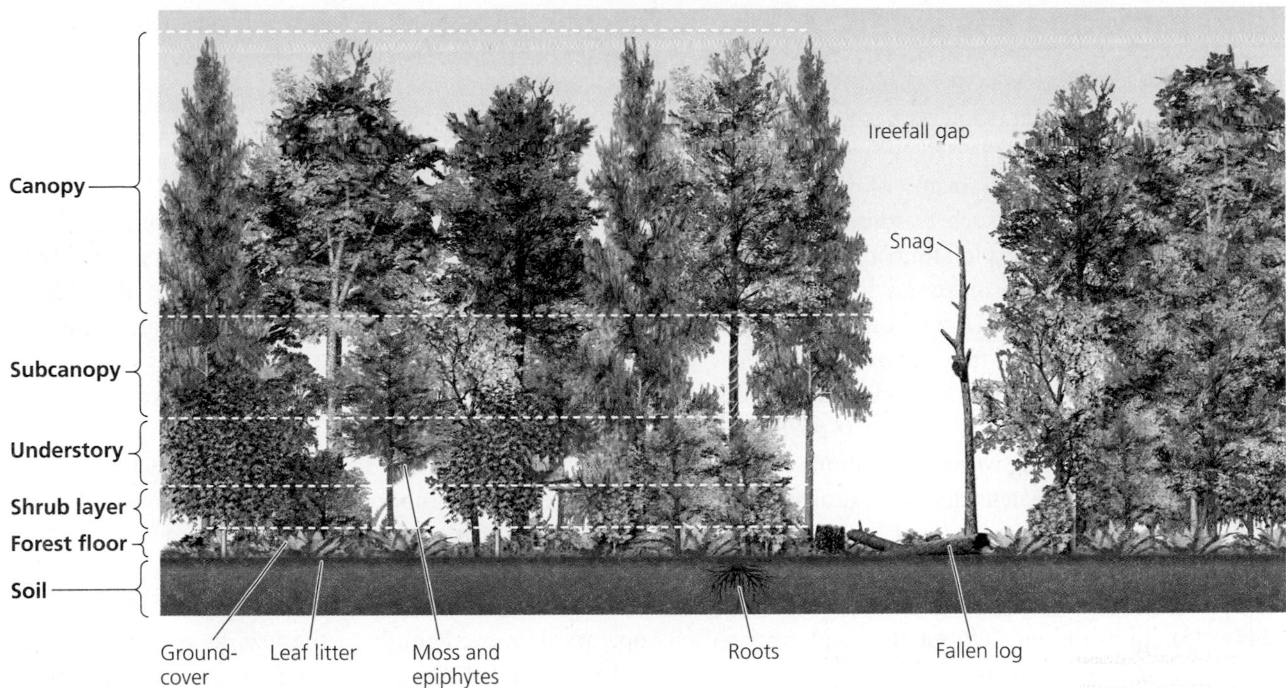


**FIGURE 12.4** Mature forests are complex ecosystems. In this cross-section of a generalized mature forest, the crowns of the largest trees form the canopy, and smaller trees beneath them form the shaded subcanopy and understory. Shrubs and groundcover grow just above the forest floor, which may be covered in leaf litter rich with invertebrate animals. Vines, mosses, lichens, and epiphytes cover portions of trees and the forest floor. Snags (standing dead trees), whose wood can be easily hollowed out, provide food and nesting and roosting sites for woodpeckers and other animals. Fallen logs nourish the soil and young plants and provide habitat for countless invertebrates as the logs decompose. Treefall gaps caused by fallen trees let light through the canopy and create small openings in the forest, allowing early successional plants to grow in patches within the mature forest.

another. It allowed us to produce paper, the medium of the first information revolution.

In recent decades, industrial harvesting has allowed the extraction of more timber than ever before, supplying all these needs of a rapidly growing human population and its expanding economy. The exploitation of forest resources has been instrumental in helping our society achieve the standard of living we enjoy today. Indeed, without industrial timber harvesting, you would not be reading this book.

Most commercial logging today takes place in Canada, Russia, and other nations that hold large expanses of boreal forest, and in tropical countries with large amounts of rainforest, such as Brazil and Indonesia. In the United States, most logging takes place in the conifer forests of the West and the pine plantations of the South, on land both private and public.

Nations maintain and use forests for all these economic and ecological reasons. An international survey in 2005 found that globally, over one-third of forests were designated primarily for timber production. Others were designated for a variety of functions, including conservation of biodiversity, protection of soil and water quality, and "social services" such as recreation, tourism, education, and conservation of culturally important sites (**Figure 12.5**).

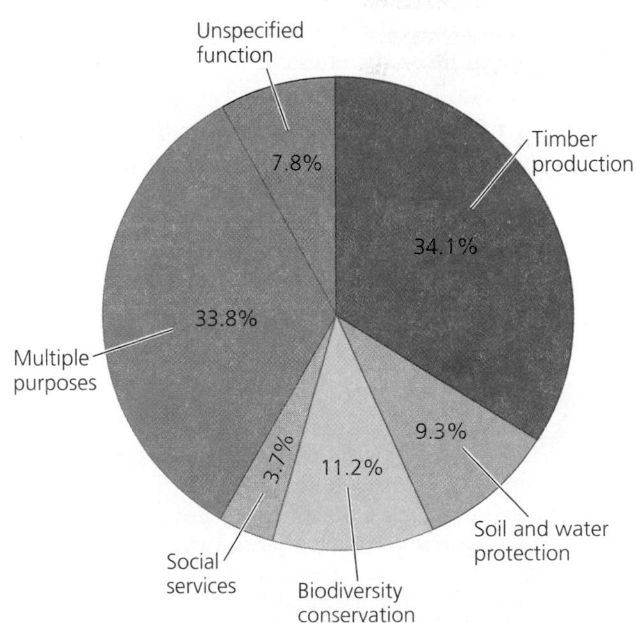


**FIGURE 12.5** Worldwide, nations designate over one-third of forests primarily for production of timber and other forest products. Smaller areas are designated for conservation of biodiversity, protection of soil and water quality, and "social services" such as recreation, tourism, education, and conservation of culturally important sites. About one-third of forests are designated for combinations of these functions. Data from U.N. Food and Agriculture Organization (FAO). 2005. *Global forest resources assessment.*

## Demand for wood has led to deforestation

We all depend in some way on wood, from the subsistence herder in Nepal cutting trees for firewood to the American student consuming reams upon reams of paper in the course of getting a degree. For such reasons, people have cleared forests for millennia to exploit timber resources. Still more forest has been cleared to make way for agriculture.

**Deforestation**, the clearing and loss of forests, has altered the landscapes and ecosystems of much of our planet. It has caused soil degradation, population declines, and species extinctions, and, as we saw with Easter Island (• pp. 8–9), it has in some cases helped bring whole civilizations to ruin. Continued deforestation threatens dire ecological and economic consequences. Impacts are greatest in tropical areas because of the potentially massive loss of biodiversity, and in arid regions because of the vulnerability to desertification (• pp. 244–245). In addition, deforestation adds carbon dioxide ($CO_2$) to the atmosphere: $CO_2$ is released when plant matter is burned or decomposed, and thereafter less vegetation remains to soak up $CO_2$. Deforestation is thereby one contributor to global climate change (Chapter 18).

Today forests are being felled at the fastest rates in the tropical rainforests of Latin America and Africa (**Figure 12.6**). Developing countries in these regions are striving to expand areas of settlement for their burgeoning populations and to boost their economies by extracting natural resources and selling them abroad. Moreover, many people in these societies cut trees for fuelwood for their daily cooking and heating needs (• p. 589). In contrast, areas of Europe and eastern North America are slowly gaining forest cover as they recover from severe deforestation of past decades and centuries. Overall, the world is losing its forests, but some advances are being made toward sustainable forest management (see "The Science behind the Story," • pp. 336–337).

## The growth of the United States and Canada was fed by deforestation

Deforestation for timber and farmland propelled the growth of the United States and Canada throughout their phenomenal expansion westward across the North American continent over the past 400 years. The vast deciduous forests of the East were virtually stripped of their trees by the mid-19th century, making way for countless small farms. Timber from these forests built the cities of the Atlantic seaboard. Later, cities such as Chicago were constructed with timber felled in the vast pine and hardwood forests of Wisconsin and Michigan.

As a farming economy shifted to an industrial one, wood was used to stoke the furnaces of industry. Logging operations moved south to the Ozarks of Missouri and Arkansas, and the vast pine woodlands and bottomland hardwood forests were logged and converted to pine plantations. Once most mature trees were removed from these areas, timber companies moved west, cutting the continent's biggest trees in the Rocky Mountains, the Sierra Nevada, the Cascade Mountains, and the Pacific Coast ranges (**Figure 12.7**).

By the early 20th century, very little **primary forest**—natural forest uncut by people—was left in the lower 48 U.S. states (**Figure 12.8**). Today, the largest oaks and maples found in eastern North America, and even most redwoods of the California coast, are merely **second-growth** trees: trees that have sprouted and grown to partial maturity after old-growth timber has been cut. The size of the gargantuan trees they replaced can be seen in the enormous stumps that remain in the more recently logged areas of the Pacific

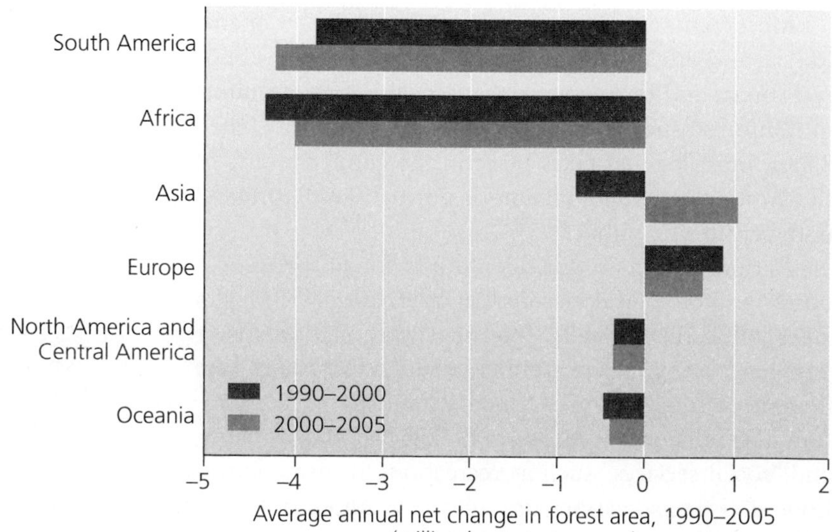


FIGURE 12.6 Nations in South America and Africa are experiencing rapid deforestation as they attempt to develop, extract resources, and provide new agricultural land for their growing populations. In Europe, meanwhile, forested area is slowly increasing as some formerly farmed areas are abandoned and allowed to grow back into forest. The data for North and Central America reflect a balance of forest regrowth in North America and forest loss in Central America. In Asia, natural forests are being lost, but the extensive planting of tree plantations (here counted as forests) in China has increased forest cover for Asia since 2000. Go to GRAPHIt! at www.aw-bc.com/withgott or on the student CD-ROM. Data from U.N. Food and Agriculture Organization (FAO). 2005. *Global forest resources assessment.*

FIGURE 12.7 A mule team drags logs of Douglas fir from a clear-cut in Mason County, Washington, in 1901. Early timber harvesting practices in North America caused significant environmental impacts and removed virtually all the virgin timber from one region after another.

coast. The scarcity of old-growth trees on the North American continent today explains the concern that scientists have for old-growth ecosystems and the passion with which environmental advocates have fought to preserve ancient forests in areas such as Clayoquot Sound.

The fortunes of loggers have risen and fallen with the availability of big trees. As each region was denuded of timber, the industry declined, and the timber companies moved on while local loggers lost their jobs. If the remaining ancient trees of North America—most in British Columbia and Alaska—are cut, many U.S. and Canadian loggers will likely be out of jobs once again. Their employers will move on to nations of the developing world, as many already have.

## Deforestation is proceeding rapidly in many developing nations

Uncut tropical forests still remain in many developing countries, and these nations are in the position the United States and Canada faced a century or two ago: having a vast frontier that they can develop for human use. Today's

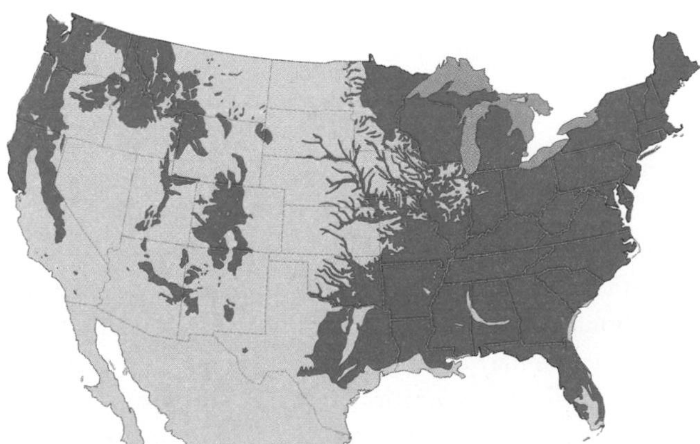

(a) 1620: Areas of uncut forest

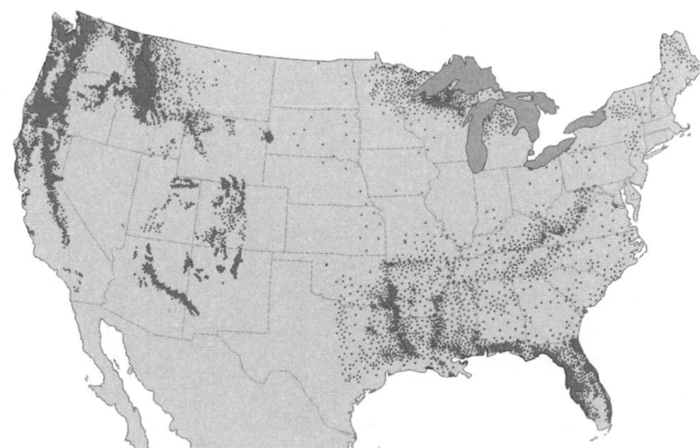

(b) 1920: Areas of uncut forest

FIGURE 12.8 When Europeans first were colonizing North America, the entire eastern half of the continent and substantial portions of the western half were covered in forest, shown in green (**a**). By the early 20th century, the vast majority of this forest had been cut, replaced mostly by agriculture and other human land uses (**b**). Since that time, most of the remaining original forest has been cut, but much of the landscape has also become reforested with second-growth forest. *Source:* Williams, M. 1989. *Americans and their forests.* Cambridge: Cambridge University Press. As adapted by Goudie, A., 2000. *The human impact.* Cambridge, MA: MIT Press.

THE SCIENCE BEHIND THE STORY

## Surveying Earth's Forests

*Old growth forest in British Columbia*

In the time it takes you to read this sentence, 2 hectares (5 acres) of tropical forest will have been cleared.

We have all heard pronouncements like this before, but where do such numbers come from? How do we know how much forest our planet is losing—or even how much forest there is to begin with?

Every 5–10 years since 1948, the United Nations Food and Agriculture Organization (FAO) has conducted a global inventory of forest resources. FAO researchers have asked the world's national governments to respond to questionnaires and provide data on forest area, types of forest, deforestation, regrowth, and other parameters. But it is risky to rely on questionnaires. Nations may not keep accurate data or may not care to share it.

In recent years, satellite technology has provided a way to confirm information from national surveys. With remote sensing data from satellites, we can now measure and map forest cover from space.

Satellites that observe Earth's surface measure wavelengths of energy being emitted. The data is then processed by software that interprets this information to infer what is on the surface. Plants absorb most wavelengths of light but reflect infrared radiation and green light (which is why plants appear green). When satellite data shows green and infrared wavelengths reflected from the surface but an absence of other wavelengths such as red and blue, then the software infers the presence of vegetation.

Researchers use data from a number of satellites (for instance, the Land Remote Sensing Satellite [Landsat], the Advanced Very High Resolution Radiometer [AVHRR], and the Moderate Resolution Imaging Spectroradiometer [MODIS]) to infer vegetative cover. By quantifying plant cover with indices such as the Normalized Difference Vegetation Index, researchers can compare one site to another and can measure how sites change across seasons or years. (Turn back to Figure 7.7b, • p. 182, for an example of satellite measurements of net primary productivity from plants.)

A 2001 study of global forest cover by the United Nations Environment Programme (UNEP) used AVHRR data and loaded this into a geographic information system, or GIS (• p. 185). Researchers then added data on human population distribution, political boundaries, and land protected against development. The study found that forests in densely populated nations such as India and Indonesia are under pressure from expanding human settlement and require urgent conservation efforts.

In 2005, the FAO released its latest global accounting of forests—one that superceded all previous efforts. In its *Global Forest Resources Assessment*, researchers combined remote sensing data, questionnaire responses, analysis from forestry experts, and statistical modeling, all to form a comprehensive picture of the state of the world's forests.

Major findings of the FAO assessment included the following:

- Forests cover 30% of the world's land area.
- Just 10 nations account for two-thirds of all forests.

advanced technology, however, has allowed these countries to exploit their resources and push back their frontiers even faster than occurred in North America. As a result, deforestation is rapid in places such as Brazil and Indonesia.

Developing nations are often desperate enough for economic development that they impose few or no restrictions on logging. Often their timber is extracted by foreign multinational corporations, which have paid fees to the developing nation's government for a *concession,* or right to extract the resource. In such cases, the foreign corporation has little or no incentive to manage forest resources sustainably. Many of the short-term economic benefits are reaped not by local residents but by the corporations that log the timber and export it elsewhere. Local people may or may not receive temporary employment from the corporation, but once the timber is harvested they no longer have the forest and the ecosystem services it once provided.

In Sarawak, the Malaysian portion of the island of Borneo, foreign corporations that were granted logging concessions have deforested several million hectares of tropical rainforest since 1963. The clearing of this forest—one of the world's richest, hosting such organisms as orangutans and

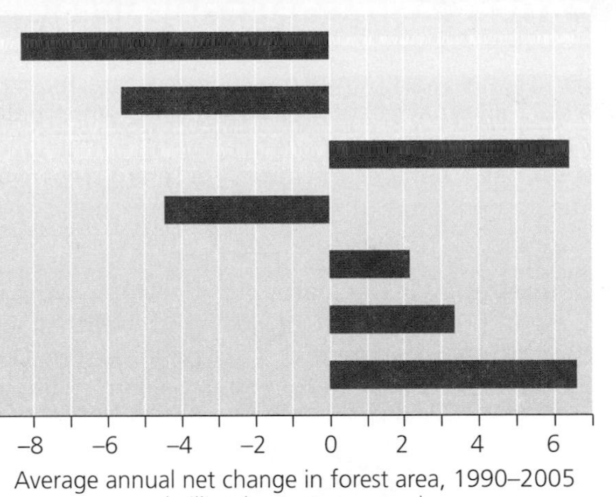


**Progress toward sustainable forestry is mixed, as shown by 7 of 21 indicators assessed by the Food and Agriculture Organization (FAO). Forest uses are shifting toward conservation and away from production, but forest area is still shrinking.** Data from U.N. Food and Agriculture Organization (FAO), 2005. *Global forest resources assessment.*

- 34% of forests are designated primarily for wood production.
- Forests store 283 billion metric tons of carbon in living tissue, and more overall than the atmosphere.
- Rare tree species valued for wood (such as teak and mahogany) are often in danger of vanishing.
- 84% of forests are publicly owned, but private ownership is increasing.
- Forestry employs 10 million people.

By comparing the results from 2005 with those of previous assessments, the FAO could also reveal trends through time. Major findings included the following:

- We deforest 13 million hectares per year.
- 5.7 million hectares per year support regrowing forests.
- Thus, net annual forest loss is 7.3 million hectares—about the area of West Virginia or South Carolina.
- Regions vary in rates of forest loss or gain (see Figure 12.6, • p. 334).
- Primary forests are being lost more quickly than second-growth forests.
- Forest plantations make up only 3.8% of forests but are increasing.
- The world's forests hold 1.1 billion fewer tons of carbon each year.

Finally, the study examined how our society was progressing toward sustainable management of forests. Researchers examined a number of indicators (see the figure). Among other trends, they found that:

- More and more forests are being managed for multiple uses.
- 11% of forests are designated for biodiversity conservation.
- 9% of forests are designated for soil and water conservation.
- Use of forests for recreation and education is increasing rapidly.

These trends, however, are occurring against a backdrop of continued net loss of forests, especially primary forests. Overall, the report concludes that the outlook is mixed: "There are many good signs and positive trends, but many negative trends remain. While intensive forest plantation and conservation efforts are on the rise, primary forests continue to become degraded or converted to agriculture at alarming rates in some regions."

the world's largest flower, *Rafflesia arnoldii*—has had direct impacts on the 22 tribes of people who live as hunter-gatherers in Sarawak's rainforest. The Malaysian government did not consult the tribes about the logging, which decreased the wild game on which these people depended. Oil palm agriculture was established afterward, leading to pesticide and fertilizer runoff that killed fish in local streams. The tribes protested peacefully and finally began blockading logging roads. The government, which at first jailed them, now is negotiating, but it insists on converting the tribes to a farming way of life.

### Weighing THE Issues | Logging Here or There

Imagine you are an environmental activist protesting a logging operation that is cutting old-growth trees near your hometown. Now let's say you know that if the protest is successful, the company will move to a developing country and cut its primary forest instead. Would you still protest the logging in your hometown? Would you pursue any other approaches?

## CAUSES AND CONSEQUENCES

Although we are making some strides toward sustainable forestry, we are still deforesting 13 million hectares (32 million acres) globally each year. When balanced against forest regrowth, this **deforestation** gives us an annual net loss of 7.3 million hectares (18 million acres). While deforestation has a number of causes, there are steps we can take to reduce the loss of forests and the consequences that deforestation brings.

Write in two causes of deforestation in the spaces provided. Then write in two consequences (impacts on the environment, human health, or quality of life) that result from deforestation. Finally, offer two solutions to this issue and its consequences. One cause, one consequence, and one solution have been filled in for you, providing examples.

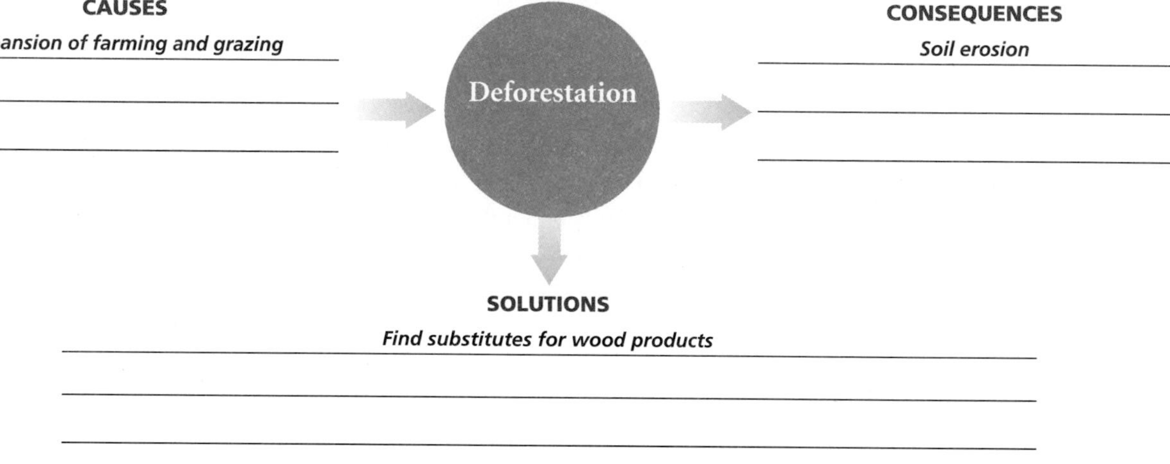


Sometimes solutions can have unintended consequences. Can you think of an undesired consequence that one solution to this issue might have? How might we then deal with *that* consequence?

## Fear of a "timber famine" spurred establishment of national forests

In the United States, the depletion of the eastern forests and widespread fear of a "timber famine" spurred the formation of a system of forest reserves: public lands set aside to grow trees, produce timber, protect watersheds, and serve as insurance against future scarcities of lumber. Today the U.S. **national forest** system consists of 77 million ha (191 million acres) spread across all but a few states (**Figure 12.9**). The system, managed by the U.S. Forest Service, covers over 8% of the nation's land area.

The U.S. Forest Service was established in 1905 under the leadership of Gifford Pinchot (• pp. 33–34). Pinchot and others developed the concepts of resource management and conservation during the Progressive Era, a time of social reform when people urged that science be applied to public policy to improve society. In line with Pinchot's conservation ethic (• pp. 33–34), the Forest Service aimed to manage the forests for "the greatest good of the greatest number in the long run." Pinchot believed the nation should extract and use resources from its public lands, so timber harvesting was, from the start, a goal behind establishing the national forests. But conservation meant planting trees as well as harvesting them, and the Forest Service intended to pursue wise management of timber resources.

Management goals are similar for the Canadian Forest Service and the provincial forestry ministries. Of Canada's 310 million ha (765 million acres) of forested land, 77% belongs to the provinces, and only 16% is federally owned and 7% privately owned. About one-third of the nation's forests are in British Columbia, and 38% are in Quebec and Ontario.

## Timber is extracted from public and private lands

Timber is extracted from publicly held forests in the United States and Canada not by the governments of these nations, but by private timber companies. In the United States, Forest Service employees plan and manage timber sales and build roads to provide access for logging companies, which sell the timber they harvest for profit.

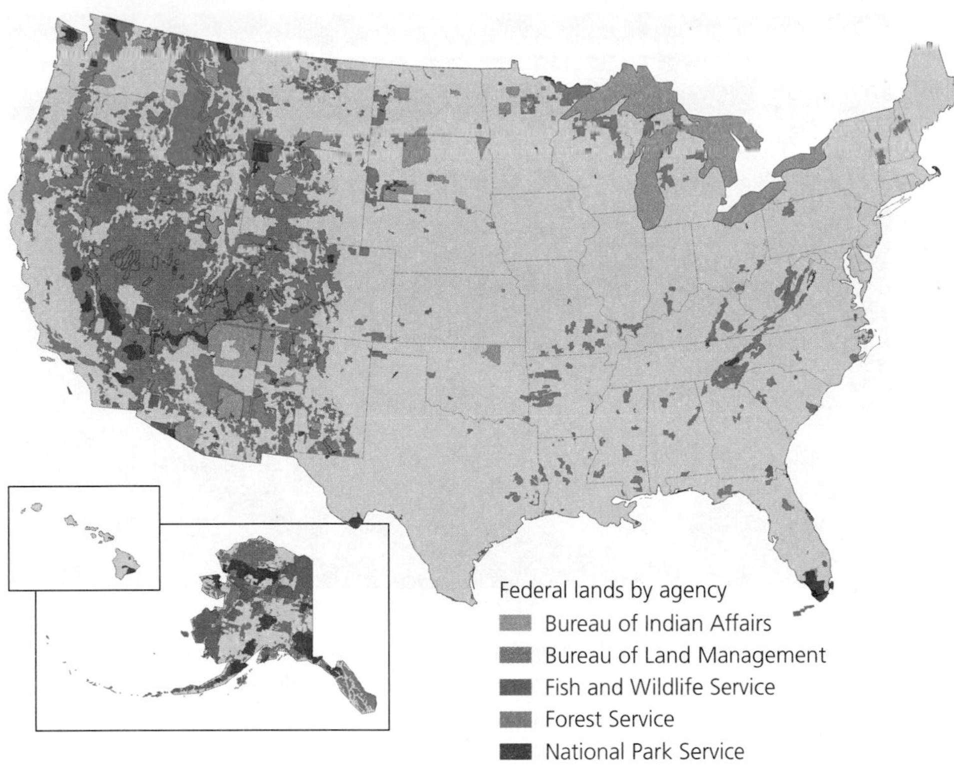


FIGURE 12.9 Federal agencies own and manage well over 250 million ha (600 million acres) of land in the United States, particularly in the western states. These include national forests, national parks, national wildlife refuges, Native American reservations, and Bureau of Land Management lands. Data from United States Geological Survey.

However, most timber harvesting in the United States these days takes place on private land, including land owned by timber companies. In 2001, the most recent year for which comprehensive data are available, timber companies extracted 31.7 million $m^3$ (340 million $ft^3$) of live timber from national forests. Although this is a large amount, it is considerably less than the amount cut from private lands and other public forests (**Figure 12.10**). Timber harvesting declined on national forests during the 1980s and 1990s, and in 2001 tree regrowth outpaced tree removal on these lands by nearly 12 to 1. Overall, timber harvesting in the United States and other developed countries has remained roughly stable for the past 40 years. Meanwhile, it has more than doubled in developing countries.

The equivalent rates of growth and removal on private lands in Figure 12.10 reflect attempts by timber companies to manage their resources in accordance with the maximum sustainable yield approach, so that they can obtain maximal profits over many years for their owners and investors. On public lands, rates of growth and removal reflect not only economic forces but also social and political ones, and these have changed over time. From the U.S. national forests,

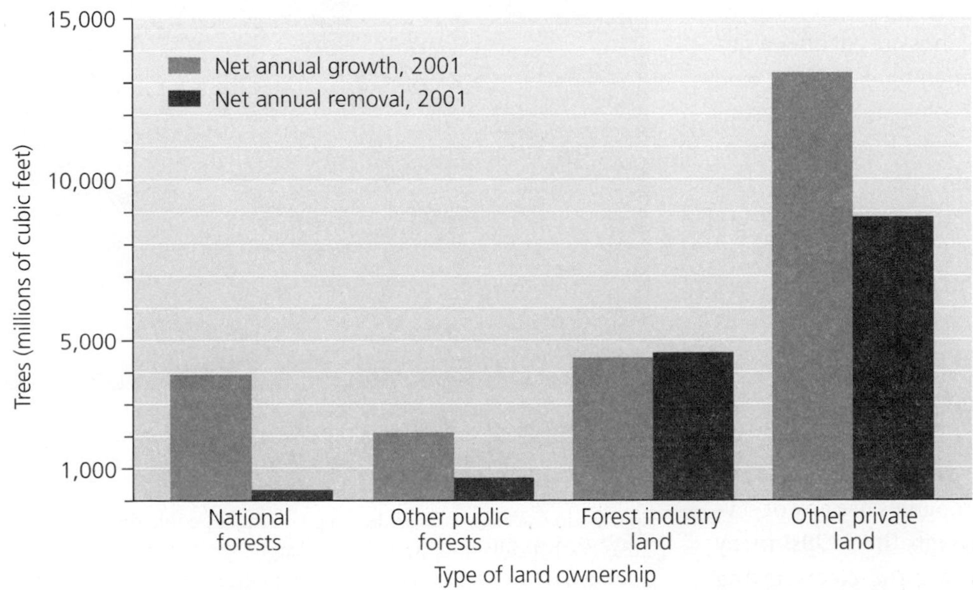


FIGURE 12.10 Forest Service data indicate that in the United States, trees (measured in cubic feet of wood biomass) are growing at a faster rate than they are being removed. The exception is on land privately owned by the timber industry, where extraction is narrowly outpacing growth. Go to GRAPHit! at www.aw-bc.com/withgott or on the student CD-ROM. Data are for 2001, from United States Forest Service, 2001.

FIGURE 12.11 Even-aged tree stand management is practiced on tree plantations where all trees are of equal age, as seen in the stand in the foreground that is regrowing after clear-cutting. In uneven-aged tree stand management, harvests are designed to maintain a mix of tree ages, as seen in the more mature forest in the background. The increased structural diversity of uneven-aged stands provides superior habitat for most wild species and makes these stands more akin to ecologically functional forests.

private timber extraction began to increase in the 1950s as the country experienced a postwar economic boom, consumption of paper products rose, and the population expanded into newly built suburban homes. More recently, harvests from national forests decreased as economic trends shifted, public concern over clear-cutting grew, and forest management philosophy evolved.

Note, however, that even when regrowth outpaces removal, the character of forests may still change. In North America and worldwide, primary forest continues to be lost and to be replaced by younger second-growth forest.

## Plantation forestry has grown

Today the North American timber industry focuses on production in the Northwest and the South from plantations of fast-growing tree species that are single-species monocultures (• pp. 267–268). Because all trees in a given stand are planted at the same time, the stands are **even-aged**, with all trees the same age (**Figure 12.11**). Stands are cut after a certain number of years (called the *rotation time*), and the land is replanted with seedlings. Most ecologists and foresters view these plantations more as crop agriculture than as ecologically functional forests. Because there are few tree species and little variation in tree age, plantations do not offer many forest organisms the habitat they need. However, some harvesting methods aim to maintain **uneven-aged** stands, where a mix of ages (and often a mix of tree species) makes the stand more similar to a natural forest.

## Timber is harvested by several methods

When they harvest trees, timber companies use any of several methods. From the 1950s through the 1970s, many timber harvests were conducted using the **clear-cutting** method, in which all trees in an area are cut, leaving only stumps. Clear-cutting is generally the most cost-efficient method in the short term, but it has the greatest impacts on forest ecosystems (**Figure 12.12**). In the best-case scenario, clear-cutting may mimic natural disturbance events such as fires, tornadoes, or windstorms that knock down trees across large areas. In the worst-case scenario, entire communities of organisms are destroyed or displaced, soil erodes, and the penetration of sunlight to ground level changes microclimatic conditions such that new types of plants replace those that had comprised the native forest. Essentially, clear-cutting sets in motion an artificially driven process of succession (• p. 155) in which the resulting

FIGURE 12.12 Clear-cutting is the most cost-efficient method for timber companies, but it can have severe ecological consequences, including soil erosion and species turnover. Although certain species do use clear-cuts as they regrow, most people find these areas aesthetically unappealing, and public reaction to clear-cutting has driven changes in forestry methods.

climax community may turn out to be quite different from the original climax community.

Clear-cutting occurred widely across North America at a time when public awareness of environmental problems was blossoming. The combination produced public outrage toward the timber industry and public forest managers. Eventually the industry integrated other harvesting methods (**Figure 12.13**). Clear-cutting (**Figure 12.13a**) is still widely practiced, but other methods involve cutting some trees and leaving some standing. In the *seed-tree* approach (**Figure 12.13b**), small numbers of mature and vigorous seed-producing trees are left standing so that they can reseed the logged area. In the *shelterwood* approach (also Figure 12.13b), small numbers of mature trees are left in place to provide shelter for seedlings as they grow. These three methods all lead to even-aged stands of trees.

Selection systems, in contrast, allow uneven-aged stand management. In selection systems (**Figure 12.13c**), only some trees in a forest are cut at any one time. The stand's overall rotation time may be the same as in an even-aged approach, because multiple harvests are made, but the stand remains mostly intact between harvests. Selection systems include single-tree selection, in which widely spaced trees are cut one at a time, and group selection, in which small patches of trees are cut.

It was a form of selection harvesting that MacMillan Bloedel and other timber companies pursued at

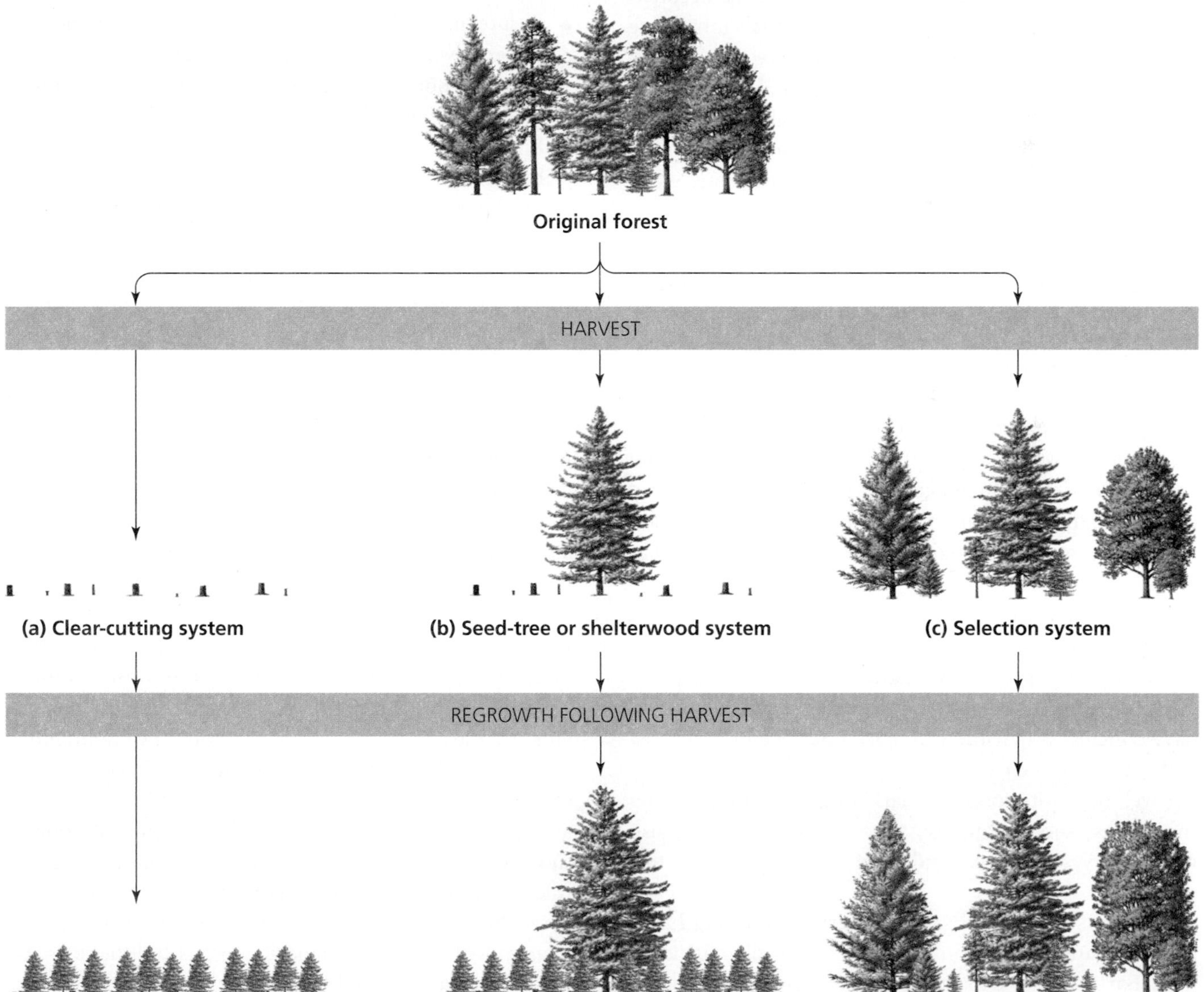


FIGURE 12.13 Foresters and timber companies have devised various methods to harvest timber from forests. In clear-cutting **(a)**, all trees in an area are cut, extracting a great deal of timber inexpensively but leaving a vastly altered landscape. In seed-tree systems and shelterwood systems **(b)**, small numbers of large trees are left in clear-cuts to help reseed the area or provide shelter for growing seedlings. In selection systems **(c)**, a minority of trees is removed at any one time, while most are left standing. These latter methods involve less environmental impact than clear-cutting, but all methods can cause significant changes to the structure and function of natural forest communities.

Clayoquot Sound after old-growth advocates applied pressure and the scientific panel published its guidelines. Not wanting to bring a complete end to logging when so many local people depended on the industry for work, these activists and scientists instead promoted what they considered a more environmentally friendly method of timber removal.

However, selection systems are by no means ecologically harmless. Moving trucks and machinery over an extensive network of roads and trails to access individual trees compacts the soil and disturbs the forest floor. Selection methods are also unpopular with timber companies because they are expensive, and loggers dislike them because they are more dangerous than clear-cutting.

All methods of logging result in habitat disturbance, which invariably affects the plants and animals inhabiting an area. All methods change forest structure and composition. Most methods increase soil erosion, leading to siltation of waterways, which can degrade habitat and affect drinking water quality. Most methods also speed runoff, sometimes causing flooding. In extreme cases, as when steep hillsides are clear-cut, landslides can result.

## Public forests may be managed for recreation and ecosystems

In recent decades, increased awareness of these problems has prompted many citizens to protest the way public forests are managed in the United States and Canada. These citizens have urged that the national, state, and provincial forests be managed for recreation, wildlife, and ecosystem integrity, rather than for timber. They want forests managed as ecologically functional entities, not as cropland for trees.

Critics of the U.S. Forest Service have also protested the fact that taxpayers' money is used to subsidize the extraction of publicly held resources by private corporations. Scientists who have analyzed government subsidies have concluded that the U.S. Forest Service loses at least $100 million of taxpayers' money each year by selling timber well below its costs for marketing and administering the harvest and for building access roads. Subsidies also inflate harvest levels beyond what would occur in a free market.

In one sense, the U.S. Forest Service has long had a policy of attending to interests besides timber production. For the past half century, forest management has nominally been guided by the policy of **multiple use**, meaning that the national forests were to be managed for recreation, wildlife habitat, mineral extraction, and various other uses. In reality, however, timber production was most often the primary use.

In 1976 the U.S. Congress passed the **National Forest Management Act**, which mandated that plans for renewable resource management be drawn up for every national forest. These plans were to be explicitly based on the concepts of multiple use and sustained yield, and they were to be subject to broad public participation under the National Environmental Policy Act (• pp. 66–67). Guidelines specified that these plans:

- Consider both economic and environmental factors.
- Provide for diversity of plant and animal communities and preserve the regional diversity of tree species.
- Ensure research and monitoring of management practices.
- Permit increases in harvest levels only if sustainable.
- Ensure that timber is harvested only where soils and wetlands will not be irreversibly damaged, lands can be restocked quickly, and economic return alone does not guide the choice of harvest method.
- Ensure that logging is conducted only where impacts have been assessed; cuts are shaped to the terrain; maximum size limits are established; and "cuts are carried out in a manner consistent with the protection of soil, watershed, fish, wildlife, recreation, and aesthetic resources, and the regeneration of the timber resource."

Over the years following passage of the National Forest Management Act, the U.S. Forest Service developed new programs to manage wildlife, nongame animals, and endangered species. It pushed for ecosystem-based management and ran programs of ecological restoration, attempting to recover plant and animal communities that had been lost or degraded. Timber harvesting methods were brought more in line with ecosystem-based management goals. A set of approaches dubbed **new forestry** called for timber cuts that came closer to mimicking natural disturbances. For instance, "sloppy clear-cuts" that leave a variety of trees standing were intended to mimic the changes a forest might experience if hit by a severe windstorm.

In 2004, however, the George W. Bush administration issued new regulations that bucked these trends. These new rules freed local forest managers from requirements imposed by the National Forest Management Act, granting them more flexibility in managing forests, but the rules also loosened environmental protections and restricted public oversight.

Then in 2005, the Bush administration repealed the Clinton administration's roadless rule, by which 23.7 million ha (58.5 million acres)—31% of national forest land and 2% of total U.S. land—were in 2001 put off limits to

FIGURE 12.14 Forest fires are natural phenomena to which many plants are adapted and which maintain many ecosystems. The suppression of fire by people over the past century has led to a buildup of leaf litter, woody debris, and young trees, which serve as fuel to increase the severity of fires when they do occur. As a result, catastrophic wildfires (such as this one in Yellowstone National Park in 1988) have become more common in recent years. To avoid these unnaturally severe fires, most fire ecologists suggest allowing natural fires to burn when possible and instituting controlled burns to reduce fuel loads and restore forest ecosystems.

further road construction or maintenance. Although the roadless rule had been supported by a record 4.2 million public comments, the Bush administration overturned it and required state governors to petition the federal government if they want to keep areas in their states roadless. The states of California, Oregon, and New Mexico responded by suing the federal government, asking that the roadless rule be reinstated.

## Fire policy has also stirred controversy

Some ecosystem management efforts, ironically, run counter to the U.S. Forest Service's best-known symbol, Smokey Bear. The cartoon bear wearing a ranger's hat who advises us to fight forest fires is widely recognized, but many scientists assert that Smokey's message has done great harm to American forests.

For over a century, the Forest Service and other land management agencies suppressed fire whenever and wherever it broke out. Yet ecological research now clearly shows that many ecosystems depend on fire. Certain plants have seeds that germinate only in response to fire, and researchers studying tree rings have documented that many ecosystems historically experienced frequent fire. Burn marks in a tree's rings reveal past fires, giving scientists an accurate history of fire events extending back hundreds or even thousands of years. Researchers have found that North America's grasslands and open pine woodlands burned regularly. Ecosystems dependent on fire are adversely affected by its suppression; pine woodlands become cluttered with hardwood understory that ordinarily would be cleared away by fire, for instance, and animal diversity and abundance decline.

In the long term, fire suppression can lead to catastrophic fires that truly do damage forests, destroy human property, and threaten human lives. Fire suppression allows limbs, logs, sticks, and leaf litter to accumulate on the forest floor, effectively producing kindling for a catastrophic fire. Such fuel buildup helped cause the 1988 fires in Yellowstone National Park (**Figure 12.14**), the 2003 fires in southern California, the 2003 fires in British Columbia, and thousands of other wildfires across the continent. Fire suppression and fuel buildup have made catastrophic fires significantly greater problems than they were in the past. Now, global climate change is bringing drier weather to much of the American West (• pp. 418, 520), further worsening wildfire risk. At the same time, increasing residential development on the edges of forested land is placing more homes in fire-prone situations.

To reduce fuel load and improve the health and safety of forests, the Forest Service and other land management agencies have in recent years been burning areas of forest under carefully controlled conditions. These **prescribed burns**, or **controlled burns**, have worked effectively, but they have been implemented on only a relatively small amount of land. And every once in a while, a prescribed burn may get out of control, as happened in 2000 when homes and government labs were destroyed at Los Alamos, New Mexico. All too often, these worthy efforts have been impeded by public misunderstanding and by interference from politicians who have not taken time to understand the science behind the approach.

In the wake of the 2003 California fires, the U.S. Congress, intending to make forests less fire-prone,

THE SCIENCE BEHIND THE STORY

## Fighting over Fire and Forests

*OSU graduate student Daniel Donato testifies in a Congressional hearing*

It's not often that a single scientific paper throws an entire college into turmoil and lands a graduate student in a Congressional hearing to face hostile questioning from federal lawmakers. But such is the political sensitivity of salvage logging.

When a fire burns a forest, should the killed trees be cut and sold for timber? Proponents of salvage logging say yes: we should not let economically valuable wood go to waste.

Opponents of post-fire logging say the burned wood is more valuable left in place—for erosion control, wildlife habitat (snags provide holes for cavity-dwelling animals and food for insects and birds), and organic material to enhance the soil and nurse the growth of future trees.

Proponents of salvage logging argue that forests regenerate best after a fire if they are logged and replanted with seedlings. Moreover, they maintain, salvage logging reduces future fire risk by removing woody debris that could serve as fuel for the next fire. When the gigantic Biscuit Fire consumed 200,000 ha (500,000 acres) in Oregon in 2002, foresters from the College of Forestry at Oregon State University (OSU) made these arguments in support of plans to log portions of the burned area.

Meanwhile, however, OSU forestry graduate student Daniel Donato and five other OSU researchers were setting up research plots in areas burned by the Biscuit Fire to test whether salvage logging really does reduce fire risk and help seedlings regenerate. They measured seedling growth and survival and the amount of woody debris in a number of study plots on burned land before (2004) and after (2005) salvage logging took place, and on burned land that was not logged.

The researchers found that conifer seedlings sprouted naturally in the burned areas at densities exceeding what foresters aim for when they replant sites manually. This suggested that artificial planting of seedlings may be unnecessary. In contrast, natural seedling densities in logged areas were only 29% as high (see the first figure). This indicated that salvage logging was hindering seedling survival, presumably because logging disturbed the soil and crushed many seedlings.

Donato's team also found that salvage logging more than tripled the amount of woody debris on the ground relative to unlogged sites (see the second figure). The research team suggested that the best strategy after fire may be to leave the site alone and leave dead trees standing, so that seedlings regenerate safely.

These conclusions directly contradicted what some OSU forestry professors had argued following the 2002 fire. When these colleagues learned that the prestigious journal *Science* had accepted the research team's paper for publication, they took the unusual step of asking the journal's editors to reconsider their decision. Claiming that *Science*'s peer review process had failed, professor John Sessions and others tried through back channels to stop publication of the paper—actions that were widely

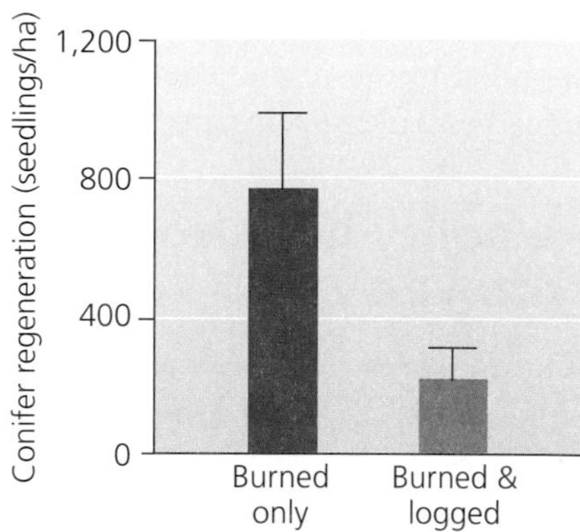


In the wake of the Biscuit Fire, natural growth of conifer seedlings was lower in areas (orange bar) that underwent salvage logging. Data from Donato, D. C., et al., 2006. Post-wildfire logging hinders regeneration and increases fire risk. *Science* 311:352.

passed the Bush administration's *Healthy Forests Restoration Act.* Although this legislation encourages some prescribed burning, it primarily promotes the physical removal of small trees, underbrush, and dead trees by timber companies. The removal of dead trees, or snags, following a natural disturbance is called **salvage logging**. From an economic standpoint, salvage logging may seem to make good sense. However, ecologically, snags have immense value; the insects that decay them provide food for wildlife, and many birds, mammals, and reptiles depend on holes in snags for nesting and roosting sites. Removing timber from recently burned condemned as an attempt at censor ship once they became known.

The paper's publication in January 2006 unleashed a torrent of bizarre events. U.S. Congressmen Greg Walden of Oregon and Brian Baird of Washington felt that the paper threatened legislation they had sponsored to accelerate salvage logging. Walden and Baird called on Donato and others to appear before a hearing of the House of Representatives' Committee on Resources and grilled them before a hometown crowd that packed City Hall chambers in Medford, Oregon. There the 29-year-old graduate student stood up for his team's findings under harsh questioning.

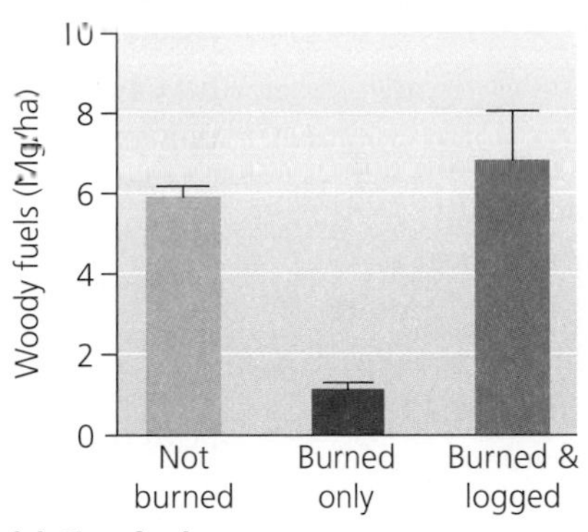


**(a) Fine fuels**

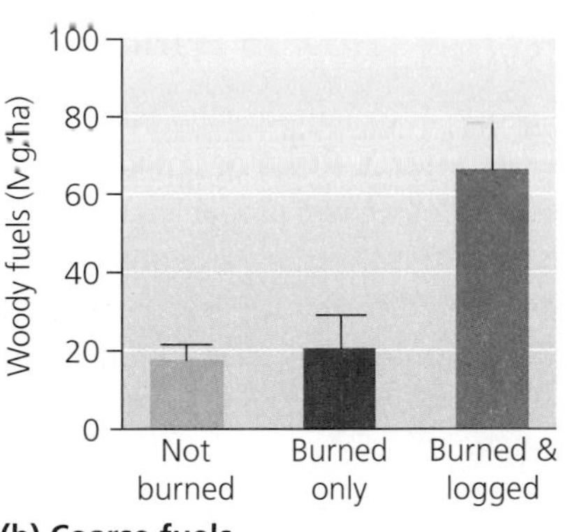


**(b) Coarse fuels**

Burned sites that were salvage-logged (orange bar) contained more fine woody debris **(a)** and coarse woody debris **(b)** than unlogged burned sites (red bar) or than sites that did not burn (yellow bar). Data from Donato, D. C., et al., 2006. Post-wildfire logging hinders regeneration and increases fire risk. *Science* 311:352.

The Bureau of Land Management suspended the team's research grant, in what many viewed as a response to pressure from congressmen or the Bush administration. This highly unusual action drew media attention, and the BLM quickly reinstated the funding.

A heated debate roiled for months in the OSU College of Forestry. The college receives 12% of its funding from taxes on timber sales, leading many to suggest that the college is open to influence from industry. E-mail correspondence was subpoenaed, and some of it showed the college's dean, Hal Salwasser, collaborating with timber industry representatives to refute the paper. As negative publicity built, the college's reputation suffered, graduate admissions declined, and a faculty committee on academic freedom criticized Salwasser for "significant failures of leadership." The dean admitted mistakes, survived a no-confidence vote of the faculty, and pledged to make reforms.

In the pages of *Science* and elsewhere, scientific criticisms of the study's methods were largely rebutted, but many felt that the study's conclusions stretched beyond what its short-term data could support. Scientists on both sides of the debate agreed that long-term research was needed to truly assess the effects of salvage logging on forest regeneration and fire risk.

In 2007, two studies by different OSU forestry scientists provided the first such long-term data. Jeffrey Shatford and colleagues documented widespread natural regrowth of conifers across areas of Oregon and northern California that had burned 9–19 years earlier. And Jonathan Thompson and colleagues examined satellite data, aerial photography, and government agency records for regions within the Biscuit Fire area that had also burned in a previous fire 15 years earlier. They found that of the regions burned in that 1987 fire, those that were salvage-logged burned more severely in 2002 than regions that were not logged. In a paper in the *Proceedings of the National Academy of Sciences*, Thompson and his colleagues concluded that salvage logging increases the risk of severe fires, even when debris is removed and seedlings are manually planted.

More long-term studies on the impacts of salvage logging are forthcoming. In the end, these should help us to better understand and manage our forests in an era of frequent wildfire.

land can also cause severe erosion and soil damage. Moreover, it may impede forest regeneration and promote further wildfires (see "The Science behind the Story," above).

Many scientists and environmental advocates have criticized the Healthy Forests Restoration Act, saying it increases commercial logging in national forests while doing little to reduce catastrophic fires near populated areas. By streamlining procedures for timber removal on public lands, the law also decreases oversight and public participation in enforcing environmental regulations, critics contend.

**Weighing THE Issues | How to Handle Fire?**

A century of fire suppression has left vast swaths of forested lands in North America in danger of catastrophic wildfires. Yet we will probably never have adequate resources to conduct careful prescribed burning over all these lands. Can you suggest any solutions to help protect people's homes near forests while improving the ecological condition of some forested lands? Do you think people should be allowed to develop homes in fire-prone areas?

### Sustainable forestry is gaining ground

Any company can claim that its timber harvesting practices are sustainable, but how is the purchaser of wood products to know whether they really are? In the last several years, a consumer movement has grown that is making informed consumer choice possible. Several organizations now examine the practices of timber companies and offer **sustainable forestry certification** to products produced using methods they consider sustainable (**Figure 12.15**).

Organizations such as the International Organization for Standardization (ISO), the Sustainable Forestry Initiative (SFI) program, and the Forest Stewardship Council (FSC) have varying standards for certification. Consumers can look for the logos of these organizations on forest products they purchase. The FSC is widely perceived to have the strictest certification standards. In 2001, Iisaak, the Native-run timber company at Clayoquot Sound, became the first tree farm license holder in British Columbia to receive FSC certification.

FIGURE 12.15 A Brazilian woodcutter taking inventory marks timber harvested from a forest certified for sustainable management in Amazonian Brazil. A consumer movement centered on independent certification of sustainable wood products is allowing consumer choice to promote sustainable forestry practices.

Consumer demand for sustainable wood has been great enough that Home Depot and other major retail businesses have begun selling sustainable wood. The decisions of such retailers are influencing the logging practices of many timber companies. In British Columbia, 70% of the province's annual harvest now is certified or meets ISO requirements.

Sustainable forestry is more costly for the timber industry, but if certification standards can be kept adequately strong, then consumer choice in the marketplace can be a powerful driver for good forestry practices for the future.

## Agricultural Land Use

Having replaced many forests, agriculture now covers more of the planet's surface than does forest. Thirty-eight percent of Earth's terrestrial surface is devoted to agriculture—more than the area of North America and Africa combined. Of this land, 26% supports pasture, and 12% consists of crops and arable land. Agriculture is the most widespread type of human land use and causes tremendous impacts on land and ecosystems. Although agricultural methods such as organic farming and no-till farming can be sustainable, the majority of the world's cropland hosts either intensive traditional agriculture or monocultural industrial agriculture, involving heavy use of fertilizers, pesticides, and irrigation, and often producing soil erosion, salinization, and desertification (Chapters 9 and 10).

In theory, the marketplace should discourage people from farming with intensive methods that degrade land they own if such practices are not profitable. But agriculture in many countries is supported by billions of dollars in government subsidies. Roughly one-fifth of the income of the average U.S. or Canadian farmer comes from subsidies. Proponents of such subsidies stress that the vagaries of weather make profits and losses from farming unpredictable from year to year. To persist, these proponents say, an agricultural system needs some way to compensate farmers for bad years. Opponents of subsidies argue that farmers can buy insurance to protect themselves against crop failures and that subsidizing environmentally destructive agricultural practices is unsustainable.

### Wetlands have been drained for farming

Many of today's crops grow on the sites of former wetlands (• pp. 414–415)—swamps, marshes, bogs, and river floodplains—that people drained and filled in (**Figure 12.16**). Throughout recent history, governments have encouraged laborious efforts to drain wetlands. To promote settlement and farming, the United States passed a series of laws known as the Swamp Lands Acts in 1849, 1850, and 1860, which

FIGURE 12.16 Most of North America's wetlands have been drained and filled, and the land converted to agricultural use. The northern Great Plains region of Canada and the United States was pockmarked with thousands of "prairie potholes," water-filled depressions that served as nesting sites for most of the continent's waterfowl. Today many of these wetlands have been lost; shown are farmlands encroaching on prairie potholes in North Dakota.

encouraged draining wetlands for agriculture. The government transferred over 24 million ha (60 million acres) of wetlands to state ownership (and eventually to private hands) to stimulate drainage, conversion, and flood control.

In the Mississippi River valley, the Midwest, and a handful of states from Florida to Oregon, these transfers eradicated malaria (because mosquito vectors breed in wetlands) and created over 10 million ha (25 million acres) of new farmland. A U.S. Department of Agriculture (USDA) program in 1940 provided monetary aid and technical assistance to farmers draining wetlands on their property, resulting in the conversion of almost 23 million ha (57 million acres).

Today, less than half the original wetlands in the lower 48 U.S. states and southern Canada remain. However, many people now have a new view of wetlands. Rather than viewing them as worthless swamps, science has made clear that they are valuable ecosystems. This scientific knowledge, along with a preservation ethic, has induced policymakers to develop regulations to safeguard remaining wetlands. Yet, because of loopholes, differing state laws, development pressures—and even debate over the legal definition of wetlands—many of these vital ecosystems are still being lost.

Financial incentives are also being used as a tool to influence agricultural land use in the United States. The Conservation Reserve Program begun in 1985 was a landmark initiative that provided farmers with a different kind of subsidy—it paid them to take highly erodible lands out of production and instead encouraged them to make the areas more habitable for wildlife. Now, under the Wetland Reserve Program, the U.S. government is offering subsidies to landowners who refrain from developing wetland areas.

**Weighing THE Issues | Subsidies, Soil, and Wetlands**

Do you think that subsidy programs such as the Conservation Reserve Program and the Wetland Reserve Program are a good use of taxpayers' money? Are financial incentives such as these better tools than government regulation for promoting certain land use goals?

## Livestock graze one-fourth of Earth's land

Cropland agriculture uses less than half the land taken up by livestock grazing, which covers a quarter of the world's land surface. Human use of rangeland, however, does not necessarily exclude its use by wildlife or its continued functioning as a grassland ecosystem. Grazing can be sustainable if done carefully and at low intensity. In the American West, ranching proponents claim that cattle are merely taking the place of the vast herds of bison that once roamed the plains. Indeed, most of the world's grasslands have historically been home to large herds of grass-eating mammals, and grasses have adapted to herbivory. Nonetheless, poorly managed grazing, as we have seen (• pp. 254–257), can have adverse impacts on soil and grassland ecosystems.

Most U.S. rangeland is federally owned and managed by the **Bureau of Land Management (BLM)**. The BLM is the nation's single largest landowner; its 106 million ha (261 million acres) are spread across 12 western states (see Figure 12.9). Ranchers are allowed to graze cattle on BLM lands for inexpensive fees, a practice that many public lands advocates say encourages overgrazing. Thus ranchers and environmentalists have traditionally been at loggerheads. In the past several years, however, ranchers and environmentalists have been finding common ground (• pp. 256–257), teaming up to preserve ranchland against what each of them views in common as a threat—the encroaching housing developments of suburban sprawl (• pp. 361–366). Although developers often pay high prices for ranchland, many ranchers do not want to see the loss of the wide open spaces and the ranching lifestyle that they cherish.

### Land use in the American West might have been better managed

Land uses such as grazing, farming, and timber harvesting need not have strongly adverse impacts. It is not these activities per se that cause environmental problems, but rather the overexploitation of resources beyond what ecosystems can handle. In the American West, a great deal of damage was done to the land by poor farming practices, overgrazing, and attempts to farm arid lands that were more suitable for grazing or preservation.

Most land to the west of the 100th meridian, the longitudinal line slicing through the Great Plains from Manitoba south through Texas, receives less than 50 cm (20 in.) of rain per year, making it too arid for unirrigated agriculture. One man in U.S. history recognized this fact and attempted to reorient policy so the West could be settled in a way that allowed farmers to succeed. John Wesley Powell, an extraordinary individual who explored the raging Colorado River by boat despite having lost an arm in the Civil War, undertook vast surveys of the Western lands for the U.S. government in the late 19th century (**Figure 12.17**). A pioneer in calling for government agencies to base their policies on science, Powell maintained that lands beyond the 100th meridian were too dry to support farming on the 65-ha (160-acre) plots the government parceled out through the Homestead Act (• p. 64). Plots in the West, he said, would have to be 16 times as large and would require irrigation. Moreover, to prevent individuals or corporations from monopolizing scarce water resources, he urged the government to organize farmers into cooperative irrigation districts, with each district encompassing a watershed.

Powell's ideas, based on science and close study of the land, were too revolutionary for the entrenched political interests and prevailing misconceptions of his time, which held that the West was a utopia for frontier settlement. The ideas in Powell's 1878 *Report on the Lands of the Arid Region of the United States* were, for the most part, never implemented. Instead, less sustainable land use policies were put in place and contributed to failures such as the Dust Bowl of the 1930s (• pp. 245–246).

For agriculture and forestry alike, debates continue today over how best to use land and manage resources.

FIGURE 12.17 John Wesley Powell, 19th-century Western explorer and government scientist, tried to shift U.S. land use policy to take account of the aridity of Western lands.

Resource extraction from public lands in the United States and Canada has helped propel the economies of these countries. But as resources dwindle, as forests and soils are degraded, and as the landscape fills with more people, the arguments for conservation of resources—for their sustainable use—have grown stronger. Also growing stronger is the argument for preservation of land—setting aside tracts of relatively undisturbed land intended to remain forever undeveloped.

## Parks and Reserves

Preservation has been part of the American psyche ever since John Muir rallied support for saving scenic lands in the Sierras (• p. 33). For ethical reasons as well as pragmatic ecological and economic ones, U.S. citizens and many other people worldwide have chosen to set aside tracts of land in perpetuity to be preserved and protected from development.

### Why have we created parks and reserves?

The historian Alfred Runte has cited four traditional reasons that parks and protected areas have been established:

1. Enormous, beautiful, or unusual features such as the Grand Canyon, Mount Rainier, or Yosemite Valley inspire people to protect them—an impulse termed *monumentalism* (**Figure 12.18**).
2. Protected areas offer recreational value to tourists, hikers, fishers, hunters, and others.

FIGURE 12.18 The awe-inspiring beauty of some regions of the western United States was one reason for the establishment of national parks. Images of scenic vistas such as this one of Bridal Veil Falls in Yosemite National Park, portrayed by the landscape painter Albert Bierstadt, inspired millions of people from North America and abroad to visit these parks.

3. Protected areas offer utilitarian benefits. For example, undeveloped watersheds provide cities with clean drinking water and a buffer against floods.
4. Parks make use of sites lacking economically valuable material resources or that are hard to develop; land that holds little monetary value is easy to set aside.

To these four traditional reasons, a fifth has been added in recent years: the preservation of biodiversity. As we saw in Chapter 11, human impact alters habitats and has led to countless population declines and species extinctions. A park or reserve is widely viewed as a kind of Noah's Ark, an island of habitat that can, scientists hope, maintain species that might otherwise disappear.

## Federal parks and reserves began in the United States

The striking scenery of the American West impelled the U.S. government to create the world's first **national parks**, publicly held lands protected from resource extraction and development but open to nature appreciation and various forms of recreation. In 1872, Yellowstone National Park was established as "a public park or pleasuring-ground for the benefit and enjoyment of the people." Yosemite, General Grant, Sequoia, and Mount Rainier National Parks followed after 1890. The Antiquities Act of 1906 gave the president authority to declare selected public lands as national monuments, which can be an interim step to national park status. Presidents from Theodore Roosevelt to Bill Clinton have used this authority to expand the nation's system of protected lands.

The National Park Service (NPS) was created in 1916 to administer the growing system of parks and monuments, which today numbers 388 sites totaling 32 million ha (79 million acres) and includes national historic sites, national recreation areas, national wild and scenic rivers, and other types of areas (see Figure 12.9). This most widely used park system in the world received 273 million reported recreation visits in 2006—almost as many visits as there are U.S. residents. The high visitation rates signal success for the park system, but they also create overcrowded conditions at some parks, suggesting to many observers that there is a pressing need to expand the system.

Many other nations now have national park systems. Canada's system covers 28 million ha (69 million acres) and receives 16 million visits yearly. At Clayoquot Sound, Pacific Rim National Park Reserve is a protected area designated for future national park status. The Clayoquot Sound region also encompasses several provincial parks. Provincial parks in Canada cover more area than national parks. In the United States, state parks are numerous and tend to be more oriented toward recreation than are national parks.

Another type of federal protected area in the United States is the **national wildlife refuge**. The system of national wildlife refuges, begun in 1903 by President Theodore Roosevelt, now totals 37 million ha (91 million acres) spread over 541 sites (see Figure 12.9). The U.S. Fish and Wildlife Service (USFWS) administers the refuges with management ranging "from preservation to active manipulation of habitats and populations." Indeed, these refuges not only serve as havens for wildlife, but also in many cases encourage hunting, fishing, wildlife observation, photography, environmental education, and other public uses.

Some wildlife advocates find it ironic that hunting is allowed at many refuges, but hunters have long been in the forefront of the conservation movement and have traditionally supplied the bulk of funding for land acquisition and habitat management for the refuges. Many refuges are managed for waterfowl, but the USFWS increasingly considers nongame species as well as game species. The USFWS manages at the habitat and ecosystem levels, engaging in ecological restoration of marshes and grasslands, for example.

## Wilderness areas have been established on federal lands

In response to the public's desire for undeveloped areas of land, in 1964 the U.S. Congress passed the Wilderness Act, which allowed areas of existing federal lands to be designated as **wilderness areas**. These areas are off-limits to development of any kind, but they are open to public recreation such as hiking, nature study, and other activities that have minimal impact on the land.

Congress declared that wilderness areas were necessary "to assure that an increasing population, accompanied by expanding settlement and growing mechanization, does not occupy and modify all areas within the United States and its possessions, leaving no lands designated for preservation and protection in their natural condition." Wilderness areas have been established within portions of national forests, national parks, national wildlife refuges, and BLM land, and they are overseen by the agencies that administer those areas (**Figure 12.19**). Some preexisting extractive land uses, such as grazing and mining, were allowed to continue within wilderness areas as a political compromise so the act could be passed.

## Not everyone supports land set-asides

The restriction of activities in wilderness areas has helped generate opposition to U.S. land protection policies. Sources of such opposition include the governments of some western states, where large portions of land are federally owned. When those states came into existence, the federal government retained ownership of much of the acreage inside their borders. Idaho, Oregon, and Utah own less than 50% of the land within their borders, and in Nevada 80% of the land is federally owned. Western state governments have traditionally sought to obtain land from the federal government and encourage resource extraction and development on it.

The drive to extract more resources, secure local control of lands, and expand recreational access to public lands is epitomized by the **wise-use movement**, a loose confederation of individuals and groups that coalesced in the 1980s and 1990s in response to the increasing success of environmental advocacy. Wise-use advocates are dedicated to protecting private property rights; opposing government regulation; transferring federal lands to state, local, or private hands; and promoting motorized recreation on public lands. Wise-use advocates include farmers, ranchers, trappers, and mineral prospectors at the grassroots level who live off the land, as well as groups representing the industries that extract timber, mineral, and fossil fuel resources.

Debate between mainstream environmental groups and wise-use spokespeople has been vitriolic. Each side claims to represent the will of the people and paints the other as the oppressive establishment. Wise-use advocates have played key roles in ongoing debates over national park policy, such as whether recreational activities that disturb wildlife, such as snowmobiles and jet-skis, should be allowed. Under the George W. Bush administration, wilderness protection policies have been weakened, and federal agencies have generally shifted policies and enforcement away from preservation and conservation and toward recreation and extractive uses.

FIGURE 12.19 Wilderness areas were designated on various federally managed lands in the United States following the 1964 Wilderness Act. These include areas little disturbed by human activities, such as the Selway-Bitterroot Wilderness in Idaho, shown here.

## Nonfederal entities also protect land

Efforts to set aside land—and the debates over such efforts—at the federal level are paralleled at regional and local levels. Each U.S. state and Canadian province has agencies that manage resources on state or provincial lands, as do many counties and municipalities. As just one example, New York State in the 19th century created Adirondack State Park in a mountainous area whose streams converge to form the headwaters of the Hudson River, which flows south past Albany to New York City. Seeing the need for river water to power industries, keep canals filled, and provide drinking water, the state set the land aside, a farsighted decision that has paid dividends through the years.

Private nonprofit groups also preserve land. **Land trusts** are local or regional organizations that purchase land with the aim of preserving it in its natural condition. The Nature Conservancy can be considered the world's largest land trust, but smaller ones are springing up throughout North America. Nearly 1,700 local and state land trusts in the United States together own 690,000 ha (1.7 million acres) and have helped preserve an additional 4.1 million ha (10.2 million acres), including well-known scenic areas such as Big Sur on the California coast, Jackson Hole in Wyoming, and Maine's Mount Desert Island.

## Parks and reserves are increasing internationally

Many nations have established national park systems and are benefiting from ecotourism as a result—from Costa Rica (Chapter 5) to Ecuador to Thailand to Tanzania. The total worldwide area in protected parks and reserves increased more than fourfold from 1970 to 2000, and in 2003 the world's 38,536 protected areas covered 1.3 billion ha (3.2 billion acres), or 9.6% of the planet's land area. However, parks in developing countries do not always receive the funding they need to manage resources, provide for recreation, and protect wildlife from poaching and timber from logging. Thus many of the world's protected areas are merely *paper parks*—protected on paper but not in reality.

Some types of protected areas fall under national sovereignty but are designated or partly managed internationally by the United Nations. *World heritage sites* are an example; currently over 830 sites across 184 countries are listed for their natural or cultural value. One such site is Australia's Kakadu National Park, discussed in Chapter 2. Another is the mountain gorilla reserve shared by three African countries.

The gorilla reserve, which integrates national parklands of Rwanda, Uganda, and the Democratic Republic of Congo, is also an example of a *transboundary park*, an area of protected land overlapping national borders. Transboundary parks can be quite large, and they account for 10% of protected areas worldwide, involving over 100 countries. A North American example is Waterton-Glacier National Parks on the Canadian–U.S. border.

Some transboundary reserves function as *peace parks*, helping ease tensions by acting as buffers between nations that have quarreled over boundary disputes. This is the case with Peru and Ecuador as well as Costa Rica and Panama, and many people hope that peace parks can also help resolve conflicts between Israel and its neighbors.

**Biosphere reserves** are tracts of land with exceptional biodiversity that couple preservation with sustainable development to benefit local people. They are designated by UNESCO (the United Nations Educational, Scientific, and Cultural Organization) following application by local stakeholders. Each biosphere reserve consists of (1) a core area that preserves biodiversity, (2) a buffer zone that allows local activities and limited development that do not hinder the core area's function, and (3) an outer transition zone in which agriculture, human settlement, and other land uses can be pursued in a sustainable way (**Figure 12.20**).

Clayoquot Sound was designated as Canada's 12th biosphere reserve in 2000, in an attempt to help build cooperation among environmentalists, timber companies, Native people, and local residents and businesses. The core area consists of provincial parks and Pacific Rim National Park Reserve. Environmentalists hoped the designation would help promote stronger land preservation efforts. Local residents supported it because outside money was being offered for local development efforts. The timber industry did not stand in the way once it was clear that harvesting operations would not be affected. The designation has brought Clayoquot Sound more international attention, but it has not created new protected areas and has not altered land use policies.

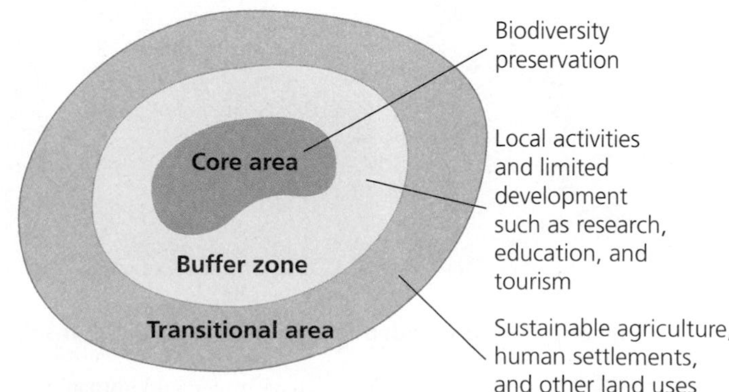


FIGURE 12.20 Biosphere reserves are international efforts that couple preservation with sustainable development to benefit local residents. Each reserve includes a core area that preserves biodiversity, a buffer zone that allows limited development, and a transition zone that permits various uses.

## The design of parks and reserves has consequences for biodiversity

Often it is not outright destruction of habitat that threatens species, but rather the fragmentation of habitat (• pp. 316–317). Expanding agriculture, spreading cities, highways, logging, and many other impacts have chopped up large contiguous expanses of habitat into small disconnected ones (**Figure 12.21a** and **12.21c**). When this happens, many species suffer. Bears, mountain lions, and other animals that need large ranges in which to roam may disappear. Bird species that thrive in the interior of forests may fail to reproduce when forced near the edge of a fragment (**Figure 12.21b**). Their nests often are attacked by predators and parasites that favor open habitats surrounding the fragment or that travel along habitat edges. Avian ecologists judge forest fragmentation to be a main reason why populations of many songbirds of eastern North America are declining.

Because habitat fragmentation is such a central issue in biodiversity conservation, and because there are limits on how much land can be set aside, conservation biologists have argued heatedly about whether it is better to make reserves large in size and few in number, or many in number but small in size. Nicknamed the **SLOSS dilemma**, for "**s**ingle **l**arge **o**r **s**everal **s**mall," this debate is ongoing and complex, but it seems clear that large species that roam great distances, such as the Siberian tiger (Chapter 11), benefit more from the "single large" approach to reserve design. In contrast, creatures such as insects that live as larvae in small areas may do just fine in a number of small isolated reserves, if they can disperse as adults by flying from one reserve to another.

**(a) Mount Hood National Forest, Oregon**

**(b) Wood thrush**

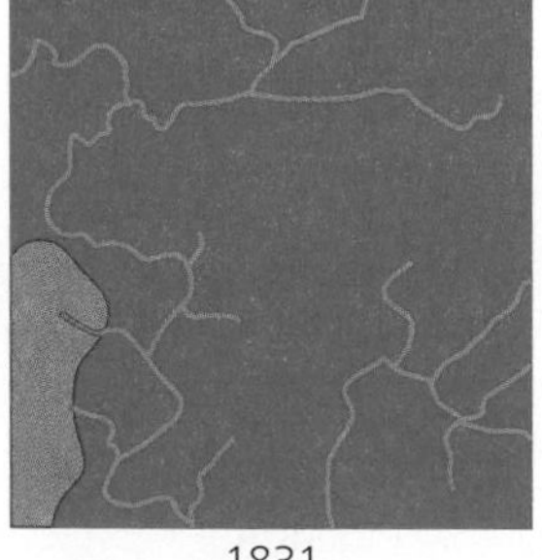

1831

1882

1902

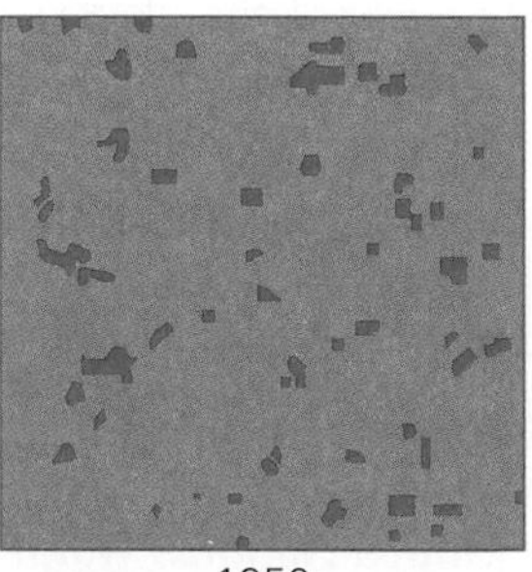

1950

**(c) Fragmentation of wooded area (green) in Cadiz Township, Wisconsin**

FIGURE 12.21 As human populations have grown and human impacts have increased, most large expanses of natural habitat have become fragmented into smaller disconnected areas. Forest fragmentation from timber harvesting, for example, is evident on the Mount Hood National Forest, Oregon (**a**). Fragmentation has significant impacts on forest-dwelling species such as the wood thrush, *Hylocichla mustelina* (**b**), a migrant songbird of eastern North America. In forest fragments, wood thrush nests are parasitized by cowbirds that thrive in open country and edge habitats. Forest fragmentation has been extreme in the eastern and midwestern United States; shown in (**c**) are historical changes in forested area in Cadiz Township, Wisconsin, between 1831 and 1950. *Source for (c):* Curtis, J. T. 1956. The modification of mid-latitude grasslands and forests by man. In Thomas, W. L. Jr., ed., *Man's role in changing the face of the earth.* Chicago: University of Chicago Press.

A related issue is whether **corridors** of protected land are important for allowing animals to travel between islands of protected habitat. In theory, connections between fragments provide animals access to more habitat and help enable gene flow to maintain populations in the long term. Many land management agencies and environmental groups try, when possible, to join new reserves to existing reserves for these reasons. It is clear that we will need to think on the landscape level if we are to preserve a great deal of our natural heritage.

## Conclusion

Managing natural resources sustainably is important for resources such as timber, which otherwise can be carelessly exploited and overharvested. The United States, Canada, and other nations have established various federal and regional agencies to oversee and manage publicly held land and the natural resources that are extracted from public land.

Forest management in North America reflects trends in land and resource management in general. Early emphasis on resource extraction evolved into policies on sustained yield and multiple use, a shift that occurred as land and resource availability declined and as the public became more aware of environmental degradation. Public forests today are managed not only for timber production, but also for recreation, wildlife habitat, and ecosystem integrity.

Meanwhile, public support for preservation of natural lands has resulted in parks, wilderness areas, and other reserves, both in North America and abroad. These trends are positive ones, because the preservation and conservation of land and resources is essential if we wish our society to be sustainable and to thrive in the future.

## REVIEWING OBJECTIVES

**You should now be able to:**

**Identify the principles, goals, and approaches of resource management**

- Resource management enables us to sustain natural resources that are renewable if we are careful not to deplete them. (pp. 329–330)
- Resource managers have increasingly focused not only on extraction, but also on sustaining the ecological systems that make resources available. (pp. 330–331)
- Resource managers have long managed for maximum sustainable yield and are beginning to implement ecosystem-based management and adaptive management. (pp. 330–331)

**Summarize the ecological roles and economic contributions of forests, and outline the history and scale of forest loss**

- Forests not only provide us economically important timber, but also support biodiversity and contribute ecosystem services. (pp. 331–333)
- Developed nations deforested much of their land as settlement, farming, and industrialization proceeded. Today deforestation is taking place most rapidly in developing nations. (pp. 334–337)

**Explain the fundamentals of forest management and describe the major methods of harvesting timber**

- The U.S. national forests were established to conserve timber and allow its sustainable extraction. (p. 338)
- Harvesting methods include clear-cutting and other even-aged techniques, as well as selection strategies that maintain uneven-aged stands that more closely resemble natural forest. (pp. 340–342)
- Foresters are beginning to manage for recreation, wildlife habitat, and ecosystem integrity, as well as timber production. (p. 342)
- Fire policy has been politically controversial, but scientists agree that we need to address the impacts of a century of fire suppression. (pp. 343–345)
- Certification of sustainable forest products allows consumer choice in the marketplace to influence forestry techniques. (p. 346)

**Analyze the scale and impacts of agricultural land use**

- Agriculture has contributed greatly to deforestation and has had enormous impacts on landscapes and ecosystems worldwide. (pp. 346–348)

**Identify major federal land management agencies and the lands they manage**

- The U.S. Forest Service, National Park Service, Fish and Wildlife Service, and Bureau of Land Management manage U.S. national forests, national parks, national wildlife refuges, and BLM land. (pp. 339, 349–350)

**Recognize types of parks and reserves, and evaluate issues involved in their design**

- Public demand for preservation and recreation has led to the creation of parks, reserves, and wilderness areas in North America and across the world. (pp. 348–349, 351)
- Biosphere reserves are one of several types of internationally managed protected lands. (p. 351)
- Because habitat fragmentation affects wildlife, conservation biologists are working on how best to design parks and reserves. (pp. 352–353)

## TESTING YOUR COMPREHENSION

1. How do minerals differ from timber when it comes to resource management?
2. Compare and contrast maximum sustainable yield, adaptive management, and ecosystem-based management. Why may pursuing maximum sustainable yield sometimes conflict with what is ecologically desirable?
3. Name several major causes of deforestation. Where is deforestation most severe today?
4. Compare and contrast the major methods of timber harvesting.
5. Describe several ecological effects of logging. How has the U.S. Forest Service responded to public concern over the ecological effects of logging?
6. Are forest fires a bad thing? Explain your answer.
7. Approximately what percentage of Earth's land is used for agriculture? What policies have caused conversion of wetlands for agriculture in the United States?
8. Name five reasons that people have created parks and reserves. Why did the U.S. Congress determine in 1964 that wilderness areas were necessary? How do these areas differ from national parks? From national wildlife refuges?
9. Why do some people in the United States oppose federal land protection?
10. Roughly what percentage of Earth's land is protected? What types of protected areas have been established in countries outside North America?

## SEEKING SOLUTIONS

1. Do you think maximum sustainable yield represents an appropriate policy for resource managers to follow? Why or why not?
2. People in developed countries are fond of warning people in developing countries to stop destroying rainforest. People of developing countries often respond that this is hypocritical, because the developed nations became wealthy by deforesting their land and exploiting its resources in the past. What would you say to the president of a developing nation, such as Brazil, that is seeking to clear much of its forest?
3. Can you think of a land use conflict that has occurred in your region? How was it resolved? If it is unresolved, then how could it be resolved?
4. What are some ecological effects of agricultural subsidies? Propose arguments for and against subsidies from an ecological point of view.
5. **THINK IT THROUGH** You have just become the supervisor of a national forest. Timber companies are requesting to cut as many trees as you will let them, and environmentalists want no logging at all. Ten percent of your forest is old-growth primary forest, and the remaining 90% is secondary forest. Your forest managers are split among preferring maximum sustainable yield, ecosystem-based management, and adaptive management. What management approach(es) will you take? Will you allow logging of all, none, or some old-growth trees? Will you allow logging of secondary forest? If so, what harvesting strategies will you encourage? What would you ask your scientists before deciding on policies on fire management and salvage logging?
6. **THINK IT THROUGH** You have just been elected mayor of a town on Clayoquot Sound. A timber company that employs 20% of your town's residents wants to log a hillside above the town, and the provincial government is supportive of the harvest. But owners of ecotourism businesses that run whale-watching excursions and rent kayaks to out-of-town visitors are complaining that the logging would destroy the area's aesthetic appeal and devastate their businesses—and these businesses provide 40% of the tax base for your town. Greenpeace is organizing a demonstration in your town soon, and news reporters are beginning to call your office, asking what you will do. How will you proceed?

## INTERPRETING GRAPHS AND DATA

The invention of the movable-type printing press by Johannes Gutenberg in 1450 stimulated a demand for paper that has only increased as the world population has grown. The 20th-century invention of the xerographic printing process used in photocopiers and laser printers has accelerated our demand for paper, with most raw fiber for paper production coming from wood pulp from forest trees.

1. How many millions of tons of paper and paperboard were consumed worldwide in 1970? 1980? 1990? 2000?
2. By what percentage did worldwide consumption of paper and paperboard increase from 1970 to 1980? From 1980 to 1990? From 1990 to 2000?
3. Name three steps that your school could take to reduce its paper consumption.

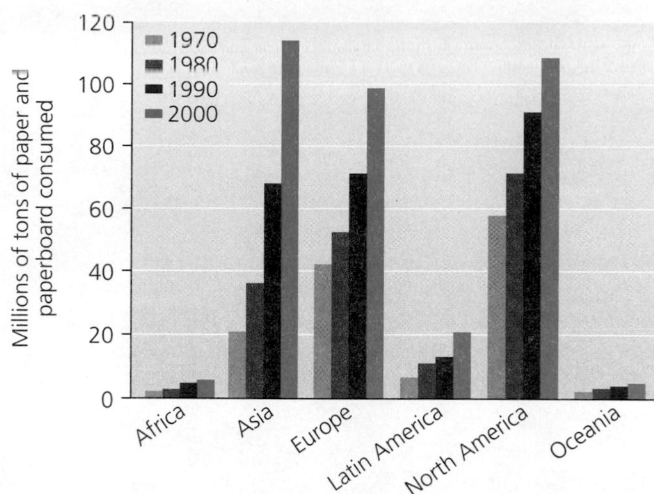


Global consumption of paper and paperboard, 1970–2000.
Data from the Food and Agriculture Organization of the United Nations.

## CALCULATING ECOLOGICAL FOOTPRINTS

The average North American uses over 300 kg (660 lb) of paper and paperboard per year. Using the estimates of paper and paperboard consumption for each region of the world for the year 2000—as shown in the figure in the "Interpreting Graphs and Data" section—calculate the per capita consumption of paper and paperboard for each region of the world using the population data in the table.

1. How much paper would North Americans save each year if we consumed paper at the rate of Europeans?
2. How much paper would be consumed if everyone in the world used as much paper as the average European? As the average North American?
3. Why do you think people in other regions consume less paper, per capita, than North Americans? Name three things you could do to reduce your paper consumption.

| | Population in 2000 (millions)* | Total paper consumed in 2000 (millions of tons) | Per capita paper consumed in 2000 (pounds) |
|---|---|---|---|
| **Africa** | 840 | 6 | 14 |
| **Asia** | 3,766 | | |
| **Europe** | 728 | | |
| **Latin America** | 531 | | |
| **North America** | 319 | | |
| **Oceania** | 32 | | |
| **World** | 6,216 | | ~114 |

**Data source:* Population Reference Bureau.

### Take It Further

Go to www.aw-bc.com/withgott or the student CD-ROM, where you'll find:

- Suggested answers to end-of-chapter questions
- Quizzes, animations, and flashcards to help you study
- *Research Navigator*™ database of credible and reliable sources to assist you with your research projects
- **GRAPHit!** Tutorials to help you interpret graphs
- **INVESTIGATEit!** Current news articles that link the topics that you study to case studies from your region to around the world

CHAPTER

# 13 Urbanization and Creating Livable Cities

Intersection near Shibuya station, Tokyo, Japan

## Upon completing this chapter, you will be able to:

- Describe the scale of urbanization
- Assess urban and suburban sprawl
- Outline city and regional planning and land use strategies
- Evaluate transportation options
- Describe the roles of urban parks
- Analyze environmental impacts and advantages of urban centers
- Assess the pursuit of sustainable cities

Pedestrians and joggers use Portland, Oregon's Tom McCall Waterfront Park

CENTRAL CASE

# Managing Growth in Portland, Oregon

**"Sagebrush subdivisions, coastal condomania, and the ravenous rampage of suburbia in the Willamette Valley all threaten to mock Oregon's status as the environmental model for the nation."**
—Oregon governor Tom McCall, 1973

**"We have planning boards. We have zoning regulations. We have urban growth boundaries and 'smart growth' and sprawl conferences. And we still have sprawl."**
—Environmental scientist Donella Meadows, 1999

With the fighting words above, Oregon governor Tom McCall challenged his state's legislature in 1973 to take action against runaway sprawling development, which many Oregon residents feared would ruin the communities and landscapes they had come to love. McCall echoed the growing concerns of state residents that farms, forests, and open space were being gobbled up for development, including housing for people moving in from California and elsewhere. Foreseeing a future of subdivisions, strip malls, and traffic jams engulfing the pastoral Willamette Valley, Oregon acted. With Senate Bill 100, the state legislature in 1973 passed a sweeping land use law that would become the focus of acclaim, criticism, and careful study for years afterward by other states and communities trying to manage their own urban and suburban growth.

Oregon's law required every city and county to draw up a comprehensive land use plan in line with statewide guidelines that had gained popular support from the state's electorate. As part of each land use plan, each metropolitan area had to establish an **urban growth boundary (UGB)**, a line on a map intended to separate areas desired to be urban from areas desired to remain rural. Development for housing, commerce, and industry would be encouraged within these urban growth boundaries but severely restricted beyond them. The intent was to revitalize city centers, prevent suburban sprawl, and protect farmland, forests, and open landscapes around the edges of urbanized areas.

Residents of the area around Portland, the state's largest city, established a new regional entity to help plan how land would be apportioned in their region. The Metropolitan Service District, or Metro, represents 25 municipalities and three counties. Metro adopted the Portland-area urban growth boundary in 1979 and has tried to focus growth on existing urban centers and to build communities where people can walk or take mass transit between home, work, and shopping. These policies have largely worked as intended; Portland's downtown and older neighborhoods have thrived, regional urban centers are becoming denser and more community oriented, mass transit has expanded, and development has been limited on land beyond the UGB. Portland began attracting international attention for its "livability."

To many Portlanders today, the UGB remains the key to maintaining quality of life in city and countryside alike. In the view of its critics, however, the "Great Wall of Portland" is an elitist and intrusive government regulatory tool. Ironically, the Portland area's successes may one day prove its undoing. A continuing influx of people has meant rapid development and rising housing prices. Still, most citizens have supported Oregon's land use rules for the past 30 years, and the system survived three state referenda and many legal challenges.

In November 2004, however, Oregon voters approved a ballot measure that threatens to eviscerate the very land use reforms they had backed for three decades. Ballot Measure 37 requires the state to compensate certain landowners if government regulation has decreased the value of their land. For example, regulations prevent landowners outside UGBs from subdividing their lots and selling them for housing development. Under Ballot Measure 37, the state now has to pay these landowners to make up for theoretically lost income or else allow them to ignore the regulations. Because the state does not have enough money to pay such claims, the measure could effectively gut Oregon's zoning, planning, and land use rules.

Once the measure cleared months' of challenges in the courts, over 7,000 landowners filed claims for payments or waivers. The state legislature, under pressure from Measure 37 opponents to revoke or revise the law and from Measure 37 supporters to maintain it, settled on a compromise: A new ballot measure revisiting the issue will be sent to the voters in November 2007. As developments unfold, events in Oregon over the next few years could tell us much about how our cities and landscapes may change in the future.

# Our Urbanizing World

We live at a turning point. Beginning about the year 2007, for the first time in human history, more people will live in urban areas than in rural areas. This shift from the countryside into towns and cities, or **urbanization**, is arguably the single greatest change our society has undergone since its transition from a nomadic hunter-gatherer lifestyle to a sedentary agricultural one.

## Industrialization has driven the move to urban centers

Since 1950, the world's urban population has more than quadrupled. Urban populations are growing for two reasons: (1) the human population overall is growing (Chapter 8), and (2) more people are moving from farms to cities than are moving from cities to farms.

The shift from country to city began long ago. Agricultural harvests that produced surplus food freed a proportion of citizens from farm life and allowed the rise of specialized manufacturing professions, class structure, political hierarchies, and urban centers (• p. 236). The industrial revolution (• pp. 4–5) spawned technological innovations that created jobs and opportunities in urban centers for people who were no longer needed on farms. Industrialization and urbanization bred further technological advances that increased production efficiencies, both on the farm and in the city. This process of positive feedback continues today.

Worldwide, the proportion of population that is urban rose from 30% half a century ago to 49% in 2005. Between 1950 and 2005, the global urban population increased by 2.65% each year, whereas the rural population rose only by 1.12% annually. From 2005 to 2030, the United Nations projects that the urban population will grow by 1.78% annually, whereas the rural population will decline by 0.03% each year.

Trends differ between developed and developing nations, however (**Figure 13.1**). In developed nations such as the United States and Canada, urbanization has slowed, because roughly three of every four people already live in cities, towns, and **suburbs**, the smaller communities that ring cities. In 1850, the U.S. Census Bureau classified only 15% of U.S. citizens as urban dwellers. That percentage passed 50% shortly before 1920 and now stands at 80%. Most U.S. urban dwellers reside in suburbs; fully 50% of the U.S. population today is suburban.

In contrast, today's developing nations, where many people still reside on farms, are urbanizing rapidly. In China, India, and Nigeria, and other nations, rural people are streaming to cities in search of jobs and urban

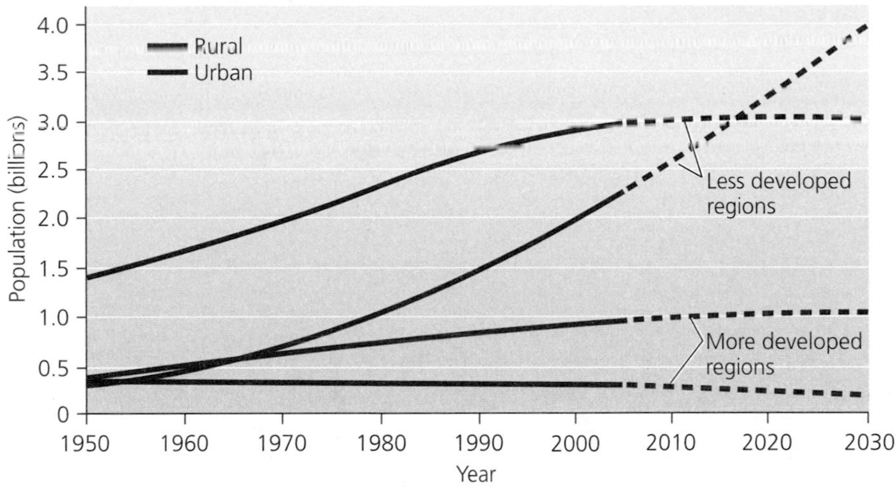


**FIGURE 13.1** In developing countries today, urban populations are growing quickly, whereas rural populations are leveling off and may soon begin to decline. Developed countries are already largely urbanized, so in these countries urban populations are growing more slowly, whereas rural populations are falling. Solid lines in the graph indicate past data, and dashed lines indicate projections of future trends. Data from United Nations Population Division (UNPD). 2007. *World urbanization prospects: The 2005 revision.* New York: UNPD.

lifestyles. U.N. demographers estimate that virtually all the world's population growth over the next 25 years will be absorbed by urban areas of developing nations.

## Today's urban centers are unprecedented in scale

Cities in themselves are nothing novel. Urban centers where population—and political power—is concentrated have been part of human culture for several thousand years. Ancient Mediterranean civilizations, the great Chinese dynasties, and the Mayan and Incan empires all featured sophisticated and powerful urban centers.

What is new is the sheer scale of today's metropolitan areas. Human population growth (Chapter 8) has placed greater numbers of people in towns and cities than ever before. Today, 20 cities are home to over 10 million residents (**Table 13.1**). The metropolitan area of the world's most populous city, Tokyo, Japan, is home to 35 million people. North America's largest metropolises, Mexico City and New York City, each hold about 19 million. However, the majority of urban dwellers live in smaller cities, such as Portland, Omaha, Winnipeg, Raleigh, Austin, and their still-smaller suburbs.

## Urban growth has often been rapid

Many American cities have followed a similar trajectory: rapid population expansion due to increased trade as the nation expanded westward, followed by stagnation in the mid-20th century as many city residents moved outward into the growing suburbs. Portland provides a typical example: The city grew as it shipped farm products from the fertile Willamette Valley overseas and accepted products shipped in from other North American ports and from Asia. Then, as in many U.S. cities, Portland's population growth stalled in the 1950s to 1970s as crowding and deteriorating economic conditions drove city dwellers to the suburbs. Subsequently, policies undertaken to improve the city center's attractiveness helped restart Portland's growth (**Figure 13.2**).

In recent years, many cities in the southern and western United States have undergone growth spurts as people (particularly retirees) from northern and eastern states

**TABLE 13.1 Metropolitan Areas with 10 Million Inhabitants or More, as of 2005**

| City, Country | Millions of people |
|---|---|
| Tokyo, Japan | 35.2 |
| Mexico City, Mexico | 19.4 |
| New York–Newark, United States | 18.7 |
| Sao Paulo, Brazil | 18.3 |
| Bombay, India | 18.2 |
| Delhi, India | 15.0 |
| Shanghai, China | 14.5 |
| Calcutta, India | 14.3 |
| Jakarta, Indonesia | 13.2 |
| Buenos Aires, Argentina | 12.6 |
| Dhaka, Bangladesh | 12.4 |
| Los Angeles–Long Beach–Santa Ana, United States | 12.3 |
| Karachi, Pakistan | 11.6 |
| Rio de Janeiro, Brazil | 11.5 |
| Osaka-Kobe, Japan | 11.3 |
| Cairo, Egypt | 11.1 |
| Lagos, Nigeria | 10.9 |
| Beijing, China | 10.7 |
| Manila, Philippines | 10.7 |
| Moscow, Russian Federation | 10.7 |

*Source:* United Nations Population Division. 2007. *World urbanization prospects: The 2005 revision.* New York: UNPD.

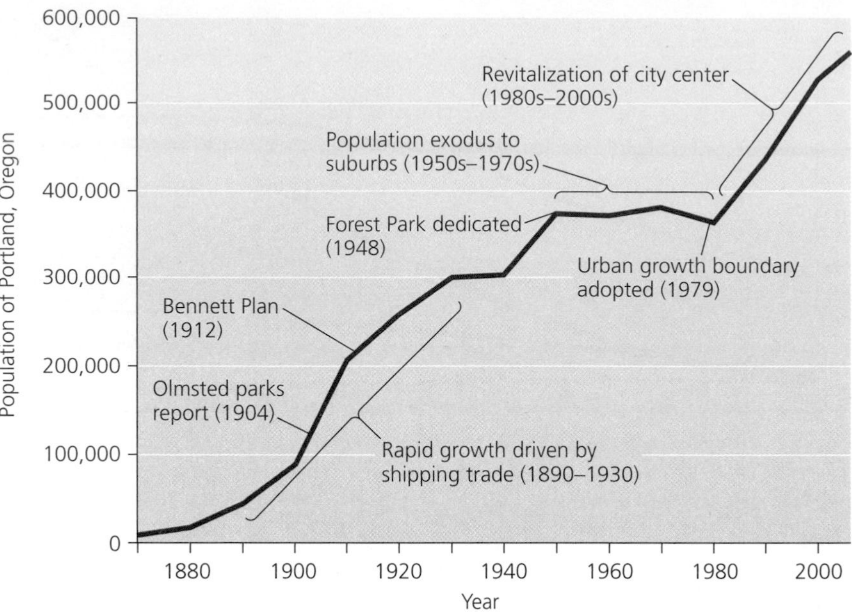


FIGURE 13.2 Once Portland established its position as a strategically located port, international shipping trade provided jobs and boosted its economy, and the city's population grew rapidly. City residents began leaving for the suburbs in the 1950s to 1970s, but policies designed to enhance the city center then revitalized the city's growth.

have moved south and west in search of warmer weather or more space. Between 1990 and 2000, for instance, the population of the Atlanta metropolitan area grew by 39%; that of the Phoenix region grew by 45%; and that of the Las Vegas metropolitan area grew by 83%.

Internationally, most fast-growing cities today are in the developing world, because industrialization is decreasing the need for farm labor and is increasing commerce and jobs in cities. Sadly, another reason is that wars, conflict, and ecological degradation are driving millions of people out of the countryside and into cities. Cities like Mumbai (Bombay), India; Lagos, Nigeria; and Cairo, Egypt are growing in population even faster than American cities did. All too often, they are doing so without the economic growth to match their population growth. As a result, many of these cities are facing overcrowding, pollution, and poverty. Nearly three of every four governments of developing nations have by now enacted policies to discourage the movement of people from the countryside into cities.

## Various factors influence the geography of urban areas

Real estate agents have long used the saying, "Location, location, location," to stress how much a home's whereabouts determines its value. Location is vitally important for urban centers, as well. Environmental variables such as climate, topography, and the configuration of waterways go a long way toward determining whether a small settlement will become a large city—and successful cities tend to be located in places that give them economic advantages. Think of any major city, and chances are it's situated along a major river, seacoast, railroad, or highway—some corridor for trade that has driven economic growth (**Figure 13.3**).

Many well-located cities have acted as linchpins in trading networks, funneling in resources from agricultural regions, processing them and manufacturing products, and shipping those products to other markets. Portland got its start in the mid-19th century as pioneers arriving by the Oregon Trail settled where the Willamette River flowed into the Columbia River. Situated at the juncture of these two major rivers, Portland had a strategic advantage in trade. Chicago grew with extraordinary speed as railroads funneled through it the resources of the vast lands to the west on their way to the cities and consumers to the east. Chicago became a center for grain processing, livestock slaughtering, meatpacking, and much else.

All cities, from ancient times to the present day, have supported themselves by drawing in resources from outlying rural areas through trade, persuasion, or conquest. In turn, cities have historically influenced how people use land in surrounding areas. Although city life and country life may seem very different, cities and the rural regions surrounding them have always been linked by tight economic relationships.

### Weighing THE Issues | What Made Your City?

Consider the town or city in which you live, or the major urban center located nearest you. Why do you think it developed into an urban area? What physical, social, or environmental factors may have aided its growth?

Spatial patterns of urbanization can change, however, with changing times. Today several factors are causing population centers to decentralize in developed nations. For one thing, people now are globally interconnected to

**(a) St. Louis, Missouri**

**(b) Fort Worth, Texas**

FIGURE 13.3 St. Louis (**a**) is situated on the Mississippi River near its confluence with the Missouri River—and this excellent strategic location for river trade drove its growth in the 19th and early 20th centuries. Fort Worth, Texas (**b**) grew most in the late 20th century as a result of linkages by the interstate highway system and a major international airport.

an unprecedented degree. Being located on a river or seacoast is no longer as vital to a city's success in our age of global commerce, jet travel, diplomacy, television, cell phones, and the Internet. Globalization has connected distant societies, and businesses and individuals can more easily communicate from locations away from major city centers. Moreover, fossil fuels have enabled the outward spread of cities. By easing long-distance transport, fossil fuels and the proliferation of highway networks have made it easier to commute into and out of cities and to import and export resources, goods, and waste. Such factors have enabled a shift of population from cities to suburbs, particularly in the United States and Canada.

### People have moved to suburbs

By the mid-20th century, many cities in the United States, Canada, and other developed nations had accumulated more people than these cities had jobs to offer. Unemployment rose, and crowded inner-city areas began to suffer increasing poverty and crime. As inner cities declined economically from the 1950s onward, many affluent city dwellers chose to move outward to the cleaner, less crowded, and more parklike suburban communities beginning to surround the cities. These people were pursuing more space, better economic opportunities, cheaper real estate, less crime, and better schools for their children.

Government policy also encouraged the population shift to the suburbs; the development of the interstate highway system in the United States allowed millions of people to commute by car to their downtown workplaces from new homes in suburban "bedroom communities." The exodus to the suburbs further hastened the economic decline of central cities. Chicago's population declined to 80% of its peak because so many residents moved to its suburbs. Philadelphia's population fell to 76% of its peak, and Detroit's to just 55%.

In most ways, suburbs have delivered the qualities people sought in them. The wide spacing of houses, with each house on its own plot of land, gives families room and privacy. However, by allotting more space to each person, suburban growth has spread human impact across the landscape. Natural areas have disappeared as housing developments are constructed. We have built extensive road networks to ease travel, but suburbanites now find themselves needing to drive everywhere. They commute longer distances to work and spend more time in congested traffic. The expanding rings of suburbs surrounding cities have grown larger than the cities themselves, and towns are running into one another. These aspects of suburban growth have inspired a new term: *sprawl*.

## Sprawl

The term *sprawl* has become laden with meanings and connotes different things to different people. To some, sprawl is aesthetically ugly, environmentally harmful, and economically inefficient. To others, it is the collective outgrowth of reasonable individual desires and decisions in a world of growing human population. We can begin our discussion by giving **sprawl** a simple, nonjudgmental definition: the spread of low-density urban or suburban development outward from an urban center.

### Today's urban areas spread outward

As urban and suburban areas have grown in population, they have also grown spatially. This growth is obvious from maps and satellite images of rapidly spreading cities such as Las Vegas (**Figure 13.4**). Houses and roads supplant over 1 million ha (2.5 million acres) of U.S. rural land each year—over 2,700 ha (6,700 acres) every day.

(a) Las Vegas, Nevada, 1972

(b) Las Vegas, Nevada, 2002

FIGURE 13.4 Satellite images show the type of rapid urban and suburban expansion that many people have dubbed *sprawl.* Las Vegas, Nevada, is currently one of the fastest-growing cities in North America. Between 1972 (**a**) and 2002 (**b**), the population increased more than fivefold, and the developed area rose more than threefold.

Because suburban growth entails allotting more space per person than does city dwelling, in most cases this outward spatial growth across the landscape has outpaced the growth in numbers of people. In fact, many researchers define *sprawl* as the physical spread of development at a rate greater than the rate of population growth. For instance, Phoenix grew from 105,000 residents spread over 44 km$^2$ (17 mi$^2$) in 1950 to 1.3 million residents spread over 1,200 km$^2$ (470 mi$^2$) in 2002; its land area grew 27 times larger, whereas the population grew 12 times larger. Between 1950 and 1990, the population of 58 major U.S. metropolitan areas rose by 80%, but the land area they covered rose by 305%. Even in 11 metro areas where population declined between 1970 and 1990 (for instance, rust-belt cities such as Detroit, Cleveland, and Pittsburgh), the amount of land covered increased.

Several types of development approaches can lead to sprawl (**Figure 13.5**). As a result, Chicago's metropolitan area now consists of more than 9.5 million people spread over 23,000 km$^2$ (9,000 mi$^2$)—an area 40 times the size of the city proper. Each person in the suburban region takes up an average of 11 times as much space as do residents of the city proper. As for Portland, today its growing suburbs hold more people than does the city itself.

## Sprawl has several causes

There are two main components of sprawl. One is human population growth—there are simply more people alive each year. The other is per capita land consumption—each person takes up more land. The amount of sprawl is a function of the number of people added to an area times the amount of land the average person occupies.

A study of the 100 major metropolitan areas of the United States between 1970 and 1990 found that, on average, each of these two factors contributes about equally to sprawl. Cities varied, however, in which factor was more important. The Los Angeles metro area increased in population density by 9% between 1970 and 1990, becoming the nation's most densely populated metro area. Increasing density should be a good recipe for preventing sprawl. Yet L.A. grew in size by a whopping 1,021 km$^2$ (394 mi$^2$). This spatial growth, despite the increase in density, clearly resulted from an overwhelming influx of new people.

In contrast, the Detroit metro area lost 7% of its population between 1970 and 1990, yet it expanded in area by 28%. Clearly population growth was not the issue here; rather, sprawl was caused solely by increased per capita land consumption.

We discussed reasons for human population growth in Chapter 8. As for the increase in per capita land consumption, there are numerous reasons. The interstate highways and technologies such as telecommunications and the Internet have fostered movement away from city centers because they free businesses from dependence on the centralized infrastructure a major city provides, and they give workers greater flexibility to live wherever they

(a) Uncentered commercial strip development

(b) Low-density single-use development

(c) Scattered, or leapfrog, development

(d) Sparse street network

FIGURE 13.5 Several standard approaches to development can result in sprawl. In uncentered commercial strip development (**a**), businesses are arrayed in a long strip along a roadway, and no attempt is made to create a centralized community with easy access for consumers. In low-density, single-use residential development (**b**), homes are located on large lots in residential tracts far away from commercial amenities. In scattered or leapfrog development (**c**), developments are created at great distances from a city center and are not integrated. In developments with a sparse street network (**d**), roads are far enough apart that moderate-sized areas go undeveloped, but not far enough apart for these areas to function as natural areas or sites for recreation. All these development approaches necessitate frequent automobile use.

desire. The primary reasons for greater per capita land consumption, however, are that most people simply like having some space and privacy and dislike congestion. Furthermore, in the consumption-oriented American lifestyle that promotes bigger houses, bigger cars, and bigger TVs, having more space to house one's possessions becomes important. Unless there are overriding economic or social disadvantages, most people prefer living in a less congested, more spacious, more affluent community.

Economists, politicians, and city boosters have almost universally encouraged the unbridled spatial expansion of cities and suburbs. The conventional assumption has been that growth is good and that attracting business, industry, and residents will unfailingly increase a community's economic well-being, political power, and cultural influence. Today, however, this assumption is increasingly being challenged. As the negative effects of sprawl on citizens' lifestyles accumulate, growing numbers of people have begun to question the mantra that all growth is good.

## What is wrong with sprawl?

*Sprawl* means different things to different people. To some, the word evokes strip malls, homogenous commercial development, and tracts of cookie-cutter houses encroaching on farmland and ranchland. It may suggest traffic jams, destruction of wildlife habitat, and loss of natural land around cities. However, for other people, sprawl represents the collective result of choices made by millions of well-meaning individuals trying to make a better life for themselves and their families. In this view, those who decry sprawl are being elitist and fail to appreciate the good things about suburban life. Let us try, then, to leave the emotional debate aside and assess the impacts of sprawl (see "The Science behind the Story," • pp. 364–365).

THE SCIENCE BEHIND THE STORY

## Measuring the Impacts of Sprawl

*Dr. Reid Ewing, Rutgers University*

Critics of sprawl have blamed it for so many societal ills that it can make a person feel guilty just for being born in the suburbs or shopping at a mall. But what does scientific research tell us are the actual consequences of sprawl?

When Reid Ewing of Rutgers University and his team set out to measure the impacts of sprawl, they discovered that researchers studying sprawl have been hard-pressed even to agree on a definition of the term or on how to measure it. Surveying the literature, Ewing's team found that researchers using different criteria ranked cities in very different ways. For instance, in most studies Los Angeles was deemed more sprawling than Portland; even so, in some studies Portland was judged more sprawling than L.A.

So Ewing, Rolf Pendall of Cornell University, and Don Chen of the nonprofit group Smart Growth America tried to define *sprawl* as simply as possible, without mixing sprawl's consequences into the definition. They decided that sprawl occurs when the spread of development across the landscape far outpaces population growth.

Ewing, Pendall, and Chen then devised four criteria by which to rank 83 of the largest U.S. metropolitan areas in terms of sprawl. Sprawling cities would show:

- Low residential density
- Distant separation of homes, employment, shopping, and schools
- Lack of "centeredness," that is, lack of activity in community centers and downtown areas
- Street networks that make many streets hard to access

For each criterion, Ewing's team measured multiple factors—22 variables in all. They then devised a way to analyze the variables and arrive at a cumulative index of sprawl. Finally, they obtained data from municipalities throughout the country.

Traffic congestion is one of the most recognized impacts of sprawl.

The research team's rankings showed that the nation's most sprawling area was the Riverside–San Bernardino, California, region, which has expanded quickly in recent decades. The area with the least sprawl was New York City, whose historically dense population and vibrant neighborhoods kept it geographically compact, relative to its number of inhabitants. The accompanying tables list the 10 most- and least-sprawling areas.

The researchers next correlated their sprawl scores with a number of transportation variables. They found that people in the 10 most-sprawling metros owned more cars (180 per 100 households) than people in the 10 least-sprawling metros (162 per 100 households). They found that residents of the most-sprawling metros drove an average of 43 km (27 mi) per day, whereas those of the least-sprawling metros drove only 34 km (21 mi). And they determined that people in the most-sprawling metros used public transit far less and suffered 67% more traffic fatalities than those in the least-sprawling metros.

Strikingly, the study found no significant difference in commute time

**Transportation** Most studies show that sprawl constrains transportation options, essentially forcing people to drive cars. These constraints include the need to own a vehicle and to drive it most places, the need to drive greater distances or to spend more time in vehicles, a lack of mass transit options, and more traffic accidents. Across the United States, during the 1980s and 1990s the average length of work trips rose by 36%, and total vehicle miles driven increased at three times the rate of population growth. An automobile-oriented culture also increases dependence on nonrenewable petroleum, with its attendant economic and environmental consequences (• pp. 559–565).

**Pollution** Sprawl's effects on transportation give rise to increased pollution. Carbon dioxide emissions from vehicles cause global climate change (Chapter 18) while nitrogen- and sulfur-containing air pollutants contribute to tropospheric ozone, urban smog, and acid precipitation (• pp. 491–495). Motor oil and road salt from roads and parking lots pollute waterways, posing risks to from home to work for people of sprawling versus less-sprawling metro areas. Critics of sprawl have long blamed sprawl for traffic congestion and commute delays. Advocates of suburban spread have argued that more streets ease commutes and that regions can sprawl their way out of congestion. The Ewing team's results seem to suggest that each side may have a point and that the effects may cancel one another out.

Because vehicle emissions cause air pollution, the researchers also measured levels of tropospheric ozone (• p. 482). Sprawling areas had worse air pollution, they found; ozone levels were 40% higher in the most-sprawling metros.

None of these results prove that sprawl *causes* these impacts, because statistical correlation alone does not imply causation. However, taken together, the results suggest that spatial patterns of development may influence people's options, impacts, and behavior relative to transportation. The results were published in 2003 in the *Transportation Research Record* and in a 2002 report published by Smart Growth America.

Ewing and his colleagues are continuing to examine other impacts of sprawl. For instance, a study in 2004 went beyond ozone to look at other health correlates of sprawl. Because studies had found that residents of sprawling areas depend more on cars and walk less, the researchers hypothesized that they would find more obesity and poorer health in people living in sprawling areas. Indeed, they found that people from sprawling metros are on average heavier for their height and show increased instances of high blood pressure (although not more diabetes or cardiovascular disease).

As more studies on the impacts of sprawl accumulate, we will have a better idea of the benefits and costs of our choices in urban design.

**The 10 Most-Sprawling American Urban Areas**

| Rank | Metropolitan Region |
|---|---|
| 1 | Riverside–San Bernardino, CA |
| 2 | Greensboro–Winston-Salem–High Point, NC |
| 3 | Raleigh–Durham, NC |
| 4 | Atlanta, GA |
| 5 | Greenville–Spartanburg, SC |
| 6 | West Palm Beach–Boca Raton–Delray Beach, FL |
| 7 | Bridgeport–Stamford–Norwalk–Danbury, CT |
| 8 | Knoxville, TN |
| 9 | Oxnard–Ventura, CA |
| 10 | Fort Worth–Arlington, TX |

1 = most-sprawling, 10 = less-sprawling.
*Source:* Ewing, R., et.al. 2002. *Measuring sprawl and its impact.* Washington, DC: Smart Growth America.

**The 10 Least-Sprawling American Urban Areas**

| Rank | Metropolitan Region |
|---|---|
| 83 | New York, NY |
| 82 | Jersey City, NJ |
| 81 | Providence–Pawtucket–Woonsocket, RI |
| 80 | San Francisco, CA |
| 79 | Honolulu, HI |
| 78 | Omaha, NE-IA |
| 77 | Boston–Lawrence–Salem–Lowell–Brockton, MA |
| 76 | Portland, OR |
| 75 | Miami–Hialeah, FL |
| 74 | New Orleans, LA |

83 = least-sprawling, 74 = more-sprawling.
*Source:* Ewing, R., et al. 2002. *Measuring sprawl and its impact.* Washington, DC: Smart Growth America.

ecosystems and human health. Runoff of polluted water from paved areas is estimated to be about 16 times greater than from naturally vegetated areas.

**Health** Aside from the health impacts of pollution, some research suggests that sprawl promotes physical inactivity because driving cars largely takes the place of walking during daily errands. Physical inactivity increases obesity and high blood pressure, which can in turn lead to other ailments. A 2003 study found that people from the most-sprawling U.S. counties weigh 2.7 kg (6 lb) more for their height than people from the least-sprawling U.S. counties and that slightly more people from the most-sprawling counties show high blood pressure.

**Land use** The spread of low-density development means that more land is developed while less is left as forests, fields, farmland, or ranchland. Of the estimated 1 million ha (2.5 million acres) of U.S. land converted each year, roughly 60% is agricultural land and 40% is forest.

These lands provide vital resource production, aesthetic beauty, habitat for wildlife, cleansing of water, places for recreation, and many other ecosystem services (• p. 38, 40). Sprawl generally diminishes all these amenities.

**Economics** Sprawl drains tax dollars from existing communities and funnels them into infrastructure for new development on the fringes of those communities. Money that could be spent maintaining and improving downtown centers is instead spent on extending the road system, water and sewer system, electricity grid, telephone lines, police and fire service, schools, and libraries. For instance, one study calculated that sprawling development at Virginia Beach, Virginia, would require 81% more in infrastructure costs and would drain 3.7 times more from the community's general fund each year than compact urban development. Advocates for sprawling development argue that taxes on new development eventually pay back the investment made in infrastructure, but studies have found that in most cases taxpayers continue to subsidize new development if municipalities do not pass on infrastructure costs to developers.

**Weighing THE Issues | Sprawl Near You**

Is there sprawl in the area where you live? Are you bothered by it, or not? Has development in your area had any of the impacts described above? Do you think your city or town should use its resources to encourage outward growth?

**(a) Chicago lakefront in 1890**

**(b) Architectural drawing from the *Plan of Chicago***

**(c) Modern-day Chicago**

**FIGURE 13.6** Daniel Burnham's 1909 *Plan of Chicago* helped Chicagoans reclaim the city's abused lakefront (**a**) and beautify it for public access in the form of Grant Park. The *Plan*'s architectural drawings (**b**) included parks, greenways, and efficient transportation routes. Burnham's plan served as a model for later city planners, and today Chicago (**c**) boasts a thriving and livable downtown.

# Creating Livable Cities

To respond to the challenges that urban and suburban sprawl present, architects, planners, developers, and policymakers across North America today are trying to restore the vitality of city centers and to plan and manage how urbanizing areas develop.

## City and regional planning are means for creating livable urban areas

**City planning** is the professional pursuit that attempts to design cities so as to maximize their efficiency, functionality, and beauty. City planners advise policymakers on development options, transportation needs, public parks, and other matters.

City planning in North America came into its own in the early 20th century. Landscape architect Daniel Burnham's 1909 *Plan of Chicago* (**Figure 13.6**) represented the first thorough plan for an American city, and it was largely implemented over the following years and decades. This plan expanded city parks and playgrounds, improved neighborhood living conditions, streamlined traffic systems, and cleared industry and railroads from the shore of Lake Michigan to provide public access to the lake.

Portland gained its own comprehensive plan just 3 years later. Edward Bennett's *Greater Portland Plan* recommended rebuilding the harbor; dredging the river channel; constructing new docks, bridges, tunnels, and a waterfront railroad; superimposing wide radial boulevards on the old city street grid; establishing civic centers downtown; and greatly expanding the number of parks. Voters approved the plan by a 2:1 margin, but they defeated a bond issue that would have paid for park development. As the century progressed, several other major planning efforts were conducted, and some ideas, such as establishing a downtown public square, came to fruition.

City planning grew in importance throughout the 20th century as urban populations expanded, inner cities decayed, and wealthier residents fled to the suburbs. In today's world of sprawling metropolitan areas, **regional planning** has become just as important (see, for example, • pp. 186–187). Regional planners deal with the same issues as city planners, but they work on broader geographic scales and must coordinate their work with multiple municipal governments. In some places, regional planning has been institutionalized in formal governmental bodies; the Portland area's Metro is the epitome of such a regional planning entity.

**Weighing the Issues** | **Your Urban Area**

Think of your favorite parts of the city you know best. What aspects do you like about them? What do you dislike about some of your least favorite parts of the city? What could this city do to improve the quality of life for its inhabitants?

## Zoning is a key tool for planning

One tool that planners use is **zoning**, the practice of classifying areas for different types of development and land use (**Figure 13.7**). For instance, to preserve the cleanliness and tranquility of residential neighborhoods, industrial plants may be kept out of districts zoned for residential use. The specification of zones for different types of development gives planners a powerful means of guiding what gets built where. Zoning can restrict areas to a single use, as is often done with suburban residential tracts in so-called bedroom communities. Or, zoning can allow the type of mixed use—residential and commercial, for instance—that some planners say can reinvigorate urban neighborhoods. Zoning also gives home buyers and business owners security; they know in advance what types of development can and cannot be located nearby.

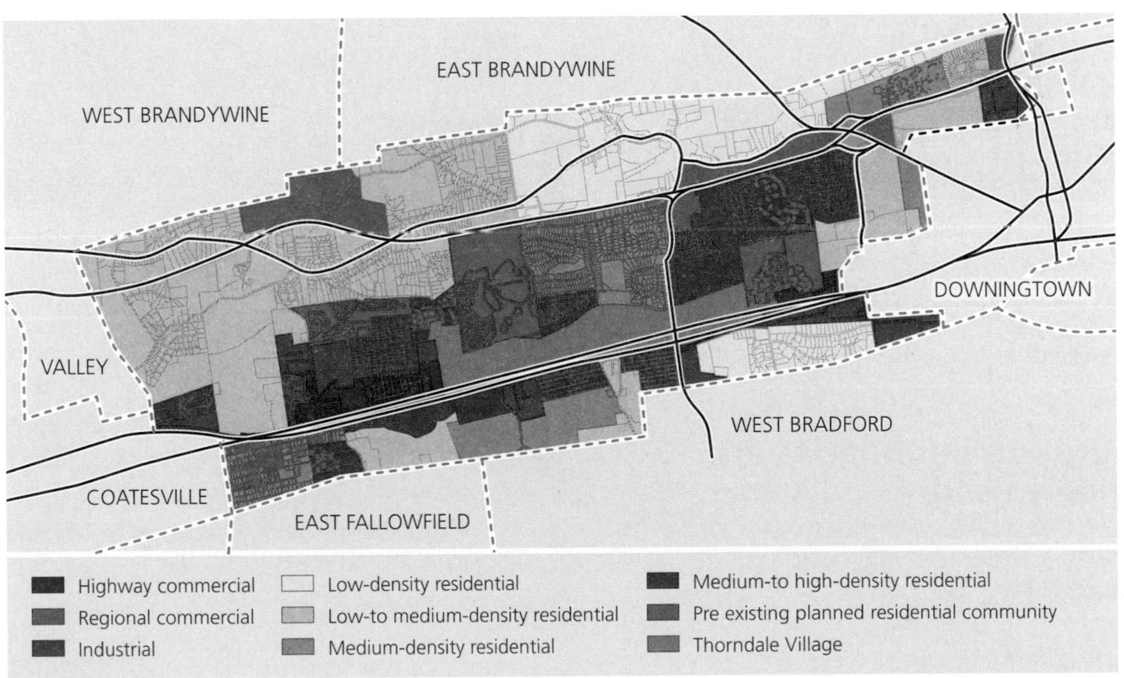


**FIGURE 13.7** By zoning areas for different uses, planners guide how a community develops. Zoning restricts what landowners can do with their land, but it is intended to maximize prosperity, efficiency, and quality of life for the community. This zoning map for Caln Township, Pennsylvania, shows several patterns common to modern zoning practice. Public and institutional uses are clustered together in a downtown area. Industrial uses are clustered together, away from most residential areas. Commercial uses are clustered along major roadways, and residential zones generally are higher in density toward the center of town.

Zoning involves government restriction on the use of private land and represents a top-down constraint on personal property rights. For this reason, some people consider zoning a regulatory taking (• p. 63) that violates individual freedoms. Many others defend zoning, saying that government has a proper role in setting certain limitations on property rights for the good of the community. Similar debates arise with endangered species management (• pp. 317–321) and other environmental issues.

Oregon voters sided with private property rights when in 2004 they passed Ballot Measure 37, which shackled government's ability to enforce zoning regulations with landowners who had owned their land before the regulations were enacted. The passage of Oregon's measure spawned similar efforts in other U.S. states. In the 2006 elections, copycat ballot measures were passed in Arizona, but defeated in California, Idaho, and Washington.

Opponents of Measure 37 in Oregon predicted that many proponents will change their minds once they begin seeing new development they do not wish to occur. For the most part, people have supported zoning over the years because the common good it produces for communities is widely felt to outweigh the restrictions on private use.

### Weighing THE Issues | Zoning and Development

Imagine you own a 10-acre parcel of land that you want to sell for housing development—but the local zoning board rezones the land so as to prohibit the development. How would you respond?

Now imagine that you live next to someone else's undeveloped 10-acre parcel. You enjoy the privacy it provides—but the local zoning board rezones the land so that it can be developed into a dense housing subdivision. How would you respond?

What factors do you think members of a zoning board should take into consideration when deciding how to zone or rezone land in a community?

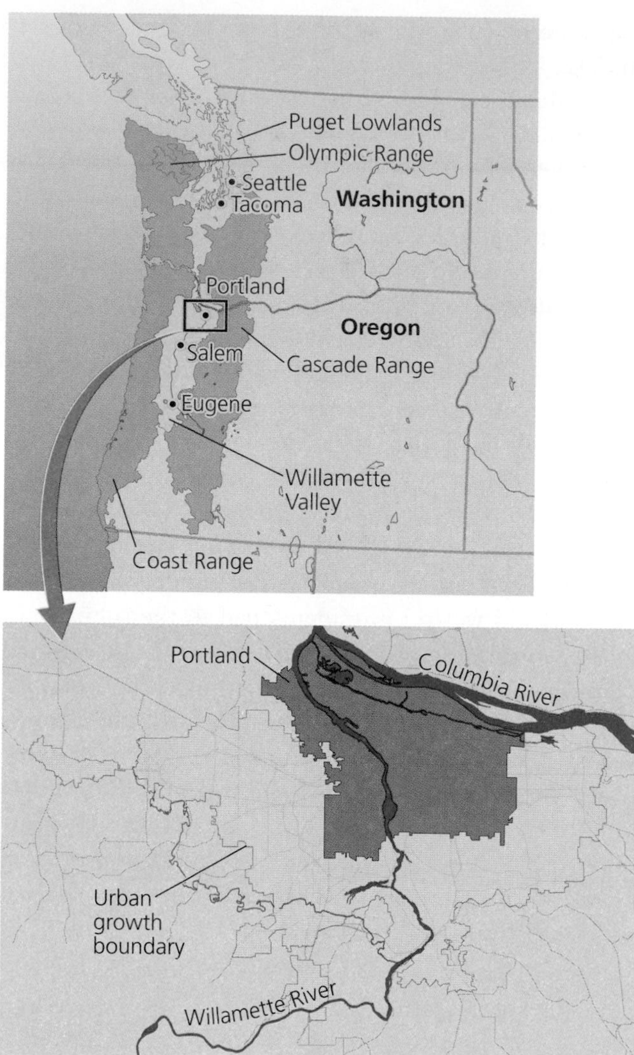


FIGURE 13.8 Oregon's urban growth boundaries (UGBs) were a response to fears that suburban sprawl might one day stretch in a great megalopolis from Eugene, Oregon, up the Willamette Valley to Portland, and up the Puget Lowlands to Seattle, Washington. The Portland area's 956 km$^2$ (369 mi$^2$) UGB encompasses Portland (dark gray) and portions of 24 other communities and three counties (light gray). Its jagged edge separates areas planners have earmarked for urban development from areas they have chosen to protect from urban development.

## Urban growth boundaries are now widely used

Planners intended Oregon's urban growth boundaries to limit sprawl by containing future growth largely within existing urbanized areas. The UGBs aimed to revitalize downtowns; protect farms, forests, and their industries; and ensure urban dwellers some access to open space near cities. Among Willamette Valley towns, the long-term goal was to head off the growth of a potential megalopolis stretching from Eugene to Seattle (**Figure 13.8**).

Since Oregon began its experiment, a number of other states, regions, and cities have adopted UGBs—from Boulder, Colorado, to Lancaster, Pennsylvania, to many California communities. In their own ways, all the UGBs aim to concentrate development, prevent sprawl, and preserve working farms, orchards, ranches, and forests. UGBs also appear to reduce the amounts municipalities have to pay for infrastructure, compared to sprawl. The best estimate nationally is that UGBs save taxpayers about 20% on infrastructure costs. However, UGBs also seem to increase housing prices within their boundaries. In the Portland area, housing is becoming less affordable, but in most other ways its UGB (see Figure 13.8) is working as intended. It has lowered prices for land outside the UGB while increasing prices within it. It has restricted development outside the

UGB. It has increased the density of new housing inside the UGB by over 50% as homes are built on smaller lots and as multistory apartments fulfill a vision of "building up, not out." Downtown employment rose by 73% between 1970 and 1995 as businesses and residents alike invested in the central city. And Portland has been able to absorb considerable immigration while avoiding rampant sprawl.

However, urbanized area still increased by 101 $km^2$ (39 $mi^2$) in the decade after Portland's UGB was established, because 146,000 people were added to the population. This fact suggests that relentless population growth may thwart even the best antisprawl efforts. Indeed, Metro has enlarged the Portland-area UGB three dozen times, and population projections for the region suggest there will be pressure for still more expansion. Many other locations that have instituted UGBs have chosen to expand them later. For instance, a group of 48 local governments from cities, towns, and counties along the Colorado Front Range in 1997 agreed to a UGB that allowed a 30% expansion of urbanized area by the year 2020. Soon, however, this limit was raised three times, finally to 36% growth by the year 2030. Given the reality of UGB expansions, it must be asked whether UGBs will truly limit sprawl in the long run.

## "Smart growth" aims to counter sprawl

As more people have begun to feel negative effects of sprawl on their everyday lives, efforts to control growth have sprung up throughout North America. Oregon's Senate Bill 100 was one of the first, and since then dozens of states, regions, and cities have adopted similar land use policies. Urban growth boundaries and many other ideas from these policies have coalesced under the concept of **smart growth** (Table 13.2).

**TABLE 13.2 Ten Principles of "Smart Growth"**

- Mix land uses
- Take advantage of compact building design
- Create a range of housing opportunities and choices
- Create walkable neighborhoods
- Foster distinctive, attractive communities with a strong sense of place
- Preserve open space, farmland, natural beauty, and critical environmental areas
- Strengthen and direct development toward existing communities
- Provide a variety of transportation choices
- Make development decisions predictable, fair, and cost-effective
- Encourage community and stakeholder collaboration in development decisions

*Source:* U.S. Environmental Protection Agency, 2005.

Proponents of smart growth want municipalities to manage the rate, placement, and style of development so as to promote healthy neighborhoods and communities, jobs and economic development, transportation options, and environmental quality. They aim to rejuvenate the older existing communities that so often are drained and impoverished by sprawl. Smart growth means "building up, not out"—focusing development and economic investment in existing urban centers and favoring multistory shop-houses and high-rises.

People have pursued the goals of smart growth in various ways. Several hundred referenda and ballot measures in dozens of U.S. states in the past decade have proposed various solutions to slow sprawl and rechannel development into more community-friendly paths. Many of these ballot measures have won, and a number of experiments are going on right now across North America.

## The "new urbanism" is now in vogue

A related movement among many architects, planners, and developers is labeled the **new urbanism**. This approach seeks to design neighborhoods on a walkable scale, with homes, businesses, schools, and other amenities all close together for convenience. The aim is to create functional neighborhoods in which most of a family's needs can be met close to home without the use of a car. Greenspaces, trees, a mix of architectural styles, and creative street layouts add to the visual interest and pleasantness of new urbanist developments (**Figure 13.9**). These developments mimic the traditional urban neighborhoods that existed until the advent of suburbs.

New urbanist neighborhoods are generally connected to public transit systems. In *transit-oriented development*, compact communities in the new urbanist style are arrayed around stops on a major rail transit line, enabling people to travel most places they need to go by train and foot alone. Several lines of the Washington, D.C., Metro system are developed in this manner.

Over 600 communities in the new urbanist style across North America are in planning or construction, and many are now complete. Among them are Seaside, Florida; Kentlands in Gaithersburg, Maryland; Addison Circle in Addison, Texas; Mashpee Commons in Mashpee, Massachusetts; Harbor Town in Memphis, Tennessee; Celebration in Orlando, Florida; and Orenco Station, west of Portland.

To develop urban centers in these ways, zoning rules must allow it. Traditionally, many zoning rules have limited the density of development. Although well-intentioned, such rules encourage sprawl, so some are now being rethought in communities that desire new urbanism and smart growth.

FIGURE 13.9 This plaza at Mizner Park in Boca Raton, Florida, is part of a planned community built in the style of the "new urbanism." Homes, schools, and businesses are mixed close together in a centered neighborhood so that most amenities are within walking distance.

## Transportation options are vital to livable cities

A key ingredient in any planner's recipe for improving the quality of urban life is making multiple transportation options available to citizens. These options include public buses, trains and subways, and light rail (smaller rail systems powered by electricity). As long as an urban center is large enough to support the infrastructure necessary, these mass transit options are cheaper, more energy-efficient, and cleaner than roadways choked with cars (**Figure 13.10**). They also ease traffic congestion by carrying passengers who would otherwise be driving cars. An average Portland bus is estimated to keep about 250 cars off the road each day. Mass transit rail systems take up less space than road networks and emit less pollution than cars. The fuel and productivity lost on roadways to traffic jams have been estimated to cost the U.S. economy $74 billion each year.

The nation's most-used train systems are the extensive heavy rail systems in America's largest cities, such as New York's subways, Washington, D.C.'s Metro, the T in Boston, and the San Francisco Bay area's BART, each of which carries more than one-fourth of each city's daily commuters. In Canada, rail systems in Montreal and Toronto enjoy similarly heavy use. Major cities internationally from Moscow to Beijing to Paris to Tokyo also have large and heavily utilized subway systems.

Some cities with severe traffic problems, such as Bangkok, Thailand, and Athens, Greece, have recently opened new rail systems that carry hundreds of thousands of commuters a day. Light rail use is increasing in Europe, and U.S. ridership is now rising faster than is the rate of new car drivers. Most countries have bus systems

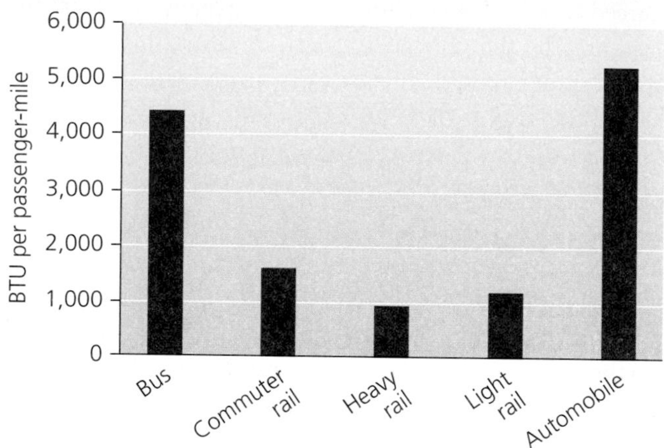


**(a) Energy consumption for different modes of transit**

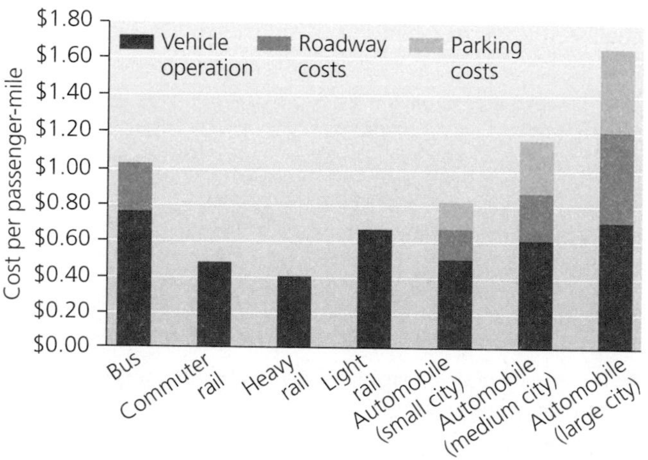


**(b) Operating costs for different modes of transit**

FIGURE 13.10 Rail transit consumes far less energy per passenger mile (**a**) than bus or automobile transit. Rail transit involves fewer costs per passenger mile (**b**) than bus or automobile transit. Data from Litman, T. 2005. *Rail transit in America: A comprehensive evaluation of benefits.* Victoria, BC: Victoria Transport Policy Institute.

that are far more accessible to citizens than are those of the United States.

One city famous for its efficient transportation system is Curitiba, Brazil. Faced with a heavy influx of immigrants from outlying farms in the 1970s, visionary city leaders led by Mayor Jaime Lerner decided to pursue an aggressive planning process so that they could direct growth rather than being overwhelmed by it. They reconfigured Curitiba's road system to maximize the efficiency of a large fleet of public buses. Today this metropolis of 2.5 million people has an outstanding bus system that is used each day by three-quarters of the population. The 340 bus routes, 250 terminals, and 1,900 buses accompany measures to encourage bicycles and pedestrians. All of this has resulted in a steep drop in car use, despite the city's rapidly growing population.

Portland's well-organized bus system carries 63 million riders per year. In 1986 Portland introduced a light rail system called MAX, and Metro policy encouraged the development of self-sufficient neighborhood communities in the new urbanist style along the rail lines. Light rail ridership has steadily increased as the system has expanded (**Figure 13.11**).

Establishing mass transit is not always easy, however. Once a road system has been developed and businesses and homes are built alongside roads, it can be difficult and expensive to replace or complement the road system with a mass transit system. In addition, modes of mass transit differ in their effectiveness (see "The Science behind the Story", • pp. 372–373), depending on city size, size of the transit system, and other factors.

To make urban transportation more efficient, governments can also raise fuel taxes, tax inefficient modes of transport, reward carpoolers with carpool lanes, encourage bicycle use and bus ridership, and charge trucks for road damage. They can choose to minimize investment in infrastructure that encourages sprawl and to stimulate investment in renewed urban centers.

**(a) MAX light rail train**

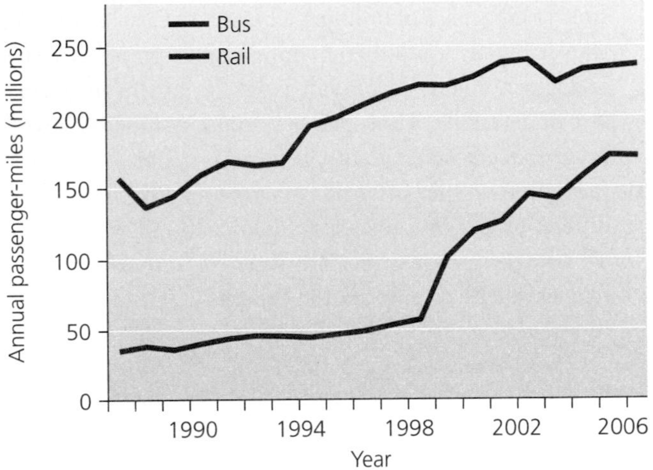


**(b) Portland transit ridership trends**

FIGURE 13.11 Portland's light rail system (**a**) is one component of an urban planning strategy that has helped make it one of North America's most livable cities. In Portland, bus ridership has been increasing slowly, whereas ridership on the MAX light rail system is growing quickly (**b**) as the system expands.

## Parks and open space are key elements of livable cities

City dwellers often desire some sense of escape from the noise, commotion, and stress of urban life. Natural lands, public parks, and open space provide greenery, scenic beauty, freedom of movement, and places for recreation. These lands also keep ecological processes functioning by regulating climate, producing oxygen, filtering air and water pollutants, and providing habitat for wildlife. The animals and plants of urban parks and natural lands also serve to satisfy biophilia (• p. 314), our natural affinity for contact with other organisms.

Protecting natural lands and establishing public parks become more important as our societies become more urbanized, because many urban dwellers come to feel increasingly isolated and disconnected from nature. In the wake of urbanization and sprawl, people of every industrialized society in the world today have, to some degree, chosen to set aside land in public parks.

## City parks were widely established at the turn of the last century

At the turn of the 20th century in urban America, civic improvement was garnering interest and support as politicians and citizens alike yearned for ways to make their crowded and dirty cities more livable. In the late 1800s, public parks began to be established in eastern U.S. cities, using aesthetic ideals borrowed from European

THE SCIENCE BEHIND THE STORY

## Assessing Benefits of Rail Transit

*Commuters on a New York City subway line*

Most urban planning experts see benefits in public mass transit and in making transit systems readily accessible to citizens. But building a mass transit system—and then maintaining and expanding it—can be an expensive undertaking. Thus, planners and policymakers value quantitative information on the costs and benefits of different types of transit systems and on how extensive a given transit system should be to produce the benefits they desire.

For this reason, researcher Todd Litman of the nonprofit Victoria Transport Policy Institute conducted a comprehensive evaluation of the benefits of rail transit in the United States, published in 2005 by the Victoria Transport Policy Institute, with support from the American Public Transport Association.

Litman first divided American cities into three categories. "Bus-only" cities have bus service but no rail service. "Small-rail" cities have rail systems serving fewer than 12% of daily commuters; these include 16 cities ranging from Atlanta to St. Louis to Denver to Salt Lake City. "Large-rail" cities feature rail service as a major component of the transportation system; these include New York, Washington, D.C., Boston, San Francisco, Chicago, Philadelphia, and Baltimore, whose rail systems serve 12–48% of daily commuters.

Litman compared these three groups of cities for a number of variables. Several key results are summarized in the accompanying figure. Compared with bus-only cities, large-rail cities had 36% fewer per capita traffic deaths each year. Residents of large-rail cities drove 21% fewer miles yearly than those of bus-only cities. Large-rail city residents also saved money on transportation, spending 14% less than bus-only city residents on transportation—equivalent to annual savings of $448. Transportation costs took up only 12% of their household budgets, compared to 14.9% of those of bus-only city residents.

The analysis also revealed benefits of large-rail systems for municipalities running the systems. Per passenger mile, large-rail cities paid only 42 cents in operating costs, versus 63 cents paid by bus-only cities. In addition, large-rail cities recovered far more of their costs than bus-only cities (38% cost recovery versus 24%).

Small-rail cities were intermediate in traffic deaths and vehicle mileage, as might be expected. However, they performed no better than bus-only cities in terms of operating costs per passenger mile or transit cost recovery. And residents of small-rail cities fared slightly worse than residents of

parks, gardens, and royal hunting grounds. The lawns, shaded groves, curved pathways, and pastoral vistas we see today in many American city parks and cemeteries originated with these European ideals, as interpreted by the leading American landscape architect, Frederick Law Olmsted. Olmsted designed New York's Central Park in 1853 and a host of urban park systems afterwards (**Figure 13.12**).

Two sometimes conflicting goals motivated the establishment and design of early city parks. On the one hand, the parks were meant to be "pleasure grounds" for the wealthy, who helped support their establishment financially and who would ride the parks' winding roadways in carriages. On the other hand, the parks were meant to alleviate congestion for poverty-stricken immigrants, and these park users were more interested in active recreation, such as ballgames, than in carriage rides. At times the aesthetic interests of the

**FIGURE 13.12** City parks were developed in many urban areas in the late 19th century to provide citizens aesthetic pleasure, recreation, and relief from the stresses of the city. Manhattan's Central Park, shown here, was one of the first.

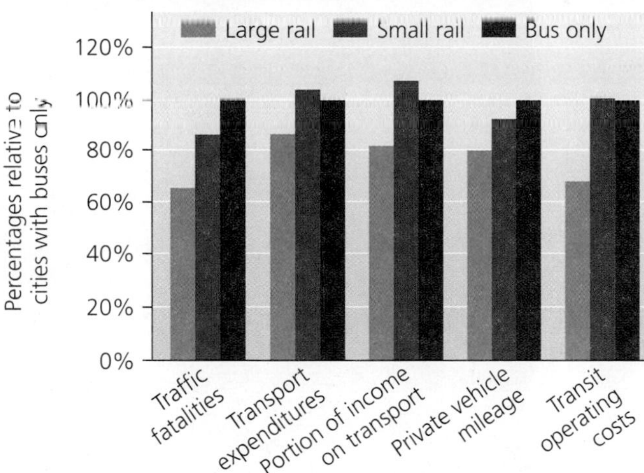


**Large-rail cities outperform bus-only cities in transport-oriented variables, but the benefits of rail are not so clear-cut for small-rail cities.** Adapted from Litman, T. 2005. *Rail transit in America: A comprehensive evaluation of benefits.* Victoria, BC: Victoria Transport Policy Institute.

bus-only cities in annual transport expenditures and proportion of income spent on transportation. These results suggest that not all benefits of rail transit begin to accrue with small systems. Rather, for many factors, a rail system has to be large enough or accommodate enough riders to gain the economy of scale needed to provide benefits.

A similar pattern was found for annual per capita costs due to traffic congestion. For bus-only and small-rail cities, congestion costs increased with city size. But for large-rail cities, congestion costs did not vary with size and were lower than in comparably sized cities.

Litman went on to quantify and assess benefits and costs of rail transit overall. He found that each year governments spend $12.5 billion on rail transit systems that they do not get back in revenues from fares. That means governments are subsidizing rail transit by $140 per resident of cities with rail systems.

Although these numbers are considerable, they are surpassed by the monetary benefits of rail systems. Each year, rail systems are estimated to save $19.4 billion in congestion costs, $22.6 billion in consumer transportation costs, $8.0 billion in roadway costs, $12.1 billion in parking costs, and $5.6 billion in accident costs, for a total of $67.7 billion in total annual savings. And that figure does not even account for indirect savings due to enhanced environmental quality.

A number of studies have critiqued rail transit and portrayed it as ineffective. The Litman study indicates that although not all rail systems are cost-effective, rail systems become more beneficial as they become larger and carry a greater proportion of a city's commuters.

educated elite and the recreational interests of the laboring class came into conflict—a friction that survives today in debates over recreation in city and national parks.

East Coast cities, such as New York, Boston, and Philadelphia, developed parks early on, but cities further west were not far behind. In Chicago, civic boosters striving to overcome the city's reputation as a rough-and-ready frontier town of slaughterhouses and meat-packing plants were glad to have Daniel Burnham and other planners design parks, playgrounds, and a greenbelt of forest preserves circling the city's outskirts (**Figure 13.13**). Further west, in San Francisco, William Hammond Hall transformed 2,500 ha (1,000 acres) of the peninsula's natural landscape of dunes into a verdant playground of lawns, trees, gardens, and sports fields. Golden Gate Park remains today one of the world's foremost city parks.

Portland's quest for urban parks began in 1900, when city leaders created a parks commission and then hired Frederick Law Olmsted's son, John Olmsted, to design a citywide park system. His 1904 plan recommended acquiring land to ring the city generously with parks, but no action was taken. A full 44 years later, citizen pressure resulted in the creation of Forest Park along a large forested ridge on the northwest side of the city. At 11 km (7 mi) long, it is today the largest city park in the United States.

## Smaller public spaces are also important

Large city parks are a key component of a healthy urban environment, but even small spaces can make a big difference. Playgrounds provide places where children can be active outdoors and interact with their peers. Community gardens

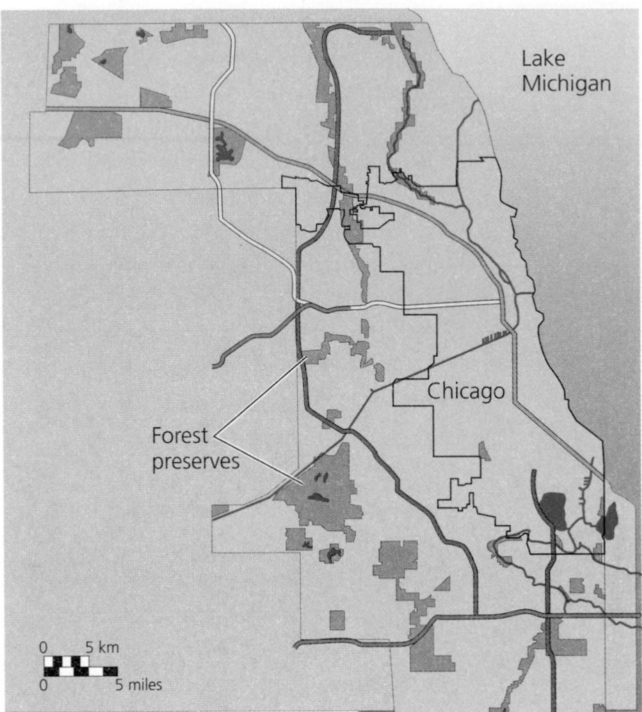


FIGURE 13.13 The forest preserves of Cook County, DuPage County, Lake County, and Will County, Illinois, wind through the suburbs surrounding the city of Chicago. This regional system features 40,000 ha (100,000 acres) of woodlands, fields, marshes, and prairies, comprising the largest holding of locally owned public conservation land in the United States.

FIGURE 13.14 Urban community gardens like this one in Seattle provide city residents with a place to grow vegetables and also serve as greenspaces that beautify cities.

allow people to grow their own vegetables and flowers in a neighborhood setting (**Figure 13.14**). Portland, Seattle, Baltimore, Boston, and many other cities feature thriving community gardens.

*Greenways*, strips of land that connect parks or neighborhoods, are often located along rivers, streams, or canals, and they may provide access to networks of walking trails. They can protect water quality, boost property values, and serve as corridors for the movement of birds and wildlife. The Rails-to-Trails Conservancy has spearheaded the conversion of abandoned railroad rights-of-way into trails for walking, jogging, and biking. To date, nearly 24,000 km (15,000 mi) of 1,200 rail lines have been converted across North America.

Besides creating new types of urban spaces, many cities are working to enhance the "naturalness" of their parks through ecological restoration (• pp. 160–161), the practice of restoring native communities. In Portland parks, volunteer teams remove English ivy, an invasive plant that covers trees and smothers native plants on the forest floor. At some Chicago-area forest preserves, scientists and volunteers use prescribed burns (• p. 343) to restore prairie native to the region. In San Francisco's Presidio, areas are being restored to the native dune communities that were displaced by urban development.

# Urban Sustainability

Urbanization and urban centers exert both positive and negative environmental impacts. These impacts depend strongly on how we utilize resources, produce goods, transport materials, and deal with waste.

## Urban resource consumption brings a mix of environmental impacts

Most of us might guess that urban living has a greater environmental impact than rural living. However, the picture is not so simple; instead, urbanization brings a complex mix of consequences.

**Resource sinks** Cities and towns are sinks for resources, having to import from beyond their borders nearly everything they need to feed, clothe, and house their inhabitants. Urban and suburban areas rely on large expanses of land elsewhere to supply food and other crops, as well as natural resources such as water, timber, metal ores, and mined fuels. Urban centers also need areas of natural land to provide ecosystem services, including purification of water and air, nutrient cycling, and waste treatment (• pp. 38, 40). Major cities such as New York, Boston, San Francisco, and Los Angeles depend for their day-to-day survival on water they pump in from faraway watersheds (**Figure 13.15**).

As cities have grown, and as the material wealth of most societies has risen, the inexorable pull of resources from the countryside to the cities has become stronger. And as urban areas extend their reach, it becomes increasingly hard for urban residents isolated from natural lands to have a tangible sense of the environmental impacts of their choices.

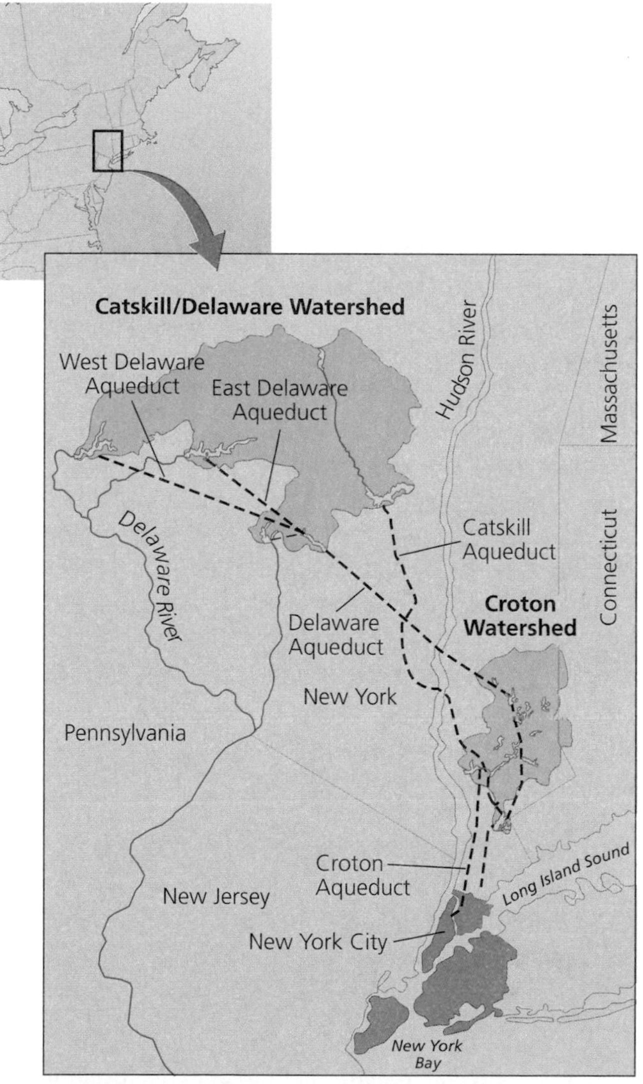


**FIGURE 13.15** New York City pipes in its drinking water from upstate reservoirs in the Croton and Catskill/Delaware Watersheds. The city has gone to lengths to acquire, protect, and manage watershed land to minimize pollution of these water sources. When New York City was confronted in 1989 with an order by the Environmental Protection Agency to build a $6 billion filtration plant to protect its citizens against waterborne disease, the city opted instead to purchase and better protect watershed land—for a fraction of the cost.

The long-distance transportation of resources and goods requires a great deal of fossil fuel use, which has significant environmental impacts (Chapter 19). For this reason, the centralization of resource use that urbanization entails may seem a bad thing for the environment. However, imagine that all the world's 3.3 billion urban residents were instead spread evenly across the landscape. What would the transportation requirements be, then, to move all those resources and goods around to all those people? A world without cities would likely require *more* transportation to provide people the same level of access to resources and goods.

**Efficiency** Once resources have arrived at an urban center where people are densely concentrated, cities should be able to minimize per capita consumption by maximizing the efficiency of resource use and delivery of goods and services. For instance, providing electricity from a power plant for urban houses close together is more efficient than providing electricity to far-flung homes in the countryside. The density of cities facilitates the provision of many social services that improve quality of life, including medical services, education, water and sewer systems, waste disposal, and public transportation.

**More consumption** Because cities draw resources from afar, their ecological footprints are much greater than their actual land areas. For instance, urban scholar Herbert Girardet calculated that the ecological footprint of London, England, extends 125 times larger than the city's actual area. By another estimate, cities take up only 2% of the world's land surface but consume over 75% of its resources.

However, the ecological footprint concept is most meaningful when used on a per capita basis. So, in asking whether urbanization causes increased resource consumption, we must ask whether the average urban dweller has a larger footprint than the average rural dweller. The answer is yes, but urban and suburban residents also tend to be wealthier than rural residents, and wealth correlates with resource consumption. Thus, although urban citizens tend to consume more than rural ones, the reason could be simply that they tend to be wealthier.

## Urban centers preserve land

The ecological footprints of urban areas are large, but because people are packed densely together in cities, more land outside cities is left undeveloped. Indeed, this is the idea behind urban growth boundaries. If cities did not exist, and if instead all 6.7 billion of us were evenly spread across the planet's land area, we would have much less room for agriculture, wilderness, biodiversity, or privacy.

There would be no large blocks of land left uninhabited by people or of habitat unfragmented for wildlife. The fact that half the human population is concentrated in discrete locations helps allow room for natural ecosystems to maintain themselves, continue functioning, and provide the ecosystem services on which all of us, urban and rural, depend.

## Urban centers suffer and export pollution

Just as cities import resources, they export wastes, either passively through pollution or actively through trade. In so doing, urban centers transfer the costs of their activities to other regions—and mask the costs from their own residents. Citizens of Toronto may not recognize that pollution from coal-fired power plants in their region worsens acid precipitation hundreds of miles to the east. Citizens of New York City may not realize how much garbage their city produces if it is shipped to other states or nations for disposal.

However, not all waste and pollution leaves the city. Urban residents are exposed to heavy metals, industrial compounds, and chemicals from manufactured products that accumulate in soil and water. Airborne pollutants cause photochemical smog, industrial smog, and acid precipitation. Fossil fuel combustion releases carbon dioxide and other pollutants, leading to climate change.

Urban residents also suffer noise pollution and light pollution. *Noise pollution* consists of undesired ambient sound. Excess noise degrades one's surroundings aesthetically, can induce stress, and at intense levels (such as with prolonged exposure to the sounds of leaf blowers, lawn mowers, and jackhammers) can harm hearing. *Light pollution* describes the way that city lights obscure the night sky, impairing the visibility of stars.

These various forms of pollution and the health threats they pose are not evenly shared among urban residents. Those who receive the brunt of the pollution are often those who are too poor to live in cleaner areas. Environmental justice concerns (• pp. 35–37) center on the fact that a disproportionate number of people living near, downstream from, or downwind from factories, power plants, and other polluting facilities are people who are poor and, often, people of racial minorities.

## Urban centers foster innovation

One of the greatest impacts of urbanization on environmental quality is also one of the most indirect and intangible. Cities promote a flourishing cultural life and, by mixing together diverse people and influences, spark innovation and creativity. The urban environment can promote education and scientific research, and cities have long been viewed as engines of technological and artistic inventiveness. This inventiveness can lead to solutions to societal problems, including ways to reduce environmental impacts.

FIGURE 13.16 Urban and suburban children take part in an environmental education program in a Chicago-area forest preserve. Although natural land is rare in urbanized areas, the education and innovation that urbanization often promotes can lead to solutions that reduce environmental impact.

For instance, research into renewable energy sources is helping us develop ways to replace fossil fuels. Technological advances have helped us reduce pollution. Wealthy and educated urban populations provide markets for low-impact goods, such as organic produce. Recycling programs help reduce the solid waste stream. Environmental education (**Figure 13.16**) is helping people choose their own ways to live cleaner, healthier, lower-impact lives. All these phenomena grow from the education, innovation, science, and technology that are part of urban culture.

## Some seek sustainability for cities

Modern cities that import all their resources and export all their wastes have a linear, one-way metabolism. Such linear models of production and consumption tend to destabilize environmental systems and are not sustainable. Proponents of sustainability for cities stress the need to develop circular systems, akin to systems found in nature, which recycle materials and use renewable sources of energy.

Researchers in the field of **urban ecology** hold that cities can be viewed explicitly as ecosystems and that the fundamentals of ecosystem ecology and systems science (Chapter 7) apply to urban areas. Major urban ecology projects are ongoing in Baltimore and Phoenix, where researchers funded by the National Science Foundation's Long Term Ecological Research (LTER) program are studying these cities explicitly as ecosystems. The researchers are examining such topics as nutrient cycling of carbon and nitrogen, biodiversity patterns, and air and water quality. They are also exploring ways in which humans perceive their environment and react to environmental health threats.

To help cities improve their standards of living while reducing their environmental impacts, urban sustainability advocates suggest that cities:

- Maximize efficient use of resources.
- Recycle as much as possible (• pp. 641–642).
- Develop environmentally friendly technologies.
- Account fully for external costs (• p. 42).
- Offer tax incentives to encourage sustainable practices.
- Use locally produced resources.
- Use organic waste and wastewater to restore soil fertility.
- Encourage urban agriculture.

More and more cities are adopting these strategies. For instance, urban agriculture is a growing pursuit in many urban areas, from Portland to Cuba (• pp. 287, 289) to Japan. Singapore produces all its meat and 25% of its vegetable needs within its city limits. In Berlin, Germany, 80,000 people grow food in community gardens, and 16,000 more are on waiting lists.

Curitiba, Brazil, shows the kind of success that can result when a city invests in well-planned infrastructure. Besides the highly effective bus transportation network described earlier, the city provides recycling, environmental education, job training for the poor, and free health care. Surveys show that its citizens are unusually happy and better off economically than people living in other Brazilian cities.

In 2007, New York City unveiled an extensive new plan that Mayor Michael Bloomberg hoped would make New York "the first environmentally sustainable 21st-century city." Bloomberg's "PlaNYC" is a 127-item program that aims to reduce energy use and greenhouse gas emissions, improve mass transit, plant trees, clean up polluted land and rivers, and enhance access to parkland, all in order to make New York City a better place to live even as it gains 1 million more people by 2030. A proposal to charge drivers for driving into downtown Manhattan (• p. 474) is the most controversial aspect of the program, which will need the support of the state government as well as city residents.

In general, experts advise that developed countries should invest in resource-efficient technologies to reduce their impacts and enhance their economies, whereas developing countries should invest in basic infrastructure to improve health and living conditions. Successes in places from Portland to Curitiba suggest that cities need not be unsustainable. Indeed, because they affect the environment in some positive ways and have the potential for efficient resource use, cities can and should be a key element in achieving progress toward global sustainability.

## Conclusion

As half the human population has shifted from rural to urban lifestyles, the nature of our impact on the environment has changed. As urban and suburban dwellers, our impacts are less direct but often more far-reaching. Resources must be delivered to us over long distances, requiring the use of still more resources. Limiting the waste of those resources by making our urban and suburban areas more sustainable will be vital for the future. Fortunately, the innovative cultural environment that cities foster has helped us develop solutions to alleviate impact and promote sustainability.

Part of seeking urban sustainability lies in making urban areas better places to live. One key component of these efforts involves expanding transportation options to relieve congestion and make cities run more efficiently. Another lies in ensuring access to adequate park lands and greenspaces near and within our urban centers, to keep us from becoming wholly isolated from nature. Accomplishments in city and regional planning have made many American cities more livable than they once were, and we should be encouraged about such progress. Proponents of smart growth and the new urbanism believe they have solutions to the challenges posed by urban and suburban sprawl. Continuing experimentation in cities from Portland, Oregon, to Curitiba, Brazil, will help us determine how best to ensure that urban growth improves our quality of life and does not degrade the quality of our environment.

## REVIEWING OBJECTIVES

**You should now be able to:**

**Describe the scale of urbanization**

- The world's population is becoming predominantly urban. (p. 358)
- The shift from rural to urban living is driven largely by industrialization and is proceeding fastest now in the developing world. (pp. 358–360)
- Nearly all future population growth will be in cities of the developing world. (pp. 359–360)
- The geography of urban areas is changing as cities decentralize and suburbs grow and expand. (pp. 360–361)

**Assess urban and suburban sprawl**

- Sprawl covers large areas of land with low-density development. Both population growth and increased per capita land use contribute to sprawl. (pp. 361–363)
- Sprawl has resulted from the home-buying choices of individuals who prefer suburbs to cities, and it has been facilitated by government policy and technological developments. (pp. 362–363)
- Sprawl may lead to negative impacts involving transportation, pollution, health, land use, natural habitat, and economics. (pp. 363–366)

**Outline city and regional planning and land use strategies**

- City and regional planning and zoning are key tools for improving the quality of urban life. (pp. 366–368)
- "Smart growth," urban growth boundaries, and the "new urbanism" attempt to re-create compact and vibrant urban spaces. (pp. 368–370)

**Evaluate transportation options**

- Mass transit systems can enhance the efficiency of urban areas, but bus and train systems of different sizes bring different benefits. (pp. 370–373)

**Describe the roles of urban parks**

- Urban park lands are vital for active recreation, soothing the stress of urban life, and keeping people in touch with natural areas. (pp. 371–374)

**Analyze environmental impacts and advantages of urban centers**

- Cities are resource sinks with high per capita resource consumption. However, cities also allow natural lands to be preserved. (pp. 374–376)
- Urban centers can maximize efficiency and help foster innovation that can lead to solutions for environmental problems. (pp. 375–376)

**Assess the pursuit of sustainable cities**

- The linear mode of consumption and production is unsustainable, and more circular modes will be needed to create sustainable cities. (pp. 376–377)

## TESTING YOUR COMPREHENSION

1. What factors lie behind the shift of population from rural areas to urban areas? What types of cities and countries are experiencing the fastest urban growth today, and why?
2. Why have so many city dwellers in the United States, Canada, and other nations moved into suburbs?
3. Give two definitions of *sprawl*. Describe five negative impacts that have been suggested to result from sprawl.
4. What are city planning and regional planning? Contrast planning with zoning. Give examples of some of the suggestions made by early planners, such as Daniel Burnham and Edward Bennett.
5. How are some people trying to prevent or slow sprawl? Describe some key elements of "smart growth." What effects, positive and negative, do urban growth boundaries tend to have?
6. Describe several apparent benefits of rail transit systems. What is a potential drawback?
7. How are city parks thought to make urban areas more livable? What types of smaller spaces in cities can serve some of the functions of parks?
8. Why do urban dwellers tend to consume more resources per capita than rural dwellers?
9. Describe the connection between urban ecology and sustainable cities. List three actions a city can take to enhance its sustainability.
10. Name two positive effects of urban centers on the natural environment.

## SEEKING SOLUTIONS

1. Assess the reasons why urban populations are rising and why rural populations are stable or falling. Do you think these trends will continue in the future, or might they change for some reason?
2. Evaluate the causes of the spread of suburbs and of the environmental, social, and economic impacts of sprawl. Overall, do you think the spread of urban and suburban development that many people label *sprawl* is predominantly a good thing or a bad thing? Do you think it is inevitable? Give reasons for your answers.
3. Would you personally want to live in a neighborhood developed in the style of the new urbanism? Would you like to live in a city or region with an urban growth boundary? Why or why not?
4. All things considered, do you feel that cities are a positive thing or a negative thing for environmental quality? How much do you feel we may be able to improve the sustainability of our urban areas?
5. **THINK IT THROUGH** You are a person who aims to live in the most ecologically sustainable way you can. Which of the following places would you choose to live: in a high-rise apartment in a big city or on a 40-acre ranch abutting a national forest? Why? What considerations will you factor into your decision?
6. **THINK IT THROUGH** After you earn your college degree, you are offered three equally desirable jobs, in three very different locations. If you take the first, you will live in the midst of a densely populated city. If you accept the second, you will live in a suburb where you have more space but where development and sprawl may soon surround you for many miles. If you select the third, you will live in a rural area with plenty of space but few cultural amenities. Where would you choose to live? Why?

## INTERPRETING GRAPHS AND DATA

In the accompanying graph, urban population density is used as an indicator of sprawl (lower density = more sprawl), and carbon emissions per capita provide some measure of the environmental impact of the transportation system or preferences for each of the cities represented.

1. Describe the relationship between urban density and carbon emissions, as shown in the graph.
2. Assuming that the standard of living is similar in these cities, to what might you attribute the relationship described in your answer to question 1?
3. If zoning ordinances slowed urban sprawl and resulted in a doubling of urban population density in a city like Houston, Texas, how would you predict that carbon emissions per capita in that city might change?

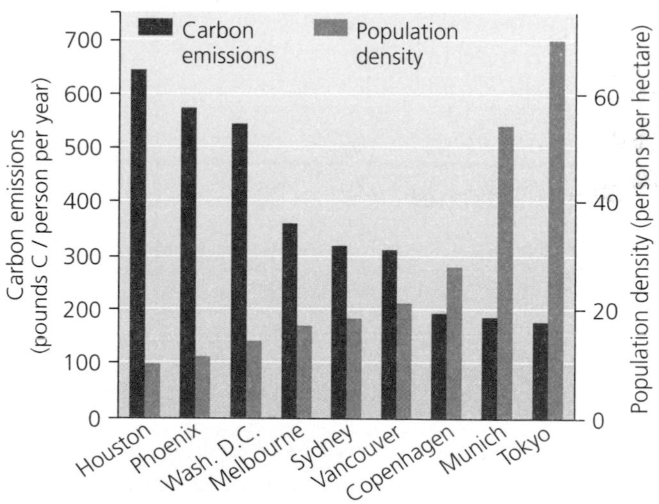


**Population density versus carbon emissions from transportation in 1990.** Data from Kenworthy, J., et al. 1999. *An international sourcebook of automobile dependence in cities.* Boulder, CO: University Press of Colorado, as cited by Sheehan, M. O. 2002. *What will it take to halt sprawl?* Washington DC: Worldwatch Institute.

## CALCULATING ECOLOGICAL FOOTPRINTS

One way of altering your ecological footprint is to consider transportation alternatives. Each gallon of gasoline is converted to approximately 20 lb of carbon dioxide ($CO_2$) during combustion, and this $CO_2$ is then released into the atmosphere. The table below lists typical amounts of $CO_2$ released for each person per mile, through various forms of transportation, assuming typical fuel efficiencies.

For an average North American person who travels 12,000 miles per year, calculate and record in the table the $CO_2$ emitted yearly for each transportation option, and the reduction in $CO_2$ emission that one could achieve by relying solely on each option.

| | $CO_2$ per person per mile | $CO_2$ per person per year | $CO_2$ emission reduction | Your estimated mileage per year | Your $CO_2$ emissions per year |
|---|---|---|---|---|---|
| **Automobile (driver only)** | 0.825 lb | 9,900 lb | 0 | | |
| **Automobile (2 persons)** | 0.413 lb | | | | |
| **Automobile (4 persons)** | 0.206 lb | | | | |
| **Vanpool (8 persons)** | 0.103 lb | | | | |
| **Bus** | 0.261 lb | | | | |
| **Walking** | 0.082 lb | | | | |
| **Bicycle** | 0.049 lb | | | | |
| | | | | **Total** = 12,000 | |

1. What transportation option will give you the most miles traveled per unit of carbon dioxide emitted?
2. Clearly, it is unlikely that any of us will walk or bicycle 12,000 miles per year or travel only in vanpools of eight people. In the last two columns, estimate what proportion of the 12,000 annual miles you think that you actually travel by each method, and then calculate the $CO_2$ emissions that you are responsible for generating over the course of a year. Which transportation option accounts for the most emissions for you?
3. How could you reduce your $CO_2$ emissions? How many pounds of emissions do you think you could realistically eliminate over the course of the next year?

## Take It Further

Go to www.aw-bc.com/withgott or the student CD-ROM, where you'll find:

- Suggested answers to end-of-chapter questions
- Quizzes, animations, and flashcards to help you study
- *Research Navigator*™ database of credible and reliable sources to assist you with your research projects
- **GRAPHIt!** Tutorials to help you interpret graphs
- **INVESTIGATEIt!** Current news articles that link the topics that you study to case studies from your region to around the world

CHAPTER

# 14 Environmental Health and Toxicology

Alligator hatchling from Lake Apopka, Florida

## Upon completing this chapter, you will be able to:

- ▶ Identify the major types of environmental health hazards and explain the goals of environmental health
- ▶ Describe the types, abundance, distribution, and movement of toxicants in the environment
- ▶ Discuss the study of hazards and their effects, including case histories, epidemiology, animal testing, and dose-response analysis
- ▶ Assess risk assessment and risk management
- ▶ Compare philosophical approaches to risk
- ▶ Describe policy and regulation in the United States and internationally

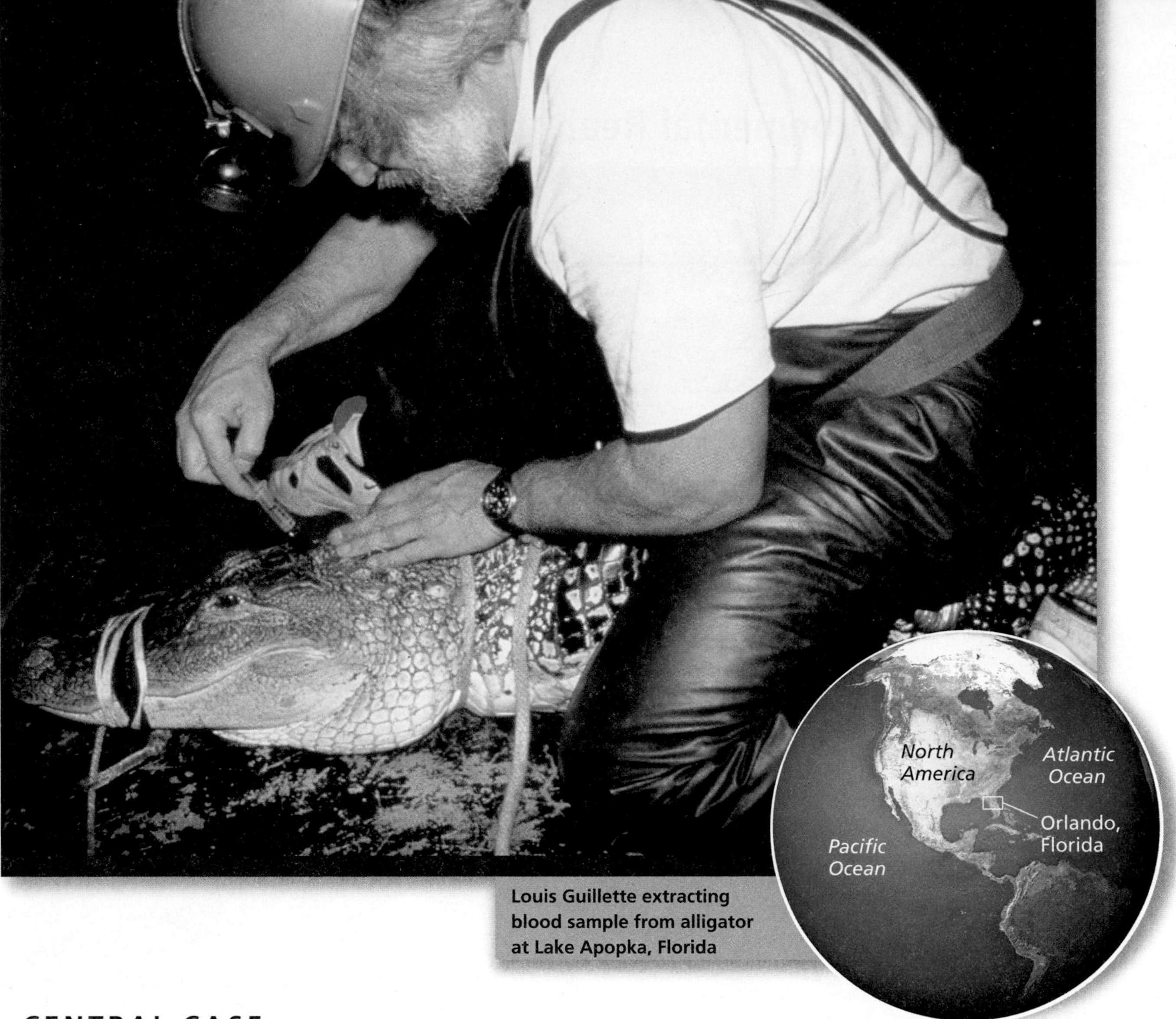


Louis Guillette extracting blood sample from alligator at Lake Apopka, Florida

## CENTRAL CASE

# Alligators and Endocrine Disruptors at Lake Apopka, Florida

**"Over the past 50 years we have all been unwitting participants in a vast, uncontrolled, worldwide chemistry experiment involving the oceans, air, soils, plants, animals, and human beings."**
—United Nations Environment Programme, in a guide to the Stockholm Convention on Persistent Organic Pollutants

**"Every man in this room is half the man his grandfather was."**
—Reproductive biologist Louis Guillette to a U.S. Congressional committee in 1995, regarding declining sperm counts in men worldwide over the previous half century

When biologist Louis Guillette began studying the reproductive biology of alligators in Florida lakes in 1985, he soon discovered that not all was well. Alligators from one lake in particular, Lake Apopka northwest of Orlando, showed a number of bizarre reproductive problems.

Females were having trouble producing viable eggs. Male hatchling alligators had severely depressed levels of the male sex hormone testosterone, and female hatchlings showed greatly elevated levels of the female sex hormone estrogen. The young animals had abnormal gonads, too. Testes of the males produced sperm at a premature age, and ovaries of the females contained multiple eggs per follicle instead of the expected one egg per follicle.

As Guillette and his co-workers tested and rejected hypotheses for the reproductive abnormalities one by one, they grew to suspect that environmental contaminants might be somehow responsible. Lake Apopka had suffered a major spill of the pesticides dicofol and DDT in 1980, and yearly surveys thereafter showed a precipitous decline in the number of juvenile alligators in the lake. In addition, the lake received high levels of chemical runoff from agriculture and was experiencing eutrophication (• p. 179 and • pp. 429–430) from nutrient input from fertilizers.

The puzzle pieces fell together when a colleague of Guillette's told him that recent research was showing that some environmental contaminants mimic hormones and interfere with the functioning of animal endocrine (hormone) systems. At very low doses, these **endocrine disruptors** can affect animals, mimicking estrogen and feminizing males. Guillette realized that his alligator observations were similar to results from tests in which rodents were exposed to estrogen during embryonic development. He formulated the hypothesis that certain chemical contaminants in Lake Apopka were disrupting the endocrine systems of alligators during development in the egg.

To test his hypothesis, Guillette and his co-workers compared alligators from heavily polluted Lake Apopka with those from cleaner lakes nearby. They found that Lake Apopka alligators had abnormally low hatching rates in the years after the pesticide spill and, even as hatching rates recovered in the 1990s, continued to show aberrant hormone levels and bizarre gonad abnormalities. In addition, the penises of male alligators from Lake Apopka were 25% smaller than those of male alligators from surrounding lakes.

Similar problems began cropping up in other lakes that experienced runoff of chemical pesticides. In the lab, researchers found that several contaminants detected in alligator eggs and young could bind to receptors for estrogen and exist in concentrations great enough to cause sex reversal of male embryos. One chemical in particular, atrazine—the most widely used herbicide in the United States—appeared to disrupt hormones by inducing production of aromatase, an enzyme that converts testosterone to estrogen. In 2003, Guillette reported preliminary findings that nitrate from fertilizer runoff may also act as an endocrine disruptor; when nitrate concentrations in lakes are above the standard for drinking water, juvenile male alligators have smaller penises and 50% lower testosterone levels.

Guillette's results have raised concern not only for alligator health but also for human health. Because alligators and humans share many of the same hormones, many scientists suspect that chemical contaminants could be affecting people, just as they have affected alligators.

# Environmental Health

Examining impacts of human-made chemicals on wildlife and people is one aspect of the broad field of environmental health. The study and practice of **environmental health** assesses environmental factors that influence human health and quality of life. Those environmental factors include wholly natural aspects of the environment over which we have little or no control, as well as anthropogenic (human-caused) factors that we ourselves are responsible for. Practitioners of environmental health seek to prevent adverse effects on human health and on the ecological systems that are essential to environmental quality and human well-being.

## Environmental hazards can be physical, chemical, biological, or cultural

There are innumerable environmental health threats, or hazards, in the world around us. For ease of understanding, we can categorize them into four main types: physical, biological, chemical, and cultural. For each of these four types of hazards, there is some amount of risk that we cannot avoid—but there is also some amount of risk that we *can* avoid by taking precautions. Much of environmental health consists of taking steps to minimize the impacts of hazards and the risks of encountering them.

**Physical hazards** *Physical* processes that occur naturally in our environment and pose health hazards include discrete events such as earthquakes, volcanic eruptions, fires, floods, blizzards, landslides, hurricanes, and droughts. We can do little to predict the timing of a natural disaster such as an earthquake, and nothing to prevent one. However, scientists can map geologic faults to determine areas at risk of earthquakes, engineers can design buildings in ways that help them resist damage, and citizens and governments can take steps to prepare for the aftermath of a severe quake. Physical processes also include ongoing natural phenomena, such as ultraviolet (UV) radiation from sunlight (**Figure 14.1a**). Excessive exposure to UV radiation damages DNA and has been tied to skin cancer, cataracts, and immune suppression in humans.

Some common practices increase our vulnerability to certain physical hazards. Deforesting slopes makes landslides more likely, for instance, and channelizing rivers makes flooding more likely in some areas while preventing flooding in others. We can reduce risk from such hazards by improving our forestry and flood control practices and by choosing not to build in areas prone to floods, landslides, fires, and coastal waves. For hazards such as exposure to UV light, we can reduce our exposure and risk by using clothing and sunscreen to shield our skin from intense sunlight.

**Chemical hazards** *Chemical* hazards include many of the synthetic chemicals that our society produces, such as disinfectants, pesticides (**Figure 14.1b**), and the compounds that contributed to reproductive problems for the alligators at Lake Apopka. Chemicals produced naturally

FIGURE 14.1 Environmental health hazards can be divided into four types. The sun's ultraviolet radiation is an example of a physical hazard (**a**). Excessive exposure increases the risk of skin cancer. Chemical hazards (**b**) include both artificial and natural chemicals. Much of our exposure comes from household chemical products, such as pesticides that some people apply to their lawns. Biological hazards (**c**) include other organisms that transmit disease. Some mosquitoes are vectors for certain pathogenic microbes, including those that cause malaria. Cultural or lifestyle hazards (**d**) include the decisions we make about how to behave, as well as the constraints forced on us by socioeconomic factors. Smoking is a lifestyle choice that raises one's risk of lung cancer and other diseases considerably.

**(a) Physical hazard**

**(b) Chemical hazard**

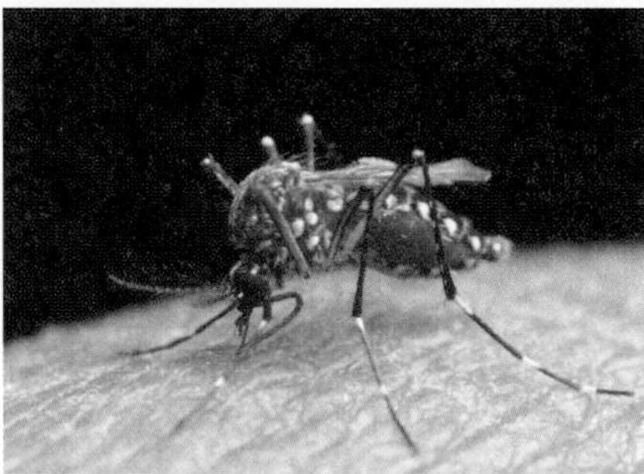

**(c) Biological hazard**

**(d) Cultural hazard**

by organisms also can be hazardous. Following our overview of environmental health, much of our chapter will focus on chemical health hazards and the ways people study and regulate them.

**Biological hazards** *Biological* hazards result from ecological interactions among organisms (**Figure 14.1c**). When we become sick from a virus, bacterial infection, or other pathogen, we are suffering parasitism by other species that are simply fulfilling their ecological roles. This is **infectious disease**, also called *communicable* or *transmissible disease*. Infectious diseases such as malaria, cholera, tuberculosis, and influenza (flu) all are considered environmental health hazards. As with physical and chemical hazards, it is impossible for us to avoid risk from biological agents completely, but we can take steps to reduce the likelihood of infection.

**Cultural hazards** Hazards that result from the place we live, our socioeconomic status, our occupation, or our behavioral choices can be thought of as *cultural* or *lifestyle* hazards. For instance, choosing to smoke cigarettes, or living or working with people who do, greatly increases our risk of lung cancer (**Figure 14.1d**). Choosing to smoke is a personal behavioral decision, but exposure to secondhand smoke in the home or workplace may not be under one's control. Much the same might be said for drug use, diet and nutrition, crime, and mode of transportation. As advocates of environmental justice (• pp. 35–37) argue, such health factors as living in proximity to toxic waste sites or working unprotected with pesticides are often correlated with socioeconomic deprivation.

## Disease is a major focus of environmental health

Among the hazards people face, disease stands preeminent. Despite all our technological advances, we still find ourselves battling disease, which causes the vast majority of human deaths worldwide (**Figure 14.2a**). Many major killers such as cancer, heart disease, and respiratory disorders have genetic bases but are also influenced by environmental factors. For instance, whether a person develops asthma is influenced not only by the genes, but also by environmental conditions. Pollutants from fossil fuel combustion worsen asthma, and children raised on farms suffer less asthma than children raised in cities, studies

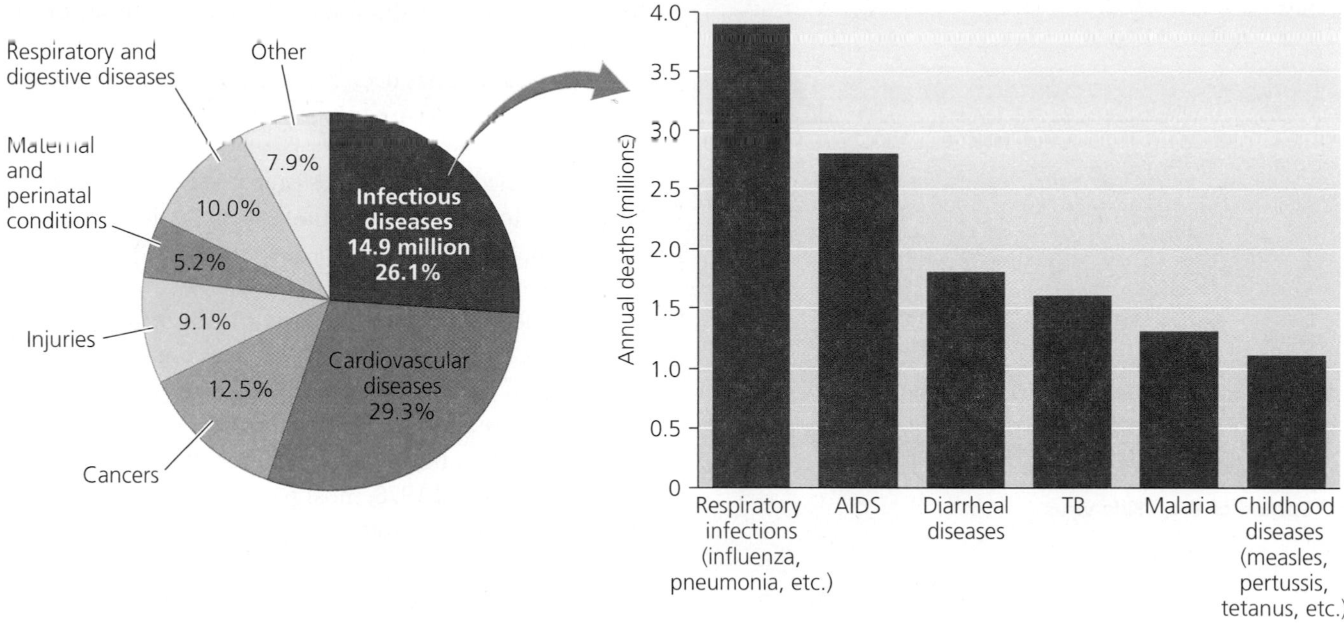


**FIGURE 14.2** Infectious diseases are the second-leading cause of death worldwide, accounting for over one-quarter of all deaths per year (**a**). Six types of diseases—respiratory infections, AIDS, diarrhea, tuberculosis (TB), malaria, and childhood diseases such as measles—account for 80% of all deaths from infectious disease (**b**). Data are for 2004, from World Health Organization.

have shown. Malnutrition (• pp. 265–266) can foster a wide variety of illnesses, as can poverty and poor hygiene. Moreover, lifestyle choices can affect risks of acquiring some noninfectious diseases: Smoking can lead to lung cancer, and lack of exercise to heart disease, for example.

Infectious diseases account for 26% of deaths that occur worldwide each year—nearly 15 million people (**Figure 14.2b**). Some pathogenic microbes attack us directly, and sometimes infection occurs through a *vector*, an organism that transfers the pathogen to the host. Infectious diseases account for close to half of all deaths in developing countries, but for very few deaths in developed nations. This discrepancy is due to differences in hygiene conditions and access to medicine, which are tightly correlated with wealth. Public health efforts have lessened the impact of infectious disease in developed nations and even have eradicated some diseases. Nevertheless, other diseases, among them tuberculosis, acquired immunodeficiency syndrome (AIDS), and West Nile virus, are increasing (**Figure 14.3**). Still others, such as the avian flu that began spreading worldwide in 2005–2006, remain as threats for a possible global epidemic.

Many diseases are spreading due to the mobility we have achieved in our era of globalization; a virus for influenza or West Nile can now hop continents in a matter of hours by airplane in its human host. Other diseases, such as tuberculosis and strains of malaria, are evolving resistance to our antibiotics, in the same way as pests evolve resistance to our pesticides (• p. 269). Tropical diseases such as malaria, dengue, cholera, and yellow fever threaten to expand into the temperate zones with climate change. And habitat alteration can affect the abundance, distribution, and movement of certain disease vectors.

To predict and prevent infectious disease, environmental health experts deal with the often complicated interrelationships among technology, land use, and ecology. One of the world's leading infectious diseases, malaria (which takes nearly 1.3 million lives each year), provides an example. The microscopic protists (four species of *Plasmodium*) that cause malaria depend on mosquitoes as a vector. These protists can sexually reproduce only within a mosquito, and it is the mosquito that injects the protists into a human or other host. Thus the primary mode of malaria control has been to use insecticides such as DDT to kill mosquitoes. Large-scale eradication projects involving insecticide use and draining of wetlands have removed malaria from large areas of the temperate world where it used to occur, such as throughout the southern United States. However, types of land use that create pools of standing water in formerly well-drained areas can boost mosquito populations and allow malaria to reinvade.

## Environmental health hazards exist indoors as well as outdoors

Outdoor hazards are generally more familiar to us, but we spend most of our lives indoors. Therefore, we must consider the spaces inside our homes and workplaces to be

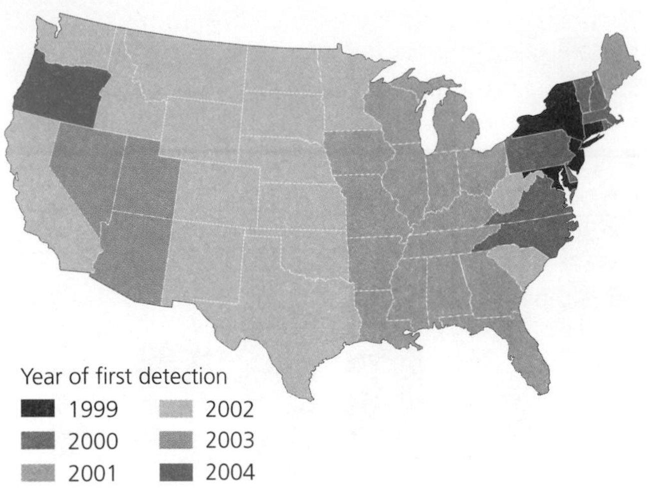


(a) Spread of West Nile virus

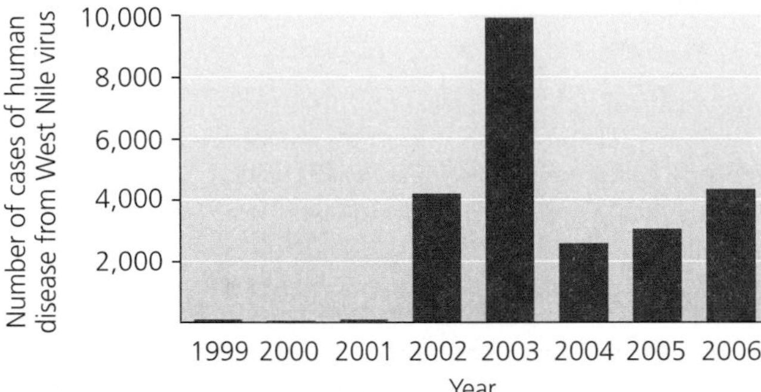


(b) Human cases of West Nile virus, 1999–2006

**FIGURE 14.3** Modern public health efforts have greatly diminished the burden of disease in developed nations, but some infectious diseases are currently on the rise. One example is West Nile virus, an Old World virus that spread rapidly through the Western Hemisphere after first being detected in the New York City area in 1999. Birds are most affected by this mosquito-transmitted virus, but more than 22,000 human disease cases, including over 800 deaths, had been recorded in the United States through the end of 2006. In the map (**a**), colors show the year that West Nile virus was first detected within each state. The disease spread westward across the country in less than 5 years. The graph (**b**) shows the number of cases reported nationwide in each year. Data from Centers for Disease Control and Prevention.

part of our environment and, as such, the sources of potential environmental hazards (Table 14.1). Radon is one indoor hazard (• p. 497). *Radon* is a highly toxic radioactive gas that is colorless and undetectable without specialized kits. Radon seeps up from the ground in areas with certain types of bedrock and can build up inside basements and homes with poor air circulation. The U.S. Environmental Protection Agency (EPA) estimates that slightly less than 1 person in 1,000 may contract lung cancer as a result of a lifetime of radon exposure at average levels for U.S. homes.

*Lead poisoning* represents another indoor health hazard. When ingested, lead, a heavy metal, can cause damage to the brain, liver, kidney, and stomach; learning problems and behavioral abnormalities; anemia; hearing loss; and even death. It has been suggested that the downfall of ancient Rome was caused in part by chronic lead poisoning. The elites of the power structure regularly drank wine sweetened with mixtures prepared in leaden vessels. Several historical personalities, such as the composer Beethoven, suffered maladies that some historians have attributed to lead poisoning.

Today lead poisoning can result from drinking water that has passed through the lead pipes common in older houses. Even in newer pipes, lead solder was widely used into the 1980s and is still sold in stores. Lead is perhaps most dangerous, however, to children through its presence in paint. Until 1978, most paints contained lead, and interiors in most houses were painted with lead-based paint. Babies and young children often take an interest in peeling paint from walls and may ingest or inhale some of it. Today lead poisoning is thought to affect 3–4 million young children, fully one in six children under age 6.

Another indoor hazard is asbestos. There are several types of *asbestos,* each of which is a mineral that forms long, thin microscopic fibers. This structure allows asbestos to insulate heat, muffle sound, and resist fire. Because of these qualities, asbestos was used widely as insulation in buildings, as well as in many products. Unfortunately, its fibrous structure also makes asbestos

**TABLE 14.1 Selected Environmental Hazards**

**Air**

- Smoking and secondhand smoke
- Chemicals from automotive exhaust
- Chemicals from industrial pollution
- Tropospheric ozone
- Pesticide drift
- Dust and particulate matter

**Water**

- Pesticide and herbicide runoff
- Nitrates and fertilizer runoff
- Mercury, arsenic, and other heavy metals in groundwater and surface water

**Food**

- Natural toxins
- Pesticide and herbicide residues

**Indoors**

- Asbestos
- Radon
- Lead in paint and pipes
- Toxicants in plastics and consumer products

dangerous when inhaled. When lodged in lung tissue, asbestos induces the body to produce acid to combat it. The acid scars the lung tissue but does little to dislodge or dissolve the asbestos. Consequently, within a few decades the scarred lungs may cease to function, a disorder called *asbestosis*. Asbestos can also cause types of lung cancer. Because of these risks, asbestos has been removed from many schools and offices (**Figure 14.4**), although the dangers of exposure from asbestos removal may sometimes exceed the dangers of leaving it in place.

One recently recognized hazard is a group of chemicals known as *polybrominated diphenyl ethers (PBDEs)*. These compounds provide fire-retardant properties and are used in a diverse array of consumer products, including computers, televisions, plastics, and furniture. They appear to be released during production and disposal of products and also to evaporate at very slow rates throughout the lifetime of products. These chemicals persist and accumulate in living tissue, and their abundance in the environment and in people in the United States is doubling every few years.

FIGURE 14.4 Asbestos was widely used in insulation and other products. A cause of lung cancer and asbestosis, the substance has now been removed from many buildings in which it was used. Its removal poses risks as well, however, and removal workers must wear protective clothing and respirators.

Like the chemicals affecting Louis Guillette's alligators, PBDEs appear to be endocrine disruptors; lab testing with animals shows them to affect thyroid hormones. Animal testing also shows limited evidence that PBDEs may affect brain and nervous system development and might possibly cause cancer. Concern about PBDEs rose after a study showed that concentrations in the breast milk of Swedish mothers had increased exponentially from 1972 to 1997. The European Union decided in 2003 to ban PBDEs, and industries in Europe had already begun phasing them out. As a result, concentrations in breast milk of European mothers have fallen substantially. In the United States, however, there has so far been little movement to address the issue.

## Toxicology is the study of poisonous substances

Studying the health effects of chemical agents suspected to be harmful, such as PBDEs, is the focus of the field of **toxicology**, the science that examines the effects of poisonous substances on humans and other organisms. Toxicologists assess and compare substances to determine their *toxicity*, the degree of harm a chemical substance can inflict. The concept of toxicity among chemical hazards is analogous to that of *pathogenicity* or *virulence* of the biological hazards that spread infectious disease. Just as types of microbes differ in their ability to cause disease, chemical hazards differ in their capacity to endanger us.

However, any chemical substance may exert negative effects if it is ingested in great enough quantities or if exposure is extensive enough. Conversely, a toxic agent, or **toxicant**, in a minute enough quantity may pose no health risk at all. These facts are often summarized in the catchphrase, "The dose makes the poison." In other words, a substance's toxicity depends not only on its chemical identity, but also on its quantity.

During the past century, our ability to produce new chemicals has expanded, concentrations of chemical contaminants in the environment have increased, and public concern for health and the environment have grown. These trends have driven the rise of **environmental toxicology**, which deals specifically with toxic substances that come from or are discharged into the environment. Environmental toxicology includes the study of health effects on humans, other animals, and ecosystems, and it represents one approach within the broader scope of environmental health.

Toxicologists generally focus on human health, using other organisms as models and test subjects. In environmental toxicology, animals are also studied out of concern for their welfare and because—like canaries in a coal mine—animals can serve as indicators of health threats that could soon affect humans. For example, the reproductive abnormalities of Lake Apopka alligators seem analogous to those seen in some Taiwanese boys. Studies showed that these boys were born to mothers who used cooking oil contaminated with toxicants called polychlorinated biphenyls (PCBs), which are by-products of chemicals used in transformers and other electrical equipment.

As we review the sometimes poisonous effects of human-made chemicals throughout this chapter, it is important to keep in mind that artificially produced chemicals have played a crucial role in giving us the standard of living we enjoy today. These chemicals have helped create the industrial agriculture that produces our food, the medical advances that protect our health and prolong our lives, and many of the modern materials and conveniences we use every day. It is important to remember these benefits as we examine some of the unfortunate side effects of these advances and as we search for better alternatives.

## Toxic Agents in the Environment

The environment contains countless natural chemical substances that may pose health risks. These substances include oil oozing naturally from the ground; radon gas seeping up from bedrock; and toxic chemicals stored or manufactured in the tissues of living organisms—for example, *toxins* that plants use to ward off herbivores and toxins that insects use to defend themselves from predators. In addition, we are exposed to many synthetic (artificial, or human-made) chemicals.

### Synthetic chemicals are ubiquitous in our environment

Synthetic chemicals are all around us in our daily lives (**Figure 14.5**). Thousands of synthetic chemicals have been manufactured (Table 14.2), and many have found their way into soil, air, and water. A 2002 study by the U.S. Geological Survey found that 80% of U.S. streams contain at least trace amounts of 82 wastewater contaminants, including antibiotics, detergents, drugs, steroids, plasticizers, disinfectants, solvents, perfumes, and other substances. The pesticides we use to kill insects and weeds (• pp. 268–269) on farms, lawns, and golf courses are some of the most widespread synthetic chemicals. As a result of all this exposure, every one of us carries traces of numerous industrial chemicals in our bodies.

This should not *necessarily* be cause for alarm. Not all synthetic chemicals pose health risks, and relatively few are known with certainty to be toxicants. However, of the roughly 100,000 synthetic chemicals on the market today, very few have been thoroughly tested for harmful effects.

FIGURE 14.5 Synthetic chemicals, such as those in household products, are everywhere around us in our everyday lives. Some of these compounds may potentially pose environmental or human health risks.

For the vast majority, we simply do not know what effects, if any, they may have.

Why are there so many synthetic chemicals around us? Let's consider pesticides and herbicides, made widespread by advances in chemistry and production capacity during and following World War II. As material prosperity grew in Westernized nations in the decades following the war, people began using pesticides not only for agriculture, but also to improve the look of their lawns and golf

**TABLE 14.2 Estimated Numbers of Chemicals in Commercial Substances during the 1990s**

| Type of chemical | Estimated number |
|---|---|
| Chemicals in commerce | 100,000 |
| Industrial chemicals | 72,000 |
| New chemicals introduced per year | 2,000 |
| Pesticides (21,000 products) | 600 |
| Food additives | 8,700 |
| Cosmetic ingredients (40,000 products) | 7,500 |
| Human pharmaceuticals | 3,300 |

Data from Harrison, P., and F. Pearce. 2000. *AAAS Atlas of Population and Environment*. Berkeley, CA: University of California Press.

courses and to fight termites, ants, and other insects inside their homes and offices. Pesticides were viewed as means toward a better quality of life.

It was not until the 1960s that people began to learn about the risks of exposure to pesticides. The key event was the publication of Rachel Carson's 1962 book *Silent Spring* (• pp. 65–66), which brought the pesticide dichloro-diphenyl-trichloroethane (DDT) to the attention of the public.

## *Silent Spring* began the public debate over synthetic chemicals

Rachel Carson was a naturalist, author, and government scientist. In *Silent Spring*, she brought together a diverse collection of scientific studies, medical case histories, and other data that no one had previously synthesized and presented to the general public. Her message was that DDT in particular and artificial pesticides in general were hazardous to people's health, the health of wildlife, and the well-being of ecosystems. Carson wrote at a time when large amounts of pesticides virtually untested for health effects were indiscriminately sprayed over residential neighborhoods and public areas, on the assumption that the chemicals would do no harm to people (**Figure 14.6**). Most consumers had no idea that the store-bought chemicals they used in their houses, gardens, and crops might be toxic.

Although challenged vigorously by spokespeople for the chemical industry, who attempted to discredit both the author's science and her personal reputation, Carson's book was a best-seller. Carson suffered from cancer as she finished *Silent Spring*, and she lived only briefly after its publication. However, the book helped generate significant social change in views and actions toward the environment. The use of DDT is now illegal in a number of nations and was banned in the United States in 1973.

The United States still manufactures and exports DDT to countries that do use it, however. Many developing countries with tropical climates use DDT to control human disease vectors, such as mosquitoes that transmit malaria. In these countries, malaria represents a far greater health threat than do the toxic effects of the pesticide.

### Weighing the Issues | The Circle of Poison

It has been called the "circle of poison." Although the United States has banned the use of DDT, U.S. companies still manufacture and export the compound to many developing nations. Thus, it is possible that pesticide-laden food can be imported back into the United States. How do you feel about this? Is it unethical for one country to sell to others a substance that it has deemed toxic? Are there factors or circumstances that might change the view you take?

## Toxicants come in several different types

Toxicants, whether they are natural or synthetic, can be classified into different types based on their particular effects on health. The best-known are **carcinogens**, which are chemicals or types of radiation that cause cancer. In cancer, malignant cells grow uncontrollably, creating tumors, damaging the body's functioning, and often leading to death. In our society today, the greatest number of cancer cases is thought to result from carcinogens contained in cigarette smoke. Carcinogens can be difficult to identify because there may be a long lag time between exposure to the agent and the detectable onset of cancer. Historically, much toxicological work focused on carcinogens. Now, however, we know that toxicants can produce

**FIGURE 14.6** Before the 1960s, the environmental and health effects of potent pesticides such as DDT were not widely studied or publicly known. Public areas such as parks, neighborhoods, and beaches were regularly sprayed for insect control without safeguards against excessive human exposure. Here children on a Long Island, New York, beach are fogged with DDT from a pesticide spray machine being tested in 1945.

many different types of effects, so scientists have many more endpoints, or health impacts, to look for.

**Mutagens** are chemicals that cause mutations in the DNA of organisms (• p. 97). Although most mutations have little or no effect, some can lead to severe problems, including cancer and other disorders. If mutations occur in an individual's sperm or egg cells, then the individual's offspring suffer the effects.

Chemicals that cause harm to the unborn are called **teratogens**. Teratogens• that affect the development of human embryos in the womb can cause birth defects. One example involves the drug thalidomide, developed in the 1950s as a sleeping pill and to prevent nausea during pregnancy. Tragically, the drug turned out to be a powerful teratogen, and its use caused birth defects in thousands of babies (**Figure 14.7**). Even a single dose during pregnancy could result in limb deformities and organ defects. Thalidomide was banned in the 1960s once scientists recognized its connection with birth defects. Ironically, today the drug shows promise in treating a wide range of diseases, including Alzheimer's disease, AIDS, and various types of cancer.

The human immune system protects our bodies from disease. Some toxicants weaken the immune system, reducing the body's ability to defend itself against bacteria, viruses, allergy-causing agents, and other attackers. Others, called **allergens**, overactivate the immune system, causing an immune response when one is not necessary. One hypothesis for the increase in asthma in recent years is that allergenic synthetic chemicals are more prevalent in our environment.

FIGURE 14.7 The drug thalidomide turned out to be a potent teratogen. It was banned in the 1960s, but not before causing thousands of birth defects in babies born to mothers who took the product to relieve nausea during pregnancy. Butch Lumpkin was an exceptional "thalidomide baby" who learned to overcome his short arms and deformed fingers, becoming a professional tennis instructor in Georgia.

Still other chemical toxicants, **neurotoxins**, assault the nervous system. Neurotoxins include various heavy metals such as lead, mercury, and cadmium, as well as pesticides and some chemical weapons developed for use in war. A famous case of neurotoxin poisoning occurred in Japan, where a chemical factory dumped mercury waste into Minamata Bay between the 1930s and 1960s. Thousands of people in and around the town on the bay were poisoned by eating fish contaminated with the mercury. First the town's cats began convulsing and dying, and then people began to show odd symptoms, including slurred speech, loss of muscle control, sudden fits of laughter, and in some cases death. The company and the government eventually paid out about $5,000 in compensation to each affected resident.

Most recently, scientists have recognized the importance of **endocrine disruptors**, toxicants that interfere with the *endocrine system*, or hormone system. The endocrine system consists of chemical messengers *(hormones)* that travel through the body. Sent through the bloodstream at extremely low concentrations, these messenger molecules have many vital functions. They stimulate growth, development, and sexual maturity, and they regulate brain function, appetite, sex drive, and many other aspects of our physiology and behavior. Hormone-disrupting toxicants can affect an animal's endocrine system by blocking the action of hormones or accelerating their breakdown. Many endocrine disruptors are so similar to some hormones in their molecular structure and chemistry that they "mimic" the hormone by interacting with receptor molecules just as the actual hormone would (**Figure 14.8**). One common type of endocrine disruption involves the feminization of male animals, as Louis Guillette found with alligators, and as other studies have indicated with fish, frogs, and other organisms. Feminization may be widespread because a number of chemicals appear to mimic the female sex hormone estrogen and bind to estrogen receptors.

## Endocrine disruption may be widespread

Scientists first noted endocrine-disrupting effects as far back as the 1960s with the pesticide DDT. However, the idea that synthetic chemicals might be altering the hormones of animals was not widely appreciated until the 1996 publication of the book *Our Stolen Future*, by Theo

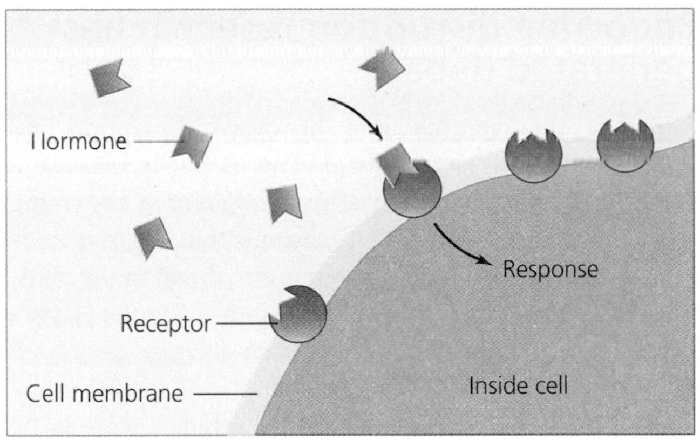


**(a) Normal hormone binding**

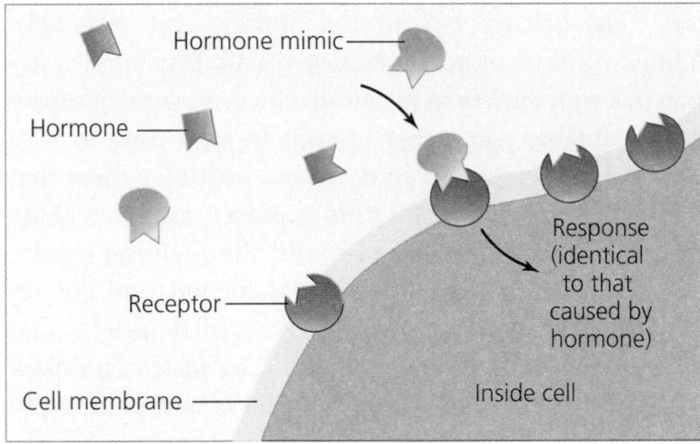


**(b) Hormone mimicry**

**FIGURE 14.8** Many endocrine-disrupting substances mimic the structure of hormone molecules. Like a key similar enough to fit into another key's lock, the hormone mimic binds to a cellular receptor for the hormone, causing the cell to react as though it had encountered the hormone.

Colburn, Dianne Dumanoski, and J. P. Myers. Colburn, a scientist with the World Wildlife Fund, had spent years bringing toxicologists, medical doctors, and wildlife biologists together to share their findings. She and her coauthors then presented the data and stories in *Our Stolen Future* (and on a Web page that provides updates on new research). Like *Silent Spring,* this book integrated scientific work from various fields and presented a unified picture that shocked many readers—and brought criticism from some scientists and from the chemical industry.

One line of research that has sparked debate is that of scientist Tyrone Hayes of the University of California at Berkeley (**Figure 14.9**), who has found in frogs gonadal abnormalities similar to those of Guillette's alligators and has attributed them to atrazine. In lab experiments, male frogs raised in water containing very low doses of the herbicide became feminized and hermaphroditic, developing both testes and ovaries. In field studies, leopard frogs across North America showed similar problems in areas of atrazine usage. Hayes found these effects at atrazine concentrations well below EPA guidelines for human health and at concentrations lower than commonly exist in U.S. drinking water. Thus, he suggests, people may be at risk.

To date, endocrine effects have been found most widely in nonhuman animals, but many scientists attribute the striking drop in sperm counts among men worldwide to endocrine disruptors. In the most noted study, Danish researchers reported in 1992 that the number and motility of sperm in men's semen had declined by 50% since 1938 (**Figure 14.10a**). The research involved a review of 61 studies that included 15,000 men from 20 nations on six continents. Subsequent studies by other researchers—including some who set out to disprove the findings—have largely confirmed the results using other methods and other populations (although there is tremendous geographic variation that remains unexplained).

Some researchers have also voiced concerns about rising rates of testicular cancer (**Figure 14.10b**), undescended testicles, and genital birth defects in men. Other scientists have proposed that the rise in breast cancer rates (one in eight U.S. women today develops breast cancer) may also be due to hormone disruption, because an

**FIGURE 14.9** According to research by Tyrone Hayes of the University of California at Berkeley, frogs may suffer reproductive abnormalities from exposure to the best-selling herbicide atrazine. Industry-backed scientists have disputed the findings.

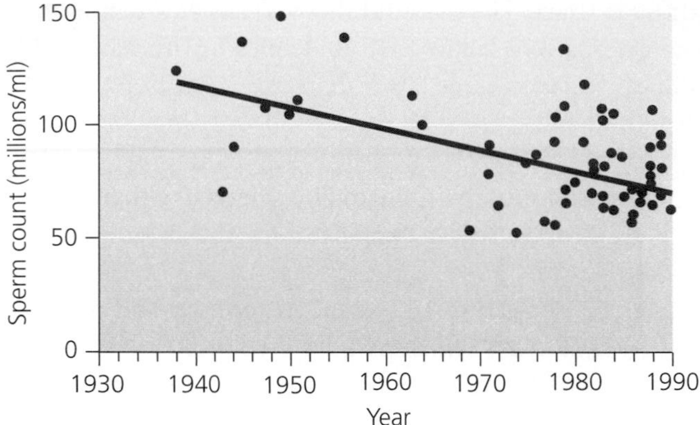


**(a) Declining sperm count in men, based on 61 studies**

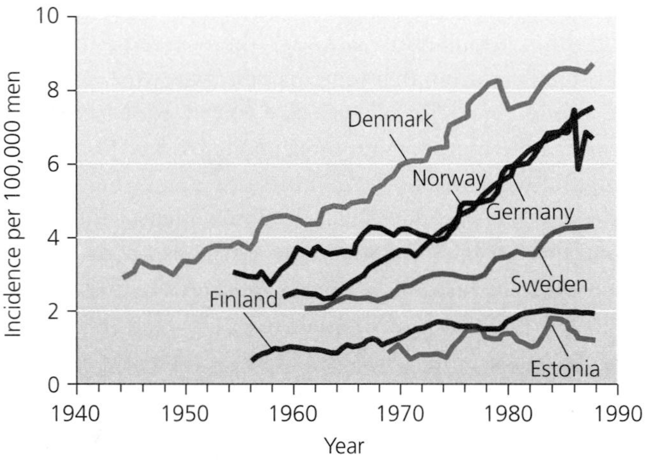


**(b) Increasing incidence of testicular cancer**

**FIGURE 14.10** Research in 1992 synthesized the results of 61 studies that had reported sperm counts in men from various localities since 1938. The data were highly variable but showed a significant decrease in human sperm counts over time (**a**). Many scientists have hypothesized that this decrease may result from exposure to endocrine-disrupting chemicals in the environment. Some have also hypothesized that endocrine disruptors could be behind an increased incidence of testicular cancer (**b**). (a) Data from Carlsen, E., et al. 1992. Evidence for decreasing quality of semen during the last 50 years. *British Medical Journal* 305: 609–613, as adapted by Toppari, J., et al. 1996. Male reproductive health and environmental xenoestrogens. *Environmental Health Perspectives* 104 (Suppl 4): 741–803. (b) Adami, H.O., et al. 1994. Testicular cancer in nine northern European countries. *International Journal of Cancer* 50: 33–38.

excess of estrogen appears to feed tumor development in older women.

Endocrine disruptors can affect more than just the reproductive system. Some impair the brain and nervous system. North American studies have shown neurological problems associated with PCB contamination. In one study in Michigan, mothers who ate Great Lakes fish contaminated with PCBs had babies with lower birth weights and smaller heads, compared to mothers who did not eat fish. These babies grew into children who showed weak and jerky reflexes and tested poorly in intelligence tests.

## Endocrine disruption research has generated debate

Much of the research into hormone disruption has brought about strident debate. This is partly because a great deal of scientific uncertainty is inherent in any young and developing field. Another reason is that negative findings about chemicals pose an economic threat to the manufacturers of those chemicals. For instance, Tyrone Hayes's work has met with fierce criticism from scientists associated with atrazine's manufacturer, which stands to lose many millions of dollars if its top-selling herbicide were to be banned or restricted in the United States.

Indeed, our society has invested heavily in some chemicals that now are suspect. One example is bisphenol-A (see "The Science behind the Story," • pp. 394–395). A building block of polycarbonate plastic, bisphenol-A occurs in a wide variety of plastic products we use daily, from drink containers to eating utensils to auto parts to CDs and DVDs. It is used in epoxy resins, including those that coat metal food and drink cans. It is even used in making dental sealants. An estrogen mimic, the chemical leaches out of many of these products into water and food, and recent experimental evidence ties it to birth defects and other abnormalities in lab animals. The plastics industry vehemently protests that the chemical is safe, pointing to other research backing its contention. However, by one count through the end of 2006, 151 of the 178 published studies with lab animals report harm from low doses of bisphenol-A. Almost without exception, the studies reporting harm received public government funding, whereas those reporting no harm were industry-funded.

Research results with bisphenol-A and mice, and those with atrazine and frogs, each have shown effects at extremely low levels of the chemical. This is also the case with research on other known or purported endocrine disruptors. The apparent reason is that the endocrine system is geared to respond to minute concentrations of substances (normally, hormones in the bloodstream). Because the endocrine system responds to minuscule amounts of chemicals, it is especially vulnerable to effects from environmental contaminants that are dispersed and diluted through the environment and that reach our bodies in very low concentrations.

## Toxicants may concentrate in water

Toxicants are not evenly distributed in the environment, and they move about in specific ways (**Figure 14.11**). For instance, water, in the form of runoff, often carries toxicants from large areas of land and concentrates them in small volumes of surface water. The U.S. Geological Survey findings regarding surface waterways (• p. 388)

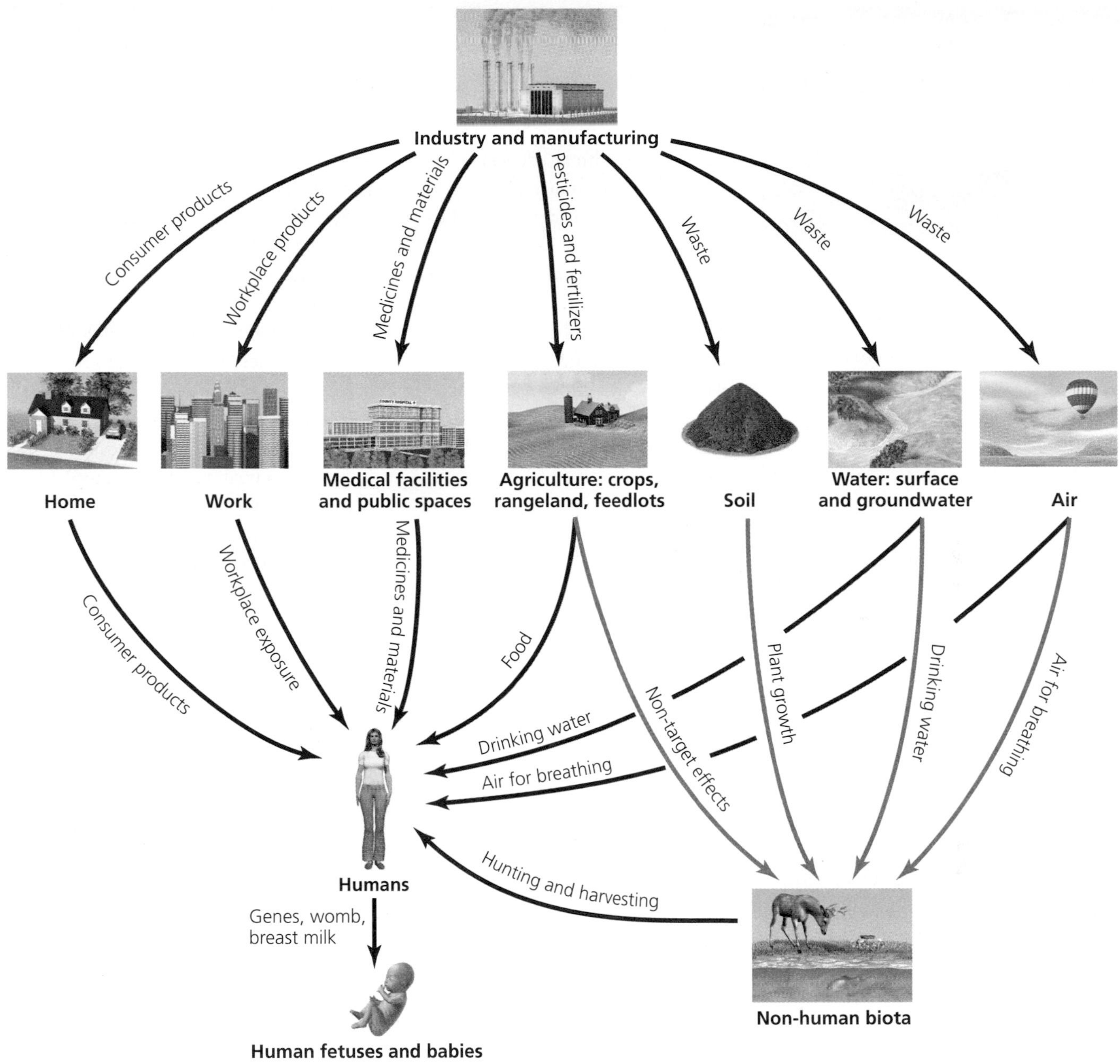


FIGURE 14.11 Synthetic chemicals take many routes in traveling through the environment. Although humans take in only a tiny proportion of these compounds, and although many compounds are harmless, humans—particularly babies—receive small amounts of toxicants from many sources.

demonstrated this concentrating effect. If chemicals persist in soil, they can leach into groundwater and contaminate drinking water supplies.

Many chemicals are soluble in water and enter organisms' tissues through drinking or absorption. For this reason, aquatic animals such as fish, frogs, and stream invertebrates are effective indicators of pollution. When aquatic organisms become sick, we can take it as an early warning that something is amiss. This is why many scientists see findings that show impacts of low concentrations of pesticides on frogs, fish, and invertebrates as a warning that humans could be next. The contaminants that wash into streams and rivers also flow and seep into the water we drink and drift through the air we breathe.

## Airborne toxicants can travel widely

Because many chemical substances can be transported by air (Chapter 17), the toxicological effects of chemical use can occur far from the site of direct chemical use. For instance, airborne transport of pesticides is sometimes termed *pesticide drift.* The Central Valley of California is widely considered the most productive agricultural region in the world. But because it is a naturally arid area,

THE SCIENCE BEHIND THE STORY

## Is Bisphenol-A Safe?

Dr. Patricia Hunt, Case Western Reserve University

Can a compound in everyday plastic products damage the most basic biological processes necessary for healthy pregnancies and births? A scientist in Ohio found evidence that it does—all because one of her lab assistants reached for the wrong soap.

At a laboratory at Case Western Reserve University in 1998, geneticist Patricia Hunt was making a routine check of her female lab mice, which included extracting and examining developing eggs from the ovaries. The results made her wonder what had gone wrong. About 40% of the eggs showed problems with their chromosomes, and 12% had irregular amounts of genetic material, a dangerous condition called aneuploidy, which can lead to miscarriages or birth defects in mice and people alike.

A bit of sleuthing revealed that a lab assistant had mistakenly washed the lab's plastic mouse cages and water bottles with an especially harsh soap. The soap damaged the cages so badly that parts of them seemed to have melted.

The cages were made from polycarbonate plastic, which contains the chemical bisphenol-A (BPA). BPA is used to manufacture thousands of plastic products, from food containers to sports equipment to baby bottles. The plastics industry produces over 2.5 billion kg (5.6 billion lb) of BPA every year—nearly a pound for every person on the planet.

Bisphenol-A, however, is an estrogen mimic and a long-suspected endocrine disruptor. Hunt knew at the time that some studies had linked the chemical to reproductive abnormalities in mice, such as low sperm counts and early sexual development. Other research indicated that BPA can leach out of plastic into water and food when the plastic is treated with extreme heat, acidity, or harsh soap. One such study conducted by researchers for the magazine *Consumer Reports* found that BPA seeped out of the plastic walls of heated baby bottles into infant formula.

The plastics industry insisted that BPA was safe and fought back by funding its own studies. Government health officials, faced with conflicting evidence, had so far declined to strengthen regulations on BPA.

Hunt, however, thought the chemical might be adversely affecting the mice in her lab. Deciding to recreate experimentally the accidental BPA exposure from the cage-washing incident, Hunt instructed researchers in her lab to wash polycarbonate cages and water bottles using varying levels of the harsh soap. They then compared mice kept in damaged cages with plastic water bottles to mice kept in undamaged cages with glass water bottles.

The developing eggs of mice exposed to BPA through the deliberately damaged plastic showed significant problems during meiosis, the division of chromosomes during egg formation—just as they had in the original incident (see the photo). In contrast, the eggs of mice in the control cages were normal.

In an additional round of tests, three sets of female mice were given daily oral doses of BPA over 3, 5, and

food production depends on intensive use of irrigation, fertilizers, and pesticides. Roughly 143 million kg (315 million lb) of pesticide active ingredients are used in California each year, mostly in the Central Valley. The region's frequent winds often blow the airborne spray—and dust particles containing pesticide residue—for long distances. In the mountains of the Sierra Nevada, research has associated pesticide drift from the Central Valley with population declines in four species of frogs. Families living in towns in the Central Valley have suffered health impacts, and activists for farmworkers maintain that hundreds of thousands of the state's residents are at risk.

Synthetic chemical contaminants are ubiquitous worldwide. Despite being manufactured and applied mainly in the temperate and tropical zones, contaminants appear in substantial quantities in the tissues of Arctic polar bears, Antarctic penguins, and people living in Greenland. Scientists can travel to the most remote and seemingly pristine alpine lakes in British Columbia and find them contaminated with foreign toxicants, such as PCBs. The surprisingly high concentrations in polar regions result from patterns of global atmospheric circulation that move airborne chemicals systematically toward the poles (**Figure 14.12**).

### Some toxicants persist for a long time

Once toxic agents arrive somewhere, they may degrade quickly and become harmless, or they may remain unaltered and persist for many months, years, or decades. Large numbers of persistent synthetic chemicals exist in

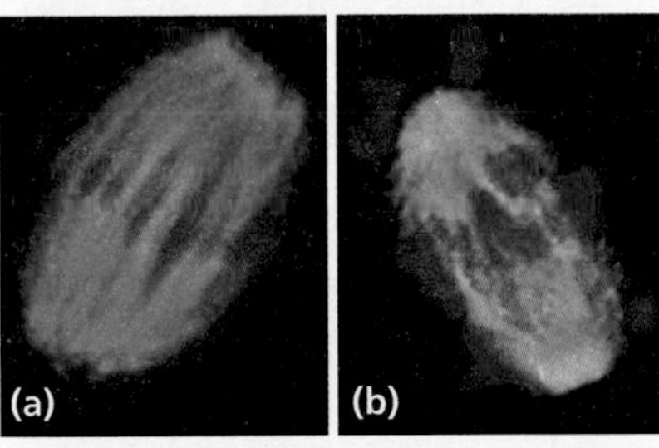


During normal cell division **(a)**, chromosomes align properly. Exposure to bisphenol-A causes abnormal cell division **(b)**, whereby chromosomes scatter and are distributed improperly and unevenly between daughter cells.

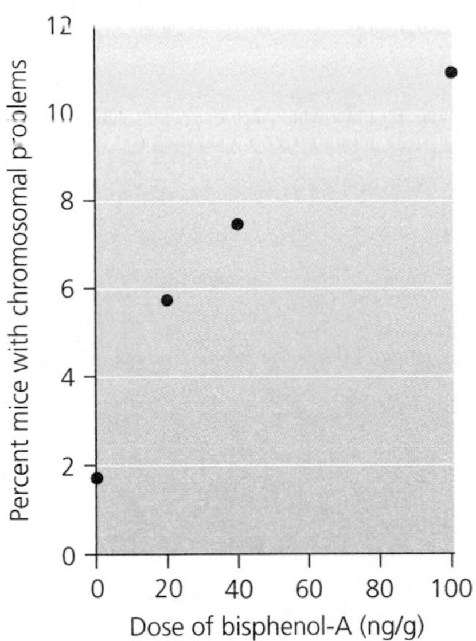


In this dose-response experiment, the percentage of mice showing chromosomal problems during cell division rose with increasing dose of bisphenol-A. In the United States and Europe, regulators have set safe intake levels for people at doses of 50 ng/g of body weight per day. Data from Hunt, P. A., et al. 2003. Bisphenol A exposure causes meiotic aneuploidy in the female mouse. *Current Biology* 13: 546–553.

7 days, and the same meiotic abnormalities were observed, although at lower levels (see the graph). The mice given BPA for 7 days were most severely affected.

Published in 2003 in the journal *Current Biology*, Hunt's findings set off a new wave of concern over the safety of bisphenol-A. Although Hunt's research dealt with mice, not humans, the findings were disturbing because sex cells of mice and people divide and function in similar ways. "We have observed meiotic defects in mice at exposure levels close to or even below those considered 'safe' for humans," the research paper stated. "Clearly, the possibility that BPA exposure increases the likelihood of genetically abnormal offspring is too serious to be dismissed without extensive further study."

Since that time, dozens of other studies have come out, and most have shown harmful effects in lab animals. A diversity of reproductive and other effects has been shown, and findings with obese mice have led some to suggest that bisphenol-A may be contributing to the industrialized world's current obesity epidemic. In 2005, a survey found bisphenol-A in urine samples from 95% of a sample of U.S. adults.

Scientist Frederick vom Saal, whose research in 1997 showed the first evidence for BPA's effects, said in 2007, "This chemical is harming snails, insects, lobsters, fish, frogs, reptiles, birds, and rats, and the chemical industry is telling people that because you're human, unless there's human data, you can feel completely safe."

Regulators in Europe recently lowered the level of BPA intake they consider safe. In the United States, some scientists are pushing for more research and federal scrutiny into the safety of BPA, and governments at various levels are considering new restrictions. In 2007, a billion-dollar class action lawsuit was filed against baby-bottle manufacturers on behalf of California babies who drank from plastic bottles. With a chemical that is so widely present in our lives, the scientific and social debate over bisphenol-A seems set to continue building.

our environment today because we design them to persist. The synthetic chemicals used in plastics, for instance, are used precisely because they resist breakdown.

The rate at which chemicals degrade depends on factors such as temperature, moisture, and sun exposure and on how these factors interact with the chemistry of the toxicant. Toxicants that persist in the environment have the greatest potential to harm many organisms over long periods of time. A major reason people have been so concerned about toxic chemicals such as DDT and PCBs is that they have long persistence times. In contrast, the Bt toxin used in biocontrol and in genetically modified (GM) crops (• p. 270; Figure 10.13, • pp. 274–275) has a very short persistence time. The herbicide atrazine is variable in its persistence, depending on environmental conditions.

Most toxicants eventually degrade into simpler compounds called *breakdown products.* Often these are less harmful than the original substance, but sometimes they are just as toxic as the original chemical, or more so. For instance, DDT breaks down into DDE, a highly persistent and toxic compound in its own right. Atrazine produces a large number of breakdown products whose effects have not been fully studied.

## Toxicants may accumulate and move up the food chain

Of the toxicants that organisms absorb, breathe, or consume, some are quickly excreted, and some are degraded into harmless breakdown products. However, others remain intact in the body. Toxicants that are fat soluble or

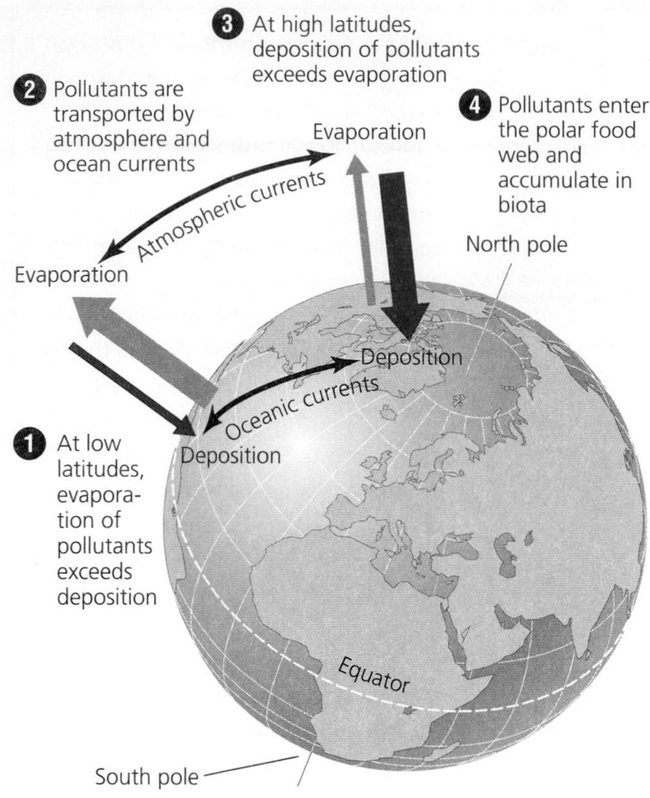


FIGURE 14.12 In a process called *global distillation,* pollutants that evaporate and rise high into the atmosphere at lower latitudes, or are deposited in the ocean, are carried toward the poles by atmospheric currents of air and oceanic currents of water. For this reason, polar organisms take in more than their share of toxicants, despite the fact that relatively few synthetic chemicals are manufactured or used near the poles.

oil soluble (often organic compounds such as DDT and DDE) are absorbed and stored in fatty tissues. Others, such as methyl mercury, may be stored in muscle tissue. Such toxicants may build up in an animal in a process termed **bioaccumulation**.

Toxicants that bioaccumulate in the tissues of one organism may be transferred to other organisms as predators consume prey. When one organism consumes another, it takes in any stored toxicants and stores them itself, along with the toxicants it has received from eating other prey. Thus with each step up the food chain, from producer to primary consumer to secondary consumer and so on, concentrations of toxicants can be greatly magnified. This process, called **biomagnification**, occurred most famously with DDT. Top predators, such as birds of prey, ended up with high concentrations of the pesticide because concentrations were magnified as DDT moved from water to algae to plankton to small fish to bigger fish and finally to fish-eating birds (**Figure 14.13**).

Biomagnification caused populations of many North American birds of prey to decline precipitously from the 1950s to the 1970s. The peregrine falcon was almost totally wiped out in the eastern United States, and the bald eagle, the U.S. national bird, was virtually eliminated from the lower 48 states. The brown pelican vanished from its Atlantic Coast range, remaining only in Florida, and the osprey and other hawks saw substantial population declines. Eventually scientists determined that DDT was causing these birds' eggshells to grow thinner, so that eggs were breaking while in the nest.

All these birds' populations have rebounded since the United States banned DDT, but such scenarios are by no means a thing of the past. The polar bears of Svalbard Island in Arctic Norway are at the top of the food chain and feed on seals that have biomagnified toxicants. Despite their remote Arctic location, Svalbard Island's polar bears show some of the highest levels of PCB contamination of any wild animals tested, as a result of biomagnification and the process of global distillation shown in Figure 14.12. The contaminants are likely responsible

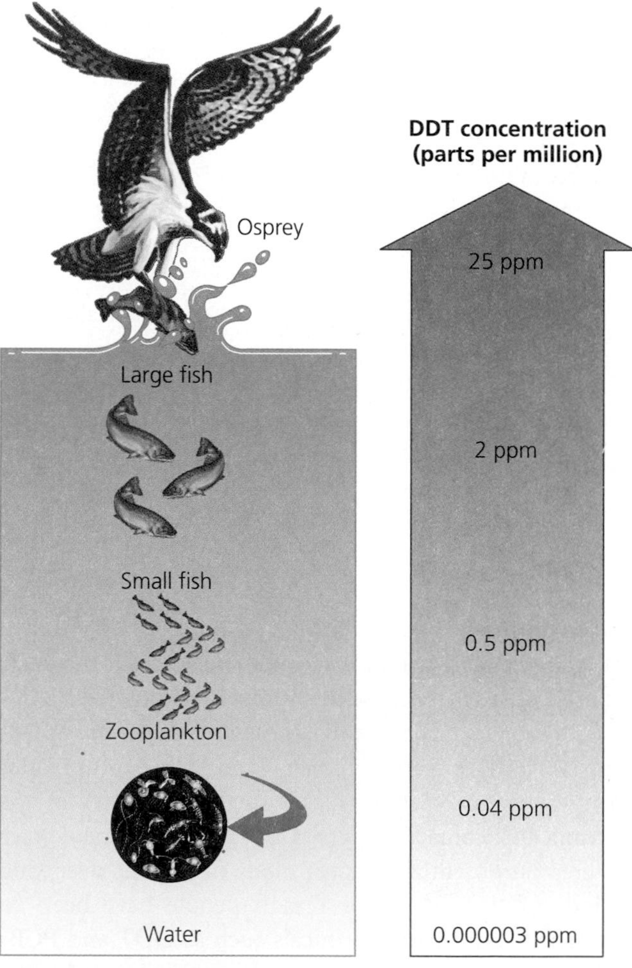


FIGURE 14.13 Fat-soluble compounds such as DDT bioaccumulate in the tissues of organisms. As animals at higher trophic levels eat organisms lower on the food chain, their load of toxicants passes up to each consumer. DDT moves from zooplankton through various types of fish, finally becoming highly concentrated in fish-eating birds such as ospreys.

for the immune suppression, hormone disruption, and high cub mortality that the bears seem to be suffering. Cubs that survive receive PCBs in their mothers' milk, so that contamination persists and accumulates over generations.

### Not all toxicants are synthetic

Although we have focused on synthetic chemicals thus far, chemical toxicants also exist naturally in the environment around us and in the foods we eat. We have good reason as citizens and consumers to insist on being informed about risks synthetic chemicals may pose, but it is a mistake to assume that all artificial chemicals are unhealthy and that all natural chemicals are healthy. In fact, the plants and animals we eat contain many chemicals that can cause us harm. Recall that plants produce toxins to ward off animals that eat them. In domesticating crop plants, we have selected for strains with reduced toxin content, but we have not eliminated these dangers. Furthermore, when we consume animal meat, we take in toxins the animals have ingested from plants or animals they have eaten.

Scientists are actively debating just how much risk natural toxicants pose. Biochemist Bruce Ames of the University of California at Berkeley maintains that the amounts of synthetic chemicals in our food from pesticide residues are dwarfed by the quantities of natural toxicants. Ames also holds that the natural defenses against toxins that our bodies have evolved are largely effective against synthetic toxicants. Therefore, he reasons, synthetic chemicals are a relatively minor worry. Moreover, he contends, if fear of pesticide residues or increased cost to remove them causes people to eat fewer fruits and vegetables, then people will be at greater risk for disease, because a balanced diet that includes fruits and vegetables is important for health.

A respected senior scientist, Ames was a hero of environmentalists for his early work. The *Ames test*, a bacterial assay for mutagens, allows easy screening of suspected toxins. Today Ames is a target of criticism from many environmental advocates. His critics say that natural toxicants are usually more readily metabolized and excreted by the body than synthetic ones, that synthetic toxicants persist and accumulate in the environment, and that synthetic chemicals expose people (such as farmworkers and factory workers) to risks in ways other than the ingestion of food. What is clear is that more research is required in this area.

## Studying Effects of Hazards

Determining health effects of particular environmental hazards is a challenging job, especially because any given person or organism has a complex history of exposure to many hazards throughout life. Scientists rely on several different methods, including correlative surveys and manipulative experiments (• pp. 13–14).

### Wildlife studies use careful observations in the field and lab

When scientists were zeroing in on the impacts of DDT, one key piece of evidence came from museum collections of wild birds' eggs from the decades before synthetic pesticides were manufactured. Eggs from museum collections had measurably thicker shells than the eggs scientists were studying in the field from present-day birds. Scientists have pieced together the puzzle of toxicant effects on alligators by taking measurements from animals in the wild, then doing controlled experiments in the lab to test hypotheses. With frogs and atrazine, scientists first measured toxicological effects in lab experiments, then sought to demonstrate correlations with herbicide use in the wild.

Often the study of wildlife advances in the wake of some conspicuous mortality event. Off the California coast in 1998–2001, populations of sea otters fell noticeably, and many dead otters washed ashore. Field biologists documented the population decline, and specialists went to work in the lab performing autopsies to determine causes of death. The most common cause of death was found to be infection with the protist parasite *Toxoplasma*, which killed otters directly and also made them vulnerable to shark attack. *Toxoplasma* occurs in the feces of cats, so scientists hypothesized that sewage runoff containing waste from litter boxes was entering the ocean from urban areas and infecting the otters.

### Human studies rely on case histories, epidemiology, and animal testing

In studies of human health, we gain much knowledge by studying sickened individuals directly. Medical professionals have long treated victims of poisonings, so the effects of common poisons are well known. Autopsies help us understand what constitutes a lethal dose. This process of observation and analysis of individual patients is known as a *case history* approach. Case histories have advanced our understanding of human illness, but they do not always help us infer the effects of rare hazards, newly manufactured compounds, or chemicals that exist at low environmental concentrations and exert minor long-term effects. Case histories also tell us little about probability and risk, such as how many extra deaths we might expect in a population due to a particular cause.

For such situations, which are common in environmental toxicology, epidemiological studies are necessary.

**Epidemiological studies** involve large-scale comparisons among groups of people, usually contrasting a group known to have been exposed to some hazard and a group that has not. Epidemiologists track the fate of all people in the study, generally for a long period of time (often years or decades) and measure the rate at which deaths, cancers, or other health problems occur in each group. The epidemiologist then analyzes the data, looking for observable differences between the groups, and statistically tests hypotheses accounting for differences. When a group exposed to a hazard shows a significantly greater degree of harm, it suggests that the hazard may be responsible. This process is akin to a natural experiment (• pp. 13–14), in which the experimenter takes advantage of the presence of groups of subjects made possible by some event that has already occurred. A slightly different type of natural experiment was conducted by anthropologist Elizabeth Guillette (see "The Science behind the Story," • pp. 400–401).

The advantages of epidemiological studies are their realism and their ability to yield relatively accurate predictions about risk. The drawbacks include the need to wait a long time for results and the inability to address future effects of new products just coming to market. In addition, participants in epidemiological studies encounter many factors that affect their health besides the one under study. Epidemiological studies measure a statistical association between a health hazard and an effect, but they do not confirm that the hazard *causes* the effect.

Manipulative experiments are needed to establish causation. However, subjecting people to massive doses of toxicants in a lab experiment would clearly be unethical. So researchers have traditionally used other animals as subjects to test toxicity. Foremost among these animal models have been laboratory strains of rats, mice, and other mammals. Because of our shared evolutionary history, the bodies of other mammals function similarly to ours. The extent to which results from animal lab tests apply to humans is always somewhat uncertain, and it can be expected to vary from one study to the next.

Some people feel the use of rats and mice for testing is unethical, but animal testing enables scientific and medical advances that would be impossible or far more difficult otherwise. However, new techniques (with human cell cultures, bacteria, or tissue from chicken eggs) are being devised that may one day replace some live-animal testing.

## Dose-response analysis is a mainstay of toxicology

The standard method of testing with lab animals in toxicology is dose-response analysis. Scientists quantify the toxicity of a given substance by measuring how much effect a toxicant produces at different doses or how many animals are affected by different doses of the toxic agent. The *dose* is the amount of toxicant the test animal receives, and the *response* is the type or magnitude of negative effects the animal exhibits as a result. The response is generally quantified by measuring the proportion of animals exhibiting negative effects. The data are plotted on a graph, with dose on the *x* axis and response on the *y* axis (**Figure 14.14a**). The resulting curve is called a **dose-response curve**.

Once they have plotted a dose-response curve, toxicologists can calculate a convenient shorthand gauge of a substance's toxicity: the amount of toxicant it takes to kill half the population of study animals used. This lethal

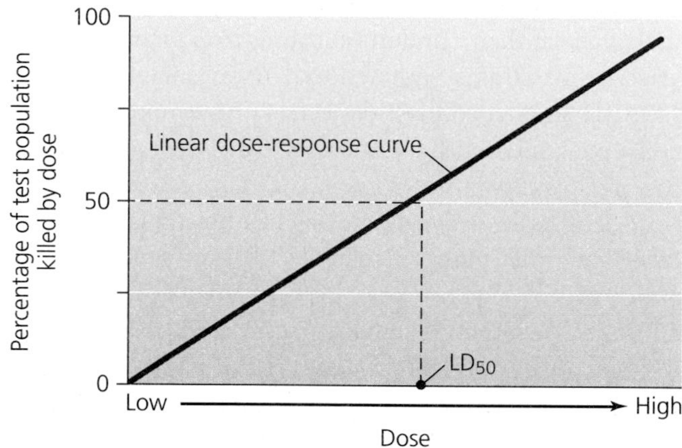


**(a) Linear dose-response curve**

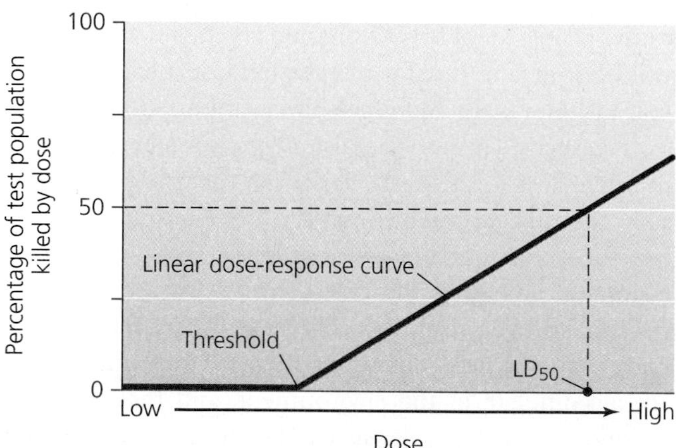


**(b) Dose-response curve with threshold**

FIGURE 14.14 In a classic linear dose-response curve (**a**), the percentage of animals killed or otherwise affected by a substance rises with the dose. The point at which 50% of the animals are killed is labeled the lethal-dose-50, or $LD_{50}$. For some toxic agents, a threshold dose (**b**) exists, below which doses have no measurable effect. Go to **GRAPHit!** at www.aw-bc.com/withgott or on the student CD-ROM.

dose for 50% of individuals is termed the **$LD_{50}$**. A high $LD_{50}$ indicates low toxicity, and a low $LD_{50}$ indicates high toxicity.

If the experimenter is instead interested in nonlethal health effects, he or she may want to document the level of toxicant at which 50% of a population of test animals is affected in some other way (for instance, what level of toxicant causes 50% of lab mice to lose their hair?). Such a level is called the effective-dose-50%, or **$ED_{50}$**.

Sometimes responses occur only above a certain dose. Such a **threshold** dose (**Figure 14.14b**) might be expected if the body's organs can fully metabolize or excrete a toxicant at low doses but become overwhelmed at higher concentrations. It might also occur if cells can repair damage to their DNA only up to a certain point.

Sometimes responses *decrease* with increased dose. Toxicologists are finding that some dose-response curves are U-shaped, J-shaped, or shaped like an inverted U. Such counterintuitive curves often occur with endocrine disruptors, likely because the hormone system is geared to function with extremely low concentrations of hormones and so is vulnerable to disruption by toxicants at extremely low concentrations. Inverted dose-response curves present a challenge for policymakers attempting to set safe environmental levels for toxicants. We may have underestimated the dangers of compounds that behave in these ways, because many such chemicals exist in very low concentrations over wide areas.

Knowing the shape of dose-response curves is crucial if one is planning to extrapolate from them to predict responses at doses below those that have been tested. Scientists generally give lab animals much higher doses relative to body mass than humans would receive in the environment. This is so that the response is great enough to be measured, and so that differences between the effects of small and large doses are evident. Data from a range of doses help give shape to the dose-response curve. Once the data from animal tests are plotted, scientists generally extrapolate downward to estimate the effect of still-lower doses on a hypothetically large population of animals. This way, they can come up with an estimate of, say, what dose causes cancer in 1 mouse in 1 million. A second extrapolation is then required to estimate the effect on humans, with our greater body mass. Because these two extrapolations go beyond the actual data obtained, they introduce uncertainty into the interpretation of what doses are acceptable for humans. As a result, to be on the safe side, regulatory agencies set standards for maximum allowable levels of toxicants that are well below the minimum toxicity levels estimated from lab studies.

## Individuals vary in their responses to hazards

Different individuals may respond quite differently to identical exposures to hazards. These differences can be genetically based or can be due to a person's current condition. People in poorer health are often more sensitive to biological and chemical hazards. Sensitivity also can vary with sex, age, and weight. Because of their smaller size and rapidly developing organ systems, fetuses, infants, and young children tend to be much more sensitive to toxicants than are adults. Regulatory agencies such as the U.S. Environmental Protection Agency (EPA) traditionally set standards for adults and extrapolated downward for infants and children. However, they have subsequently found that in many cases their linear extrapolations did not lower standards enough to protect babies adequately. Many critics today contend that despite improvements, regulatory agencies still do not account explicitly enough for risks to fetuses, infants, and children.

## The type of exposure can affect the response

The risk posed by a hazard often varies according to whether a person experiences high exposure for short periods of time, known as **acute exposure**, or lower exposure over long periods of time, known as **chronic exposure**. Incidences of acute exposure are easier to recognize, because they often stem from discrete events, such as accidental ingestion, an oil spill, a chemical spill, or a nuclear accident. Lab tests and $LD_{50}$ values generally reflect acute toxicity effects. However, chronic exposure is more common—and more difficult to detect and diagnose. Chronic exposure often affects organs gradually, as when smoking causes lung cancer, or when alcohol abuse induces liver or kidney damage. Pesticide residues on food or low levels of arsenic in drinking water (• pp. 434–435) also pose chronic risk. Because of the long time periods involved, relationships between cause and effect may not be readily apparent.

## Mixes may be more than the sum of their parts

It is difficult enough to determine the impact of a single hazard on an organism, but the task becomes astronomically more difficult when multiple hazards interact. For instance, chemical substances, when mixed, may act in concert in ways that cannot be predicted from the effects

THE SCIENCE BEHIND THE STORY

## Pesticides and Child Development in Mexico's Yaqui Valley

*A Yaqui mother with her child*

With spindly arms and big, round eyes, one set of pictures shows the sorts of stick figures drawn by young children everywhere. Next to them is another group of drawings, mostly disconnected squiggles and lines, resembling nothing. Both sets of pictures are intended to depict people. The main difference identified between the two groups of young artists: long-term pesticide exposure.

Children's drawings are not a typical tool of toxicology, but Elizabeth Guillette, an anthropologist married to Louis Guillette, wanted to try new methods. Guillette was interested in the effects of pesticides on children. She devised tests to measure childhood development based on techniques from anthropology and medicine. Searching for a study site, Guillette found the Yaqui Valley region of northwestern Mexico.

The Yaqui Valley is farming country, worked for generations by the indigenous group that gives the region its name. Synthetic pesticides arrived in the area in the 1940s. Some Yaqui embraced the agricultural innovations, spraying their farms in the valley to increase their yields. Yaqui farmers in the surrounding foothills, however, generally chose to bypass the chemicals and to continue following more traditional farming practices. Although differing in farming techniques, Yaqui in the valley and foothills continued to share the same culture, diet, education system, income levels, and family structure.

At the time of the study, in 1994, valley farmers planted crops twice a year, applying pesticides up to 45 times from planting to harvest. A previous study conducted in the valley in 1990, focusing on areas with the largest farms, had indicated high levels of multiple pesticides in the breast milk of mothers and in the umbilical cord blood of newborn babies. In contrast, foothill families avoided chemical pesticides in their gardens and homes.

To understand how pesticide exposure affects childhood development, Guillette and fellow researchers studied 50 preschoolers aged 4 to 5, of whom 33 were from the valley and 17 from the foothills. Each child underwent a half-hour exam, during which researchers showed a red balloon, promising to give the balloon later as a gift, and using the promise to evaluate long-term memory. Each child was then put through a series of physical and mental tests:

- Catching a ball from distances of up to 3 m (10 ft) away, to test overall coordination
- Jumping in place for as long as possible, to assess endurance
- Drawing a picture of a person, as a measure of perception
- Repeating a short string of numbers, to test short-term memory
- Dropping raisins into a bottle cap from a height of about 13 cm (5 in.), to gauge fine-motor skills

The researchers also measured each child's height and weight but, because of lack of time and money, stopped short of taking blood or tissue samples to check for pesticides or other toxins. When all tests were completed, each child was asked what he or she had been promised and received a red balloon.

Although the two groups of children were not significantly different in height and weight, they differed markedly in other areas of development. Valley children were far behind the foothill children developmentally in coordination, physical endurance, long-term memory, and fine motor skills:

- From a distance of 3 m (10 ft), valley children had great difficulty catching the ball.
- Valley children could jump for an average of 52 seconds, compared to 88 seconds for foothill children.

of each in isolation. Mixed toxicants may sum each other's effects, cancel out each other's effects, or multiply each other's effects. Whole new types of impacts may arise when toxicants are mixed together. Such interactive impacts—those that are more than or different from the simple sum of their constituent effects—are called **synergistic effects**.

With Florida's alligators, lab experiments have indicated that DDE can either help cause or inhibit sex reversal, depending on the presence of other chemicals. Mice exposed to a mixture of nitrate, atrazine, and aldicarb have been found to show immune, hormone, and nervous system effects that were not evident from exposure to each of these chemicals alone. Wood frogs

**Drawings by children in the foothills**

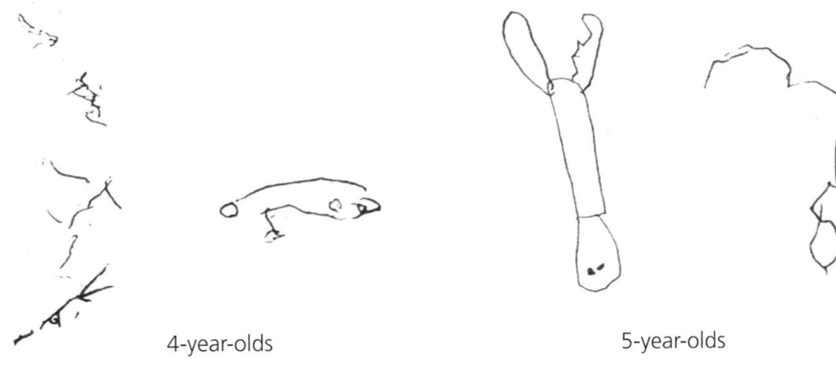


**Drawings by children in the valley**

Elizabeth Guillette's study in Mexico's Yaqui Valley offers a startling example of apparent neurological effects of pesticide poisoning. Young children from foothills areas where pesticides were not commonly used drew recognizable figures of people. Children the same age from valley areas where pesticides were used heavily in industrialized agriculture could draw only scribbles. Adapted from Guillette, E. A., et al. 1998. *Environmental Health Perspectives* 106: 347–353.

- Most valley children missed the bottle cap when dropping their raisins, whereas foothill children dropped them into the caps far more often.
- Each group did fairly well repeating numbers, but valley children showed poor long-term memory. At the end of the test, all but one of the foothill children remembered that they had been promised a balloon, and 59% remembered it was red. However, of the valley children only 27% remembered the color of the balloon, only 55% remembered they'd be getting a balloon, and 18% were unable to remember anything about the balloon.

It was the children's drawings, however, that exhibited the most dramatic difference between valley and foothill children (see the figure). The researchers determined each drawing could earn 5 points, with 1 point each for a recognizable feature: head, body, arms, legs, and facial features. The foothill children drew pictures that looked like people, averaging about 4.5 points per drawing. The valley children, in contrast, averaged 1.6 points per drawing; their scribbles resembled little that looked like a person. By the standards of developmental medicine, the 4- and 5-year-old valley children drew at the level of a 2-year-old.

Some scientists greeted Guillette's study skeptically, pointing out that its sample size was too small to be meaningful. Others said that factors the researchers missed, such as different parenting styles or unknown health problems, could be to blame. Prominent toxicologists argued that without blood or tissue tests on the children, the study results couldn't be tied to agricultural chemicals. Regardless of these criticisms, Guillette maintains that her findings show that nontraditional study methods are a valid way to track the effects of environmental toxins and that pesticides present a complex long-term risk to human growth and health.

in the wild are increasingly suffering limb deformities, apparently the result of being parasitized by trematode flatworms. Being near an agricultural field with pesticide runoff increases the rate of parasitic infection, because, as lab studies have shown, pesticides suppress the frog's immune response, making it more vulnerable to parasites.

Traditionally, environmental health has tackled effects of single hazards one at a time. In toxicology, the complex experimental designs required to test interactions, and the sheer number of chemical combinations, have meant that single-substance tests have received priority. This approach is changing, but scientists in environmental health and toxicology will never be able to test all possible combinations. There are simply too many hazards in the environment.

## Risk Assessment and Risk Management

Policy decisions on whether to ban chemicals or restrict their use generally follow years of rigorous testing for toxicity. Likewise, strategies for combating disease and other health threats are often based on extensive research. Policy and management decisions reach beyond the scientific results on health to incorporate considerations about economics and ethics. And all too often, they are influenced by political pressure from powerful interests. The steps between the collection and interpretation of scientific data and the formulation of policy involve assessing and managing risk.

### Risk is expressed in terms of probability

Exposure to an environmental health threat does not invariably produce some given effect. Rather, it causes some probability of harm, some statistical chance that damage will result. To understand the impact of a health threat, a scientist must know more than just its identity and strength. He or she must also know the chance that an organism will encounter it, the frequency at which the organism may encounter it, the amount of substance or degree of threat to which the organism is exposed, and the organism's sensitivity to the threat. Such factors help determine the overall risk posed. Risk can be measured in terms of *probability*, a quantitative description of the likelihood of a certain outcome. The probability that some harmful outcome (for instance, injury, death, environmental damage, or economic loss) will result from a given action, event, or substance expresses the **risk** posed by that phenomenon.

### Our perception of risk may not match reality

Every action we take and every decision we make involves some element of risk, some (generally small) probability that things will go wrong. We try in everyday life to behave in ways that minimize risk, but our perceptions of risk do not always match statistical reality (**Figure 14.15**). People often worry unduly about negligibly small risks but happily engage in other activities that pose high risks. For instance, most people perceive flying in an airplane as a riskier activity than driving a car, but driving a car is statistically far more dangerous.

Psychologists agree that this difference between risk perception and reality stems from the fact that we feel more at risk when we are not controlling a situation and more safe when we are "at the wheel"—regardless of the actual risk involved. When we drive a car, we feel we are in control, even though statistics show we are at greater risk

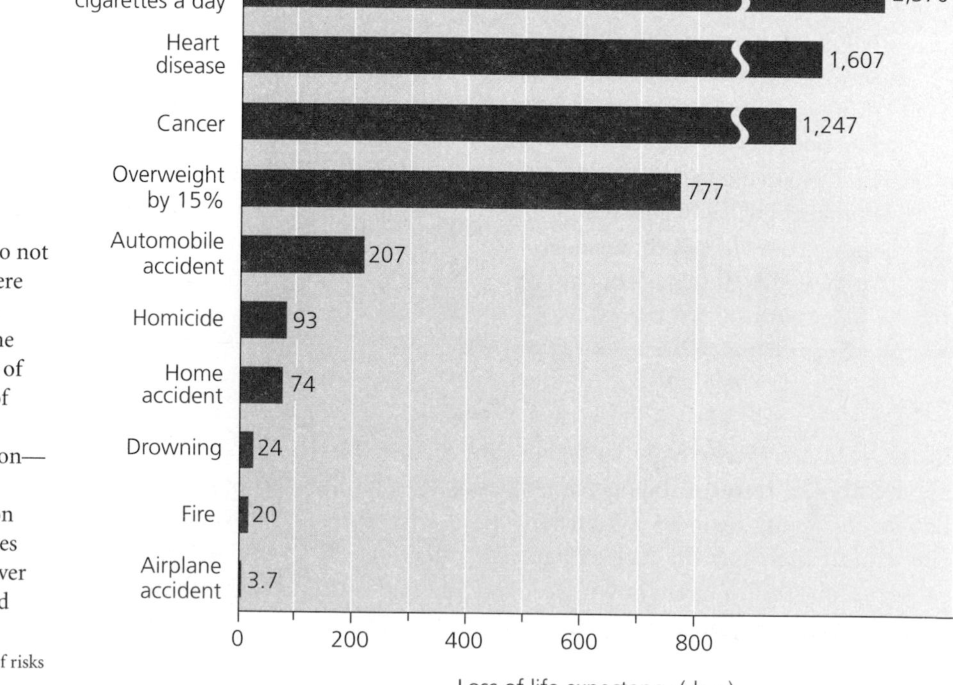


FIGURE 14.15 Our perceptions of risk do not always match the reality of risk. Listed here are several leading causes of death in the United States, along with a measure of the risk each poses. Risk is measured in days of lost life expectancy, that is, the number of days of life lost by people suffering the hazard, spread across the entire population—a measure commonly used by insurance companies. By this measure, one common source of anxiety, airplane accidents, poses 20 times less risk than home accidents, over 50 times less risk than auto accidents, and over 200 times less risk than being overweight. Data from Cohen, B. 1991. Catalog of risks extended and updated. *Health Physics* 61: 317–335.

than as a passenger in an airplane. This psychology can account for people's great fear of nuclear power, toxic waste, and pesticide residues on foods—environmental hazards that are invisible or little understood and whose presence in their lives is largely outside their personal control. In contrast, people are more ready to accept and ignore the risks of smoking cigarettes, overeating, and not exercising, all voluntary activities statistically shown to pose far greater risks to health.

## Risk assessment analyzes risk quantitatively

The quantitative measurement of risk and the comparison of risks involved in different activities or substances together are termed **risk assessment**. Risk assessment is a way of identifying and outlining problems. In environmental health, it helps ascertain which substances and activities pose health threats to people or wildlife and which are largely safe.

Assessing risk for a chemical substance involves several steps. The first steps involve the scientific study of toxicity outlined above—determining whether a given substance has toxic effects and, through dose-response analysis, measuring how effects on an organism vary with the degree of toxicant exposure. Subsequent steps involve assessing the individual's or population's likely extent of exposure to the substance, including the frequency of contact, the concentrations likely encountered, and the length of time the substance is expected to be encountered. Risk assessment studies are often performed by scientists associated with the industries that manufacture toxicants, which in many people's minds can undermine the objectivity of the process.

## Risk management combines science and other social factors

Accurate risk assessment is a vital step toward effective **risk management**, which consists of decisions and strategies to minimize risk (**Figure 14.16**). In most developed nations, risk management is handled largely by federal agencies, such as the EPA, the Centers for Disease Control and Prevention (CDC), and the Food and Drug Administration (FDA) in the United States. In risk management, scientific assessments of risk are considered in light of economic, social, and political needs and values. The costs and benefits of addressing risk in various ways are assessed with regard to both scientific and nonscientific concerns. Decisions whether to reduce or eliminate risk are then made.

In environmental health and toxicology, comparing costs and benefits (• pp. 40–41) can be difficult because the benefits are often economic whereas the costs often pertain to health. Moreover, economic benefits are generally known, easily quantified, and of a discrete and stable amount, whereas health risks are hard-to-measure probabilities, often involving a very small percentage of people

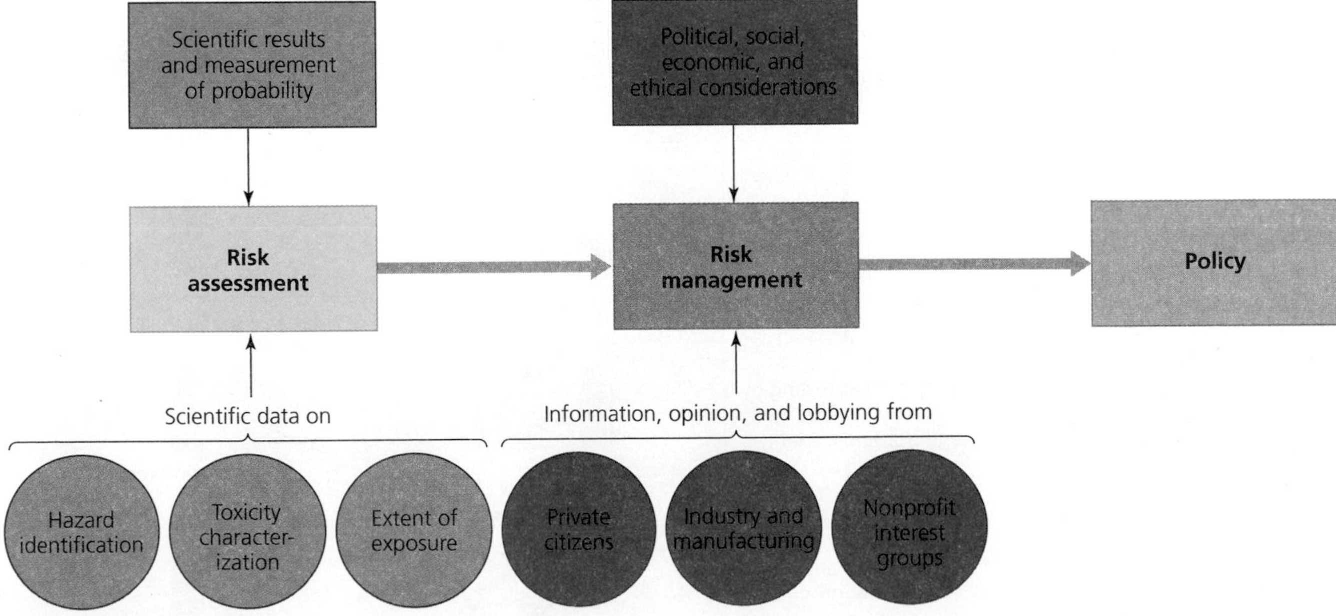


FIGURE 14.16 The first step in addressing the risk of an environmental hazard is risk assessment, a process of quantifying the risk of the hazard and comparing it to other risks. Once science identifies and measures risks, then risk management can proceed. In this process, economic, political, social, and ethical issues are considered in light of the scientific data from risk assessment. The consideration of all these types of information is designed to result in policy decisions that minimize the risk of the environmental hazard.

likely to suffer greatly and a large majority likely to experience little effect. When a government agency bans a pesticide, it may mean measurable economic loss for the manufacturer and the farmer, whereas the benefits accrue less predictably over the long term to some percentage of factory workers, farmers, and the general public. Because of the lack of equivalence in the way costs and benefits are measured, risk management frequently tends to stir up debate.

# Philosophical and Policy Approaches

Because we cannot know a substance's toxicity until we measure and test it, and because there are so many untested chemicals and combinations, science will never eliminate the many uncertainties that accompany risk assessment. In such a world of uncertainty, there are two basic philosophical approaches to categorizing substances as safe or dangerous (**Figure 14.17**).

## Two approaches exist for determining safety

One approach is to assume that substances are harmless until shown to be harmful. We might nickname this the *innocent-until-proven-guilty approach.* Because thoroughly testing every existing substance (and combination of substances) for its effects is a hopelessly long, complicated, and expensive pursuit, the innocent-until-proven-guilty approach has the benefit of not slowing down technological innovation and economic advancement. However, it has the disadvantage of putting into wide use some substances that may later turn out to be dangerous.

The other approach is to assume that substances are harmful until they are shown to be harmless. This approach follows the precautionary principle (• p. 277). This more cautious approach should enable us to identify troublesome toxicants before they are released into the environment, but it may also significantly impede the pace of technological and economic advance.

These two approaches are actually two ends of a continuum of possible approaches. The two endpoints differ

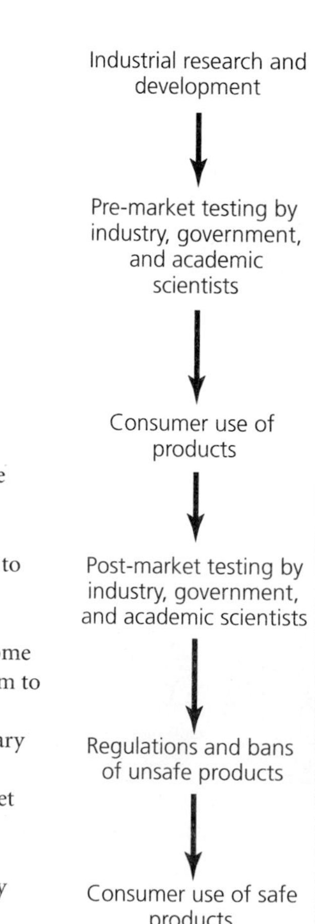


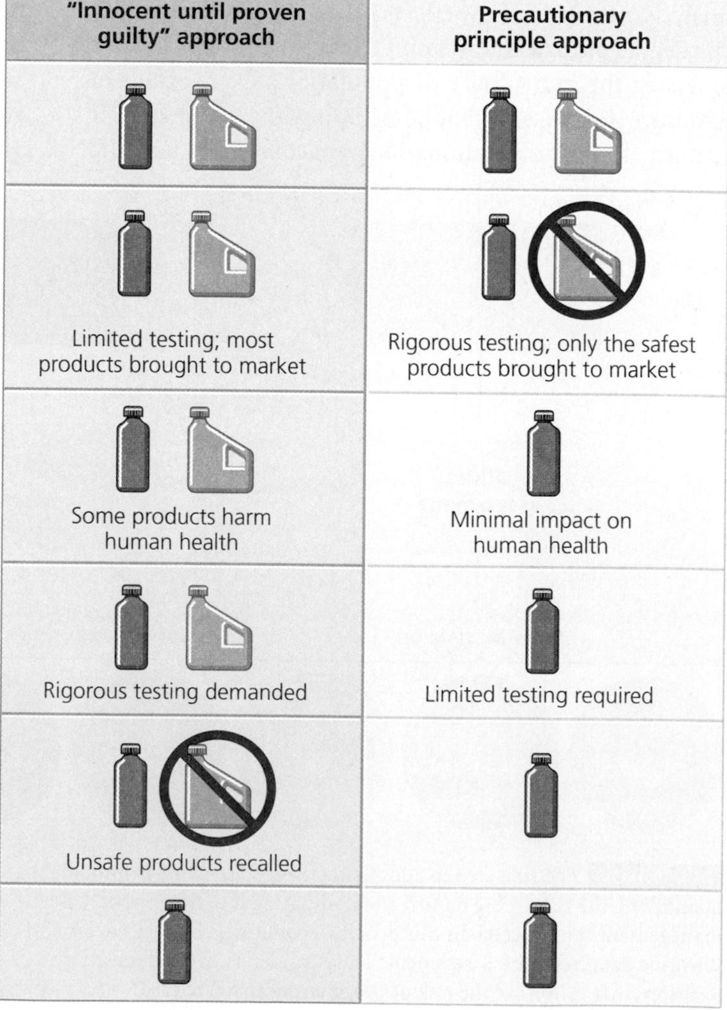


**FIGURE 14.17** Two main approaches can be taken to introduce new substances to the market. In one approach, substances are innocent until proven guilty; they are brought to market relatively quickly after limited testing. Products reach consumers more quickly, but some fraction of them may cause harm to some fraction of people. In the other approach, the precautionary principle is adopted, and substances are brought to market cautiously, only after extensive testing. Products that reach the market should be safe, but many perfectly safe products will be delayed in reaching consumers.

mainly in where they lay the burden of proof—specifically, whether product manufacturers are required to prove safety or whether government, scientists, or citizens are required to prove danger.

**Weighing THE Issues** | **The Precautionary Principle**

Industry's critics say chemical manufacturers should be made to bear the burden of proof for the safety of their products before they hit the market. Industry's supporters say that mandating more safety research will hamper the introduction of products that consumers want, increase the price of products as research costs are passed on to consumers, and cause companies to move to nations where standards are more lax. What do you think? Should government follow the precautionary principle and require proof of safety prior to a chemical's introduction into the market?

## Philosophical approaches are reflected in policy

Because of the health and environmental consequences of hazardous substances, governments of developed nations have viewed regulatory oversight over manufactured substances as one solution to environmental health threats (see "Causes and Consequences," • p. 406). One's philosophical approach has implications for policy, affecting what materials are allowed into our environment. Most nations follow a blend of the two approaches, but there is marked variation among countries. At the present time, European nations are embarking on a new policy course that largely incorporates the precautionary principle regarding the regulation of synthetic chemicals, whereas the United States is not. Although industry frequently complains that government regulation is cumbersome, environmental and consumer advocates criticize U.S. policies for largely following the innocent-until-proven-guilty approach. For instance, compounds involved in cosmetics require no FDA review or approval before being sold to the public.

In the United States, several federal agencies apportion responsibility for tracking and regulating synthetic chemicals. The FDA, under the Food, Drug, and Cosmetic Act of 1938 and its subsequent amendments, regulates foods and food additives, cosmetics, drugs, and medical devices. The EPA regulates pesticides under the Federal Insecticide, Fungicide, and Rodenticide Act of 1947 (FIFRA) and its amendments. The Occupational Safety and Health Administration (OSHA) regulates workplace hazards under a 1970 act. Several other agencies regulate other substances. Synthetic chemicals not covered by other laws are regulated by the EPA under the 1976 Toxic Substances Control Act (TSCA).

## The EPA regulates pesticides and other substances

FIFRA was enacted as the post–World War II U.S. chemical industry was expanding, but before the environmental activism of the 1960s and 1970s gave rise to many environmental laws. As such, FIFRA was primarily intended not to protect public health or the environment, but to assure consumers that products actually worked as their manufacturers claimed. Subsequent amendments shifted the focus somewhat toward protecting health and safety and charged the EPA with "registering" each new pesticide that manufacturers propose to bring to market.

The registration process involves risk assessment and risk management. The EPA first asks the pesticide manufacturer to provide information, including the results of safety assessments the company has performed according to EPA guidelines. The EPA examines the company's research and all other relevant scientific research. It examines the product's ingredients and how the product will be used and evaluates whether the chemical poses risks to humans, other organisms, or water or air quality. The EPA then approves, denies, or sets limits on the chemical's sale and use. It also must approve the language used on the product's label.

Because the registration process takes economic considerations into account, critics say it allows hazardous chemicals to be approved if the economic benefits outweigh the hazards. Here the challenges of weighing intangible risks involving human health and environmental quality against the tangible and quantitative numbers of economics become apparent.

TSCA directed the EPA to monitor some 75,000 industrial chemicals manufactured in or imported into the United States, including PCBs and other compounds involved in plastics. The act gave the agency power to regulate these chemicals and ban them if they are found to pose excessive risk.

However, many public health and environmental advocates view TSCA as being far too weak. They note that the screening required of industry is minimal and that to mandate more extensive and meaningful testing, the EPA must show proof of the chemical's toxicity. In other words, the agency is trapped in a Catch-22: To push for studies looking for toxicity, it must have proof of toxicity already. The result, these advocates say, is that most synthetic chemicals are not thoroughly tested before being put on the market. Of those that fall under TSCA, only 10% have been thoroughly tested for toxicity; only 2% have been screened for carcinogenicity, mutagenicity, or teratogenicity; fewer than 1% are government regulated; and almost none have been tested for endocrine, nervous, or immune system damage, according to the U.S. National Academy of Sciences.

## CAUSES AND CONSEQUENCES

In the wake of industrialization, our society has manufactured many thousands of products, materials, and chemical substances. These have helped bring us higher standards of living and more comfortable lives, but they have also introduced many new **toxic substances in the environment**. There is plenty we can do, however, to minimize concentrations of artificial toxic substances in our environment and to limit our exposure to them.

Write in two causes of toxic substances in the environment in the spaces provided. Then write in two consequences (impacts on the environment, human health, or quality of life) that result. Finally, offer two solutions to this issue and its consequences. One cause, one consequence, and one solution have been filled in for you, providing examples.

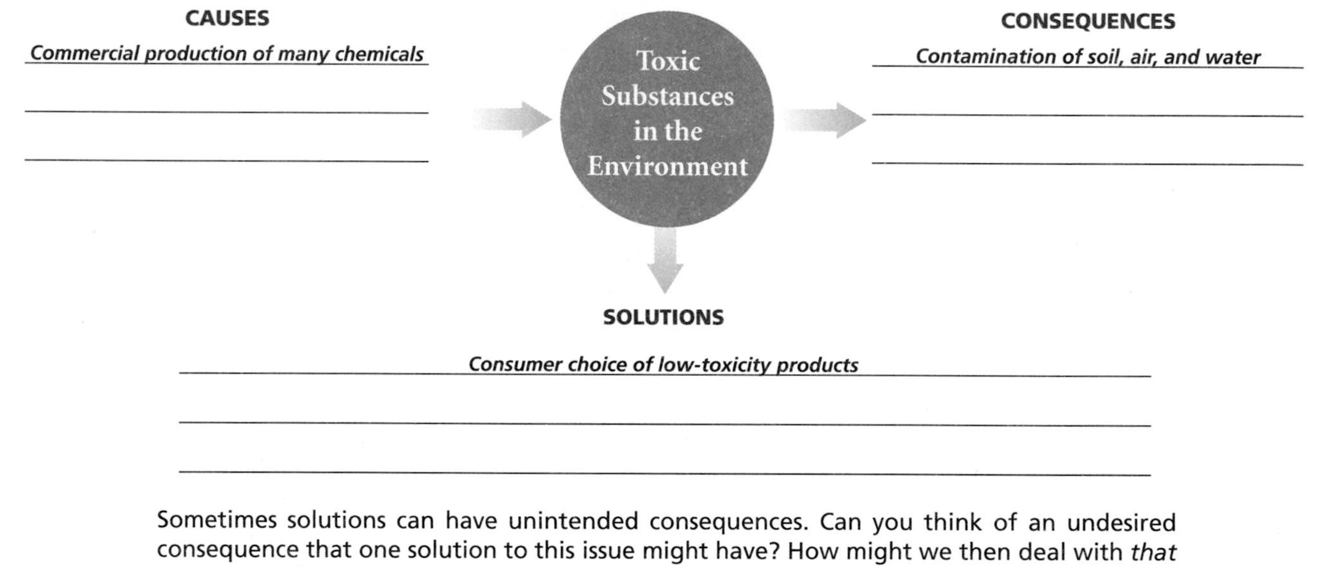


Sometimes solutions can have unintended consequences. Can you think of an undesired consequence that one solution to this issue might have? How might we then deal with *that* consequence?

## Toxicants are regulated internationally

Nations have sought to address chemical pollution with international treaties. For instance, the *Stockholm Convention on Persistent Organic Pollutants (POPs)* came into force in 2004 and has been ratified by roughly 140 nations. POPs are toxic chemicals that persist in the environment, bioaccumulate in the food chain, and often can travel long distances. The PCBs and other contaminants found in polar bears are a prime example. Because contaminants often cross international boundaries, an international treaty seemed the best way of dealing fairly with such transboundary pollution. The Stockholm Convention aims first to end the use and release of 12 of the POPs shown to be most dangerous, a group nicknamed the "dirty dozen" (Table 14.3). It sets guidelines for phasing out these chemicals and encourages transition to safer alternatives.

The European Union is taking the world's boldest step toward testing and regulating manufactured chemicals. In 2007, the EU's **REACH** program went into effect (*REACH* stands for Registration, Evaluation, Authorization, and Restriction of Chemicals). REACH shifts the burden of proof for testing chemical safety from national governments to industry and requires that chemical substances produced or imported in amounts of over 1 metric ton per year be registered with a new European Chemicals Agency. This agency will evaluate industry research and decide whether the chemical seems safe and should be approved, whether it is unsafe and should be restricted, or whether more testing is needed.

Previous policy had required industry to test chemicals brought to market after 1981 (there are 4,300 such chemicals) but required no such testing for chemicals already on the market in 1981 (which number over 100,000). It is expected that REACH will require 30,000 substances to be registered.

The REACH policy also aims to help industry—by giving it a single streamlined regulatory system and by exempting it from having to file paperwork on substances under 1 metric ton. By requiring stricter review of major chemicals already in use, exempting chemicals made only in small

TABLE 14.3 The "Dirty Dozen" Persistent Organic Pollutants (POPs) Targeted by the Stockholm Convention

| Toxicant | Type | Description |
|---|---|---|
| Aldrin | Pesticide | Kills termites, grasshoppers, corn rootworm, and other soil insects |
| Chlordane | Pesticide | Kills termites and is a broad-spectrum insecticide on various crops |
| DDT | Pesticide | Widely used in the past to protect against insect-spread diseases; continues to be applied in several countries to control malaria |
| Dieldrin | Pesticide | Controls termites and textile pests; also used against insect-borne diseases and insects in agricultural soil |
| Dioxins | Unintentional by-product | Produced by incomplete combustion and in chemical manufacturing; released in some kinds of metal recycling, pulp and paper bleaching, automobile exhaust, tobacco smoke, and wood and coal smoke |
| Endrin | Pesticide | Kills insects on cotton and grains; also used against rodents |
| Furans | Unintentional by-product | Result from the same processes that release dioxins; also are found in commercial mixtures of PCBs |
| Heptachlor | Pesticide | Kills soil insects, termites, cotton insects, grasshoppers, and mosquitoes |
| Hexachlorobenzene | Fungicide; unintentional by-product | Kills fungi that affect crops; released during chemical manufacture and from processes that give rise to dioxins and furans |
| Mirex | Pesticide | Combats ants and termites; also is a fire retardant in plastics, rubber, and electronics |
| PCBs | Industrial chemical | Used in industry as heat-exchange fluids, in electrical transformers and capacitors, and as additives in paint, sealants, and plastics |
| Toxaphene | Pesticide | Kills insects on crops; kills ticks and mites on livestock |

Data from United Nations Environment Programme (UNEP), 2001.

amounts, and providing financial incentives for innovating new chemicals, the EU hopes to help European industries research and develop safer new chemicals and products while safeguarding human health and the environment.

The EU expects that roughly 1,500 substances may be judged harmful enough to be replaced with safer substances. These include carcinogens, mutagens, teratogens, persistent substances, and chemicals that bioaccumulate. (EU commissioners held off on including endocrine disruptors.) The EU also expects that 1–2% of substances may cease to be manufactured because their production will no longer be profitable.

In an impacts assessment in 2003, EU commissioners estimated that REACH will cost the chemical industry and chemical users 2.8–5.2 billion euros (US $3.8–7.0 billion) over 11 years, but that the health benefits to the public would be roughly 50 billion euros (US $67 billion) over 30 years. Changes in the program since then have made the predicted cost:benefit ratio even better.

## Conclusion

International agreements such as REACH and the Stockholm Convention represent a hopeful sign that governments will act to protect the world's people, wildlife, and ecosystems from toxic chemicals and other environmental hazards. At the same time, solutions often come more easily when they do not arise from government regulation alone. To many minds, consumer choice, exercised through the market, may be the best way to influence industry's decision making. Consumers of products can make decisions that influence industry when they have full information from scientific research regarding the risks involved. Once scientific results are in, a society's philosophical approach to risk management will determine what policy decisions are made.

Whether the burden of proof is laid at the door of industry or of government, it is important to realize that we will never attain complete scientific knowledge of any risk. Rather, we must make choices based on the information available. Synthetic chemicals have brought us innumerable modern conveniences, a larger food supply, and medical advances that save and extend human lives. Human society would be very different without them. Yet a safer and happier future, one that safeguards the well-being of both humans and the environment, depends on knowing the risks that some hazards pose and on having means in place to phase out harmful substances and replace them with safer ones.

## REVIEWING OBJECTIVES

**You should now be able to:**

**Identify the major types of environmental health hazards and explain the goals of environmental health**

- Environmental health seeks to assess and mitigate environmental factors that adversely affect human health and ecological systems. (p. 383)
- Environmental health threats include physical, chemical, biological, and cultural hazards. (pp. 383–384)
- Disease is a major focus of environmental health. (pp. 384–385)
- Environmental hazards exist indoors as well as outdoors. (pp. 385–387)
- Toxicology is the study of poisonous substances. (pp. 387–388)

**Describe the types, abundance, distribution, and movement of toxicants in the environment**

- Thousands of potentially toxic substances exist around us. (pp. 388–389)
- Toxicants may be of human or natural origin. They include carcinogens, mutagens, teratogens, allergens, neurotoxins, and endocrine disruptors. (pp. 389–390)
- Toxicants may enter and move through surface and groundwater reservoirs, or they may travel long distances through the atmosphere. (pp. 392–394)
- Some chemicals break down very slowly and thus persist in the environment. (pp. 394–395)
- Some organic poisons bioaccumulate and move up the food chain, poisoning consumers at high trophic levels through the process of biomagnification. (pp. 395–397)

**Discuss the study of hazards and their effects, including case histories, epidemiology, animal testing, and dose-response analysis**

- In case histories, researchers study health problems in individual people. (p. 397)
- Epidemiology involves gathering data from large groups of people over long periods of time and comparing groups with and without exposure to the environmental health threat being assessed. (p. 398)
- In dose-response analysis, scientists measure the response of test animals to various doses of the suspected toxicant. (pp. 398–399)
- Toxicity or strength of response may be influenced by the dose or amount of exposure, the nature of exposure (acute or chronic), individual variation, and synergistic interactions with other hazards. (pp. 399–401)

**Assess risk assessment and risk management**

- Risk assessment involves quantifying and comparing risks involved in different activities or substances. (p. 403)
- Risk management integrates science with political, social, and economic concerns, in order to design strategies to minimize risk. (pp. 403–404)

**Compare philosophical approaches to risk**

- An innocent-until-proven-guilty approach assumes that a substance is not harmful unless it is shown to be so. (pp. 404–405)
- A precautionary approach entails assuming that a substance may be harmful unless proven otherwise. (pp. 404–405)

**Describe policy and regulation in the United States and internationally**

- The EPA, CDC, FDA, and OSHA are responsible for regulating environmental health threats under U.S policy. (p. 405)
- European nations take a more precautionary approach than does the United States when it comes to testing chemical products. (pp. 406–407)

## TESTING YOUR COMPREHENSION

1. What four major types of health hazards does research in the field of environmental health encompass?
2. In what way is disease the greatest hazard that people face? What kinds of interrelationships must environmental health experts study to learn about how diseases affect human health?
3. Where does most exposure to lead, asbestos, radon, and PBDEs occur? How has each been addressed?
4. When did concern over the effects of pesticides start to grow in the United States? Describe the argument presented by Rachel Carson in *Silent Spring*. What policy resulted from the book's publication? Is DDT still used?
5. List and describe the six types or general categories of toxicants described in this chapter.
6. How do toxicants travel through the environment, and where are they most likely to be found? What are the life spans of toxic agents? Describe the processes of bioaccumulation and biomagnification.
7. What are epidemiological studies, and how are they most often conducted?

8. Why are animals used in laboratory experiments in toxicology? Explain the dose-response curve. Why is a substance with a high $LD_{50}$ considered safer than one with a low $LD_{50}$ ?
9. What factors may affect an individual's response to a toxic substance? Why is chronic exposure to toxic agents often more difficult to measure and diagnose than acute exposure? What are synergistic effects, and why are they difficult to measure and diagnose?
10. How do scientists identify and assess risks from substances or activities that may pose health threats?

## SEEKING SOLUTIONS

1. Describe some environmental hazards that you think you may be living with indoors. How do you think you may have been affected by indoor or outdoor environmental hazards in the past? What philosophical approach do you plan to take in dealing with these toxicants in your own life?
2. Why is it that research on endocrine disruption has spurred so much debate? What steps do you think could be taken to help establish more consensus among scientists, industry, regulators, policymakers, and the public?
3. Do you feel that laboratory-bred animals should be used in experiments in toxicology? Why or why not?
4. Describe differences in the policies of the United States and the European Union toward the study and management of the risks of synthetic chemicals. Which do you believe is better, the policies of the United States or the European Union, and why?
5. **THINK IT THROUGH** You are the parent of two young children, and you want to minimize the environmental health risks to your family as your kids grow up. Name five steps that you could take in your household and in your daily life that would reduce your children's exposure to environmental health hazards.
6. **THINK IT THROUGH** You have just been hired as the office manager for a high-tech startup company that employs bright and motivated young people but is located in an old, dilapidated building. Despite their youth and vigor, the company's employees seem perpetually sick with colds, headaches, respiratory ailments, and other unexplained illnesses. Looking into the building's history, you discover that the water pipes and ventilation system are many decades old, that there have been repeated termite infestations, and that part of the building was remodeled just before your company moved in but there are no records of what was done in the remodel. Your company has all the latest furniture, computers, and other electronics. Most windows are sealed shut. There is an indoor smoking lounge.

   You want to figure out what is making the employees sick, and you want to convince your boss to give you a budget to hire professionals to examine the building for hazards. What hazards might you expect? What will you tell your boss to interest him in funding inspections? What questions will you ask employees to help focus and prioritize any inspections you are granted?

## INTERPRETING GRAPHS AND DATA

To minimize their exposure to ultraviolet (UV) radiation and thus their risk of skin cancer, people have increased their use of sunscreen lotions in recent decades. Recently, however, some research has shown that chemicals in sunscreens may themselves pose some risk to human health. The compounds most commonly used as UV protectants are fat soluble, environmentally persistent, and prone to bioaccumulation. Moreover, they exhibit estrogenic effects in laboratory rats (see Schlumpf, et al., 2001, as cited in the source note to the graph). Although the benefits of sunscreen use are substantial, the possible risks are not yet well understood. A hypothetical trade-off between the risk factors of UV exposure and sunscreen use illustrates the balancing act known as risk management.

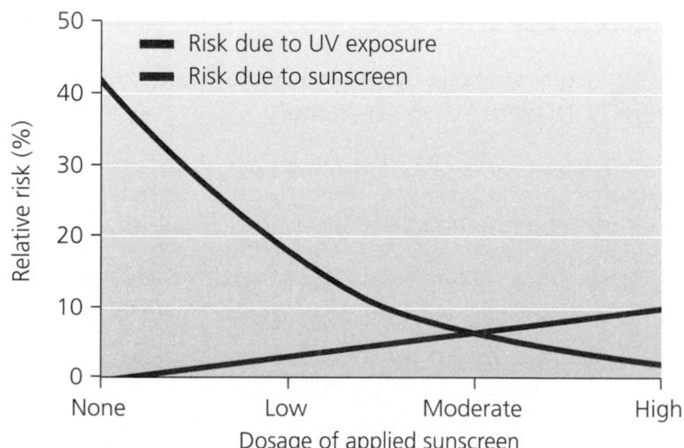


**Hypothetical risk distributions for individuals using an estrogenic sunscreen to prevent skin cancer.** Schlumpf, M., et al. 2001. *In vitro* and *in vivo* estrogenicity of UV screens. *Environmental Health Perspectives* 109: 239–244.

1. What dosage of applied sunscreen on the graph corresponds to the greatest risk due to UV exposure? What

dosage corresponds to the greatest risk due to chemicals in the sunscreen? Which of these two points on the graph is associated with the greater risk?

2. What dosage of applied sunscreen on the graph corresponds to the least risk due to UV exposure? What dosage corresponds to the least risk due to chemicals in the sunscreen? Which of these two points is associated with the greater risk?
3. The total risk to the individual is the sum of the two individual risks. What point on the graph corresponds to the greatest total risk? What sunscreen dosage corresponds to the least total risk? Based on the data shown here, how much sunscreen would you choose to apply the next time you go to the beach? Is there any other information you'd like to know before you change the way you use sunscreen? Can you think of any other cases that illustrate this sort of trade-off between dose-dependent risk factors?

## CALCULATING ECOLOGICAL FOOTPRINTS

In 2001, the population of the United States was approximately 285 million, and the world's population totaled 6.16 billion. In that same year, pesticide use in the United States was approximately 1.20 billion pounds of active ingredient, and world pesticide use totaled 5.05 billion pounds of active ingredient. Pesticides include hundreds of chemicals used as insecticides, fungicides, herbicides, rodenticides, repellants, and disinfectants. They are used by farmers, governments, industries, and individuals. In the table, calculate your share of pesticide use as a U.S. citizen in 2001 and the amount used by (or on behalf of) the average citizen of the world.

| | Annual pesticide use (pounds of active ingredient) |
|---|---|
| You | 4.21 |
| Your class | |
| Your state | |
| United States | |
| World (total) | |
| World (per capita) | |

1. What is the ratio of your annual pesticide use to the world's per capita average?
2. Refer to the "Calculating Ecological Footprints" section in Chapter 1 (• p. 24), and find the ecological footprints of the average U.S. citizen and the average world citizen. Compare the ratio of pesticide usage with the ratio of the overall ecological footprints. What is the difference, and how would you account for it?
3. Does the figure for per capita pesticide use for you as a U.S. citizen seem reasonable for you personally? Why or why not? Do you find this figure alarming or of little concern? What else would you like to know to assess the risk associated with this level of pesticide use?

## Take It Further

Go to www.aw-bc.com/withgott or the student CD-ROM, where you'll find:

- Suggested answers to end-of-chapter questions
- Quizzes, animations, and flashcards to help you study
- *Research Navigator*™ database of credible and reliable sources to assist you with your research projects
- GRAPHit! Tutorials to help you interpret graphs
- INVESTIGATEit! Current news articles that link the topics that you study to case studies from your region to around the world

CHAPTER

# 15 Freshwater Resources: Natural Systems, Human Impact, and Conservation

Hoover Dam on the Colorado River

## Upon completing this chapter, you will be able to:

- Explain the importance of water and the hydrologic cycle to ecosystems, human health, and economic pursuits
- Delineate the distribution of fresh water on Earth
- Describe major types of freshwater ecosystems
- Discuss how we use water and alter freshwater systems
- Assess problems of water supply and propose solutions to address depletion of fresh water
- Assess problems of water quality and propose solutions to address water pollution
- Explain how wastewater is treated

Colorado River delta

CENTRAL CASE

# Plumbing the Colorado River

**"Water promises to be to the 21st century what oil was to the 20th century: the precious commodity that determines the wealth of nations."**
—FORTUNE MAGAZINE, MAY 2000

**"The wars of the 21st century will be fought over water."**
—WORLD WATER COMMISSION CHAIRMAN ISMAIL SERAGELDIN

The Colorado River begins in the high peaks of the Rocky Mountains, charges through the Grand Canyon, crosses into Mexico, and empties into the Gulf of California, draining 637,000 km$^2$ (246,000 mi$^2$) of southwestern North America. Its raging waters chiseled through thousands of feet of bedrock, creating the Grand Canyon and leaving extraordinary scenery along its 2,330-km (1,450-mi) length.

Today, however, only a trickle reaches the river's mouth. Instead, the waters of the Colorado River are impounded by massive dams to provide flood control, recreation, and hydroelectric power. Water is siphoned off for irrigation to make agriculture possible in this dry region. It is piped to cities in canals and aqueducts to quench the thirst of 30 million people. It keeps golf courses green in the desert, fills swimming pools in southern California backyards, and gushes through fountains in Las Vegas casinos. Water is the lifeline for any civilization in an arid environment. Without water, our society in the western United States would simply not exist.

Since 1922, the seven states along the Colorado have divided the river's water among themselves, guided by the Colorado River Compact they signed that year. In this treaty, Arizona, California, Colorado, Nevada, New Mexico, Utah, and Wyoming apportioned water according to the needs and negotiating power each state had at the time.

As years passed, California was permitted to exceed its allotted portion because the other six states did not need their full allotments. However, the populations of these states are booming today, so the U.S. Interior Department pressured California to reduce its withdrawals—and cut off 15% of California's water when an irrigation district tried to back out of the agreement in 2003.

At the Interior Department's insistence, the seven states sat down to negotiate how to share dwindling water resources in the face of growing demand. These negotiations turned frantic as the region fell into the grip of a multiyear drought. The Colorado River's flow was below average for 7 of 8 years leading up to 2007, leaving reservoirs at half capacity.

Worse still, new scientific data paint a grim picture for the region: Tree-ring studies show that megadroughts were common in the past, and climate modelers predict that global climate change will bring still more drought.

In spring 2007, the states worked out a plan that allowed upper-basin states to withhold more water when needed and that encouraged lower-basin states to develop supplies elsewhere. In particular, the plan backed Las Vegas's controversial proposal to mine groundwater from rural areas of eastern Nevada. Barely populated in 1922, Nevada had won only 4% of the river's water in the Colorado River Compact, but today rapidly growing Las Vegas needs more water, fast. The city already treats wastewater and returns it to the river so it can legally take more than its share, and it pays its residents to rip up their lawns to reduce water use. But in a city adding tens of thousands of people each year, such steps may not be enough.

The other states fear that if Las Vegas is not allowed adequate water from somewhere, it may back out of the Colorado River Compact or challenge it in court. To safeguard their own shares of Colorado River water, the states are giving Las Vegas the go-ahead to pump groundwater from beneath a scenic area of Great Basin desert, even though this threatens the area's ecology and people. The contentious plan is currently awaiting approval by Nevada's state engineer. It is just the latest episode in the colorful history of western water politics among the states jockeying for rights to what was once the West's wildest river.

## Freshwater Systems

"Water, water, everywhere, nor any drop to drink." The well-known line from Coleridge's poem *The Rime of the Ancient Mariner* describes the situation on our planet quite well. Water may seem abundant to us, but water that we can drink is actually quite rare and limited (**Figure 15.1**). Roughly 97.5% of Earth's water resides in the oceans and is too salty to drink or use to water crops. Only 2.5% is considered **fresh water**, water that is relatively pure, with few dissolved salts. Because most fresh water is tied up in glaciers, icecaps, and underground aquifers, just over 1 part in 10,000 of Earth's water is easily accessible for human use.

Water is constantly moving among the reservoirs specified in Figure 15.1 via the *hydrologic cycle* (Figure 7.17, • p. 196). As water moves, it redistributes heat, erodes mountain ranges, builds river deltas, maintains organisms

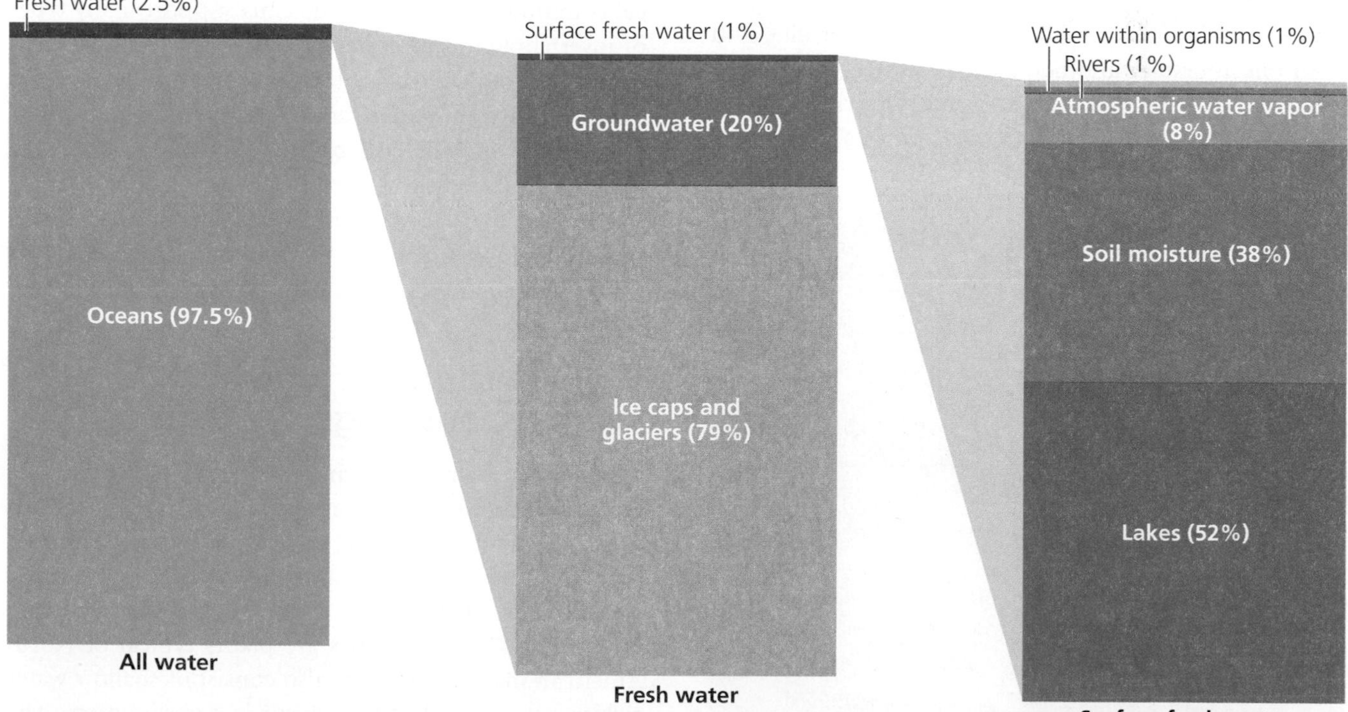


**FIGURE 15.1** Only 2.5% of Earth's water is fresh water. Of that 2.5%, most is tied up in glaciers and ice caps. Of the 1% that is surface water, most is in lakes and soil moisture. Data from United Nations Environment Programme (UNEP) and World Resources Institute.

and ecosystems, shapes civilizations, and gives rise to political conflicts. Let's first examine the portions of the hydrologic cycle that are most conspicuous to us—surface water bodies—and take stock of the ecological systems they support.

## Rivers and streams wind through landscapes

Water from rain, snowmelt, or springs runs downhill and converges where the land dips lowest, forming streams, creeks, or brooks. These watercourses merge into rivers, whose water eventually reaches the ocean (or sometimes ends in a landlocked water body). A smaller river flowing into a larger one is a *tributary,* and the area of land drained by a river and all its tributaries is that river's *watershed* (• p. 58).

Rivers shape the landscape through which they run. The force of water rounding a river's bend gradually eats away at the outer shore, eroding soil from the bank. Meanwhile, sediment is deposited along the inside of the bend, where water currents are weaker. In this way, over time, river bends become exaggerated in shape (**Figure 15.2**). Eventually, a bend may become such an extreme loop (called an *oxbow*) that water erodes a shortcut from one end of the loop to the other, pursuing a direct course. The bend is cut off and remains as an isolated, U-shaped water body called an *oxbow lake.*

Over thousands or millions of years, a river may shift from one course to another, back and forth over a large area, carving out a flat valley. Areas nearest a river's course that are flooded periodically are said to be within the river's **floodplain**. Frequent deposition of silt from flooding makes floodplain soils especially fertile. As a result, agriculture thrives in floodplains, and *riparian* (riverside) forests are productive and species-rich.

FIGURE 15.2 Rivers and streams flow downhill, shaping landscapes, as shown by an oxbow of this meandering river in Colorado.

FIGURE 15.3 Shallow water bodies with ample vegetation are called wetlands and include swamps, bogs, and marshes such as this one in Botswana, Africa.

The water of rivers and streams hosts diverse ecological communities. Algae and detritus support many types of invertebrates, from water beetles to crayfish. Insects as diverse as dragonflies, mayflies, and mosquitoes develop as larvae in streams and rivers before maturing into adults that take to the air. Fish consume aquatic insects, and birds such as kingfishers, herons, and ospreys dine on fish. Many amphibians spend their larval stages in streams, and some live their entire lives in streams. Salmon migrate from oceans up rivers and streams to spawn.

## Wetlands include marshes, swamps, and bogs

Systems that combine elements of fresh water and dry land are enormously rich and productive. Often lumped under the term **wetlands**, such areas include different types of systems. In *freshwater marshes* (**Figure 15.3**), shallow water allows plants to grow above the water surface. Cattails and bulrushes are plants typical of North American marshes. *Swamps* also consist of shallow water rich in vegetation, but they occur in forested areas. The cypress swamps of the southeastern United States, where cypress trees grow in standing water, are an example. Swamps are also created when beavers build dams across streams with limbs from trees they have cut, flooding wooded areas upstream. *Bogs* are ponds thoroughly covered with thick floating mats of vegetation, and can represent a stage in aquatic succession (• p. 157).

Wetlands are extremely valuable as habitat for wildlife. They also provide important ecosystem services by slowing runoff, reducing flooding, recharging aquifers, and filtering pollutants. Despite these vital roles, people have drained and filled wetlands extensively, largely for agriculture (• pp. 346–347). It is estimated that southern Canada and the United States have lost well over half their wetlands since European colonization. In 2006, the U.S Supreme Court was asked to decide whether the Clean Water Act protected small isolated wetlands (in the case *Rapanos v. United States*, dealing with a Michigan developer who filled wetlands on his property). The Court's split decision required the Army Corps of Engineers to devise guidelines to determine when such wetlands are ecologically valuable enough to protect by law.

## Lakes and ponds are ecologically diverse systems

Lakes and ponds are bodies of open standing water. Their physical conditions and the types of life within them vary with depth and the distance from shore. As a result, scientists have described several zones typical of lakes and ponds (**Figure 15.4**).

The region ringing the edge of a water body is named the *littoral zone.* Here the water is shallow enough that aquatic plants grow from the mud and reach above the water's surface. The nutrients and productive plant growth of the littoral zone make it rich in invertebrates—such as insect larvae, snails, and crayfish—on which fish, birds, turtles, and amphibians feed. The *benthic zone* extends along the bottom of the entire water body, from shore to the deepest point. Many invertebrates live in the mud on the bottom, feeding on detritus or preying on one another.

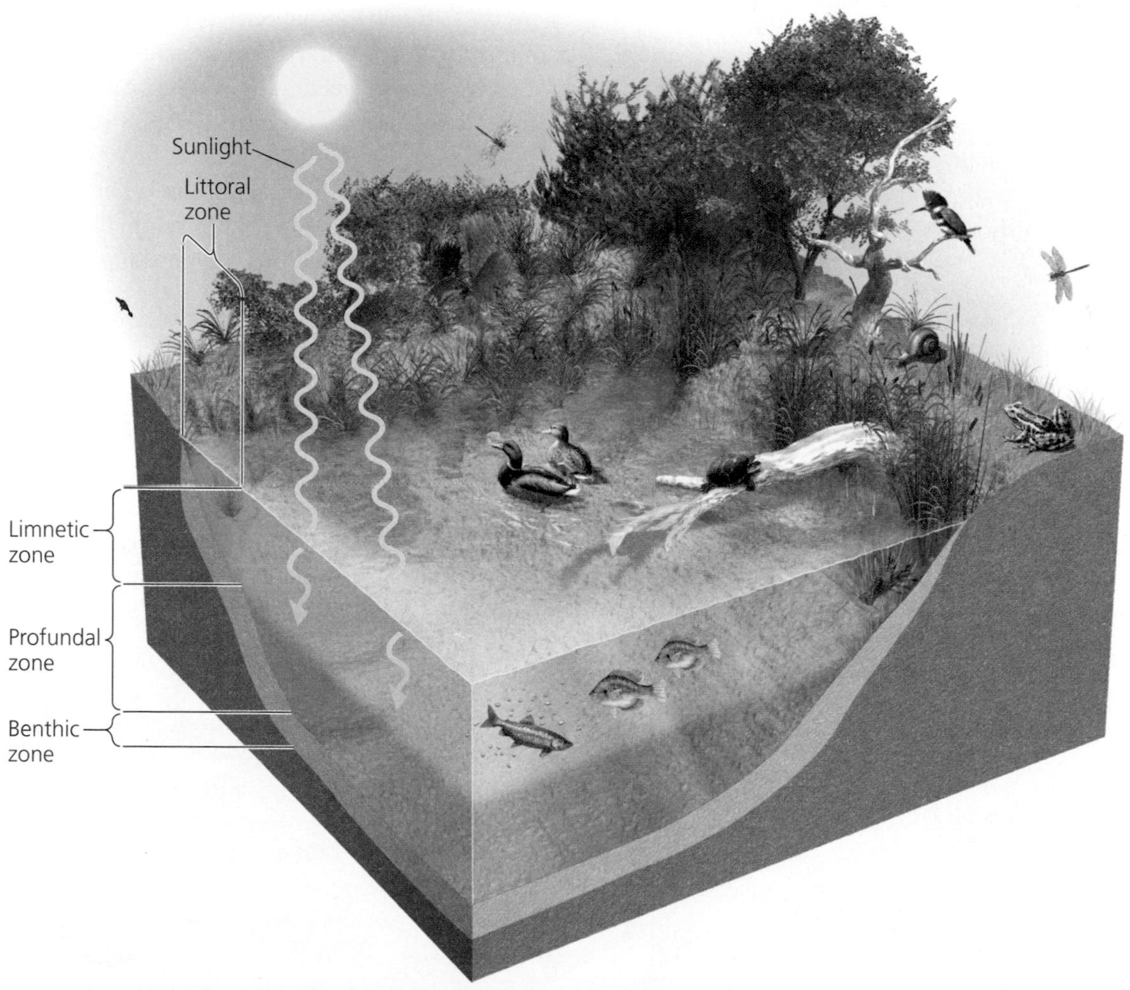


**FIGURE 15.4** In lakes and ponds, emergent plants grow around the shoreline in the littoral zone. The limnetic zone is the layer of open, sunlit water, where photosynthesis takes place. Sunlight does not reach the deeper profundal zone. The benthic zone, which is the bottom of the water body, often is muddy, rich in detritus and nutrients, and low in oxygen.

In the open portion of a lake or pond, away from shore, sunlight penetrates shallow waters of the *limnetic zone.* Because light enables photosynthesis and plant growth, the limnetic zone supports phytoplankton, which in turn support zooplankton, both of which are eaten by fish. Within the limnetic zone, sunlight intensity (and therefore water temperature) decreases with depth. The water's turbidity affects the depth of this zone; water that is clear allows sunlight to penetrate deeply, whereas turbid water does not. Below the limnetic zone is the *profundal zone,* the volume of open water that sunlight does not reach. This zone lacks plant life and thus is lower in dissolved oxygen and supports fewer animals. Aquatic animals rely on dissolved oxygen, and its concentration depends on the amount released by photosynthesis and the amount removed by animal and microbial respiration, among other factors.

Ponds and lakes change over time naturally as streams and runoff bring them sediment and nutrients. *Oligotrophic* lakes and ponds, which have low-nutrient and high-oxygen conditions, may slowly give way to the high-nutrient, low-oxygen conditions of *eutrophic* water bodies (jump ahead to see Figure 15.20, • p. 430). Eventually, water bodies may fill in completely by the process of aquatic succession. As lakes or ponds change over time, species of fish, plants, and invertebrates adapted to oligotrophic conditions may give way to those that thrive under eutrophic conditions.

Some lakes are so large that they differ substantially in their characteristics from small lakes. These large lakes are sometimes known as inland seas. North America's Great Lakes are prime examples. Because they hold so much water, most of their biota is adapted to open water. Major fish species of the Great Lakes include lake sturgeon, lake whitefish, northern pike, alewife, bass, walleye, and perch. Lake Baikal in Asia is the world's deepest lake, at 1,637 m (5,370 ft) deep, and the Caspian Sea is the world's largest freshwater body, at 371,000 km$^2$ (143,000 mi$^2$).

## Groundwater plays key roles in the hydrologic cycle

Any precipitation reaching Earth's land surface that does not evaporate, flow into waterways, or get taken up by organisms infiltrates the surface. Most percolates downward through the soil to become **groundwater** (**Figure 15.5**). Groundwater makes up one-fifth of Earth's fresh water supply and plays a key role in meeting human water needs.

As we first saw in Chapter 7 (• p. 197), groundwater is contained within **aquifers**: porous, spongelike formations of rock, sand, or gravel that hold water. An aquifer's upper layer, or *zone of aeration,* contains pore spaces partly filled with water. In the lower layer, or *zone of saturation,* the spaces are completely filled with water. The boundary between these two zones is the **water table**. Picture a sponge resting partly submerged in a tray of water; the lower part

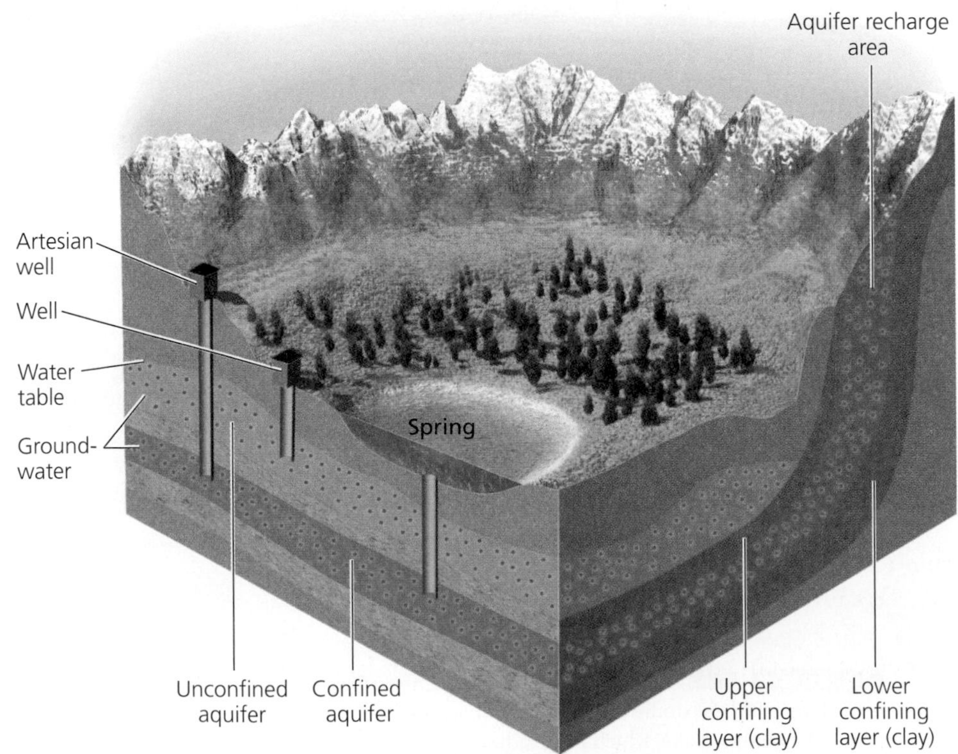


**FIGURE 15.5** Groundwater may occur in unconfined aquifers above impermeable layers or in confined aquifers under pressure between impermeable layers. Water may rise naturally to the surface at springs and through the wells we dig. Artesian wells tap into confined aquifers to mine water under pressure.

of the sponge is completely saturated, whereas the upper portion may be moist but contains plenty of air in its pores. Any area where water infiltrates Earth's surface and reaches an aquifer below is known as an *aquifer recharge zone.*

There are two broad categories of aquifers. A **confined aquifer**, or **artesian aquifer**, exists when a water-bearing porous layer of rock, sand, or gravel is trapped between upper and lower layers of less permeable substrate (often clay). In such a situation, the water is under great pressure. In contrast, an **unconfined aquifer** has no such upper layer to confine it, so its water is under less pressure and can be readily recharged by surface water.

Just as surface water becomes groundwater by infiltration and percolation, groundwater becomes surface water through springs (and human-drilled wells), sometimes keeping streams flowing when surface conditions are otherwise dry. Groundwater flows downhill and from areas of high pressure to areas of low pressure. A typical rate of groundwater flow might be only about 1 m (3 ft) per day, so groundwater may remain in an aquifer for a long time. In fact, groundwater can be ancient. The average age of groundwater has been estimated at 1,400 years, and some is tens of thousands of years old.

Nonetheless, volumes of groundwater are large enough that each day in the United States alone, aquifers release 1.9 trillion L (492 billion gal) of groundwater into bodies of surface water—nearly as much as the daily flow of the Mississippi River. The world's largest known aquifer is the Ogallala Aquifer, which underlies the Great Plains of the United States (**Figure 15.6**). Water from this massive aquifer has enabled American farmers to create the most bountiful grain-producing region in the world.

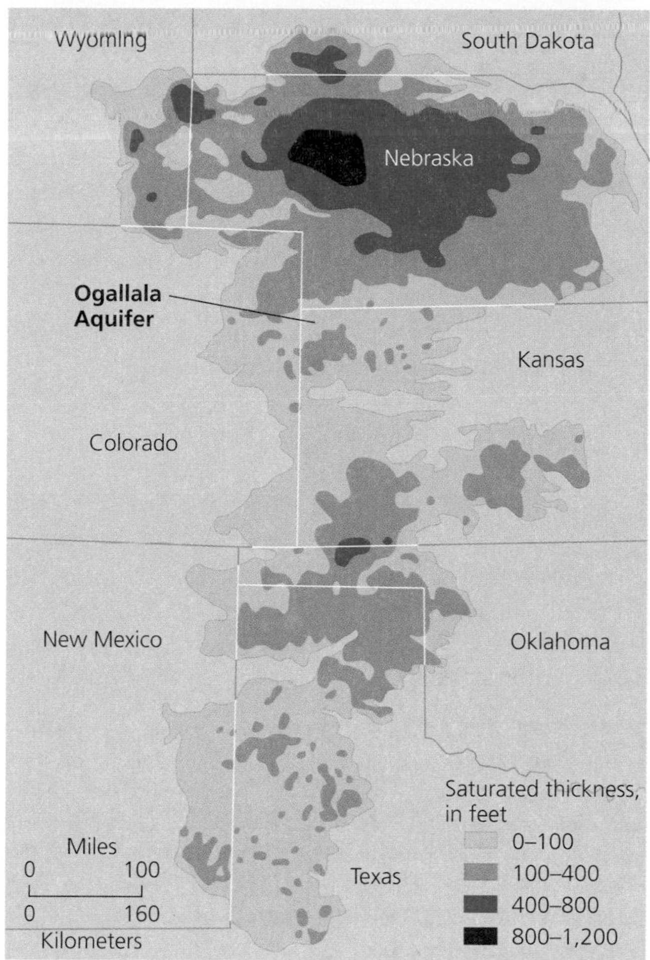


**FIGURE 15.6** The Ogallala Aquifer is the world's largest aquifer, and it held 3,700 km$^3$ (881 mi$^3$) of water before pumping began. This aquifer underlies 453,000 km$^2$ (175,000 mi$^2$) of the Great Plains beneath eight U.S. states from South Dakota to Texas. Overpumping for irrigation is currently reducing the volume and extent of this aquifer.

## Water is unequally distributed across Earth's surface

The Great Plains and its farmlands boast the massive Ogallala Aquifer as a source of fresh water, but many other areas are not so endowed. Different regions possess vastly different amounts of groundwater, surface water, and precipitation. Precipitation ranges from about 1,200 cm (470 in.) per year at Mount Waialeale on the Hawaiian island of Kauai to virtually zero in Chile's Atacama Desert.

People are not distributed across the globe in accordance with water availability. Many areas with high population density are water-poor, leading to inequalities in per capita water resources among and within nations (**Figure 15.7**). For example, Canada has 20 times more water for each of its citizens than does China. The Amazon River carries 15% of the world's runoff, but its watershed holds less than half a percent of the world's human population. Many densely populated nations, such as Pakistan, Iran, and Egypt, face serious water shortages. Asia possesses the most water of any continent but has the least water available per person, whereas Australia, with the least amount of water, boasts the most water available per person. Because of this mismatched distribution of water and population, one challenge for our societies has always been to transport fresh water from its source to where people need it.

Fresh water is distributed unevenly in time as well as space. India's monsoon season brings concentrated storms in which half of a region's annual rain may fall in a few hours. Northwest China receives three-fifths of its annual precipitation during 3 months when crops do not need it. The uneven distribution of water across time is one reason people have erected dams to store water, so that it may be distributed when needed.

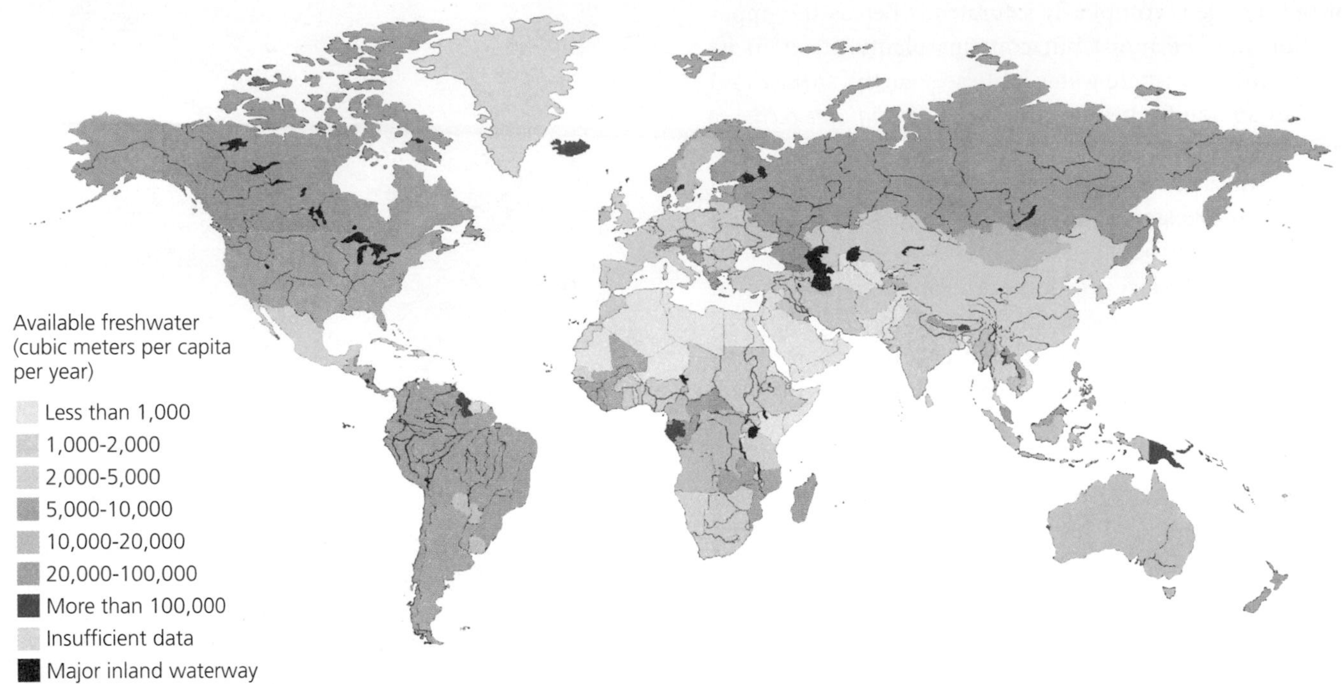


FIGURE 15.7 Nations vary tremendously in the amount of fresh water per capita available to their citizens. For example, Iceland, Papua New Guinea, Gabon, and Guyana (dark blue in this map) each have over 100 times more water per person than do many Middle Eastern and North African countries. Go to GRAPHIt! at www.aw-bc.com/withgott or on the student CD-ROM. Data from U.N. Environment Programme and World Resources Institute, as presented by Harrison, P., and F. Pearce. 2000. *AAAS atlas of population and the environment.* Berkeley, CA: University of California Press.

## Climate change will cause water shortages

As if the existing mismatches between water availability and human need were not enough, global climate change (Chapter 18) will worsen conditions in many regions by altering precipitation patterns, melting glaciers, causing early-season runoff, and intensifying droughts and flooding. The Colorado River basin is experiencing the effects already (**Figure 15.8**). A drought has hung over the area for most of the past decade, and climate models (• pp. 513–516) predict a drier future for the American Southwest. The prospect of having still less water for more people has added new urgency for the western states, and they are responding; four states have committed \$2.5 billion to water supply projects now underway or planned.

FIGURE 15.8 Water dropped to abnormally low levels at Lake Mead in 2006, as shown by the "bathtub ring" of exposed shoreline in this photograph. This was the result of a multiyear drought, but if global climate change makes the region more arid, then scenes more extreme than this could become commonplace.

# How We Use Water

In our attempts to harness fresh water sources for countless purposes and pursuits, we have achieved impressive engineering accomplishments. In so doing, we also have altered many environmental systems. It is estimated that 60% of the world's largest 227 rivers (and 77% of those in North America and Europe) have been strongly or moderately affected by artificial dams, canals, and diversions.

We are also using too much water. Data indicate that at present our consumption of fresh water in much of the world is unsustainable, and we are depleting many sources of surface water and groundwater. One-third of the world's people are already affected by water scarcity (with less than 1,000 $m^3$ [35,000 $ft^3$] of water per person

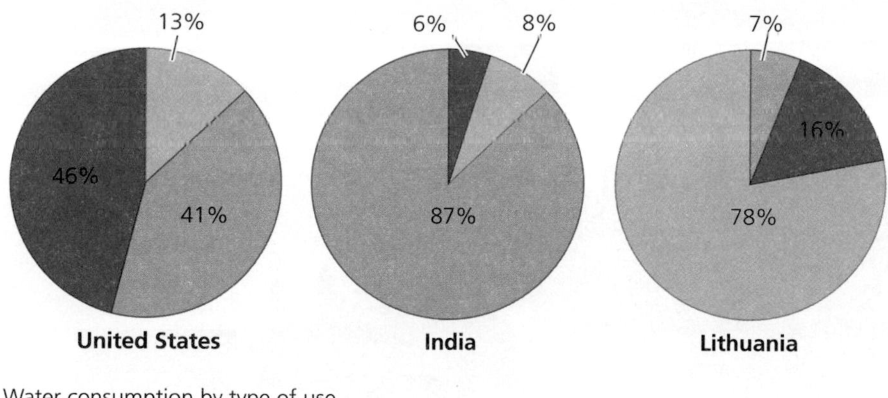


**FIGURE 15.9** Nations apportion their fresh water consumption differently. Industry consumes most water used in the United States, agriculture uses the most in India, and most water in Lithuania goes toward domestic use. Data are for 2000, the most recently available year, from U.N. Food and Agriculture Organization (FAO).

per year), according to a comprehensive global assessment presented in 2006.

## Water supplies our households, agriculture, and industry

We all use water at home for drinking, cooking, and cleaning. Farmers and ranchers use water to irrigate crops and water livestock. Most manufacturing and industrial processes require water. The proportions of each of these three types of use—residential/municipal, agricultural, and industrial—vary dramatically among nations (**Figure 15.9**). Nations with arid climates tend to use more fresh water for agriculture, and heavily industrialized nations use a great deal for industry. Globally, we spend about 70% of our annual fresh water allotment on agriculture. Industry accounts for roughly 20%, and residential and municipal uses for only 10%.

When we remove water from an aquifer or surface water body and do not return it, this is called **consumptive use**. A large portion of agricultural irrigation and of many industrial and residential uses is consumptive. **Nonconsumptive use** of water does not remove, or only temporarily removes, water from an aquifer or surface water body. Using water to generate electricity at hydroelectric dams is an example of nonconsumptive use; water is taken in, passed through dam machinery to turn turbines, and released downstream.

## We have erected thousands of dams

A **dam** is any obstruction placed in a river or stream to block the flow of water so that water can be stored in a reservoir. We build dams to prevent floods, provide drinking water, facilitate irrigation, and generate electricity (**Figure 15.10**; **Table 15.1**). Power generation with hydroelectric dams is discussed in Chapter 20 (• pp. 595–598).

Worldwide, we have erected more than 45,000 large dams (greater than 15 m, or 49 ft, high) across rivers in over 140 nations. We have built tens of thousands of smaller dams. Only a few major rivers in the world remain undammed and free-flowing, and these run through the tundra and taiga of Canada, Alaska, and Russia and in remote regions of Latin America and Africa.

Our largest dams comprise some of humanity's greatest engineering feats. The two behemoths of the Colorado River, Hoover Dam and Glen Canyon Dam, stand 221 m (726 ft) and 216 m (710 ft) high, respectively, stretch for 379 m (1,244 ft) and 476 m (1,560 ft) across, and consist of 6.6 and 7.3 million tons of concrete and steel. Hoover Dam holds back 35.2 $km^3$ (8.4 $mi^3$) of water in a reservoir that stretches 177 km (110 mi) long and reaches 152 m (500 ft) deep. Glen Canyon Dam holds back 33.3 $km^3$ (7.9 $mi^3$) of water in a reservoir that stretches 300 km (186 mi) long and reaches 171 m (560 ft) deep. Together these reservoirs store four times more water than flows down the river in an entire year.

## China's Three Gorges Dam is the world's largest

The complex mix of benefits and costs that dams produce is exemplified by the world's largest dam project. The Three Gorges Dam on China's Yangtze River, 186 m (610 ft) high and 2 km (1.3 mi) wide, was completed in 2003 (**Figure 15.11a**). When filled in 2009, its reservoir should stretch for 616 km (385 mi; as long as Lake Superior) and hold over 38 trillion L (10 trillion gal) of water. This project will enable boats and barges to travel farther upstream, provide flood control, and generate enough hydroelectric power to replace dozens of large coal or nuclear plants.

However, the Three Gorges Dam has cost $25 billion to build, and its reservoir is flooding 22 cities and the homes of 1.13 million people, requiring the largest resettlement project in China's history (**Figure 15.11b**). The reservoir is also submerging 10,000-year-old archaeological sites, productive farmlands, and wildlife habitat. Moreover, the reservoir slows the river's flow so that suspended sediment settles behind the dam. Indeed, the reservoir began

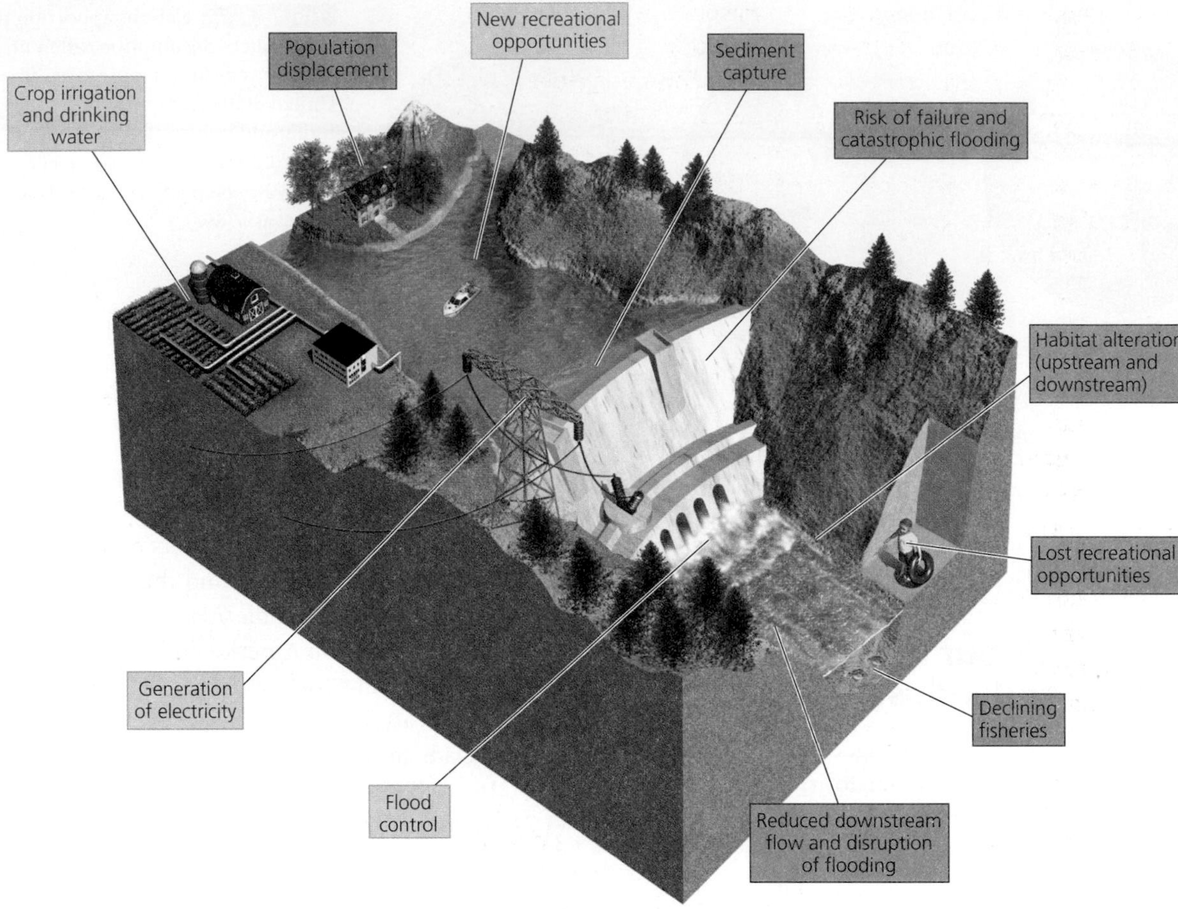


FIGURE 15.10 Damming rivers has diverse consequences for people and the environment. The generation of clean and renewable electricity is one of several major benefits (green boxes) of hydroelectric dams. Habitat alteration is one of several negative impacts (red boxes).

**TABLE 15.1 Major Benefits and Costs of Dams**

| Benefits | Costs |
|---|---|
| ▶ **Power generation.** Hydroelectric dams (• pp. 595–598) provide inexpensive electricity. | ▶ **Habitat alteration.** Reservoirs flood riparian habitats and displace or kill riparian species. Dams modify rivers downstream. Shallow warm water downstream from a dam is periodically flushed with cold reservoir water, stressing or killing many fish. |
| ▶ **Emissions reduction.** Hydroelectric power produces no greenhouse gases in its operation (although some are produced during construction and maintenance of infrastructure). By replacing fossil fuel combustion as an electricity source, hydropower reduces air pollution, climate change, and their health and environmental consequences. | ▶ **Fisheries declines.** Salmon and other fish that migrate up rivers to spawn encounter dams as a barrier. Although "fish ladders" at many dams allow passage, most fish do not make it. |
| ▶ **Crop irrigation.** Reservoirs can release irrigation water when farmers most need it and can buffer regions against drought. | ▶ **Population displacement.** Reservoirs generally flood fertile farmland and have flooded many human settlements. An estimated 40–80 million people globally have been displaced by dam projects over the past half century. |
| ▶ **Drinking water.** Many reservoirs store plentiful, reliable, and clean water for municipal drinking water supplies, provided that watershed lands draining into the reservoir are not developed or polluted. | ▶ **Sediment capture.** Sediment settles behind dams. Downstream floodplains and estuaries are no longer nourished, and reservoirs fill with silt. |
| ▶ **Flood control.** Dams can prevent floods by storing seasonal surges, such as those following snowmelt or heavy rain. | ▶ **Disruption of flooding.** Floods create productive farmland by depositing rich sediment. Without flooding, topsoil is lost, and farmland deteriorates. |
| ▶ **Shipping.** By replacing rocky river beds with deep placid pools, dams enable ships to transport goods over longer distances. | ▶ **Risk of failure.** There is always risk that a dam could fail, causing massive property damage, ecological damage, and loss of life. |
| ▶ **New recreational opportunities.** People can fish from boats and use personal watercraft on reservoirs in regions where such recreation was not possible before. | ▶ **Lost recreational opportunities.** Tubing, whitewater rafting, fly-fishing and kayaking opportunities are lost. |

(a) The Three Gorges Dam in Yichang, China

(b) Displaced people in Sichuan Province, China

FIGURE 15.11 China's Three Gorges Dam, completed in 2003, is the world's largest dam (**a**). Well over a million people were displaced and whole cities were leveled for its construction, as shown here in Sichuan Province (**b**). The reservoir began filling in 2003 and will continue filling for several years.

accumulating sediment as soon as the dam was completed, and because the river downstream is deprived of sediment, the tidal marshes at the Yangtze's mouth are eroding away, leaving the city of Shanghai with a degraded coastal environment and less coastal land to develop. Many scientists worry that the Yangtze's many pollutants will also be trapped in the reservoir, making the water undrinkable. In fact, high levels of bacteria were found in the water as it began building up behind the dam. The Chinese government plans to sink $5 billion into building hundreds of sewage treatment and waste disposal facilities.

## Some dams are being removed

People who feel that the costs of some dams have outweighed their benefits are pushing for such dams to be dismantled. By removing dams and letting rivers flow free, these people say, we can restore riparian ecosystems, reestablish economically valuable fisheries, and revive river recreation such as fly-fishing and rafting. Increasingly, private dam owners and the Federal Energy Regulatory Commission (FERC), the U.S. government agency charged with renewing licenses for dams, have agreed. Roughly 500 dams have been removed in the United States in recent years. One reason is that many aging dams are in need of costly repairs or have outlived their economic usefulness.

The drive to remove dams first gathered steam in 1999 with the dismantling of the Edwards Dam on Maine's Kennebec River, which resulted from a FERC determination that the environmental benefits of removing the dam outweighed the economic benefits of relicensing it. Within just a year after the 7.3-m (24-ft) high, 279-m (917-ft) long dam was removed, large numbers of 10 species of migratory fish, including salmon, sturgeon, shad, herring, alewife, and bass, ventured upstream and began using the 27-km (17-mi) stretch of river above the dam site. Some property owners along the former reservoir who had opposed the dam's removal had a change of heart once they saw the healthy and vibrant river that now ran past their property. More small dams—and perhaps some large ones—will come down in the years ahead as over 500 FERC licenses come up for renewal in the next decade.

## Dikes and levees are meant to control floods

Flood prevention ranks high among reasons we control the movement of fresh water. People have always been attracted to riverbanks for their water supply and for the flat topography and fertile soil of floodplains. Flooding is a normal, natural process due to snowmelt or heavy rain, and floodwaters spread nutrient-rich sediments over large areas, benefiting both natural systems and human agriculture.

In the short term, however, floods can do tremendous damage to the farms, homes, and property of people who choose to live in floodplains. To protect against floods, individuals and governments have built *dikes* and *levees* (long raised mounds of earth) along the banks of rivers to hold rising water in main channels. Many dikes are small and locally built, but in the United States the Army Corps of Engineers has constructed thousands of miles of massive levees along the banks of major waterways (those that failed in New Orleans after Hurricane Katrina are examples). Although these structures prevent flooding at most times and places, they can sometimes worsen flooding because they force water to stay in channels and accumulate, building up enormous energy and leading to occasional catastrophic overflow events.

## We divert—and deplete—surface water to suit our needs

People have long diverted water from rivers, streams, lakes, and ponds to farm fields, homes, and cities. The Colorado River's water is heavily diverted and utilized (**Figure 15.12**). Early on in its course, some Colorado River water is piped through a mountain tunnel and down the

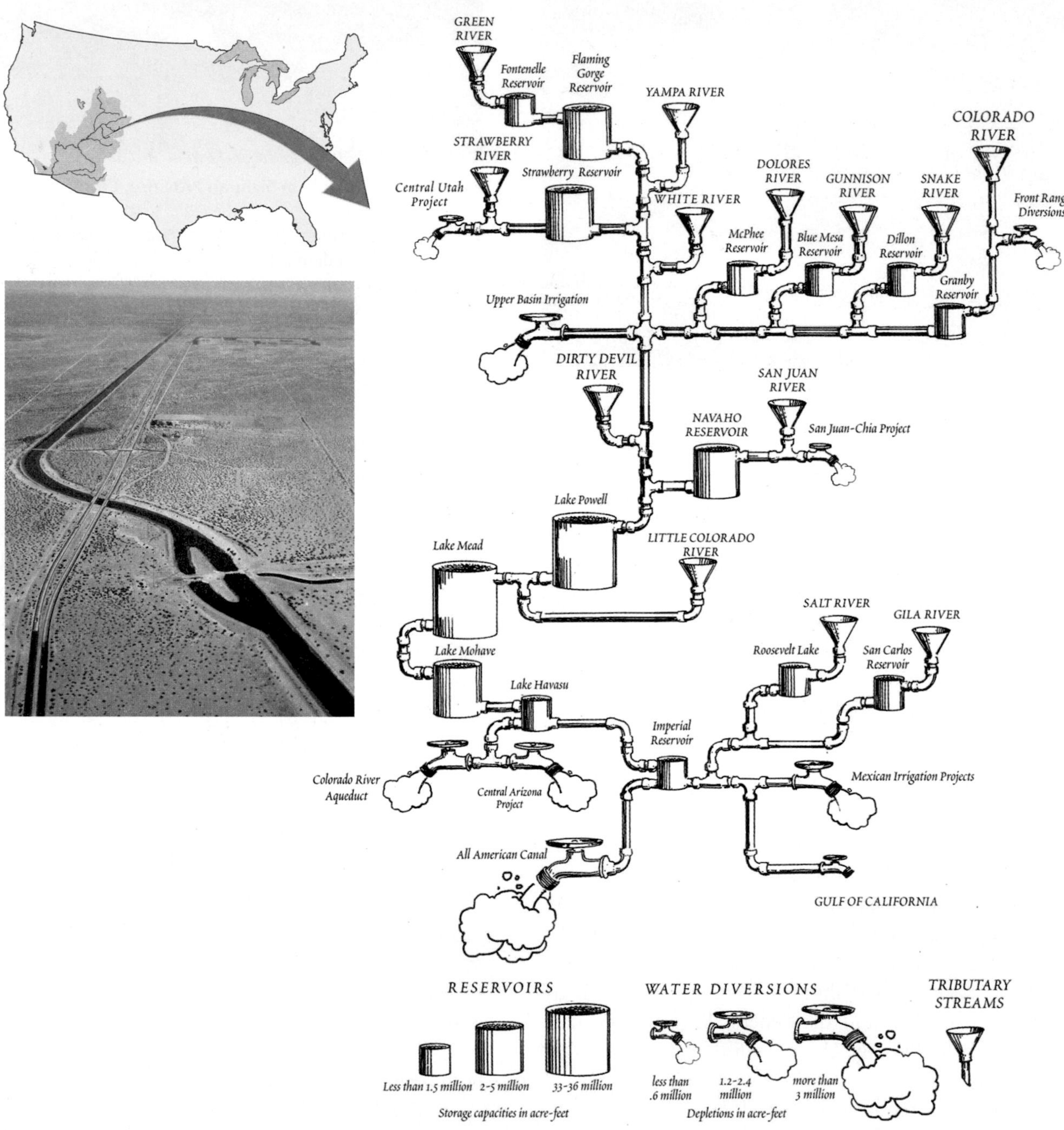


**FIGURE 15.12** The 2,330-km (1,450-mile) Colorado River drains a watershed of 637,000 km$^2$ (246,000 mi$^2$). The degree to which this once-wild river has been dammed and diverted led cartoonist Lester Dore to portray the Colorado and its tributaries as an immense plumbing system. Thirteen major dams pool water in enormous reservoirs along the lengths of the Colorado River and its tributaries. Several major canals (such as the All American Canal shown winding through the California desert in the photo) divert water from the lower portion of the river, mostly to irrigate farmland in desert regions. *Source: High Country News,* 10 November 1997.

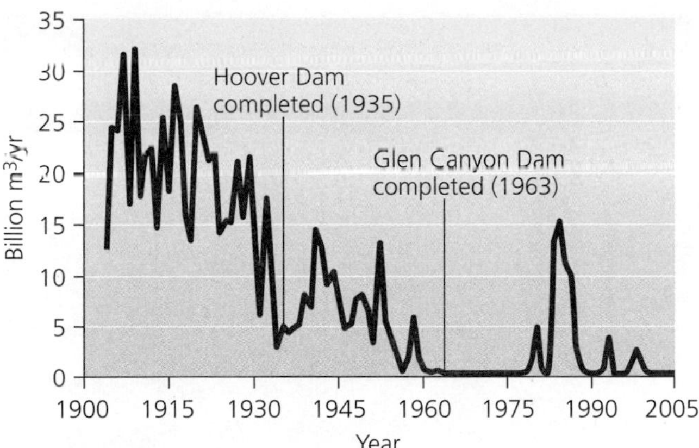


**FIGURE 15.13** Flow at the mouth of the Colorado River has greatly decreased over the past century as a result of withdrawals, mostly for agriculture. The river now often runs dry at its mouth. Data from Postel, S. 2005. *Liquid assets: The critical need to safeguard freshwater ecosystems.* Worldwatch Paper 170. Washington, D.C.: Worldwatch Institute.

Rockies' eastern slope to supply the city of Denver. More is removed for Las Vegas and other cities and for farmland as the water proceeds downriver. When it reaches Parker Dam on the California–Arizona state line, large amounts are diverted into the Colorado River Aqueduct, which brings water to millions of people in the Los Angeles and San Diego areas via a long open-air canal. From Parker Dam, Arizona also draws water, transporting it in the large canals of the Central Arizona Project. Further south at Imperial Dam, water is diverted into the Coachella and All-American Canals, destined for agriculture, mostly in California's Imperial Valley. To make this desert bloom, Imperial Valley farmers soak the soil with subsidized water for which they pay one penny per 795 L (210 gal).

What water is left in the Colorado River after all the diversions comprises just a trickle making its way to the Gulf of California. On some days, water does not reach the Gulf at all. This reduction in flow (**Figure 15.13**) has drastically altered the ecology of the lower river and the once-rich delta, changing plant communities, wiping out populations of fish and invertebrates, and devastating fisheries.

The Colorado's plight is not unique. Several hundred miles to the east, the Rio Grande also frequently runs dry, the victim of over-extraction by both Mexican and U.S. farmers in times of drought. The situation is even worse for China's Yellow River, so the Chinese government wants to build a massive aqueduct to supplement its flow with water from the Yangtze River. Even the river that has nurtured human civilization as long as any other, the Nile, now peters out before reaching its mouth.

Nowhere are the effects of surface water depletion so evident as at the Aral Sea. Once the fourth-largest lake on Earth, just larger than Lake Huron, it has lost over four-fifths of its volume in just 45 years (**Figure 15.14**). This dying inland sea, on the border of present-day Uzbekistan and Kazakhstan, is the victim of irrigation practices. The former Soviet Union instituted large-scale cotton farming in this region by flooding the dry land with water from the two rivers leading into the Aral Sea. For a few decades this action boosted Soviet cotton production, but it led the Aral Sea to shrink, and the irrigated soil became waterlogged and salinized. Today 60,000 fishing jobs are gone, winds blow pesticide-laden dust up from the dry lake bed, and what cotton grows on the blighted soil cannot bring the regional economy back. However, all may not be lost: Scientists, engineers, and local people struggling to save the northern portion of the Aral Sea and its damaged ecosystems may now have finally begun reversing its decline.

### Weighing THE Issues | The Klamath Crisis

In 2001, angry farmers of Klamath County, Oregon, disobeyed a federal order to divert irrigation waters downstream to save two endangered species of fish. During that bone-dry year, there simply wasn't enough water to irrigate farmers' fields and also keep endangered salmon and suckerfish alive. The government had long ago begun overallocating water here: Homesteaders enjoyed federal incentives to settle and farm the region, but their descendants are now subject to a conflicting federal mandate—the Endangered Species Act. Can you think of ways to reconcile these needs and produce solutions to the Klamath crisis? Are there similar conflicts in your region? If so, how are they being addressed?

## Inefficient irrigation wastes water

The green revolution (• pp. 237, 266–268) required significant increases in irrigation, and 70% more water is withdrawn for irrigation today than in 1960. During this period, the amount of land under irrigation has doubled (**Figure 15.15**). Expansion of irrigated agriculture has kept pace with population growth; irrigated area per capita has remained stable for at least four decades at around 460 $m^2$ (4,900 $ft^2$).

Irrigation can more than double crop yields by allowing farmers to apply water when and where it is needed. The world's 274 million ha (677 million acres) of irrigated cropland make up only 18% of world farmland but yield fully 40% of world agricultural produce, including 60% of the global grain crop. Still, most irrigation remains highly inefficient. Only about 45% of the fresh water we use for irrigation actually is taken up by crops.

(a) Ships stranded by the Aral Sea's fast-receding waters

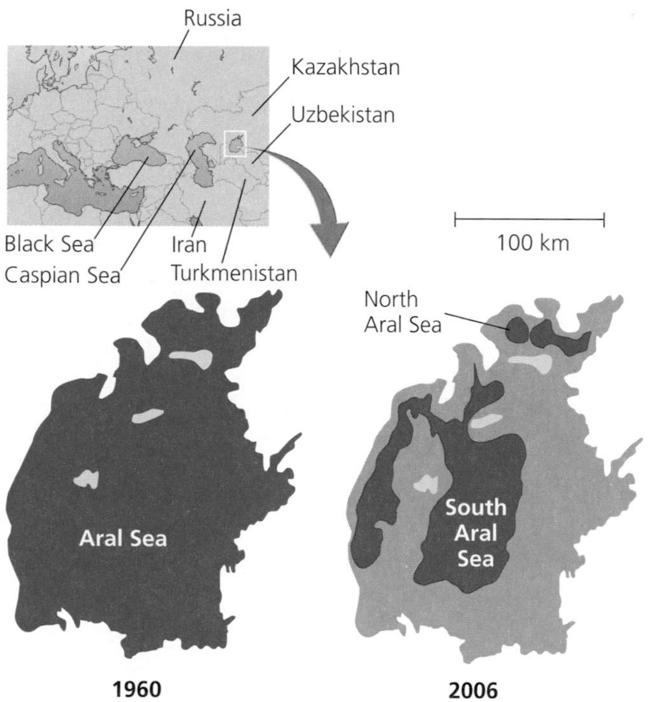


(c) The shrinking Aral Sea, then and now

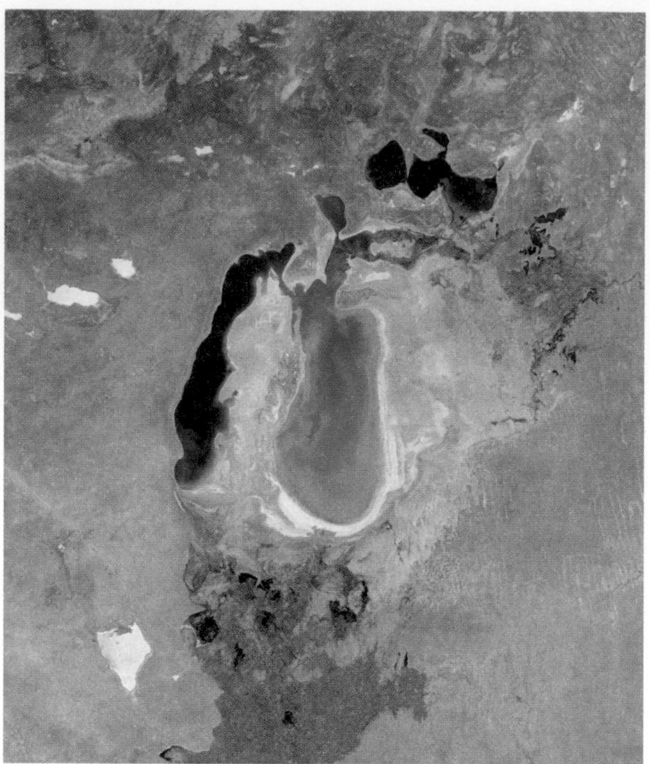

(b) Satellite view of Aral Sea, 2002

**FIGURE 15.14** Ships lie stranded in the sand (**a**) because the waters of Central Asia's Aral Sea have receded so far and so quickly (**b**). The Aral Sea was once the world's fourth-largest lake. However, it has been shrinking for the past four decades (**c**) because of overwithdrawal of water to irrigate cotton crops. Today restoration efforts are beginning to reverse the decline in the northern portion of the sea, and waters there are slowly rising.

Inefficient "flood and furrow" irrigation, in which fields are liberally flooded with water that may evaporate from standing pools, accounts for 90% of irrigation worldwide. Overirrigation leads to waterlogging and salinization, which affect one-fifth of farmland today and reduce world farming income by $11 billion (• pp. 251–252).

Many national governments have subsidized irrigation to promote agricultural self-sufficiency. Unfortunately, inefficient irrigation methods in arid regions such as the Middle East are using up huge amounts of groundwater for little gain. Worldwide, roughly 15–35% of water withdrawals for irrigation are thought to be unsustainable.

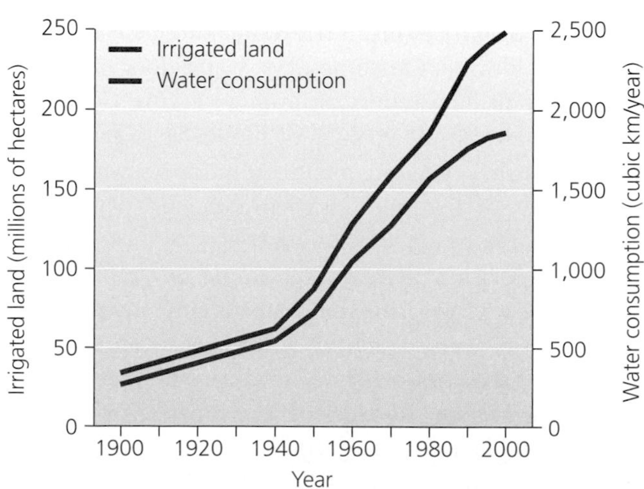


**FIGURE 15.15** Throughout the 20th century, overall global water consumption rose in tandem with the area of land irrigated for agriculture. Data from United Nations Sustainable Development Commission on Freshwater, 2002.

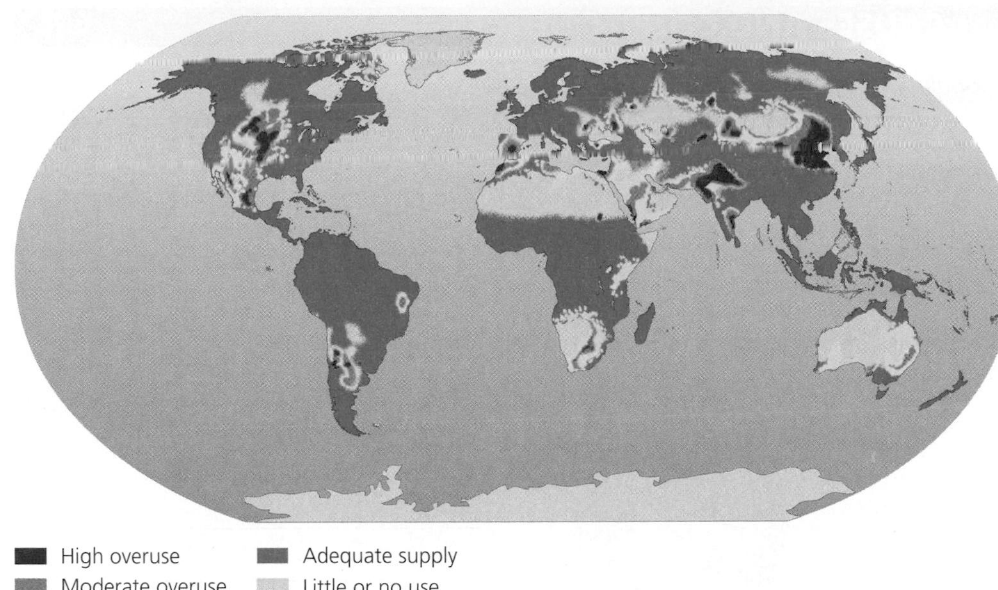


FIGURE 15.16 Irrigation for agriculture is the main contributor to unsustainable water use. Mapped are regions where overall use of fresh water (for agriculture, industry, and domestic use) exceeds the available supply, requiring groundwater depletion or diversion of water from other regions. The map understates the problem, because it does not reflect seasonal shortages. Data from UNESCO. 2006. *Water: A shared responsibility.* World Water Development Report 2. UNESCO and Berghahn Books, Paris and New York.

In areas where agriculture is demanding more fresh water than can be sustainably supplied, *water mining*—withdrawing water faster than it can be replenished—is taking place (**Figure 15.16**). In these areas, aquifers are being depleted or surface water is being piped in from other regions.

## We are depleting groundwater

Groundwater is more easily depleted than surface water because most aquifers recharge very slowly. If we compare an aquifer to a bank account, we are making more withdrawals than deposits, and the balance is shrinking. Today we are extracting 160 km$^3$ (5.65 trillion ft$^3$) more water each year than is finding its way back into the ground. This is a major problem because one-third of Earth's human population—including 99% of the rural population of the United States—relies on groundwater for its needs.

As aquifers are depleted, water tables drop. Groundwater becomes more difficult and expensive to extract, and eventually it may run out. In parts of Mexico, India, China, and other Asian and Middle Eastern nations, water tables are falling 1–3 m (3–10 ft) per year. In the United States, by the late 1990s overpumping had drawn the Ogallala Aquifer down by 325 km$^3$ (11.5 trillion ft$^3$). This volume is equal to the yearly flow of 18 Colorado Rivers.

When groundwater is overpumped in coastal areas, salt water can intrude into aquifers, making water undrinkable. This has occurred widely in the Middle East and in localities as varied as Florida, Turkey, and Bangkok. Moreover, as aquifers lose water, their substrate can become weaker and less capable of supporting overlying strata, and the land surface above may subside. For this reason, cities from Venice to Bangkok to Beijing are slowly sinking. Mexico City's downtown has sunk over 10 m (33 ft) since the time of Spanish arrival; streets are buckled, old buildings lean at angles, and underground pipes break so often that 30% of the system's water is lost to leaks.

Sometimes land subsides suddenly in the form of **sinkholes**, areas where the ground gives way with little warning, occasionally swallowing people's homes (**Figure 15.17**). Once the ground subsides, soil becomes compacted, losing the porosity that enabled it to hold water. Recharging a depleted aquifer may thereafter become more difficult. Estimates suggest that compacted

FIGURE 15.17 When too much groundwater is withdrawn too quickly, the land above it may collapse in sinkholes, sometimes bringing buildings down with it.

## CAUSES AND CONSEQUENCES

As the world's population and our demand for water rise, we are tapping groundwater and surface water sources, often at unsustainable rates. The resulting **fresh water depletion** is having ecological and social impacts in many areas. However, we have at our disposal many proven strategies for avoiding water shortages and making our water use more sustainable.

Write in two causes of fresh water depletion in the spaces provided. Then write in two consequences (impacts on the environment, human health, or quality of life) that result from fresh water depletion. Finally, offer two solutions to this issue and its consequences. One cause, one consequence, and one solution have been filled in for you, providing examples.

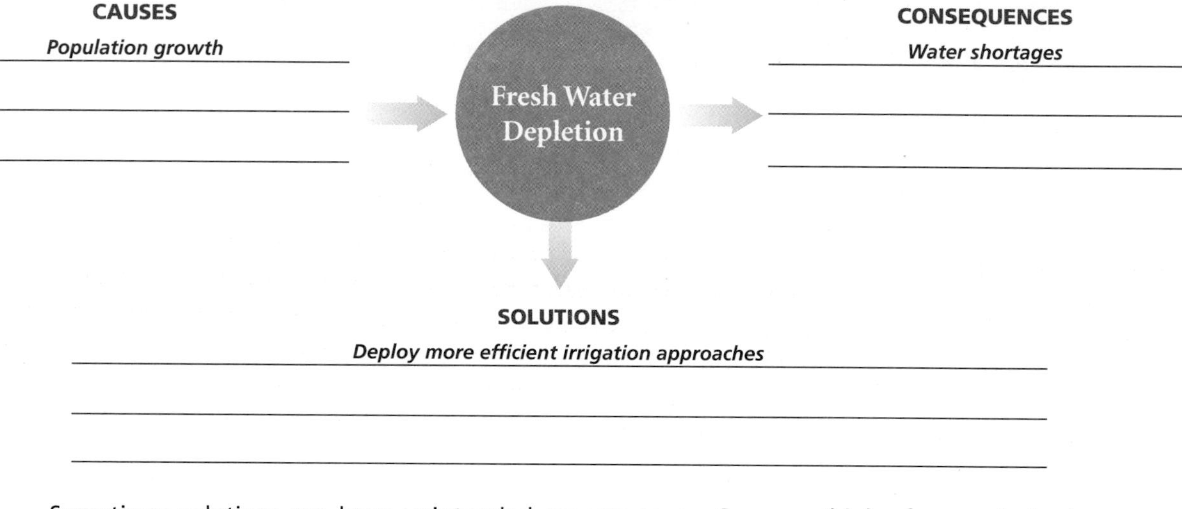


Sometimes solutions can have unintended consequences. Can you think of an undesired consequence that one solution to this issue might have? How might we then deal with *that* consequence?

aquifers under California's Central Valley have lost storage capacity equal to that of 40% of the state's artificial surface reservoirs.

Falling water tables also do vast ecological harm. Permanent wetlands exist where water tables are high enough to reach the surface, so when water tables drop, wetland ecosystems dry up. In Jordan, the Azraq Oasis covered 7,500 ha (18,500 acres) and enabled migratory birds and other animals to find water in the desert. The water table beneath this oasis dropped from 2.5 m (8.2 ft) to 7 m (23 ft) during the 1980s because of increased well use by the city of Amman. As a result, the oasis dried up altogether during the 1990s. Today international donors are collaborating with the Jordanian government to try to find alternative sources of water and restore this oasis.

## Will we see a future of water wars?

Depletion of fresh water leads to shortages, and resource scarcity can lead to conflict. On the Colorado River in 1933, the governor of Arizona foresaw that California's water diversion might endanger Arizona's future allotment, so he sent the state's National Guard to threaten the construction of Parker Dam. After a long standoff, the U.S. interior secretary halted the project to avoid hostilities while the issue was mediated in court. Arizona won the court case, but California persuaded Congress to authorize the dam, and Arizona chose not to tackle the U.S. Army troops sent to protect the dam's construction.

Many predict that water's role in regional conflicts will increase as human population continues to grow in water-poor areas and as climate change alters regional patterns of precipitation. A total of 261 major rivers, whose watersheds cover 45% of the world's land area, cross national borders, and transboundary disagreements are common. Water is already a key element in the hostilities among Israel, the Palestinian people, and neighboring nations.

On the positive side, many nations have cooperated with neighbors to resolve water disputes. India has struck cooperative agreements over management of transboundary rivers with Pakistan, Bangladesh, Bhutan, and

Nepal. In Europe, international conventions have been signed by multiple nations along the Rhine and the Danube rivers. Such progress gives reason to hope that future water wars will be few and far between.

# Solutions to Depletion of Fresh Water

Human population growth, expansion of irrigated agriculture, and industrial development doubled our annual fresh water use between 1960 and 2000. We now use an amount equal to 10% of total global runoff. The hydrologic cycle makes fresh water a renewable resource, but if our usage exceeds what a lake, river, or aquifer can provide, we must either reduce our use, find another water source, or be prepared to run out of water.

## Solutions can address supply or demand

To address depletion of fresh water, we can aim either to increase supply or to reduce demand. Strategies for reducing demand include conservation and efficiency measures. Lowering demand is more difficult politically in the short term but may be necessary in the long term. In the developing world, international aid agencies are increasingly funding demand-based solutions over supply-based solutions, because demand-based solutions offer better economic returns and cause less ecological and social damage.

To increase supply in a given area, people have transported water through pipes and aqueducts from areas where it is more plentiful or accessible. In many instances, water-poor regions have forcibly appropriated water from communities too weak to keep it for themselves. For instance, Los Angeles grew by using water it appropriated from the Owens Valley, Mono Lake, and other less-inhabited regions of California. In so doing, it desertified the environments of these areas, creating dustbowls and destroying rural economies. Today Las Vegas is trying to win approval for a 450-km (280-mi) pipeline to import groundwater from sparsely populated eastern Nevada, where local residents and wildlife advocates oppose the diversion plan.

### Weighing the Issues | Reaching for Water

In 1941, the burgeoning metropolis of Los Angeles needed water and decided to divert streams feeding into Mono Lake, over 565 km (350 mi) away in northern California. As the lake level fell 14 m (45 ft) over 40 years, salt concentrations doubled and aquatic communities suffered. Other desert cities—such as Las Vegas, Phoenix, and Denver—are expected to double in population in the coming decades. What challenges might they face in trying to pipe in water from distant sources? How will people living in these source areas be affected? Do you think such diversions are justified? How else could these cities meet their future water needs? Is your region facing any of these issues?

## Desalination "makes" more water

Another supply-side strategy is to develop technologies to find or "make" more water. The best-known technological approach to generate fresh water is **desalination**, or *desalinization,* the removal of salt from seawater or other water of marginal quality. One method of desalination mimics the hydrologic cycle by hastening evaporation from allotments of ocean water with heat and then condensing the vapor—essentially *distilling* fresh water. Another method involves forcing water through membranes to filter out salts; the most common process of this type is called *reverse osmosis.*

Over 7,500 desalination facilities are operating worldwide, most in the arid Middle East and some in small island nations that lack groundwater. The largest plant, in Saudi Arabia, produces 485 million L (128 million gal) of fresh water every day. However, desalination is expensive, requires large inputs of fossil fuel energy, and generates concentrated salty waste. In 1992, the world's largest reverse osmosis plant was completed along the Colorado River near Yuma, Arizona (**Figure 15.18**). It was intended to reduce the salinity of irrigation runoff reentering the river, but it proved too expensive to operate and closed after only 8 months. In 2007 it was reopened experimentally for 3 months, and engineers are now evaluating whether to reopen the plant for good to provide water for Yuma.

## Agricultural demand can be reduced

Because most water is used for agriculture, it makes sense to look first to agriculture for ways to decrease demand. Farmers can improve efficiency by lining irrigation canals to prevent leaks, leveling fields to minimize runoff, and adopting efficient irrigation methods. Low-pressure spray irrigation sprays water downward toward plants, and drip irrigation systems target individual plants and introduce water directly onto the soil (see Figure 9.18c, • p. 252). Both methods reduce water lost to evaporation and surface runoff. Low-pressure precision sprinklers in Texas

FIGURE 15.18 The Yuma Desalting Plant along the Colorado River in Arizona was shut down years ago because it was not cost-effective. Today, however, the region's supply of fresh water is lower and the demand is higher, so engineers are testing the plant in preparation for its possible reopening. Here, U.S. Bureau of Reclamation area manager Jim Cherry fills a bottle with desalinized water from a faucet at the plant during its 2007 trial run.

have efficiencies of 80–95% and have resulted in water savings of 25–37%. Experts have estimated that drip irrigation, which has efficiencies as high as 90%, could cut water use in half while raising yields by 20–90% and giving developing-world farmers $3 billion in extra annual income.

Choosing crops to match the land and climate in which they are being farmed can save huge amounts of water. Currently, crops that require a great deal of water, such as cotton, rice, and alfalfa, are often planted in arid areas with government-subsidized irrigation. As a result of the subsidies, the true cost of water is not part of the costs of growing the crop. Eliminating subsidies and growing crops in climates with adequate rainfall could greatly reduce water use. Finally, selective breeding (• pp. 115–117) and genetic modification (• pp. 273–276) can result in crop varieties that require less water.

## We can lessen residential and industrial water use in many ways

We can each help reduce agricultural water use by decreasing the amount of meat we eat, because producing meat requires far greater water inputs than producing grain or vegetables (• p. 283). In households, we can reduce water use by installing low-flow faucets, showerheads, washing machines, and toilets. Automatic dishwashers, studies show, use less water than does washing dishes by hand. If your home has a lawn, it is best to water it at night, when water loss from evaporation is minimal. Better yet, you can replace a water-intensive lawn with native plants adapted to your region's natural precipitation patterns. *Xeriscaping*, landscaping using plants adapted to arid conditions, has become popular in much of the U.S. Southwest (**Figure 15.19**).

Industry and municipalities can take water-saving steps as well. Manufacturers are shifting to processes that use less water and in doing so are reducing their costs. St. Petersburg, Florida, and many other cities are recycling municipal wastewater for irrigation and industrial uses. Governments in Arizona and in England are capturing excess surface runoff during their rainy seasons and pumping it into aquifers. Finding and patching leaks in pipes has saved some cities and companies large amounts of water—and money. Boston and its suburbs reduced water demand by 31% between 1987 and 2004 by patching leaks, retrofitting homes with efficient plumbing, auditing industry, and promoting conservation to the public. This program enabled Massachusetts to avoid an unpopular $500-million river diversion scheme.

FIGURE 15.19 Homeowners and businesses can reduce water consumption by xeriscaping their grounds—landscaping them with drought-tolerant plants. This xeriscaped yard is in Boulder City, Nevada, near Lake Mead.

## Economic approaches to water conservation are being debated

Economists who want to use market-based strategies to achieve sustainable water use have suggested ending government subsidies of inefficient practices and letting water become a commodity whose price reflects the true costs of its extraction. Others worry that making water a fully priced commodity would make it less available to the world's poor and increase the gap between rich and poor. Because industrial use of water can be 70 times more profitable than agricultural use, market forces alone might favor uses that would benefit wealthy and industrialized people, companies, and nations at the expense of the poor and less industrialized.

Similar concerns surround another potential solution, the privatization of water supplies. During the 1990s, many public water systems were partially or wholly privatized, their construction, maintenance, management, or ownership transferred to private companies. This was done in hope of increasing the systems' efficiency, but many firms have little incentive to allow equitable access to water for rich and poor alike. Already in some developing countries, rural residents without access to public water supplies, who are forced to buy water from private vendors, end up paying on average 12 times more than those connected to public supplies.

Other experiences indicate that decentralization of control over water, from the national level to the local level, may help conserve water. In Mexico, the effectiveness of irrigation systems improved dramatically once they were transferred from public ownership to the control of 386 local water user associations.

Regardless of how demand is addressed, the ongoing shift from supply-side to demand-side solutions is beginning to pay dividends. In Europe, a new focus on demand (through government mandates and public education) has decreased public water consumption, and industries are becoming more water-efficient. The United States decreased its water consumption by 10% from 1980 to 1995, thanks to conservation measures, even while its population grew 16%.

# Fresh Water Pollution and Its Control

The quantity and distribution of fresh water poses one set of environmental and social challenges. Safeguarding the *quality* of water involves another collection of environmental and human health dilemmas. To be safe for human consumption and other organisms, water must be relatively free of disease-causing organisms and toxic substances.

Although developed nations have made admirable advances in cleaning up water pollution over the past few decades, the World Commission on Water recently concluded that over half the world's major rivers are "seriously depleted and polluted, degrading and poisoning the surrounding ecosystems, threatening the health and livelihood of people who depend on them." The largely invisible pollution of groundwater, meanwhile, has been termed a "covert crisis."

## Water pollution takes many forms

The term **pollution** describes the release of matter or energy into the environment that causes undesirable impacts on the health and well-being of humans or other organisms. Pollution can be physical, chemical, or biological and can affect water, air, or soil.

Water pollution comes in many forms and can cause diverse impacts on aquatic ecosystems and human health. We can categorize pollution into several types, including nutrient pollution, biological pollution by disease-causing organisms, toxic chemical pollution, physical pollution by sediment, and thermal pollution.

**Nutrient pollution** We saw in Chapter 7 with the Gulf of Mexico's dead zone how nutrient pollution from fertilizers and other sources can lead to eutrophication and hypoxia in coastal marine areas (see Figure 7.5, • p. 179). Eutrophication proceeds in a similar fashion in freshwater systems, where phosphorus is usually the nutrient that spurs growth (• pp. 181, 183). When excess phosphorus enters surface waters, it fertilizes algae and aquatic plants, boosting their growth rates and populations. Although such growth provides oxygen and food for other organisms, algae can cover the water's surface, depriving deeper-water plants of sunlight. As algae die off, they provide food for decomposing bacteria. Decomposition requires oxygen, so the increased bacterial activity drives down levels of dissolved oxygen. These levels can drop too low to support fish and shellfish, leading to dramatic changes in aquatic ecosystems.

Eutrophication (**Figure 15.20**) is a natural process, but excess nutrient input from runoff from farms, golf courses, lawns, and sewage can dramatically increase the rate at which it occurs. We can reduce nutrient pollution by treating wastewater, reducing fertilizer application, planting vegetation to increase nutrient uptake, and purchasing phosphate-free detergents.

**Pathogens and waterborne diseases** Disease-causing organisms (pathogenic viruses, protists, and bacteria) can enter drinking water supplies when these are

**(a) Oligotrophic water body**

**(b) Eutrophic water body**

FIGURE 15.20 An oligotrophic water body (**a**) with clear water and low nutrient content may eventually become a eutrophic water body (**b**) with abundant algae and high nutrient content. Pollution of freshwater bodies by excess nutrients accelerates the process of eutrophication.

contaminated with human waste from inadequately treated sewage or with animal waste from feedlots (• pp. 281–282). Specialists monitoring water quality can tell when water has been contaminated by waste when they detect fecal coliform bacteria, which live in the intestinal tracts of people and other vertebrates. These bacteria are usually not pathogenic themselves, but they serve as indicators of fecal contamination, which may mean that the water holds other pathogens that can cause ailments such as giardiasis, typhoid, or hepatitis A.

Biological pollution by pathogens causes more human health problems than any other type of water pollution. A study of global water supply and sanitation issues by the World Health Organization (WHO) and the United Nations Children's Fund (UNICEF) in 2000 showed that despite advances in many parts of the world, major problems still existed. On the positive side, 4.9 billion people (82% of the population) had access to safe water as a result of some form of improvement in their water supply—an increase from 4.1 billion (79% of the population) in 1990. However, over 1.1 billion people were still without safe water supplies. In addition, 2.4 billion people had no sewer or sanitation facilities. Most of these people were Asians and Africans, and four-fifths of the people without sanitation lived in rural areas. These conditions contribute to widespread health impacts and 5 million deaths per year.

Treating sewage (• pp. 435–438) constitutes one approach for reducing the risks that waterborne pathogens pose. Another is using chemical or other means to disinfect drinking water. Others include hygienic measures such as public education to encourage personal hygiene and government enforcement of regulations to ensure the cleanliness of food production, processing, and distribution.

**Toxic chemicals** Our waterways have become polluted with toxic organic substances of our own making, including pesticides, petroleum products, and other synthetic chemicals. Many of these can poison animals and plants, alter aquatic ecosystems, and cause a wide array of human health problems, including cancer. In addition, toxic metals, such as arsenic, lead, and mercury, and acids from acid precipitation and from acid drainage from mining sites (• p. 560) also cause negative impacts on human health and the environment. Health impacts of toxic chemicals are discussed in Chapter 14.

Legislating and enforcing more stringent regulations of industry can help reduce releases of these toxic inorganic chemicals. Better yet, we can modify our industrial processes and our purchasing decisions to rely less on these substances.

**Sediment** Although floods build fertile farmland, sediment that rivers transport can also impair aquatic ecosystems. Mining, clear-cutting, land clearing for housing development, and careless cultivation of farm fields all expose soil to wind and water erosion (• pp. 242–245). Some water bodies, such as the Colorado River and China's Yellow River, are naturally sediment-rich, but many others are not. When a clear-water river receives a heavy influx of eroded sediment, aquatic habitat can change dramatically, and fish adapted to clear-water environments may not be able to adjust. We can reduce sediment pollution by better managing farms and forests and avoiding large-scale disturbance of vegetation.

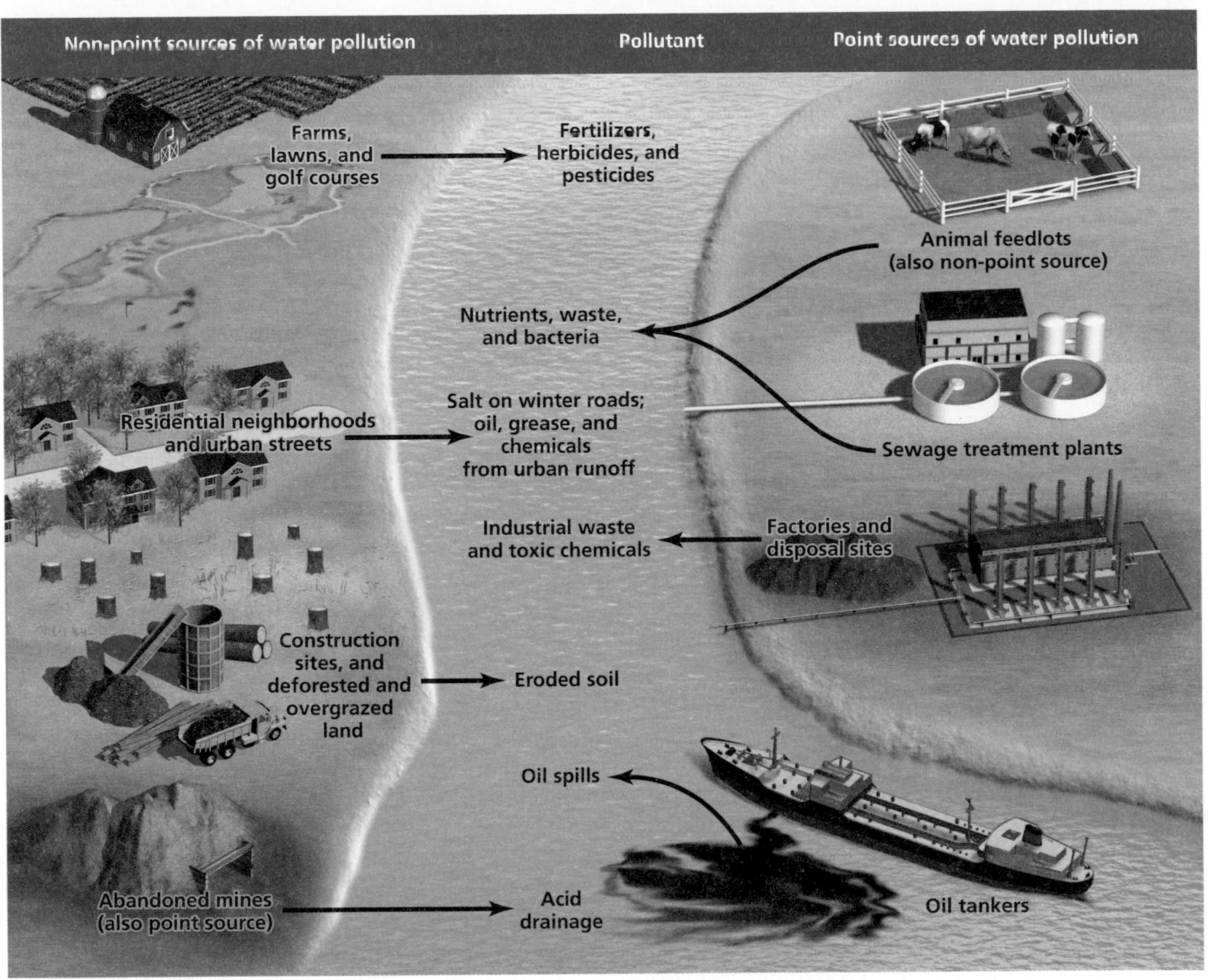


FIGURE 15.21 Point-source pollution comes from discrete facilities or locations, usually from single outflow pipes. Non-point-source pollution (such as runoff from streets, residential neighborhoods, lawns, and farms) originates from numerous sources spread over large areas.

**Thermal pollution** Water's ability to hold dissolved oxygen decreases as temperature rises, so some aquatic organisms may not survive when human activities raise water temperatures. When we withdraw water from a river and use it to cool an industrial facility, we transfer heat energy from the facility back into the river where the water is returned. People also raise surface water temperatures by removing streamside vegetation that shades water.

Too little heat can also cause problems. On the Colorado and many other dammed rivers, water at the bottoms of reservoirs is colder than water at the surface. When dam operators release water from the depths of a reservoir, downstream water temperatures drop suddenly. In the Colorado River system, these low water temperatures have favored cold-loving invasive trout over an endangered native species of suckerfish.

## Water pollution comes from point and non-point sources

Some water pollution is emitted from **point sources**—discrete locations, such as a factory or sewer pipe. In contrast, **non-point-source** pollution arises from multiple cumulative inputs over larger areas, such as farms, city streets, and residential neighborhoods (**Figure 15.21**). The U.S. Clean Water Act (• p. 68), by targeting industrial discharges, addressed point-source pollution with some success, such that non-point-source pollution exerts a greater impact on water quality in the United States today. Many common activities give rise to non-point-source water pollution, such as applying fertilizers and pesticides to lawns, applying salt to roads in winter, and changing automobile oil. To minimize non-point-source pollution of drinking water, governments limit development on watershed land surrounding reservoirs.

## Scientists use several indicators of water quality

Most forms of water pollution are not very visible to the human eye, so scientists and technicians measure certain physical, chemical, and biological properties of water to characterize its quality (**Figure 15.22**). Biological properties include the presence of fecal coliform bacteria and other disease-causing organisms, as discussed above. Algae and aquatic invertebrates are also commonly used as biological indicators of water quality.

Chemical properties include nutrient concentrations, pH (• pp. 94–96), taste and odor, and hardness. Hard water contains high concentrations of calcium and magnesium ions, prevents soap from lathering, and leaves chalky deposits behind when heated or boiled. An important chemical characteristic is dissolved oxygen content. Dissolved oxygen is an indicator of aquatic ecosystem health because surface waters low in dissolved oxygen are less capable of supporting aquatic life.

Among physical characteristics, turbidity measures the density of suspended particles in a water sample. Fast-moving rivers that cut through arid or eroded landscapes, such as the Colorado and Yellow Rivers, carry a great deal of sediment and are turbid and muddy-looking as a result. Water color can reveal particular substances present in a sample. Some forest streams run the color of iced tea because of chemicals called tannins that occur naturally in decomposing leaf litter. Finally, temperature can be used to assess water quality. High temperatures can interfere with some biological processes, and warmer water holds less dissolved oxygen. If scientists can measure only one parameter, they will often choose turbidity, because it tends to correlate with many others and is thereby a good indicator of overall water quality.

**FIGURE 15.22** Scientists and technicians characterize water quality by measuring certain physical, chemical, and biological properties. Here a researcher collects water samples from a lagoon near a landfill in England.

## Groundwater pollution is a serious problem

Most efforts at pollution control have focused on surface water. Yet increasingly, groundwater sources once assumed to be pristine have been contaminated by pollution from industrial and agricultural practices. Groundwater pollution is largely hidden from view and is extremely difficult to monitor; it can be out-of-sight, out-of-mind for decades until widespread contamination of drinking supplies is discovered.

Groundwater pollution is also more difficult to manage than surface water pollution. Rivers flush their pollutants fairly quickly, but groundwater retains its contaminants until they decompose, which in the case of persistent pollutants can be many years or decades. The long-lived pesticide DDT, for instance, is still found widely in U.S. aquifers even though it was banned 35 years ago. Moreover, chemicals break down much more slowly in aquifers than in surface water or soils. Groundwater generally contains less dissolved oxygen, microbes, minerals, and organic matter, so decomposition is slower. For instance, concentrations of the herbicide alachlor decline by half after 20 days in soil, but in groundwater this takes almost 4 years.

## There are many sources of groundwater pollution

Some chemicals that are toxic at high concentrations, including aluminum, fluoride, nitrates, and sulfates, occur naturally in groundwater. The poisoning of Bangladesh's wells by arsenic is one case of natural contamination (see "The Science behind the Story," • pp. 434–435). However, groundwater pollution from human activity is widespread. Industrial, agricultural, and urban wastes—from heavy metals to petroleum products to industrial solvents to pesticides—can leach through soil and seep into aquifers. Pathogens and other pollutants can enter groundwater through improperly designed wells and from the pumping of liquid hazardous waste below ground (• pp. 651–652). A 17-year study of volatile organic compounds (VOCs; • p. 484) published in 2006 detected VOCs from manufactured products and industrial processes in most U.S. aquifers and in 20% of wells sampled. Chemicals detected included chloroform, MTBE (• p. 85), and 40 other VOCs.

FIGURE 15.23 Leaky underground storage tanks are a major source of groundwater pollution. Under an EPA program, hundreds of thousands of these tanks are being unearthed and repaired.

Leakage from underground septic tanks, tanks of industrial chemicals, and tanks of oil and gas also pollute groundwater. Across the United States, the Environmental Protection Agency (EPA) has embarked on a nationwide cleanup program to unearth and repair leaky tanks before they do further damage to soil and groundwater quality (**Figure 15.23**). After more than a decade of work, the EPA by March 2007 had confirmed leaks from 468,000 tanks, had initiated cleanups on 439,000 of them, and had completed cleanups of 357,000. Intercepting carcinogenic or otherwise toxic pollutants such as chlorinated solvents and gasoline before they reach aquifers is vital because once an aquifer is contaminated, it is extremely difficult to remediate.

Agriculture also contributes to groundwater pollution. Nitrate from fertilizers has leached into groundwater in Canada and in 49 U.S. states. Pesticides were detected in over half of the shallow aquifer sites tested in the United States in the mid-1990s, although generally below the standards set by the EPA for drinking water. Nitrate in drinking water has been linked to cancers, miscarriages, and "blue-baby" syndrome, which reduces the oxygen-carrying capacity of infants' blood. Agriculture can also contribute pathogens; in 2000, the groundwater supply of Walkerton, Ontario, became contaminated with the bacterium *Escherichia coli,* or *E. coli.* Two thousand people became ill, and seven died.

Manufacturing industries and military sites have been heavy polluters through the years. Although legislation has forced them to reduce discharges, groundwater can bear a toxic legacy long afterwards. During World War II, the U.S. Army operated the world's largest facility to produce trinitrotoluene (TNT) near St. Louis. Nitroaromatic byproducts seeped into the drinking water for miles around. The site was prioritized to be cleaned up under the 1980 Superfund legislation (• pp. 652–653). At another Superfund site, the Hanford Nuclear Reservation in Washington State, groundwater has been contaminated with radioactive waste, some of it with a half-life of a quarter-million years.

## Legislative and regulatory efforts have helped reduce pollution

As numerous as our freshwater pollution problems may seem, it is important to remember that many of them were worse a few decades ago, when the Cuyahoga River would catch fire (• p. 66). Citizen activism and government response during the 1960s and 1970s in the United States resulted in legislation such as the Federal Water Pollution Control Act of 1972 (later amended and renamed the Clean Water Act in 1977). These acts made it illegal to discharge pollution from a point source without a permit, set standards for industrial wastewater, set standards for contaminant levels in surface waters, and funded construction of sewage treatment plants. Thanks to such legislation, point-source pollution in the United States has been reduced, and rivers and lakes are cleaner than they have been in decades.

Other developed nations have also reduced pollution. In Japan, Singapore, China, and South Korea, legislation, regulation, enforcement, and investment in wastewater treatment have brought striking water quality improvements. However, non-point-source pollution, eutrophication, and acid precipitation remain major challenges.

The Great Lakes of Canada and the United States represent a success story in fighting water pollution. In the 1970s these lakes, which hold 18% of the world's surface fresh water, were badly polluted with wastewater, fertilizers, and toxic chemicals. Algal blooms occurred along beaches, and Lake Erie was pronounced dead. Today, efforts of the Canadian and U.S. governments have paid off. According to Environment Canada, releases of seven toxic chemicals are down by 71%, municipal phosphorus has decreased by 80%, and chlorinated pollutants from paper mills are down by 82%. Levels of PCBs and DDE are down by 78% and 91%, respectively. Bird populations are rebounding, and Lake Erie is now home to the world's largest walleye fishery. The Great Lakes' troubles are by no means over—sediment pollution is still heavy, PCBs and mercury still settle on the lakes from the air, and fish are not always safe to eat. However, the progress so far shows how conditions can improve when citizens push their governments to take action.

## We treat our drinking water

Technological advances have also improved our ability to control pollution. The treatment of drinking water and the treatment of wastewater are mainstream practices in

THE SCIENCE BEHIND THE STORY

## Arsenic in the Waters of Bangladesh

*Skin lesions caused by arsenic poisoning in Bangladesh*

In the 1970s, UNICEF, with the help of environmental scientists at the British Geological Survey, launched a campaign to improve access to fresh water in Bangladesh. By digging thousands of small artesian wells, the designers of the program hoped to reduce Bangladeshis' dependence on disease-ridden surface waters. In the mid-1990s, however, scientists began to suspect that the wells dug to improve Bangladeshis' health were contaminated with arsenic, a poison that, if ingested frequently, can cause serious skin disorders and other illnesses, including cancer.

A medical doctor sounded the first alarm. In 1983, dermatologist K. C. Saha of the School of Tropical Medicine in Calcutta, India, saw the first of many patients from West Bengal, an area of India just west of Bangladesh, who showed signs of arsenic poisoning. Through a process of elimination, contaminated well water was identified as the likely cause of the poisoning. The hypothesis was confirmed by groundwater testing and by the work of epidemiologists, among them Dipankar Chakraborti of Calcutta's Jadavpur University.

However, it was not until the late 1990s that large-scale testing of Bangladesh's wells began. By 2001, when the British Geological Survey and the government of Bangladesh published their final report, 3,524 wells had been tested. Of the shallow wells, those less than 150 m (490 ft) deep, 46% exceeded the World Health Organization's maximum recommended level of 10 μg/L ($1.33 \times 10^{-6}$ oz/gal) of arsenic. Extrapolating across all of Bangladesh, the scientists estimated that as many as 2.5 million wells serving 57 million people were contaminated. As the figure shows, arsenic contamination is most prevalent in southern Bangladesh, but localized hot spots are found in northern regions of the country.

Scientists have not yet reached consensus on the chemical processes by which Bangladesh's shallow aquifers became contaminated. All agree that the arsenic is of natural origin; what remains unclear is how the low levels of arsenic naturally present in soils were dissolved in the aquifers in elemental and highly toxic form. One initial explanation, suggested by Chakraborti and his colleagues, placed most of the blame on agricultural irrigation. By drawing large amounts of water out of aquifers during Bangladesh's dry season, they argued, irrigation had permitted oxygen to enter the aquifers and prompted the release of arsenic from pyrite, a common mineral.

Other scientists contend that pyrite oxidation cannot explain most cases of arsenic contamination. In a 1998 paper in the journal *Nature*, Ross Nickson of the British Geological Survey and his colleagues suggested that arsenic was being released from

developed nations today. The U.S. EPA sets standards for over 80 drinking water contaminants, which local governments and private water suppliers are obligated to meet. Before being sent to your tap, water from a reservoir or aquifer is treated with chemicals to remove particulate matter; passed through filters of sand, gravel, and charcoal; and/or disinfected with small amounts of an agent such as chlorine.

### It is better to prevent pollution than to mitigate it after it occurs

In many cases, solutions to pollution will need to involve prevention, not simply "end-of-pipe" treatment and cleanup. For instance, with groundwater contamination, preventing pollution in the first place would seem to be the best strategy when one considers the other options for dealing with the problem:

- Filtering groundwater before distributing it can be expensive; facilities in the U.S. Midwest by one estimate spend $400 million annually just to remove the herbicide atrazine from water supplies.
- Pumping water out of an aquifer, treating it, then injecting it back in, repeatedly, takes an impracticably long time. Superfund sites are using this method, but work has begun on just over 1% of all sites with heavily polluted groundwater, and the eventual cleanup bill has been estimated at $1 trillion.
- Restricting pollutants on lands above selected aquifers would alleviate contamination in those aquifers—but could simply shift pollution elsewhere.

There are many things ordinary people can do to help minimize freshwater pollution. One is to exercise the power of consumer choice in the marketplace by purchasing

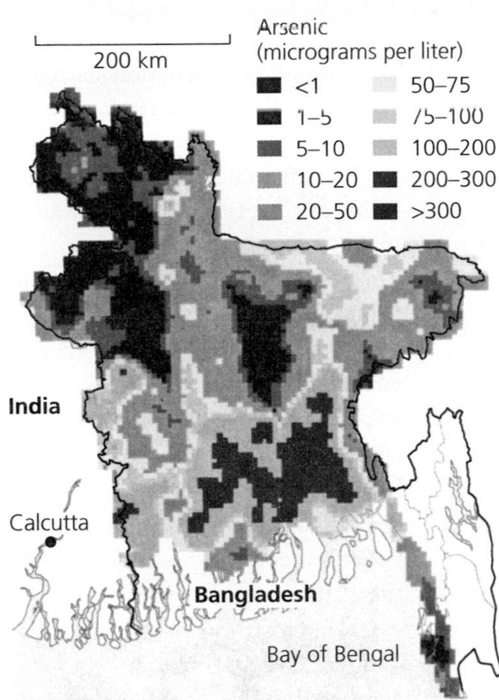


In a modern tragedy, thousands of wells dug for drinking water in Bangladesh at the urging of international aid workers turned out to be laced with arsenic. This map shows that arsenic concentrations are highest in the southern portion of the country. *Source:* Kinniburgh, D. G., and P. L. Smedley, eds. 2001. Arsenic contamination of groundwater in Bangladesh. Department of Public Health Engineering Bangladesh, British Geological Survey Report.

iron oxides carried into Bangladesh by the Ganges River. They pointed to results of a hydrochemical survey that measured the chemical composition of aquifers throughout Bangladesh. Contrary to the predictions of the pyrite oxidation hypothesis, the survey found that arsenic concentrations tended to increase with aquifer depth and to be inversely correlated with concentrations of sulfur, a component of pyrite. Nickson and his colleagues concluded that highly reducing chemical conditions created by buried organic matter, such as peat, had probably leached arsenic from iron oxides over thousands of years.

Recently, Massachusetts Institute of Technology hydrologist Charles Harvey and colleagues suggested that irrigation may contribute to the arsenic problem after all, but not because of pyrite oxidation. In a 2002 paper in the journal *Science,* they described an experiment in which more than a dozen wells were dug near the capital city of Dhaka. Contrary to the pyrite oxidation hypothesis, and in agreement with Nickson and his colleagues, they found little evidence of a connection between sulfur or oxygen and arsenic.

However, they also found that they could increase arsenic concentrations by injecting organic matter, such as molasses, into their experimental wells. In the process of being metabolized by microbes, the molasses appeared to be freeing arsenic from iron oxides. A similar process might take place naturally, Harvey's team argued, when runoff from rice paddies, ponds, and rivers recharges aquifers that have been depleted by heavy pumping for irrigation. In support of this hypothesis, they found that much of the carbon in the shallow wells was of recent origin. Other scientists, however, have found arsenic in much older waters. This finding suggests that Bangladesh's arsenic problem may be caused by multiple hydrological and geological factors.

phosphorus-free detergents and other "environmentally friendly" products. Another is to become involved in protecting local waterways. Locally based "riverwatch" groups or watershed associations enlist volunteers to collect data and help state and federal agencies safeguard the health of rivers and other water bodies. Such programs are proliferating as citizens and policymakers increasingly demand clean water.

# Wastewater and Its Treatment

**Wastewater** refers to water that has been used by people in some way. It includes water carrying sewage; water from showers, sinks, washing machines, and dishwashers; water used in manufacturing or cleaning processes by businesses and industries; and stormwater runoff.

Although natural systems can process moderate amounts of wastewater, the large and concentrated amounts generated by our densely populated areas can harm ecosystems and pose threats to human health. Thus, attempts are now widely made to treat wastewater before releasing it into the environment.

## Municipal wastewater treatment involves several steps

In rural areas, **septic systems** are the most popular method of wastewater disposal. In a septic system, wastewater runs from the house to an underground septic tank, inside which solids and oils separate from water. The clarified water proceeds downhill to a drain field of perforated pipes laid horizontally in gravel-filled trenches underground. Microbes decompose the

THE SCIENCE BEHIND THE STORY

## Using Nature to Treat Wastewater

*An Arcata city worker checks levels of microbes on marsh plants*

On a stretch of northern California's scenic Redwood Coast, a beautiful marsh hugs the waterfront lining the town of Arcata. Bulrushes, cattails, and wildflowers wave in the breeze. Locals come to jog or walk their dogs. Thousands of birds forage for food in the shallow waters. Few visitors would guess the whole thing is built on sewage.

About 450 km (280 mi) up the coast from San Francisco, the Arcata Marsh and Wildlife Sanctuary is a 121-ha (300-acre) site where science and nature merge to clean up pollution. The marsh is a human-engineered wetland created to filter wastewater. Aquatic plants and microbes perform secondary wastewater treatment, cleaning partially treated sewage water.

Arcata, a town of 16,000 people that sits on Humboldt Bay, built its treatment wetland after requirements of the Federal Water Pollution Control Act took effect in 1974. Arcata's first plan to meet the new requirements called for building a larger treatment plant that would cost more than $50 million and pipe wastewater far out into the ocean. Local residents opposed the plan, pushing for less expensive options more in step with the local environment. Two environmental scientists from nearby Humboldt State University, Robert Gearheart and George Allen, came forward, saying wetlands might provide the treatment Arcata needed.

Gearheart's and Allen's Marsh Pilot Project began in 1979 with 10 small constructed wetlands that together handled 10% of Arcata's wastewater. After 2 years, the pilot study revealed that wetlands could clean wastewater and also restore coastal wetland habitat. Arcata's civic leaders voted to use marshland to solve their sewage treatment problem. Wetland construction started in 1981, and the Arcata marsh system opened in 1986.

Treatment begins with heavy filtering at a conventional wastewater plant. To remove nitrogen compounds, pathogens, and suspended solids that initial sewage treatment fails to catch, Gearheart and Allen engineered a system of plants and microbes in the 20 ha (49 acres) of aeration ponds, where oxygen-breathing bacteria break down sewage, and in the six treatment and enhancement marshes, which receive the water for further filtering.

The marshes are lined with a mix of aquatic plants, including duckweed, pennywort, and a native species, hardstem bulrush. The roots and stems of these plants form a dense network of underwater vegetation, hosting a wide variety of single-celled microbes and fungi. An especially productive zone lies in the plants' rhizome network, where tiny root hairs form a thick web that shelters algae, bacteria, and other microscopic sewage-treaters.

wastewater these pipes emit. Periodically, solid waste needs to be pumped from the septic tank and taken to a landfill.

In more densely populated areas, municipal sewer systems carry wastewater from homes and businesses to centralized treatment locations. There, pollutants in wastewater are removed by physical, chemical, and biological means (**Figure 15.24**).

At a treatment facility, **primary treatment**, the physical removal of contaminants in settling tanks or clarifiers, generally removes about 60% of suspended solids from wastewater. Wastewater then proceeds to **secondary treatment,** in which water is stirred and aerated so that aerobic bacteria degrade organic pollutants. Roughly 90% of suspended solids may be removed after secondary treatment. Finally, the clarified water is treated with chlorine, and sometimes ultraviolet light, to kill bacteria. Most often, the treated water, called *effluent*, is piped into rivers or the ocean following primary and secondary treatment. Sometimes, however, "reclaimed" water is used for lawns and golf courses, for irrigation, or for industrial purposes such as cooling water in power plants.

As water is purified throughout the treatment process, the solid material that is removed is termed *sludge*. Sludge is sent to digesting vats, where microorganisms decompose much of the matter. The result, a wet solution of "biosolids," is then dried and either disposed of in a landfill, incinerated, or used as fertilizer on cropland. Methane- rich gas created by the decomposition process is sometimes burned to generate electricity, helping to offset

The Arcata Marsh and Wildlife Sanctuary is the site of an artificially constructed wetland that helps treat this northern California city's wastewater.

Some of the roots and microbes in this network trap suspended solids, holding them until the plants can use them as fertilizer. Other microbes break down nitrogen compounds such as nitrate through denitrification (• p. 192). Still other microbes consume coliform bacteria, which may also die when exposed to sunlight in the marsh's shallow waters. Scientists who monitor the marshes know the cleaning system is working when the roots and stems of marsh plants are slimy and slippery: evidence that cleansing microbes are growing and feeding and that the rhizome network is trapping suspended solids. After 2 months of treatment, the wastewater is released into Humboldt Bay.

The marshes are monitored carefully to avoid overwhelming the wetland's cleaning abilities. Too much nitrogen can lead to algal blooms. If initial sewage treatment does not catch contaminants such as lead, these toxins can accumulate and harm microbes or wildlife. Dissolved oxygen, temperature, and nitrogen content are measured regularly, and marsh plants are thinned every few years.

The system's low maintenance needs and simplicity have won awards and saved money. The cost of creating the marshes was approximately $7 million, far less than that of the conventional sewage plant that was first proposed.

The marsh treatment system has also brought the Arcata waterfront back to life. More than 100,000 people visit the marsh each year, and more than 250 species of birds have been observed there.

Other cities have used the Arcata Marsh and Wildlife Sanctuary as a model for their own wastewater treatment wetlands, and the marsh is a site for ongoing research on wetlands and solutions to water pollution problems.

the cost of the treatment facility. Each year about 6 million dry tons of sludge are generated in the United States.

### Weighing THE Issues | Sludge on the Farm

It is estimated that anywhere from 38% to over half of the sewage sludge, or biosolids, produced each year is used as fertilizer on agricultural lands. This practice makes productive use of the sludge, increases crop output, and conserves landfill space, but many people have voiced concern over accumulation of toxic metals, proliferation of dangerous pathogens, and odors. Do you feel this practice represents an efficient use of resources or an unnecessary risk? What further information would you want to know to inform your decision?

## Artificial wetlands can aid treatment

Natural wetlands already perform the ecosystem service of water purification, and wastewater treatment engineers are now manipulating wetlands and even constructing wetlands *de novo* to employ them as tools to cleanse wastewater (see "The Science behind the Story," above). Generally in this approach, wastewater that has gone through primary treatment at a conventional facility is pumped into the wetland, where microbes living amid the algae and aquatic plants decompose the remaining pollutants. Water cleansed in the wetland can then be released into waterways or allowed to percolate underground.

Constructed wetlands also serve as havens for wildlife and areas for human recreation. For example, at Sweetwater

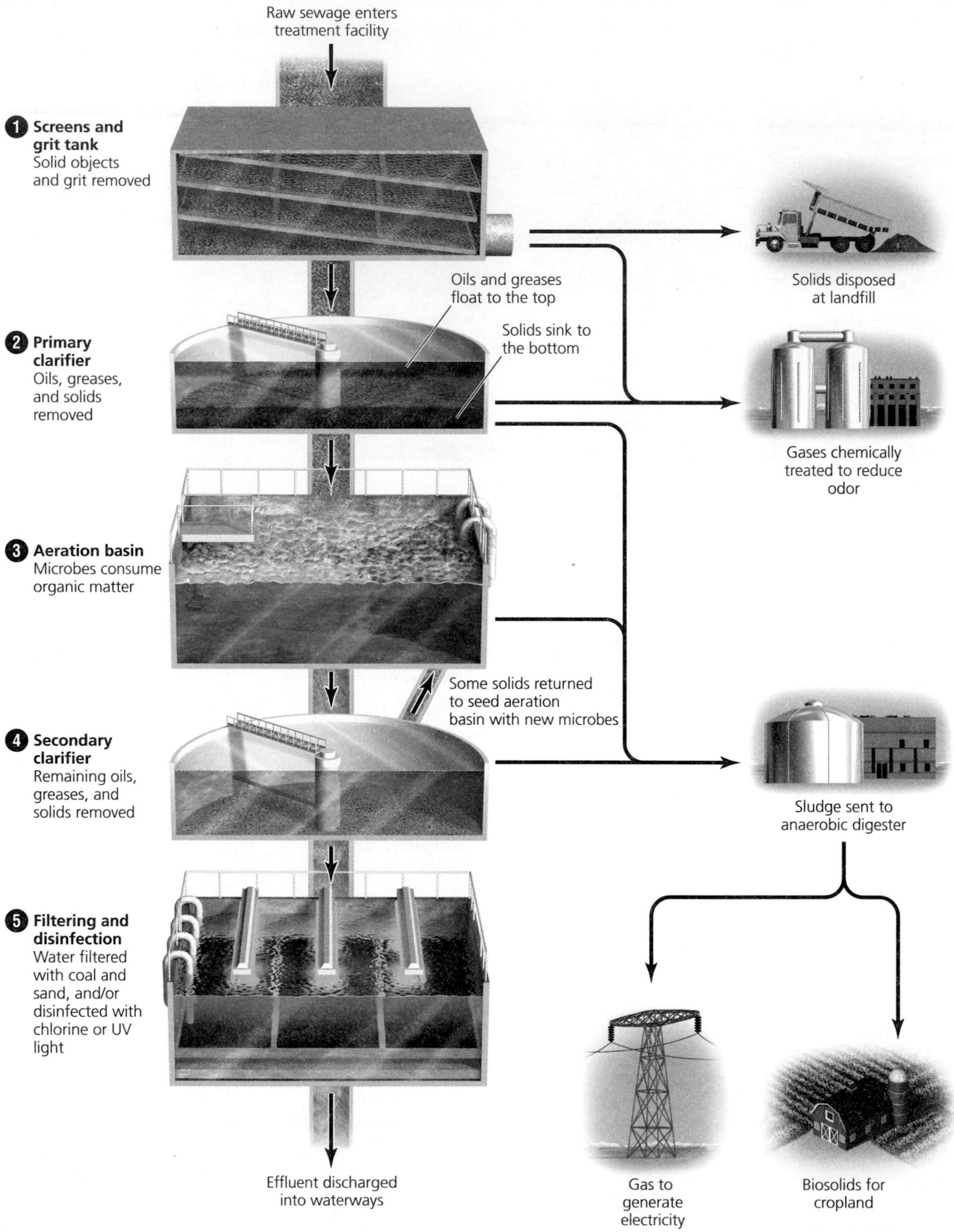

FIGURE 15.24 Shown here is a generalized process from a modern, environmentally sensitive wastewater treatment facility. Wastewater initially passes through screens to remove large debris and into grit tanks to let grit settle (1). It then enters tanks called primary clarifiers (2), in which solids settle to the bottom and oils and greases float to the top for removal. Clarified water then proceeds to aeration basins (3) that oxygenate the water to encourage decomposition by aerobic bacteria. Water then passes into secondary clarifier tanks (4) for removal of further solids and oils. Next, the water may be purified by chemical treatment with chlorine, passage through carbon filters, and /or exposure to ultraviolet light (5). The treated water (called *effluent*) may then be piped into natural water bodies, used for urban irrigation, flowed through an artificial wetland, or used to recharge groundwater. In addition, most treatment facilities control odor in the early steps and use anaerobic bacteria to digest sludge removed from the wastewater. Sludge from digesters may be sent to farm fields as fertilizer, and gas from digestion may be used to generate electric power.

Wetlands in Tucson, Arizona, the artificial marshes constructed in 1996 comprise a wetland oasis for birds and wildlife in this desert region while helping to recharge a depleted aquifer. The practice of treating wastewater with artificial wetlands is growing fast; today over 500 artificially constructed or restored wetlands in the United States are performing this service.

## Conclusion

Citizen action, government legislation and regulation, new technologies, economic incentives, and public education are all enabling us to confront what will surely be one of the great environmental challenges of the new century: ensuring adequate quantity and quality of fresh water for ourselves and for the planet's ecosystems.

Accessible fresh water comprises only a minuscule percentage of the hydrosphere, but we generally take it for granted. With our expanding population and increasing water usage, we are approaching conditions of widespread scarcity. Water depletion and water pollution are already taking a toll on the health, economies, and societies of the developing world, and they are beginning to do so in arid areas of the developed world. There is reason to hope that we may yet attain sustainability in our water usage, however. Potential solutions are numerous, and the issue is too important to ignore.

## REVIEWING OBJECTIVES

**You should now be able to:**

**Explain the importance of water and the hydrologic cycle to ecosystems, human health, and economic pursuits**

- We depend utterly on drinkable water, and a functioning hydrologic cycle is vital to maintaining ecosystems and our civilization. (pp. 413–414, 416–417)

**Delineate the distribution of fresh water on Earth**

- Of all the water on Earth, only about 1% is readily available for our use. (p. 413)
- Water availability varies in space and time, and regions vary greatly in the amounts they possess. (pp. 417–418)

**Describe major types of freshwater ecosystems**

- The main types of freshwater ecosystems include rivers and streams, wetlands, and lakes and ponds. (pp. 414–416)

**Discuss how we use water and alter freshwater systems**

- We use water for agriculture, industry, and residential use. The ratio of these uses varies among societies, but globally 70% is used for agriculture. (pp. 418–419)
- Most of the world's rivers are dammed. Dams bring a diverse set of benefits and costs. Increasingly, people are proposing dam removal. (pp. 419–421)
- We divert water with canals and irrigation ditches and attempt to control floods with dikes and levees. (pp. 421–423)
- We pump water from aquifers and surface water bodies, sometimes at unsustainable rates. (pp. 424–426)

**Assess problems of water supply and propose solutions to address depletion of fresh water**

- Water tables are dropping worldwide from unsustainable groundwater extraction. Surface water extraction has caused rivers to run dry and water bodies to shrink. (pp. 424–426)
- Unequal water distribution amid shrinking supplies may heighten political tensions over water in the future. (p. 426)
- Solutions to expand supply, such as desalination, are worth pursuing, but not to the exclusion of finding ways to decrease demand. (pp. 427–428)
- Solutions to reduce demand include technology, approaches, and consumer products that increase efficiency in agriculture, industry, and the home. (pp. 428)

**Assess problems of water quality and propose solutions to address water pollution**

- Water pollutants include excessive nutrients, microbial pathogens, toxic chemicals, sediment, and thermal pollution. (pp. 429–431)
- Water pollution stems from point sources and nonpoint sources. (p. 431)
- Scientists who monitor water quality use biological, chemical, and physical indicators. (p. 432)
- Groundwater pollution can be more persistent than surface water pollution. (pp. 432–433)
- Legislation and regulation have improved water quality in developed nations in recent decades. (p. 433)
- Preventing water pollution is better than mitigation. (pp. 434–435)

**Explain how wastewater is treated**

- Septic systems are used to treat wastewater in rural areas. (pp. 435–436)
- Wastewater is treated physically, biologically, and chemically in a series of steps at municipal wastewater treatment facilities. (pp. 436, 438)
- Artificial wetlands enhance wastewater treatment while restoring habitat for wildlife. (pp. 436–437, 439)

## TESTING YOUR COMPREHENSION

1. Define *groundwater*. What role does groundwater play in the hydrologic cycle?
2. Why are sources of fresh water unreliable for some people and plentiful for others?
3. Describe three benefits of damming rivers, and three costs. What particular environmental, health, and social concerns has China's Three Gorges Dam and its reservoir raised?
4. Why do the Colorado, Rio Grande, Nile, and Yellow rivers now slow to a trickle or run dry before reaching their deltas?
5. Why are water tables dropping around the world? What are some environmental costs of falling water tables?
6. Name three major types of water pollutants, and provide an example of each. List three properties of water that scientists use to determine water quality.
7. Why do many scientists consider groundwater pollution a greater problem than surface water pollution?
8. What are some anthropogenic (human) sources of groundwater pollution?
9. Describe how drinking water is treated. How does a septic system work?
10. Describe and explain the major steps in the process of wastewater treatment. How can artificial wetlands aid such treatment?

## SEEKING SOLUTIONS

1. Discuss possible strategies for equalizing distribution of water throughout the world. Consider supply and transport issues. Have our methods of drawing, distributing, and storing water changed very much throughout history? How is the scale of our efforts affecting the availability of water supplies?
2. How can we lessen agricultural demand for water? Describe some ways in which we can reduce household water use. How can industrial uses of water be reduced?
3. Have the provisions of the Clean Water Act been effective? Discuss some of the methods we can adopt, in addition to "end-of-pipe" solutions, to prevent contamination and ensure "water security."
4. How might desalination technology help "make" more water? Describe two methods of desalination. Where is this technology being used?
5. **THINK IT THROUGH** Your state's governor has put you in charge of water policy for the state. The aquifer beneath your state has been overpumped, and many wells have already run dry. Agricultural production last year decreased for the first time in a generation, and farmers are clamoring for you to do something. Meanwhile, the state's largest city is growing so fast that more water is needed for its burgeoning urban population. What policies would you consider to restore your state's water supply? Would you try to take steps to increase supply, to decrease demand, or both? Explain why you would choose such policies.
6. **THINK IT THROUGH** Having solved the water depletion problem in your state, your next task is to deal with pollution of the groundwater that provides your state's drinking water supply. Recent studies have shown that one-third of the state's groundwater has levels of pollutants that violate EPA standards for human health. The federal government is threatening enforcement, and citizens are fearful for their safety. What steps would you consider taking to safeguard the quality of your state's groundwater supply, and why?

## INTERPRETING GRAPHS AND DATA

Close to 75% of the fresh water used by people is used in agriculture, and about 1 of every 14 people live where water is scarce, according to a review by hydrologist J. S. Wallace. By the year 2050, scientists project that two-thirds of the world's population will live in water-scarce areas, including most of Africa, the Middle East, India, and China. How much water is required to feed over 6 billion people a basic dietary requirement of 2,700 calories per day? The answer depends on the efficiency with which we use water in agricultural production and on the type of diet we consume.

1. How many liters of water are needed to produce 2,300 calories of vegetable food? How many liters of water are needed to produce 400 calories of animal food? How many liters of water are needed daily to provide this diet? Annually?
2. How many liters of water would be saved daily, compared to the diet in the graph, if the 2,700 calories were provided entirely by vegetables? Annually?
3. Reflect on one of the quotes at the beginning of this chapter: "Water promises to be to the 21st century what oil was to the 20th century: the precious commodity that determines the wealth of nations." How do you think the demographic pressure on the water supply could affect world trade, particularly trade of agricultural products? Do you think it could affect prospects for peace and stability in and among nations? How so?

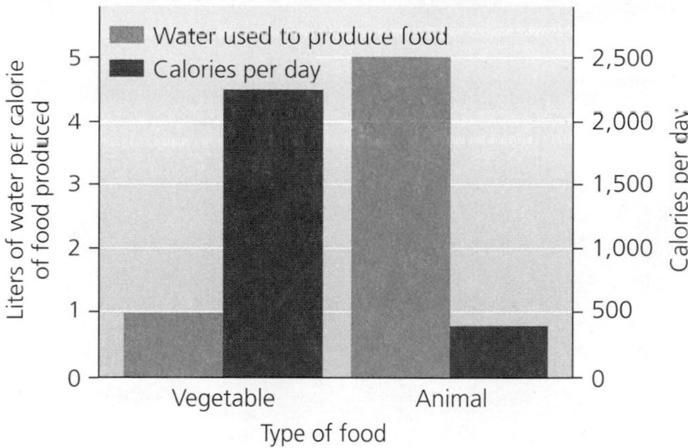


Amount of water needed to produce vegetable and animal food (orange), and global average calories per day consumed of vegetable and animal food (red). Data from Wallace, J. S. 2000. Increasing agricultural water use efficiency to meet future food production. *Agriculture, Ecosystems and Environment* 82: 105–119.

## CALCULATING ECOLOGICAL FOOTPRINTS

In the United States, the EPA estimates that household water use averages 750 liters per person per day. One of the single greatest personal uses of water is for showering. Standard showerheads dispense 15 liters of water per minute, but so-called low-flow showerheads dispense only 9 liters per minute. Given an average daily shower time of 10 minutes, calculate the amounts of water used and saved over the course of a year with standard versus low-flow showerheads, and record your results in the table below.

| | Annual water use with standard showerheads (liters) | Annual water use with low-flow showerheads (liters) | Annual water savings with low-flow showerheads (liters) |
|---|---|---|---|
| **You** | 54,750 | 32,850 | 21,900 |
| **Your class** | | | |
| **Your state** | | | |
| **United States** | | | |

Data from U.S. EPA. 1995. *Cleaner water through conservation: Chapter 1—How we use water in the United States.* EPA 841-B-95-002.

1. What percentage of personal water consumption would you calculate is used for showering?
2. How much additional water would you be able to save by shortening your average shower time from 10 minutes to 8 minutes? To 5 minutes?
3. Can you think of any factors that are not being considered in this scenario of water savings? Explain.

## Take It Further

Go to www.aw-bc.com/withgott or the student CD-ROM, where you'll find:

- Suggested answers to end-of-chapter questions
- Quizzes, animations, and flashcards to help you study
- *Research Navigator*™ database of credible and reliable sources to assist you with your research projects
- **GRAPHIt!** Tutorials to help you interpret graphs
- **INVESTIGATEIt!** Current news articles that link the topics that you study to case studies from your region to around the world

CHAPTER

# 16 Marine and Coastal Systems: Resources, Impacts, and Conservation

Schooling marine fish

## Upon completing this chapter, you will be able to:

- Identify physical, geographical, chemical, and biological aspects of the marine environment
- Describe major types of marine ecosystems
- Outline historic and current human uses of marine resources
- Assess human impacts on marine environments
- Review the current state of ocean fisheries and reasons for their decline
- Evaluate marine protected areas and reserves as innovative solutions

Massachusetts cod fishermen haul in a dwindling catch

CENTRAL CASE

# Collapse of the Cod Fisheries

**"All of a sudden they just crashed."**
—DONALD PAUL, NEWFOUNDLAND FISHERMAN, IN 1997

**"Either we have sustainable fisheries, or we have no fishery."**
—CANADIAN FISHERIES MINISTER DAVID ANDERSON, IN 1998

No fish has had more impact on human civilization than the Atlantic cod. Europeans exploring the coasts of North America 500 years ago discovered that they could catch these abundant fish merely by dipping baskets over the railings of their ships, and the race that ensued to harvest this resource helped lead to the colonization of the New World. Starting in the early 1500s, schooners captured countless millions of cod, and the fish became a dietary staple in cultures on both sides of the Atlantic.

Since then, cod fishing has been the economic engine for hundreds of communities in coastal New England and eastern Canada. Massachusetts honored the fish by naming Cape Cod after it and by erecting a carved wooden cod statue in its statehouse. In many Canadian coastal villages, cod fishing has been a way of life for generations. So it came as a shock when the cod all but disappeared, and governments had to step in and close the fisheries.

The Atlantic cod (*Gadus morhua*) is a type of *groundfish*, a name given to any fish that lives or feeds along the bottom. People have long coveted groundfish such as halibut, pollock, haddock, and flounder. Adult cod eat smaller fish and invertebrates, commonly grow 60–70 cm long, and can live 20 years. A mature female cod can produce several million eggs. Atlantic cod inhabit cool ocean waters on both sides of the North Atlantic and occur in 24 discrete populations, commonly called *stocks*. One stock inhabits the Grand Banks off the Newfoundland coast, and another lives on Georges Bank off Massachusetts (**Figure 16.1**).

The Grand Banks provided ample fish for centuries. With advancing technology, however, ships became larger and more effective at finding fish. By the 1960s, massive industrial trawlers from Europe were vacuuming up unprecedented numbers of groundfish. In 1977, Canada exercised its legal right to the waters 200 nautical miles from shore, kicked out foreign fleets, and claimed most of the Grand Banks for itself. Canada then developed the same industrial technologies and revved up its fishing industry like never before.

Then came the crash. Catches dwindled throughout the 1980s because too many fish had been taken and because trawling (• p. 459–460) had destroyed so much underwater habitat. By 1992 the situation was dire: scientists reported that mature cod were at just 10% of their long-term abundance. On July 2, Fisheries Minister John Crosbie announced a 2-year ban on commercial cod fishing off Labrador and Newfoundland, where the $700 million fishery supplied income to 16% of the province's workforce. To compensate fishers, the government offered 10 weekly payments of $225, along with training for new job skills and incentives for early retirement. Over the next 2 years, 40,000 fishers and processing plant workers lost their jobs. Some coastal communities faced economic ruin; in this bleak resource-poor land, fishing was their only reason for being.

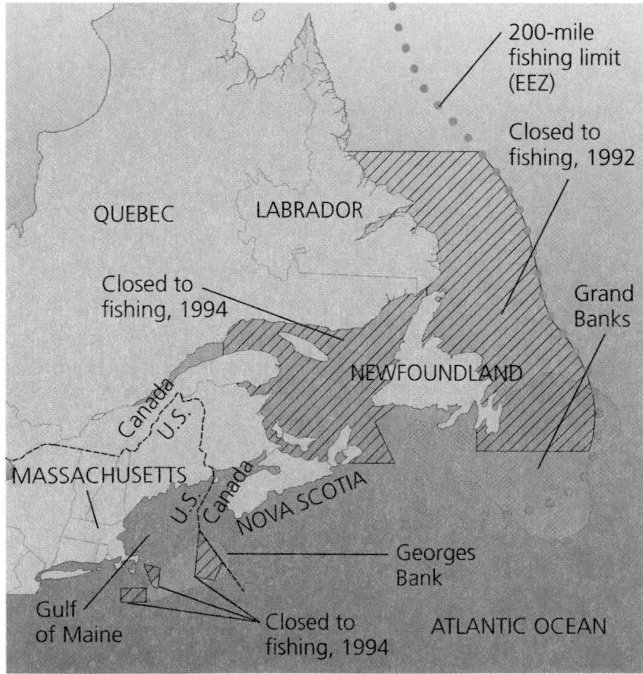


**FIGURE 16.1** Ten stocks of Atlantic cod inhabit areas of the northwestern Atlantic Ocean, including the Grand Banks and Georges Bank, regions of shallow water that are especially productive for groundfish. Portions of these and other areas have been closed to fishing in recent years because cod populations have collapsed after being overfished.

Cod stocks did not rebound by 1994, so the government extended the moratorium, enacted bans on all other major cod fisheries, and scrambled to offer more compensation, eventually spending over $4 billion.

In 1997–1998, Canada partially reopened some fisheries, but data soon confirmed that the stocks were not recovering. In April 2003, the cod fisheries were closed indefinitely—to recreational fishing as well as commercial fishing. Today it is illegal for Canadians in these areas to catch even one cod for their family's dinner. Fishers challenging the ban have been arrested, fined, and jailed.

Across the border in U.S. waters, cod stocks were collapsing in the Gulf of Maine and on Georges Bank. In 1994, the National Marine Fisheries Service (NMFS) closed three prime fishing areas on Georges Bank. Over the next several years, NMFS designed a number of regulations, but these steps were too little, too late. A 2005 report revealed that the cod were not recovering as hoped, and further restrictions were enacted. Today scientists are struggling to explain why cod are not recovering. Research so far suggests that once mature cod were eliminated, the species they preyed upon proliferated, and now those species compete with and prey on young cod, preventing the population from rebuilding.

There is good news, however: The Georges Bank closures have helped some other species to rebound. Seafloor invertebrates have begun to recover in the absence of trawling, spawning stock of haddock and yellowtail flounder have risen, and sea scallops have increased in biomass 14-fold. Such recoveries in no-fishing areas are showing scientists, fishers, and policymakers that protecting areas of ocean can help save dwindling marine populations.

## The Oceans

It's been said that our planet Earth should more properly be named "Ocean." After all, ocean water covers the vast majority of our planet's surface. Moreover, the oceans strongly influence how our planet's systems work. They influence global climate, teem with biodiversity, facilitate transportation and commerce, and provide us resources. Even landlocked areas far from the coasts are affected.

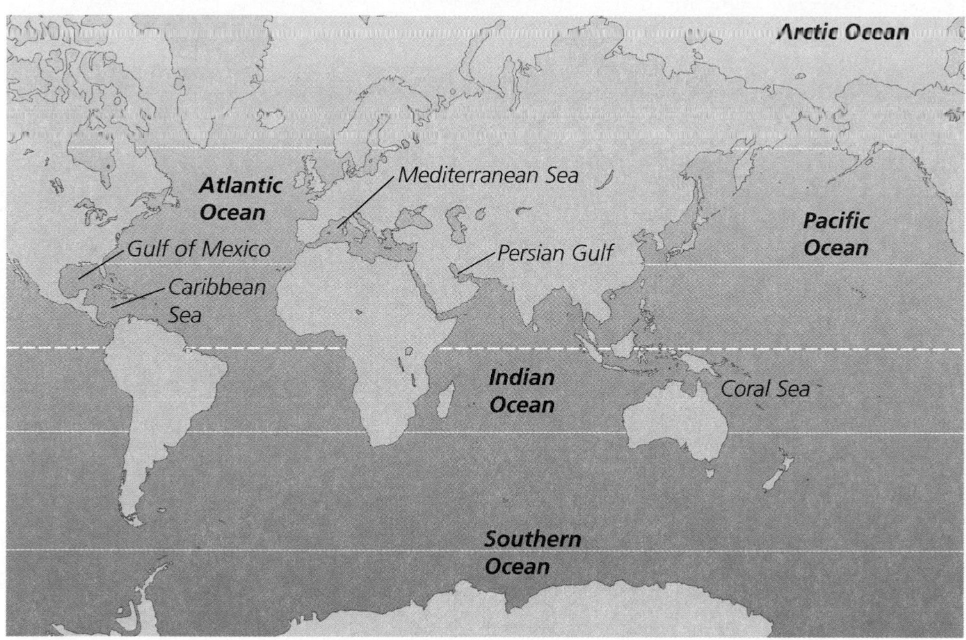


FIGURE 16.2 The world's oceans are connected in a single vast body of water but are given different names. The Pacific Ocean is the largest and, like the Atlantic and Indian Oceans, includes both tropical and temperate waters. The smaller Arctic and Southern Oceans include the waters in the north and south polar regions, respectively. Many smaller bodies of water are named as seas or gulfs; a selected few are shown here.

The oceans provide fish for people to eat in Iowa, supply oil to power cars in Ontario, and influence the weather in Nebraska.

## Oceans cover most of Earth's surface

Although we generally speak of the world's oceans (**Figure 16.2**) in the plural, giving each major basin a name—Pacific, Atlantic, Indian, Arctic, and Southern—all these oceans are connected, comprising a single vast body of water. This one "world ocean" covers 71% of Earth's surface and contains 97.2% of its surface water. The oceans take up most of the hydrosphere, influence the atmosphere and lithosphere, and encompass much of the biosphere, including at least 250,000 species. The world's oceans touch and are touched by virtually every environmental system and every human endeavor.

## The oceans contain more than water

Ocean water contains approximately 96.5% $H_2O$ by mass; most of the remainder consists of ions from dissolved salts (**Figure 16.3**). Ocean water is salty primarily because ocean basins are the final repositories for water that runs off the land. Rivers carry sediment and dissolved salts from the continents into the ocean, as do winds. Evaporation from the ocean surface then removes pure water, leaving a higher concentration of salts. If we were able to evaporate all the water from the oceans, the empty basins would be covered with a layer of dried salt 63 m (207 ft) thick.

The salinity of ocean water generally ranges from 33 to 37 parts per thousand (ppt), varying from place to place because of differences in evaporation, precipitation, and

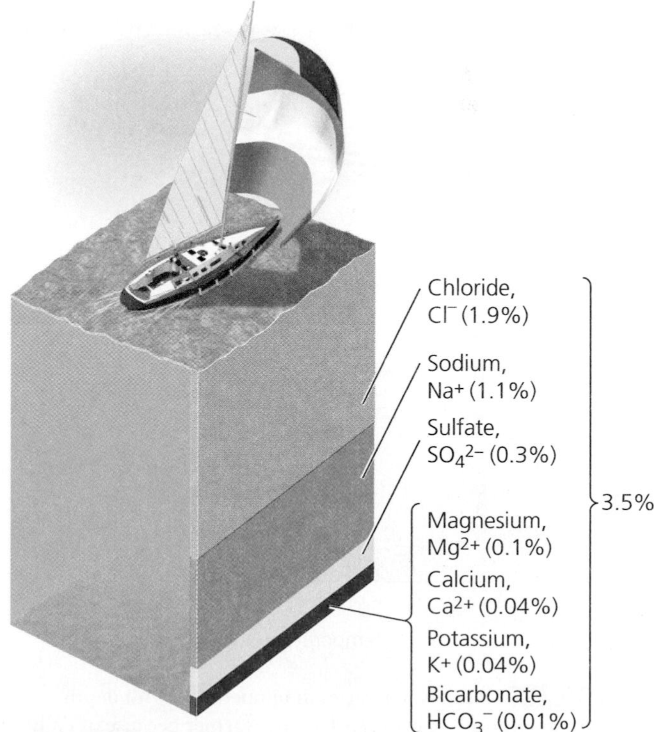


FIGURE 16.3 Ocean water consists of 3.5% salt, by mass, as shown by the proportionally thin colored slices of the cube in this diagram. Most of this salt is NaCl in solution, so sodium and chloride ions are abundant. A number of other ions and trace elements are also present.

freshwater runoff from land and glaciers. Salinity near the equator is low because this region has a great deal of precipitation, which is relatively salt-free. In contrast, surface salinity is high at latitudes roughly 30–35 degrees north and south, where evaporation exceeds precipitation.

Besides the dissolved salts shown in Figure 16.3, nutrients such as nitrogen and phosphorus occur in seawater in trace amounts (well under 1 part per million) and play essential roles in nutrient cycling (• pp. 186–198) in marine ecosystems. Another aspect of ocean chemistry is dissolved gas content. Roughly 36% of the gas dissolved in seawater is oxygen, which is produced by photosynthetic plants, bacteria, and phytoplankton (• p. 142), and by diffusion from the atmosphere. Marine animals depend on this oxygen, and oxygen concentrations are highest in the upper layer of the ocean, reaching 13 ml/L of water.

## Ocean water is vertically structured

Surface waters in tropical regions receive more solar radiation and therefore are warmer than surface waters in temperate or polar regions. In all regions, however, temperature declines with depth (**Figure 16.4**). Water density increases as salinity rises and as temperature falls. These relationships give rise to different layers of water; heavier (colder and saltier) water sinks, and lighter (warmer and less salty) water remains nearer the surface. Waters of the surface zone are heated by sunlight each day and are stirred by wind such that they are of similar density throughout, down to a depth of approximately 150 m (490 ft). Below the zone of surface water lies the *pycnocline,* a region in which density increases rapidly with depth. The pycnocline contains about 18% of ocean water by volume, compared to the surface zone's 2%. The remaining 80% resides in the deep zone beneath the pycnocline. The dense water in this zone is sluggish and unaffected by winds, storms, sunlight, and daily temperature fluctuations.

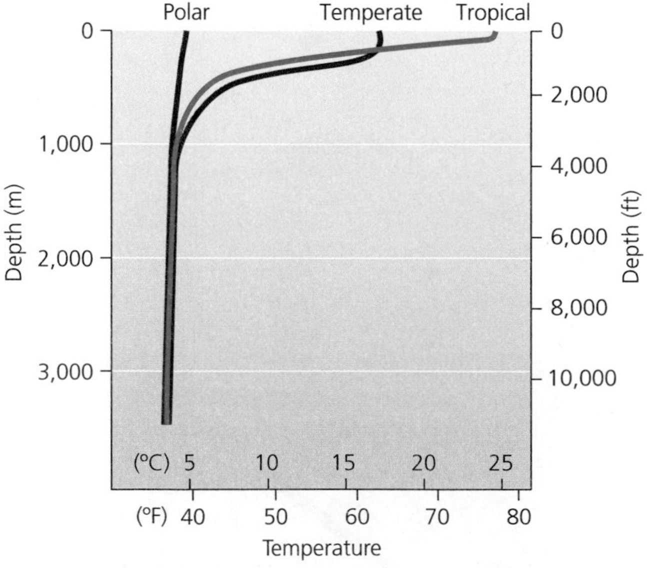


FIGURE 16.4 Ocean water varies in temperature with depth. Water temperatures near the surface are warmer because of daily heating by the sun, and within the top 1,000 m (3,300 ft) they become rapidly colder with depth. This temperature differential is greatest in the tropics because of intense solar heating and is least in the polar regions. Deep water at all latitudes is equivalent in temperature. *Source:* Garrison, T. 2005. *Oceanography,* 5th ed. Belmont, CA: Brooks/Cole.

Despite the daily heating and cooling of surface waters, ocean temperatures are much more stable than temperatures on land. Midlatitude oceans experience yearly temperature variation of only around 10° C (18° F), and tropical and polar oceans are still more stable. The reason for this stability is that water has a very high *heat capacity,* a measure of the heat required to increase temperature by a given amount. It takes much more heat energy to increase the temperature of water than it does to increase the temperature of air by the same amount. High heat capacity enables the oceans to absorb a tremendous amount of heat from the atmosphere. In fact, the heat content of the entire atmosphere is equal to that of just the top 2.6 m (8.5 ft) of the oceans. By absorbing heat and releasing it to the atmosphere, the oceans help regulate Earth's climate (Chapter 18). Also influencing climate is the ocean's surface circulation, a system of currents that move in the pycnocline and the surface zone.

## Ocean water flows horizontally in currents

Far from being a static pool of water, Earth's ocean is composed of vast riverlike flows (**Figure 16.5**) driven by density differences, heating and cooling, gravity, and wind. These surface **currents** flow within the upper 400 m (1,300 ft) of water, horizontally and for great distances. These long-lasting patterns influence global climate and play key roles in the phenomena known as El Niño and La Niña (• pp. 510–511). They also have been crucial in navigation and human history; currents helped carry Polynesians to Easter Island, Darwin to the Galapagos, and Europeans to the New World. Currents transport heat, nutrients, pollution, and the larvae of cod and many other marine species from place to place.

Some currents are very slow. Others, like the Gulf Stream, are rapid and powerful. From the Gulf of Mexico, the Gulf Stream moves up the U.S. Atlantic coast and flows past the eastern edges of Georges Bank and the Grand Banks at a rate of 160 km per day (nearly 2 m/sec, or over 4.1 mi/hr). Averaging 70 km (43 mi) across, the Gulf Stream

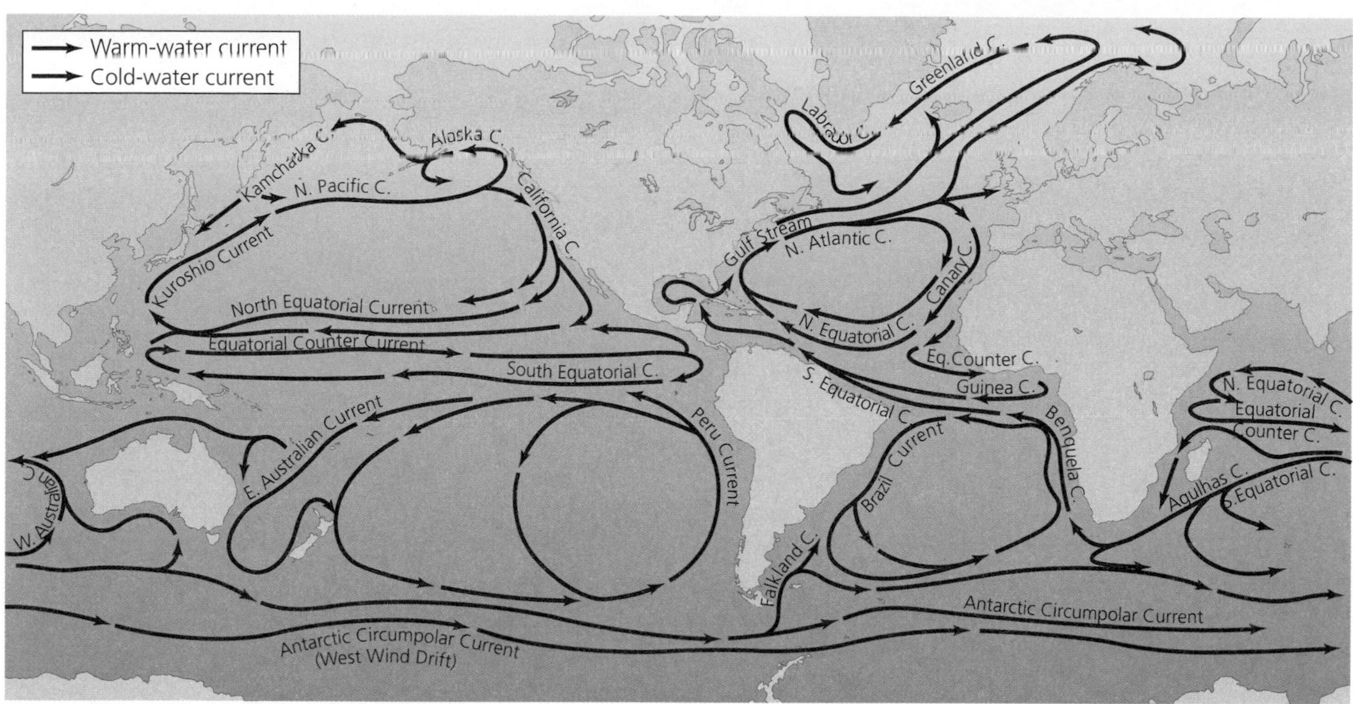


**FIGURE 16.5** The upper waters of the oceans flow in currents, which are long-lasting and predictable global patterns of water movement. Warm- and cold-water currents interact with the planet's climate system, and people have used them for centuries to navigate the oceans.
*Source:* Garrison, T. 2005. *Oceanography,* 5th ed. Belmont, CA: Brooks/Cole.

continues across the North Atlantic, bringing warm water to Europe and moderating that continent's climate (• p. 511), which otherwise would be much colder.

## Vertical movement of water affects marine ecosystems

Surface winds and heating also create vertical currents in seawater. **Upwelling,** the vertical flow of cold, deep water toward the surface, occurs where horizontal currents diverge, or flow away from one another. Because upwelled water is rich in nutrients from the bottom, upwellings are often sites of high primary productivity (• pp. 181–182) and lucrative fisheries. Upwellings also occur where strong winds blow away from or parallel to coastlines (**Figure 16.6**). An example is the California coast, where north winds and the Coriolis effect (• p. 480) move surface waters away from shore, raising nutrient-rich water from below and creating a biologically rich region. The cold water also chills the air along the coast, giving San Francisco its famous fog and cool summers.

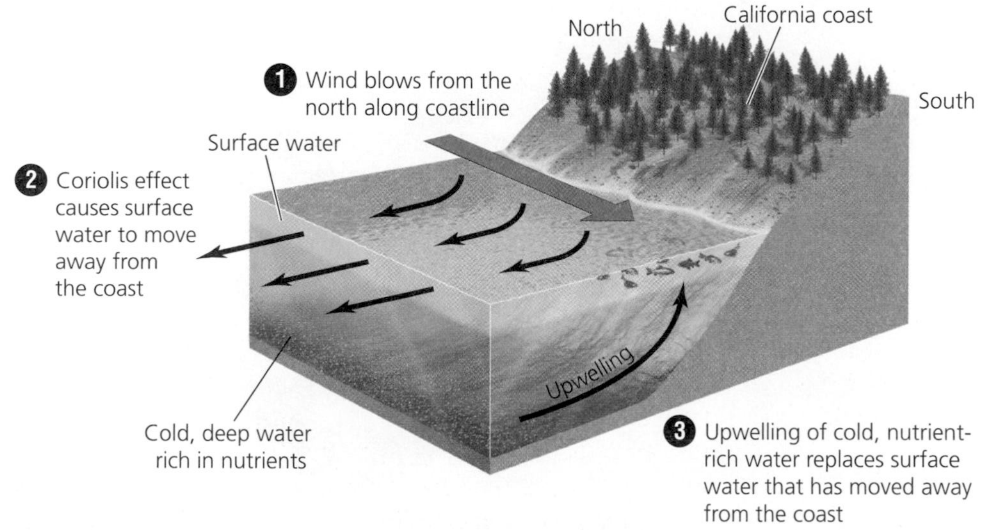


**FIGURE 16.6** Upwelling is the movement of bottom waters upward. This type of vertical current often brings nutrients up to the surface, creating rich areas for marine life. For example, north winds blow along the California coastline (1), while the Coriolis effect (• p. 480) draws wind and water away from the coast (2). Water is then pulled up from the bottom (3) to replace the water that moves away from shore.

In areas where surface currents converge, or come together, surface water sinks, a process called **downwelling**. Downwelling transports warm water rich in dissolved gases, providing an influx of oxygen for deep-water life. Vertical currents also occur in the deep zone, where differences in water density can lead to rising and falling convection currents, such as those seen in molten rock (•pp. 200–201) and in air (• pp. 447, 479–480). The thermohaline circulation (• p. 511) is an example of such circulation that has far-reaching effects on global climate.

## Seafloor topography can be rugged and complex

Although oceans are depicted on most maps and globes as smooth, blue swaths, portions of the ocean floor are complex. Underwater volcanoes shoot forth enough magma to build islands above sea level, such as the Hawaiian Islands. Steep canyons similar in scale to Arizona's Grand Canyon lie just offshore of some continents. The lowest spot in the oceans—the Mariana Trench in the South Pacific—is deeper than Mount Everest is high, by over 2.1 km (1.3 mi). Our planet's longest mountain range is under water—the Mid-Atlantic Ridge runs the length of the Atlantic Ocean (**Figure 16.7**).

Georges Bank and the Grand Banks are essentially huge underwater mounds formed from the debris dumped by glaciers at their southernmost extent. As climate warmed and the glaciers retreated, sea level rose and this hilly terrain became submerged.

FIGURE 16.7 The seafloor can be rugged. The spreading margin between tectonic plates at the Mid-Atlantic Ridge gives rise to a vast underwater volcanic mountain chain.

We can gain an understanding of underwater geographic features by examining a stylized map (**Figure 16.8**) that reflects *bathymetry* (the measurement of ocean depths) and *topography* (physical geography, or the shape and arrangement of landforms). In bathymetric profile, gently sloping **continental shelves** underlie the shallow waters bordering the continents. Continental shelves vary in width from 100 m (330 ft) to 1,300 km (800 mi), averaging 70 km (43 mi) wide, with an average slope of 1.9 m/km (10 ft/mi). These shelves drop off with relative suddenness at the *shelf-slope break*. The *continental slope*

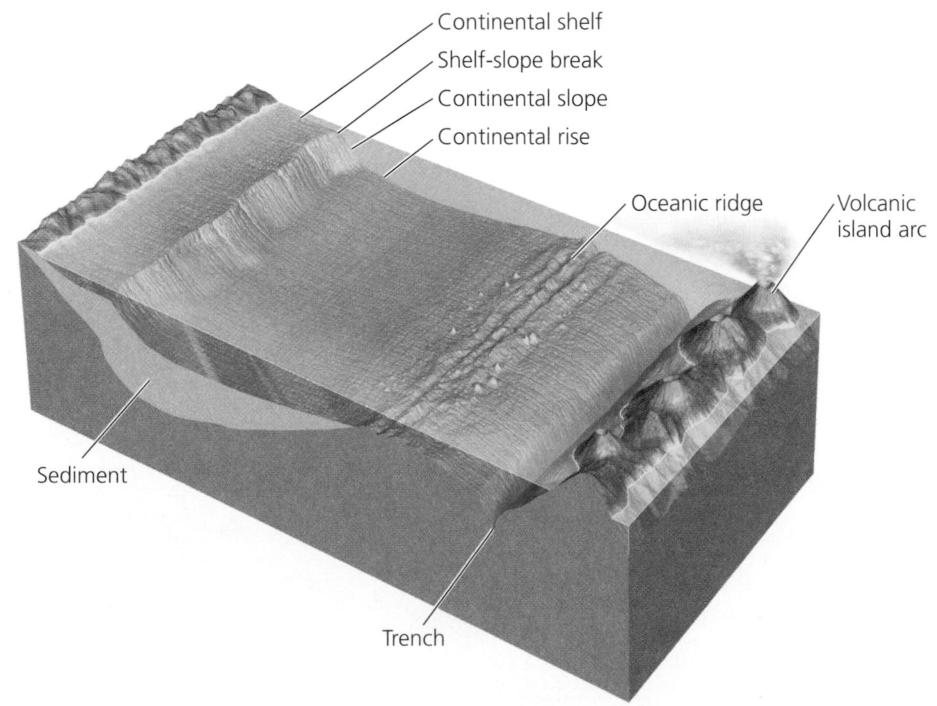


FIGURE 16.8 A stylized bathymetric profile shows key geologic features of the submarine environment. Shallow regions of water exist around the edges of continents over the continental shelf, which drops off at the shelf-slope break. The relatively steep dropoff called the continental slope gives way to the more gradual continental rise, all of which are underlain by sediments from the continents. Vast areas of seafloor are flat abyssal plain. Seafloor spreading occurs at oceanic ridges, and oceanic crust is subducted in trenches. Volcanic activity along trenches often gives rise to island chains such as the Aleutian Islands. Features on the left side of this diagram are more characteristic of the Atlantic Ocean, and features on the right side of the diagram are more characteristic of the Pacific Ocean. Adapted from Thurman, H. V. 1990. *Essentials of oceanography,* 4th ed. New York: Macmillan.

angles somewhat steeply downward, connecting the continental shelf to the deep ocean basin below.

Most of the seafloor is flat, but volcanic peaks that rise above the ocean floor provide physical structure for marine animals and are often the site of productive fishing grounds. Some island chains, such as the Florida Keys, are formed by the development of reefs (• pp. 450–451) and lie atop the continental shelf. Others, such as the Aleutian Islands, which curve across the North Pacific from Alaska toward Russia, are volcanic in origin. The Aleutians are also the site of a deep trench that, like the Mariana Trench, formed at a convergent tectonic plate boundary, where one slab of crust dives beneath another in the process of subduction (• p. 201).

Regions of ocean water differ greatly, and some zones support more life than others. The uppermost 10 m (33 ft) of water absorbs 80% of the solar energy that reaches its surface. For this reason, nearly all of the oceans' primary productivity occurs in the well-lighted top layer, or **photic zone**. Generally, the warm, shallow waters of continental shelves are most biologically productive and support the greatest species diversity. Habitats and ecosystems occurring between the ocean's surface and floor are termed **pelagic**, whereas those that occur on the ocean floor are called **benthic**.

## Marine and Coastal Ecosystems

With their variation in topography, temperature, salinity, nutrients, and sunlight, marine and coastal environments feature a variety of ecosystems. Most marine and coastal ecosystems are powered by solar energy, with sunlight driving photosynthesis by phytoplankton in the photic zone. Yet even the darkest ocean depths host life.

### Open-ocean ecosystems vary in their biological diversity

Biological diversity in pelagic regions of the open ocean is highly variable in its distribution. Primary productivity (• p. 181) and animal life near the surface are concentrated in regions of nutrient-rich upwelling. Microscopic phytoplankton constitute the base of the marine food chain in the pelagic zone. These photosynthetic algae, protists, and cyanobacteria feed zooplankton (• p. 142), which in turn become food for fish, jellyfish, whales, and other free-swimming animals (**Figure 16.9**). Predators at higher trophic levels include larger fish, sea turtles, and sharks. Many fish-eating birds such as puffins, petrels, and shearwaters feed at the surface of the open ocean, returning periodically to nesting sites on islands and coastlines.

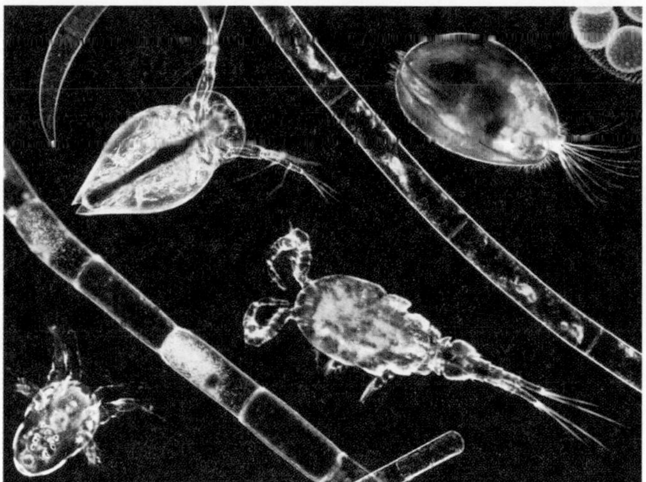

FIGURE 16.9 The uppermost reaches of ocean water contain billions upon billions of phytoplankton—tiny photosynthetic algae, protists, and bacteria that form the base of the marine food chain—as well as zooplankton, small animals and protists that dine on phytoplankton and comprise the next trophic level.

In recent years biologists have been learning more about animals of the deep ocean, although tantalizing questions remain. In deep-water ecosystems, animals have adapted to deal with extreme water pressures and to live in the dark without food from plants. Some of these often bizarre-looking creatures scavenge carcasses or organic detritus that falls from above. Others are predators, and still others attain food from symbiotic mutualistic (• p. 147) bacteria. Some species carry bacteria that produce light chemically by bioluminescence (**Figure 16.10**).

As we saw in Chapter 4 (• pp. 103–104), some ecosystems form around hydrothermal vents, where heated

FIGURE 16.10 Life is scarce in the dark depths of the deep ocean, but the creatures that do live there can appear bizarre. The anglerfish lures prey toward its mouth with a bioluminescent (glowing) organ that protrudes from the front of its head.

water spurts from the seafloor, often carrying minerals that precipitate to form large rocky structures. Tubeworms, shrimp, and other creatures in these recently discovered systems use symbiotic bacteria to derive their energy from chemicals in the heated water rather than from sunlight. They manage to thrive within amazingly narrow zones between scalding-hot and icy cold water.

## Kelp forests harbor many organisms

Large brown algae, or **kelp,** grow from the floor of continental shelves, reaching upward toward the sunlit surface. Some kelp reaches 60 m (200 ft) in height and can grow 45 cm (18 in.) in a single day. Dense stands of kelp form underwater "forests" along many temperate coasts (**Figure 16.11**). Kelp forests supply shelter and food for invertebrates and fish, which in turn provide food for higher-trophic-level predators, such as seals and sharks. Indeed, kelp forests were the setting for our discussion of keystone species in Chapter 6 (• pp. 151–154, 156–157). Recall that sea otters control sea urchin populations, and when otters disappear, urchins overgraze the kelp, destroying the forests and creating "urchin barrens." Kelp forests also absorb wave energy and protect shorelines from erosion. People in Asian cultures eat some types of kelp, and kelp provides compounds known as alginates, which serve as thickeners in a wide range of consumer products, including cosmetics, paints, paper, and soaps.

## Coral reefs are treasure troves of biodiversity

In shallow subtropical and tropical waters, coral reefs occur. A reef is an underwater outcrop of rock, sand, or other material. A **coral reef** is a mass of calcium carbonate composed of the skeletons of tiny colonial marine organisms known as *corals.* A coral reef may occur as an extension of a shoreline, along a *barrier island* paralleling a shoreline, or as an *atoll,* a ring around a submerged island.

FIGURE 16.11 "Forests" of tall brown algae known as *kelp* grow from the floor of the continental shelf. Numerous fish and other creatures eat kelp or find refuge among its fronds.

Corals are tiny invertebrate animals related to sea anemones and jellyfish. They remain attached to rock or existing reef and capture passing food with stinging tentacles. Corals also derive nourishment from symbiotic algae, known as *zooxanthellae,* which inhabit their bodies and produce food through photosynthesis. Most corals are colonial, and the colorful surface of a coral reef consists of millions of densely packed individuals. As corals die, their skeletons remain part of the reef while new corals grow atop them, increasing the reef's size.

Like kelp forests, coral reefs protect shorelines by absorbing wave energy. They also host tremendous biodiversity (**Figure 16.12a**). The likely reason is that coral reefs provide complex physical structure (and thus many habitats) in shallow nearshore waters, which are regions of high primary productivity. Besides the staggering diversity of anemones, sponges, hydroids, tubeworms, and other sessile (stationary) invertebrates, innumerable molluscs, flatworms, sea stars, and urchins patrol the reefs, while thousands of fish species find food and shelter in reef nooks and crannies.

Coral reefs are experiencing worldwide declines, however. Many have undergone "coral bleaching," a process that occurs when zooxanthellae leave the coral, depriving it of nutrition. Corals lacking zooxanthellae lose color and frequently die, leaving behind ghostly white patches in the reef (**Figure 16.12b**). Coral bleaching is thought to result from increased sea surface temperatures associated with global climate change, from the influx of pollutants, from unknown natural causes, or from some combination of these factors. Nutrient pollution in coastal waters also promotes the growth of algae, which are blanketing reefs in the Florida Keys and many other regions. In addition, coral reefs sustain damage when divers stun fish with cyanide to capture them for food or for the pet trade, a common practice in waters of Indonesia and the Philippines. Finally, as global climate change proceeds, the oceans are becoming more acidic, as excess carbon dioxide from the atmosphere reacts with seawater to form carbonic acid. Acidification threatens to deprive corals of the carbonate ions they need to produce their structural parts.

A few coral species thrive in waters outside the tropics and build reefs on the ocean floor at depths of 200–500 m (650–1,650 ft). These little-known reefs, which occur in cold-water areas off the coasts of Norway, Spain, the British Isles, and elsewhere, are only now beginning to be studied by scientists. Already, however, many have

**(a) Coral reef community**

**(b) Bleached coral**

FIGURE 16.12 Corals reefs provide food and shelter for a tremendous diversity (a) of fish and other creatures. However, these reefs face multiple environmental stresses from human impacts, and many corals have died as a result of coral bleaching (b), in which corals lose their zooxanthellae. Such bleaching is evident in the whitened portion of this coral.

been badly damaged by trawling (• pp. 459–460)—the same practice that has so degraded the benthic habitats of groundfish such as the Atlantic cod. Norway and other countries are now beginning to protect some of these deep-water reefs.

## Intertidal zones undergo constant change

Where the ocean meets the land, **intertidal**, or **littoral**, ecosystems (**Figure 16.13**) spread between the uppermost reach of the high tide and the lowest limit of the low tide. **Tides** are the periodic rising and falling of the ocean's height at a given location, caused by the gravitational pull of the moon and sun. High and low tides occur roughly 6 hours apart, so intertidal organisms spend part of each day submerged in water, part of the day exposed to the air and sun, and part of the day being lashed by waves. Subject to tremendous extremes in temperature, moisture, sun exposure, and salinity, these creatures must also protect themselves from marine predators at high tide and terrestrial predators at low tide.

The intertidal environment is a tough place to make a living, but it is home to a remarkable diversity of organisms. Rocky shorelines can be full of life among the crevices, which provide shelter and pools of water (tidepools) during low tides. Sessile animals such as anemones, mussels, and barnacles live attached to rocks, filter-feeding on plankton in the water that washes over them. Urchins, sea slugs, chitons, and limpets eat intertidal algae or scrape food from the rocks. Sea stars (starfish) creep slowly along, preying on the filter-feeders and herbivores at high tide. Crabs clamber around the rocks, scavenging detritus.

The rocky intertidal zone is so diverse because environmental conditions such as temperature, salinity, and moisture change dramatically from the high to the low reaches. This environmental variation gives rise to horizontal bands formed by dominant organisms as they array themselves according to their habitat needs. Sandy intertidal areas, such as those of Cape Cod, host less biodiversity, yet plenty of organisms burrow into the sand at low tide to await the return of high tide, when they emerge to feed.

## Salt marshes occur widely along temperate shorelines

Along many of the world's coasts at temperate latitudes, **salt marshes** occur where the tides wash over gently sloping sandy or silty substrates. Rising and falling tides flow into and out of channels called *tidal creeks* and at highest tide spill over onto elevated marsh flats (**Figure 16.14**). Marsh flats grow thick with grasses, rushes, shrubs, and other herbaceous plants. Grasses such as those in the genera *Spartina* and *Distichlis* comprise the dominant vegetation in most salt marshes.

Salt marshes boast very high primary productivity and provide critical habitat for shorebirds, waterfowl, and the adults and young of many commercially important fish and shellfish species. Salt marshes also filter out pollution and stabilize shorelines against storm surges. However, people desire to live along coasts, and coastal sites are desirable for commerce. As a result, people have altered or destroyed vast expanses of salt marshes worldwide to make way for coastal development. When salt marshes are

(a) Tidal zones

(b) Tidepools at low tide

FIGURE 16.13 The rocky intertidal zone is the swath of a rocky shoreline between the lowest and highest reaches of the tides (a). This is an ecosystem rich in biodiversity, typically containing large invertebrates such as seastars (starfish), barnacles, crabs, sea anemones, corals, bryozoans, snails, limpets, chitons, mussels, nudibranchs (sea slugs), and sea urchins. Fish swim in tidal pools (b), and many types of algae cover the rocks. Areas higher on the shoreline are exposed to the air more frequently and for longer periods, so organisms that can tolerate exposure best specialize in the upper intertidal zone. The lower intertidal zone is exposed less frequently and for shorter periods, so organisms less able to tolerate exposure thrive in this zone.

destroyed, we lose the ecosystem services they provide. When Hurricane Katrina struck the Gulf Coast, for instance, the flooding was made worse because vast areas of salt marshes had vanished due to development, subsidence from oil and gas drilling, and dams that had held back marsh-building sediment.

## Mangrove forests line coasts in the tropics and subtropics

In tropical and subtropical latitudes, mangrove forests replace salt marshes along gently sloping sandy and silty coasts. **Mangroves** are trees with unique types of roots

FIGURE 16.14 Salt marshes occur in temperate intertidal zones where the substrate is muddy, allowing salt-adapted grasses to grow. Tidal waters generally flow through marshes in channels called *tidal creeks*, amid flat areas called *benches*, sometimes partially submerging the grasses.

that curve upward like snorkels to attain oxygen lacking in the mud or that curve downward like stilts to support the tree in changing water levels (**Figure 16.15**). Fish, shellfish, crabs, snakes, and other organisms thrive among the root networks, and birds feed and nest in the dense foliage of these coastal forests. Besides serving as nurseries for fish and shellfish that people harvest, mangroves also provide materials that people use for food, medicine, tools, and construction.

From Florida to Mexico to the Philippines, mangrove forests have been destroyed as people have developed coastal areas for residential, commercial, and recreational uses. Shrimp farming in particular has driven the conversion of large areas of mangroves. We have eliminated half the world's mangrove forests, and their area continues to decline by 2–8% per year. When mangroves are removed, coastal areas lose the ability to slow runoff, filter pollutants, and retain soil. As a result, offshore systems such as eelgrass beds and coral reefs are more readily degraded. Moreover, mangrove forests protect coastal communities against storm surges and tsunamis (tidal waves), as was shown when the 2004 Indian Ocean tsunami devastated areas where mangroves had been removed but caused less damage where mangroves were intact. Despite these important ecosystem services, we have granted only about 1% of the world's remaining mangroves protection against development.

## Fresh water meets salt water in estuaries

Many salt marshes and mangrove forests occur in or near **estuaries,** water bodies where rivers flow into the ocean, mixing fresh water with salt water. Estuaries are biologically productive ecosystems that experience fluctuations in salinity as tides and freshwater runoff vary daily and seasonally. For shorebirds and for many commercially important shellfish species, estuaries provide critical habitat. For anadromous fishes (fishes, such as salmon, that spawn in fresh water and mature in salt water), estuaries provide a transitional zone where young fish make the passage from fresh water to salt water.

Estuaries around the world have been affected by urban and coastal development, water pollution, habitat alteration, and overfishing. Florida Bay, where fresh water from the Everglades system mixes with salt water, provides

FIGURE 16.15 Mangrove forests are important ecosystems along tropical and subtropical coastlines throughout the world. Mangrove trees, such as these at Lizard Island, Australia, show specialized adaptations for growing in salt water and provide habitat for many types of fish, birds, crabs, and other animals.

one example. This estuary has suffered pollution and a reduction in fresh water flow due to irrigation and fertilizer use by sugarcane farmers, housing development, septic tank leakage, and other human impacts. Coastal ecosystems have borne the brunt of human impact because two-thirds of Earth's people choose to live within 160 km (100 mi) of the ocean.

### Weighing THE Issues | Coastal Development

A developer wants to build a large marina on an estuary in your coastal town. The marina would boost the town's economy but eliminate its salt marshes. As a homeowner living adjacent to the marshes, how would you respond? Do you think that developers or town officials should offer homeowners insurance against damage from storm surges in such a situation?

FIGURE 16.16 Crude oil and natural gas from beneath the seafloor are two of the economically valuable resources that we take from the oceans. Offshore petroleum drilling at platforms such as these off Santa Barbara, California, creates some of the many human impacts on the marine environment.

# Human Use and Impact

Our species has a long history of interacting with the oceans. We have long traveled across their waters, clustered our settlements along coastlines, and been fascinated by the beauty, power, and vastness of the seas. We have also left our mark upon them by exploiting oceans for their resources and polluting them with our waste.

## Oceans provide transportation routes

We have used the oceans for transportation for thousands of years, and the oceans continue to provide affordable means of moving people and products over vast distances. Ocean shipping has accelerated the global reach of some cultures and has promoted interaction among long-isolated peoples. It has had substantial impacts on the environment as well. The thousands of ships plying the world's oceans today carry everything from cod to cargo containers to crude oil. Ships transport ballast water (• p. 159), which, when discharged at ports of destination, may transplant aquatic organisms picked up at ports of departure. Some of these species—such as the zebra mussel (Chapter 6)—establish themselves and become invasive.

## We extract energy and minerals

We mine the oceans for sources of commercially valuable energy. Worldwide, about 30% of our crude oil and nearly half of our natural gas come from seafloor deposits (**Figure 16.16**). Most offshore oil and gas is concentrated in petroleum-rich regions such as the North Sea and the Gulf of Mexico, but energy companies extract smaller amounts of oil and gas from diverse locations, among them the Grand Banks and adjacent Canadian waters. Proposals to drill for oil and gas in Georges Bank and the Gulf of Maine, however, have been stalled by both the U.S. and Canadian governments, in large part because any spilled oil could damage the region's valuable fisheries.

Ocean sediments also contain a novel potential source of fossil fuel energy. **Methane hydrate** (• p. 558–559) is an ice-like solid consisting of molecules of methane ($CH_4$, the main component of natural gas) embedded in a crystal lattice of water molecules. Methane hydrates are stable at temperature and pressure conditions found in many sediments on the Arctic seafloor and the continental shelves. The U.S. Geological Survey estimates that the world's deposits of methane hydrates may hold twice as much carbon as all known deposits of oil, coal, and natural gas combined.

Could methane hydrates be developed as an energy source to power our civilization through the 21st century and beyond? Perhaps, but a great deal of research remains before scientists and engineers can be sure how to extract these energy sources safely. Destabilizing a methane hydrate deposit could lead to a catastrophic release of gas. This could cause a massive landslide and tsunami and would also release huge amounts of methane, a potent greenhouse gas, into the atmosphere, exacerbating global climate change.

Fortunately, the oceans also hold potential for providing renewable energy sources that do not emit greenhouse gases. Engineers have developed ways of harnessing energy from waves, tides, and the heat of ocean water (•pp. 620–621). These promising energy sources await further research, development, and investment.

We extract minerals from the ocean floor, as well. Using large vacuum-cleaner-like hydraulic dredges, miners collect sand and gravel from beneath the sea. Also extracted are sulfur from salt deposits in the Gulf of Mexico and phosphorite from offshore areas near the California coast and elsewhere. Other valuable minerals found on or beneath the seafloor include calcium carbonate (used in making cement) and silica (used as fire-resistant insulation and in manufacturing glass), as well as rich deposits of copper, zinc, silver, and gold ore. Many minerals are concentrated in manganese nodules, small ball-shaped accretions that are scattered across parts of the ocean floor. Over 1.5 trillion tons of manganese nodules may exist in the Pacific Ocean alone, and their reserves of metal may exceed all terrestrial reserves. The logistical difficulty of mining them, however, has kept their extraction uneconomical so far.

FIGURE 16.17 This northern fur seal became entangled in a discarded fishing net. Each year many thousands of marine mammals, birds, and sea turtles are killed by plastic debris, abandoned nets, and other trash that people have dumped in the ocean.

## Marine pollution threatens resources

People have long made the oceans a sink for waste and pollution. Even into the mid-20th century, it was common for coastal U.S. cities to dump trash and untreated sewage along their shores. Fort Bragg, a bustling town on the northern California coast, boasts of its Glass Beach, an area where beachcombers collect sea glass, the colorful surf-polished glass sometimes found on beaches after storms. Glass Beach is in fact the site of the former town dump, and besides well-polished glass, the perceptive visitor may also spot old batteries, rusting car frames, and other trash protruding from the bluffs above the beach.

Oil, plastic, industrial chemicals, and excess nutrients all eventually make their way from land into the oceans. Raw sewage and trash from cruise ships and abandoned fishing gear from fishing boats add to the input. The scope of trash in the sea can be gauged by the amount picked up each September by volunteers who trek beaches in the Ocean Conservancy's annual International Coastal Cleanup. In this nonprofit organization's 2006 cleanup, 359,000 people from 66 nations picked up 3.2 million kg (7.0 million lb) of trash from 55,600 km (34,600 mi) of shoreline.

## Nets and plastic debris endanger marine life

Plastic bags and bottles, discarded fishing nets, gloves, fishing line, buckets, floats, abandoned cargo, and much else that people transport on the sea or deposit into it can harm marine organisms. Because most plastic is not biodegradable, it can drift for decades before washing up on beaches. Marine mammals, seabirds, fish, and sea turtles may mistake floating plastic debris for food and can die as a result of ingesting material they cannot digest or expel. Fishing nets that are lost or intentionally discarded can continue snaring animals for decades (**Figure 16.17**).

Of 115 marine mammal species, 49 are known to have eaten or become entangled in marine debris, and 111 of 312 species of seabirds are known to ingest plastic. All five species of sea turtle in the Gulf of Mexico have died from consuming or contacting marine debris. Marine debris affects people, as well. A survey of fishers off the Oregon coast indicated that more than half had encountered equipment damage or other problems from plastic debris, and debris has caused over $50 million dollars in insurance payments. In December 2006, the U.S. Congress responded to these threats and passed the Marine Debris Research, Prevention, and Reduction Act.

## Oil pollution comes from spills of all sizes

Major oil spills, such as the *Exxon Valdez* spill in Prince William Sound, Alaska (Chapter 4), make headlines and cause serious environmental problems. Yet it is important to put such accidents into perspective. The majority of oil pollution in the oceans comes not from large spills in a few particular locations, but from the accumulation of innumerable, widely spread small sources (non-point sources, • p. 431), including leakage from small boats and runoff from human activities on land. Moreover, the amount of petroleum spilled into the oceans in recent years is equaled by the amount that seeps into the water from naturally occurring seafloor deposits (**Figure 16.18a**).

Nonetheless, minimizing the amount of oil we release into coastal waters is important, because petroleum pollution is detrimental to marine life and to our economies. Petroleum can physically coat and kill marine organisms and can poison them when ingested (• p. 88). In response to headline-grabbing oil spills, governments worldwide have begun to implement more stringent safety standards for tankers, such as requiring industry to pay for tugboat escorts in sensitive and hazardous coastal waters and to develop prevention and response plans for major spills.

The U.S. Oil Pollution Act of 1990 created a $1 billion prevention and cleanup fund and required that by 2015 all oil tankers in U.S. waters be equipped with double hulls as a precaution against puncture. The oil industry has resisted many such safeguards, and today the ship that oiled Prince William Sound is still plying the world's oceans, renamed the *Sea River Mediterranean* and still featuring only a single hull. However, over the past three decades, the amount of oil spilled in U.S. and global waters has decreased (**Figure 16.18b**), in part because of an increased emphasis on spill prevention and response.

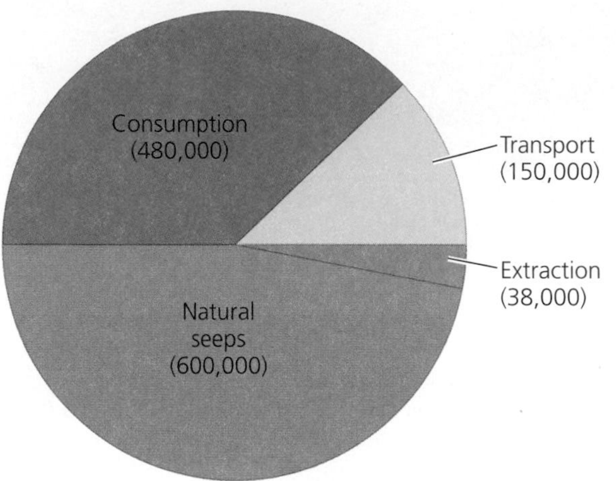


**(a) Sources of petroleum input into oceans (metric tons)**

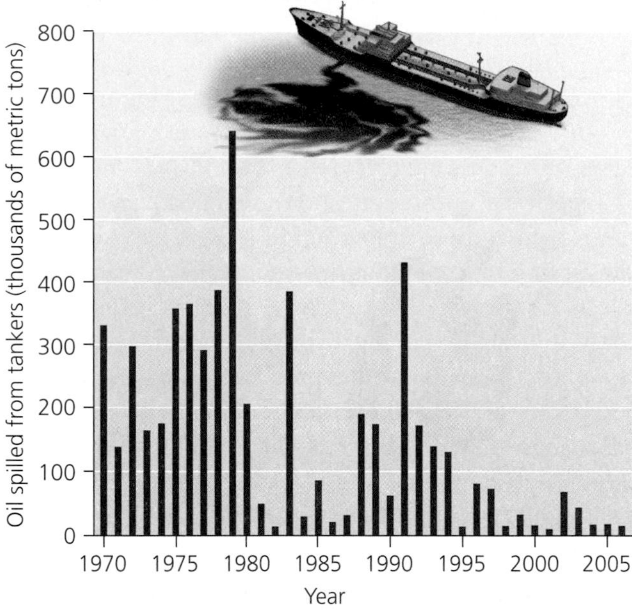


**(b) Quantity of petroleum spilled from tankers, 1970–2006**

**FIGURE 16.18** Of the 1.3 million metric tons of petroleum entering the world's oceans each year, nearly half is from natural seeps. Petroleum consumption by people accounts for 38% of total input, and this includes numerous diffuse non-point sources, especially runoff from rivers and coastal communities and leakage from two-stroke engines. Spills during petroleum transport account for 12%, and leakage during petroleum extraction accounts for 3% (**a**). Less oil is being spilled into ocean waters today in large tanker spills, thanks in part to regulations on the oil shipping industry and improved spill response techniques (**b**). The figure shows cumulative quantities of oil spilled worldwide from nonmilitary spills over 7 metric tons. Data from: National Research Council. 2003. *Oil in the sea III. Inputs, fates, and effects*. Washington, DC: National Academies Press (a); and International Tanker Owners Pollution Federation Ltd. (ITOPF). 2006. (b).

## Toxic pollutants can contaminate seafood

Marine pollution can make some fish and shellfish unsafe for people to eat. One prime concern today is mercury contamination. Mercury is a toxic heavy metal (• pp. 386, 647–648) that is emitted in coal combustion (• pp. 548–549) and from other sources. After settling onto land and water, mercury bioaccumulates in animals' tissues and biomagnifies as it makes its way up the food chain (• pp. 395–397). As a result, fish and shellfish at high trophic levels can contain substantial levels of mercury. Eating seafood high in mercury is particularly dangerous for young children and for pregnant or nursing mothers, because the fetus, baby, or child can suffer neurological damage as a result.

Because seafood is a vital part of a healthy diet, nutritionists do not advocate avoiding seafood entirely. However, people in at-risk groups should avoid fish high in mercury (such as swordfish, shark, and albacore tuna) while continuing to eat seafood low in mercury (such as catfish, salmon, and canned light tuna). We should also be careful not to eat seafood from local areas where health advisories have been issued.

## Excess nutrients cause algal blooms

Pollution from fertilizer runoff or other nutrient inputs can have dire effects on marine ecosystems, as we saw with the Gulf of Mexico's dead zone in Chapter 7. The release of excess nutrients into surface waters can spur unusually rapid growth of phytoplankton, causing eutrophication (• pp. 179, 429–430) in freshwater and saltwater systems.

Excessive nutrient concentrations sometimes give rise to population explosions among several species of marine algae that produce powerful toxins that attack the nervous systems of vertebrates. Blooms of these algae are known as **harmful algal blooms.** Some algal species produce reddish

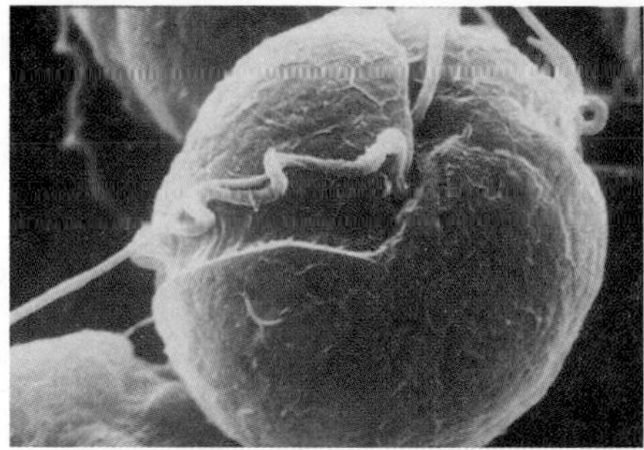

(a) Dinoflagellate (*Gymnodinium*)

(b) Red tide, Gulf of Carpentaria, Australia

FIGURE 16.19 In a harmful algal bloom, certain types of algae multiply to great densities in surface waters, producing toxins that can bioaccumulate and harm organisms. Red tides are a type of harmful algal bloom in which the algae, such as dinoflagellates of the genus *Gymnodinium* (a), produce pigment that turns the water red (b).

pigments that discolor surface waters, and blooms of these species are nicknamed **red tides** (**Figure 16.19**). Harmful algal blooms can cause illness and death among zooplankton, birds, fish, marine mammals, and people as their toxins are passed up the food chain. They also cause economic loss for communities dependent on fishing or beach tourism. Reducing nutrient runoff into coastal waters can lessen the frequency of these outbreaks. When they occur, we can minimize their health impacts by monitoring to prevent human consumption of affected organisms.

As severe as the impacts of marine pollution can be, however, most marine scientists concur that the more worrisome dilemma is overharvesting. Unfortunately, the old cliché that "there are always more fish in the sea" is not true; the oceans today have been overfished, and like the groundfish of the Northwest Atlantic, many stocks have been largely depleted.

# Emptying the Oceans

The oceans and their biological resources have provided for human needs for thousands of years, but today we are placing unprecedented pressure on marine resources. Half the world's marine fish populations are fully exploited, meaning that we cannot harvest them more intensively without depleting them, according to the U.N. Food and Agriculture Organization (FAO). An additional 25% of marine fish populations are overexploited and already being driven toward extinction. Thus only one-quarter of the world's marine fish populations can yield more than they are already yielding without being driven into decline.

Total global fisheries catch, after decades of increases, leveled off after about 1988 (**Figure 16.20**), despite increased fishing effort. Fishery collapses such as those off Newfoundland and New England are ecologically devastating and also take a severe economic toll on human communities that depend on fishing. If current trends continue, a comprehensive 2006 study in the journal *Science* predicted, populations of *all* ocean species that we fish for today will collapse by the year 2048.

As our population grows, we will become even more dependent on the oceans' bounty. This makes it vital, many scientists and fisheries managers say, that we turn immediately to more sustainable fishing practices.

## We have long overfished

People have always harvested fish, shellfish, turtles, seals, and other animals from the oceans. Although much of this harvesting was sustainable, scientists are learning that people began depleting some marine species centuries or millennia ago. Overfishing then accelerated during the colonial period of European expansion and intensified further in the 20th century.

A recent synthesis of historical evidence by marine biologist Jeremy Jackson and others revealed that ancient overharvesting likely affected ecosystems in astounding ways we only partially understand today. Several large animals, including the Caribbean monk seal, Steller's sea cow, and Atlantic gray whale, were hunted to extinction prior to the 20th century—before scientists were able to study them or the ecological roles they played. Overharvesting of the vast oyster beds of Chesapeake Bay led to the collapse of its oyster fishery in the late 19th century. Eutrophication and hypoxia similar to that of the Gulf of Mexico (Chapter 7) resulted, because there are no longer oysters to filter algae and bacteria from the water.

## CAUSES AND CONSEQUENCES

Our oceans and coastlines suffer many forms of pollution from human activities on land and on the water. This **marine pollution** has impacts on human health, human economies, fish and wildlife, natural systems, and ecosystem services. Fortunately, we have many solutions at hand to address the causes and mitigate the consequences of marine pollution.

Write in two causes of marine pollution in the spaces provided. Then write in two consequences (impacts on the environment, human health, or quality of life) that result from marine pollution. Finally, offer two solutions to this issue and its consequences. One cause, one consequence, and one solution have been filled in for you, providing examples.

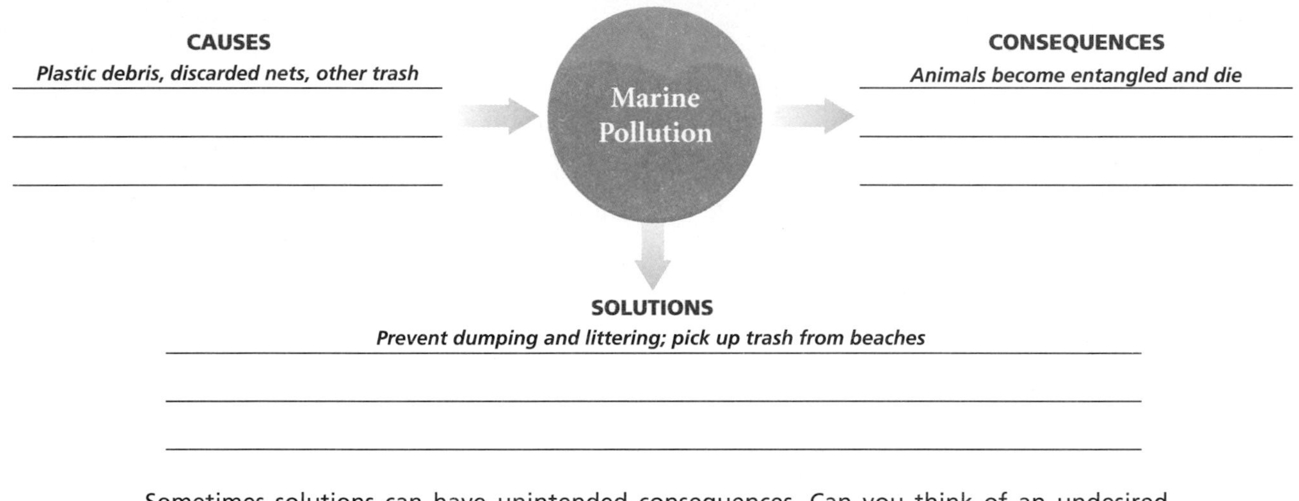


Sometimes solutions can have unintended consequences. Can you think of an undesired consequence that one solution to this issue might have? How might we then deal with *that* consequence?

In the Caribbean, green sea turtles ate sea grass and likely kept it cropped low, like a lawn. But with today's turtle population a fraction of what it was, sea grass grows thickly, dies, and rots, giving rise to disease such as sea grass wasting disease, which ravaged Florida Bay sea grass in the 1980s. The best-known case of historical overharvesting is the near-extinction of many species of whales. This resulted from commercial whaling that began centuries ago and was curtailed only in 1986. Since then, some species have been recovering, but others have not.

Groundfish in the Northwest Atlantic historically were so abundant that the people who harvested them never imagined they could be depleted. Yet careful historical analysis of fishing records has revealed that even in the 19th century, fishers repeatedly experienced locally dwindling catches, and each time needed to introduce some new approach or technology to extend their reach and restore their catch rate.

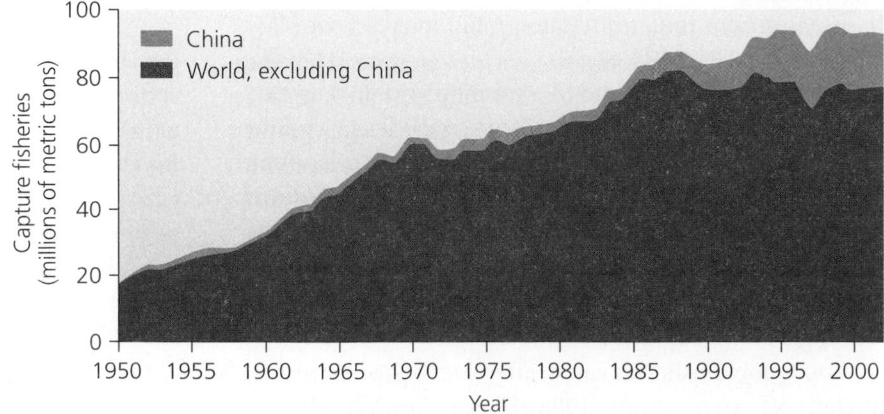


FIGURE 16.20 The total global fisheries catch has increased over the past half century, but in recent years growth has stalled, and many fear that a global catch decline is imminent if conservation measures are not taken soon. The figure shows trends with and without China's data (see "The Science behind the Story," • p. 463). With China's data, global catch has leveled off since the mid-1990s. Without China's data, catch has decreased slightly since 1988. Go to GRAPHIt! at www.aw-bc.com/withgott or on the student CD-ROM. Data from U.N. Food and Agricultural Organization (FAO).

## Fishing has industrialized

Today's commercial fishing fleets are highly industrialized, employing fossil fuels, huge vessels, and powerful new technologies to capture fish in volumes never dreamed of by 19th-century mariners. So-called *factory fishing* vessels even process and freeze their catches while at sea. The global reach of today's fleets makes our impacts much more rapid and intensive than in past centuries.

The modern fishing industry uses a number of methods to capture fish at sea. Some vessels set out long *driftnets* that span large expanses of water (**Figure 16.21a**). These strings of nets are arrayed strategically to drift with currents so as to capture passing fish, and are held vertical by floats at the top and weights at the bottom. Driftnetting usually targets species that traverse the open water in immense schools (flocks), such as herring, sardines, and mackerel. Specialized forms of driftnetting are used for sharks, shrimp, and other animals.

*Longline fishing* (**Figure 16.21b**) involves setting out extremely long lines with up to several thousand baited hooks spaced along their lengths. Tuna and swordfish are among the species targeted by longline fishing.

*Trawling* entails dragging immense cone-shaped nets through the water, with weights at the bottom and floats at the top to keep the nets open. Trawling in open water captures pelagic fish, whereas *bottom-trawling* (**Figure 16.21c**) involves dragging weighted nets across the floor of the continental shelf to catch groundfish and other benthic organisms, such as scallops.

**(a) Driftnetting**

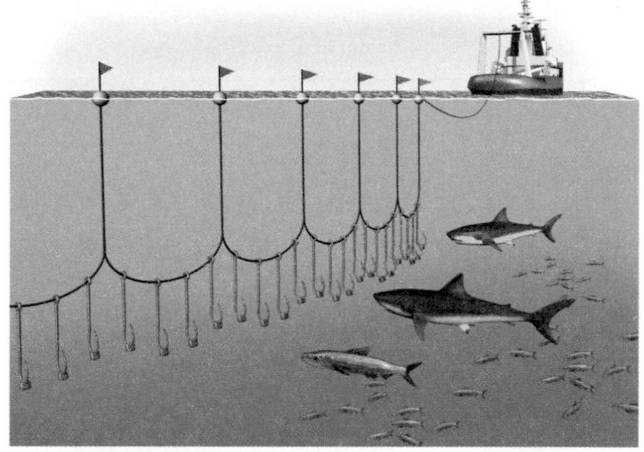

**(b) Longlining**

**(c) Bottom-trawling**

**FIGURE 16.21** Commercial fishing fleets use several main methods of capture. In driftnetting (**a**), huge nets are set out to drift through the open water to capture schools of fish. In longlining (**b**), lines with numerous baited hooks are set out in open water. In bottom-trawling (**c**), weighted nets are dragged along the floor of the continental shelf. All methods result in large amounts of by-catch, the capture of nontarget animals. The illustrations above are schematic for clarity and do not capture the immense scale that these technologies can attain; for instance, industrial trawling nets can be large enough to engulf multiple Boeing 747 jumbo jets.

## Fishing practices kill nontarget animals and damage ecosystems

Unfortunately, these fishing practices catch more than just the species they target. **By-catch** refers to the accidental capture of animals, and it accounts for the deaths of many thousands of fish, sharks, marine mammals, and birds each year.

Driftnetting captures substantial numbers of dolphins, seals, and sea turtles, as well as countless nontarget fish. Most of these end up drowning (mammals and turtles need to surface to breathe) or dying from air exposure on deck (fish breathe through gills in the water). Many nations have banned or restricted driftnetting because of excessive by-catch. The widespread death of dolphins in driftnets motivated consumer efforts to label tuna as "dolphin-safe" if its capture uses methods designed to avoid dolphin by-catch. Such measures helped reduce dolphin deaths from an estimated 133,000 per year in 1986 to less than 2,000 per year since 1998.

Similar by-catch problems exist with longline fishing, which kills turtles, sharks, and albatrosses, magnificent seabirds with wingspans up to 3.6 m (12 ft). Several methods are being developed to limit by-catch from longline fishing, but an estimated 300,000 seabirds of various species die each year when they become caught on hooks while trying to ingest bait.

Bottom-trawling can destroy entire communities and ecosystems. The weighted nets crush organisms in their path and leave long swaths of damaged sea bottom. Trawling is especially destructive to structurally complex areas, such as reefs, that provide shelter and habitat for many animals. In recent years, underwater photography has begun to reveal the extent of structural and ecological disturbance done by trawling (**Figure 16.22**). Trawling is often likened to clear-cutting (• pp. 340–341) and strip-mining (• p. 547), and in heavily fished areas, the bottom may be damaged more than once. At Georges Bank, it is estimated that the average expanse of bottom has been trawled three times.

**(a) Before trawling**

**(b) After trawling**

**FIGURE 16.22** Bottom-trawling causes severe structural damage to reefs and benthic habitats, and it can decimate underwater communities and ecosystems. A photo of an untrawled location (a) on the seafloor of Indonesia shows a vibrant and diverse coral reef community. A photo of a trawled location (b) nearby shows a flattened and lifeless expanse of broken coral draped with an abandoned stretch of trawling net.

## Modern fishing fleets deplete marine life rapidly

We can see the effects of large-scale industrialized fishing in the catch records of groundfish from the Northwest Atlantic. Although cod had been harvested since the 1500s on the Grand Banks, catches more than doubled once immense industrial trawlers from Europe, Japan, and the United States appeared in the 1960s (**Figure 16.23a**). These record-high catches lasted only a decade; the industrialized approach removed so many fish that the stock has not recovered. Likewise, on Georges Bank, cod catches rose greatly in the 1960s, remained high for 30 years, then collapsed (**Figure 16.23b**).

Throughout the world's oceans, today's industrialized fishing fleets are depleting marine populations quickly. In a 2003 study, Canadian fisheries biologists Ransom Myers and Boris Worm analyzed fisheries data from FAO archives, looking for changes in the catch rates of fish in various regions of ocean since they were first exploited by industrialized fishing. For one region after another, they found the same pattern: Catch rates dropped precipitously, with 90% of large-bodied fish and sharks eliminated within only a decade (**Figure 16.24**). Following that, populations stabilized at 10% of their former levels. This means, Myers and Worm concluded, that the oceans today contain only one-tenth of the large-bodied animals they once did.

As we have seen (• pp. 151–157), when animals at high trophic levels are removed from a food web, the proliferation of their prey can alter the nature of the entire community. Many scientists now conclude that most marine communities may have been very different prior to industrial fishing.

## Several factors mask declines

Although industrialized fishing has depleted fish stocks in region after region, the overall global catch has remained roughly stable for two decades (see Figure 16.20). How could this be? The seeming stability of the total global catch can be explained by several factors that mask population declines. One is that fishing fleets have been traveling longer distances to reach less-fished portions of the ocean. They also have been fishing in deeper waters; average depth of catches was 150 m (495 ft) in 1970 and 250 m (820 ft) in 2000. Moreover, fishing fleets have been spending more time fishing and have been setting out more nets and lines—expending increasing effort just to catch the same number of fish.

Improved technology also helps explain large catches despite declining stocks. Today's Japanese, European, Canadian, and U.S. fleets can reach almost any spot on the globe with vessels that attain speeds of 80 kph (50 mph).

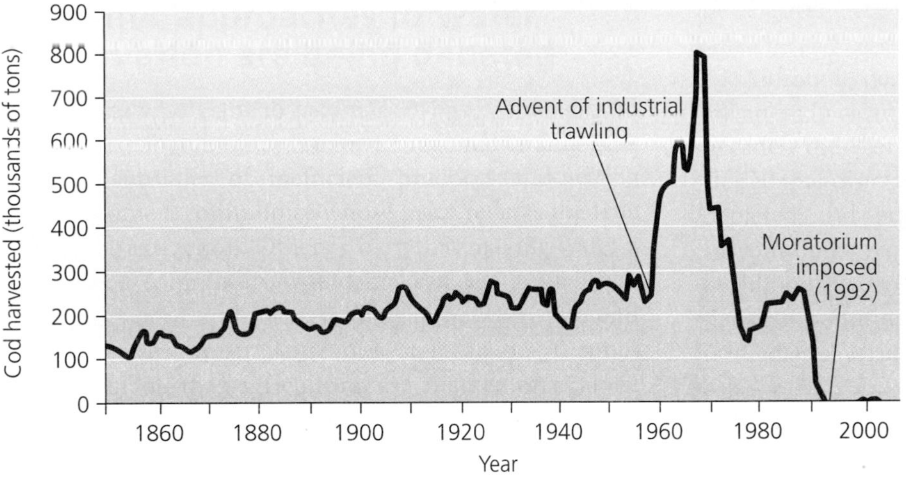


(a) Cod harvested from Grand Banks

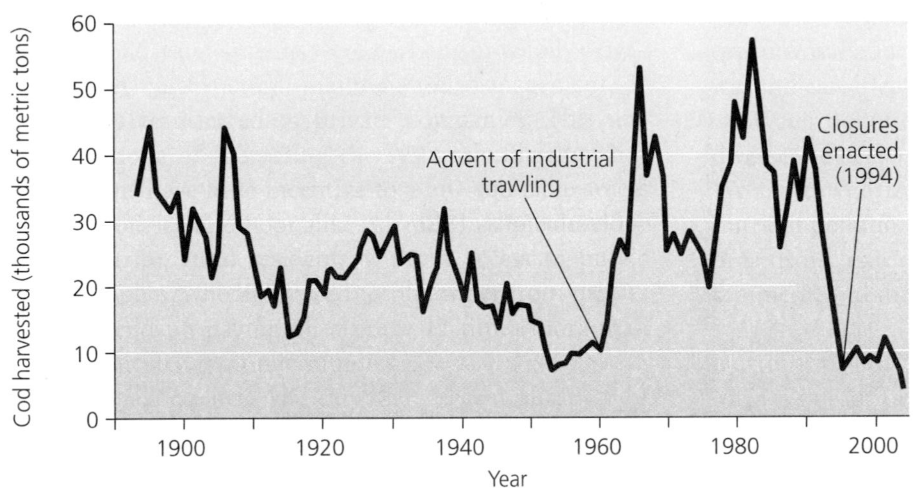


(b) Cod harvested from Georges Bank

FIGURE 16.23 In the North Atlantic off the coast of Newfoundland, commercial catches of Atlantic cod (**a**) increased with intensified fishing by industrial trawlers in the 1960s and 1970s. The fishery subsequently crashed, and moratoria imposed in 1992 and 2003 have not brought it back. A similar pattern is seen in the cod catches at Georges Bank (**b**); industrial fishing produced 30 years of high catches, followed by a collapse and the closure of some areas to fishing. Note also that in each case, there is one peak before 1977 and one after 1977. The first peak and decline resulted from foreign fishing fleets, whereas the second peak and decline resulted from Canadian and U.S. fleets, respectively, after they laid claim to their 200-mile exclusive economic zones. Data from Millennium Ecosystem Assessment. 2005 (a); and O'Brien, et al. 2005. Georges Bank Atlantic Cod. In Mayo, R. K. and M. Terceiro, eds. *Assessment of 19 Northeast groundfish stocks through 2004*. Northeast Fisheries Science Center, Woods Hole, MA (b).

They have access to an array of technologies that militaries have developed for spying and for chasing enemy submarines, including advanced sonar mapping equipment, satellite navigation, and thermal sensing systems. Some fleets rely on aerial spotters to find schools of commercially valuable fish such as bluefin tuna.

Lastly, one cause of misleading stability in global catch numbers is that not all data supplied to international monitoring agencies may be accurate (see "The Science behind the Story," • p. 463).

## We are "fishing down the food chain"

Overall figures on total global catch tell only part of the story, because they do not relate the species, age, and size of fish harvested. Careful analyses of fisheries data have revealed in case after case that as fishing increases, the size and age of fish caught decline. Cod caught in the Northwest Atlantic today are on average much smaller than they were decades ago, and it is now rare to find a cod over 10 years of age, although cod of this age formerly were common.

In addition, as particular species become too rare to fish profitably, fleets begin targeting other species that are in greater abundance. Generally this means shifting from large, desirable species to smaller, less desirable ones. Fleets have time and again depleted popular food fish (such as cod) and shifted to species of lower value (such as capelin, smaller fish that cod eat). Because this often entails catching species at lower trophic levels, this phenomenon has been termed "fishing down the food chain."

## Consumer choice can influence fishing practices

To most of us, marine fishing practices may seem a distant phenomenon over which we have no control. Yet by exercising careful choice when we buy seafood, consumers

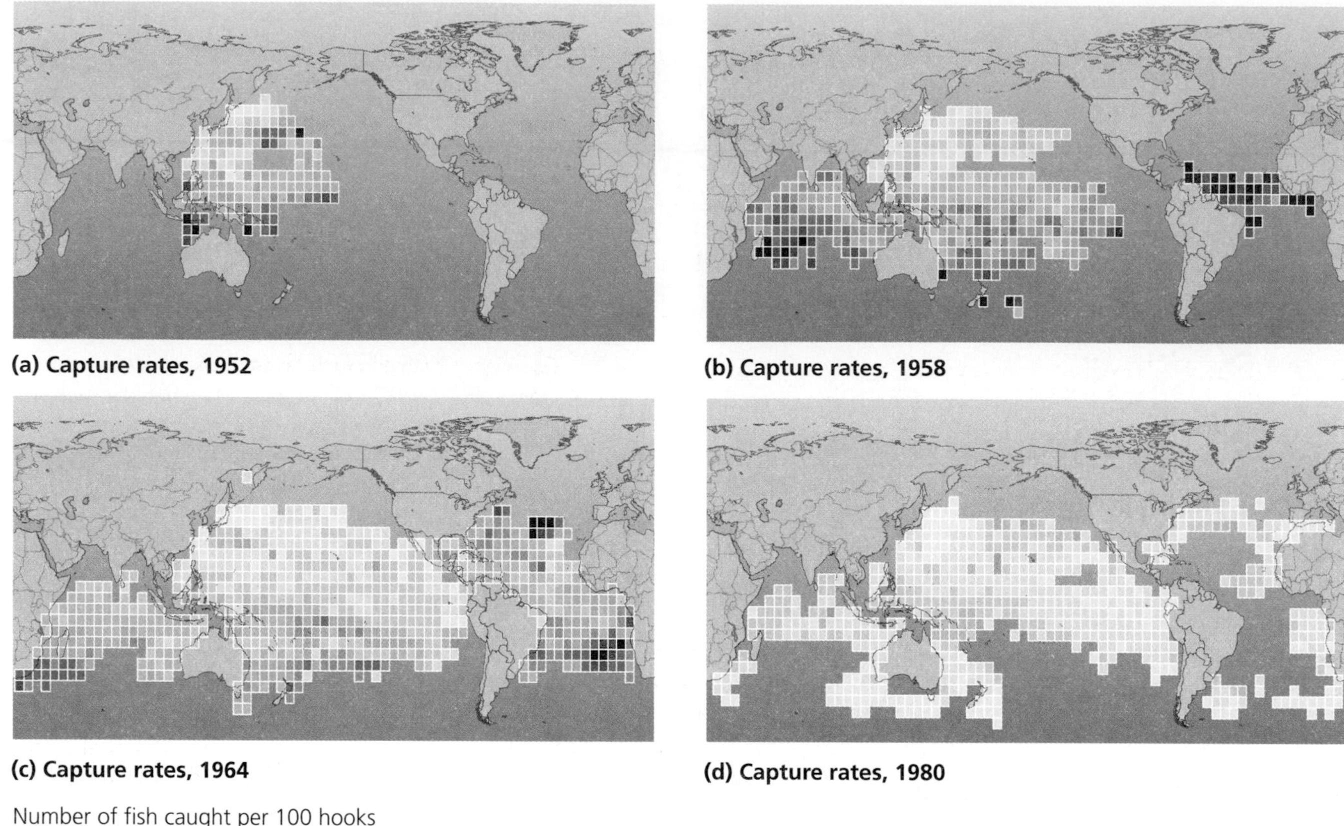


FIGURE 16.24 As industrial fishing fleets reached each new region of the world's oceans, capture rates of large predatory fishes were initially high and then within a decade declined markedly. In the figure, darker colors signify high capture rates, and lighter colors signify low capture rates. High capture rates in the southwestern Pacific in 1952 (**a**) gave way to low ones in later years. Excellent fishing success in the tropical Atlantic and Indian Oceans in 1958 (**b**) had turned mediocre by 1964 (**c**) and poor by 1980 (**d**). High capture rates in the north and south Atlantic in 1964 (**c**) gave way to low capture rates there in 1980 (**d**). Adapted from Myers, R. A. and B. Worm. 2003. Rapid worldwide depletion of predatory fish communities. *Nature* 423: 280–283.

can influence the ways in which fisheries function (**Figure 16.25**). Purchasing ecolabeled seafood, such as dolphin-safe tuna, is one way to exercise choice, but in most cases consumers have no readily available information about how their seafood was caught. Thus, several nonprofit organizations have recently devised concise guides to help consumers make informed choices. These guides differentiate fish and shellfish that are overfished or whose capture is ecologically damaging from those that are harvested more sustainably. For instance, the Monterey Bay Aquarium provides a wealth of this information on its Web site, and Table 16.1 includes a few examples.

### Weighing THE Issues | Eating Seafood

After reading this chapter, do you plan to alter your decisions about eating seafood in any way? If so, how? If not, why not? Do you think consumer buying choices can exert an influence on fishing practices? On mercury contamination in seafood?

## Marine biodiversity loss erodes ecosystem services

Overfishing, pollution, habitat change, and other factors that deplete biodiversity can threaten the ecosystem services we derive from the oceans. In the 2006 study that predicted global fisheries collapse by 2048 (• p. 457), the study's 14 authors analyzed all existing scientific literature to summarize the effects of biodiversity loss on ecosystem function and ecosystem services. They found that across 32 different controlled experiments conducted by various researchers, systems with less species diversity or genetic diversity showed less primary and secondary production and were less able to withstand disturbance.

The team also found that when biodiversity was reduced, so were habitats that serve as nurseries for fish and shellfish. Moreover, biodiversity loss was correlated with reduced filtering and detoxification (as from wetland vegetation and oyster beds), which can lead to harmful algal blooms, dead zones, fish kills, and beach closures.

FIGURE 16.25 You can discourage overfishing and support sustainable fisheries by exercising choice as a consumer when buying seafood. Here, a choosy customer shops for sustainably harvested fish at a fish market in Mahon, on the island of Menorca, Spain.

TABLE 16.1 Seafood Choices for Consumers

| Best choices* | Seafood to avoid† |
|---|---|
| Catfish (farmed) | Atlantic cod |
| Caviar (farmed) | Caviar (wild-caught) |
| Sardines | Chilean seabass/toothfish |
| Snow crab (Canada) | King crab (imported) |
| Pacific halibut | Atlantic halibut |
| Striped bass (farmed) | Orange roughy |
| Salmon (wild-caught from Alaska) | Salmon (farmed, including Atlantic) |
| Oysters (farmed) | Sharks |
| Shrimp (trap-caught) | Shrimp (imported, farmed, or trawl-caught) |
| Sturgeon (farmed) | Sturgeon (imported wild-caught) |
| Pollock (U.S.-caught from Alaska) | Swordfish (imported) |
| Tuna: albacore (troll/pole-caught) | Tuna: bluefin |

* Fish or shellfish that are abundant, well-managed, and fished or farmed in environmentally friendly ways.

† Fish or shellfish from sources that are overfished and/or fished or farmed in ways that harm other marine life or the environment.

*Source:* Monterey Bay Aquarium. *Seafood watch national seafood guide.* June 2005.

# Marine Conservation

Because we bear responsibility and stand to lose a great deal if valuable ecological systems collapse, marine scientists have been working to develop solutions to the problems that threaten the oceans. Many have begun by taking a hard look at the strategies used traditionally in fisheries management.

## Fisheries management has been based on maximum sustainable yield

Fisheries managers conduct surveys, study fish population biology, and monitor catches. They then use that knowledge to regulate the timing of harvests, the techniques used to catch fish, and the scale of harvests. The goal is to allow for maximal harvests of particular populations while keeping fish available for the future—the concept of *maximum sustainable yield* (• pp. 330–331). If data indicate that current yields are unsustainable, managers might limit the number or total biomass of that species that can be harvested, or they might restrict the type of gear fishers can use.

Despite such efforts, a number of fish and shellfish stocks have plummeted, and many scientists and managers now feel it is time to rethink fisheries management. One key change these reformers suggest is to shift the focus away from individual species and toward viewing marine resources as elements of larger ecological systems. This means considering the impacts of fishing practices on habitat quality, on species interactions, and on other factors that may have indirect or long-term effects on populations. One key aspect of such an *ecosystem-based management* approach (• p. 331) is to set aside areas of ocean where systems can function without human interference.

## We can protect areas in the ocean

Hundreds of **marine protected areas (MPAs)** have been established, most of them along the coastlines of developed countries. The United States is now inventorying areas for inclusion in a national network of MPAs, as called for under a Clinton administration executive order. However, despite their name, marine protected areas do not necessarily protect all their natural resources, because nearly all MPAs allow fishing or other extractive activities. As a recent report from an environmental advocacy group put it, even national marine sanctuaries "are dredged, trawled, mowed for kelp, crisscrossed with oil pipelines and fiber-optic cables, and swept through with fishing nets."

Because of the lack of true refuges from fishing pressure, many scientists—and some fishers—want to establish areas where fishing is prohibited. Such "no-take" areas have come to be called **marine reserves**. Designed to preserve entire ecosystems intact without human interference, marine reserves are also intended to improve fisheries. Scientists argue that marine reserves can act as production factories for fish for surrounding areas, because fish larvae produced inside reserves will disperse outside and stock other parts of the ocean (see "The Science behind the

THE SCIENCE BEHIND THE STORY

# China's Fisheries Data

*Dr. Daniel Pauly, University of British Columbia*

China is responsible for a larger share of the world's fisheries catch than any other nation, according to the United Nations Food and Agriculture Organization (FAO). Thus, China's catch data has a major impact on the FAO's attempts to assess the health of global fisheries.

In 2001, two fisheries scientists published a paper in the journal *Nature* suggesting that China had exaggerated its catch data by as much as 100% during the 1990s. Because local officials are rewarded for meeting production targets set by the government, they have economic incentive to falsely inflate their catch numbers, these researchers argued, and China's national data ends up inflated as a result. The high catch data had led to complacency among the world's fisheries managers and policymakers, they maintained, because it led the FAO to overestimate the true amount of fish left in the ocean.

The authors of the paper, University of British Columbia researchers Reg Watson and Daniel Pauly, had become suspicious of China's data for several reasons. First, China's coastal fisheries had been overexploited for decades, yet reported catches continued to increase. Second, China's "catch per unit effort" remained unchanged from 1980 to 1995, even though the abundance of fish had decreased. Finally, China's catch statistics suggested that its waters were far more productive than ecologically similar areas fished by other nations.

Using a number of databases, including FAO catch data collected since the 1950s, Watson and Pauly built a statistical model to predict global fisheries catch. They divided the world's oceans into approximately 180,000 "cells," each spanning half a degree longitudinally and latitudinally. The cells were characterized by factors that affect catch size, such as depth, primary productivity, fishing rights, and species distribution. The researchers used these factors to predict annual catch for each cell.

For most cells, the catch predicted by the model was similar to the catch reported by fleets that fished those waters. In large swaths of China's coastal waters, however, predicted catches were far smaller than reported catches. For instance, the model predicted that in 1999 China's total catch would be 5.5 million metric tons, whereas the figure reported by Chinese fisheries authorities was 10.1 million metric tons.

When Watson and Pauly analyzed global fisheries trends using the model rather than China's official numbers, they found that the annual global catch had decreased by 360,000 metric tons since 1988, instead of increasing by 330,000 metric tons, as reported by the FAO. Their finding suggested that the global catch, rather than remaining stable through the 1990s, actually had begun to decline in the 1980s.

Fisheries managers and policymakers depend on the FAO for information about the state of the world's fisheries. By using China's apparently inflated numbers, Watson later wrote, the FAO had encouraged a global "mopping-up operation" of the last of the world's fish.

The Chinese government and the FAO both criticized the study. China's reported catch was "basically correct," said Yang Jian, director-general of the Chinese Bureau of Fisheries. Watson and Pauly's model, he suggested, failed to take into account unique aspects of China's fisheries, such as its large catch of crabs and jellyfish.

The FAO noted that it was already treating China's statistics with caution. In its recent publications, global fisheries catches were being reported with and without China's data (see Figure 16.20). The FAO had also held meetings with Chinese officials to discuss ways of reducing overreporting, which Yang agreed did exist.

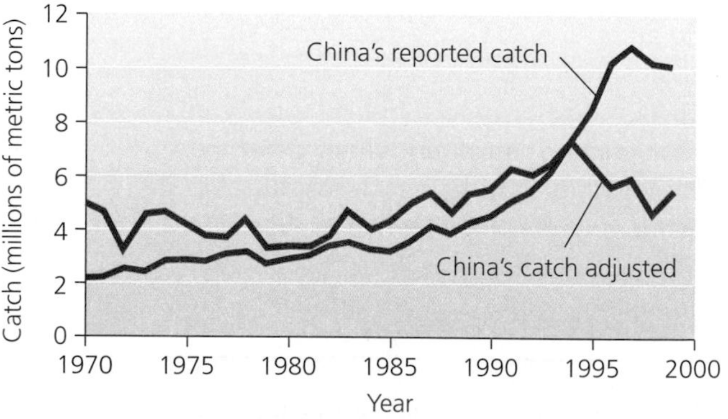


**Models estimated China's fish harvests to be much lower than reported.** Data from Watson, R., and D. Pauly. 2001. Systematic distortions in world fisheries catch trends. *Nature* 414: 536–538.

Story," • pp. 466–467). Proponents maintain that by serving both purposes, marine reserves are a win-win proposition for environmentalists and fishers alike.

**Weighing THE Issues** | **Preservation on Land and at Sea**

Almost 4% of U.S. land area is designated as wilderness, yet far less than 1% of coastal waters are protected in reserves. Why do you think it is taking so long for the preservation ethic to make the leap to the oceans?

Many fishers dislike the idea of no-take reserves, however, just as most were opposed to the Canadian groundfish moratoria and the Georges Bank closures. Nearly every marine reserve that has been established or proposed has met with pockets of intense opposition from people and businesses who use the area for fishing or recreation. Opposition comes from commercial fishing fleets as well as from individuals who fish recreationally. Both types of fishers are concerned that marine reserves will simply put more areas off-limits to fishing. In some parts of the world, protests have become violent. For instance, to protest fishing restrictions, fishermen in the Galapagos Islands destroyed offices at Galapagos National Park and threatened researchers and park managers with death.

## Reserves can work for both fish and fishers

In the past decade, data synthesized from marine reserves around the world have been indicating that reserves *do* work as win-win solutions that benefit ecosystems, fish populations, and fishing economies. In 2001, 161 prominent marine scientists signed a "consensus statement" summarizing the effects of marine reserves. Besides boosting fish biomass, total catch, and record-sized fish, the report stated, marine reserves yield several benefits. Within reserve boundaries, they

- Produce rapid and long-term increases in abundance, diversity, and productivity of marine organisms.
- Decrease mortality and habitat destruction.
- Lessen the likelihood of extirpation of species.

Outside reserve boundaries, marine reserves

- Can create a "spillover effect" when individuals of protected species spread outside reserves.
- Allow larvae of species protected within reserves to "seed the seas" outside reserves.

The consensus statement was backed up by research into reserves worldwide. At Apo Island in the Philippines, biomass of large predators increased eightfold inside a marine reserve, and fishing improved outside the reserve. At two coral reef sites in Kenya, commercially fished and keystone species were up to 10 times more abundant in the protected area as in the fished area. At Leigh Marine Reserve in New Zealand, snapper increased 40-fold, and spiny lobsters were increasing by 5–11% yearly. Spillover from this reserve improved fishing and ecotourism, and—as in Florida—local residents who once opposed the reserve now support it.

The review of data from existing marine reserves as of 2001 revealed that just 1–2 years after their establishment, marine reserves

- Increased densities of organisms on average by 91%.
- Increased biomass of organisms on average by 192%.
- Increased average size of organisms by 31%.
- Increased species diversity by 23%.

Since that time, further research has shown that reserves create a fourfold increase in catch per unit effort in fished areas surrounding reserves, and that they can greatly increase ecotourism by divers and snorkelers.

On Georges Bank, once commercial trawling was halted in 1994, populations of many organisms began to recover. As benthic invertebrates began to come back, numbers of groundfish such as haddock and yellowtail flounder rose inside the closed areas, and scallops increased by 14 times.

Moreover, fish from the closure areas appear to be spilling over into adjacent waters, because fishers have been catching more and more groundfish from Georges Bank as a whole since the late 1990s. From these and other datasets, increasing numbers of scientists, fishers, and policymakers are advocating the establishment of fully protected marine reserves as a central management tool.

## How should reserves be designed?

If marine reserves work in principle, the question becomes how best to design reserves and arrange them into networks. Scientists today are asking how large reserves need to be, how many there need to be, and where they need to be placed. Involving fishers directly in the planning process is crucial for coming up with answers to such questions.

Of several dozen studies that have estimated how much area of the ocean should be protected in no-take reserves, estimates range from 10% to 65%, with most falling between 20% and 50%. Other studies are modeling how to optimize the size and spacing of individual reserves so that ecosystems are protected, fisheries are sustained, and people are not overly excluded from marine areas (**Figure 16.26**). If marine reserves are designed strategically to take advantage of ocean currents, many

THE SCIENCE BEHIND THE STORY

## Do Marine Reserves Work?

*Dr. Callum M. Roberts, University of York, monitoring reef fish in Belize*

In November 2001, a team of fisheries scientists published a paper in the journal *Science*, providing some of the first clear evidence that marine reserves can benefit nearby fisheries. The team, led by York University researcher Callum Roberts, focused on reserves off the coasts of Florida and the Caribbean island of St. Lucia.

Following the establishment in 1995 of the Soufrière Marine Management Area (SMMA), a network of reserves intended to help restore St. Lucia's severely depleted coral reef fishery, Roberts and his colleague, Julie Hawkins, conducted annual visual surveys of fish abundance in the reserves and nearby areas. Within 3 years, they found that the biomass of five commercially important families of fish—surgeonfishes, parrot fishes, groupers, grunts, and snappers—had tripled inside the reserves and doubled outside them (see the first figure).

Roberts and Hawkins also interviewed local fishers and found that those with large traps were catching 46% more fish per trip in 2000–2001 than they had in 1995–1996 and that fishers with small traps were catching 90% more (see the second figure). Roberts and his colleagues concluded that "in 5 years, reserves have led to improvement in the SMMA fishery, despite the 35% decrease in area of fishing grounds."

Roberts and his coworkers also studied the oldest fully protected marine reserve in the United States, the Merritt Island National Wildlife Refuge (MINWR), established in 1962 as a buffer around what is today the Kennedy Space Center on Cape Canaveral, Florida. In a previous study, Darlene Johnson and James Bohnsack of the National Oceanic and Atmospheric Administration and Nicholas Funicelli of the United States Geological Service had found that the reserve contained more and larger fish than did nearby unprotected areas. This team also found that some of the reserve's fish appeared to be migrating to nearby fishing areas.

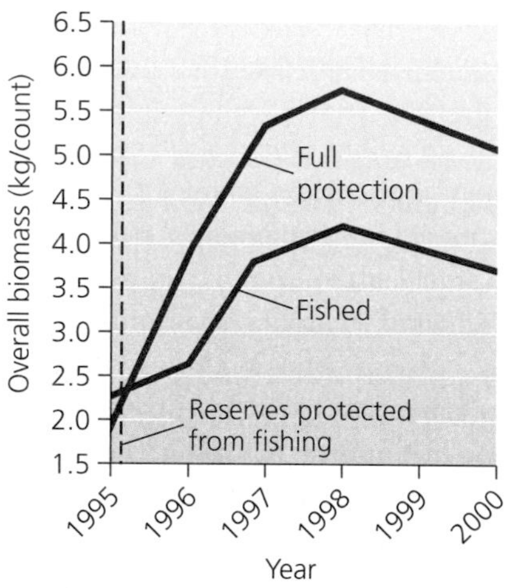


Established in 1995, the Soufrière Marine Management Area (SMMA) along the coast of St. Lucia had a rapid impact. By 1998, fish biomass within the five reserves tripled, and in adjacent fished areas it doubled. Data from Roberts, C., et al. 2001. Effects of marine reserves on adjacent fisheries. *Science* 294: 1920–1923.

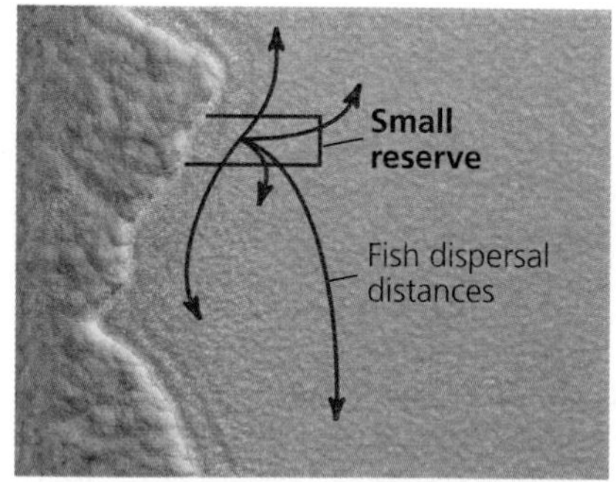


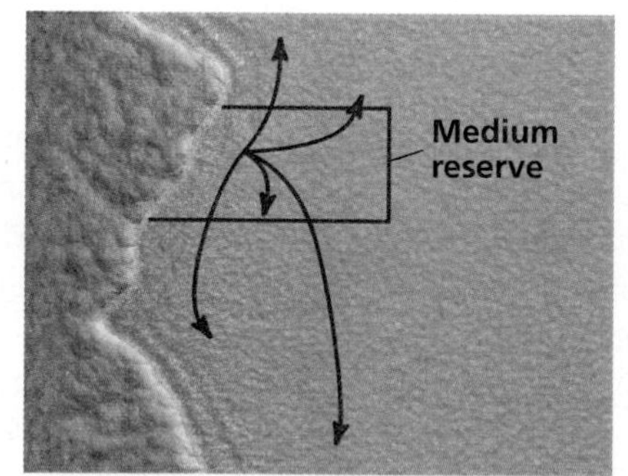


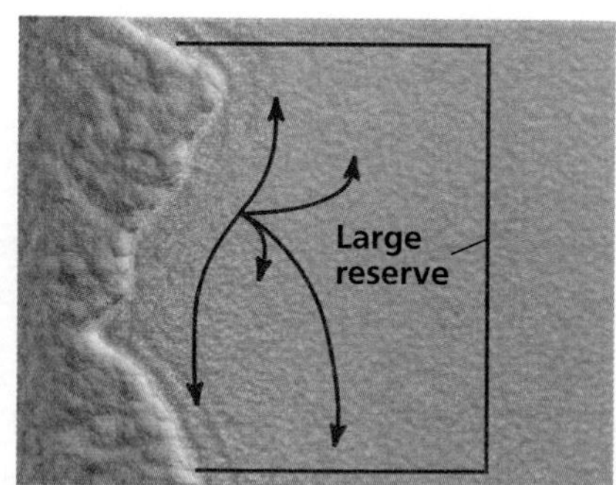


**FIGURE 16.26** Marine reserves of different sizes may have varying effects on ecological communities and fisheries. Young and adult fish and shellfish of different species can disperse different distances, as indicated by the red arrows in the figure. A small reserve (left panel) may fail to protect animals because too many disperse out of the reserve. A large reserve (right panel) may protect fish and shellfish very well but will provide relatively less "spillover" into areas where people can legally fish. Thus medium-sized reserves (middle panel) may offer the best hope of preserving species and ecological communities while also providing adequate fish to fishers and human communities. *Source:* Halpern, B. S., and R. R. Warner. 2003. Matching marine reserve design to reserve objectives. *Proceedings of the Royal Society of London B* 270: 1871–1878.

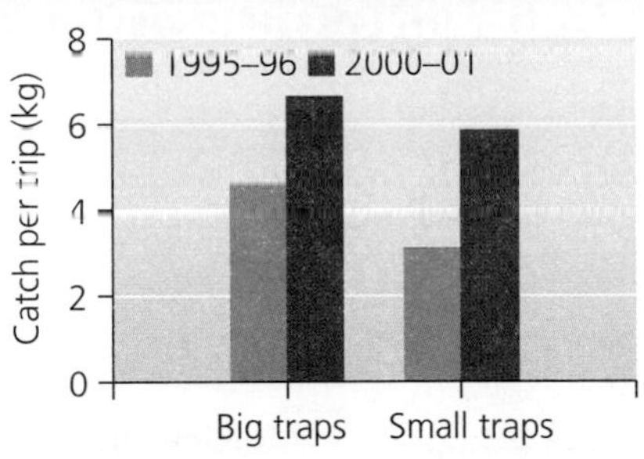


**(a) Catch per trip**

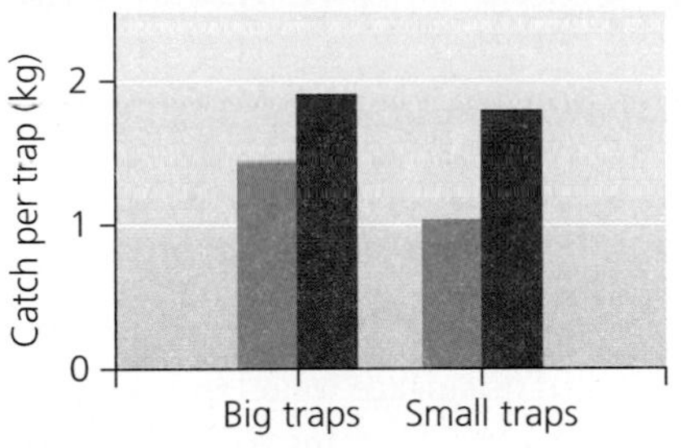


**(b) Catch per trap**

Roberts and his colleagues studied biomass of fish caught at the SMMA over two 5-month periods in 1995–1996 and 2000–2001. Per fishing trip **(a)**, the catch for fishers with big traps increased by 46%, and the catch for fishers with small traps increased by 90%. Per trap **(b)**, fishers with big traps caught 36% more fish, and fishers with big traps caught 80% more fish. Data from Roberts, C., et al. 2001. Effects of marine reserves on adjacent fisheries. *Science* 294: 1920–1923.

Bohnsack, Roberts, and their colleagues corroborated the evidence for migration by analyzing trophy records from the International Game Fish Association. They found that the proportion of Florida's record-sized fish caught near Merritt Island increased significantly after 1962. Nine years after the refuge was established, for instance, the number of spotted sea trout records from the Merritt Island area jumped dramatically. Bohnsack, Roberts, and their colleagues hypothesized that the reserve was providing a protected zone in which fish could grow to trophy size before migrating to nearby areas, where they were caught by recreational fishers.

Not everyone saw the St. Lucia and Merritt Island cases as proof that marine reserves could rescue depleted fisheries. In February 2002, several alternative interpretations of the evidence were published as letters in *Science.* Mark Tupper, a fisheries scientist at the University of Guam, suggested that the St. Lucia results were relevant only to coral reef fisheries in developing nations, whereas Florida's boost in fish populations was due primarily to limits on recreational fishing. Karl Wickstrom, editor-in-chief of *Florida Sportsman* magazine, suggested that the increase in trophy fish near MINWR was caused by commercial fishing regulations and changes in how trophies were recorded and promoted. And Ray Hilborn, a fisheries scientist at the University of Washington, challenged the study's scientific methods. In the St. Lucia case, he pointed out, there had been no control condition.

In response, Roberts and his colleagues reaffirmed the validity of their results while acknowledging some limitations. They agreed with Tupper that marine reserves are not always effective and often need to be complemented by other management tools, such as size limits. "We agree that inadequately protected reserves are useless," they wrote, "but our study shows that well-enforced reserves can be extremely effective and can play a critical role in achieving sustainable fisheries."

scientists say, then they may well seed the seas and help lead us toward solutions to one of our most pressing environmental problems.

## Conclusion

Oceans cover most of our planet and contain diverse topography and ecosystems, some of which we are only now beginning to explore and understand. We are learning more about the oceans and coastal environments while we are intensifying our use of their resources and causing these areas more severe impacts. In so doing, we are coming to understand better how to use these resources without depleting them or causing undue ecological harm to the marine and coastal systems on which we depend.

Today, scientists are demonstrating that setting aside protected areas of the ocean can serve to maintain natural systems and also to enhance fisheries. This is vital at a time when we are depleting many of the world's marine fish stocks. As historical studies reveal more information on how much biodiversity our oceans formerly contained and have now lost, we may increasingly look beyond simply making fisheries stable and instead consider restoring the ecological systems that once flourished in our waters.

## REVIEWING OBJECTIVES

**You should now be able to:**

**Identify physical, geographical, chemical, and biological aspects of the marine environment**

- Oceans cover 71% of Earth's surface and contain over 97% of its surface water. (p. 445)
- Ocean water contains 96.5% $H_2O$ by mass and various dissolved salts. (pp. 445–446)
- Colder, saltier water is denser and sinks. Water temperatures vary with latitude, and temperature variation is greater in surface layers. (p. 446)
- Persistent currents move horizontally through the oceans, driven by density differences, sunlight, and wind. (pp. 446–447)
- Vertical water movement includes upwelling and downwelling, which affect the distribution of nutrients and life. (pp. 447–448)
- Seafloor topography can be complex. (pp. 448–449)

**Describe major types of marine ecosystems**

- Major types of marine and coastal ecosystems include pelagic and deep-water open ocean systems, kelp forests, coral reefs, intertidal zones, salt marshes, mangrove forests, and estuaries. (pp. 449–454)
- Many of these systems are highly productive and rich in biodiversity. Many also suffer heavy impacts from human influence. (pp. 449–454)

**Outline historic and current human uses of marine resources**

- For millennia, people have fished the oceans and used ocean waters for transportation. (p. 454)
- Today we extract energy and minerals from the oceans. (pp. 454–455)

**Assess human impacts on marine environments**

- People pollute ocean waters with trash, including plastic and nets that harm marine life. (p. 455)
- Marine oil pollution results from non-point sources on land as well as from tanker spills at sea. (pp. 455–456)
- Heavy metal contaminants in seafood affect human health, and nutrient pollution can lead to harmful algal blooms. (pp. 456–457)
- Overharvesting is perhaps the major human impact on marine systems. (p. 457)

**Review the current state of ocean fisheries and reasons for their decline**

- Half the world's marine fish populations are fully exploited, 25% are already overexploited, and only 25% can yield more without declining. (p. 457)
- Global fish catches have stopped growing since the late 1980s, despite increased fishing effort and improved technologies. (pp. 457–458)
- People began depleting marine resources long ago, but impacts have intensified in recent decades. (pp. 457–461)
- Commercial fishing practices include driftnetting, long-line fishing, and trawling, all of which capture nontarget organisms, called by-catch. (pp. 459–460)
- Today's oceans hold only one-tenth the number of large animals that they did before the advent of industrialized commercial fishing. (pp. 460, 462)
- As fishing intensity increases, the fish available become smaller. (p. 461)
- Marine biodiversity loss affects ecosystem services. (p. 462)
- Consumers can encourage good fishery practices by shopping for sustainable seafood. (pp. 462, 464)
- Traditional fisheries management has not stopped declines, so many scientists feel that ecosystem-based management is needed. (p. 464)

**Evaluate marine protected areas and reserves as innovative solutions**

- We have established fewer protected areas in the oceans than we have on land, and most marine protected areas allow many extractive activities. (pp. 464–465)
- No-take marine reserves can protect ecosystems while also boosting fish populations and making fisheries sustainable. (pp. 464–467)

## TESTING YOUR COMPREHENSION

1. What proportion of Earth's surface do oceans cover? What is the average salinity of ocean water? How are density, salinity, and temperature related in each layer of ocean water?
2. What factors drive the system of ocean currents? In what ways do these movements affect conditions for life in the oceans?
3. Where in the oceans are productive areas of biological activity likely to be found?
4. Describe three kinds of ecosystems found near coastal areas and the kinds of life they support.
5. Why are coral reefs biologically valuable? How are they being degraded by human impact? What is causing the disappearance of mangrove forests and salt marshes?
6. Discuss three ways in which people are combating pollution in the oceans and on our coasts.
7. Describe an example of how overfishing can lead to ecological damage and fishery collapse.
8. Explain the conclusion of the Myers and Worm study of 2003 (see Figure 16.24).
9. Name three industrial fishing practices, and explain how they create by-catch and harm marine life.
10. How does a marine reserve differ from a marine protected area? Why do many fishers oppose marine reserves? Explain why many scientists say no-take reserves will be good for fishers.

## SEEKING SOLUTIONS

1. What benefits do you derive from the oceans? How does your behavior affect the oceans? Give specific examples.
2. We have been able to reduce the amount of oil we spill into the oceans, but petroleum-based products such as plastic continue to litter our oceans and shorelines. Discuss some ways that we can reduce this threat to the marine environment.
3. Describe the trends in global fish capture over the past 50 years, and explain several factors that account for these trends.
4. Consider what you know about biological productivity in the oceans, about the scientific data on marine reserves, and about the social and political issues surrounding the establishment of marine reserves. What ocean regions do you think it would be particularly appropriate to establish as marine reserves? Why?
5. **THINK IT THROUGH** You make your living fishing on the ocean, just as your father and grandfather did, and as most of your neighbors do in your small coastal village. Your region's fishery has just collapsed, however, and everyone is blaming it on overfishing. The government has closed the fishery for 3 years, and scientists are pushing for a permanent marine reserve to be established on your former fishing grounds. You have no desire to move away from your village, so what steps will you take now? Will you protest the closure? What compensation will you ask of the government if it prevents you from fishing? Will you work with scientists to establish a reserve that improves fishing in the future, or will you oppose their attempts to create a reserve? What data and what assurances will you ask of them?
6. **THINK IT THROUGH** You are mayor of a coastal town where some residents are employed as commercial fishers and others make a living serving ecotourists who come to snorkel and scuba-dive at the nearby coral reef. In recent years, several fish stocks have crashed, and ecotourism is dropping off as fish disappear from the increasingly degraded reef. Scientists are urging you to help establish a marine reserve around portions of the reef, but most commercial and recreational fishers are opposed to this idea. What steps would you take to restore your community's economy and environment?

## INTERPRETING GRAPHS AND DATA

The accompanying graph presents trends in the status of North Atlantic swordfish, a highly migratory species managed directly by the National Marine Fishery Service. The solid red line shows the mortality rate from fishing. The solid blue line indicates the biomass of the stock. The dotted lines of corresponding colors indicate the reference levels used to determine whether the stock is overfished or recovered. The graph also indicates the date when an international recovery plan was implemented.

1. Describe the trends in swordfish stocks (1) before the United States adopted an international management plan and (2) since the plan was adopted. Describe the interactions between fishing mortality and biomass as illustrated by the graph.
2. Based on the data in the graph, predict the likely trend in swordfish production over the next 10 years, assuming no change to the status quo.
3. This graph illustrates an effort that is succeeding, but not all rebuilding plans lead to stock recovery. Beyond the existence of a plan, what actions might play a role in supporting stock recovery efforts?

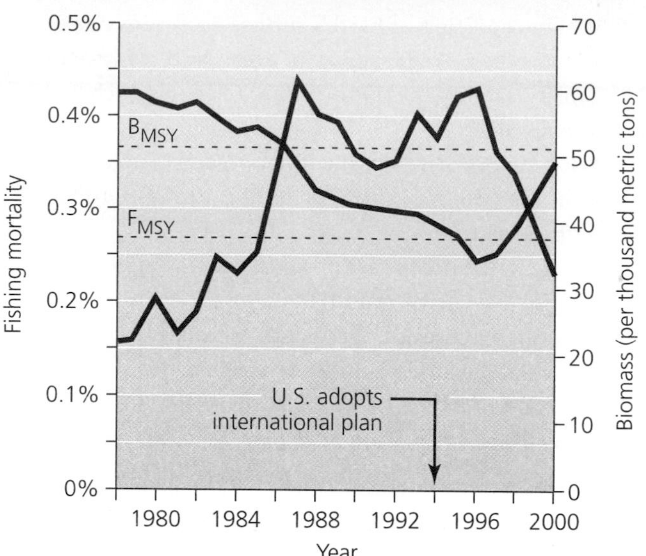


**Trends in fishing mortality and stock biomass for North Atlantic swordfish, 1978–2000.** Adapted from Rosenberg, A., et al. 2006. Rebuilding U.S. fisheries: Progress and problems. *Frontiers in Ecology and the Environment* 4(6), August 2006.

## CALCULATING ECOLOGICAL FOOTPRINTS

The relationship between the ecological goods and services used by individuals and the amount of *land* area needed to provide those goods and services is relatively well developed. People also use goods and services from Earth's oceans, where the concept of *area* is less useful. It is clear, however, that our removal of fish from the oceans has an impact, or an ecological footprint.

The table shows data on the mean annual per capita consumption from ocean fisheries for North America, China, and the world as a whole. Using the data provided, calculate the amount of fish each consumer group would consume per year, given the annual per capita consumption rates, for each of these three regions. Record your results in the table.

| | **Annual Consumption** | | |
|---|---|---|---|
| **Consumer group** | **North America (21.6 kg per capita)** | **China (27.7 kg per capita)** | **World (16.2 kg per capita)** |
| **You** | | | |
| **Your class** | | | |
| **Your state** | | | |
| **United States** | $6.48 \times 10^9$ kg | $8.31 \times 10^9$ kg | $4.86 \times 10^9$ kg |
| **World** | | | |

Data from U.N. Food and Agriculture Organization (FAO), Fisheries Department. 2004. *The state of world fisheries and aquaculture: 2004.* Data are for 2002, the most recent year for which comparative data are available.

1. Calculate the ratio of North America's per capita fish consumption rate to that of the world. Compare this ratio to the ratio of the per capita ecological footprints for the United States, Canada, and Mexico (see Figure 1.12, • p.17) versus the world average footprint of 2.2 ha/person/year. Can you account for similarities and differences between these ratios?
2. The population of China has grown at an annual rate of 1.1% since 1987, while over the same period fish consumption in China has grown at an annual rate of 8.9%. Speculate on the reasons behind China's rapidly increasing consumption of fish.
3. What ecological concerns do the combined trends of human population growth and increasing per capita fish consumption raise for you? What role might you play in contributing to these concerns or to their solutions?

## Take It Further

Go to www.aw-bc.com/withgott or the student CD-ROM, where you'll find:

- Suggested answers to end-of-chapter questions
- Quizzes, animations, and flashcards to help you study
- *Research Navigator*™ database of credible and reliable sources to assist you with your research projects
- GRAPHIt! Tutorials to help you interpret graphs
- INVESTIGATEIt! Current news articles that link the topics that you study to case studies from your region to around the world

CHAPTER

# 17 Atmospheric Science and Air Pollution

View of Earth's atmosphere from space

## Upon completing this chapter, you will be able to:

- Describe the composition, structure, and function of Earth's atmosphere
- Outline the scope of outdoor air pollution and assess potential solutions
- Explain stratospheric ozone depletion and identify steps taken to address it
- Define acidic deposition and illustrate its consequences
- Characterize the scope of indoor air pollution and assess potential solutions

Bus leaves London's congestion charging area

CENTRAL CASE

# Charging toward Cleaner Air in London

**"We are encouraging people to take into account the impact of their choices . . . on the environment."**
—Ken Livingstone, Mayor of London, November 2006

**"This is a bad idea. It's [the government] trying to make yet more money from honest, hard-working people."**
—London Gardener Roger Grover, December 2006

No city can claim a longer or more notorious history of air pollution than London, England. As far back as the 13th century, pollution from coal burning in London was so bad that King Edward I decreed that anyone caught burning a certain dirty form of coal would be beheaded. One unfortunate citizen reputedly paid the price.

As the United Kingdom led the world into the industrial revolution in the 18th and 19th centuries, London expanded its use of coal, firing the nation's industries, heating people's homes—and polluting the city's air. In 1952, tragedy struck as weather conditions trapped pollutants over the city for days, creating a thick and blinding "killer smog" that killed 4,000 people—and by some estimates up to 12,000. Although similar events had occurred in 1813, 1873, 1880, 1891, and 1948, the severity of the 1952 catastrophe helped Great Britain and other developed nations down the road to pollution control.

Today, Londoners breathe much cleaner air, but they find themselves battling a new source of pollution: smog from the emissions of congested urban traffic. Roughly 250,000 vehicles commute into London's downtown every workday, spewing a cocktail of pollutants including carbon dioxide, carbon monoxide, nitrogen oxides, and particulate matter as they creep along at an average speed of 8 mph.

Enter Ken Livingstone, a populist mayor with a daring plan: If London were to charge drivers a fee for entering the central city, it could relieve the traffic gridlock, encourage mass transit options, and cut down on pollution.

After 20 months of consulting with residents, businesses, and interest groups, Livingstone pushed through his "congestion-charging" program in February 2003, and today people driving into central London during weekdays are required to pay 8 pounds (about $15) per day. Drivers can pay online, by phone, or in shops, and fines are steep for those who do not. The money—expected to amount to 1.3 billion pounds ($2.5 billion) after 10 years—goes to enhance bus service and encourage transport by rail, taxi, bicycle, and foot.

Many citizens were outraged. Daily commuters fumed that paying 1,900 pounds ($3,600) per year, on top of existing taxes and parking fees, was simply too much. Others argued that the system discriminated against poorer people who could not afford the fees. Still others complained that the system was not working or that the promised improvements in public transport were slow in coming.

However, many other Londoners supported the program, and business support grew as the benefits became clearer. Traffic congestion decreased by nearly 30% as fewer cars entered the charging zone and driving speeds rose. Londoners in the zone suffered 40–70 fewer injuries from traffic accidents per year. Business and economic indicators remained strong. And the air became cleaner.

In just the first year within the charging zone, particulate matter declined by 15.5%, nitrogen oxide emissions decreased by 13.4%, and carbon dioxide emissions fell by 16.4%. Experts judged that nearly all the carbon dioxide reduction and about half the cuts in particulate matter and nitrogen oxides were due to reduced congestion, with the rest of the reductions due to improved vehicle technology.

In November 2006, Livingstone announced a proposal to charge highly polluting vehicles more and to exempt low-emission vehicles. At the same time, plans to expand the charging zone proceeded.

London's congestion-charging program is being studied by other cities, including New York City and San Francisco. Similar programs have already been implemented in Singapore, Toronto, Melbourne, Riga, and several Swedish cities, helping to improve air quality in these cities of the developed world. Meanwhile, air pollution remains a major problem in much of the developing world as poorer nations industrialize.

# The Atmosphere

Every breath we take reaffirms our connection to the **atmosphere,** the thin layer of gases that surrounds Earth. We live at the bottom of this layer, which provides us oxygen, absorbs hazardous solar radiation, burns up incoming meteors, transports and recycles water and nutrients, and moderates climate.

The atmosphere consists of roughly 78% nitrogen gas ($N_2$) and 21% oxygen gas ($O_2$). The remaining 1% is composed of argon gas (Ar) and minute concentrations of several other gases (**Figure 17.1**). These include *permanent gases* that remain at stable concentrations and *variable gases* that vary in concentration from time to time or place to place as a result of natural processes or human activities.

Over Earth's long history, the atmosphere's chemical composition has changed. Oxygen gas began to build up in an atmosphere dominated by carbon dioxide ($CO_2$), nitrogen, carbon monoxide (CO), and hydrogen ($H_2$) about 2.7 billion years ago, with the emergence of autotrophic microbes that emitted oxygen as a by-product of photosynthesis (• pp. 102–103). Today, human activity is altering the quantities of some atmospheric gases, such as carbon dioxide, methane ($CH_4$), and ozone ($O_3$). In this chapter and in Chapter 18, we will explore the atmospheric changes brought about by artificial pollutants, but we must first begin with an overview of Earth's atmosphere.

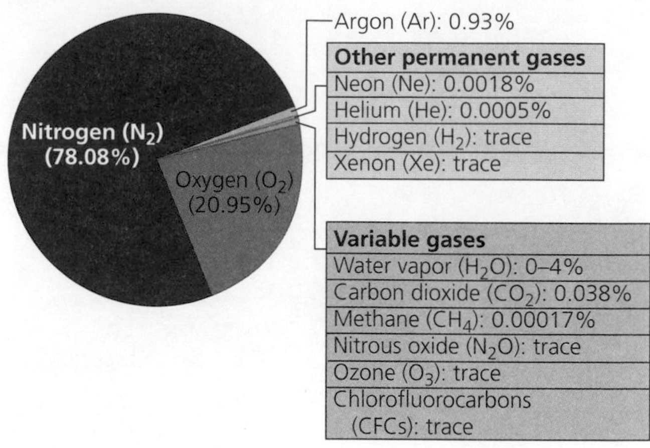


**FIGURE 17.1** Earth's atmosphere consists mostly of nitrogen, secondarily of oxygen, and lastly of a mix of gases at dilute concentrations. Permanent gases are fixed in concentration. Variable gases vary in concentration as a result of either natural processes or human activities. Data from Ahrens, C. D. 2007. *Meteorology today,* 8th ed. Belmont, CA: Brooks/Cole.

## The atmosphere consists of layers

The atmosphere that stretches so high above us and seems so vast is actually just a thin coating about 1/100th of Earth's diameter, like the fuzzy skin of a peach. This coating consists of four layers that atmospheric scientists recognize by measuring differences in temperature, density, and composition (**Figure 17.2**).

The bottommost layer, the **troposphere,** blankets Earth's surface and provides us the air we need to live. The movement of air within the troposphere is also largely responsible for the planet's weather. Although it is thin (averaging 11 km [7 mi] high) relative to the atmosphere's other layers, the troposphere contains three-quarters of the atmosphere's mass, because air is denser near Earth's surface. On average, tropospheric air temperature declines by about 6° C for each kilometer in altitude (or 3.5° F per 1,000 ft), dropping to roughly −52° C (−62° F) at its highest point. At the top of the troposphere, however, temperatures cease to decline with altitude, marking a boundary

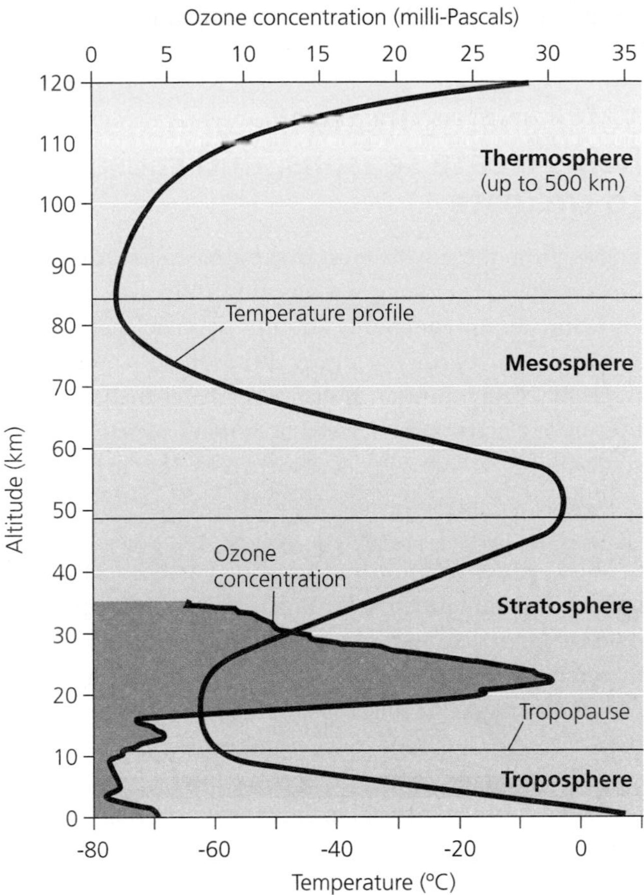


**FIGURE 17.2** Temperature drops with altitude in the troposphere, rises with altitude in the stratosphere, drops in the mesosphere, then rises again in the thermosphere. The tropopause separates the troposphere from the stratosphere. Ozone reaches a peak in a portion of the stratosphere, giving rise to the term *ozone layer.* Adapted from Jacobson, M. Z. 2002. *Atmospheric pollution: History, science, and regulation*. Cambridge: Cambridge University Press; Parson, E. A. 2003. *Protecting the ozone layer: Science and strategy*. Oxford: Oxford University Press.

called the *tropopause.* The tropopause acts like a cap, limiting mixing between the troposphere and the atmospheric layer above it, the stratosphere.

The **stratosphere** extends 11–50 km (7–31 mi) above sea level. Similar in composition to the troposphere, the stratosphere is 1,000 times drier and less dense. Its gases experience little vertical mixing, so once substances (including pollutants) enter it, they tend to remain for a long time. The stratosphere attains a maximum temperature of −3° C (27° F) at its highest altitude but is colder in its lower reaches. The reason is that ozone and oxygen absorb and scatter the sun's ultraviolet (UV) radiation (• pp. 101–102), so that much of the UV radiation penetrating the upper stratosphere fails to reach the lower stratosphere. Most of the atmosphere's minute amount of ozone concentrates in a portion of the stratosphere roughly 17–30 km (10–19 mi) above sea level, a region that has come to be called Earth's **ozone layer.** The ozone layer greatly reduces the amount of UV radiation that reaches Earth's surface. Because UV light can damage living tissue and induce mutations in DNA, the ozone layer's protective effects are vital for life on Earth.

Above the stratosphere lies the *mesosphere,* which extends 50–80 km (31–56 mi) above sea level. Air pressure is extremely low here, and temperatures decrease with altitude, reaching their lowest point at the top of the mesosphere. From here, the *thermosphere,* our atmosphere's top layer, extends upward to an altitude of 500 km (300 mi).

## Atmospheric properties include temperature, pressure, and humidity

Although the lower atmosphere is stable in its chemical composition, it is dynamic in its movement; air movement within it is due to differences in the physical properties of air masses. Among these properties are pressure and density, relative humidity, and temperature.

Gravity pulls gas molecules toward Earth's surface, causing air to be most dense near the surface and less so as altitude increases. **Atmospheric pressure,** which measures the force per unit area produced by a column of air, also decreases with altitude, because at higher altitudes fewer molecules are pulled down by gravity (**Figure 17.3**). At sea level, atmospheric pressure is 14.7 lb/in.$^2$ or 1,013 millibars (mb). Mountain climbers trekking to Mount

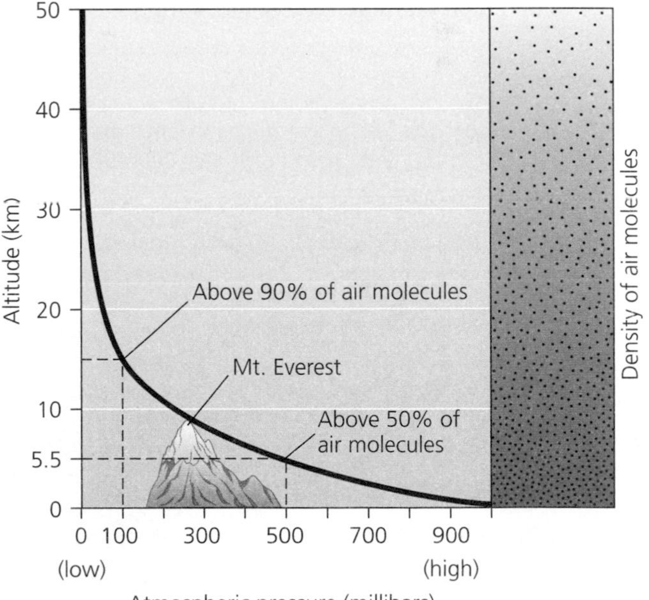


**FIGURE 17.3** As one climbs higher through the atmosphere, gas molecules become less densely packed. As density decreases, so does atmospheric pressure. Because most air molecules lie low in the atmosphere, one needs to be only 5.5 km (3.4 mi) high to be above half the planet's air molecules. Adapted from Ahrens, C. D. 2007. *Meteorology today,* 8th ed. Belmont, CA: Brooks/Cole.

Everest, the world's highest mountain, can look up and view their destination from Kala Patthar, a nearby peak, at roughly 5.5 km (18,000 ft). At this altitude, pressure is 500 mb, and half the atmosphere's air molecules are above the climber, and half are below. A climber who reaches Everest's peak (8.85 km [29,035 ft]), where the "thin air" is just over 300 mb, stands above two-thirds of the molecules in the atmosphere. When we fly on a commercial jet airliner, at a typical cruising altitude of 11 km (36,000 ft), we are above roughly 80% of the atmosphere's molecules.

Another property of air is **relative humidity,** the ratio of water vapor a given volume of air contains to the maximum amount it *could* contain at a given temperature. Average daytime relative humidity in June in the desert at Phoenix, Arizona, is only 31% (meaning that the air contains less than a third of the water vapor it possibly can at its temperature), whereas on the tropical island of Guam, relative humidity rarely drops below 88%. People are sensitive to changes in relative humidity because we perspire to cool our bodies. When humidity is high, the air is already holding nearly as much water vapor as it can, so sweat evaporates slowly and the body cannot cool itself efficiently. This is why high humidity makes it feel hotter than it really is. Low humidity speeds evaporation and makes it feel cooler.

The temperature of air also varies with location and time. At the global scale, temperature varies over Earth's surface because the sun's rays strike some areas more directly than others. At more local scales, temperature varies because of topography, plant cover, proximity of land to water, and many other factors.

## Solar energy heats the atmosphere, helps create seasons, and causes air to circulate

Energy from the sun heats air in the atmosphere, drives air movement, helps create seasons, and influences weather and climate. An enormous amount of solar energy continuously bombards the upper atmosphere—over 1,000 watts/m$^2$, many thousands of times greater than the total output of electricity generated by human society. Of that solar energy, about 70% is absorbed by the atmosphere and planetary surface, while the rest is reflected back into space (see Figure 18.1, • p. 507).

The spatial relationship between Earth and the sun determines how much solar radiation strikes each point on Earth's surface. Sunlight is most intense when it shines directly overhead and meets the planet's surface at a perpendicular angle. At this angle, sunlight passes through a minimum of energy-absorbing atmosphere, and Earth's surface receives a maximum of solar energy per unit surface area. Conversely, solar energy that approaches Earth's surface at an oblique angle loses intensity as it traverses a longer distance through the atmosphere, and it is less intense when it reaches the surface. This is why, on average, solar radiation intensity is highest near the equator and weakest near the poles (**Figure 17.4**).

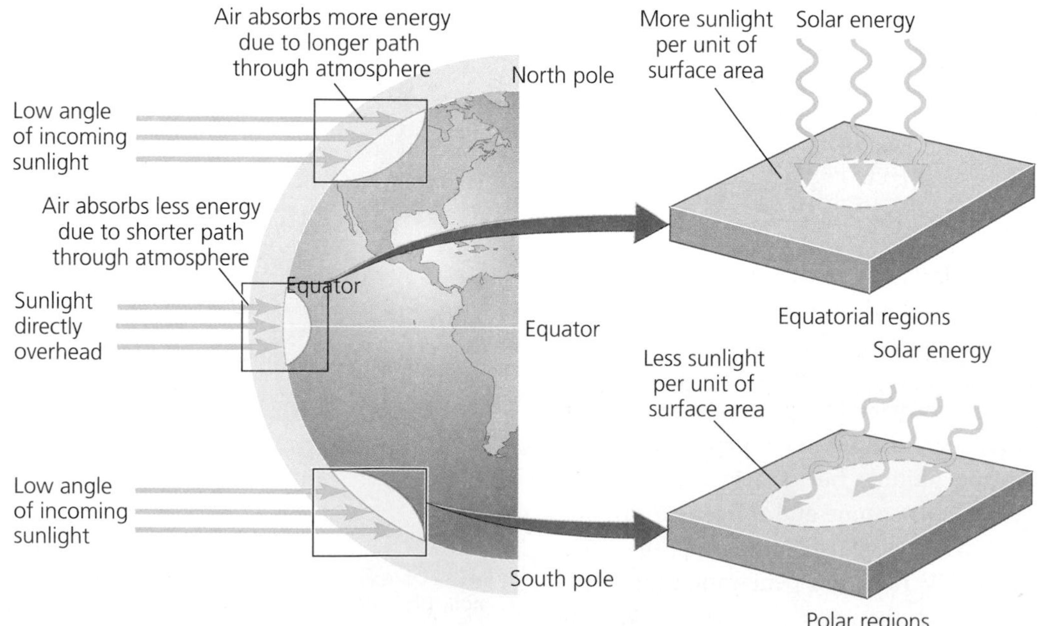


FIGURE 17.4 Because of Earth's curvature, polar regions receive on average less solar energy than equatorial regions. One reason is that sunlight gets spread over a larger area when striking the surface at an angle. Another reason is that sunlight approaching at a lower angle near the poles must traverse a longer distance through the atmosphere, during which more energy is absorbed or reflected. These patterns represent year-round averages; the latitude at which radiation approaches the surface perpendicularly varies with the seasons (see Figure 17.5).

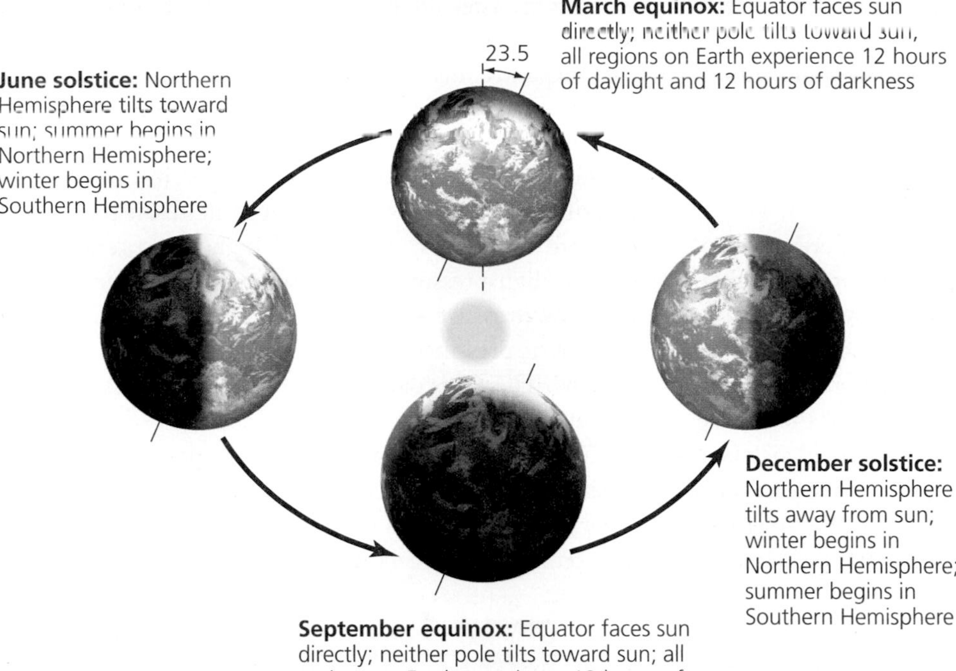


**FIGURE 17.5** The seasons occur because Earth is tilted on its axis by 23.5 degrees. As Earth revolves around the sun, the Northern Hemisphere tilts toward the sun for one half of the year, and the Southern Hemisphere tilts toward the sun for the other half of the year. In each hemisphere, summer occurs during the period in which the hemisphere receives the most solar energy because of its tilt toward the sun.

Because Earth is tilted on its axis (an imaginary line connecting the poles, running perpendicular to the equator) by about 23.5 degrees, the Northern and Southern Hemispheres each tilt toward the sun for half the year, resulting in the change in seasons (**Figure 17.5**). Regions near the equator are largely unaffected by this tilt; they experience about 12 hours each of sunlight and darkness every day throughout the year. Near the poles, however, the effect is strong, and seasonality is pronounced.

Land and surface water absorb solar energy, radiating some heat and causing some water to evaporate. Air near Earth's surface therefore tends to be warmer and moister than air at higher altitudes. These differences set into motion a process of **convective circulation** (**Figure 17.6**). Warm air, being less dense, rises and creates vertical currents. As air rises into regions of lower atmospheric pressure, it expands and cools. Once the air cools, it descends and becomes denser, replacing warm air that is rising. The air picks up heat and moisture near ground level and prepares to rise again, continuing the process. Similar convective circulation patterns occur in ocean waters (• pp. 447–448, 511), in magma beneath Earth's surface (• pp. 200–201), and even in a simmering pot of soup. Convective circulation influences both weather and climate.

## The atmosphere drives weather and climate

Weather and climate each involve the physical properties of the troposphere, such as temperature, pressure, humidity, cloudiness, and wind. **Weather** specifies atmospheric conditions over short time periods, typically hours or days, and within relatively small geographic areas. **Climate,** in contrast, describes the pattern of atmospheric conditions found across large geographic regions over long periods of time, typically seasons, years, or millennia. Mark Twain once noted the distinction between climate and weather by

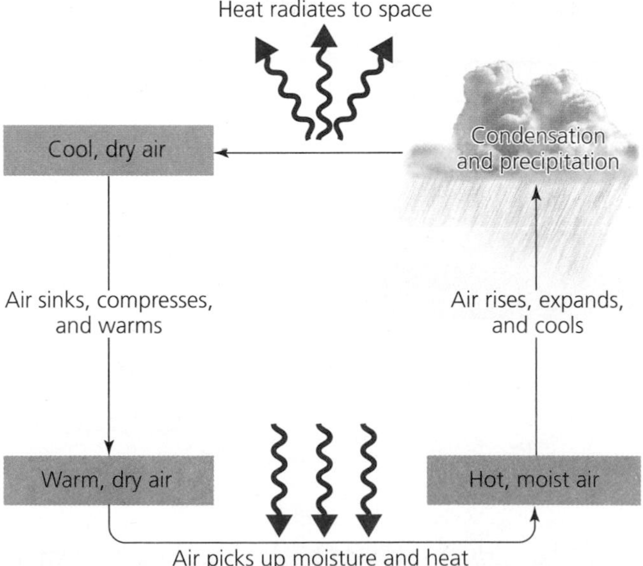


**FIGURE 17.6** Weather is driven in part by the convective circulation of air in the atmosphere. Air being heated near Earth's surface picks up moisture and rises. Once aloft, this air cools, and moisture condenses, forming clouds and precipitation. Cool, drying air begins to descend, compressing and warming in the process. Warm, dry air near the surface begins the cycle anew.

saying, "Climate is what we expect; weather is what we get." For example, London has a moist, temperate climate, but occasional days in summer can turn hot, dry, and sunny.

## Air masses interact to produce weather

Weather can change when air masses with different physical properties meet. The boundary between air masses that differ in temperature and moisture (and therefore density) is called a *front.* The boundary along which a mass of warmer, moister air replaces a mass of colder, drier air is termed a **warm front** (**Figure 17.7a**). Some of the warm, moist air behind a warm front rises over the cold air mass and then cools and condenses to form clouds that may produce light rain. A **cold front** (**Figure 17.7b**) is the boundary along which a colder, drier air mass displaces a warmer, moister air mass. The colder air, being denser, tends to wedge beneath the warmer air. The warmer air rises, expands, then cools to form clouds that can produce thunderstorms. Once a cold front passes through, the sky usually clears, and the temperature and humidity drop.

Adjacent air masses may also differ in atmospheric pressure. A **high-pressure system** contains air that moves outward away from a center of high pressure as it descends. High-pressure systems typically bring fair weather. In a **low-pressure system,** air moves toward the low atmospheric pressure at the center of the system and spirals upward. The air expands and cools, and clouds and precipitation often result.

Under most conditions, air in the troposphere decreases in temperature as altitude increases. Because warm air rises, vertical mixing results. Occasionally, however, a layer of cool air occurs beneath a layer of warmer air. This departure

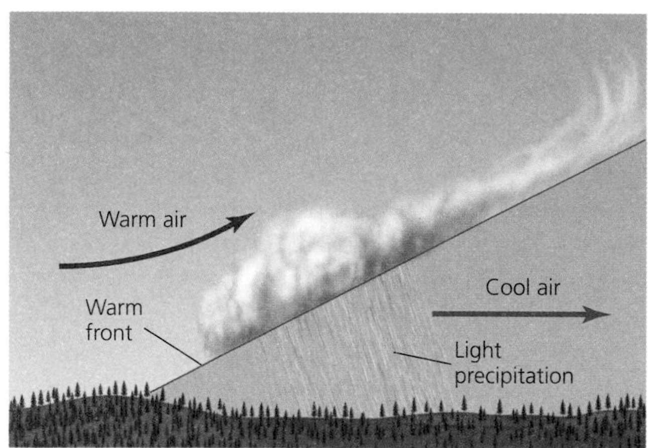


**(a) Warm front**

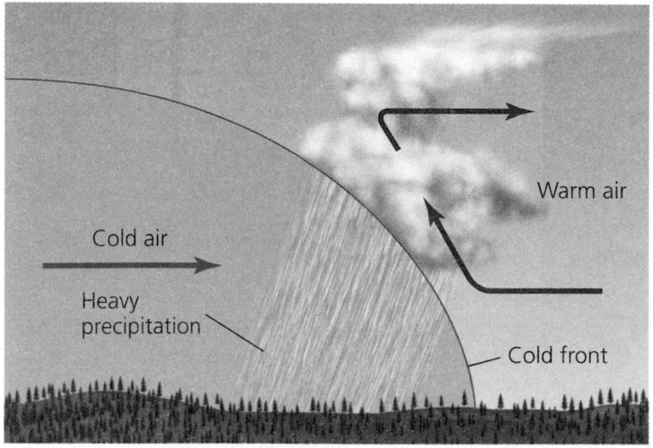


**(b) Cold front**

FIGURE 17.7 When a warm front approaches (**a**), warmer air rises over cooler air, causing light or moderate precipitation as moisture in the warmer air condenses. When a cold front approaches (**b**), colder air pushes beneath warmer air, and the warmer air rises, resulting in condensation and heavy precipitation.

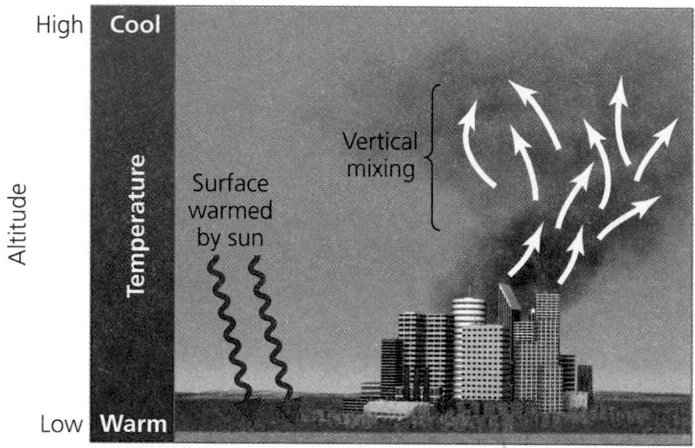


**(a) Normal conditions**

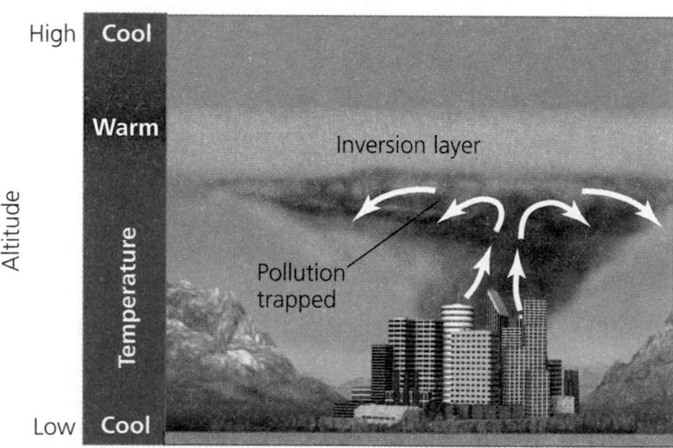


**(b) Thermal inversion**

FIGURE 17.8 A thermal inversion is a natural atmospheric occurrence that can worsen air pollution locally. Under normal conditions (**a**), tropospheric temperature decreases with altitude, and air of different altitudes mixes, dispersing pollutants upward and outward from their sources. During a thermal inversion (**b**), cool air remains near the ground underneath an "inversion layer" of air that warms with altitude. Little mixing occurs, and pollutants are trapped near the surface.

from the normal temperature profile is known as a **temperature inversion,** or **thermal inversion** (Figure 17.8). The band of air in which temperature rises with altitude is called an **inversion layer** (because the normal direction of temperature change is inverted). The cooler air at the bottom of the inversion layer is denser than the warmer air at the top of the inversion layer, so it resists vertical mixing and remains stable. Thermal inversions can occur in different ways, sometimes involving cool air at ground level and sometimes producing an inversion layer higher above the ground (as shown in Figure 17.8b). One common type of inversion occurs in mountain valleys where slopes block morning sunlight, keeping ground-level air within the valley shaded and cool.

Whereas vertical mixing normally allows air pollution to be diluted upward, thermal inversions trap pollutants near the ground. It was a thermal inversion that sparked London's "killer smog" of 1952. A high-pressure system settled over the city, acting like a cap on the pollution and keeping it in place. Inversions regularly cause smog buildup in large metropolitan areas in valleys ringed by mountains, such as Los Angeles; Mexico City; Seoul, Korea; and Rio de Janeiro and São Paulo, Brazil.

## Large-scale circulation systems produce global climate patterns

At larger geographic scales, convective air currents contribute to broad climatic patterns (**Figure 17.9a**). Near the equator, solar radiation sets in motion a pair of convective cells known as **Hadley cells.** Here, where sunlight is most intense, surface air warms, rises, and expands. As it does so, it releases moisture, producing the heavy rainfall that gives rise to tropical rainforests near the equator. After releasing much of its moisture, this air diverges and moves in currents heading northward and southward.

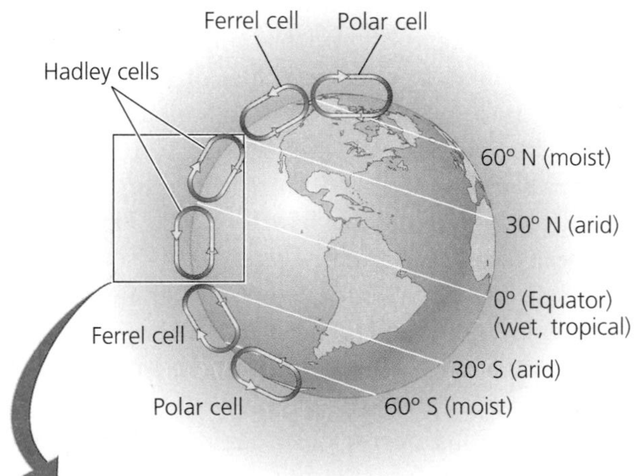


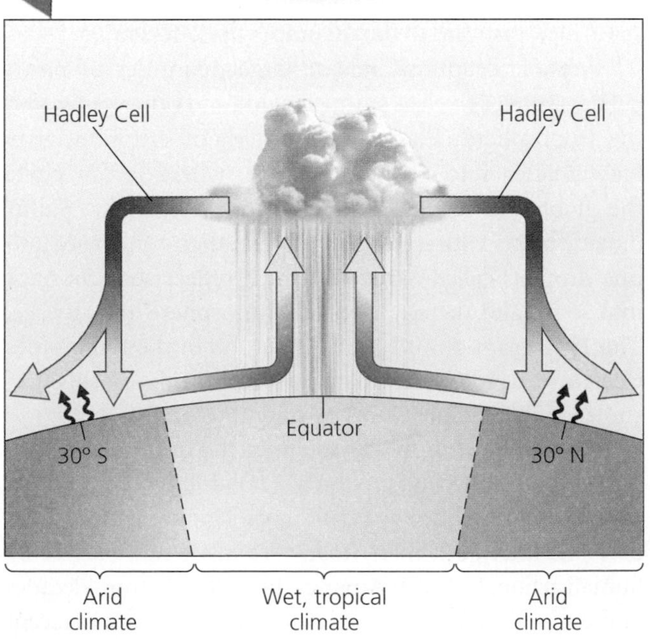


(a) Convection currents

**FIGURE 17.9** A series of large-scale convective cells (**a**) helps determine global patterns of humidity and aridity. Warm air near the equator rises, expands, and cools, and moisture condenses, giving rise to a wet climate in tropical regions. Air travels toward the poles and descends around 30 degrees latitude. This air, which loses its moisture in the tropics, causes regions around 30 degrees latitude to be arid. This convective circulation, a Hadley cell, occurs on both sides of the equator. Between roughly 30 and 60 degrees latitude north and south, Ferrel cells occur; and between 60 and 90 degrees latitude, polar cells occur. As a result, air rises around 60 degrees latitude, creating a moist climate, and falls around 90 degrees, creating a dry climate. Global wind currents (**b**) show latitudinal patterns as well. Trade winds between the equator and 30 degrees latitude blow westward, whereas westerlies between 30 and 60 degrees latitude blow eastward.

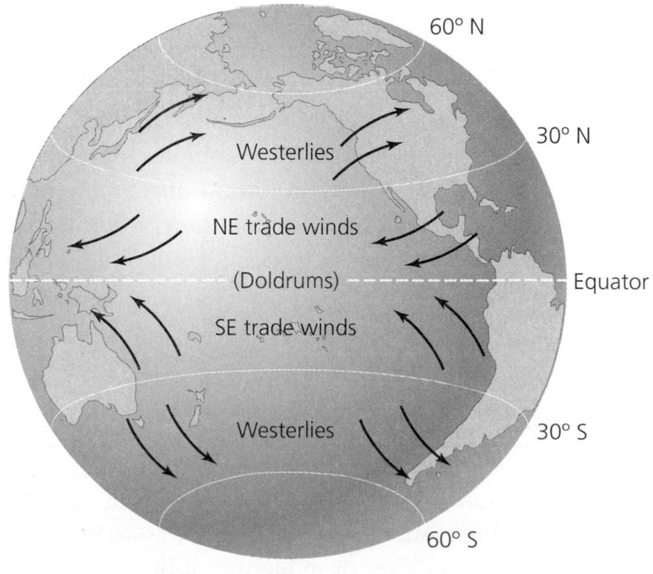


(b) Global wind patterns

The air in these currents cools and descends back to Earth at about 30 degrees latitude north and south. Because the descending air has low relative humidity, the regions around 30 degrees latitude are quite arid, giving rise to deserts. Two pairs of similar but less intense convective cells, called **Ferrel cells** and **polar cells,** lift air and create precipitation around 60 degrees latitude north and south and cause air to descend at around 30 degrees latitude and in the polar regions.

These three pairs of cells account for the latitudinal distribution of moisture across Earth's surface: wet climates near the equator; arid climates near 30 degrees latitude; moist regions near 60 degrees latitude; and dry conditions near the poles. These patterns, combined with temperature variation, help explain why biomes tend to be arrayed in latitudinal bands (Figure 6.18, • p. 162).

The Hadley, Ferrel, and polar cells interact with Earth's rotation to produce the global wind patterns shown in **Figure 17.9b**. As Earth rotates on its axis, locations on the equator spin faster than locations near the poles. As a result, the north-south air currents of the convective cells appear to be deflected from a straight path as some portions of the globe move beneath them more quickly than others. This apparent deflection is called the **Coriolis effect,** and it results in the curving global wind patterns evident in Figure 17.9b. Near the equator lies a region with few winds known as the *doldrums.* Between the equator and 30 degrees latitude, the *trade winds* blow from east to west. From 30 to 60 degrees latitude, the *westerlies* originate from the west and blow east.

People used these global circulation patterns for centuries to facilitate ocean travel by wind-powered sailing ships. Moreover, the atmosphere interacts with the oceans to affect weather, climate, and the distribution of biomes. For instance, winds and convective circulation in ocean water together maintain ocean currents (• pp. 446–447), and trade winds weaken periodically, leading to El Niño conditions (• pp. 510–511). The atmosphere's interactions with other systems of the planet are complex, but even a basic understanding of how the atmosphere functions can help us comprehend how our pollution of the atmosphere can affect ecological systems, economies, and human health.

# Outdoor Air Pollution

Throughout human history, we have made the atmosphere a dumping ground for our airborne wastes. Whether from primitive wood fires or modern coal-burning power plants, people have generated **air pollutants,** gases and particulate material added to the atmosphere that can affect climate or harm people or other organisms. **Air pollution** refers to the release of air pollutants. In recent decades, government policy and improved technologies have helped us diminish **outdoor air pollution** (often called **ambient air pollution**) substantially in countries of the developed world. However, outdoor air pollution remains a problem, particularly in developing nations and in urban areas.

## Natural sources can pollute

When we think of outdoor air pollution, we tend to envision smokestacks belching black smoke from industrial plants. However, natural processes produce a great deal of the world's air pollution. Some of these natural impacts can be exacerbated by human activity and land-use policies.

Winds sweeping over arid terrain can send huge amounts of dust aloft. In 2001, strong westerlies lifted soil from deserts in Mongolia and China. The dust blanketed Chinese towns, spread to Japan and Korea, traveled eastward across the Pacific Ocean to the United States, then crossed the Atlantic and left evidence atop the French Alps. Every year, hundreds of millions of tons of dust are blown westward by trade winds across the Atlantic Ocean from northern Africa to the Americas (**Figure 17.10a**). These dust storms bring nutrients to the Amazon basin, as well as fungal and bacterial spores that have been linked to die-offs in Caribbean coral reef systems.

Although dust storms are natural, the immense scale of these events results from unsustainable farming and grazing practices that strip vegetation from the soil, promote wind erosion, and lead to desertification (• pp. 244–245). Continental-scale dust storms took place in the United States in the 1930s, when soil from the drought-stricken Dust Bowl states blew eastward to the Atlantic (• pp. 245–246).

Volcanic eruptions release large quantities of particulate matter, as well as sulfur dioxide and other gases, into the troposphere (**Figure 17.10b**). Major eruptions may blow matter into the stratosphere, where it can circle the globe and remain aloft for months or years. Sulfur dioxide reacts with water and oxygen, then condenses into fine droplets called **aerosols,** which reflect sunlight back into space and thereby cool the atmosphere and surface. The 1991 eruption of Mount Pinatubo in the Philippines ejected nearly 20 million tons of ash and aerosols and cooled global temperatures by roughly 0.5° C.

Burning vegetation also pollutes the atmosphere with soot and gases. Over 60 million ha (150 million acres) of forest and grassland burn in a typical year (**Figure 17.10c**). Fires occur naturally, but many today are made more severe by human action. In North America, fuel buildup from decades of fire suppression has caused damaging forest fires in recent

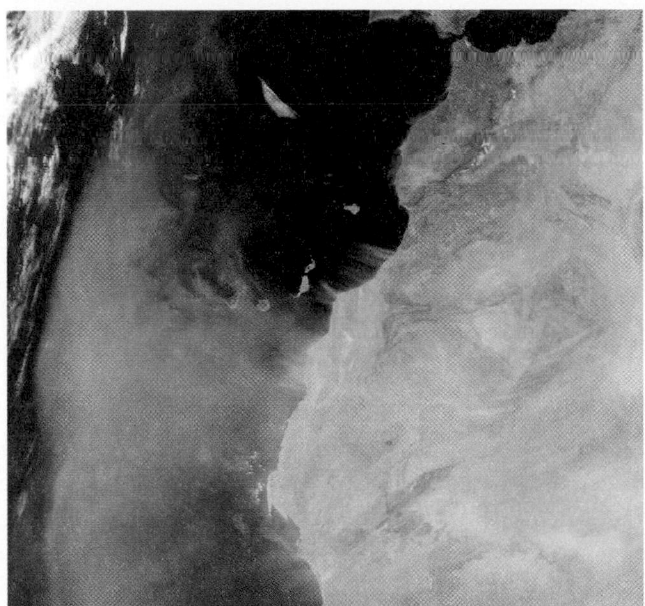

(a) Dust storm off west coast of Africa

(b) Mount Saint Helens eruption, 1980

(c) Natural fire in California

FIGURE 17.10 Massive dust storms, such as this one blowing across the Atlantic Ocean from Africa to the Americas **(a)**, are one type of natural air pollution. Volcanoes are another, as shown by Mount Saint Helens **(b)**, which erupted in the state of Washington in 1980. A third cause is fires in forests and grasslands **(c)**. Often, pollution from natural sources is made worse by human influence, such as when poor agricultural practices lead to soil erosion by wind, or when fire suppression leads to more devastating fires.

years (• p. 343). In the tropics, many fires result as farmers clear forest for farming and grazing using a "slash-and-burn" approach (• pp. 241–242). In 1997, a severe drought brought on by the 20th century's strongest El Niño event caused forest fires in Indonesia to rage out of control. Their smoke sickened 20 million Indonesians and caused a plane to crash and ships to collide. Along with tens of thousands of fires in drought-plagued Mexico, Central America, and Africa, these fires released more carbon monoxide into the atmosphere during 1997–1998 than did our worldwide combustion of fossil fuels.

## We create various types of outdoor air pollution

Since the onset of industrialization, human activity has introduced a variety of sources of air pollution. As with water pollution, air pollution can emanate from mobile or stationary sources, and from *point sources* or *non-point sources* (• p. 431). A point source describes a specific spot where large quantities of pollutants are discharged. Non-point sources are more diffuse, often consisting of many small sources. In London today, power plants and factories act as stationary point sources, whereas millions of automobiles on the roadways together comprise a mobile non-point source.

Once pollutants are in the atmosphere at sufficient concentrations, they may do harm directly, or they may induce chemical reactions that produce harmful compounds. **Primary pollutants,** such as soot and carbon monoxide, are pollutants emitted into the troposphere in a form that can be directly harmful or that can react to form harmful substances. Harmful substances produced when primary pollutants interact or react with constituents of the atmosphere are called **secondary pollutants.** Secondary pollutants include tropospheric ozone, sulfuric acid, and other examples we will explore below.

Arguably the greatest human-induced air pollution problem today is our emission of greenhouse gases that contribute to global climate change. Addressing our release of excess carbon dioxide, methane, and other gases that warm the atmosphere stands as one of our civilization's primary challenges. We will discuss this issue separately and in depth in Chapter 18.

## Clean Air Act legislation addresses pollution in the United States

To address air pollution in the United States, Congress has passed a series of laws, beginning with the Air Pollution Control Act of 1955. The Clean Air Act of 1963 funded research into pollution control and encouraged emissions standards for automobiles and stationary point sources, such as industrial plants. Subsequent amendments expanded the legislation's scope and established a nationwide air quality monitoring system.

In 1970, Congress thoroughly revised the law in what came to be known as the **Clean Air Act of 1970.** This legislation set stricter standards for air quality, imposed limits on emissions from new stationary and mobile sources, provided new funds for pollution-control research, and enabled citizens to sue parties violating the standards. Some of these goals came to be viewed as too ambitious, so amendments in 1977 altered some standards and extended some deadlines for compliance.

The **Clean Air Act of 1990** sought to strengthen regulations pertaining to air quality standards, auto emissions, toxic air pollution, acidic deposition, and stratospheric ozone depletion. It also introduced an emissions trading program for sulfur dioxide (• pp. 79–82). Beginning in 1995, businesses and utilities were allocated permits for emitting this pollutant and could then buy, sell, or trade these allowances with one another. Each year the overall amount of allowed pollution was decreased. This market-based incentive program has helped reduce sulfur dioxide emissions. It has also spawned similar programs for other pollutants, including greenhouse gases (• p. 535).

As a result of Clean Air Act legislation, the U.S. Environmental Protection Agency (EPA) sets nationwide standards for emissions of pollutants and for concentrations of pollutants in ambient air throughout the nation. It is largely up to the states to monitor air quality and develop, implement, and enforce regulations within their borders. States submit implementation plans to the EPA for approval, and if a state's plans are not adequate, the EPA can take over enforcement in that state.

## The EPA sets air quality standards for "criteria pollutants"

The EPA and the states focus on six **criteria pollutants,** pollutants judged to pose especially great threats to human health—carbon monoxide (CO), sulfur dioxide ($SO_2$), nitrogen dioxide ($NO_2$), tropospheric ozone ($O_3$), particulate matter, and lead (Pb). For these, the EPA has established *national ambient air quality standards (NAAQS),* which are maximum allowable concentrations of these pollutants in ambient outdoor air.

**Carbon monoxide** **Carbon monoxide** is a colorless, odorless gas produced primarily by the incomplete combustion of fuel. Vehicles and engines account for about 78% of CO emissions in the United States; other sources include industrial processes, combustion of waste, and residential wood burning. Carbon monoxide poses risk to humans and other animals, even in small concentrations. It can bind irreversibly to hemoglobin in red blood cells, preventing the hemoglobin from binding with oxygen.

**Sulfur dioxide** Like CO, **sulfur dioxide** is a colorless gas, but unlike CO, it has a pungent odor. The vast majority of $SO_2$ pollution results from the combustion of coal for electricity generation and industry. During combustion, elemental sulfur (S) in coal reacts with oxygen gas ($O_2$) to form $SO_2$. Once in the atmosphere, $SO_2$ may react to form sulfur trioxide ($SO_3$) and sulfuric acid ($H_2SO_4$), which may then fall back to Earth in the form of acid precipitation.

**Nitrogen dioxide** **Nitrogen dioxide** is a highly reactive, foul-smelling reddish brown gas that contributes to smog and acid precipitation. Along with nitric oxide (NO), $NO_2$ belongs to a family of compounds called **nitrogen oxides** ($NO_x$). Nitrogen oxides result when atmospheric nitrogen and oxygen react at the high temperatures created by combustion engines. Over half of U.S. $NO_x$ emissions result from combustion in motor vehicle engines. Electrical utility and industrial combustion account for most of the rest.

**Tropospheric ozone** Although ozone in the stratosphere shields us from the dangers of UV radiation, $O_3$ from human activity forms and accumulates low in the troposphere and acts as a pollutant. In the troposphere, this colorless gas with an objectionable odor results from the interaction of sunlight, heat, nitrogen oxides, and volatile carbon-containing chemicals. **Tropospheric ozone** is therefore categorized as a secondary pollutant. A major component of smog, $O_3$ can pose health risks as a result of its instability as a molecule; this triplet of oxygen atoms will readily release one of its threesome, leaving a molecule of oxygen gas and a free oxygen atom. The free oxygen atom may then participate in reactions that can injure living tissues and cause respiratory problems. Tropospheric

$O_3$ is the pollutant that most frequently exceeds its EPA air quality standard.

**Particulate matter** **Particulate matter** is composed of solid or liquid particles small enough to be suspended in the atmosphere. Particulate matter includes primary pollutants such as dust and soot, as well as secondary pollutants such as sulfates and nitrates. Particulate matter can damage respiratory tissues when inhaled. Most particulate matter (about 60%) in the atmosphere is wind-blown dust; human activity accounts for much of the rest. Along with sulfur dioxide, it was largely the emission of particulate matter from industrial and residential coal combustion that produced London's 1952 killer smog and the deaths resulting from that episode.

**Lead** **Lead** is a heavy metal that enters the atmosphere as a particulate pollutant. The lead-containing compounds tetraethyl lead and tetramethyl lead, when added to gasoline, improve engine performance. However, the exhaust from the combustion of leaded gasoline emits lead into the atmosphere, from which it can be inhaled or can be deposited on land and water. Lead can enter the food chain, accumulate within body tissues, and cause central nervous system malfunction and many other ailments. Once people recognized the dangers of lead, leaded gasoline was phased out in the United States and other industrialized nations, beginning in the 1970s (• pp. 8–9). Today the greatest source of atmospheric lead pollution in developed nations is industrial metal smelting. However, many developing nations still add lead to gasoline and experience significant lead pollution.

EPA monitoring finds that large numbers of Americans live in areas where concentrations of these pollutants regularly reach unhealthy levels. In 2005, 122 million Americans lived in counties that violated the national ambient air quality standards for at least one of the six criteria pollutants (**Figure 17.11**).

### Weighing the Issues | Your County's Air Quality

Locate where you live on the map in Figure 17.11. What is the status of your county's air quality, and how does your county compare to the rest of the nation? What factors do you think account for the quality of its air? Can you propose any solutions for reducing air pollution in your county?

## Agencies monitor pollutants that affect air quality

Besides measuring concentrations of the six criteria pollutants in ambient air, state and local agencies also monitor, calculate, and report to the EPA the emissions of certain

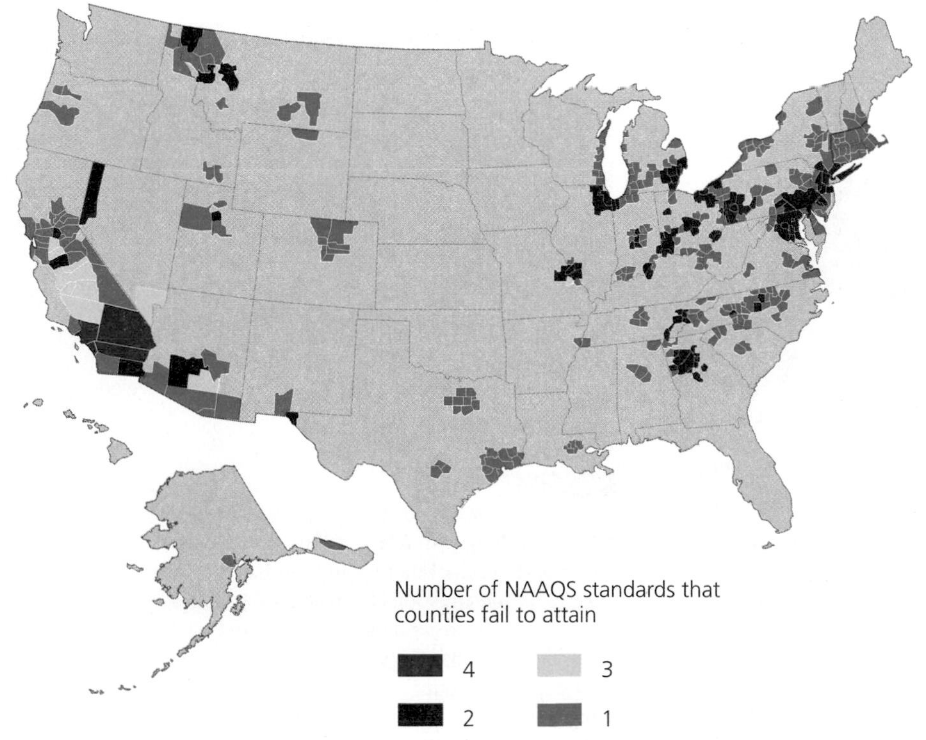


FIGURE 17.11 Roughly 122 million Americans live in counties that in 2005 failed to meet the EPA's national ambient air quality standards (NAAQS) for at least one criteria pollutant. This map shows counties that failed to attain the standards for one (green) through four (red) of the six criteria pollutants. Data from U.S. EPA.

major pollutants that affect ambient concentrations of the criteria pollutants. The major pollutants whose emissions are measured include the four criteria pollutants that are primary pollutants (carbon monoxide, sulfur dioxide, particulate matter, and lead), as well as all nitrogen oxides (because NO reacts readily in the atmosphere to form $NO_2$, which is both a primary and secondary pollutant). Tropospheric ozone is a secondary pollutant only, so there are no emissions to monitor. Instead, agencies monitor emissions of volatile organic compounds, which can react to produce ozone and other secondary pollutants.

**Volatile organic compounds (VOCs)** are carbon-containing chemicals used in and emitted by vehicle engines and a wide variety of solvents and industrial processes, as well as by many household chemicals and consumer items. One group of VOCs consists of hydrocarbons (• pp. 95–96) such as methane ($CH_4$, the primary component of natural gas), propane ($C_3H_8$, used as a portable fuel), butane ($C_4H_{10}$, found in cigarette lighters), and octane ($C_8H_{18}$, a component of gasoline). Human activities account for about half the VOC emissions in the United States, and the remainder comes from natural sources. For example, plants produce isoprene ($C_5H_8$) and terpene ($C_{10}H_{15}$).

In the United States in 2006, human activity polluted the air with 137 million tons of the six monitored pollutants. Carbon monoxide was the most abundant pollutant by mass, followed by $NO_x$, VOCs, and $SO_2$ (**Figure 17.12**).

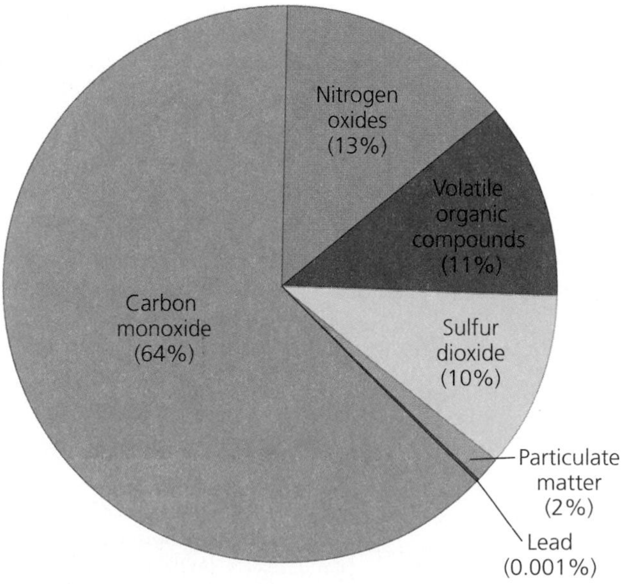


FIGURE 17.12 In 2006, the United States emitted 137 million tons of the six major pollutants whose emissions are monitored by the EPA and state and local agencies. Carbon monoxide accounted for most of these emissions, by mass. Data from U.S. EPA.

## Air pollution has decreased markedly since 1970

Since the Clean Air Act of 1970, emissions of each of the six monitored pollutants have decreased, and total emissions of the six together have declined by 53% (**Figure 17.13a**). These dramatic reductions in emissions have occurred despite substantial increases in the nation's population, energy consumption, miles traveled by vehicle, and gross domestic product (**Figure 17.13b**).

There are several reasons for these declines. Cleaner-burning motor vehicle engines and automotive technologies such as catalytic converters (• p. 672) have played a large part, decreasing the emissions of carbon monoxide and several other pollutants. The sulfur dioxide permit-trading program (• pp. 79–82) and clean coal technologies (• pp. 548–549) have reduced $SO_2$ emissions. Technologies such as baghouse filters, electrostatic precipitators, and **scrubbers** (**Figure 17.14**) that chemically convert or physically

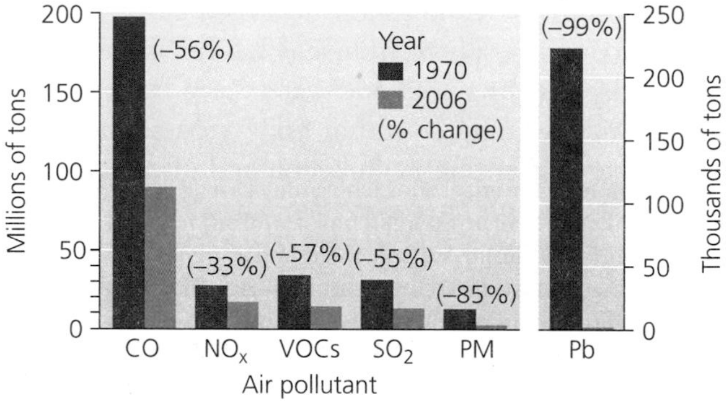


**(a) Declines in six major pollutants**

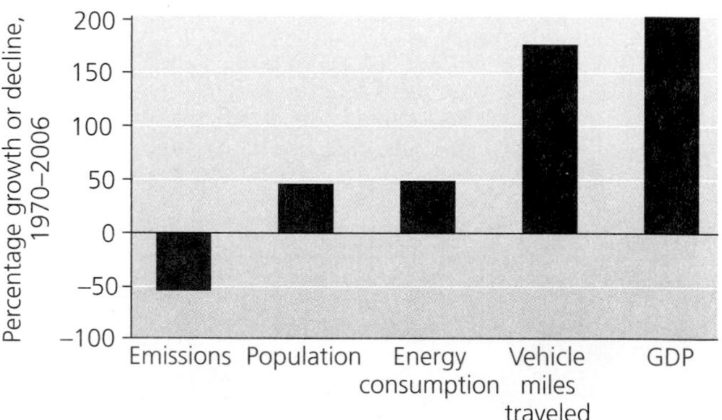


**(b) Trends in major indicators**

FIGURE 17.13 The EPA tracks emissions of several major pollutants into ambient air. Shown are emissions of four of the six criteria pollutants, along with nitrogen oxides and volatile organic compounds. Each of these pollutants has shown substantial declines since 1970, and emissions from all six together have declined by 54% **(a)**. This decrease in emissions has occurred despite increases in U.S. population, energy consumption, vehicle miles traveled, and gross domestic product **(b)**. Go to GRAPHIt! at www.aw-bc.com/withgott or on the student CD-ROM. Data from U.S. EPA.

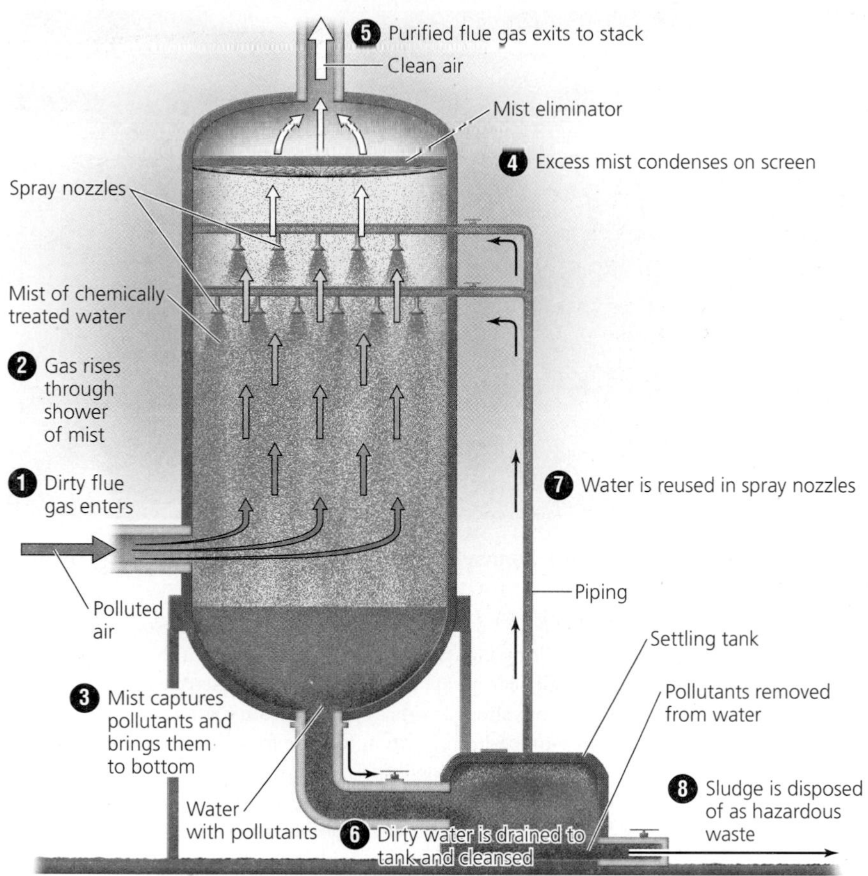


FIGURE 17.14 In this spray-tower wet scrubber, polluted air (1) rises through a chamber while arrays of nozzles spray a mist of water mixed with lime or other active chemicals (2). The falling mist captures pollutants and carries them to the bottom of the chamber (3), essentially washing them out of the air. Excess mist is captured on a screen (4), and air emitted from the scrubber has largely been cleansed (5). Periodically, the dirty water is drained from the chamber (6), cleansed in a settling tank, and recirculated (7) through the spray nozzles. The resulting sludge must be disposed of (8) as hazardous waste (• pp. 646–653). Scrubbers and other pollution-control devices come in many designs; the type shown here typically removes at least 90% of particulate matter and gases such as sulfur dioxide.

remove airborne pollutants before they are emitted from smokestacks have allowed factories, power plants, and refineries to decrease emissions of several pollutants. And the leaded gasoline phaseout caused U.S. lead emissions to plummet by 93% from 1980 to 1990 alone. Overall, the reduction of outdoor air pollution since 1970 represents one of the United States's greatest accomplishments in safeguarding human health and environmental quality.

Other industrialized nations have also succeeded in reducing emissions and improving air quality, thanks to improved technologies and targeted federal policies. England is among them. Between 1996 and 2005, London achieved a 56% reduction in CO emissions, a 41% drop in $NO_X$ emissions, a 28% decline in particulate matter release, and an impressive 73% decrease in $SO_2$ emissions. Only tropospheric ozone showed an increase, rising 33%. With the congestion-charging program and related transportation initiatives, Mayor Livingstone hopes to help turn around the trend in ozone pollution as well.

## Toxic substances also pollute

The 1990 Clean Air Act identifies 188 different **toxic air pollutants.** These include substances known to cause cancer, reproductive defects, or neurological, developmental, immune system, or respiratory problems in people, as well as substances that affect the health of animals and plants. Some toxic air pollutants are produced naturally. For example, hydrogen sulfide gas ($H_2S$) gives the mud of swamps and bogs the odor of rotten eggs. However, most toxic air pollutants are produced by human activities, such as metal smelting, sewage treatment, and industrial processes. Toxic air pollutants range from the heavy metal mercury (from coal-burning power plant emissions and other sources) to VOCs such as benzene (a component of gasoline) and methylene chloride (found in paint stripper). Among the 188 pollutants are 21 from mobile sources (such as diesel exhaust) and 33 "urban hazardous" pollutants judged to pose the greatest health risks in urban areas.

State and federal agencies do not monitor toxic air pollutants as extensively as they do the six criteria pollutants, but so far 300 monitoring sites are operating, and coverage is improving. The EPA estimates that because of Clean Air Act regulations, emissions of toxic air pollutants decreased by 30% during the 1990s.

## Recent policy proposals have been contentious

Despite the gains in air quality made under the Clean Air Act, the George W. Bush administration has vigorously pushed proposals that would overturn key aspects of the

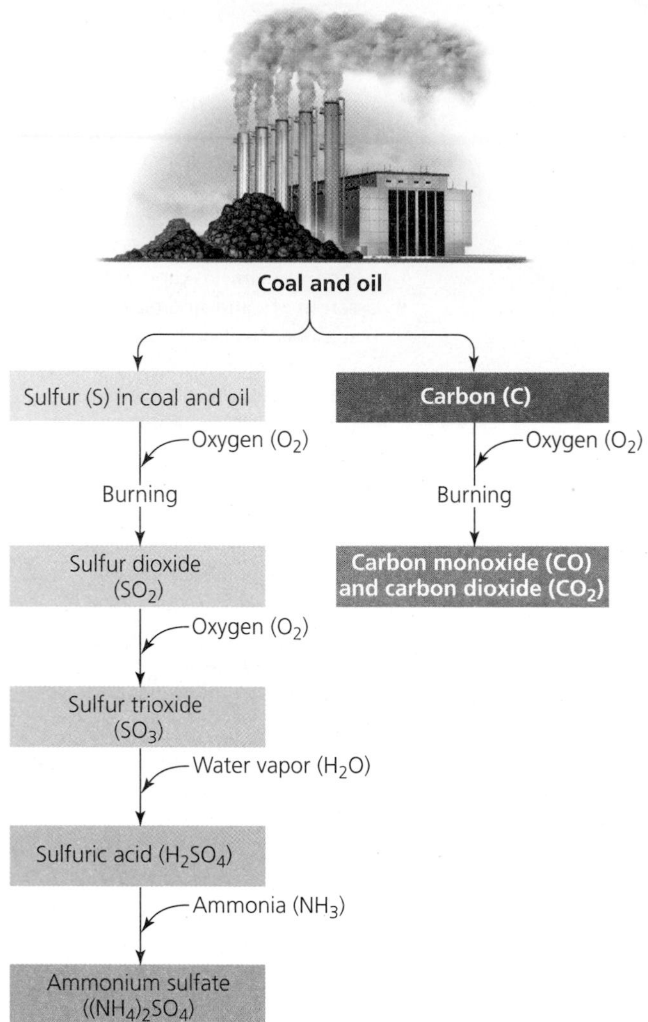


**(a) Burning sulfur-rich oil or coal without adequate pollution control technologies**

**(b) Donora, Pennsylvania, at midday in the 1948 smog event**

FIGURE 17.15 Emissions from the combustion of coal and oil in manufacturing plants and utilities without pollution-control technologies can create industrial smog. Industrial smog consists primarily of sulfur dioxide and particulate matter, as well as carbon monoxide and carbon dioxide from the carbon component of fossil fuels. When fossil fuels are combusted, sulfur contaminants give rise to sulfur dioxide, which in the presence of other chemicals in the atmosphere can produce several other sulfur compounds (**a**). Under certain weather conditions, industrial smog can blanket whole towns or regions, as it did in Donora, Pennsylvania, shown here in the daytime during its deadly 1948 smog episode (**b**).

legislation. The EPA under Bush proposed to eliminate *new source review.* This policy had been established in 1977, when utilities protested that meeting new pollution targets by retrofitting aging plants with new technologies would be too expensive. Congress came up with a compromise: Old plants would be exempt from the new pollution requirements, as long as they installed the "best available" current technology for pollution control if they ever upgraded their plants in the future. Some plants failed to do this, and the Clinton administration eventually sued 51 plants for violating the law. The Bush administration not only dropped these lawsuits, but also proposed abolishing new source review altogether. In 2006, federal courts struck down the EPA's attempts to eliminate new source review.

The Bush administration also pushed its *Clear Skies* initiative, aimed to abandon a command-and-control policy approach and establish a market-based cap-and-trade program for sulfur dioxide, nitrogen oxides, and mercury. The Clear Skies legislation was stopped in the Senate in 2005 by senators who concluded that it would increase pollution, relative to existing Clean Air Act policy. An independent Congressional Research Service analysis at the end of that year judged that the EPA had skewed its analysis of costs and benefits to promote the legislation.

## Burning fossil fuels produces industrial smog

In response to the increasing incidence of fogs polluted by the smoke of Britain's industrial revolution, an early British scientist coined the term *smog*. Today the term is used worldwide to describe unhealthy mixtures of air pollutants that often form over urban areas.

The smog that enveloped London in 1952 was what we today call **industrial smog,** or gray-air smog. When coal or oil is burned, some portion is completely combusted, forming $CO_2$; some is partially combusted, producing CO; and some remains unburned and is released as soot, or particles of carbon. Moreover, coal contains varying amounts of contaminants, including mercury and sulfur. Sulfur reacts with oxygen to form sulfur dioxide, which can undergo a series of reactions to form sulfuric acid and ammonium sulfate (**Figure 17.15a**). These chemicals and

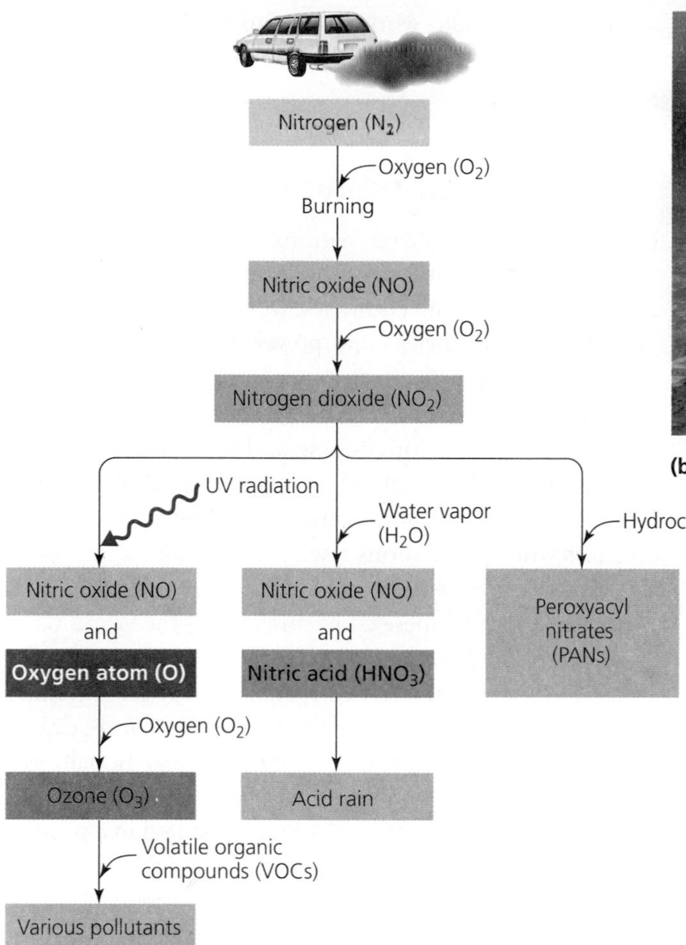


**(a) Formation of photochemical smog**

**(b) Photochemical smog over Mexico City**

FIGURE 17.16 Nitric oxide, a key element of photochemical smog, can start a chemical chain reaction **(a)** that results in the production of other compounds, including nitrogen dioxide, nitric acid, ozone, and peroxyacyl nitrates (PANs). PANs can induce further reactions that damage living tissues in animals and plants. Nitric acid contributes to acidic deposition as well as photochemical smog. Photochemical smog is common today over many urban areas, especially those with hilly topography or frequent inversion layers. Mexico City **(b)** is one city that frequently experiences photochemical smog.

others produced by further reactions, along with soot, are the main components of industrial smog and give the smog its characteristic gray color.

Industrial smog is far less common today in developed nations than it was 50–100 years ago. In the wake of the 1952 London episode and others, the governments of most developed nations began regulating industrial emissions to minimize the external costs (• p. 42) they impose on citizens. However, in regions that are industrializing today, such as China, India, and Eastern Europe, heavy reliance on coal burning (by industry and by citizens heating and cooking in their homes), combined with lax pollution controls, produces industrial smog that poses significant health risks in many areas.

Although coal combustion supplies the chemical constituents for industrial smog, weather also plays a role, as it did in London in 1952. A similar event occurred 4 years earlier in Donora, Pennsylvania. Here, air near the ground cooled during the night, and because Donora is located in hilly terrain, too little morning sun reached the valley floor to warm and disperse the cold air. The resulting thermal inversion trapped smog containing particulate matter emissions from a steel and wire factory. Twenty-one people were killed, and over 6,000 people—nearly half the town—became ill (**Figure 17.15b**).

Hilly topography such as Donora's is a factor in the air pollution of many other cities where surrounding mountains trap air and create inversions. This is true for the Los Angeles basin, which has long symbolized smog problems in American popular culture. Modern-day Los Angeles, however, suffers from a different type of smog, one called photochemical smog.

## Photochemical smog is produced by a complex series of reactions

A photochemical process is one whose activation requires light. **Photochemical smog,** or brown-air smog, is formed through light-driven chemical reactions of primary pollutants and normal atmospheric compounds that produce a mix of over 100 different chemicals, tropospheric ozone often being the most abundant among them (**Figure 17.16a**). High levels of $NO_2$ cause photochemical smog to form a brownish haze over cities (**Figure 17.16b**). Hot, sunny, windless days in urban areas provide perfect conditions for the formation of photochemical smog. Exhaust from morning traffic releases large amounts of NO and VOCs into a city's air. Sunlight then promotes the production of ozone and other constituents of photochemical smog. Levels of photochemical pollutants in urban areas typically peak in midafternoon and can irritate people's eyes, noses, and throats.

London's congestion-charging scheme should cut down on photochemical smog by reducing traffic exhaust. So far, a reduction has been hard to prove, however, because during the first year of the program unusual weather resulted in optimal conditions for photochemical reactions, and smog increased. Indeed, the cities most afflicted by photochemical smog are those with weather and topography that promote it. In Athens, Greece, the problem was bad enough that the city government provided incentives to replace aging vehicles. It also mandated that autos with odd-numbered license plates be driven only on odd-numbered days, and those with even-numbered plates only on even-numbered days. According to Greek officials, smog was reduced by 30% as a result.

Some U.S. states have cut emissions leading to photochemical smog by 30% as well, through vehicle inspection programs. Drivers in 34 states are required to have their vehicle exhaust inspected periodically at check stations to maintain their registrations. Although a failed "smog check" means inconvenience for the auto owner, these programs help maintain vehicle condition and make the air measurably cleaner for all of us.

### Weighing THE Issues | Congestion Charging

Does the city you live in, or the nearest major city to you, suffer from air pollution? Do you think this city should adopt a congestion charging program like London's? What benefits would your city enjoy from such a program, and what problems might it bring? What other steps should this city take to tackle pollution?

## Air quality is a rural issue, too

Air quality is not only an urban issue. In rural areas, people suffer from drift of airborne pesticides from farms, as well as industrial pollutants that drift far from cities, factories, and power plants. A great deal of rural air pollution emanates from feedlots (• pp. 281–282), where cattle, hogs, or chickens are raised in dense concentrations. The huge numbers of animals at feedlots and the voluminous amounts of waste they produce release dust as well as methane, hydrogen sulfide, and ammonia. These gases create objectionable odors, and the ammonia contributes to nitrogen deposition across wide areas. Studies have shown that people working at and living near feedlots have high rates of respiratory problems.

## Industrializing nations are suffering increasing air pollution

Although industrialized nations have been improving their air quality, outdoor air pollution is growing worse in many industrializing countries. In these societies, rapidly proliferating factories and power plants are releasing emissions with little effort to control pollution, and citizens continue to burn traditional sources of fuel such as wood and charcoal for cooking and home heating. Thus, just as occurred in England during its period of industrialization, new pollution sources are added to traditional sources while populations rise.

China suffers among the world's worst air pollution. Four out of five Chinese cities surveyed by the World Bank in 2000 suffered $SO_2$ or $NO_2$ emissions above the threshold set by the World Health Organization. Together, China and India suffer 58% of the 1.8 million premature deaths that the World Bank estimates occur each year globally as a result of outdoor air pollution.

Pollution from southern Asia has resulted in a persistent 2-mile-thick layer of pollution that hangs over the subcontinent throughout the dry season each December through April. Dubbed the *Asian Brown Cloud*, this massive layer of pollution is thought to reduce the sunlight reaching Earth's surface in that region by 10–15%, influence climate, decrease rice productivity by 5–10%, and account for many thousands of deaths each year.

## Synthetic chemicals deplete stratospheric ozone

A pollutant in smog in the troposphere, ozone is a highly beneficial gas at altitudes centering around 25 km (15 mi) in the lower stratosphere, where it is concentrated in the so-called *ozone layer* (see Figure 17.2). Here, concentrations of ozone are only about 12 parts per million. However, ozone molecules are so effective at absorbing incoming ultraviolet radiation from the sun that this concentration helps to protect life on Earth's surface from the damaging effects of ultraviolet (UV) radiation.

In the 1960s, atmospheric scientists began wondering why their measurements of ozone were lower than theoretical models predicted. Researchers hypothesizing that natural or artificial chemicals were depleting ozone finally pinpointed a group of human-made compounds derived from simple hydrocarbons, such as ethane and methane, in which hydrogen atoms are replaced by chlorine, bromine, or fluorine. One class of such compounds, **chlorofluorocarbons (CFCs),** was being mass-produced by industry at a rate of a million metric tons per year in the early 1970s, and this rate was growing by 20% a year.

Soon researchers showed that CFCs could deplete stratospheric ozone by releasing chlorine atoms that split ozone molecules, creating from each of them an $O_2$ molecule and a ClO molecule (see "The Science behind the Story," • pp. 490–491). Then in 1985, scientists announced that stratospheric ozone levels over Antarctica had declined by 40–60% in the previous decade, leaving a thinned ozone concentration that was soon dubbed the *ozone hole* (**Figure 17.17**). Research over the next few years confirmed the link between CFCs and ozone loss in the Antarctic and indicated that depletion was also occurring in the Arctic, and perhaps globally. Already concerned that increased UV radiation would lead to more skin cancer, scientists were becoming anxious over possible ecological effects as well, including harm to crops and to the productivity of ocean phytoplankton, the base of the marine food chain.

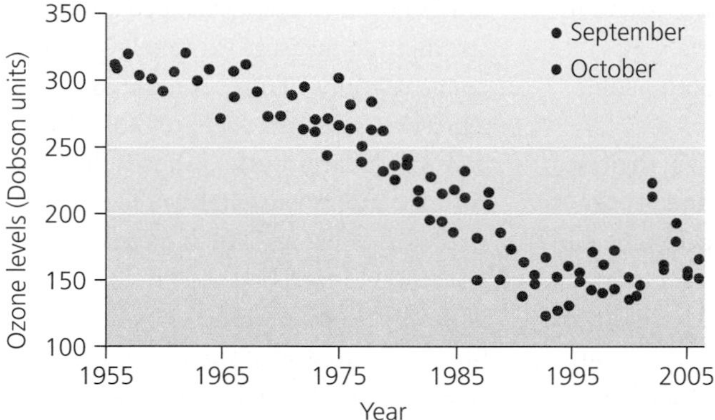


**(a) Monthly mean ozone levels at Halley, Antarctica**

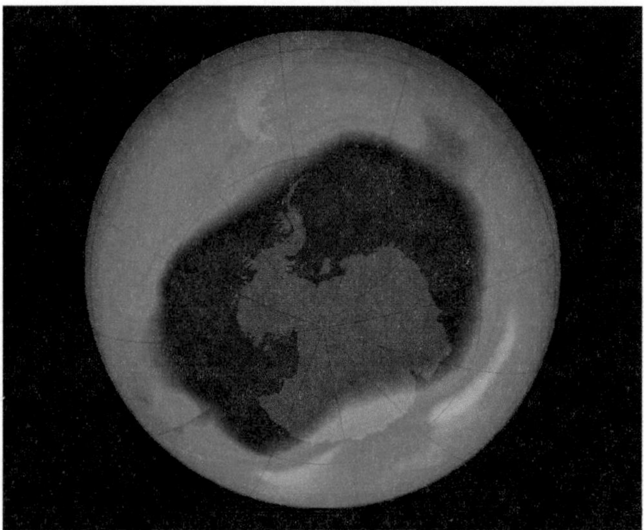

**(b) The "ozone hole" over Antarctica, September 24, 2006**

**FIGURE 17.17** The "ozone hole" consists of a region of thinned ozone density in the stratosphere over Antarctica and the southernmost ocean regions. It has reappeared seasonally each September in recent decades. Data from Halley, Antarctica (**a**), show a steady decrease in stratospheric ozone concentrations from the 1960s to 1990. Ozone-depleting CFCs began to be regulated under the Montreal Protocol in 1987, and ozone concentrations stopped declining. Colorized satellite imagery from September 24, 2006 (**b**), shows the "ozone hole" (blue) at its maximal recorded extent to date. Data in (a) from British Antarctic Survey.

## The Montreal Protocol addressed ozone depletion

In response to the scientific concerns, international policy efforts to restrict CFC production finally bore fruit in 1987 with the **Montreal Protocol.** In this treaty, signatory nations (eventually numbering 180) agreed to cut CFC production in half. Five follow-up agreements strengthened the pact by deepening the cuts, advancing timetables for compliance, and addressing related ozone-depleting chemicals. Today the production and use of ozone-depleting compounds has fallen by 95% since the late 1980s, and scientists can discern the beginnings of long-term recovery of the ozone layer. Industry was able to shift to alternative chemicals that have largely turned out to be cheaper and more efficient.

There are still challenges to overcome. Much of the 5 billion kg (11 billion lb) of CFCs emitted into the troposphere has yet to diffuse up into the stratosphere, and CFCs are slow to dissipate or break down. Thus, we can expect a considerable lag time between implementation of policy and the desired environmental effect. (This is one reason scientists often argue for proactive policy guided by the precautionary principle, rather than reactive policy that may respond too late.) Moreover, nations can plead for some ozone-depleting chemicals to be exempted from the ban, and the United States recently was allowed to continue using methyl bromide, a fumigant used to control pests on strawberries. However, despite the remaining challenges, the Montreal Protocol and its follow-up amendments are widely considered the biggest success story so far in addressing any global environmental problem.

Environmental scientists have attributed this success primarily to two factors:

1. Policymakers engaged industry in helping to solve the problem, and government and industry worked together on developing replacement chemicals. This cooperation reduced the battles that typically erupt between environmentalists and industry.
2. Implementation of the Montreal Protocol after 1987 followed an adaptive management approach (• p. 331), altering strategies midstream in response to new scientific data, technological advances, or economic figures.

THE SCIENCE BEHIND THE STORY

## Identifying CFCs as the Main Cause of Ozone Depletion

*Drs. F. Sherwood Rowland (left), Mario Molina (center), and Paul Crutzen (right), at a press conference after receiving their Nobel Prize*

In their attempts to understand how chlorofluorocarbons (CFCs) influence stratospheric ozone, scientists have relied on a wide variety of data sources, including historical records, field observations, laboratory experiments, and computer models.

Ozone was discovered in 1839, and its presence in the upper atmosphere was first proposed in the 1880s. In 1924, British scientist G. M. B. Dobson built an instrument that could measure ozone concentrations by sampling incoming sunlight at ground level and comparing the intensities of wavelengths that ozone does and does not absorb. By the 1970s, the Dobson ozone spectrophotometer was being used by a global network of observation stations.

Research on CFCs also had a long history. First invented in 1928, CFCs were useful as refrigerants, fire extinguishers, and propellants for aerosol spray cans. Starting in the 1960s, CFCs also found wide use as cleaners for electronics and as a part of the process of manufacturing rigid polystyrene foams. Research on the chemical properties of CFCs showed that they were almost completely inert; that is, they rarely reacted with other chemicals. Therefore, scientists surmised that, at trace levels, CFCs would be harmless to both people and the environment.

However, in June 1974, chemists F. Sherwood Rowland and Mario Molina published a paper in the journal *Nature,* arguing that the inertness that made CFCs so ideal for industrial purposes could also have disastrous consequences for the ozone layer.

Whereas more-reactive chemicals are broken down in the lower atmosphere, CFCs reach the stratosphere unchanged. Once CFCs reach the stratosphere, intense ultraviolet radiation from the sun breaks them into their constituent chlorine and carbon atoms. In a two-step chemical reaction (see the figure), a chlorine atom can split an ozone molecule, and then ready itself to split another one. Over its lifetime, each free chlorine atom, it was calculated, can catalyze the destruction of as many as 100,000 ozone molecules.

Rowland and Molina were the first to assemble a complete picture of the threat posed by CFCs, but they could not have reached their conclusions without the contributions of other scientists. British researcher James Lovelock had developed an instrument to measure extremely low concentrations of atmospheric gases. American scientists Richard Stolarski and Ralph Cicerone had shown that chlorine atoms can catalyze the destruction of ozone. Dutch meteorologist Paul Crutzen had shown that naturally produced nitrous oxide breaks down ozone. And American researcher James McDonald had predicted that ozone loss, by allowing more UV radiation to reach the surface, would result in thousands more skin cancer cases each year.

Rowland and Molina's analysis earned them the 1995 Nobel Prize in chemistry jointly with Crutzen. It also helped spark discussion among scientists, policymakers, and industry leaders over limits on CFC production.

Because of its success in addressing ozone depletion, the Montreal Protocol is widely seen as a model for international cooperation in addressing other pressing global problems, such as persistent organic pollutants (• pp. 406–407), climate change (• p. 534), and biodiversity loss (• p. 321).

### Acidic deposition is another transboundary pollution problem

Just as stratospheric ozone depletion crosses political boundaries, so does another atmospheric pollution concern—acidic deposition. **Acidic deposition** refers to the deposition of acidic (• pp. 94–96) or acid-forming pollutants from the atmosphere onto Earth's surface. This can take place either by precipitation (commonly referred to as *acid rain,* but also including acid snow, sleet, and hail), by fog, by gases, or by the deposition of dry particles. Acidic deposition is one type of **atmospheric deposition,** which refers more broadly to the wet or dry deposition on land of a wide variety of pollutants, including mercury, nitrates, organochlorines, and others.

Acidic deposition originates primarily with the emission of sulfur dioxide and nitrogen oxides, largely through fossil fuel combustion by automobiles, electric

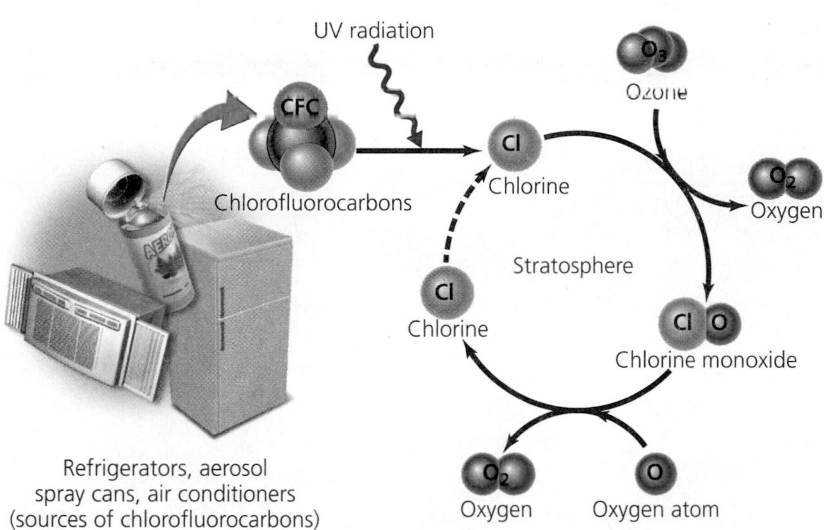


A chlorine atom released from a CFC molecule in the presence of UV radiation reacts with an ozone molecule, forming one molecule of oxygen gas and one chlorine monoxide (ClO) molecule. The oxygen atom in the ClO molecule will then bind with a stray oxygen atom to form oxygen gas, leaving the chlorine atom to begin the destructive cycle anew. In this way, any given chlorine atom may destroy up to 100,000 ozone molecules.

As a result, the United States and several other nations banned the use of CFCs in aerosol spray cans in 1979. Other uses continued, however, and by the early 1980s global production of CFCs was increasing.

Then, a new finding shocked scientists and spurred the international community to take further action. Scientists at a British research station in Antarctica had been recording ozone concentrations since the 1950s. In May 1985, Joseph Farman and colleagues reported in *Nature* that Antarctic ozone concentrations had declined dramatically since the 1970s. The decline exceeded even the worst-case predictions.

To determine what was causing the "ozone hole" over Antarctica, expeditions were mounted in 1986 and 1987 to measure trace amounts of atmospheric gases using ground stations and high-altitude balloons and aircraft. Together with other researchers, Crutzen analyzed data collected on the expeditions and concluded that the ozone hole resulted from a combination of Antarctic weather conditions and human-made chemicals.

In the frigid Antarctic winter, high-altitude clouds, or polar stratospheric clouds, form. In the spring, those clouds provide ideal conditions for CFC-derived chlorine and other chemicals to catalyze the destruction of massive amounts of ozone. The problem is made worse by the fact that prevailing air currents largely isolate Antarctica's atmosphere from the rest of Earth's atmosphere.

In subsequent years, scientists used data from ground stations and satellites to show that ozone levels were declining globally. In 1987, those findings helped convince the world's nations to agree on the Montreal Protocol, which aimed to cut CFC production in half by 1998. Within 2 years, however, further scientific evidence and computer modeling showed that more drastic measures would be needed if serious damage to the ozone layer were to be avoided. In 1990, the Montreal Protocol was strengthened to include a complete phaseout of CFCs by 2000. By 1998, the amount of chlorine in the atmosphere appeared to be leveling off.

utilities, and industrial facilities. Once airborne, these pollutants can react with water, oxygen, and oxidants to produce compounds of low pH (• pp. 94–96), primarily sulfuric acid and nitric acid. Suspended in the troposphere, droplets of these acids may travel days or weeks, sometimes covering hundreds or thousands of kilometers before falling in precipitation (**Figure 17.18**).

Acidic deposition can have wide-ranging, cumulative detrimental effects on ecosystems and on our built environment (**Table 17.1**). Acids leach nutrients such as calcium, magnesium, and potassium from the topsoil, altering soil chemistry and harming plants and soil organisms. This occurs because hydrogen ions from acidic precipitation take the place of calcium, magnesium, and potassium ions in soil compounds, and these valuable nutrients leach into the subsoil, where they become inaccessible to plant roots.

Acidic precipitation also "mobilizes" toxic metal ions such as aluminum, zinc, mercury, and copper by chemically converting them from insoluble forms to soluble forms. Elevated soil concentrations of metals such as aluminum hinder water and nutrient uptake by plants. In some regions of Britain and the United States, acid fog with a pH of 2.3 (equivalent to vinegar) can envelop forests for extended periods, leading to widespread tree mortality.

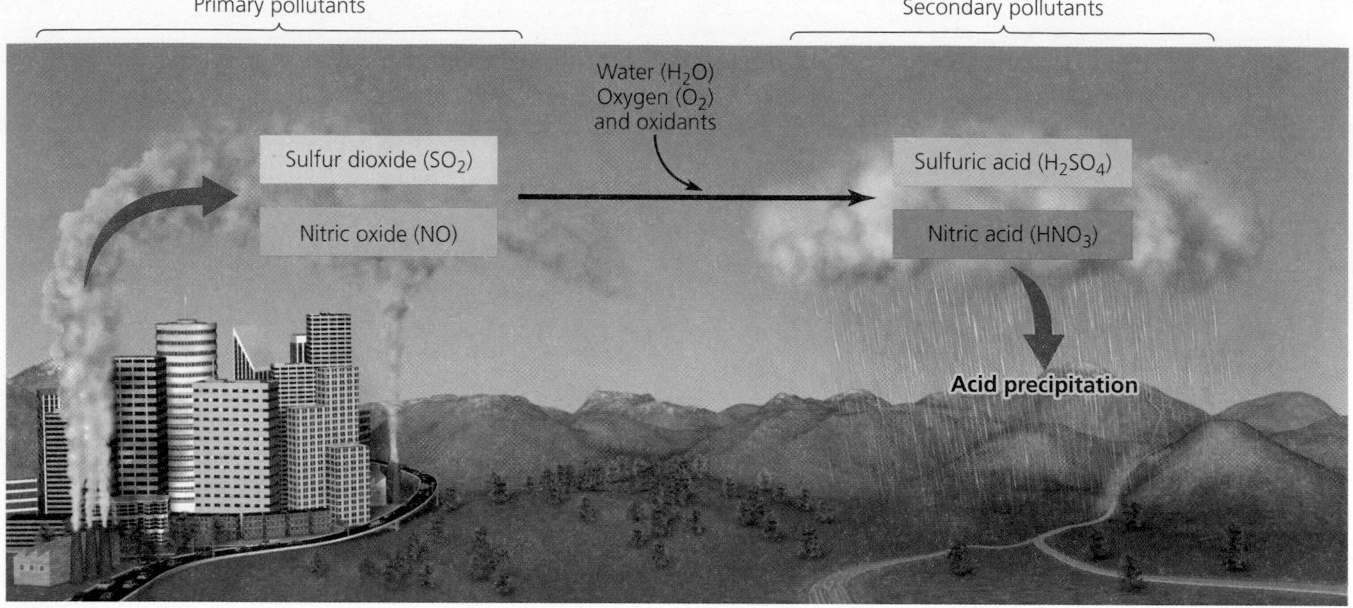


FIGURE 17.18 Acidic deposition can have consequences many miles downwind from its source. Sulfur dioxide and nitric oxide emitted by industries and utilities can be transformed into sulfuric acid and nitric acid through chemical reactions in the atmosphere. These acidic compounds then descend to Earth's surface in rain, snow, fog, and dry deposition.

When acidic water runs off from land, it affects streams, rivers, and lakes. In fact, thousands of lakes in Canada, Scandinavia, the United States, and elsewhere now have lost their fish, because acid precipitation leaches aluminum out of soil and rock and into waterways in a form that can be deadly to aquatic life. Aluminum can kill fish by damaging their gills and disrupting their salt balance, water balance, breathing, and circulation.

**TABLE 17.1 Effects of Acidic Deposition on Ecosystems in the Northeastern United States**

**Acidic deposition in northeastern forests has . . .**

- Accelerated leaching of base cations (ions that counteract acidic deposition) from soil
- Allowed sulfur and nitrogen to accumulate in soil
- Increased dissolved inorganic aluminum in soil, hindering plant uptake of water and nutrients
- Leached calcium from needles of red spruce, leading to tree mortality from wintertime freezing
- Increased mortality of sugar maples due to leaching of base cations from soil and leaves
- Acidified 41% of Adirondack, New York, lakes and 15% of New England lakes
- Lowered lakes' capacity to neutralize further acids
- Elevated aluminum levels in surface waters
- Reduced species diversity and abundance of aquatic life, and negatively affected entire food webs

*Source:* Adapted from Driscoll, C. T., et al. 2001. *Acid rain revisited.* Hubbard Brook Research Foundation.

Besides altering natural ecosystems, acid precipitation also damages agricultural crops. Moreover, it erodes stone buildings, corrodes cars, and erases the writing from tombstones. Ancient cathedrals in Europe, monuments in Washington, D.C., temples in Asia, and stone statues in London are experiencing billions of dollars of damage as their features gradually wear away (**Figure 17.19**).

Because the pollutants leading to acid deposition can travel long distances, their effects may be felt far from their sources—a situation that has led to political bickering among the leaders of states and nations. For instance, much of the pollution from power plants and factories in Pennsylvania, Ohio, and Illinois falls out in states to their east, including New York, Vermont, and New Hampshire, as well as in regions of Canada to the north. As **Figure 17.20** shows, many regions of greatest acidification are downwind of major source areas of pollution.

## Acid deposition has not been reduced as much as scientists had hoped

Reducing acid precipitation involves reducing the pollution that contributes to it. New technologies such as scrubbers (• p. 485) have helped. As a result of declining

**(a) Before acid rain damage**

**(b) After acid rain damage**

FIGURE 17.19 Acidic deposition can corrode statues and buildings. Shown is an Egyptian obelisk known as Cleopatra's Needle, in Central Park, New York City, before **(a)** and after **(b)** the onset of significant acidic deposition.

emissions of $SO_2$, average sulfate precipitation in 1996–2000 was 10% lower than in 1990–1994 across the United States and 15% lower in the eastern states. However, because of increasing $NO_x$ emissions, average nitrate precipitation increased nationally by 3% between these time periods.

A recent report by scientists at New Hampshire's Hubbard Brook research forest, where acidic deposition's effects were first demonstrated in the United States, disputed the notion that the problem of acid deposition is being solved (see "The Science behind the Story," • next page). Instead, the report said, the effects are worse

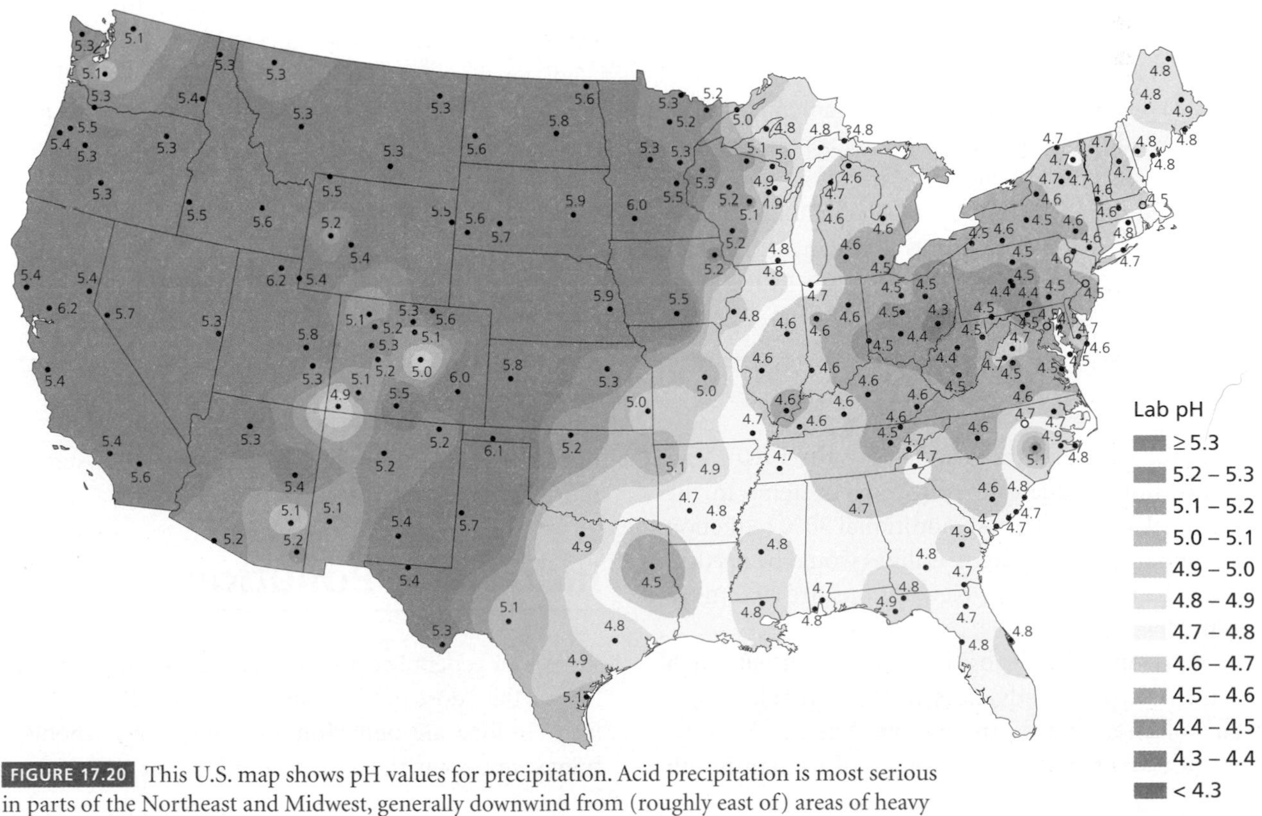


FIGURE 17.20 This U.S. map shows pH values for precipitation. Acid precipitation is most serious in parts of the Northeast and Midwest, generally downwind from (roughly east of) areas of heavy industrial development. Data are for 2005, from the National Atmospheric Deposition Program.

THE SCIENCE BEHIND THE STORY

## Acid Rain at Hubbard Brook Research Forest

*Dr. Gene E. Likens, Institute of Ecosystem Studies*

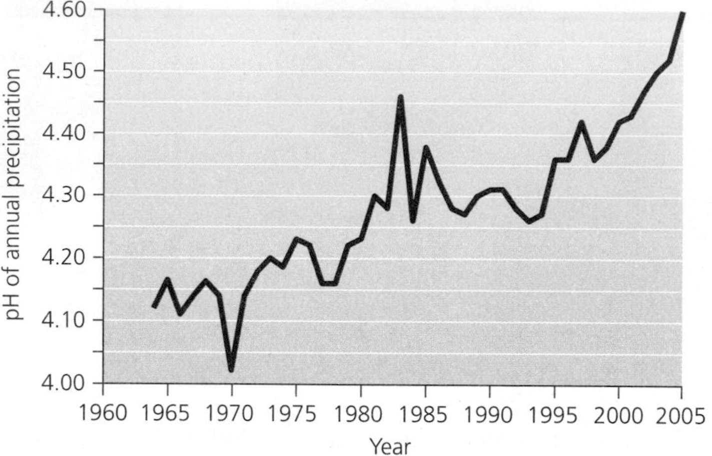


**Over the past 40 years, precipitation at the Hubbard Brook Experimental Forest has become slightly less acidic. However, it is still far more acidic than is natural precipitation.** Data from Likens, G. E. 2004. *Ecology* 85: 2355–2362.

Acidic deposition involves subtle and incremental changes in pH levels that take place over long periods of time, so no single experiment can give us a complete picture of acidic deposition's effects. Nonetheless, one long-term study conducted in the Hubbard Brook Experimental Forest in New Hampshire's White Mountains has been critically important to our understanding of acidic deposition in the United States.

Established by the U.S. Forest Service in 1955, Hubbard Brook was initially devoted to research on hydrology, the study of water flow through forests and streams. In 1963, in collaboration with scientists at Dartmouth University, Hubbard Brook researchers broadened their focus to include a long-term study of nutrient cycling in forest ecosystems. Since then, they have collected and analyzed weekly samples of precipitation. The measurements make up the longest-running North American record of acid precipitation and have helped shape U.S. policy on sulfur and nitrogen emissions.

Throughout Hubbard Brook's 3,160 ha (7,800 acres), small plastic collecting funnels channel precipitation into clean bottles, which researchers retrieve and replace each week. Hubbard Brook's laboratory measures acidity and conductivity, which indicates the amount of salts and other electrolytic contaminants dissolved in the water. Concentrations of sulfuric acid, nitrates, ammonia, and other compounds are measured elsewhere.

By the late 1960s, ecologists Gene Likens, F. Herbert Bormann, and others had found that precipitation at Hubbard Brook was several hundred times more acidic than natural rainwater. By the early 1970s, a number of other studies had corroborated their findings. Together, these studies indicated that precipitation from Pennsylvania to Maine had pH values averaging around 4 and that individual rainstorms showed values as low as 2.1—almost 10,000 times more acidic than ordinary rainwater.

In 1978, the National Atmospheric Deposition Program was launched to monitor precipitation and dry deposition across the United States. Initially

than first predicted, and the mandates of the 1990 Clean Air Act are not adequate to restore ecosystems in the northeastern United States. An additional 80% reduction in sulfur emissions from electric utilities would be needed, the report estimates, to allow New Hampshire streams to recover in 20–25 years.

The data on acid deposition show that although there have been many advances in the control of outdoor air pollution, more can be done. The same can be said for indoor air pollution, a source of human health threats that is less familiar to most of us, but statistically more dangerous.

## Indoor Air Pollution

Indoor air generally contains higher concentrations of pollutants than does outdoor air. As a result, the health effects from **indoor air pollution** in workplaces, schools, and homes outweigh those from outdoor air pollution. One consisting of 22 sites, including Hubbard Brook, the program now consists of more than 200, each of which gathers weekly data on acidic deposition and deposition of other substances. By the late 1980s, this program had produced a nationwide map of pH values (see Figure 17.20). The most severe problems were found to be in the Northeast, where prevailing west-to-east winds were blowing emissions from fossil-fuel-burning power plants in the Midwest. Scientists hypothesized that when sulfur dioxide, nitrogen oxides, and other pollutants arrived in the Northeast, they were absorbed by water droplets in clouds, converted to acidic compounds such as sulfuric acid, and deposited on farms, forests, and cities in rain or snow.

Acidic deposition killed these trees on Mount Mitchell in western North Carolina.

The Clean Air Act of 1970 helped reduce acidic deposition in the Northeast. At an area of Hubbard Brook known as Watershed 6, average pH increased slightly between 1965 and 1995, from about 4.15 to about 4.35 (see the first figure). In 1990, as a consequence of the Hubbard Brook study and the nationwide research that followed, the Clean Air Act of 1970 was amended to further restrict emissions of sulfur dioxide and other acid-forming compounds. Nonetheless, acidic deposition continues to be a serious problem in the Northeast.

Some long-term consequences of acidic deposition are now becoming clear. In 1996, researchers reported that approximately 50% of the calcium and magnesium in Hubbard Brook's soils had leached out. Meanwhile, acidic deposition had increased the concentration of aluminum in the soil, which can prevent tree roots from absorbing nutrients. The resulting nutrient deficiency slows forest growth and weakens trees, making them more vulnerable to drought and insects (see the second figure). It also reduces the ability of soil and water to neutralize acidity, making the ecosystem increasingly vulnerable to further inputs of acid.

In October 1999, researchers used a helicopter to distribute 50 tons of a calcium-containing mineral called wollastonite over one of Hubbard Brook's watersheds. Their objective was to raise the concentration of base cations to estimated historical levels. Over the next 50 years, scientists plan to evaluate the impact of calcium addition on the watershed's soil, water, and life. By providing a comparison to watersheds in which calcium remains depleted, the results should provide new insights into the consequences of acid rain and the possibilities for reversing its negative effects.

estimate, from the U.N. Development Programme in 1998, attributed 2.2 million deaths worldwide to indoor air pollution and 500,000 deaths to outdoor air pollution. Indoor air pollution alone, then, takes roughly 6,000 lives each day.

If the impact of indoor air pollution seems surprising, consider that the average U.S. citizen spends at least 90% of his or her time indoors. Then consider that in the past half century a dizzying array of consumer products has been manufactured and sold, many of which we keep in our homes and offices and have come to play major roles in our daily lives. Many of these products are made of synthetic materials, and, as we saw in Chapter 14, novel synthetic substances are not comprehensively tested for health effects before being brought to market. Products as diverse as insecticides, cleaning fluids, plastics, and chemically treated wood can all exude volatile chemicals into the air.

In an ironic twist, some attempts to be environmentally prudent during the "energy crisis" of 1973–1974 (•pp. 562–563) worsened indoor air pollution in developed countries. To reduce heat loss and improve energy efficiency, building managers sealed off most ventilation in existing buildings, and building designers constructed new buildings with limited ventilation and with windows that did not open. These steps may have saved energy, but they also worsened indoor air pollution by trapping stable, unmixed air—and its pollutants—inside.

## Indoor air pollution in the developing world arises from fuelwood burning

Indoor air pollution has the greatest impact in the developing world. Millions of people in developing nations burn wood, charcoal, animal dung, or crop waste inside their homes for cooking and heating with little or no ventilation (**Figure 17.21**). In the process, they inhale dangerous amounts of soot and carbon monoxide. In the air of such homes, concentrations of particulate matter are commonly 20 times above U.S. EPA standards, the World Health Organization (WHO) has found. Poverty forces fully half the population and 90% of rural residents of developing countries to heat and cook with indoor fires.

Indoor air pollution from fuelwood burning, the WHO estimates, kills 1.6 million people each year, comprising over 5% of all deaths in some developing nations and 2.7% of the entire global disease burden. Many people who tend indoor fires are not aware of the health risks. They do not have access to the statistics showing that chemicals and soot released by burning coal, plastic, and other materials indoors can increase risks of pneumonia, bronchitis, allergies, sinus infections, cataracts, asthma, emphysema, heart disease, cancer, and premature death. Many who are aware of the health risks are too poor to have viable alternatives.

Even in the developed world, recognizing indoor air pollution as a problem is still quite novel. Fortunately, scientists have identified the most deadly indoor threats. Particulate matter and chemicals from wood and charcoal smoke are the primary health risks in the developing world. In developed nations, the top risks are cigarette smoke and radon, a naturally occurring radioactive gas.

## Tobacco smoke and radon are the most dangerous indoor pollutants in the developed world

The health effects of smoking cigarettes are well known, but only recently have scientists quantified the risks of inhaling secondhand smoke. Secondhand smoke, or environmental tobacco smoke, is smoke inhaled by a nonsmoker who is nearby or shares an enclosed airspace with a smoker. Secondhand smoke has been found to cause many of the same problems as directly inhaled cigarette smoke, ranging from irritation of the eyes, nose, and throat, to exacerbation of asthma and other respiratory ailments, to lung cancer. This hardly seems surprising when one considers that environmental tobacco smoke consists of a brew of over 4,000 chemical compounds, many of which are known or suspected to be toxic or carcinogenic.

FIGURE 17.21 In the developing world, many people build fires inside their homes for cooking and heating, as seen here in a South African kitchen. Indoor fires expose family members to particulate matter and carbon monoxide. In most regions of the developing world, indoor air pollution is estimated to cause upwards of 3% of all health risks.

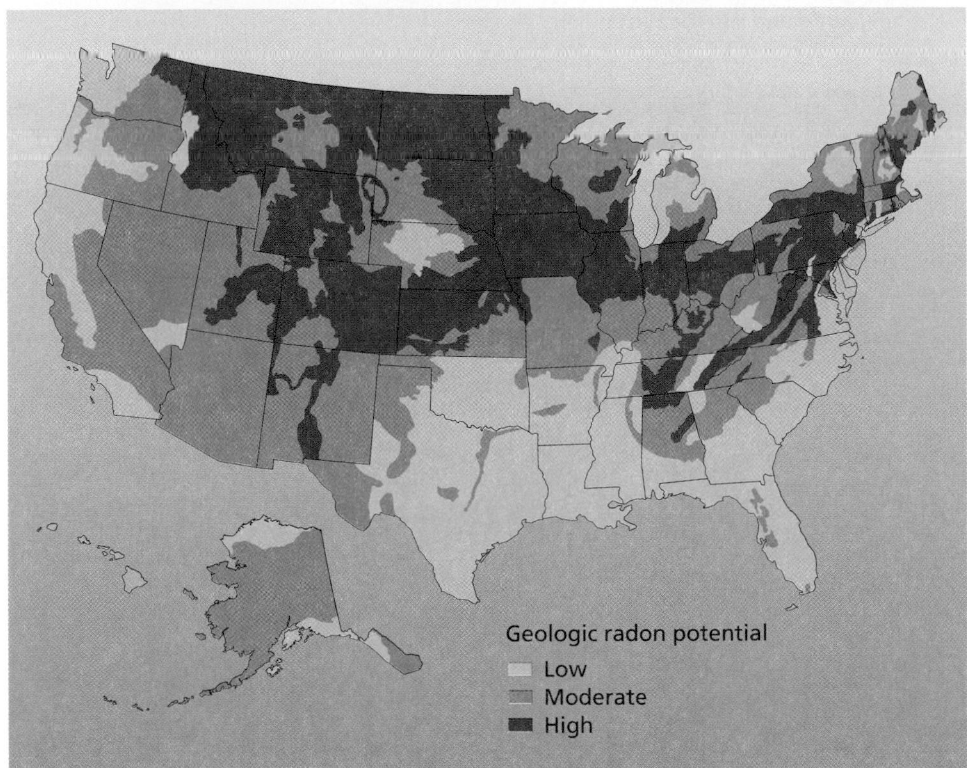


FIGURE 17.22 One's risk from radon depends largely on underground geology. This map shows relative levels of risk from radon across the United States, but there is much fine-scale geographic variation from place to place not evident on the map. Testing your home for radon is the surest way to determine whether this colorless, odorless gas could be a problem in your home. Data from U.S. Geological Survey. 1993. *Generalized geological radon potential of the United States, 1993.*

Fortunately, the popularity of smoking has declined greatly in developed nations in recent years, and many public and private venues now ban smoking. In London, people are engaged in an ongoing civic debate as to whether smoking should be banned in pubs. The health benefits of such a ban are clear, but pubs are such a traditional feature of British life that any change in custom is bound to be controversial.

After cigarette smoke, radon gas is the second-leading cause of lung cancer in the developed world, responsible for an estimated 20,000 deaths per year in the United States alone. The WHO estimates that radon may account for 15% of lung cancer cases worldwide. As we saw in Chapter 14 (• p. 386), radon is a radioactive gas resulting from the natural decay of uranium in soil, rock, or water, which seeps up from the ground and can infiltrate buildings. Radon is colorless and odorless, and it can be impossible to predict where it will occur without knowing details of an area's underlying geology (**Figure 17.22**). As a result, the only way to determine whether radon is entering a building is to measure radon with a test kit. Testing in 1991 led the EPA to estimate that 6% of U.S. homes exceeded the EPA's maximum recommended level for radon. Since the mid-1980s, 18 million U.S. homes have been tested for radon, and 700,000 have undergone radon mitigation. New homes are now being built with radon-resistant features—over a million such homes since 1990.

## Many VOCs pollute indoor air

In our daily lives at home, we are exposed to many indoor air pollutants (**Figure 17.23**). The most diverse indoor pollutants are volatile organic compounds. These airborne carbon-containing compounds are released by everything from plastics to oils to perfumes to paints to cleaning fluids to adhesives to pesticides. VOCs evaporate from furnishings, building materials, color film, carpets, laser printers, fax machines, and sheets of paper. Some products, such as chemically treated furniture, release large amounts of VOCs when new and progressively less as they age. Other items, such as photocopying machines, emit VOCs each time they are used.

Although we are surrounded by products that emit VOCs, they are released in very small amounts. EPA and European surveys have both found overall levels of VOCs in buildings to be nearly always less than 0.1 part per million. This is, however, a substantially greater concentration than is generally found outdoors. When the EPA surveyed U.S. office buildings between 1994 and 1998, it found 34 of the 48 VOCs it tested for, and it found at least one type in 81% of buildings. All of these compounds existed at higher levels indoors than outdoors, suggesting that they originated from indoor sources.

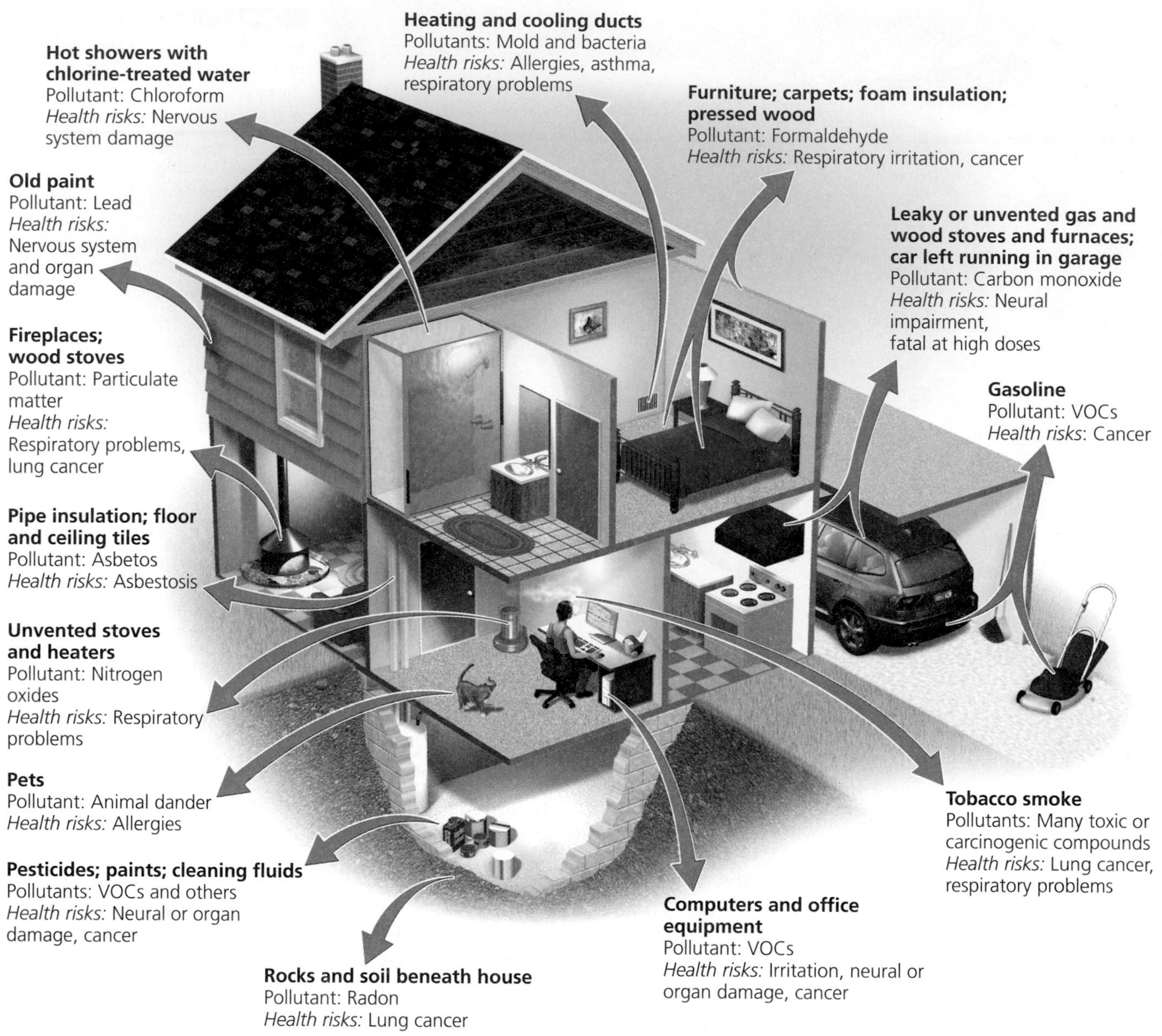


FIGURE 17.23 The typical U.S. home contains a variety of potential sources of indoor air pollution. Shown are some of the most common sources, the major pollutants they emit, and some of the health risks they pose.

The implications for human health of chronic exposure to VOCs are far from clear. Because they exist in such low concentrations and because individuals regularly are exposed to mixtures of many different types, it is extremely difficult to study the effects of any one pollutant. An exception is formaldehyde, which does have clear and known health impacts. This VOC, one of the most common synthetically produced chemicals, irritates mucous membranes, induces skin allergies, and causes other ailments. Formaldehyde is used in numerous products, but health complaints have mainly resulted from its leakage from pressed wood and insulation. The use of plywood has decreased in the last decade because of health concerns over formaldehyde.

VOCs also include pesticides, which we examined in Chapters 10 and 14. Three-quarters of U.S. homes use at least one pesticide indoors during an average year, but most are used outdoors. Thus it may seem surprising that the EPA found in a 1990 study that 90% of people's pesticide exposure came from indoor sources. Households that the agency tested had multiple pesticide volatiles in their air, at levels 10 times above levels measured outside. Some of the pesticides had apparently been used years earlier against termites and then seeped into the houses

## CAUSES AND CONSEQUENCES

**Indoor air pollution** receives less attention than outdoor air pollution, yet it exerts severe health impacts on millions of people. Poverty worsens indoor air pollution in the developing world; high-technology products and chemicals contribute to it in the developed world; and certain lifestyle choices can affect all of us. Fortunately, many solutions are within reach to address the causes and lessen the consequences of this environmental health risk.

Write two causes of indoor air pollution in the spaces provided. Then write two consequences (impacts on the environment, human health, or quality of life) that result from indoor air pollution. Finally, offer two solutions to this issue and its consequences. One cause, one consequence, and one solution have been filled in for you as examples.

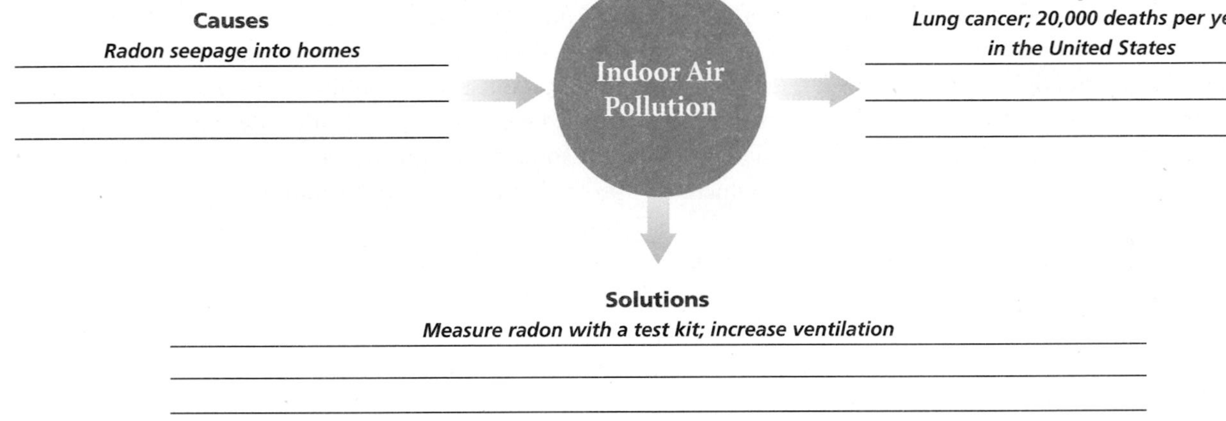


Sometimes solutions can have unintended consequences. Can you think of an undesired consequence that one solution to this issue might have? How might we then deal with *that* consequence?

through floors and walls. DDT, banned 15 years before the study, was found in five of eight homes, probably having been brought in from outdoors on the soles of occupants' shoes.

## Living organisms can pollute indoor spaces

Tiny living organisms can pollute. In fact, they may be the most widespread source of indoor air pollution in the developed world. Dust mites and animal dander can worsen asthma in children. Some fungi, mold, and mildew (in particular, their airborne spores) can cause severe health problems, including allergies, asthma, and other respiratory ailments. Some airborne bacteria can cause infectious disease. One example is the bacterium that causes Legionnaires' disease. Of the estimated 10,000–15,000 annual U.S. cases of Legionnaires' disease, 5–15% are fatal. Heating and cooling systems in buildings make ideal breeding grounds for microbes, providing moisture, dust, and foam insulation as substrates, as well as air currents to carry the organisms aloft.

Microbes that induce allergic responses are thought to be a major cause of building-related illness, a sickness produced by indoor pollution. When the cause of such an illness is a mystery, and when symptoms are general and nonspecific, the illness is often called *sick-building syndrome.* The U.S. Occupational Safety and Health Administration (OSHA) has estimated that 30–70 million Americans have suffered ailments due to the environment of the building in which they live. We can reduce the prevalence of sick-building syndrome by using low-toxicity construction materials and ensuring that buildings are well ventilated.

### Weighing the Issues | How Safe Is Your Indoor Environment?

Think about the amount of time you spend indoors. Name the potential indoor air quality hazards in your home, work, or school environment. Are these spaces well ventilated? What could you do to make the indoor spaces you use safer?

## We can reduce indoor air pollution

Using low-toxicity material, monitoring air quality, keeping rooms clean, and providing adequate ventilation are the keys to alleviating indoor air pollution in most situations. In the developed world, we can try to limit our use of plastics and treated wood where possible and to limit our exposure to pesticides, cleaning fluids, and other known toxicants by keeping them in a garage or outdoor shed rather than in the house. The EPA recommends that we test our homes and offices for radon. Because carbon monoxide is so deadly and so hard to detect, many homes are equipped with detectors that sound an alarm if incomplete combustion produces dangerous levels of CO. In addition, keeping rooms and air ducts clean and free of mildew and other biological pollutants will reduce potential irritants and allergens. Finally, it is important to keep our indoor spaces as well ventilated as possible to minimize concentrations of the pollutants among which we live.

Remedies for fuelwood pollution in the developing world include drying wood before burning (which reduces the amount of smoke produced), cooking outside, shifting to less-polluting fuels (such as natural gas), and replacing inefficient fires with cleaner stoves that burn fuel more efficiently. For example, the Chinese government has invested in a program that has placed more fuel-efficient stoves in millions of homes in China. Installing hoods, chimneys, or cooking windows can increase ventilation for little cost, alleviating the majority of indoor smoke pollution.

# Conclusion

Indoor air pollution is a potentially serious health threat. However, by keeping informed of the latest scientific findings and taking appropriate precautions, we as individuals can significantly minimize the risks to our families and ourselves. Outdoor air pollution has been addressed more effectively by government legislation and regulation. In fact, reductions in outdoor air pollution in the United States, Great Britain, and other developed nations represent some of the greatest strides made in environmental protection to date. Much room for improvement remains, however, particularly in reducing acidic deposition and the photochemical smog that results from urban congestion. Fortunately, developed nations no longer experience the type of pollution that Londoners suffered in their 1952 killer smog. Nevertheless, avoiding such high pollutant levels in the developing world will continue to pose a challenge as less-wealthy nations industrialize.

## REVIEWING OBJECTIVES

**You should now be able to:**

**Describe the composition, structure, and function of Earth's atmosphere**

- The atmosphere consists of 78% nitrogen gas, 21% oxygen gas, and a variety of other gases in minute concentrations. (p. 474)
- The atmosphere includes four principal layers: the troposphere, stratosphere, mesosphere, and thermosphere. Temperature and other characteristics vary across these layers. Ozone is concentrated in the stratosphere. (pp. 474–475)
- The sun's energy heats the atmosphere, drives air circulation, and helps determine weather, climate, and the seasons. (pp. 476–477)
- Weather is a short-term phenomenon, whereas climate is a long-term phenomenon. Fronts, pressure systems, and the interactions among air masses influence weather. (pp. 477–479)
- Global convective cells called Hadley, Ferrel, and polar cells create latitudinal climate zones. (pp. 479–480)

**Outline the scope of outdoor air pollution and assess potential solutions**

- Natural sources such as windblown dust, volcanoes, and fires account for much atmospheric pollution, but human activity can worsen some of these phenomena. (pp. 480–481)
- Human-emitted pollutants include primary and secondary pollutants from point and non-point sources. (pp. 481–482)
- To safeguard public health, the U.S. EPA monitors six criteria pollutants (carbon monoxide, lead, nitrogen dioxide, tropospheric ozone, sulfur dioxide, and particulate matter), as well as volatile organic compounds and 188 toxic air pollutants. (pp. 482–486)
- Emissions in the United States have decreased substantially since the Clean Air Act of 1970, and ambient air quality is much improved. (pp. 484–485)
- Recent proposals by the Bush administration to revise Clean Air Act policy proved controversial and so far have been blocked. (p. 486)

- Industrial smog like that which blanketed 1952 London is produced by fossil fuel combustion and is still a problem in urban and industrial areas of many developing nations. (pp. 486–487)
- Photochemical smog like that which affects London and other cities today is created by chemical reactions of pollutants in the presence of sunlight. It impairs visibility and human health in urban areas. (pp. 486–488)
- Rural areas suffer air pollution from feedlots and other sources. (p. 488)
- Industrializing nations such as China and India are experiencing some of the world's worst air pollution today. (p. 488)

**Explain stratospheric ozone depletion and identify steps taken to address it**

- CFCs destroy stratospheric ozone, and thinning ozone concentrations pose dangers to life because they allow more ultraviolet radiation to reach Earth's surface. (pp. 489–491)
- The Montreal Protocol and its follow-up agreements have proven remarkably successful in reducing emissions of ozone-depleting compounds. (pp. 489–491)
- The long residence time of CFCs in the atmosphere means a time lag between the protocol and the actual restoration of stratospheric ozone. (pp. 489–490)

**Define acidic deposition and illustrate its consequences**

- Acidic deposition results when pollutants such as $SO_2$ and NO react in the atmosphere to produce strong acids that are deposited on Earth's surface. (pp. 491–492)
- Acidic deposition may occur a long distance from the source of pollution. (pp. 492–493)
- Water bodies, soils, trees, animals, and ecosystems all experience negative impacts from acidic deposition. (pp. 492–495)

**Characterize the scope of indoor air pollution and assess potential solutions**

- Indoor air pollution causes far more deaths and health problems worldwide than outdoor air pollution. (pp. 495–496)
- Indoor burning of fuelwood is the developing world's primary indoor air pollution risk. (p. 496)
- Tobacco smoke and radon are the deadliest indoor pollutants in the developed world. (pp. 496–497)
- Volatile organic compounds and living organisms can pollute indoor air. (pp. 497–499)
- Using low-toxicity building materials, keeping spaces clean, monitoring air quality, and maximizing ventilation are some of the steps we can take to reduce indoor air pollution. (pp. 499–500)

## TESTING YOUR COMPREHENSION

1. About how thick is Earth's atmosphere? Name one characteristic of each of the four atmospheric layers.
2. Where is the "ozone layer" located? How and why is stratospheric ozone beneficial for people, and tropospheric ozone harmful?
3. How does solar energy influence weather and climate? How do Hadley, Ferrel, and polar cells help to determine long-term climatic patterns and the location of biomes?
4. What factors led to the deadly smog in London in 1952? Describe a thermal inversion.
5. Name three natural sources of outdoor air pollution and three sources caused by human activity.
6. What is the difference between a primary and a secondary pollutant? Give an example of each.
7. What is smog? How is smog formation influenced by the weather? By topography? How does photochemical smog differ from industrial smog?
8. How do chlorofluorocarbons (CFCs) deplete stratospheric ozone? Why is this depletion considered a long-term international problem? What was done to address this problem?
9. Why are the effects of acidic deposition often felt in areas far from where the primary pollutants are produced? List three impacts of acidic deposition.
10. Name five common sources of indoor pollution. For each, describe one way to reduce one's exposure to this source.

## SEEKING SOLUTIONS

1. Consider London's "killer smog" of 1952 and its modern-day pollution by photochemical smog. Describe several factors that make it difficult to study causes of air pollution and to develop solutions. How has London tried to improve its air quality?
2. Name one type of natural air pollution, and discuss how human activity can sometimes worsen it. What potential solutions can you think of to minimize this human impact?
3. Describe how and why emissions of major pollutants have been reduced by over 50% in the United States since 1970, despite increases in population and economic activity.
4. International regulatory action has produced reductions in CFCs, but other transboundary pollution issues, including acidic deposition, have not yet been addressed as effectively. What types of actions do you feel are appropriate for pollutants that cross political boundaries?
5. **THINK IT THROUGH** You have just been elected mayor of the largest city in your state. Your city's residents are complaining about photochemical smog and traffic congestion. Traffic engineers and city planners project that population and traffic will grow by 20% in the next decade. A citizen's group is urging you to implement a congestion-charging program like London's, but businesses are fearful of losing money if shoppers are discouraged from visiting. Consider the particulars of your city, and then decide whether you will pursue a congestion-charging program. If so, how would you do it? If not, why not, and what other steps would you take to address your city's problems?
6. **THINK IT THROUGH** You have just become the head of your county health department, and the EPA has informed you that your county has failed to meet the national ambient air quality standards for ozone, sulfur dioxide, and nitrogen dioxide. Your county is partly rural but is home to a city of 200,000 people and 10 sprawling suburbs. There are several large and aging coal-fired power plants, a number of factories with advanced pollution-control technology, and no public transportation system. What steps would you urge the county government to take to meet the air quality standards? Explain how you would prioritize these steps.

## INTERPRETING GRAPHS AND DATA

Since the Clean Air Act of 1970, total emissions of carbon monoxide, sulfur dioxide, nitrogen oxides, VOCs, lead, and particulate matter have dropped by over 50% while U.S. population, energy consumption, and economic productivity have all increased. As you learned from the "Interpreting Graphs and Data" feature in Chapter 3 (• pp. 84–85), lead emissions resulted mostly from the combustion of leaded gasoline and dropped precipitously once leaded gasoline was phased out. Consider the other pollutants listed above as you interpret the data in the graph.

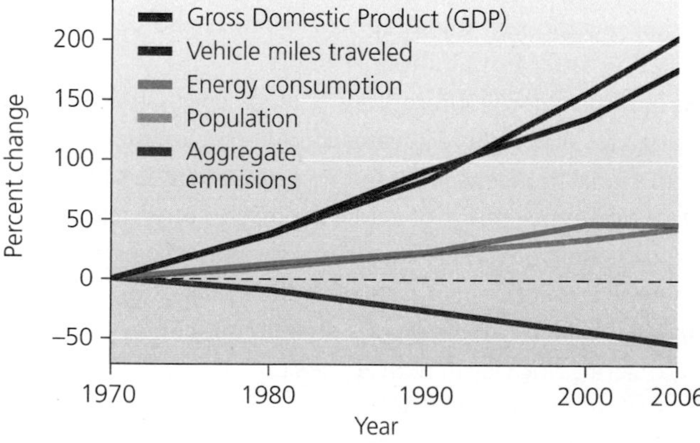


**Trends from 1970 to 2006 in economic production, vehicle miles, energy consumption, population, and aggregate emissions of the six principal air pollutants monitored by the EPA.** Data from U.S. EPA.

1. What are the percentage changes from 1970 to 2006 for each of the five variables graphed? Now calculate the percentage changes in the per capita values of each variable over this time period.
2. Fossil fuel combustion accounts for most of our energy consumption and is the major source of most of the pollutants named above. What is the percentage change in aggregate emissions of these six principal pollutants per unit of energy consumed in 2006, compared to 1970? What do you think accounts for this change?
3. Do you think that additional reductions in emissions are likely to result primarily from changing technology (e.g., hybrid cars, advanced catalytic converters) or from changing behavior (e.g., driving fewer miles per person per year)? Use the data in the graph to support your position.

## CALCULATING ECOLOGICAL FOOTPRINTS

According to EPA data, emissions of nitrogen oxides in the United States in 2002 totaled 21,102,000 tons. Nitrogen oxides come from fuel combustion in motor vehicles, power plants, and other industrial, commercial, and residential sources, but fully 11,452,000 tons of the 2002 total came from the transportation sector. Of this amount, 7,365,000 tons came from on-road vehicles, with 3,583,000 tons of this total coming from light-duty cars and trucks. The U.S. Census Bureau estimated the nation's population to be 288,368,698 at mid-year in 2002 and projects that it will reach 308,936,000 by mid-2010. Considering these data, calculate the missing values in the table below (1 ton = 2,000 lb).

| | Total $NO_x$ emissions (lb) | $NO_x$ emissions due to light-duty vehicles (lb) |
|---|---|---|
| You | | |
| Your class | | |
| Your state | | |
| United States | | |

Data from U.S. EPA.

1. By what percentage is the U.S. population projected to increase between 2002 and 2010? Do you think that $NO_x$ emissions will increase, decrease, or remain the same over that period of time? Why? (You may want to refer to Figure 17.13a.)
2. Assuming you are an average American driver, how many pounds of $NO_x$ emissions would you prevent if you were to reduce by half the vehicle miles you travel? What percentage of your total $NO_x$ emissions would that be?
3. How might you reduce your vehicle miles traveled by 50%? What other steps could you take to reduce the $NO_x$ emissions for which you are responsible?

### Take It Further

Go to www.aw-bc.com/withgott or the student CD-ROM, where you'll find:

- Suggested answers to end-of-chapter questions
- Quizzes, animations, and flashcards to help you study
- *Research Navigator*™ database of credible and reliable sources to assist you with your research projects
- GRAPHIt! Tutorials to help you interpret graphs
- INVESTIGATEIt! Current news articles that link the topics that you study to case studies from your region to around the world

CHAPTER

# 18 Global Climate Change

Muratti Island of the Maldives

## Upon completing this chapter, you will be able to:

- Describe Earth's climate system and explain the many factors influencing global climate
- Characterize human influences on the atmosphere and global climate
- Summarize modern methods of climate research
- Outline current and future trends and impacts of global climate change
- Suggest ways we may respond to climate change

Fishermen of the Maldives

## CENTRAL CASE

# Rising Seas May Flood the Maldives

**"The impact of global warming and climate change can effectively kill us off, make us refugees. . . ."**
—Ismail Shafeeu, Minister of Environment, Maldives, 2000

**"We simply must do everything we can in our power to slow down global warming before it is too late. The science is clear. The global warming debate is over."**
—Arnold Schwarzenegger, Governor of California, September 2006

A nation of low-lying islands in the Indian Ocean, the Maldives is known for its spectacular tropical setting, colorful coral reefs, and sun-drenched beaches. For visiting tourists it seems to be paradise, and for 370,000 Maldives residents it is home. But residents and tourists alike now fear that the Maldives could soon be submerged by the rising seas that are accompanying global climate change.

Nearly 80% of the Maldives' land area lies less than 1 m (39 in.) above sea level. In a nation of 1,200 islands whose highest point is just 2.4 m (8 ft) above sea level, rising seas are a matter of life or death. The world's oceans rose 10–20 cm (4–8 in.) during the 20th century as warming temperatures expanded ocean water and as melting polar ice discharged water into the ocean. According to current projections, sea level will rise another 18–59 cm (7–23 in.) by the year 2100.

Higher seas are expected to flood large areas of the Maldives and cause salt water to contaminate drinking water supplies. Storms intensified by warmer water temperatures will erode beaches, cause flooding, and damage the coral reefs that are so vital to the tourism and fishing industries that drive the nation's economy. Because of such concerns, the Maldives government recently evacuated residents from several of the lowest-lying islands.

On December 26, 2004, the nation got a taste of what could be in store in the future, when a massive *tsunami*, or tidal wave, devastated coastal areas throughout the Indian Ocean. The tsunami killed 100 Maldives residents and left 20,000 homeless. Schools, boats, tourist resorts, hospitals, and transportation and communication infrastructure were damaged or destroyed. The World Bank estimates that direct damage in the Maldives totaled $470 million, an astounding 62% of the nation's gross domestic product (GDP). Soil erosion, saltwater contamination of aquifers, and other environmental damage will result in still greater long-term economic losses. The tsunami was caused *not* by climate change, but by an earthquake. Yet as sea levels rise, the damage that such natural events—or ordinary storm waves—can inflict increases considerably.

Maldives islanders are not alone in their predicament. Other island nations, from the Galapagos to Fiji to the Seychelles, also fear a future of constant vigilance against encroaching seawater. Mainland coastal areas of the world, such as the hurricane-battered coasts of Florida, Louisiana, Texas, the Carolinas, and other states, will face similar challenges. In one way or another, global climate change will affect each and every one of us for the remainder of our lifetimes.

# Our Dynamic Climate

Climate influences virtually everything around us, from the day's weather to major storms, from crop success to human health, and from national security to the ecosystems that support our economies. If you are a student in your teens or twenties, climate change may well be *the* major event of your lifetime and the phenomenon that most shapes your future.

As this chapter was being prepared, the world's people—particularly those of the United States—were beginning to undergo a revolution in awareness and concern. The 2007 release of the *Fourth Assessment Report* of the Intergovernmental Panel on Climate Change (IPCC; • p. 516) made clear to the world the scientific consensus that climate is changing, that we are the cause, and that climate change is already exerting impacts that will become increasingly severe if we do not take action. In 2006–2007, millions of people watched former U.S. Vice President Al Gore's Oscar-winning film on climate change, *An Inconvenient Truth*, or read his best-selling book of the same title. In April 2007, thousands of U.S. students and citizens staged the first nationwide rallies for action against climate change.

Climate change is the fastest-moving area of environmental science today. New scientific papers that refine our understanding of climate are published every week, and policymakers and businesspeople make headlines with decisions and announcements just as quickly. By the time you read this chapter, some of its information will already be out of date. We urge you to explore further, with your instructor and on your own, the most recent information on climate change and the impacts it will have on your future.

## What is climate change?

**Climate** describes an area's long-term atmospheric conditions, including temperature, moisture content, wind, precipitation, barometric pressure, solar radiation, and other characteristics. As we have learned (• pp. 477–478), *climate* differs from *weather* in that weather specifies conditions at localized sites over hours or days, whereas climate describes conditions across broader regions over seasons, years, or millennia. **Global climate change** describes trends and variation in Earth's climate, involving aspects such as temperature, precipitation, and storm frequency and intensity. People often use the term **global warming** synonymously in casual conversation, but *global warming* refers specifically to an increase in Earth's average surface temperature. It is thus only one aspect of global climate change, although warming does in turn drive other components of climate change.

Our planet's climate varies naturally through time, but the climatic changes taking place today are unfolding at an exceedingly rapid rate. Moreover, scientists agree that human activities, notably fossil fuel combustion and deforestation, are largely responsible. Understanding how and why climate is changing requires understanding how our planet's climate functions. Thus, we first will survey the fundamentals of Earth's climate system—a complex and finely tuned system that has nurtured life for billions of years.

## The sun and atmosphere keep Earth warm

Three factors exert more influence on Earth's climate than all others combined. The first is the sun. Without it, Earth would be dark and frozen. The second is the atmosphere. Without it, Earth would be as much as 33° C (91° F) colder on average, and temperature differences between night and day would be far greater than they are. The third is the oceans, which shape climate by storing and transporting heat and moisture.

The sun supplies most of our planet's energy. Earth's atmosphere, clouds, land, ice, and water together absorb about 70% of incoming solar radiation, and reflect the remaining 30% back into space (**Figure 18.1**).

## Greenhouse gases warm the lower atmosphere

As Earth's surface absorbs solar radiation, the surface increases in temperature and emits infrared radiation (• pp. 101–102), radiation with longer wavelengths than visible light. Some atmospheric gases absorb this infrared radiation very effectively. These include water vapor, ozone ($O_3$), carbon dioxide ($CO_2$), nitrous oxide ($N_2O$), and methane ($CH_4$), as well as halocarbons, a diverse group that

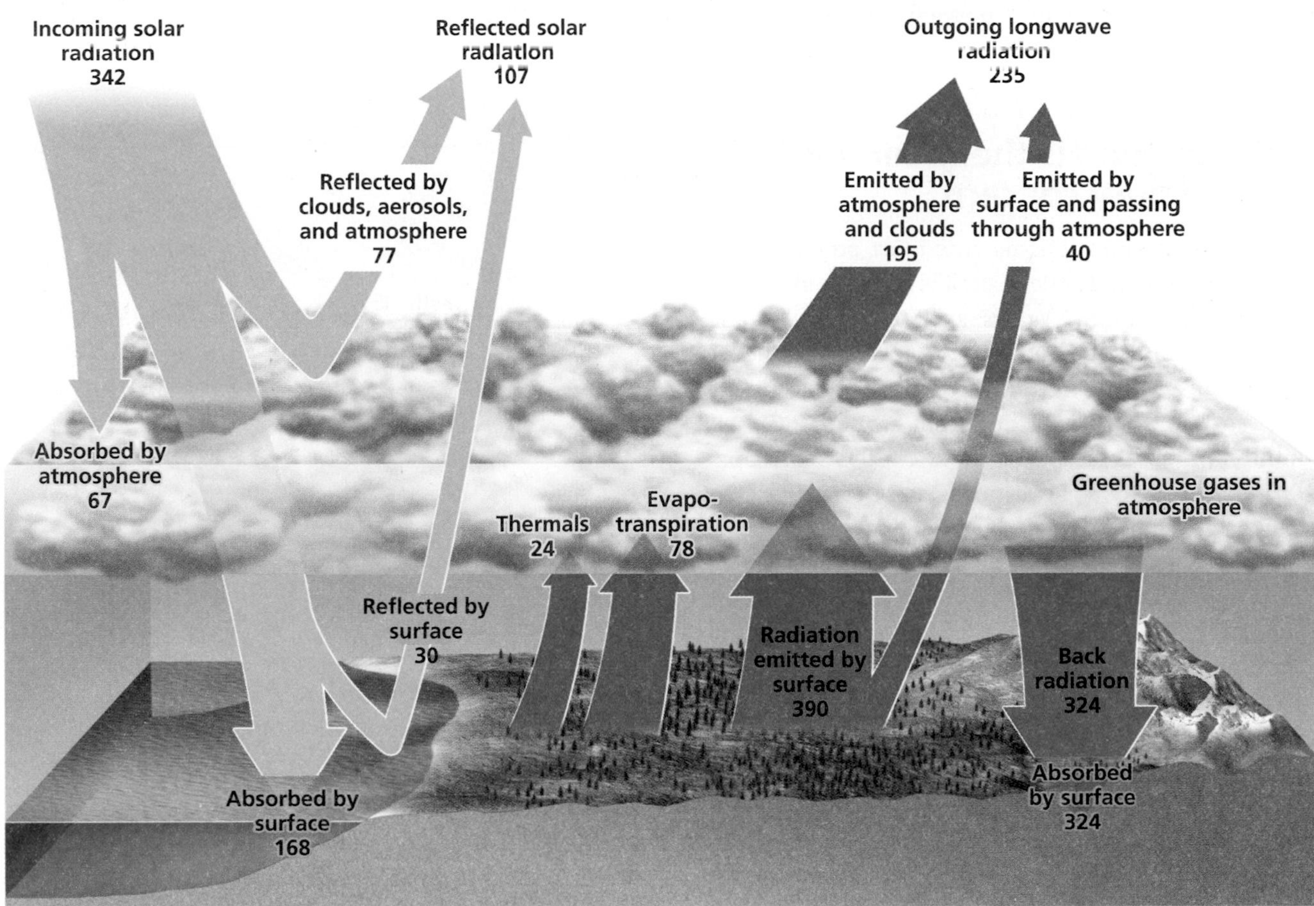


FIGURE 18.1 Our planet absorbs nearly 70% of the solar radiation it receives from the sun, and it reflects the rest back into space (yellow arrows). Absorbed radiation is then re-emitted (orange arrows) as heat. Greenhouse gases in the atmosphere absorb some of this radiation and re-emit it again, sending some downward to warm the atmosphere and the surface by the greenhouse effect. Overall, Earth's climate system before human impact was in rough equilibrium; this illustration, showing major pathways of energy flow in watts per square meter, indicates that our planet naturally emits and reflects 342 watts/$m^2$, the same amount it receives from the sun. Arrow thicknesses in the diagram are proportional to flows of energy in each pathway. Data from Kiehl, J. T., and K. E. Trenberth. 1997. Earth's annual global mean energy budget. *Bulletin of the American Meteorological Society* 78: 197–208.

includes chlorofluorocarbons (CFCs; • pp. 489–491). Such gases are known as **greenhouse gases.** After absorbing radiation emitted from the surface, greenhouse gases subsequently re-emit infrared energy of slightly different wavelengths. Some of this re-emitted energy is lost to space, but some travels back downward, warming the atmosphere (specifically the *troposphere;* • pp. 474–475) and the planet's surface in a phenomenon known as the **greenhouse effect.**

The greenhouse effect is a natural phenomenon, and greenhouse gases (with the exception of the anthropogenic halocarbons) have been present in our atmosphere for billions of years. However, human activities have increased the concentrations of many greenhouse gases in the past 250–300 years, thereby enhancing the greenhouse effect.

Greenhouse gases differ in their ability to warm the troposphere and surface. *Global warming potential* refers to the relative ability of one molecule of a given greenhouse gas to contribute to warming. Table 18.1 shows the global warming potentials for several greenhouse gases. Values

TABLE 18.1 Global Warming Potentials of Four Greenhouse Gases

| Greenhouse gas | Relative heat-trapping ability (in $CO_2$ equivalents) |
|---|---|
| Carbon dioxide | 1 |
| Methane | 23 |
| Nitrous oxide | 296 |
| Hydrochlorofluorocarbon HFC-23 | 12,000 |

Data from Intergovernmental Panel on Climate Change. 2001. *Third assessment report. Climate change 2001: The scientific basis.*

are expressed in relation to carbon dioxide, which is assigned a global warming potential of 1. Thus, a molecule of methane is 23 times as potent as a molecule of carbon dioxide, and a molecule of nitrous oxide is 296 times as potent as a $CO_2$ molecule.

## Carbon dioxide is the greenhouse gas of primary concern

Although carbon dioxide is not the most potent greenhouse gas on a per-molecule basis, it is far more abundant in the atmosphere than gases such as methane and nitrous oxide, so it contributes more to the greenhouse effect. Human activities have boosted Earth's atmospheric concentration of carbon dioxide from around 280 parts per million (ppm) as recently as the late 1700s to 383 ppm in 2007 (**Figure 18.2**). The atmospheric $CO_2$ concentration is now, by far, at its highest level in over 650,000 years, and likely the highest in the last 20 million years.

What has caused levels of this greenhouse gas to increase so rapidly? As you may recall from our discussion of the carbon cycle in Chapter 7 (• pp. 188–189), and as we will see further in Chapter 19 (• p. 544), most carbon is stored for long periods in the upper layers of the lithosphere. The deposition, partial decay, and compression of organic matter (mostly plants) that grew in wetland or marine areas during the Carboniferous Period (290–354 million years ago; Appendix D) led to the formation of coal, oil, and natural gas in sediments from that time. In the absence of human activity, these carbon reservoirs would be practically permanent. However, over the past two centuries we have burned fossil fuels in our homes, factories, and automobiles, transferring large amounts of carbon from one reservoir (the underground deposits that stored the carbon for millions of years) to another (the atmosphere). This flux of carbon from lithospheric reservoirs into the atmosphere is the main reason atmospheric carbon dioxide concentrations have increased so dramatically.

At the same time, people have cleared and burned forests to make room for crops, pastures, villages, and cities. Forests serve as sinks (• p. 187) for recently active carbon, and their removal reduces the biosphere's ability to absorb carbon dioxide from the atmosphere. In this way, deforestation has contributed to rising atmospheric $CO_2$ concentrations.

## Other greenhouse gases add to warming

Carbon dioxide is not the only greenhouse gas increasing in the atmosphere because of our activities. We release methane by tapping into fossil fuel deposits, raising livestock that emit methane as a metabolic waste product, disposing of organic matter in landfills, and growing certain crops such as rice. Since 1750, atmospheric methane concentrations have risen 2.5-fold (see Figure 18.2), and today's concentration is the highest by far in over 650,000 years.

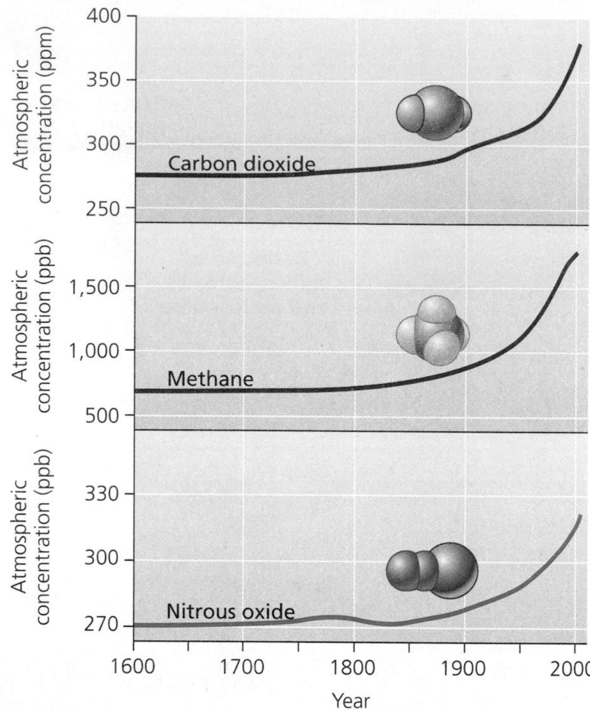


FIGURE 18.2 Since the start of the industrial revolution, global atmospheric concentrations of carbon dioxide (top panel), methane (middle panel), and nitrous oxide (bottom panel) have increased markedly. Data from Intergovernmental Panel on Climate Change. 2007. *Fourth assessment report.*

Human activities have also enhanced atmospheric concentrations of nitrous oxide. This greenhouse gas, a by-product of feedlots, chemical manufacturing plants, auto emissions, and synthetic nitrogen fertilizers, has risen by 18% since 1750 (see Figure 18.2). Ozone concentrations in the troposphere have risen roughly 36% since 1750 because of photochemical smog (• pp. 486–488). The contribution of halocarbon gases to global warming has begun to slow because of the Montreal Protocol and subsequent controls (• pp. 489–491).

Emissions of greenhouse gases from human activity in the United States consist mostly of carbon dioxide. Even after accounting for the greater global warming potential of molecules of other gases, carbon dioxide's abundance in our emissions makes it the major contributor to global warming (**Figure 18.3**).

Water vapor is the most abundant greenhouse gas in our atmosphere and contributes most to the greenhouse effect. Its concentrations vary locally, but its global concentration has not changed over recent centuries, so it is

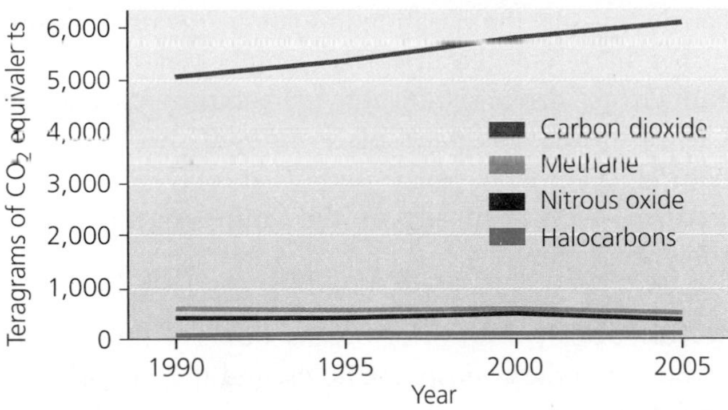


FIGURE 18.3 Emissions of the main greenhouse gases in the United States increased by 16.3% between 1990 and 2005. Lines on this graph show contributions of each gas to global warming, after accounting for global warming potentials. Despite the greater global warming potentials of molecules of methane, nitrous oxide, and halocarbons, carbon dioxide is so abundant that it accounts for nearly 85% of warming. Data from U.S. Environmental Protection Agency (EPA). 2007. *Inventory of U.S. greenhouse gas emissions and sinks: 1990–2005.* Washington, D.C.: U.S. EPA.

not viewed as having driven industrial-age climate change. However, as tropospheric temperatures continue to increase, Earth's water bodies should transfer more water vapor into the atmosphere. Such a positive feedback mechanism (• p. 176) could amplify the greenhouse effect. Alternatively, more water vapor could give rise to increased cloudiness, which might, in a negative feedback loop (• pp. 175–176), slow global warming by reflecting more solar radiation back into space. Depending on whether low- or high-elevation clouds resulted, they might either shade and cool Earth (negative feedback) or else contribute to warming and accelerate evaporation and further cloud formation (positive feedback). Because of feedback loops (see Figure 7.1, • p. 176), minor modifications of components of the atmosphere can potentially lead to major effects on climate.

## Aerosols may exert a cooling effect

Whereas greenhouse gases exert a warming effect on the atmosphere, **aerosols** (• p. 480), microscopic droplets and particles, can have either a warming or cooling effect. Generally speaking, soot, or black carbon aerosols, can cause warming by absorbing solar energy, but most tropospheric aerosols cool the atmosphere by reflecting the sun's rays. Sulfate aerosols produced by fossil fuel combustion may slow global warming, at least in the short term. When sulfur dioxide enters the atmosphere, it undergoes various reactions, some of which lead to acid precipitation (• pp. 491–495). These reactions, along with volcanic eruptions, can also contribute to the formation of a sulfur-rich aerosol haze in the upper atmosphere that reduces the amount of sunlight that reaches Earth's surface. Aerosols released by major volcanic eruptions can exert short-term cooling effects on Earth's climate over periods of up to several years (• p. 480).

## Radiative forcing expresses change in energy input over time

Scientists have made quantitative estimates of the degree of influence that aerosols, greenhouse gases, and other factors exert over Earth's energy balance (**Figure 18.4**). The amount of change in energy that a given factor causes is called its **radiative forcing**. Positive forcing warms the surface, whereas negative forcing cools it. Scientists' best estimate is that Earth today compared with the pre-industrial Earth of 1750 is experiencing overall radiative forcing of about 1.6 watts/m$^2$. For context, look back at Figure 18.1 and note that Earth is estimated to receive and give off 342 watts/m$^2$ of energy. Although 1.6 may seem like a small proportion of 342, it is enough to alter climate significantly.

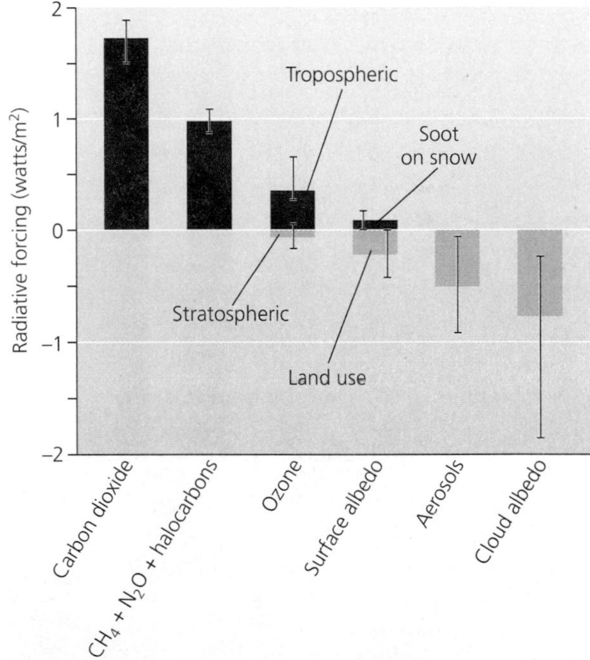


FIGURE 18.4 For each emitted gas or other human impact on the atmosphere since the industrial revolution, we can estimate the warming or cooling effect this has had on Earth's climate. We express this as *radiative forcing*, which in this graph is shown as the amount of influence on climate today relative to 1750, in watts per square meter. Red bars indicate positive forcing (warming), and blue bars indicate negative forcing (cooling). *Albedo* (• p. 521) refers to the reflectivity of a surface. A number of more minor influences are not shown. In total, scientists estimate that human impacts on the atmosphere exert a cumulative radiative forcing of 1.6 watts/m$^2$. Data from Intergovernmental Panel on Climate Change. 2007. *Fourth assessment report.*

## The atmosphere is not the only factor that influences climate

Our climate is influenced by factors other than atmospheric composition. Among these are cyclic changes in Earth's rotation and orbit, variation in energy released by the sun, absorption of carbon dioxide by the oceans, and oceanic circulation patterns.

**Milankovitch cycles** During the 1920s, Serbian mathematician Milutin Milankovitch described three types of periodic changes in Earth's rotation and orbit around the sun. These variations, now known as **Milankovitch cycles**, alter the way solar radiation is distributed over Earth's surface (**Figure 18.5**). By modifying patterns of atmospheric heating, these cycles trigger long-term climate variation, such as periodic episodes of glaciation.

**Solar output** The sun varies in the amount of radiation it emits, over both short and long time scales. For example, at each peak of its 11–year sunspot cycle the sun may emit solar flares, bursts of energy strong enough to disrupt satellite communications. However, scientists are concluding that the variation in solar energy reaching our planet in recent centuries has simply not been great enough to drive significant temperature change on Earth's surface. Estimates place the radiative forcing of natural changes in solar output at only about 0.12 watts/m$^2$—less than any of the anthropogenic causes shown in Figure 18.4.

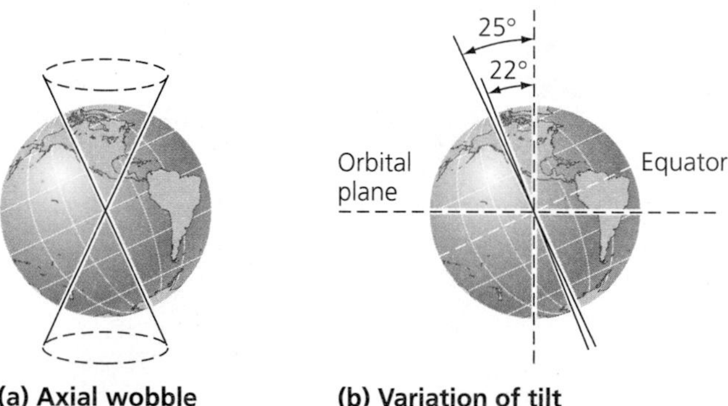


(a) Axial wobble (b) Variation of tilt

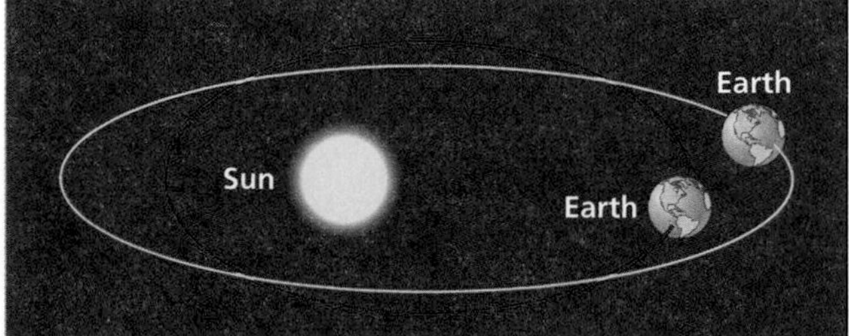


(c) Variation of orbit

FIGURE 18.5 There are three types of Milankovitch cycles. The first is an axial wobble (**a**) that occurs on a 19,000- to 23,000-year cycle. The second is a 3-degree shift in the tilt of Earth's axis (**b**) that occurs on a 41,000-year cycle. The third is a variation in Earth's orbit from almost circular to more elliptical (**c**), which repeats itself every 100,000 years. These variations affect the intensity of solar radiation that reaches portions of Earth at different times, contributing to long-term changes in global climate.

**Ocean absorption** The oceans hold 50 times more carbon than the atmosphere holds, and they absorb carbon dioxide from the atmosphere, both through direct solubility of gas in water, and through uptake by marine phytoplankton for photosynthesis. However, the oceans absorb $CO_2$ more slowly than we are adding $CO_2$ to the atmosphere (see Figure 7.12, • p. 188). Thus, carbon absorption by the oceans is slowing global warming but not preventing it. Moreover, recent evidence indicates that this absorption is now decreasing. As ocean water warms, it absorbs less $CO_2$ because gases are less soluble in warmer water—a positive feedback effect that accelerates warming.

**Ocean circulation** Ocean water exchanges tremendous amounts of heat with the atmosphere, and ocean currents move energy from place to place. In equatorial regions, such as the area around the Maldives, the oceans receive more heat from the sun and atmosphere than they emit. Near the poles, the oceans emit more heat than they receive. Because cooler water is denser than warmer water, the cooling water at the poles tends to sink, and the warmer surface water from the equator moves to take its place. This is one principle underlying global ocean circulation patterns (• pp. 446–447).

One interaction between ocean and atmosphere that influences climate is the **El Niño–Southern Oscillation (ENSO)**, a systematic shift in atmospheric pressure, sea surface temperature, and ocean circulation in the tropical Pacific Ocean. Under normal conditions, prevailing winds blow from east to west along the equator, from a region of high pressure in the eastern Pacific to one of low pressure in the western Pacific, forming a large-scale convective loop, or atmospheric circulation pattern (**Figure 18.6a**). The winds push surface waters westward, causing water to "pile up" in the western Pacific. As a result, water near Indonesia can be 50 cm (20 in.) higher and 8 °C warmer than water near South America. The westward-moving surface waters allow cold water to rise up from the deep in a nutrient-rich upwelling (• p. 447) along the coast of Peru and Ecuador.

**El Niño** conditions are triggered when air pressure increases in the western Pacific and decreases in the eastern Pacific, causing the equatorial winds to weaken. Without these winds, the warm water that collected in the

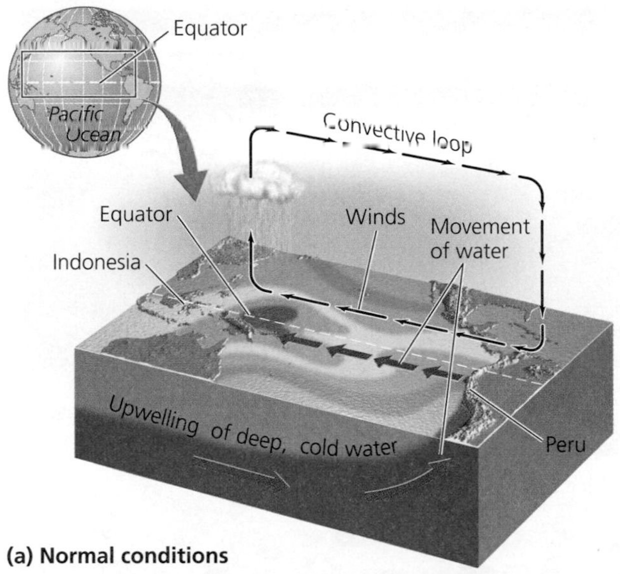


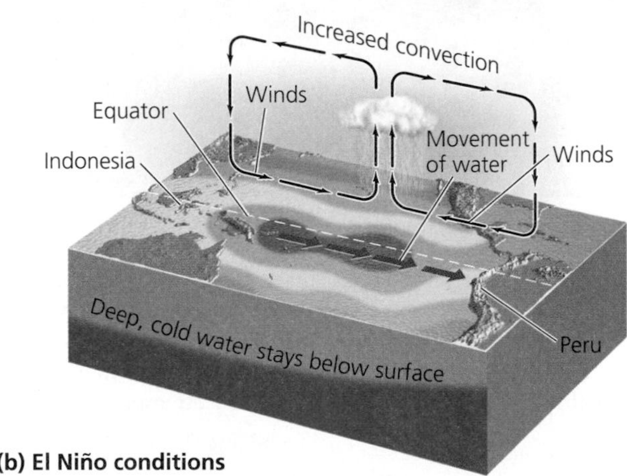


FIGURE 18.6 In these diagrams, red and orange colors denote warmer water, and blue and green colors denote colder water. Under normal conditions (**a**), prevailing winds push warm surface waters toward the western Pacific. Under El Niño conditions (**b**), the winds weaken and the warm water flows back across the Pacific toward South America, like water sloshing in a bathtub. This shuts down upwelling along the American coast and alters precipitation patterns regionally and globally. Adapted from National Oceanic and Atmospheric Administration, Tropical Atmospheric Ocean Project.

western Pacific flows eastward (**Figure 18.6b**), suppressing upwellings along the Pacific coast of South, Central, and North America, and shutting down the delivery of nutrients that support marine life and fisheries. Coastal industries such as Peru's anchovy fisheries are devastated by each El Niño event, and the 1982–1983 El Niño caused over $8 billion in economic losses worldwide. El Niño events alter weather patterns around the world, creating rainstorms and floods in areas that are generally dry (such as southern California), and causing drought and fire in regions that are typically moist (such as Indonesia).

**La Niña** events are the opposite of El Nino events; under these conditions, cold surface waters extend far westward in the equatorial Pacific, and weather patterns are affected in opposite ways.

ENSO cycles are periodic but irregular, occurring every 2–8 years. Scientists are exploring whether globally warming air and sea temperatures may be increasing the frequency and strength of these cycles.

Ocean currents and climate also interact through the **thermohaline circulation,** a worldwide current system in which warmer, fresher water moves along the surface and colder, saltier water (which is more dense) moves deep beneath the surface (**Figure 18.7**). In the Atlantic Ocean, warm surface water flows northward from the equator in the Gulf Stream (• pp. 446–447), carrying heat to high latitudes and keeping Europe warmer than it would otherwise be. As the surface water of this conveyor belt system releases heat energy and cools, it becomes denser and sinks, creating the *North Atlantic Deep Water (NADW)*.

Recently, scientists have hypothesized that interrupting the thermohaline circulation could trigger rapid climate change. If global warming causes much of Greenland's ice sheet to melt, fresh water runoff into the North Atlantic would dilute surface waters, making them less dense (because fresh water is less dense than salt water). This could potentially stop the NADW formation and shut down the northward flow of warm equatorial water, causing Europe to cool rapidly. This scenario inspired the 2004 blockbuster film *The Day After Tomorrow*, although the filmmakers chose entertainment value over science and grossly exaggerated the potential impacts. Some data suggest that the thermohaline circulation in this region is already slowing, but other researchers argue that Greenland's runoff will not be enough to cause a shutdown this century.

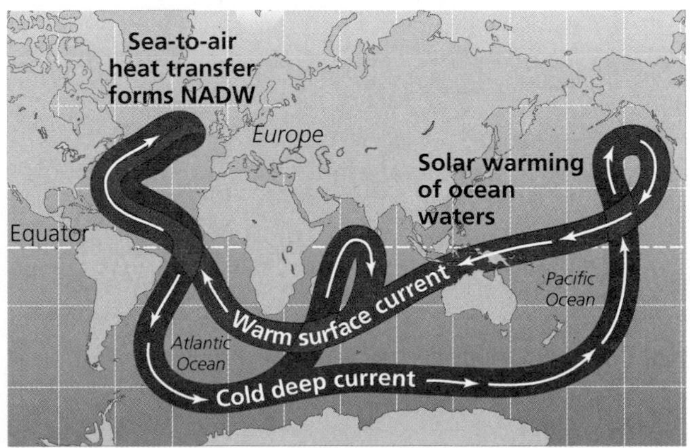


FIGURE 18.7 Warm surface currents carry heat from equatorial waters of the ocean north toward Europe and Greenland, where they release heat into the atmosphere and then cool and sink, forming the North Atlantic Deep Water. Scientists debate whether rapid melting of Greenland's ice sheet could interrupt the flow of heat from equatorial regions and cause Europe to cool dramatically.

# Studying Climate Change

To comprehend any phenomenon that is changing, we must study its past, present, and future. Climate scientists monitor present-day climatic conditions, but they also have devised clever means of inferring past change and have developed sophisticated methods to predict future change.

## Proxy indicators tell us about the past

To understand how climate is changing today, and to predict future change, scientists must learn what climatic conditions were like thousands or millions of years ago. Environmental scientists have developed a number of methods to decipher clues from the past. **Proxy indicators** are types of indirect evidence that serve as proxies, or substitutes, for direct measurement and that shed light on past climate.

For instance, Earth's ice caps, ice sheets, and glaciers hold clues to climate history. Over the ages, these huge expanses of snow and ice have accumulated to great depths, preserving within them tiny bubbles of the ancient atmosphere (**Figure 18.8**). Scientists can examine these trapped air bubbles by drilling into the ice and extracting long columns, or cores. From these ice cores, scientists can determine atmospheric composition, greenhouse gas concentrations, temperature trends, snowfall, solar activity, and even (from trapped soot particles) frequency of forest fires.

**(a) Ice core**

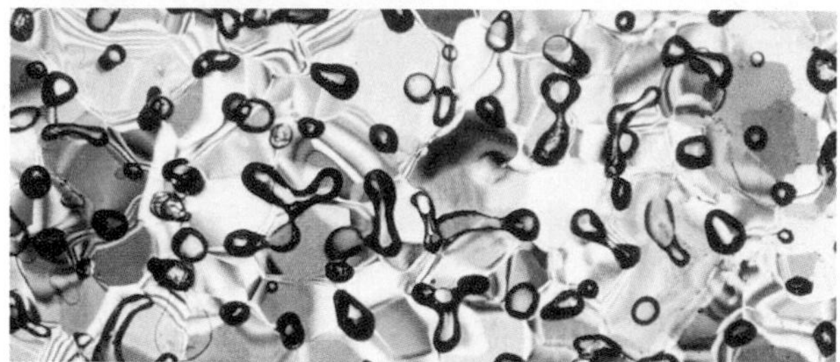

**(b) Micrograph of ice core**

FIGURE 18.8 In Greenland and Antarctica, scientists have drilled deep into ancient ice sheets and removed cores of ice like this one (**a**), held by Dr. Gerald Holdsworth of the University of Calgary, to extract information about past climates. Bubbles (black shapes) trapped in the ice (**b**) contain small samples of the ancient atmosphere.

By extracting ice cores from Antarctica, scientists have now been able to go back in time 740,000 years, reading Earth's history across eight glacial cycles (see "The Science behind the Story, • pp. 514–515.) During 2007–2009, such research is being funded and promoted as part of the *International Polar Year*, a large international scientific program coordinating research in the Arctic and Antarctic.

Researchers also drill cores into beds of sediment beneath bodies of water. Sediments often preserve pollen grains and other remnants from plants that grew in the past, and as we saw with the study of Easter Island (• pp. 8–9), analyzing these materials can illuminate the history of past vegetation. Because climate influences the types of plants that grow in an area, knowing what plants occurred in a location at a given time can tell us much about the climate at that place and time.

Tree rings provide another proxy indicator. The width of each ring of a tree trunk cut in cross-section reveals how much the tree grew in a particular growing season; a wide ring means more growth, generally indicating a wetter year. Long-lived trees such as bristlecone pines can provide records of precipitation and drought going back hundreds or thousands of years. Tree rings are also used to study fire history, since a charred ring indicates that a fire took place in that year. In arid regions such as the U.S. Southwest, packrat middens are a valuable source of climate data. Packrats are rodents that carry seeds and plant parts back to their middens, or dens, in caves and rock crevices sheltered from rain. In an arid-enough location, plant parts may preserve for centuries, allowing researchers to study the past flora of the region.

Researchers gather data on past ocean conditions from coral reefs (• pp. 450–451). Living corals take in trace elements and isotope ratios (• pp. 92–93) from ocean water as they grow, and incorporate these chemical clues to ocean conditions into growth bands in the structure of the reefs they build.

Proxy indicators often tell us information about local or regional areas, so to get a global perspective scientists need to combine multiple records from various areas. Because the number of available indicators decreases the further back in time we go, estimates of global climate conditions for the recent past tend to be more reliable than those for the distant past.

## Direct atmospheric sampling tells us about the present

Studying present-day climate is more straightforward because scientists can measure atmospheric conditions directly. The late Charles Keeling of the Scripps Institution of Oceanography in La Jolla, California, documented trends in atmospheric carbon dioxide concentrations starting in 1958 (**Figure 18.9**). Keeling collected four air samples from five towers every hour from his monitoring station at the Mauna Loa Observatory in Hawaii. These data show that atmospheric $CO_2$ concentrations have increased from 315 ppm in 1958 to 383 ppm in 2007. Today Keeling's colleagues are continuing these measurements, building upon the best long-term dataset we have of direct atmospheric sampling of any greenhouse gas.

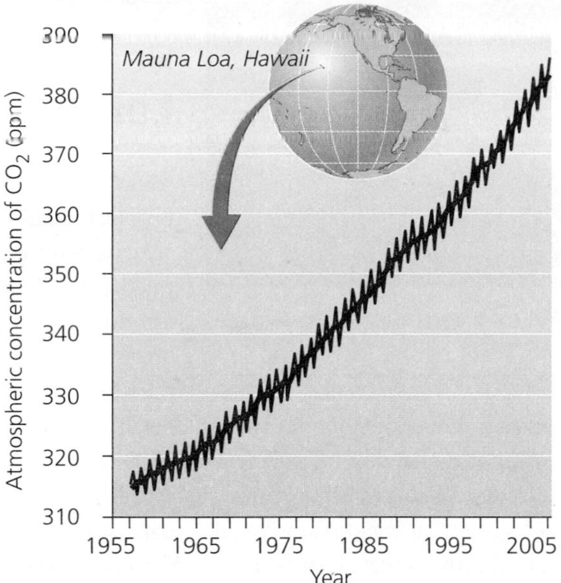


FIGURE 18.9 Atmospheric concentrations of carbon dioxide have risen steeply since 1958, when Charles Keeling began collecting these data at the Mauna Loa Observatory in Hawaii. The jaggedness of the upward trend reflects seasonal variation, which is due to the fact that the Northern Hemisphere has more land area and thus more vegetation than the Southern Hemisphere. Thus, more carbon dioxide is absorbed during the northern summer, when Northern Hemisphere plants are more photosynthetically active. Go to GRAPHIt! at www.aw-bc.com/withgott or on the student CD-ROM. Data from National Oceanic and Atmospheric Administration, Earth System Research Laboratory, Global Monitoring Division, 2007.

## Models help us understand climate

To understand how climate systems function and to predict future climate change, scientists simulate climate processes with sophisticated computer programs. **Coupled general circulation models** (often simply called **climate models**) are programs that combine what is known about atmospheric circulation, ocean circulation, atmosphere-ocean interactions, and feedback mechanisms to simulate climate processes (**Figure 18.10**). They couple, or combine, climate influences of the atmosphere and oceans in a single simulation. This requires manipulating vast

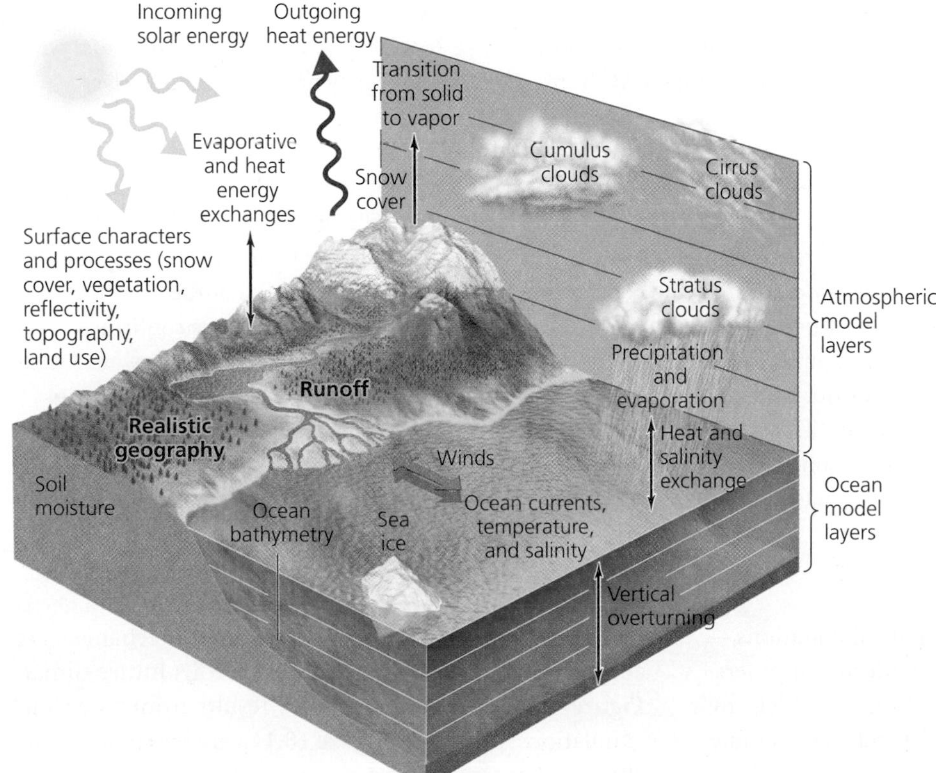


FIGURE 18.10 Modern climate models incorporate many factors, including processes involving the atmosphere, land, oceans, ice, and biosphere. Such factors are shown graphically here, but the actual models deal with them as mathematical equations in computer simulations.

THE SCIENCE BEHIND THE STORY

## Reading History in the World's Longest Ice Core

*An EPICA researcher prepares a Dome C ice core sample for analysis*

In the most frigid reaches of our planet, snow falling year after year for millennia compresses into ice and stacks up into immense sheets that scientists can mine for clues to Earth's climate history. The ice sheets of Antarctica and Greenland trap tiny air bubbles, dust particles, and other proxy indicators (•p. 512) of past conditions. By drilling boreholes and extracting ice cores, researchers can tap into these valuable archives.

Recently, researchers drilled and analyzed the deepest core ever. At a remote and pristine site in Antarctica named Dome C, they drilled down 3,270 m (10,728 ft) to bedrock and pulled out more than 800,000 years' worth of ice. The longest previous ice core (from Antarctica's Vostok station) had gone back "only" 420,000 years.

Ice near the top of these cores was laid down most recently, and ice at the bottom is oldest, so by analyzing ice at intervals along the core's length, researchers can generate a timeline of environmental change.

Dome C, a high summit of the Antarctic ice sheet, is one of the coldest spots on the planet, with an annual mean temperature of −54.5° C (−98.1° F). The Dome C ice core was drilled by the European Project for Ice Coring in Antarctica (EPICA), a consortium of researchers from 10 European nations. Antarctic operations are expensive and logistically complicated, ice-drilling requires powerful technology, and the analysis requires a diverse assemblage of experts. When the team published a research paper detailing its data in the journal *Nature* in 2004, the paper had 56 authors.

That landmark paper reported data across 740,000 years. The researchers obtained data on surface air temperature by measuring the ratio of deuterium isotopes (•pp. 91, 93) to normal hydrogen in the ice, because this ratio is temperature-dependent (see top panel of figure). And by examining the density of dust particles, they could tell when arid and/or windy climates sent more dust aloft.

In 2005, two followup papers in the journal *Science* reported analyses of greenhouse gas concentrations from the EPICA ice core. By analyzing air bubbles trapped in the ice, the researchers quantified atmospheric concentrations of carbon dioxide, methane, and nitrous oxide from across 650,000 years.

One finding from the EPICA studies was expected—yet important. The researchers documented that temperature swings in the past were tightly correlated with concentrations of carbon dioxide (see middle panel of figure), as well as methane and nitrous oxide. This bolstered the scientific consensus that greenhouse gas emissions are causing Earth to warm today.

Also clear and expected from the data was that temperature varied with swings in solar radiation due to Milankovitch cycles (• p. 510). The complex interplay of these cycles produced periodic temperature fluctuations on Earth resulting in periods of *glaciation* (when temperate regions of the planet were covered in ice) as well as warm *interglacial* periods. The Dome C ice core spanned eight glacial cycles.

In addition, the EPICA data demonstrate that by increasing greenhouse gas concentrations since the industrial revolution, we have brought them well above the highest levels they reached naturally across 650,000 years. Today's carbon dioxide concentration (383 ppm in 2007) is too recent to show up in the ice core, but is far above previous maximum values (of ~300 ppm) shown in the middle panel of the figure. Present-day concentrations of methane and nitrous oxide are likewise the highest in 650,000 years. These data show that we as a society have brought ourselves deep into uncharted territory.

Other findings from the ice core are not easily explained. Intriguingly, the earlier glacial cycles are of a different character than the more recent cycles (see figure). For the most recent cycles, the Dome C core confirmed what the Vostok data had shown: Glacial periods were long, whereas interglacial periods were brief, with a rapid rise and fall of temperature. Interglacials thus appear on a graph of

amounts of data and complex mathematical equations—a task not possible until the advent of modern computers.

Researchers test the efficacy of climate models by entering past climate data. If a model produces accurate reconstructions of past and current climate, then we have reason to believe that it simulates climate mechanisms realistically and that it may accurately predict future climate. **Figure 18.11** shows temperature results from three such simulations. Results in **Figure 18.11a** are based on natural climate-changing factors alone (such as volcanic activity

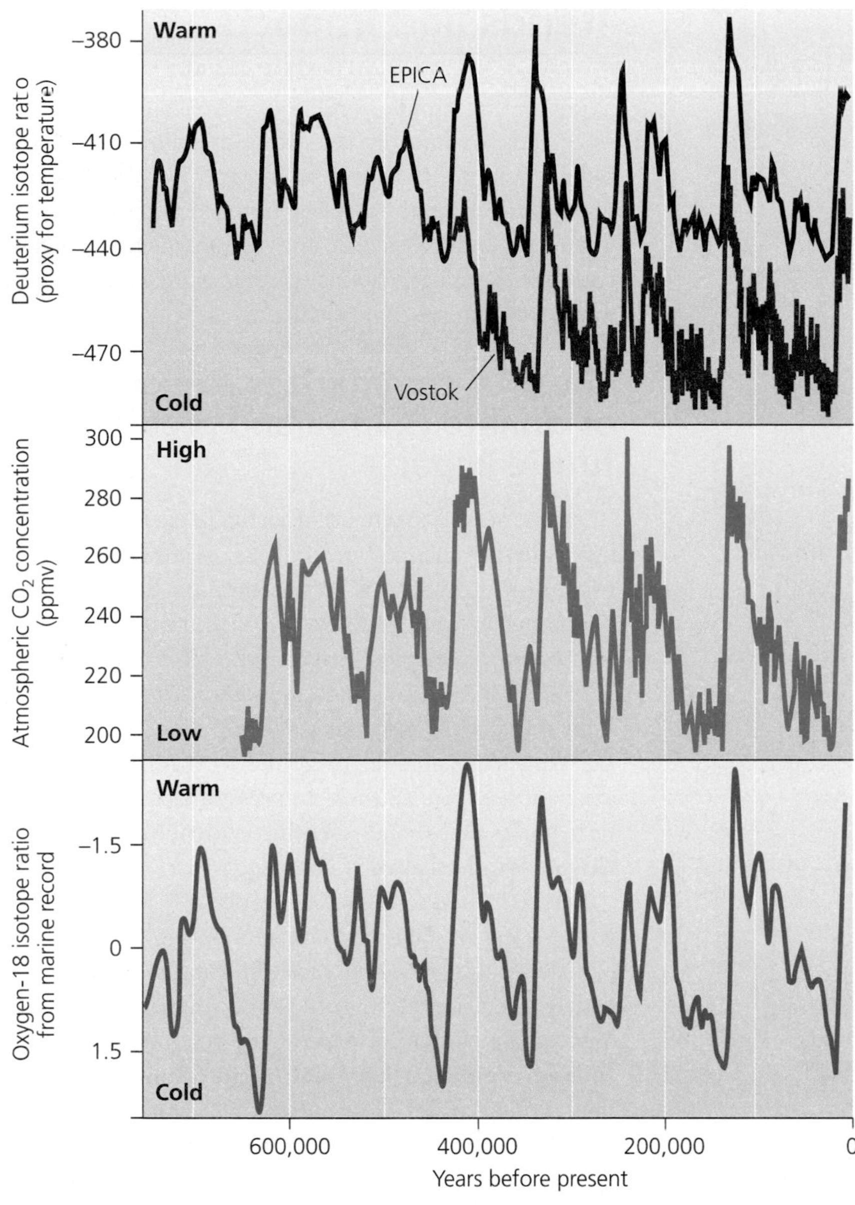


Deuterium isotope data from the EPICA ice core reveal changes in surface temperature across 740,000 years (top panel). High peaks indicate warm interglacial periods, and low troughs indicate cold glacial periods. Data from the previous Vostok ice core are shown for comparison. Atmospheric carbon dioxide concentrations (middle panel) from the EPICA ice core rise and fall in tight correlation with temperature. These datasets are consistent with oxygen-18 isotope data from the marine record (bottom panel), an independent proxy indicator for global temperature.
Adapted from EPICA community members. 2004. Eight glacial cycles from an Antarctic ice core. *Nature* 429: 623–628; and Siegenthaler, U. 2005. Stable carbon cycle-climate relationship during the late Pleistocene. *Science* 310: 1313–1317.

temperature through time as tall thin spikes. However, older glacial cycles revealed by the Dome C core look different: The glacial and interglacial periods were of more equal duration, and the warm extremes of interglacials were not as great.

This change in the nature of glacial cycles through time had been noted before by researchers working with oxygen isotope data from the fossils of marine organisms (bottom panel of figure). But why glacial cycles should be so different before and after the 450,000-year mark, no one knows.

Today polar scientists are searching for a site that might provide an ice core stretching back more than 1 million years. For at that time, data from marine isotopes tell us that glacial cycles switched from a periodicity of roughly 41,000 years (conforming to the influence of planetary tilt), to intervals of about 100,000 years (more similar to orbital changes). An ice core that captures cycles on both sides of the 1-million-year divide might help clarify the influence of Milankovitch cycles, or perhaps offer other explanations.

The intriguing patterns revealed by the Dome C ice core show that we still have plenty to learn about our complex climate history. However, the clear relationship between greenhouse gases and temperature evident in the EPICA data confirm that we would do well to address our society's greenhouse emissions.

and variation in solar energy). Results in **Figure 18.11b** are based on anthropogenic factors only (such as human emissions of greenhouse gases and sulfate aerosols). Results in **Figure 18.11c** are based on natural and anthropogenic factors combined, and this produces the closest match between predictions and actual climate.

Results such as those in Figure 18.11c support the notion that both natural and human factors contribute

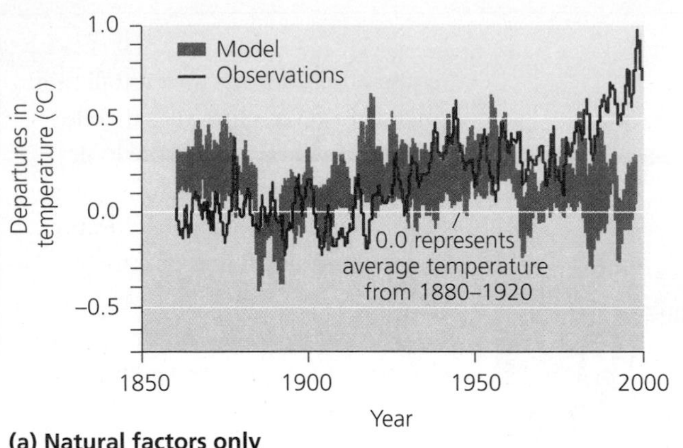


(a) Natural factors only

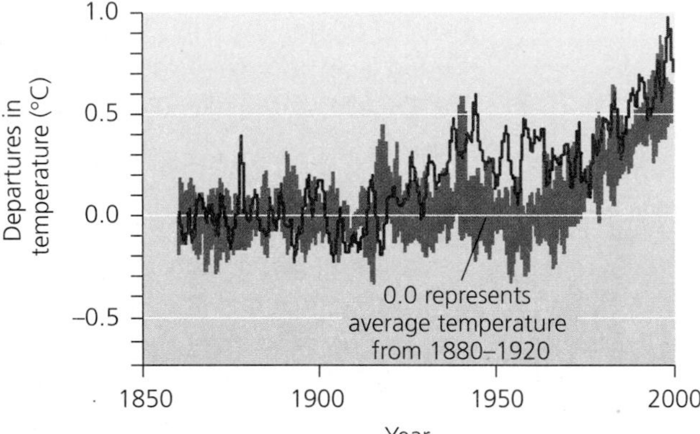


(b) Anthropogenic factors only

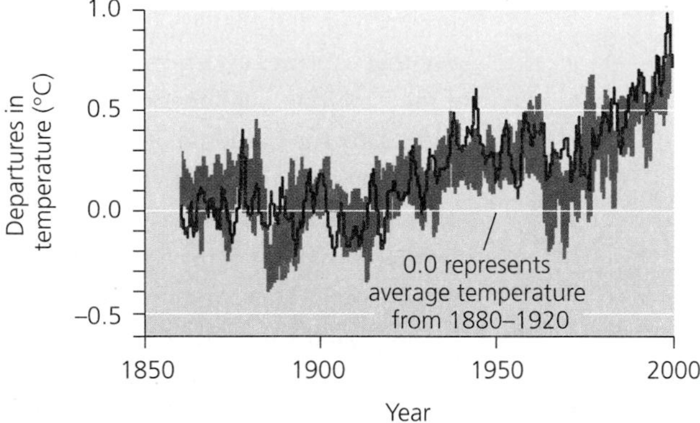


(c) All factors

FIGURE 18.11 Scientists test climate models by entering climate data from past years and comparing model predictions (blue areas) with actual observed data (red lines). Models that incorporate only natural factors (**a**) or only human-induced factors (**b**) do not predict real climate trends as well as models (**c**) that incorporate both natural and anthropogenic factors. Data from the Intergovernmental Panel on Climate Change. 2001. *Third assessment report.*

to climate dynamics, and they also indicate that global climate models can produce reliable predictions. As computing power increases and as we glean more and better data from proxy indicators, these models become increasingly reliable. They also are improving in resolution and are beginning to predict climate change region by region for various areas of the world.

# Current and Future Trends and Impacts

Evidence that climate conditions have changed worldwide since industrialization is now overwhelming and indisputable. Climate change in recent years has already had numerous effects on the physical properties of our planet, on organisms and ecosystems, and on human well-being. If we continue to emit greenhouse gases into the atmosphere, the impacts of climate change will only grow more severe.

## The IPCC summarizes evidence of climate change and predicts future impacts

In recent years, it seems that virtually everyone is detecting climatic changes around us. A fisherman in the Maldives notes the seas encroaching on his home island. A rancher in west Texas suffers a multi-year drought. A homeowner in Florida finds it impossible to obtain insurance against the hurricanes and storm surges that increasingly threaten. New Yorkers, Bostonians, Chicagoans, and Los Angelenos marvel over one freakish weather event after another. Are all these impressions part of a real pattern, and is there solid scientific evidence to confirm that climate is indeed already changing?

Over the past century, and particularly in recent years, a wide variety of data sets have shown significant trends in climate conditions. The most thoroughly reviewed and widely accepted synthesis of scientific information concerning climate change is a series of reports issued by the **Intergovernmental Panel on Climate Change (IPCC)**. This international panel of scientists and government officials was established in 1988 by the United Nations Environment Programme (UNEP) and the World Meteorological Organization.

In 2007 the IPCC released its ***Fourth Assessment Report,*** which represents the consensus of scientific climate research from around the world. This report summarizes many thousands of scientific studies, and it documents observed trends in surface temperature, precipitation patterns, snow and ice cover, sea levels, storm intensity, and other factors. It also predicts future changes in these phenomena after considering a range of potential scenarios for future greenhouse gas emissions. The report addresses impacts of current and future climate change on wildlife, ecosystems, and human societies. Finally, it discusses possible strategies we might pursue in response to climate change. **Figure 18.12** summarizes a selection of the IPCC report's major observed and predicted trends and impacts.

The IPCC report is authoritative but, like all science, deals in uncertainties. Its authors have therefore taken great

| Major Trends and Impacts of Climate Change, from IPCC Fourth Assessment Report, 2007 | |
|---|---|
| **Global physical indicators** | **Regional physical indicators** |
| Earth's average surface temperature increased 0.74° C (1.33 °F) in the past 100 years, and will rise 1.8–4.0° C (3.2–7.2° F) in the 21st century. | Arctic areas warmed fastest. Future warming will be greatest in the Arctic and greater over land than over water. |
| Eleven of the years from 1995 to 2006 were among the 12 warmest on record. | Summer Arctic sea ice thinned by 7.4% per decade since 1978. |
| Atmospheric water vapor increased since at least the 1980s. | Average Northern Hemisphere temperatures of the past 50 years were the highest in at least 1,300 years.[1] |
| Oceans absorbed >80% of heat added to the climate system, and warmed to depths of at least 3,000 m (9,800 ft). | Thawing decreased area of Arctic permafrost in spring by 15% since 1900. |
| Glaciers, snow cover, ice caps, ice sheets, and sea ice will continue melting, contributing to sea-level rise. | Precipitation increased in e. North America, e. South America, n. Europe, and n. and c. Asia since 1900. |
| Sea level rose by an average of 17 cm (7 in.) in the 20th century, and will rise 18–59 cm (7–23 in.) in the 21st century. | Precipitation decreased in the Sahel, the Mediterranean, s. Africa, and parts of s. Asia since 1900. |
| Ocean water became more acidic by about 0.1 pH unit, and will decrease in pH by 0.14–0.35 units more by century's end. | Precipitation will generally increase at high latitudes and decrease at subtropical latitudes, often making wet areas wetter and dry ones drier.[1] |
| Storm surges increased, and will increase further.[1] | Heavy precipitation events increased over most land areas.[1] |
| Sea level rise will worsen coastal erosion and degrade wetlands. | Droughts became longer, more intense, and more widespread since the 1970s, especially in the tropics and subtropics.[1] |
| Carbon uptake by terrestrial ecosystems will likely peak by the mid-21st century and then weaken or reverse, amplifying climate change.[2] | Droughts and flooding will increase, leading to agricultural losses.[2] |
| **Social indicators** | Over most land areas, cold and frost days decreased[3] and will continue to decrease[4] while hot days and heat waves increased[3] and will continue to increase.[3] |
| Farmers and foresters have had to adapt to altered growing seasons and disturbance regimes. | Hurricanes intensified in the North Atlantic since 1970[1], and will continue to intensify.[1] |
| Temperate-zone crop yields will rise until temperature warms beyond 3° C (5.4° F), but in the dry tropics and subtropics, crop productivity will fall and lead to hunger.[5] | The thermohaline circulation will slow, but will not shut down and chill Europe in the 21st century.[3] |
| Timber production may rise slightly in the near-term, but will vary by region.[5] | Antarctica will continue accumulating snow, but may also continue losing ice around its edges. |
| Impacts on biodiversity will cause losses of food, water, and other ecosystem goods and services.[2] | **Biological indicators** |
| Sea-level rise will displace people from islands and coastal regions.[3] | Species ranges are shifting toward the poles and upward in elevation, and will continue to shift. |
| Melting of mountain glaciers will reduce water supplies to millions of people.[2] | The timing of seasonal phenomena (such as migration and breeding) is shifting, and will continue to shift. |
| Economic costs will outweigh benefits as climate change worsens;[2] costs could average 1–5% of GDP globally for 4° C (7.2° F) of warming. | About 20–30% of species studed so far will face extinction risk if temperature rises more than 1.5–2.5° C (2.7–4.5° F).[5] |
| Poorer nations and communities suffer more from climate change, because they rely more on climate-sensitive resources and have less capacity to adapt.[2] | Species interactions and ecosystem structure and function could change greatly, resulting in biodiversity loss. |
| Human health will suffer as increased warm-weather health hazards outweigh decreased cold-weather health hazards.[2] | Corals will experience further mortality from bleaching and ocean acidification.[5,4] |

**FIGURE 18.12** Climate change has had numerous consequences already and is predicted to have many more. Listed here are some of the main observed and predicted trends and impacts described in the Intergovernmental Panel on Climate Change's *Fourth Assessment Report*. For simplicity, this table expresses mean estimates only; the IPCC report provides ranges of estimates as well. [1]Certainty level = 66–90% probability of being correct. [2]Certainty level = ~80% probability of being correct. [3]Certainty level = 90–99% probability of being correct. [4]Certainty level = >99% probability of being correct. [5]Certainty level = ~50% probability of being correct. Data from the Intergovernmental Panel on Climate Change (IPCC). 2007. *Fourth assessment report*.

care to assign statistical probabilities to its conclusions and predictions. In addition, its estimates regarding impacts of change on human societies are conservative, because its scientific conclusions had to be approved by representatives of the world's national governments, some of which are reluctant to move away from a fossil-fuel-based economy.

## Temperature increases will continue

The IPCC report concludes that average surface temperatures on Earth increased by an estimated 0.74° C (1.33° F) in the century from 1906 to 2005 (**Figure 18.13**), with most of this increase occurring in the last few decades. Eleven of the years from 1995 to 2006 were among the 12 warmest on record since global measurements began 150 years earlier. The numbers of extremely hot days and heat waves have increased, whereas the number of cold days has decreased.

Temperature changes are greatest in the Arctic (**Figure 18.14**). Here, ice sheets are melting, sea ice is thinning, storms are increasing, and altered conditions are posing challenges for people and wildlife. As sea ice melts earlier, freezes later, and recedes from shore, it becomes harder for Inuit people and for polar bears alike to hunt the seals they each rely on for food. Thin sea ice is dangerous for people to travel and hunt upon, and in recent years, polar bears have been dying of exhaustion and starvation as they try to swim long distances between ice floes. *Permafrost* (permanently frozen ground) is thawing in the Arctic, destabilizing countless buildings. The strong Arctic warming is contributing to sea-level rise by melting ice caps and ice sheets.

### Weighing THE Issues | Climate Change and Human Rights

In December 2005, a group representing North America's Inuit people sent a legal petition to the Inter-American Commission on Human Rights, demanding that the United States restrict its greenhouse gas emissions, which the Inuit maintained were destroying their way of life in the Arctic. After a year, the Commission dismissed the petition with a terse three-sentence letter. Do you think Arctic-living people deserve some sort of compensation from industrialized nations whose emissions have caused climate change that has disproportionately affected the Arctic? Do you think climate change can be viewed as a human rights issue? What ethical issues, if any, do you think climate change presents? How could these best be resolved?

In the future, we can expect average surface temperatures on Earth to rise roughly 0.2° C (0.4° F) per decade for the next 20 years, according to IPCC analysis. If we were to cease greenhouse gas emissions today, temperatures would still rise 0.1° C (0.2° F) per decade because of the time lag from gases already in the atmosphere that have yet to exert their full influence. At the end of the 21st century, the IPCC predicts global temperatures will be 1.8–4.0° C (3.2–7.2° F) higher than today's, depending upon the emission scenario. Unusually hot days and

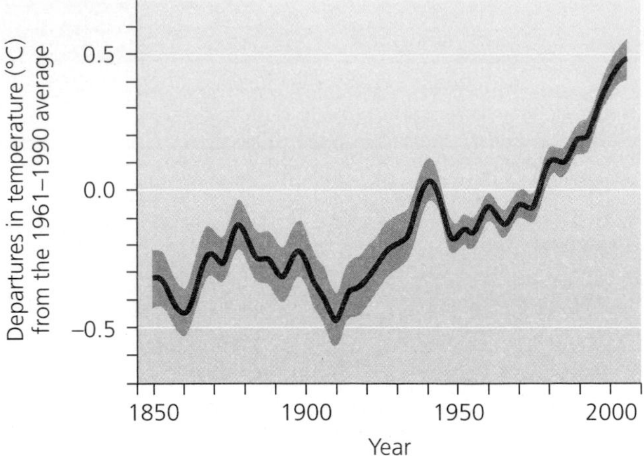


**(a) Global temperature measured since 1850**

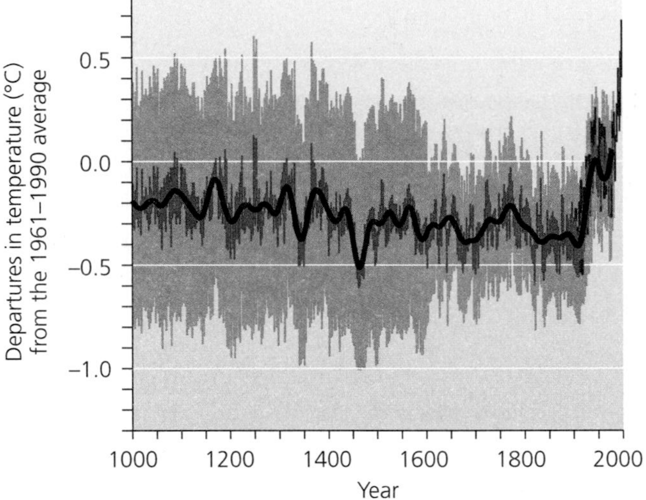


**(b) Northern Hemisphere temperature over the past 1,000 years**

FIGURE 18.13 Data from thermometers (**a**) show changes in Earth's average surface temperature since 1850. Gray shaded area indicates range of uncertainty. In (**b**), proxy indicators (blue line) and thermometer data (red line) together show average temperature changes in the Northern Hemisphere over the past 1,000 years. The gray-shaded zone represents the 95% confidence range. This record shows that 20th-century warming has eclipsed the magnitude of change during both the "Medieval Warm Period" (10th–14th centuries) and the "Little Ice Age" (15th–19th centuries). Go to GRAPHit! at www.aw-bc.com/withgott or on the student CD-ROM. Data from the Intergovernmental Panel on Climate Change, 2007. *Fourth assessment report* (a); and IPCC 2001. *Third assessment report* (b).

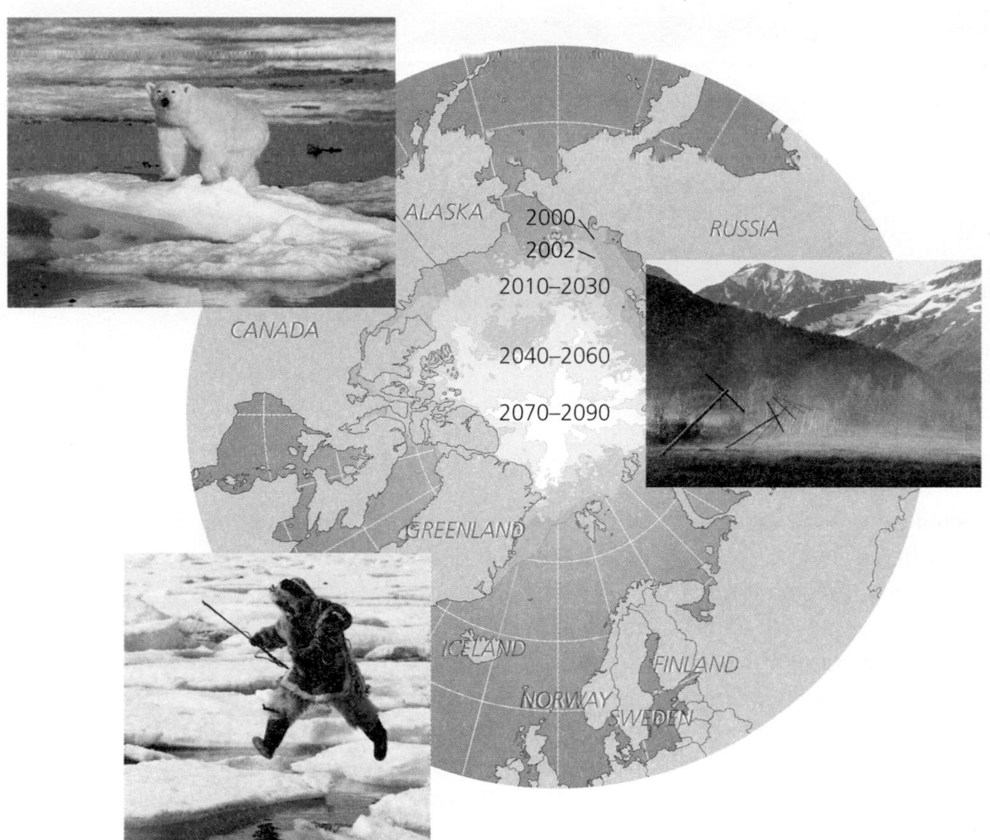


FIGURE 18.14 The Arctic has borne the brunt of climate change's impacts so far. As Arctic sea ice melts, it recedes from large areas, as shown by the map indicating the mean minimum summertime extent of sea ice for the recent past, present, and future. Inuit people find it difficult to hunt and travel in their traditional ways, and polar bears starve because they are less able to hunt seals. As permafrost thaws beneath them, human-made structures are damaged: phone poles topple and buildings can lean, buckle, crack, and fall. Map data from National Center for Atmospheric Research and National Snow and Ice Data Center.

heat waves will become more frequent. Temperature change is predicted to vary from region to region in ways that parallel regional differences already apparent (**Figure 18.15**).

Sea surface temperatures are also increasing as the oceans absorb heat added to the climate system. The record number of hurricanes and tropical storms in 2005—Katrina and 27 others—left many people wondering whether global warming was to blame. Are warmer ocean temperatures spawning more hurricanes, or hurricanes that are more powerful or long-lasting? Scientists are not yet sure, but recent analyses of storm data suggest that warmer seas may not be increasing the number of storms, but likely are increasing the power of storms, and possibly their duration.

## Changes in precipitation vary by region

A warmer atmosphere holds more water vapor, but changes in precipitation patterns have been complex, with some regions of the world receiving more precipitation than usual and others receiving less. In regions from the African Sahel to the southwestern United States, droughts have become more frequent and severe, harming agriculture, promoting soil erosion, reducing drinking water supplies, and encouraging forest fires. Meanwhile, in both dry and humid regions, heavy rain events have increased, contributing to flooding, such as the 2002 flood in the Czech Republic, which killed 16 people, left 30,000 homeless, and inflicted $2.8 billion in damage.

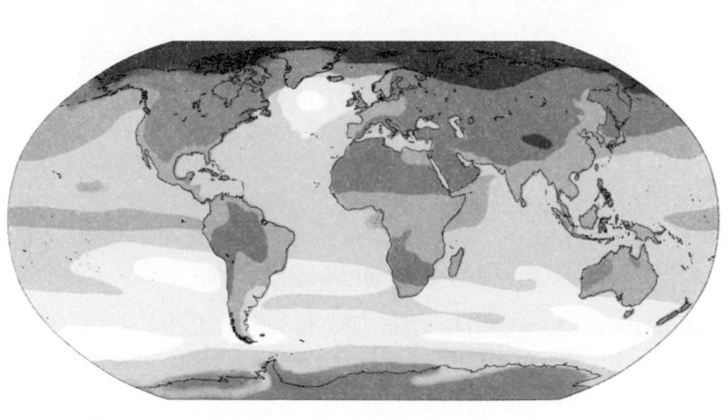


FIGURE 18.15 This map shows projected increases in surface temperature for the decade 2090–2099, relative to temperatures in 1980–1999. Land masses will warm more than oceans, and the Arctic will warm the most, up to 7.5° C (13.5° F). The IPCC uses multiple climate models when predicting regional variation in how temperature will change, and this map was generated using an emission scenario that is intermediate in its assumptions, involving an average global temperature rise of 2.8° C (5.0° F) by 2100. Data from Intergovernmental Panel on Climate Change. 2007. *Fourth assessment report.*

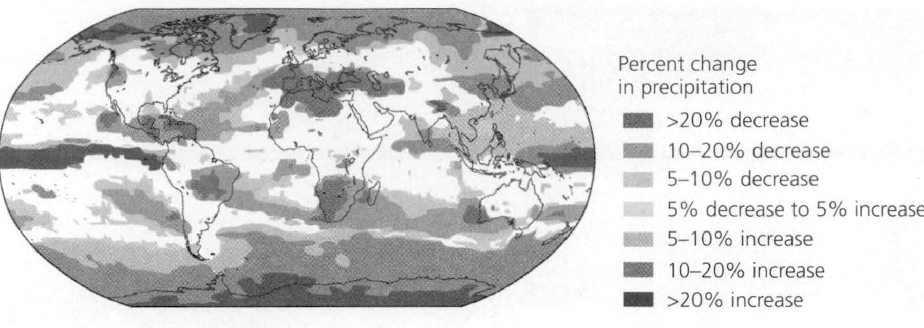


FIGURE 18.16 This map shows projected changes in June–August precipitation for the decade 2090-2099, relative to precipitation levels in 1980–1999. Browner shades indicate less precipitation, and bluer shades indicate more precipitation. White indicates areas for which models could not agree. This map was generated using an emission scenario that is intermediate in its assumptions, involving an average global temperature rise of 2.8° C (5.0° F) by 2100. Data from Intergovernmental Panel on Climate Change. 2007. *Fourth assessment report.*

Precipitation changes are predicted to vary among regions in ways that parallel regional differences seen over the past century (**Figure 18.16**). In general, precipitation will increase at high latitudes and decrease at low and middle latitudes, magnifying differences in rainfall that already exist and worsening water shortages in many developing countries of the arid subtropics. In many areas, heavy precipitation events will become more frequent, increasing the risk of flooding.

## Melting ice and snow have far-reaching effects

As the world warms, mountaintop glaciers are disappearing (**Figure 18.17**). Since 1980, the World Glacier Monitoring Service estimates, major glaciers have lost an average of 9.6 m (31.5 ft) in vertical thickness. Many glaciers on tropical mountaintops have disappeared already. In Glacier National Park in Montana, only 27 of 150 glaciers present at the park's inception remain, and scientists estimate that by 2030 even these will be gone.

Mountains accumulate snow in the winter and release meltwater gradually during the summer. Throughout high-elevation areas of the world, however, warming temperatures will continue to melt mountain glaciers, posing risks of sudden floods as ice dams burst, and reducing summertime water supplies to millions of people. Over one-sixth of the world's people live in regions supplied by mountain meltwater, and some of these people are already beginning to face water shortages. If this water vanishes during drier months, whole communities will be forced to look elsewhere for water, or to move.

Warming temperatures are also reducing snow cover at high latitudes and melting the immense ice sheets of the Arctic. Recent research reveals that melting of the Greenland ice sheet is accelerating (see "The Science behind the Story," • pp. 522–523). At the other end of the world, in Antarctica, ice shelves the size of Rhode Island

(a) Grinnell Glacier in 1938

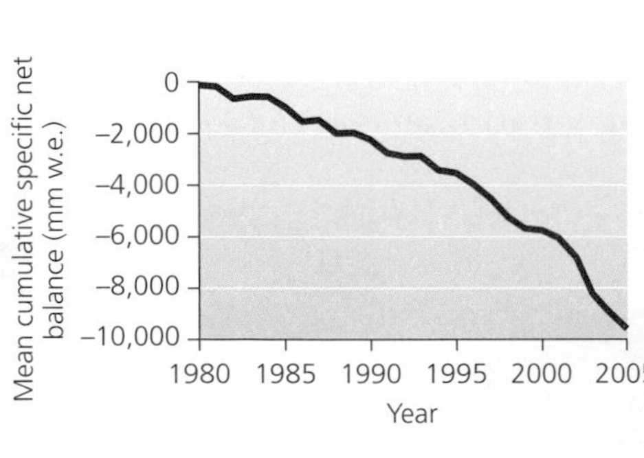

(b) Grinnell Glacier in 2005

FIGURE 18.17 Glaciers are melting rapidly around the world as global warming proceeds. The Grinnell Glacier in Glacier National Park, Montana, retreated substantially between 1938 (**a**) and 2005 (**b**). The graph shows declines in mass in 30 major glaciers monitored since 1980 by the World Glacier Monitoring Service. Data from World Glacier Monitoring Service.

have disintegrated as a result of contact with warmer ocean water, but increased precipitation has supplied the continent with enough extra snow to compensate for the loss of ice around its edges.

One reason warming is accelerating in the Arctic is that as snow and ice melt, darker, less-reflective surfaces are exposed, and Earth's *albedo*, or capacity to reflect light, decreases. As a result, more of the sun's rays are absorbed at the surface, fewer reflect back into space, and the surface warms. In a process of positive feedback, this warming causes more ice and snow to melt, which in turn causes more absorption of radiation and more warming.

Near the poles, snow cover, permafrost, and ice sheets are projected to decrease, and sea ice will continue to shrink in both the Arctic and Antarctic. Some emission scenarios show Arctic sea ice disappearing completely by the late 21st century, creating new shipping lanes for commerce (and likely a rush to exploit underwater oil and mineral reserves that may exist in Arctic waters).

## Rising sea levels will affect hundreds of millions of people

As glaciers and ice sheets melt, increased runoff into the oceans causes sea levels to rise. Sea levels also are rising because ocean water is warming, and water expands in volume as its temperature increases. In fact, recent sea-level rise has resulted primarily from the thermal expansion of seawater. Worldwide, average sea levels rose an estimated 17 cm (6.7 in.) during the 20th century (**Figure 18.18**). Seas rose by an estimated 1.8 mm/year from 1961–2003 and 3.1 mm/year from 1993–2006. Note that all these numbers represent vertical rises in water level, and on most coastlines a vertical rise of a few inches means many feet of horizontal incursion inland.

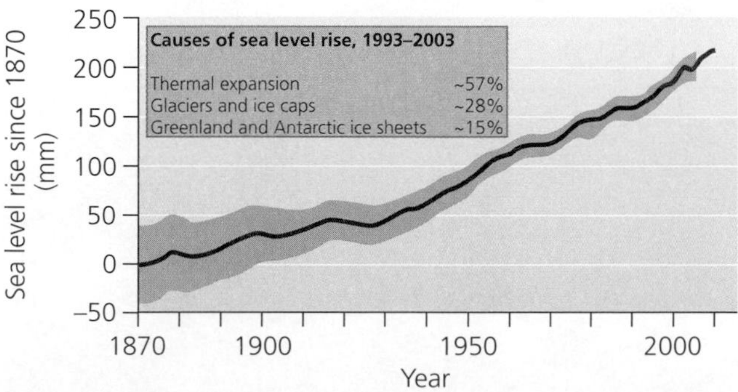


**FIGURE 18.18** Data from tide gauges (black line) and satellite observations (red line) show that global average sea level has risen about 200 mm (7.9 in.) since 1870. Gray shaded area indicates range of uncertainty. Thermal expansion of water accounts for most sea-level rise. Data from Intergovernmental Panel on Climate Change. 2007. *Fourth assessment report.*

Higher sea levels lead to beach erosion, coastal flooding, intrusion of saltwater into aquifers, and other impacts. In 1987, unusually high waves struck the Maldives and triggered a campaign to build a large seawall around Male, the nation's capital. Known as "The Great Wall of Male," the seawall is intended to protect buildings and roads by dissipating the energy of incoming waves during storm surges. A *storm surge* is a temporary and localized rise in sea level brought on by the high tides and winds associated with storms. The higher that sea level is to begin with, the further inland a destructive storm surge can reach. "With a mere 1-meter rise [in sea level]," Maldives' President Maumoon Abdul Gayoom warned in 2001, "a storm surge would be catastrophic, and possibly fatal to the nation."

The Maldives is not the only nation with such concerns. In fact, among island nations, the Maldives has fared better than many others. It saw sea level rise about 2.5 mm per year throughout the 1990s, but most Pacific islands are experiencing greater changes. Regions experience differing amounts of sea level change because land elevations may be rising or subsiding naturally, depending on local geological conditions.

In the United States, 53% of the population lives in coastal areas. Vulnerability to storm surges became tragically apparent in 2005 when Hurricane Katrina struck New Orleans and the Gulf Coast, followed shortly thereafter by Hurricane Rita (**Figure 18.19**).

The levees surrounding New Orleans that Katrina's floodwaters breached are now repaired, but large portions of the city will always remain below sea level. Areas that are now 2.1 m (7 ft) below sea level may be as much as 3.3 m (10 ft) below sea level by 2100 as the land subsides and sea level rises. Outside the city, marshes of the Mississippi River delta are being lost rapidly, as dams upriver hold back silt that used to maintain the delta, as land subsides because of petroleum extraction, and as rising seas eat away at coastal vegetation. Approximately 2.5 million ha (1 million acres) of Louisiana wetlands have become open water since 1940, and continued wetland loss will deprive New Orleans of protection against future storm surges.

At the end of the 21st century, the IPCC predicts mean sea level to be 18–59 cm (7–23 in.) higher than today's, depending upon the emission scenario. However, these estimates do not take into account findings on accelerated ice melting in Greenland, because that research (see "The Science behind the Story, • pp. 522–523) is so new that it has not yet been incorporated into climate models. If Greenland's melting continues to accelerate, then sea levels will rise more quickly.

Rising sea levels will force hundreds of millions of people to choose between moving upland or investing in costly protections against high tides and storm surges.

THE SCIENCE BEHIND THE STORY

# Timing Greenland's Glaciers as They Race to the Sea

*Outlet glaciers melting into Scoresby Sund, Greenland's largest fjord*

Scientists have known for years that the Arctic is bearing the brunt of global warming and that the massive ice sheet covering Greenland is melting around its edges. But data from 1993 to 2003 showed Greenland's ice loss accounting for only 4–12% of global sea level rise, about 0.21 mm/yr. And as authors of the IPCC's *Fourth Assessment Report* used results from climate models to predict Greenland's future contributions to sea-level rise, the models told them to expect more of the same.

However, some brand-new research hadn't made it into the models. Scientists studying how ice moves were learning that ice sheets can collapse more quickly than expected and that Greenland's ice loss is accelerating. As a result, they said, the IPCC report underestimates the likely speed and extent of future sea level rise.

Greenland's ice sheet is massive, averaging nearly a mile deep and covering as much area as Texas, California, Michigan, and Minnesota combined. If the entire ice sheet were to melt, global sea level would rise by a whopping 7 m (23 ft).

The ice sheet gains mass by accumulating snow during cold weather, which becomes packed into ice over time. It loses mass as surface ice melts in warm weather, generally at the periphery, where ice is thinnest or contacts seawater. If melting and runoff outpace accumulation, then the ice sheet shrinks.

But researchers are now learning that the internal physical dynamics of how ice moves may be more important. These dynamics can speed the flow of immense amounts of ice in *outlet glaciers* downhill toward the coast, where eroding ice sloughs off and melts into the sea.

The first good indication of this process came in 2002 when a team led by Jay Zwally of NASA's Goddard Space Flight Center in Maryland noted that ice in outlet glaciers flows more quickly during warm months, when pools of meltwater form on the surface. In a paper published in the journal *Science* in 2002, Zwally's group proposed that meltwater leaks down through crevasses and vertical tunnels called *moulins* to the bottom of the glacier. There, the water runs downhill in a layer between bedrock and ice, lubricating the bedrock surface and enabling the ice to slide downhill like a car hydroplaning on a wet road. In addition, the meltwater weakens ice on its way down and warms the base of the glacier, melting some of it to create more water (see the figure).

Other scientists proposed an additional mechanism: Warming ocean water melts ice shelves along the coast, depriving outlet glaciers of the buttressing support that holds them in place. Without a floating ice tongue at its terminus, a glacier slides into the ocean more readily.

These physical dynamics represent positive feedback, researchers said; once global warming initiates these processes, they encourage further melting. As such, they said, we should expect that Greenland's melting will accelerate.

Several research groups soon attempted to measure the rates of

**FIGURE 18.19** In a world of higher sea levels and stronger storms, rescues like this one from the floodwaters of Hurricane Katrina's storm surge in coastal Mississippi in 2005 could become more frequent.

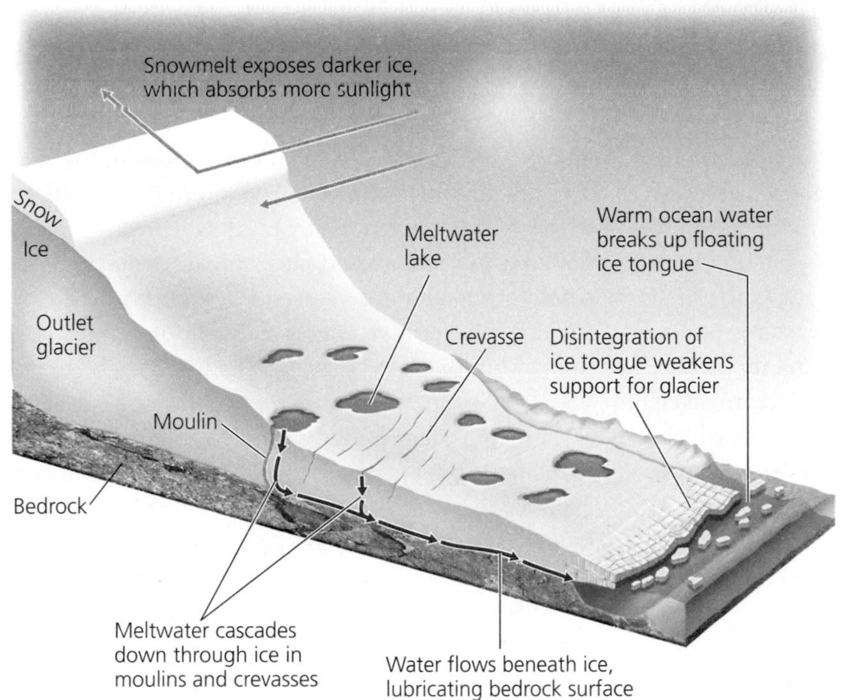


Outlet glaciers from Greenland's ice sheet are accelerating their slide into the ocean because meltwater is descending through moulins and crevasses, lubricating the bedrock surface. Moreover, melting snow exposes darker ice, which absorbs sunlight and speeds melting. Finally, breakup of floating ice tongues weakens support for the glacier.

Greenland's ice loss and account for the impact of these physical dynamics. In 2006, Eric Rignot of NASA's Jet Propulsion Laboratory in Pasadena, California, and Pannir Kanagaratnam of the University of Kansas, Lawrence, published research in *Science* showing that rates of ice loss had more than doubled between 1996 and 2005.

To measure how ice moves over such a scale and in such inhospitable terrain, Rignot and Kanagaratnam analyzed satellite data showing how fast Greenland's ice was moving, along with aircraft radar measurements giving data on ice thickness. This let them estimate how much ice was lost during this period.

The researchers found that the physical dynamics of ice movement were responsible for the loss of 56 $km^3$ of ice in 1996, 92 $km^3$ in 2000, and 167 $km^3$ in 2005. The 2005 amount is equivalent to 44 trillion gallons of water, enough to supply the New York City metro area for over a century.

They also determined that the physical dynamics of ice flow accounted for two-thirds of total ice loss. The other third was due to runoff outpacing snow accumulation, and this too was accelerating. Including these amounts showed that Greenland lost a grand total of 91 $km^3$ of ice in 1996, 138 $km^3$ in 2000, and 224 $km^3$ in 2005. This last amount exceeds all the water consumed in the United States in an entire year.

Clearly, ice losses were accelerating quickly. Rignot and Kanagaratnam calculated that Greenland's contribution to sea level rise increased from 0.23 mm/yr in 1996 to 0.57 mm/yr in 2005. Scientists will now be keeping a close eye on Greenland's ice sheet to see whether it continues to discharge water more and more quickly. If it does, climate modelers will have several years in which to incorporate into their models the new and evolving understanding of the physical dynamics of ice—just in time for the next IPCC report scheduled for 2013.

Densely populated regions on low-lying river deltas, such as Bangladesh, will be most affected. So will storm-prone regions such as Florida, coastal cities such as Houston and Charleston, and areas where land is subsiding, such as the U.S. Gulf Coast. Many Pacific islands will need to be evacuated, and some nations such as Tuvalu and the Maldives fear for their very existence. In the meantime, the Maldives is likely to suffer from shortages of fresh water because rising seas threaten to bring salt water into the nation's wells, just as the 2004 tsunami did. The contamination of groundwater and soils by seawater threatens not only island nations like the Maldives but also coastal areas such as the Tampa, Florida, region, which depend on small lenses of fresh water that float atop saline groundwater.

### Weighing THE Issues | Environmental Refugees

Citizens of the Maldives see an omen of their future in the Pacific island nation of Tuvalu, which has been losing 9 cm (3.5 in.) of elevation per decade to rising seas. Appeals from Tuvalu's 11,000 citizens were heard by New Zealand, which began accepting these environmental refugees in 2003. Do you think a national culture can survive if its entire population is relocated? Think of the tens of thousands of refugees from Hurricane Katrina. How are their lives and culture faring in the wake of that tragedy?

Maldives residents also worry about damage to the marine ecosystems that are critical for their economy, including coral reefs (• pp. 450–451). Coral reefs provide habitat for important food fish that are consumed locally and exported; offer snorkeling and scuba diving sites for tourism; and reduce wave intensity, protecting coastlines from erosion. Around the world, rising seas will eat away at the coral reefs, mangrove forests, and salt marshes that serve as barriers protecting our coasts (• pp. 450–453).

Climate change poses two additional threats to coral reefs: warmer waters are causing coral bleaching (• pp. 450–451), and enhanced $CO_2$ concentrations in the atmosphere are changing ocean chemistry. As ocean water absorbs atmospheric $CO_2$, it becomes more acidic, which impairs the growth of coral and other organisms whose exoskeletons consist of calcium carbonate. The oceans have already decreased by 0.1 pH unit (a fairly large amount; • pp. 94–96), and are predicted to decline in pH by 0.14–0.35 more units over the next 100 years.

## Climate change affects organisms and ecosystems

The many changes in Earth's physical systems have direct consequences for life on our planet. Organisms are adapted to their environments, so they are affected when those environments are altered. As global warming proceeds, it modifies temperature-dependent biological phenomena. For instance, in the spring, birds are migrating earlier, insects are hatching earlier, and animals are breeding earlier. Plants are leafing out earlier, too—an effect confirmed by satellite photography that records whole landscapes "greening up" each year.

These changes in seasonal timing are expected to continue, and they are having complex effects. For instance, European birds known as great tits time their breeding cycle so that their young hatch and grow at the time of peak caterpillar abundance. However, as plants leaf out earlier and insects emerge earlier, research shows that great tits are not breeding earlier. As a result of the mismatch in timing, fewer caterpillars are available when young birds need them, and fewer birds survive. Although some organisms will no doubt adapt to such changes in seasonal timing, research so far shows that in most cases mismatches occur.

Biologists also have recorded spatial shifts in the ranges of organisms, with plants and animals moving toward the poles or upward in elevation (i.e., toward cooler regions) as temperatures warm. As these trends continue, some organisms will not be able to cope, and as many as 20–30% of all plant and animal species could be threatened with extinction, the IPCC estimates. Trees may not be able to shift their distributions fast enough. Animals adapted to montane environments may be forced uphill until there is nowhere left to go (as we saw with Monteverde's organisms in Chapter 5). Rare species finding refuge in protected preserves may be forced out of preserves into developed areas, making such refuges far less effective tools for conservation.

These changes will greatly affect species interactions, and scientists foresee major modifications in the structure and function of communities and ecosystems. In regions where precipitation and stream flow increase, erosion and flooding will pollute and alter aquatic systems. In regions where precipitation decreases, lakes, ponds, wetlands, and streams will diminish, affecting aquatic organisms, as well as human health and well-being. Acidification of the oceans may pose major threats for corals and other marine animals. Given that corals will likely also suffer increased bleaching from thermal stress, the world's coral reefs are expected to decline substantially. This would reduce marine biodiversity significantly, because so many other organisms depend on living coral reefs for food and shelter.

Effects on plant communities comprise an important component of climate change, because by drawing in $CO_2$ for photosynthesis, plants act as sinks for carbon. If climate change increases vegetative growth, this could help mitigate carbon emissions, in a process of negative feedback. However, if climate change decreases plant growth (through drought or fire, for instance), then positive feedback could increase carbon flux to the atmosphere. The many impacts on ecological systems will reduce the ecosystem goods and services we receive from nature and that our societies depend on, from food to clean air to drinking water.

## Climate change exerts societal impacts

Human society has begun to feel impacts of climate change. Damage from drought, flooding, hurricanes, storm surges, and sea level rise, as discussed above, has already taken a toll on the lives and livelihoods of millions of people. However, climate change will have even more consequences for human populations, including impacts on agriculture, forestry, economics, and health.

**Agriculture** For farmers, earlier springs require earlier crop planting. For some crops in the temperate zones, production may increase slightly with moderate warming, because growing seasons become longer and because more carbon dioxide is available to plants for photosynthesis. However, rainfall will shift in space and time, and in areas where droughts and floods become more severe, these will cut into agricultural productivity. Overall, global crop yields are predicted to increase somewhat, but beyond a rise of 3° C (5.4° F), the IPCC expects crop yields to decline. In seasonally dry tropical and subtropical regions, growing seasons may be shortened, and harvests may be more susceptible to drought and crop failure. Thus, scientists predict that crop production will fall in these regions even with minor warming. This would worsen hunger in many of the world's developing nations.

## Weighing THE Issues | Agriculture in a Warmer World

The IPCC predicts that slight warming could shift agricultural belts toward the poles and marginally increase global crop production. Locate the nations of Russia, Canada, and Argentina on a world map, and hypothesize how such a poleward shift of agriculture might affect these nations. Now locate India, Nigeria, and Ethiopia, and hypothesize how the shift might affect them. Given that many developing nations near the equator are already suffering from food shortages and soil degradation, what effects do you think that global warming may have? If climate change magnifies inequities between developed and developing nations, what could we do to alleviate this problem?

**Forestry** Forest managers increasingly find themselves having to battle insect and disease outbreaks, invasive species, and catastrophic fires, which are mostly caused by decades of fire suppression (• p. 343) but are also promoted by longer, warmer, drier fire seasons. For timber and forest products, enriched atmospheric $CO_2$ may spur greater growth in the near term, but this will vary substantially from region to region. Other climatic effects such as drought may eliminate these gains. For instance, droughts brought about by a strong El Niño in 1997–1998 allowed immense forest fires to destroy millions of hectares of rainforest in Indonesia, Brazil, Mexico, and elsewhere.

**Health** As a result of climate change, we will face more heat waves—and heat stress can cause death (**Figure 18.20**), especially among older adults. A 1995 heat wave killed at least 465 people in Chicago, for example, and one killed 35,000 people in Europe in 2003. In addition, a warmer climate exposes us to other health problems:

- Respiratory ailments from air pollution, as hotter temperatures promote formation of photochemical smog (• pp. 486–488)
- Expansion of tropical diseases, such as dengue fever, into temperate regions as vectors of infectious disease (such as mosquitoes) move toward the poles
- Disease and sanitation problems when floods overcome sewage treatment systems
- Injuries and drowning if storms become more frequent or intense
- Hunger-related ailments as human population grows and demands on agricultural systems increase

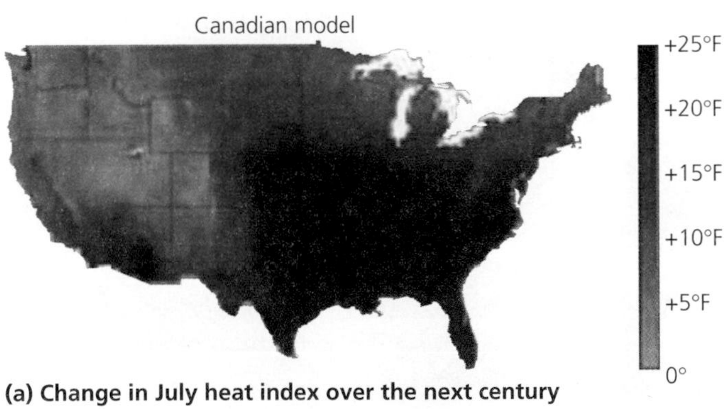


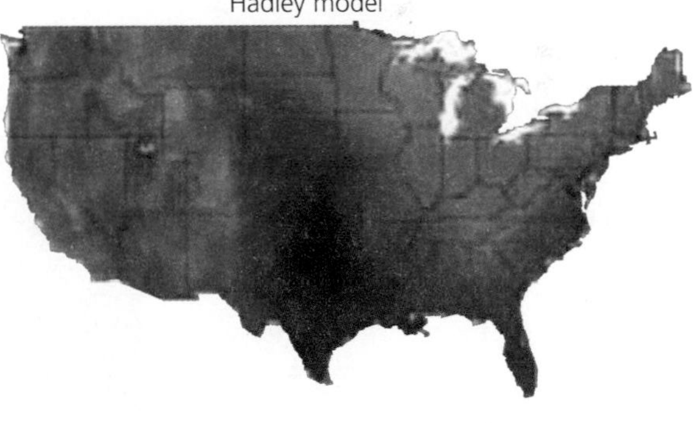


**(a) Change in July heat index over the next century**

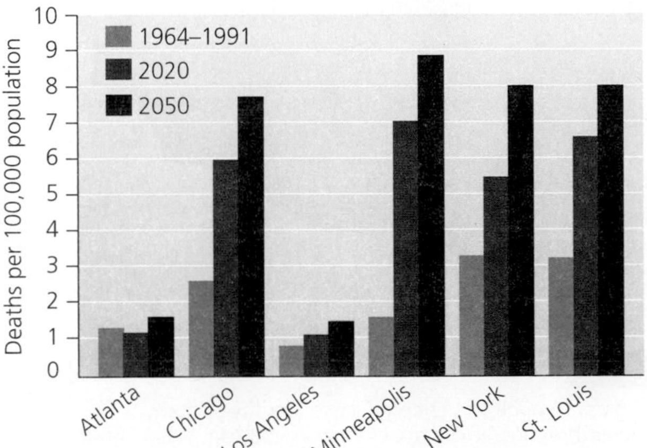


**(b) Average summer mortality rates attributed to hot weather episodes**

**FIGURE 18.20** Climate models—in this case, the Canadian and Hadley models (**a**)—provide projections of the heat index (a product of temperature and humidity) for July across the United States for the coming century. The Canadian model predicts that the July heat index will be 14° C (25° F) hotter across much of the southeastern United States. Past and projected mortality rates attributed to hot weather (**b**) are shown for several U.S. cities. Data from National Assessment Synthesis Team, U.S. Global Change Research Program (USGCRP). 2001. *Climate change impacts on the United States: The potential consequences of climate variability and change: Foundation.* Cambridge, U.K.: Cambridge University Press and USGCRP.

Health hazards from cold weather will decrease, but researchers feel that the increase from warm-weather hazards will more than offset these gains.

**Economics** People will experience a variety of economic costs and benefits from the many impacts of climate change, but on the whole researchers predict that costs will outweigh benefits, and that this gap will widen as climate change grows more severe. Climate change will also widen the gap between rich and poor, both within and among nations. Poorer people have less wealth and technology with which to adapt to climate change, and poorer people rely more on resources (such as local food and water) that are particularly sensitive to climatic conditions.

From a wide variety of economic studies, the IPCC estimated that climate change will cost 1–5% of GDP on average globally, although poor nations would lose more proportionally than rich nations. Economists trying to quantify damages from climate change by measuring the "social cost of carbon" (i.e., external costs; • p. 42) have proposed costs of anywhere from \$10 to \$350 per ton of carbon. The highest profile economic study to date has been the *Stern Review* commissioned by the British government (see "The Science behind the Story, Chapter 2, • pp. 44–45). This study maintained that climate change could cost us roughly 5–20% of GDP by the year 2200 but that investing just 1% of GDP starting now could enable us to avoid these future costs. Regardless of the precise numbers, many economists and policymakers are concluding that spending money now to mitigate climate change will save us a great deal more in the future.

All these physical, biological, and social impacts of climate change are consequences of the warming effect of our greenhouse gas emissions (**Figure 18.21**). We are

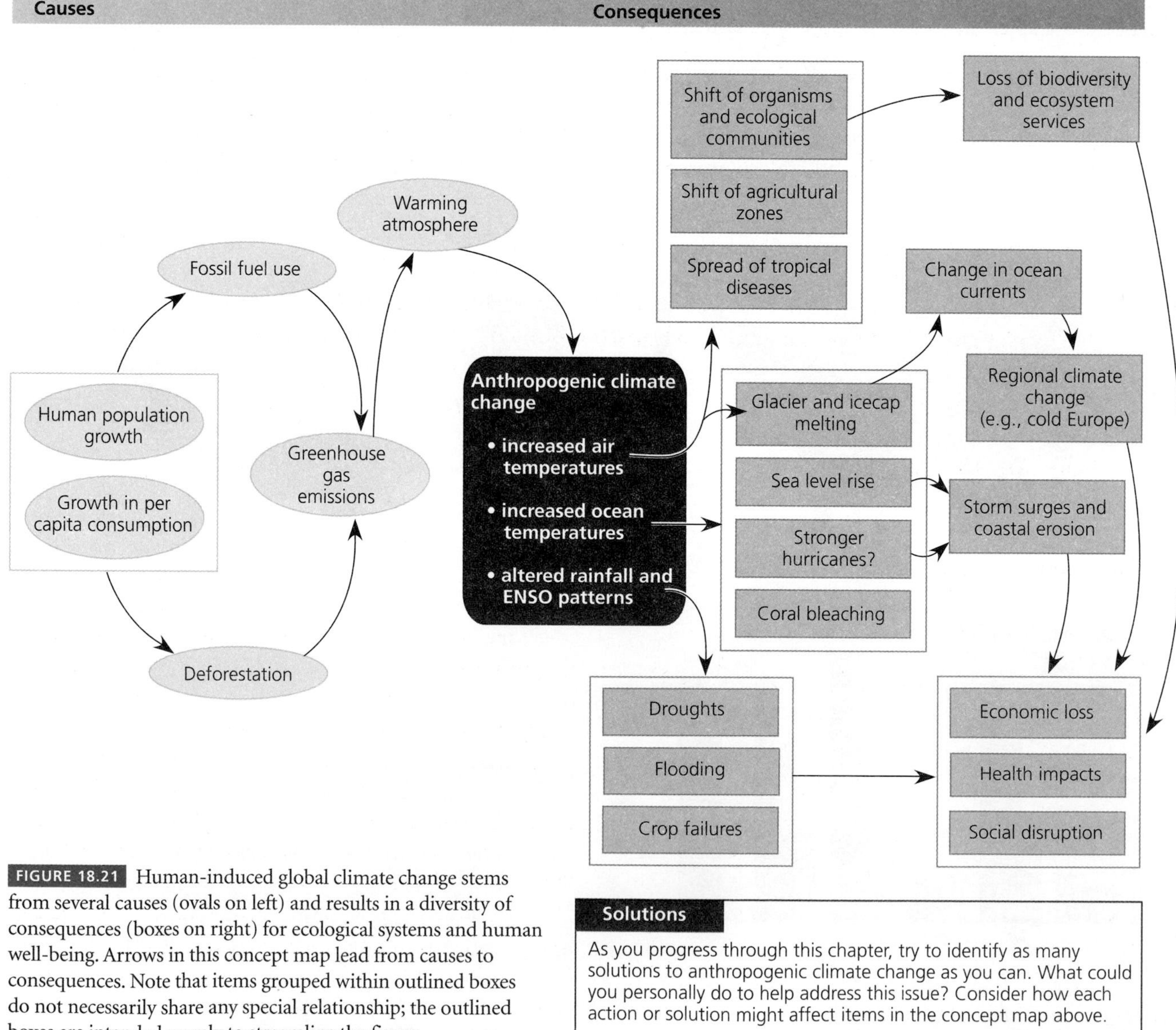


FIGURE 18.21 Human-induced global climate change stems from several causes (ovals on left) and results in a diversity of consequences (boxes on right) for ecological systems and human well-being. Arrows in this concept map lead from causes to consequences. Note that items grouped within outlined boxes do not necessarily share any special relationship; the outlined boxes are intended merely to streamline the figure.

**Solutions**

As you progress through this chapter, try to identify as many solutions to anthropogenic climate change as you can. What could you personally do to help address this issue? Consider how each action or solution might affect items in the concept map above.

bound to experience further consequences, but by addressing the root causes of anthropogenic climate change now, we can still prevent the most severe future impacts.

## Impacts will vary regionally

All future impacts of climate change are subject to regional variation, so the way each of us experiences these impacts over the coming decades will vary tremendously, depending on where we live.

For the United States, potential impacts were analyzed by the U.S. Global Change Research Program (USGCRP), which Congress created in 1990 to coordinate federal climate research. In 2000–2001, this program issued a comprehensive report highlighting the past and future effects of global climate change on the United States, where average temperatures increased by 0.6° C (1.0° F) during the 20th century.

To develop its predictions (**Table 18.2**), the report used coupled general circulation models (• p. 513). These models allowed scientists to present graphical depictions summarizing the predicted impacts of climate change on particular geographic areas (**Figure 18.22**). For instance, climate models predicted that forest communities should in general shift northward and upward in elevation. Although some forest types will likely decline, others, including oak-hickory and oak-pine forests, may expand in the eastern United States (**Figure 18.23**). The report also predicted that of 16 crops in the United States, yields of 13 would increase and only 1 (potatoes) would decrease.

The USGCRP report reached conclusions broadly similar to those of the internationally accepted IPCC assessment

**TABLE 18.2 Some Predicted Impacts of Climate Change in the United States**

- Average temperatures will rise 3 5° C (5.4 9.0° F) in 100 years.
- Droughts and flooding will worsen.
- Drought and other factors could decrease crop yields, but longer growing seasons and enhanced $CO_2$ could raise yields.
- Snowpack will decrease; water shortages will worsen.
- Greater temperature extremes will increase health problems and human mortality. Some tropical diseases will spread north.
- Forest growth may increase in the short term, but in the long term, drought, pests, and fire may alter forest ecosystems.
- Alpine ecosystems and barrier islands will begin to disappear.
- Southeastern U.S. forests will break up into savanna/grassland/forest mosaics.
- Northeastern U.S. forests will lose sugar maples.
- Sea level rise will cause loss of coastal wetlands and real estate.
- Melting permafrost will undermine Alaskan buildings and roads.

Adapted from National Assessment Synthesis Team, U.S. Global Change Research Program (USGCRP). 2000. *Climate change impacts on the United States: The potential consequences of climate variability and change: Overview.* Cambridge, U.K.: Cambridge University Press and USGCRP.

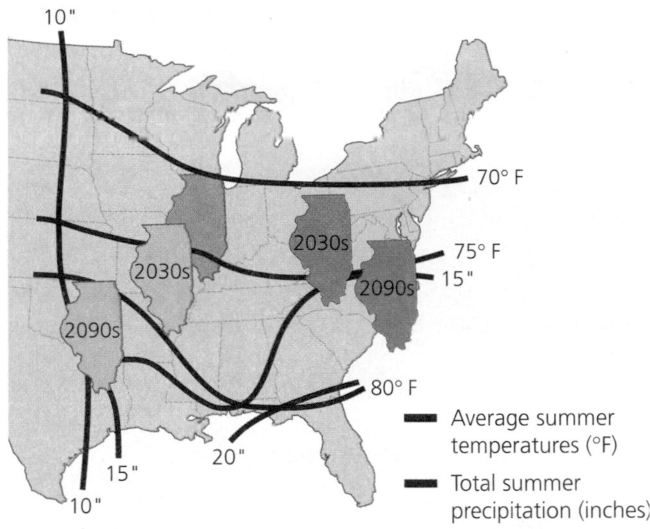


**FIGURE 18.22** Researchers used two climate models to predict the future climate of Illinois. The Canadian model (green shades for Illinois) predicts that Illinois will become warmer and drier, attaining in the 2030s a climate similar to Missouri's, and in the 2090s a climate similar to eastern Oklahoma's. The Hadley model (blue shades for Illinois) predicts that Illinois will become warmer and moister, attaining in the 2030s a climate similar to West Virginia's, and in the 2090s one similar to North Carolina's. Adapted from National Assessment Synthesis Team, U.S. Global Change Research Program (USGCRP). 2001. *Climate change impacts on the United States: The potential consequences of climate variability and change: Foundation.* Cambridge, U.K.: Cambridge University Press and USGCRP.

reports, but climate scientists say the USGCRP report was suppressed once the George W. Bush administration came to power. Climate change has been a politically divisive issue because any decision to move away from a fossil fuel economy challenges many entrenched and powerful interests.

## Are we responsible for climate change?

The IPCC's 2007 report concluded conservatively that it is greater than 90% likely that most of the global warming recorded over the past half century is due to the well-documented increase in greenhouse gas concentrations in our atmosphere. Scientists agree that this increase in greenhouse gases results primarily from our combustion of fossil fuels for energy and secondarily from land use changes, including deforestation and agriculture.

By the time the IPCC's *Fourth Assessment Report* came out, many scientists had already become concerned enough to put themselves on record urging governments to address climate change. In June 2005, as the leaders of the "G8" industrialized nations met, the national academies of science from 11 nations (Brazil, Canada, China, France, Germany, India, Italy, Japan, Russia, the United Kingdom, and the United States) issued a joint statement urging these political leaders to take action. Such a broad consensus statement from the

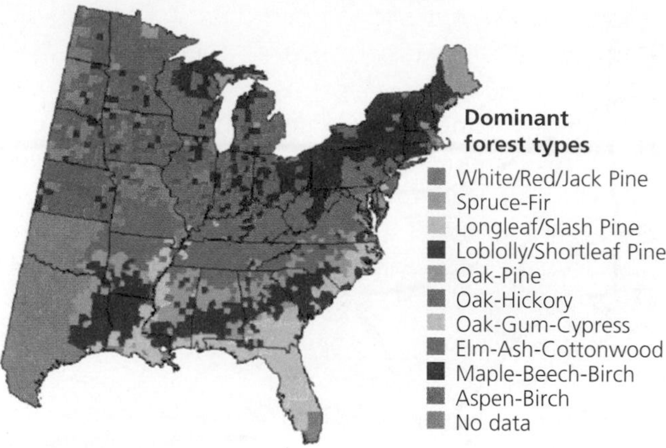


**(a) Current distribution (1960–1990)**

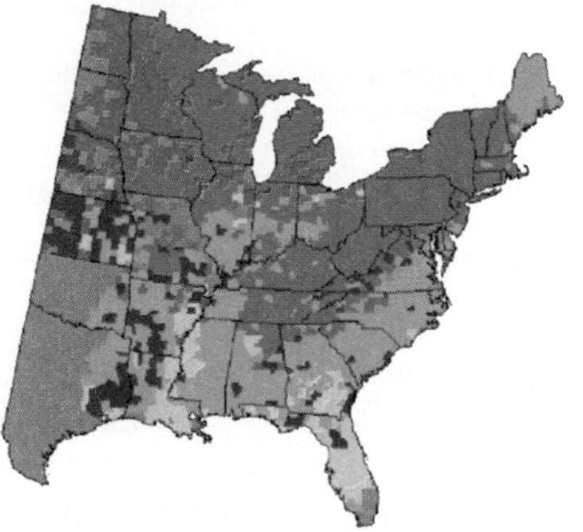

**(b) Canadian scenario (2070–2100)**

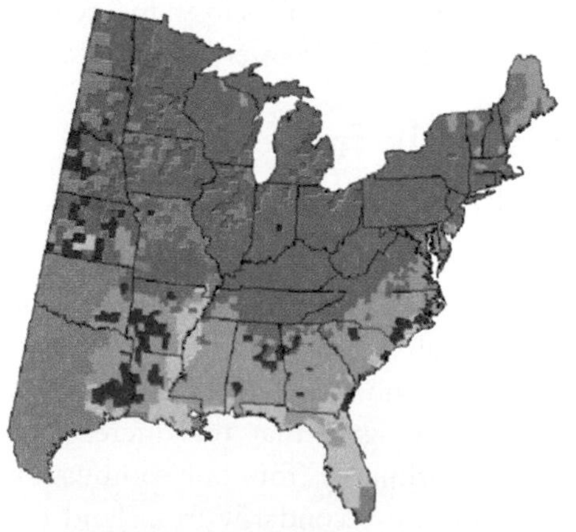

**(c) Hadley scenario (2070–2100)**

**FIGURE 18.23** Although some forest types will decline, others, including oak-hickory and oak-pine forests, may expand in the eastern United States. Shown is the current distribution of forest types (**a**) along with distributions predicted by the Canadian (**b**) and Hadley (**c**) climate models. Data from National Assessment Synthesis Team, U.S. Global Change Research Program (USGCRP). 2001. *Climate change impacts on the United States: The potential consequences of climate variability and change: Foundation*. Cambridge, U.K.: Cambridge University Press and USGCRP.

world's scientists was virtually unprecedented, on any issue. The statement read, in part:

> The scientific understanding of climate change is now sufficiently clear to justify nations taking prompt action. It is vital that all nations identify cost-effective steps that they can take now, to contribute to substantial and long-term reduction in net global greenhouse gas emissions. . . . A lack of full scientific certainty about some aspects of climate change is not a reason for delaying an immediate response that will, at a reasonable cost, prevent dangerous anthropogenic interference with the climate system.

Despite broad consensus among scientists and most national governments that climate change is a pressing issue, public discussion of climate change in the United States remained mired in an outdated debate over whether the phenomenon was real and whether humans were to blame. The debate was fanned largely by "greenhouse skeptics," a handful of scientists typically backed by funding from Exxon-Mobil and other corporations in the fossil fuel and auto industries. These people aimed to cast doubt on the scientific consensus, and their views were amplified by the American news media, which seeks to present two sides to every issue, even when the sides' arguments are not equally supported by evidence.

Today that debate is largely over, and most Americans accept that our fossil fuel consumption is altering the planet that our children will inherit. The tide turned with the 2007 release of the IPCC report, the broad popularity of Al Gore's 2006 movie and book, *An Inconvenient Truth* (**Figure 18.24**), grassroots activism, and action by political leaders at the state and local levels.

In a public opinion poll taken by the *New York Times* and CBS News in April 2007, 84% of respondents viewed human activity as contributing to global warming, and 78% of respondents supported immediate action to curb global warming. Fifteen percent of respondents rated global warming as the "most important" environmental problem, up from only 2% a decade earlier. More than half said global warming was "a very serious problem and should be one of the highest priorities for government leaders," whereas only 9% thought it was not serious or did not need addressing.

As a result of this shift in public perception, and in response to demand from their shareholders, many corporations and industries began offering support for reductions in greenhouse gas emissions. Strikingly, these advocates included some of the titans of industry in our fossil-fuel based economy who previously had worked to block action on climate change. Companies joining together in the U.S. Climate Action Partnership to ask government leaders to "quickly enact strong national

FIGURE 18.24 Former U.S. Vice President Al Gore's Oscar-winning film and best-selling book, *An Inconvenient Truth*, were based on a slideshow he had given on the lecture circuit for years. The film and book helped ignite the U.S. public's awakening to the reality of climate change in 2006–2007.

legislation to require significant reductions of greenhouse gas emissions" included Alcoa, BP America, ConocoPhillips, Dow Chemical, Duke Energy, DuPont, General Electric, General Motors, Pacific Gas and Electric, and Shell. These giants of the American economy thereby joined ranks with the insurance industry, which years earlier had grown concerned with climate change as it foresaw increased payouts for damage due to coastal storms, drought, and floods.

## Responding to Climate Change

Today we possess a newly broad consensus that climate change is a clear and present challenge to our society. How precisely we should respond to climate change is a difficult question, however, and one we will likely be wrestling with for decades.

### Shall we pursue mitigation or adaptation?

We can respond to climate change in two fundamental ways. One is to pursue actions that reduce greenhouse gas emissions, in order to lessen the severity of future climate change. This strategy is called **mitigation** because the aim is to mitigate, or alleviate, the problem. The second type of response is to accept that climate change is happening and to pursue strategies to minimize its impacts on us. This strategy is called **adaptation** because the goal is to adapt to change by finding ways to cushion oneself from its blows.

Erecting a seawall like Maldives residents did with the Great Wall of Male is an example of adaptation, using technology and engineering. The people of Tuvalu also adapted, but with a behavioral choice—some chose to leave their island and make a new life in New Zealand (see "Weighing the Issues," • p. 523). Other examples of adaptation include restricting coastal development; adjusting farming practices to cope with drought; and modifying water management practices to deal with reduced river flows, glacial outburst floods, or salt contamination of groundwater.

In contrast, mitigation strategies center on reducing greenhouse gas emissions by improving energy efficiency, switching to clean and renewable energy sources, encouraging farm practices that protect soil quality, recovering landfill gas, and preventing deforestation.

Many environmental advocates have criticized adaptation strategies, because they view them as escapist—a way of sidestepping the hard work that must be done to protect future generations from climate change. However, adaptation and mitigation are not mutually exclusive approaches; a person or a nation can pursue both.

Indeed, both approaches are necessary. Adaptation is needed because even if we were to halt all our emissions now, global warming would continue until the planet's systems reach a new equilibrium, with temperature rising an estimated 0.6° C (1.0° F) more by the end of the century. Because we will face this change no matter what we do, it is wise to develop ways to minimize its impacts.

Mitigation is necessary because if we do nothing to slow down climate change, it will eventually overwhelm any efforts at adaptation we could make. To leave a sustainable future for our civilization and to safeguard the living planet that we know, we will need to pursue mitigation. Moreover, the faster we begin reducing our emissions, the lower the level at which they will peak, and the less we will alter climate (**Figure 18.25**). We will spend the remainder of our chapter examining approaches for the mitigation of climate change.

### Electricity generation is the largest source of U.S. greenhouse gases

The generation of electricity produces the largest portion (40%) of U.S. carbon dioxide emissions (**Figure 18.26**). From cooking and heating to the clothes we wear, much of what we own and do depends on electricity. Fossil fuel combustion generates 69% of U.S. electricity, and coal alone accounts for 50%, along with most of the greenhouse gas emissions. Reducing the volume of fossil fuels we burn to generate electricity would lessen greenhouse gas emissions, as would decreasing electricity consumption. There are two ways to reduce the amount of fossil

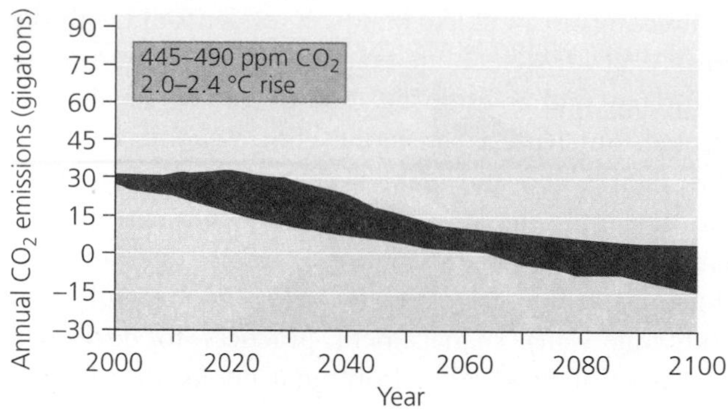


**(a) Scenario 1: Emissions peak soon**

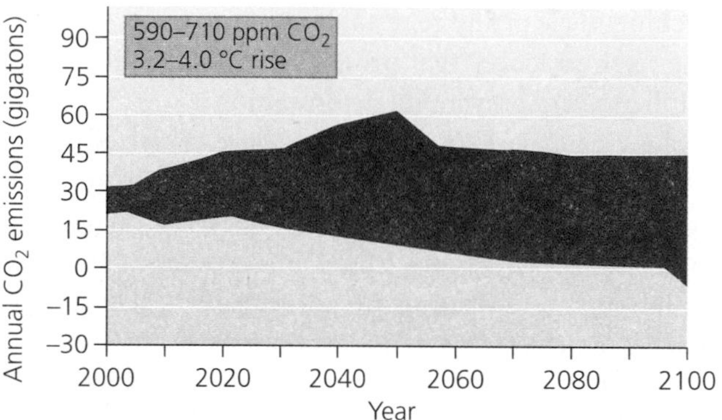


**(b) Scenario 2: Emissions peak in 2020–2060**

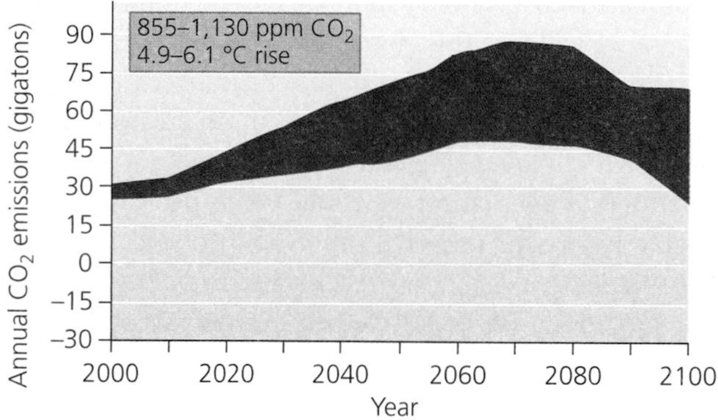


**(c) Scenario 3: Emissions peak in 2060–2090**

FIGURE 18.25 The sooner we stabilize our emissions, the less climate change we will cause. Shown are three scenarios studied by the IPCC for emissions of carbon dioxide. In (**a**), emissions peak in the next several years, leading to eventual atmospheric stabilization at 445–490 ppm $CO_2$ and a global mean temperature rise of 2.0–2.4° C. In (**b**), emissions peak in 2020-2060, bringing stabilization at 590–710 ppm $CO_2$ and a temperature rise of 3.2–4.0° C. In (**c**), emissions peak in 2060–2090, leading to stabilization at 855–1,130 ppm $CO_2$ and a temperature rise of 4.9–6.1° C. Predictions are based on a large number of climate models. Data from Intergovernmental Panel on Climate Change. 2007. *Fourth assessment report.*

FIGURE 18.26 Coal-fired electricity-generating power plants, such as this one in Maryland, are the largest contributors to U.S. greenhouse emissions.

fuels we use: (1) encouraging conservation and efficiency (• pp. 565–568) and (2) switching to cleaner and renewable energy sources (Chapters 20 and 21).

**Conservation and efficiency** Conservation and efficiency in energy use can arise from new technologies, such as high-efficiency lightbulbs and appliances, or from individual ethical choices to reduce electricity consumption. The U.S. Environmental Protection Agency (EPA) offers technological solutions through its Energy Star Program, which rates household appliances, lights, windows, fans, office equipment, and heating and cooling systems by their energy efficiency. For instance, replacing an old washing machine with an Energy Star washing machine can cut your $CO_2$ emissions by 200 kg (440 lb) annually. Replacing standard lightbulbs with compact fluorescent lights reduces energy use for lighting by 40%. Energy Star homes use highly efficient windows, duct work, insulation, and heating and cooling systems to reduce energy use and emissions by as much as 30%. Such technological solutions are popular, and they can be profitable for manufacturers while also saving consumers money.

Consumers can also opt for lifestyle choices. For nearly all of human history, people managed without the electrical appliances that most of us take for granted today. It is possible for each of us to choose to use fewer greenhouse-gas-producing appliances and technologies and to take practical steps to use electricity more efficiently.

**Sources of electricity** We can also reduce greenhouse gas emissions by altering the types of energy we use. Among fossil fuels, natural gas burns more cleanly than oil, and oil is cleaner-burning than coal. Using natural gas instead of coal produces the same amount of energy with roughly one-half the emissions (• p. 58). Moreover, approaches to boost the efficiency of fossil fuel use, such as cogeneration (• pp. 566–567) produce fewer emissions per unit energy generated.

Currently, interest in *carbon capture* and *carbon sequestration* or *storage* is intensifying. Carbon capture refers to technologies or approaches that remove carbon dioxide from power plant emissions. Successful carbon capture technology would allow plants to continue using fossil fuels while cutting greenhouse gas pollution. The carbon would then be sequestered, or stored, somewhere—perhaps underground under pressure in locations where it will not seep out. However, we are still a long way from developing adequate technology and secure storage space to accomplish this, and some experts doubt that we will ever be able to sequester enough carbon to make a dent in our emissions.

Technologies and energy sources that generate electricity without using fossil fuels represent another means of reducing greenhouse gas emissions. These include nuclear power (• pp. 575–588), hydroelectric power (• pp. 595–598), geothermal energy (• pp. 617–620), photovoltaic cells (• pp. 609–610), wind power (• pp. 612–617), and ocean energy sources (• pp. 620–621). These energy sources give off no emissions during their use (but some in the production of their infrastructure). We will examine these clean and renewable energy sources in detail in Chapters 20 and 21.

## Transportation is the second largest source of U.S. greenhouse gases

Can you imagine life without a car? Most Americans probably can't—a reason why transportation is the second-largest source of U.S. greenhouse emissions. One-third of the average American city—including roads, parking lots, garages, and gas stations—is devoted to use by the nation's 220 million registered automobiles. The average American family makes 10 trips by car each day, and governments across the nation spend $200 million per day on road construction and repairs.

Unfortunately, the typical automobile is highly inefficient. Close to 85% of the fuel you pump into your gas tank does something other than move your car down the road. According to the U.S. Department of Energy, only about 13–14% of the fuel energy actually moves the vehicle and its occupants from point A to point B (**Figure 18.27**). Although more aerodynamic designs, increased engine efficiency, and improved tire design could help reduce these losses, gasoline-fueled automobiles may always remain somewhat inefficient.

**Automotive technology** The technology exists to make our vehicles more fuel-efficient than they currently are. Indeed, the vehicles of many nations are more fuel-efficient than those of the United States. Raising fuel efficiency (• pp. 565–566) for American-made vehicles will require government mandate and/or consumer demand, and as gasoline prices rise, demand for more fuel-efficient automobiles will intensify.

Advancing technology is also bringing us alternatives to the traditional combustion-engine automobile. These include hybrid vehicles that combine electric motors and gasoline-powered engines for greater efficiency (• pp. 566–567). They also include fully electric vehicles, alternative fuels such as compressed natural gas and biodiesel (• pp. 591–592), and hydrogen fuel cells that use oxygen and hydrogen and produce only water as a waste product (• pp. 622–625).

**Driving less and using public transportation** People can also opt to make lifestyle choices that reduce their reliance on cars. For example, some people are choosing to live nearer to their workplaces. Others use mass transit such as buses, subway trains, and light rail. Still others

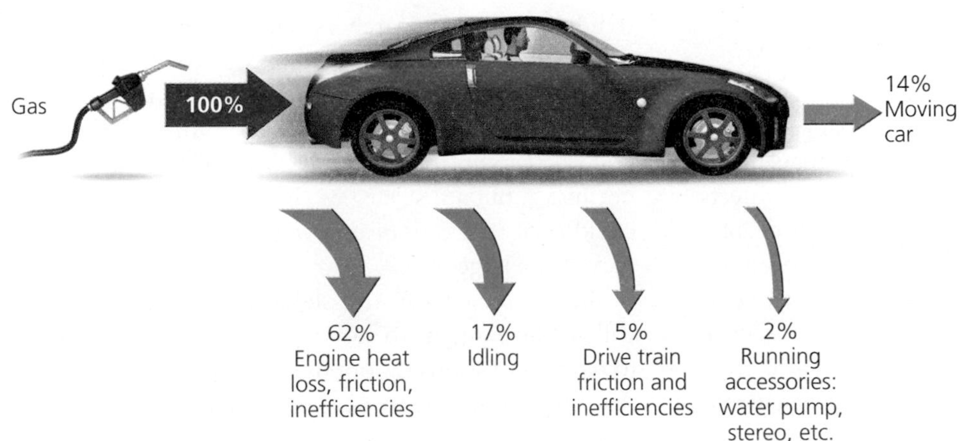


FIGURE 18.27 Conventional automobiles are extremely inefficient. Almost 85% of useful energy is lost, and only 14% actually moves the car down the road.

FIGURE 18.28 By choosing human-powered transportation methods, such as bicycles, we can greatly reduce our transportation-related greenhouse gas emissions. More people are choosing to live closer to their workplaces and to enjoy the dual benefits of exercise and reduced emissions by walking or cycling to work or school.

bike or walk to work or for their errands (**Figure 18.28**). Unfortunately, reliable and convenient public transit is not yet available in many U.S. communities. Making automobile-based cities and suburbs more friendly to pedestrian and bicycle traffic and improving people's access to public transportation stand as central challenges for city and regional planners (• pp. 370–373).

In a 2002 study, the American Public Transportation Association (APTA) concluded that increasing the use of public transportation is the single most effective strategy for conserving energy and reducing pollution. Already, public transportation in the United States reduces fossil fuel consumption by 855 million gallons of gas (45 million barrels of oil) each year, the APTA estimates. Yet according to the study, if U.S. residents increase their use of public transportation to the levels of Canadians (7% of daily travel needs) or Europeans (10% of daily travel needs), the United States could substantially cut its air pollution, its dependence on imported oil, and its contribution to climate change.

## We can reduce emissions in other ways as well

Other pathways toward mitigating climate change include advances in agriculture, forestry, and waste management. In agriculture, sustainable land management that protects the integrity of soil on cropland and rangeland enables soil to store more carbon. Techniques have also been developed to reduce the emission of methane from rice cultivation and from cattle and their manure, and to reduce nitrous oxide emissions from fertilizer. We can also grow renewable biofuels, and this is an active area of current research (• p. 590–591).

In forest management, the rapid reforestation of cleared areas helps restore forests, which act as reservoirs that pull carbon from the air. Sustainable forestry practices and preserving existing forests can help to reverse the carbon dioxide emissions resulting from deforestation.

Waste managers are doing their part to cut greenhouse emissions by recovering methane seeping from landfills (• pp. 638–639), treating wastewater (• pp. 435–438), and generating energy from waste in incinerators (• pp. 637–638). Individuals, communities, and waste haulers also help reduce emissions by encouraging recycling, composting, and the reduction and reuse of materials and products (• pp. 639–644).

## We will need to follow multiple strategies to reduce emissions

We should not expect to find a single magic bullet for mitigating climate change. Instead, reducing emissions will require many steps by many people and institutions across many sectors of our economy. The good news is that most reductions can be achieved using current technology and that we can begin implementing these changes right away.

Environmental scientists Stephen Pacala and Robert Socolow argue that "humanity already possesses the fundamental scientific, technical, and industrial know-how to solve the carbon and climate problem for the next half-century." They advise that we follow some age-old wisdom: When the job is big, break it into small parts. Pacala and Socolow propose that we adopt a portfolio of strategies, each one feasible in itself, that could together stabilize our $CO_2$ emissions at current levels (**Figure 18.29**).

Pacala and Socolow began with graphs that predicted a doubling of emissions over the next 50 years and asked: What would we need to do to hold our emissions flat instead and to avoid the additional future emissions represented by the triangular area of the graph above the flat trend line? The researchers subdivided their so-called *stabilization triangle* into 7 equal wedges, like slices of a pie. To eliminate one wedge, a strategy would need to reduce emissions equivalent to 1 billion tons of carbon per year 50 years in the future. Pacala and Socolow identify not just 7, but 15, strategies (listed in Figure 18.29) that could each take care of one wedge if developed and deployed at a large scale.

In the long term, the stabilization-wedge approach will not be enough. To stop climate change, we will need to reduce emissions (as opposed to stabilizing them), and this may require us to develop new technology, further change our lifestyles, and/or reverse our population

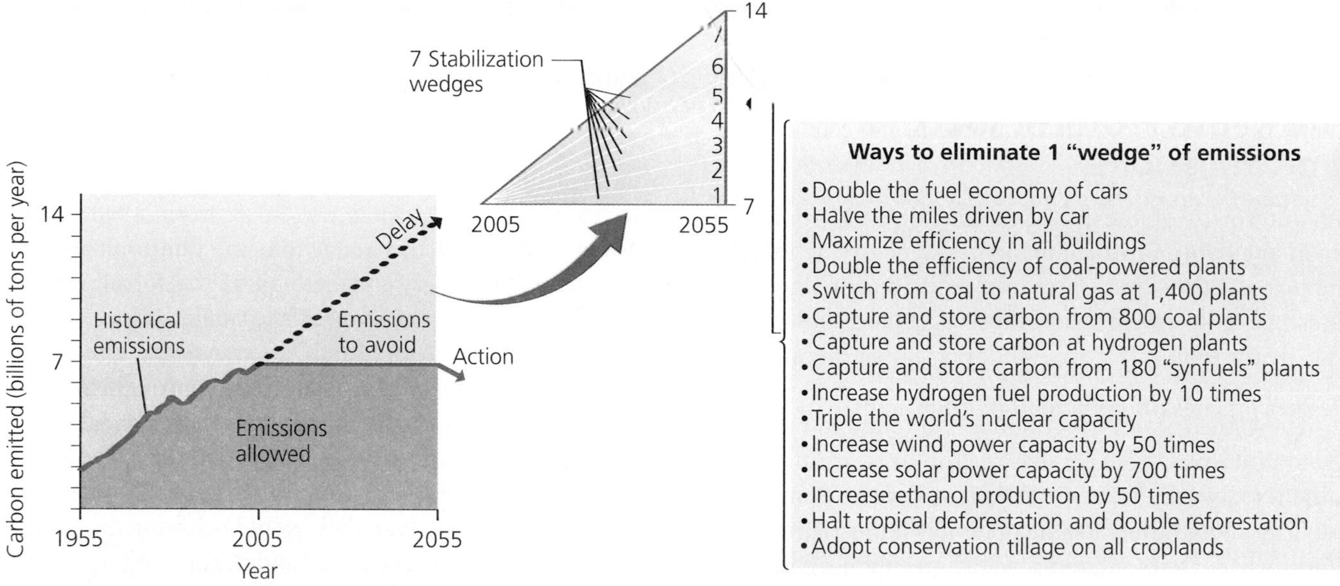


FIGURE 18.29 Pacala and Socolow began with a standard graph (left) of predicted carbon emissions from $CO_2$ showing the doubling of emissions that scientists expect to occur from 2005 to 2055. They added a flat line to represent the trend if emissions were held constant and separated the graph into emissions allowed (below the line) and emissions to be avoided (the triangular area above the flat line). They then divided this "stabilization triangle" into seven equal-sized portions, which they called "stabilization wedges" (center). Each stabilization wedge represents 1 billion tons of $CO_2$ emissions in 2055 to be avoided. Finally, they identified 15 strategies (box at right), each of which could take care of 1 wedge. If we accomplish just 7 of these 15 strategies, we could halt our growth in emissions for the next half century.
Adapted from Pacala, S. and R. Socolow. 2004. Stabilization wedges: Solving the climate problem for the next 50 years with current technologies. *Science* 305: 968–972; and www.princeton.edu/~cmi/resources/CMI_Resources_new_files/CMI_Stab_Wedges_Movie.swf.

growth. However, there is plenty we can do in the meantime to mitigate climate change simply by scaling up the technologies and approaches for emissions reductions that we already have developed.

## Shall we use government mandates or market incentives?

How quickly and successfully we translate our science and technology into practical solutions for reducing emissions depends largely on the policies we urge our leaders to pursue and on how government and the market economy interact. As we saw in Chapter 3 and in numerous instances throughout this book, governmental command-and-control policy has been vital in safeguarding environmental quality and promoting human well-being. However, government mandates are often resisted by industry, and market incentives can sometimes be more effective in driving change.

With climate change policy, we are in the midst of a dynamic period of debate and experimentation. At all levels—international, national, state, regional, and local—policymakers, industry, commerce, and citizens are searching for ways to employ government and the market to reduce emissions in ways that are fair, economically palatable, effective, and enforceable.

## We began tackling climate change by international treaty

In 1992, the United Nations convened the U.N. Conference on Environment and Development Earth Summit in Rio de Janeiro, Brazil. Nations represented at the Earth Summit signed the **U.N. Framework Convention on Climate Change (FCCC)**, which outlined a plan for reducing greenhouse gas emissions to 1990 levels by the year 2000 through a voluntary, nation-by-nation approach.

By the late 1990s, it was clear that a voluntary approach was not likely to succeed. For example, from 1990 to 2006, U.S. greenhouse emissions increased by 17.9%. However, certain other nations demonstrated that economic vitality does not require ever-increasing emissions. For instance, Germany has the third most technologically advanced economy in the world and is a leading producer of iron, steel, coal, chemicals, automobiles, machine tools, electronics, textiles, and other goods—yet it managed between 1990 and 2004 to reduce its greenhouse gas emissions by 17.2%. In the same period, the United Kingdom cut its emissions by 14.3%.

After watching the seas rise and observing the failure of most industrialized nations to cut their emissions, nations of the developing world—the Maldives among them—helped initiate an effort to create a binding international treaty that would *require* all signatory nations to reduce their emissions. This effort led to the Kyoto Protocol.

## The Kyoto Protocol seeks to limit emissions

An outgrowth of the FCCC drafted in 1997 in Kyoto, Japan, the **Kyoto Protocol** mandates signatory nations, by the period 2008–2012, to reduce emissions of six greenhouse gases to levels below those of 1990 (**Table 18.3**). The treaty took effect in 2005 after Russia became the 127th nation to ratify it.

The United States has continued to refuse to ratify the Kyoto Protocol. U.S. leaders have called the treaty unfair because it requires industrialized nations to reduce emissions but does not require the same of rapidly industrializing nations such as China and India, whose greenhouse emissions have risen over 50% in the past 15 years. Proponents of the Kyoto Protocol say the differential requirements are justified because industrialized nations created the current problem and therefore should take the lead in resolving it. The refusal of the United States to join international efforts to curb greenhouse emissions has generated resentment among its allies and has undercut the effectiveness of these efforts.

Because resource use and per capita emissions are far greater in the industrialized world, governments and industries there often feel they have more to lose economically from restrictions on emissions. Ironically, industrialized nations are also the ones most likely to gain economically, because they are best positioned to invent, develop, and market new technologies to power the world in a post-fossil-fuel era.

**TABLE 18.3 Emissions Reductions Required and Achieved**

| Nation | Required change,* 1990–2008/2012 | Observed change, 1990–2004† |
|---|---|---|
| Russia | 0.0% | −32.0%‡ |
| Germany | −21.0% | −17.2% |
| United Kingdom | −12.5% | −14.3% |
| France | 0.0% | −0.8% |
| Italy | −6.5% | +12.1% |
| Japan | −6.0% | +6.5% |
| United States | −7.0% | +15.8% |
| Canada | −6.0% | +26.6% |

*Percentage decrease in emissions (carbon-equivalents of six greenhouse gases) from 1990 to period 2008–2012, as mandated under Kyoto Protocol.

†Actual percentage change in emissions (carbon-equivalents of six greenhouse gases) from 1990 to 2004. Negative values indicate decreases; positive values indicate increases. Values do not include influences of land use and forest cover.

‡Russia's substantial decrease was due mainly to economic contraction following the breakup of the Soviet Union.

Data from U.N. Framework Convention on Climate Change, National Greenhouse Gas Inventory Reports, 2007.

As of 2004 (the most recent year with full international data), nations that signed the Kyoto Protocol had decreased their emissions by 3.3% from 1990 levels. However, much of this reduction was due to economic contraction in Russia and nations of the former Soviet Bloc following the breakup of the Soviet Union. When these nations are factored out, the remaining signatories showed an 11.0% *increase* in emissions from 1990 to 2004.

Kyoto Protocol critics and supporters alike acknowledge that even if every nation complied with the treaty, greenhouse gas emissions would continue to increase—albeit more slowly than they would without the treaty. Nations are now looking ahead and negotiating over what will come next to supercede Kyoto.

## States and cities are advancing climate change policy

In the absence of action to address climate change by the George W. Bush administration and the U.S. Congress, numerous state and local governments across the United States are responding to popular sentiment and advancing policies to limit greenhouse emissions. By mid-2007, mayors from over 600 cities from all 50 U.S. states (representing 22% of the U.S. population) had signed on to the U.S. Mayors Climate Protection Agreement, led by Seattle Mayor Greg Nickels. Under this agreement, mayors commit their cities to pursue policies to "meet or beat" Kyoto Protocol guidelines and to urge their states and the federal government to take action as well.

At the state level, the boldest action so far has come from California, where in 2006 California's legislature worked with governor Arnold Schwarzenegger to pass the Global Warming Solutions Act, which aims to cut the state's greenhouse gas emissions 25% by the year 2020. This law is the first state legislation with penalties for noncompliance, and followed earlier efforts in California to mandate higher fuel efficiency for automobiles.

Bold action was also taken by 10 northeastern states that launched the Regional Greenhouse Gas Initiative (RGGI) in 2007. In this effort, Connecticut, Delaware, Maine, Maryland, Massachusetts, New Hampshire, New Jersey, New York, Rhode Island, and Vermont have set up a cap-and-trade program for carbon emissions from power plants. This emissions trading program (• pp. 79–82) is one example of how a government-sponsored and mandated plan can engage the market economy to achieve public policy goals.

## Market mechanisms are being used to address climate change

As we first discussed in Chapter 3 (• pp. 79–82), permit trading programs represent a way to harness the economic efficiency of the free market to achieve policy goals while allowing business, industry, or utilities flexibility in how they meet those goals. Supporters of permit trading programs argue that they provide the fairest, least expensive, and most effective method of achieving emissions reductions. Polluters get to choose how to reduce their emissions and are given financial incentives for reducing emissions below the legally required amount. We will likely discover how successful these ventures are over the next decade as various carbon trading programs get up and running around the world.

As an example of how a *cap-and-trade* emissions trading program can work, consider the planned approach for the Regional Greenhouse Gas Initiative:

1. Each state decides what polluting sources it will require to participate in the program.
2. Each state sets a cap on the total $CO_2$ emissions it will allow.
3. Each state distributes to each emissions source one permit for each ton they emit, up to the amount of the cap.
4. Sources with too few permits to cover their emissions must find ways to reduce their emissions, buy permits from other sources, or pay for credits through a *carbon offset* project. Sources with excess permits may keep them or sell them.
5. Any source emitting more than its permitted amount will face penalties.

Once up and running, it is hoped that the system will be self-sustaining. The price of a permit is meant to fluctuate freely in the market, creating the same kinds of financial incentives as any other commodity that is bought and sold in our capitalist system.

The world's first emissions trading program for greenhouse gas reduction (operating since 2003) is the Chicago Climate Exchange, which now boasts over 120 corporations, institutions, and municipalities in North America and Brazil. This legally binding trading system imposes a 6% reduction on overall emissions by 2010.

The world's largest cap-and-trade program is the European Union Emission Trading Scheme, which began on January 1, 2005. All EU member states participate, and each submits for approval a national allocation plan that conforms to the nation's obligations under the Kyoto Protocol. This market got off to a successful start, and carbon prices reached 30 euros per ton in early 2006. However, when investors determined that national governments had allocated too many emissions permits to their industries, the price of carbon fell. The over-allocation gave companies little incentive to reduce emissions, so permits lost their value. By early 2007, prices in the market had tanked to below 0.30 euros. This drop is roughly equivalent to a stock valued at $40 falling to less than 40 cents.

Proponents of emissions trading chalk up the freefall in the European market as a learning experience. Europeans will have the chance to correct their allocations and revive their market beginning in 2008 as the program enters its next phase, when it expands to include more greenhouse gases, more emissions sources, and additional members.

## Carbon offsets are in vogue

Emissions trading programs have allowed participants who cannot or will not adequately reduce their own emissions to use *carbon offsets* instead. A **carbon offset** is a voluntary payment to another entity intended to enable that entity to reduce the greenhouse emissions that one is unable or unwilling to reduce oneself. The payment thus offsets one's own emissions. For instance, a coal-burning power plant could pay a reforestation project to plant trees that will soak up as much carbon as the coal plant emits. Or a university could fund the development of clean and renewable energy projects to make up for fossil fuel energy the university uses.

Carbon offsets are fast becoming popular among utilities, businesses, universities, governments, and individuals trying to achieve *carbon-neutrality*, a state in which no net carbon is emitted. For time-stressed people with enough wealth, offsets represent a simple and convenient way to reduce one's emissions without investing in efforts to change one's habits. For example, you can go to the website of the company TerraPass, calculate your emissions for travel by car or plane, or for your home or campus dorm, and purchase offsets for those emissions. Your money funds renewable energy and efficiency projects, and you can advertise your donation with bumper stickers and decals.

In principle, carbon offsets seem a great idea, but in practice they often fall short. Without rigorous oversight to make sure that the offset money actually accomplishes what it is intended for, carbon offsets risk being no more than a way for wealthy consumers to assuage a guilty conscience. Efforts to create a transparent and enforceable offset infrastructure are ongoing. If these efforts succeed, then carbon offsets could become an effective and important key to mitigating climate change.

## You can reduce your own carbon footprint

Carbon offsets, emissions trading schemes, national policies, international treaties, and technological innovations will all play roles in mitigating climate change. But in the end the most influential factor may be the collective decisions of millions of regular people. In our everyday lives, each one of us can take steps to approach a carbon-neutral lifestyle by reducing greenhouse emissions that result from our decisions and activities. Just as we each have an ecological footprint (• p. 6), we each have a **carbon footprint** that expresses the amount of carbon we are responsible for emitting.

You can apply many of the strategies discussed in this chapter in your everyday life—from deciding where to live and how to get to work to choosing appliances. You will encounter still more solutions in our discussions of energy sources, conservation, and renewable energy in Chapters 19–21 and elsewhere throughout this book.

Global climate change may be the biggest challenge facing us and our children. Fortunately, it is still early enough that, with concerted action, we can avert the most severe impacts. Taking immediate, resolute action is the most important thing that we, personally and as a society, can do.

# Conclusion

Many factors influence Earth's climate, and human activities have come to play a major role. Climate change is well underway, and further greenhouse gas emissions will increase global warming and cause increasingly severe and diverse impacts. Sea-level rise and other consequences of global climate change will affect locations worldwide from the Maldives to Bangladesh to Alaska to Florida. As scientists and policymakers come to better understand anthropogenic climate change and its environmental, economic, and social consequences, more and more of them are urging immediate action. Reducing greenhouse gas emissions and taking other actions to mitigate and adapt to climate change represents the foremost challenge for our society in the coming years.

## REVIEWING OBJECTIVES

**You should now be able to:**

**Describe Earth's climate system and explain the many factors influencing global climate**

- Earth's climate changes naturally over time, but it is now changing rapidly because of human influence. (p. 506)
- The sun provides most of Earth's energy and interacts with the atmosphere, land, and oceans to drive climate processes. (pp. 506, 510–511)
- Earth absorbs about 70% of incoming solar radiation and reflects about 30% back into space. (pp. 506–507)
- Greenhouse gases such as carbon dioxide, methane, water vapor, nitrous oxide, ozone, and halocarbons warm the atmosphere by absorbing infrared radiation and re-emitting infrared radiation of different wavelengths. (pp. 506–509)
- Milankovitch cycles influence climate in the long term. (p. 510)

**Characterize human influences on the atmosphere and global climate**

- Increased greenhouse gas emissions enhance the greenhouse effect. (pp. 507–509)
- By burning fossil fuels, clearing forests, and manufacturing halocarbons, humans are increasing atmospheric concentrations of many greenhouse gases. (pp. 508–509)
- Human input of aerosols into the atmosphere exerts a variable but slight cooling effect. (p. 509)

**Summarize modern methods of climate research**

- Geologic records, such as cores through ice or sediments, reveal information about past climatic conditions. (pp. 512, 514–515)
- Direct atmospheric sampling tells us about current composition of the atmosphere. (p. 513)
- Coupled general circulation models serve to predict future changes in climate. (pp. 513–516)

**Outline current and future trends and impacts of global climate change**

- The IPCC has comprehensively synthesized current climate research, and its periodic reports represent the consensus of the scientific community. (pp. 516–518)
- Temperatures on Earth have warmed by an average of 0.74° C (1.33° F) over the past century and are predicted to rise 1.8–4.0° C (3.2–7.2° F) over the next century. (pp. 518–519)
- Changes in precipitation vary by region. (pp. 519–520)
- Sea level has risen an average of 17 cm (7 in.) over the past century. (pp. 521–523)
- Other impacts include melting of glaciers and polar ice, frequency of extreme weather events, impacts on

agriculture, forestry, and health, and effects on plants and animals. (pp. 520–528)

- Climate change and its impacts will vary regionally. (p. 527)
- Despite some remaining uncertainties, the scientific community feels that evidence for humans' role in influencing climate is strong enough to justify governments taking action to reduce greenhouse emissions. (pp. 527–528)

**Suggest ways we may respond to climate change**

- Both adaptation and mitigation are necessary for responding to climate change. (p. 529)
- Conserving electricity, improving efficiency of energy use, and switching to clean and renewable energy sources will help reduce fossil fuel consumption and greenhouse emissions. (pp. 529–531)
- Encouraging new automotive technologies and investment in public transportation will help reduce greenhouse emissions. (pp. 531–532)
- Solving the climate problem will require the deployment of multiple strategies. (pp. 532–533)
- The Kyoto Protocol has provided a first step for nations to begin addressing climate change. (pp. 533–534)
- U.S. states and cities are taking action to address greenhouse emissions because the federal government has not. (p. 534)
- Emissions trading programs are providing a way to harness the free market and engage industry in reducing emissions. (pp. 534–535)
- Individuals are increasingly exploring carbon offsets and other means of reducing personal carbon footprints. (pp. 535–536)

## TESTING YOUR COMPREHENSION

1. What happens to solar radiation after it reaches Earth? How do greenhouse gases warm the lower atmosphere?
2. Why is carbon dioxide considered the main greenhouse gas? How could an increase in water vapor create either a positive or negative feedback effect?
3. How do scientists study the ancient atmosphere?
4. Has simulating climate change with computer programs been effective in helping us predict climate? How do these programs work?
5. List five major trends in climate that scientists have documented so far. Now list five future trends or impacts that they are predicting.
6. Describe how rising sea levels, caused by global warming, can create problems for people. How may climate change affect marine ecosystems?
7. How might a warmer climate affect agriculture? How is it affecting distributions of plants and animals? How might it affect human health?
8. What are the largest two sources of greenhouse gas emissions in the United States? In what ways can we reduce these emissions?
9. What roles have international treaties played in addressing climate change? Give two specific examples.
10. Describe one market-based approach for reducing greenhouse emissions. Explain one reason it may work well and one reason it may not work well.

## SEEKING SOLUTIONS

1. To determine to what extent current climate change is the result of human activity versus natural processes, which type(s) of scientific research do you think is (are) most helpful? Why?
2. Some people argue that we need "more proof," or "better science" before we commit to substantial changes in our energy economy. How much "science," or certainty, do you think we need before we should take action regarding climate change? How much certainty do you need in your own life before you make a major decision? Should nations and elected officials follow a different standard? Do you believe that the precautionary principle (• pp. 277, 405) is an appropriate standard in the case of global climate change? Why or why not?
3. Describe several ways in which we can reduce greenhouse gas emissions from transportation. Which approach do you think is most realistic, which approach do you think is least realistic, and why?
4. Imagine that you would like to make your own lifestyle carbon-neutral and that you aim to begin by reducing the emissions you are responsible for by 25%. What actions would you take first to achieve this reduction?
5. **THINK IT THROUGH** You have been appointed as the United States representative to an international conference to negotiate terms of a treaty to take hold after the Kyoto Protocol ends. All nations recognize that the Kyoto Protocol was not fully effective, and most are committed to creating a stronger agreement. The U.S.

government has instructed you to take a leading role in designing the new treaty and to engage constructively with other nations' representatives, while protecting your nation's economic and political interests. What type of agreement will you try to shape? Describe at least three components that you would propose or agree to, and at least one that you would oppose.

6. **THINK IT THROUGH** You have just been elected governor of a medium-sized U.S. state. Polls show that the public wants you to take bold action to reduce greenhouse gas emissions. However, polls also show that the public does not want prices of gasoline or electricity to rise very much. Carbon-emitting industries in your state are wary of emissions reductions being required of them, but are willing to explore ideas with you. Your state legislature will support you in your efforts as long as you remain popular with voters. The state to your west has just passed ambitious legislation mandating steep greenhouse gas emissions reductions. The state to your east has just joined a new regional emissions-trading consortium. What actions will you take in your first year as governor?

## INTERPRETING GRAPHS AND DATA

We burn fossil fuels to generate electricity, to power vehicles for transportation, and as primary energy sources (non-electricity uses, mostly for heating) in homes, businesses, and industry. For each of these uses, the accompanying graph shows trends in the emission of carbon dioxide from fossil fuel combustion in the United States.

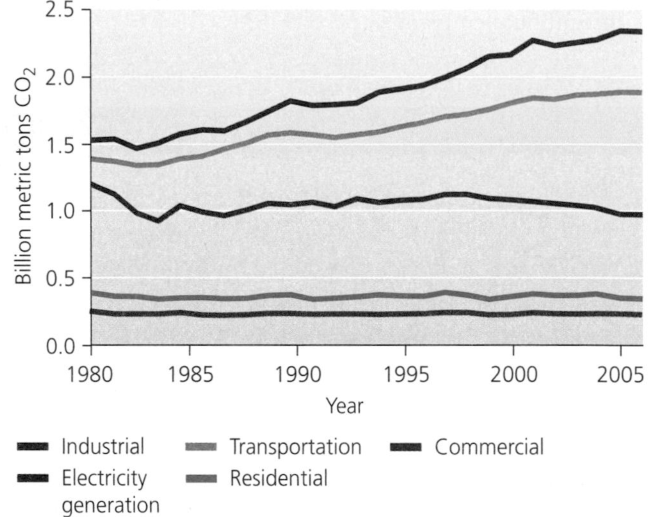


**Emissions of $CO_2$ from fossil fuel combustion by end-use sector in the United States, 1980–2006.** Data from U.S. Department of Energy, Energy Information Administration. 2007. *Annual energy review 2006.*

1. Calculate the approximate percentage changes in $CO_2$ emissions from transportation; electricity generation; and residential, commercial, and industrial primary energy use (mostly heating) between 1980 and 2006.
2. Between 1980 and 2006, U.S. population increased by 32% and the inflation-adjusted U.S. gross domestic product (GDP) more than doubled. What quantitative conclusions can you draw from these data about $CO_2$ emissions per capita? About $CO_2$ emissions per unit of total economic activity? Create a graph and sketch a trend line of $CO_2$ emissions per capita from 1980 to 2006. Now sketch a trend line of $CO_2$ emissions per unit of total economic activity from 1980 to 2006.
3. Imagine you are put in charge of designing a strategy to reduce U.S. emissions of $CO_2$ from fossil fuel combustion. Based on the data presented here, what approaches would you recommend, and how would you prioritize these? Explain your answers.

## CALCULATING ECOLOGICAL FOOTPRINTS

Global climate change is something to which we all contribute, because fossil fuel combustion plays such a large role in supporting the lifestyles we lead. Conversely, as individuals, each one of us can contribute to mitigating global climate change through personal decisions and actions that affect the way we live our lives. Several online calculators enable you to calculate your own personal *carbon footprint*, the amount of carbon emissions for which you are responsible. Go to one of these, at http://www.carbonfootprint.com, follow the link for the U.S. version, take the quiz, and enter the relevant data in the table.

| | Carbon footprint (kg per person per year) |
|---|---|
| **World average** | |
| **Average for industrialized nations** | |
| **U.S. average** | |
| **Your footprint** | |
| **Average needed to halt climate change** | |
| **Your footprint with three changes** | |

1. How does your personal carbon footprint compare to that of the average U.S. resident? How does it compare to that of the average person in the world? Why do you think your footprint differs from these in the ways it does?
2. Think of three changes you could make in your lifestyle that would lower your carbon footprint. Now take the footprint quiz again, incorporating these three changes. Enter your resulting footprint in the table. By how much did you reduce your yearly emissions?
3. Now take the quiz again, trying to make enough changes to reduce your footprint to the level at which we could halt climate change. Do you think you could achieve such a footprint? What do you think would be an admirable yet realistic goal for you to set as a target value for your own footprint? Would you choose to purchase carbon offsets to help reduce your impact? Why or why not?

## Take It Further

Go to www.aw-bc.com/withgott or the student CD-ROM, where you'll find:

- Suggested answers to end-of-chapter questions
- Quizzes, animations, and flashcards to help you study
- *Research Navigator*™ database of credible and reliable sources to assist you with your research projects
- GRAPHit! Tutorials to help you interpret graphs
- INVESTIGATEit! Current news articles that link the topics that you study to case studies from your region to around the world

CHAPTER

# 19 Fossil Fuels, Their Impacts, and Energy Conservation

Oil production facility at Prudhoe Bay, Alaska

## Upon completing this chapter, you will be able to:

- Identify the energy sources that we use
- Describe the nature and origin of coal and evaluate its extraction and use
- Describe the nature and origin of natural gas and evaluate its extraction and use
- Describe the nature and origin of petroleum and evaluate its extraction, use, and future depletion
- Describe the nature, origin, and potential of alternative fossil fuels
- Outline and assess environmental impacts of fossil fuel use
- Evaluate political, social, and economic impacts of fossil fuel use
- Specify strategies for conserving energy and enhancing efficiency

Caribou forced to cross haul road at Prudhoe Bay

## CENTRAL CASE

# Oil or Wilderness on Alaska's North Slope?

**"America is addicted to oil."**
—U.S. President George W. Bush, 2006

**"The era of easy oil is over."**
—Chevron CEO David J. O'Reilly, 2005

Above the Arctic Circle, at the top of the North American continent, the land drops steeply down from the jagged mountains of Alaska's spectacular Brooks Range and stretches north in a vast, flat expanse of tundra until it meets the icy waters of the Arctic Ocean. Few Americans have been to this remote region, yet it has come to symbolize a struggle between two values in our modern life.

For some U.S. citizens, Alaska's North Slope is the last great expanse of wilderness in their industrialized nation—a rare place that people have left untouched. For these millions of Americans, simply knowing that this wilderness still exists is of tremendous value. For millions of others, this land represents something else entirely—a source of petroleum, the natural resource that, more than any other, fuels our society and shapes our way of life. To these people, it seems wrong to leave such an important resource sitting unused in the ground. Those who propose drilling for oil here accuse wilderness preservationists of neglecting the country's economic interests. Advocates for wilderness maintain that drilling will sacrifice the nation's natural heritage for little gain.

These competing visions for Alaska's North Slope exist side by side across three regions of this vast swath of land (**Figure 19.1**). The westernmost portion of the North Slope was set aside in 1923 by the U.S. government as an emergency reserve for petroleum. This parcel of land, the size of Indiana, is called the National Petroleum Reserve–Alaska (NPR–A) and was intended to remain untapped for oil unless the nation faced an emergency. In recent years, most of this region's 9.5 million ha (23.5 million acres) have been opened for development.

East of the National Petroleum Reserve are state lands that were developed after oil was discovered at Prudhoe Bay in 1968. Since drilling began in 1977, we have extracted over 14 billion barrels (1 barrel = 159 L or

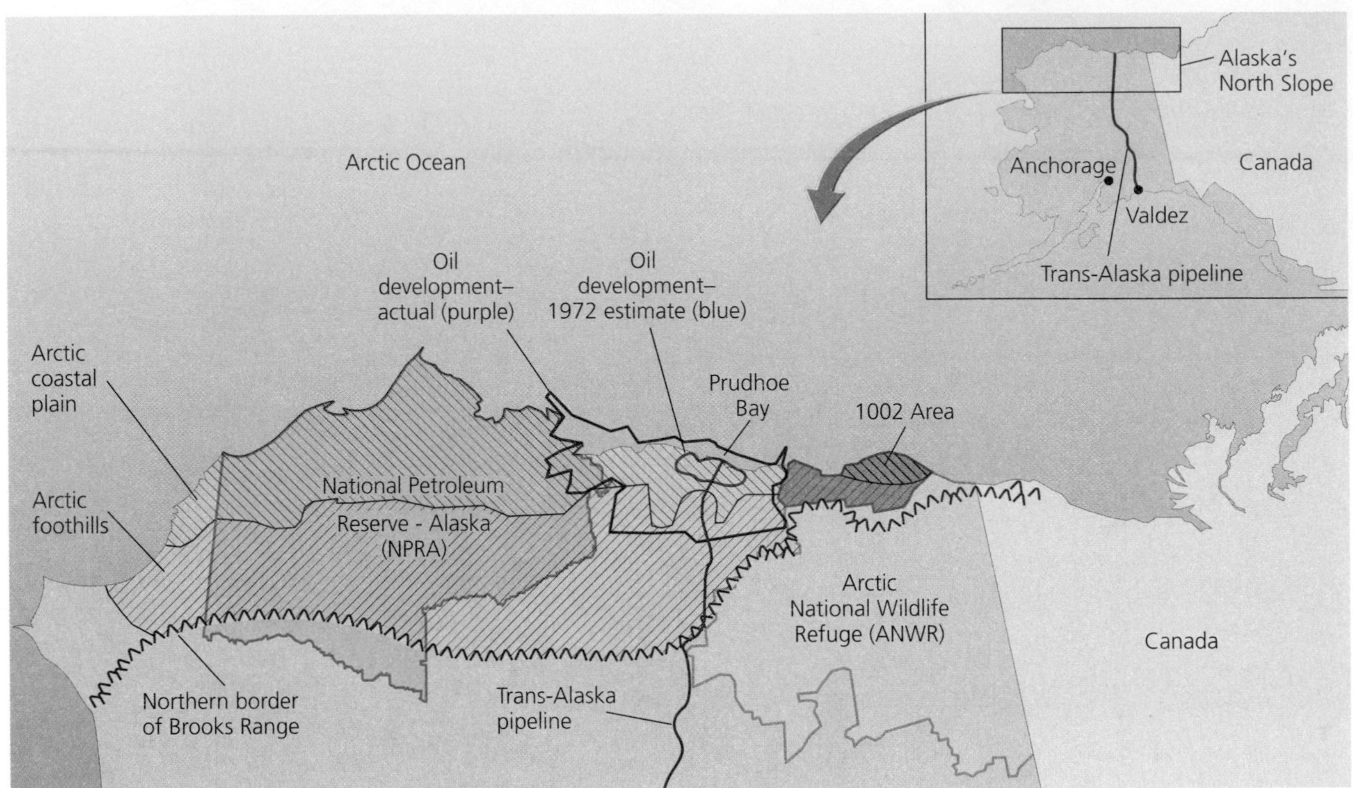


FIGURE 19.1 Alaska's North Slope is the site of both arctic wilderness and oil exploration. In the western portion of this region, the U.S. government established the National Petroleum Reserve–Alaska as an area in which to drill for oil if it is needed in an emergency. It is well explored but not yet widely developed. To the east of this area, the Prudhoe Bay region is the site of widespread oil extraction. Oil development has expanded much farther than experts estimated it would in 1972. Farther east lies the Arctic National Wildlife Refuge, home to untrammeled Arctic wilderness. Debate has focused on whether the 1002 Area of the coastal plain north of the Brooks Range should be opened to oil development.

42 gal) of crude oil from 19 oil fields spread over 160,000 ha (395,000 acres) of this region. The oil is transported across Alaska by the 1,300-km (800-mi) trans-Alaska pipeline south to the port of Valdez, where it is loaded onto tankers.

East of the Prudhoe Bay region lies the Arctic National Wildlife Refuge (ANWR), an area the size of South Carolina consisting of federal lands set aside in 1960 and 1980 mainly to protect wildlife and preserve pristine ecosystems of tundra, mountains, and seacoast. This scenic region is home to 160 nesting bird species, numerous fish and marine mammals, grizzly bears, polar bears, Arctic foxes, timber wolves, musk oxen, and other animals. In summer, thousands of caribou arrive from the south and give birth to and raise their calves. Because of the vast caribou herd and the other large mammals, ANWR has been called "the Serengeti of North America."

ANWR has been the focus of debate for decades. Advocates of oil drilling have tried to open its lands for development, and proponents of wilderness have fought for its preservation. Scientists, oil industry experts, politicians, environmental groups, citizens, and Alaska residents have all been part of the debate. So have the two Native groups in the area, the Gwich'in and the Inupiat, who disagree over whether the refuge should be opened to oil development. The Gwich'in depend on hunting caribou and fear that oil industry activity will reduce caribou herds, whereas the Inupiat see oil extraction as one of the few opportunities for economic development in the area.

In a compromise in 1980, the U.S. Congress put most of the refuge off limits to development but reserved for future decision making a 600,000-ha (1.5-million-acre) area of coastal plain. This region, called the 1002 Area (after Section 1002 of the bill that established it), can be opened for oil development by a vote of both houses of Congress. Its unsettled status has made it the center of the oil-versus-wilderness debate, and Congress has been caught between passionate feelings on both sides for a quarter of a century.

In recent years, Republican-controlled Congresses have repeatedly attempted to open the Arctic Refuge to drilling. The election of a Democratic majority to Congress in 2006 may take the issue off the table in the short term, but as our society depletes the resource it depends on most, and as gasoline prices rise, calls to drill in the refuge will come up again and again.

In addition, we can expect accelerated development of lands in the National Petroleum Reserve–Alaska, which have until now served as a *de facto* wildlife refuge. Stymied in its efforts to drill in ANWR, the George W. Bush administration in 2006—ignoring objections from biologists, environmentalists, and hunters—opened for development an ecologically sensitive area of the NPR–A around Lake Teshekpuk, a key breeding ground for caribou and for many of the continent's migratory waterfowl.

Behind the noisy policy debate over ANWR, scientists have attempted to inform the dialogue through research. Geologists have tried to ascertain how much oil lies underneath the refuge, and biologists have tried to assess the impacts of drilling on Arctic ecosystems. Moreover, scientists and nonscientists alike are debating the need for the oil beneath the refuge for the security and prosperity of the nation, at a time when oil supplies threaten to decline, climate change worsens, and renewable energy sources remain underdeveloped.

## Sources of Energy

Humanity has devised many ways to harness the renewable and nonrenewable forms of energy available on our planet (Table 19.1). We use these energy sources to heat and light our homes, power our machinery, fuel our vehicles, and provide the comforts and conveniences to which we've grown accustomed in the industrial age.

### We use a variety of energy sources

A great deal of energy emanates from Earth's core, enabling us to harness geothermal power. Energy also results from the gravitational pull of the moon and sun, and we are just beginning to harness the power from the ocean tides that these forces generate. An immense amount of energy resides within the bonds among protons and neutrons in atoms, and this energy provides us with nuclear power.

Most of our energy, however, comes from the sun. We can harness energy from the sun's radiation directly in a number of ways. Solar radiation also helps drive wind patterns and the hydrologic cycle, making possible forms of energy such as wind power and hydroelectric power. And of course, sunlight drives photosynthesis (• p. 102–103) and the growth of plants, from which we take wood and other biomass as a fuel source. Finally, when plants die and are preserved in sediments under particular conditions, they may impart their stored chemical energy to **fossil fuels**, highly combustible substances formed from the remains of organisms from past geological ages. The three fossil fuels we use widely today are oil, coal, and natural gas.

Since the industrial revolution, fossil fuels have replaced biomass as our society's dominant source of energy. Global consumption of the three main fossil fuels has risen steadily for years and is now at its highest level ever (**Figure 19.2**). The high energy content of fossil fuels makes them efficient to burn, ship, and store. Besides providing for transportation, heating, and cooking, these fuels are used to generate **electricity**, a secondary form of energy that is easier to transfer over long distances and apply to a variety of uses.

As we first noted in Chapter 1 (• pp. 3–4), energy sources such as sunlight, geothermal energy, and tidal energy are considered perpetually *renewable* because their supplies will not be depleted by our use. Other sources, such as timber, are renewable only if we do not harvest them at too great a rate. In contrast, energy sources such as oil, coal, and natural gas are considered *nonrenewable*, because at our current rates of consumption we will use up Earth's accessible store of them in a matter of decades to centuries. Nuclear power as currently harnessed through fission of

**TABLE 19.1 Energy Sources We Use Today**

| Energy source | Description | Type of energy | Chapter |
|---|---|---|---|
| Crude oil | Fossil fuel extracted from ground (liquid) | Nonrenewable | 19 |
| Natural gas | Fossil fuel extracted from ground (gas) | Nonrenewable | 19 |
| Coal | Fossil fuel extracted from ground (solid) | Nonrenewable | 19 |
| Nuclear energy | Energy from atomic nuclei of uranium | Nonrenewable | 20 |
| Biomass energy | Energy stored in plant matter from photosynthesis | Renewable | 20 |
| Hydropower | Energy from running water | Renewable | 20 |
| Solar energy | Energy from sunlight directly | Renewable | 21 |
| Wind energy | Energy from wind | Renewable | 21 |
| Geothermal energy | Earth's internal heat rising from core | Renewable | 21 |
| Tidal and wave energy | Energy from tidal forces and ocean waves | Renewable | 21 |

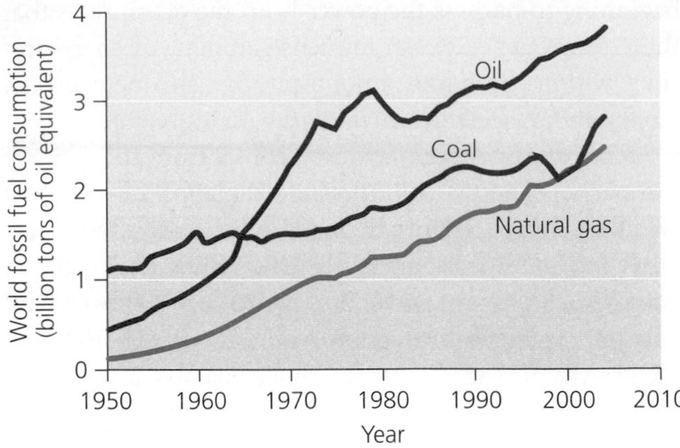


FIGURE 19.2 Global consumption of fossil fuels has risen greatly over the past half century. Oil use rose steeply during the 1960s to overtake coal, and today it remains our leading energy source. Data from Worldwatch Institute, 2006. *Vital signs 2006–2007.*

uranium (• p. 576) can be considered nonrenewable to the extent that uranium ore is in limited supply.

Although these nonrenewable fuels result from ongoing natural processes, the timescales on which they are created are so long that once the fuels are depleted, they cannot be replaced within any time span useful to our civilization. It takes a thousand years for the biosphere to generate the amount of organic matter that must be buried to produce a single day's worth of fossil fuels for our society. To replenish the fossil fuels we have depleted so far would take many millions of years. For this reason, and because fossil fuels exert severe environmental impacts, renewable energy sources increasingly are being developed as alternatives to fossil fuels, as we will see in Chapters 20 and 21.

## Fossil fuels are indeed fuels created from "fossils"

The fossil fuels we burn today in our vehicles, homes, industries, and power plants were formed from the tissues of organisms that lived 100–500 million years ago. The energy these fuels contain came originally from the sun and was converted to chemical-bond energy as a result of photosynthesis (• pp. 102–103). The chemical energy in these organisms' tissues then became concentrated as these tissues decomposed and their hydrocarbon compounds were altered and compressed (**Figure 19.3**).

Most organisms, after death, do not end up as part of a coal, gas, or oil deposit. A tree that falls and decays as a rotting log undergoes mostly **aerobic** decomposition; in the presence of air, bacteria and other organisms that use oxygen break down plant and animal remains into simpler carbon molecules that are recycled through the ecosystem. Fossil fuels are produced only when organic material is broken down in an **anaerobic** environment,

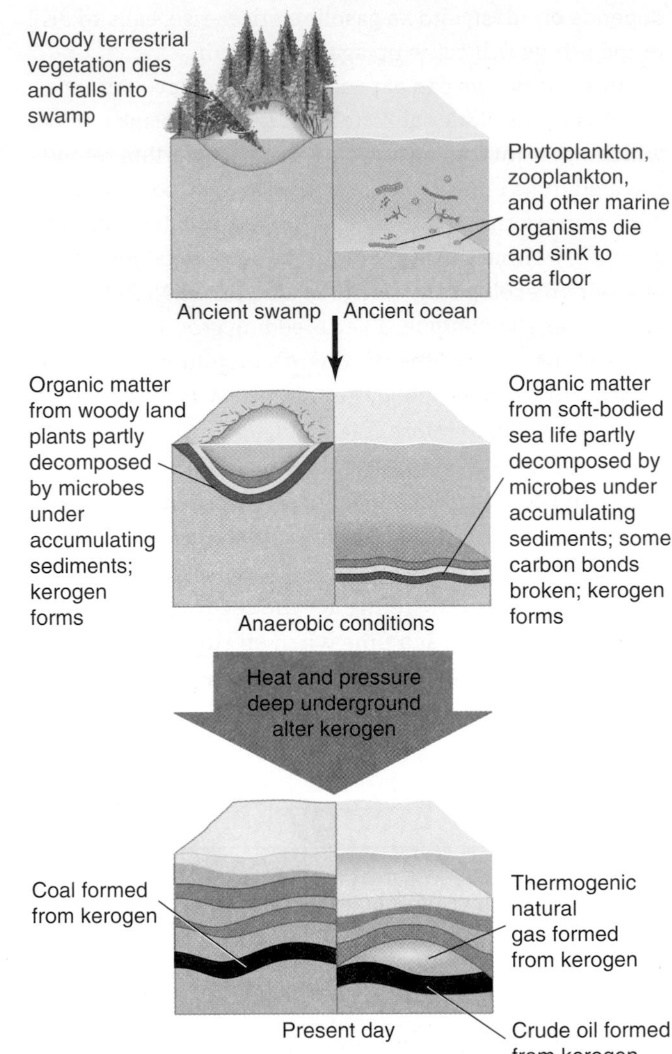


FIGURE 19.3 Fossil fuels begin to form when organisms die and end up in oxygen-poor conditions, such as when trees fall into lakes and are buried by sediment, or when phytoplankton and zooplankton drift to the seafloor and are buried (top diagram). Organic matter that undergoes slow anaerobic decomposition deep under sediments forms kerogen (middle diagram). Geothermal heating then acts on kerogen to create crude oil and natural gas (bottom diagram). Natural gas can also be produced nearer the surface by anaerobic bacterial decomposition of organic matter. Oil and gas come to reside in porous rock layers beneath dense, impervious layers. Coal is formed when plant matter is compacted so tightly that there is little decomposition.

one that has little or no oxygen. Such environments include the bottoms of deep lakes, swamps, and shallow seas. Over millions of years, organic matter that accumulates at the bottoms of such water bodies is converted into crude oil, natural gas, or coal. Which fuel forms in any given place depends on the chemical composition of the starting material, the temperatures and pressures to which the material is subjected, the presence or absence of anaerobic decomposers, and the passage of time.

## Fossil fuel reserves are unevenly distributed

Fossil fuel deposits are localized and unevenly distributed over Earth's surface, so some regions have substantial reserves of fossil fuels whereas others have very few. How long each nation's fossil fuel reserves will last depends on how much the nation extracts, how much it consumes, and how much it imports from and exports to other nations. Nearly two-thirds of the world's proven reserves of crude oil lie in the Middle East. The Middle East is also rich in natural gas, but Russia contains more than twice as much natural gas as any other country. Russia is also rich in coal, as is China, but the United States possesses more coal than any other nation (Table 19.2).

**TABLE 19.2 Nations with the Largest Proven Reserves of Fossil Fuels**

| Oil (% world reserves) | Natural gas (% world reserves) | Coal (% world reserves) |
|---|---|---|
| Saudi Arabia, 21.9 | Russia, 26.3 | United States, 27.1 |
| Iran, 11.4 | Iran, 15.5 | Russia, 17.3 |
| Iraq, 9.5 | Qatar, 14.0 | China, 12.6 |
| Kuwait, 8.4 | Saudi Arabia, 3.9 | India, 10.2 |
| United Arab Emirates, 8.1 | United Arab Emirates, 3.3 | Australia, 8.6 |
| Venezuela, 6.6 | United States, 3.3 | South Africa, 5.4 |
| Russia, 6.6 | Nigeria, 2.9 | Ukraine, 3.8 |
| Libya, 3.4 | Algeria, 2.5 | Kazakhstan, 3.4 |
| Kazakhstan, 3.3 | Venezuela, 2.4 | Poland, 1.5 |
| Nigeria, 3.0 | Iraq, 1.7 | Brazil, 1.1 |

Data from British Petroleum. 2007. *Statistical review of world energy 2007.*

## Developed nations consume more energy than developing nations

Citizens of developed regions generally consume far more energy than do those of developing regions (**Figure 19.4**). Per person, the most-industrialized nations use up to 100 times more energy than do the least-industrialized nations. Although the United States has only 4.6% of the world's population, it consumes 22.5% of its energy.

Moreover, developed and developing nations tend to apportion their energy use differently. Industrialized nations use roughly one-third of their energy on transportation, one-third on industry, and one-third on all other uses. Developing nations devote a greater proportion of their energy to subsistence activities such as agriculture,

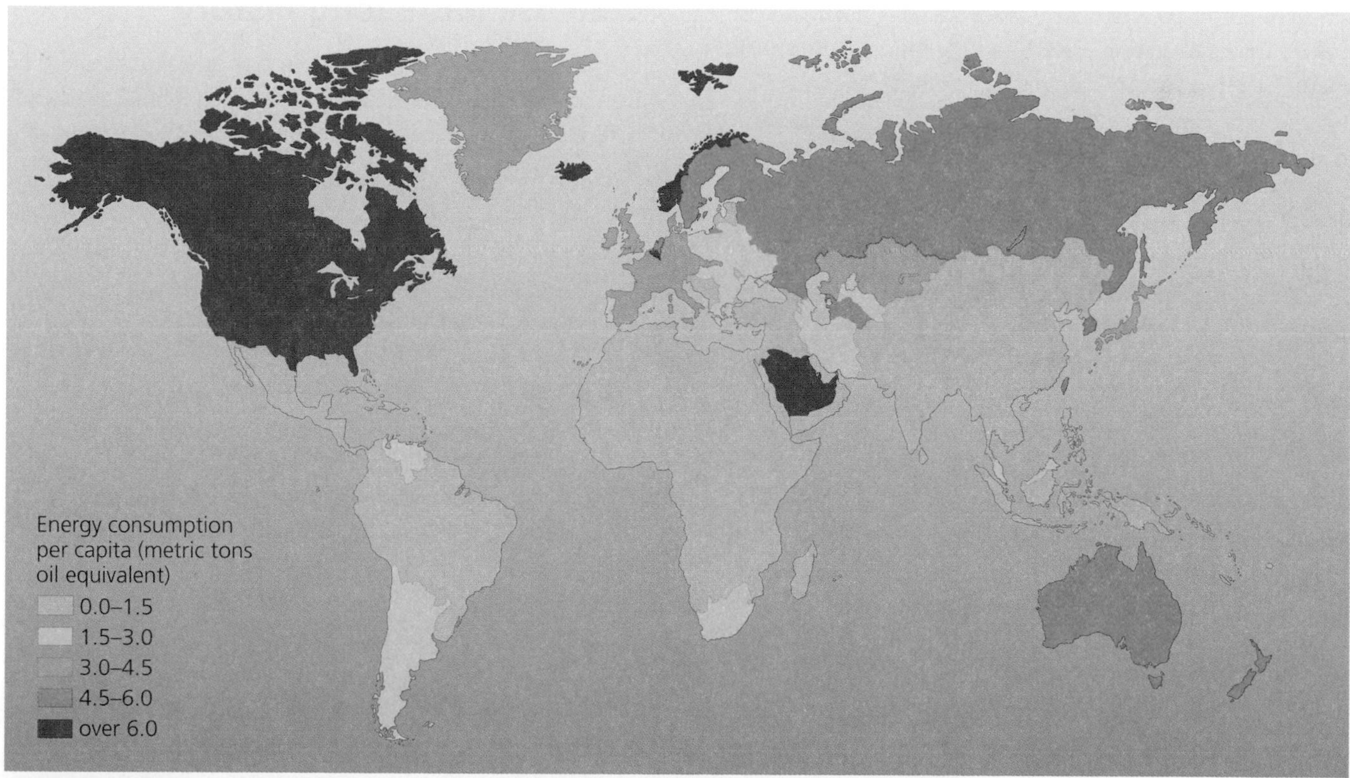


**FIGURE 19.4** Regions vary greatly in their consumption of energy per person. People in industrialized nations consume the most. This map combines all types of energy, standardized to metric tons of "oil equivalent," that is, the amount of fuel needed to produce the energy gained from combusting one metric ton of crude oil. Data from British Petroleum. 2007. *Statistical review of world energy 2007.*

food preparation, and home heating, and substantially less on transportation. In addition, people in developing countries often rely on manual or animal energy sources instead of automated ones. For instance, rice farmers in Bali plant rice by hand, but industrial rice growers in California use airplanes. Because industrialized nations rely more on equipment and technology, they use more fossil fuels. In the United States, oil, coal, and natural gas together supply 89% of energy needs.

### It takes energy to make energy

We don't simply get energy for free. To harness, extract, process, and deliver the energy that we use requires that we invest substantial inputs of energy. For instance, drilling for oil at Prudhoe Bay has required the construction of an immense infrastructure of roads, wells, vehicles, storage tanks, pipelines, housing for workers, and more—all requiring the use of energy. Piping and shipping the oil out of the North Slope, and then refining the crude oil into products we can use, requires further energy inputs. Thus, when evaluating the value of an energy source, it is important to subtract costs in energy invested from benefits in energy received. **Net energy** expresses the difference between energy returned and energy invested:

Net energy = Energy returned − Energy invested

When comparing energy sources, it is useful to use a ratio often denoted as **EROI**—*energy returned on investment.* EROI ratios are calculated as follows:

EROI = Energy returned/Energy invested

Higher ratios mean that we receive more energy from each unit of energy that we invest. Fossil fuels are widely used because their EROI ratios are high. However, EROI ratios can change over time. For instance, those for U.S. oil and natural gas declined from over 100:1 in the 1940s to about 30:1 in the 1970s, and today they hover around 5:1. The EROI ratios have declined because we extracted the easiest deposits first and now must work harder and harder to extract the remaining amounts.

## Coal

Coal is the world's most abundant fossil fuel. The proliferation 300–400 million years ago of swampy environments where organic material could be buried has resulted in substantial coal deposits throughout the world. **Coal** is organic matter (generally woody plant material) that was compressed under very high pressure to form dense, solid carbon structures (**Figure 19.5**). Coal typically results when little decomposition takes place because the material cannot be digested or appropriate decomposers are not present. One-quarter of the world's coal is located in the United States, and coal provides one-quarter of the world's commercial energy consumption.

### Coal use has a long history

People have used coal longer than any other fossil fuel. The Romans used coal for heating in the second and third centuries in Britain, as have people in parts of China for

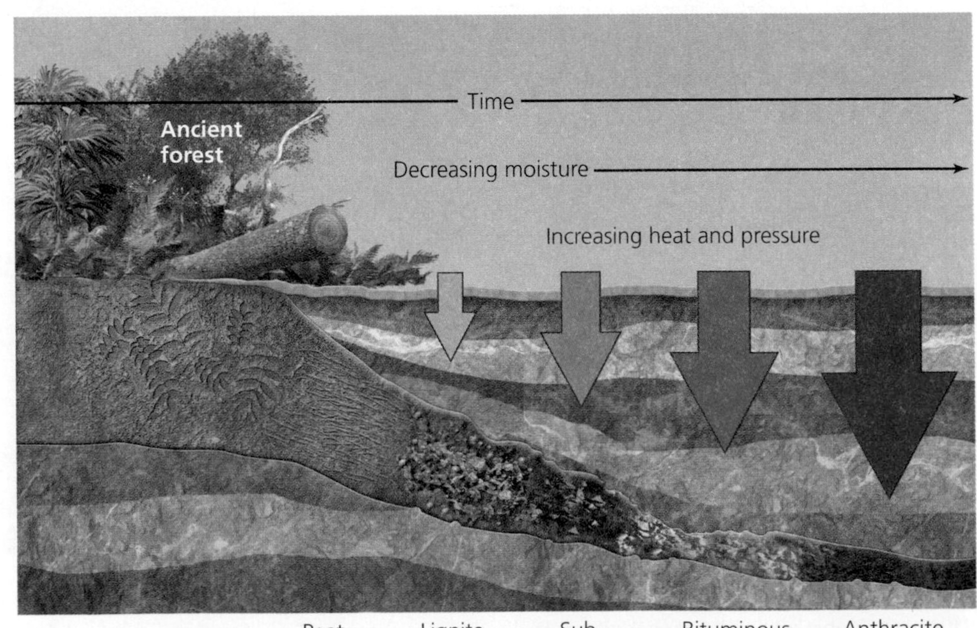


FIGURE 19.5 Coal forms as ancient plant matter is compacted underground. Scientists categorize coal into several types, depending on the amount of heat, pressure, and moisture involved in its formation. Anthracite coal is formed under greatest pressure, where temperatures are high and moisture content is low. Lignite coal is formed under conditions of much less pressure and heat, but more moisture. Peat is also part of this continuum, representing plant matter that is minimally compacted.

TABLE 19.3 Top Producers and Consumers of Coal

| Production (% world production) | Consumption (% world consumption) |
|---|---|
| China, 39.4 | China, 38.6 |
| United States, 19.3 | United States, 18.4 |
| India, 6.8 | India, 7.7 |
| Australia, 6.6 | Japan, 3.9 |
| South Africa, 4.7 | Russia, 3.6 |
| Russia, 4.7 | South Africa, 3.0 |
| Indonesia, 3.9 | Germany, 2.7 |
| Poland, 2.2 | Poland, 1.9 |
| Germany, 1.6 | South Korea, 1.8 |
| Kazakhstan, 1.6 | Australia, 1.7 |

Data from British Petroleum. 2007. *Statistical review of world energy 2007.*

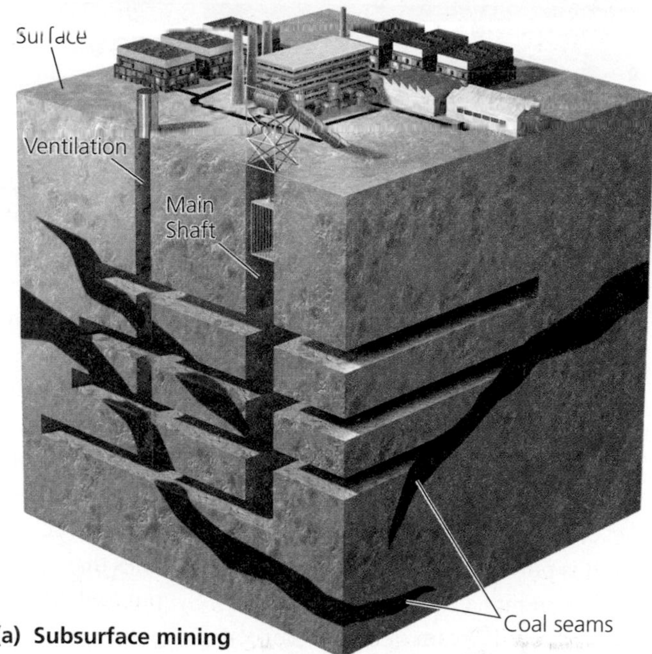

**(a) Subsurface mining**

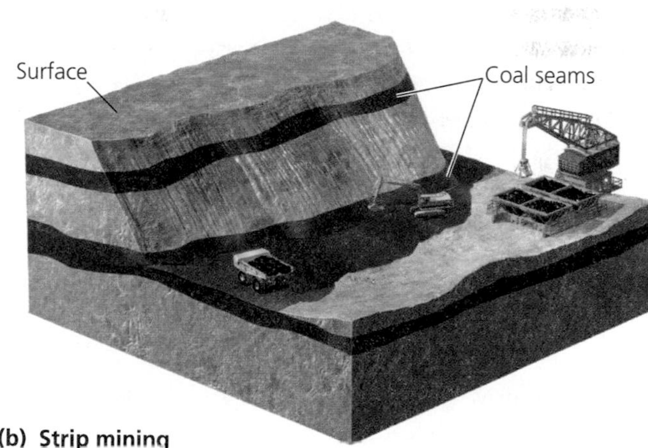

**(b) Strip mining**

FIGURE 19.6 Coal is mined in two major ways. In subsurface mining (**a**), miners work below ground in shafts and tunnels blasted through the rock; these passageways provide access to underground seams of coal. This type of mining poses dangers and long-term health risks to miners. In strip mining (**b**), soil is removed from the surface, exposing coal seams from which coal is mined. This type of mining can cause substantial environmental impact.

2,000–3,000 years. In Arizona, Native Americans of the Hopi Nation follow ancestral traditions by using coal to fire pottery, cook food, and heat their homes. Once commercial mining began in Europe in the 1700s, people began using coal widely as a heating source. Coal found an expanded market after the invention of the steam engine, because it was used to boil water to produce steam. Coal-fired steam engines helped drive the industrial revolution, powering factories, agriculture, trains, and ships. The birth of the steel industry in 1875 increased demand still further because coal fueled the furnaces used to produce steel.

In the 1880s, people began to use coal to generate electricity. In coal-fired power plants, coal combustion converts water to steam, which turns a turbine to create electricity (see "The Science behind the Story," • pp. 548–549). Today coal provides over half the electrical generating capacity of the United States. China and the United States are the primary producers and consumers of coal (**Table 19.3**).

## Coal is mined from the surface and from below ground

We extract coal using two major methods (**Figure 19.6**). We reach underground deposits with **subsurface mining**. Shafts are dug deep into the ground, and networks of tunnels are dug or blasted out to follow coal seams. The coal is removed systematically and shipped to the surface. When coal deposits are at or near the surface, strip-mining methods are used. In **strip mining**, heavy machinery removes huge amounts of earth to expose and extract the coal. The pits are subsequently refilled with the soil that had been removed. Strip-mining operations can occur on immense scales; in some cases entire mountaintops are lopped off. This environmentally destructive process, called **mountaintop removal** (• p. 560), has become more common recently in the Appalachian Mountains. We will explore some of coal's environmental impacts later in this chapter.

## Coal varies in its qualities

Coal varies from deposit to deposit in many ways, including in its water content and the amount of potential energy it contains. Organic material that is broken down

THE SCIENCE BEHIND THE STORY

## How Electricity Is Generated

*Worker checking furnace at the "Big Cajun 2" coal-fired power plant in Louisiana*

We use coal to generate electricity in a process that dates back more than a century (see the figure). Once mined, coal is hauled to power plants, where it is pulverized. The crushed coal is blown into a boiler furnace on a superheated stream of air and burned in a blaze of intense heat—typical furnace temperatures often flare at 815° C (1,500° F).

Water circulating around the boiler absorbs the heat and is converted to high-pressure steam. This steam is injected into a **turbine**, a rotary device that converts the kinetic energy of a moving substance such as steam into mechanical energy. As steam from the boiler exerts pressure on the blades of the turbine, they spin, turning the turbine's drive shaft.

The drive shaft is connected to a generator, which features a rotor that rotates and a stator that remains stationary. Generators make use of a phenomenon that you may have experimented with yourself in your high-school physics class: Moving magnets adjacent to coils of copper wire cause electrons in the copper wires to move, generating alternating electric current.

In some generators, the rotor consists of magnets and spins within a stator of coiled copper wire. As the turbine's drive shaft rotates, it causes the rotor to revolve, creating a magnetic field and causing electrons in the stator's copper wires to move, thereby creating an electrical current. This current flows into transmission lines that travel from the power plant out to the customers who use the plant's electricity.

As the United States tries to balance its growing demand for electricity with rising concerns about environmental and health impacts of coal combustion, power plants continue to rely heavily on coal while scientists work to limit the pollution that use of this fuel creates.

The national energy legislation passed by the U.S. Congress in 2005 provided $1.3 billion in tax credits to power plants that use state-of-the-art "clean coal" technologies, and it set aside $2.5 billion over 10 years for further research into such technologies. **Clean coal** technologies largely focus on approaches to rid the generation process of toxic chemicals before or after the coal is burned.

Technologies to do this include *scrubbers* (• pp. 484–485), materials based on minerals such as calcium or sodium that absorb and remove sulfur dioxide ($SO_2$) from smokestack emissions. Other approaches use chemical reactions to strip away nitrogen oxides ($NO_X$), breaking them down into elemental nitrogen and water. Multilayered filtering devices are used to capture tiny ash particles.

Research at the Niles Station power plant in Ohio found a way to remove 95% of $SO_2$ and 90% of $NO_X$ from the plant's emissions. Researchers first installed smokestacks with efficient filter bags and captured significant amounts of ash and gases. The filters also proved effective at removing dangerous substances such as mercury and selenium. The captured ash and gases were reheated, and small amounts of ammonia were added to chemically convert $NO_X$ into harmless nitrogen gas and water vapor. Further addition of oxygen to $SO_2$ in the smokestack gas converted the sulfur compounds into sulfuric acid, which the plant bottled and sold for use in fertilizer and manufacturing processes. The gas-scrubbing process worked well enough for the plant to begin using it permanently, and several power plants in Europe adopted the technology.

anaerobically but remains wet, near the surface, and not well compressed is called **peat**. A kind of precursor stage to coal, peat has been widely used as a fuel in Britain and other locations. As peat decomposes further, as it becomes buried more deeply under sediments, as pressure and heat increase, and as time passes, water is squeezed out of the material, and carbon compounds are packed more tightly together, forming coal. Scientists classify coal into four types: lignite, sub-bituminous, bituminous, and anthracite. Lignite is the least-compressed type of coal, and anthracite is the most-compressed type (see Figure 19.5). The greater the compression, the greater is the energy content per unit volume.

Most coal contains impurities, including sulfur, mercury, arsenic, and other trace metals, and coal deposits vary in the amount of impurities they contain. Sulfur content depends in part on whether the coal was formed in freshwater or saltwater sediments. Coal in what is today the eastern United States tends to be high in sulfur because it was formed in marine sediments, where sulfur from seawater was present.

When high-sulfur coal is burned, it produces sulfate air pollutants, which contribute to industrial smog and

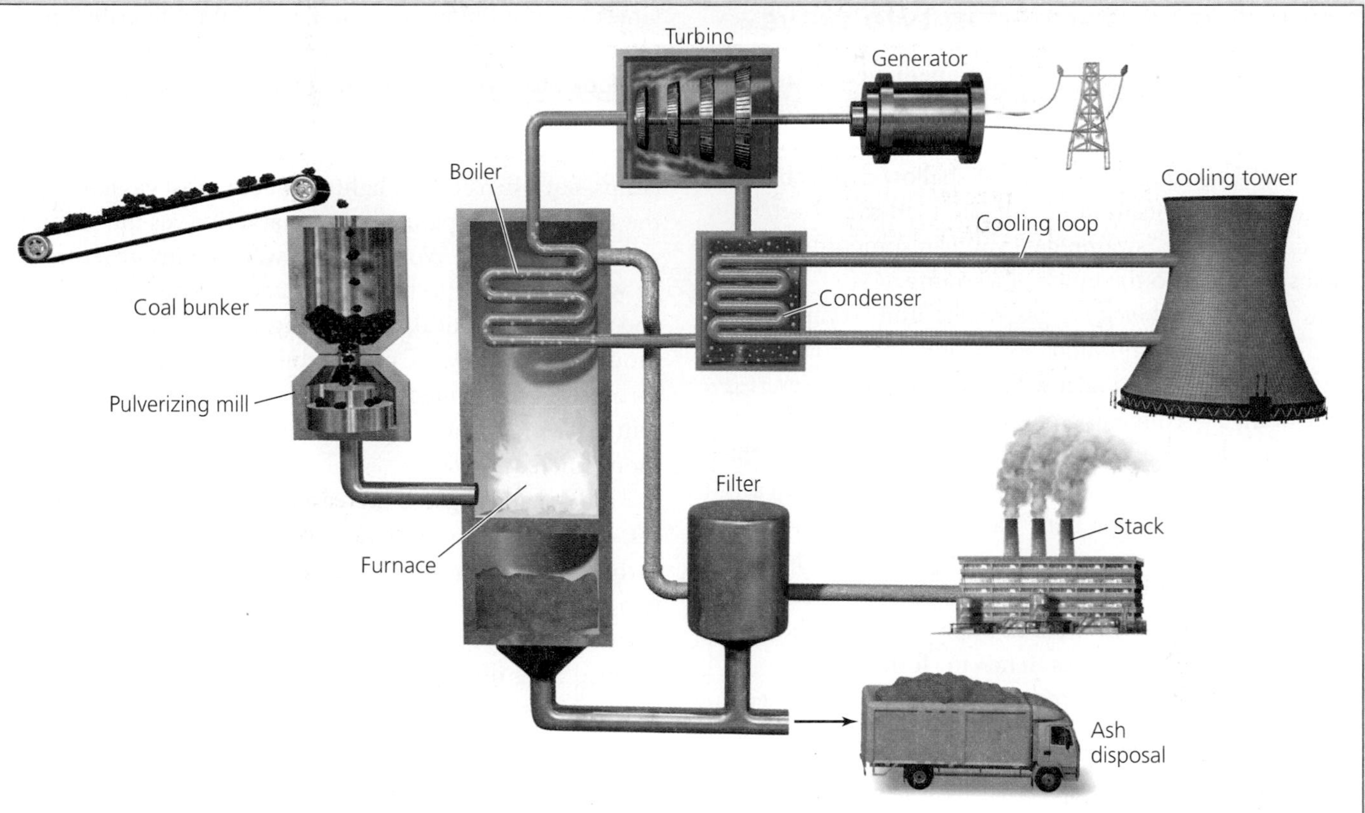


Coal is the primary fuel source used to generate electricity in the United States. Pieces of coal are pulverized and blown into a high-temperature furnace. Heat from the combustion boils water, and the resulting steam turns a turbine, generating electricity by passing magnets past copper coils. The steam is then cooled and condensed in a cooling loop and returned to the furnace. "Clean coal" technologies help filter out pollutants from the combustion process, and toxic ash residue is disposed of at hazardous waste disposal sites.

Some energy analysts and environmental advocates question a policy emphasis on clean coal. Coal, they maintain, is an inherently dirty way of generating power and should be replaced outright with cleaner energy sources.

Nonetheless, the U.S. government is forging ahead with plans to build a protoype of a zero-emissions coal-fired power plant. Working with 11 energy companies and two other national governments, the $1 billion *FutureGen* project aims to design, construct, and operate a power plant that burns coal using gasification and combined cycle generation, produces electricity and hydrogen, and captures its carbon dioxide emissions and sequesters them underground. (At this point the sequestration technology is least developed.) Four sites in Illinois and Texas are being considered for the plant, and if all goes well, it could be operating in 2012.

acidic deposition (• pp. 491–495). Combustion of coal high in mercury content emits mercury that can bioaccumulate in organisms' tissues, poisoning animals as it moves up food chains. Such pollution problems commonly occur downwind of coal-fired power plants. Scientists and engineers are seeking ways to cleanse coal of its impurities so that it can continue to be used as an energy source while minimizing impact on health and the environment (see "The Science behind the Story," above). Reducing pollution from coal is important because society's demand for this relatively abundant fossil fuel may soon rise as supplies of oil and natural gas decline.

## Natural Gas

Natural gas is the fastest-growing fossil fuel in use today and provides for one-quarter of global commercial energy consumption. It is a much cleaner-burning fuel than coal or oil, so it produces less pollution. World supplies of natural gas are projected to last perhaps 60 more years.

## Natural gas is formed in two ways

**Natural gas** consists primarily of methane, $CH_4$, and typically includes varying amounts of other volatile hydrocarbons. Natural gas can arise from either of two processes. *Biogenic* gas is created at shallow depths by the anaerobic decomposition of organic matter by bacteria. An example is the "swamp gas" you can sometimes smell when stepping into the muck of a swamp.

In contrast, *thermogenic* gas results from compression and heat deep underground. As organic matter is buried more and more deeply under sediments, the pressure exerted by the overlying sediments grows, and temperatures increase. Carbon bonds in the organic matter begin breaking, and the organic matter turns to a substance called *kerogen,* which acts as a source material for both natural gas and crude oil. Further heat and pressure act on the kerogen to degrade complex organic molecules into simpler hydrocarbon molecules. At very deep levels—below about 3 km (1.9 mi)—the high temperatures and pressures tend to form natural gas. Whereas biogenic gas is nearly pure methane, thermogenic gas contains small amounts of other gases as well as methane.

Thermogenic gas may be formed directly, along with coal or crude oil, or from coal or oil that is altered by heating. Most gas extracted commercially is thermogenic and is found above deposits of crude oil or seams of coal, so its extraction often accompanies the extraction of those fossil fuels.

All too often, however, natural gas goes to waste as it escapes from coal mines or oil wells. Methane from coal seams, called *coalbed methane*, commonly leaks to the atmosphere during mining. To avoid this waste, and because methane is a potent greenhouse gas that contributes to climate change (• pp. 507–508), mining engineers are now trying to capture more of this gas for energy. Likewise, in most remote oil-drilling areas, where the transport of natural gas remains prohibitively expensive, natural gas is flared—wasted by being simply burned off. In Alaska, gas captured during oil drilling requires too much expense to export, but now it is being reinjected into the ground for potential future extraction.

One source of biogenic natural gas is the decay process in landfills, and many landfill operators are now capturing this gas to sell as fuel (• pp. 638–639). This practice decreases energy waste, can be profitable for the operator, and helps reduce the atmospheric release of methane.

## Natural gas has only recently been widely used

Throughout history, naturally occurring seeps of natural gas would occasionally be ignited by lightning and could be seen burning in parts of what is now Iraq, inspiring the Greek essayist Plutarch around A.D. 100 to describe their "eternal fires." The first commercial extraction of natural gas took place in 1821, but until recently its use was localized because technology did not exist to pipe gas safely over long distances. Natural gas was used to fuel streetlamps, but when electric lights replaced most gas lamps in the 1890s, gas companies began marketing gas for heating and cooking. After World War II, wartime improvements in welding and pipe building made gas transport safer and more economical, and during the 1950s and 1960s, thousands of miles of underground pipelines were laid throughout the United States. If laid end to end, our current network of natural gas pipelines would extend to the moon and back twice.

Today natural gas is increasingly favored because it is versatile and clean-burning, emitting just half as much carbon dioxide per unit of energy produced as coal and two-thirds as much as oil. Converted to a liquid at low temperatures (*liquefied natural gas,* or *LNG*), it can be shipped long distances in refrigerated tankers, although this poses risks of catastrophic explosions. Natural gas deposits are greatest in Russia and the Middle East, and Russia and the United States lead the world in gas production and gas consumption, respectively (Table 19.4).

## Natural gas extraction becomes more challenging with time

To access some natural gas deposits, prospectors need only drill an opening, because pressure and low molecular weight drive the gas upward naturally. The first gas fields to be tapped were of this type. Most fields remaining today, however, require that gas be pumped to Earth's surface. In Texas, Kansas, California, and other areas of

**TABLE 19.4 Top Producers and Consumers of Natural Gas**

| Production (% world production) | Consumption (% world consumption) |
|---|---|
| Russia, 21.3 | United States, 22.0 |
| United States, 18.5 | Russia, 15.1 |
| Canada, 6.5 | Iran, 3.7 |
| Iran, 3.7 | Canada, 3.4 |
| Norway, 3.0 | United Kingdom, 3.2 |
| Algeria, 2.9 | Germany, 3.0 |
| United Kingdom, 2.8 | Japan, 3.0 |
| Indonesia, 2.6 | Italy, 2.7 |
| Saudi Arabia, 2.6 | Saudi Arabia, 2.6 |
| Netherlands, 2.2 | Ukraine, 2.3 |

Data from British Petroleum. 2007. *Statistical review of world energy 2007.*

FIGURE 19.7 Horsehead pumps are used to extract natural gas as well as oil. They are a common feature of the landscape in areas such as west Texas. The pumping motion of the machinery draws gas and oil upward from below ground.

the United States, it is common to see a device called a horsehead pump (**Figure 19.7**). This pump moves a rod in and out of a shaft, creating pressure to pull both natural gas and crude oil to the surface.

As with oil and coal, many of the most accessible natural gas reserves have already been exhausted. The average U.S. well produces just over one-quarter as much as it did in 1970. Thus, much extraction today makes use of sophisticated techniques to break into rock formations and pump gas to the surface. One such "fracturing technique" is to pump salt water under high pressure into the rocks to crack them. Sand or small glass beads are inserted to hold the cracks open once the water is withdrawn.

## Offshore drilling produces much of our gas and oil

Drilling for natural gas, as well as for oil, takes place not just on land but also in the seafloor on the continental shelves. Offshore drilling has required developing technology that can withstand the forces of wind, waves, and ocean currents. Some drilling platforms are fixed, standing platforms built with unusual strength. Others are resilient floating platforms anchored in place above the drilling site. Roughly 25% of the gas and oil extracted in the United States comes from offshore drilling sites, primarily in the Gulf of Mexico and off the southern California coast. This is why Hurricanes Katrina and Rita in 2005 caused such disruption to U.S. oil and gas supplies. By damaging offshore platforms off the coasts of Louisiana and neighboring states and by damaging refineries onshore, the storms interrupted a substantial portion of the nation's oil and gas supply, and prices rose accordingly.

# Oil

Oil has been the world's most-used fuel since the 1960s, when it eclipsed coal. It now accounts for 37% of the world's commercial energy consumption. Its use worldwide over the past decade has risen over 17%.

## Heat and pressure underground form petroleum

The sludgelike liquid we know as **crude oil**, or **petroleum**, tends to form within a window of temperature and pressure conditions often found 1.5–3 km (1–2 mi) below the surface. Crude oil is a mixture of hundreds of different types of hydrocarbon molecules characterized by carbon chains of different lengths (• pp. 95–96). A chain's length affects its chemical properties, which has consequences for human use, such as whether a given fuel burns cleanly in a car engine. Oil refineries sort the various hydrocarbons of crude oil, separating those intended for use in gasoline engines from those, such as tar and asphalt, used for other purposes.

The crude oil of Alaska's North Slope was formed when dead plant material (and small amounts of animal material) drifted down through coastal marine waters millions of years ago and was buried in sediments on the ocean floor. These organic remains were then transformed by time, heat, and pressure into the crude oil of today.

## The age of oil began in the mid-19th century

People have used solid forms of oil (such as tar and asphalt) from deposits easily accessible at Earth's surface for at least 6,000 years. The modern extraction and use of petroleum for energy began in the 1850s. In Pennsylvania, miners drilling for salt occasionally encountered oily rocks instead. At first, entrepreneurs bottled the crude oil from these deposits and sold it as a healing aid, unaware that crude oil is carcinogenic when applied to the skin and poisonous when ingested. Soon, however, a Dartmouth College scholar named George Bissell realized that this "rock oil" could be used to light lamps and lubricate machinery, and in 1854 Bissell started the Pennsylvania Rock Oil Company.

By the time the firm developed its drilling technology and struck oil, Bissell was no longer at the company. Instead, Edwin Drake is credited with drilling the world's first oil well, in Titusville, Pennsylvania, in 1859. Over the next 40 years, Pennsylvania's oil fields produced half the world's oil supply and helped establish a fossil-fuel-based economy that would hold sway for decades to come.

**TABLE 19.5 Top Producers and Consumers of Oil**

| Production (% world production) | Consumption (% world consumption) |
|---|---|
| Saudi Arabia, 13.1 | United States, 24.1 |
| Russia, 12.3 | China, 9.0 |
| United States, 8.0 | Japan, 6.0 |
| Iran, 5.4 | Russia, 3.3 |
| China, 4.7 | Germany, 3.2 |
| Mexico, 4.7 | India, 3.1 |
| Canada, 3.9 | South Korea, 2.7 |
| Venezuela, 3.7 | Canada, 2.5 |
| Kuwait, 3.4 | France, 2.4 |
| Norway, 3.3 | Saudi Arabia, 2.4 |

Data from British Petroleum. 2007. *Statistical review of world energy 2007.*

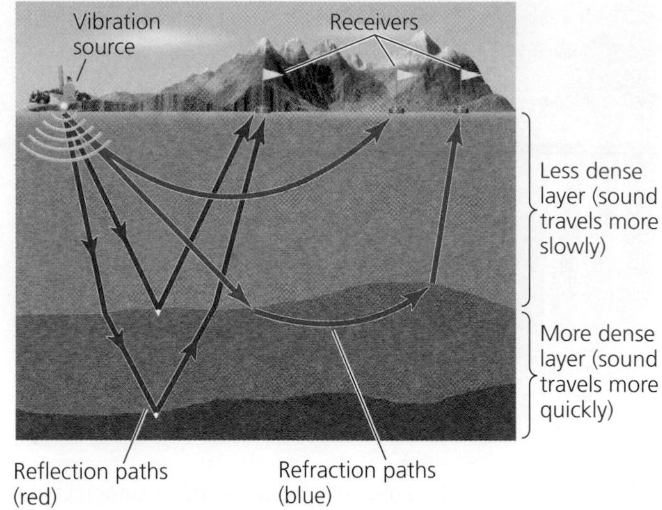


**FIGURE 19.8** Petroleum geologists use seismic surveying to locate fossil fuel deposits. One method is to create powerful vibrations (by exploding dynamite, thumping the ground with a large weight, or using an electric vibrating machine) at the surface in one location and then measure how long it takes the seismic waves to reach receivers at other surface locations. Seismic waves travel more quickly through denser layers, and density differences in the substrate cause waves to reflect off layers, refract, or bend. Scientists and engineers interpret the patterns of wave reception to infer the densities, thicknesses, and location of underlying geological layers—which in turn provide clues about the location and size of fuel deposits.

Today our global society produces and consumes nearly 750 L (200 gal) of oil each year for every man, woman, and child. The United States consumes nearly one-fourth of the world's oil. U.S. oil consumption has increased 16% in the past decade and shows little sign of abating. Table 19.5 shows the top oil-producing and oil-consuming nations.

## Petroleum geologists infer the location and size of deposits

Because petroleum forms only under certain conditions, it occurs in isolated deposits. Once geothermal heating separates hydrocarbons from their source material and produces crude oil, this liquid migrates upward through rock pores, sometimes assisted by seismic faulting. It tends to collect in porous layers beneath dense, impermeable layers.

Geologists searching for oil (or other fossil fuels) drill rock cores and conduct ground, air, and seismic surveys (**Figure 19.8**) to map underground rock formations, understand geological history, and predict where fossil fuel deposits might lie. Using such techniques, geologists from the U.S. Geological Survey (USGS) in 1998 estimated, with 95% certainty, the total amount of oil underneath ANWR's 1002 Area to be between 11.6 and 31.5 billion barrels. The geologists' average estimate of 20.7 billion represents their best guess as to the number of barrels of oil the 1002 Area holds. This amount is enough to supply the United States for 33 months at its current rate of consumption.

Some portion of oil will be impossible to extract using current technology, however, and will need to wait for future advances in extraction equipment or methods. Thus, estimates are generally made of "technically recoverable" amounts of fuels. In its 1998 estimates, the USGS calculated technically recoverable amounts of oil under the 1002 Area to be between 4.3 and 11.8 billion barrels, with an average estimate of 7.7 billion barrels (an amount equivalent to 1 year of U.S. consumption). Because some Native lands and state-owned offshore areas can be developed if the 1002 Area is developed, the USGS also surveyed these areas and estimated that 1.4–4.2 billion additional barrels likely lie beneath them. In 2002, the USGS estimated that the National Petroleum Reserve–Alaska contained 9.3 billion barrels of technically recoverable oil, spread over an area 15 times as large as ANWR.

However, oil companies will not be willing to extract these entire amounts. Some oil would be so difficult to extract that the expense of doing so would exceed the income the company would receive from the oil's sale. Thus, the amount a company chooses to drill for will be determined by the costs of extraction (and transportation), together with the current price of oil on the world market. Because the price of oil fluctuates, the portion of oil from a given deposit that is "economically recoverable" fluctuates as well. USGS scientists calculated that at a price of $30 per barrel, 3.0–10.4 billion barrels would be economically worthwhile to recover from the 1002 Area. The USGS did not present estimates for today's much higher prices, but as prices climb, economically recoverable amounts approach technically recoverable amounts.

Thus, technology sets a limit on the amount that *can* be extracted, whereas economics determines how much *will* be extracted. The amount of oil, or any other fossil fuel, in a deposit that is technologically and economically feasible to remove under current conditions is termed the **proven recoverable reserve** of that fuel.

## We drill to extract oil

Once geologists have identified an oil deposit, an oil company will typically conduct *exploratory drilling.* Holes drilled during this phase are usually small in circumference and descend to great depths. If enough oil is encountered, extraction begins. Just as you would squeeze a sponge to remove its liquid, pressure is required to extract oil from porous rock. Oil is typically already under pressure—from above by rock or trapped gas, from below by groundwater, or internally from natural gas dissolved in the oil. All these forces are held in place by surrounding rock until drilling reaches the deposit, whereupon oil will often rise to the surface of its own accord.

Once pressure is relieved, however, oil becomes more difficult to extract and may need to be pumped out. Even after pumping, a great deal of oil remains stuck to rock surfaces. As much as two-thirds of a deposit may remain in the ground after **primary extraction**, the initial drilling and pumping of available oil (**Figure 19.9a**). Companies may then begin **secondary extraction**, in which solvents are used or underground rocks are flushed with water or steam to remove additional oil (**Figure 19.9b**).

At Prudhoe Bay, seawater is piped in from the coast and pumped into wells to flush out remaining oil. Even after secondary extraction, quite a bit of oil can remain; we lack the technology to remove every last drop. Secondary extraction is more expensive than primary extraction, so many U.S. deposits did not undergo secondary extraction when they were first drilled because the price of oil was too low to make the procedure economical. When oil prices rose in the 1970s, many drilling sites were reopened for secondary extraction. Still more are being reopened today, as prices rise again.

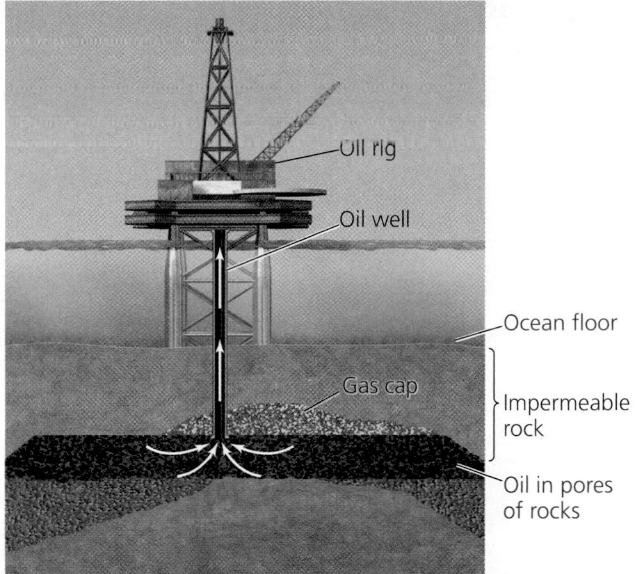


**(a) Primary extraction of oil**

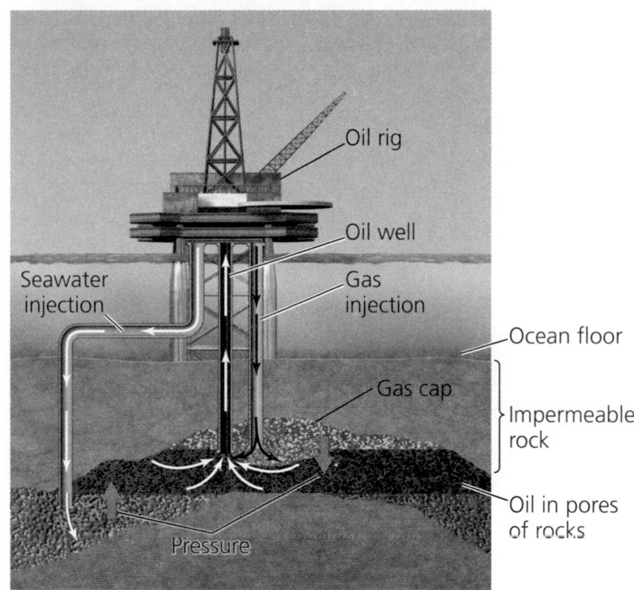


**(b) Secondary extraction of oil**

FIGURE 19.9 In primary extraction (**a**), oil is drawn up through the well by keeping pressure at the top lower than pressure at the level of the oil deposit. Once the pressure in the deposit drops, however, material must be injected into the deposit to increase the pressure. Thus, secondary extraction (**b**) involves injecting seawater beneath the oil and/or gases just above the oil to force more oil up and out of the deposit.

## Petroleum products have many uses

Once crude oil is extracted, it is put through refining processes (see "The Science behind the Story," • pp. 554–555). Because crude oil is a complex mix of hydrocarbons, we can create many types of petroleum products by separating its various components. Since the 1920s, refining techniques and chemical manufacturing have greatly expanded our uses of petroleum to include a wide array of products and applications, from lubricants to plastics to fabrics to pharmaceuticals. Today, petroleum-based products are all around us in our everyday lives (**Figure 19.10**).

THE SCIENCE BEHIND THE STORY

# How Crude Oil is Refined

*Workers at a Citgo oil refinery in Lemont, Illinois*

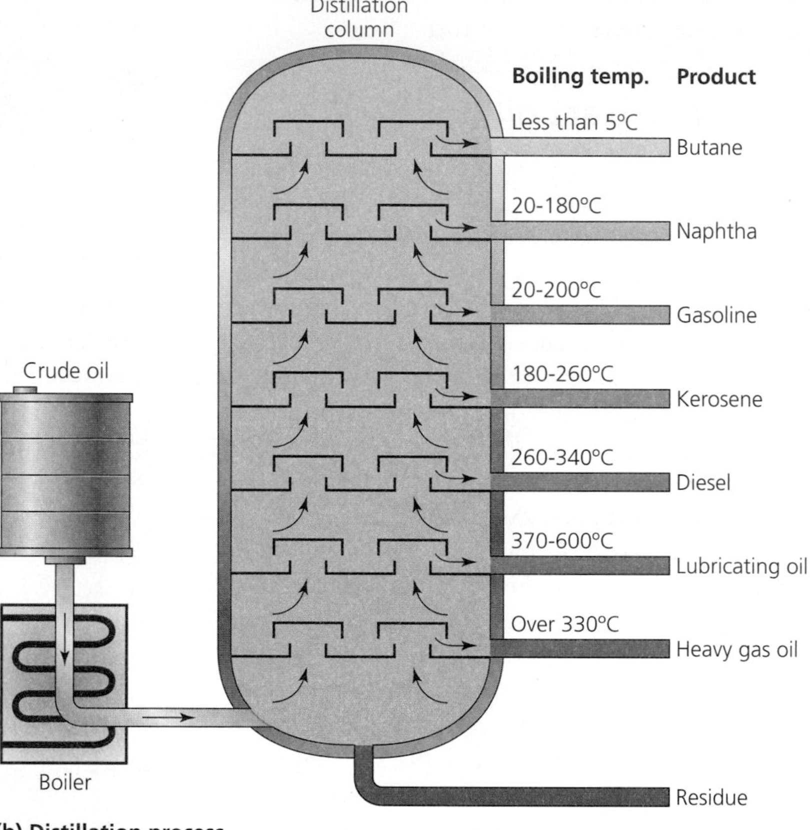


**(a) Distillation columns** **(b) Distillation process**

At crude petroleum refineries **(a)**, crude oil is boiled, causing its many hydrocarbon constituents to volatilize and proceed upward through a distillation column **(b)**. Constituents that boil at the highest temperatures and condense readily once the temperature drops will condense at low levels in the column. Constituents that volatilize at lower temperatures will continue rising through the column and condense at higher levels, where temperatures are lower. In this way, heavy oils (generally consisting of long hydrocarbon molecules) are separated from lighter oils (generally those with short hydrocarbon molecules).

Crude oil is a complex mixture of thousands of kinds of hydrocarbon molecules. Through the process of **refining**, these hydrocarbon molecules are separated into classes of different sizes and chemically transformed to create specialized fuels for heating, cooking, and transportation and to create lubricating

Because petroleum products have become so central to our lives, many fossil fuel experts today are voicing concern that oil production may soon decline as we continue to deplete the world's recoverable oil reserves.

## We may have already depleted half our oil reserves

Some scientists and oil industry analysts calculate that we have already extracted nearly half of the world's oil reserves. So far we have used up about 1.1 trillion barrels of oil, and most estimates hold that somewhat more than 1 trillion barrels remain. To estimate how long this remaining oil will last, analysts calculate the **reserves-to-production ratio**, or **R/P ratio**, by dividing the amount of total remaining reserves by the annual rate of production (i.e., extraction and processing). At current levels of production (30 billion barrels globally per year), 1.2 trillion barrels would last about 40 more years.

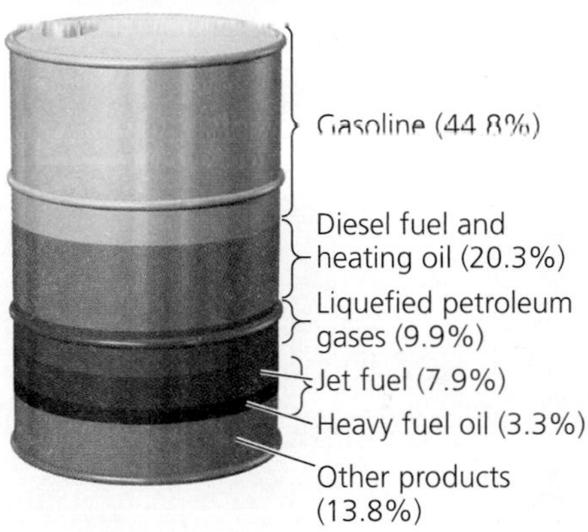


The refining process converts crude oil into a range of petroleum products. Shown are percentages of each major category of product typically produced from a barrel of crude oil. Adapted from U.S. Energy Information Administration.

oils, asphalts, and the precursors of plastics and other petrochemical products. To maximize the production of marketable products while minimizing negative environmental impacts, petroleum engineers have developed a variety of refining techniques.

The first step in processing crude oil is *distillation*, or *fractionation*. This process is based on the fact that different components of crude oil boil at different temperatures. In refineries, the distillation process takes place in tall columns filled with perforated horizontal trays (see the first figure). The columns are cooler at the top than at the bottom. When heated crude oil is introduced into the column, lighter components rise as vapor to the upper trays, condensing into liquid as they cool, while heavier components sink to the lower trays. Light gases, such as butane, boil at less than 32° C (90° F), and heavier oils, such as industrial fuel oil, boil only at temperatures above 343° C (650° F).

Since the early 20th century, light gasoline, used in automobiles, has been in much higher demand than most other derivatives of crude oil. The demand for high-performance, clean-burning gasoline has also risen. To meet these demands, refiners have developed several techniques to convert heavy hydrocarbons into gasoline.

The general name for processes that convert heavy oil into lighter oil is *cracking*. One of the simplest methods is thermal cracking, in which long-chained molecules are broken into smaller chains by heating in the absence of oxygen. (The oil would ignite if oxygen were present.)

Catalytic cracking, a related method, uses catalysts—substances that promote chemical reactions without being consumed by them—to control the cracking process. The result is an increase in the amount of a desired lighter product from a given amount of heavy oil.

Today, the most widely used form of cracking is fluidized catalytic cracking, in which a finely powdered catalyst that behaves like a fluid is fed continuously into a reaction chamber with heavy oils. The products of cracking are then fed into a distillation column.

Refiners can also change the chemical composition of oil through a process called *catalytic reforming*. Catalytic reforming uses catalysts to promote chemical reactions that transform certain hydrocarbons that are slightly heavier than gasoline so that they can be blended with gasoline to obtain higher octane ratings. The octane rating reflects the amount of compression gasoline can undergo before it spontaneously ignites.

Besides distilling crude oil and altering the chemical structure of some of its components, refineries also remove contaminants. Sulfur and nitrogen compounds, which can be harmful when released into the atmosphere, are the two most common contaminants in crude oil. Government regulations stemming from legislation such as the Clean Air Act (• p. 482) have forced refiners to develop scrubbers (• pp. 484–485) and other methods of removing such contaminants, particularly sulfur. Some methods successfully remove up to 98% of sulfur.

As a result of all these approaches, each barrel of crude oil is eventually converted into gasoline and a wide variety of other petroleum products (see the second figure).

Unfortunately, this does not mean that we have a full 40 years in which to figure out what to do once the oil runs out. A growing number of scientists and analysts insist that we will face a crisis not when the last drop of oil is pumped, but when the rate of production first begins to decline. They point out that when production declines as demand continues to increase (because of rising global population and consumption), we will experience an oil shortage immediately. Because production tends to decline once reserves are depleted halfway, most of these experts calculate that this crisis will likely begin within the next several years.

To understand the basis of these concerns, we need to turn back the clock to 1956. In that year, Shell Oil geologist M. King Hubbert calculated that U.S. oil production would peak around 1970. His prediction was ridiculed at the time, but it proved to be accurate; U.S. production peaked in that very year and has continued to fall since

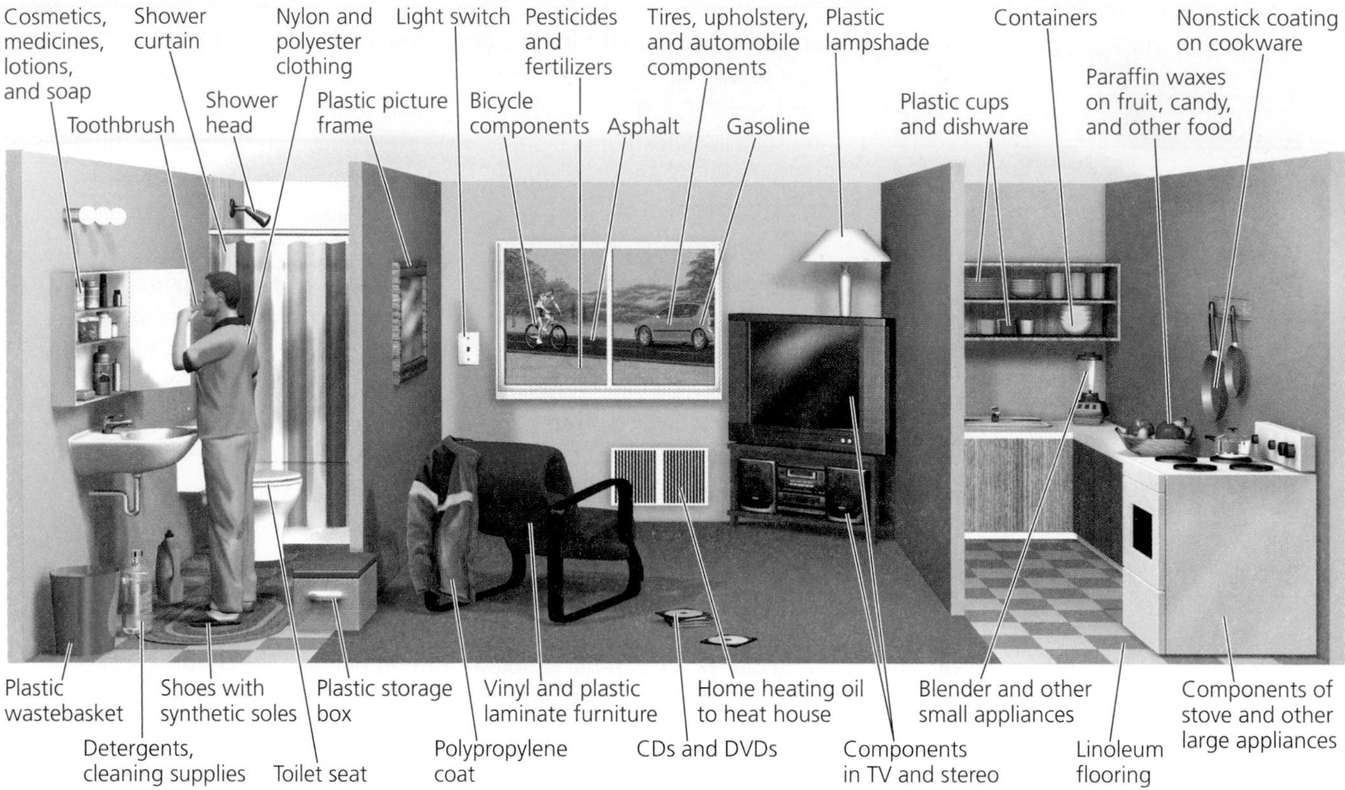


FIGURE 19.10 Petroleum products are everywhere in our daily lives. Besides the gasoline and other fuels we use for transportation and heating, petroleum products include many of the fabrics that we wear and most of the plastics that help make up countless items we use every day.

then (**Figure 19.11a**). The peak in production came to be known as **Hubbert's peak.**

Then in 1974, Hubbert analyzed data on technology, economics, and geology, predicting that global oil production would peak in 1995. It grew past 1995, but many scientists using newer, better data today predict that at some point in the coming decade, production will begin to decline (**Figure 19.11b**). Oil geologist Kenneth Deffeyes even contends that we have already passed the peak—that we did so in December of 2005. Indeed, because of year-to-year variability in production, we will be able to recognize that we have passed the peak of oil production only several years after it has happened.

Predicting an exact date for **"peak oil"** and the coming decline in production is difficult. Many companies and governments do not reveal their true data on oil reserves, and a estimates differ as to how much oil we can extract secondarily from existing deposits. Moreover, a recent U.S. Geological Survey report estimated 2 trillion barrels remaining in the world, rather than 1 trillion, and some estimates predict still greater amounts. A 2007 report by the U.S. General Accounting Office reviewed 21 studies and found that most estimates for the timing of the oil production peak ranged from now through 2040. Regardless of the exact timing, it seems certain that a peak in global oil production will occur. Discoveries of new oil fields peaked 30 years ago, and since then we have been extracting and consuming more oil than we have been discovering. Meanwhile, global demand continues to rise, particularly as China and India industrialize rapidly.

The coming divergence of demand and supply will likely have momentous economic, social, and political consequences that will profoundly affect the lives of each and every one of us. One prophet of peak oil, writer James Howard Kunstler, has sketched a frightening scenario of our post-peak world during what he calls "the long emergency": Lacking cheap oil with which to transport goods long distances, today's globalized economy would collapse and our economies would become intensely localized. Large cities could no longer be supported without urban agriculture, and even by expanding agricultural land we could only feed a fraction of the world's 6.7 billion people without petroleum-based fertilizers and pesticides.

The American suburbs would be hit particularly hard because of their utter dependence on the automobile. Kunstler argues that the suburbs will become the slums of the future, a bleak and crime-ridden landscape littered

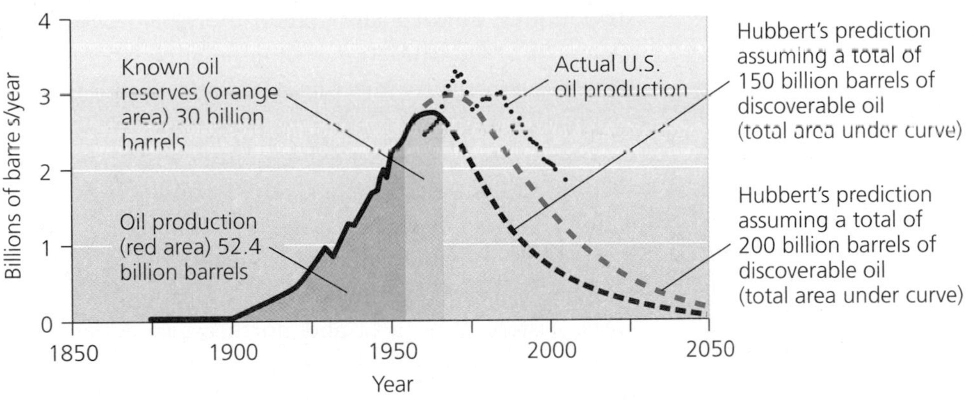


(a) Hubbert's prediction of peak in U.S. oil production, with actual data

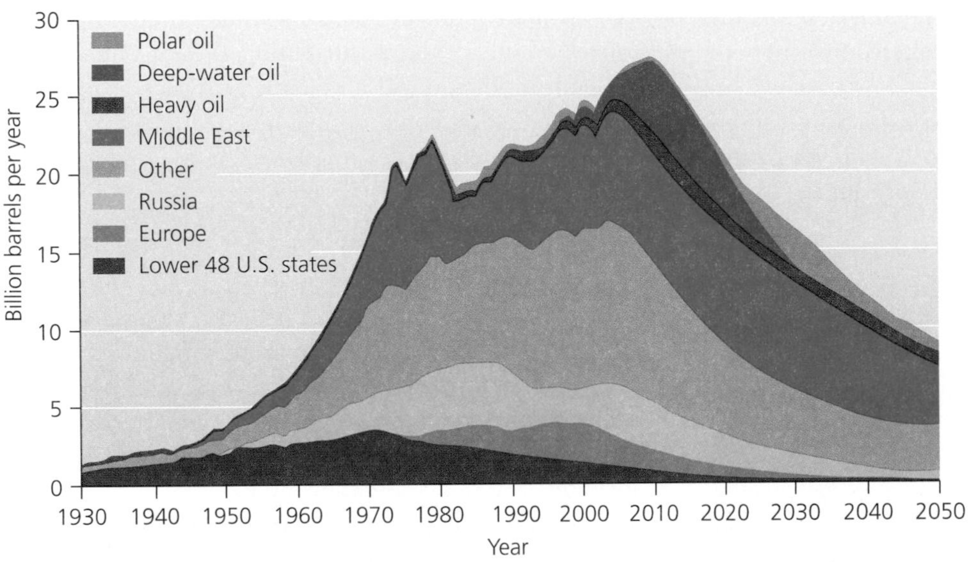


(b) Modern prediction of peak in global oil production

**FIGURE 19.11** Because fossil fuels are nonrenewable resources, supplies at some point pass the midway point of their depletion, and annual production begins to decline. U.S. oil production peaked in 1970, just as geologist M. King Hubbert predicted decades previously; this highpoint is referred to as *Hubbert's peak* (**a**). Today many analysts believe global oil production is about to peak. Shown (**b**) is the latest projection, from a 2007 analysis by scientists at the Association for the Study of Peak Oil. Go to **GRAPHit!** at www.aw-bc.com/withgott or on the student CD-ROM. Data from (a) Deffeyes, K. S. 2001. *Hubbert's peak: The impending world oil shortage*, and U.S. Energy Information Administration; (b) Campbell, C. J, and Association for the Study of Peak Oil. 2007.

with the hulls of rusted-out SUVs. One 29-year-old software consultant learning about the prospect of peak oil recently put it this way: "I realized the future was going to look a lot more like *Mad Max* than *I, Robot*."

More optimistic observers argue that as oil supplies dwindle, rising prices will create powerful incentives for businesses, governments, and individuals to conserve energy and develop alternative energy sources (Chapters 20 and 21)—and that these developments will save us from major disruptions caused by the coming oil peak.

Indeed, to achieve a sustainable society, we will need to switch to renewable energy sources. Energy conservation (• pp. 565–568) can extend the time we have in which to make this transition. However, the research and development needed to construct the infrastructure for a new energy economy depend on having cheap oil, and the time we will have to make this enormous transition will be quite limited.

### Weighing THE Issues | The End of Oil

Physicist David Goodstein has calculated that once world oil production begins to decline, the gap between rising demand and falling supply may widen by 5% per year. As a result, just 10 years after the production peak we will have only half the oil available to us that we had at the peak. He worries that we may not be able to modify our infrastructure and institutions quickly enough to bring other energy sources online before the economic impacts of an oil shortage undermine our very ability to do so.

How do you think your life would be affected if our society were to suffer a 50% decrease in oil availability over the next 10 years? What steps would you take to adapt to these changes? What steps should our society take to deal with the coming depletion of oil?

## Other Fossil Fuels

As oil production declines, we will rely more on natural gas and coal—yet these in turn will also peak and decline in future years. Are there other fossil fuels that can replace them and stave off our day of reckoning? At least three types of alternative fossil fuels exist in large amounts: oil sands, oil shale, and methane hydrates.

### Oil sands can be mined and processed

**Oil sands** (also called **tar sands**) are deposits of moist sand and clay containing 1–20% *bitumen*, a thick and heavy form of petroleum that is rich in carbon and poor in hydrogen. Oil sands represent crude oil deposits that have been degraded and chemically altered by water erosion and bacterial decomposition.

Because bitumen is too thick to extract by conventional oil drilling, oil sands are generally removed by strip mining (**Figure 19.12**), using methods similar to coal strip mining. For deposits 75 m or more below ground, a variety of *in situ* extraction techniques are being devised. Most of these involve injecting steam or chemical solvents to liquefy the bitumen so it can be extracted through conventional wells. After extraction, bitumen may be sent to specialized refineries, where several types of chemical reactions that add hydrogen or remove carbon can upgrade it into more valuable synthetic crude oil.

FIGURE 19.12 In Alberta, companies strip-mine oil sands with the world's largest dump trucks and power shovels. On average, 2 metric tons of oil sands are required to produce 1 barrel of synthetic crude oil.

Three-quarters of the world's oil sands lie in two areas: eastern Venezuela and northeastern Alberta. Oil sands in each region hold at least 175 billion barrels of oil. In Alberta, strip mining began in 1967, but as rising crude oil prices make oil sands more profitable, dozens of companies are now angling to begin 100 or more mining projects in the region. In 2005, oil sands produced 966,000 barrels of oil per day, contributing 39% of Canada's petroleum production. If all planned projects go through, then production could double by 2010 and could reach 5 million barrels per day by 2030.

### Oil shale is abundant in the U.S. West

**Oil shale** is sedimentary rock filled with kerogen (organic matter) and can be processed to produce liquid petroleum. Oil shale is formed by the same processes that form crude oil but occurs when kerogen was not buried deeply enough or subjected to enough heat and pressure to form oil.

We mine oil shale using strip mines or subsurface mines. Once mined, oil shale can be burned directly like coal, or it can be baked in the presence of hydrogen and in the absence of air to extract liquid petroleum (a process called *pyrolysis*). Currently, industry is developing *in situ* extraction processes in which rock is heated underground to liquefy and release oil into conventional wells.

The world's known deposits of oil shale may be able to produce over 600 billion barrels of oil (roughly half as much as the crude oil remaining in the world). About 40% of global oil shale reserves are in the United States, mostly on federally owned land in Colorado, Wyoming, and Utah. Low prices for crude oil have kept investors away from the more costly oil shale, but as crude prices rise, oil shale is again attracting attention.

### Methane hydrate shows potential

In Chapter 16 (• p. 454), we discussed **methane hydrate** (also called *methane clathrate* or *methane ice*). This solid consisting of molecules of methane within a crystal lattice of water ice molecules occurs underground in some Arctic locations and more widely under the seafloor on the continental shelves. Most methane in these gas hydrates was formed by bacterial decomposition in anaerobic environments, but some results from thermogenic formation deeper below the surface. Scientists believe there to be immense amounts of methane hydrate on Earth—from twice to 20 times the amount of natural gas from all other sources.

However, we still need to develop technology to extract methane hydrates safely, avoiding the risk of destabilizing these deposits during extraction. Such destabilization could lead to underwater landslides, tsunamis, and the release of large amounts of methane, a potent greenhouse gas.

### These alternative fossil fuels have downsides

These three sources of alternative fossil fuels are abundant, but they are no panacea for our energy challenges. For one thing, their net energy values are low, because they are expensive to extract and process. Thus the energy returned on energy invested (EROI) ratio is low. For instance, at least 40% of the energy content of oil shale is consumed in its production, and oil shale's EROI is only about 3:1, compared to a 5:1 or greater ratio for conventional crude oil.

Second, these fuels exert severe environmental impacts. Oil sands and oil shale require extensive strip mining, which utterly devastates landscapes over large areas and pollutes waterways running into other areas. Although most governments require mining companies to restore mined areas to their original condition, regions denuded by the very first oil sand mine in Alberta 30 years ago have still not recovered. *In situ* extraction methods exert less environmental impact, but these techniques are not yet well developed.

Besides impacts from their extraction, our combustion of alternative fossil fuels would emit at least as much carbon dioxide, methane, and other air pollutants as does our use of coal, oil, and gas. Thus, they will worsen the effects that fossil fuels are already causing, including air pollution and global climate change.

## Environmental Impacts of Fossil Fuel Use

Our society's love affair with fossil fuels and the many petrochemical products we have developed from them has boosted our material standard of living beyond what our ancestors could have dreamed, has eased constraints on travel, and has helped lengthen our life spans. It has, however, also caused harm to the environment and human health. Concern over these impacts is a prime reason many scientists, environmental advocates, businesspeople, and policymakers are increasingly looking toward renewable sources of energy that exert less impact on natural systems.

### Fossil fuel emissions cause pollution and drive climate change

When we burn fossil fuels, we alter certain flux rates in Earth's carbon cycle (• pp. 188–189). We essentially take carbon that has been retired into a long-term reservoir underground and release it into the air. This occurs as carbon from within the hydrocarbon molecules of fossil fuels unites with oxygen from the atmosphere during combustion, producing carbon dioxide ($CO_2$). Carbon dioxide is a greenhouse gas, and $CO_2$ released from fossil fuel combustion warms our planet and drives changes in global climate (Chapter 18). Because global climate change may have diverse, severe, and widespread ecological and socioeconomic impacts, carbon dioxide pollution (**Figure 19.13**) is becoming recognized as the greatest environmental impact of fossil fuel use.

Fossil fuels release more than carbon dioxide when they burn. Methane is a potent greenhouse gas, and other air pollutants resulting from fossil fuel combustion can have serious consequences for human health and the environment. Deposition of mercury and other pollutants from coal-fired power plants is increasingly recognized as a substantial health risk. The burning of oil and coal in our power plants and vehicles releases sulfur dioxide and nitrogen oxides, which contribute to industrial and photochemical smog and to acidic deposition (• pp. 486–488, 491–495). We have already employed technologies such as

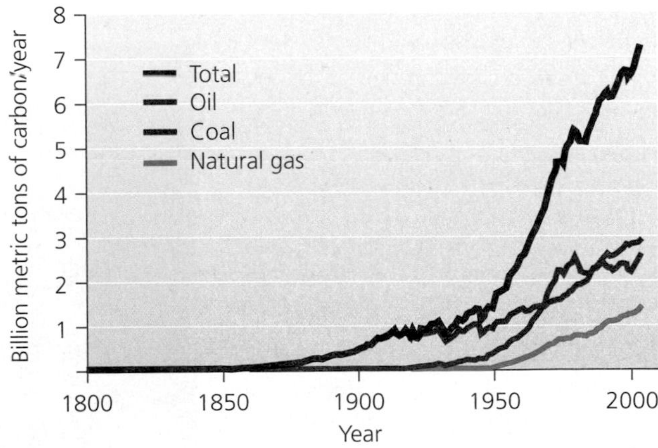


FIGURE 19.13 As industrialization has proceeded, and as population and consumption have grown, emissions from fossil fuel combustion have risen dramatically. Here, worldwide emissions of carbon from carbon dioxide are subdivided by their source (oil, coal, or natural gas). Some other minor sources (such as cement production) are not shown but are included in the graphed total. Data from Marland, G., et al. 2006. Carbon Dioxide Information Analysis Center, Oak Ridge National Laboratory, U.S. Department of Energy, Oak Ridge, TN.

catalytic converters (• p. 672) to cut down on vehicle exhaust pollution. To address pollution for power plants, the U.S. government is teaming up with 11 companies to design the world's first zero-emission coal-fired power plant (see "The Science behind the Story," • p. 549).

Gasoline combustion in automobiles releases pollutants that irritate the nose, throat, and lungs. Some hydrocarbons, such as benzene and toluene, are carcinogenic to laboratory animals and likely also to people. In addition, gases such as hydrogen sulfide can evaporate from crude oil, irritate the eyes and throat, and cause asphyxiation. Crude oil also often contains trace amounts of known poisons, such as lead and arsenic. As a result, workers at drilling operations, refineries, and other jobs that entail frequent exposure to oil or its products can develop serious health problems, including cancer.

Fossil fuels pollute water as well as air. Atmospheric deposition of pollutants exerts many impacts on freshwater ecosystems (• p. 492). Moreover, oil from non-point sources, such as industries, homes, automobiles, gas stations, and businesses, runs off roadways and enters rivers and sewage treatment facilities to be discharged eventually into the ocean (• p. 431). Although most spilled oil results from these non-point sources, large catastrophic oil spills can have significant impacts on the marine environment. Crude oil's toxicity to most plants and animals can lead to high mortality. This was the case with the *Exxon Valdez* spill in 1989 (Chapter 4), in which oil from Alaska's North Slope, piped to the port of Valdez through the trans-Alaska pipeline, exerted long-term damage to ecosystems and economies in Alaska's Prince William Sound when a ship ran aground.

Oil can also contaminate groundwater supplies, such as when leaks from oil operations penetrate deeply into soil. Of greater concern, as we saw in Chapter 15 (• p. 433), thousands of underground storage tanks containing petroleum products have leaked, threatening drinking water supplies.

## Coal mining affects the environment

The mining of coal also has substantial impacts on natural systems and human well-being. Surface strip mining can destroy large swaths of habitat and cause extensive soil erosion. It also can cause chemical runoff into waterways through the process of **acid drainage**. This occurs when sulfide minerals in newly exposed rock surfaces react with oxygen and rainwater to produce sulfuric acid. As the sulfuric acid runs off, it leaches metals from the rocks, many of which are toxic to organisms in high concentrations. Acid drainage is a natural phenomenon, but its rate accelerates greatly when mining exposes many new rock surfaces at once.

Regulations in the United States require mining companies to restore strip-mined land following mining, but impacts are severe and long-lasting just the same. Most other nations exercise less oversight.

Mountaintop removal (**Figure 19.14**) can have even greater impacts than conventional strip mining. When countless tons of rock and soil are removed from the top of a mountain, it is difficult to keep material from sliding

**FIGURE 19.14** Strip mining in some areas is taking place on massive scales, such that entire mountain peaks are leveled, as at this site in West Virginia. Such "mountaintop removal" can cause enormous amounts of erosion into waterways that flow into surrounding valleys, affecting ecosystems over large areas, as well as the people who live there.

downhill, where immense areas of habitat can be degraded or destroyed and creek beds can be polluted and clogged. Loosening of U.S. government regulations in 2002 enabled mining companies to legally dump mountaintop rock and soil into valleys and rivers below, regardless of the consequences for ecosystems, wildlife, and local residents.

Whereas mountaintop removal threatens the welfare of nearby residents, subsurface mining raises health concerns for miners. Underground coal mining is one of our society's most dangerous occupations. Besides risking injury or death from collapsing shafts and tunnels and from dynamite blasts, miners constantly inhale coal dust in the enclosed spaces of mines, which can lead to respiratory diseases, including fatal black lung disease.

The costs of alleviating all these health and environmental impacts are high, and the public eventually pays them in an inefficient manner. The reason is that the costs are generally not internalized (• p. 42) in the market prices of fossil fuels, which are kept inexpensive through government subsidies to extraction companies.

## Oil and gas extraction can alter the environment

Much more than drilling is involved in the development of an oil or gas field (see the photo that opens this chapter, • p. 540). Road networks must be constructed, and many sites may be explored in the course of prospecting. The extensive infrastructure needed to support a full-scale drilling operation typically includes housing for workers, access roads, transport pipelines, and waste piles for removed soil. Ponds may be constructed for collecting the toxic sludge that remains after the useful components of oil have been removed. At extraction sites for coalbed methane (• p. 550), groundwater is pumped out to free gas to rise, but salty groundwater dumped on the surface can contaminate soil and kill vegetation over large areas.

Many onshore North American oil reserves are located in arctic or semi-arid areas. Plants grow slowly in tundra and semi-desert ecosystems, so even minor impacts can have long-lasting repercussions. For example, tundra vegetation at Prudhoe Bay still has not fully recovered from temporary roads last used 30 years ago during the exploratory phase of development.

Whether Prudhoe Bay's oil operations have had a negative impact on the region's caribou is widely debated. Surveys show that the region's summer population of caribou has increased in the 25 years since Prudhoe Bay was developed. Many supporters of drilling have used this trend to argue that oil development does no harm to caribou. Other studies, however, show that female caribou and their calves avoid all parts of the Prudhoe Bay oil complex, including its roads, sometimes detouring miles to do so. These studies also show that the reproductive rate of female caribou in the Prudhoe Bay region is lower than for those in undeveloped areas in Alaska. As a result, although the herd near Prudhoe Bay has increased over the past 25 years, it has not increased as much as have herds in some other parts of Alaska.

There is no way of knowing how the Prudhoe Bay herd would have performed in the absence of development; that is, there is no control, as there would be for a manipulative experiment. It is difficult, therefore, to draw conclusions about the impacts of oil development on caribou. In the Arctic Refuge, the herd that visits the undeveloped 1002 Area rose from 100,000 animals in the 1970s to 178,000 in 1989 and has since declined to 123,000, all for unknown reasons.

## Many scientists anticipate negative impacts of drilling in ANWR

To predict the possible ecological effects of drilling in ANWR's 1002 Area, scientists have examined the effects of development on arctic vegetation, air quality, water quality, and wildlife (including caribou, grizzly bears, and various birds) in Alaska locales where the environment is similar to that of ANWR's coastal plain (**Figure 19.15**). Researchers have compared different areas, and they have contrasted single areas or populations before and after drilling, as described above with caribou. In addition, scientists have run small-scale manipulative experiments when possible, to study, for example, the effects of ice roads and secondary extraction methods such as seawater flushing. In one way or another they have examined the effects of road building,

FIGURE 19.15 Alaska's North Slope is home to a variety of large mammals, including grizzly bears, polar bears, wolves, Arctic foxes, and large herds of caribou. How oil development may affect these animals is a controversial issue, and scientific studies are ongoing. The caribou herd near Prudhoe Bay has increased since oil extraction began there, but less than herds in other parts of Alaska. Grizzly bears such as the ones shown here have been found near, or even walking atop, the trans-Alaska pipeline.

oil pad construction, worker presence, oil spills, accidental fires, trash buildup, permafrost melting, off-road vehicle trails, and dust from roads.

Based on these studies, many scientists anticipate damage to vegetation and wildlife if drilling takes place in ANWR. Vegetation can be killed when saltwater pumped in for flushing deposits is spilled or when plants are buried under gravel pits or roads. Air and water quality can be degraded by fumes from equipment and drilling operations, flaring of natural gas associated with oil extraction, sludge ponds, waste pits, and oil spills. Roads can fragment habitats, and prospecting and drilling can produce enough noise and disruption to affect wildlife.

Other scientists, however, contend that drilling operations in ANWR would have little environmental impact. Roads would be built of ice that will melt in the summer, they point out, and most drilling activity would be confined to the winter, when caribou are elsewhere. Moreover, drilling proponents maintain, Prudhoe Bay is not an appropriate model for ANWR because Prudhoe Bay's development is larger than that projected for ANWR. Furthermore, they say, much of the technology used at Prudhoe Bay is now outdated, and ANWR would be developed with more environmentally sensitive technology and approaches.

# Political, Social, and Economic Impacts

The political, social, and economic consequences of fossil fuel use are numerous, varied, and far-reaching. Our discussion focuses on several negative consequences of fossil fuel use and dependence, but it is important to bear in mind that their use has enabled much of the world's population to achieve a higher material standard of living than ever before. It is also important to ask in each case whether switching to more renewable sources of energy would solve existing problems.

## Nations can become dependent on foreign energy

Virtually all our modern technologies and services depend in some way on fossil fuels. Putting all of one's eggs in one basket is always a risky strategy. The fact that our economies are utterly tied to fossil fuels means that we are vulnerable to supplies' becoming suddenly unavailable or extremely costly. Nations that lack adequate fossil fuel reserves of their own are especially vulnerable. For instance, Germany, France, South Korea, and Japan consume far more energy than they produce and thus rely almost entirely on imports for their continued economic well-being (**Figure 19.16**). Since its 1970 oil production peak, the United States has relied more and more on foreign energy, and today the nation imports two-thirds of its crude oil.

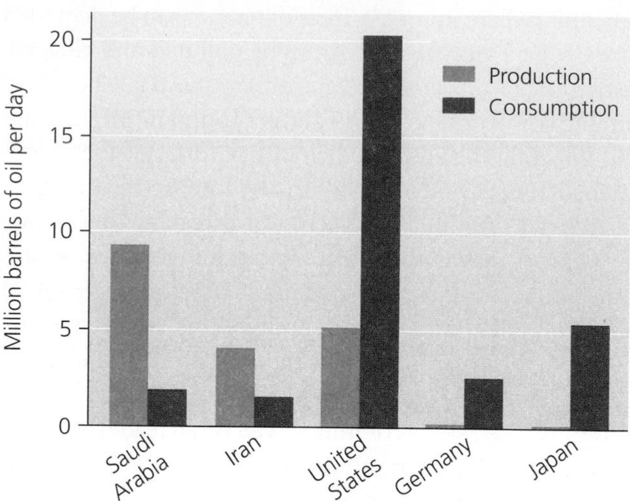


FIGURE 19.16 Japan, Germany, and the United States are among nations that consume far more oil than they produce. Iran and Saudi Arabia produce more oil than they consume and are able to export oil to high-consumption countries. Data from U.S. Energy Information Administration. 2007. *Annual energy review 2006.* Washington, D.C.

Such reliance means that seller nations can control energy prices, forcing buyer nations to pay more and more as supplies dwindle. This became clear in 1973, when the *Organization of Petroleum Exporting Countries (OPEC)* resolved to stop selling oil to the United States. The predominantly Arab nations of OPEC opposed U.S. support of Israel in the Arab-Israeli Yom Kippur War and wished to raise prices by restricting supply. The embargo created panic in the West and caused oil prices to skyrocket (**Figure 19.17**), spurring inflation. Fear of oil shortages drove American consumers to wait in long lines at gas pumps.

## Oil supply and prices affect the economies of nations

More recently, when Hurricanes Katrina and Rita slammed into the Gulf Coast in 2005, they damaged offshore platforms and refineries, causing oil and gas prices to spike significantly. The economic ripple effects served to remind us yet again how much we rely on a steady and ever-increasing supply of petroleum.

With the majority of world oil reserves located in the politically volatile Middle East, crises such as the 1973–1974 embargo, the Iranian Revolution, the Iran-Iraq War, and recent events in Iraq are a constant concern for U.S. policymakers. The United States has cultivated a close relationship with Saudi Arabia, the owner of 22% of world oil reserves, despite the fact that that country's political system allows for little of the democracy that U.S. leaders claim to cherish and promote. The world's third-largest holder of oil reserves, at

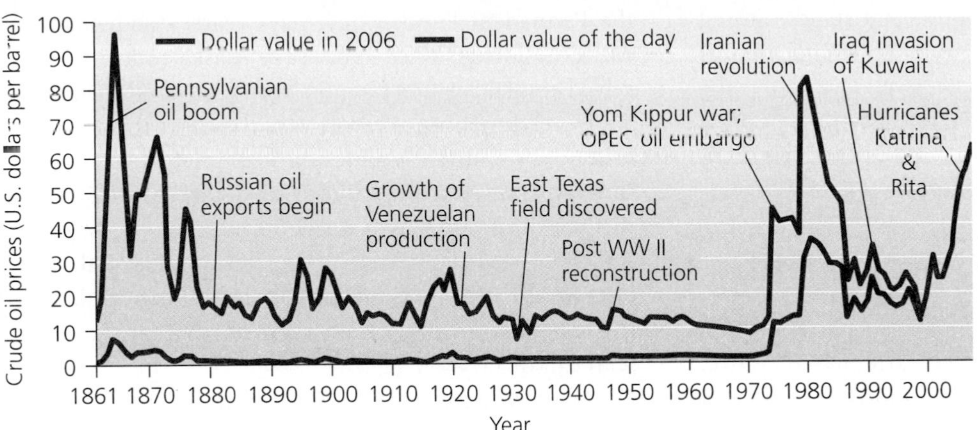


FIGURE 19.17 World oil prices have gyrated greatly over the decades, often because of political and economic events in oil-producing countries. The greatest price hikes in recent times have resulted from wars and unrest in the oil-rich Middle East. Data from U.S. Energy Information Administration.

10%, is Iraq, which is why many people around the world believe the U.S.-led invasion of that nation in 2003 was motivated primarily to secure access to oil.

In response to the 1973 embargo, the U.S. government enacted a series of policies designed to reduce reliance on foreign oil. The government called for developing additional domestic sources, such as those on Alaska's North Slope. Since then, concern over reliance on foreign oil has repeatedly driven the proposal to open ANWR to drilling, despite critics' charges that such drilling would do little to decrease the nation's dependence.

The United States also diversified its sources of petroleum and today receives most of it from non–Middle Eastern nations, including Canada, Mexico, Venezuela, and Nigeria. Major petroleum trade relations among nations and regions of the world are depicted in **Figure 19.18.**

The U.S. government also called for resuming extraction at sites shut down after primary extraction had ceased being cost-effective, capping the price that domestic producers could charge for oil, funding research into renewable energy sources, and enacting conservation measures we will discuss below. It also established a stockpile of oil

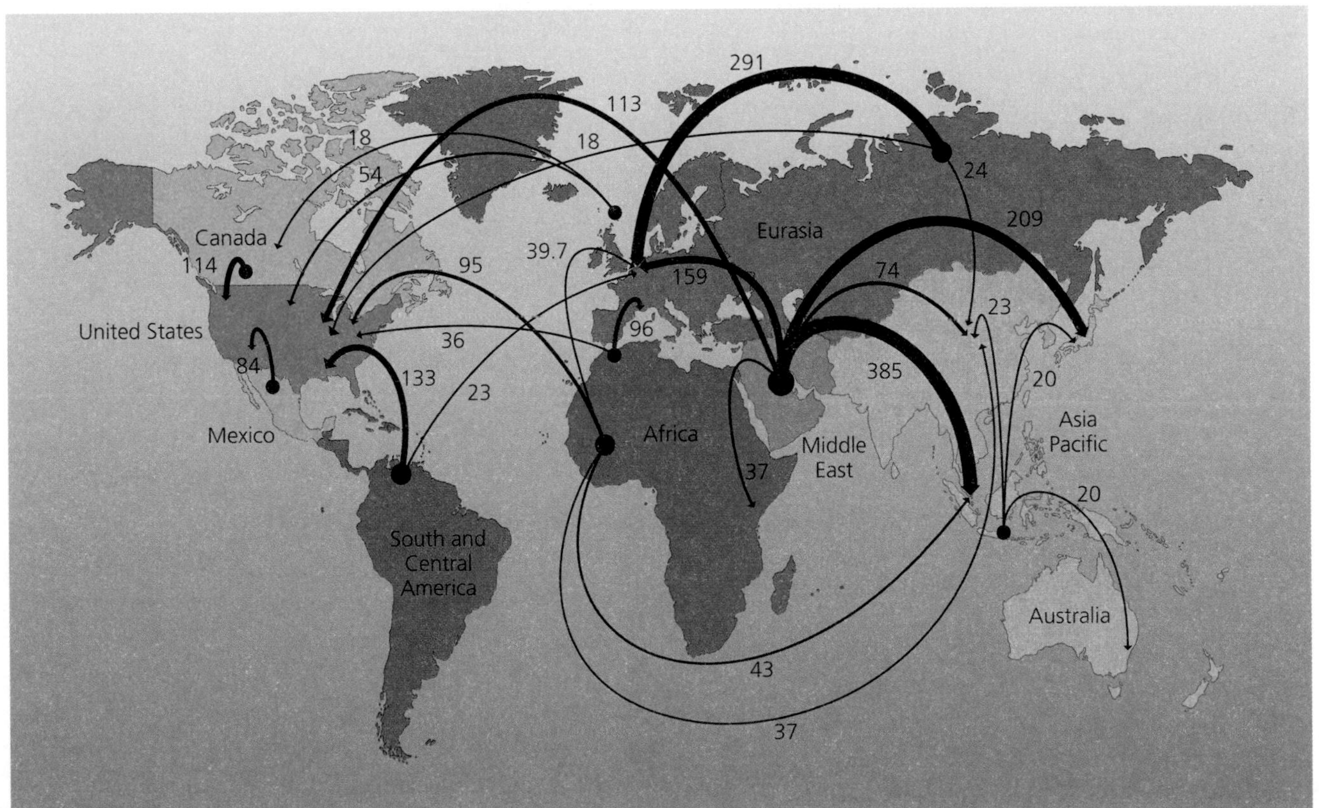


FIGURE 19.18 The global trade in oil is lopsided; relatively few nations account for most exports, and some nations are highly dependent on others for energy. The United States obtains most of its imported oil from Canada, Mexico, Saudi Arabia, Venezuela, and Nigeria. Canada imports some North Sea oil while exporting more to the United States. Numbers in the figure represent millions of metric tons. Data from British Petroleum. 2007. *Statistical review of world energy 2007.*

as a short-term buffer against future shortages. Stored deep underground in salt caverns in Louisiana, this is called the *Strategic Petroleum Reserve.* Currently the Reserve contains nearly 700 million barrels of oil; at present rates of U.S. consumption (20.7 million barrels/day), this equals just over one month's supply.

## Residents may or may not benefit from their fossil fuel reserves

The extraction of fossil fuels can be extremely lucrative; many of the world's wealthiest corporations deal in fossil fuel energy or related industries. These industries provide jobs to millions of employees and provide dividends to millions of investors. Development can potentially yield economic benefits for people who live in petroleum-bearing areas, as well. Since the construction of the trans-Alaska pipeline in the 1970s, the state of Alaska has received over $60 billion in oil revenues. Alaska's state constitution requires that one-quarter of state oil revenues be placed in the Permanent Fund, which pays yearly dividends to all Alaska residents. Since 1982, each Alaska resident has received annual payouts ranging from $331 to $1,964.

Development of ANWR would add to this fund and create jobs. Some estimates anticipate creation of many thousands of jobs and billions of dollars of income. For this reason, most Alaska residents support oil drilling in ANWR. So do many Inupiat who live on the North Slope, because the income could pay for health care, police and fire protection, and other services that are currently scarce in this remote region.

Alaska's distribution of revenue among its citizenry is unusual, however. In most parts of the world where fossil fuels have been extracted, local residents have not seen great benefits, but instead have frequently suffered. When multinational corporations have extracted oil or gas in developing countries, paying those countries' governments for access, the money often has not trickled down to residents of the regions where the extraction takes place. Moreover, oil-rich developing countries such as

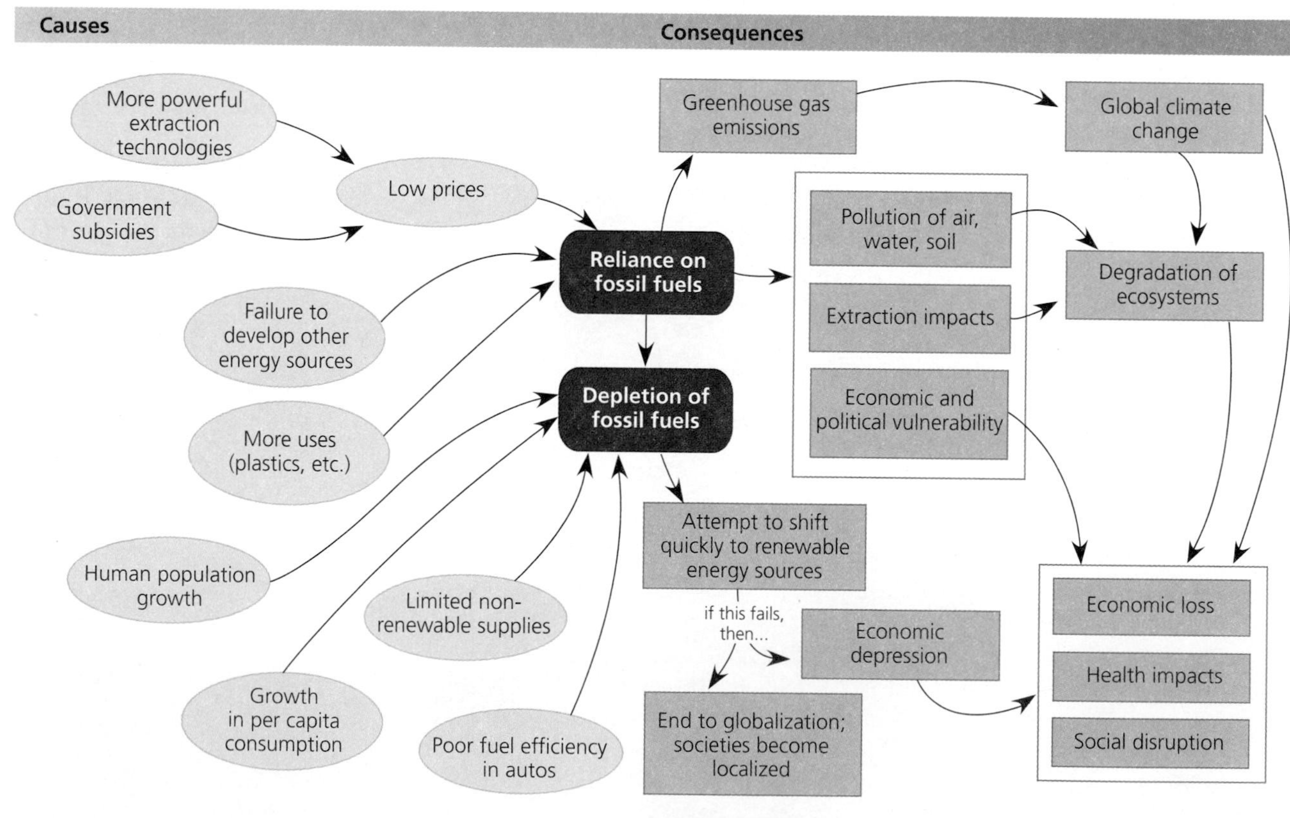


**Solutions**

As you progress through this chapter, try to identify as many solutions to our reliance on and depletion of fossil fuels as you can. What could you personally do to help address this issue? Consider how each action or solution might affect items in the concept map above.

**FIGURE 19.19** Our reliance on, and depletion of, fossil fuels have many causes (ovals on left) and many consequences (boxes on right). Arrows in this concept map lead from causes to consequences. Note that items grouped within outlined boxes do not necessarily share any special relationship; the outlined boxes are merely intended to streamline the figure.

Ecuador, Venezuela, and Nigeria tend to have few environmental regulations, and existing regulations may go unenforced if a government does not want to risk losing the large sums of money associated with oil development.

In Nigeria, oil was discovered in 1958 in the territory of the Ogoni, one of Nigeria's native peoples, and the Shell Oil Company moved in to develop oil fields. Although Shell extracted $30 billion of oil from Ogoni land over the years, the Ogoni still live in poverty, with no running water or electricity. The profits from oil extraction on Ogoni land went to Shell and to the military dictatorships of Nigeria. The development resulted in oil spills, noise, and constantly burning gas flares, all of which caused illness among people living nearby. From 1962 until his death in 1995, Ogoni activist and leader Ken Saro-Wiwa worked for fair compensation to the Ogoni for oil extraction and environmental degradation on their land. After years of persecution by the Nigerian government, Saro-Wiwa was arrested in 1994, given a trial universally regarded as a sham, and put to death by military tribunal.

## How will we convert to renewable energy?

Fossil fuel supplies are limited, and their use and our continued dependence on them have health, environmental, political, and socioeconomic consequences (**Figure 19.19**). Given this, the world's nations have several policy options for guiding future energy use. One option is to continue relying on fossil fuels until they are no longer economically practical and to develop other energy sources only after supplies have dwindled. A second option is to increase funding to develop alternative energy sources dramatically and immediately and to hasten a rapid shift to them. Third, we could steer a middle course and attempt to reduce our reliance on fossil fuels gradually.

Regardless of which course we take, it will benefit us to prolong the availability of fossil fuels as we make the transition to renewable sources. We can prolong our access to fossil fuels by instituting measures to conserve energy, primarily through lifestyle changes that reduce energy use and technological advances that improve efficiency.

# Energy Conservation

Until our society makes the transition to renewable energy sources, we will need to find ways to minimize the expenditure of energy from our dwindling fossil fuel resources. **Energy conservation** is the practice of reducing energy use to extend the lifetimes of our nonrenewable energy supplies, to be less wasteful, and to reduce our environmental impact.

## Energy conservation has followed economic need

In the United States, many people first saw the value of conserving energy following the OPEC embargo of 1973–1974. Conservation measures enacted by the U.S. government in response to that event included a mandated increase in the mile-per-gallon (mpg) fuel efficiency of automobiles and a reduction in the national speed limit to 55 miles per hour.

Over the past three decades, however, many of the conservation initiatives that followed the 1973–1974 oil crisis were abandoned. Without high market prices and an immediate threat of shortages, people lacked economic motivation to conserve. Government funding for research into alternative energy sources decreased, speed limits increased, and countless bills to raise the mandated average fuel efficiency of vehicles failed in Congress. The average fuel efficiency of new vehicles has fallen from a high of 22.1 mpg in 1988 to 21.0 mpg in 2006 (**Figure 19.20**). This decrease is due to increased sales of light trucks (averaging 18.4 mpg), including sport-utility vehicles, relative to cars (averaging 24.6 mpg).

U.S. policymakers have repeatedly failed to raise the *corporate average fuel efficiency (CAFE) standards*, which set benchmarks for auto manufacturers to meet. This is despite a 2001 National Academy of Sciences report concluding that fuel efficiency could feasibly be raised by 40% without adverse economic effect, and despite the fact that most other developed nations boast autos with considerably better fuel efficiency. In 2007, both houses of Congress were considering the Fuel Economy Reform

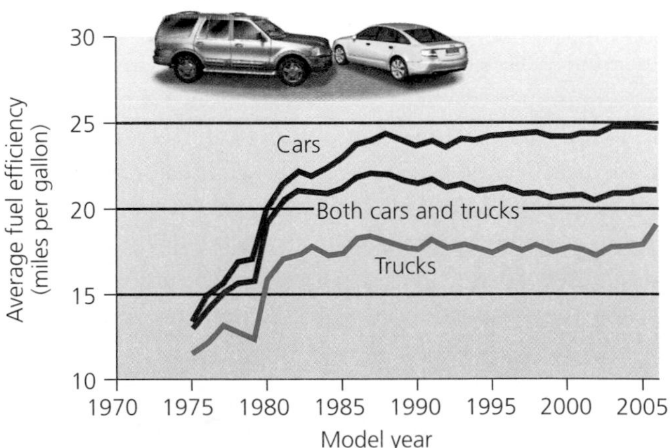


**FIGURE 19.20** Fuel efficiency for automobiles in the United States rose dramatically in the late 1970s as a result of legislative mandates, but it has declined slightly since 1988. The decline is due to a lack of further legislation for improved fuel economy and to the increased popularity in recent years of sport-utility vehicles. Data from U.S. Environmental Protection Agency. 2006. *Light-duty automotive technology and fuel economy trends: 1975 through 2006.*

Act, the first legistation in years that seemed likely to raise CAFE standards. Most fuel efficiency advocates are praising Congress's action, but say the proposed legislation does not raise standards high enough or fast enough.

Transportation accounts for two-thirds of U.S. oil use, and passenger vehicles consume over half this energy. Thus, the failure to improve vehicular fuel economy over the past 20 years, despite the existence of technology to do so, has added greatly to U.S. oil consumption. This is unfortunate because the inefficient use of gasoline in auto engines wastes oil that we could put to better use in manufacturing countless products that enhance our lives (see Figure 19.10).

The United States has also kept its taxes on gasoline extremely low, relative to other nations. Americans pay two-and-a-half times *less* per gallon of gas than drivers in many European countries, for example. Although this keeps U.S. consumers happy at the pump, it means that gasoline prices do not account for the substantial external costs (• p. 42) that oil production and consumption impose on society.

### Weighing THE Issues | More Miles, Less Gas

If you drive an automobile, what gas mileage does it get? How does it compare to the vehicle averages in Figure 19.20? If your vehicle's fuel efficiency were 10 mpg greater, and if you drove the same amount, how many gallons of gasoline would you no longer need to purchase each year? How much money would you save? Do you think that the government should raise taxes on gasoline sales as an incentive to consumers to conserve energy?

Many critics of oil drilling in the Arctic National Wildlife Refuge point out the vast amounts of oil wasted by our fuel-inefficient automobiles. They argue that a small amount of conservation would save the nation far more oil than it would ever obtain from ANWR. As mentioned earlier, the USGS's average estimate for recoverable oil in the 1002 Area, 7.7 billion barrels, represents just 1 year's supply for the United States at current consumption rates. Spread over a period of extraction of many years, the proportion of U.S. oil demand that ANWR would fulfill appears strikingly small (**Figure 19.21**).

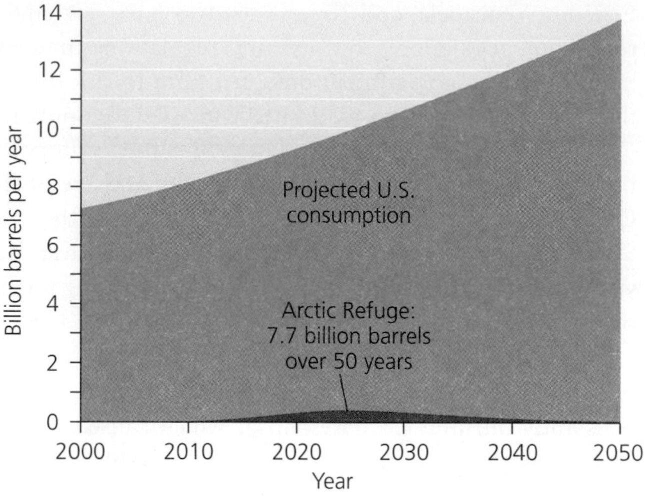


FIGURE 19.21 Opponents of oil drilling in the Arctic National Wildlife Refuge contend that its recoverable oil would make only a small contribution toward overall U.S. oil demand. In this graph, the best USGS estimate of oil from ANWR's 1002 Area is shown in red, in the context of total U.S. oil consumption (in orange), assuming that current consumption trends are extrapolated into the future and that oil production takes place over many years. Any actual ANWR contribution would depend greatly on the amount of oil actually present under ANWR, the time it would take to extract it, and future trends in consumption. Adapted from Natural Resources Defense Council. 2002. *Oil and the Arctic National Wildlife Refuge;* and U.S. Geological Survey. 2001. *Arctic National Wildlife Refuge, 1002 Area, petroleum assessment, 1998, including economic analysis.*

## Personal choice and increased efficiency are two routes to conservation

Energy conservation can be accomplished in two primary ways. As individuals, we can make conscious choices to reduce our own energy consumption. Examples include driving less, turning off lights when rooms are not being used, turning down thermostats, and investing in more efficient machines and appliances. For any given individual or business, reducing energy consumption can save money while also helping to conserve resources.

As a society, we can conserve energy by making our energy-consuming devices and processes more efficient. Currently, more than two-thirds of the fossil fuel energy we use is simply lost, as waste heat, in automobiles and power plants. The United States burns through twice as much energy per dollar of Gross Domestic Product (GDP) as do most other industrialized nations. However, the good news is that over the past three decades the United States has decreased its energy use per dollar of GDP by about 50%. Clearly, we have achieved tremendous gains in efficiency already, and we should be able to make still-greater progress in the future.

In the case of automobiles, we already possess the technology to increase fuel efficiency far above the current U.S. average of 21 mpg. We could accomplish this with more efficient gasoline engines, lightweight materials, continuously variable transmissions, alternative technology vehicles such as electric/gasoline hybrids (**Figure 19.22**), or vehicles that use hydrogen fuel cells (• pp. 622–625).

We can also vastly improve the efficiency of our power plants. One way is to use **cogeneration**, in which excess

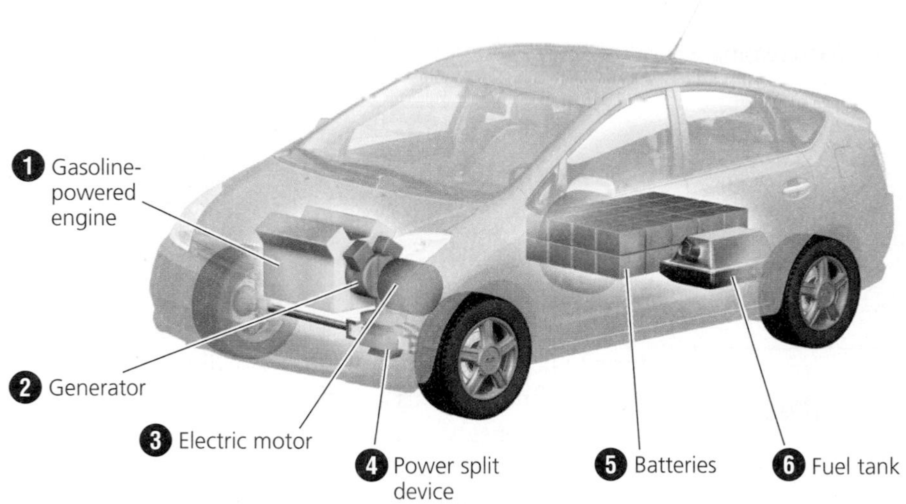


FIGURE 19.22 A hybrid car, such as the Toyota Prius diagrammed here, uses a small, clean, and efficient gasoline-powered engine (1) to produce power that the generator (2) can convert to electricity to drive the electric motor (3). The power split device (4) integrates the engine, generator, and motor, serving as a continuously variable transmission. The car automatically switches between all-electrical power, all-gas power, and a mix of the two, depending on the demands being placed on the engine. Typically, the motor provides power for low-speed city driving and adds extra power on hills. The motor and generator charge a pack of nickel-metal-hydride batteries (5), which can in turn supply power to the motor. Energy for the engine comes from gasoline carried in a typical fuel tank (6).

heat produced during the generation of electricity is captured and used to heat workplaces and homes and to produce other kinds of power. Cogeneration can almost double the efficiency of a power plant. The same is true of *coal gasification* and *combined cycle* generation. In this process, coal is treated to create hot gases that turn a gas turbine, while the hot exhaust of this turbine heats water to drive a conventional steam turbine like that shown on page 549.

In homes and public buildings, a significant amount of heat is lost in winter and gained in summer because of inadequate insulation (**Figure 19.23**). Improvements in the design of homes and offices can reduce the energy required to heat and cool them. Such design changes can involve the building's location, the color of its roof (light colors keep buildings cooler by reflecting the sun's rays), and its insulation.

Among consumer products, scores of appliances, from refrigerators to lightbulbs, have been reengineered through the years to increase energy efficiency. Energy-efficient lighting, for example, can reduce energy use by 80%, and new federal standards for energy-efficient appliances have already reduced per-person home electricity use below what it was in the 1970s. Even so, there remains room for further improvement.

While manufacturers can improve the energy efficiency of appliances, consumers need to "vote with their wallets" by purchasing these energy-efficient appliances. Decisions by consumers to purchase energy-efficient products are crucial in keeping those products commercially available. The U.S. Environmental Protection Agency (EPA) estimates that if all U.S. households purchased energy-efficient appliances, the national annual energy expenditure would be reduced by $200 billion. For the individual consumer, studies show that the slightly higher cost of buying energy-efficient washing machines is rapidly offset by savings on water and electricity bills. On the national level, France, Great Britain, and many other developed countries have standards of living equal to that of the

FIGURE 19.23 Many of our homes and offices could be made more energy-efficient. One way to determine how much heat a building is losing is to take a photograph that records energy in the infrared portion of the electromagnetic spectrum (• pp. 101–102). In such a photograph, or *thermogram* (shown here), white, yellow, and red signify hot and warm temperatures at the surface of the house, whereas blue and green shades signify cold and cool temperatures. The white, yellow, and red colors indicate areas where heat is escaping.

United States, but they use much less energy per capita. This disparity indicates that U.S. citizens could significantly reduce their energy consumption without decreasing their quality of life.

**Weighing THE Issues** | **Drilling in ANWR**

Do you think the United States should open the Arctic National Wildlife Refuge to oil extraction? Why or why not? What would be gained? What would be lost?

### Both conservation and renewable energy are needed

It is often said that reducing our energy use is equivalent to finding a new oil reserve. Some estimates hold that effective energy conservation in the United States could save 6 million barrels of oil a day. Such a savings would likely far more than offset the energy produced by any oil under the Arctic National Wildlife Refuge while also reducing the negative impacts of fossil fuel extraction and use. Indeed, conserving energy is better than finding a new reserve, because it lessens impacts on the environment while extending our access to fossil fuels.

However, energy conservation does not add to our supply of available fuel. Regardless of how much we conserve, we will still need energy, and it will need to come from somewhere. The only sustainable way of guaranteeing ourselves a reliable long-term supply of energy is to ensure sufficiently rapid development of renewable energy sources.

## Conclusion

Over the past 200 years, fossil fuels have helped us build the complex industrialized societies we enjoy today. However, we are now approaching a turning point in history: Our production of fossil fuels will begin to decline. We can respond to this new challenge in creative ways, encouraging conservation and developing alternative energy sources. Or we can continue our current dependence on fossil fuels and wait until they near depletion before we try to develop new technologies and ways of life. The path we choose will have far-reaching consequences for human health and well-being, for Earth's climate, and for our environment.

The ongoing debate over the future of the Arctic National Wildlife Refuge is a microcosm of this debate over our energy future. Fortunately, there is not simply a trade-off between benefits of energy for us and harm to the environment, climate, and health. Instead, as evidence builds that renewable energy sources are becoming increasingly feasible and economical, it becomes easier to envision giving up our reliance on fossil fuels and charting a win-win future for humanity and the environment.

## REVIEWING OBJECTIVES

**You should now be able to:**

**Identify the energy sources that we use**

- A variety of renewable and nonrenewable energy sources is available to us. (pp. 543–544)
- Since the industrial revolution, nonrenewable fossil fuels—including oil, natural gas, and coal—have become our primary sources of energy. (pp. 543–544)
- Fossil fuels are formed very slowly as buried organic matter is chemically transformed by heat, pressure, and/or anaerobic decomposition. (p. 544)
- In evaluating energy sources, it is important to compare the amount of energy obtained from them with the amount invested in their extraction and production. (p. 546)

**Describe the nature and origin of coal and evaluate its extraction and use**

- Coal is our most abundant fossil fuel. It results from organic matter that undergoes compression but little decomposition. (p. 546)
- The first fossil fuel to be widely used for heating homes and powering industry, coal is used today principally to generate electricity. (pp. 546–549)
- Coal is mined underground and strip-mined from the land surface. (p. 547)
- Coal comes in different types and varies in its composition. Combustion of coal that is high in contaminants emits toxic air pollution. (pp. 547–549)

**Describe the nature and origin of natural gas and evaluate its extraction and use**

- Natural gas consists mostly of methane and can be formed in two ways. (p. 550)
- Use of natural gas is growing rapidly, and it is cleaner-burning than coal or oil. (p. 550)
- Natural gas often occurs with oil and coal deposits, is extracted in similar ways, and becomes depleted in similar ways. (pp. 550–551)

**Describe the nature and origin of petroleum and evaluate its extraction, use, and future depletion**

- Crude oil is a thick, liquid mixture of hydrocarbons that is formed underground under certain temperature and pressure conditions. (p. 551)
- Scientists locate fossil fuel deposits by analyzing subterranean geology. Geologists estimate total reserves, as well as the technically and economically recoverable portions of those reserves. (pp. 552–553)
- Oil drilling often involves primary extraction followed by secondary extraction, in which gas or liquid is injected into the ground to help force up additional oil. (p. 553)
- Petroleum-based products, from gasoline to clothing to plastics, are everywhere in our daily lives. (pp. 553–556)
- Components of crude oil are separated in refineries to produce a wide variety of fuel types. (pp. 554–555)
- We have nearly depleted half the world's oil. Once we pass the peak and production slows, the gap between rising demand and falling supply may pose immense economic and social challenges for our society. (pp. 554–557)

**Describe the nature, origin, and potential of alternative fossil fuels**

- Oils sands can be mined and processed into synthetic oil. (p. 558)
- Oil shale is abundant in the western United States. (p. 558)
- Methane hydrate could provide a source of methane gas. (pp. 558–559)

**Outline and assess environmental impacts of fossil fuel use**

- Emissions from fossil fuel combustion pollute air, pose human health risks, and drive global climate change. (pp. 559–560)
- Oil is a major contributor to water pollution. (p. 560)
- Strip mining and mountaintop removal can devastate ecosystems locally or regionally, and acid drainage from coal mines pollutes waterways. (pp. 560–561)
- Development for oil and gas extraction exerts various environmental impacts. (pp. 561–562)

**Evaluate political, social, and economic impacts of fossil fuel use**

- Today's societies are so reliant on fossil fuel energy that sudden restrictions in oil supplies can have major economic consequences. (pp. 562–563)
- Nations that consume far more fossil fuels than they produce are especially vulnerable to supply restrictions. (pp. 562–563)
- People living in areas of fossil fuel extraction do not always benefit from their extraction. (pp. 564–565)

**Specify strategies for conserving energy and enhancing efficiency**

- Energy conservation involves both personal choices and efficient technologies. These two forces interact through the market power of consumer choice. (pp. 565–566)
- Increases in automotive fuel efficiency and efficiency in power plant combustion could help us conserve immense amounts of oil. (pp. 565–567)
- Conservation helps lengthen our access to fossil fuels and reduce environmental impact, but to build a sustainable society we will also need to shift to renewable energy sources. (p. 568)

## TESTING YOUR COMPREHENSION

1. Why are fossil fuels our most prevalent source of energy today? Why are they considered nonrenewable sources of energy?
2. How are fossil fuels formed? How do environmental conditions determine what type of fossil fuel is formed in a given location? Why are fossil fuels often concentrated in localized deposits?
3. Describe how net energy differs from energy returned on investment (EROI). Why are these concepts important when evaluating energy sources?
4. Describe how coal is used to generate electricity.
5. Why is natural gas often extracted simultaneously with other fossil fuels? What constraints on its extraction does it share with oil?
6. How have geologists estimated the total amount of oil beneath the Arctic National Wildlife Refuge (ANWR) 1002 Area? How is this amount different from the "technically recoverable" and "economically recoverable" amounts of oil?

7. How do we create petroleum products? Provide examples of several of these products.
8. What is Hubbert's peak? Why do many experts think we are about to pass the global production peak for oil? What consequences could there be for our society if we do not transition soon to renewable energy sources?
9. List three environmental impacts of fossil fuel production and consumption. Compare some of the contrasting views of scientists regarding the environmental impacts of drilling for oil in ANWR.
10. Describe two main approaches to energy conservation, and give a specific example of each.

## SEEKING SOLUTIONS

1. Roughly how much oil is left in the world, and how much longer can we expect to use it? How effective were the conservation methods adopted by the United States in response to the "energy crisis" of 1973–1974? What steps should we take to avoid energy shortages in the future?
2. Compare the effects of coal and oil consumption on the environment. Which process do you think has ultimately been more detrimental to the environment, oil extraction or coal mining, and why? What steps could governments, industries, and individuals take to reduce environmental impacts?
3. If the United States and other developed countries reduced dependence on foreign oil and on fossil fuels in general, do you think that their economies would benefit or suffer? Might your answer be different for the short term and the long term? What factors come into play in trying to make such a judgment?
4. Contrast the experiences of the Ogoni people of Nigeria with those of the citizens of Alaska. How have they been similar and different? Do you think businesses or governments should take steps to ensure that local people benefit from oil drilling operations? How could they do so?
5. **THINK IT THROUGH** You have been elected U.S. senator from the state of Alaska. Your colleague, the other senator from Alaska, has just introduced legislation to open the Arctic National Wildlife Refuge to oil drilling. Would you vote in favor of the legislation? Why or why not? Now imagine that you are instead a senator from another U.S. state. How would you vote on the bill, and why?
6. **THINK IT THROUGH** Throughout this book in these questions, we have asked you to imagine yourself in various roles. This time we ask you simply to be yourself. Given the information in this chapter on petroleum supplies, consumption, and depletion, what actions, if any, do you plan to take to prepare yourself for changes in our society that may come about as oil production declines? Describe in detail how you think your life may change, and suggest one thing you could do to help reduce negative impacts of oil depletion on our society.

## INTERPRETING GRAPHS AND DATA

The fossil fuels that we burn today were formed long ago from buried organic matter. However, only a small fraction of the original organic carbon remains in the coal, oil, or natural gas that is formed. Thus, it requires approximately 90 metric tons of ancient organic matter—so-called paleoproduction—to result in just 3.8 L (1 gal) of gasoline. The graph presents estimates of the amount of paleoproduction required to produce the fossil fuels humans have used each year over the past 250 years.

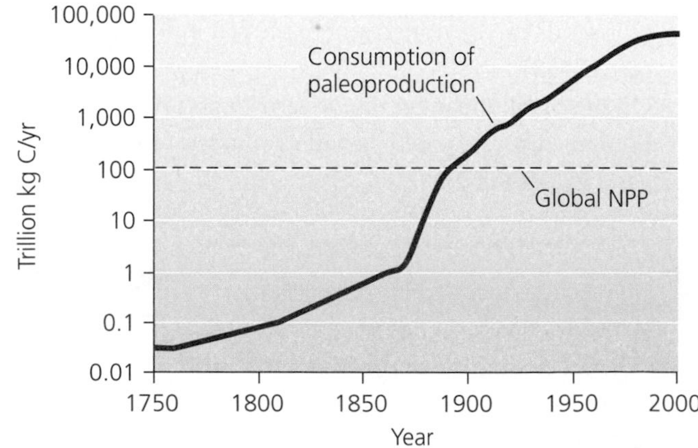


Annual human consumption of paleoproduction by fossil fuel combustion (red line), 1750–2000. The dashed line indicates current annual net primary production (NPP; • p. 181) for the entire planet. Data from Dukes, J. 2003. Burning buried sunshine: Human consumption of ancient solar energy. *Climatic Change* 61: 31–44.

1. Estimate in what year the annual consumption of paleoproduction, represented by our combustion of fossil fuels, surpassed Earth's current annual net primary production (• p. 181).
2. In 2000, approximately how many times greater than global net primary production was our consumption of paleoproduction?
3. If on average it takes 7,000 units of paleoproduction to produce 1 unit of fossil fuel, estimate the total carbon content of the fossil fuel consumed in 2000. How does this amount compare to global NPP?

## CALCULATING ECOLOGICAL FOOTPRINTS

Wackernagel and Rees calculated the energy component of our ecological footprint by estimating the amount of ecologically productive land required to absorb the carbon released from fossil fuel combustion. For the average American, this translates into 3 ha of his or her total ecological footprint. Another way to think about our footprint, however, is to estimate how much land would be needed to grow biomass with an energy content equal to that of the fossil fuel we burn.

| | Hectares of land for fuel production |
|---|---|
| **You** | 1,794 |
| **Your class** | |
| **Your state** | |
| **United States** | |

Data from Wackernagel, M., and W. Rees. 1996. *Our ecological footprint: Reducing human impact on the Earth*. Gabriola Island, British Columbia: New Society Publishers.

Assume that you are an average American who burns about 300 gigajoules of fossil fuels per year and that average terrestrial net primary productivity can be expressed as 160 megajoules/ha/year. Calculate how many hectares of land it would take to supply our fuel use by present-day photosynthetic production. A gigajoule is $10^9$ (1 billion) joules; a megajoule is $10^6$ (1 million) joules.

1. Compare the energy component of your ecological footprint calculated in this way with the 3 ha calculated using the method of Wackernagel and Rees. Explain why results from the two methods may differ.
2. Earth's total land area is approximately $1.5 \times 10^{10}$ (15 billion) ha. Compare this to the hectares of land for fuel production from the table.
3. How large a human population could Earth support at the level of consumption of the average American, if all of Earth's land were devoted to fuel production? Do you consider this realistic? Provide two reasons why or why not.

## Take It Further

Go to www.aw-bc.com/withgott or the student CD-ROM, where you'll find:

- Suggested answers to end-of-chapter questions
- Quizzes, animations, and flashcards to help you study
- *Research Navigator*™ database of credible and reliable sources to assist you with your research projects
- **GRAPHit!** Tutorials to help you interpret graphs
- **INVESTIGATEit!** Current news articles that link the topics that you study to case studies from your region to around the world

CHAPTER

# 20 Conventional Energy Alternatives

Cooling towers of nuclear power plant, Shropshire, U.K.

## Upon completing this chapter, you will be able to:

- Discuss the reasons for seeking alternatives to fossil fuels
- Summarize the contributions to world energy supplies of conventional alternatives to fossil fuels
- Describe nuclear energy and how it is harnessed
- Outline the societal debate over nuclear power
- Describe the major sources, scale, and impacts of biomass energy
- Describe the scale, methods, and impacts of hydroelectric power

Biomass power plant, Skellefteå, Sweden

## CENTRAL CASE

# Sweden's Search for Alternative Energy

**"Nowhere has the public debate over nuclear power plants been more severely contested than Sweden."**
—Writer and Analyst Michael Valenti

**"If [Sweden] phases out nuclear power, then it will be virtually impossible for the country to keep its climate-change commitments."**
—Yale University economist William Nordhaus

On the morning of April 28, 1986, workers at a nuclear power plant in Sweden detected suspiciously high radiation levels. Their concern turned to confusion when they determined that the radioactivity was coming not from their own plant, but through the atmosphere from the direction of the Soviet Union.

They had, in fact, discovered evidence of the disaster at Chernobyl, more than 1,200 km (750 mi) away in what is now the nation of Ukraine. Chernobyl's nuclear reactor had exploded 2 days earlier, but the Soviet government had not yet admitted it to the world.

As low levels of radioactive fallout rained down on the Swedish countryside in the days ahead, contaminating crops and cows' milk, many Swedes felt more certain than ever about the decision they had made collectively 6 years earlier. In a 1980 referendum, Sweden's electorate had voted to phase out their country's nuclear power program, shutting down all nuclear plants by the year 2010.

But trying to phase out nuclear power has proven difficult. Nuclear power today provides Sweden with one-third of its overall energy supply and nearly half its electricity. If nuclear plants are shut down, something will have to take their place. Aware of the environmental impacts of fossil fuels, Sweden's government and citizens do not favor expanding fossil fuel use. In fact, Sweden is one of the few nations that have managed to decrease use of fossil fuels since the 1970s—and it has done so largely by replacing them with nuclear power.

To fill the gap that would be left by a nuclear phaseout, Sweden's government has promoted research and development of renewable energy sources. Hydroelectric power from running water was already supplying most of the other half of the nation's electricity, but it could not be expanded much more. The government hoped that energy from biomass sources and wind power could fill the gap.

Sweden has made itself an international leader in renewable energy alternatives, but because renewables have taken longer to develop than hoped, the government has repeatedly postponed the nuclear phaseout. Only one of the 12 reactors operating in 1980 has been shut down so far, and efforts to close a second one have generated sustained controversy.

Proponents of nuclear power say it would be fiscally and socially irresponsible to dismantle the nation's nuclear program without a ready replacement. And environmental advocates worry that if nuclear power is simply replaced by fossil fuel combustion, or if converting to biomass energy means cutting down more forests, the nuclear phaseout would be bad news for the environment. Moreover, Sweden has international obligations to hold down its carbon emissions under the Kyoto Protocol (• p. 534), so its incentive to keep nuclear power is strong. Nuclear energy is free of atmospheric pollution and seems, to many, the most effective way to minimize carbon emissions in the short term.

In 2003, a poll showed 55% of the Swedish public in favor of maintaining or increasing nuclear power, and 41% in favor of abandoning it. But despite the mixed feelings over nuclear power, Swedes have little desire to return to an energy economy dominated by fossil fuels. A concurrent poll showed that 80% of Swedes supported boosting research on renewable energy sources—a higher percentage than in any other European country.

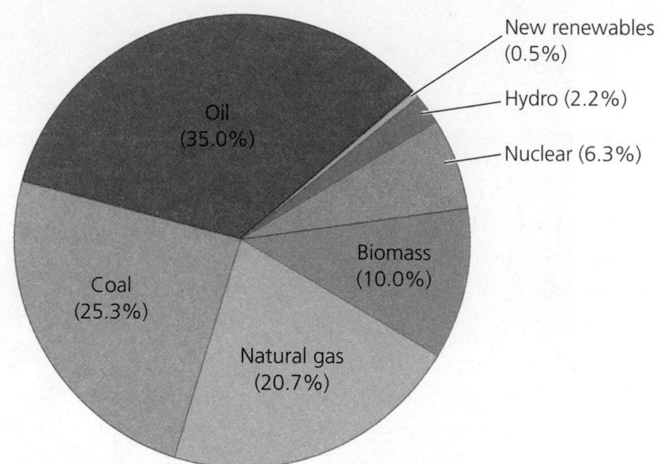


**(a) World energy production, by source**

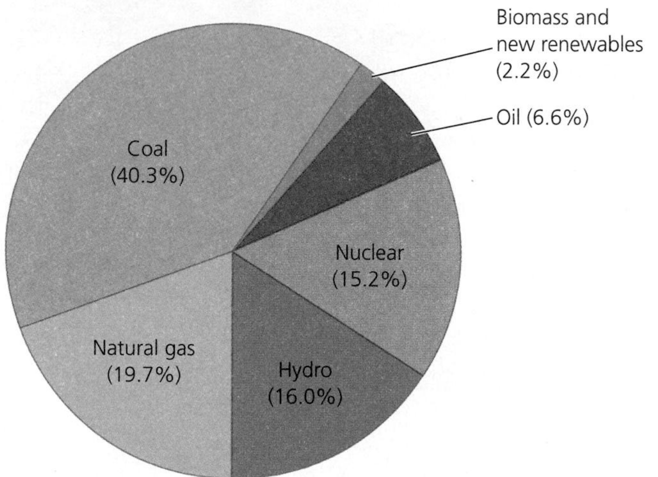


**(b) World electricity generation, by source**

**FIGURE 20.1** Fossil fuels account for 80% of the world's energy production (**a**). Nuclear and hydroelectric power contribute substantially to global electricity generation, but fossil fuels still power two-thirds of our electricity (**b**). Data are for 2005, from the International Energy Agency (IEA). 2007. *Key world energy statistics 2007.* Paris: IEA.

# Alternatives to Fossil Fuels

Fossil fuels helped to drive the industrial revolution and to create the unprecedented material prosperity we enjoy today. Our global economy is largely powered by fossil fuels; 80% of our energy comes from oil, coal, and natural gas (**Figure 20.1a**), and these three fuels also power two-thirds of the world's electricity generation (**Figure 20.1b**). However, these nonrenewable energy sources will not last forever. As we saw in Chapter 19, oil production is thought to be peaking, and easily extractable supplies of oil and natural gas may not last half a century more (• pp. 554–557). Moreover, the use of coal, oil, and natural gas entails substantial environmental impacts, as described in Chapters 17, 18, and 19.

For these reasons, most scientists and energy experts, as well as many economists and policymakers, accept that we will need to shift from fossil fuels to energy sources that are less easily depleted and gentler on our environment. Developing alternatives to fossil fuels has the added benefit of helping to diversify an economy's mix of energy, thus lessening price volatility and dependence on foreign fuel imports.

We have developed a range of alternatives to fossil fuels. Most of these energy sources are renewable, and most have less impact on the environment than oil, coal, or natural gas. However, at this time most remain more costly than fossil fuels, at least in the short term, and external costs (• p. 42) are not included in market prices. Moreover, many depend on technologies that are not yet fully developed.

## Nuclear power, biomass energy, and hydropower are conventional alternatives

Three alternative energy sources are currently the most developed and most widely used: nuclear energy, hydroelectric power, and energy from biomass. Each of these well-established energy sources plays substantial roles in the energy and electricity budgets of nations today. We can therefore call nuclear energy, hydropower, and biomass energy "conventional alternatives" to fossil fuels.

In some respects, this trio of conventional energy alternatives makes for an odd collection. They are generally considered to exert less environmental impact than fossil fuels, but more impact than the "new renewable" alternatives we will discuss in Chapter 21. Yet, as we will see, they each involve a unique and complex mix of benefits and drawbacks for human well-being and the environment. Nuclear energy is commonly termed a nonrenewable energy source, and hydropower and biomass are generally described as renewable; the reality, however, is more complicated. They are perhaps best viewed as intermediates along a continuum of renewability.

## Conventional alternatives provide some of our energy and much of our electricity

Fuelwood and other biomass sources provide 10.0% of the world's energy, nuclear power provides 6.3%, and hydropower provides 2.2%. The less established renewable energy sources account for only 0.5% (see Figure 20.1a). Although their global contributions to overall energy supply are still minor, alternatives to fossil fuels do contribute greatly to our generation of electricity. Nuclear energy and hydropower each account for nearly one-sixth of the world's electricity generation (see Figure 20.1b).

Energy consumption patterns in the United States (**Figure 20.2a**) are similar to those globally, except that the United States relies less on fuelwood and more on fossil fuels and nuclear power than most other countries. A graph showing trends in energy consumption in the United States over the past half century (**Figure 20.2b**) reveals two things. First, conventional alternatives play minor yet substantial roles in overall energy use. Second, use of conventional alternatives has been growing more slowly than use of fossil fuels.

Sweden and some other nations, however, have shown that it is possible to replace fossil fuels gradually with alternative sources. Since 1970, Sweden has decreased its fossil fuel use by 36%, and today nuclear power, biomass, and hydropower together provide Sweden with 63% of its total energy and virtually all of its electricity.

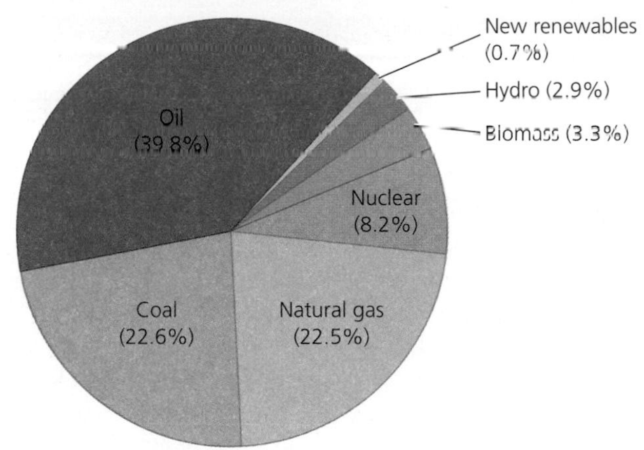


**(a) U.S. energy consumption, by source**

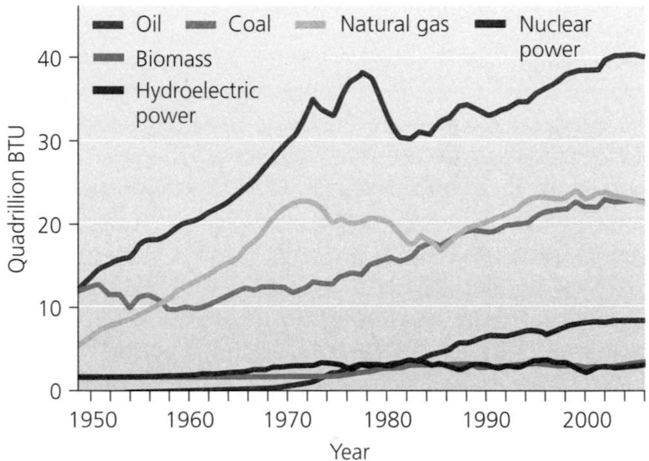


**(b) U.S. energy consumption, 1949–2006**

FIGURE 20.2 In the United States, fossil fuels account for 85% of energy consumption, nuclear power for 8.2%, biomass for 3.3%, and hydropower for 2.9% (**a**). U.S. consumption of the three fossil fuels has grown faster than that of biomass or hydropower over the past half century (**b**). Nuclear power grew considerably between 1970 and 2000. Data are for 2006, from Energy Information Administration. 2007. *Annual energy review 2006*. U.S. Department of Energy.

# Nuclear Power

Nuclear power occupies an odd and conflicted position in our modern debate over energy. It is free of the air pollution produced by fossil fuel combustion, so it has long been put forth as an environmentally friendly alternative to fossil fuels. Yet nuclear power's great promise has been clouded by nuclear weaponry, the dilemma of radioactive waste disposal, and the long shadow of Chernobyl and other power plant accidents. As such, public safety concerns and the costs of addressing them have constrained the development and spread of nuclear power in the United States, Sweden, and many other nations.

TABLE 20.1 Top Producers of Nuclear Power

| Nation | Nuclear power produced* | Number of plants† | Percentage of electricity from nuclear power‡ |
|---|---|---|---|
| United States | 98.4 | 103 | 19.4 |
| France | 63.4 | 59 | 78.1 |
| Japan | 47.6 | 55 | 30.0 |
| Russia | 21.7 | 31 | 15.9 |
| Germany | 20.4 | 17 | 31.8 |
| South Korea | 17.5 | 20 | 38.6 |
| Ukraine | 13.1 | 15 | 47.5 |
| Canada | 12.6 | 18 | 15.8 |
| United Kingdom | 11.0 | 19 | 18.4 |
| Sweden | 9.1 | 10 | 48.0 |
| Rest of world | 54.2 | 89 | 8.0 |

* In gigawatts, 2007 data, from the European Nuclear Society.
† 2007 data, from the International Atomic Energy Agency.
‡ 2006 data, from the International Atomic Energy Agency.

First developed commercially in the 1950s, nuclear power has expanded 15-fold worldwide since 1970, experiencing most of its growth during the 1970s and 1980s. Of all nations, the United States generates the most electricity from nuclear power—nearly a third of the world's production—and is followed by France and Japan. However, only 20% of U.S. electricity comes from nuclear sources. A number of other nations rely more heavily on nuclear power (Table 20.1). France leads the list, receiving 78% of its electricity from nuclear power.

## Fission releases nuclear energy

Strictly defined, **nuclear energy** is the energy that holds together protons and neutrons within the nucleus of an atom. We harness this energy by converting it to thermal energy, which can then be used to generate electricity. Several processes can convert the energy within an atom's nucleus into thermal energy, releasing it and making it available for use. Each process involves transforming isotopes (• pp. 90–91) of one element into isotopes of other elements by the addition or loss of neutrons.

The reaction that drives the release of nuclear energy in power plants is **nuclear fission**, the splitting apart of atomic nuclei (**Figure 20.3**). In fission, the nuclei of large, heavy atoms, such as uranium or plutonium, are bombarded with neutrons. Ordinarily neutrons move too quickly to split nuclei when they collide with them, but if neutrons are slowed down they can break apart nuclei. Each split nucleus emits heat, radiation, and multiple neutrons. These neutrons (two to three in the case of fissile isotopes of uranium-235) can in turn bombard other nearby uranium-235 ($^{235}U$) atoms, resulting in a self-sustaining chain reaction.

If not controlled, this chain reaction becomes a runaway process of positive feedback that releases enormous amounts of energy. It is this process that creates the explosive power of a nuclear bomb. Inside a nuclear power plant, however, fission is controlled so that on average only one of the two or three neutrons emitted with each fission event goes on to induce another fission event. In this way, the chain reaction maintains a constant output of energy at a controlled rate.

## Nuclear energy comes from processed and enriched uranium

We generate electricity from nuclear power by controlling fission in **nuclear reactors**, facilities contained within nuclear power plants. But this is just one step in a longer process sometimes called the *nuclear fuel cycle*. This process begins when the naturally occurring element uranium is mined from underground deposits, as we saw with the mines on Australian Aboriginal land in Chapter 2.

Uranium is an uncommon mineral, and uranium ore is in finite supply, which is why nuclear power is generally considered a nonrenewable energy source. Uranium is used for nuclear power because it is radioactive. Radioactive isotopes, or *radioisotopes* (• p. 91), emit subatomic particles and high-energy radiation as they decay into lighter radioisotopes until they ultimately become stable isotopes.

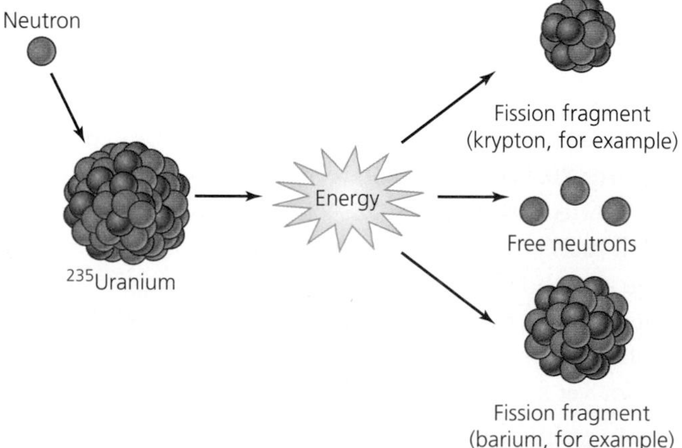


FIGURE 20.3 In nuclear fission, an atom of uranium-235 is bombarded with a neutron. The collision splits the uranium atom into smaller atoms and releases two or three neutrons, along with energy and radiation. The neutrons can continue to split other uranium atoms and set in motion a runaway chain reaction, so engineers at nuclear plants must use control rods to absorb excess neutrons and thereby regulate the rate of the reaction.

The isotope uranium-235 decays into a series of daughter isotopes, eventually forming lead-207. Each radioisotope decays at a rate determined by that isotope's *half-life* (• p. 91), the time it takes for half of the atoms to give off radiation and decay. The half-life of $^{235}U$ is about 700 million years.

Over 99% of the uranium in nature occurs as the isotope uranium-238. Uranium-235 (with three fewer neutrons) makes up less than 1% of the total. Because $^{238}U$ does not emit enough neutrons to maintain a chain reaction when fissioned, we use $^{235}U$ for commercial nuclear power. Therefore, mined uranium ore must be processed to enrich the concentration of $^{235}U$ to at least 3%. The enriched uranium is formed into pellets of uranium dioxide ($UO_2$), which are incorporated into metallic tubes called *fuel rods* (**Figure 20.4**) that are used in nuclear reactors. After several years in a reactor, enough uranium has decayed so that the fuel cannot generate adequate energy, and it must be replaced with new fuel. In some countries, the spent fuel is reprocessed to recover what usable energy may be left. Most spent fuel, however, is disposed of as radioactive waste.

FIGURE 20.4 Enriched uranium fuel is packaged into fuel rods, which are encased in metal and used to power fission inside the cores of nuclear reactors. In this photo, the fuel rods are visible, arrayed in a circle within the blue water.

## Fission in reactors generates electricity in nuclear power plants

For fission to begin in a nuclear reactor, the neutrons bombarding uranium are slowed down with a substance called a *moderator*, most often water or graphite. As fission proceeds, it becomes necessary to soak up the excess neutrons produced when uranium nuclei divide, so that on average only a single neutron from each nucleus goes on to split another nucleus. For this purpose, *control rods*, made of a metallic alloy that absorbs neutrons, are placed into the reactor among the water-bathed fuel rods. Engineers move these control rods into and out of the water to maintain the fission reaction at the desired rate.

All this takes place within the reactor core and is the first step in the electricity-generating process of a nuclear power plant (**Figure 20.5**). The reactor core is housed within a reactor vessel, and the vessel, steam generator, and associated plumbing are protected within a *containment building*. Containment buildings, with their meter-thick concrete and steel walls, are constructed to prevent leaks of radioactivity due to accidents or natural catastrophes such as earthquakes.

## Breeder reactors make better use of fuel, but have raised safety concerns

Using $^{235}U$ as fuel for fission is only one potential way to harness nuclear energy. **Breeder reactors** make use of $^{238}U$, which in conventional fission goes unused as a waste product. In breeder reactors, the addition of a neutron to $^{238}U$ and its subsequent electron loss forms plutonium ($^{239}Pu$). When plutonium is bombarded by a neutron, it splits into fission products and releases more neutrons, which convert more of the remaining $^{238}U$ fuel into $^{239}Pu$, continuing the process. Because 99% of all uranium is $^{238}U$, fission in breeder reactors makes much better use of fuel, generates far more power, and produces far less waste.

However, breeder reactors are more dangerous than conventional reactors because highly reactive liquid sodium, rather than water, is used as a coolant, raising the risk of explosive accidents. Breeder reactors also are more expensive than conventional reactors. Finally, breeder reactors can be used to supply plutonium to nuclear weapons programs. As a result, all but a handful of the world's breeder reactors have now been shut down.

## Fusion remains a dream

For as long as scientists and engineers have generated power from nuclear fission, they have tried to figure out how they might use nuclear fusion instead. **Nuclear fusion**—the process that drives our sun's vast output of energy, and the

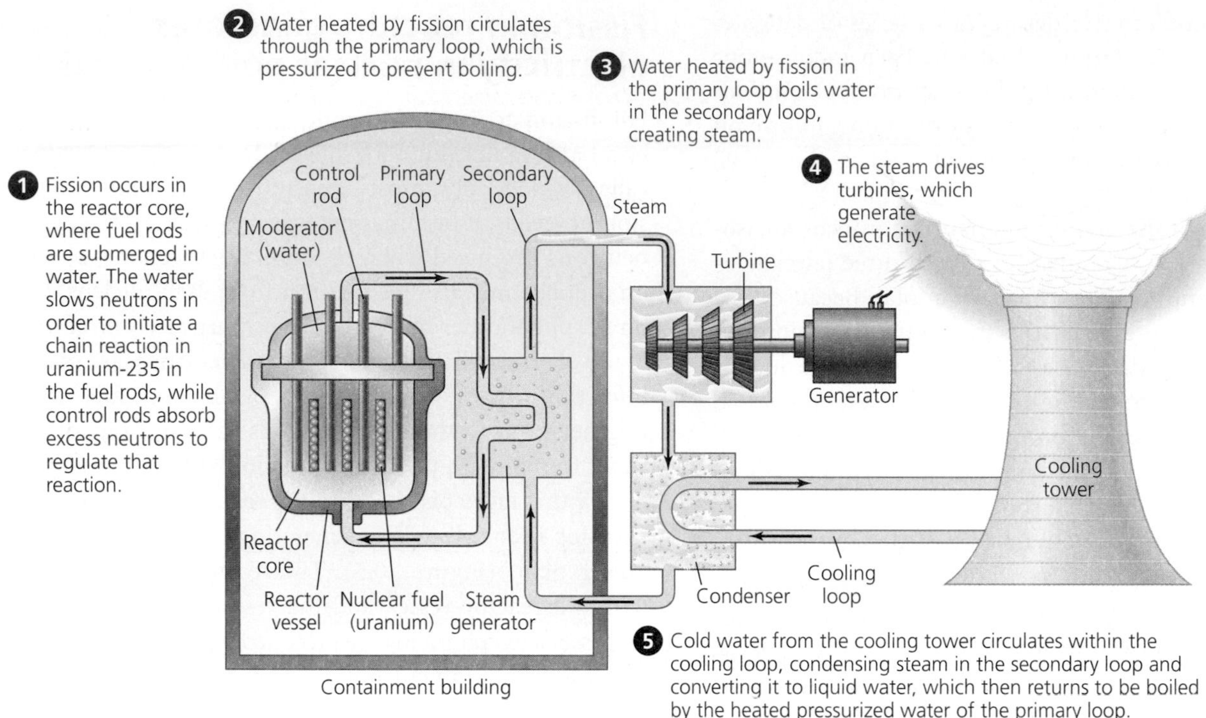


FIGURE 20.5 In a pressurized light water reactor, the most common type of nuclear reactor, uranium fuel rods are placed in water, which slows neutrons so that fission can occur (1). Control rods that can be moved into and out of the reactor core absorb excess neutrons to regulate the chain reaction. Water heated by fission circulates through the primary loop (2) and warms water in the secondary loop, which turns to steam (3). Steam drives turbines, which generate electricity (4). The steam is then cooled in the cooling tower by water from an adjacent river or lake and returns to the containment building (5), to be heated again by heat from the primary loop.

force behind hydrogen or thermonuclear bombs—involves forcing together the small nuclei of lightweight elements under extremely high temperature and pressure. The hydrogen isotopes deuterium and tritium can be fused together to create helium, releasing a neutron and a tremendous amount of energy (**Figure 20.6**).

Overcoming the mutually repulsive forces of protons in a controlled manner is difficult, and fusion requires temperatures of many millions of degrees Celsius. Thus, researchers have not yet developed this process for commercial power

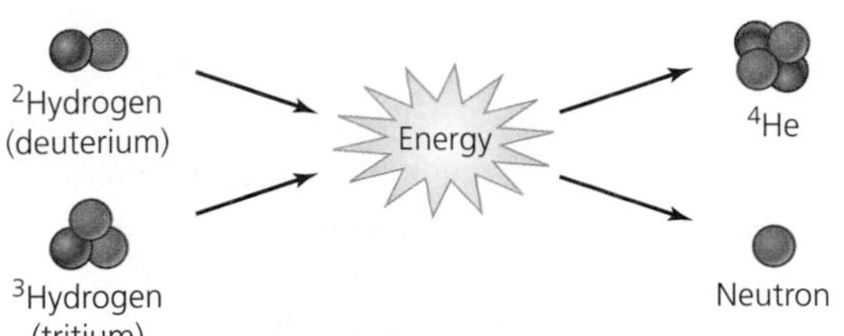


FIGURE 20.6 In nuclear fusion, two small atoms, such as the hydrogen isotopes deuterium and tritium, are fused together, releasing energy along with a helium nucleus and a free neutron. So far, however, scientists have not been able to fuse atoms without supplying far more energy than the reaction produces, so this process is not used commercially.

generation. Despite billions of dollars of funding and decades of research, fusion experiments in the lab still require scientists to input more energy than they produce from the process. Fusion's potentially huge payoffs, though, make many scientists eager to keep trying.

If one day we were to find a way to control fusion in a reactor, we could produce vast amounts of energy using water as a fuel. The process would create only low-level radioactive wastes, without pollutant emissions or the risk of dangerous accidents, sabotage, or weapons proliferation. A consortium of industrialized nations is collaboratingto build a prototype fusion reactor called the International Thermonuclear Experimental Reactor (ITER) in southern France. Even if this multibillion-dollar effort succeeds, however, power from fusion seems likely to remain many years in the future.

## Nuclear power delivers energy more cleanly than fossil fuels

Using fission, nuclear power plants generate electricity without creating air pollution from stack emissions. In contrast, combusting coal, oil, or natural gas emits sulfur dioxide that contributes to acidic deposition, particulate matter that threatens human health, and carbon dioxide and other greenhouse gases that drive global climate change. Even considering all the steps involved in building plants and generating power, researchers from the International Atomic Energy Agency (IAEA) have calculated that nuclear power reduces emissions 4–150 times below fossil fuel combustion (see "The Science behind the Story," • pp. 580–581). IAEA scientists estimate that at current global levels of use, nuclear power helps us avoid emitting 600 million metric tons of carbon each year, equivalent to 8% of global greenhouse gas emissions.

Nuclear power has additional environmental advantages over fossil fuels—coal in particular. Because uranium generates far more power than coal by weight or volume, less of it needs to be mined, so uranium mining causes less damage to landscapes and generates less solid waste than coal mining. Moreover, in the course of normal operation, nuclear power plants are safer for workers than coal-fired plants.

Nuclear power also has drawbacks. One is that the waste it produces is radioactive. Radioactive waste must be handled with great care and must be disposed of in a way that minimizes danger to present and future generations. The second main drawback is that if an accident occurs at a power plant, or if a plant is sabotaged, the consequences can potentially be catastrophic.

Given this mix of advantages and disadvantages (**Figure 20.7**), many governments (although not necessarily most citizens) have judged the good to outweigh the bad, and today the world has 436 operating nuclear plants in 30 nations.

**Weighing THE Issues** | **Choose Your Risk**

Given the choice of living next to a nuclear power plant or living next to a coal-fired power plant, which would you choose? What would concern you most about each option?

## Nuclear power poses small risks of large accidents

Although scientists calculate that nuclear power poses fewer chronic health risks than does fossil fuel combustion, the possibility of catastrophic accidents has spawned a great deal of public anxiety over nuclear power. Two

**Environmental Impacts of Coal-fired and Nuclear Power**

| Type of Impact | Coal | Nuclear |
|---|---|---|
| Land and ecosystem disturbance from mining | Extensive, on surface or underground | Less extensive |
| Greenhouse gas emissions | Considerable emissions | None from plant operation; much less than coal over the entire life cycle |
| Other air pollutants | Sulfur dioxide, nitrogen oxides, particulate matter, and other pollutants | No pollutant emissions |
| Radioactive emissions | No appreciable emissions | No appreciable emissions during normal operation; possibility of emissions during severe accident |
| Occupational health among workers | More known health problems and fatalities | Fewer known health problems and fatalities |
| Health impacts on nearby residents | Air pollution impairs health | No appreciable known health impacts under normal operation |
| Effects of accident or sabotage | No widespread effects | Potentially catastrophic widespread effects |
| Solid waste | More generated | Less generated |
| Radioactive waste | None | Radioactive waste generated |
| Fuel supplies remaining | Should last several hundred more years | Uncertain; supplies could last longer or shorter than coal supplies |

**FIGURE 20.7** Coal-fired power plants and nuclear power plants pose very different risks and impacts to human health and the environment. This chart compares the major impacts of each mode of electricity generation. The more severe impacts are indicated by red boxes.

THE SCIENCE BEHIND THE STORY

## Assessing Emissions from Power Sources

*Coal-fired power plant in Hebei Province, China*

Combusting coal, oil, or natural gas emits carbon dioxide and other greenhouse gases into the atmosphere, where they contribute to global climate change (Chapter 18). Reducing greenhouse emissions is one of the main reasons so many people want to replace fossil fuels with alternative energy sources.

But determining how different energy alternatives compare in emissions is a complex process. A number of studies have tried to quantify and compare emission rates of different energy types, but the varied methods used have made it hard to synthesize this information into a coherent picture.

Researchers from the International Atomic Energy Agency (IAEA) attempted such a synthesis for the generation of electricity. Experts met at six meetings between 1994 and 1998 and reviewed the scientific literature, together with data from industry and government. Their goal was to come up with a range of estimates of greenhouse gas emissions for nuclear energy, each major fossil fuel type, and each major renewable energy source. IAEA scientists Joseph Spadaro, Lucille Langlois, and Bruce Hamilton then published the results in the *IAEA Bulletin* in 2000.

The researchers had to decide how much of the total life cycle of electric power production to include in their estimates. Simply comparing the rotation of turbines at a wind farm to the operation of a coal-fired power plant might not be fair, because it would not reveal that greenhouse gases were emitted as a result of manufacturing the turbines, transporting them to the site, and erecting them there. Similarly, because uranium mining is part of the nuclear fuel cycle, and because we use oil-fueled machinery to mine uranium, perhaps emissions from this process should be included in the estimate for nuclear power.

The researchers decided to conduct a "cradle-to-grave" analysis and include all sources of emissions throughout the entire life cycle of each energy source. This included not just power generation, but also the mining of fuel, preparation and transport of fuel, manufacturing of equipment, construction of power plants, disposal of wastes, and decommissioning of plants. They did, however, separate stack emissions from all other sources of emissions in the chain of steps so that these data could be analyzed independently.

Different greenhouse gases were then standardized to a unit of "carbon equivalence" according to their global warming potential (• pp. 507–508). For instance, because methane is 21 times as powerful a greenhouse gas as carbon, each unit of methane emitted was counted as 21 units of carbon-equivalence. The researchers then calculated rates of emission per unit of power produced. They presented figures in grams of carbon-equivalent emitted per kilowatt-hour ($gC_{eq}$/kWh) of electric power produced.

The overall pattern they found was clear: Fossil fuels produce much higher emission rates than renewable energy sources and nuclear energy (see the figure). The maximum emission rate for fossil fuels (357 $gC_{eq}$/kWh for coal) was 4.7 times higher than the maximum emission rate for any renewable energy source (76.4 $gC_{eq}$/kWh for solar power). The minimum emission rate for any fossil

events were influential in shaping public opinion about nuclear energy.

The first took place at the **Three Mile Island** plant in Pennsylvania (**Figure 20.8**), where in 1979 the United States experienced its most serious nuclear power plant accident. Through a combination of mechanical failure and human error, coolant water drained from the reactor vessel, temperatures rose inside the reactor core, and metal surrounding the uranium fuel rods began to melt, releasing radiation. This process is termed a **meltdown**, and it proceeded through half of one reactor core at Three Mile Island. Area residents stood ready to be evacuated as the nation held its breath, but fortunately most radiation remained trapped inside the containment building.

The accident was brought under control within days, the damaged reactor was shut down, and multibillion-dollar cleanup efforts stretched on for years. Three Mile Island is best regarded as a near-miss; the emergency could have been far worse had the meltdown proceeded through the entire stock of uranium fuel or had the containment building not contained the radiation. Although residents have shown no significant health impacts in the years since, the event put safety concerns squarely on the map for U.S. citizens and policymakers.

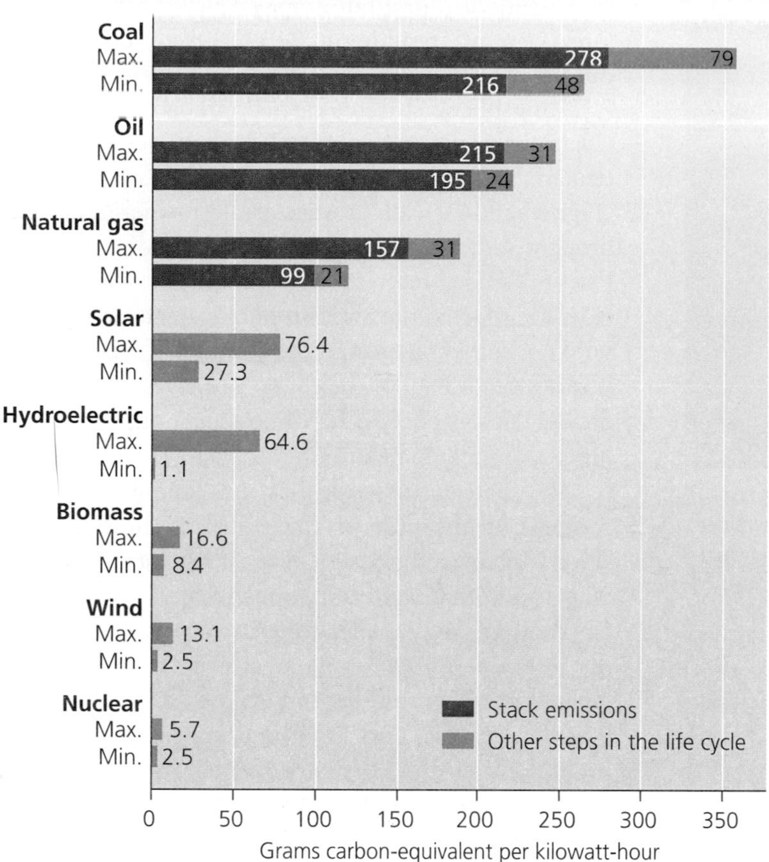


Coal, oil, and natural gas emit far more greenhouse gases than do renewable energy sources and nuclear energy. Red portions of bars represent stack emissions, and orange portions show emissions from other steps in the life cycle. Maximum and minimum values are given for each energy source.

Data from Spadaro, J. V., et al. 2000. Greenhouse gas emissions of electricity generation chains: Assessing the difference. *IAEA Bulletin* 42(2).

fuel (120 $gC_{eq}$/kWh for natural gas) was nearly 100 times greater than the minimum rate for renewables (1.1 $gC_{eq}$/kWh for one form of hydropower). Overall, emissions decreased in the following order: coal, oil, natural gas, photovoltaic solar, hydroelectric, biomass, wind, and nuclear.

Most fossil fuel emissions were stack emissions directly from power generation, and the amounts due to other steps in the life cycle were roughly comparable to those from renewable sources.

Within each category, emissions values varied considerably. This variation was due to many factors, including the type of technology used, geographic location and transport costs, carbon content of the fuel, and the efficiency with which fuel was converted to electricity.

However, technology was expected to improve in the future, creating greater fuel-to-electricity conversion efficiency and lowering emissions rates. Thus, the researchers devised separate emissions estimates for newer technologies expected between 2005 and 2020. These estimates suggested that fossil fuels will improve but still will not approach the cleanliness of nuclear energy and most renewable sources.

Because the IAEA is charged with promoting nuclear energy, critics point out that the agency has clear motivation for conducting a study that shows nuclear power in a favorable light. However, few experts would quibble with the overall trend in the study's data: Nuclear and renewable energy sources are demonstrably cleaner than fossil fuels.

## Chernobyl saw the worst accident yet

In 1986 an explosion at the **Chernobyl** plant in Ukraine (part of the Soviet Union at the time) caused the most severe nuclear power plant accident the world has yet seen. Engineers had turned off safety systems to conduct tests, and human error, combined with unsafe reactor design, led to explosions that destroyed the reactor and sent clouds of radioactive debris billowing into the atmosphere. For 10 days radiation escaped from the plant while emergency crews risked their lives (some later died from radiation exposure) putting out fires. Most residents of the surrounding countryside remained at home for these 10 days, exposed to radiation, before the Soviet government belatedly began evacuating more than 100,000 people.

In the months and years afterwards, workers erected a gigantic concrete sarcophagus around the demolished reactor, scrubbed buildings and roads, and removed irradiated materials (**Figure 20.9**). However, the landscape for at least 30 km (19 mi) around the plant remains contaminated today, and an international team plans to build a larger sarcophagus around the original one, which is deteriorating.

The accident killed 31 people directly and sickened or caused cancer in thousands more. Exact numbers are

FIGURE 20.8 The Three Mile Island nuclear power plant near Harrisburg, Pennsylvania, was the site of a partial meltdown in 1979. This emergency was a "near-miss"—radiation was released but was mostly contained, and no health impacts were confirmed. The incident put the world on notice, however, that a major accident could potentially occur.

uncertain because of inadequate data and the difficulty of determining long-term radiation effects (see "The Science behind the Story," • pp. 584–585). Health authorities estimate that most of the over 4,000 cases of thyroid cancer diagnosed in people who were children at the time resulted from radioactive iodine spread by the accident. Estimates for the total number of cancer cases attributable to Chernobyl, past and future, vary widely, but an international consensus effort 20 years after the event estimated an increase in the cancer rate among exposed people of "up to a few percent," resulting in "up to several thousand fatal cancers."

Atmospheric currents carried radioactive fallout from Chernobyl across much of the Northern Hemisphere, particularly Ukraine, Belarus, and parts of Russia and Europe (**Figure 20.10**). Fallout was greatest where rainstorms brought radioisotopes down from the radioactive cloud. Parts of Sweden received high amounts of fallout. The accident reinforced the Swedish public's fears about nuclear power. A survey taken after the event asked, "Do you think it was good or bad for the country to invest in nuclear energy?" The proportion of respondents answering "bad" jumped from 25% before Chernobyl to 47% afterward.

Fortunately, the world has not experienced another accident on the scale of Chernobyl in the two decades since. Moreover, the design of most reactors in the United States and other western nations is far safer than that of Chernobyl's. Yet, smaller-scale incidents have occurred; for instance, a 1999 accident at a plant in Tokaimura, Japan, killed two workers and exposed over 400 others to leaked radiation. And Sweden experienced a near-miss in 2006, when the Forsmark plant north of Stockholm narrowly avoided a meltdown after only two of four generators started up following a power outage.

As plants around the world age, they require more maintenance and are therefore less safe. New concerns have also surfaced. The September 11, 2001, terrorist attacks raised fears that similar airplane attacks could be

**(a) The Chernobyl sarcophagus**

**(b) Technicians measuring radiation**

FIGURE 20.9 The world's worst nuclear power plant accident unfolded in 1986 at Chernobyl, in present-day Ukraine (then part of the Soviet Union). As part of the extensive cleanup operation, the destroyed reactor was encased in a massive concrete sarcophagus (**a**) to contain further radiation leakage. Technicians scoured the landscape surrounding the plant (**b**), measuring radiation levels, removing soil, and scrubbing roads and buildings.

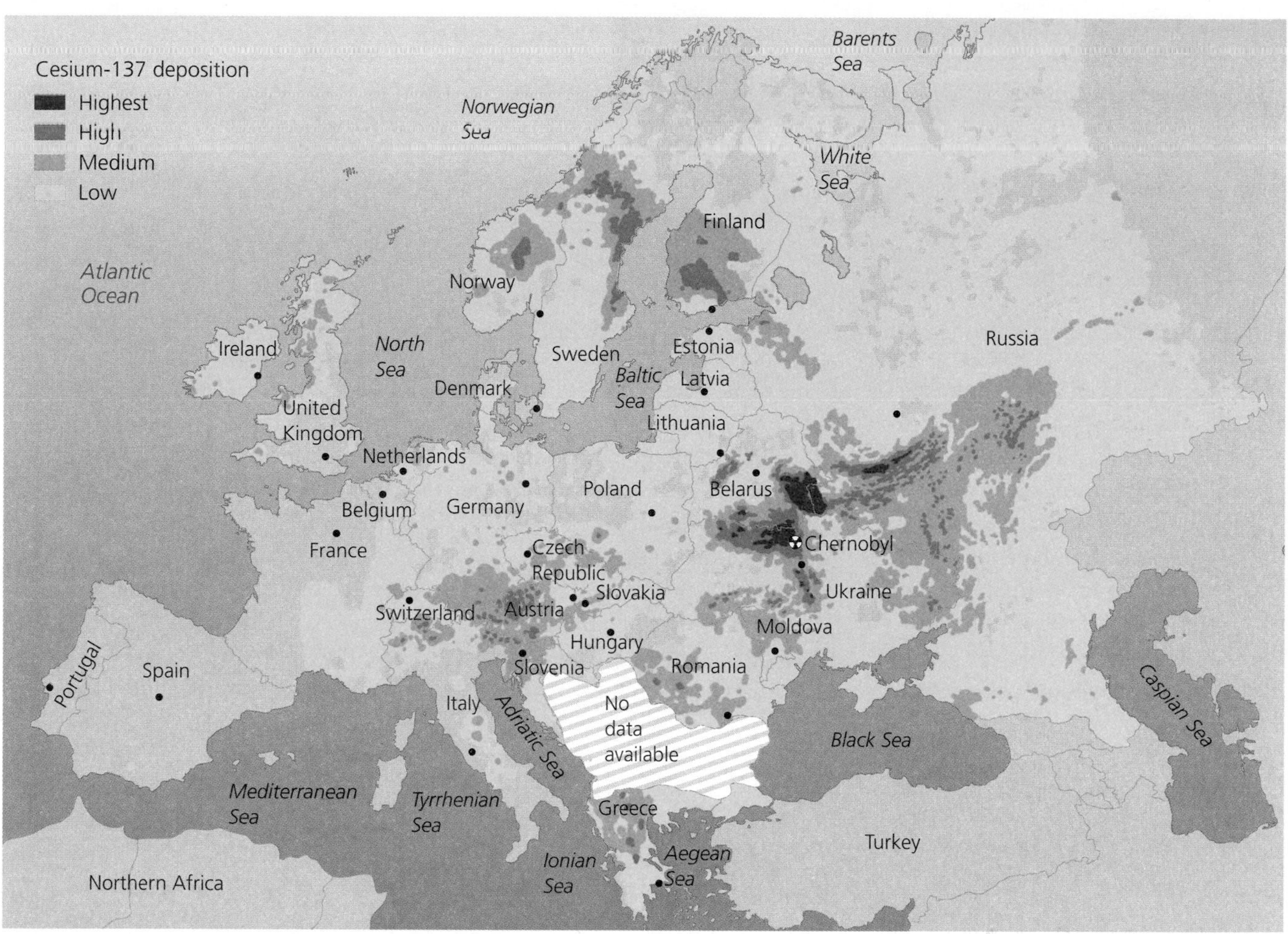


**FIGURE 20.10** Radioactive fallout from the Chernobyl disaster was deposited across Europe in complex patterns resulting from atmospheric currents and rainstorms in the days following the accident. Darker colors in this map of cesium–137 deposition indicate higher levels of radioactivity. Although Chernobyl produced 100 times more fallout than the U.S. bombs dropped on Hiroshima and Nagasaki in World War II, it was distributed over a much wider area. Thus, levels of contamination in any given place outside of Ukraine, Belarus, and western Russia were relatively low; the average European received less than the amount of radiation a person receives naturally in a year. Data from chernobyl.info, Swiss Agency for Development and Cooperation, Bern, 2005.

carried out against nuclear plants. Moreover, radioactive material could be stolen from plants and used in terrorist attacks. This possibility is especially worrisome in the cash-strapped nations of the former Soviet Union, where hundreds of former nuclear sites have gone without adequate security for years. In a cooperative international agreement, the U.S. government, through the "megatons to megawatts" program, has been buying up some of this material and diverting it to peaceful use in power generation.

## Waste disposal remains a problem

Even if nuclear power generation could be made completely safe, we would still be left with the conundrum of what to do with spent fuel rods and other radioactive waste. Recall that fission utilizes $^{235}U$ as fuel, leaving as waste the 97% of uranium that is $^{238}U$. This $^{238}U$, as well as all irradiated material and equipment that is no longer being used, must be disposed of in a location where radiation will not escape. Because the half-lives of uranium, plutonium, and many other radioisotopes are far longer than human lifetimes, this waste will continue emitting radiation for thousands of years. Thus, radioactive waste must be placed in unusually stable and secure locations where radioactivity will not harm future generations.

Currently, nuclear waste from power generation is being held in temporary storage at nuclear power plants across the United States and the world. Spent fuel rods are sunken in pools of cooling water to minimize radiation leakage (**Figure 20.11a**). However, the U.S. Department of Energy (DOE) estimates that by 2010, three-fourths of

THE SCIENCE BEHIND THE STORY

## Health Impacts of Chernobyl

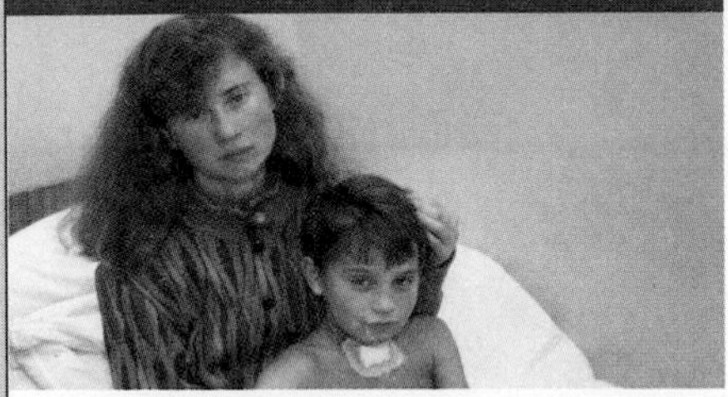

*Chernobyl survivors*

In the wake of the nuclear power plant accident at Chernobyl in 1986, medical scientists from around the world rushed to study how the release of radiation would affect human health. Yet determining long-term health impacts of an event is difficult, so it is not surprising that the hundreds of researchers trying to pin down Chernobyl's impacts sometimes came up with very different conclusions.

In an effort to reach consensus, the World Health Organization (WHO) engaged 100 experts from various nations to review all studies through 2006 and issue a report summarizing what scientists had learned in the 20 years since the accident. The WHO report was part of a collaborative 20-year review by an array of international agencies and the governments of Ukraine, Russia, and Belarus.

Doctors documented the most severe effects among emergency workers who battled to contain the incident in its initial hours and days. Medical staff treated and recorded the progress of 134 workers hospitalized with acute radiation sickness (ARS). Radiation destroys cells in the body, and if the destruction outpaces the body's abilities to repair the damage, the person will soon die. Symptoms of ARS include vomiting, fever, diarrhea, thermal burns, mucous membrane damage, and weakening of the immune system by the depletion of white blood cells. In total, 28 people died from acute effects soon after the accident, and those dying had the greatest estimated exposure to radiation.

The major health impact of Chernobyl's radiation, however, has been thyroid cancer. Studies have documented a clear excess of cases among Chernobyl-area residents, particularly children. The thyroid gland is where the human body concentrates iodine, and one of the most common radioactive isotopes released early in the disaster was iodine-131 ($^{131}$I). Children have large and active thyroid glands, so they are especially vulnerable to thyroid cancer induced by radioisotopes of iodine.

Realizing that thyroid cancer might be a problem, medical workers took measurements of iodine activity from the thyroid glands of several hundred thousand people in Russia, Ukraine, and Belarus in the months following the accident. They also measured food contamination and had people fill out questionnaires on their food consumption. These data showed that drinking milk from cows that had grazed on contaminated grass was the main route of exposure to $^{131}$I for most people, although fresh vegetables also contributed.

As doctors had feared, rates of thyroid cancer began rising among children in the regions of highest exposure (see the graph). The yearly number of thyroid cancer cases in the 1990s, particularly in Belarus, far exceeded numbers from years before Chernobyl. Multiple studies found linear dose-response relationships (• p. 398) in data from Ukraine and Belarus.

Fortunately, treatment of thyroid cancer has a high success rate, and as of 2002, only 15 of the 4,000 children diagnosed with thyroid cancer had died from it. By 2006, medical professionals were estimating that the number of cases had increased to 5,000 and was still rising.

Critics pointed out that any targeted search tends to turn up more of whatever medical problem is being looked for. But experts now agree that the increase in childhood thyroid cancer was undeniably attributable to Chernobyl. Furthermore, thyroid cancer also appears to have risen markedly in adults in Belarus and the most contaminated regions of Russia.

U.S. plants will have no room left for this type of storage. Many plants are now expanding their storage capacity by storing waste in thick casks of steel, lead, and concrete (**Figure 20.11b**).

In total, U.S. power plants are storing over 56,000 metric tons of high-level radioactive waste, enough to fill a football field to the depth of 5.5 m (18 ft)—as well as much more low-level radioactive waste. This waste is held at 125 sites spread across 39 states (**Figure 20.12**). A 2005 National Academy of Sciences report judged that most of these sites were vulnerable to terrorist attacks. The DOE estimates that over 161 million U.S. citizens live within 125 km (75 mi) of temporarily stored waste.

Because storing waste at many dispersed sites creates a large number of potential hazards, nuclear waste managers

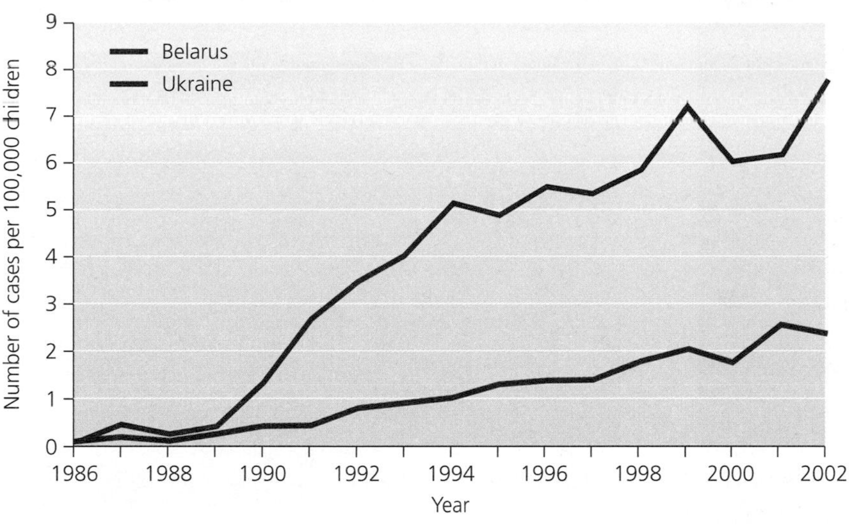


**(a) Rates of child thyroid cancer**

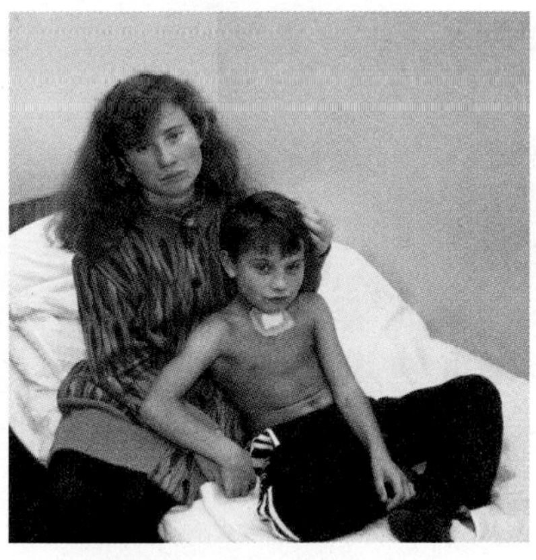

**(b) Cancer patient with mother after surgery**

The incidence of thyroid cancer **(a)** jumped in Belarus and Ukraine starting 4 years after the Chernobyl accident released radioactive iodine isotopes. Many babies and young children at the time of the accident developed thyroid cancer in later years. Most, like this boy shown with his mother after surgery **(b)**, have undergone treatment and survived. Data (a) from The Chernobyl Forum. 2006. *Chernobyl's legacy: Health, environmental, and socio-economic impacts.* The Chernobyl Forum: 2003–2005, 2nd rev. version. IAEA, Vienna.

Studies addressing other health impacts have turned up varying results. Some research has shown an increase in cataracts due to radiation. Other data has revealed that there was apparently no rise in reproductive problems. Studies have also shown psychological effects of the accident among exposed people, including heightened anxiety, particularly about health. However, the WHO report noted that much of this anxiety may result from constant portrayals of exposed people as victims and sufferers rather than as survivors—causing those people to think of themselves in that way.

Researchers have conducted many studies on leukemia and other cancers, but the WHO report concluded that there is "no convincing evidence" so far that rates of any cancer (aside from thyroid cancer) have increased among people exposed to Chernobyl's radiation.

Based on previous studies on survivors of the atomic bombs dropped on Hiroshima and Nagasaki, researchers had expected increases in cancer mortality among the 600,000 people with significant exposure from Chernobyl to be "up to a few percent." This would translate into adding perhaps 4,000 cancer fatalities to a normally expected 100,000 cancer deaths in the population from all other causes. However, conducting epidemiological studies (• pp. 397–398) with enough statistical power to detect such minor increases in already-rare events is difficult, requiring that huge numbers of people be observed over many years.

Thus, the jury is still out on leukemia and other cancers. Moreover, many cancers do not generally appear for at least 10–15 years after exposure, so it is possible that most illnesses have yet to arise. For all these reasons, the WHO recommends continued monitoring. As future cancer cases accumulate, continued research will be needed to measure the full scope of health effects from Chernobyl.

want to send all waste to a central repository that can be heavily guarded. In Sweden, that nation's nuclear industry has established a single repository for low-level waste near one power plant and is searching for a single disposal site deep within bedrock for spent fuel rods and other high-level waste. In the United States, the multiyear search homed in on Yucca Mountain, a remote site in the desert of southern Nevada, 160 km (100 mi) from Las Vegas (**Figure 20.13a**). Choice of this site followed extensive study by government scientists (**Figure 20.13b**), but Nevadans were not happy about the choice and fought against it. In 2002 the site was recommended by the president and approved by Congress, and the Department of Energy is now awaiting approval from the Nuclear Regulatory Commission. If given final approval, Yucca Mountain is expected to begin receiving waste from

(a) Wet storage

(b) Dry storage

FIGURE 20.11 Spent uranium fuel rods are currently stored at nuclear power plants and will likely remain at these scattered sites until a central repository for commercial radioactive waste is fully developed. Spent fuel rods are most often kept in "wet storage" in pools of water (**a**), which keep them cool and reduce radiation release. Alternatively, the rods may be kept in "dry storage" in thick-walled casks layered with lead, concrete, and steel (**b**).

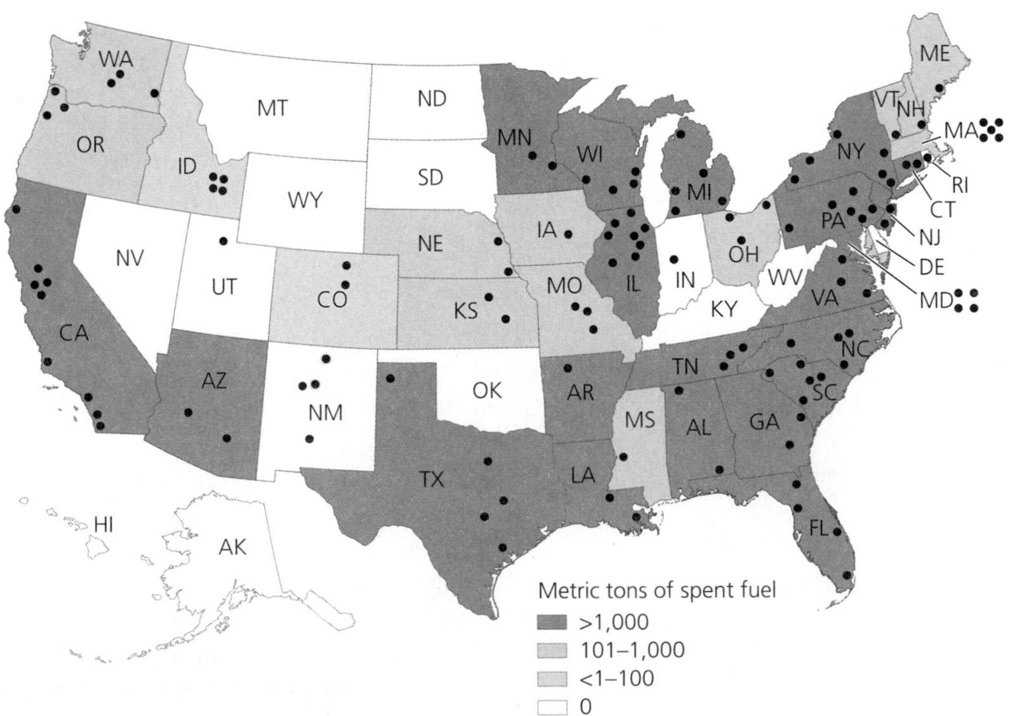


FIGURE 20.12 High-level radioactive waste from civilian reactors is currently stored at over 120 sites in 39 states across the United States. In this map, dots indicate storage sites, and the four shades of color indicate the total amount of waste stored in each state. Slightly different classifications of waste mean that some states shaded white show storage sites for certain types of waste. Data from Office of Civilian Radioactive Waste Management, U.S. Department of Energy; and Nuclear Energy Institute, Washington, D.C.

(a) Yucca Mountain

(b) Scientific testing

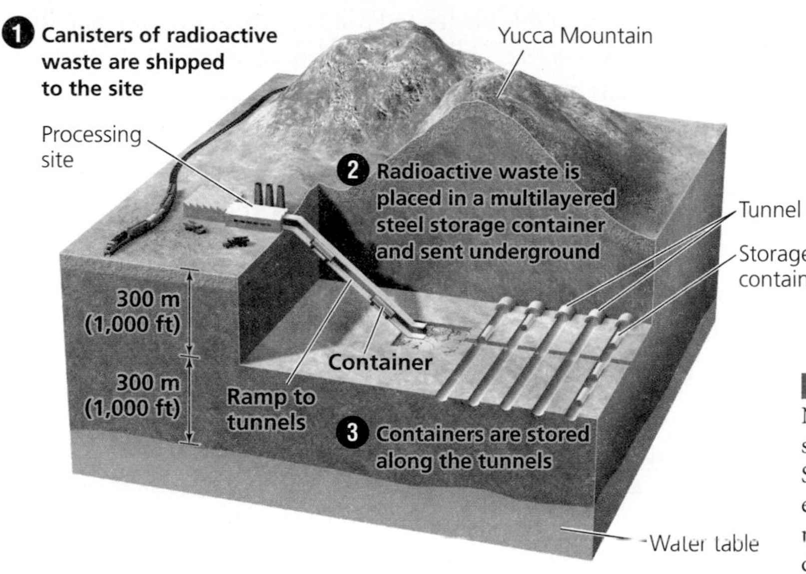

(c) Proposed design

FIGURE 20.13 Yucca Mountain (**a**), in a remote part of Nevada, awaits development as the central repository site for all the commercial nuclear waste in the United States. Here (**b**), technicians are testing the effects of extreme heat from radioactive decay on the stability of rock. Waste would be buried in a network of tunnels deep underground yet still high above the water table (**c**).

nuclear reactors, as well as high-level radioactive waste from military installations, in 2017. According to the design, waste would be stored in a network of tunnels 300 m (1,000 ft) underground, yet 300 m (1,000 ft) above the water table (**Figure 20.13c**). Scientists and policymakers chose the Yucca Mountain site because they determined that:

- It is unpopulated, lying 23 km (14 mi) from the nearest year-round residences.
- It has stable geology, with minimal risk of earthquakes that could damage the tunnels and release radioactivity.
- Its dry climate should minimize water infiltration, reducing chances of groundwater contamination.
- The water table is deep underground, making groundwater contamination less likely.
- The pool of groundwater does not connect with groundwater elsewhere, so any contamination would be contained.
- The location, on federal land, can feasibly be protected from sabotage.

Some scientists, antinuclear activists, and concerned Nevadans have challenged these conclusions. For instance, they argue that earthquakes or volcanic activity could destabilize the site's geology. They also fear that fissures in the mountain's rock could allow rainwater to seep into the caverns.

Another concern is that nuclear waste will need to be transported to Yucca Mountain from the 120-some current storage areas and from current and future nuclear plants and military installations. Because this would involve many thousands of shipments by rail and truck across hundreds of public highways through almost every state of the union, many people worry that the risk of an accident or of sabotage is unacceptably high.

**Weighing THE Issues** How to Store Waste?

Which do you think is a better option—to transport nuclear waste cross-country to a single repository or to store it permanently at numerous power plants and military bases scattered across the nation? Would your opinion be affected if you lived near the repository site? Near a power plant? On a highway route along which waste is transported?

## Multiple dilemmas have slowed nuclear power's growth

Dogged by concerns over waste disposal, safety, and expensive cost overruns, nuclear power's growth has slowed. Since the late 1980s, nuclear power has grown by 2.5% per year, about the same rate as electricity generation overall. Public anxiety in the wake of Chernobyl made utilities less willing to invest in new plants. So did the enormous expense of building, maintaining, operating, and ensuring the safety of nuclear facilities. Almost every nuclear plant has turned out to be more expensive than expected. In addition, plants have aged more quickly than expected because of problems that were underestimated, such as corrosion in coolant pipes. The plants that have been shut down—well over 100 around the world to date—have served on average less than half their expected lifetimes. Moreover, shutting down, or decommissioning, a plant can sometimes be more expensive than the original construction.

As a result of these economic issues, electricity from nuclear power today remains more expensive than electricity from coal and other sources. Governments are still subsidizing nuclear power to keep consumer costs down, but many private investors lost interest long ago. Nonetheless, nuclear power remains one of the few currently viable alternatives to fossil fuels with which we can generate large amounts of electricity in short order.

Many experts predict nuclear power will decrease because three-quarters of Western Europe's capacity is scheduled to be retired by 2030. In Western Europe, not a single reactor is under construction today, and Germany and Belgium, like Sweden, have declared an intention to phase out nuclear power altogether.

Asian nations, in contrast, are adding nuclear capacity. China, India, and South Korea are expanding their nuclear programs to help power their rapidly growing economies. Japan is so reliant on imported oil that it is eager to diversify its energy options. Altogether, Asia hosts two-thirds of the most recent nuclear plants to go into operation and 15 of the 26 plants now under construction.

In the United States, the nuclear industry stopped building plants following Three Mile Island, and public opposition scuttled many that were under construction. The $5.5-billion Shoreham Nuclear Power Plant on New York's Long Island was shut down just 2 months after being licensed because officials determined that evacuation would be impossible in this densely populated area should an accident ever occur. Of the 259 U.S. nuclear plants ordered since 1957, nearly half have been cancelled. At its peak in 1990, the United States had 112 operable plants; today it has 103. However, expanding U.S. nuclear capacity would decrease reliance on fossil fuels, and nuclear proponents point out that engineers are planning a new generation of reactors designed to be safer and less expensive.

**Weighing THE Issues** More Nuclear Power?

Do you think the United States should expand its nuclear power program? Why or why not?

With little or no growth predicted for nuclear power, and with fossil fuels in limited supply, where will our growing human population turn for additional energy? Increasingly, people are turning to renewable sources of energy: energy sources that cannot be depleted by our use. Although many renewable sources are still early in their stages of development, two of them—biomass energy and hydroelectric power—are already well developed and widely used.

# Biomass Energy

When people use the term *biomass energy* these days, they can mean very different things. To a poor farmer in Africa, biomass energy means cutting wood from trees or collecting livestock manure by hand and burning it to heat and cook for her family. To an industrialized farmer in Iowa, biomass energy means shipping his grain to a hi-tech refinery that converts it to liquid fuel to run automobiles. The diversity of sources and approaches involved in biomass energy (Table 20.2) provides great potential for addressing our energy challenges, but it also means we must be careful in selecting the paths we follow.

*Biomass* (• p. 150) consists of the organic material that makes up living organisms. People harness **biomass energy** from many types of plant matter, including wood from trees, charcoal from burned wood, and matter from agricultural crops, as well as from combustible animal waste products such as cattle manure. Fossil fuels are not considered biomass energy sources because their organic matter has not been part of living organisms for millions of years and has undergone considerable chemical alteration.

**TABLE 20.2 Major Sources of Biomass Energy**

| Major Sources of Biomass Energy |
|---|
| **Direct combustion for heating** |
| ▶ Wood cut from trees (fuelwood) |
| ▶ Charcoal |
| ▶ Manure from farm animals |
| **Biofuels for powering vehicles** |
| ▶ Corn grown for ethanol |
| ▶ *Bagasse* (sugarcane residue) grown for ethanol |
| ▶ Soybeans, rapeseed, and other crops grown for biodiesel |
| ▶ Used cooking oil for biodiesel |
| ▶ Plant matter treated with enzymes to produce cellulosic ethanol |
| **Biopower for generating electricity** |
| ▶ Crop residues (such as cornstalks) burned at power plants |
| ▶ Forestry residues (such as wood waste from logging) burned at power plants |
| ▶ Processing wastes (such as solid or liquid waste from sawmills, pulp mills, and paper mills) burned at power plants |
| ▶ "Landfill gas" burned at power plants |
| ▶ Livestock waste from feedlots for gas from anaerobic digesters |
| ▶ Organic components of municipal solid waste from landfills |

## Traditional biomass sources are widely used in the developing world

Over 1 billion people still use wood from trees as their principal energy source. In developing nations, especially in rural areas, families gather fuelwood to burn in their homes for heating, cooking, and lighting (**Figure 20.14**; also see Figure 17.21, • p. 496). In these nations, fuelwood, charcoal, and manure account for fully 35% of energy use—in the poorest nations, up to 90%.

Fuelwood and other traditional biomass sources constitute nearly 80% of all renewable energy used worldwide. However, considering what we have learned about the loss of forests (• pp. 334–337), it is fair to ask whether biomass should truly be considered a renewable resource. In reality, biomass is renewable only if it is not overharvested. At moderate rates of use, trees and other plants can replenish themselves over months to decades. However, when forests are cut too quickly, or when overharvesting leads to soil erosion and forests fail to grow back, then biomass is not effectively replenished. The potential for deforestation makes biomass energy less sustainable than other renewable sources, particularly as human population continues to increase.

As developing nations industrialize, fossil fuels are replacing traditional energy sources (**Figure 20.15**). As a result, biomass use is growing more slowly worldwide than overall energy use.

**FIGURE 20.14** Well over a billion people in developing countries rely on fuelwood for heating and cooking. Wood cut from trees remains the major source of biomass energy used in the world today. In theory, biomass is renewable, but in practice it may not be if forests are overharvested.

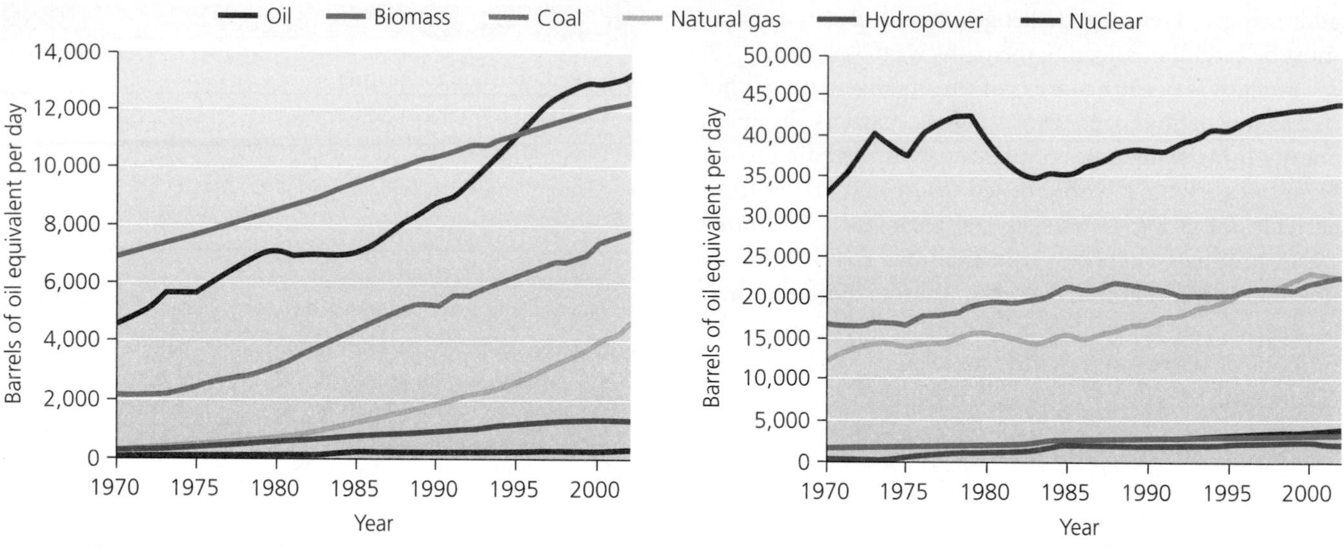


**(a) Energy consumption in developing nations**

**(b) Energy consumption in industrialized nations**

FIGURE 20.15 Energy consumption patterns vary greatly between developing nations (**a**) and industrialized nations, here represented by nations of the Organisation for Economic Cooperation and Development (OECD) (**b**). Note the large role that biomass (primarily fuelwood) plays in supplying energy to developing countries. Note also that the *y* axes differ; people in developing nations consume far less energy than those in industrialized nations. Data from Energy Information Administration, U.S. Department of Energy.

## New biomass strategies are being developed in industrialized countries

Besides the fuelwood, charcoal, and manure traditionally used in direct combustion for heating, biomass energy sources in today's world include a variety of materials that can be made into several innovative types of energy (see Table 20.2). Some of these sources can be burned in power plants to produce **biopower**, generating heat and electricity in the same way that coal is burned for power (• pp. 548–549). Other new biomass sources can be converted into fuels used primarily to power automobiles; these are termed **biofuels**. Because many of these novel biofuels and biopower strategies depend on technologies resulting from extensive research and development, they are being developed primarily in wealthier industrialized nations, such as Sweden and the United States.

## Biofuels can power automobiles

Liquid fuels from biomass sources are helping to power millions of vehicles on today's roads. The two primary biofuels developed so far are ethanol (for gasoline engines) and biodiesel (for diesel engines).

**Ethanol** is the alcohol in beer, wine, and liquor. It is produced as a biofuel by fermenting biomass, generally from carbohydrate-rich crops, in a process similar to brewing beer. In fermentation, carbohydrates are converted to sugars and then to ethanol. Spurred by the 1990 Clean Air Act amendments and generous government subsidies, ethanol is widely added to gasoline in the United States to reduce automotive emissions. In 2006 in the United States, 18.4 billion L (4.86 billion gal) of ethanol were produced, mostly from corn (**Figure 20.16**). This amount is growing rapidly, and the number of U.S. ethanol plants has soared past 100, with 70–80 more now under construction.

Any vehicle with a gasoline engine runs well on gasoline blended with up to 10% ethanol, but more and more vehicles are being produced that can run primarily on ethanol. Sweden has many such public buses, and the U.S. "big three" automakers are now producing *flexible fuel vehicles* that run on E-85, a mix of 85% ethanol and 15% gasoline. Over 5 million such cars are on U.S. roads today, but so few gas stations offer E-85 (**Figure 20.17**) that drivers generally are forced to fill these cars with conventional gasoline. However, increasing infrastructure for ethanol will change this.

In Brazil, sugarcane residue is crushed to make *bagasse*, a material that is then used to make ethanol. Half of all new Brazilian cars are flexible-fuel vehicles, and ethanol from sugarcane accounts for 40% of all automotive fuel that Brazil's drivers use.

Because growing crops to produce ethanol may not be sustainable, as we shall see shortly, researchers are refining techniques to use enzymes to produce ethanol from the cellulose that gives structure to all plant material. If we can produce *cellulosic ethanol* in commercially feasible ways, then ethanol could be made from low-value crop waste (residues such as corn stalks and husks), rather than from high-value crops.

**(a) Corn grown for ethanol**

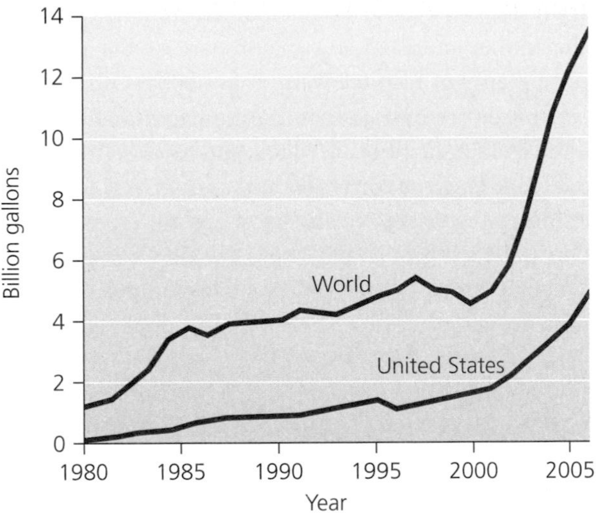


**(b) Ethanol production, 1980–2006**

**FIGURE 20.16** As much as 20% of the U.S. corn crop (**a**) is used to produce ethanol, a biofuel that is widely added to gasoline in the United States. Brazil produces most of the rest of the world's ethanol, from *bagasse* (sugarcane residue). Ethanol production (**b**) has grown rapidly in the last several years.

Vehicles with diesel engines have their own biofuel that is growing even faster than ethanol. **Biodiesel** is produced from vegetable oils. Oil is mixed with small amounts of ethanol or methanol (wood alcohol) in the presence of a chemical catalyst. In Europe, where most biodiesel is used, rapeseed oil is the oil of choice, whereas U.S. biodiesel producers use mostly soybean oil. In addition, biodiesel producers can utilize animal fats and used grease and cooking oil from restaurants.

Vehicles with diesel engines can run on 100% biodiesel. In fact, when Rudolf Diesel invented the diesel engine in 1895, he designed it to run on a variety of fuels, and he showcased his invention at the 1900 World's Fair using peanut oil. Since that time, we have mainly used petroleum- based fuel *(petrodiesel)* because it is cheaper. Today's diesel engines are designed to work with petrodiesel, and although biodiesel will also power the vehicle, some engine parts wear out more quickly with its use. Most frequently, biodiesel is mixed with conventional petrodiesel; a 20% biodiesel mix (called B20) is common today.

To run on straight vegetable oil, a diesel engine needs to be modified. Extra parts need to be added, so that there are tanks for both the oil and for petrodiesel, which is often needed to start the engine in cooler weather. Although these parts can be bought for as little as $800, it remains to be seen whether using straight vegetable oil might entail further costs, such as reduced longevity or greater engine maintenance.

Biodiesel cuts down on emissions compared with petrodiesel (**Figure 20.18**). Its fuel economy is almost as good, and it costs just slightly more at today's oil prices. It is also nontoxic and biodegradable. Increasing numbers of environmentally conscious individuals in North America and Europe are fueling their cars with biodiesel from waste oils, and some buses and recycling trucks are now running on biodiesel. Governments are encouraging its use, too; Minnesota, for instance, mandates that all

**FIGURE 20.17** Few gas stations offer biofuels, but that is changing quickly. This station in New Mexico sells a 20% biodiesel blend (left pump), an 85% ethanol blend (center pump), and a standard 10% ethanol blend (right pump).

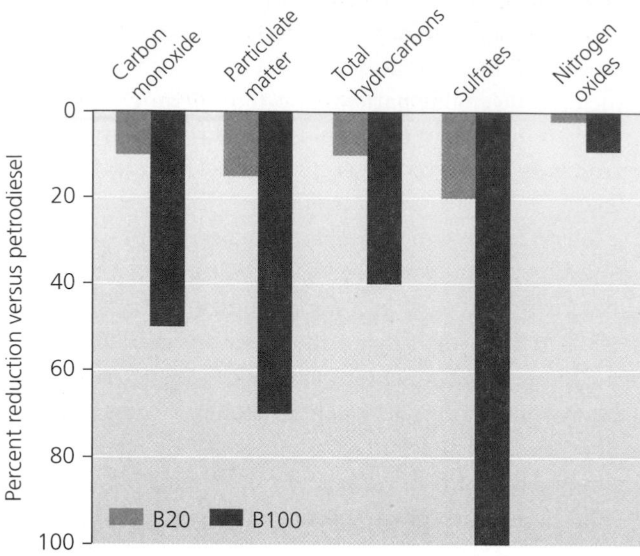


FIGURE 20.18 Burning biodiesel in a diesel engine emits less pollution than burning conventional petroleum-based diesel. Shown are the percentage reductions in several major automotive pollutants that one can attain by using B20 (a mix of 20% biodiesel and 80% petrodiesel) and B100 (pure biodiesel). Data from U.S. Environmental Protection Agency.

diesel sold must include a 2% biodiesel component, and already over 40 state and federal fleets are using biodiesel blends.

Some enthusiasts have taken biofuel use further. Eliminating the processing step that biodiesel requires, they use straight vegetable oil in their diesel engines. One notable effort is the BIO Tour, in which a group of students, environmentalists, and artists drives across North America in a bus fueled entirely by waste oil from restaurants (**Figure 20.19**). Each summer since 2003, a group has gone on tour with the bus, hosting festive events that combine music and dancing with seminars on environmental sustainability and spreading the word about nonpetroleum fuels.

## With biopower we generate electricity from biomass

We harness biopower by combusting biomass to generate electricity. This can be done using a variety of sources and techniques. Many of the sources used for biopower are the waste products of existing industries or processes. For instance, the forest products industry generates large amounts of woody debris in logging operations and at sawmills, pulp mills, and paper mills (**Figure 20.20**). Sweden's efforts to promote biomass energy have focused largely on using forestry residues. Because so much of the nation is forested and the timber industry is a major part of the national economy, plenty of forestry waste is available. Other waste sources include organic waste from municipal landfills, animal waste from agricultural feedlots, and residue from agricultural crops (such as cornstalks and corn husks).

Besides using waste, we also grow certain crops specifically to produce biopower. These include fast-growing trees, such as specially bred willows and poplars, and various fast-growing grasses, such as bamboo, fescue, and switchgrass (**Figure 20.21**).

Power plants built to combust biomass operate similarly to those fired by fossil fuels; the combustion heats water, creating steam to turn turbines and generators, thereby generating electricity. Much of the biopower produced so far comes from power plants that generate both electricity and heating through cogeneration (•pp. 566–567).

FIGURE 20.19 Each summer a group of alternative-fuel advocates goes on tour in the "BioBus," a bus fueled entirely on used vegetable oil from restaurants. Their "Bio Tours" sponsor events across North America that include music, dancing, and seminars on environmental sustainability and alternative fuels. Their motto: "Solar-powered sound, veggie-powered bus."

FIGURE 20.20 Forestry residues (here from a Swedish logging operation) are a major source material for biopower in some regions.

These plants are often located where they can take advantage of forestry waste.

Biomass is also increasingly being combined with coal in coal-fired power plants in a process called *co-firing*. Wood chips, wood pellets, or other biomass is introduced with coal into a high-efficiency boiler that uses one of several technologies. We can substitute biomass for up to 15% of the coal with only minor equipment modification and no appreciable loss of efficiency. Co-firing can be a relatively easy and inexpensive way for fossil-fuel-based utilities to expand their use of renewable energy.

The decomposition of biomass by microbes produces gas that can be used to generate electricity. The anaerobic bacterial breakdown of waste in landfills produces methane and other components, and this "landfill gas" is now being captured at many solid waste landfills and sold as fuel (• pp. 550, 638–639). Methane and other gases can also be

FIGURE 20.21 Switchgrass is one of several fast-growing plants that are being grown as fuel for biopower.

produced in a more controlled way in anaerobic digestion facilities. This "biogas" can then be burned in a power plant's boiler to generate electricity.

We also harness biopower through *gasification*, a process in which biomass is vaporized at extremely high temperatures in the absence of oxygen, creating a gaseous mixture including hydrogen, carbon monoxide, carbon dioxide, and methane. This mixture can generate electricity when used in power plants to turn a gas turbine to propel a generator. Gas from gasification can also be treated in various ways to produce methanol, synthesize a type of diesel fuel, or isolate hydrogen for use in hydrogen fuel cells (• pp. 622–625). An alternative method of heating biomass in the absence of oxygen results in *pyrolysis*, which produces a mix of solids, gases, and liquids. This includes a liquid fuel called pyrolysis oil, which can be burned to generate electricity.

At small scales, farmers, ranchers, or villages can operate modular biopower systems that use livestock manure to generate electricity. Small household biodigesters now provide portable and decentralized energy production for remote rural areas.

At large scales, industries such as the forest products industry are using their waste to generate power, and industrialized farmers are growing crops for biopower. In Sweden, one-sixth of the nation's energy supply now comes from biomass, and biomass provides more fuel for electricity generation than coal, oil, or natural gas. Pulp mill liquors are the main source, but solid wood waste, municipal solid waste, and biogas from digestion are all increasingly used. In the United States, several dozen biomass-fueled power plants are now operating, and several dozen coal-fired plants are experimenting with co-firing.

## Biomass energy brings environmental and economic benefits

Biomass energy has one overarching environmental benefit: It is essentially carbon-neutral, releasing no net carbon into the atmosphere. Although burning biomass emits plenty of carbon, the carbon released is simply the carbon that photosynthesis had pulled from the atmosphere to create the biomass in the first place. That is, the carbon that biomass combustion emits is balanced by the carbon that photosynthesis had sequestered within the biomass just years, months, or weeks before. Therefore, when we replace fossil fuels with bioenergy, we reduce net carbon flux to the atmosphere, helping to mitigate global climate change.

However, this holds only if biomass sources are not overharvested. Deforestation will increase carbon flux to the atmosphere because less vegetation means less carbon uptake by plants for photosynthesis. In addition, biomass energy use is not carbon-neutral if we need to input fossil fuel energy to produce the biomass (for instance, by driving tractors, using fertilizers, and applying pesticides to produce crops used for biofuel).

Biofuels can reduce greenhouse gas emissions that contribute to climate change in additional ways. Capturing landfill gas reduces emissions of methane, a potent greenhouse gas. And biofuels such as ethanol and biodiesel contain oxygen, which when added to gasoline or diesel helps those fuels to combust more completely, reducing pollution.

Shifting from fossil fuels to biomass energy also can have economic benefits. As a resource, biomass tends to be well spread geographically, so using it should help support rural economies and reduce many nations' dependence on imported fuels. Biomass also tends to be the least expensive type of fuel for burning in power plants, and improved energy efficiency brings lower prices for consumers.

By enhancing energy efficiency and recycling waste products, using biomass energy helps move our industrial systems toward greater sustainability. The U.S. forest products industry now obtains over half its energy by combusting the waste it recycles, including woody waste and liquor from pulp mill processing.

Relative to fossil fuels, biomass also benefits human health. By replacing coal in co-firing and direct combustion, biopower reduces emissions of sulfur dioxide because plant matter, unlike coal, contains no appreciable sulfur content. By replacing gasoline and petrodiesel and by burning more cleanly in vehicles, biofuels reduce emissions of nitrogen oxides and other air pollutants.

## Biomass energy also brings drawbacks

Biomass energy has negative environmental impacts, too. Burning fuelwood and other biomass in traditional ways for cooking and heating leads to health hazards from indoor air pollution (• p. 496). Harvesting fuelwood at an unsustainably rapid rate leads to deforestation, soil erosion, and desertification, damaging landscapes, diminishing biodiversity, and impoverishing human societies. In arid regions that are heavily populated and that support meager woodlands, fuelwood harvesting can have enormous impacts. Such is the case with many regions of Africa and Asia. In contrast, moister, well-forested areas with lower population densities, such as Sweden, stand less chance of deforestation.

Growing crops to produce biofuels exerts tremendous impacts on ecosystems. Although crops grown for energy typically receive lower inputs of pesticides and fertilizers than those grown for food, cultivating biofuel crops is land-intensive and brings with it all the impacts of monocultural agriculture (• pp. 237, 267–268). Biofuel crops take up precious land that might otherwise be left in its natural condition or developed for other purposes (**Figure 20.22**). If we were to try to produce all the automotive fuel currently used in the United States with ethanol from U.S. corn, the nation would need to expand its already immense corn acreage by more than 60%, with no loss of productivity and without producing any corn for food. Even at our current level of production, biofuel is already competing with food production. As farmers shifted more corn crops to ethanol in 2006 and 2007, corn supplies for food dropped, and international corn prices skyrocketed. In Mexico, where corn tortillas are emblematic of the national cuisine, average citizens found themselves struggling to buy their staple food, and protests erupted across the country over the inflated prices.

Growing bioenergy crops also requires substantial inputs of energy. We currently operate farm equipment using fossil fuels, and farmers apply petroleum-based pesticides and fertilizers to increase yields. Moreover, fossil fuels are used in refineries to heat water so that we can distill pure ethanol. Thus, shifting from gasoline to ethanol for our transportation needs would not eliminate our reliance on fossil fuels.

Furthermore, growing corn for ethanol yields only a modest amount of energy relative to the energy that needs to be input. Recall our discussion of *energy returned on investment* (EROI; • p. 546). The EROI, or ratio of energy returned to energy invested, for corn-based ethanol is controversial, but the best recent estimates place it around 1.5:1. This means that to gain 1.5 units of energy from ethanol, we need to expend 1 unit of energy. The EROI of Brazilian *bagasse* ethanol is much higher, but the

**FIGURE 20.22** These hybrid poplars are specially bred for fast growth in dense plantations, and they are harvested for use in biopower. This is renewable energy, but it also has environmental impacts: The vast monocultural plantations, which may replace natural systems over large areas, do not function ecologically as forests.

low ratio for corn-based ethanol makes this fuel quite inefficient. For this reason many critics do not view ethanol as an effective path to sustainable energy use.

Future advances in cellulosic ethanol may ease the environmental impacts of biofuel crops considerably, and researchers are studying how to obtain biofuel from algae, which could take up far less space than growing corn. Although biomass energy in industrialized nations currently revolves around a few easily grown crops and the efficient use of waste products, the U.S. government envisions a future of specialized crops serving as the basis for a wide variety of fuels and products.

Meanwhile, however, use of fuelwood in the developing world is expected to increase, and the International Energy Agency estimates that in the year 2030, 2.6 billion people will be using traditional fuels for heating and cooking in unsustainable ways. Like nuclear power and like hydropower, biomass energy use involves a complex mix of advantages and disadvantages for the environment and human society.

**Weighing THE Issues | Ethanol**

Do you think producing and using ethanol from corn or other crops is a good idea? Do the benefits outweigh the drawbacks? Can you suggest ways of using biofuels that would minimize environmental impacts?

# Hydroelectric Power

Next to biomass, we draw more renewable energy from the motion of water than from any other resource. In **hydroelectric power**, or **hydropower**, we use the kinetic energy of moving water to turn turbines and generate electricity. We examined hydropower and its environmental impacts in our discussion of freshwater resources in Chapter 15 (• pp. 419–421). Here we will take a closer look at hydropower as an energy source.

## Modern hydropower uses two approaches

Just as people have long burned fuelwood, we have long harnessed the power of moving water. Waterwheels spun by river water powered mills in past centuries. Today we harness water's kinetic energy in two major ways: with storage impoundments and by using "run-of-river" approaches.

Most of our hydroelectric power today comes from impounding water in reservoirs behind concrete dams that block the flow of river water and then letting that water pass through the dam. Because immense amounts of water are stored behind dams, this is called the **storage** technique. If you have ever seen Hoover Dam on the Colorado River (• p. 411), or any other large dam, you have witnessed an impressive example of the storage approach to hydroelectric power.

As reservoir water passes through a dam, it turns the blades of turbines, which cause a generator to generate electricity (**Figure 20.23**). Electricity generated in the powerhouse of a dam is transmitted to the electric grid by transmission lines, and the water is allowed to flow into the riverbed below the dam to continue downriver. The amount of power generated depends on the distance the water falls and the volume of water released. By storing water in reservoirs, dam operators can ensure a steady and predictable supply of electricity at all times, even during seasons of naturally low river flow.

An alternative approach is the **run-of-river** approach, which generates electricity without greatly disrupting the flow of river water. This approach sacrifices the reliability of water flow across seasons that the storage approach guarantees, but it minimizes many of the impacts of large dams. The run-of-river approach can use several methods, one of which is to divert a portion of a river's flow through a pipe or channel, passing it through a powerhouse and then returning it to the river (**Figure 20.24**). This can be done with or without a small reservoir that pools water temporarily, and the pipe or channel can be run along the surface or underground. Another method is to flow river water over a dam small enough not to impede fish passage, siphoning off water to turn turbines, and then returning the water to the river. Run-of-river systems are particularly useful in areas remote from established electrical grids and in regions without the economic resources to build and maintain large dams.

## Hydroelectric power is widely used

Hydropower accounts for 2.2% of the world's energy supply and 16.0% of the world's electricity production (see Figure 20.1). For nations with large amounts of river water and the economic resources to build dams, hydroelectric power has been a keystone of their development and wealth. Canada, Brazil, Norway, Austria, Switzerland, and many other nations today obtain large amounts of their energy from hydropower (**Table 20.3**).

Sweden receives 11.5% of its total energy and nearly half its electricity from hydropower. In the wake of the nation's decision to phase out nuclear power, many people had hoped hydropower could play a still larger role and compensate for the electrical capacity that would be lost. However, Sweden has already dammed so many of its rivers that it cannot gain much additional hydropower by erecting more dams. Moreover, Swedish citizens have

(a) Ice Harbor Dam, Snake River, Washington

(b) Turbine generator inside McNary Dam, Columbia River

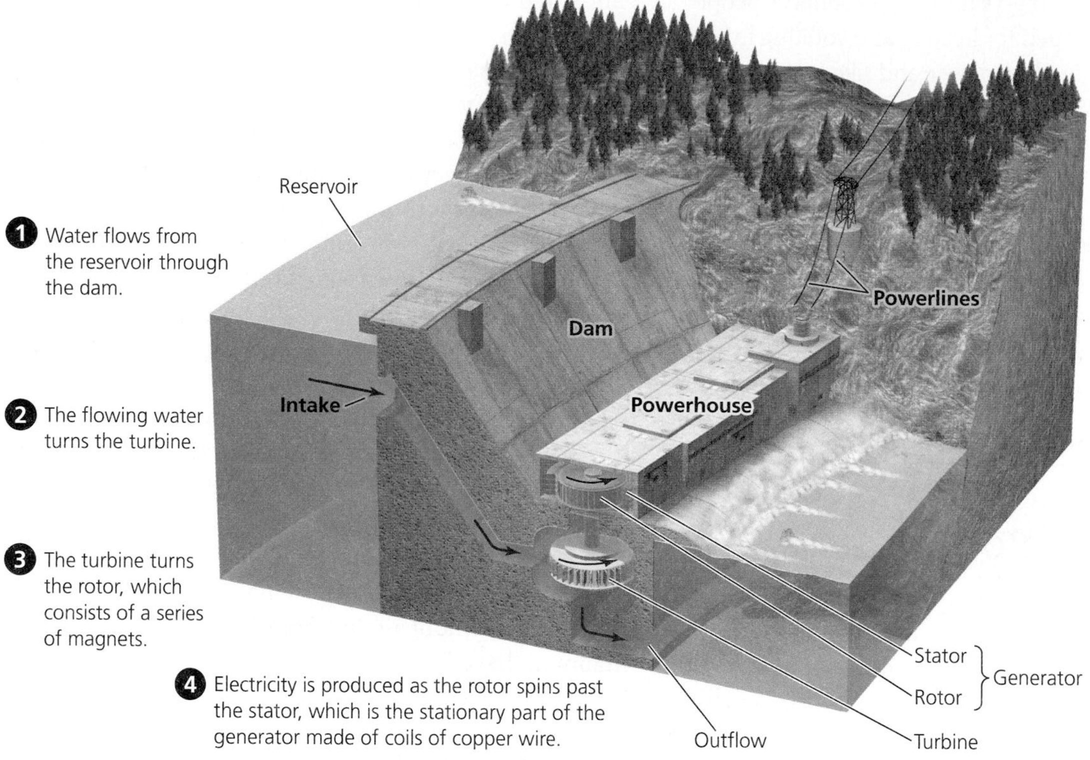


(c) Hydroelectric power

**FIGURE 20.23** Large dams, such as the Ice Harbor Dam on the Snake River in Washington **(a)**, generate substantial amounts of hydroelectric power. Inside these dams, flowing water is used to turn turbines **(b)** and generate electricity. Water is funneled from the reservoir through a portion of the dam **(c)** to rotate turbines, which turn rotors containing magnets. The spinning rotors generate electricity as their magnets pass coils of copper wire. Electrical current is transmitted away through power lines, and the river's water flows out through the base of the dam.

made it clear that they want some rivers to remain undammed, preserved in their natural state.

In the United States, the great age of dam building for hydroelectric power (as well as for flood control and irrigation) began in the 1930s, when the federal government constructed dams as public projects, partly to employ people and help end the economic depression of the time. U.S. dam construction peaked in 1960, when 3,123 dams were completed in a single year.

## Hydropower is clean and renewable

For producing electricity, hydropower has two clear advantages over fossil fuels. First, it is renewable; as long as precipitation falls from the sky and fills rivers and reservoirs, we can use water to turn turbines.

The second advantage of hydropower over fossil fuels is its cleanliness. No carbon compounds are burned in the production of hydropower, so no carbon dioxide

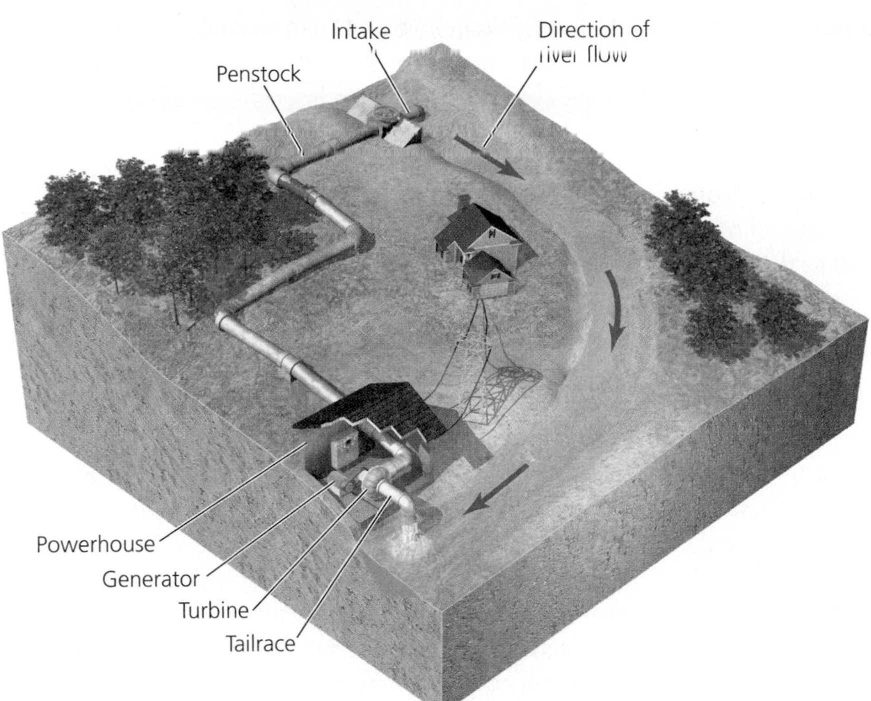


**FIGURE 20.24** Run-of-river systems divert a portion of a river's water and can be designed in various ways. Some designs involve piping water downhill through a powerhouse and releasing it downriver, and some involve using water as it flows over shallow dams.

or other pollutants are emitted into the atmosphere, which helps safeguard air quality, climate, and human health. Of course, fossil fuels *are* used in constructing and maintaining dams—and recent evidence indicates that large reservoirs release the greenhouse gas methane as a result of anaerobic decay in deep water. But overall, hydropower accounts for only a small fraction of the greenhouse gas emissions typical of fossil fuel combustion.

In addition, hydropower is efficient, for it is thought to have an EROI of 10:1 or more, at least as high as any other modern-day energy source. Fossil fuels had higher EROI values in the past, but as it has become more expensive to reach remaining deposits, their EROI values have dipped below that of hydropower.

**TABLE 20.3 Top Producers of Hydropower**

| Nation | Hydropower produced (terawatt-hours) | Percentage of electricity generation from hydropower |
|---|---|---|
| China | 397 | 15.9 |
| Canada | 364 | 57.9 |
| Brazil | 337 | 83.7 |
| United States | 290 | 6.8 |
| Russia | 175 | 18.3 |
| Norway | 137 | 98.9 |
| India | 100 | 14.3 |
| Japan | 86 | 7.8 |
| Venezuela | 75 | 73.9 |
| Sweden | 73 | 46.0 |
| Rest of world | 960 | 13.9 |

Data are for 2005, from the International Energy Agency.

## Hydropower has negative impacts

Although it is renewable, efficient, and produces little air pollution, hydropower does create other impacts. Damming rivers destroys habitat for wildlife as riparian areas above dam sites are submerged and those below dam sites often are starved of water. Because water discharge is regulated to optimize electricity generation, the natural flooding cycles of rivers are disrupted. Suppressing flooding prevents river floodplains from receiving fresh nutrient-laden sediments. Instead, sediments become trapped behind dams, where they begin filling the reservoir.

Dams also cause thermal pollution, because water downstream may become unusually warm if water levels are kept unnaturally shallow. Moreover, periodic flushes of cold water occur from the release of reservoir water. Such thermal shocks, together with habitat alteration, have diminished or eliminated many native fish populations in dammed waterways throughout the world. In addition, dams generally block the passage of fish and other aquatic creatures, effectively fragmenting the river and reducing biodiversity in each stretch.

All these ecological impacts generally translate into negative social and economic impacts on local communities. We discussed the environmental, economic, and social impacts of dams, and their advantages and disadvantages, more fully in Chapter 15 (• pp. 419–421).

## Hydropower may not expand much more

Today the world is witnessing some gargantuan hydroelectric projects. China's recently completed Three Gorges Dam (• pp. 419, 421) is the world's largest. Its reservoir displaced over 1 million people, and the dam should soon be generating as much electricity as dozens of coal-fired or nuclear plants.

However, unlike other renewable energy sources, hydropower is not likely to expand much more. One reason is that, as in Sweden, most of the world's large rivers that offer excellent opportunities for hydropower are already dammed. Another reason is that people have grown more aware of the ecological impacts of dams, and in some regions residents are resisting dam construction. Indeed, in Sweden, hydropower's contribution to the national energy budget has remained virtually unchanged for over 35 years. In the United States, 98% of rivers appropriate for dam construction already are dammed, many of the remaining 2% are protected under the Wild and Scenic Rivers Act, and many people now want to dismantle some dams and restore river habitats (• p. 421).

Overall, hydropower will likely continue to increase in developing nations that have yet to dam their rivers, but in developed nations hydropower growth will likely slow or stop. The International Energy Agency (IEA) forecasts that hydropower's share of electricity generation will decline between now and 2030, whereas the share of other renewable energy sources will triple, from 2% to 6%.

# Conclusion

Given limited supplies of fossil fuels and their considerable environmental impacts, many nations have sought to diversify their energy portfolios with alternative energy sources. The three most developed and widely used alternatives so far are nuclear power, biomass energy, and hydropower.

Nuclear power showed promise at the outset to be a pollution-free and highly efficient form of energy. But high costs and public fears over safety in the wake of accidents at Chernobyl and Three Mile Island stalled its growth, and some nations are attempting to phase it out completely. Biomass energy sources include traditional fuelwood, as well as newer biofuels and various means of generating biopower. These sources can be carbon-neutral but are not all strictly renewable. Hydropower is a renewable, pollution-free alternative, but it is nearing its maximal use and can involve substantial ecological impacts. Although some nations, such as Sweden, already rely heavily on these three conventional alternatives, it appears that we will need further renewable sources of energy.

## REVIEWING OBJECTIVES

**You should now be able to:**

**Discuss the reasons for seeking energy alternatives to fossil fuels**

- Fossil fuels are nonrenewable resources, and we are gradually depleting them. (p. 574)
- Fossil fuel combustion causes air pollution that results in many environmental and health impacts and contributes to global climate change. (p. 574)

**Summarize the contributions to world energy supplies of conventional alternatives to fossil fuels**

- Biomass provides 10.0% of global primary energy use, nuclear power provides 6.3%, and hydropower provides 2.2%. (p. 574)
- Nuclear power generates 15.2% of the world's electricity, and hydropower generates 16.0%. (p. 574)
- "Conventional energy alternatives" are the alternatives to fossil fuels that are most widely used. They include biomass energy, nuclear energy, and hydroelectric power. In their renewability and environmental impact, these sources fall between fossil fuels and the less widely used "new renewable" sources. (p. 575)

**Describe nuclear energy and how it is harnessed**

- Nuclear power comes from converting the energy of subatomic bonds into thermal energy, using uranium isotopes. (p. 576)
- Uranium is mined, enriched, processed into pellets and fuel rods, and used in nuclear reactors. (pp. 576–578)
- By controlling the reaction rate of nuclear fission, nuclear power plant engineers produce heat that powers electricity generation. (pp. 577–578)

**Outline the societal debate over nuclear power**

- Many advocates of "clean" energy support nuclear power because it lacks the pollutant emissions of fossil fuels. (pp. 578–581)
- For many people, the risk of a major power plant accident, such as the one at Chernobyl, outweighs the benefits of clean energy. (pp. 579–585)

- The disposal of nuclear waste remains a major dilemma. Temporary storage and single-repository plans each involve health, security, and environmental risks. (pp. 583–588)
- Economic factors and cost overruns have slowed the nuclear industry's growth. (p. 588)

**Describe the major sources, scale, and impacts of biomass energy**

- Fuelwood remains the major source of biomass energy today, especially in developing nations. (p. 589)
- Biofuels, including ethanol and biodiesel, are used to power automobiles. Some crops are grown specifically for this purpose, and waste oils are also used. (pp. 590–592)
- We use biomass to generate electrical power (biopower) in several ways. Sources include special crops as well as waste products from agriculture and forestry. (pp. 592–593)
- Biomass energy theoretically adds no net carbon to the atmosphere and can use waste efficiently. But overharvesting of wood can lead to deforestation, and growing crops solely for fuel production is inefficient and ecologically damaging. (pp. 593–595)

**Describe the scale, methods, and impacts of hydroelectric power**

- Hydroelectric power is generated when water from a river runs through a powerhouse and turns turbines. (pp. 595–596)
- Dams and reservoirs and run-of-river systems are alternative approaches. (pp. 595–597)
- Hydropower produces little air pollution, but dams and reservoirs can greatly alter riverine ecology and local economies. (pp. 596–597)

## TESTING YOUR COMPREHENSION

1. How much of our global energy supply do nuclear power, biomass energy, and hydroelectric power contribute? How much of our global electricity do these three conventional energy alternatives generate?
2. Describe how nuclear fission works. How do nuclear plant engineers control fission and prevent a runaway chain reaction?
3. In terms of greenhouse gas emissions, how does nuclear power compare to coal, oil, and natural gas? How do hydropower and biomass energy compare?
4. In what ways did the incident at Three Mile Island differ from that at Chernobyl? What consequences resulted from each of these incidents?
5. What has been done so far about disposing of radioactive waste?
6. List five sources of biomass energy. What is the world's most-used source of biomass energy? How does biomass energy use differ between developed and developing nations?
7. Describe two biofuels, where each comes from, and how each is used.
8. Evaluate two potential benefits and two potential drawbacks of biomass energy.
9. Contrast two major approaches to generating hydroelectric power.
10. Assess two benefits and two negative environmental impacts of hydroelectric power.

## SEEKING SOLUTIONS

1. Given what you learned about fossil fuels in Chapter 19 and about some of the conventional alternatives discussed in this chapter, do you think it is important for us to minimize our use of fossil fuels and maximize our use of alternatives? If such a shift were to require massive public subsidies, how much should we subsidize this? What challenges or obstacles would we need to overcome to shift to such alternatives?
2. Nuclear power has by now been widely used for over three decades, and the world has experienced only one major accident (Chernobyl) responsible for any significant number of deaths. Would you call this a good safety record? Should we maintain, decrease, or increase our reliance on nuclear power? Why might safety at nuclear power plants be better in the future? Why might it be worse?
3. How serious a problem do you think the disposal of radioactive waste represents? How would you like to see this issue addressed?
4. There are many different sources of biomass and many ways of harnessing energy from biomass. Discuss one that seems particularly beneficial to you, and one with which you see problems. What biomass energy sources and strategies do you think our society should focus on investing in?

5. **THINK IT THROUGH** Imagine that you are the head of the national department of energy in a country that has just experienced a minor accident at one of its nuclear plants. A partial meltdown released radiation, but the radiation was fully contained inside the containment building, and there were no health impacts on area residents. However, citizens are terrified, and the media is playing up the dangers of nuclear power. Your country relies on its 10 nuclear plants for 25% of its energy and 50% of its electricity needs. It has no fossil fuel deposits and recently began a promising but still-young program to develop renewable energy options. What will you tell the public at your next press conference, and what policy steps will you recommend taking to ensure a safe and reliable national energy supply?

6. **THINK IT THROUGH** You are an investor and would like to invest in alternative energy. You are considering buying stock in companies that (1) construct nuclear reactors, (2) construct turbines for hydroelectric dams, (3) operate pulpmills, (4) build ethanol refineries, and (5) supply farm waste to co-fired power plants. What questions would you research about each of these companies before deciding how to invest your money? How do you expect you might apportion your investments, and why? Do you expect there may be ways in which you will feel torn between doing what seems financially wise and what seems best for the environment or for energy security? Explain.

## INTERPRETING GRAPHS AND DATA

Recently, national security issues have received increased attention in energy policy discussions. Although nuclear reactors are potential targets for terrorism, they do help diversify our energy portfolio and lessen our dependence on petroleum imports. The United States produces reactor fuel but also relies on imports of uranium oxide from abroad, particularly from Canada, Australia, Russia, Kazakhstan, Uzbekistan, South Africa, and Namibia.

1. How much uranium oxide did the United States produce in 2005? How much did it export that year? How much did it import?
2. What was the net amount of uranium imported in 2005? How does this amount compare to domestic production for that year? How do these amounts compare to the amounts in 1980?
3. What national security concerns do these data suggest to you? How might such concerns be alleviated?

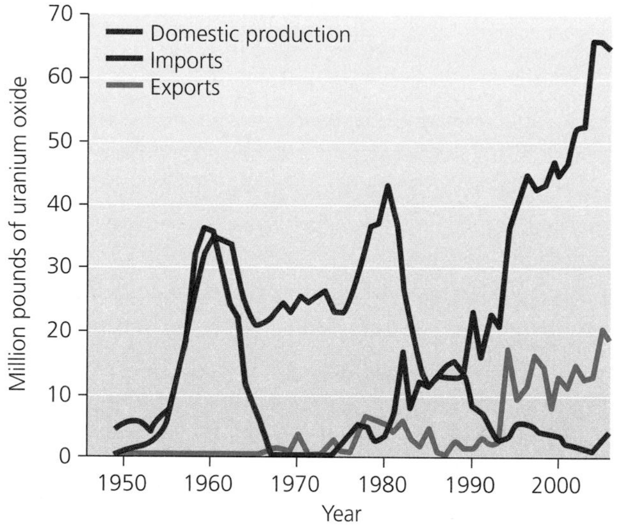


**Production and trade of uranium oxide by the United States, 1949–2006.** Data from Energy Information Administration. 2007. *Annual energy review 2006*. U.S. Department of Energy.

## CALCULATING ECOLOGICAL FOOTPRINTS

Each of the conventional energy alternatives releases considerably less net carbon dioxide ($CO_2$) to the atmosphere than do any of the fossil fuels.

For each fuel source in the table, calculate the net greenhouse gas emissions it produces while providing electricity to a typical U.S. household that uses 30 kilowatt-hours per day. Use data on emissions rates from each of the energy sources as provided in the figure from "The Science behind the Story" on • p. 581. For each energy source, use the average of the maximum and minimum values given.

1. What is the ratio of greenhouse emissions from fossil fuels, on average, to greenhouse emissions from alternative energy sources, on average?
2. Why are there significant emissions from hydroelectric power, which is commonly touted as being a nonpolluting energy source?
3. Which energy source has the lowest emission rate? Would you advocate that your nation further develop that source? Why or why not?

| Energy source | Greenhouse gas emission rate (g $C_{eq}$/ kW-hr) | Time period | | |
|---|---|---|---|---|
| | | 1 day | 1 year | 30 years |
| Coal | 311 | 9,330 | 3,405,450 | 102,163,500 |
| Oil | | | | |
| Natural gas | | | | |
| Photovoltaic solar | | | | |
| Hydroelectric | | | | |
| Biomass | | | | |
| Wind | | | | |
| Nuclear | | | | |

## Take it Further

Go to www.aw-bc.com/withgott or the student CD-ROM, where you'll find:

- Suggested answers to end-of-chapter questions
- Quizzes, animations, and flashcards to help you study
- *Research Navigator*™ database of credible and reliable sources to assist you with your research projects
- GRAPHIt! Tutorials to help you interpret graphs
- INVESTIGATEIt! Current news articles that link the topics that you study to case studies from your region to around the world

CHAPTER

# 21 New Renewable Energy Alternatives

Iceland's Blue Lagoon and Svartsengi geothermal power plant

## Upon completing this chapter, you will be able to:

- Outline the major sources of renewable energy and assess their potential for growth
- Describe solar energy and the ways it is harnessed, and evaluate its advantages and disadvantages
- Describe wind energy and the ways it is harnessed, and evaluate its advantages and disadvantages
- Describe geothermal energy and the ways it is harnessed, and evaluate its advantages and disadvantages
- Describe ocean energy sources and the ways they could be harnessed
- Explain hydrogen fuel cells and assess future options for energy storage and transportation

Iceland's hydrogen buses arrive in Reykjavik

## CENTRAL CASE

# Iceland Moves toward a Hydrogen Economy

**"I believe that water will one day be employed as fuel, that hydrogen and oxygen which constitute it, used singly or together, will furnish an inexhaustible source of heat and light. . . . Water will be the coal of the future."**
—Jules Verne, in *The Mysterious Island*, 1874

**"Our long-term vision is of a hydrogen economy."**
—Robert Purcell, Jr., executive director, General Motors, 2000

The Viking explorers who first set foot centuries ago on the remote island of Iceland were trailblazers. Today the citizens of the nation of Iceland are blazing a bold new path, one they believe the rest of the world may follow. Iceland aims to become the first nation to leave fossil fuels behind and convert to an economy based completely on renewable energy.

Iceland is essentially a hunk of lava the size of Kentucky that has risen out of the North Atlantic from the rift between tectonic plates known as the Mid-Atlantic Ridge (• pp. 200, 448). Most of Iceland's 300,000 people live in the capital city of Reykjavik, leaving much of the island an unpeopled and starkly beautiful landscape of volcanoes, hot springs, and glaciers.

The magma that gave birth to the island heats its groundwater in many places, giving Iceland some of the world's best sources of geothermal energy. Iceland has also dammed some of its many rivers for hydropower. Together these two renewable energy sources provide 73% of the country's energy supply and virtually all its electricity generation.

Yet the nation, like most others, also depends on fossil fuels. Oil powers its automobiles, some of its factories, and its economically vital fishing fleet. Because it possesses no fossil fuel reserves, all fossil fuels must be imported—a weak link in an otherwise robust economy that has given Iceland's citizens one of the highest per capita incomes in the world.

Enter Bragi Árnason, a University of Iceland professor popularly known as "Dr. Hydrogen." Árnason began arguing in the 1970s that Iceland could break its reliance on fossil fuel imports, boost its economy, and serve as a model to the world by converting from fossil fuels to a renewable energy economy based on hydrogen. He suggested zapping water with Iceland's cheap and renewable electricity in a chemical reaction called *electrolysis*, splitting the hydrogen from the oxygen and then using the hydrogen to power fuel cells that would produce and store electricity. The process is clean; nothing is combusted, and the only waste product is water.

In the late 1990s Árnason's countrymen began to listen. The nation's leaders decided to embark on a grand experiment to test the efficacy of switching to a "hydrogen economy." By setting an example for the rest of the world to follow, and by getting a head start at producing and exporting hydrogen fuel, these leaders hoped Iceland could become a "Kuwait of the North" and get rich by exporting hydrogen to an energy-hungry world, as Kuwait has done by exporting oil.

The leaders planning the shift to a hydrogen economy sketched a stepped transition in which fossil fuels would be phased out over 30–50 years. Conversion of the Reykjavik bus fleet to run on hydrogen fuel is the first step. After that, Iceland's 180,000 private cars would be powered by fuel cells, and then the fishing fleet would be converted to hydrogen. The final stage would be the export of hydrogen fuel to mainland Europe.

To make this happen, Icelanders in 1999 teamed up with corporate partners looking to develop technology for the future. Auto company Daimler-Chrysler is producing hydrogen buses, oil company Royal Dutch Shell is running hydrogen filling stations to fuel the buses, and Norsk Hydro is providing electrolysis technology.

In 2003, the world's first commercial hydrogen filling station opened in Reykjavik, and three hydrogen-fueled buses began operation. The public-private consortium, *Icelandic New Energy (INE)*, monitored the technology's effectiveness and the costs of developing infrastructure. In 2008 improved buses will be introduced, along with prototype cars and boats, and Iceland will continue determining how to switch its fuel-intensive fishing fleet to hydrogen. Iceland's citizens are behind the effort, with one poll showing 93% support.

Meanwhile, Daimler-Chrysler has introduced trios of hydrogen-fueled buses to a number of other European cities. Hydrogen buses are also being developed by other companies and run in cities in Europe and throughout the world, from Tokyo to Chicago to Perth to Winnipeg. Hydrogen refueling stations are being demonstrated in Japan, Singapore, and the United States, and fuel-cell passenger automobiles are being tested in Japan and California. A global hydrogen economy could be closer than we suspect.

# "New" Renewable Energy Sources

Iceland's bold drive toward a hydrogen economy is just one facet of a global move toward renewable energy. With the economic costs, security risks, and environmental impacts of fossil fuel dependence all growing, nations across the world are searching for ways to move away from fossil fuels while ensuring a continued supply of energy for their economies.

In Chapter 20 we explored the two renewable energy sources that are most developed and widely used: biomass energy, the energy from combustion of wood and other plant matter, and hydropower, the energy from running water. These "conventional" alternatives to fossil fuels are renewable energy sources that can be depleted with overuse and that exert some undesirable environmental impacts.

In this chapter we explore a group of alternative energy sources that are often called "new renewables." These include energy from the sun, from wind, from Earth's geothermal heat, and from the movement of ocean water. These energy sources are not truly new. In fact, they are as old as our planet, and people have used them for millennia. They are commonly referred to as "new" because (1) they are not yet used on a wide scale in our modern industrial society; (2) they are harnessed using technologies that are still in a rapid phase of development; and (3) they will likely come to play a much larger role in our energy use in the future.

## The new renewables currently provide little of our power

We obtain only half of one percent of our global energy from the new renewable energy sources. Fossil fuels provide 81% of the world's energy, nuclear power provides 6.3%, and renewable energy sources account for 12.7%, nearly all of which is provided by biomass and hydropower (see Figure 20.1a, • p. 574). The new renewables make a similarly small contribution to our global generation of electricity (see Figure 20.1b, • p. 574). Only 18% of our

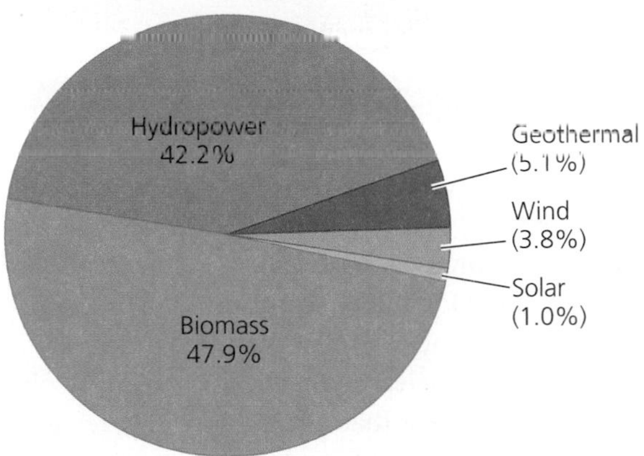


**(a) U.S. consumption of renewable energy, by source**

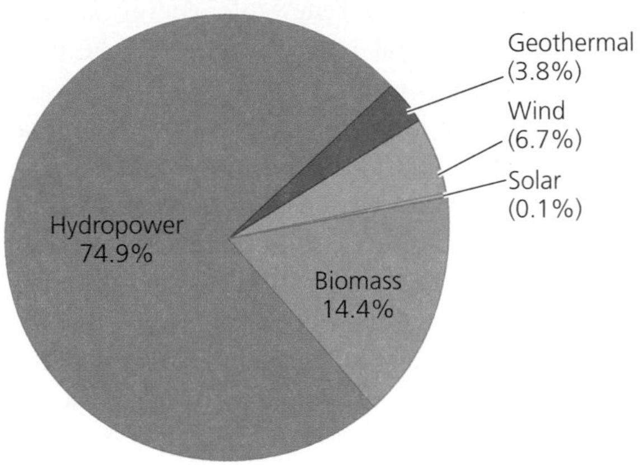


**(b) U.S. electricity generation from renewable sources**

**FIGURE 21.1** Only 6% of the energy consumed in the United States each year comes from renewable sources. Of this amount, most derives from hydropower and biomass energy (**a**). Geothermal, wind, and solar energy together account for less than 10% of this amount. Similarly, only 9% of electricity generated in the United States comes from renewable energy sources, predominantly hydropower and biomass (**b**). Data are for 2006, from Energy Information Administration, U.S. Department of Energy. 2007. *Annual energy review, 2006.*

electricity comes from renewable energy, and of this amount, hydropower accounts for nearly 90%.

Nations and regions vary in the renewable sources they use. In the United States, most renewable energy comes from hydropower and biomass, in nearly equal proportions. As of 2006, geothermal energy accounted for 5.1%, wind energy for 3.8%, and solar energy for 1.0% (**Figure 21.1a**). Of electricity generated in the United States from renewables, hydropower accounts for nearly 75%, and biomass 14.4%. As of 2006, wind power accounted for 6.7%, geothermal for 3.8%, and solar for just 0.1% (**Figure 21.1b**).

## The new renewables are growing fast

Although they comprise only a minuscule proportion of our energy budget, the new renewable energy sources are growing at much faster rates than are conventional energy sources. Over the past three decades, solar, wind, and geothermal energy sources have grown far faster than has the overall energy supply (**Figure 21.2**). The leader in growth is wind power, which has expanded by nearly 50% *each year* since the 1970s. Because these sources started from such low levels of use, however, it will take them some time to catch up to conventional sources. For instance, the absolute amount of energy added by a 50% increase in wind power is still far less than the amount added by just a 1% increase in oil, coal, or natural gas.

Use of new renewables has been expanding quickly because of growing concerns over diminishing fossil fuel supplies (which cause both high prices and national security concerns) and the environmental impacts of fossil fuel combustion (Chapter 19). Advances in technology are also making it easier and less expensive to harness renewable energy sources. The new renewables promise several benefits over fossil fuels. For one, they help alleviate air pollution and the greenhouse gas emissions that drive global climate change (Chapter 18). Unlike fossil fuels, renewable sources are inexhaustible on time scales relevant to human societies. Developing renewables can also diversify an economy's mix of energy, lowering price volatility and protecting against supply restrictions such

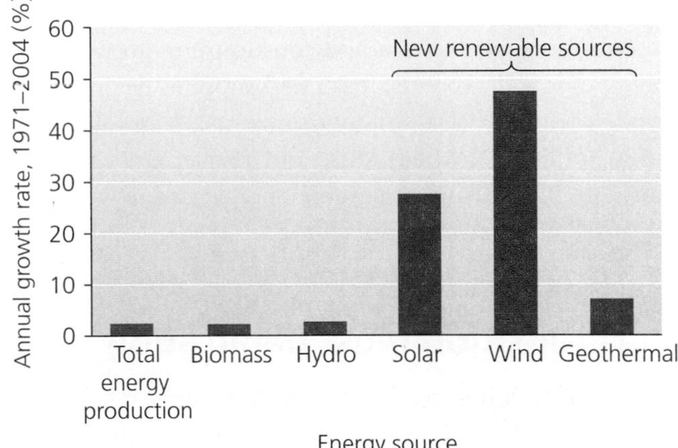


**FIGURE 21.2** Globally, the "new renewable" energy sources are growing substantially faster than the total primary energy supply. Solar power has grown by 28% each year since 1971, and wind power has grown by 48% each year. Because these sources began from such low starting levels, however, their overall contribution to our energy supply is still small. Go to GRAPHit! at www.aw-bc.com/ withgott or on the student CD-ROM. Data from International Energy Agency Statistics, 2007.

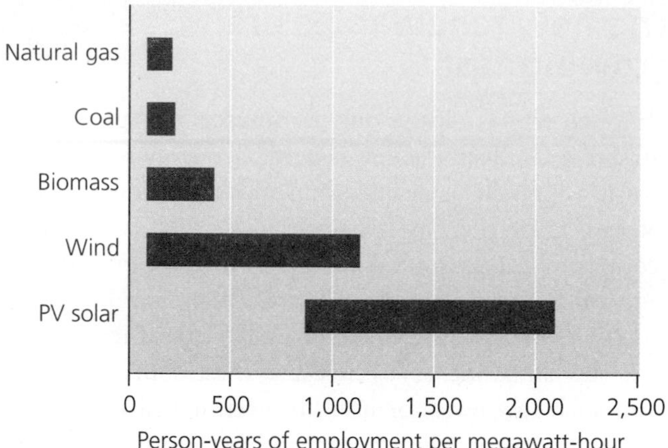


FIGURE 21.3 Manufacturing, installing, and servicing young renewable energy technologies is more labor-intensive than for established fossil fuel technologies, so a shift to renewable energy should increase employment. Renewable energy sources such as photovoltaic (PV) solar, wind, and biomass tend to support more jobs than natural gas and coal. Data from Worldwatch Institute and Center for American Progress. 2006. *American energy: The renewable path to energy security.* Washington, D.C.

as those caused by the 1973 oil embargo or by Hurricane Katrina in 2005 (• pp. 551, 562–563). New energy sources also can create new employment opportunities and sources of income and property tax for communities, especially in rural areas passed over by other economic development. New and developing technologies generally require more labor per unit of energy output than do established technologies, so shifting to renewable energy should generate more employment than remaining with a fossil fuel economy (**Figure 21.3**).

Rapid growth in renewable energy sectors seems likely to continue as population and consumption grow, global energy demand expands, fossil fuel supplies decline, and people demand cleaner environments. More governments, utilities, corporations, and consumers are now promoting and using renewable energy, and, as a result, the prices of renewables are falling.

## Our transition must begin soon

If our civilization is to persist in the long term, it will need to shift to renewable energy sources. A key question is whether we will be able to shift soon enough and smoothly enough to avoid widespread war, economic depression, social unrest, and further damage to the environment. The answer to this question will largely determine the quality of life for all of us in the coming decades.

We cannot switch completely to renewable energy sources overnight, because there are technological and economic barriers. Currently, most renewables lack adequate technological development and lack infrastructure to transfer power on the required scale. However, rapid advances in science, technology, and infrastructure in recent years suggest that most remaining barriers are political. Renewable energy sources have received far less in subsidies, tax breaks, and other incentives from governments than have conventional sources (**Figure 21.4**). By one estimate, of the $150 billion in subsidies the U.S. government provided to nuclear, solar, and wind power in the past half century, the nuclear industry received 96%, solar received 3%, and wind less than 1%. For decades, research and development of renewable sources have gone underfunded because of the continuing availability of fossil fuels made inexpensive in part by government policy.

Failing to transition to renewable energy sources would have severe societal repercussions (see "Causes and Consequences," • p. 607). Most corporations in the fossil fuel and automobile industries understand that they will eventually need to switch to renewable sources. They also know that when the time comes, they will need to act fast to stay ahead of their competitors. However, in light of continuing short-term profits and unclear policy signals, companies have not been eager to invest in the transition from fossil fuels to renewable energy. Under these circumstances, our best hope may be for a gradual shift driven largely by economic supply and demand. However, if the transition proceeds too slowly—if we wait solely for the market to do its work, without government

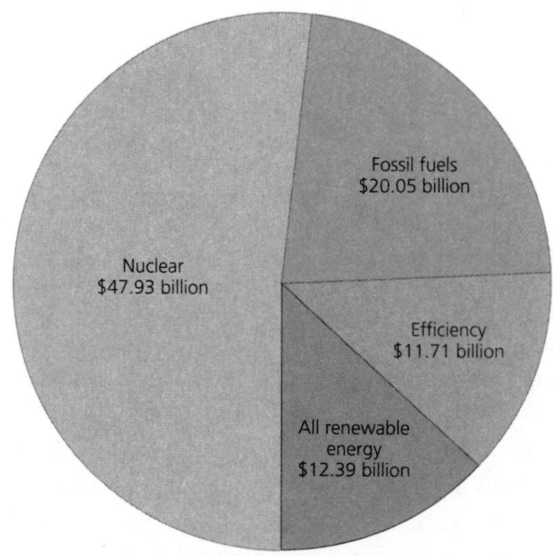


FIGURE 21.4 Most U.S. research and development funding for energy has gone toward nuclear power and fossil fuels. Between 1974 and 2005, as data in this figure show, only 13% went toward all renewable energy sources combined. Data from International Energy Agency.

## CAUSES AND CONSEQUENCES

Despite the appeal of renewable energy sources, they currently comprise less than 14% of our global energy supply and generate less than 19% of our electricity. Moreover, the "new renewables" account for only a very small fraction of these numbers. The fact that **renewables provide little of our energy** is a dilemma with multiple causes and consequences. However, most renewable sources are now expanding quickly, and there are many ways in which we can encourage their development and spread.

Write in two causes of the fact that renewables provide little of our energy in the spaces provided. Then write in two consequences (impacts on the environment, human health, or quality of life) that result. Finally, offer two solutions to this issue and its consequences. One cause, one consequence, and one solution have been filled in for you, providing examples.

**CAUSES**

*Fossil fuels are rich in energy*

______

______

Renewables provide little of our energy

**CONSEQUENCES**

*Greenhouse emissions spur climate change*

______

______

**SOLUTIONS**

*Increase research and development funding for renewables*

______

______

Sometimes solutions can have unintended consequences. Can you think of an undesired consequence that one solution to this issue might have? How might we then deal with *that* consequence?

encouragement—then fossil fuel supplies could dwindle faster than we are able to develop new sources, and we could find our economies disrupted and our environment highly degraded. Thus, encouraging the speedy development of renewable energy alternatives holds promise for bringing us a vigorous and sustainable energy economy without the environmental impacts of fossil fuels.

## Solar Energy

The sun provides energy for almost all biological activity on Earth (• pp. 102–103) by converting hydrogen to helium in nuclear fusion (• pp. 577–578). On average, each square meter of Earth's surface receives about 1 kilowatt of solar energy—17 times the energy of a lightbulb. As a result, a typical house has enough roof area to generate all its power needs with rooftop panels that harness solar energy. The amount of energy Earth receives from the sun each day, if it could be collected in full for our use, would be enough to power human consumption for a quarter of a century. Clearly, the potential for using sunlight to meet our energy needs is tremendous. However, we are still in the process of developing solar technologies and learning the most effective and cost-efficient ways to put the sun's energy to use.

The most commonly used way to harness solar energy is through **passive solar** energy collection. In this approach, buildings are designed and building materials are chosen to maximize direct absorption of sunlight in winter, even as they keep the interior cool in the heat of summer. This approach contrasts with **active solar** energy collection, which makes use of technological devices to focus, move, or store solar energy.

We tend to think of using solar energy as a novel phenomenon, but people have chosen and designed their living sites with passive solar collection in mind for millennia. Moreover, solar energy was first harnessed with active solar technology in 1767, when Swiss scientist Horace de Saussure built a thermal solar collector to heat water and cook food. In 1891, U.S. inventor Clarence

Kemp claimed the first commercial patent for a solar-powered water heater. Two California entrepreneurs bought the patent rights and outfitted one-third of the homes in Pasadena, California, with solar water heaters by 1897.

## Passive solar heating is simple and effective

One passive solar design technique involves installing low, south-facing windows to maximize sunlight capture in the winter (in the Northern Hemisphere; north-facing windows are used in the Southern Hemisphere). Overhangs block light from above, shading these windows in the summer, when the sun is high in the sky and when cooling, not heating, is desired. Passive solar techniques also include the use of heat-absorbing construction materials (often called *thermal mass*) that absorb heat, store it, and release it later. Thermal mass (of straw, brick, concrete, or other materials) most often makes up floors, roofs, and walls, but it also can comprise portable blocks.

Thermal mass may be strategically located to capture sunlight in cold weather and radiate heat in the interior of the building. In warm weather, the mass should be located away from sunlight so that it absorbs warmed air in the interior to cool the building. Passive solar design can also involve planting vegetation in particular locations around a building. By heating buildings in cold weather and cooling them in warm weather, passive solar methods conserve energy and reduce energy costs.

## Active solar energy collection can heat air and water in buildings

One active method for harnessing solar energy involves using *solar panels* or *flat-plate solar collectors*, most often installed on rooftops. These panels generally consist of dark-colored, heat-absorbing metal plates mounted in flat boxes covered with glass panes. Water, air, or antifreeze solutions are run through tubes that pass through the collectors, transferring heat throughout a building. Heated water can be pumped to tanks to store the heat for later use and through pipes designed to release the heat into the building. Such systems have proven especially effective for heating water for residences.

Over 1.5 million homes and businesses in the United States heat their water with solar panels, but most of this goes toward swimming pools. Active solar heating is used more in China and Europe, even in isolated locations. In Gaviotas, a remote town in the high plains of Colombia far from any electrical grid, active solar technology is used for heating, cooling, and water purification (**Figure 21.5**). Engineers developed inexpensive solar panels that harvest enough solar energy, even under cloudy skies, to boil drinking water for a family of four. They also designed, built, and installed a unique solar refrigerator in a rural hospital, along with a large-scale solar collector to boil and sterilize water. Their innovations show that solar power does not need to be expensive or confined to regions that are always sunny.

FIGURE 21.5 Engineers in Gaviotas, a remote highland town in Colombia, developed inexpensive solar panels that provide residences and businesses active solar power for heating, cooling, and water purification.

## Concentrating solar rays magnifies energy

We can magnify the strength of solar energy by gathering sunlight from a wide area and focusing it on a single point. This is the principle behind *solar cookers*, simple portable ovens that use reflectors to focus sunlight onto food and cook it (see Figure 3.13, • p. 69). Such cookers are proving extremely useful in parts of the developing world.

Utilities have put the solar cooker principle to work in large-scale, high-tech approaches to generating electricity. In one approach, mirrors concentrate sunlight onto a receiver atop a tall "power-tower" (**Figure 21.6**). From the receiver, heat is transported by fluids (often molten salts) piped to a steam-driven generator to create electricity. These solar power plants can harness light from large mirrors spread across many hectares of land. The world's largest such plant so far—a collaboration among government, industry, and utility companies in the California desert—produces power for 10,000 households. Another approach is the use of solar-trough collection systems, which consist of mirrors that gather sunlight and focus it on oil in troughs. The superheated oil creates steam that drives turbines, as in conventional power plants.

FIGURE 21.6 At the Solar Two facility operated by Southern California Edison in the desert of southern California, mirrors are spread across wide expanses of land to concentrate sunlight onto a receiver atop a "power-tower." Heat is then transported through fluid-filled pipes to a steam-driven generator that produces electricity.

## Photovoltaic cells generate electricity directly

A more direct approach to producing electricity from sunlight involves photovoltaic (PV) systems. **Photovoltaic (PV) cells** collect sunlight and convert it to electrical energy by making use of the *photovoltaic effect*, or *photoelectric effect*. This effect occurs when light strikes one of a pair of metal plates in a PV cell, causing the release of electrons, which are attracted by electrostatic forces to the opposing plate. The flow of electrons from one plate to the other creates an electrical current (direct current, DC), which can be converted into alternating current (AC) and used for residential and commercial electrical power (**Figure 21.7**).

The plates of a typical PV cell are made primarily of silicon, a semiconductor that conducts electricity. One silicon plate (the *n-type layer*) is enriched with phosphorus and is rich in electrons. The other plate (the *p-type layer*) is enriched with boron and is electron-poor. When sunlight strikes the PV cell, it knocks electrons loose from some of the silicon atoms, and they tend to flow toward the n-type layer. Connecting the two plates with wires generates electricity as electrons flow from the n-type layer back to the p-type layer. Photovoltaic cells can be connected to batteries that store the accumulated charge until it is needed.

You may be familiar with small PV cells that power your watch or your calculator. Atop the roofs of homes and other buildings, PV cells are arranged in modules, which comprise panels, which can be gathered together in arrays. Increasingly, PV roofing tiles are being used instead of these arrays. In some remote areas, such as Xcalak, Mexico, PV systems are used in combination with wind turbines (• pp. 612–613) and a diesel generator to power entire villages.

## Solar power is little used but fast growing

Although active solar technology dates from the 18th century, it was pushed to the sidelines as fossil fuels came to dominate our energy economy. Funding for research and development of solar technology has been erratic. After the 1973 oil embargo (• pp. 562–563), the U.S. Department of Energy funded the installation and testing of over 3,000 PV systems, providing a boost to companies in the solar industry. As oil prices declined, however, so did government support for solar power, and funding for solar has remained far below that for fossil fuels.

Largely because of the lack of investment, solar energy contributes only a minuscule portion of today's energy production. In 2006, solar accounted for only 0.07%—7 parts in 10,000—of the U.S. energy supply, and less than 0.02% of U.S. electricity generation. However, solar energy use has grown by 28% annually worldwide since 1971, a growth rate second only to that of wind power. Solar power is proving especially attractive in developing countries, many of which are rich in sun but poor in power infrastructure, and where hundreds of millions of people are still without electricity. Some multinational corporations that built themselves on fossil fuels are now investing in alternative energy as well. BP Solar, British Petroleum's solar energy wing, recently completed $30 million projects in the Philippines and Indonesia and is working on a $48 million project to supply electricity to 400,000 people in 150 villages.

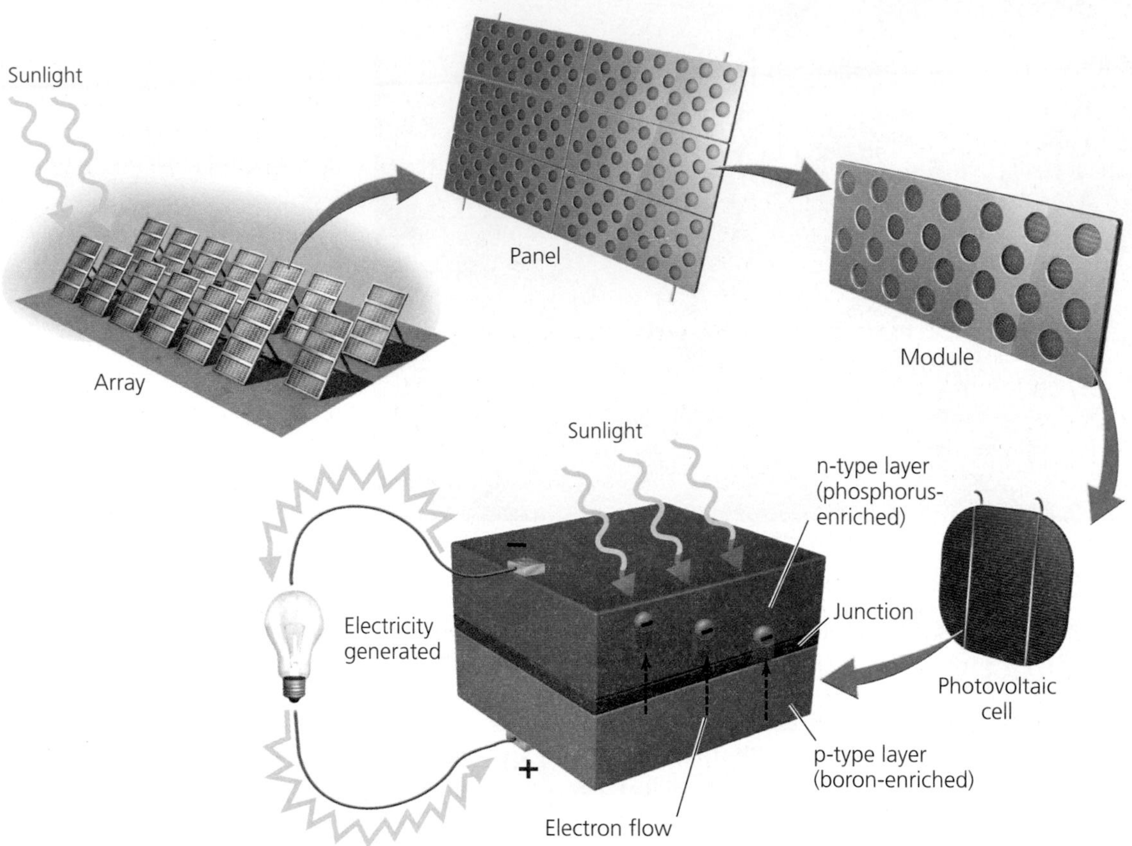


FIGURE 21.7 A photovoltaic (PV) cell converts sunlight to electrical energy. When sunlight hits the silicon layers of the cell, electrons are knocked loose and tend to move from the boron-enriched p-type layer toward the phosphorus-enriched n-type layer. Connecting the two layers with wiring remedies this imbalance as electrical current flows from the n-type layer back to the p-type layer. This direct current (DC) is converted to alternating current (AC) to produce usable electricity. PV cells are grouped in modules, which comprise panels, which can be erected in arrays.

Use of PV cells is growing fast (**Figure 21.8**), with Japan leading the world in production and Germany leading the world in installation. A federal tax credit enacted in 2005 and initiatives by some states (led by California's $3.35 billion program in 2006) may help the United States recover the leadership it lost to other nations in this technology. Use of solar technology should continue to increase as prices fall, technologies improve, and governments enact economic incentives to spur investment.

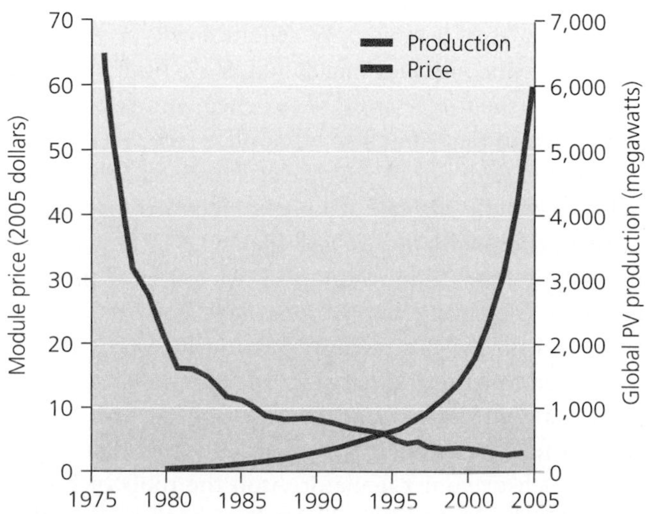


FIGURE 21.8 Production of PV cells grew sixfold just between 2000 and 2005, and prices have fallen rapidly. Data from Worldwatch Institute and Center for American Progress. 2006. *American energy: The renewable path to energy security*. Washington, D.C.

## Solar power offers many benefits

The fact that the sun will continue burning for another 4–5 billion years makes it practically inexhaustible as an energy source for human civilization. Moreover, the amount of solar energy reaching Earth's surface should be enough to power our civilization once solar technology is adequately developed. These primary benefits of solar energy are clear, but the technologies themselves also provide benefits. PV cells and other solar technologies use no fuel, are quiet and safe, contain no moving parts, require little maintenance, and do not even require a turbine or generator to create electricity. An average unit can produce energy for 20–30 years.

Another advantage of solar systems is that they allow for local, decentralized control over power. Homes, businesses, and isolated communities can use solar power to produce their own electricity without being near a power plant or connected to the grid of a city.

In developing nations, solar cookers enable families to cook food without gathering fuelwood; as a result, they lessen people's daily workload and help reduce deforestation. In locations such as refugee camps, solar cookers are helping relieve social and environmental stress. The low cost of solar cookers—many can be built locally for $2 to $10 each—has made them available for purchase or donation in many impoverished areas.

In the developed world, most PV systems are connected to the regional electric grid. This may enable owners of houses with PV systems to sell their excess solar energy to their local power utility. In this process, called *net metering*, the value of the power the consumer sells to the utility is subtracted from the consumer's monthly utility bill. Nearly 40 U.S. states currently permit some sort of net metering.

Another advantage of solar power is that its development is producing many new jobs. Currently, among major energy sources, PV technology employs the most people per unit energy output (see Figure 21.3, • p. 606).

Finally, a major advantage of solar power over fossil fuels is that it does not pollute the air with greenhouse gas emissions and other air pollutants. The manufacture of photovoltaic cells *does* currently require fossil fuel use, but once a PV system is up and running, it produces no emissions. Consumers can access online calculators to calculate the economic and environmental impacts of installing PV solar cells. For example, a calculator offered by the U.S. Department of Energy and two solar power associations estimated at the time of this writing that installing a 5-kilowatt PV system in a home in Dallas, Texas, to provide 50% of annual power needs could save the homeowners $869 per year on energy bills and would prevent the emission of 10,800 lb of carbon dioxide per year. Even in overcast Seattle, Washington, a 5-kilowatt system producing half of a home's energy needs could save $416 per year and prevent 8,160 lb of carbon dioxide emissions.

## Location and cost can be drawbacks

Solar energy currently has two major disadvantages. One is that not all regions are sunny enough to provide adequate power, given current technology. Although Earth as a whole receives vast amounts of sunlight, not every location on Earth does (**Figure 21.9**). People in cities such as

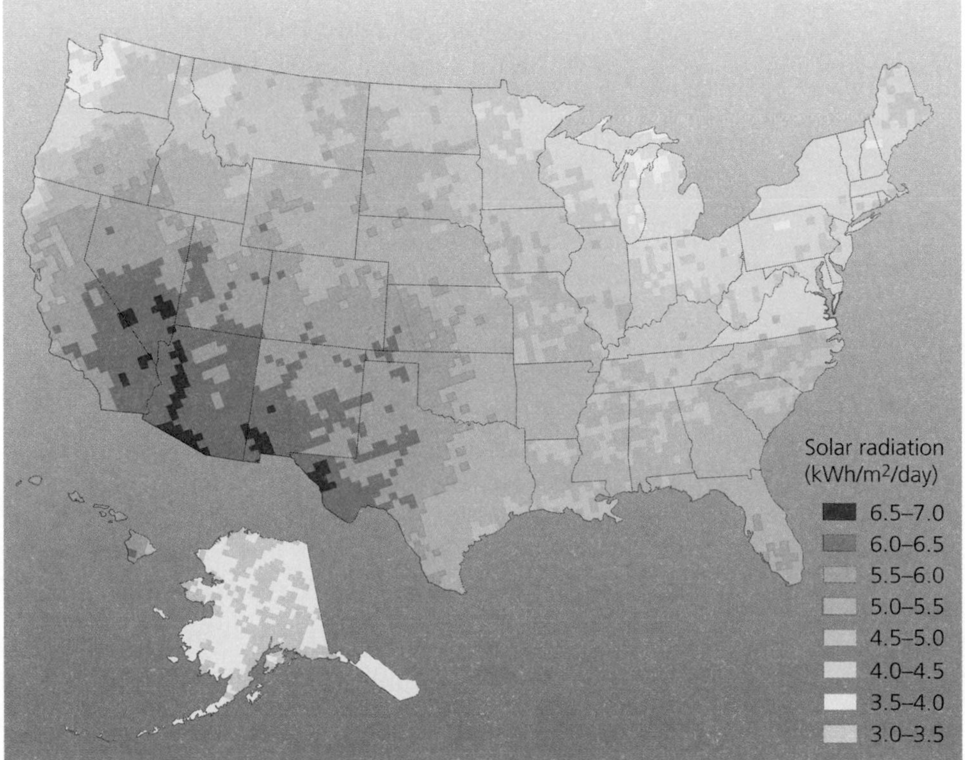


**FIGURE 21.9** Because some locations receive more sunlight than others, harnessing solar energy is more profitable in some areas than in others. In the United States, many areas of Alaska and the Pacific Northwest receive only 3–4 kilowatt-hours per square meter per day, whereas most areas of the Southwest receive 6–7 kilowatt-hours per square meter per day. Data from National Renewable Energy Laboratory, U.S. Department of Energy.

Seattle might find it difficult to harness enough sunlight most of the year to rely on solar power. Daily or seasonal variation in sunlight can also pose problems for stand-alone solar systems if storage capacity in batteries or fuel cells is not adequate or if backup power is not available from a municipal electricity grid.

The primary disadvantage of current solar technology is the up-front cost of investing in the equipment. The investment cost for solar is higher than that for fossil fuels, and indeed, solar power remains the most expensive way to produce electricity (**Figure 21.10**). For instance, for the Dallas home mentioned above, it would take an estimated 23 years for the homeowners to break even on their investment (and 30 years if we factor out the increase in their home's property value). Proponents of solar power argue that decades of government promotion of fossil fuels and nuclear power—which have received many financial breaks that solar power has not—have made solar power unable to compete. Because the external costs (• p. 42) of nonrenewable energy are not included in market prices, these energy sources have remained relatively cheap. Governments, businesses, and consumers thus have had little economic incentive to switch to solar and other renewables.

However, decreases in price and improvements in energy efficiency of solar technologies so far are encouraging, even in the absence of significant financial commitment from government and industry. At their advent in the 1950s, solar technologies had efficiencies of around 6% while costing $600 per watt. Recent single-crystal silicon PV cells are showing 15% efficiency commercially and 24% efficiency in lab research, suggesting that future solar technologies may be more efficient than any energy technologies we have today. Solar systems have become much less expensive over the years and now can often pay for themselves in 10–20 years. After that time, they provide energy virtually for free as long as the equipment lasts. With future technological advances, some experts believe that the time to recoup investment could fall to 1–3 years.

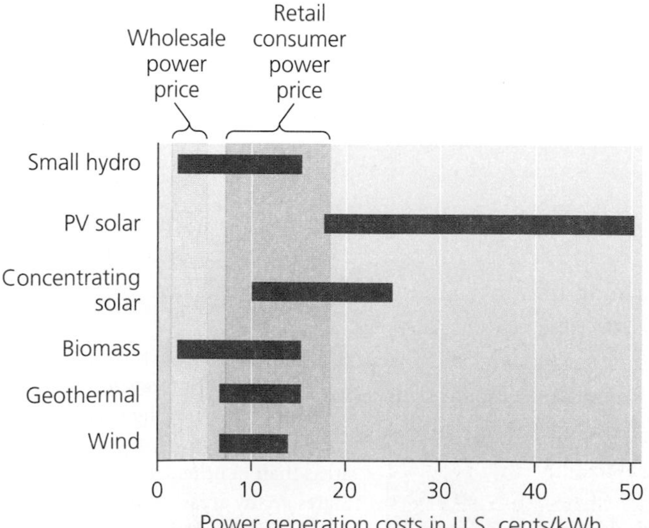


**FIGURE 21.10** Although electricity prices from many renewable energy sources are becoming competitive with those from fossil fuels, costs of solar energy have remained higher. However, these costs are coming down with advancing technology and increased production. Data from Worldwatch Institute and Center for American Progress. 2006. *American energy: The renewable path to energy security*. Washington, D.C.

# Wind Energy

Wind energy can be thought of as an indirect form of solar energy, because it is the sun's differential heating of air masses on Earth that causes wind to blow. We can harness power from wind by using devices called **wind turbines**, mechanical assemblies that convert wind's kinetic energy (• p. 100), or energy of motion, into electrical energy.

## Wind has long been used for energy

Today's wind turbines have their historical roots in Europe, where wooden windmills have been used for 800 years to pump water to drain wetlands and irrigate crops and to grind grain into flour. In each application, wind causes a windmill's blades to turn, driving a shaft connected to several cogs that turn wheels, which either grind grain or pull buckets from a well. In North America, countless ranches in the West and the Great Plains use windmills to draw groundwater up for thirsty cattle.

The first wind turbine built to generate electricity was constructed in the late 1800s in Cleveland, Ohio, by inventor Charles Brush, who designed a turbine 17 m (50 ft) tall with 144 rotor blades made of cedar wood. But it was not until after the 1973 oil embargo that governments in North America and Europe began funding research and development for wind power. This moderate infusion of funding boosted technological progress, and the cost of wind power was cut in half in less than 10 years. Today wind power at favorable locations generates electricity for nearly as little cost per kilowatt-hour as do conventional sources, and modern wind turbines look more like airplane propellers or sleek new helicopters than romantic old Dutch paintings.

## Modern wind turbines convert kinetic energy to electrical energy

Wind blowing into a turbine turns the blades of the rotor, which rotate machinery inside a compartment called a *nacelle*, which sits atop a tall tower (**Figure 21.11**). Inside the

FIGURE 21.11 A wind turbine converts wind's energy of motion into electrical energy. Wind causes the blades of a wind turbine to spin, turning a shaft that extends into the nacelle that is perched atop the tower. Inside the nacelle, a gearbox converts the rotational speed of the blades, which can be up to 20 revolutions per minute (rpm) or more, into much higher rotational speeds (over 1500 rpm). These high speeds provide adequate motion for a generator inside the nacelle to produce electricity.

nacelle are a gearbox and a generator, as well as equipment to monitor and control the turbine's activity. Most of today's towers range from 40 to 100 m (from 131 to 328 ft) tall, so the largest are taller than a football field is long. Higher is generally better, to minimize turbulence (and potential damage) and to maximize wind speed. Most rotors consist of three blades and measure 42–80 m (138–262 ft) across. Turbines are designed to yaw, or rotate back and forth in response to changes in wind direction, ensuring that the motor faces into the wind at all times. Turbines can be erected singly, but they are most often erected in groups called wind parks or *wind farms*. The world's largest wind farms contain hundreds of turbines spread across the landscape.

Engineers have designed turbines to begin turning at specific wind speeds to harness wind energy as efficiently as possible. Some turbines create low levels of electricity by turning in light breezes. Others are programmed to rotate only in strong winds, operating less frequently but generating large amounts of electricity in short time periods. Slight differences in wind speed yield substantial differences in power output, for two reasons. First, the energy content of a given amount of wind increases as the square of its velocity; thus if wind velocity doubles, energy quadruples. Second, an increase in wind speed causes more air molecules to pass through the wind turbine per unit time, making power output equal to wind velocity cubed. Thus a doubled wind velocity actually results in an eightfold increase in power output.

## Wind power is the fastest-growing energy sector

Like solar energy, wind provides only a minuscule proportion of the world's power needs, but wind power is growing fast—26% per year globally between 2000 and 2005. Just five nations account for 80% of the world's wind power output (**Figure 21.12**). Denmark is a leader in wind power; there, a series of wind farms supplies over 20% of the nation's electricity needs. California and Texas account for close to half of the wind power generated in the United States. Growth in the United States has been haphazard because Congress has not yet committed to a long-term federal tax credit for wind development, but instead has passed a series of short-term renewals, leaving the industry uncertain about how much to invest. However, experts agree that wind power's rapid growth will continue, because only a very small portion of this resource is currently being tapped. Meteorological evidence suggests that wind power could be expanded in the United States to meet the electrical needs of the entire country (• p. 617).

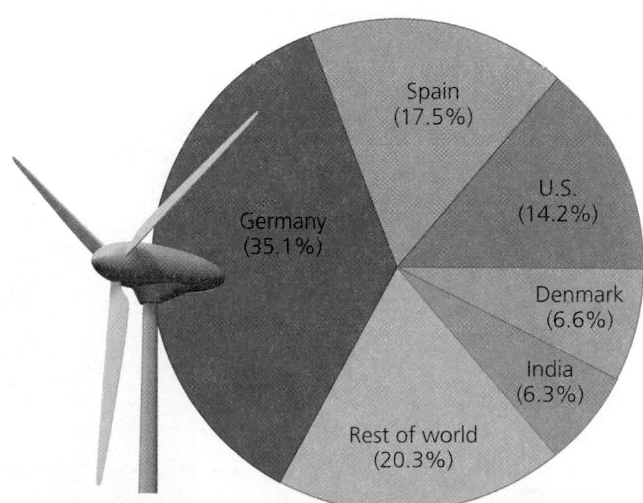


FIGURE 21.12 Most of the world's fast-growing wind-power-generating capacity is concentrated in a handful of countries. Tiny Denmark obtains the highest percentage of its energy needs from wind, but the larger nations of Germany, Spain, and the United States have so far developed more total wind capacity. Data from American Wind Energy Association (AWEA). 2005. *Global wind energy market report.* AWEA.

## Offshore sites can be promising

Wind speeds on average are roughly 20% greater over water than over land. There is also less air turbulence over water than land. For these reasons, offshore wind turbines are becoming popular (**Figure 21.13**). Costs to erect and maintain turbines in water are higher, but the stronger, less turbulent winds produce more power and make offshore wind potentially more profitable. Currently, offshore wind farms are limited to shallow water, where towers are sunk into sediments singly or using a tripod configuration to stabilize them. However, in the future, towers may be placed on floating pads anchored to the seafloor in deep water. At great distances from land, it may be best to store the generated electricity in hydrogen fuel and then ship or pipe this to land (instead of building submarine cables to carry electricity to shore), but further research is needed.

Denmark erected the first offshore wind farm in 1991. Over the next decade, nine more came into operation across northern Europe, where the North and Baltic Seas offer strong winds. The power output of these farms increased by 43% annually as larger turbines were erected. In Iceland, wind advocates are considering developing 240 offshore turbines in the nation's waters to meet future electricity demand for its hydrogen economy.

## Wind power has many benefits

Like solar power, wind produces no emissions once the necessary equipment is manufactured and installed. As a replacement for fossil fuel combustion in the average U.S. utility generator, the U.S. Environmental Protection Agency (EPA) has calculated that running a 1-megawatt wind turbine for 1 year prevents the release of more than 1,500 tons of carbon dioxide, 6.5 tons of sulfur dioxide, 3.2 tons of nitrogen oxides, and 60 lb of mercury. The amount of carbon pollution that all U.S. wind turbines together prevent from entering the atmosphere is greater than the cargo of a 50-car freight train, with each car holding 100 tons of solid carbon, each and every day.

Wind power, under optimal conditions, appears considerably more efficient than conventional power sources in its energy returned on investment (EROI; • p. 546). One recent study found that wind turbines produce 23 times more energy than they consume. For nuclear energy, the ratio was 16:1; for coal it was 11:1; and for natural gas it was 5:1. Wind farms also use less water than do conventional power plants. Wind turbine technology can be used on many scales, from a single tower for local use to fields of hundreds that supply large regions. Small-scale turbine development can help make local areas more self-sufficient, just as solar energy can. For instance, the Rosebud Sioux Tribe of Native Americans in 2003 set up a single turbine on its reservation in South Dakota. The turbine produces electricity for 200 homes and brings the tribe an estimated $15,000 per year in revenue. Wind resources are rich in this region, and the tribe is developing a 30-megawatt wind farm nearby in coming years.

Another societal benefit of wind power is that farmers and ranchers can lease their land for wind development, which provides them extra revenue while also increasing property tax income for rural communities. A single large turbine can bring in $2,000 to $4,500 in

FIGURE 21.13 More and more wind farms are being developed offshore, because offshore winds tend to be stronger yet less turbulent. Denmark is a world leader in wind power, and much of it comes from offshore turbines. This Danish wind farm is one of several that provide over 20% of the nation's electricity.

THE SCIENCE BEHIND THE STORY

# Idaho's Wind Prospectors

*Installing a wind turbine*

Where does wind translate into energy? In Idaho, where resource planners decided that the state's power future lies in generating electricity from wind. Now scores of Idaho residents, from small farmers to Native American tribes, have joined in the search for gusts with energy potential.

By handing out wind-measuring devices to interested landowners, Idaho turned its citizens into "wind prospectors" who pinpoint potential areas for wind farms. Idaho first launched the public wind prospecting program in 2001, after joining several other northwestern states in a regional research effort. People who join the program must collect data on wind speed and direction, share that data with the state, and agree to make it public.

A promising wind farm site requires some infrastructure, such as roads for erecting wind turbines and transmission lines for sending out power the turbines generate. But the most important factor is the speed and frequency of the wind. Effective commercial wind farms have a steady flow of wind just above ground level, with regular gusts of at least 21 km/hr (13 mi/hr) at a height of about 50 m (164 ft).

People analyze existing data to determine whether a site merits further study. In many parts of the developed world, these data include decades of weather information compiled into computerized wind maps that indicate general wind conditions.

In Idaho, energy planners provide prospectors with starter maps that divide the state into seven wind "classes" and reveal which areas might have enough wind to make a wind farm worthwhile. Areas listed as "Class 3" or higher, with wind speeds of about 23 km (14.3 mi) per hour at 50 m (164 ft) above ground, offer the best possibilities.

Such maps, however, may not provide enough detail about a specific location. A piece of property, for example, may sit in a Class 3 area but be sheltered by a small hill that blocks the wind. Knowing that kind of detail requires site-specific on-the-ground research.

To make such research possible, Idaho lends landowners in areas listed as Class 3 or higher devices called *anemometers* (see the figure), which measure wind speed and direction. The Idaho program uses a common cup anemometer, with an array of three or four hollow cups set to catch the wind and rotate around a vertical rod. The greater the wind speed, the faster the cups rotate. Wind direction is measured by a vane that turns on a vertical axis pointing directly into the wind. The cup wheel and wind vane are connected electrically to speed and direction dials, which relay wind data.

To determine where to build wind farms, Idaho's wind prospectors use anemometers, which collect and relay wind data. Cup wheels rotate to indicate wind speed, and a vane turns to reveal wind direction.

More than 80 landowners borrowed anemometers from the state in the Idaho program's first year and sent data to the state every 60 days for review by energy planners and for subsequent posting online. The studies have generated new funding and wind farm plans in the state. In the fall of 2003, one farm near Idaho Falls won a $500,000 federal grant to help build a 1.5-megawatt wind farm that could supply power for approximately 500 homes.

In eastern Idaho, five anemometers set up on Shoshone-Bannock tribal lands have revealed good prospects for a commercial wind farm on two Native American reservations. The research effort has shown that the lands are "world-class sites" for wind power, according to a state energy official. With average wind speeds in the 29 km/hr (18 mi/hr) range, further study of the tribal lands revealed possible sites for large-scale commercial wind farms, which could mean jobs and revenue for the reservations.

Similar wind prospecting programs are now underway on other reservations as well as in other states, including Utah, Oregon, Virginia, and Missouri.

annual royalties while occupying just a quarter-acre of land. Because each turbine takes up only a small area, most of the land can still be used for farming, ranching, or other uses.

Financially, wind energy involves up-front costs for erecting turbines and expanding infrastructure to allow electricity distribution, but over the lifetime of a project it requires only maintenance costs. Unlike fossil fuel power plants, wind turbines incur no ongoing fuel costs. Startup costs of wind farms generally are higher than those of fossil-fuel-driven plants, but wind farms incur fewer expenses once they are up and running. Moreover, advancing technology is driving down the costs of wind farm construction; as large wind farms become more efficient, the cost of each unit of electricity produced is dropping.

## Wind power has some downsides

Wind is an intermittent resource; we have no control over when wind will occur, and this unpredictability is a major limitation in relying on it as an electricity source. However, this poses little problem if wind is only one of several sources contributing to a utility's power generation. Moreover, several technologies can store energy generated by wind and release it later when needed; for example, batteries or hydrogen fuel.

Just as wind varies from time to time, it varies from place to place. Some areas are simply windier than others. Global wind patterns combine with local topography—mountains, hills, water bodies, forests, cities—to create local wind patterns, and companies study these patterns closely before investing in a wind farm. Meteorological research has given us information with which to judge prime areas for locating wind farms. A map of average wind speeds across the United States (**Figure 21.14a**) shows that mountainous regions are best, along with areas of the Great Plains. Based on such information, the young wind power industry has located much of its generating capacity in states with high wind speeds (**Figure 21.14b**) and is seeking to expand in the Great Plains and mountain states. Provided that wind farms are strategically erected in optimal locations, an estimated 15% of U.S. energy demand could be met using only 43,000 km$^2$ (16,600 mi$^2$) of land (with less than 5% of this land area actually occupied by turbines, equipment, and access roads).

Good wind resources, however, are not always near population centers that need the energy. Most of North America's people live near the coasts, far from the Great Plains and mountain regions that have the best wind resources. Thus, transmission networks would need to be greatly expanded to get wind power to where people live.

Furthermore, when wind farms *are* proposed near population centers, local residents often oppose them. Turbines are generally located in exposed, conspicuous sites, and many people object to wind farms for aesthetic reasons, feeling that the structures clutter the landscape. Although polls show wide public approval of existing wind projects and of the concept of wind power in general, newly proposed wind projects often elicit the so-called *not-in-my-backyard (NIMBY)* syndrome among people living nearby. For instance, a proposal for North America's first offshore wind farm, in Nantucket Sound between Cape Cod and the islands of Nantucket and Martha's Vineyard, has faced stiff opposition from wealthy area residents, even though many of these residents like to think of themselves as progressive environmentalists.

Wind turbines also pose a threat to birds and bats, which can be killed when they fly into the rotating blades. At California's Altamont Pass wind farm, turbines killed many golden eagles and other raptors during the 1990s. Studies since then at other sites have suggested that bird deaths may be a less severe problem than was initially feared, but uncertainty remains. For instance, one European study indicated that migrating seabirds fly past offshore turbines without problem, but other data show that resident seabird densities decline near turbines. On land, an estimated one to two birds are killed per turbine per year—far fewer than the millions already being killed annually by television, radio, and cell phone towers; tall buildings; automobiles; pesticides; and domestic cats. Bat mortality appears to be a more severe problem, but more research is needed. The key for protecting birds and bats may be selecting sites that are not on flyways or in the midst of prime habitat for species that are likely to fly into the blades.

### Weighing THE Issues | Wind and NIMBY

If you could choose to get your electricity from a wind farm or a coal-fired power plant, which would you choose? How would you react if the electric utility proposed to build the wind farm that would generate your electricity atop a ridge running in back of your neighborhood, such that the turbines would be clearly visible from your living room window? Would you support or oppose the development? Why? If you would oppose it, where would you suggest the farm be located? Do you think anyone might oppose it in that location?

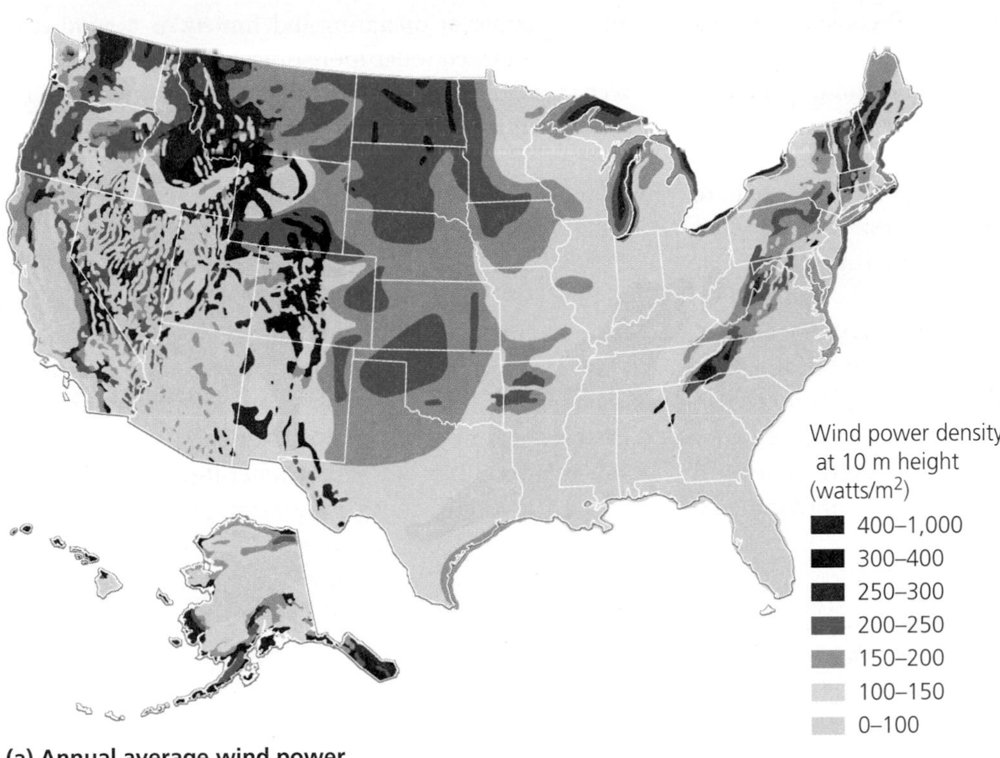


(a) Annual average wind power

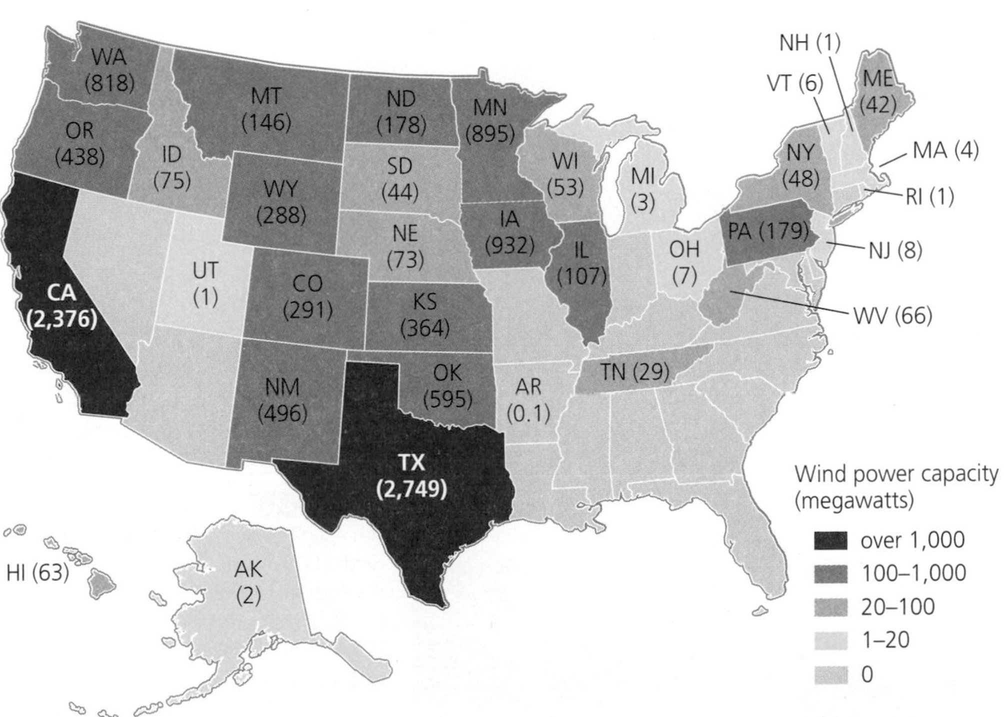


(b) Wind generating capacity, 2007

**FIGURE 21.14** Wind's capacity to generate power varies according to wind speed. Meteorologists have measured wind speed to calculate the potential generating capacity from wind in different areas. The map in **(a)** shows average wind power in watts per square meter at a height of 10 m (33 ft) above ground across the United States. Such maps are used to help guide placement of wind farms. The development of U.S. wind power so far is summarized in **(b)**, which shows the megawatts of generating capacity developed in each state through mid-2007. *Sources:* (a) Elliott, D. L., et al. 1987. *Wind energy resource atlas of the United States.* Golden, CO: Solar Energy Research Institute; (b) American Wind Energy Association.

## Geothermal Energy

Geothermal energy is one form of renewable energy that does not originate from the sun. Instead, it is generated deep within Earth. The radioactive decay of elements amid the extremely high pressures deep in the interior of our planet generates heat that rises to the surface through magma (molten rock, • p. 198) and through fissures and cracks. Where this energy heats groundwater, spurts of heated water and steam are sent up from below. Terrestrial geysers and submarine hydrothermal vents (• pp. 103–104) are the surface manifestations of these processes. Iceland is built from magma that extruded above the ocean's surface and cooled—magma from the Mid-Atlantic Ridge (• pp. 200, 448), the area of volcanic activity along the spreading boundary of two tectonic plates. Because of the geothermal heat in this region, volcanoes and geysers are numerous in Iceland. In fact, the word *geyser* originated from the Icelandic *Geysir*, the name for the island's largest geyser, which recently resumed periodic eruptions after many years in dormancy.

Geothermal power plants use the energy of naturally heated underground water and steam for direct heating and to turn turbines and generate electricity (**Figure 21.15**). Geothermal energy is renewable in principle (its use does not affect the amount of heat produced in Earth's interior), but the power plants we build to use this energy may not all be capable of operating indefinitely. If a geothermal plant uses heated water more quickly than groundwater is recharged, the plant will eventually run out of water. This is occurring at The Geysers, in Napa Valley, California, where the first generator was built in 1960. In response, operators have begun injecting municipal wastewater into the ground to replenish the supply. More and more geothermal power plants throughout the world are now injecting water, after it is used, back into aquifers to help maintain pressure and thereby sustain the resource.

A second reason geothermal energy may not always be renewable is that patterns of geothermal activity in Earth's crust shift naturally over time. This means that an area that produces hot groundwater now may not always do so.

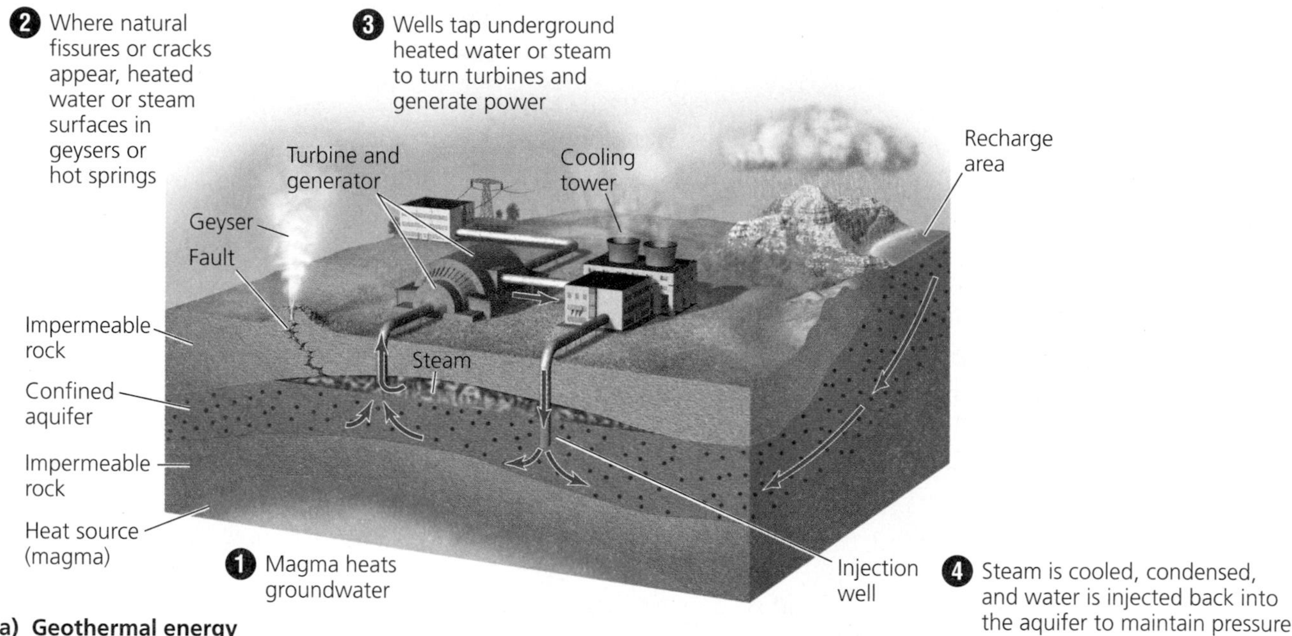


**(a) Geothermal energy**

**(b) Nesjavellir geothermal power station, Iceland**

FIGURE 21.15 With geothermal energy (**a**), magma heats groundwater deep in the earth (1), some of which is let off naturally through surface vents such as geysers (2). Geothermal facilities tap into heated water below ground and channel steam through turbines in buildings to generate electricity (3). After being used, the steam is often condensed, and the water is pumped back into the aquifer to maintain pressure (4). At the Nesjavellir geothermal power station in Iceland (**b**), steam is piped from four wells to a condenser at the plant, where cold water pumped from lakeshore wells 6 km (3.7 mi) away is heated. The water, heated to 83° C (181° F), is sent through an insulated 270-km (170-mi) pipeline to Reykjavik and environs, where residents use it for washing and space heating.

## We harness geothermal energy for heating and electricity

Geothermal energy can be harnessed directly from geysers at the surface, but most often wells must be drilled down hundreds or thousands of meters toward heated groundwater. Generally, water at temperatures of 150–370° C (300–700° F) or more is brought to the surface and converted to steam by lowering the pressure in specialized compartments. The steam is then employed in turning turbines to generate electricity.

Hot groundwater can also be used directly for heating homes, offices, and greenhouses; for driving industrial processes; and for drying crops. Iceland heats most of its homes through direct heating with piped hot water. Iceland began putting geothermal energy to use in the 1940s, and today 30 municipal district heating systems and 200 small private rural networks supply heat to 86% of the nation's residences. Other locales are benefiting in similar ways; the Oregon Institute of Technology heats its buildings with geothermal energy for a fraction of the cost it would take to heat them with natural gas. Such direct use of naturally heated water is cheap and efficient, but it is feasible only in areas such as Iceland or parts of Oregon, where geothermal energy sources are available and near where the heat must be transported (**Figure 21.16**).

Geothermal *ground source heat pumps* (GSHPs) use thermal energy from near-surface sources of earth and water. Soil varies in temperature from season to season less than air does, so the pumps heat buildings in the winter by transferring heat from the ground into buildings, and they cool buildings in the summer by transferring heat from buildings into the ground. Both types of heat transfer are accomplished by a single network of underground plastic pipes that circulate water. Because heat is simply moved from place to place rather than being produced using outside energy inputs, heat pumps can be highly energy-efficient.

More than 600,000 GSHPs are already used to heat U.S. residences. Compared to conventional electric heating and cooling systems, GSHPs heat spaces 50–70% more efficiently, cool them 20–40% more efficiently, can reduce electricity use by 25%–60%, and can reduce emissions by up to 70%.

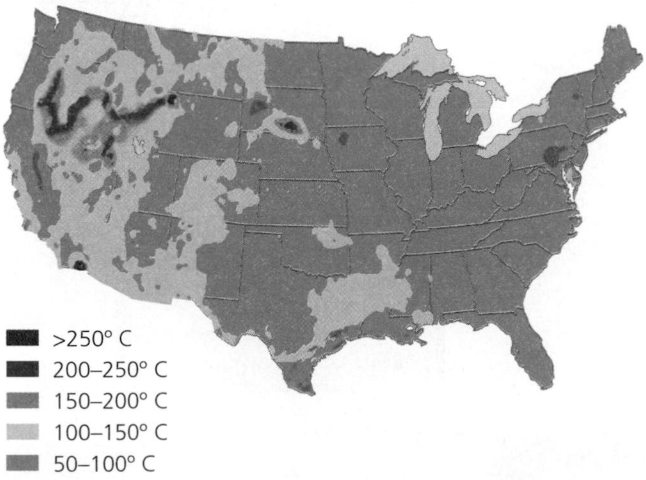


FIGURE 21.16 Geothermal resources in the United States are greatest in the western states. This map shows water temperatures 3 km (1.9 mi) below ground. Although deep subterranean temperatures are greatest in the West, ground-source heat pumps can be used anywhere in the country. Data from Idaho National Laboratory, 2007.

## Use of geothermal power is growing

Geothermal energy provides less than 0.5% of the total energy used worldwide. It provides more power than solar and wind combined, but only a small fraction of the power from hydropower and biomass. Geothermal energy in the United States provides enough power to supply electricity to over 4 million people. At the world's largest geothermal power plants, The Geysers in northern California, generating capacity has declined by more than 50% since 1989 as steam pressure has declined, but The Geysers still provide enough electricity to supply 750,000 homes. Currently Japan, China, and the United States lead the world in use of geothermal power.

## Geothermal power has benefits and limitations

Like other renewable sources, geothermal power greatly reduces emissions relative to fossil fuel combustion. Geothermal sources can release variable amounts of gases dissolved in their water, including carbon dioxide, methane, ammonia, and hydrogen sulfide. However, these gases are generally in very small quantities, and geothermal facilities using the latest filtering technologies produce even fewer emissions. By one estimate, each megawatt of geothermal power prevents the emission of 7.0 million kg (15.5 million lb) of carbon dioxide emissions each year.

On the negative side, geothermal sources, as we have seen, may not always be truly sustainable. In addition, the water of many hot springs is laced with salts and minerals that corrode equipment and pollute the air. These factors may shorten the lifetime of plants, increase maintenance costs, and add to pollution.

Moreover, use of geothermal energy is limited to areas where the energy can be tapped. Unless technology is developed to penetrate far more deeply into the ground, geothermal energy use will remain more localized than solar, wind, biomass, or hydropower. Places such as Iceland are rich in geothermal sources, but most of the world is not. Nonetheless, many hydrothermal resources remain unexploited, awaiting improved technology and governmental support for their development.

# Ocean Energy Sources

The oceans are home to several underexploited energy sources. Each involves continuous natural processes that could potentially provide us sustainable energy predictably through time and in substantial amounts. Of the four approaches being developed, three involve motion, and one involves temperature.

## We can harness energy from tides, waves, and currents

Just as dams on rivers use flowing fresh water to generate hydroelectric power, some scientists, engineers, businesses, and governments are developing ways to use the kinetic energy from the natural motion of ocean water to generate electrical power.

The rising and falling of ocean tides (• p. 451) twice each day at coastal sites throughout the world moves large amounts of water past any given point on the world's coastlines. Differences in height between low and high tides are especially great in long, narrow bays such as Alaska's Cook Inlet or the Bay of Fundy between New Brunswick and Nova Scotia. Such locations are best for harnessing *tidal energy*, which is accomplished by erecting dams across the outlets of tidal basins. The incoming tide flows through sluices past the dam, and as the outgoing tide passes through the dam, it turns turbines to generate electricity (**Figure 21.17**). Some designs allow for generating electricity from water moving in both directions.

The world's largest tidal generating facility is the La Rance facility in France (see Figure 21.17, inset), which has operated for over 40 years. Smaller facilities operate in China, Russia, and Canada. Tidal stations release few or no pollutant emissions, but they can have impacts on the ecology of estuaries and tidal basins.

*Wave energy* could be developed at a greater variety of sites than could tidal energy. The principle is to harness the motion of wind-driven waves at the ocean's surface and convert this mechanical energy into electricity. Many designs for machinery to harness wave energy exist, but few have been adequately tested. Some designs are for offshore facilities and involve floating devices that move up and down with the waves. Wave energy is greater at deep-ocean sites, but transmitting the electricity produced to shore would be expensive.

Other designs are for coastal onshore facilities. Some of these designs funnel waves from large areas into narrow channels and elevated reservoirs, from which water is then allowed to flow out, generating electricity as hydroelectric dams do. Other coastal designs use rising and

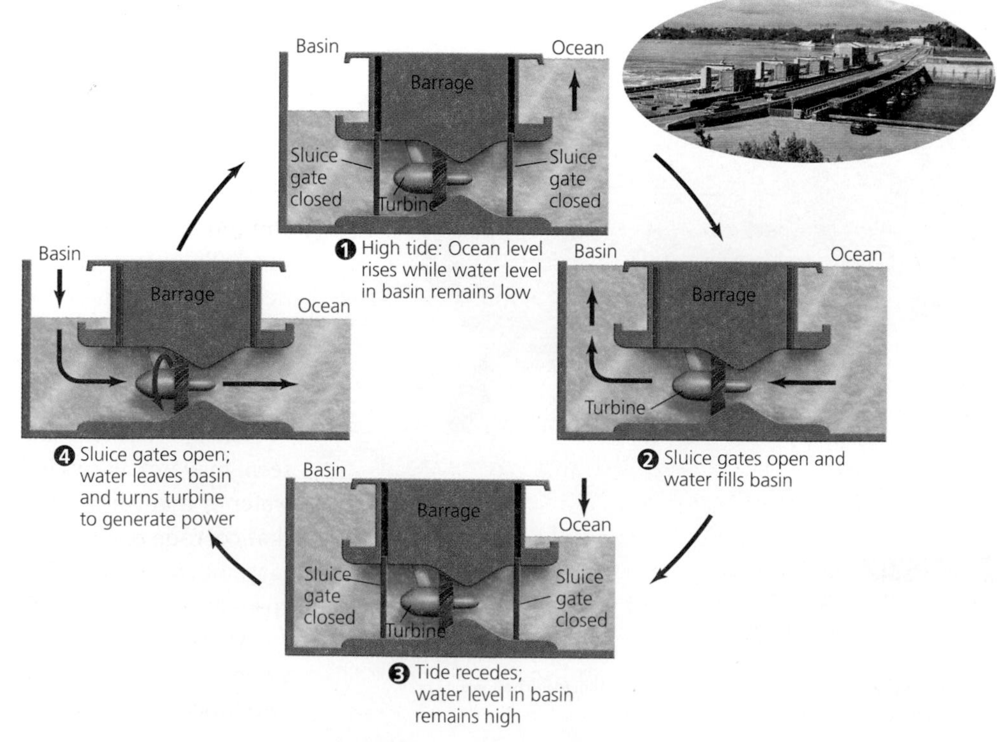


**FIGURE 21.17** Energy can be extracted from the movement of the tides at coastal sites where tidal flux is great enough. One way of doing so involves using bulb turbines in concert with the outgoing tide. At high tide (1), ocean water is let through the sluice gates, filling an interior basin (2). At low tide (3), the basin water is let out into the ocean, spinning turbines to generate electricity (4). This technology has been used at the La Rance facility in France (photo) for over four decades.

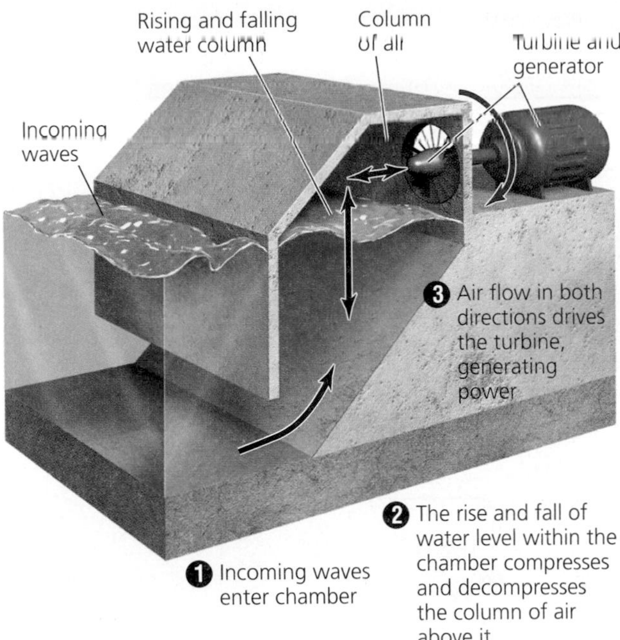


FIGURE 21.18 Coastal facilities can make use of energy from the motion of ocean waves. As waves are let into and out of a tightly sealed chamber, the air inside is alternately compressed and decompressed, creating air flow that rotates turbines to generate electricity.

falling waves to push air into and out of chambers, turning turbines to generate electricity (**Figure 21.18**). No commercial wave energy facilities are operating yet, but demonstration projects exist in Europe and Japan.

A third way of harnessing marine kinetic energy is to use the motion of ocean currents (• pp. 446–447), such as the Gulf Stream. Devices that look essentially like underwater wind turbines have been erected in European waters to test this idea.

## The ocean stores thermal energy

Each day the tropical oceans absorb an amount of solar radiation equivalent to the heat content of 250 billion barrels of oil—enough to provide 20,000 times the electricity used daily in the United States. The ocean's sun-warmed surface is higher in temperature than its deep water, and **ocean thermal energy conversion (OTEC)** is based on this gradient in temperature.

In the *closed cycle* approach, warm surface water is piped into a facility to evaporate chemicals, such as ammonia, that boil at low temperatures. These evaporated gases spin turbines to generate electricity. Cold water piped in from ocean depths then condenses the gases so they can be reused. In the *open cycle* approach, warm surface water is evaporated in a vacuum, and its steam turns turbines and then is condensed by cold water. Because ocean water loses its salts as it evaporates, the water can be recovered, condensed, and sold as desalinized fresh water for drinking or agriculture. Research on OTEC systems has been conducted in Hawaii and Japan, but costs remain high, and as of yet no facility is commercially operational.

### Weighing THE Issues | Your Island's Energy?

Imagine that you have been elected the president of an island nation the size of Iceland and that your nation's congress is calling on you to propose a national energy policy. Unlike Iceland, your country is located in equatorial waters. Your geologists do not yet know whether there are fossil fuel deposits or geothermal resources under your land, but your country gets a lot of sunlight and a fair amount of wind, and broad, shallow shelf regions surround its coasts. Your island's population is moderately wealthy but is growing fast, and importing fossil fuels from mainland nations is becoming increasingly expensive.

What approaches would you propose in your energy policy? What specific steps would you urge your congress to fund immediately? What trade relationships would you seek to establish with other countries? What questions would you ask of your economic advisors? What questions would you fund your country's scientists to research?

# Hydrogen

All the renewable energy sources we have discussed can be used to generate electricity more cleanly than can fossil fuels. As useful as electricity is to us, however, it cannot be stored easily in large quantities for use when and where it is needed. This is why vehicles rely on fossil fuels for power. The development of fuel cells and hydrogen fuel shows promise to store energy conveniently and in considerable quantities and to produce electricity at least as cleanly and efficiently as renewable energy sources.

In the "hydrogen economy" that Iceland's leaders and many energy experts worldwide envision, hydrogen fuel, together with electricity, would serve as the basis for a clean, safe, and efficient energy system. This system would use as a fuel the universe's simplest and most abundant element. In this system, electricity generated from renewable sources that are intermittent, such as wind or solar energy, could be used to produce hydrogen. Fuel cells could then employ hydrogen to produce electrical energy as needed to power vehicles, computers, cell phones, home heating, and countless other applications.

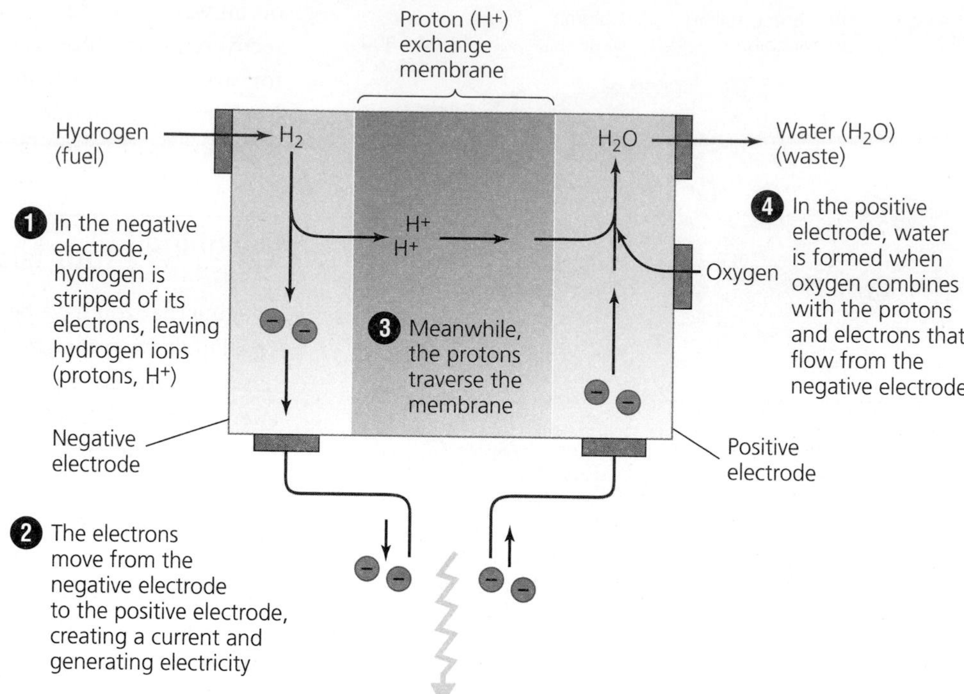


FIGURE 21.19 Hydrogen fuel drives electricity generation in a fuel cell, creating water as a waste product. Atoms of hydrogen are first stripped of their electrons (1). The electrons move from a negative electrode to a positive one, creating a current and generating electricity (2). Meanwhile, the hydrogen ions pass through a proton exchange membrane (3) and combine with oxygen to form water molecules (4).

NASA's space flight programs have used fuel cell technology (**Figure 21.19**) since the 1960s. Basing an energy system on hydrogen could alleviate dependence on foreign fuels and help fight climate change. For these reasons, governments are funding research into hydrogen and fuel cell technology, and automobile companies are investing in research and development to produce vehicles that run on hydrogen (**Figure 21.20**).

## Hydrogen fuel may be produced from water or from other matter

Hydrogen gas ($H_2$) does not tend to exist freely on Earth; rather, hydrogen atoms bind to other molecules, becoming incorporated in everything from water to organic molecules. To obtain hydrogen gas for fuel, we must force these substances to release their hydrogen atoms, and this requires an input of energy. Several potential ways of producing hydrogen are being studied (see "The Science behind the Story," • p. 624). In **electrolysis**, the process being pursued by Iceland, electricity is input to split hydrogen atoms from the oxygen atoms of water molecules:

$$2H_2O \rightarrow 2H_2 + O_2$$

Electrolysis produces pure hydrogen, and it does so without emitting the carbon- or nitrogen-based pollutants of fossil fuel combustion. However, whether this strategy for producing hydrogen will cause pollution over its entire life cycle depends on the source of the electricity used for the electrolysis. If coal is burned to create the electricity, then the entire process will not reduce emissions compared with reliance on fossil fuels. If, however, the electricity is produced by some less-polluting renewable source, then hydrogen production by electrolysis would create much less pollution and greenhouse warming than reliance on fossil fuels. The "cleanliness" of a future hydrogen economy in Iceland or anywhere else would, therefore, depend largely on the source of electricity used in electrolysis.

The environmental impact of hydrogen production will also depend on the source material for the hydrogen. Besides water, hydrogen can be obtained from biomass and fossil fuels. Obtaining hydrogen from these sources generally requires less energy input but results in emissions of carbon-based pollutants. For instance, extracting hydrogen from the methane ($CH_4$) in natural gas entails producing one molecule of the greenhouse gas carbon dioxide for every four molecules of hydrogen gas:

$$CH_4 + 2H_2O \rightarrow 4H_2 + CO_2$$

Thus, whether a hydrogen-based energy system is environmentally cleaner than a fossil fuel system depends on how the hydrogen is extracted.

In addition, some new research suggests that leakage of hydrogen from its production, transport, and use could potentially deplete stratospheric ozone and lengthen the atmospheric lifetime of the greenhouse gas methane. Research into these questions is ongoing, because scientists do not want society to switch from fossil fuels to hydrogen without first knowing the possible risks from hydrogen.

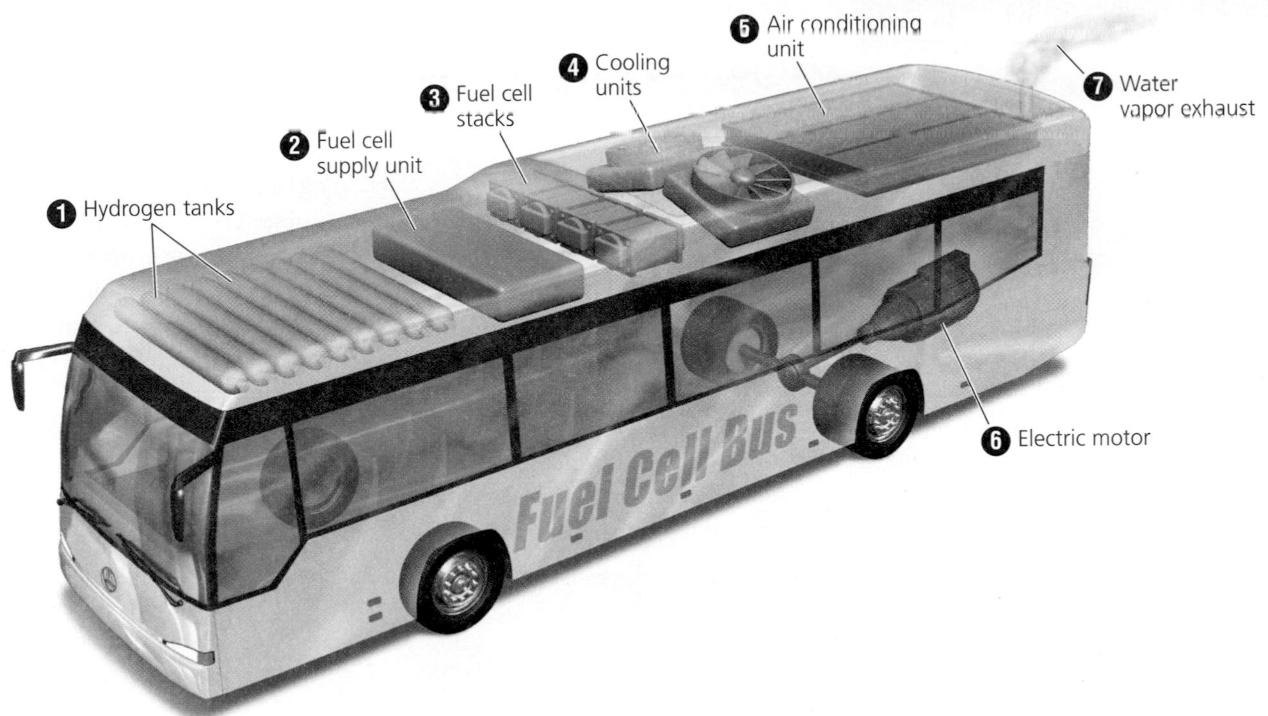


FIGURE 21.20 In the hydrogen-fueled Citaro buses operating in Reykjavik and other European capitals, hydrogen is stored in nine fuel tanks (1). The fuel cell supply unit (2) controls the flow of hydrogen, air, and cooling water into the fuel cell stacks (3). Cooling units (4) and the air conditioning unit (5) dissipate waste heat produced by the fuel cells. Electricity generated by the fuel cells is changed from direct current (DC) to alternating current (AC) by an inverter, and it is transmitted to the electric motor (6), which powers the operation of the bus. The vehicle's exhaust (7) consists simply of water vapor.

**Weighing THE Issues | Precaution over Hydrogen?**

Some environmental scientists have recently warned that we do not yet know enough about the environmental consequences of replacing fossil fuels with hydrogen fuel. An increase in tropospheric hydrogen gas would deplete hydroxyl (OH) radicals, they hypothesize, possibly leading to stratospheric ozone depletion and global warming from increased concentrations of methane. Some scientists say such effects will be small; others say there could be further effects that are currently unknown. Do you think we should apply the precautionary principle to the development of hydrogen fuel and fuel cells? Or should we embark on pursuing a hydrogen economy before knowing all the scientific answers? What factors inform your view?

## Fuel cells produce electricity by joining hydrogen and oxygen

Once isolated, hydrogen gas can be used as a fuel to produce electricity within fuel cells. The chemical reaction involved in a fuel cell is simply the reverse of that shown for electrolysis; an oxygen molecule and two hydrogen molecules each split so that their atoms can bind and form two water molecules:

$$2H_2 + O_2 \rightarrow 2H_2O$$

The way this occurs within one common type of fuel cell is shown in Figure 21.19. Hydrogen gas (usually compressed and stored in an attached fuel tank) is allowed into one side of the cell, whose middle consists of two electrodes that sandwich a membrane that only protons (hydrogen ions) can move across. One electrode, helped by a chemical catalyst, strips the hydrogen gas of its electrons, creating two hydrogen ions that begin moving across the membrane. Meanwhile, on the other side of the cell, oxygen molecules from the open air are split into their component atoms along the other electrode. These oxygen ions soon bind to pairs of hydrogen ions traveling across the membrane, forming molecules of water that are expelled as waste, along with heat. While this is occurring, the electrons from the hydrogen atoms have traveled to a device that completes an electric current between the two electrodes. The movement of the hydrogen's electrons from one electrode to the other creates the output of electricity.

THE SCIENCE BEHIND THE STORY

# Algae as a Hydrogen Fuel Source

*The green alga Chlamydomonas reinhardtii.*

As scientists search for new ways to generate energy, some are looking to an unlikely power source—pond scum. Algae are being studied as an innovative way to generate large amounts of hydrogen to move society toward a more sustainable energy future.

Hydrogen's benefits hinge on how hydrogen fuel is produced. Some methods release substantial amounts of carbon dioxide, and other, nonpolluting, processes can be costly. These drawbacks have kept scientists searching for new hydrogen sources.

At the University of California at Berkeley, biologist Anastasios Melis thought one possible hydrogen source might be a single-celled aquatic plant. The alga *Chlamydomonas reinhardtii* was known to emit small amounts of hydrogen for brief periods of time when deprived of light.

So Melis set up an experiment with energy experts at the National Renewable Energy Laboratory in Colorado, aiming to develop ways to tweak the alga's basic biological functions so that the plant produced greater quantities of hydrogen.

Green algae photosynthesize, absorbing energy from light that converts carbon dioxide and water into food, and then expelling oxygen as a waste product. Additional nutrients from soil or water, and catalysts called *enzymes* within the plant, keep this process running smoothly. To conduct photosynthesis effectively, *Chlamydomonas reinhardtii* needs the element sulfur as a nutrient. The alga also contains an enzyme called *hydrogenase*, which can trigger the alga to stop producing oxygen as a metabolic by-product and start releasing hydrogen instead.

Hydrogenase is normally active only after *Chlamydomonas reinhardtii* has been deprived of light. When deprived of light, the light-dependent reactions of photosynthesis ebb, little oxygen is produced, and hydrogenase is activated. When light returns and the alga begins producing oxygen again, hydrogenase is deactivated, and the hydrogen release stops.

Dr. Anastasios Melis inspects flasks of algae in his laboratory.

Melis's team wanted to activate hydrogenase so that more hydrogen would be produced. But simply keeping the algae in the dark would not escalate hydrogen production because the alga's metabolic functions slowed without light.

The researchers decided to try putting the algae on a sulfur-free, bright-light regimen. The lack of sulfur would hinder photosynthesis, limiting oxygen output enough to activate hydrogenase and trigger hydrogen production. The presence of light would keep the algae metabolically active and releasing large amounts of by-products.

The researchers cultured large quantities of the algae in bottles in labs. Then they deprived the cultures of sulfur but kept the algae exposed to light for long periods of time—up to 150 hours. After the sustained light exposure, gas and liquids were extracted from the bottles and analyzed.

The analysis supported the team's hypothesis. Without sulfur or photosynthesis, the algae were not producing oxygen. This low-oxygen, or anaerobic, environment had induced hydrogenase, which spurred the algae to begin splitting water molecules and releasing hydrogen gas. Hydrogen dominated the alga's emissions—in gas collection analysis, approximately 87% of the gas was hydrogen, 1% was carbon dioxide, and the remaining 12% was nitrogen with traces of oxygen. The research teams published their findings in the journal *Plant Physiology* in 2000.

Since then, Melis and his collaborators have genetically engineered a strain of algae to give it enhanced energy efficiency. With further advances, they envision farms of algae pumping out hydrogen for our use in the future.

Questions remain about how much fuel can be harvested continuously using this *photobiological* process. But the research results so far are helping to fuel the momentum of a future hydrogen economy.

## Hydrogen and fuel cells have many benefits

As a fuel, hydrogen offers a number of benefits. We will never run out of hydrogen; it is the most abundant element in the universe. It can be clean and nontoxic to use, and—depending on the source of the hydrogen and the source of electricity for its extraction—it may produce few greenhouse gases and other pollutants. Pure water and heat may be the only waste products from a hydrogen fuel cell, along with negligible traces of other compounds. In terms of safety for transport and storage, hydrogen can catch fire, but if it is kept under pressure, it is probably no more dangerous than gasoline in tanks.

Hydrogen fuel cells are energy-efficient. Depending on the type of fuel cell, 35% to 70% of the energy released in the reaction can be used. If the system is designed to capture heat as well as electricity, then the energy efficiency of fuel cells can rise to 90%. These rates are comparable or superior to most nonrenewable alternatives.

Fuel cells are also silent and nonpolluting. Unlike batteries (which also produce electricity through chemical reactions), fuel cells will generate electricity whenever hydrogen fuel is supplied, without ever needing recharging. For all these reasons, hydrogen fuel cells are being used to power vehicles, including the buses now operating on the streets of Reykjavik and many other European, American, and Asian cities.

# Conclusion

The coming decline of fossil fuel supplies and the increasing concern over air pollution and global climate change have convinced many people that we will need to shift to renewable energy sources that will not run out and will pollute far less. Renewable sources with promise for sustaining our civilization far into the future without greatly degrading our environment include solar energy, wind energy, geothermal energy, and ocean energy sources. Moreover, by using electricity from renewable sources to produce hydrogen fuel, we may be able to use fuel cells to produce electricity when and where it is needed, helping to convert our transportation sector to a nonpolluting, renewable basis.

Most renewable energy sources have been held back by inadequate funding for research and development and by artificially cheap market prices for nonrenewable resources that do not include external costs. Despite these obstacles, renewable technologies have progressed far enough to offer hope that we can shift from fossil fuels to renewable energy with a minimum of economic and social disruption. Whether we can also limit environmental impact will depend on how soon and how quickly we make the transition and to what extent we put efficiency and conservation measures into place.

## REVIEWING OBJECTIVES

**You should now be able to:**

**Outline the major sources of renewable energy and assess their potential for growth**

- The "new renewable" energy sources include solar, wind, geothermal, and ocean energy sources. They are not truly "new," but rather are in a stage of rapid development. (p. 604)
- The new renewables currently provide far less energy and electricity than we obtain from fossil fuels or other conventional energy sources. (pp. 604–605)
- Use of new renewables is growing quickly, and this growth is expected to continue as people seek to move away from fossil fuels. (pp. 605–606)

**Describe solar energy and the ways it is harnessed, and evaluate its advantages and disadvantages**

- Energy from the sun's radiation can be harnessed using passive methods or by active methods involving powered technology. (p. 608)
- Solar technologies include solar panels for heating, mirrors to concentrate solar rays, and photovoltaic cells to generate electricity. (pp. 608–610)
- Solar energy is perpetually renewable, creates no emissions, and enables decentralized power. (pp. 610–611)
- Solar radiation varies in intensity from place to place and time to time, and harnessing solar energy remains expensive. (pp. 611–612)

**Describe wind energy and the ways it is harnessed, and evaluate its advantages and disadvantages**

- Energy from wind is harnessed using wind turbines mounted on towers. (pp. 612–613)
- Turbines are often erected in arrays at wind farms located on land or offshore, in locations with optimal wind conditions. (pp. 613–615, 617)
- Wind energy is renewable, turbine operation creates no emissions, wind farms can generate economic benefits, and the cost of wind power is competitive with that of electricity from fossil fuels. (pp. 614, 616)

- Wind is an intermittent resource and is adequate only in some locations. Turbines kill some birds and bats, and wind farms can face opposition from local residents. (p. 616)

**Describe geothermal energy and the ways it is harnessed, and evaluate its advantages and disadvantages**

- Energy from radioactive decay in Earth's core rises toward the surface and heats groundwater. This energy is harnessed at the surface or by drilling at geothermal power plants. (pp. 617–618)
- Use of geothermal energy for direct heating of water, electricity generation, and in heat pumps can be efficient, clean, and renewable. (pp. 618–619)
- Geothermal sources occur only in certain areas and may be exhausted if water is overpumped. (pp. 619–620)

**Describe ocean energy sources and the ways they could be harnessed**

- Major ocean energy sources include the motion of tides, waves, and currents, and the thermal heat of ocean water. (pp. 620–621)
- Ocean energy is perpetually renewable and holds much promise, but so far technologies have seen only limited development. (pp. 620–621)

**Explain hydrogen fuel cells and assess future options for energy storage and transportation**

- Hydrogen can serve as a fuel to store and transport energy, so that electricity generated by renewable sources can be made portable and used to power vehicles. (pp. 621–623)
- Hydrogen can be produced through electrolysis, but also by using fossil fuels—in which case its environmental benefits are reduced. (p. 622)
- There is concern that releasing excess hydrogen could have negative impacts on the atmosphere. (pp. 622–623)
- Fuel cells create electricity by controlling an interaction between hydrogen and oxygen, and they produce only water as a waste product. (pp. 622–623)
- Hydrogen can be clean, safe, and efficient. Fuel cells are silent, are nonpolluting, and do not need recharging. (p. 625)

## TESTING YOUR COMPREHENSION

1. About how much of our energy now comes from renewable sources? What is the most prevalent form of renewable energy we use? What form of renewable energy is most used to generate electricity?
2. What factors and concerns are causing renewable energy sectors to expand? Which renewable source is experiencing the most rapid growth?
3. Contrast passive and active solar heating. Describe how each works, and give examples of each.
4. Define the photoelectric effect. Explain how photovoltaic (PV) cells function and are used.
5. What are the environmental and economic advantages of solar power? What are its disadvantages?
6. How do modern wind turbines generate electricity? How does wind speed affect the process? What factors affect where wind turbines are placed?
7. What are the environmental and economic benefits of wind power? What are its drawbacks?
8. Define *geothermal energy*, and explain how it is obtained and used. In what ways is it renewable, and in what way is it not renewable?
9. List and describe four approaches to obtaining energy from ocean water.
10. How is hydrogen fuel produced? Is this a clean process? What factors determine the amount of pollutants hydrogen production will emit?

## SEEKING SOLUTIONS

1. Why might a hydrogen economy be closer than we think? Why might it instead not come to pass? Do you think water could be "the coal of the future"? Why or why not?
2. For each source of renewable energy discussed in this chapter, what factors are standing in the way of an expedient transition from fossil fuel use? What could be done in each case to ease a shift to these renewable sources?
3. Do you think we can develop and implement renewable energy resources to replace fossil fuels without great social, economic, and environmental disruption? What steps would we need to take? Will market forces alone suffice to bring about this transition? Do you think such a shift will be good for our economy?
4. Iceland is giving itself many years to phase in its planned hydrogen economy. Do you think the United States could transition to a hydrogen economy more quickly, less quickly, or not at all? Why? What steps could the United States take to accelerate such a transition?

5. **THINK IT THROUGH** You have just graduated from college, gotten married, landed a good job, and purchased your first home. You and your spouse plan to stay in this home for the foreseeable future and are considering installing solar panels and/or PV tiles on your roof. What factors will you consider, and what questions will you ask before deciding whether to make the investment in solar energy?
6. **THINK IT THROUGH** You are the CEO of a company that develops wind farms. Your staff is presenting you with three options, listed below, for sites for your next development. Describe at least one likely advantage and at least one likely disadvantage you would expect to encounter with each option. What further information would you like to know before deciding which to pursue?
   - Option A: A remote rural site in North Dakota
   - Option B: A ridge-top site among the suburbs of Philadelphia
   - Option C: An offshore site off the Florida coast

## INTERPRETING GRAPHS AND DATA

Of the new renewable energy alternatives discussed in this chapter, photovoltaic conversion of solar energy is the one that most areas of the United States could most easily adopt. The influx of solar radiation varies with time of day, time of year, and location, so all areas are not equally well suited. Today's photovoltaic technology is approximately 10% efficient at converting the energy of sunlight into electricity, but new technologies under development may increase that efficiency to as much as 40%.

1. Given a 10% efficiency for photovoltaic conversion of solar energy, approximately how many square meters of photovoltaic cells would be needed to supply one person's residential electrical needs for a year, based on the yearly average values? How many square meters would be needed if efficiency were improved to 40%?
2. Given the same 10% conversion efficiency, approximately how many square meters of photovoltaic cells would be required to supply one person's residential electrical needs during the month of April? During July? How many square meters would be required to supply the average U.S. household of four people for each of those months?
3. Commercially available photovoltaic systems of this capacity cost approximately $20,000. The average cost of electricity in the United States is approximately 9¢ per kilowatt-hour. At these prices, how long would it take for the PV system to generate $20,000 worth of electricity? Calculate a combination of PV system cost and electricity cost at which the system would pay for itself in 10 years.

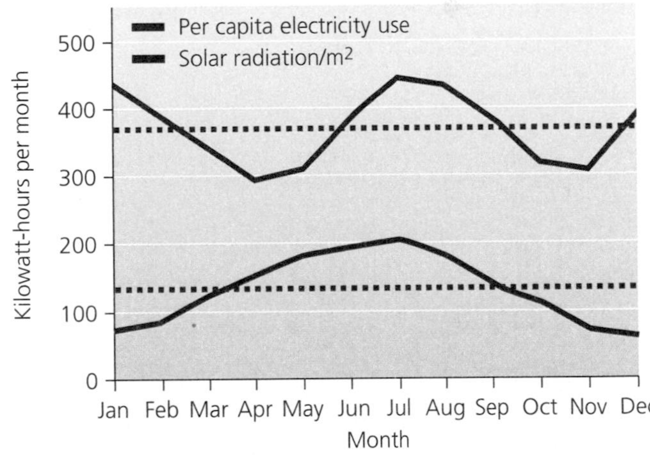


Average per capita residential use of electricity in the United States in 2004 (red line) and average influx of solar radiation per square meter for Topeka, Kansas (blue line). The dashed lines represent the yearly average values for each. Data from Renewable Resource Data Center, National Renewable Energy Laboratory, U.S. Department of Energy (DOE); and Energy Information Administration. 2005. *Annual energy review 2004.* DOE.

## CALCULATING ECOLOGICAL FOOTPRINTS

Assume that average per capita residential consumption of electricity is 12 kilowatt-hours per day, that photovoltaic cells have an electrical output of 10% incident solar radiation, and that PV cells cost $800 per square meter. Now refer to Figure 21.9 on page 611, and estimate the area and cost of the PV cells needed to provide all of the residential electricity used by each group in the table.

| | Area of photovoltaic cells | Cost of photovoltaic cells |
|---|---|---|
| **You** | 25 | $20,000 |
| **Your class** | | |
| **Your state** | | |
| **United States** | | |

1. What additional information do you need to increase the accuracy of your estimates for the areas in the table above?
2. Considering the distribution of solar radiation in the United States, where do you think it will be most feasible to greatly increase the percentage of electricity generated from photovoltaic solar cells?
3. The purchase price of a photovoltaic system is considerable. What other costs and benefits should you consider, in addition to the purchase price, when contemplating "going solar"?

## Take It Further

Go to www.aw-bc.com/withgott or the student CD-ROM, where you'll find:

- Suggested answers to end-of-chapter questions
- Quizzes, animations, and flashcards to help you study
- *Research Navigator*™ database of credible and reliable sources to assist you with your research projects
- GRAPHit! Tutorials to help you interpret graphs
- INVESTIGATEit! Current news articles that link the topics that you study to case studies from your region to around the world

CHAPTER

# 22 Waste Management

Containers en route to a recycling facility

## Upon completing this chapter, you will be able to:

- Summarize and compare the types of waste we generate
- List the major approaches to managing waste
- Delineate the scale of the waste dilemma
- Describe conventional waste disposal methods: landfills and incineration
- Evaluate approaches for reducing waste: source reduction, reuse, composting, and recycling
- Discuss industrial solid waste management and principles of industrial ecology
- Assess issues in managing hazardous waste

Fresh Kills Landfill,
Staten Island, New York

## CENTRAL CASE

# Transforming New York's Fresh Kills Landfill

**"An extraterrestrial observer might conclude that conversion of raw materials to wastes is the real purpose of human economic activity."**
—Gary Gardner and Payal Sampat, Worldwatch Institute

**"Recycling is one of the best environmental success stories of the late 20th century."**
—U.S. Environmental Protection Agency

The closure of a landfill is not the kind of event that normally draws politicians and the press, but the Fresh Kills Landfill was no ordinary dump. The largest landfill in the world, Fresh Kills was the primary repository of New York City's garbage for half a century. On March 22, 2001, New York City Mayor Rudolph Giuliani and New York Governor George Pataki were on hand to celebrate as a barge arrived on the western shore of New York City's Staten Island and dumped the final load of 650 tons of trash at Fresh Kills.

The landfill's closure was a welcome event for Staten Island's 450,000 residents, who had long viewed the landfill as a bad-smelling eyesore, health threat, and civic blemish. The 890-ha (2,200-acre) landfill featured six gigantic mounds of trash and soil. The highest, at 69 m (225 ft), was higher than the nearby Statue of Liberty.

New York City had grandiose plans for the site. It planned to transform the old landfill into a world-class public park—a verdant landscape of rolling hills and wetlands teeming with wildlife, a mecca for recreation for New York's residents. The site certainly had potential. It was two-and-a-half times bigger than Manhattan's Central Park. It was the region's largest remaining complex of saltwater tidal marshes and freshwater creeks and wetlands, which still attracted birds and wildlife. And the mounds offered panoramic views of the Manhattan skyline and the rest of the region. The city sponsored an international competition to select a landscape architecture firm to design plans for the new park.

Meanwhile, with its only landfill closed, New York City began exporting its waste. The city began plans to develop an efficient network of stations to package and transfer the waste and ship it outward by barge and railroad. However, these plans soon fell apart amid neighborhood opposition, economic misjudgments, and accusations of political favoritism and mob influence, New York City instead found itself paying contractors exorbitant prices to haul its garbage away inefficiently, one truckload at a time. In the years following the Fresh Kills closure, trucks full of trash rumbled through neighborhood streets, carrying 12,000 tons of waste each day bound for 26 different landfills and incinerators in New York, New Jersey, Virginia, Pennsylvania, and Ohio. The city sanitation department's budget nearly doubled, and budget woes caused the city to scale back its recycling program. Some New Yorkers suggested reopening Fresh Kills.

The landfill *was* reopened, but not for a reason anyone could have foreseen. After the September 11, 2001, terrorist attacks, the 1.8 million tons of rubble from the collapsed World Trade Center towers, including unrecoverable human remains, was taken by barge to Fresh Kills, where it was sorted and buried. A monument will be erected at the site as part of the new park.

Today, plans for the park are forging ahead. A draft master plan that incorporated suggestions from the public was released in 2006, and an environmental impact statement was scheduled for release in late 2007. The master plan involves everything from ecological restoration of the wetlands to construction of roads, ball fields, sculptures, and roller-blading rinks. People will be able to bicycle on trails paralleling tidal creeks of the region's largest estuary and reach stunning vistas atop the hills. Public tours of the site are now being offered, and the first parcel to be developed, Owl Hollow Fields, will open for recreation in 2009.

This immense undertaking—one of the largest public works projects in the world—will not be completed overnight. Designers and city officials expect to develop and open the new park one portion at a time over the next 30 years. In the end, they aim to transform a longtime symbol of waste into a world-class center for recreation and urban ecological restoration.

# Approaches to Waste Management

As the world's human population rises, and as we produce and consume more material goods, we generate more waste. **Waste** refers to any unwanted material or substance that results from a human activity or process.

For management purposes, waste is divided into several main categories. **Municipal solid waste** is nonliquid waste that comes from homes, institutions, and small businesses. **Industrial solid waste** includes waste from production of consumer goods, mining, agriculture, and petroleum extraction and refining. **Hazardous waste** refers to solid or liquid waste that is toxic, chemically reactive, flammable, or corrosive. It can include everything from paint and household cleaners to medical waste to industrial solvents. Another type of waste is *wastewater*, water we use in our households, businesses, industries, or public facilities and drain or flush down our pipes, as well as the polluted runoff from our streets and storm drains. We discussed wastewater in Chapter 15 (• pp. 435–438).

## We have several aims in managing waste

Waste can degrade water quality, soil quality, and air quality, thereby degrading human health and the environment. Waste is also a measure of inefficiency, so reducing waste can potentially save industry, municipalities, and consumers both money and resources. Waste is also unpleasant aesthetically. For these and other reasons, waste management has become a vital pursuit.

There are three main components of **waste management**:

- Minimizing the amount of waste we generate
- Recovering waste materials and finding ways to recycle them
- Disposing of waste safely and effectively

Minimizing waste at its source—called *source reduction*—is the preferred approach. There are several ways to reduce the amount of waste that enters the **waste stream**, the flow of waste as it moves from its sources toward disposal destinations (**Figure 22.1**). Manufacturers can use materials more efficiently. Consumers can buy fewer goods, buy goods with less packaging, and use those goods longer. Reusing goods you already own, purchasing used items, and donating your used items for others also help reduce the amount of material entering the waste stream.

*Recovery* (*recycling* and *composting*) is widely viewed as the next best strategy in waste management. Recycling involves sending used goods to facilities that extract and reprocess raw materials to manufacture new goods. Newspapers, white paper, cardboard, glass, metal cans, appliances, and some plastic containers have all become increasingly recyclable as new technologies have been developed and as markets for recycled materials have grown. Organic waste can be recovered through *composting*, or biological decomposition. Recycling is not a concept that humans invented; recall that all materials are recycled in ecosystems (• p. 180). Recycling is a fundamental feature of the way natural systems function.

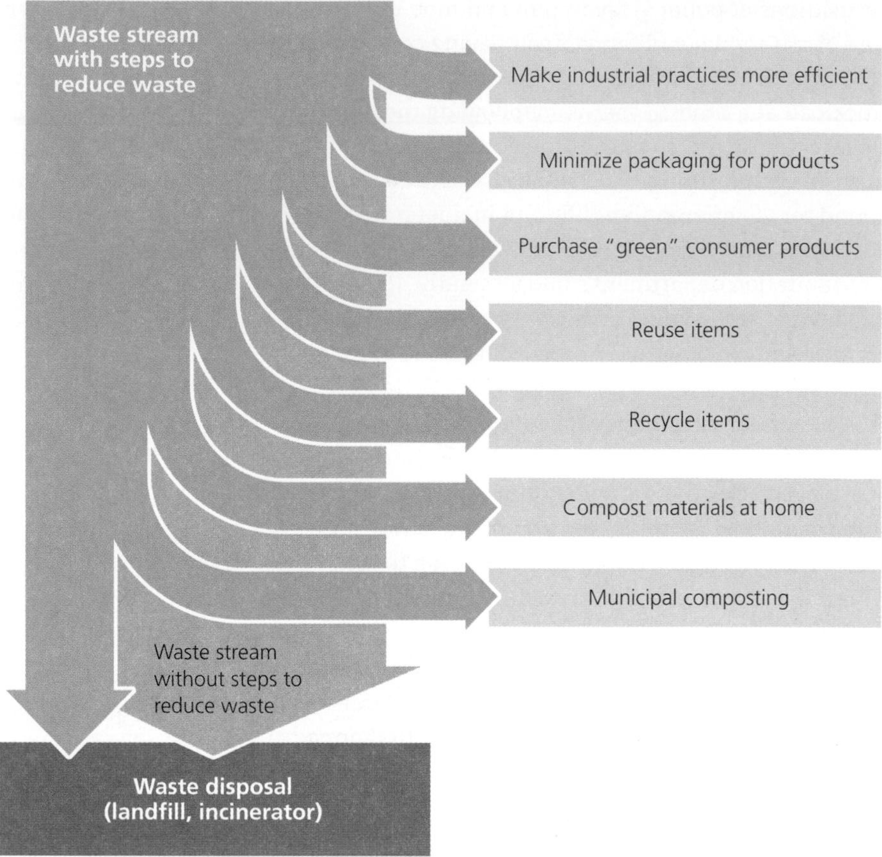


FIGURE 22.1 The most effective way to manage waste is to minimize the amount of material that enters the waste stream. To do this, manufacturers can increase efficiency, and consumers can buy "green" products that have minimal packaging or are produced in ways that minimize waste. Individuals can compost food scraps and yard waste at home and can reuse items rather than buying new ones. Many of us can recycle materials and compost yard waste through municipal recycling and composting programs. For all remaining waste, waste managers attempt to find disposal methods that minimize impact to human health and environmental quality.

Regardless of how effectively we reduce our waste stream, there will likely always be some waste left to dispose of. Disposal methods include burying waste in landfills and burning waste in incinerators. In this chapter we first examine how these approaches are used to manage municipal solid waste, and then we address industrial solid waste and hazardous waste.

## Municipal Solid Waste

Municipal solid waste is waste produced by consumers, public facilities, and small businesses. It is what we commonly refer to as "trash" or "garbage." Everything from paper to food scraps to roadside litter to old appliances and furniture is considered municipal solid waste.

### Patterns in the municipal solid waste stream vary from place to place

In the United States, paper, yard debris, food scraps, and plastics are the principal components of municipal solid waste, together accounting for 71% of the waste stream (**Figure 22.2**). Even after recycling, paper is the largest component of U.S. municipal solid waste. Patterns differ in developing countries; there, food scraps are often the primary contributor to solid waste, and paper makes up a smaller proportion.

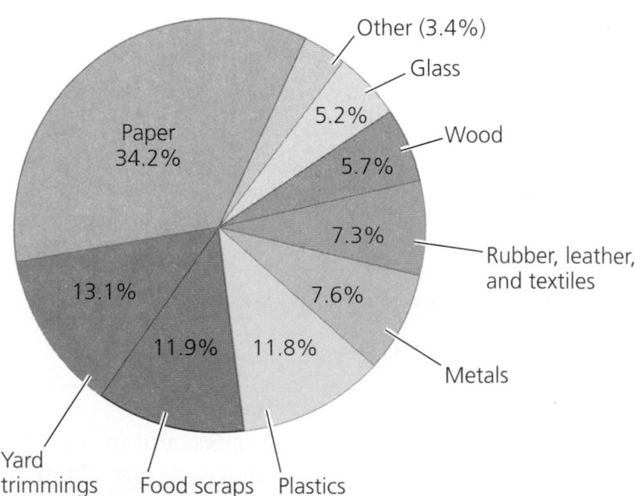


FIGURE 22.2 Paper products comprise the largest component of the municipal solid waste stream in the United States, followed by yard trimmings, food scraps, and plastics. In total, each U.S. citizen generates close to 1 ton of solid waste each year. Data are for 2005, from U.S. Environmental Protection Agency, 2006. *Municipal solid waste generation, recycling, and disposal in the United States: Facts and figures for 2005.* EPA-530-F-06-039, Washington, D.C.

Most municipal solid waste comes from packaging and nondurable goods (products meant to be discarded after a short period of use). In addition, consumers throw away old durable goods and outdated equipment as they purchase new products. As we acquire more goods, we generate more waste. In 2005, U.S. citizens produced 246 million tons of municipal solid waste, almost 1 ton per person. The average American generates over 2.0 kg (4.5 lb) of trash per day.

Following the United States in per capita solid waste generation are Canada, with 1.7 kg (3.75 lb) per day, and the Netherlands, with 1.4 kg (3 lb) per day. Among developed nations, Germany and Sweden produce the least waste per capita, generating just under 0.9 kg (2.0 lb) per day. Differences among nations result in part from differences in the cost of waste disposal; where disposal is expensive, people have incentive to waste less. The relative wastefulness of the U.S. lifestyle, with its excess packaging and reliance on nondurable goods, has caused critics to label the United States "the throwaway society."

In developing nations, people consume less and generate considerably less waste. One study (see "Interpreting Graphs and Data," • p. 655) found that people of high-income nations waste more than twice as much as people of low-income nations. However, wealthier nations also invest more in waste collection and disposal, so they are often better able to manage their waste proliferation and minimize impacts on human health and the environment.

## Waste generation is rising in all nations

In the United States since 1960, waste generation has increased (**Figure 22.3**) by 2.8 times, and per capita waste generation has risen by 69%. Plastics, which came into wide consumer use only after 1970, have accounted for the greatest relative increase in the waste stream during the last several decades.

The intensive consumption that has long characterized the United States and other wealthy nations is now increasing rapidly in developing nations. To some extent, this trend reflects rising material standards of living, but an increase in packaging is also to blame. Items made for temporary use and poor-quality goods designed to be inexpensive wear out and pile up quickly as trash, littering the landscapes of countries from Mexico to Kenya to Indonesia. Over the past three decades, per capita waste generation rates have more than doubled in Latin American nations and have increased more than fivefold in the Middle East. Like U.S. consumers in the "throwaway society," wealthy consumers in developing nations often discard items that can still be used. At many dumps and landfills in the developing world, in fact, poor people support themselves by selling items they scavenge (**Figure 22.4**).

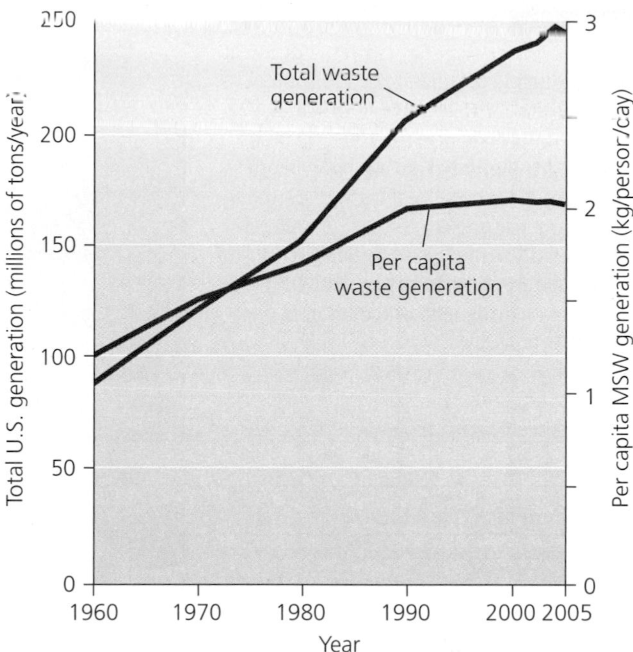


**FIGURE 22.3** U.S. waste generation has increased by 2.8 times since 1960, and U.S. per capita waste generation has risen by 69%. Per capita waste generation has leveled off in recent years largely because of recycling and source-reduction efforts. Data from U.S. Environmental Protection Agency, 2006. *Municipal solid waste generation, recycling, and disposal in the United States: Facts and figures for 2005.* EPA-530-F-06-039, Washington, D.C.

In many industrialized nations, per capita generation rates have leveled off or decreased in recent years. For instance, note in Figure 22.3 that per capita waste production in the United States has been essentially stable since 1990. This is due largely to the increased popularity of recycling, composting, reduction, and reuse. We will examine these nondisposal approaches to waste management shortly, but let's first assess how we dispose of waste.

## Open dumping of the past has given way to improved disposal methods

Historically, people dumped their garbage wherever it suited them. For example, until the mid-19th century, New York City's official method of garbage disposal was to dump it off piers into the East River. As population densities increased, municipalities took on the task of consolidating trash into open dumps at specified locations to keep other areas clean. To decrease the volume of trash, these dumps would be burned from time to time. Open dumping and burning still occur throughout much of the world.

FIGURE 22.4 Tens of thousands of people used to scavenge each day from the dump at Payatas, outside Manila in the Philippines, finding items for themselves and selling material to junk dealers for 100–200 pesos (U.S. $2–$4) per day. That so many people could support themselves this way testifies to the immense amount of usable material needlessly discarded by wealthier portions of the population. The dump was closed in 2000 after an avalanche of trash killed hundreds of people.

As population and consumption rose in developed nations, waste increased, and dumps grew larger. At the same time, expanding cities and suburbs forced more people into the vicinity of dumps and exposed them to the noxious smoke of dump burning. Reacting to opposition from residents living near dumps, and to a rising awareness of health and environmental threats posed by unregulated dumping and burning, many nations improved their methods of waste disposal. Most industrialized nations now bury waste in lined and covered landfills and burn waste in incineration facilities.

In the 1980s in the United States, waste generation increased while incineration was restricted, and recycling was neither economically feasible nor widely popular. As a result, landfill space became limited, and there was much talk of a solid waste "crisis." New York and other East Coast urban areas felt this situation most acutely; Fresh Kills Landfill accepted its all-time annual peak of trash, 29,000 tons, in 1986–1987. Since the late 1980s, however, recovery of materials for recycling has expanded, decreasing the pressure on landfills (**Figure 22.5**). As of 2005, U.S. waste managers were landfilling 54.3% of

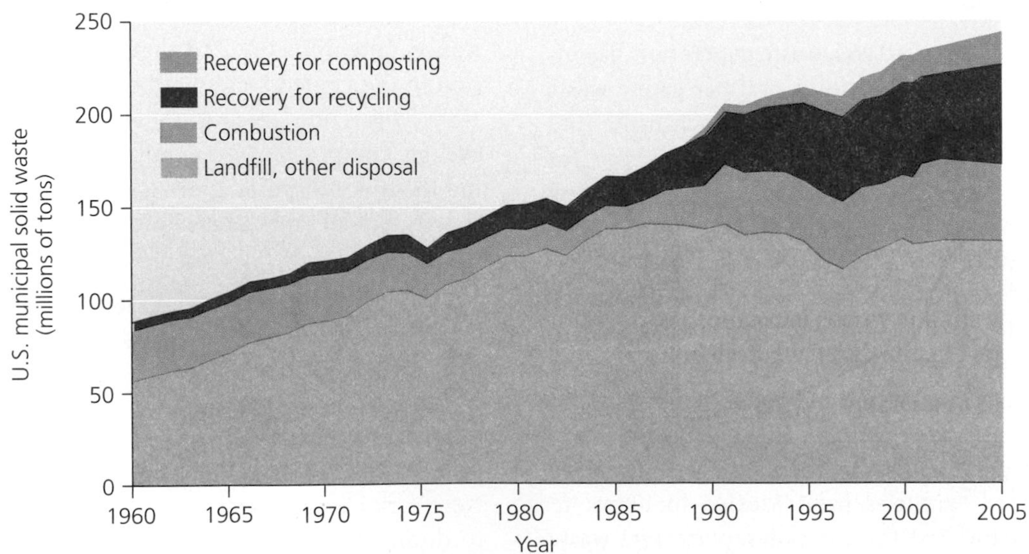


FIGURE 22.5 Since the 1980s, recycling and composting have grown in the United States, allowing a smaller proportion of waste to go to landfills. As of 2005, 54.3% of U.S. municipal solid waste was going to landfills and 13.6% to incinerators, whereas 32.1% was being recovered for composting and recycling. Go to GRAPHIt! at www.aw-bc.com/withgott or on the student CD-ROM. Data from U.S. Environmental Protection Agency, 2006. *Municipal solid waste generation, recycling, and disposal in the United States: Facts and figures for 2005*. EPA-530-F-06-039, Washington, D.C.

municipal solid waste, incinerating 13.6%, and recovering 32.1% for composting and recycling.

## Sanitary landfills are regulated by health and environmental guidelines

In modern **sanitary landfills**, waste is buried in the ground or piled up in large, carefully engineered mounds. In contrast to open dumps, sanitary landfills are designed to prevent waste from contaminating the environment and threatening public health (**Figure 22.6**). Although most municipal landfills in the United States are regulated locally or by the states, they must meet national standards set by the U.S. Environmental Protection Agency (EPA), under the federal **Resource Conservation and Recovery Act (RCRA)**, enacted in 1976 and amended in 1984.

In a sanitary landfill, waste is partially decomposed by bacteria and compresses under its own weight to take up less space. Waste is layered along with soil, a method that speeds decomposition, reduces odor, and reduces infestation by pests. Limited infiltration of rainwater allows for biodegradation by aerobic and anaerobic bacteria.

To protect against environmental contamination, U.S. regulations require that landfills be located away from wetlands and earthquake-prone faults and be at least 6 m (20 ft) above the water table. The bottoms and sides of sanitary landfills must be lined with heavy-duty plastic and 60 to 120 cm (2 to 4 ft) of impermeable clay to help prevent contaminants from seeping into aquifers. Sanitary landfills also have systems of pipes, collection ponds, and treatment facilities to collect and treat **leachate**, liquid that results when substances from the trash dissolve in water as rainwater percolates downward. Landfill managers are required to maintain leachate collection systems for 30 years after a landfill has closed. Regulations also require that area groundwater be monitored regularly for contamination.

After a landfill is closed, it is capped with an engineered cover that must be maintained. This cap consists of a hydraulic barrier of plastic that prevents water from seeping down and gas from seeping up; a gravel layer above the hydraulic barrier that drains water, lessening pressure on the hydraulic barrier; a soil barrier of at least 60 cm (24 in.) that stores water and protects the hydraulic layer from weather extremes; and a topsoil layer of at least 15 cm (6 in.) that encourages plant growth, helping to prevent erosion.

Although it was considered a model for advanced landfill technology at the time of its construction, the Fresh Kills Landfill predated most of the EPA guidelines. As a result, it caused some environmental contamination. However, engineers retrofitted the landfill with clay liners and a sophisticated leachate collection system. Three of the

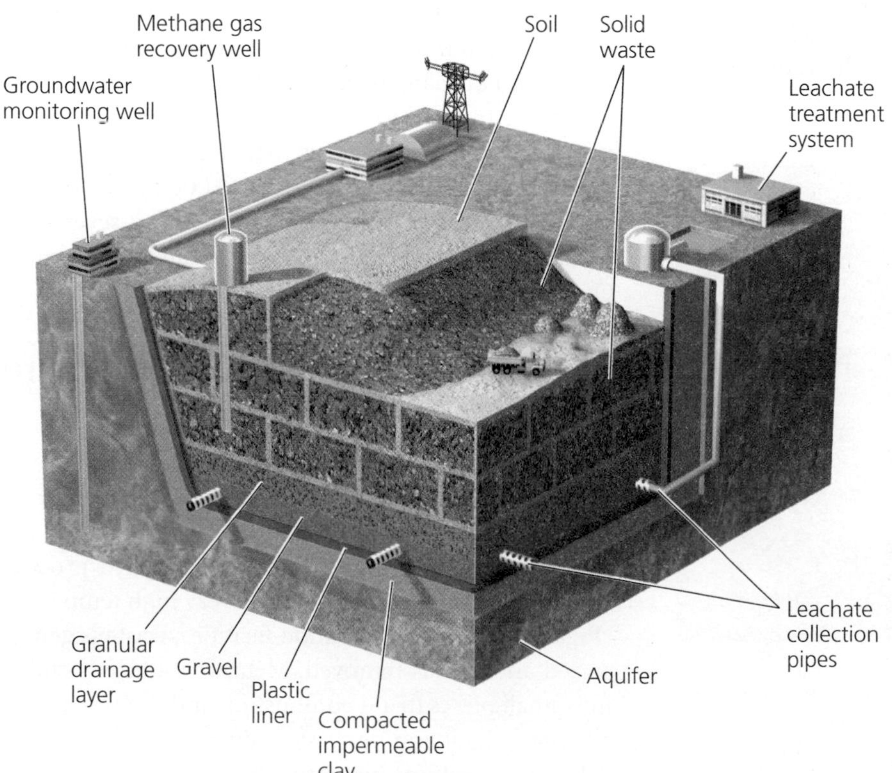


FIGURE 22.6 Sanitary landfills are engineered to prevent waste from contaminating soil and groundwater. Waste is laid in a large depression lined with plastic and impervious clay designed to prevent liquids from leaching out. Pipes of a leachate collection system draw out these liquids from the bottom of the landfill. Waste is layered along with soil until the depression is filled, and it continues to be built up until the landfill is capped. Landfill gas produced by anaerobic bacteria may be recovered, and waste managers monitor groundwater for contamination.

six mounds have been capped with a "final cover," and the remaining mounds will soon be capped. Indeed, the city's investment of several hundred million dollars in an existing landfill to bring environmental protection measures up to modern standards was unprecedented. Because these safeguards need to be maintained and monitored for 30 years after closure, designs for a public park at Fresh Kills have had to work around these constraints.

## Landfills can be transformed after closure

Today thousands of landfills lie abandoned. One reason is that waste managers have closed many smaller landfills and consolidated the trash stream into fewer, much larger, landfills. In 1988 the United States had nearly 8,000 landfills, but today it has fewer than 1,700.

Growing numbers of cities have converted closed landfills into public parks (**Figure 22.7**). The Fresh Kills redevelopment endeavor will be the world's largest landfill conversion project, but such efforts date back at least to 1938, when an ash landfill at Flushing Meadows, in Queens, was redeveloped for the 1939 World's Fair. The site subsequently hosted the United Nations and the 1964–1965 World's Fair. Designated a park in 1967, today the site hosts Shea Stadium, the Queens Museum of Art, the New York Hall of Science, and the Queens Botanical Garden, as well as playgrounds, wetlands, and festival events.

## Landfills have drawbacks

Despite improvements in liner technology and landfill siting, many experts believe that leachate will eventually escape even from well-lined landfills. Liners can be punctured, and leachate collection systems eventually cease to be maintained. Moreover, landfills are kept dry to reduce leachate, but the bacteria that break down material thrive in wet conditions. Dryness, therefore, slows waste decomposition. In fact, it is surprising how slowly some materials biodegrade when they are tightly compressed in a landfill. Innovative archaeological research has revealed that landfills often contain food that has not decomposed and 40-year-old newspapers that are still legible (see "The Science behind the Story," • pp. 638–639).

FIGURE 22.7 Old landfills, once properly capped, can serve other purposes. A number of them, such as Cesar Chavez Park in Berkeley, California, shown here, have been developed into areas for human recreation.

Another problem is finding suitable areas to locate landfills, because most communities do not want them nearby. This not-in-my-backyard (NIMBY) reaction (• p. 616) is one reason why New York decided to export its waste and why residents of states receiving that waste are increasingly protesting. As a result of the NIMBY syndrome, landfills are rarely sited in neighborhoods that are home to wealthy and educated people with the political clout to keep them out. Instead, they are disproportionately sited in poor and minority communities, as environmental justice advocates have frequently made clear.

The unwillingness of most communities to accept waste became apparent with the famed case of the "garbage barge." In Islip, New York, in 1987, the town's landfills were full, prompting town administrators to ship waste by barge to a methane production plant in North Carolina. Prior to the barge's arrival, however, it became known that the shipment was contaminated with 16 bags of medical waste, including syringes, hospital gowns, and diapers. Because of the medical waste, the methane plant rejected the entire load. The barge sat in a North Carolina harbor for 11 days before heading for Louisiana. However, Louisiana would not permit the barge to dock. The barge traveled toward Mexico, but the Mexican navy prevented it from entering that nation's waters. In the end, the barge traveled 9,700 km (6,000 mi) before eventually returning to New York, where, after several court battles, the waste was finally incinerated at a facility in Queens.

## Incinerating trash reduces pressure on landfills

Just as sanitary landfills are an improvement over open dumping, incineration in specially constructed facilities can be an improvement over open-air burning of trash. **Incineration**, or combustion, is a controlled process in which mixed garbage is burned at very high temperatures (**Figure 22.8**). At incineration facilities, waste is generally sorted and metals removed. Metal-free waste is chopped into small pieces to aid combustion and then is burned in a furnace. Incinerating waste reduces its weight by up to 75% and its volume by up to 90%.

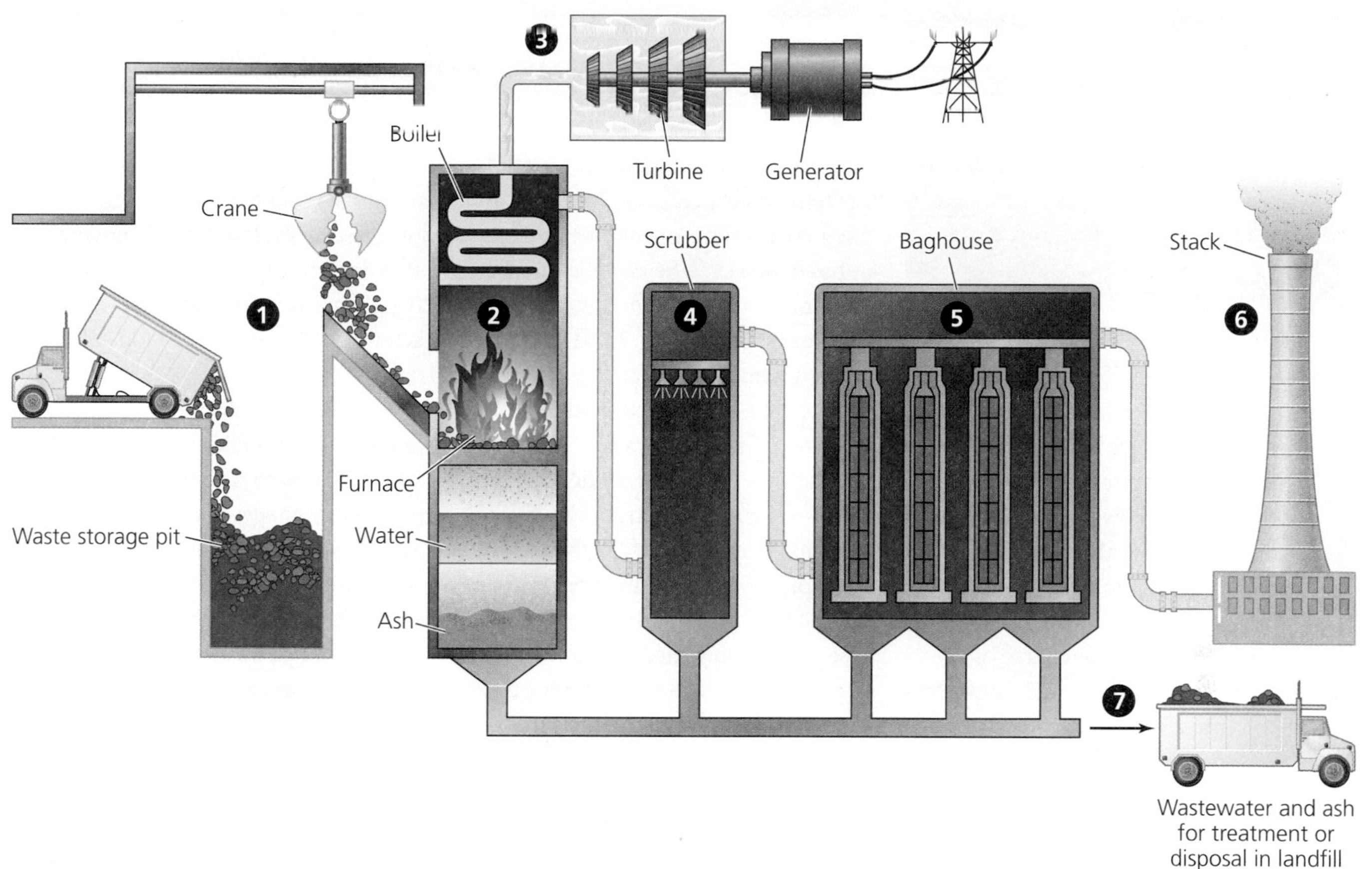


FIGURE 22.8 Incinerators reduce the volume of solid waste by burning it but may emit toxic compounds into the air. Many incinerators are waste-to-energy (WTE) facilities that use the heat of combustion to generate electricity. In a WTE facility, solid waste (1) is burned at extremely high temperatures (2), heating water, which turns to steam. The steam turns a turbine (3), which powers a generator to create electricity. In an incinerator outfitted with pollution-control technology, toxic gases produced by combustion are mitigated chemically by a scrubber (4), and airborne particulate matter is filtered physically in a baghouse (5) before air is emitted from the stack (6). Ash remaining from the combustion process is disposed of (7) in a landfill.

However, simply reducing the volume and weight does not rid trash of components that are toxic. The ash remaining after trash is incinerated therefore must be disposed of in hazardous waste landfills (• p. 650). Moreover, when trash is burned, hazardous chemicals—including dioxins, heavy metals, and PCBs (Chapter 14)—can be created and released into the atmosphere. Such releases caused a backlash against incineration from citizens concerned about health hazards. Most developed nations now regulate incinerator emissions, and some have banned incineration outright.

As a result of real and perceived health threats from incinerator emissions—and of community opposition to these plants—engineers have developed several technologies to mitigate emissions. *Scrubbers* chemically treat the gases produced in combustion to remove hazardous components and neutralize acidic gases, such as sulfur dioxide and hydrochloric acid, turning them into water and salt. Scrubbers generally do this either by spraying liquids formulated to neutralize the gases or by passing the gases through dry lime. A liquid-spraying scrubber was illustrated in Chapter 17 as an example of technology to mitigate air pollution (see Figure 17.14, • p. 485).

Particulate matter is physically removed from incinerator emissions in a system of huge filters known as a *baghouse.* These tiny particles, called fly ash, often contain some of the worst dioxin and heavy metal pollutants. In addition, burning garbage at especially high temperatures can destroy certain pollutants, such as PCBs. Even all these measures, however, do not fully eliminate toxic emissions.

## Weighing the Issues | Environmental Justice?

Do you know where your trash goes? Where is your landfill or incinerator located? Are the people who live closest to the facility wealthy, poor, or middle class? What race or ethnicity are they? Do you know whether the people of this neighborhood protested against the introduction of the landfill or incinerator?

THE SCIENCE BEHIND THE STORY

## Digging Garbage: The Archaeology of Solid Waste

*Fresh Kills Landfill, before its closure*

*Garbage* and *knowledge* are two words rarely put together. But when scientist William Rathje dons trash-flecked clothes and burrows into a city dump, he gleans valuable information about how modern Americans live.

By pulling tons of trash out of disposal sites over the course of decades, Rathje has turned dumpster-diving into a noteworthy field of scientific inquiry that he calls *garbology*. An archaeologist by training who has been called "the Indiana Jones of solid waste," Rathje has brought exacting archaeological techniques to the contents of trash cans.

As a professor at the University of Arizona in the early 1970s, Rathje wanted his students to learn a technique common among archaeologists—sorting through ancient trash mounds to understand the lives of past cultures. With few ancient civilizations or their trash close at hand, however, he arranged for his students to dig through their neighbors' garbage. In 1973, he gave that effort a name, "The Garbage Project," and began a methodical study of the contents of modern trash.

Rathje asked communities to divert some of their garbage trucks and bring trash to his study teams. With rakes and notebooks, the researchers sorted, weighed, itemized, and analyzed the refuse. They then visited the homes of the people who had generated the trash and asked residents about their shopping and consumption habits.

Then in 1987, amid growing debates about how quickly U.S. landfills were filling up, Rathje decided to see what was taking up space in them. The Garbage Project headed to landfills with a truck-mounted bucket auger—a large drill commonly employed by geologists and construction crews to handle everything from excavating soil samples to creating new water wells. Rathje and his researchers dug into landfills around North America, boring as far as 30 m (100 ft) down, often drilling approximately 15 to 20 garbage "wells" at each site, with each well yielding up to 25 tons of trash.

Once excavated, landfill contents were sorted, weighed, and identified. Rathje's teams sometimes froze the trash before they worked with it to make the garbage easier to separate and to limit odor and flies. Smaller bits of trash were put through sieves and sometimes washed with water to make them easier to label. Rathje has excavated at least 21 dumps, including sites in Tucson, San Francisco, Chicago, Phoenix, New York City, and Philadelphia, uncovering a host of interesting facts in the process:

- *Not much rot.* Trash doesn't decay much in closed landfills, Rathje has found. In the low-oxygen conditions inside most closed dumps, trash turns into a sort of time capsule. Rathje's teams have

### Many incinerators burn waste to create energy

Incineration was initially practiced simply to reduce the volume of waste, but today it often serves to generate electricity as well. Most North American incinerators today are **waste-to-energy (WTE)** facilities that use the heat produced by waste combustion to boil water, creating steam that drives electricity generation or that fuels heating systems. When burned, waste generates approximately 35% of the energy generated by burning coal. Over 100 WTE facilities are operating across the United States (mostly in the Northeast and South), with a total capacity to process nearly 100,000 tons of waste per day.

Revenues from power generation, however, are usually not enough to offset the considerable financial cost of building and running incinerators. Because it can take many years for a WTE facility to become profitable, many companies that build and operate these facilities require communities contracting with them to guarantee the facility a minimum amount of garbage. In a number of cases, such long-term commitments have interfered with communities' later efforts to reduce their waste through recycling and other waste-reduction strategies.

### Landfills can produce gas for energy

Combustion in WTE plants is not the only way to gain energy from waste. Deep inside landfills, bacteria decompose waste in an oxygen-deficient environment. This anaerobic decomposition produces *landfill gas*, a mix of gases that consists of roughly half methane (• pp. 95–96, 508). Landfill gas can be collected, processed, and used in the same way as natural gas (• pp. 549–551).

"Garbologist" William Rathje has pioneered the study of our culture through the waste we generate.

found whole hot dogs in most digs, intact pastries that are decades old, and grass clippings that are still green. Decades-old newspapers are legible and can be used to date layers of trash.

- *Paper rules.* Paper-based products make up more than 40% of most landfill content, and construction debris makes up about 20%. Newspapers are often a high-volume item, averaging about 14% of landfill space.
- *Plastic packaging no problem.* Rathje says plastic packaging is not the landfill problem many believe it to be. Plastic packaging makes up only about 4.5% of landfill content, and that figure has not increased substantially since the 1970s, Rathje reported in 1997. Fast-food packaging, polystyrene foam, and disposable diapers also aren't a major problem, making up only about 3% of landfill content. If all plastic packaging were to be replaced by containers made of glass, paper, steel, or similar materials, Rathje maintains, the packaging discarded by U.S. households would more than double.
- *Poison in small bottles.* Toxic waste comes in all sizes. If nail polish were sold in 55-gallon drums, its chemical composition would make it illegal to throw out in a regular dump. Nail polish, however, is tossed in small bottles. In 1991, after studying Tucson dumps, Rathje calculated that about 350,000 bottles were getting thrown away each year. Luckily, however, the potentially toxic ingredients in nail polish don't always spread far, he found. Paper, diapers, and other nontoxic garbage often absorb toxic materials in landfills and keep the poisons from leaching out.

Through garbology, Rathje has gleaned unique insights into how we can change our often-wasteful habits. Now a consulting professor at Stanford University, Rathje has emerged as a leading expert on how to reduce waste.

Today more than 330 operational projects collect landfill gas in the United States. Other countries take advantage of this resource as well. In Chile, four facilities in Valparaiso and Santiago supply 40% of the region's demand for natural gas. At Fresh Kills, landfill gas collection wells pull gas upward through a network of pipes by vacuum pressure. Landfill gas collected from Fresh Kills should soon provide enough energy for 25,000 homes. Where gas is not collected for commercial use, it is burned off in flares to reduce odors and greenhouse emissions.

## Reducing waste is a better option

Reducing the amount of material entering the waste stream avoids costs of disposal and recycling, helps conserve resources, minimizes pollution, and can often save consumers and businesses money. Preventing waste generation in this way is known as **source reduction**.

Much of our waste stream consists of materials used to package goods. Packaging serves worthwhile purposes—preserving freshness, preventing breakage, protecting against tampering, and providing information—but much packaging is extraneous. Consumers can give manufacturers incentive to reduce packaging by choosing minimally packaged goods, buying unwrapped fruit and vegetables, and buying food in bulk. In addition, manufacturers can use packaging that is more recyclable. They can also reduce the size or weight of goods and materials, as they already have with many items, such as aluminum cans, plastic soft drink bottles, and personal computers.

### Weighing THE Issues | Reducing Packaging: Is It a Wrap?

Reducing packaging cuts down on the waste stream, but how, when, and how much should we reduce? Packaging can serve very worthwhile purposes, such as safeguarding consumer health and safety. Can you think of three products for which you would *not* want to see less packaging? Why? Can you name three products for which packaging could easily be reduced without ill effect to the consumer? Would you be any more or less likely to buy these products if they had less packaging?

Some governments have recently taken aim at a major source of waste and litter—plastic grocery bags. These lightweight polyethylene bags can persist for centuries in the environment, choking and entangling wildlife and littering the landscape—yet Americans discard 100 billion of them each year. Several nations have now banned their use. When Ireland began taxing these bags, their use dropped 90%. The Ikea company began charging for them and saw similar drops in usage. San Francisco has now become the first U.S. city to begin banning nonbiodegradable plastic bags.

Increasing the longevity of goods also helps reduce waste. Consumers generally choose goods that last longer, all else being equal. To maximize sales, however, companies often produce short-lived goods that need to be replaced frequently. Thus, increasing the longevity of goods is largely up to the consumer. If demand is great enough, manufacturers will respond.

## Reuse is one main strategy for waste reduction

To reduce waste, you can save items to use again or substitute disposable goods with durable ones. Habits as simple as bringing your own coffee cup to coffee shops or bringing sturdy reusable cloth bags to the grocery store can, over time, have substantial impact. You can also donate unwanted items and shop for used items yourself at yard sales and resale centers. Over 6,000 reuse centers exist in the United States, including stores run by organizations that resell donated items, such as Goodwill Industries and the Salvation Army. Besides doing good for the environment, reusing items is often economically advantageous. Used items are quite often every bit as functional as new ones, and much cheaper. Table 22.1 presents a sampling of actions that we all can take to reduce the waste we generate.

**TABLE 22.1 Some Everyday Things You Can Do to Reduce and Reuse**

- Donate used items to charity
- Reuse boxes, paper, plastic wrap, plastic containers, aluminum foil, bags, wrapping paper, fabric, packing material, etc.
- Rent or borrow items instead of buying them, when possible . . . and lend your items to friends
- Buy groceries in bulk
- Decline bags at stores when you don't need them
- Bring reusable cloth bags shopping
- Make double-sided photocopies
- Bring your own coffee cup to coffee shops
- Pay a bit extra for durable, long-lasting, reusable goods rather than disposable ones
- Buy rechargeable batteries
- Select goods with less packaging
- Compost kitchen and yard wastes in a compost bin or worm bin (often available from your community or waste hauler)
- Buy clothing and other items at resale stores and garage sales
- Use cloth napkins and rags rather than paper napkins and towels
- Write to companies to tell them what you think about their packaging and products
- When solid waste policy is being debated, let your government representatives know your thoughts
- Support organizations that promote waste reduction

## Composting recovers organic waste

**Composting** is the conversion of organic waste into mulch or humus (• pp. 238–239) through natural biological processes of decomposition. The compost can then be used to enrich soil. Householders can place waste in compost piles, underground pits, or specially constructed containers. As wastes are added, heat from microbial action builds in the interior, and decomposition proceeds. Banana peels, coffee grounds, grass clippings, autumn leaves, and countless other organic items can be converted into rich, high-quality compost through the actions of earthworms, bacteria, soil mites, sow bugs, and other detritivores and decomposers. Home composting is a prime example of how we can live more sustainably by mimicking natural cycles and incorporating them into our daily lives.

Municipal composting programs—3,800 across the United States at last count—divert food and yard waste from the waste stream to central composting facilities, where they decompose into mulch that community residents can use for gardens and landscaping. Nearly half of U.S. states now ban yard waste from the municipal waste stream, helping accelerate the drive toward composting. Approximately one-fifth of the U.S. waste stream is made up of materials that can easily be composted. Composting reduces landfill waste, enriches soil and helps it resist erosion, encourages soil biodiversity, makes for healthier plants and more pleasing gardens, and reduces the need for chemical fertilizers.

**FIGURE 22.9** The familiar recycling symbol consists of three arrows to represent the three components of a sustainable recycling strategy: collection and processing of recyclable materials, use of the materials in making new products, and consumer purchase of these products.

## Recycling consists of three steps

Recycling, too, offers many benefits. **Recycling** consists of collecting materials that can be broken down and reprocessed to manufacture new items. Recycling diverted 58 million tons of materials away from incinerators and landfills in the United States in 2005.

The recycling loop contains three basic steps (**Figure 22.9**). The first step is collecting and processing used recyclable goods and materials. Communities may designate locations where residents can drop off recyclables or receive money for them. Many of these have now been replaced by the more convenient option of curbside recycling, in which trucks pick up recyclable items in front of houses, usually in conjunction with municipal trash pickup. Curbside recycling has grown rapidly, and its convenience has helped boost recycling rates. About 9,000 curbside recycling programs across all 50 U.S. states now serve nearly half of all Americans.

Items collected are taken to **materials recovery facilities (MRFs)**, where workers and machines sort items, using automated processes including magnetic pulleys, optical sensors, water currents, and air classifiers that separate items by weight and size. The facilities clean the materials, shred them, and prepare them for reprocessing.

Once readied, these materials are used in manufacturing new goods. Newspapers and many other paper products use recycled paper, many glass and metal containers are now made from recycled materials, and some plastic containers are of recycled origin. Some large objects, such as benches and bridges in city parks, are now made from recycled plastics, and glass is sometimes mixed with asphalt (creating "glassphalt") for paving roads and paths. The pages in this textbook are made from recycled paper that is up to 20% post-consumer waste.

If the recycling loop is to function, consumers and businesses must complete the third step in the cycle by purchasing products made from recycled materials. Buying recycled goods provides economic incentive for industries to recycle materials and for new recycling facilities to open or existing ones to expand. In this arena, individual consumers have power to encourage environmentally friendly options through the free market. Many businesses now advertise their use of recycled materials, a widespread instance of ecolabeling (• p. 49). As markets for products made with recycled materials expand, prices continue to fall.

## Recycling has grown rapidly and can expand further

The thousands of curbside recycling programs and the 500 MRFs in operation today have sprung up only in the last 20 years. Recycling in the United States has risen from 6.4% of the waste stream in 1960 to 23.8% in 2005 (and 32.1% if you include composting), according to EPA data (**Figure 22.10**). The EPA calls the growth of

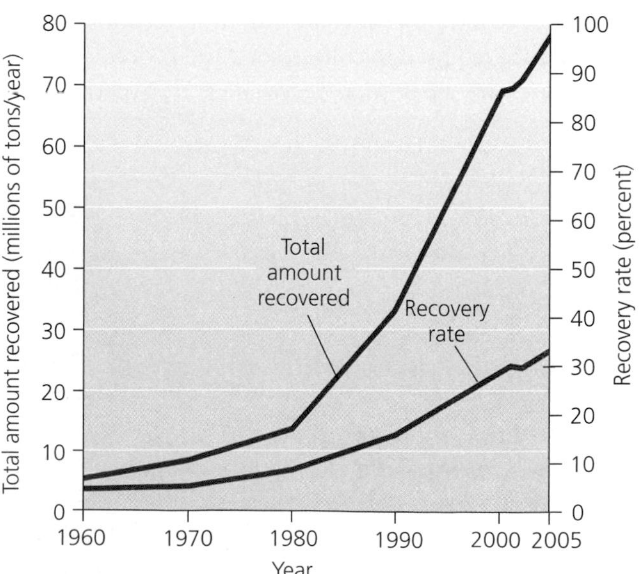


**FIGURE 22.10** Recovery has risen sharply in the United States over the past 45 years. Today nearly 80 million tons of material is recovered (58 million tons through recycling and 21 million tons through municipal composting), comprising more than 32% of the waste stream. Data from U.S. EPA, 2006.

**TABLE 22.2 Recovery Rates for Various Materials in the United States**

| Material | Percentage that is recycled or composted |
|---|---|
| Major appliances | 67 |
| Steel packaging | 63 |
| Yard trimmings | 62 |
| Paper and paperboard | 50 |
| Aluminum cans | 45 |
| Plastic soft drink bottles | 34 |
| Glass containers | 22 |
| Total plastics | 6 |

Data are for 2005, the most recent year available, from U.S. Environmental Protection Agency.

recycling "one of the best environmental success stories of the late 20th century."

Recycling rates vary greatly from one product or material type to another and from one location to another. Rates for different types of materials and products range from nearly zero to almost 100% (Table 22.2). Recycling rates among U.S. states also vary greatly, from 2% to 46% (**Figure 22.11**).

Recycling's growth has been propelled in part by economic forces as established businesses see opportunities to save money and as entrepreneurs see opportunities to start new businesses. It has also been driven by the desire of municipalities to reduce waste and by the satisfaction people take in recycling. These two forces have driven recycling's rise even though it has often not been financially profitable. In fact, many of the increasingly popular municipal recycling programs are run at an economic loss. The expense required to collect, sort, and process recycled goods is often more than recyclables are worth in the market. Furthermore, the more people recycle, the more glass, paper, and plastic is available to manufacturers for purchase, driving down prices.

Recycling advocates, however, point out that market prices do not take into account external costs (• p. 42)—in particular, the environmental and health impacts of *not* recycling. For instance, it has been estimated that globally, recycling saves enough energy to power 6 million households per year. And recycling aluminum cans saves 95% of the energy required to make the same amount of aluminum from mined virgin bauxite, its source material.

As more manufacturers use recycled products and as more technologies and methods are developed to use recycled materials in new ways, markets should continue to expand, and new business opportunities may arise. We are still at an early stage in the shift from an economy that moves linearly from raw materials to products to waste, to an economy that moves circularly, using waste products as raw materials for new manufacturing processes. The steps we have taken in recycling so far are central to this transition, which many analysts view as key to building a sustainable economy.

### Weighing THE Issues | Costs of Recycling and Not Recycling

Should recycling programs be subsidized by governments even if they are run at an economic loss? What types of external costs—costs not reflected in market prices—do you think would be involved in not recycling, say, aluminum cans? Do you feel these costs justify sponsoring recycling programs even when they are not financially self-supporting? Why or why not?

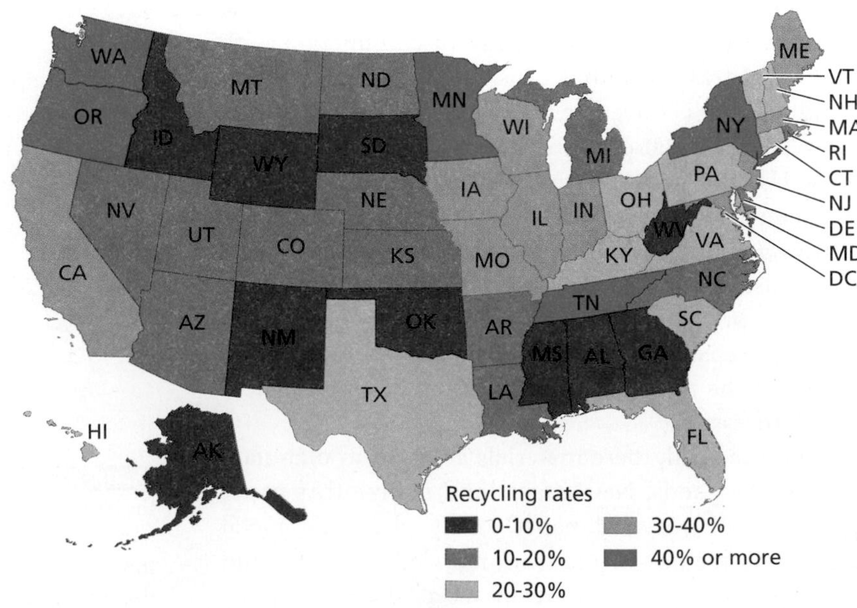


**FIGURE 22.11** U.S. states vary greatly in the rates at which their citizens recycle. Data are for 2004 (with earlier data for several states), from Simmons, P., et al. 2006. The state of garbage in America. *BioCycle* 47: 26.

## Financial incentives can help address waste

Waste managers have employed economic incentives to reduce the waste stream. The "pay-as-you-throw" approach to garbage collection uses a financial incentive to influence consumer behavior. In these programs, municipalities charge residents for home trash pickup according to the amount of trash they put out. The less waste the household generates, the less the resident has to pay. Over 7,000 of these programs now exist in the United States, serving more than one out of every four communities.

"Bottle bills" represent another approach that hinges on financial incentive. Eleven U.S. states have these laws, which allow consumers to return bottles and cans to stores after use and receive a refund—generally 5 cents per bottle or can. The first bottle bills were passed in the 1970s to cut down on litter, but they have also served to decrease the waste stream. In states where they have been enacted, these laws have proved profoundly effective and resoundingly popular; they are recognized as among the most successful state legislation of recent decades (**Figure 22.12**). States with bottle bills have reported that their beverage container litter has decreased by 69–84%, their total litter has decreased by 30–64%, and their per capita container recycling rate has risen 2.6-fold.

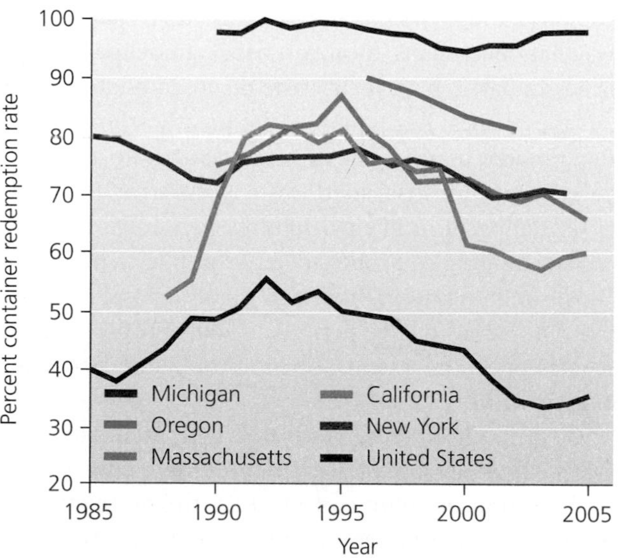


FIGURE 22.12 Data suggest that bottle bills increase recycling rates and that higher redemption amounts boost recycling rates further. The five states with bottle bills shown in this graph all have higher recycling rates than the United States as a whole. Michigan, the only state with a 10-cent deposit, has the highest recycling rate of all. Data from Container Recycling Institute, Arlington, VA, 2007.

As of 2007, 5 of the 11 bottle bill states were seeking to expand their programs, and six other states were considering establishing programs. It is a testament to the lobbying influence of the beverage industries and grocery retailers, which have traditionally opposed passage of bottle bills, that more states do not have such legislation.

States with bottle bills now face two challenges. One is to amend these laws to include new kinds of containers. In New York State, where polls show 70% of the public favors expanding their state's law to include more types of containers, the legislature in 2007 considered a "bigger better bottle bill" that would add plastic containers for water, fruit juice, and sports drinks. The version of the bill being pushed by Governor Eliot Spitzer would funnel the tens of millions of dollars in unclaimed deposit money to the state to fund environmental protection programs, but so far the legislature has failed to pass the bill.

The second challenge for bottle-bill states is to adjust refunds for inflation. In the 36 years since Oregon passed the nation's first bottle bill, the value of a nickel has dropped such that today, the refund would need to be 26 cents to reflect the refund's original intended value. Proponents argue that increasing refund amounts will raise return rates, and available data support this view (see Figure 22.12).

## One Canadian city showcases reduction and recycling

Edmonton, Alberta, has created one of the world's most advanced waste management programs. As recently as 1998, fully 85% of the city's waste was being landfilled, and space was running out. Today, just 35% goes to the new sanitary landfill, whereas 15% is recycled, and an impressive 50% is composted. Edmonton's citizens are proud of the program, and 81% of them participate in its curbside recycling program.

When Edmonton's residents put out their trash, city trucks take it to their new co-composting plant—at the size of eight football fields, the largest in North America (**Figure 22.13a**). The waste is dumped on the floor of the facility, and large items, such as furniture, are removed and landfilled. The bulk of the waste is mixed with dried sewage sludge for 1–2 days in five large rotating drums, each the length of six buses. The resulting mix travels on a conveyor to a screen that removes nonbiodegradable items. It is aerated for several weeks in the largest stainless steel building in North America (**Figure 22.13b**). The mix is then passed through a finer screen and finally is left outside for 4–6 months. The resulting compost—80,000 tons annually—is made available to area farmers and residents. The facility even filters the air it emits with a

**(a) Composting facility, Edmonton, Alberta**

**(b) Aeration building, Edmonton composting facility**

FIGURE 22.13 Edmonton, Alberta, boasts one of North America's most successful waste management programs. Edmonton's gigantic composting facility (**a**) is the size of eight football fields. Inside the aeration building (**b**), which is the size of 14 professional hockey rinks, mixtures of solid waste and sewage sludge are exposed to oxygen and composted for 14–21 days.

1-m (3.3-ft) layer of compost, bark, and wood chips, which eliminates the release of unpleasant odors into the community.

Edmonton's program also includes a state-of-the-art MRF that handles 30,000–40,000 tons of waste annually, a leachate treatment plant, a research center, public education programs, and a wetland and landfill revegetation program. In addition, 100 pipes collect enough landfill gas to power 4,000 homes, bringing thousands of dollars to the city and helping power the new waste management center. Five area businesses reprocess the city's recycled items. Newsprint and magazines are turned into new newsprint and cellulose insulation, and cardboard and paper are converted into building paper and shingles. Household metal is made into rebar and blades for tractors and graders, and recycled glass is used for reflective paint and signs.

# Industrial Solid Waste

Each year, U.S. industrial facilities generate about 7.6 billion tons of waste, according to the EPA, about 97% of which is wastewater. Thus, very roughly, 228 million or so tons of solid waste are generated by 60,000 facilities each year—an amount about equal to that of municipal solid waste. In the United States, industrial solid waste is defined as solid waste that is considered neither municipal solid waste nor hazardous waste under the Resource Conservation and Recovery Act.

Whereas the federal government regulates municipal solid waste, state or local governments regulate industrial solid waste (although with federal guidance). Industrial waste includes waste from factories, mining activities, agriculture, petroleum extraction, and more. Waste is generated at various points along the process from raw materials extraction to manufacturing to sale and distribution (**Figure 22.14**).

## Regulation and economics each influence industrial waste generation

Most methods and strategies of waste disposal, reduction, and recycling by industry are similar to those for municipal solid waste. For instance, businesses that manage their own waste on site most often dispose of it in landfills, and companies must design and manage their landfills in ways that meet state, local, or tribal guidelines. Other businesses pay to have their waste disposed of at municipal disposal sites. Regulation varies greatly from state to state and area to area, but in most cases, state and local regulation of industrial solid waste is less strict than federal regulation of municipal solid waste. This is one reason that industry continues to suffer public criticism for its waste and pollution impacts. In many areas, industries are not required to have permits, install landfill liners or leachate collection systems, or monitor groundwater for contamination.

The amount of waste generated by a manufacturing process is one measure of its efficiency; the less waste produced per unit or volume of product, the more efficient that process is, from a physical standpoint. However, physical efficiency is not always equivalent to economic efficiency. Often it is cheaper for industry to manufacture its products or perform its services quickly but messily. That is, it can be cheaper to generate waste than to avoid generating waste. In such cases, economic efficiency is maximized, but physical efficiency is not. The frequent

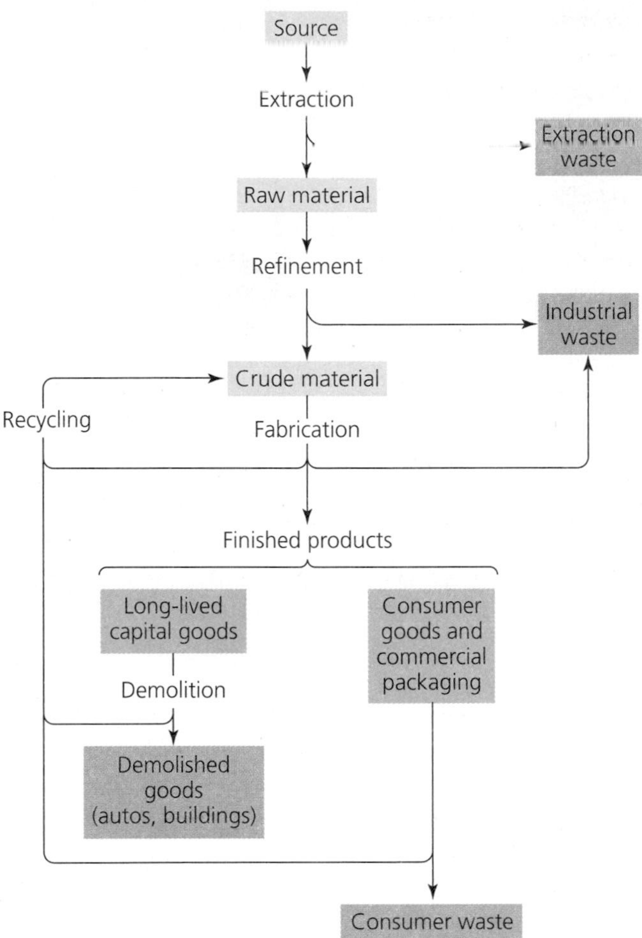


FIGURE 22.14 Industrial and municipal waste is generated at a number of stages throughout the life cycles of products. Waste is first generated when raw materials needed for production are extracted. Further industrial waste is produced as raw materials are processed and as products are manufactured. Waste results from the demolition or disposal of products once they are used by businesses and individuals. At each stage, there often are opportunities for efficiency improvements, waste reduction, or recycling.

mismatch between these two types of efficiency is a major reason why the output of industrial waste is so great.

Rising costs of waste disposal, however, enhance the financial incentive to decrease waste and increase physical efficiency. Once either government or the market makes the physically efficient use of raw materials also economically efficient, businesses have financial incentives to reduce their own waste.

## Industrial ecology seeks to make industry more sustainable

To reduce waste, growing numbers of industries today are experimenting with industrial ecology. A holistic approach that integrates principles from engineering, chemistry, ecology, and economics, **industrial ecology** seeks to redesign industrial systems to reduce resource inputs and to minimize physical inefficiency while maximizing economic efficiency. Industrial ecologists would reshape industry so that nearly everything produced in a manufacturing process is used, either within that process or in a different one.

The larger idea behind industrial ecology is that industrial systems should function more like ecological systems, in which almost everything produced is used by some organism, with very little being wasted. This principle brings industry closer to the ideal of ecological economists, in which human economies attain sustainability by functioning in a circular fashion rather than a linear one (• pp. 45, 671–673).

Industrial ecologists pursue their goals in several ways. For one, they examine the entire life cycle of a given product—from its origins in raw materials, through its manufacturing, to its use, and finally its disposal—and look for ways to make the process more ecologically efficient. This strategy is called **life-cycle analysis**.

Industrial ecologists also try to identify how waste products from one manufacturing process can be used as raw materials for a different process. For instance, used plastic beverage containers cannot be refilled because of the potential for contamination, but they can be shredded and reprocessed to make other plastic items, such as benches, tables, and decks. In addition, industrial ecologists examine industrial processes with an eye toward eliminating environmentally harmful products and materials. Finally, they study the flow of materials through industrial systems to look for ways to create products that are more durable, recyclable, or reusable. Goods that are currently thrown away when they become obsolete, such as computers, automobiles, and some appliances, could be designed to be more easily disassembled and their component parts reused or recycled.

## Businesses are adopting industrial ecology

Businesses are taking advantage of the insights of industrial ecology to reduce waste and lessen their impact on health and the environment while saving money. For example, American Airlines switched from hazardous to nonhazardous materials in its Chicago facility, decreasing its need to secure permits from the EPA. The company used over 50,000 reusable plastic containers to ship goods, reducing packaging waste by 90%. Its Dallas–Fort Worth headquarters recycled enough aluminum cans and white paper in 5 years to save $205,000 and recycled 3,000 broken baggage containers into lawn furniture. A program to gather suggestions from employees brought over 700 ideas to reduce

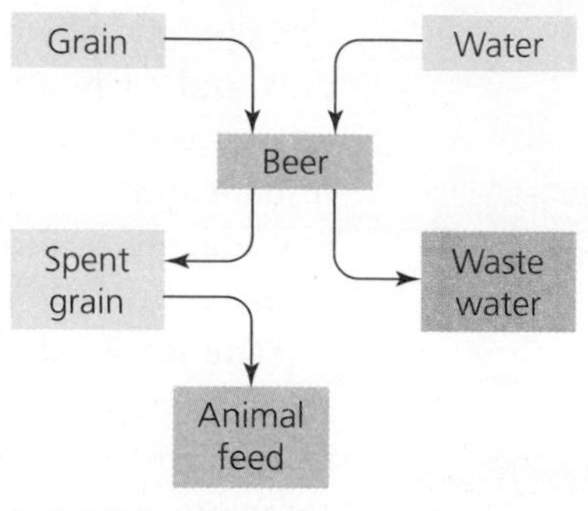


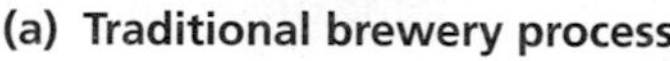
(a) Traditional brewery process

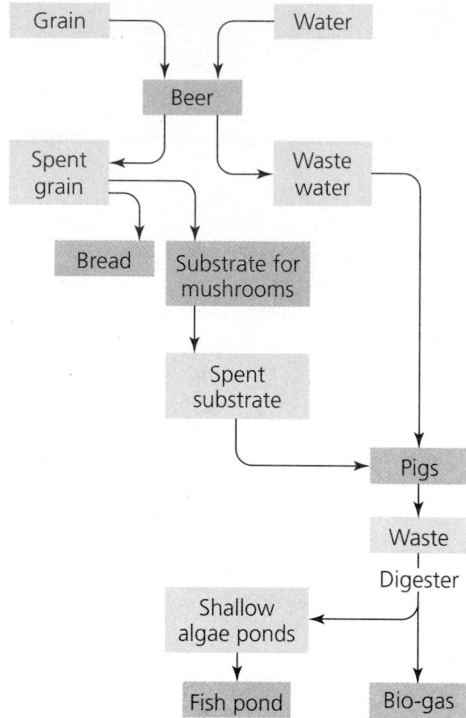

(b) ZERI brewery process

FIGURE 22.15 Traditional breweries (**a**) produce only beer while generating much waste, some of which goes toward animal feed. ZERI-sponsored breweries (**b**) use their waste grain to make bread and to farm mushrooms. Waste from the mushroom farming, along with brewery wastewater, goes to feed pigs. The pigs' waste is digested in containers that capture natural gas and collect nutrients used to nourish algae for growing fish in fish farms. The brewer derives income from bread, mushrooms, pigs, gas, and fish, as well as beer.

waste—and 15 of these ideas saved the company over $8 million in the first year of implementation.

Few businesses have taken industrial ecology to heart as much as the carpet tile company Interface, which founder Ray Anderson set on the road to sustainability over a decade ago (• p. 51). Interface asks customers to return used tiles for recycling and for reuse as backing for new carpet. It modified its tile design and its production methods to reduce waste. It adapted its boilers to use landfill gas for its energy needs. Through such steps, Anderson's company has cut its waste generation by 80%, its fossil fuel use by 45%, and its water use by 70%—all while saving $30 million per year, holding prices steady for its customers, and raising profits by 49%.

The Swiss Zero Emissions Research and Initiatives (ZERI) Foundation sponsors dozens of innovative projects worldwide that attempt to create goods and services without generating waste. One example involves breweries, currently being pursued in Canada, Sweden, Japan, and Namibia (**Figure 22.15**). Brewers in these projects take waste from the beer-brewing process and use it to fuel other processes. As a result, the brewer can make money from bread, mushrooms, pigs, gas, and fish, as well as beer, all while producing little waste. Although most ZERI projects are not fully closed-loop systems, they attempt to approach this ideal. In so doing, they cut down on waste while increasing output and income, and often they generate new jobs as well.

For businesses, governments, and individuals alike, there are plenty of ways to reduce waste and mitigate the impacts of our waste generation (see "Causes and Consequences," • p. 647)—and quite often, doing so brings economic benefits. This is true both for solid waste and for hazardous waste.

# Hazardous Waste

Hazardous wastes are diverse in their chemical composition and may be liquid, solid, or gaseous. By EPA definition, **hazardous waste** is waste that is one of the following:

- *Ignitable.* Substances that easily catch fire (for example, natural gas or alcohol).
- *Corrosive.* Substances that corrode metals in storage tanks or equipment.
- *Reactive.* Substances that are chemically unstable and readily react with other compounds, often explosively or by producing noxious fumes.
- *Toxic.* Substances that harm human health when they are inhaled, are ingested, or contact human skin.

Materials with these characteristics can harm human health and environmental quality. Flammable and explosive materials can cause ecological damage and atmospheric pollution. For instance, fires at large tire dumps in California's Central Valley have caused air pollution and highway closures. Toxic wastes in lakes and rivers have caused fish die-offs and closed important domestic fisheries, such as those in Chesapeake Bay.

## CAUSES AND CONSEQUENCES

Until we can achieve truly sustainable closed-loop systems, our activities will continue to generate large amounts of municipal solid waste, industrial solid waste, and hazardous waste. **Excessive waste generation** results from several causes and exerts various impacts on human health and ecological systems. Fortunately, there are many steps we can take to reduce our generation of waste.

Write in two causes of excessive waste generation in the spaces provided. Then write in two consequences (impacts on the environment, human health, or quality of life) that result. Finally, offer two solutions to this issue and its consequences. One cause, one consequence, and one solution have been filled in for you, providing examples.

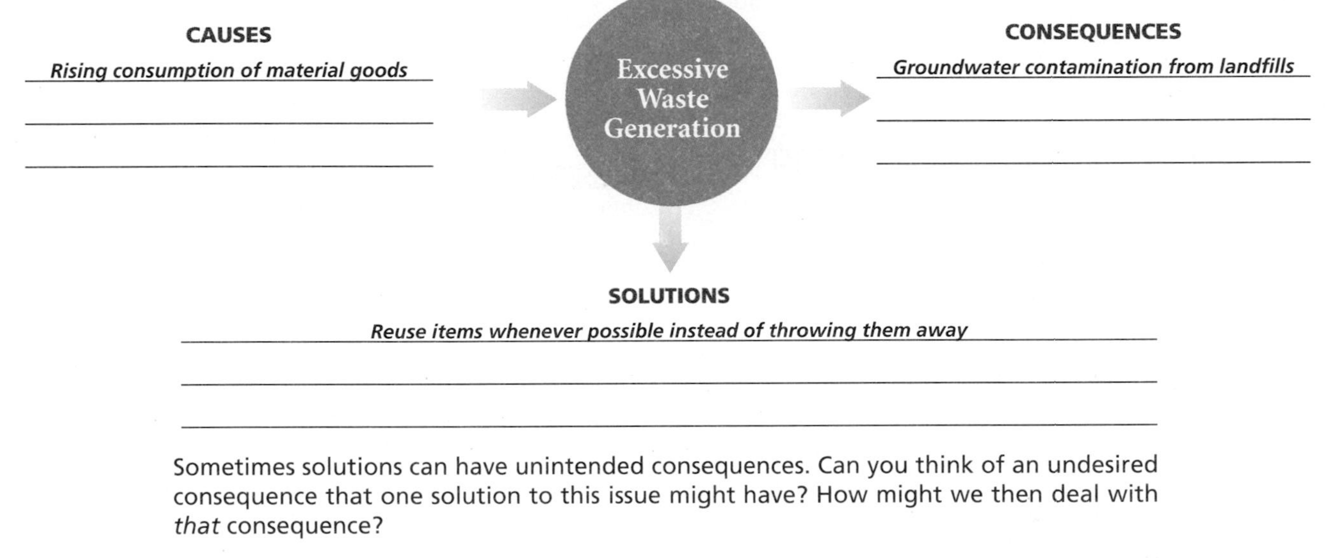


Sometimes solutions can have unintended consequences. Can you think of an undesired consequence that one solution to this issue might have? How might we then deal with *that* consequence?

## Hazardous wastes have diverse sources

Industry, mining, households, small businesses, agriculture, utilities, and building demolition all create hazardous waste. Industry produces the largest amounts of hazardous waste, but in most developed nations industrial waste generation and disposal is highly regulated. This regulation has reduced the amount of hazardous waste entering the environment from industrial activities. As a result, households currently are the largest source of unregulated hazardous waste.

Household hazardous waste includes a wide range of items, such as paints, batteries, oils, solvents, cleaning agents, lubricants, and pesticides. U.S. citizens generate 1.6 million tons of household hazardous waste annually, and the average home contains close to 45 kg (100 lb) of it in sheds, basements, closets, and garages.

Although many hazardous substances become less hazardous over time as they degrade chemically, two classes of chemicals are particularly hazardous because their toxicity persists over time: organic compounds and heavy metals.

## Organic compounds and heavy metals can be hazardous

In our day-to-day lives, we rely on the capacity of synthetic organic compounds and petroleum-derived compounds to resist bacterial, fungal, and insect activity. Items such as plastic containers, rubber tires, pesticides, solvents, and wood preservatives are useful to us precisely because they resist decomposition. We use these substances to protect our buildings from decay, kill pests that attack crops, and keep stored goods intact. However, the resistance of these compounds to decay is a double-edged sword, for it also makes them persistent pollutants. Many synthetic organic compounds are toxic because they can be readily absorbed through the skin of humans and other animals and can act as mutagens, carcinogens, teratogens, and endocrine disruptors (• pp. 389–390).

Heavy metals such as lead, chromium, mercury, arsenic, cadmium, tin, and copper are used widely in industry for wiring, electronics, metal plating, metal fabrication, pigments, and dyes. Heavy metals enter the environment when paints, electronic devices, batteries, and other materials are disposed of improperly. Lead from fishing weights and from hunters' lead shot has accumulated in many rivers, lakes, and forests. In older homes, lead from pipes contaminates drinking water, and lead paint remains a problem, especially for infants. Heavy metals that are fat soluble and break down slowly are prone to bioaccumulate (• pp. 395–397). In California's Coast Range, for instance, mercury washed downstream from abandoned mercury mines enters low-elevation lakes and rivers, is consumed by bacteria and invertebrates, and accumulates in increasingly larger quantities up the food chain, poisoning organisms at higher trophic levels and making fish unsafe to eat.

## "E-waste" is a new and growing problem

When we first began to conduct much of our business, learning, and communication with computers and other electronic devices, many people predicted that our paper waste would decrease. Instead, the proliferation of computers, printers, VCRs, fax machines, cell phones, GPSs, MP3 players, and other gadgets has created a substantial new source of waste. These products have short lifetimes before people judge them obsolete, and most are discarded after only a few years.

The amount of this **electronic waste**—often called **e-waste**—is growing rapidly. Over 2 billion electronic devices have been sold in the United States since 1980. Of these, about half are still being used (or reused), 9% are in storage, and nearly 40% have been disposed of. The Consumer Electronics Association estimates that U.S. households discarded 304 million electronic devices in 2005—two-thirds of them still in working order.

Most e-waste is disposed of in landfills as conventional solid waste. However, most electronic products contain heavy metals and toxic flame retardants, and recent research suggests that e-waste should instead be treated as hazardous waste (see "The Science behind the Story," (• pp. 650–651). The EPA and a number of states are now moving toward keeping e-waste out of conventional sanitary landfills.

In addition, more and more electronics are being recycled. The devices are taken apart, and parts are either reused or disposed of more safely. Roughly one-fifth of the 1.5–1.9 million tons of electronics discarded in 2005 in the United States were recycled, the EPA estimates. As an example, **Figure 22.16** shows numbers of desktop computers sold, removed from useful service, and recycled in recent years.

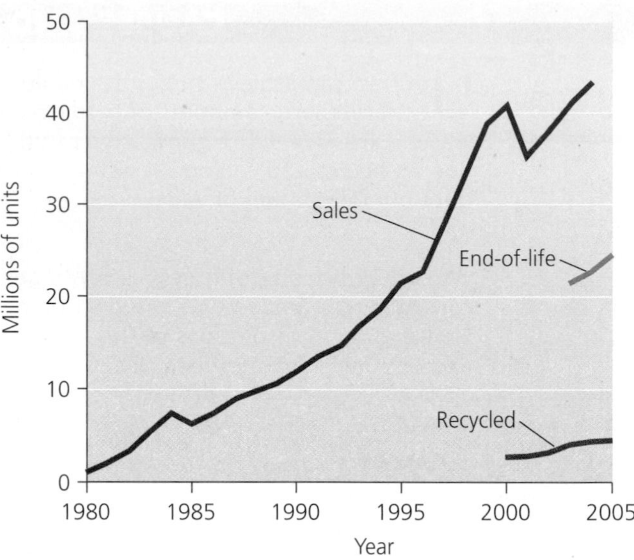


FIGURE 22.16 Although 5 million desktop computers are recycled from the United States each year (bottom line), this is only a small proportion of the number removed from service (middle line) and sold (top line). All these numbers are rising quickly as electronic waste becomes a larger component of our waste stream.

There are serious concerns about the health risks that recycling may pose to workers doing the disassembly, and wealthy nations ship much of their e-waste to developing countries, where the disassembly is done by poor workers with minimal safety regulations. These environmental justice concerns need to be resolved, but if electronics recycling can be done responsibly, it seems likely to be the way of the future.

In many North American cities, businesses, nonprofit organizations, or municipal services now collect used electronics for reuse or recycling. So next time you upgrade to a new computer, TV, DVD player, VCR, or cell phone, find out what opportunities exist in your area to recycle your old ones.

### Weighing the Issues | Toxic Computers?

The cathode ray tubes in televisions and computer screens can hold up to 5 kg (8 lb) of heavy metals, such as lead and cadmium. These represent the second-largest source of lead in U.S. landfills today, behind auto batteries. With more computer screens being purchased, the transition to high-definition television, and the rapid turnover of computers, what future waste problems and environmental health issues might you expect? How do you think we should handle the reuse, recycling, and disposal of these products?

## Several steps precede the disposal of hazardous waste

For many years we discarded hazardous waste without special treatment. In many cases, people did not know that certain substances were harmful to human health. In other cases, the danger posed by these substances was known or suspected, but it was assumed that the substances would disappear or be sufficiently diluted in the environment. The resurfacing of toxic chemicals in a residential area years after their burial at Love Canal (• p. 652) in upstate New York demonstrated to the public that hazardous waste deserves special attention and treatment.

Since the 1980s, many communities have designated sites or special collection days to gather household hazardous waste, or have designated facilities for the exchange and reuse of substances (**Figure 22.17**). Once consolidated in such sites, the waste is transported for treatment and ultimate disposal.

Under the Resource Conservation and Recovery Act, the EPA sets standards by which states are to manage hazardous waste. RCRA also requires large generators of hazardous waste to obtain permits and mandates that hazardous materials be tracked "from cradle to grave." As hazardous waste is generated, transported, and disposed of, the producer, carrier, and disposal facility must each report to the EPA the type and amount of material generated; its location, origin, and destination; and the way it is being handled. This process is intended to prevent illegal dumping and to encourage the use of reputable waste carriers and disposal facilities.

FIGURE 22.17 Many communities designate collection sites or collection days for household hazardous waste. Here, workers handle waste from an Earth Day collection event near Los Angeles.

FIGURE 22.18 Unscrupulous individuals or businesses sometimes dump hazardous waste illegally to avoid disposal costs.

Because current U.S. law makes disposing of hazardous waste quite costly, irresponsible companies sometimes illegally and anonymously dump waste, creating health risks for residents and financial headaches for local governments forced to deal with the mess (**Figure 22.18**).

Hazardous waste from industrialized nations is also sometimes dumped illegally in developing nations—a major environmental justice issue (• p. 37). This practice occurs despite the Basel Convention, an international treaty to prevent such instances. In 2006, a ship secretly dumped toxic wastes in Abidjan, the capital of the Ivory Coast, after being told by Dutch authorities that the Netherlands would charge it money to dispose of the waste in Amsterdam. The waste caused several deaths and thousands of illnesses in Abidjan, and street protests forced the government to resign over the scandal. Because of the difficulty of tracking deliveries in the international shipping industry, the responsible parties have not yet been brought to justice.

Fortunately, high costs of disposal have also encouraged conscientious businesses to invest in reducing their hazardous waste. Many biologically hazardous materials can be broken down by incineration at high temperatures in cement kilns. Some hazardous materials can be treated by exposure to bacteria that break down harmful components and synthesize them into new compounds. Besides bacterial bioremediation, phytoremediation (• pp. 90–91) is also used. Various plants have now been bred or engineered to take up specific contaminants from soil and then break down organic contaminants into safer compounds or concentrate heavy metals in their tissues. The plants are eventually harvested and disposed of.

THE SCIENCE BEHIND THE STORY

## Testing the Toxicity of "E-Waste"

*Dr. Brajesh Dubey (L) and Dr. Timothy Townsend (R) preparing the TCLP test*

Most electronic waste, or "e-waste," is disposed of in conventional sanitary landfills. However, most electronic appliances contain heavy metals that can cause environmental contamination and public health risks. For instance, over 6% of a typical computer is composed of lead.

The EPA funded Timothy Townsend's lab at the University of Florida at Gainesville to determine whether e-waste is toxic enough to be classified as hazardous waste under the Resource Conservation and Recovery Act.

With students and colleagues, Townsend determined in 1999–2000 that cathode ray tubes (CRTs) from computer monitors and color televisions leach an average of 18.5 mg/L of lead, far above the regulatory threshold of 5 mg/L. Following this research, the EPA proposed classifying CRTs as hazardous waste, and several U.S. states banned these items from conventional landfills.

Then in 2004, Townsend's lab group completed experiments on 12 other types of electronic devices. To measure their toxicity, Townsend's group used the EPA's standard test, the Toxicity Characteristic Leaching Procedure (TCLP), designed to mimic the process by which chemicals leach out of solid waste in landfills. In the TCLP, waste is ground up into fine pieces, and 100 g (3.5 oz) of it is put in a container with 2 L (0.53 gal) of an acidic leaching fluid. The container is rotated for 18 hours, after which the leachate is analyzed for its chemical content. Researchers look for eight heavy metals—arsenic, barium, cadmium, chromium, lead, mercury, selenium, and silver—and determine for each whether their concentration in the leachate exceeds that allowed by EPA regulations. Of these eight elements, electronic devices contain notable amounts of four: cadmium, chromium, lead, and mercury.

Discarded electronic waste can leach heavy metals and should be considered hazardous waste, researchers say.

To conduct the standard TCLP, Townsend's team ground up the central processing units (CPUs) of personal computers, creating a mix made up by weight of 15.8% circuit board, 7.5% plastic, 68.2% ferrous metal, 5.4% nonferrous metal, and 3.1% wire and cable. However, grinding up a computer into small bits is no easy task, and it is hard to obtain a sample that accurately represents all components and materials. So the researchers also designed a modified TCLP test in which they placed whole CPUs—with the parts disassembled but not ground up—in a rotating 55-gallon drum full of leaching liquid. Then they tested their 12 types of

## We have three disposal methods for hazardous waste

We have developed three primary means of hazardous waste disposal: landfills, surface impoundments, and injection wells. These do nothing to lessen the hazards of the substances, but they do help keep the waste isolated from people, wildlife, and ecosystems. Design and construction standards for landfills that receive hazardous waste are stricter than those for ordinary sanitary landfills. Hazardous waste landfills must have several impervious liners and leachate removal systems and must be located far from aquifers. Dumping of hazardous waste in ordinary landfills has long been a problem. In New York City, Fresh Kills largely managed to keep hazardous waste out, but most of the city's older landfills were declared to be hazardous sites because of past toxic waste dumping.

Liquid hazardous waste, or waste in dissolved form, may be stored in ponds or **surface impoundments**, shallow depressions lined with plastic and an impervious material, such as clay. Water containing dilute hazardous waste is placed in the pond and allowed to evaporate, leaving a residue of solid hazardous waste on the bottom (**Figure 22.19**). This process is repeated until the dry material is removed and transported elsewhere for permanent disposal. Impoundments are not ideal. The

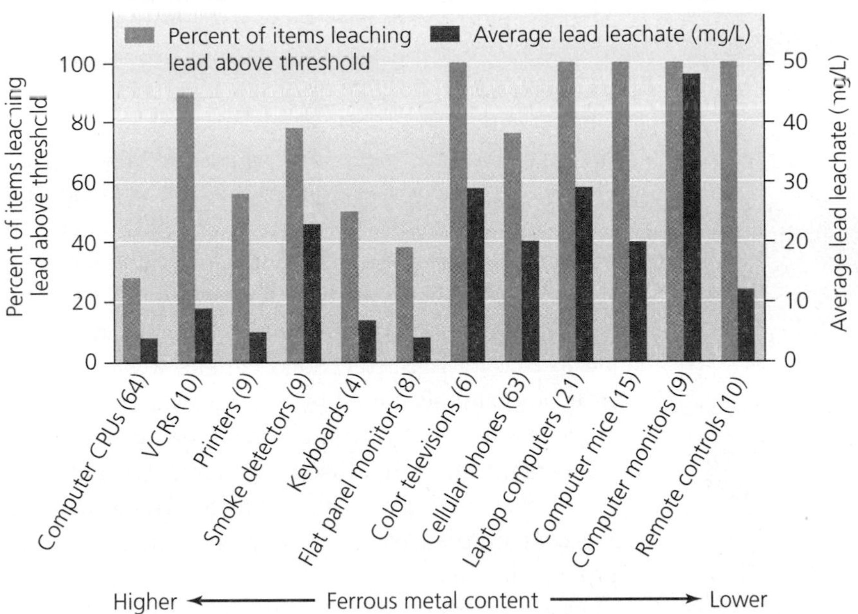


**Some proportion of all 12 devices tested exceeded the EPA regulatory standard for lead leachate. Devices with higher ferrous metal content tended to leach less lead. Where both standard and modified TCLPs were used, results are averaged.** Data from Townsend, T.G., et al. 2004. *RCRA toxicity characterization of computer CPUs and other discarded electronic devices.* July 15, 2004, report to the U.S. EPA.

devices using a combination of the standard and modified TCLP methods.

The team's results are summarized in the accompanying bar chart. Lead was the only heavy metal found to exceed the EPA's regulatory threshold, but this threshold (5 mg/L) was exceeded in the majority of trials. Computer monitors leached the most lead (47.7 mg/L on average), as expected, because monitors include the cathode ray tubes already known to be a problem. However, laptops, color TVs, smoke detectors, cell phones, and computer mice also leached high levels of lead. Next came remote controls, VCRs, keyboards, and printers, all of which leached more lead on average than the EPA threshold, and did so in 50% or more of the trials. Whole CPUs and flat panel monitors were the only devices to leach less than 5 mg/L of lead on average, but even these exceeded the threshold more than one-quarter of the time.

The researchers found that items containing more ferrous metals (such as iron) tended to leach less lead. For instance, CPUs contain 68% ferrous metals (compared to only 7% in laptops), and laptops leached seven times as much lead as CPUs. Further experiments confirmed that ferrous metals were chemically reacting with lead and stopping it from leaching.

Townsend says the work suggests that many electronic devices have the potential to be classified as hazardous waste because they frequently surpass the toxicity criterion for lead. However, EPA scientists must decide how to judge results from the modified TCLP methods, and must evaluate other research, before determining whether to alter regulatory standards.

Furthermore, lab tests may or may not accurately reflect what actually happens in landfills. So Townsend's team is filling columns measuring 24 cm (2 ft) wide by 4.9 m (16 ft) long with e-waste and municipal solid waste, burying them in a Florida landfill, and then testing the leachate that results. The results from such research should help regulators decide how best to dispose of the e-waste that is not reused or recycled.

underlying layer can crack and leak waste. Some material may evaporate or blow into surrounding areas. Rainstorms may cause waste to overflow and contaminate nearby areas. For these reasons, surface impoundments are used only for temporary storage.

The third method is intended for long-term disposal. In **deep-well injection**, a well is drilled deep beneath the water table into porous rock, and wastes are injected into it (**Figure 22.20**). The waste is meant to remain deep underground, isolated from groundwater and human contact. This idea seems attractive in principle, but in practice wells become corroded and can leak wastes into soil, allowing them to enter aquifers. Roughly 34 billion L (9 billion gal) of hazardous waste continue to be placed in U.S. injection wells each year.

## Radioactive waste is especially hazardous

Radioactive waste is particularly dangerous to human health and is persistent in the environment. The dilemma of disposal has dogged the nuclear energy industry and the U.S. military for decades. As we saw in our discussion of radioactive waste disposal in Chapter 20 (• pp. 583–588), Yucca Mountain in Nevada is now designated as the single-site repository for all U.S. nuclear waste.

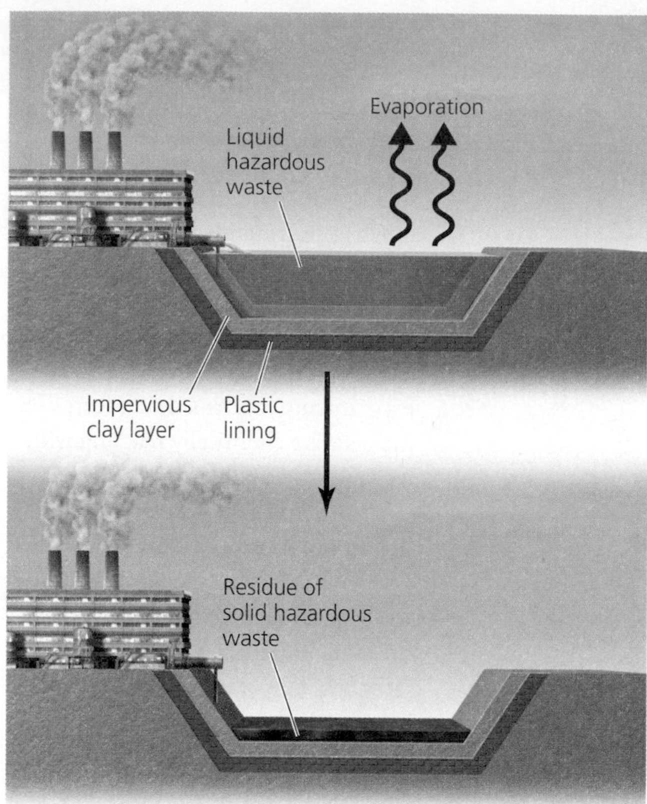


FIGURE 22.19 Surface impoundments are a strategy for temporarily disposing of liquid hazardous waste. The waste, mixed with water, is poured into a shallow depression lined with plastic and clay to prevent leakage. When the water evaporates, leaving a crust of the hazardous substance, new liquid is poured in and the process repeated. This method alone is not satisfactory, because waste can potentially leak, overflow, evaporate, or blow away.

Currently, a site in the Chihuahuan Desert in southeastern New Mexico serves as a permanent disposal site for radioactive waste. The Waste Isolation Pilot Plant (WIPP) is the world's first underground repository for transuranic waste from nuclear weapons development. The mined caverns holding the waste are located 655 m (2,150 ft) below ground in a huge salt formation thought to be geologically stable. Twenty years in the planning, WIPP became operational in 1999 and is receiving thousands of shipments of waste from 23 other locations over the next three decades.

## Contaminated sites are being cleaned up, slowly

Many thousands of former military and industrial sites remain contaminated with hazardous waste in the United States and virtually every other nation on Earth. For most nations, dealing with these messes is simply too difficult, time-consuming, and expensive. In 1980, however, the U.S. Congress passed the Comprehensive Environmental Response Compensation and Liability Act (CERCLA). This legislation established a federal program to clean up U.S. sites polluted with hazardous waste from past activities. The EPA administers this cleanup program, called the **Superfund**. Under EPA auspices, experts identify sites polluted with hazardous chemicals, take action to protect groundwater near these sites, and clean up the pollution. Later laws also charged the EPA with cleaning up *brownfields*, lands whose reuse or development are complicated by the presence of hazardous materials.

Two well-publicized events spurred creation of the Superfund legislation. In *Love Canal*, a residential neighborhood in Niagara Falls, New York, families were evacuated after toxic chemicals buried by a company and the city in past decades rose to the surface, contaminating homes and an elementary school. In Missouri, the entire town of *Times Beach* was evacuated and its buildings demolished after being contaminated by dioxin (• p. 407) from waste oil sprayed on its roads.

Once a Superfund site is identified, EPA scientists evaluate how close the site is to human habitation, whether wastes are currently confined or likely to spread, and whether the site threatens drinking water supplies. Sites that appear harmful are placed on the EPA's National Priority List, ranked according to the level of risk to human health that they pose. Cleanup proceeds on a site-by-site basis as funds are available. Throughout the process, the EPA is required to hold public hearings to inform area residents of its findings and to receive feedback.

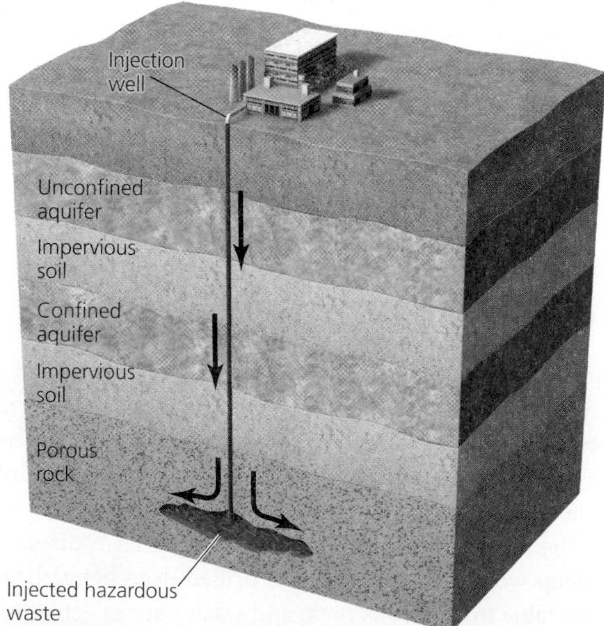


FIGURE 22.20 Liquid hazardous waste may be pumped deep underground, by deep-well injection. The well must be drilled below any aquifers, into porous rock separated by impervious clay. The technique is expensive, and waste may leak from the well shaft into groundwater.

The objective of CERCLA was to charge the polluting parties for cleanup of their sites, according to the *polluter-pays principle*. For many sites, however, the responsible parties cannot be found or held liable, and in such cases—roughly one of four so far—Superfund activities have been covered by taxpayers' funds and from a trust fund established by a federal tax on industries producing petroleum and chemical raw materials. However, Congress let the tax expire, the trust fund went bankrupt in 2004, and neither the Bush administration nor Congress has moved to restore it, so taxpayers are now shouldering the entire burden of the program. As funding dwindles and the remaining cleanup jobs become more expensive, fewer cleanups are being completed.

As of mid-2007, 1,013 of the 1,562 Superfund sites on the National Priority List had been cleaned up. The average cleanup has cost $25 million and has taken 12–15 years. Many sites are contaminated with hazardous chemicals we have no effective way to deal with. In such cases, cleanups simply involve trying to isolate waste from human contact, either by building trenches and clay or concrete barriers around a site or by excavating contaminated material, placing it in industrial-strength containers, and shipping it to a hazardous waste disposal facility. For all these reasons, the current emphasis in the United States and elsewhere is on preventing hazardous waste contamination in the first place.

## Conclusion

Our societies have made great strides in addressing our waste problems. Modern methods of waste management are far safer for people and gentler on the environment than past practices of open dumping and open burning. In many countries, recycling and composting efforts are making rapid strides. The United States has gone in a few decades from a country that did virtually no recycling to a nation in which nearly one-third of all solid waste is diverted from disposal. The continuing growth of recycling, driven by market forces, government policy, and consumer behavior, shows potential to further alleviate our waste problems.

Despite these advances, our prodigious consumption habits have created more waste than ever before. Our waste management efforts are marked by a number of difficult dilemmas, including the cleanup of Superfund sites, safe disposal of hazardous and radioactive waste, and frequent local opposition to disposal sites. These dilemmas make clear that the best solution to our waste problem is to reduce our generation of waste. Finding ways to reduce, reuse, and efficiently recycle the materials and goods that we use stands as a key challenge for the new century.

## REVIEWING OBJECTIVES

**You should now be able to:**

**Summarize and compare the types of waste we generate**

- Municipal and industrial solid waste, hazardous waste, and wastewater are major types of waste. (p. 631)

**List the major approaches to managing waste**

- Source reduction, recovery, and disposal are the three main components of waste management. (pp. 631–632)

**Delineate the scale of the waste dilemma**

- Developed nations generate far more waste than developing nations. (p. 633)
- Waste everywhere is increasing as a result of growth in population and consumption. (pp. 633–634)

**Describe conventional waste disposal methods: landfills and incineration**

- Sanitary landfills guard against contamination of groundwater, air, and soil. Nonetheless, such contamination can occur. (pp. 635–636)
- Incinerators reduce waste volume by burning it. Pollution control technology removes most pollutants from emissions, but some escape, and highly toxic ash needs to be disposed of in landfills. (pp. 636–637)
- We are harnessing energy from landfill gas and generating electricity from incineration. (pp. 638–639)

**Evaluate approaches for reducing waste: source reduction, reuse, composting, and recycling**

- Reducing waste before it is generated is the best waste management approach. Recovery comprises the next-best option. (pp. 639–640)
- Consumers can take an array of simple steps to reduce their waste output. (pp. 639–640)
- Composting reduces waste while creating organic matter for gardening and agriculture. (p. 640)
- Recycling has grown in recent years and now removes nearly 24% of the U.S. waste stream. (pp. 641–643)

**Discuss industrial solid waste management and principles of industrial ecology**

- ▶ Regulations differ, but industrial waste management is similar to that for municipal solid waste. (p. 644)
- ▶ Industrial ecology urges industrial systems to mimic ecological systems and provides ways for industry to increase its efficiency. (pp. 645–646)

**Assess issues in managing hazardous waste**

- ▶ Hazardous waste is ignitable, corrosive, reactive, or toxic. (p. 646)
- ▶ Electronic waste may be considered hazardous. (pp. 648, 650–651)
- ▶ Hazardous waste is regulated and monitored, yet illegal dumping remains a problem. (p. 649)
- ▶ No fully satisfactory method of disposing of hazardous waste has yet been devised. (pp. 649–652)
- ▶ Cleanup of hazardous waste sites is a long and expensive process. (pp. 652–653)

## TESTING YOUR COMPREHENSION

1. Describe five major methods of managing waste. Why do we practice waste management?
2. Why have some people labeled the United States "the throwaway society"? How much solid waste do Americans generate, and how does this amount compare to that of people from other countries?
3. Name several guidelines by which sanitary landfills are regulated. Describe three problems with landfills.
4. Describe the process of incineration or combustion. What happens to the resulting ash? What is one drawback of incineration?
5. What is composting, and how does it help reduce input to the waste stream?
6. What are the three elements of a sustainable process of recycling?
7. What are the goals of industrial ecology?
8. What four criteria are used to define hazardous waste? Why are heavy metals and synthetic organic compounds particularly hazardous?
9. What are the largest sources of hazardous waste? Describe three ways to dispose of hazardous waste.
10. What is the Superfund program? How does it work?

## SEEKING SOLUTIONS

1. How much waste do you generate? Look into your waste bin at the end of the day, and categorize and measure the waste there. List all other waste you may have generated in other places throughout the day. How much of this waste could you have avoided generating? How much could have been reused or recycled?
2. Some people have criticized current waste management practices as merely moving waste from one medium to another. How might this criticism apply to the methods now in practice? What are some potential solutions?
3. Of the various waste management approaches covered in this chapter, which ones are your community or campus pursuing, and which are they not pursuing? Would you suggest that your community or campus start pursuing any new approaches? If so, which ones, and why?
4. Can manufacturers and businesses benefit from source reduction if consumers were to buy fewer products as a result? How? Given what you know about industrial ecology, what do you think the future of sustainable manufacturing may look like?
5. **THINK IT THROUGH** You are the CEO of a major corporation that produces containers for soft drinks and a wide variety of other consumer products. Your company's shareholders are asking that you improve the company's image—while not cutting into profits—by taking steps to reduce waste. What steps would you consider taking?
6. **THINK IT THROUGH** You are the president of your college or university. Your trustees want you to engage with local businesses and industries in ways that benefit both the school and the community. Your faculty and students want you to make the school a leader in waste reduction and industrial ecology. Consider the industries and businesses in your community and the ways they interact with facilities on your campus. Bearing in mind the principles of industrial ecology, can you think of any novel ways that your school and local businesses might mutually benefit from one another's services, products, or waste materials? Are there waste products from one business, industry, or campus facility that another might put to good use? Can you design an eco-industrial park that might work on your campus? What steps would you propose to take as president?

## INTERPRETING GRAPHS AND DATA

Using 1990 data from 149 countries, David Beede, an economist at the U.S. Department of Commerce, and David Bloom, a professor of economics at Columbia University, examined global patterns in the generation and management of municipal solid waste (MSW). Beede and Bloom were particularly interested in the relationships among wealth, population size, and per capita generation of MSW. Their results are presented in the accompanying table.

1. Create a bar chart with the four income categories of nations as entries on the *x* axis, and with percentages from 0 to 60 on the *y* axis. For each category of nation, plot as paired bars the values of MSW generation and population size as percentages of the world total.
2. Using the data for total MSW generation and for population size (and remembering that there are 2,000 pounds in a ton), calculate the pounds of MSW per capita per day for each category of nation, and enter these values in the table.
3. Now add a second *y* axis to your graph, on the right side, ranging from 0 to 3. Plot the values you calculated for per capita waste generation for each of the four income categories of nations, placing them as data points connected by a line. Describe in general terms the relationship between wealth and per capita generation of MSW. Can you offer at least one possible reason for the trend that you see?
4. Do you think it's possible for wealthy nations to reduce their per capita MSW generation to the rates of poorer nations? Why or why not?

| Income category of nation | Total MSW generation | | Population size | | Pounds MSW per capita per day |
|---|---|---|---|---|---|
| | Millions of tons/year | % of world total | Millions of people | % of world total | |
| Low | 658 | 46.3 | 3,091 | 58.5 | 1.17 |
| Low-middle | 160 | 11.2 | 629 | 11.9 | |
| Upper-middle | 212 | 14.9 | 748 | 14.2 | |
| High | 393 | 27.6 | 816 | 15.4 | |
| **All economies** | **1,422** | **100.0** | **5,284** | **100.0** | |

Data from Beede, D. N. and D. E. Bloom. 1995. The economics of municipal solid waste. *World Bank Research Observer* 10: 113–150.

## CALCULATING ECOLOGICAL FOOTPRINTS

The 15th biennial "State of Garbage in America" survey documents the ability of U.S. residents to generate prodigious amounts of municipal solid waste (MSW). According to the survey, on a per capita basis, South Dakotans generate the least MSW (3.82 lb/day), and Indiana residents generate the most (11.37 lb/day). The average for the entire country is 7.25 lb MSW per person per day. Compare this number to the data in the "Interpreting Graphs and Data" table. Now calculate the amount of MSW generated in 1 day and in 1 year by each of the groups indicated, at each of the rates shown in the accompanying table.

1. Suppose your town of 50,000 people has just approved construction of a landfill nearby. Estimates are that it will accommodate 1 million tons of MSW. Assuming the landfill is serving only your town, for how many years will it accept waste before filling up? How much longer would a landfill of the same capacity serve a town of the same size in another industrialized ("high-income") country?
2. Why do you think U.S. residents generate so much more MSW than people in other "high-income" countries, when standards of living in those countries are comparable?

| Groups generating municipal solid waste | Per capita MSW generation rates | | | | | | | |
|---|---|---|---|---|---|---|---|---|
| | U.S. average (7.25 lb/day) | | Indiana (11.37 lb/day) | | "High- income" countries (2.64 lb/day) | | World average (1.47 lb/day) | |
| | Day | Year | Day | Year | Day | Year | Day | Year |
| You | 7.25 | 2,647 | | | | | | |
| Your class | | | | | | | | |
| Your town | | | | | | | | |
| Your state | | | | | | | | |
| United States | | | | | | | | |
| World | | | | | | | | |

Data from Simmons, P., et al. 2006. The state of garbage in America. *BioCycle* 47: 26.

## Take It Further

Go to www.aw-bc.com/withgott or the student CD-ROM, where you'll find:

- Suggested answers to end-of-chapter questions
- Quizzes, animations, and flashcards to help you study
- *Research Navigator*™ database of credible and reliable sources to assist you with your research projects
- **GRAPHit!** Tutorials to help you interpret graphs
- **INVESTIGATEit!** Current news articles that link the topics that you study to case studies from your region to around the world

CHAPTER

# 23 Sustainable Solutions

Sri Lankan children planting tree seedlings on deforested hillsides around their village

## Upon completing this chapter, you will be able to:

- List and describe approaches being taken on college and university campuses to promote sustainability
- Explain the concept of sustainable development
- Discuss how protecting the environment can be compatible with promoting economic welfare
- Describe and assess key approaches to designing sustainable solutions
- Explain how time is limited but how human potential to solve problems is tremendous

Ball State University

CENTRAL CASE

# Ball State University Aims for Campus Sustainability

**"We solemnly pledge to the peoples of the world and the generations that will surely inherit this Earth that we are determined to ensure that our collective hope for sustainable development is realized."**

—Declaration signed by 193 nations at the World Summit, Johannesburg, South Africa, 2002

**"We hope—surely we must believe—that our species will emerge from the environmental bottleneck in better condition than when we entered . . . taking as much of the rest of life with us as possible."**

—Edward O. Wilson, *Consilience*, 1998

At Ball State University, "BSU" stands for more than the institution's name. It also stands for "Becoming a Sustainable University." The efforts of students, faculty, staff, and administrators at this 18,000-student campus in Muncie, Indiana, have made Ball State a leader in the international movement for campus sustainability.

Proponents of campus sustainability view colleges and universities as microcosms of society at large, noting that institutions of learning consume resources, emit pollution, and exert other environmental impacts. These proponents seek to change the ways these institutions operate and to reduce their ecological footprints.

Many colleges and universities have student-run environmental organizations, recycling programs, and courses or majors in environmental science or environmental studies. Ball State began taking steps beyond these back in 1994, when it initiated a program to promote environmental literacy among its faculty. Its "Green for Green" program pays faculty to take seminars on sustainability and then integrate what they've learned into their teaching, promote changes in their departments, or conduct community outreach. To date, over 200 instructors from nearly all of the university's 47 departments have participated—over one-fifth of the BSU faculty.

Ball State accelerated its sustainability efforts after 1999, once the university's president signed the **Talloires Declaration**, a commitment to pursue sustainability. This document was composed in Talloires, France, in

1990 and has been signed by over 350 university presidents and chancellors from 50 nations.

To plan green initiatives at Ball State, faculty, staff, students, and community members formed the Council on the Environment (COTE). COTE wrote a guide for future goals and actions, meets monthly to plan initiatives, and works with BSU administrators to achieve its goals. A Center for Energy Research/Education/Service coordinates and implements COTE's initiatives.

In 2003, COTE persuaded the university to replace its virgin stock paper with paper of 30% post-consumer recycled content. It also made 100% post-consumer recycled paper available, which a number of departments adopted. COTE estimates that, besides reducing landfill waste, switching to 30% recycled-content paper saves 3,000 trees each year and prevents the emission of 45,000 kg (100,000 lb) of $CO_2$ equivalents and 90 kg (200 lb) of nitrogen oxides.

Sustainability advocates are also greening Ball State's transportation system. Three gas-electric hybrid vehicles were added to the vehicle fleet in 2003. The next year, Ball State began running the entire bus fleet, as well as certain equipment, on B20 biodiesel fuel. In 2005, Ball State added a hybrid bus to its fleet.

Other recent initiatives include composting, Earth Day events, and reducing mercury use on campus. Most recently, the university agreed that two new buildings would be built using certified sustainable design principles. COTE also sponsors annual awards honoring people at Ball State and in the community who contribute to sustainability efforts.

Many of these sustainability initiatives have cost money, and their proponents have had to argue their case to penny-conscious administrators time after time. But the university's administration has recognized that many short-term costs are actually investments that will save money in the long term.

Beyond serving as a model for other schools, Ball State has launched the *Greening of the Campus* Conference Series. At each of these international conferences, people involved in sustainability efforts from well over 100 colleges and universities have gathered to share their experiences. The university hosted the seventh of these biennial conferences in September 2007.

# Sustainability on Campus

If we are to attain a sustainable civilization, we will need to make efforts at every level, from the individual to the household to the community to the nation to the world. Governments, corporations, and organizations must all encourage and pursue sustainable practices. Among the institutions that can contribute to sustainability efforts are colleges and universities.

We tend to think of colleges and universities as enlightened and progressive institutions that generate benefits for society. However, colleges and universities are also centers of lavish resource consumption. Institutions of higher education feature extensive infrastructure, including classrooms, offices, research labs, and residential housing. Most also have dining establishments, sports arenas, vehicle fleets, and road networks. The 4,100 campuses in the United States interact with many thousands of businesses and spend over $200 billion each year on products and services. The ecological footprint of a typical college or university is substantial.

Reducing the size of this footprint is challenging. Colleges and universities tend to be bastions of tradition, where institutional habits are deeply ingrained and where bureaucratic inertia can often block the best intentions for positive change. Nonetheless, faculty, staff, administrators, and especially students are progressing on a variety of fronts to make the operations of educational institutions more sustainable.

## Why strive for campus sustainability?

You enrolled at your college or university to gain an education, not to transform the institution. Why, then, are increasing numbers of students promoting sustainable practices on their campuses? First, reducing the ecological footprint of a campus really can make a difference. The consumptive impact of educating, feeding, and housing hundreds or thousands of students is immense. Second, campus sustainability efforts make students aware of the need to address environmental problems, and students who act to promote campus sustainability serve as models for their peers. Finally, the student who engages in sustainability efforts learns and grows as a result. The challenges, successes, and failures that you encounter can serve as valuable preparation for similar efforts in transforming inertia-bound institutions in the broader society.

Support from faculty, staff, and administrators is crucial for success, but students are often the ones who initiate change. Students often feel freer than faculty or staff to express themselves. Students also arrive on campus with new ideas and perspectives, and they generally are less attached to traditional ways of doing things.

## Campus efforts may begin with an audit

Campus sustainability efforts often begin with a quantitative assessment of the institution's operations. Such audits provide baseline information on what an institution is doing or how much it is consuming, and audits help set priorities and goals. For instance, Harford Community College in Maryland hired engineers and specialists to audit its energy use and pollutant emissions, which then served as the basis for setting reduction goals. Students at Franklin Pierce College in New Hampshire conducted a campuswide audit that gave them data and direction on how to address energy use, food operations, waste management, and more. At the University of Vermont, graduate student Erika Swahn gathered data on heating, transportation, electricity, waste, food, and water use to calculate her school's ecological footprint—over 59,000 acres annually, or about 4.5 acres for each student, instructor, and staff member.

It is most useful in an audit to target items that can lead directly to specific recommendations. For instance, an audit should quantify the performance of individual appliances currently in operation so that decision makers can identify particular ones to replace. Once changes are implemented, the institution can monitor progress by comparing future measurements to the audit's baseline data.

**FIGURE 23.1** In "landfill on the lawn" events, piles of waste are dumped onto a campus open space. Volunteers sort through the rubbish, separating out recyclables, as these students at the University of North Carolina at Greensboro are doing. These events can dramatically demonstrate just how many recyclable items are needlessly thrown away.

## Recycling and waste reduction are the most common campus efforts

Campus sustainability efforts most frequently involve waste reduction, recycling, or composting. Depending on the type of material, 46–85% of schools had recycling services in place, as of the first comprehensive survey of campus sustainability efforts in 2001 (**Table 23.1**). Waste management initiatives are relatively easy to start and maintain because they offer many opportunities for small-scale improvements and because people generally enjoy recycling and reducing waste.

**TABLE 23.1 Frequency of Campus Sustainability Efforts**

| Activity | Percentage of schools performing activity* |
|---|---|
| Water efficiency upgrades | 72 |
| Recycling | 65† |
| Energy efficiency and conservation | 63 |
| Research and service opportunities for students | 58 |
| Professional development for faculty | 50 |
| Setting and reviewing goals | 45 |
| Sustainable landscaping | 43 |
| Staffing environmental programs with a coordinator | 29 |
| Managing transportation demand | 24 |

*Based on voluntary responses from 891 schools.
†46–85%, depending on type of material.
*Source:* McIntosh, M., et al. 2001. *State of the campus environment: A national report card on environmental performance and sustainability in higher education.* National Wildlife Federation Campus Ecology; survey conducted by Princeton Survey Research Associates.

Students at Ohio University and Miami University in Ohio kicked off *RecycleMania,* a 10-week competition among schools to see which can recycle the most. The annual competition involved 201 schools in its seventh year in 2007, when California State University at San Marcos won the overall award, recycling 59.94% of its waste. Students at Columbia University's Lamont-Doherty Earth Observatory won the title for per capita recycling, with 101.12 lb (45.9 kg) of material recycled per student. Other top-performing schools in 2007 included Rutgers University, Kalamazoo College, University of San Francisco, Sheldon Jackson College, University of Texas at Austin, and West Los Angeles Community College.

"Trash audits" or "landfill on the lawn" events involve tipping dumpsters onto a campus open space and sorting out recyclable items (**Figure 23.1**). When students at Ashland University in Ohio audited their waste, they found that 70% was recyclable and used this data to press their administration to support recycling programs. Composting is becoming popular as well; Ball State University composts bulky wood waste, shredding surplus furniture and wood pallets and making them into mulch to nourish campus plantings.

Students at some campuses run events to promote the reuse of items. The University of North Carolina–Charlotte ran a "Take It or Leave It Tour" that enabled over

600 students to swap 55 kg (122 lb) of unwanted notebooks, file folders, posters, CDs, and other items with one another. Clemson University in South Carolina donated to local charities 5,220 kg (11,500 lb) of food, clothing, and other items that might otherwise have been thrown out.

Administrators are more easily convinced to enact institutional changes if they save money. At Ithaca College in New York, about 44% of the food waste generated annually on campus is composted. Disposal fees at the local landfill are $60 per ton, so composting saves the college $11,500 each year. The compost is used on campus plantings, and experiments showed that the plantings grew better with the compost mix than with chemical soil amendments.

## Green building design is a key to sustainable campuses

Dozens of campuses now boast "green" buildings that are constructed from sustainable building materials and whose design and technologies encourage energy efficiency, water efficiency, renewable energy, and the reduction of pollution. As with any type of ecolabeling, there need to be agreed-upon standards, and for sustainable buildings these are the *Leadership in Energy and Environmental Design (LEED)* standards. Developed and maintained by the nonprofit U.S. Green Building Council, LEED standards guide the design and certification of new construction and the renovation of existing structures on campuses and elsewhere.

One of the first green buildings on a college campus was the Adam Joseph Lewis Center for Environmental Studies at Oberlin College in Ohio (**Figure 23.2a**). This building was constructed using materials that were recycled or reused, took little energy to produce, or were locally harvested, produced, or distributed. Some materials, such as carpeting, are leased and then returned to the company for recycling when they wear out. The Lewis Center contains energy-efficient lighting, heating, and appliances, and it maximizes indoor air quality with a state-of-the-art ventilation system as well as paints, adhesives, and carpeting that emit few volatile organic compounds. The structure is powered largely by solar energy from PV panels on the roof, active solar heating, and passive solar heating from south-facing walls of glass and a tiled slate floor that acts as a thermal mass (• p. 608). Over 150 sensors throughout the building monitor conditions such as temperature and air quality.

A few years after the Lewis Center was constructed, Bren Hall at the University of Santa Barbara in California (**Figure 23.2b**) became the only laboratory building in the nation to achieve a "platinum" LEED ranking (the highest possible). Bren Hall uses an innovative and efficient heating and cooling system integrated with solar panels and white roofing material to reflect sunlight. The structural steel and other materials in the building are composed of 40% recycled content, and over 90% of construction waste was recycled. The building contains no formaldehyde, asbestos, or CFCs, and other toxic substances were kept to a minimum. Bren Hall conserves water with low-flow fixtures, waterless urinals, reclaimed water for toilets, and automatic sensors. Only 2% of the $26 million construction cost was due to sustainable materials and approaches, and this cost is easily being recovered through energy savings.

**(a) Lewis Center at Oberlin**

**(b) Bren Hall at UC Santa Barbara**

**FIGURE 23.2** Oberlin College's Adam Joseph Lewis Center for Environmental Studies (**a**) and Bren Hall at the University of Santa Barbara, California (**b**), are two of the best-known green buildings on American campuses.

The movement for "green buildings" continues to grow. The University of Florida has built or started construction on 18 green buildings since 2003. The new Kirsch Center for Environmental Studies at DeAnza College in California is energy-efficient, water-efficient, and climate-responsive; is built with recycled, renewable, and nontoxic materials; is solar-powered; and includes outdoor learning spaces and labs. At Catawba College in North Carolina, students convinced the board of trustees to commit to construct only sustainable buildings. Catawba is now working on a major renovation and addition to its library, in which students are involved in the greening process. A recently completed green building at Catawba, which houses its Center for the Environment, is next to 76 ha (189 acres) of reclaimed farmland that is now a wildlife preserve used for teaching purposes.

As Catawba's preserve suggests, sustainable architecture doesn't stop at the walls of a building. The landscaping around Bren Hall includes drought-tolerant native plants that require little watering—and the water used for irrigation is reclaimed from other uses. Oberlin's Lewis Center is set among orchards, gardens, and a restored wetland that helps filter wastewater, and with urban agriculture and lawns of grass specially bred to require less chemical care. Careful design of campus landscaping can create livable spaces that promote social interaction and where plantings supply shade, prevent soil erosion, create attractive settings, and provide wildlife habitat. It has been said, in fact, that groundskeepers are more vital to colleges' recruiting efforts than are vice presidents.

FIGURE 23.3 University of Arizona students re-engineer the landscape on their campus to prevent flooding and harvest rainwater.

## Efficient water use is important

Managing water efficiently is a key element of sustainable campuses—especially in arid regions, as students at the University of Arizona in Tucson know. Despite Tucson's dry desert climate, rain falls in torrents during the late-summer monsoon season, when water runs off paved surfaces and down riverbeds, causing erosion, carrying pollution, and flowing too swiftly to sink into the ground and recharge aquifers. So UA students sought to redirect these floodwaters and put them to use. With the help of an $11,000 grant written by student Chet Phillips and Dr. James Riley, the group created an independent study course to design and implement a rainwater harvesting project on campus (**Figure 23.3**). Students surveyed the site, researched its hydrology, and worked with staff to design and engineer channels, dams, berms, and basins to slow the water down and direct it into swales, where it can nourish plants and sink in to recharge the aquifer. The group is now pursuing further such projects.

Water conservation is just as important indoors. To promote efficiency, sustainability advocates at Manhattanville College in New York persuaded 217 people to reduce the length of their showers one day. On average, people cut their showers short by 30%, saving a total of 5,173 gallons of water. Water-saving technologies such as waterless urinals and "living machines" to treat wastewater are being installed at a number of campuses. The University of British Columbia in Vancouver, Canada, sank $35 million into retrofitting 300 campus buildings with water- and energy-efficient upgrades that now save the school $2.6 million annually. The upgrades reduce water use by 30% each year—enough water for 11,785 homes.

## Energy conservation is achievable

Students are finding many ways to conserve energy. A student energy audit at Daemen College in New York led the college to turn down thermostats, require sleep-mode settings for computers, and shut off unused lights. Students at SUNY-Purchase in New York saved the school $86,000 per year simply by turning down hot water temperatures by 5° F (2.8° C). The University of British Columbia's extensive retrofits are reducing energy use by 20% and greenhouse emissions by 80%—equal to taking 2,500 cars off the road each year. At Middlebury College in Vermont and Williams College in Massachusetts, students distributed thousands of energy-efficient compact fluorescent lightbulbs to their peers and to community members, in exchange for incandescent bulbs.

Williams College students also made energy conservation into a fun campus event. Their "Do It in the Dark" competition pitted residential houses against one another to reduce energy consumption. The initial month's competition resulted in just a 3% decrease in energy use, but the second try produced a 13% cut as students got caught up in the fun.

FIGURE 23.4 In October 2005, 18 teams from colleges and universities across the United States and the world converged on the Mall in Washington, D.C., for the second Solar Decathlon. Each team erected an entire house, of the students' own design, fully powered by solar energy. The University of Colorado won the 2005 competition, edging out Cornell University and California Polytechnic State University.

Campuses can harness large energy savings simply by not powering unused buildings. How many times have you walked through a classroom building at night or in the summer when it was totally empty, yet found that all the lights were on and the heating or air-conditioning was running full-blast? Large buildings are expensive to heat, cool, and light, so powering them down when not in use saves a great deal of energy, money, and greenhouse gas emissions. Central Florida Community College booked its summer classes into a minimum number of efficient buildings so that other buildings could be shut down. Students at Central Florida also surveyed employees, asking whether they could do with less lighting, and then removed unneeded lightbulbs. The college also installed motion detectors to turn off lights when people are not in rooms.

## Students can promote renewable energy

Campuses can reduce energy consumption and greenhouse emissions by altering the type of energy they use. Ball State replaced its coal-fired heating plant with cleaner and more efficient fluidized bed combustion technology. Middlebury is seeking to switch its central plant from fuel oil to carbon-neutral wood chips.

Student initiative on campus can also influence the type of energy we use in our society. In October 2005, teams of students from 18 universities competed in the second-ever Solar Decathlon. In this remarkable event, teams of students travel to the National Mall in Washington, D.C., bringing material for the solar-powered homes they have spent months designing. They erect their homes on the Mall, where they stand for 3 weeks (**Figure 23.4**). The homes are judged on ten criteria, and prizes are awarded to winners in each category. One of the 2005 teams, from the University of Massachusetts at Dartmouth, donated its home to Habitat for Humanity for a Washington, D.C., family to live in, and is partnering with Habitat for Humanity to build several other homes with solar technology. The next Solar Decathlon takes place in fall 2007.

Students learn about bioenergy at the University of Florida at Gainesville, where their faculty sponsor, Ann Wilkie, has developed an innovative system to use livestock manure to produce bioenergy while recovering water and nutrients for fertilizer and minimizing waste and greenhouse emissions. The students study sustainable energy production through lectures and field trips, and promote bioenergy across campus.

Solar and wind power play roles on many campuses. At the University of Vermont, PV panels provide enough electricity for nine desktop computers or 95 energy-efficient lightbulbs for 10 hours each day. Middlebury College and Macalester College in Minnesota have each installed wind turbines to help meet their energy needs. For students with less inclination to erect renewable energy technology themselves, any educational institution can invest in renewable energy by purchasing "green tags," or carbon offsets (• p. 535), that subsidize renewable energy sources. Both Western Washington University and College of the Atlantic in Maine offset 100% of their greenhouse gas emissions by buying green tags for renewable energy.

## Carbon-neutrality is a new goal

Now that global climate change has vaulted to the forefront of society's concerns, reducing greenhouse gas emissions from fossil fuel combustion has become a top priority for campus sustainability proponents. Today many campuses are aiming to become carbon-neutral.

Students at Lewis and Clark College in Oregon began the trend in 2002, when they made their school the first in the nation to comply with the Kyoto Protocol (• p. 534). After conducting a campus audit, student leaders reduced greenhouse emissions by the percentage required under the Protocol, largely by purchasing carbon offsets from a nonprofit that funds energy efficiency and revegetation projects. Although U.S. leaders have repeatedly cited economic expense in refusing to slow emissions at the national level, Lewis and Clark students found that becoming Kyoto-compliant on their campus cost only $10 per student per year. Similarly, Western Washington University's carbon offset program costs just $10.50 per student per year—an amount for which students voted overwhelmingly to tax themselves.

Since then, student pressure and petitions at many campuses have nudged administrators and trustees to set targets for reducing greenhouse emissions. As of fall 2007, over 350 university presidents had signed onto the American College and University Presidents Climate Commitment. This pledge to reduce campus greenhouse emissions commits presidents to inventory emissions, set target dates and milestones for becoming carbon-neutral, and take immediate steps to lower emissions with short-term actions, while also integrating sustainability into the curriculum.

On January 31, 2008, campuses across the United States will join together in *Focus the Nation*, a national "teach-in" on solutions to global climate change. This effort will involve thousands of students and aims to engage millions of Americans in a national conversation on climate change. It could be the biggest and most meaningful mass event on college campuses since the first Earth Day.

## Dining services and campus gardens let students eat sustainably

Campus food service operations can promote sustainable practices by buying organic produce, composting food scraps, and purchasing food in bulk or with less packaging. Buying locally grown or produced food supports local economies and cuts down on fuel use from long-distance transportation. Some college campuses even have gardens where students can grow food (**Figure 23.5**).

At Sterling College in Vermont, many foods are organic, grown by local farmers, or produced by Vermont-based companies. Some foods are grown and breads are baked right on the Sterling campus, and food shipped in is purchased in bulk to reduce packaging. Dish soap is biodegradable, and kitchen scraps are composted along with recycled unbleached paper products. By reducing unnecessary equipment use, the college saved money and slashed its electricity use by 23%.

**FIGURE 23.5** A number of campuses now include gardens where students can grow organic vegetables that are used for meals in dining halls. Here, a student works in the garden at Middlebury College in Vermont.

Some such strategies are difficult to scale up to large state universities, but these schools can take their own approaches. At the University of Massachusetts at Boston, the campus took a step toward "zero-waste dining" when it replaced plastic dinnerware in its food courts with compostable dinnerware, eliminating 45,000 kg (100,000 lb) of plastic waste.

Students at Northwest Indian College (NWIC) in Washington promote healthy, local, organic foods that are also important in traditional Native American cultures of the Pacific Northwest. At their main campus they planted camas lilies, a vital traditional food plant, and distributed camas seeds to other campuses and institutions in the community. Along with students from nearby Western Washington University, NWIC volunteers also help maintain a garden for elderly community members that grows traditional food plants.

## Institutional purchasing matters

The kinds of purchasing decisions made in dining halls favoring local food, organic food, and biodegradable products can be applied across the entire spectrum of a campus's needs. When campus purchasing departments buy recycled paper, certified sustainable wood, energy-efficient appliances, goods with less packaging, and other ecolabeled products, they send signals to manufacturers and increase the demand for such items. Ball State recently pioneered a new avenue in this vein. Students are working with the Barnes and Noble–run campus bookstore to carry, promote, and sell more environmentally and socially sustainable books, paper products, and school supplies.

At Chatham College in Pennsylvania, students chose to honor their school's best-known alumnus, Rachel Carson (• pp. 65–66, 389), by seeking to eliminate toxic chemicals on campus. Administrators agreed to this effort, provided that alternative products to replace the toxic ones worked just as well and were not more expensive. Students brought in the CEO of a company that produces nontoxic cleaning products, who demonstrated to the janitorial staff that his company's products were superior. The university switched to the nontoxic products, which were also cheaper, and proceeded to save $10,000 per year. Chatham students then found a company offering paint without volatile organic compounds and negotiated with it for a free paint job and discounted prices on later purchases. Students also worked with grounds staff to eliminate herbicides and fertilizers used on campus lawns and to find alternative treatments.

## Transportation alternatives are many

Many campuses struggle with traffic congestion, parking shortages, commuting delays, and pollution from vehicle exhaust. Some are addressing these issues by establishing or expanding bus and shuttle systems; encouraging bicycling, walking, and carpooling; and introducing alternative vehicles to university fleets (**Figure 23.6**). The State University of New York College of Environmental Science and Forestry in Syracuse acquired six electric vehicles for on-campus use, a gas-electric hybrid car, and a delivery van that runs on compressed natural gas. At the same time, it converted its buses to biodiesel.

**FIGURE 23.6** Most buses at the University of California at Davis run on compressed natural gas, and one runs on a novel mixture of natural gas and hydrogen. Several of these low-emission vehicles are shown here lined up in front of Freeborn Hall during a press conference celebrating their purchase.

A number of other campuses are using biodiesel, including Ball State, Hobart and William Smith Colleges in New York, the University of South Carolina, and the University of Vermont. Middlebury College students began Project Bio Bus, which crisscrosses North America each summer in a biodiesel bus spreading the gospel of this alternative fuel. Biofuel proponents at Rice University in Houston are now taking biodiesel initiatives to the next level—they are producing biodiesel. With the help of faculty sponsors, facilities managers, and departmental equipment, students are collecting waste cooking oil from dining halls, setting up a pilot plant with a 70-gallon reactor, and testing the biodiesel they brew—all while getting course credit.

The University of British Columbia is a leader in transportation efforts. For $20 a month, UBC's "U-Pass" program provides students with biking programs and facilities, expanded campus bus and shuttle services, unlimited use of some city transit systems, rides home in emergencies, merchant discounts, and priority parking spaces and ride-matching services for carpoolers. The program has boosted transit ridership to campus by 53% and has decreased single-person car use by 20%.

## Campuses are restoring native plants, habitats, and landscapes

No campus sustainability program would be complete without some effort to enhance the campus's natural environment. Such efforts remove invasive species, restore native plants and communities, improve habitat for wildlife, enhance soil and water quality, reduce pesticide use, and create healthier, more attractive surroundings (**Figure 23.7**).

At the University of West Alabama, a student crew conducted controlled burns (• p. 343) to restore prairie on 7 ha (17 acres) and then planted native plants in experimental plots to study ecological patterns in the wake of the restoration. At Warren Wilson College in North Carolina, students and the landscaping supervisor built a greenhouse, expanded an arboretum, and are propagating local species of grasses and wildflowers. Some schools, such as California's DeAnza College, feature demonstration gardens of native plant communities. Others, such as Loyola College in Maryland and the University of Wisconsin at Milwaukee, focus on removing invasive species. Still others, such as Ohio State University and the New College of California in San Francisco, have rooftop gardens.

At Northland College in Wisconsin, sustainability advocates relandscaped half the campus to enhance biodiversity and water quality. They replaced invasive plants with native ones and planted meadows that capture

FIGURE 23.7 Many schools have embarked on habitat restoration projects to beautify their campuses, provide wildlife habitat, restore native plants, and filter water runoff. Here, students at Catawba College in North Carolina plant a red maple tree on the central campus.

stormwater runoff and filter pollution. At North Hennepin Community College in Minnesota, students are helping to transform 2.8 ha (7 acres) of lawn and channelized ponds into areas of marsh, prairie/savanna, and native forest. So far students have grown and planted over 35,000 seedlings. Besides providing wildlife habitat, the restored area will reduce runoff, erosion, and maintenance costs and will provide new opportunities for biological and environmental education.

## Sustainability efforts include curricular changes

Campus sustainability activities provide students active, hands-on ways to influence how their campuses function. Yet sustainability concerns are also transforming academic curricula and course offerings. The course for which you are using this book right now likely did not exist a generation ago. As our society comes to appreciate the looming challenge of sustainability, colleges and universities are attempting to train students to confront this challenge more effectively.

Besides offering new courses, schools are integrating sustainability issues into established courses across many disciplines. Several of Ball State's recent "green initiative" awards have gone to faculty in the Family and Consumer Sciences Department for incorporating sustainability concepts into their classes. One college that has generated a strong and innovative green curriculum is Miami-Dade College in Florida. Besides creating a number of novel courses, workshops, and outdoor immersion programs for students, it also trains hundreds of faculty in sustainability workshops, courses, conferences, and outdoor programs. Like the trained faculty at Ball State, these instructors go on to "green their curricula" in courses ranging from American literature to interior design.

At many schools, activities and teaching are linked. For instance, students at Philadelphia University in Philadelphia remove invasive plants and plant native ones as part of a for-credit course focused on those topics. In the best tradition of colleges and universities, classroom learning and real-world learning go hand in hand.

### Weighing THE Issues | Sustainability on Your Campus

Find out what sustainability efforts are being made on your campus. What results have these achieved so far? What further efforts would you like to see pursued on your campus? Do you foresee any obstacles to these efforts? How could these obstacles be overcome? How could you become involved?

## Organizations are available to assist campus efforts

Many campus sustainability initiatives are supported by organizations such as University Leaders for a Sustainable Future, the Association for the Advancement of Sustainability in Higher Education, and the National Wildlife Federation's Campus Ecology program. These organizations act as information clearinghouses for campus sustainability efforts. Each year the NWF program recognizes the most successful campus sustainability initiatives, and Ball State University and dozens of other schools have received Campus Ecology awards. With Ball State's Greening of the Campus conferences and the assistance of these organizations, it is easier than ever to start sustainability efforts on your own campus and obtain the support to carry them through to completion.

# Sustainability and Sustainable Development

Efforts toward sustainability on college and university campuses parallel efforts in the world at large. As more people come to appreciate Earth's limited capacity to accommodate our rising population and consumption, they are voicing concern that we will need to modify our behaviors, institutions, and technologies if we wish to sustain our civilization and the natural environment on which it depends. In the quest for sustainability, the strategies pursued on campuses reflect those pursued in the wider society, and they also can serve as models.

When people speak of *sustainability*, what precisely do they mean to sustain? Generally they mean to sustain human institutions in a healthy and functional state—and also to sustain ecological systems in a healthy and functional state. The contributions of biodiversity (•pp. 310–314) and ecosystem goods and services (•pp. 38, 40, 50–51) to human welfare are tremendous. Indeed, they are so fundamental (some would say infinitely valuable, thus literally priceless) that we have long taken them for granted.

## Sustainable development aims to achieve a triple bottom line

We first explored sustainable development in our opening chapter (• p. 21), and offered the United Nations' definition: "Development that meets the needs of the present without compromising the ability of future generations to meet their own needs." Today, it is widely recognized that sustainability does not mean simply protecting the environment against the ravages of human development. Instead, it means finding ways to promote social justice, economic well-being, and environmental quality at the same time. Meeting this *triple bottom line* (• p. 21) is the goal of modern sustainable development. Achieving this goal is most pressing in nations of the developing world, but it is a vital need everywhere. It is our primary challenge for this century and likely for the rest of our species' time on Earth.

"Environmental sustainability" is one of the United Nations' *Millennium Development Goals* set by the international community at the turn of the century, and the drive for sustainability overall meshes with all eight goals. In the past few years, the Millennium Project and the Millennium Ecosystem Assessment (•p. 18) each have determined that:

- Environmental degradation is a major barrier to achieving the Millennium Development Goals.
- Investing in environmental assets and management is vital to relieving poverty, hunger, and disease.
- Reaching environmental goals requires progress in eradicating poverty.

The many actions being taken today at campuses such as Ball State's and by governments, businesses, industries, organizations, and individuals across the globe are giving people optimism that achieving these goals and developing in sustainable ways is within reach.

## Environmental protection can enhance economic opportunity

Reducing resource consumption and waste often saves money, as many colleges and universities discover when they embark on sustainability initiatives. Sometimes savings accrue immediately, and other times an up-front investment brings long-term savings.

FIGURE 23.8 The northern spotted owl (*Strix occidentalis occidentalis*) has become a symbol of the "jobs-versus-environment" debate. This bird of the Pacific Northwest rainforest is considered endangered because of the logging of mature forests. Proponents of logging argue that laws protecting endangered species cause economic harm and job loss. Advocates of endangered species protection argue that unsustainable logging practices pose a larger risk of job loss.

For society as a whole, attention to environmental quality can enhance economic opportunity by providing new types of employment. This reality contrasts with the common perception that environmental protection hurts the economy by costing people jobs. In the controversy over logging of old-growth forest in the Pacific Northwest (Chapter 12), protection for the northern spotted owl (**Figure 23.8**) under the U.S. Endangered Species Act (• p. 317) set limits on timber extraction. Proponents of logging claimed that the restrictions cost local loggers their jobs. Some jobs were indeed lost, but loggers' jobs are far more at risk when timber companies cut trees at unsustainable rates and then leave a region, seeking mature forests elsewhere, as has happened in region after region throughout U.S. history (• pp. 334–335). This is playing out in the Pacific Northwest now, as North American timber companies move operations abroad.

The jobs-versus-environment debate frequently overlooks the fact that as some industries decline, others spring up to take their place. As jobs in logging, mining, and manufacturing have disappeared in developed nations over the past few decades, jobs have proliferated in service occupations and high-technology sectors. As we decrease our dependence on fossil fuels, jobs and investment opportunities are opening up in renewable energy sectors, such as wind power and fuel cell technology (Chapter 21).

Thus, environmental protection need not lead to economic stagnation, but instead is likely to enhance economic opportunity. A recent U.S. government review concluded that the economic benefits of environmental regulations greatly exceed their economic costs (see "The Science behind the Story," Chapter 3, • pp. 80–81). This connection is also suggested by the fact that the U.S. and global economies have each expanded rapidly in the past 30 years, the very period during which environmental protection measures have proliferated.

Moreover, if we look beyond conventional economic accounting (which measures only private economic gain and loss) and instead include external costs and benefits (• p. 42) that affect people at large, then environmental protection becomes still more valuable. Take several studies reviewed by the Millennium Ecosystem Assessment (• p. 18): They each show how overall economic value is maximized by conserving natural resources rather than exploiting them for short-term private gain (**Figure 23.9**).

Although people desire private monetary gain, they also desire to live in areas that have clean air, clean water, intact forests, and parks and open space. Environmental protection increases a region's attractiveness, drawing more residents and increasing property values and the tax revenues that help fund social services. As a result, those regions that act to protect their environments are generally the ones that retain and increase their wealth and quality of life.

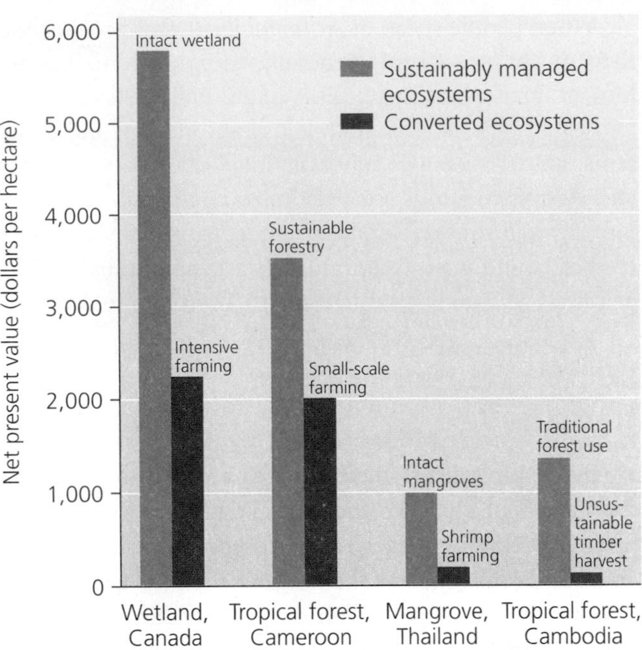


**FIGURE 23.9** Once external costs and benefits (• p. 42) are factored in, the economic value of sustainably managed ecosystems generally exceeds the economic value of ecosystems that have been converted for intensive private resource harvesting. Shown are land values calculated by researchers in four such comparisons from sites around the world. Data from Millennium Ecosystem Assessment, 2005.

## What accounts for the perceived economy-versus-environment divide?

If environmental protection and economic development are mutually reinforcing, what, then, accounts for the common view that we cannot simultaneously protect the environment and provide for people's needs? One proximate explanation lies in the fact that economic development since the industrial revolution has so clearly diminished biodiversity, decreased habitat, and degraded ecological systems. A second proximate explanation lies in the fact that many people believe command-and-control environmental policy (• pp. 77–78) poses excessive costs for industry and restricts rights of private citizens.

An ultimate explanation may lie in our species' long history. For the thousands of years that we lived as nomadic hunter-gatherers, population densities were low, and consumption and environmental impact were limited. With natural resources in little danger of running out, people were free to exploit them limitlessly and had little reason to adopt a conservation ethic. Our establishment of sedentary agricultural societies, followed by urbanization, has increased our impact while also causing us to overlook the connections between our economies and our environments. It is common to hear "humans and the environment" or "people and nature" being set in contrast, as though they were separate. Some philosophers venture to say that the perceived dichotomy between humans and nature is the root of all our environmental problems.

## Humans are not separate from the environment

On a day-to-day basis, it is easy to feel disconnected from the natural environment, particularly in industrialized nations and large cities. We live inside houses, work in shuttered buildings, travel in enclosed vehicles, and generally know little about the plants and animals around us. Millions of urban citizens have never set foot in an undeveloped area. Just a few centuries or even decades ago, most of the world's people were able to name and describe the habits of the plants and animals that lived nearby. They knew exactly where their food, water, and clothing came from. Today it seems that water comes from the faucet, clothing from the mall, and food from the grocery store. It's little wonder we have lost track of the connections that tie us to our natural environment.

**FIGURE 23.10** A banana split eaten at an ice cream shop in Tulsa, Denver, or Des Moines consists of ingredients from around the world, whose production has impacts on the environments of many far-away locations. Ice cream requires milk from dairy cows that graze pastures or are raised in feedlots on grain grown in industrial monocultures. Ice cream is sweetened with sugar from sugar beet farms or sugarcane plantations. The banana was shipped thousands of miles by oil-fueled transport from a tropical country, where it grew on a plantation that displaced rainforest and where it was liberally treated with fertilizers and fungicides. Fruits and nuts grown in California's Central Valley were irrigated generously with water piped in from the Sierras. The spoon originated with metal ores mined along with thousands of tons of soil and processed into stainless steel using energy from fossil fuels.

However, this doesn't make our connections to the environment any less real. Consider a thoroughly un-"natural" (yet delicious!) invention of the human species: the banana split (**Figure 23.10**). Even in this triumph of human creation, seemingly concocted *de novo* at an ice cream shop, each and every element has ties to the resources of the natural environment, and each exerts environmental impacts.

Once we learn to consider where the things we use and value each day actually come from, it becomes easier to see how people are part of the environment. And once we reestablish this connection, it becomes readily apparent that our own interests are best served by preservation or responsible stewardship of the natural systems around us. Because what is good for the environment can also be good for people, win-win solutions are very much within reach, if we learn from what science can teach us, think creatively, and act on our ideas.

# Strategies for Sustainability

Sustainable solutions to environmental problems are numerous, and we have seen specific examples throughout this book. The challenges lie in being imaginative enough to think of solutions and being shrewd and dogged enough to overcome political or economic obstacles that may lie in the path of their implementation. We will now summarize several broad strategies or approaches that can spawn sustainable solutions (**Table 23.2**).

**TABLE 23.2 Some Major Approaches to Sustainability**

- Refine our ideas about economic growth and quality of life
- Reduce unnecessary consumption
- Limit population growth
- Encourage green technologies
- Mimic natural systems by promoting closed-loop industrial processes
- Think in the long term
- Enhance local self-sufficiency, and embrace some aspects of globalization
- Be politically active
- Vote with our wallets
- Promote research and education

## We can refine our ideas about economic growth and quality of life

It is conventional among economists and the policymakers who heed their advice to speak of economic growth as an ultimate goal. Many politicians view nurturing an expanding economy as their prime responsibility while in office. Yet economic growth is merely a tool with which we try to attain the real goal of maximizing human happiness. Thus, if economic growth depends on an ever-increasing consumption and depletion of nonrenewable resources, then we will not be able to attain long-term happiness by endlessly expanding the size of our economy (• pp. 43–47).

FIGURE 23.11 Citizens of the United States consume more than the people of any other nation. Unless we find ways to increase the sustainability of our manufacturing processes, this rate of consumption cannot be sustained in the long run.

We may also want to incorporate external costs (• p. 42) into the market prices of goods and services. Currently, goods and services are priced as though pollution and resource extraction involved no costs to society. If we can make our accounting practices reflect indirect consequences to the public, then we can provide a clearer view of the full costs and benefits of any given action or product. In that case, the free market could become the optimal tool for improving environmental quality, our economy, and our quality of life. Moreover, implementing green taxes (• p. 79) and phasing out harmful subsidies could hasten our attainment of prosperous and sustainable economies. The political obstacles to this are considerable, and such changes will require educated citizens to push for them and courageous policymakers to implement them.

## We can consume less

Economic growth is largely driven by consumption: the purchase of material goods and services (and thus the use of resources involved in their manufacture) by consumers (**Figure 23.11**). Our tendency to believe that more, bigger, and faster are always better is reinforced by advertisers seeking to sell more goods more quickly. Consumption has grown tremendously, with the wealthiest nations leading the way. The United States, with less than 5% of the world's population, consumes 30% of world energy resources and 40% of total global resources. U.S. houses are larger than ever, sports-utility vehicles are among the most popular automobiles, and many citizens have more material belongings than they know what to do with. We think nothing today of having home computers with high-speed Internet access, let alone the televisions, telephones, refrigerators, and dishwashers that were marvels just decades ago.

Because many of Earth's natural resources are limited and nonrenewable, consumption cannot continue growing forever. Eventually, if we do not shift to sustainable resource use, per capita consumption will drop for rich and poor alike as resources dwindle. Cornucopian critics often scoff at the notion that resources are limited, but we must remember that our perspective in time is limited and that our consumption is taking place within an extraordinarily brief slice of time in the long course of history (**Figure 23.12**). Our lavishly consumptive lifestyles

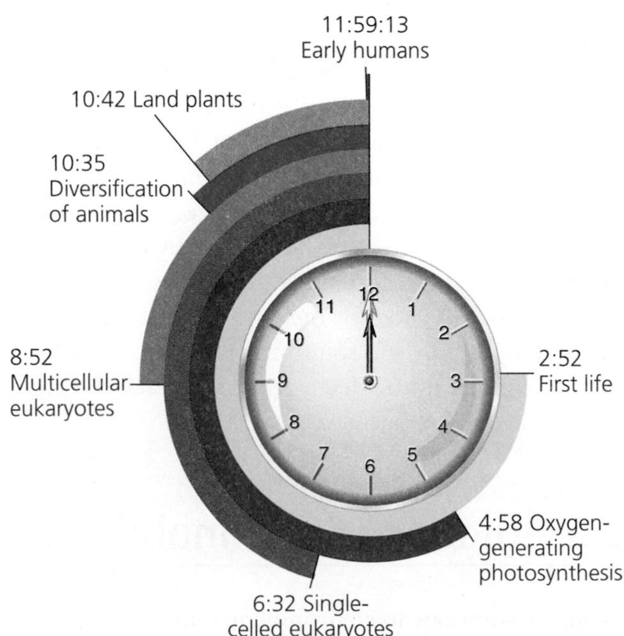


FIGURE 23.12 By viewing Earth's 4.5-billion-year history as a 12-hour clock, we can gain a better understanding of relative time scales across the immense span of geological time. *Homo sapiens,* as a species, has come into existence only during the final 1 or 2 seconds, around 11:59:59. The agricultural and industrial revolutions that have increased our environmental impacts have taken up only a minuscule fraction of a second.

are a brand-new phenomenon on Earth. We are enjoying the greatest material prosperity in all of history, but if we do not find ways to make our wealth sustainable, the party may not last much longer.

Fortunately, material consumption alone does not reflect a person's quality of life. For many people in industrialized nations, the accumulation of possessions has not brought contentment. Observing how affluent people often fail to find happiness in their material wealth, social critics have given this phenomenon a name like a dread disease: *affluenza* (• p. 44). Whether or not it is a disease, scientific research does back up the contention that money cannot buy nearly as much happiness as people typically believe (**Figure 23.13**). Although economic growth is generally equated with "progress," true progress consists of an increase in human happiness. In the end we are, one would hope, more than just the sum of what we buy.

We can reduce our consumption while enhancing our quality of life—squeezing more from less—in at least three ways. One way is to improve the technology of materials and the efficiency of manufacturing processes, so that industry produces goods using fewer natural resources. Another way is to develop a sustainable manufacturing system—one that is circular and based on recycling, in which the waste from a process becomes raw material for input into that process or others (• pp. 645–646). A third way is to modify our behavior, attitudes, and lifestyles to minimize consumption. At the outset, such choices may seem like sacrifices, but people who have slowed down the pace of their busy lives and freed themselves of an attachment to material possessions say it can feel tremendously liberating.

## Population growth must eventually cease

Just as continued growth in consumption is not sustainable, neither is growth in the human population. We have seen (• pp. 129–131) that populations may grow exponentially for a time but eventually encounter limiting factors and decline or level off. We have used technology to increase Earth's carrying capacity for our species, but our population cannot continue growing forever; sooner or later, human population growth will end. The question is how: through war, plagues, and famine, or through voluntary means as a result of wealth and education?

The demographic transition (• pp. 219–221) is already far along in many industrialized nations thanks to urbanization, wealth, education, and the empowerment of women. If today's developing nations also pass through a demographic transition, then there is hope that humanity may halt its population growth while creating a more prosperous and equitable society.

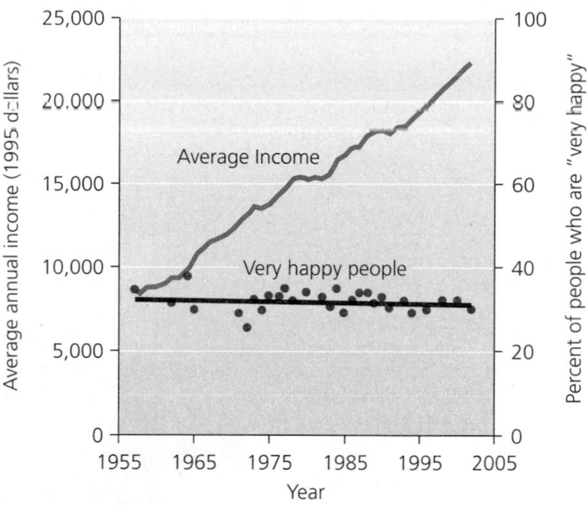


**(a) Happiness versus income, through time**

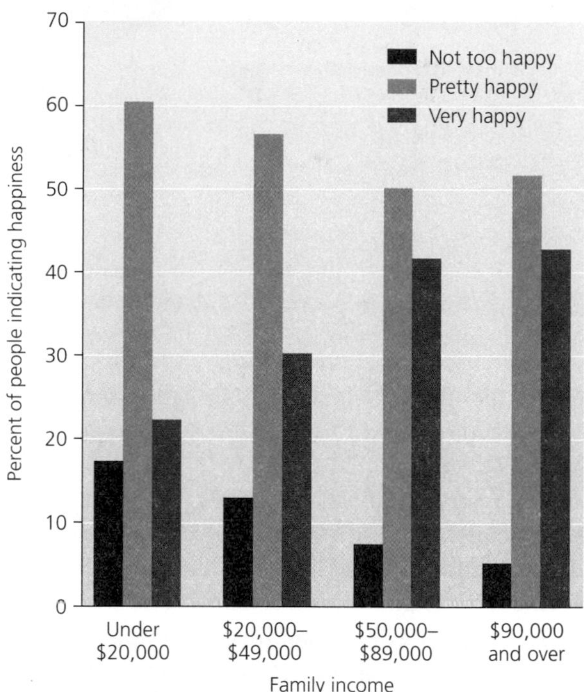


**(b) Happiness versus family income**

FIGURE 23.13 Although average income in the United States has risen steadily in the past half-century, the percentage of people reporting themselves as being "very happy" has remained stable or declined slightly **(a)**. A different study found that Americans in higher income brackets were more likely to report themselves as "very happy" and less likely to report themselves as "not too happy." **(b)**. However, the difference was far less than people generally expect, and the researchers showed this, too. For instance, when asked how much a fivefold increase in income improves a person's mood day to day, respondents guessed on average that such an increase would improve mood by 32%, but the actual improvement, from respondents' self-reporting, was only 12%. Data in (a) from Myers, D. G. 2000. *The American paradox: Spiritual hunger in an age of plenty*. New Haven, CT: Yale University Press; and Gardner, G., and E. Assadourian, 2004. Rethinking the good life, pp. 164–179 in *State of the World 2004*, Worldwatch Institute. Data in (b) from Kahneman, D., et al. 2006, Would you be happier if you were richer? A focusing illusion. *Science* 312: 1908–1910.

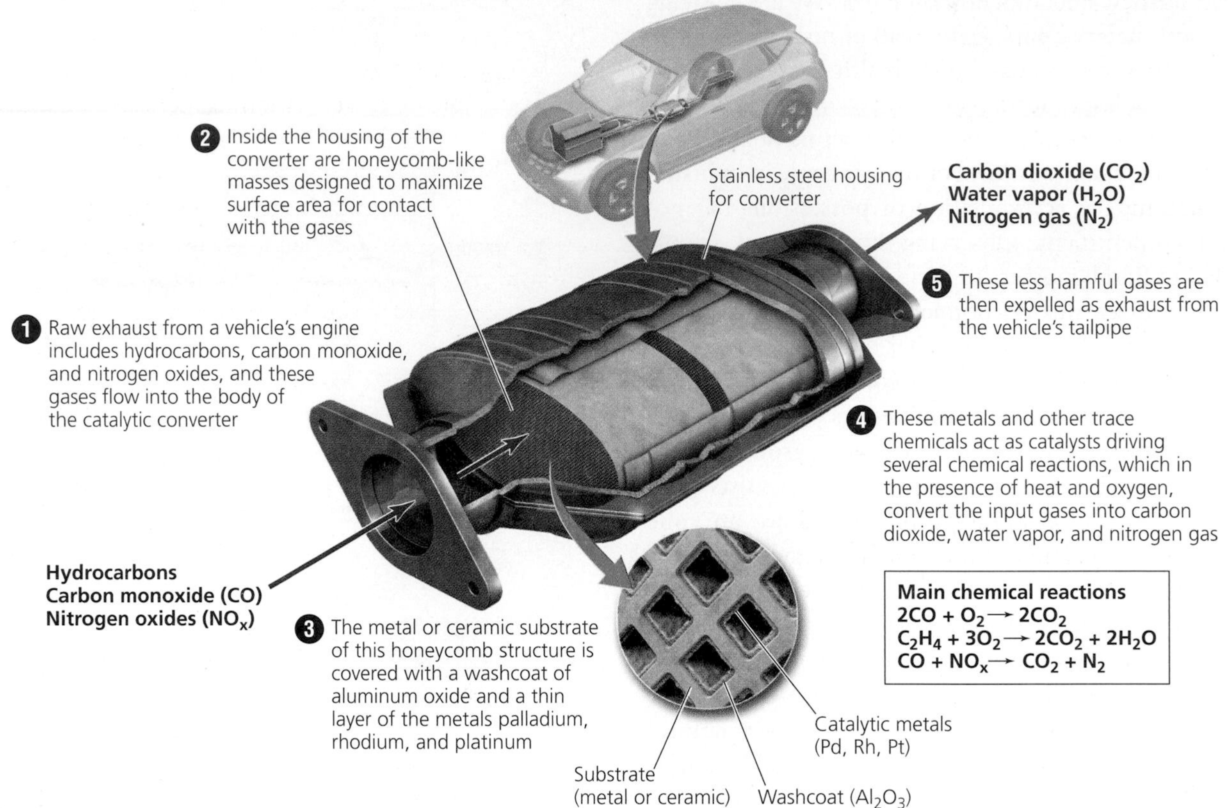


FIGURE 23.14 The catalytic converter is a classic example of green technology. This device filters air pollutants from vehicle exhaust and has helped bring about the improvement of air quality in the United States and other nations.

## Technology can help us toward sustainability

It is largely technology—developed with the agricultural revolution, the industrial revolution, and advances in medicine and health—that has spurred our population increase. Technology has magnified our impact on Earth's environmental systems, yet it can also give us ways to reduce our impact. Recall the I=PAT equation (• p. 211), which summarizes human environmental impact (I) as the interaction of population (P), consumption or affluence (A), and technology (T). Technology can exert either a positive or negative value in this equation. The short-sighted use of technology may have gotten us into this mess, but wiser use of environmentally friendly, or "green," technology can help get us out.

In recent years, technology has intensified environmental impact in developing countries as industrial technologies from the developed world have been exported to poorer nations eager to industrialize. In developed nations, meanwhile, green technologies have begun mitigating our environmental impact. Catalytic converters on cars have reduced emissions (**Figure 23.14**), as have scrubbers on industrial smokestacks (Figure 17.14, • p. 485). Recycling technology and advances in wastewater treatment are helping reduce our waste output. Solar, wind, and geothermal energy technologies are producing cleaner renewable energy. Countless technological advances such as these are one reason that people of the United States and western Europe today enjoy cleaner environments—although they consume far more—than people of eastern Europe or rapidly industrializing nations such as China.

## Industry can mimic natural systems

As industries seek to develop green technologies and sustainable practices, they have an excellent model: nature itself. As we saw in Chapter 7 and throughout this book, environmental systems tend to operate in cycles consisting of feedback loops and the circular flow of materials. In natural systems, output is recycled into input. In contrast, human manufacturing processes have run on a linear model in which raw materials are input and processed to create a product, while by-products and waste are generated and discarded. Some forward-thinking industrialists are making their processes more sustainable by transforming linear pathways into circular ones, in which waste is recycled and reused (• p. 645–646). For instance, several companies now produce carpets that can be retrieved from the consumer when they wear out, and these materials are then recycled to create new carpeting (• p. 646). Some automobile manufacturers are planning cars that can be disassembled and recycled into new cars. Proponents of this industrial model see little reason why virtually all appliances and other products cannot be recycled, given the right technology. Their ultimate vision is to create truly closed-loop industrial processes, generating no waste.

## We can think in the long term

To be sustainable, a solution must work in the long term. Often the best long-term solution is not the best short-term solution, which explains why much of what we currently do is not sustainable. Policymakers in democracies often act for short-term good because they aim to produce immediate, positive results so that they will be reelected. This poses a major hurdle for addressing environmental dilemmas, many of which are cumulative, worsen gradually, and can be resolved only over long periods. Often the costs of addressing an environmental problem are short term, whereas the benefits are long term, giving politicians little incentive to tackle the problem. In such a situation, citizen pressure on policymakers is especially vital, because policy is an essential tool for pursuing sustainability (see "The Science behind the Story," • pp. 674–675).

Businesses may act according either to long-term or short-term interests. A business committed to operating in a particular community for a long time has incentive to sustain environmental quality. However, a business merely attempting to make a profit and move on has little incentive to invest in environmental protection measures that involve short-term costs.

## We can promote local self-sufficiency and embrace some aspects of globalization

As our societies become more globally interconnected, we experience a diversity of impacts, positive and negative. To many people, encouraging local self-sufficiency is an important element of building sustainable societies. When people feel closely tied to the area in which they live, they tend to value the area and seek to sustain its environment and human communities. This line of reasoning is frequently made in relation to locally based organic or sustainable agriculture (• pp. 285–289).

Many advocates of local self-sufficiency criticize globalization. However, as the ecological economist Herman Daly has explained, globalization means different things to different people. Those who view it as a positive phenomenon generally focus on how people of the world's diverse cultures are increasingly communicating and learning about one another. Books, airplanes, television, and the Internet have made us more aware of one another's cultures and more likely to respect and celebrate, rather than fear, differences among cultures.

Those who view globalization in a negative light generally cite the homogenization of the world's cultures, by which a few cultures and worldviews displace many others. For instance, the world's many languages are going extinct with astonishing speed. Traditional ways of life in many areas are being abandoned as more people take up the material and cultural trappings of a few dominant cultures, particularly that of the United States.

In recent years, many people have reacted against this homogenization and the growing power of large multinational corporations. In France, farm activist Jose Bove became a popular hero when he wrecked a McDonald's restaurant with his tractor to protest what many French farmers view as a threat to local French cuisine. In Seattle in 1999, thousands of protestors picketed a meeting of the World Trade Organization, which they viewed as a symbol of Western market capitalism (**Figure 23.15**). Since then, protestors have picketed every WTO meeting.

**FIGURE 23.15** Thousands of protesters picketed the World Trade Organization's meeting in Seattle in 1999, criticizing the homogenizing effects of globalization, as well as relaxations in labor and environmental protections brought about by free trade.

THE SCIENCE BEHIND THE STORY

*Dr. Daniel Esty, Yale University*

## Rating the Environmental Performance of Nations

How effectively is our world moving toward sustainability? That depends on where in the world you are, because the governments of some nations are performing better than others.

To measure the progress of nations toward environmental sustainability, researchers devised the *Environmental Performance Index (EPI)*, which rates countries using data from 16 indicators of environmental conditions for which governments can be held accountable. A report detailing the results was produced and published online in 2006 by Daniel Esty and five colleagues at the Yale Center for Environmental Law and Policy and Columbia University's Center for International Earth Science Information Network, in collaboration with the World Economic Forum and the Joint Research Centre of the European Commission.

Esty and his colleagues pointed out that the U.N. Millennium Development Goals (• p. 667) did not define how to quantify progress on environmental measures. The lack of quantitative measures, Esty's team said, had stymied progress on how to implement effective environmental policy.

To address this problem, these researchers aimed to track the performance of environmental policy with the same quantitative rigor as statistics are tracked for health, poverty reduction, and other development goals. Giving nations scores and ranking them would reveal "leaders and laggards," showing which nations are on the right track and which are not.

The researchers gathered internationally available data from U.N. and other sources on 16 indicators, which funneled into six categories: environmental health, air quality, water resources, productive natural resources, sustainable energy, and biodiversity and habitat. The latter five categories were considered components of "ecosystem vitality." Esty's team scored nations on their performance in each category and gave them overall scores based 50% on their environmental health score and 50% on the five ecosystem vitality scores. Scores for each category ranged from zero to 100, with 100 representing a target value established by international consensus. The researchers included 133 nations in their 2006 report (60 nations could not be included because of lack of data).

The main pattern in the results was conspicuous: Nations with the highest scores (see the first table) are

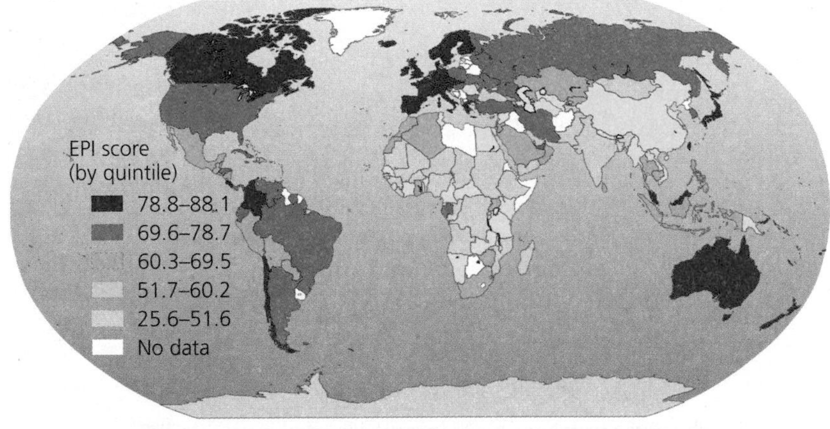


Researchers scored nations for their performance in approaching environmental sustainability goals. Higher Environmental Performance Index values (red is highest in this map) indicate better performance. Data from Esty, D. C., et al., 2006. *Pilot 2006 Environmental Performance Index*. New Haven, CT: Yale Center for Environmental Law and Policy.

Daly and others argue that globalization entails a process in which multinational corporations attain greater and greater power over global trade while governments retain less and less. Most critics of globalization consider corporations less likely than governments to support environmental protection, so they feel that globalization will hinder progress toward sustainability. Moreover, the U.S. model of market capitalism promotes a high-consumption lifestyle, which does indeed threaten efforts to attain sustainable solutions.

On the positive side, globalization may foster sustainability because Western democracy, as imperfect as it is, serves as a model and beacon for people living under repressive governments. Open societies allow for entrepreneurship and the flowering of creativity in business, research, and academia. Millions of free minds thinking about issues are more likely to come up with sustainable solutions than the minds of a few holding authoritarian power.

**Top-Ranked Nations**

| Nation | EPI Score |
|---|---|
| New Zealand | 88.0 |
| Sweden | 87.8 |
| Finland | 87.0 |
| Czech Republic | 86.0 |
| United Kingdom | 85.6 |
| Austria | 85.2 |
| Denmark | 84.2 |
| Canada | 84.0 |
| Malaysia | 83.3 |
| Ireland | 83.3 |

Data from Esty, D. C., et al., 2006. *Pilot 2006 Environmental Performance Index*. New Haven, CT: Yale Center for Environmental Law and Policy.

**Bottom-Ranked Nations**

| Nation | EPI Score |
|---|---|
| Sudan | 44.0 |
| Bangladesh | 43.5 |
| Burkina Faso | 43.2 |
| Pakistan | 41.1 |
| Angola | 39.3 |
| Ethiopia | 36.7 |
| Mali | 33.9 |
| Mauritania | 32.0 |
| Chad | 30.5 |
| Niger | 25.7 |

Data from Esty, D. C., et al., 2006. *Pilot 2006 Environmental Performance Index*. New Haven, CT: Yale Center for Environmental Law and Policy.

wealthy, industrialized nations with the capacity to commit substantial resources toward environmental protection. Nations with the lowest scores (see the second table) are largely developing nations with few resources to invest. They include countries with dense populations and stressed ecosystems that are trying to industrialize, those with arid environments and limited natural resources, and those facing extreme poverty.

This correlation between economic vitality and environmental protection was expected, but the researchers were more interested in the variation around this trend. For any given level of income or development, some nations vastly outperformed others. The researchers say this demonstrates that other factors are at work—namely, the political choices national leaders make. Indeed, their data show a strong correlation between EPI scores and "good governance," which involves aspects such as rule of law, open political debate, and lack of corruption.

These interpretations have touched sensitive nerves in nations that did not rank as highly as they would have liked. The United States placed 28th, for instance—far below most of its economic peers.

Some critics say the rankings have little meaning because only certain data sets were measured, and these were then weighted subjectively. Supporters counter that although no one will ever agree on exact numbers, the EPI provides a useful tool for evaluating actions and investments, highlighting policies that work, and identifying future priorities.

Whereas controversy generated by the overall rankings drew media attention, other trends were apparent for those who examined the data more deeply. For instance, wealthy industrialized nations tended to score highest in environmental health and lowest in ecosystem vitality indicators, especially biodiversity and habitat. This is presumably because they possess money to invest in protecting the health of their citizens but achieved economic development by exploiting their natural resources and degrading their natural environment.

Conversely, poorer nations tended to score low on environmental health and more highly on indicators that reflected the availability of remaining unexploited natural resources. Overall across all nations, performance was worst for the indicators of renewable energy and wilderness protection.

The EPI scores are only as solid as the data that go into them, and not every nation provides reliably accurate data on every issue. Moreover, Esty's group acknowledges that there are plenty of ways in which their formulas might be improved in the future. Nonetheless, the Environmental Performance Index provides a unique way to assess what is working in environmental policy. As the methodology is refined year by year, the EPI promises to become a more accurate performance indicator and a more helpful policy tool.

### Weighing THE Issues | Globalization

From your own experience, what advantages and disadvantages do you see in globalization? Have you personally benefited or been hurt by it in any way? In what ways might promoting local self-sufficiency be helpful for the pursuit of global sustainability? In what ways might it not?

Now consider this: Many people enjoy eating at Vietnamese restaurants in the United States, yet many who criticize globalization would frown on the presence of McDonald's restaurants in Vietnam. Do you think this represents a double standard, or are there reasons the two are not comparable?

## Citizens exert political influence

Politically open democracies offer a compelling route for pursuing sustainability: the power of the vote. Many of the changes needed to attain sustainable solutions require policymaking, and individually and collectively we can guide our political leaders to enact policies for sustainability. Policymakers respond to whoever exerts influence. Corporations and interest groups employ lobbyists to influence politicians all the time. Citizens in a democratic republic have the same power, *if* they choose to exercise it. You can exercise your power at the ballot box, by attending public hearings, by donating to advocacy groups that promote ideas you favor, and by writing letters and making phone calls to office-holders. You may be surprised how little input policymakers receive from the public; sometimes a single letter or phone call can, in fact, make a difference.

Today's environmental laws came about because citizens pressured their governmental representatives to tackle environmental problems. The raft of legislation enacted in the 1960s and 1970s in the United States and other nations might never have come about had ordinary citizens not stepped up and demanded action. We owe it to our children to be engaged and to act responsibly now so that they have a better world in which to live. The words of anthropologist Margaret Mead are worth repeating: "Never doubt that a small group of thoughtful, committed people can change the world. Indeed, it's the only thing that ever has."

### Consumers vote with their wallets

Expressing one's preferences through the political system is vital, but we also wield influence in the choices we make as consumers. When products produced sustainably are ecolabeled (• p. 49), consumers can "vote with their wallets" by purchasing these products. Consumer choice has helped drive sales of everything from recycled paper to organic produce to "dolphin-safe" tuna.

Individuals can multiply their own influence by promoting "green" purchasing habits at their school or workplace. We saw how purchasing power at colleges and universities has spurred sales of certified sustainable wood, organic food, energy-efficient appliances, and more. Employees in businesses and government agencies can often promote change within those institutions by voicing their preferences in purchasing decisions.

### Promoting research and education is vital

None of these approaches will succeed fully if the public is not aware of their importance. An individual's decisions to reduce consumption, purchase ecolabeled products, or vote for candidates who support sustainable approaches will have limited impact unless many others do the same. Individuals can influence large numbers of people by educating others and by serving as role models through their actions. The campus sustainability efforts at Ball State University and so many other colleges and universities accomplish both approaches. Moreover, the discipline of environmental science plays a key role in providing information that people can use to make wise decisions about environmental issues. By promoting scientific research and by educating the public about environmental science, we can all assist in the pursuit of sustainable solutions.

## Precious Time

The pace of our lives is getting faster, and life's commotion can make it hard to give attention to problems we don't need to deal with on a daily basis. The world's sheer load of environmental dilemmas can feel overwhelming, and even the best-intentioned among us may feel we have little time to devote to saving the planet.

However, the natural systems we depend on are changing quickly. Many human impacts continue to intensify, including deforestation, overfishing, land clearing, wetlands draining, and resource extraction. Our window of opportunity for turning some of these trends around is getting short. Even if we can visualize sustainable solutions to our many problems, how can we possibly find the time to implement them before we do irreparable damage to our environment and our own future?

### We need to reach again for the moon

On May 25, 1961, U.S. President John F. Kennedy announced that within the decade the United States would be "landing a man on the moon and returning him safely to the Earth" (**Figure 23.16**). It was a bold and astonishing statement; the technology to achieve this unprecedented, almost unimaginable, feat did not yet exist. Kennedy's directive had powerful motivation behind it, however. The United States was dueling the Soviet Union in the Cold War. In this competition for global hegemony, the two nations, held mutually at bay with nuclear weapons, tried to prove their mettle by other means. The race for dominance in space became the centerpiece of the rivalry. Early victories in the "space race" went to the Soviet Union, which sent humanity's first satellite into orbit, followed it with others, and then sent the first human being into space. The United States, meanwhile, was crashing rockets.

FIGURE 23.16 In 1961, U.S. President John F. Kennedy called on Congress to fund a space program to send men to the moon before 1970. Addressing our environmental problems and shifting our political, economic, and social institutions to a paradigm of sustainable development will require still more vision, resolve, and commitment. The fact that astronauts reached the moon just 8 years after Kennedy's speech demonstrates the power of human ingenuity in meeting a challenge, and provides hope that we will be able to meet the larger challenge of living sustainably on Earth.

The prospect of "losing" the space race prompted Kennedy's administration to set a national goal on an ambitious timeline. Congress supplied funding, NASA performed the science and engineering, and in 1969 astronauts walked on the moon. The United States accomplished this milestone in human history by building public support for a goal and giving its scientists and engineers the wherewithal to develop technology and strategies to meet the goal. Similarly great challenges were met when the United States confronted the Great Depression, threw itself into World War II, and conducted the Marshall Plan after the war to help rebuild western Europe.

Today humanity faces a challenge more important than any previous one—the challenge to achieve sustainability. Attaining sustainability is a larger and more complex process than traveling to the moon. However, it is one to which every single person on Earth can contribute; in which government, industry, and citizens can all cooperate; and toward which all nations can work together. If America was able to reach the moon in a mere 8 years, then certainly humanity can begin down the road to sustainability with comparable speed. Human ingenuity is capable of it; we merely need to rally public resolve and engage our governments, institutions, and entrepreneurs in the race.

## We must pass through the environmental bottleneck

Human ingenuity and compassion give us reason to hope that we may achieve sustainability before doing too much damage to our planet and our prospects, but we must be realistic about the challenges that lie ahead. As we deplete the natural capital we can draw on, we give ourselves, and the rest of the world's creatures, less room to maneuver. Until we implement sustainable solutions, we will be squeezing ourselves through a progressively tighter space, like being forced through the neck of a bottle. The key question for the future of our species and our planet is whether we can make it safely through this bottleneck. Biologist Edward O. Wilson (• pp.314, 316) has written eloquently of this view:

> At best, an environmental bottleneck is coming in the twenty-first century. It will cause the unfolding of a new kind of history driven by environmental change. Or perhaps an unfolding on a global scale of more of the old kind of history, which saw the collapse of regional civilizations [in] Mesopotamia, and subsequently Egypt, then the Mayan and many others. . . . Somehow humanity must find a way to squeeze through the bottleneck without destroying the environments on which the rest of life depends.

## We must think of Earth as an island

We began this book with the vision of Earth as an island, and indeed that is what it is (**Figure 23.17**). Islands can be paradise, as Easter Island (• pp. 8–9) likely was when the

FIGURE 23.17 This photo of Earth, taken by astronauts orbiting the moon, shows our planet as it truly is—an island in space. Everything we know, need, love, and value comes from and resides on this small sphere, so we had best treat it well.

Polynesians first reached it. But when Europeans arrived at Easter Island, they witnessed the aftermath of a civilization that had depleted its island's resources, degraded its environment, and collapsed as a result. For the few people who remained of the once-mighty culture, life was difficult and unrewarding. They had lost even the knowledge of the history of their ancestors, who had cut trees unsustainably, kicking the base out from beneath their prosperous civilization.

As Easter Island's trees disappeared, some individuals must have spoken out for conservation and for finding ways to live sustainably amid dwindling resources. And likely others ignored those calls and went on extracting more than the land could bear, assuming that somehow things would turn out all right. Indeed, whoever cut the last tree atop the most remote mountaintop could have looked out across the island and seen that it was the last tree. And yet that person cut it down.

It would be tragic folly to let such a fate occur to our planet as a whole. By recognizing this, by deciding to shift our individual behavior and our cultural institutions in ways that encourage sustainable practices, and by employing science to help us achieve these ends, we may yet be able to live happily and sustainably on our wondrous island, Earth.

## Conclusion

In any society facing dwindling resources and environmental degradation, there will be those who raise alarms and those who ignore them. Fortunately, in our global society today we have many thousands of scientists who study Earth's processes and resources. For this reason, we are amassing a detailed knowledge and an ever-developing understanding of our dynamic planet, what it offers us, and what impacts it can bear. The challenge for our global society today, our one-world island of humanity, is to support that science so that we may judge false alarms from real problems and distinguish legitimate concerns from thoughtless denial. This science, this study of Earth and of ourselves, offers us hope for our future.

## REVIEWING OBJECTIVES

**You should now be able to:**

**List and describe approaches being taken on college and university campuses to promote sustainability**

- Audits produce baseline data on how much a campus consumes and pollutes. (pp. 659–660)
- The most common campus sustainability efforts involve recycling and waste reduction. (pp. 660–661)
- Green buildings are being constructed on a growing number of campuses. (pp. 661–662)
- There are many ways to reduce water use, and these efforts often save money. (p. 662)
- Students have many feasible ways to conserve energy and promote renewable energy sources. (pp. 662–663)
- A current drive is to make campuses "carbon-neutral." (pp. 663–664)
- Dining services can help sustainability efforts by providing local food and reducing waste. (p. 664)
- Colleges and universities can favor sustainable products in institutional purchasing. (pp. 664–665)
- Campuses can use alternative fuels and vehicles and encourage bicycling, walking, and public transportation. (p. 665)
- Habitat restoration is one of the most popular campus sustainability activities. (pp. 665–666)
- Curricula are adding issues of sustainability. (p. 666)

**Explain the concept of sustainable development**

- Sustainable development entails environmental protection, economic development, and social justice. (p. 667)
- Proponents of sustainable development feel that economic development and environmental quality can enhance one another. (pp. 667–668)

**Discuss how protecting the environment can be compatible with promoting economic welfare**

- Environmental protection and green technologies and industries can create rich sources of new jobs. (pp. 667–668)
- Protecting environmental quality enhances a community's desirability and economy. (p. 668)

**Describe and assess key approaches to designing sustainable solutions**

- We discuss ten general approaches that can inspire specific sustainable solutions. (pp. 669–676)
- Growth in population and per capita consumption will likely need to be halted if we are to create a sustainable society. (pp. 670–671)
- Technology has traditionally increased environmental impact, but new "green" technologies can help reduce impact. (p. 672)

**Explain how time is limited but how human potential to solve problems is tremendous**

- Time for turning around our increasing environmental impacts is running short. (p. 676)
- The United States and other nations have met tremendous challenges before, so we have reason to hope that we will be able to attain a sustainable society. (pp. 676–677)

## TESTING YOUR COMPREHENSION

1. In what ways are campus sustainability efforts relevant to sustainability efforts in the broader society?
2. Name one way in which campus sustainability proponents have addressed each of the following areas: (1) recycling and waste reduction, (2) "green" building, (3) water conservation, (4) energy efficiency, (5) renewable energy, and (6) global climate change.
3. Name one way in which campus sustainability proponents have addressed each of the following areas: (1) dining services, (2) institutional purchasing, (3) transportation, (4) habitat restoration, and (5) curricula.
4. What do environmental scientists mean by *sustainable development*?
5. Describe three ways in which environmental protection can enhance economic well-being.
6. Why are many people now living at the highest level of material prosperity in history? Is this level of consumption sustainable? How can it feel good to consume less?
7. In what ways can technology help us achieve sustainability? How do natural processes provide good models of sustainability for manufacturing? Provide examples.
8. Why do many people feel that local self-sufficiency is important? What consequences of globalization may threaten sustainability? How can open democratic societies help to promote sustainability?
9. Explain Edward O. Wilson's metaphor of the "environmental bottleneck."
10. How can thinking of Earth as an island help prevent us from repeating the mistakes of previous civilizations?

## SEEKING SOLUTIONS

1. What sustainability initiatives would you like to see attempted on your campus? If you were to take the lead in promoting such initiatives, how would you go about it? What obstacles would you expect to face, and how would you deal with them?
2. Choose one item or product that you enjoy, and consider how it came to be. Think of as many components of the item or product as you can, and determine how each of them was obtained or created. Now refer to Figure 23.10. What steps were involved in creating your item's components, and where did the raw materials come from? How was your item manufactured? How was it delivered to you?
3. Do you think that we can increase our quality of life through development while also protecting the integrity of the environment? Discuss examples from your course or from other chapters of this book that illustrate possible win-win solutions. Are you familiar with any cases in your community or at your college that bear on this issue? Describe such a case, and state what lessons you would draw from it.
4. Reflect on the experiences of prior human civilizations and how they came to an end. What is your prognosis for our current human civilization? Do you see a vast world of independent cultures all individually responsible for themselves, or do you consider human civilization to be one great entity? If we accept that all people depend on the same environmental systems for sustenance, what resources and strategies do we have to ensure that the actions of a few do not determine the outcome for all and that sustainable solutions are a common global goal?
5. **THINK IT THROUGH** You have been elected president of your college class, and your school's administrators promise to be responsive to student concerns. Many of your fellow students are asking you to promote sustainability initiatives on your campus. Consider the many approaches and activities pursued by the colleges and universities mentioned in this chapter, and now think about your own school. Which of these approaches and activities are most needed at your school? Which might be most effective? What ideas would you prioritize and promote during your term as president of your class?
6. **THINK IT THROUGH** In our final "Think It Through" question, you are . . . *you!* In this chapter and throughout this book, you have encountered a diversity of ideas for sustainable solutions to environmental problems. Many of these are approaches you can pursue in your own life. Name at least five ways in which you think you can make a difference—and would most like to make a difference—in helping to attain a more sustainable society. For each approach, describe one specific thing you could do today or tomorrow or next week to begin.

## INTERPRETING GRAPHS AND DATA

An undergraduate class at Pennsylvania State University conducted an ecological assessment of one of their biology laboratory buildings in response to the question: "How is this building like an ecosystem?" The result of their assessment was a 52-page report outlining ways to reduce the ecological footprint of the Mueller Laboratory Building in the areas of energy use, water use, communications/computing, furnishings/renovation, maintenance, and food. The students found ways to save an estimated $45,500 per year in the cost of energy alone for a building occupied by 123 scientists and support staff. Their data on the current use and potential savings in the energy component of the ecological footprint are shown in the graph.

1. From the graph, estimate the amount of potential savings for each of the five areas identified in the Mueller Report. Approximately what percentage of the total electrical energy use of the building do these savings together represent?
2. What was the approximate cost of electricity in cents per kilowatt-hour used to calculate the savings of $45,500? Do you think this cost may have changed since the report was issued in 2001? How and why?
3. How is the building where you take your course like an ecosystem? How is it not like an ecosystem? Is your building operating sustainably? What improvements do you think could be made to it?

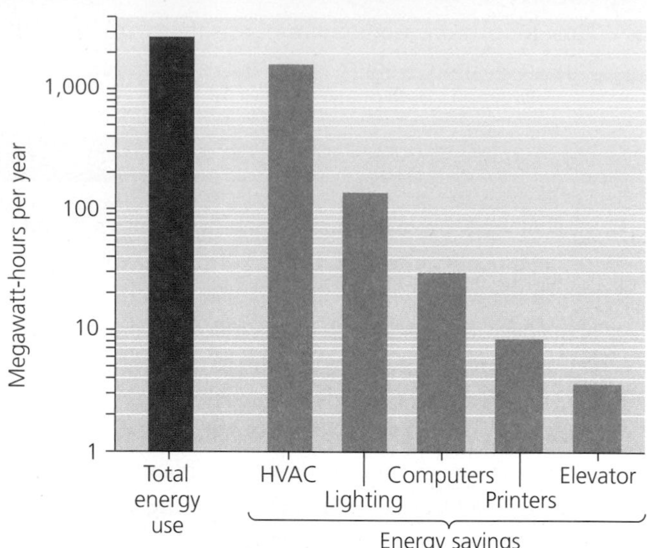


Total annual electrical energy use (red bar) and potential energy savings in five areas (orange bars) at Penn State University's Mueller Laboratory Building. The *y* axis is logarithmic, with each tick mark representing an increase equal to the value indicated below it on the axis. Data from Penn State Green Destiny Council. 2001. *The Mueller Report: Moving beyond sustainability indicators to sustainability action at Penn State.*

## CALCULATING ECOLOGICAL FOOTPRINTS

As we have seen throughout this book, individuals can contribute to sustainable solutions for our society and our planet in many ways. Some of these involve advocating for change at high levels of government or business or academia. But plenty of others involve the countless small choices we make in how we live our lives day to day. Where we live, what we buy, how we travel—these types of choices we each make as citizens, consumers, and human beings determine how we affect the environment and the people around us. As you know, such personal choices are summarized (crudely, but usefully) in an ecological footprint.

Turn back to the "Calculating Ecological Footprints" exercise in Chapter 1 (• p. 24), and recover the numerical value of your own personal ecological footprint that you calculated at the beginning of your course. Enter it in the table to the right. Now return to the same online ecological footprint calculator that you used for Chapter 1's "Calculating Ecological Footprints" exercise. For many of you, this will have been www.myfootprint.org or http://ecofoot.org. Take the footprint quiz again, and calculate your current footprint.

| | Footprint value (hectares per person) |
|---|---|
| **World average** | 2.23 |
| **U.S. average** | 9.6 |
| **Your footprint from Chapter 1** | |
| **Your footprint now** | |
| **Your footprint with three more changes** | |

1. Enter your current footprint, as determined by the online calculator, in the table. How does this value compare to your footprint at the beginning of your course? By what percentage did your footprint decrease or increase? If it changed, why do you think it changed? What changes have you made in your lifestyle since beginning this course that influence your environmental impact?
2. How does your personal footprint compare to the average footprint of a U.S. resident? How does it compare to that of the average person in the world? What do you think would be an admirable yet

realistic goal for you to set as a target value for your own footprint?

3. Now think of three changes in your lifestyle that would lower your footprint. These should be changes that you would like to make and that you believe you could reasonably make. Take the footprint quiz again, incorporating these three changes. Enter the resulting footprint in the table.

4. Now set as a goal reducing your footprint by 25%, and experiment by changing various answers in your footprint quiz. What changes would allow you to attain a 25% reduction in your footprint? What changes would be needed to reduce your footprint to the hypothetical target value you set in Question 2?

## Take It Further

Go to www.aw-bc.com/withgott or the student CD-ROM, where you'll find:

- Suggested answers to end-of-chapter questions
- Quizzes, animations, and flashcards to help you study
- *Research Navigator*™ database of credible and reliable sources to assist you with your research projects
- **GRAPHit!** Tutorials to help you interpret graphs
- **INVESTIGATEit!** Current news articles that link the topics that you study to case studies from your region to around the world

# APPENDICES

# Appendix A Some Basics on Graphs

Presenting data in ways that help make trends and patterns visually apparent is a vital part of the scientific endeavor. For scientists, businesspeople, and others, the primary tool for expressing patterns in data is the graph. Thus, the ability to interpret graphs is a skill that you will want to cultivate. This appendix guides you in how to read graphs, introduces a few vital conceptual points, and surveys the most common types of graphs, giving rationales for their use.

## Navigating a Graph

A graph is a diagram that shows relationships among *variables*, which are factors that can change in value. The most common types of graphs relate values of a *dependent variable* to those of an *independent variable*. As explained in Chapter 1 (› pp. 12–13), a dependent variable is so named because its values "depend on" the values of an independent variable. In other words, as the values of an independent variable change, the values of the dependent variable change in response. In a manipulative experiment (› p. 13), changes that a researcher specifies in the value of the independent variable *cause* changes in the value of the dependent variable. In observational studies, there may be no causal relationship, and scientists may plot a correlation (› pp. 13–14). In either case, the values of the independent variable are known or specified, and the values of the dependent variable are unknown and are what we are interested in observing or measuring.

By convention, independent variables are generally represented on the horizontal axis, or *x axis*, of a graph, while dependent variables are represented on the vertical axis, or *y axis*. Numerical values of variables generally become larger as one proceeds rightward on the *x* axis or upward on the *y* axis. In many cases, independent variables are not numerical at all, but categorical. For example, in a graph presenting population sizes of several nations, the nations comprise a categorical independent variable, whereas population size is a numerical dependent variable.

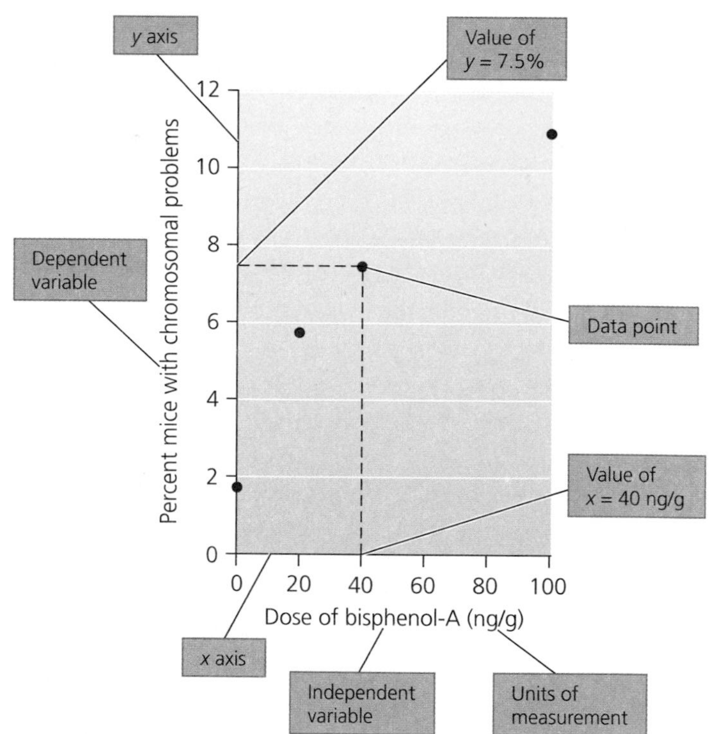


**FIGURE A.1** Frequency of chromosomal problems relative to dose of bisphenol-A. (Figure 14.SBS, p. 395)

As a simple example, **Figure A.1** shows data from a scientist who ran an experiment to test the effects on laboratory mice of the chemical bisphenol-A. The *x* axis shows values of her independent variable, the dose of bisphenol-A given to the mice. The values are expressed in units of nanograms per gram of water (ng/g). The researcher was interested in what proportion of mice develop chromosomal problems as a result. Thus, her dependent variable, presented on the *y* axis, is the percentage of mice found to develop chromosomal problems. For each of four doses of bisphenol-A she supplied in her experiment, a data point on the graph is plotted to show the corresponding percentage of mice with chromosomal problems.

Now that you're familiar with the basic building blocks of a graph, let's survey the most common types of graph you'll see, and examine a few vital concepts in graphing.

## GRAPH TYPE: Line Graph

A line graph is drawn when a data set involves a sequence of some kind, such as a series of values that occur one by one and change through time or across distance (**Figure A.2**). Line graphs are most appropriate when the *y* axis expresses a continuous numerical variable, and the *x* axis expresses either continuous numerical data or discrete sequential categories (such as years).

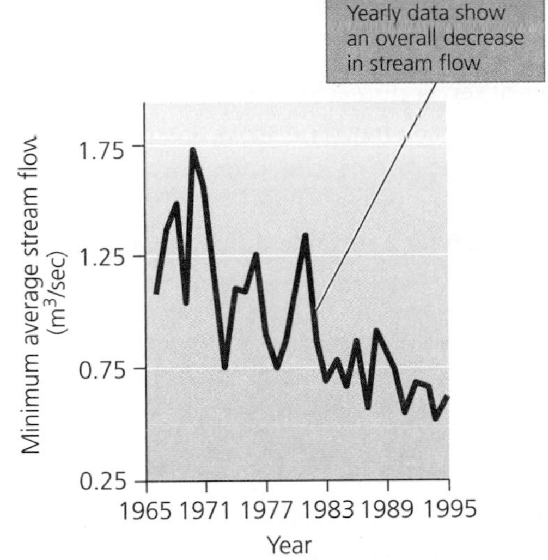


FIGURE A.2 Minimum stream flow. (Figure 5.SBS, p. 134)

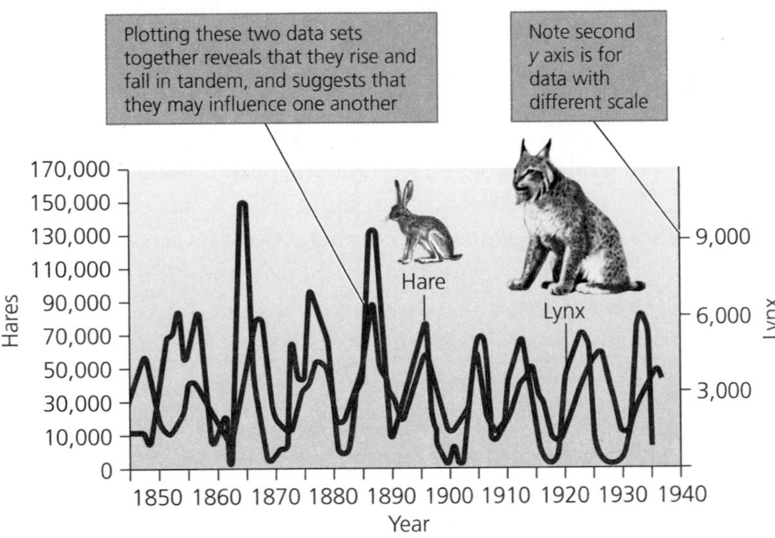


FIGURE A.3 Population fluctuations in hare and lynx. (Figure 6.5, p. 145)

One useful technique is to plot two or more data sets together on the same graph (**Figure A.3**). This allows us to compare trends in the data sets to see whether and how they may be related.

## KEY CONCEPT: Projections

Besides showing observed data, we can use graphs to show data that is predicted for the future, based on models, simulations, or extrapolations from past data. Often, projected future data on a line graph are shown with dashed lines, as in **Figure A.4,** to indicate that they are less certain than data that has already been observed.

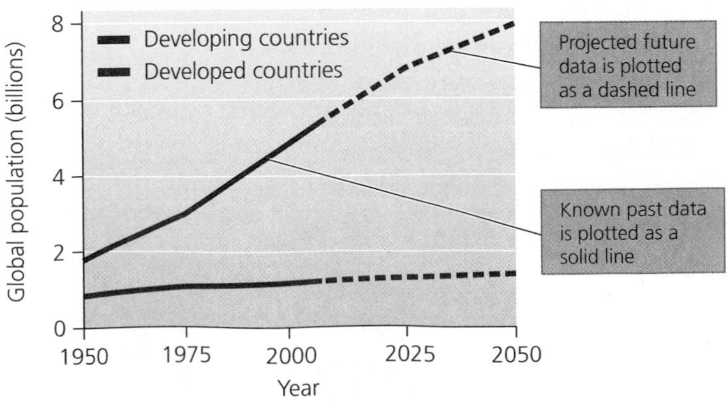


FIGURE A.4 Past and projected population growth for developing and developed countries. (Figure 8.19, p. 225)

## GRAPH TYPE: Bar Chart

A bar chart is most often used when one variable is categorical and the other is numerical. In such a chart, the height (or length) of each bar represents the quantitative value of a given category of the categorical variable; longer bars mean larger values (**Figure A.5**). Bar charts allow us to visualize how a variable differs quantitatively among categories.

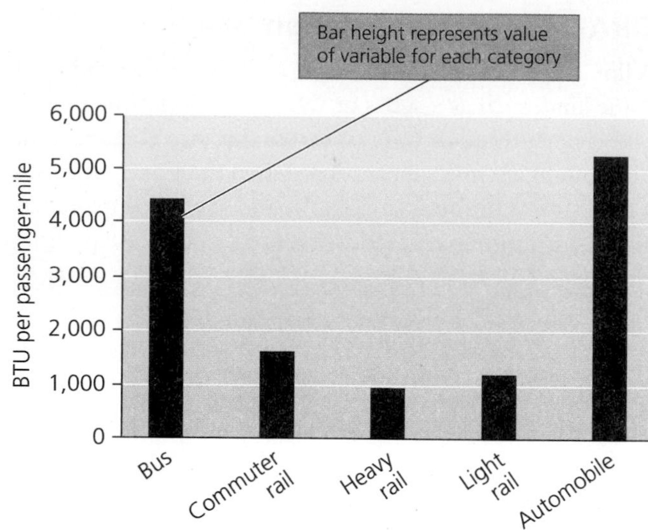


FIGURE A.5 Energy consumption for different modes of transit. (Figure 13.10a, p. 370)

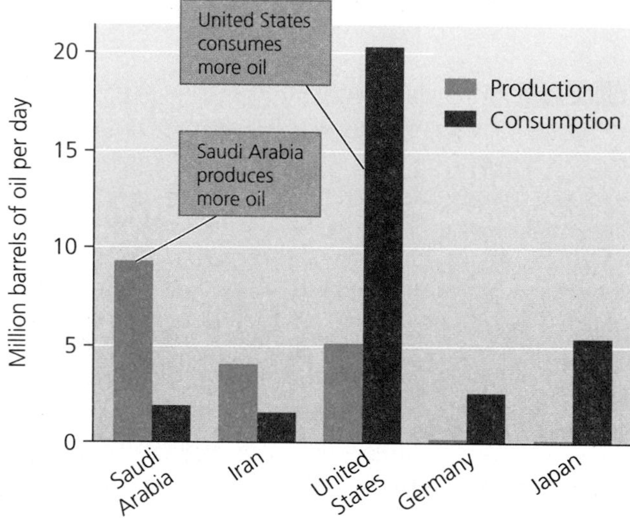
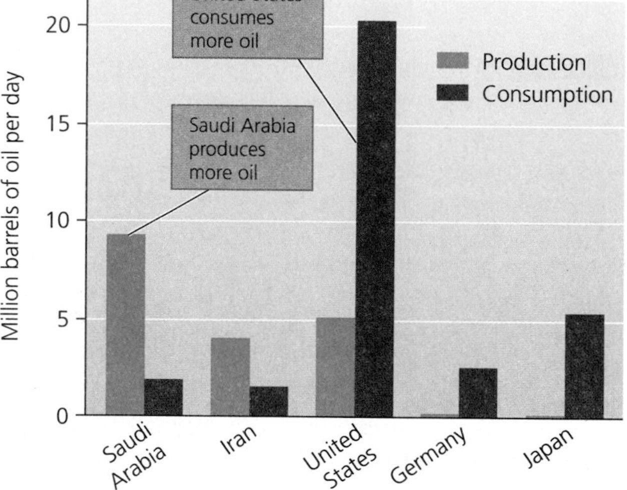


FIGURE A.6 Oil production and consumption by selected nations. (Figure 19.16, p. 562)

It is often instructive to graph two or more data sets together to reveal patterns and relationships. A bar chart such as **Figure A.6** allows us to compare two data sets (oil production and oil consumption) both within and among nations. A graph that does double duty in this way allows for higher-level analysis (in this case, suggesting which nations depend on others for petroleum imports). Most bar charts in this book illustrate multiple types of information at once in this manner.

**Figure A.7** illustrates two ways in which a bar chart can be modified. First, note that the orientation of bars is horizontal instead of vertical. In this configuration, the categorical variable is along the *y* axis and the numerical variable is along the *x* axis. Second, the bars can extend in either direction from a central *x*-axis value of zero, representing either positive (right) or negative (left) values. Depending on the nature of one's data and the points one wants to make, sometimes such arrangements can make for a clearer presentation.

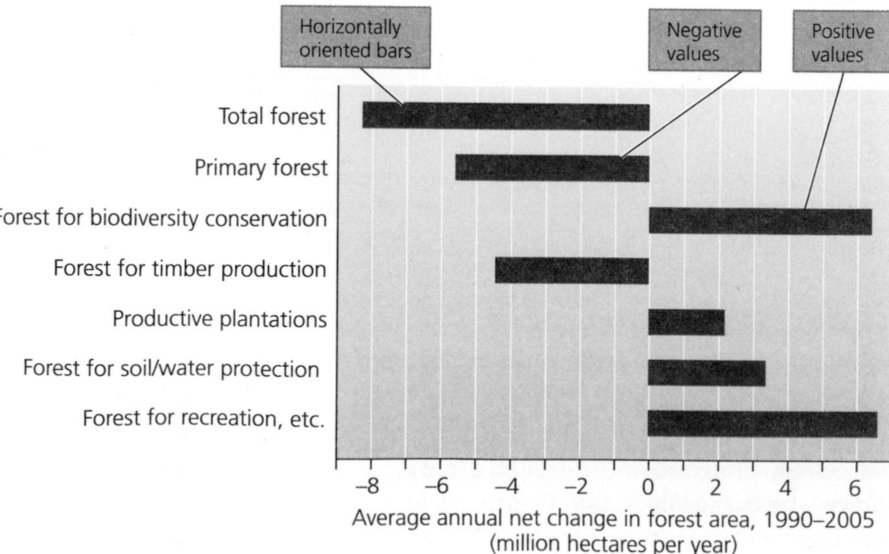


FIGURE A.7 Loss or gain of forests, by type. (Figure 12.SBS, p. 337)

One special type of horizontally oriented bar chart is the age pyramid used by demographers (**Figure A.8**). Age categories are displayed on the *y* axis, with bars representing the population size of each age group varying in their horizontal length.

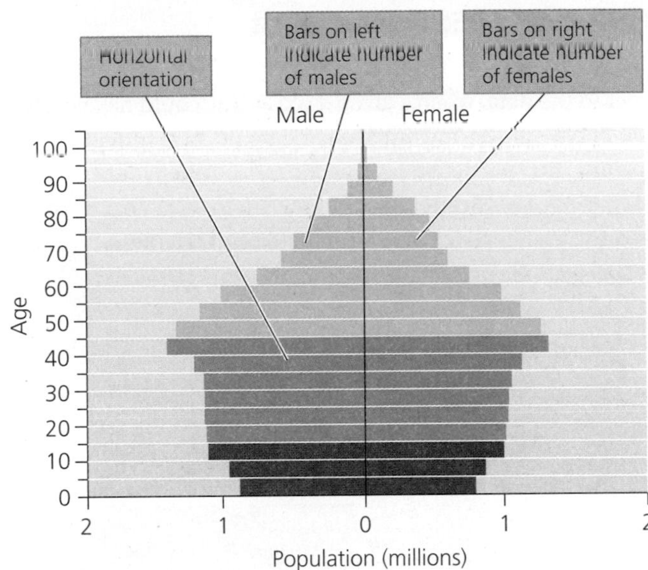


FIGURE A.8 Age structure of Canada, 2005. (Figure 8.10a, p. 216)

## KEY CONCEPT: Statistical Uncertainty

Most data sets involve some degree of uncertainty. Sometimes exact measurements are impossible, so the researcher estimates the likely range of measurement error around the data point. Other times, data points represent the *mean* (average) of many measurements, and the researcher may want to show the degree to which the data vary around this mean. Mathematical techniques are used to obtain precise statistical probabilities for degrees of variation. Results from such analyses are represented in bar charts such as **Figure A.9** as thin black lines called *error bars*. These bars extend past and within the end of each bar, representing the degree of variation around the bar's value. Longer error bars indicate more uncertainty or variation, whereas short error bars mean we can have high confidence in the value.

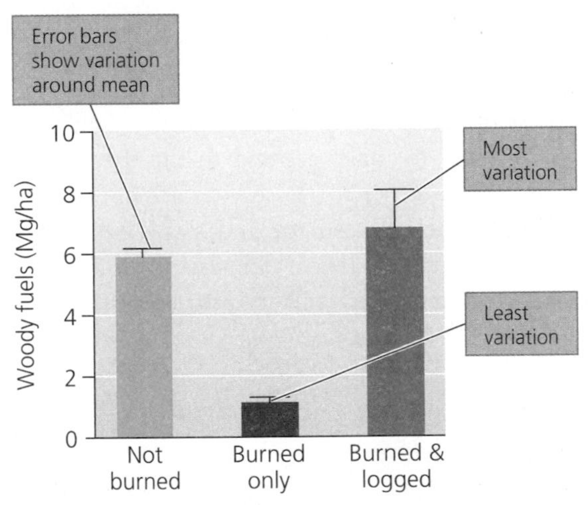


FIGURE A.9 Fine-scale woody debris left after treatments in salvage logging study. (Figure 12.SBS, p. 345)

The statistical analysis of data is critically important in science. In this book we provide a broad and streamlined introduction to many topics, so we often omit error bars from our graphs and details of statistical significance from our discussions. This is for clarity of presentation only; the research we discuss analyzes its data in far more depth than any textbook can possibly cover.

## KEY CONCEPT: Logarithmic Scales

When data span a large range of values, it can help to change the scale on a graph from the standard linear scale to a logarithmic scale. In a logarithmic scale, each equal unit of distance on an axis corresponds to a ratio of values rather than an additive increase in values. Most often, logarithmic scales advance by factors of ten. **Figure A.10** uses a logarithmic scale on its *y* axis for a graph in which we also show a linear scale using white horizontal lines crossing the graph space. The choice of scale does not affect data values, but does drastically affect the appearance of lines or bars on a graph. For further discussion, see the "Interpreting Graphs and Data" feature for Chapter 1 (pp. 23–24).

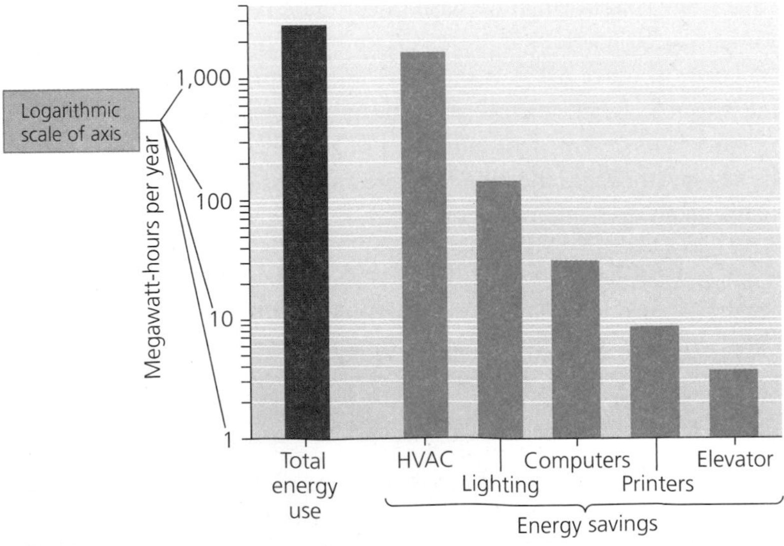


FIGURE A.10 Annual energy use and savings from a campus sustainability study. (Figure 23.IGD, p. 680)

## GRAPH TYPE: Scatter Plot

A scatter plot is often used when there is no sequential aspect to the data, when a given $x$-axis value could have multiple $y$-axis values, and when each data point is independent, having no particular connection to other data points (**Figure A.11**). Scatter plots allow us to visualize a broad positive or negative correlation between variables.

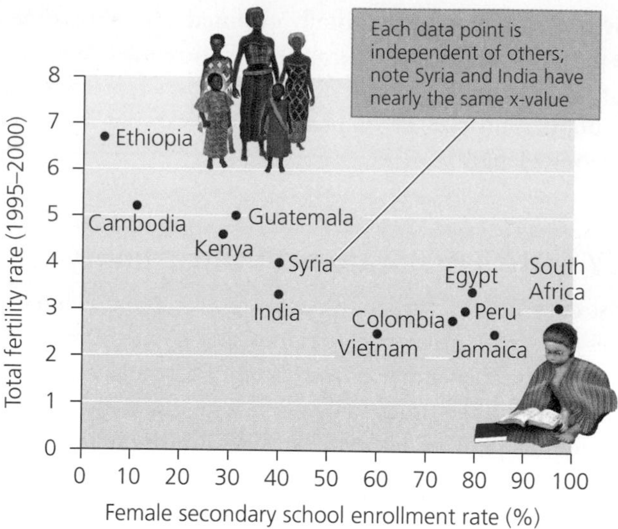


**FIGURE A.11** Fertility rate and female education. (Figure 8.17, p. 221)

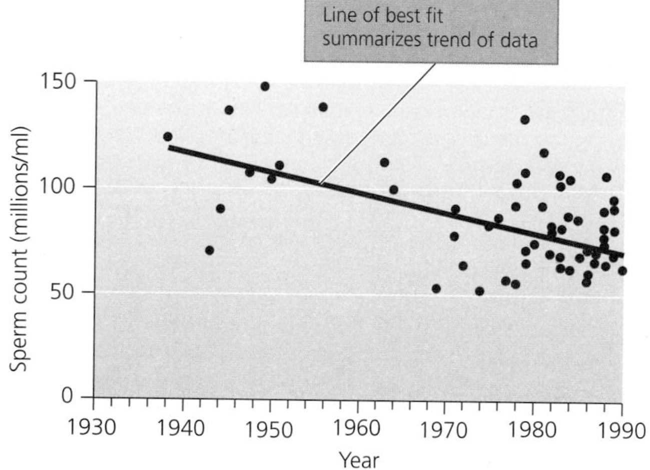


**FIGURE A.12** Declining sperm count in men across the world. (Figure 14.10a, p. 392)

A "line of best fit" may be drawn through the data of a scatter plot in order to make a trend in the data more clear to the eye (**Figure A.12**). These lines are not drawn casually, however; their placement and slope are determined by precise mathematical analysis of the data through a statistical technique called linear regression.

## GRAPH TYPE: Pie Chart

A pie chart is used when we wish to compare the proportions of some whole that are taken up by each of several categories (**Figure A.13**). A pie chart is appropriate when one variable is categorical and one is numerical. Each category is represented visually like a slice from a pie, with the size of the slice reflecting the percentage of the whole that is taken up by that category.

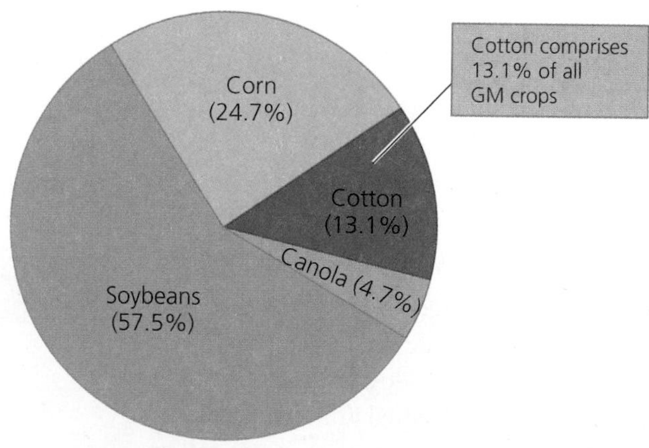


**FIGURE A.13** Genetically modified crops grown worldwide, by type. (Figure 10.14a, p. 276)

But don't stop here. Take advantage of the GRAPHit! tutorials on the CD-ROM included with this text, or go to the Withgott/Brennan Companion Website at www.aw-bc.com/withgott. The GRAPHit! tutorials allow you to plot your own data, and help you further expand your comprehension of graphs.

# Appendix B Metric System

| Measurement | Unit and Abbreviation | Metric Equivalent | Metric to English Conversion Factor | English to Metric Conversion Factor |
|---|---|---|---|---|
| Length | 1 kilometer (km) | = 1,000 ($10^3$) meters | 1 km = 0.62 mile | 1 mile = 1.61 km |
| | 1 meter (m) | = 100 ($10^2$) centimeters | 1 m = 1.09 yards | 1 yard = 0.914 m |
| | | = 1,000 millimeters | 1 m = 3.28 feet | 1 foot = 0.305 m |
| | | | 1 m = 39.37 inches | |
| | 1 centimeter (cm) | = 0.01 ($10^{-2}$) meter | 1 cm = 0.394 inch | 1 foot = 30.5 cm |
| | | | | 1 inch = 2.54 cm |
| | 1 millimeter (mm) | = 0.001 ($10^{-3}$) meter | 1 mm = 0.039 inch | |
| Area | 1 square meter ($m^2$) | = 10,000 square centimeters | 1 $m^2$ = 1.1960 square yards | 1 square yard = 0.8361 $m^2$ |
| | | | 1 $m^2$ = 10.764 square feet | 1 square foot = 0.0929 $m^2$ |
| | 1 square centimeter ($cm^2$) | = 100 square millimeters | 1 $cm^2$ = 0.155 square inch | 1 square inch = 6.4516 $cm^2$ |
| Mass | 1 metric ton (t) | = 1,000 kilograms | 1 t = 1.103 ton | 1 ton = 0.907 t |
| | 1 kilogram (kg) | = 1,000 grams | 1 kg = 2.205 pounds | 1 pound = 0.4536 kg |
| | 1 gram (g) | = 1,000 milligrams | 1 g = 0.0353 ounce | 1 ounce = 28.35 g |
| | 1 milligram (mg) | = 0.001 gram | | |
| Volume (solids) | 1 cubic meter ($m^3$) | = 1,000,000 cubic centimeters | 1 $m^3$ = 1.3080 cubic yards | 1 cubic yard = 0.7646 $m^3$ |
| | | | 1 $m^3$ = 35.315 cubic feet | 1 cubic foot = 0.0283 $m^3$ |
| | 1 cubic centimeter ($cm^3$ or cc) | = 0.000001 cubic meter | 1 $cm^3$ = 0.0610 cubic inch | 1 cubic inch = 16.387 $cm^3$ |
| | | = 1 milliliter | | |
| | 1 cubic millimeter ($mm^3$) | = 0.000000001 cubic meter | | |
| Volume (liquids and gases) | 1 kiloliter (kl or kL) | = 1,000 liters | 1 kL = 264.17 gallons | 1 gallon = 3.785 L |
| | 1 liter (l or L) | = 1,000 milliliters | 1 L = 0.264 gallons | 1 quart = 0.946 L |
| | | | 1 L = 1.057 quarts | |
| | 1 milliliter (ml or mL) | = 0.001 liter | 1 ml = 0.034 fluid ounce | 1 quart = 946 ml |
| | | = 1 cubic centimeter | 1 ml = approximately $\frac{1}{4}$ teaspoon | 1 pint = 473 ml |
| | | | | 1 fluid ounce = 29.57 ml |
| | | | | 1 teaspoon = approx. 5 ml |
| Time | 1 millisecond (ms) | = 0.001 second | | |
| Temperature | Degrees Celsius (°C) | | $°C = \frac{5}{9}(°F - 32)$ | $°F = \frac{9}{5}°C + 32$ |
| Energy and Power | 1 kilowatt-hour | = 34,113 BTUs = 860,421 calories | | |
| | 1 watt | = 3.413 BTU/hr | | |
| | | = 14.34 calorie/min | | |
| | 1 calorie | = the amount of heat necessary to raise the temperature of 1 gram (1 $cm^3$) of water 1 degree Celsius | | |
| | 1 horsepower | = 7.457 × 102 watts | | |
| | 1 joule | = 9.481 × $10^{-4}$) BTU | | |
| | | = 0.239 cal | | |
| | | = 2.778 × $10^{-7}$ kilowatt-hour | | |
| Pressure | 1 pound per square inch (psi) | = 6894.757 pascal (Pa) | | |
| | | = 0.068045961 atmosphere (atm) | | |
| | | = 51.71493 millimeters of mercury (mm hg = Torr) | | |
| | | = 68.94757 millibars (mbar) | | |
| | | = 6.894757 kilopascal (kPa) | | |
| | 1 atmosphere (atm) | = 101.325 kilopascal (kPa) | | |

# Appendix C Periodic Table of the Elements

| | Representative (main group) elements | | Transition metals | | | | | | | | | | Representative (main group) elements | | | | | |
|---|---|---|---|---|---|---|---|---|---|---|---|---|---|---|---|---|---|---|
| | IA | IIA | IIIB | IVB | VB | VIB | VIIB | VIIIB | VIIIB | VIIIB | IB | IIB | IIIA | IVA | VA | VIA | VIIA | VIIIA |
| 1 | 1 H 1.0079 Hydrogen | | | | | | | | | | | | | | | | | 2 He 4.003 Helium |
| 2 | 3 Li 6.941 Lithium | 4 Be 9.012 Beryllium | | | | | | | | | | | 5 B 10.811 Boron | 6 C 12.011 Carbon | 7 N 14.007 Nitrogen | 8 O 15.999 Oxygen | 9 F 18.998 Fluorine | 10 Ne 20.180 Neon |
| 3 | 11 Na 22.990 Sodium | 12 Mg 24.305 Magnesium | | | | | | | | | | | 13 Al 26.982 Aluminum | 14 Si 28.086 Silicon | 15 P 30.974 Phosphorus | 16 S 32.066 Sulfur | 17 Cl 35.453 Chlorine | 18 Ar 39.948 Argon |
| 4 | 19 K 39.098 Potassium | 20 Ca 40.078 Calcium | 21 Sc 44.956 Scandium | 22 Ti 47.88 Titanium | 23 V 50.942 Vanadium | 24 Cr 51.996 Chromium | 25 Mn 54.938 Manganese | 26 Fe 55.845 Iron | 27 Co 58.933 Cobalt | 28 Ni 58.69 Nickel | 29 Cu 63.546 Copper | 30 Zn 65.39 Zinc | 31 Ga 69.723 Gallium | 32 Ge 72.61 Germanium | 33 As 74.922 Arsenic | 34 Se 78.96 Selenium | 35 Br 79.904 Bromine | 36 Kr 83.8 Krypton |
| 5 | 37 Rb 85.468 Rubidium | 38 Sr 87.62 Strontium | 39 Y 88.906 Yttrium | 40 Zr 91.224 Zirconium | 41 Nb 92.906 Niobium | 42 Mo 95.94 Molybdenum | 43 Tc 98 Technetium | 44 Ru 101.07 Ruthenium | 45 Rh 102.906 Rhodium | 46 Pd 106.42 Palladium | 47 Ag 107.868 Silver | 48 Cd 112.411 Cadmium | 49 In 114.82 Indium | 50 Sn 118.71 Tin | 51 Sb 121.76 Antimony | 52 Te 127.60 Tellurium | 53 I 126.905 Iodine | 54 Xe 131.29 Xenon |
| 6 | 55 Cs 132.905 Cesium | 56 Ba 137.327 Barium | 57 La 138.906 Lanthanum | 72 Hf 178.49 Hafnium | 73 Ta 180.948 Tantalum | 74 W 183.84 Tungsten | 75 Re 186.207 Rhenium | 76 Os 190.23 Osmium | 77 Ir 192.22 Iridium | 78 Pt 195.08 Platinum | 79 Au 196.967 Gold | 80 Hg 200.59 Mercury | 81 Tl 204.383 Thallium | 82 Pb 207.2 Lead | 83 Bi 208.980 Bismuth | 84 Po 209 Polonium | 85 At 210 Astatine | 86 Rn 222 Radon |
| 7 | 87 Fr 223 Francium | 88 Ra 226.025 Radium | 89 Ac 227.028 Actinium | 104 Rf 261 Unnilquadium | 105 Db 262 Unnilpentium | 106 Sg 263 Unnilhexium | 107 Bh 262 Unnilseptium | 108 Hs 265 Unniloctium | 109 Mt 266 Unnilennium | 110 Uun 269 Ununnilium | 111 Uuu 272 Unununium | 112 Uub 277 Ununbium | | 114 | | 116 | | |

Rare earth elements

| | | | | | | | | | | | | | | |
|---|---|---|---|---|---|---|---|---|---|---|---|---|---|---|
| Lanthanides | 58 Ce 140.115 Cerium | 59 Pr 140.908 Praseodymium | 60 Nd 144.24 Neodymium | 61 Pm 145 Promethium | 62 Sm 150.36 Samarium | 63 Eu 151.964 Europium | 64 Gd 157.25 Gadolinium | 65 Tb 158.925 Terbium | 66 Dy 162.5 Dysprosium | 67 Ho 164.93 Holmium | 68 Er 167.26 Erbium | 69 Tm 168.934 Thulium | 70 Yb 173.04 Ytterbium | 71 Lu 174.967 Lutetium |
| Actinides | 90 Th 232.038 Thorium | 91 Pa 231.036 Protactinium | 92 U 238.029 Uranium | 93 Np 237.048 Neptunium | 94 Pu 244 Plutonium | 95 Am 243 Americium | 96 Cm 247 Curium | 97 Bk 247 Berkelium | 98 Cf 251 Californium | 99 Es 252 Einsteinium | 100 Fm 257 Fermium | 101 Md 258 Mendelevium | 102 No 259 Nobelium | 103 Lr 262 Lawrencium |

The periodic table arranges elements according to atomic number and atomic weight into horizontal rows called periods and vertical columns called groups.

Elements of each group in Class A have similar chemical and physical properties. This reflects the fact that members of a particular group have the same number of valence shell electrons, which is indicated by the group's number. For example, group IA elements have one valence shell electron, group IIA elements have two, and group VA elements have five. In contrast, as you progress across a period from left to right, properties of the elements change, varying from the very metallic properties of groups IA and IIA to the nonmetallic properties of group VIIA to the inert elements (noble gases) in group VIIIA. This reflects changes in the number of valence shell electrons.

Class B elements, or transition elements, are metals, and generally have one or two valence shell electrons. In these elements, some electrons occupy more distant electron shells before the deeper shells are filled.

In this periodic table, elements with symbols printed in black exist as solids under standard conditions (25 °C and 1 atmosphere of pressure), while elements in red exist as gases, and those in dark blue as liquids. Elements with symbols in green do not exist in nature and must be created by some type of nuclear reaction.

# Appendix D Geologic Timescale

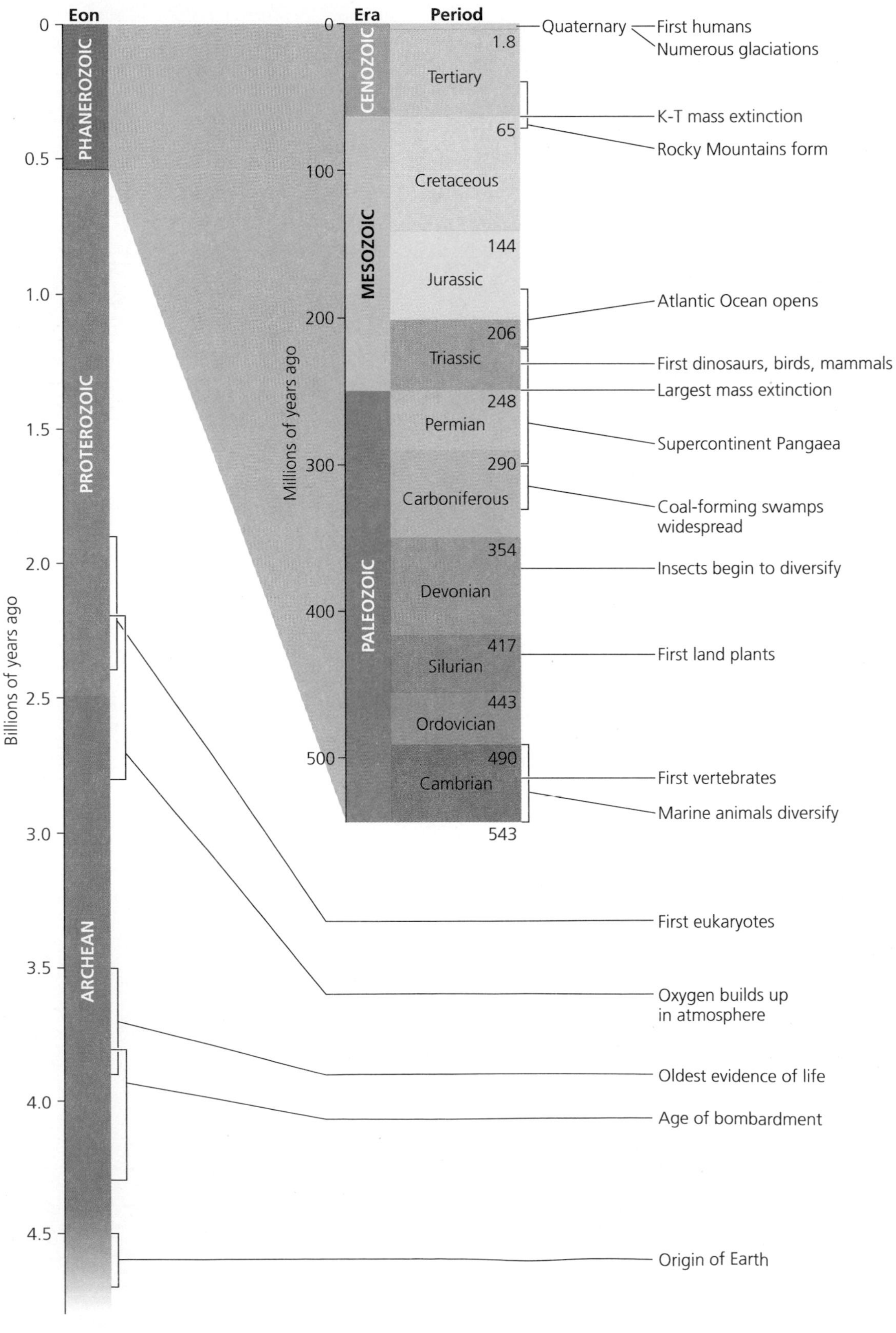

# Glossary

**acid drainage** A process in which sulfide minerals in newly exposed rock surfaces react with *oxygen* and rainwater to produce sulfuric acid, which causes chemical *runoff* as it *leaches* metals from the rocks. Although acid drainage is a natural phenomenon, mining can greatly accelerate its rate by exposing many new rock surfaces at once.
**acidic** The property of a *solution* in which the concentration of *hydrogen* ($H^+$) *ions* is greater than the concentration of hydroxide ($OH^-$) ions. Compare *basic.*
**acidic deposition** The settling of acidic or acid-forming pollutants from the *atmosphere* onto Earth's surface. This can take place by precipitation, fog, gases, or the deposition of dry particles. Compare *acid rain; acidic precipitation.*
**acidic precipitation** The deposition of *acidic* or acid-forming pollutants from the *atmosphere* onto Earth's surface by *precipitation.* Compare *acid rain.*
**acid rain** *Acidic deposition* that takes place through rain. Compare *acidic precipitation.*
**active solar energy collection** An approach in which technological devices are used to focus, move, or store solar energy. Compare *passive solar energy collection.*
**acute exposure** Exposure to a *toxicant* occurring in high amounts for short periods of time. Compare *chronic exposure.*
**adaptive management** The systematic testing of different management approaches to improve methods over time.
**adaptive radiation** A burst of *species* formation due to natural selection.
**adaptive trait (adaptation)** A trait that confers greater likelihood that an individual will reproduce.
**aerobic** Occurring in an *environment* where *oxygen* is present. For example, the decay of a rotting log proceeds by aerobic decomposition. Compare *anaerobic.*
**aerosols** Very fine liquid droplets or solid particles aloft in the atmosphere.
**affluenza** Term coined by social critics to describe the failure of material goods to bring happiness to people who have the financial means to afford them.
**age distribution** The relative numbers of organisms of each age within a *population.* Age distributions can have a strong effect on rates of population growth or decline and are often expressed as a ratio of age classes, consisting of organisms (1) not yet mature enough to reproduce, (2) capable of reproduction, and (3) beyond their reproductive years.
**age structure** See *age distribution.*
**age structure diagram (age pyramid)** A diagram demographers use to show the age structure of a population. The width of each horizontal bar represents the relative number of individuals in each age class.
**agricultural revolution** The shift around 10,000 years ago from a hunter-gatherer lifestyle to an agricultural way of life in which people began to grow their own crops and raise domestic animals. Compare *industrial revolution.*
**agriculture** The practice of cultivating *soil,* producing crops, and raising livestock for human use and consumption.
**agroforestry** The practice of growing trees and agricultural crops together. See also *alley cropping.*
**A horizon** A layer of *soil* found in a typical *soil profile.* It forms the top layer or lies below the *O horizon* (if one exists). It consists of mostly inorganic mineral components such as *weathered* substrate, with some organic matter and *humus* from above mixed in. The A horizon is often referred to as *topsoil.* Compare *B horizon; C horizon; E horizon; R horizon.*
**air pollutants** Gases and particulate material added to the atmosphere that can affect *climate* or harm people or other organisms.
**air pollution** The act of polluting the air, or the condition of being polluted by *air pollutants.*
**albedo** The capacity of a surface to reflect light. Higher albedo values refer to greater reflectivity.
**allelopathy** A phenomenon whereby certain plants release poisonous chemicals that harm others nearby.
**allergen** A *toxicant* that overactivates the immune system, causing an immune response when one is not necessary.
**alley cropping** A type of *agroforestry* in which fields planted in rows of mixed crops are surrounded by or interspersed with rows of trees that provide fruit, wood, or protection from wind. Compare *intercropping; shelterbelt.*
**allopatric speciation** Species formation due to the physical separation of populations over some geographic distance. Compare *sympatric speciation.*
**alpine tundra** *Tundra* that occurs at the tops of mountains.
**ambient air pollution** See *outdoor air pollution.*
**amensalism** A relationship between members of different *species* in which one organism is harmed and the other is unaffected. Compare *commensalism.*
**amino acids** Organic molecules that join in long chains to form *proteins.*
**anaerobic** Occurring in an *environment* that has little or no *oxygen.* The conversion of organic matter to *fossil fuels (crude oil, coal, natural gas)* at the bottom of a deep lake, swamp, or shallow sea is an example of anaerobic decomposition. Compare *aerobic.*
**anemometer** A device that measures wind speed and direction.
**anthropocentrism** A human-centered view of our relationship with the *environment.*
**application** An applied use of science, such as a new technology, policy decision, or resource management strategy.
**aquaculture** The raising of aquatic organisms for food in controlled *environments.*
**aquifer** An underground water reservoir.
**aquifer recharge zone** An area where water infiltrates Earth's surface and reaches an aquifer below.
**area effect** In *island biogeography theory,* the pattern that large islands host more species than smaller islands, because larger islands provide larger targets for immigration and because extinction rates are reduced.
**artesian aquifer** See *confined aquifer.*
**artificial selection** *Natural selection* conducted under human direction. Examples include the *selective breeding* of crop plants, pets, and livestock.
**asbestos** Any of several types of mineral that form long, thin microscopic fibers—a structure that allows asbestos to insulate buildings for heat, muffle sound, and resist fire. When inhaled and lodged in lung tissue, asbestos scars the tissue and may eventually lead to lung cancer or *asbestosis.*
**asbestosis** A disorder resulting from lung tissue scarred by acid following prolonged inhalation of *asbestos.*
**Asian Brown Cloud** A persistent 2-mile-thick layer of *air pollution* from southern Asia that hangs over the Indian subcontinent throughout the dry season, each December through April. It is estimated to reduce sunlight reaching Earth's surface in that region by 10–15%, decrease rice productivity by 5–10%, and cause thousands of deaths each year.
**atmosphere** The thin layer of gases surrounding planet Earth. Compare *biosphere; hydrosphere; lithosphere.*
**atmospheric deposition** The wet or dry deposition on land of a wide variety of pollutants, including mercury, nitrates, organochlorines, and others. *Acidic deposition* is one type of atmospheric deposition.
**atmospheric pressure** The weight per unit area produced by a column of air.
**atoll** A ring-shaped island (generally of *coral reef*) surrounding an older submerged island area.
**atom** The smallest component of an *element* that maintains the chemical properties of that element.

**atomic number** The number of *protons* in a given *atom.*
**autotroph (primary producer)** An organism that can use the energy from sunlight to produce its own food. Includes green plants, algae, and cyanobacteria.

***Bacillus thuringiensis* (Bt)** A naturally occurring *soil* bacterium that produces a protein that kills many pests, including caterpillars and the larvae of some flies and beetles.
**background rate of extinction** The average rate of *extinction* that occurred before the appearance of humans. For example, the *fossil record* indicates that for both birds and mammals, one *species* in the world typically became extinct every 500–1,000 years. Compare *mass extinction event.*
**bagasse** Crushed sugarcane residue, whose sugars are used in Brazil to make *ethanol* that helps powers millions of vehicles.
**baghouse** A system of large filters that physically removes *particulate matter* from *incinerator emissions.*
**barrier island** A long thin island that parallels a shoreline. Generally of sand or *coral reef,* barrier islands protect coasts from storms.
**basic** The property of a *solution* in which the concentration of hydroxide ($OH^-$) *ions* is greater than the concentration of *hydrogen* ($H^+$) ions. Compare *acidic.*
**bathymetry** The study of ocean depths.
**bedrock** The continuous mass of solid rock that makes up Earth's *crust.*
**benthic** Of, relating to, or living on the bottom of a water body. Compare *pelagic.*
**benthic zone** The bottom layer of water body. Compare *littoral zone; limnetic zone; profundal zone.*
**B horizon** The layer of *soil* that lies below the *E horizon* and above the *C horizon.* Minerals that leach out of the E horizon are carried down into the B horizon (or subsoil) and accumulate there. Sometimes called the "zone of accumulation" or "zone of deposition." Compare *A horizon; O horizon; R horizon.*
**bioaccumulation** The buildup of *toxicants* in the tissues of an animal.
**biocentrism** A philosophy that ascribes relative values to actions, entities, or properties on the basis of their effects on all living things or on the integrity of the *biotic* realm in general. The biocentrist evaluates an action in terms of its overall impact on living things, including—but not exclusively focusing on—human beings.
**biodiesel** Diesel fuel produced by mixing vegetable oil, used cooking grease, or animal fat with small amounts of *ethanol* or methanol (wood alcohol) in the presence of a chemical catalyst.
**biodiversity (biological diversity)** The sum total of all organisms in an area, taking into account the diversity of *species,* their *genes,* their *populations,* and their *communities.*
**biodiversity hotspot** An area that supports an especially great diversity of *species,* particularly species that are *endemic* to the area.
**biofuel** Fuel produced from *biomass energy* sources and used primarily to power automobiles.
**biogenic** Type of *natural gas* created at shallow depths by the anaerobic decomposition of organic matter by bacteria. Consists of nearly pure *methane.* Compare *thermogenic.*
**biogeochemical cycle** See *nutrient cycle.*
**biological control (biocontrol)** The attempt to battle pests and weeds with organisms that prey on or parasitize them, rather than by using *pesticides.*
**biological diversity** See *biodiversity.*
**biological hazard** Human health hazards that result from ecological interactions among organisms. These include *parasitism* by viruses, bacteria, or other *pathogens.* Compare *infectious disease; chemical hazard; cultural hazard; physical hazard.*
**biological weathering** *Weathering* that occurs when living things break down *parent material* by physical or chemical means. Compare *chemical weathering; physical weathering.*
**biomagnification** The magnification of the concentration of *toxicants* in an organism caused by its consumption of other organisms in which toxicants have *bioaccumulated.*
**biomass** Organic material that makes up living organisms.
**biomass energy** *Energy* harnessed from plant and animal matter, including wood from trees, charcoal from burned wood, and combustible animal waste products, such as cattle manure. *Fossil fuels* are not considered biomass energy sources because their organic matter has not been part of living organisms for millions of years and has undergone considerable chemical alteration since that time.
**biome** A major regional complex of similar plant *communities;* a large *ecological* unit defined by its dominant plant type and vegetation structure.
**biophilia** A hypothetical phenomenon that E. O. Wilson defined as "the connections that human beings subconsciously seek with the rest of life."
**biopower** The burning of *biomass energy* sources to generate electricity.
**bioremediation** The attempt to clean up *pollution* by enhancing natural processes of biodegradation by living organisms.
**biosphere** The sum total of all the planet's living organisms and the *abiotic* portions of the *environment* with which they interact.
**biosphere reserve** A tract of land with exceptional *biodiversity* that couples preservation with *sustainable development* to benefit local people. Biosphere reserves are designated by UNESCO (the *United Nations* Educational, Scientific, and Cultural Organization) following application by local stakeholders.
**biotechnology** The material application of biological *science* to create products derived from organisms. The creation of *transgenic* organisms is one type of biotechnology.
**biotic potential** An organism's capacity to produce offspring.
**bitumen** A thick and heavy form of *petroleum* rich in *carbon* and poor in *hydrogen.*
**bog** A type of *wetland* in which a pond is thoroughly covered with a thick floating mat of vegetation. Compare *freshwater marsh; swamp.*
**boreal forest** A *biome* of northern coniferous forest that stretches in a broad band across much of Canada, Alaska, Russia, and Scandinavia. Also known as taiga, boreal forest consists of a limited number of *species* of evergreen trees, such as black spruce, that dominate large regions of forests interspersed with occasional bogs and lakes.
**bottleneck** A step in a process that limits the progress of the overall process.
**bottom-trawling** Fishing practice that involves dragging weighted nets across the seafloor to catch *benthic* organisms. Trawling crushes many organisms in its path and leaves long swaths of damaged sea bottom.
**breakdown product** *A compound* that results from the degradation of a toxicant.
**breeder reactor** A nuclear reactor that creates more fissile material than it consumes, and uses primarily uranium-238 and plutonium-239. Although breeder reactors make better use of fuel, generate more power, and produce less waste than conventional reactors, most have been closed because of concerns over nuclear weapons proliferation.
**brownfield** An area of land whose redevelopment or reuse is complicated by the presence or potential presence of hazardous material.
**Bureau of Land Management (BLM)** Federal agency that owns and manages most U.S. rangelands. The BLM is the nation's single largest landowner; its 106 million ha (261 million acres) are spread across 12 western states.
**by-catch** That portion of a commercial fishing catch consisting of animals caught unintentionally. By-catch kills many thousands of fish, sharks, marine mammals, and birds each year.

**Calvin cycle** In *photosynthesis,* a series of chemical reactions in which *carbon atoms* from *carbon dioxide* are linked together to manufacture sugars.
**cap-and-trade** A *permit-trading* system in which government determines an acceptable level of *pollution* and then issues polluting parties permits to pollute. A company receives credit for amounts it does not emit and can then sell this credit to other companies. Compare *emissions trading system.*
**capitalist market economy** An *economy* in which buyers and sellers interact to determine which *goods* and *services* to produce, how much of them to produce, and how to distribute them. Compare *centrally planned economy.*
**captive breeding** The practice of capturing members of threatened and endangered *species* so that their young can be bred and raised in controlled *environments* and subsequently reintroduced into the wild.
**carbohydrate** An *organic compound* consisting of *atoms* of *carbon, hydrogen,* and *oxygen.*

**carbon** The chemical *element* with six protons and six neutrons. A key element in *organic compounds.*
**carbon capture** Technologies or approaches that remove *carbon dioxide* from power plant or other emissions, in an effort to mitigate *global climate change.*
**carbon cycle** A major *nutrient cycle* consisting of the routes that *carbon atoms* take through the nested networks of environmental *systems.*
**carbon dioxide ($CO_2$)** A colorless gas used by plants for *photosynthesis,* given off by *respiration,* and released by burning *fossil fuels.* A primary *greenhouse gas* whose buildup contributes to *global climate change.*
**carbon footprint** The cumulative amount of carbon, or *carbon dioxide,* that a person or institution emits, and is indirectly responsible for emitting, into the *atmosphere,* contributing to *global climate change.* Compare *ecological footprint.*
**carbon monoxide (CO)** A colorless, odorless gas produced primarily by the incomplete combustion of fuel. An EPA *criteria pollutant.*
**carbon neutrality** The state in which an individual, business, or institution emits no net carbon to the atmosphere. This may be achieved by reducing carbon emissions and/or employing *carbon offsets* to offset emissions.
**carbon offset** A voluntary payment to another entity intended to enable that entity to reduce the *greenhouse gas* emissions that one is unable or unwilling to reduce oneself. The payment thus offsets one's own emissions.
**carbon sequestration** Technologies or approaches to sequester, or store, *carbon dioxide* from industrial emissions, e.g., underground under pressure in locations where it will not seep out, in an effort to mitigate *global climate change.* We are still a long way from developing adequate technology and secure storage space to accomplish this.
**carbon storage** See *carbon sequestration.*
**carcinogen** A chemical or type of radiation that causes cancer.
**carnivore** An organism that consumes animals. Compare *herbivore; omnivore.*
**carrying capacity** The maximum *population size* that a given *environment* can sustain.
**case history** Medical approach involving the observation and analysis of individual patients.
**case law** A body of law made up of cumulative decisions rendered by courts.
**Cassandra** A person who predicts doom and disaster for the world because of our impact upon it. Compare *Cornucopian.*
**catalytic reforming** A process in oil refining in which catalysts are used to promote chemical reactions that transform certain hydrocarbons heavier than gasoline so that they can be blended with gasoline to obtain higher octane ratings.
**categorical imperative** An *ethical standard* described by Immanuel Kant, which roughly approximates Christianity's "golden rule": to treat others as you would prefer to be treated yourself.
**cation exchange** Process by which plants' roots donate *hydrogen ions* to the soil in exchange for cations (positively charged ions) such as those of calcium, magnesium, and potassium, which plants use as *nutrients.* The soil particles then replenish these cations by exchange with soil water.
**cation exchange capacity** A soil's ability to hold cations, preventing them from *leaching* and thus making them available to plants. A useful measure of soil fertility.
**cell** The most basic organizational unit of organisms.
**cellular respiration** The process by which a *cell* uses the chemical reactivity of *oxygen* to split glucose into its constituent parts, water and *carbon dioxide,* and thereby release *chemical energy* that can be used to form chemical bonds or to perform other tasks within the cell. Compare *photosynthesis.*
**cellulosic ethanol** *Ethanol* produced from the cellulose in plant tissues, by treating it with enzymes. Techniques for producing cellulosic ethanol are under development because of the desire to make ethanol from low-value crop waste (residues such as corn stalks and husks), rather than from the sugars of high-value crops.
**centrally planned economy** An *economy* in which a nation's government determines how to allocate resources in a top-down manner. Also called a "state socialist economy." Compare *capitalist market economy.*
**chaparral** A *biome* consisting mostly of densely thicketed evergreen shrubs occurring in limited small patches. Its "Mediterranean" *climate* of mild, wet winters and warm, dry summers is induced by oceanic influences. In addition to ringing the Mediterranean Sea, chaparral occurs along the coasts of California, Chile, and southern Australia.
**character displacement** A phenomenon resulting from *competition* among *species* in which competing species evolve characteristics that better adapt them to specialize on the portion of the resource they use. The species essentially become more different from one another, reducing their competition.
**chemical energy** *Potential energy* held in the bonds between atoms.
**chemical hazard** Chemicals that pose human health hazards. These include *toxins* produced naturally, as well as many of the disinfectants, *pesticides,* and other synthetic chemicals that our society produces. Compare *biological hazard; cultural hazard; physical hazard.*
**chemical weathering** *Weathering* that results when water or other substances chemically interact with *parent material.* Compare *biological weathering; physical weathering.*
**chemosynthesis** The process by which bacteria in *hydrothermal vents* use the *chemical energy* of hydrogen sulfide ($H_2S$) to transform inorganic *carbon* into *organic compounds.* Compare *photosynthesis.*
**Chernobyl** Site of a nuclear power plant in Ukraine (then part of the Soviet Union), where in 1986 an explosion caused the most severe *nuclear reactor* accident the world has yet seen. As with *Three Mile Island,* the term is often used to denote the accident itself.
**chlorofluorocarbon (CFC)** One of a group of human-made *organic compounds* derived from simple *hydrocarbons,* such as ethane and methane, in which *hydrogen atoms* are replaced by chlorine, bromine, or fluorine. CFCs deplete the protective *ozone layer* in the *stratosphere.*
**chlorophyll** The light-absorbing pigment that enables *photosynthesis* and makes plants green.
**chloroplast** A cell organelle containing *chlorophyll* in which *photosynthesis* occurs.
**C horizon** The layer of *soil* that lies below the *B horizon* and above the *R horizon.* It contains rock particles that are larger and less *weathered* than the layers above. It consists of *parent material* that has been altered only slightly or not at all by the process of *soil* formation. Compare *A horizon; E horizon; O horizon.*
**chronic exposure** Exposure for long periods of time to a *toxicant* occurring in low amounts. Compare *acute exposure.*
**city planning** The professional pursuit that attempts to design cities in such a way as to maximize their efficiency, functionality, and beauty.
**classical economics** Founded by *Adam Smith,* the study of the behavior of buyers and sellers in a free-market *economy.* Holds that individuals acting in their own self-interest may benefit society, provided that their behavior is constrained by the rule of law and by private property rights and operates within competitive markets. See also *neoclassical economics.*
**clay** *Sediment* consisting of particles less than 0.002 mm in diameter. Compare *sand; silt.*
**Clean Air Act of 1970** Revision of prior Congressional *legislation* to control *air pollution* that set stricter standards for air quality, imposed limits on emissions from new stationary and mobile sources, provided new funds for *pollution*-control research, and enabled citizens to sue parties violating the standards.
**Clean Air Act of 1990** Congressional *legislation* that strengthened *regulations* pertaining to air quality standards, auto emissions, toxic *air pollution, acidic deposition,* and depletion of the *ozone layer,* while also introducing market-based incentives to reduce *pollution.*
**clean coal** Term used to describe technologies and approaches that seek to reduce the generation and release of sulfur and other pollutants before, during, or after coal is burned for power.
**clear-cutting** The harvesting of timber by cutting all the trees in an area, leaving only stumps. Although it is the most cost-efficient method, clear-cutting is also the most damaging to the *environment.*
**Clear Skies** A George W. Bush administration initiative that aimed to abandon a command-and-control policy approach to air pollution and establish a market-based cap-and-trade program for sulfur dioxide, nitrogen oxides, and mercury. The Clear Skies legislation was stopped in 2005 by senators who concluded that it would increase pollution, relative to existing Clean Air Act policy.
**climate** The pattern of atmospheric conditions found across large geographic regions over long periods of time. Compare *weather.*

**climate diagram (climatograph)** A visual representation of a region's average monthly temperature and *precipitation.*
**climate model** See *coupled general circulation model.*
**climax community** In the traditional view of ecological *succession*, a *community* that remains in place with little modification until disturbance restarts the successional process. Today, ecologists recognize that community change is more variable and less predictable than originally thought, and that assemblages of species may instead form complex mosaics in space and time.
**closed cycle** An approach in *ocean thermal energy conversion* in which warm surface water is used to evaporate chemicals that boil at low temperatures. These evaporated gases spin turbines to generate electricity. Cold water piped in from ocean depths then condenses the gases so they can be reused.
**cloud forests** Moist forests, generally at high elevations in the tropics and subtropics, that derive much of their moisture from low-moving clouds.
**clumped distribution** Distribution pattern in which organisms arrange themselves in patches, generally according to the availability of the resources they need.
**coal** A *fossil fuel* composed of organic matter that was compressed under very high pressure to form a dense, solid *carbon* structure.
**coal gasification** The chemical breakdown of *coal* by treatment with steam and heat to form a mix of gases.
**coalbed methane** *Methane* that emanates from *coal* seams, which commonly leaks to the *atmosphere* during coal mining. To avoid this waste and reduce methane emissions, engineers are trying to capture more of this gas for energy.
**coevolution** Process by which two or more species evolve in response to one another. Parasites and hosts may coevolve, as may flowers and their pollinators.
**co-firing** A process in which *biomass* is combined with *coal* in coal-fired power plants. Can be a relatively easy and inexpensive way for *fossil-fuel*-based utilities to expand their use of *renewable energy.*
**cogeneration** A practice in which the extra heat generated in the production of electricity is captured and put to use heating workplaces and homes, as well as producing other kinds of power.
**cold front** The boundary where a mass of cold air displaces a mass of warmer air. Compare *warm front.*
**combined cycle** A process in which coal is treated to create hot gases that turn a gas turbine, while the hot exhaust of this turbine heats water to drive a conventional steam turbine.
**command and control** An approach to protecting the *environment* that sets strict legal limits and threatens punishment for violations of those limits.
**commensalism** A relationship between members of different *species* in which one organism benefits and the other is unaffected. Compare *amensalism.*
**communicable disease** See *infectious disease.*
**community** An assemblage of *populations* of organisms that live in the same place at the same time.
**community-based conservation** The practice of engaging local people to protect land and wildlife in their own region.
**community ecologist** A scientist who studies *community ecology.*
**community ecology** The study of the interactions among *species*, from one-to-one interactions to complex interrelationships involving entire *communities.*
**community-supported agriculture** A practice in which consumers pay farmers in advance for a share of their yield, usually in the form of weekly deliveries of produce.
**competition** A relationship in which multiple organisms seek the same limited resource.
**competitive exclusion** An outcome of *interspecific competition* in which one *species* excludes another species from resource use entirely.
**compost** a mixture produced when decomposers break down organic matter, including food and crop waste, in a controlled environment.
**composting** The conversion of organic *waste* into mulch or *humus* by encouraging, in a controlled manner, the natural biological processes of decomposition.
**compound** A *molecule* whose *atoms* are composed of two or more *elements.*
**concentrated animal feeding operation** See *feedlot.*
**concession** The right to extract a resource, granted by a government to a corporation.
**confined (artesian) aquifer** A water-bearing, porous layer of rock, *sand*, or gravel that is trapped between an upper and lower layer of less permeable substrate, such as *clay*. The water in a confined aquifer is under pressure because it is trapped between two impermeable layers. Compare *unconfined aquifer.*
**conservation biology** A scientific discipline devoted to understanding the factors, forces, and processes that influence the loss, protection, and restoration of *biological diversity* within and among *ecosystems.*
**conservation district** One of many county-based entities created by the Soil Conservation Service (now the Natural Resources Conservation Service) to promote practices that conserve *soil.*
**conservation ethic** An *ethic* holding that humans should put *natural resources* to use but also have a responsibility to manage them wisely. Compare *preservation ethic.*
**conservation geneticist** A scientist who studies genetic attributes of organisms, generally to infer the status of their *populations* in order to help conserve them.
**consumer** See *heterotroph.*
**consumptive use** *Freshwater* use in which water is removed from a particular *aquifer* or surface water body and is not returned to it. *Irrigation* for *agriculture* is an example of consumptive use. Compare *nonconsumptive use.*
**containment building** A building with thick concrete and steel walls that encloses a nuclear reactor, designed to prevent leaks of radioactivity in case of an accident.
**continental shelf** The gently sloping underwater edge of a continent, varying in width from 100 m (330 ft) to 1,300 km (800 mi), with an average slope of 1.9 m/km (10 ft/mi).
**continental slope** The portion of the ocean floor that angles somewhat steeply downward, connecting the *continental shelf* to the deep ocean basin below.
**contingent valuation** A technique that uses surveys to determine how much people would be willing to pay to protect a resource or to restore it after damage has been done. Compare *expressed preference.*
**contour farming** The practice of plowing furrows sideways across a hillside, perpendicular to its slope, to help prevent the formation of rills and gullies. The technique is so named because the furrows follow the natural contours of the land.
**control** The portion of an *experiment* in which a *variable* has been left unmanipulated, to serve as a point of comparison with the *treatment.*
**control rods** Rods made of a metallic alloy that absorbs *neutrons*, which are placed in a *nuclear reactor* among the water-bathed *fuel rods* of uranium. Engineers move these control rods into and out of the water to maintain the *fission* reaction at the desired rate.
**controlled burn** See *prescribed burn.*
**controlled experiment** An *experiment* in which the effects of all *variables* are held constant, except the one whose effect is being tested by comparison of *treatment* and *control* conditions.
**convective circulation** A circular *current* (of air, water, magma, etc.) driven by temperature differences. In the atmosphere, warm air rises into regions of lower *atmospheric pressure*, where it expands and cools and then descends and becomes denser, replacing warm air that is rising. The air picks up heat and moisture near ground level and prepares to rise again, continuing the process.
**convention** A *treaty* or binding agreement among national governments.
**conventional law** International law that arises from *conventions*, or treaties, that nations agree to enter into. Compare *customary law.*
**Convention on Biological Diversity** An international treaty that aims to conserve *biodiversity*, use biodiversity in a *sustainable* manner, and ensure the fair distribution of biodiversity's benefits. Although many nations have agreed to the treaty (as of 2007, 188 nations had become parties to it), several others, including the United States, have not.
**Convention on International Trade in Endangered Species of Wild Fauna and Flora (CITES)** A 1973 treaty facilitated by the *United Nations* that protects endangered *species* by banning the international transport of their body parts.

**convergent plate boundary** Area where tectonic plates collide. Can result in *subduction* or mountain range formation.
**coral** Tiny marine animals that build *coral reefs*. Corals attach to rock or existing reef and capture passing food with stinging tentacles. They also derive nourishment from photosynthetic symbiotic algae known as *zooxanthellae*.
**coral reef** A mass of calcium carbonate composed of the skeletons of tiny colonial marine organisms called *corals*.
**core** The innermost part of the Earth, made up mostly of iron, that lies beneath the *crust* and *mantle*.
**Coriolis effect** The apparent deflection of north-south air *currents* to a partly east-west direction, caused by the faster spin of regions near the equator than of regions near the poles as a result of Earth's rotation.
**Cornucopian** A person holding the view that we will find ways to make Earth's natural resources meet all of our needs indefinitely and that human ingenuity will see us through any difficulty. Term is also used to describe this view. Compare *Cassandra*.
**corporate average fuel efficiency (CAFE) standards** Miles-per-gallon fuel efficiency standards set by the U.S. Congress for auto manufacturers to meet, by a sales-weighted average of all models of the manufacturer's fleet.
**correlation** A relationship among *variables*.
**corridor** A passageway of protected land established to allow animals to travel between islands of protected *habitat*.
**corrosive** Able to corrode metals. One criterion for defining *hazardous waste*.
**cost-benefit analysis** A method commonly used by *neoclassical economists*, in which estimated costs for a proposed action are totaled and then compared to the sum of benefits estimated to result from the action.
**coupled general circulation model** A computer program that combines what is known about weather patterns, atmospheric circulation, atmosphere-ocean interactions, and feedback mechanisms to simulate *climate* processes.
**covalent bond** A chemical bond in which the uncharged *atoms* in a *molecule* share *electrons*. For example, the uncharged atoms of *carbon* and *oxygen* in *carbon dioxide* form a covalent bond. Compare *ionic bond*.
**cracking** A process in oil refining that converts heavy oil into lighter oil.
**criteria pollutants** Six *air pollutants—carbon monoxide, sulfur dioxide, nitrogen dioxide, tropospheric ozone, particulate matter*, and *lead*—for which the *Environmental Protection Agency* has established maximum allowable concentrations in ambient outdoor air because of the threats they pose to human health.
**cropland** Land that humans use to raise plants for food and fiber.
**crop rotation** The practice of alternating the kind of crop grown in a particular field from one season or year to the next.
**crude birth rate** The number of births per 1,000 individuals for a given time period.
**crude death rate** The number of deaths per 1,000 individuals for a given time period.
**crude oil (petroleum)** A *fossil fuel* produced by the conversion of *organic compounds* by heat and pressure. Crude oil is a mixture of hundreds of different types of *hydrocarbon* molecules characterized by *carbon* chains of different length.
**crust** The lightweight outer layer of the Earth, consisting of rock that floats atop the malleable *mantle*, which in turn surrounds a mostly iron *core*.
**cultivar** A cultivated variety of an agricultural crop. Compare *landrace*.
**cultural hazard** Human health hazards that result from the place we live, our socioeconomic status, our occupation, or our behavioral choices. These include choosing to smoke cigarettes, or living or working with people who do. Compare *biological hazard; chemical hazard; physical hazard*.
**culture** The overall ensemble of knowledge, beliefs, values, and learned ways of life shared by a group of people.
**current** The flow of a liquid or gas in a certain direction.
**customary law** International law that arises from long-standing practices, or customs, held in common by most *cultures*. Compare *conventional law*.

**dam** Any obstruction placed in a river or stream to block the flow of water so that water can be stored in a reservoir. Dams are built to prevent floods, provide drinking water, facilitate *irrigation*, and generate electricity.
**Darwin, Charles (1809–1882)** English naturalist who proposed the concept of *natural selection* as a mechanism for *evolution* and as a way to explain the great variety of living things. See also *Wallace, Alfred Russell*.
**data** Information, generally quantitative information.
**deciduous** Term describing trees that lose their leaves each fall and remain dormant during winter, when hard freezes would endanger leaves.
**decomposer** An organism, such as a fungus or bacterium, that breaks down leaf litter and other nonliving matter into simple constituents that can be taken up and used by plants. Compare *detritivore*.
**deep ecology** A philosophy established in the 1970s based on principles of self-realization (the awareness that humans are inseparable from nature) and *biocentric* equality (the precept that all living beings have equal value). Holds that because we are truly inseparable from our *environment*, we must protect all other living things as we would protect ourselves.
**deep-well injection** A *hazardous waste* disposal method in which a well is drilled deep beneath an area's *water table* into porous rock below an impervious *soil* layer. Wastes are then injected into the well, so that they will be absorbed into the porous rock and remain deep underground, isolated from *groundwater* and human contact. Compare *surface impoundment*.
**deforestation** The clearing and loss of forests.
**demand** The amount of a product people will buy at a given price if free to do so.
**demographer** A scientist who studies human populations.
**demographic fatigue** An inability on the part of governments to address problems caused by overwhelming challenges related to population growth.
**demographic transition** A theoretical model of economic and cultural change that explains the declining death rates and birth rates that occurred in Western nations as they became industrialized. The model holds that industrialization caused these rates to fall naturally by decreasing mortality and by lessening the need for large families. Parents would thereafter choose to invest in quality of life rather than quantity of children.
**demography** A *social science* that applies the principles of *population ecology* to the study of statistical change in human *populations*.
**denitrifying bacteria** Bacteria that convert the nitrates in *soil* or water to gaseous *nitrogen* and release it back into the *atmosphere*.
**density-dependent factor** A *limiting factor* whose effects on a *population* increase or decrease depending on the *population density*. Compare *density-independent factor*.
**density-independent factor** A *limiting factor* whose effects on a *population* are constant regardless of *population density*. Compare *density-dependent factor*.
**deoxyribonucleic acid** See *DNA*.
**dependent variable** The *variable* that is affected by manipulation of the *independent variable*.
**deposition** The arrival of eroded *soil* at a new location. Compare *erosion*.
**desalination (desalinization)** The removal of salt from seawater.
**desert** The driest *biome* on Earth, with annual *precipitation* of less than 25 cm. Because deserts have relatively little vegetation to insulate them from temperature extremes, sunlight readily heats them in the daytime, but daytime heat is quickly lost at night, so temperatures vary widely from day to night and in different seasons.
**desertification** A loss of more than 10% of a land's productivity due to *erosion, soil* compaction, forest removal, *overgrazing*, drought, *salinization, climate* change, depletion of water sources, or other factors. Severe desertification can result in the actual expansion of desert areas or creation of new ones in areas that once supported fertile land.
**detritivore** An organism, such as a millipede or soil insect, that scavenges the waste products or dead bodies of other community members. Compare *decomposer*.
**development** The use of natural resources for economic advancement (as opposed to simple subsistence, or survival).
**dike** A long raised mound of earth erected along a river bank to protect against floods by holding rising water in the main channel.
**directional selection** Mode of *natural selection* in which selection drives a feature in one direction rather than another—for example, toward larger or smaller, or faster or slower. Compare *disruptive selection; stabilizing selection*.

**disruptive selection** Mode of *natural selection* in which a trait diverges from its starting condition in two or more directions. Compare *directional selection; stabilizing selection.*

**distance effect** In *island biogeography theory*, the pattern that islands far from a mainland host fewer *species* because fewer species tend to find and colonize it.

**distillation** The process of separating components of crude oil by heating. Components with different molecular weights boil and condense at different temperatures.

**divergent plate boundary** Area where *magma* surging upward to the surface divides tectonic plates and pushes them apart, creating new *crust* as it cools and spreads. A prime example is the Mid-Atlantic ridge. Compare *transform plate boundary* and *convergent plate boundary.*

**DNA (deoxyribonucleic acid)** A double-stranded *nucleic acid* composed of four nucleotides, each of which contains a sugar (deoxyribose), a phosphate group, and a nitrogenous base. DNA carries the hereditary information for living organisms and is responsible for passing traits from parents to offspring. Compare *RNA.*

**doldrums** A region near the equator with little wind activity.

**dose** The amount of *toxicant* a test animal receives in a dose-response test. Compare *response.*

**dose–response curve** A curve that plots the response of test animals to different doses of a *toxicant.* The response is generally quantified by measuring the proportion of animals exhibiting negative effects.

**downwelling** In the ocean, the flow of warm surface water toward the ocean floor. Downwelling occurs where surface *currents* converge. Compare *upwelling.*

**driftnet** Fishing net that spans large expanses of water, arrayed strategically to drift with currents so as to capture passing fish, and held vertical by floats at the top and weights at the bottom. Driftnetting captures substantial *by-catch* of dolphins, seals, sea turtles, and nontarget fish.

**Dust Bowl** An area that loses huge amounts of *topsoil* to wind *erosion* as a result of drought and/or human impact; first used to name the region in the North American Great Plains severely affected by drought and topsoil loss in the 1930s. The term is now also used to describe that historical event and others like it.

**dynamic equilibrium** The state reached when processes within a *system* are moving in opposing directions at equivalent rates so that their effects balance out.

**ecocentrism** A philosophy that considers actions in terms of their damage or benefit to the integrity of whole ecological *systems*, including both *biotic* and *abiotic* elements. For an ecocentrist, the well-being of an individual organism—human or otherwise—is less important than the long-term well-being of a larger integrated ecological system.

**ecofeminism** A philosophy holding that the patriarchal (male-dominated) structure of society is a root cause of both social and environmental problems. Ecofeminists hold that a *worldview* traditionally associated with women, which interprets the world in terms of interrelationships and cooperation, is more in tune with nature than a worldview traditionally associated with men, which interprets the world in terms of hierarchies and competition.

**ecolabeling** The practice of designating on a product's label how the product was grown, harvested, or manufactured, so that consumers buying it are aware of the processes involved and can differentiate between brands that use processes believed to be *environmentally* beneficial (or less harmful than others) and those that do not.

**ecological economics** A developing school of *economics* that applies the principles of *ecology* and *systems* thinking to the description and analysis of *economies.* Compare *environmental economics; neoclassical economics.*

**ecological footprint** The cumulative amount of land and water required to provide the raw materials a person or *population* consumes and to dispose of or *recycle* the *waste* that is produced.

**ecological restoration** Efforts to reverse the effects of human disruption of *ecological systems* and to restore *communities* to their "natural" state.

**ecology** The *science* that deals with the distribution and abundance of organisms, the interactions among them, and the interactions between organisms and their *abiotic environments.*

**economically recoverable** Extractable such that income from a resource's sale exceeds the costs of extracting it. Applied to *fossil fuel* deposits.

**economics** The study of how we decide to use scarce resources to satisfy the demand for *goods* and *services.*

**economy** A social *system* that converts resources into *goods* and *services.*

**ecosystem** All organisms and nonliving entities that occur and interact in a particular area at the same time.

**ecosystem-based management** The attempt to manage the harvesting of resources in ways that minimize impact on the *ecosystems* and ecological processes that provide the resources.

**ecosystem diversity** The number and variety of ecosystems in a particular area. One way to express *biodiversity.* Related concepts consider the geographic arrangement of *habitats, communities,* or *ecosystems* at the landscape level, including the sizes, shapes, and interconnectedness of patches of these entities.

**ecosystem ecology** The study of how the living and nonliving components of *ecosystems* interact.

**ecosystem service** An essential service an *ecosystem* provides that supports life and makes *economic* activity possible. For example, ecosystems naturally purify air and water, cycle *nutrients*, provide for plants to be *pollinated* by animals, and serve as receptacles and *recycling* systems for the *waste* generated by our economic activity.

**ecotone** A transitional zone where *ecosystems* meet.

**ecotourism** Visitation of natural areas for tourism and recreation. Most often involves tourism by more-affluent people, which may generate *economic* benefits for less-affluent communities near natural areas and thus provide economic incentives for conservation of natural areas.

**$ED_{50}$ (effective dose–50%)** The amount of a *toxicant* it takes to affect 50% of a *population* of test animals. Compare *threshold dose; $LD_{50}$.*

**effluent** Water that flows out of a facility such as a wastewater treatment plant or power plant.

**E horizon** The layer of *soil* that lies below the *A horizon* and above the *B horizon.* The letter "E" stands for "eluviation," meaning "loss," and the E horizon is characterized by the loss of certain minerals through *leaching.* It is sometimes called the "zone of leaching." Compare *C horizon; O horizon; R horizon.*

**electricity** A secondary form of energy that can be transferred over long distances and applied for a variety of uses.

**electrolysis** A process in which electrical current is passed through a *compound* to release *ions.* Electrolysis offers one way to produce *hydrogen* for use as fuel: Electrical current is passed through water, splitting the water *molecules* into hydrogen and *oxygen atoms.*

**electron** A negatively charged particle that surrounds the nucleus of an *atom.*

**electronic waste (e-waste)** Discarded electronic products such as computers, monitors, printers, DVD players, cell phones, and other devices. *Heavy metals* in these products mean that this waste may be judged hazardous.

**element** A fundamental type of matter; a chemical substance with a given set of properties, which cannot be broken down into substances with other properties. Chemists currently recognize 92 elements that occur in nature, as well as more than 20 others that have been artificially created.

**El Niño** The exceptionally strong warming of the eastern Pacific Ocean that occurs every 2 to 7 years and depresses local fish and bird *populations* by altering the marine *food web* in the area. Originally, the name that Spanish-speaking fishermen gave to an unusually warm surface *current* that sometimes arrived near the Pacific coast of South America around Christmas time. Compare *La Niña.*

**El Niño–Southern Oscillation (ENSO)** A systematic shift in atmospheric pressure, sea-surface temperature, and ocean circulation in the tropical Pacific Ocean. ENSO cycles give rise to *El Niño* and *La Niña* conditions.

**emergent property** A characteristic that is not evident in a *system*'s components.

**Emerson, Ralph Waldo (1803–1882)** American author, poet, and philosopher who espoused transcendentalism, a philosophy that views nature as a direct manifestation of the divine, and who promoted a holistic view of nature among the public.

**emigration** The departure of individuals from a *population.*

**emissions trading system** A *permit-trading* system for emissions in which a government issues marketable emissions permits to conduct environmentally

harmful activities. Under a *cap-and-trade* system, the government determines an acceptable level of *pollution* and then issues permits to pollute. A company receives credit for amounts it does not emit and can then sell this credit to other companies. Compare *cap-and-trade.*
**Endangered Species Act (ESA)** The primary *legislation,* enacted in 1973, for protecting *biodiversity* in the United States. It forbids the government and private citizens from taking actions (such as developing land) that would destroy endangered *species* or their *habitats,* and it prohibits trade in products made from endangered species.
**endemic** Native or restricted to a particular geographic region. An endemic species occurs in one area and nowhere else on Earth.
**endocrine disruptor** A *toxicant* that interferes with the *endocrine (hormone) system.*
**endocrine system** The body's *hormone* system.
**energy** An intangible phenomenon that can change the position, physical composition, or temperature of matter.
**energy conservation** The practice of reducing *energy* use as a way of extending the lifetime of our *fossil fuel* supplies, of being less wasteful, and of reducing our impact on the *environment.*
**energy conversion efficiency** The ratio of useful output of *energy* to the amount that needs to be input. See also *EROI; net energy.*
**entropy** The degree of disorder in a substance, *system,* or process. See *second law of thermodynamics.*
**environment** The sum total of our surroundings, including all of the living things and nonliving things with which we interact.
**environmental economics** A developing school of *economics* that modifies the principles of *neoclassical economics* to address environmental challenges. An environmental economist believes that we can attain *sustainability* within our current economic *systems.* Whereas ecological economists call for revolution, environmental economists call for reform. Compare *ecological economics; neoclassical economics.*
**environmental ethics** The application of *ethical standards* to environmental questions.
**environmental health** Environmental factors that influence human health and quality of life and the health of *ecological* systems essential to environmental quality and long-term human well-being.
**environmental impact statement (EIS)** A report of results from detailed studies that assess the potential effects on the *environment* that would likely result from development projects or other actions undertaken by the government.
**environmental justice** A movement based on a moral sense of fairness and equality that seeks to expand society's domain of ethical concern from men to women, from humans to nonhumans, from rich to poor, and from majority races and ethnic groups to minority ones.
**Environmental Performance Index (EPI)** An index that rates nations on their progress toward environmental sustainability, using data from 16 indicators of environmental conditions for which governments can be held accountable.
**environmental policy** *Public policy* that pertains to human interactions with the *environment.* It generally aims to regulate resource use or reduce *pollution* to promote human welfare and/or protect natural systems.
**Environmental Protection Agency (EPA)** An administrative agency created by executive order in 1970. The EPA is charged with conducting and evaluating research, monitoring environmental quality, setting standards, enforcing those standards, assisting the states in meeting standards and goals, and educating the public.
**environmental resistance** The collective force of limiting factors, which together stabilize a population size at its carrying capacity.
**environmental science** The study of how the natural world functions and how humans and the *environment* interact.
**environmental studies** An academic *environmental science* program that heavily incorporates the social sciences as well as the natural sciences.
**environmental toxicology** The study of *toxicants* that come from or are discharged into the *environment,* including the study of health effects on humans, other animals, and *ecosystems.*
**environmentalism** A social movement dedicated to protecting the natural world.
**enzyme** A chemical that catalyzes a chemical reaction.
**epidemiological study** A study that involves large-scale comparisons among groups of people, usually contrasting a group known to have been exposed to some *toxicant* and a group that has not.
**equilibrium theory of island biogeography** A *theory* that was initially applied to oceanic islands to explain how *species* come to be distributed among them. Since its development, researchers have increasingly applied the theory to islands of *habitat* (patches of one type of habitat isolated within vast "seas" of others). Aspects of the theory include *immigration* and *extinction* rates, the effect of island size, and the effect of distance from the mainland.
**EROI (Energy Returned on Investment)** The ratio determined by dividing the quantity of *energy* returned from a process by the quantity of energy invested in the process. Higher EROI ratios mean that more energy is produced from each unit of energy invested. See also *energy conversion efficiency; net energy.*
**erosion** The removal of material from one place and its transport to another by the action of wind or water.
**estuary** An area where a river flows into the ocean, mixing *fresh water* with salt water.
**ethanol** The alcohol in beer, wine, and liquor, produced as a *biofuel* by fermenting biomass, generally from *carbohydrate*-rich crops such as corn.
**ethical standard** A criterion that helps differentiate right from wrong.
**ethics** The study of good and bad, right and wrong. The term can also refer to a person's or group's set of moral principles or values.
**eukaryote** A multicellular organism. The *cells* of eukaryotic organisms consist of a membrane-enclosed nucleus that houses *DNA,* an outer membrane of lipids, and an inner fluid-filled chamber containing *organelles.* Compare *prokaryote.*
**European Union (EU)** Political and economic organization formed after World War II to promote Europe's economic and social progress. As of 2007, the EU consisted of 27 member nations.
**eutrophic** Term describing a water body that has high-nutrient and low-oxygen conditions. Compare *oligotrophic.*
**eutrophication** The process of *nutrient* enrichment, increased production of organic matter, and subsequent *ecosystem* degradation in a water body.
**evaporation** The conversion of a substance from a liquid to a gaseous form.
**even-aged** Condition of timber plantations—generally *monocultures* of a single *species*—in which all trees are of the same age. Most *ecologists* view plantations of even-aged stands more as crop *agriculture* than as ecologically functional forests. Compare *uneven-aged.*
**evenness** See *relative abundance.*
**evolution** Genetically based change in the appearance, functioning, and/or behavior of organisms across generations, often by the process of *natural selection.*
**evolutionary arms race** A duel of escalating adaptations between species. Like rival nations racing to stay ahead of one another in military technology, *host* and *parasite* may repeatedly evolve new responses to the other's latest advance.
**e-waste** See *electronic waste.*
**executive branch** The branch of the U.S. government that is headed by the president and that includes administrative agencies. Among other powers, the president may approve (enact) or reject (veto) legislation and issue executive orders. Compare *judicial branch; legislative branch.*
**executive order** A specific legal instruction for government agencies ordered by the U.S. president.
**exotic** Non-native to an area (as, an exotic organism).
**experiment** An activity designed to test the validity of a *hypothesis* by manipulating *variables.* See *manipulative experiment* and *natural experiment.*
**exploitative interaction** A species interaction in which one participant benefits while another is harmed; that is, one species exploits the other. Such interactions include *predation, parasitism,* and *herbivory.*
**exploratory drilling** Drilling that takes place after a *fossil fuel* deposit has been identified, in order to gauge how much of the fuel exists and whether extraction will prove worthwhile. Involves drilling small holes that descend to great depths.
**exponential growth** The increase of a *population* (or of anything) by a fixed percentage each year.

**expressed preference** An approach in environmental economics that quantifies nonmarket values by asking people how much they would pay for specified items. *Contingent valuation* is an example. Compare *revealed preference.*
**external cost** A negative *externality;* a cost borne by someone not involved in an economic transaction. Examples include harm to citizens from water *pollution* or *air pollution* discharged by nearby factories.
**externality** A cost or benefit of a transaction that affects people other than the buyer or seller.
**extinction** The disappearance of an entire *species* from the face of the Earth. Compare *extirpation.*
**extirpation** The disappearance of a particular *population* from a given area, but not the entire *species* globally. Compare *extinction.*
**extrusive** Term for igneous rock formed when magma is ejected from a volcano and cools quickly (e.g., basalt).

**facilitation** A process by which the presence or action of one *species* enables another species to establish itself or grow more effectively. Most often refers to interactions among plants.
**factory farm** See *feedlot.*
**factory fishing** A highly industrialized approach to commercial fishing, employing fossil fuels, huge vessels, and powerful new technologies to capture fish in immense volumes. Factory fishing vessels even process and freeze their catches while at sea.
**fat** A type of *lipid*, along with oil, that stores energy and releases it when burned, especially for mobile animals.
**feedback loop** A circular process in which a *system*'s output serves as input to that same system. See *negative feedback loop; positive feedback loop.*
**feedlot** A huge barn or outdoor pen designed to deliver *energy*-rich food to animals living at extremely high densities. Also called a factory farm or concentrated animal feeding operation (CAFO).
**Ferrel cell** One of a pair of cells of *convective circulation* between 30° and 60° north and south latitude that influence global *climate* patterns. Compare *Hadley cell; polar cell.*
**fertilizer** A substance that promotes plant growth by supplying essential *nutrients* such as *nitrogen* or *phosphorus.*
**first law of thermodynamics** Physical law stating that *energy* can change from one form to another but cannot be created or lost. The total energy in the universe remains constant and is said to be conserved.
**flagship species** A *species* that has wide appeal with the public and that can be used to promote conservation efforts that also benefit other less charismatic species.
**flat-plate solar collectors** See *solar panels.*
**flexible fuel vehicle** A vehicle that runs on fuel that is a mixture of *ethanol* and gasoline, such as E-85, a mix of 85% ethanol and 15% gasoline.
**floodplain** The region of land over which a river has historically wandered and periodically floods.
**flux** The movement of nutrients among pools or reservoirs in a *nutrient cycle.*
**food chain** A linear series of feeding relationships. As organisms feed on one another, energy is transferred from lower to higher *trophic levels.* Compare *food web.*
**food security** An adequate, reliable, and available food supply to all people at all times.
**food web** A visual representation of feeding interactions within an *ecological community* that shows an array of relationships between organisms at different *trophic levels.* Compare *food chain.*
**forestry** The professional management of forests.
**fossil** The remains, impression, or trace of an animal or plant of past geological ages that has been preserved in rock or *sediments.*
**fossil fuel** A *nonrenewable natural resource,* such as *crude oil, natural gas,* or *coal,* produced by the decomposition and compression of organic matter from ancient life.
**fossil record** The cumulative body of *fossils* worldwide, which paleontologists study to infer the history of past life on Earth.
**Fourth Assessment Report** A 2007 report from the *Intergovernmental Panel on Climate Change (IPCC)* that summarizes thousands of scientific studies, documenting observed trends in surface temperature, precipitation patterns, snow and ice cover, sea levels, storm intensity, and other factors. It also predicts future changes in these phenomena under a range of emission scenarios; addresses impacts of *climate change* on wildlife, ecosystems, and human societies; and discusses strategies we might pursue in response. The Fourth Assessment represents the consensus of scientific climate research from around the world.
**fractionation** See *distillation.*
**Framework Convention on Climate Change (FCCC)** International agreement to reduce *greenhouse gas* emissions to 1990 levels by the year 2000, signed by nations represented at the 1992 Earth Summit convened in Rio de Janeiro by the *United Nations.* The FCCC called for a voluntary, nation-by-nation approach, but by the late 1990s it had become apparent that it would not succeed. Its imminent failure sparked introduction of the *Kyoto Protocol.*
**free rider** A party that fails to invest in controlling *pollution* or carrying out other *environmentally* responsible activities and instead relies on the efforts of other parties to do so. For example, a factory that fails to control its emissions gets a "free ride" on the efforts of other factories that do make the sacrifices necessary to reduce emissions.
**fresh water** Water that is relatively pure, holding very few dissolved salts.
**freshwater marsh** A type of *wetland* in which shallow water allows plants such as cattails to grow above the water surface. Compare *swamp; bog.*
**front** The boundary between air masses that differ in temperature and moisture (and therefore density). See *warm front; cold front.*
**fuel rods** Rods of uranium that supply the fuel for nuclear *fission,* and are kept bathed in a *moderator* in a *nuclear reactor.*
**fundamental niche** The full *niche* of a *species.* Compare *realized niche.*
**fungicide** A type of chemical *pesticide* that kills fungi.

**gasification** A process in which *biomass* is vaporized at extremely high temperatures in the absence of *oxygen,* creating a gaseous mixture including *hydrogen, carbon monoxide,* and *methane,* in order to produce *biopower* or *biofuels.*
**gene** A stretch of *DNA* that represents a unit of hereditary information.
**genera** Plural of *genus.*
**General Land Ordinances of 1785 and 1787** Laws that gave the U.S. government the right to manage Western lands and created a grid system for surveying them and readying them for private ownership.
**generalist** A *species* that can survive in a wide array of *habitats* or use a wide array of resources. Compare *specialist.*
**genetically modified (GM) organism** An organism that has been *genetically engineered* using a technique called *recombinant DNA* technology.
**genetic diversity** A measurement of the differences in *DNA* composition among individuals within a given *species.*
**genetic engineering** Any process scientists use to manipulate an organism's genetic material in the lab by adding, deleting, or changing segments of its *DNA.*
**genome** The entirety of all an organism's genes.
**Genuine Progress Indicator (GPI)** An *economic* indicator introduced in 1995 that attempts to differentiate between desirable and undesirable economic activity. The GPI accounts for benefits such as volunteerism and for costs such as *environmental* degradation and social upheaval. Compare *Gross Domestic Product (GDP).*
**genus** A taxonomic level in the Linnaean classification system that is above *species* and below family. A genus is made up of one or more closely related species.
**geographic information system (GIS)** Computer software that takes multiple types of data (for instance, on geology, hydrology, vegetation, animal species, and human development) and overlays them on a common set of geographic coordinates. The idea is to create a complete picture of a landscape and to analyze how elements of the different datasets are arrayed spatially and how they may be correlated. A common tool of geographers, landscape ecologists, resource managers, and conservation biologists.
**geothermal energy** Renewable *energy* that is generated deep within Earth. The radioactive decay of elements amid the extremely high pressures and temperatures at depth generate heat that rises to the surface in magma and through fissures and cracks. Where this energy heats *groundwater,* natural eruptions of heated water and steam are sent up from below.

**geyser** A natural spurt of heated groundwater and steam sent up from belowground that erupts through the surface.
**glaciation** The extension of ice sheets from the polar regions far into Earth's temperate zones during cold periods of Earth's history.
**global climate change** Any change in aspects of Earth's *climate,* such as temperature, *precipitation,* and storm intensity. Generally refers today to the current warming trend in global temperatures and associated climatic changes. Compare *global warming.*
**global warming** An increase in Earth's average surface temperature. The term is most frequently used in reference to the pronounced warming trend of recent years and decades. Global warming is one aspect of *global climate change,* and in turn drives other components of climate change.
**global warming potential** A quantity that specifies the ability of one *molecule* of a given *greenhouse gas* to contribute to atmospheric warming, relative to *carbon dioxide.*
**good** A material commodity manufactured for and bought by individuals and businesses.
**greenhouse effect** The warming of Earth's surface and *atmosphere* (especially the *troposphere*) caused by the *energy* emitted by *greenhouse gases.*
**greenhouse gas** A gas that absorbs infrared radiation released by Earth's surface and then warms the surface and *troposphere* by emitting *energy,* thus giving rise to the *greenhouse effect.* Greenhouse gases include *carbon dioxide* ($CO_2$), water vapor, ozone ($O_3$), nitrous oxide ($N_2O$), halocarbon gases, and *methane* ($CH_4$).
**green manure** Organic *fertilizer* comprised of freshly dead plant material.
**green revolution** An intensification of the industrialization of *agriculture* in the developing world in the latter half of the 20th century that has dramatically increased crop yields produced per unit area of farmland. Practices include devoting large areas to *monocultures* of crops specially bred for high yields and rapid growth; heavy use of *fertilizers, pesticides,* and *irrigation* water; and sowing and harvesting on the same piece of land more than once per year or per season.
**green tax** A levy on *environmentally* harmful activities and products aimed at providing a market-based incentive to correct for *market failure.* Compare *subsidy.*
**greenwashing** A public-relations effort by a corporation or business to mislead customers or the public into thinking it is acting more sustainably than it actually is.
**greenway** A strip of park land that connects parks or neighborhoods; often located along rivers, streams, or canals.
**Gross Domestic Product (GDP)** The total monetary value of final *goods* and *services* produced in a country each year. The GDP sums all *economic* activity, whether good or bad, and does not account for benefits such as volunteerism or for *external costs* such as *environmental* degradation and social upheaval. Compare *Genuine Progress Indicator (GPI).*
**gross primary production** The *energy* that results when *autotrophs* convert solar energy (sunlight) to energy of chemical bonds in sugars through *photosynthesis.* Autotrophs use a portion of this production to power their own metabolism, which entails oxidizing *organic compounds* by *cellular respiration.* Compare *net primary production.*
**groundfish** A name given to *benthic* fish that live or eat along the bottom, such as cod, halibut, pollock, haddock, and flounder.
**ground source heat pump** A pump that harnesses *geothermal energy* from near-surface sources of earth and water, and that can help heat residences.
**groundwater** Water held in aquifers underground.
**growth rate** The net change in a *population*'s size, per 1,000 individuals. Calculated by adding the *crude birth rate* to the *immigration* rate and then subtracting the *crude death rate* and the *emigration* rate, each expressed as the number per 1,000 individuals per year.

**Haber-Bosch process** A process to synthesize ammonia on an industrial scale. Developed by German chemists Fritz Haber and Carl Bosch, the process has enabled humans to double the natural rate of *nitrogen fixation* on Earth and thereby increase *agricultural* productivity, but it has also dramatically altered the *nitrogen cycle.*
**habitat** The specific *environment* in which an organism lives, including both *biotic* and *abiotic factors.*
**habitat conservation plan** A cooperative agreement that allows landowners to harm threatened or endangered species in some ways if they voluntarily improve habitat for the species in others.
**habitat heterogeneity** The diversity of habitats in a given area.
**habitat selection** The process by which organisms select *habitats* from among the range of options they encounter.
**habitat use** The process by which organisms use *habitats* from among the range of options they encounter.
**Hadley cell** One of a pair of cells of *convective circulation* between the equator and 30° north and south latitude that influence global *climate* patterns. Compare *Ferrel cell; polar cell.*
**half-life** The amount of time it takes for one-half the atoms of a *radioisotope* to emit radiation and decay. Different radioisotopes have different half-lives, ranging from fractions of a second to billions of years.
**harmful algal bloom** A *population* explosion of toxic algae caused by excessive *nutrient* concentrations.
**hazardous waste** *Waste* that is toxic, chemically reactive, flammable, or corrosive. Compare *industrial solid waste; municipal solid waste.*
**Healthy Forests Restoration Act** Controversial legislation initiated by the George W. Bush administration and passed by Congress in 2003 that promotes the physical removal of small trees, underbrush, and dead trees on *national forests* by timber companies in order to reduce the chance of fire. Critics contend that it focuses too little on *prescribed burning,* encourages *salvage logging,* subsidizes private companies, and reduces citizen participation in forest management decisions.
**heat capacity** A measure of the heat energy required to increase the temperature of a given substance by a given amount.
**herbicide** A type of chemical *pesticide* that kills plants.
**herbivore** An organism that consumes plants. Compare *carnivore; omnivore.*
**herbivory** The consumption of plants by animals.
**heterotroph (consumer)** An organism that consumes other organisms. Includes most animals, as well as fungi and microbes that decompose organic matter.
**high-pressure system** An air mass with elevated *atmospheric pressure,* containing air that descends, typically bringing fair *weather.* Compare *low-pressure system.*
**historical sciences** Sciences, such as cosmology or paleontology, that study phenomena that change across historical time.
**homeostasis** The tendency of a *system* to maintain constant or stable internal conditions.
**horizon** A distinct layer of *soil.* See *A horizon; B horizon; C horizon; E horizon; O horizon; R horizon.*
**hormone** A chemical messenger that travels though the bloodstream to stimulate growth, development, and sexual maturity; and regulate brain function, appetite, sexual drive, and many other aspects of physiology and behavior.
**host** The organism in a parasitic relationship that suffers harm while providing the *parasite* nourishment or some other benefit.
**Hubbert's peak** The peak in production of *crude oil* in the United States, which occurred in 1970 just as Shell Oil geologist M. King Hubbert had predicted in 1956.
**humus** A dark, spongy, crumbly mass of material made up of complex organic compounds, resulting from the partial decomposition of organic matter.
**hydrocarbon** An *organic compound* consisting solely of *hydrogen* and *carbon atoms.*
**hydroelectric power (hydropower)** The generation of electricity using the *kinetic energy* of moving water.
**hydrogen** The chemical *element* with 1 proton. The most abundant element in the universe. Also a possible fuel for our future economy.
**hydrogen bond** A special type of interaction in which the oxygen atom of one water molecule is weakly attracted to one or two hydrogen atoms of another.
**hydrogenase** An enzyme that can trigger algae to stop producing *oxygen* as a metabolic by-product and start releasing *hydrogen* instead.
**hydrologic cycle** The flow of water—in liquid, gaseous, and solid forms—through our *biotic* and *abiotic environment.*

**hydropower** See *hydroelectric power.*
**hydrosphere** All water—salt or fresh, liquid, ice, or vapor—in surface bodies, underground, and in the *atmosphere.* Compare *biosphere; lithosphere.*
**hydrothermal vent** Location in the deep ocean where heated water spurts from the seafloor, carrying minerals that precipitate to form rocky structures. Unique and recently discovered *ecosystems* cluster around these vents; tubeworms, shrimp, and other creatures here use symbiotic bacteria to derive their energy from chemicals in the heated water rather than from sunlight.
**hypothesis** An educated guess that explains a phenomenon or answers a *scientific* question. Compare *theory.*
**hypoxia** The condition of extremely low dissolved *oxygen* concentrations in a body of water.

**igneous rock** One of the three main categories of rock. Formed from cooling *magma.* Granite and basalt are examples of igneous rock. Compare *metamorphic rock; sedimentary rock.*
**ignitable** Easily able to catch fire. One criterion for defining *hazardous waste.*
**immigration** The arrival of individuals from outside a *population.*
**inbreeding depression** A state that occurs in a *population* when genetically similar parents mate and produce weak or defective offspring as a result.
**incineration** A controlled process of burning solid waste for disposal in which mixed garbage is combusted at very high temperatures. Compare *sanitary landfill.*
**independent variable** The *variable* that the scientist manipulates in a *manipulative experiment.*
**indoor air pollution** *Air pollution* that occurs indoors.
**industrial ecology** A holistic approach to industry that integrates principles from engineering, chemistry, *ecology, economics,* and other disciplines and seeks to redesign industrial *systems* in order to reduce resource inputs and minimize inefficiency.
**industrial revolution** The shift in the mid-1700s from rural life, animal-powered agriculture, and manufacturing by craftsmen to an urban society powered by *fossil fuels* such as *coal* and *crude oil.* Compare *agricultural revolution.*
**industrial smog** Gray-air smog caused by the incomplete combustion of *coal* or oil when burned. Compare *photochemical smog.*
**industrial solid waste** Nonliquid *waste* that is not especially hazardous and that comes from production of consumer goods, mining, *petroleum* extraction and *refining,* and *agriculture.* Compare *hazardous waste; municipal solid waste.*
**industrial stage** The third stage of the *demographic transition* model, characterized by falling birth rates that close the gap with falling death rates and reduce the rate of *population* growth. Compare *pre-industrial stage; post-industrial stage; transitional stage.*
**industrialized agriculture** A form of *agriculture* that uses large-scale mechanization and *fossil fuel* combustion, enabling farmers to replace horses and oxen with faster and more powerful means of cultivating, harvesting, transporting, and processing crops. Other aspects include *irrigation* and the use of *inorganic fertilizers.* Use of chemical herbicides and *pesticides* reduces *competition* from weeds and *herbivory* by insects. Compare *traditional agriculture.*
**infectious disease** A disease in which a pathogen attacks a host.
**inorganic fertilizer** A *fertilizer* that consists of mined or synthetically manufactured mineral supplements. Inorganic fertilizers are generally more susceptible than *organic fertilizers* to *leaching* and *runoff* and may be more likely to cause unintended off-site impacts.
**insecticide** A type of chemical *pesticide* that kills insects.
**integrated pest management (IPM)** The use of multiple techniques in combination to achieve long-term suppression of pests, including *biocontrol,* use of *pesticides,* close monitoring of *populations, habitat* alteration, *crop rotation, transgenic* crops, alternative tillage methods, and mechanical pest removal.
**intensive traditional agriculture** A form of *traditional agriculture* that aims to produce excess food to sell at market, using teams of laboring animals and employing significant quantities of irrigation water and fertilizer, but stopping short of using fossil fuels. Compare *subsistence agriculture.*
**intercropping** Planting different types of crops in alternating bands or other spatially mixed arrangements.
**interdisciplinary field** A field that borrows techniques from several more traditional fields of study and brings together research results from these fields into a broad synthesis.
**interest group** A small group of people seeking private gain that may work against the larger public interest.
**interglacial** Term describing relatively warm periods in Earth history that occur between periods of *glaciation.*
**Intergovernmental Panel on Climate Change (IPCC)** An international panel of *climate* scientists and government officials established in 1988 by the *United Nations Environment Programme* and the World Meteorological Organization. The IPCC's mission is to assess and synthesize scientific research on *global climate change* and to offer guidance to the world's policymakers. The IPCC's 2007 *Fourth Assessment Report* summarizes current and projected future global trends in climate, and represents the consensus of climate scientists around the world.
**International Polar Year** A large international scientific program coordinating research in the Arctic and Antarctic in 2007–2009.
**interspecific competition** *Competition* that takes place among members of two or more different *species.* Compare *intraspecific competition.*
**intertidal** Of, relating to, or living along shorelines between the highest reach of the highest *tide* and the lowest reach of the lowest tide.
**intraspecific competition** *Competition* that takes place among members of the same *species.* Compare *interspecific competition.*
**intrusive** Term for igneous rock formed when magma cools slowly while it is well below Earth's surface (e.g., granite).
**invasive** Spreading widely, quickly, and becoming dominant in a *community* (as, an *invasive species*).
**invasive species** A *species* that spreads widely and rapidly becomes dominant in a *community,* interfering with the community's normal functioning.
**inversion layer** In a *temperature inversion,* the band of air in which temperature rises with altitude (instead of falling with altitude, as temperature does normally).
**ion** An electrically charged *atom* or combination of atoms.
**ionic bond** A chemical bond in which oppositely charged *ions* are held together by electrical attraction. Compare *covalent bond.*
**ionic compound (salt)** An association of *ions* that are bonded electrically in an *ionic bond.*
**IPAT model** A formula that represents how humans' total impact (I) on the *environment* results from the interaction among three factors: *population* (P), affluence (A), and technology (T).
**irrigation** The artificial provision of water to support *agriculture.*
**isotope** One of several forms of an *element* having differing numbers of *neutrons* in the nucleus of its *atoms.* Chemically, isotopes of an element behave almost identically, but they have different physical properties because they differ in mass.

**judicial branch** The branch of the U.S. government, consisting of the Supreme Court and various lower courts, that is charged with interpreting the law. Compare *executive branch; legislative branch.*

**kelp** Large brown algae or seaweed that can form underwater "forests," providing habitat for marine organisms.
**kerogen** A substance derived from deeply buried organic matter that acts as a source material for both *natural gas* and *crude oil.*
**keystone species** A *species* that has an especially far-reaching effect on a *community.*
**kinetic energy** *Energy* of motion. Compare *potential energy.*
**K–selected** Term denoting a *species* with low *biotic potential* whose members produce a small number of offspring and take a long time to gestate and raise each of their young, but invest heavily in promoting the survival and growth of these few offspring. *Populations* of K–selected species are generally regulated by *density-dependent factors.* Compare *r–selected.*
**kwashiorkor** A form of *malnutrition* that results from a high-*starch* diet with inadequate *protein* or *amino acids.* In children, causes bloating of the abdomen, deterioration and discoloration of hair, mental disability, immune suppression, developmental delays, anemia, and reduced growth.

**Kyoto Protocol** An agreement drafted in 1997 that calls for reducing, by 2012, emissions of six *greenhouse gases* to levels lower than their levels in 1990. Although the United States has refused to ratify the protocol, it came into force in 2005 when Russia ratified it, the 127th nation to do so.

**La Niña** An exceptionally strong cooling of surface water in the equatorial Pacific Ocean that occurs every 2 to 7 years and has widespread climatic consequences. Compare *El Niño.*
**landfill gas** A mix of gases that consists of roughly half *methane* produced by anaerobic decomposition deep inside *landfills.*
**landrace** A variety of an agricultural crop native to a particular area. Compare *cultivar.*
**landscape ecology** The study of how landscape structure affects the abundance, distribution, and interaction of organisms. This approach to the study of organisms and their *environments* at the landscape scale focuses on broad geographical areas that include multiple *ecosystems.*
**land trust** Local or regional organization that preserves lands valued by its members. In most cases, land trusts purchase land outright with the aim of preserving it in its natural condition. The Nature Conservancy may be considered the world's largest land trust.
**latitudinal gradient** The increase in *species richness* as one approaches the equator. This pattern of variation with latitude has been one of the most obvious patterns in *ecology,* but one of the most difficult ones for scientists to explain.
**lava** *Magma* that is released from the *lithosphere* and flows or spatters across Earth's surface.
**law of conservation of matter** Physical law stating that *matter* may be transformed from one type of substance into others, but that it cannot be created or destroyed.
**$LD_{50}$ (lethal dose–50%)** The amount of a *toxicant* it takes to kill 50% of a *population* of test animals. Compare *$ED_{50}$; threshold dose.*
**leachate** Liquids that seep through liners of a *sanitary landfill* and leach into the *soil* underneath.
**leaching** The process by which solid materials such as minerals are dissolved in a liquid (usually water) and transported to another location.
**lead** A heavy metal that may be ingested through water or paint, or that may enter the *atmosphere* as a particulate pollutant through combustion of leaded gasoline or other processes. Atmospheric lead deposited on land and water can enter the *food chain,* accumulate within body tissues, and cause *lead poisoning* in animals and people. An EPA *criteria pollutant.*
**Leadership in Energy and Environmental Design (LEED)** The leading set of standards for sustainable building.
**lead poisoning** Poisoning by ingestion or inhalation of the heavy metal *lead,* causing an array of maladies including damage to the brain, liver, kidney, and stomach; learning problems and behavioral abnormalities; anemia; hearing loss; and even death. Lead poisoning can result from drinking water that passes through old lead pipes or ingesting dust or chips of old lead-based paint.
**legislation** Statutory law.
**legislative branch** The branch of the U.S. government that passes laws; it consists of Congress, which includes the House of Representatives and the Senate. Compare *executive branch; judicial branch.*
**Leopold, Aldo (1887–1949)** American scientist, scholar, philosopher, and author. His book *The Land Ethic* argued that humans should view themselves and the land itself as members of the same *community* and that humans are obligated to treat the land *ethically.*
**levee** See *dike.*
**lichen** A mutualistic aggregate of fungi and algae in which the algal component provides food and energy via photosynthesis while the fungal component takes a firm hold on rock and captures moisture. Often helps kickstart *primary succession.*
**life-cycle analysis** In *industrial ecology,* the examination of the entire life cycle of a given product—from its origins in raw materials, through its manufacturing, to its use, and finally its disposal—in an attempt to identify ways to make the process more *ecologically* efficient.
**life expectancy** The average number of years that individuals in particular age groups are likely to continue to live.
**lifestyle hazard** See *cultural hazard.*
**light pollution** Pollution from city lights that obscures the night sky, impairing people's visibility of stars.
**light reactions** In *photosynthesis,* a series of chemical reactions in which water molecules are split and react to form *hydrogen ions* ($H^+$), molecular *oxygen* ($O_2$), and small, high-energy molecules used to fuel the *Calvin cycle.*
**limiting factor** A physical, chemical, or biological characteristic of the *environment* that restrains *population* growth.
**limnetic zone** In a water body, the layer of open water through which sunlight penetrates. Compare *littoral zone; benthic zone; profundal zone.*
**lipid** One of a chemically diverse group of *macromolecules* that are classified together because they do not dissolve in water. Lipids include *fats* and *oils, phospholipids, waxes,* and *steroids.*
**liquefied natural gas (LNG)** *Natural gas* that has been converted to a liquid at low temperatures and that can be shipped long distances in refrigerated tankers.
**lithification** The formation of rock through the processes of compaction, binding, and crystallization.
**lithosphere** The solid part of the Earth, including the rocks, *sediment,* and *soil* at the surface and extending down many miles underground. Compare *atmosphere; biosphere; hydrosphere.*
**littoral** See *intertidal.*
**littoral zone** The region ringing the edge of a water body. Compare *benthic zone; limnetic zone; profundal zone.*
**Living Planet Index** A metric that summarizes trends in the *populations* of over 1,100 *species* that are well enough monitored to provide reliable data. Developed by scientists at the World Wildlife Fund and the *United Nations Environment Programme* to give an overall idea of how natural populations are faring. Between 1970 and 2003, this index fell by roughly 30%.
**loam** *Soil* with a relatively even mixture of *clay-, silt-,* and *sand*-sized particles.
**lobbying** The expenditure of time or money in an attempt to influence an elected official.
**logistic growth curve** A plot that shows how the initial *exponential growth* of a *population* is slowed and finally brought to a standstill by *limiting factors.*
**longline fishing** Fishing practice that involves setting out extremely long lines with up to several thousand baited hooks spaced along their lengths. Kills turtles, sharks, and an estimated 300,000 seabirds each year in *by-catch.*
**Love Canal** A residential neighborhood in Niagara Falls, New York, from which families were evacuated after buried toxic chemicals rose to the surface, contaminating homes and an elementary school. The well-publicized event helped spark passage of the *Superfund* legislation. Compare *Times Beach.*
**low-input agriculture** *Agriculture* that uses smaller amounts of *pesticides, fertilizers,* growth hormones, water, and *fossil fuel* energy than are used in *industrial agriculture.*
**low-pressure system** An air mass in which the air moves toward the low *atmospheric pressure* at the center of the system and spirals upward, typically bringing clouds and *precipitation.* Compare *high-pressure system.*

**macromolecule** A very large molecule, such as a *protein, nucleic acid, carbohydrate,* or *lipid.*
**macronutrient** A *nutrient* that organisms require in relatively large amounts. Compare *micronutrient.*
**magma** Molten, liquid rock.
**malnutrition** The condition of lacking *nutrients* the body needs, including a complete complement of vitamins and minerals.
**Malthus, Thomas (1766–1834)** British economist who maintained that increasing human *population* would eventually deplete the available food supply until starvation, war, or disease arose and reduced the population.
**mangrove** A tree with a unique type of roots that curve upward to obtain *oxygen,* which is lacking in the mud in which they grow, and that serve as stilts to support the tree in changing water levels. Mangrove forests grow on the coastlines of the tropics and subtropics.
**manipulative experiment** An *experiment* in which the researcher actively chooses and manipulates the *independent variable.* Compare *natural experiment.*

**mantle** The malleable layer of rock that lies beneath Earth's *crust* and surrounds a mostly iron *core.*
***maquiladora*** A U.S.-owned factory on the Mexican side of the U.S.-Mexico border.
**marasmus** A form of *malnutrition* that results from *protein* deficiency together with a lack of calories, causing wasting or shriveling among millions of children in the developing world.
**marine protected area (MPA)** An area of the ocean set aside to protect marine life from fishing pressures. An MPA may be protected from some human activities but be open to others. Compare *marine reserve.*
**marine reserve** An area of the ocean designated as a "no-fishing" zone, allowing no extractive activities. Compare *marine protected area.*
**market failure** The failure of markets to take into account the *environment's* positive effects on *economies* (for example, *ecosystem services*) or to reflect the negative effects of economic activity on the environment and thereby on people (*external costs*).
**mass extinction event** The extinction of a large proportion of the world's *species* in a very short time period due to some extreme and rapid change or catastrophic event. Earth has seen five mass extinction events in the past half-billion years.
**mass number** The combined number of *protons* and *neutrons* in an *atom.*
**materials recovery facility (MRF)** A *recycling* facility where items are sorted, cleaned, shredded, and prepared for reprocessing into new items.
**matter** All material in the universe that has mass and occupies space. See *law of conservation of matter.*
**maximum sustainable yield** The maximal harvest of a particular *renewable natural resource* that can be accomplished while still keeping the resource available for the future.
**mechanical weathering** See *physical weathering.*
**meltdown** The accidental melting of the uranium fuel rods inside the core of a *nuclear reactor*, causing the release of radiation.
**mesosphere** The atmospheric layer above the *stratosphere*, extending 50–80 km (31–56 mi) above sea level.
**metamorphic rock** One of the three main categories of rock. Formed by great heat and/or pressure that reshapes crystals within the rock and changes its appearance and physical properties. Common metamorphic rocks include marble and slate. Compare *igneous rock; sedimentary rock.*
**metapopulation** A network of subpopulations, most of whose members stay within their respective landscape *patches*, but some of whom move among patches or mate with members of other patches.
**methane ($CH_4$)** A colorless gas produced primarily by *anaerobic* decomposition. The major constituent of *natural gas*, and a *greenhouse gas* that is molecule-for-molecule more potent than *carbon dioxide.*
**methane hydrate (methane clathrate** or **methane ice)** An ice-like solid consisting of molecules of *methane* embedded in a crystal lattice of water molecules. Methane hydrates are being investigated as a potential new source of *energy* from *fossil fuels.*
**micronutrient** A *nutrient* that organisms require in relatively small amounts. Compare *macronutrient.*
**Milankovitch cycle** One of three types of variations in Earth's rotation and orbit around the sun that result in slight changes in the relative amount of solar radiation reaching Earth's surface at different latitudes. As the cycles proceed, they change the way solar radiation is distributed over Earth's surface and contribute to changes in *atmospheric* heating and circulation that have triggered the ice ages and other *climate* changes.
**Mill, John Stuart (1806–1873)** British philosopher who believed that as resources become harder to find and extract, *economic* growth will slow and eventually stabilize into a *steady-state economy.*
**Millennium Development Goals** A program of targets for *sustainable development* set by the international community through the *United Nations* at the turn of this century.
**Millennium Ecosystem Assessment** The most comprehensive scientific assessment of the present condition of the world's ecological systems and their ability to continue supporting our civilization. Prepared by over 2,000 of the world's leading environmental scientists from nearly 100 nations, and completed in 2005.
**minimum viable population size** The *population* size for a given population of a *species* that should assure its continued existence indefinitely into the future. Calculated and used by *conservation geneticists*, population biologists, and wildlife managers.
**mixed economy** An economy that combines elements of a capitalist market economy and state socialism.
**moderator** Within a *nuclear reactor*, a substance, most often water or graphite, that slows the *neutrons* bombarding uranium so that *fission* can begin.
**molecule** A combination of two or more *atoms.*
**monoculture** The uniform planting of a single crop over a large area. Characterizes *industrialized agriculture.* Compare *polyculture.*
**Montreal Protocol** International treaty ratified in 1987 in which 180 signatory nations agreed to restrict production of *chlorofluorocarbons (CFCs)* in order to forestall stratospheric ozone depletion. Because of its effectiveness in decreasing global CFC emissions, the Montreal Protocol is considered the most successful effort to date in addressing a global *environmental* problem.
**monumentalism** The impulse to protect enormous, unusual, or beautiful natural features (as in national parks).
**mortality** Rate of death within a *population.*
**mosaic** In *landscape ecology*, a spatial configuration of *patches* arrayed across a landscape.
**moulin** A tunnel extending downward from the top surface of a glacier, created by meltwater and through which water flows.
**mountaintop removal** A large-scale form of *coal* mining in which entire mountaintops are leveled. The technique exerts extreme environmental impact on surrounding ecosystems and human residents.
**Muir, John (1838–1914)** Scottish immigrant to the United States who eventually settled in California and made the Yosemite Valley his wilderness home. Today, he is most strongly associated with the *preservation ethic.* He argued that nature deserved protection for its own inherent values (an *ecocentrist* argument) but also claimed that nature played a large role in human happiness and fulfillment (an *anthropocentrist* argument).
**multiple use** A principle that has nominally guided management policy for national forests over the past half century. The multiple use principle specifies that the forests be managed for recreation, wildlife habitat, mineral extraction, and various other uses.
**municipal solid waste** Nonliquid *waste* that is not especially hazardous and that comes from homes, institutions, and small businesses. Compare *hazardous waste; industrial solid waste.*
**mutagen** A *toxicant* that causes *mutations* in the *DNA* of organisms.
**mutation** An accidental change in *DNA* that may range in magnitude from the deletion, substitution, or addition of a single nucleotide to a change affecting entire sets of chromosomes. Mutations provide the raw material for evolutionary change.
**mutualism** A relationship in which all participating organisms benefit from their interaction. Compare *parasitism.*

**nacelle** Compartment in a *wind turbine* containing machinery for generating power.
**natality** Rate of birth within a *population.*
**national ambient air quality standards** Maximum allowable concentrations of *criteria pollutants* in ambient outdoor air, set by the U.S. *EPA.*
**National Environmental Policy Act (NEPA)** A U.S. law enacted on January 1, 1970, that created an agency called the Council on Environmental Quality and required that an *environmental impact statement* be prepared for any major federal action.
**national forest** Public lands consisting of 191 million acres (more than 8% of the nation's land area) in many tracts spread across all but a few states.
**National Forest Management Act** *Legislation* passed by the U.S. Congress in 1976, mandating that plans for renewable resource management be drawn up for every national forest. These plans were to be explicitly based on the concepts of *multiple use* and *sustainable development* and be subject to broad public participation.
**national park** A scenic area set aside for recreation and enjoyment by the public. The national park system today numbers 388 sites totaling 78.8 million acres and includes national historic sites, national recreation areas, national wild and scenic rivers, and other types of areas.

**national wildlife refuge** An area set aside to serve as a haven for wildlife and also sometimes to encourage hunting, fishing, wildlife observation, photography, environmental education, and other public uses.
**natural experiment** An *experiment* in which the researcher cannot directly manipulate the *variables* and therefore must observe nature, comparing conditions in which variables differ, and interpret the results. Compare *manipulative experiment.*
**natural gas** A *fossil fuel* composed primarily of methane ($CH_4$), produced as a by-product when bacteria decompose organic material under *anaerobic* conditions.
**natural rate of population change** The rate of change in a *population's* size resulting from birth and death rates alone, excluding migration.
**natural resource** Any of the various substances and *energy* sources we need in order to survive.
**Natural Resources Conservation Service** U.S. agency that promotes soil conservation, as well as water quality protection and pollution control. Prior to 1994, known as the Soil Conservation Service.
**natural science** An academic discipline that studies the natural world. Compare *social science.*
**natural selection** The process by which traits that enhance survival and reproduction are passed on more frequently to future generations of organisms than those that do not, thus altering the genetic makeup of populations through time. Natural selection acts on genetic variation and is a primary driver of *evolution.*
**negative feedback loop** A *feedback loop* in which output of one type acts as input that moves the *system* in the opposite direction. The input and output essentially neutralize each other's effects, stabilizing the system. Compare *positive feedback loop.*
**neoclassical economics** A *theory* of *economics* that explains market prices in terms of consumer preferences for units of particular commodities. Buyers desire the lowest possible price, whereas sellers desire the highest possible price. This conflict between buyers and sellers results in a compromise price being reached and the "right" quantity of commodities being bought and sold. Compare *ecological economics; environmental economics.*
**net energy** The quantitative difference between *energy* returned from a process and energy invested in the process. Positive net energy values mean that a process produces more energy than is invested. See also *energy conversion efficiency; EROI.*
**net metering** Process by which owners of houses with *photovoltaic* systems can sell their excess *solar energy* to their local power utility.
**net primary production** The *energy* or biomass that remains in an ecosystem after *autotrophs* have metabolized enough for their own maintenance through *cellular respiration.* Net primary production is the energy or biomass available for consumption by *heterotrophs.* Compare *gross primary production; secondary production.*
**net primary productivity** The rate at which *net primary production* is produced. See *productivity; gross primary production; net primary production; secondary production.*
**neurotoxin** A *toxicant* that assaults the nervous system. Neurotoxins include heavy metals, *pesticides,* and some chemical weapons developed for use in war.
**neutron** An electrically neutral (uncharged) particle in the nucleus of an *atom.*
**new forestry** A set of *ecosystem-based management* approaches for harvesting timber that explicitly mimic natural disturbances. For instance, "sloppy clear-cuts" that leave a variety of trees standing mimic the changes a forest might experience if hit by a severe windstorm.
**new source review** A policy whereby old polluting facilities were exempted from pollution requirements introduced in 1977, as long as they installed the "best available" current technology for pollution control when upgrading their plants in the future.
**new urbanism** A school of thought among architects, planners, and developers that seeks to design neighborhoods in which homes, businesses, schools, and other amenities are within walking distance of one another. In a direct rebuttal to *sprawl,* proponents of new urbanism aim to create functional neighborhoods in which families can meet most of their needs close to home without the use of a car.
**niche** The functional role of a *species* in a *community.* See *fundamental niche; realized niche.*
**NIMBY** See *not-in-my-backyard.*
**nitrification** The conversion by bacteria of ammonium ions ($NH_4^+$) first into nitrite ions ($NO_2^-$) and then into nitrate ions ($NO_3^-$).
**nitrogen** The chemical *element* with seven *protons* and seven *neutrons.* The most abundant element in the *atmosphere,* a key element in *macromolecules,* and a crucial plant *nutrient.*
**nitrogen cycle** A major *nutrient cycle* consisting of the routes that *nitrogen atoms* take through the nested networks of environmental *systems.*
**nitrogen dioxide ($NO_2$)** A foul-smelling reddish brown gas that contributes to *smog* and *acidic deposition.* It results when atmospheric *nitrogen* and *oxygen* react at the high temperatures created by combustion engines. An EPA *criteria pollutant.*
**nitrogen fixation** The process by which inert *nitrogen* gas combines with *hydrogen* to form ammonium ions ($NH_4^+$), which are chemically and biologically active and can be taken up by plants.
**nitrogen-fixing** Term describing bacteria that live in a *mutualistic* relationship with many types of plants and provide *nutrients* to the plants by converting *nitrogen* to a usable form.
**noise pollution** Undesired ambient sound.
**nonconsumptive use** *Fresh water* use in which the water from a particular *aquifer* or surface water body either is not removed or is removed only temporarily and then returned. The use of water to generate electricity in hydroelectric *dams* is an example. Compare *consumptive use.*
**nongovernmental organization (NGO)** An organization unaffiliated with any government or corporation that exists to promote an issue or agenda. Many operate internationally, and some exert influence over environmental policy.
**nonmarket value** A value that is not usually included in the price of a *good* or *service.*
**non-point source** A diffuse source of *pollutants,* often consisting of many small sources. Compare *point source.*
**nonrenewable natural resource** A *natural resource* that is in limited supply and is formed much more slowly than we use it. Compare *renewable natural resource.*
**normative** Prescribing norms or standards, as in *ethics.*
**North Atlantic Deep Water (NADW)** The deep portion of the *thermohaline circulation* in the northern Atlantic Ocean.
**Northwest Forest Plan** A 1994 plan developed by the Clinton administration to allow logging of forests of western Washington, Oregon, and northwestern California with increased protection for *species* and *ecosystems.* The Northwest Forest Plan represented one of the first large-scale applications of *adaptive management.*
**no-till** See *no-tillage.*
**no-tillage** *Agriculture* that does not involve tilling (plowing, disking, harrowing, or chiseling) the *soil.*
**not-in-my-backyard (NIMBY)** Syndrome in which people do not want something (e.g., a polluting facility) near where they live, even if they may want or need the thing to exist somewhere.
**n-type layer** The silicon layer in a *photovoltaic cell* that is enriched with *phosphorus* and is rich in *electrons.* Compare *p-type layer.*
**nuclear energy** The *energy* that holds together *protons* and *neutrons* within the nucleus of an *atom.* Several processes, each of which involves transforming *isotopes* of one *element* into isotopes of other elements, can convert nuclear energy into thermal energy, which is then used to generate electricity. See also *nuclear fission; nuclear reactor.*
**nuclear fission** The conversion of the *energy* within an *atom's* nucleus to usable thermal energy by splitting apart atomic nuclei. Compare *nuclear fusion.*
**nuclear fuel cycle** The process from the mining of *uranium* through its use as fuel in a *nuclear reactor* through its disposal as *waste.*
**nuclear fusion** The conversion of the *energy* within an *atom's* nucleus to usable thermal energy by forcing together the small nuclei of lightweight *elements* under extremely high temperature and pressure. Developing a commercially viable method of nuclear fusion remains an elusive goal.
**nuclear reactor** A facility within a nuclear power plant that initiates and controls the process of *nuclear fission* in order to generate electricity.

**nucleic acid** A *macromolecule* that directs the production of *proteins.* Includes *DNA* and *RNA.*
**nutrient** An *element* or *compound* that organisms consume and require for survival.
**nutrient cycle** The comprehensive set of cyclical pathways by which a given *nutrient* moves through the *environment.*

**oceanography** The study of the physics, chemistry, biology, and geology of the oceans.
**ocean thermal energy conversion (OTEC)** A potential *energy* source that involves harnessing the solar radiation absorbed by tropical oceans in the tropics. See *closed cycle*; *open cycle.*
**O horizon** The top layer of *soil* in some *soil profiles,* made up of organic matter, such as decomposing branches, leaves, crop residue, and animal waste. Compare *A horizon; B horizon; C horizon; E horizon; R horizon.*
**oil** (1) *Petroleum.* (2) A type of *lipid,* along with *fat,* that stores *energy* and releases it when burned, especially for mobile animals.
**oil sands** Deposits that can be mined from the ground, consisting of moist sand and clay containing 1–20% *bitumen,* that some envision as a replacement for *crude oil* as this resource is depleted. Oil sands represent crude oil deposits that have been degraded and chemically altered by water *erosion* and bacterial decomposition.
**oil shale** *Sedimentary rock* filled with *kerogen* that can be processed to produce liquid *petroleum.* Oil shale is formed by the same processes that form *crude oil,* but occurs when kerogen was not buried deeply enough or subjected to enough heat and pressure to form oil.
**oligotrophic** Term describing a water body that has low-nutrient and high-oxygen conditions. Compare *eutrophic.*
**omnivore** An organism that consumes both plants and animals. Compare *carnivore*; *herbivore.*
**open cycle** An approach in *ocean thermal energy conversion* whereby warm surface water is evaporated in a vacuum, its steam turns turbines, and it is then condensed by cold water.
**organ** In an organism, a collection of *tissues,* which performs a particular bodily function.
**organ system** In an organism, a system of *organs* coordinated to perform a particular set of bodily functions.
**organelle** A structure, such as a ribosome or mitochondrion, inside the eukaryotic *cell* that performs specific functions.
**organic agriculture** *Agriculture* that uses no synthetic *fertilizers* or *pesticides* but instead relies on biological approaches such as *composting* and *biocontrol.*
**organic compound** A *compound* made up of *carbon atoms* (and, generally, *hydrogen* atoms) joined by *covalent bonds* and sometimes including other *elements,* such as *nitrogen, oxygen,* sulfur, or *phosphorus.* The unusual ability of carbon to build elaborate *molecules* has resulted in millions of different organic compounds showing various degrees of complexity.
**organic fertilizer** A *fertilizer* made up of natural materials (largely the remains or *wastes* of organisms), including animal manure, crop residues, fresh vegetation, and *compost.* Compare *inorganic fertilizer.*
**Organization of Petroleum Exporting Countries (OPEC)** Cartel of predominantly Arab nations that in 1973 embargoed oil shipments to the United States and other nations supporting Israel, setting off the nation's first oil shortage.
**outdoor air pollution** *Air pollution* that occurs outdoors.
**outlet glacier** A glacier, or slow-moving stretch of ice, that extends outward from the edge of an ice sheet or ice cap, and through which water from melting ice flows.
**overgrazing** The consumption by too many animals of plant cover, impeding plant regrowth and the replacement of *biomass.* Overgrazing can exacerbate damage to *soils,* natural *communities,* and the land's productivity for further grazing.
**overnutrition** A condition of excessive food intake in which people receive more than their daily caloric needs.
**overshoot** The amount by which humanity has surpassed Earth's long-term carrying capacity for our species.
**oxbow** A U-shaped loop in a river.
**oxbow lake** A U-shaped waterbody that becomes isolated from a river when river water erodes a shortcut from one end of an *oxbow* to the other, so that the river pursues a direct course.
**oxygen** The chemical *element* with eight *protons* and eight *neutrons.* A key element in the *atmosphere* that is produced by *photosynthesis.*
**ozone** A *molecule* consisting of three atoms of *oxygen.* Absorbs ultraviolet radiation in the *stratosphere.* Compare *ozone layer*; *tropospheric ozone.*
**ozone hole** Term popularly used to describe the thinning of the stratospheric *ozone layer* that occurs over Antarctica each year, as a result of *chlorofluorocarbons* and other *ozone*-destroying *pollutants.*
**ozone layer** A portion of the *stratosphere,* roughly 17–30 km (10–19 mi) above sea level, that contains most of the *ozone* in the *atmosphere.*

**Pangaea** A "supercontinent" 225 million years ago consisting of all Earth's landmasses joined together.
**paradigm** A dominant philosophical and theoretical framework within a scientific discipline.
**parasite** The organism in a parasitic relationship that extracts nourishment or some other benefit from the *host.*
**parasitism** A relationship in which one organism, the *parasite,* depends on another, the *host,* for nourishment or some other benefit while simultaneously doing the host harm. Compare *mutualism.*
**parasitoid** An insect that parasitizes other insects, generally causing eventual death of the *host.* Compare *parasite.*
**parent material** The base geological material in a particular location.
**particulate matter** Solid or liquid particles small enough to be suspended in the *atmosphere* and able to damage respiratory tissues when inhaled. Includes *primary pollutants* such as dust and soot as well as *secondary pollutants* such as sulfates and nitrates. An EPA *criteria pollutant.*
**passive solar energy collection** An approach in which buildings are designed and building materials are chosen to maximize their direct absorption of sunlight in winter, even as they keep the interior cool in the summer. Compare *active solar energy collection.*
**patch** In *landscape ecology,* spatial areas within a landscape. Depending on a researcher's perspective, patches may consist of habitat for a particular organism, or communities, or ecosystems. An array of patches forms a *mosaic.*
**pathogen** A microbe that causes disease.
**pathogenicity** The intensity with which a *pathogen* can cause harm.
**peace park** A *transboundary park* intended to help ease tensions by acting as a buffer between nations that have quarreled over boundary disputes.
**peak oil** Term used to describe the point of maximum production of petroleum in the world (or for a given nation), after which oil production declines. This is also expected to be roughly the midway point of extraction of the world's oil supplies. The term is generally used in contexts suggesting that our society will face tremendous challenges once the peak is past. Compare *Hubbert's peak.*
**peat** A kind of precursor stage to *coal,* produced when organic material that is broken down by *anaerobic* decomposition remains wet, near the surface, and poorly compressed.
**peer review** The process by which a manuscript submitted for publication in an academic journal is examined by other specialists in the field, who provide comments and criticism (generally anonymously), and judge whether the work merits publication in the journal.
**pelagic** Of, relating to, or living between the surface and floor of the ocean. Compare *benthic.*
**periodic table of the elements** (see Appendix C) Standard table in chemistry that summarizes information on the *elements* in a comprehensive and elegant way.
**permafrost** In *tundra,* underground soil that remains more or less permanently frozen.
**permanent gas** A gas that remains at a stable concentration in the *atmosphere.*
**permit-trading** The practice of buying and selling government-issued marketable emissions permits to conduct environmentally harmful activities. Under a *cap-and-trade* system, the government determines an acceptable level of *pollution* and then issues permits to pollute. A company receives credit for

amounts it does not emit and can then sell this credit to other companies. Compare *emissions trading system.*
**peroxyacyl nitrate** A chemical created by the reaction of $NO_2$ with hydrocarbons that can induce further reactions that damage living tissues in animals and plants.
**pest** A pejorative term for any organism that damages crops that are valuable to us. The term is subjective and defined by our own economic interest, and is not biologically meaningful. Compare *weed.*
**pesticide** An artificial chemical used to kill insects (insecticide), plants (herbicide), or fungi (fungicide).
**pesticide drift** Airborne transport of *pesticides.*
**petroleum** See *crude oil.*
**pH** A measure of the concentration of *hydrogen ions* in a *solution.* The pH scale ranges from 0 to 14: A solution with a pH of 7 is neutral; solutions with a pH below 7 are *acidic,* and those with a pH higher than 7 are *basic.* Because the pH scale is logarithmic, each step on the scale represents a tenfold difference in hydrogen ion concentration.
**phospholipid** A type of *lipid* similar to *fats* but consisting of one water-repellant side and one water-attracting side. When arranged in a double layer, they make up the primary component of cell membranes.
**phosphorus** The chemical *element* with 15 *protons* and 15 *neutrons.* An abundant element in the *lithosphere,* a key element in *macromolecules,* and a crucial plant *nutrient.*
**phosphorus cycle** A major *nutrient cycle* consisting of the routes that *phosphorus atoms* take through the nested networks of environmental *systems.*
**photic zone** In the ocean or a freshwater body, the well-lit top layer of water where *photosynthesis* occurs.
**photobiological** Term describing a process that produces fuel by utilizing light and living organisms.
**photochemical smog** Brown-air smog caused by light-driven reactions of *primary pollutants* with normal atmospheric *compounds* that produce a mix of over 100 different chemicals, ground-level ozone often being the most abundant among them. Compare *industrial smog.*
**photoelectric effect** Effect that occurs when light strikes one of a pair of metal plates in a *photovoltaic cell,* causing the release of *electrons,* which are attracted by electrostatic forces to the opposing plate. The flow of electrons from one plate to the other creates an electrical current.
**photosynthesis** The process by which *autotrophs* produce their own food. Sunlight powers a series of chemical reactions that convert *carbon dioxide* and water into sugar (glucose), thus transforming low-quality *energy* from the sun into high-quality energy the organism can use. Compare *cellular respiration.*
**photovoltaic (PV) cell** A device designed to collect sunlight and directly convert it to electrical *energy* by making use of the *photoelectric effect.*
**photovoltaic effect** See *photoelectric effect.*
**phylogenetic tree** A treelike diagram that represents the history of divergence of *species* or other taxonomic groups of organisms.
**physical hazard** Physical processes that occur naturally in our environment and pose human health hazards. These include discrete events such as earthquakes, volcanic eruptions, fires, floods, blizzards, landslides, hurricanes, and droughts, as well as ongoing natural phenomena such as ultraviolet radiation from sunlight. Compare *biological hazard; chemical hazard; cultural hazard.*
**physical weathering** *Weathering* that breaks rocks down without triggering a chemical change in the *parent material.* Wind and rain are two main forces. Compare *biological weathering; chemical weathering.*
**phytoplankton** Microscopic photosynthetic algae, protists, and cyanobacteria that drift near the surface of water bodies and generally form the first *trophic level* in an aquatic *food chain.* Compare *zooplankton.*
**Pinchot, Gifford (1865–1946)** The first professionally trained American *forester,* Pinchot helped establish the U.S. Forest Service. Today, he is the person most closely associated with the *conservation ethic.*
**pioneer species** A *species* that arrives earliest, beginning the ecological process of *succession* in a terrestrial or aquatic *community.*
**plastics** Synthetic (human-made) *polymers* used in numerous manufactured products.
**plate tectonics** The process by which Earth's surface is shaped by the extremely slow movement of tectonic plates, or sections of *crust.* Earth's surface includes about 15 major tectonic plates. Their interaction gives rise to processes that build mountains, cause earthquakes, and otherwise influence the landscape.
**plowpan** A layer of soil hardened by too much tilling that resists the infiltration of water and the penetration of roots.
**point source** A specific spot—such as a factory's smokestacks—where large quantities of *pollutants* are discharged. Compare *non-point source.*
**polar** A *covalent bond,* or a *molecule* with such a bond, in which *electrons* are shared unequally, with one *atom* exerting a greater pull.
**polar cell** One of a pair of cells of *convective circulation* between the poles and 60° north and south latitude that influence global *climate* patterns. Compare *Ferrel cell; Hadley cell.*
**policy** A rule or guideline that directs individual, organizational, or societal behavior.
**political action committee (PAC)** A group that raises money and distributes it to political campaigns. Corporations and industries may not legally make direct campaign contributions, but may establish PACs.
**pollination** An interaction in which one organism (for example, bees) transfers pollen (male sex cells) from one flower to the ova (female cells) of another, fertilizing the female flower, which subsequently grows into a fruit.
**polluter pays principle** Principle in which the party that produces *pollution* pays the costs of cleaning up or mitigating the pollution.
**pollution** Any matter or *energy* released into the *environment* that causes undesirable impacts on the health and well-being of humans or other organisms. Pollution can be physical, chemical, or biological, and can affect water, air, or soil.
**polybrominated diphenyl ethers (PBDEs)** Synthetic compounds that provide fire-retardant properties and are used in a diverse array of consumer products, including computers, televisions, plastics, and furniture. Released during production, disposal, and use of products, these chemicals persist and accumulate in living tissue, and appear to be *endocrine disruptors.*
**polyculture** The planting of multiple crops in a mixed arrangement or in close proximity. An example is some traditional Native American farming that mixed maize, beans, squash, and peppers. Compare *monoculture.*
**polymer** A chemical *compound* or mixture of compounds consisting of long chains of repeated *molecules.* Some polymers play key roles in the building blocks of life.
**pool** A location in which nutrients in a *biogeochemical cycle* remain for a period of time before moving to another pool. Can be living or nonliving entities. Compare *flux; residence time.*
**population** A group of organisms of the same *species* that live in the same area. Species are often composed of multiple populations.
**population density** The number of individuals within a *population* per unit area. Compare *population size.*
**population dispersion** See *population distribution.*
**population distribution** The spatial arrangement of organisms within a particular area.
**population ecology** Study of the quantitative dynamics of how individuals within a *species* interact with one another—in particular, why *populations* of some species decline while others increase.
**population size** The number of individual organisms present at a given time.
**positive feedback loop** A *feedback loop* in which output of one type acts as input that moves the *system* in the same direction. The input and output drive the system further toward one extreme or another. Compare *negative feedback loop.*
**post-industrial stage** The fourth and final stage of the *demographic transition* model, in which both birth and death rates have fallen to a low level and remain stable there, and *populations* may even decline slightly. Compare *industrial stage; pre-industrial stage; transition stage.*
**potential energy** *Energy* of position. Compare *kinetic energy.*
**precautionary principle** The idea that one should not undertake a new action until the ramifications of that action are well understood.
**precedent** A legal ruling that serves as a guide for later cases, steering judicial decisions through time.
**precipitation** Water that condenses out of the *atmosphere* and falls to Earth in droplets or crystals.
**predation** The process in which one *species* (the *predator*) hunts, tracks, captures, and ultimately kills its *prey.*

**predator** An organism that hunts, captures, kills, and consumes individuals of another species, the *prey*.
**prediction** A specific statement, generally arising from a *hypothesis*, that can be tested directly and unequivocally.
**pre-industrial stage** The first stage of the *demographic transition* model, characterized by conditions that defined most of human history. In pre-industrial societies, both death rates and birth rates are high. Compare *industrial stage; post-industrial stage; transitional stage.*
**prescribed (controlled) burns** The practice of burning areas of forest or grassland under carefully controlled conditions to improve the health of *ecosystems*, return them to a more natural state, and help prevent uncontrolled catastrophic fires.
**prescriptive** Guiding or directing how one should behave, as in *ethics*.
**preservation ethic** An ethic holding that we should protect the natural *environment* in a pristine, unaltered state. Compare *conservation ethic.*
**prey** An organism that is killed and consumed by a *predator*.
**primary consumer** An organism that consumes *producers* and feeds at the second *trophic level.*
**primary extraction** The initial drilling and pumping of the most easily accessible *crude oil.* Compare *secondary extraction.*
**primary forest** Forest uncut by people. Compare *second-growth.*
**primary pollutant** A hazardous substance, such as soot or *carbon monoxide*, that is emitted into the *troposphere* in a form that is directly harmful. Compare *secondary pollutant.*
**primary producer** See *autotroph.*
**primary production** The conversion of solar energy to the energy of chemical bonds in sugars during *photosynthesis*, performed by *autotrophs*. Compare *secondary production.*
**primary succession** A stereotypical series of changes as an *ecological community* develops over time, beginning with a lifeless substrate. In terrestrial *systems*, primary succession begins when a bare expanse of rock, *sand*, or *sediment* becomes newly exposed to the atmosphere and *pioneer species* arrive. Compare *secondary succession.*
**primary treatment** A stage of *wastewater* treatment in which contaminants are physically removed. Wastewater flows into tanks in which sewage solids, grit, and particulate matter settle to the bottom. Greases and oils float to the surface and can be skimmed off. Compare *secondary treatment.*
**probability** A quantitative description of the likelihood of a certain outcome.
**productivity** The rate at which plants convert solar *energy* (sunlight) to *biomass. Ecosystems* whose plants convert solar energy to biomass rapidly are said to have high productivity. See *net primary productivity; gross primary production; net primary production.*
**profundal zone** In a water body, the volume of open water that sunlight does not reach. Compare *littoral zone; benthic zone; limnetic zone.*
**prokaryote** A typically unicellular organism. The *cells* of prokaryotic organisms lack *organelles* and a nucleus. All bacteria and archaeans are prokaryotes. Compare *eukaryote.*
**protein** A *macromolecule* made up of long chains of amino acids.
**proton** A positively charged particle in the nucleus of an *atom.*
**proven recoverable reserve** The amount of a given *fossil fuel* in a deposit that is technologically and economically feasible to remove under current conditions.
**proxy indicator** A type of indirect evidence that serves as a proxy, or substitute, for direct measurement, and that sheds light on conditions of the past. For example, pollen from *sediment* cores and air bubbles from ice cores provide data on the past *climate.*
**p-type layer** The silicon layer in a *photovoltaic cell* that is enriched with boron and is *electron*-poor. Compare *n-type layer.*
**public policy** *Policy* that is made by governments, including those at the local, state, federal, and international levels; it consists of *legislation*, *regulations*, orders, incentives, and practices intended to advance societal welfare. See also *environmental policy.*
**pycnocline** A zone of the ocean beneath the surface in which density increases rapidly with depth.
**pyrolysis** The chemical breakdown of organic matter (such as *biomass, oil shale*, and so on) by heating in the absence of *oxygen*, which often produces materials that can be more easily converted to useable energy. A number of variations on this process exist.

**qualitative** Not expressible in numbers.
**quantitative** Expressible in numbers.

**radiative forcing** The amount of change in *energy* that a given factor (such as *aerosols, albedo*, or *greenhouse gases*) exerts over Earth's energy balance. Positive radiative forcing warms the surface, whereas negative radiative forcing cools it.
**radioactive** The quality by which some *isotopes* "decay," changing their chemical identity as they shed atomic particles and emit high-energy radiation.
**radiocarbon dating** A technique to date preserved organic materials by measuring the percentage of carbon that is the *radioisotope* carbon-14, and matching this value against carbon-14's clocklike progression of decay.
**radioisotope** A radioactive *isotope* that emits subatomic particles and high-*energy* radiation as it "decays" into progressively lighter isotopes until becoming a stable isotope.
**radon** A highly *toxic*, radioactive, colorless gas that seeps up from the ground in areas with certain types of bedrock and can build up inside basements and homes with poor air circulation.
**random distribution** Distribution pattern in which individuals are located haphazardly in space in no particular pattern (often when needed resources are spread throughout an area and other organisms do not strongly influence where individuals settle).
**rangeland** Land used for grazing livestock.
**REACH** Program of the European Union that shifts the burden of proof for testing chemical safety from national governments to industry, and requires that chemical substances produced or imported in amounts of over 1 metric ton per year be registered with a new European Chemicals Agency. *REACH* stands for Registration, Evaluation, Authorisation and Restriction of Chemicals, and went into effect in 2007.
**reactive** Chemically unstable and readily able to react with other compounds, often explosively or by producing noxious fumes. One criterion for defining *hazardous waste.*
**realized niche** The portion of the *fundamental niche* that is fully realized (used) by a *species.*
**recombinant DNA** *DNA* that has been patched together from the DNA of multiple organisms in an attempt to produce desirable traits (such as rapid growth, disease and pest resistance, or higher nutritional content) in organisms lacking those traits.
**recombination** Process by which organisms mix, or recombine, their genetic material (as during sexual reproduction), so that some of each parent's *genes* are included in the genes of the offspring. Recombination produces novel combinations of genes, generating variation among individuals.
**recovery** Waste management strategy composed of *recycling* and *composting.*
**recycling** The collection of materials that can be broken down and reprocessed to manufacture new items.
**Red List** An updated list of *species* facing unusually high risks of *extinction.* The list is maintained by the World Conservation Union.
**red tide** A *harmful algal bloom* consisting of algae that produce reddish pigments that discolor surface waters.
**refining** Process of separating the *molecules* of the various *hydrocarbons* in *crude oil* into different-sized classes and transforming them into various fuels and other petrochemical products.
**regional planning** *City planning* done on broader geographic scales, generally involving multiple municipal governments.
**regulation** A specific rule issued by an administrative agency, based on the more broadly written statutory law passed by Congress and enacted by the president.
**regulatory taking** The deprivation of a property's owner, by means of a law or *regulation*, of most or all economic uses of that property.
**relative abundance** The extent to which numbers of individuals of different species are equal or skewed. One way to express species diversity. See *evenness*; compare *species richness.*
**relative humidity** The ratio of the water vapor contained in a given volume of air to the maximum amount the air could contain, for a given temperature.

**relativist** An ethicist who maintains that *ethics* do and should vary with social context. Compare *universalist*

**renewable natural resource** A *natural resource* that is virtually unlimited or that is replenished by the *environment* over relatively short periods of hours to weeks to years. Compare *nonrenewable natural resource.*

**replacement fertility** The *total fertility rate (TFR)* that maintains a stable *population* size.

**replicate** (as verb): To stage multiple tests of the same experiment. (as noun): One of multiple tests of the same experiment.

**reserves-to-production ratio (R/P ratio)** The total remaining reserves of a *fossil fuel* divided by the annual rate of production (extraction and processing).

**reservoir** See *pool.*

**residence time** In a biogeochemical cycle, the amount of time a nutrient remains in a given pool or reservoir before moving to another. Compare *flux; pool.*

**resilience** The ability of an ecological *community* to change in response to disturbance but later return to its original state. Compare *resistance.*

**resistance** The ability of an ecological *community* to remain stable in the presence of a disturbance. Compare *resilience.*

**Resource Conservation and Recovery Act (RCRA)** Congressional *legislation* (enacted in 1976 and amended in 1984) that specifies, among other things, how to manage *sanitary landfills* to protect against environmental contamination.

**resource management** Strategic decision making about who should extract resources and in what ways, so that resources are used wisely and not wasted.

**resource partitioning** The process by which *species* adapt to *competition* by evolving to use slightly different resources, or to use their shared resources in different ways, thus minimizing interference with one another.

**response** The type or magnitude of negative effects an animal exhibits in response to a *dose* of *toxicant* in a dose-response test. Compare *dose.*

**restoration ecology** The study of the historical conditions of *ecological communities* as they existed before humans altered them.

**revealed preference** An approach in environmental economics that quantifies nonmarket values by measuring people's preferences as revealed by data on their actual behavior. Compare *expressed preference.*

**revolving door** The movement of powerful officials between the private sector and government agencies.

**R horizon** The bottommost layer of *soil* in a typical *soil profile.* Also called *bedrock.* Compare *A horizon; B horizon; C horizon; E horizon; O horizon.*

**ribonucleic acid** See *RNA.*

**riparian** Relating to a river or the area along a river.

**risk** The mathematical probability that some harmful outcome (for instance, injury, death, *environmental* damage, or *economic* loss) will result from a given action, event, or substance.

**risk assessment** The quantitative measurement of *risk,* together with the comparison of risks involved in different activities or substances.

**risk management** The process of considering information from scientific *risk assessment* in light of economic, social, and political needs and values, in order to make decisions and design strategies to minimize *risk.*

**RNA (ribonucleic acid)** A usually single-stranded *nucleic acid* composed of four nucleotides, each of which contains a sugar (ribose), a phosphate group, and a nitrogenous base. RNA carries the hereditary information for living organisms and is responsible for passing traits from parents to offspring. Compare *DNA.*

**rock cycle** The very slow process in which rocks and the minerals that make them up are heated, melted, cooled, broken, and reassembled, forming *igneous, sedimentary,* and *metamorphic* rocks.

**rotation time** The number of years that pass between the time a forest stand is cut for timber and the next time it is cut.

**r–selected** Term denoting a *species* with high *biotic potential* whose members produce a large number of offspring in a relatively short time but do not care for their young after birth. *Populations* of r–selected species are generally regulated by *density-independent factors.* Compare *K–selected.*

**runoff** The water from *precipitation* that flows into streams, rivers, lakes, and ponds, and (in many cases) eventually to the ocean.

**run-of-river** Any of several methods used to generate *hydroelectric power* without greatly disrupting the flow of river water. Run-of-river approaches eliminate much of the *environmental* impact of large *dams.* Compare *storage.*

**Ruskin, John (1819–1900)** British art critic, poet, and writer who criticized industrialized cities and their *pollution,* and who believed that people no longer appreciated the *environment*'s spiritual or aesthetic benefits.

**safe harbor agreement** A cooperative agreement that allows landowners to harm threatened or endangered species in some ways if they voluntarily improve habitat for the species in others.

**salinization** The buildup of salts in surface *soil* layers.

**salt** See *ionic compound.*

**salt marsh** Flat land that is intermittently flooded by the ocean where the *tide* reaches inland. Salt marshes occur along temperate coastlines and are thickly vegetated with grasses, rushes, shrubs, and other herbaceous plants.

**salvage logging** The removal of dead trees following a natural disturbance. Although it may be economically beneficial, salvage logging can be ecologically destructive, because the dead trees provide food and shelter for a variety of insects and wildlife and because removing timber from recently burned land can cause severe *erosion* and damage to *soil.*

**sand** *Sediment* consisting of particles 0.005–2.0 mm in diameter. Compare *clay; silt.*

**sanitary landfill** A site at which solid waste is buried in the ground or piled up in large mounds for disposal, designed to prevent the waste from contaminating the *environment.* Compare *incineration.*

**savanna** A *biome* characterized by grassland interspersed with clusters of acacias and other trees. Savanna is found across parts of Africa (where it was the ancestral home of our *species*), South America, Australia, India, and other dry tropical regions.

**science** A systematic process for learning about the world and testing our understanding of it.

**scientific method** A formalized method for testing ideas with observations that involves several assumptions and a more or less consistent series of interrelated steps.

**scrubber** Technology to chemically treat gases produced in combustion to remove hazardous components and neutralize acidic gases, such as *sulfur dioxide* and hydrochloric acid, turning them into water and salt, in order to reduce smokestack emissions.

**secondary consumer** An organism that consumes *primary consumers* and feeds at the third *trophic level.*

**secondary extraction** The extraction of *crude oil* remaining after *primary extraction* by using solvents or by flushing underground rocks with water or steam. Compare *primary extraction.*

**secondary pollutant** A hazardous substance produced through the reaction of substances added to the *atmosphere* with chemicals normally found in the atmosphere. Compare *primary pollutant.*

**secondary production** The total *biomass* that *heterotrophs* generate by consuming *autotrophs.* Compare *primary production.*

**secondary succession** A stereotypical series of changes as an *ecological community* develops over time, beginning when some event disrupts or dramatically alters an existing community. Compare *primary succession.*

**secondary treatment** A stage of *wastewater* treatment in which biological means are used to remove contaminants remaining after *primary treatment.* Wastewater is stirred up in the presence of *aerobic* bacteria, which degrade organic pollutants in the water. The wastewater then passes to another settling tank, where remaining solids drift to the bottom. Compare *primary treatment.*

**second-growth** Term describing trees that have sprouted and grown to partial maturity after virgin timber has been cut.

**second law of thermodynamics** Physical law stating that the nature of *energy* tends to change from a more-ordered state to a less-ordered state; that is, *entropy* increases.

**sediment** The eroded remains of rocks.

**sedimentary rock** One of the three main categories of rock. Formed when dissolved minerals seep through *sediment* layers and act as a kind of glue, crystallizing and binding sediment particles together. Sandstone and shale are examples of sedimentary rock. Compare *igneous rock; metamorphic rock.*

**seed bank** A storehouse for samples of the world's crop diversity.
**seed-tree** Timber harvesting approach that leaves small numbers of mature and vigorous seed-producing trees standing so that they can reseed a logged area.
**selective breeding** See *artificial selection.*
**septic system** *A wastewater* disposal method, common in rural areas, consisting of an underground tank and series of drainpipes. Wastewater runs from the house to the tank, where solids precipitate out. The water proceeds downhill to a drain field of perforated pipes laid horizontally in gravel-filled trenches, where microbes decompose the remaining waste.
**service** Work done for others as a form of business.
**sex ratio** The proportion of males to females in a *population.*
**shale oil** *Sedimentary rock* filled with organic matter that was not buried deeply enough to form *oil.* Some envision this as a replacement for current *petroleum* as this resource declines.
**shelf-slope break** The portion of the ocean floor where the *continental shelf* drops off with relative suddenness.
**shelterbelt** A row of trees or other tall perennial plants that are planted along the edges of farm fields to break the wind and thereby minimize wind *erosion.*
**shelterwood** Timber harvesting approach that leaves small numbers of mature trees in place to provide shelter for seedlings as they grow.
**sick-building syndrome** An illness produced by indoor *pollution* in which the specific cause is not identifiable.
**silt** *Sediment* consisting of particles 0.002–0.005 mm in diameter. Compare *clay; sand.*
**sink** In a *nutrient cycle,* a *pool* that accepts more *nutrients* than it releases.
**sinkhole** An area where the ground has given way with little warning as a result of subsidence caused by depletion of water from an *aquifer.*
**SLOSS (Single Large or Several Small) dilemma** The debate over whether it is better to make reserves large in size and few in number or many in number but small in size.
**sludge** Solid material that is removed during the *wastewater* treatment process.
**smart growth** A *city planning* concept in which a community's growth is managed in ways that limit *sprawl* and maintain or improve residents' quality of life. It involves guiding the rate, placement, and style of development such that it serves the *environment,* the *economy,* and the community.
**Smith, Adam (1723–1790)** Scottish philosopher known today as the father of *classical economics.* He believed that when people are free to pursue their own economic self-interest in a competitive marketplace, the marketplace will behave as if guided by "an invisible hand" that ensures that their actions will benefit society as a whole.
**smog** Term popularly used to describe unhealthy mixtures of air *pollutants* that often form over urban areas. See *industrial smog; photochemical smog.*
**social science** An academic discipline that studies human interactions and institutions. Compare *natural science.*
**socially responsible investing** Investing in companies that have met certain criteria for environmental or social sustainability.
**soil** A complex plant-supporting *system* consisting of disintegrated rock, organic matter, air, water, *nutrients,* and microorganisms.
**soil porosity** A measure of the size of spaces between soil particles.
**soil profile** The cross-section of a *soil* as a whole, from the surface to the *bedrock.*
**solar cooker** A simple portable oven that uses reflectors to focus sunlight onto food and cook it.
**solar energy** Energy from the sun. It is perpetually renewable and may be harnessed in several ways.
**solar panels** Panels generally consisting of dark-colored, heat-absorbing metal plates mounted in flat boxes covered with glass panes, often installed on rooftops to harness *solar energy.*
**solution** A homogenous mixture of substances in which elements, molecules, or compounds come together without chemically bonding. Most often liquids, but sometimes gases or solids.
**source** In a *nutrient cycle,* a *pool* that releases more *nutrients* than it accepts.
**source reduction** The reduction of the amount of material that enters the *waste stream* to avoid the costs of disposal and *recycling,* help conserve resources, minimize *pollution,* and save consumers and businesses money.
**specialist** A *species* that can survive only in a narrow range of *habitats* that contain very specific resources. Compare *generalist.*
**speciation** The process by which new *species* are generated.
**species** A *population* or group of populations of a particular type of organism whose members share certain characteristics and can breed freely with one another and produce fertile offspring. Different biologists may have different approaches to diagnosing species boundaries.
**species-area curve** A graph showing how number of *species* varies with the geographic area of a landmass or waterbody. *Species richness* commonly doubles as area increases tenfold.
**Species at Risk Act (SARA)** Canada's endangered species protection law, enacted in 2002.
**species coexistence** An outcome of *interspecific competition* in which no competing *species* fully excludes others, and the species continue to live side by side.
**species diversity** The number and variety of *species* in the world or in a particular region.
**species richness** The number of species in a particular region. One way to express species diversity. Compare *evenness; relative abundance.*
**sprawl** The unrestrained spread of urban or *suburban* development outward from a city center and across the landscape. Sometimes specified as growth in which the area of development outpaces *population* growth.
**stabilizing selection** Mode of *natural selection* in which selection produces intermediate traits, in essence preserving the status quo. Compare *directional selection; disruptive selection.*
**stable isotope** An *isotope* that is not *radioactive.*
**steady-state economy** An *economy* that does not grow or shrink but remains stable.
**steroid** A type of *lipid* used in animal *cell* membranes and in the production of *hormones,* including estrogen and androgen, vital to sexual maturation.
**stock** In fisheries management, a *population* of a particular *species* of fish.
**storage** Technique used to generate *hydroelectric power,* in which large amounts of water are impounded in a reservoir behind a concrete *dam* and then passed through the dam to turn *turbine*s that generate electricity. Compare *run-of-river.*
**storm surge** A temporary and localized rise in sea level brought on by the high tides and winds associated with storms.
**Strategic Petroleum Reserve** A stockpile of *oil* stored deep underground in salt caverns in Louisiana by the U.S. government as a short-term buffer against future shortages.
**stratosphere** The layer of the *atmosphere* above the *troposphere* and below the *mesosphere*; it extends from 11 km (7 mi) to 50 km (31 mi) above sea level.
**strip-mining** The use of heavy machinery to remove huge amounts of earth to expose *coal* or minerals, which are mined out directly. Compare *subsurface mining.*
**subduction** The *plate tectonic* process by which denser ocean *crust* slides beneath lighter continental crust at a *convergent plate boundary.*
**subsidy** A government incentive (a giveaway of cash or publicly owned resources, or a *tax break*) intended to encourage a particular activity. Compare *green tax.*
**subsistence agriculture** The oldest form of *traditional agriculture,* in which farming families produce only enough food for themselves. Compare *intensive traditional agriculture.*
**subsistence economy** A survival *economy,* one in which people meet most or all of their daily needs directly from nature and do not purchase or trade for most of life's necessities.
**subspecies** *Populations* of a *species* that occur in different geographic areas and vary from one another in some characteristics. Subspecies are formed by the same processes that drive *speciation,* but result when divergence does not proceed far enough to create separate species.
**subsurface mining** Method of mining underground *coal* deposits, in which shafts are dug deeply into the ground and networks of tunnels are dug or blasted out to follow coal seams. Compare *strip-mining.*
**suburb** A smaller community that rings a city.
**succession** A stereotypical series of changes in the composition and structure of an *ecological community* through time. See *primary succession; secondary succession.*

**sulfur dioxide ($SO_2$)** A colorless gas resulting in part from the combustion of *coal.* In the *atmosphere,* it may react to form sulfur trioxide and sulfuric acid, which may return to Earth in *acidic deposition.* An EPA *criteria pollutant.*
**Superfund** A program administered by the *Environmental Protection Agency* in which experts identify sites polluted with hazardous chemicals, protect *groundwater* near these sites, and clean up the *pollution.*
**supply** The amount of a product offered for sale at a given price.
**surface impoundment** A *hazardous waste* disposal method in which a shallow depression is dug and lined with impervious material, such as *clay.* Water containing small amounts of hazardous waste is placed in the pond and allowed to evaporate, leaving a residue of solid hazardous waste on the bottom. Compare *deep-well injection.*
**survivorship curve** A graph that shows how the likelihood of death for members of a *population* varies with age.
**sustainability** A guiding principle of *environmental science* that requires us to live in such a way as to maintain Earth's systems and its *natural resources* for the foreseeable future.
**sustainable agriculture** *Agriculture* that does not deplete *soils* faster than they form.
**sustainable development** Development that satisfies our current needs without compromising the future availability of *natural resources* or our future quality of life.
**sustainable forestry certification** A form of *ecolabeling* that identifies timber products that have been produced using *sustainable* methods. Several organizations issue such certification.
**swamp** A type of *wetland* consisting of shallow water rich with vegetation, occurring in a forested area. Compare *bog; freshwater marsh.*
**swidden** The traditional form of *agriculture* in tropical forested areas, in which the farmer cultivates a plot for 1 to a few years and then moves on to clear another plot, leaving the first to grow back to forest. When the forest is burned this may be called "slash-and-burn" agriculture.
**symbiosis** A *parasitic* or *mutualistic* relationship between different *species* of organisms that live in close physical proximity.
**sympatric speciation** *Species* formation that occurs when *populations* become reproductively isolated within the same geographic area. Compare *allopatric speciation.*
**synergistic effect** An interactive effect (as of *toxicants)* that is more than or different from the simple sum of their constituent effects.
**system** A network of relationships among a group of parts, elements, or components that interact with and influence one another through the exchange of *energy,* matter, and/or information.

**taiga** See *boreal forest.*
**takings clause** A clause from the Fifth Amendment to the U.S. Constitution that ensures, in part, that private property shall not "be taken for public use without just compensation." Courts have interpreted this clause to ban not only the literal taking of private property but also *regulatory takings.*
**Talloires Declaration** A document composed in Talloires, France, in 1990 that commits university leaders to pursue *sustainability* on their campuses. It has been signed by over 300 university presidents and chancellors from more than 40 nations.
**tar sands** See *oil sands.*
**tax break** Governmental reduction or elimination of taxes required of a business or an individual, for the purpose of promoting industries or activities deemed desirable.
**taxonomist** A scientist who classifies species, using an organism's physical appearance and/or genetic makeup, and who groups species by their similarity into a hierarchy of categories meant to reflect evolutionary relationships.
**technically recoverable** Extractable using current technology. Applied to *fossil fuel* deposits.
**temperate deciduous forest** A *biome* consisting of midlatitude forests characterized by broad-leafed trees that lose their leaves each fall and remain dormant during winter. These forests occur in areas where *precipitation* is spread relatively evenly throughout the year: much of Europe, eastern China, and eastern North America.
**temperate grassland** A *biome* whose vegetation is dominated by grasses and features more extreme temperature differences between winter and summer and less *precipitation* than *temperate deciduous forests.*
**temperate rainforest** A *biome* consisting of tall coniferous trees, cooler and less *species*-rich than *tropical rainforest* and milder and wetter than *temperate deciduous forest.*
**temperature (thermal) inversion** A departure from the normal temperature distribution in the *atmosphere,* in which a pocket of relatively cold air occurs near the ground, with warmer air above it. The cold air, denser than the air above it, traps *pollutants* near the ground and causes a buildup of *smog.*
**teratogen** A *toxicant* that causes harm to the unborn, resulting in birth defects.
**terracing** The cutting of level platforms, sometimes with raised edges, into steep hillsides to contain water from *irrigation* and *precipitation.* Terracing transforms slopes into series of steps like a staircase, enabling farmers to cultivate hilly land while minimizing their loss of *soil* to water *erosion.*
**tertiary consumer** An organism that consumes *secondary consumers* and feeds at the fourth *trophic level.*
**theory** A widely accepted, well-tested explanation of one or more cause-and-effect relationships that has been extensively validated by a great amount of research. Compare *hypothesis.*
**thermal inversion** See *temperature inversion.*
**thermal mass** Construction materials that absorb heat, store it, and release it later, for use in *passive solar energy* approaches.
**thermogenic** Type of *natural gas* created by compression and heat deep underground. Contains *methane* and small amounts of other *hydrocarbon* gases. Compare *biogenic.*
**thermohaline circulation** A worldwide system of ocean currents in which warmer, fresher water moves along the surface and colder, saltier water (which is more dense) moves deep beneath the surface.
**thermosphere** The *atmosphere*'s top layer, extending upward to an altitude of 500 km (300 mi).
**Thoreau, Henry David (1817–1862)** American transcendentalist author, poet, and philosopher. His book *Walden,* recording his observations and thoughts while he lived at Walden Pond away from the bustle of urban Massachusetts, remains a classic of American literature.
**Three Mile Island** Nuclear power plant in Pennsylvania that in 1979 experienced a partial *meltdown.* The term is often used to denote the accident itself, the most serious *nuclear reactor* malfunction that the United States has thus far experienced.
**threshold dose** The amount of a *toxicant* at which it begins to affect a *population* of test animals. Compare $ED_{50}$; $LD_{50}$.
**tidal creek** A channel in a *salt marsh* through which the *tide* flows in and out.
**tidal energy** *Energy* harnessed by erecting a *dam* across the outlet of a tidal basin. Water flowing with the incoming or outgoing *tide* through sluices in the dam turns turbines to generate *electricity.*
**tide** The periodic rise and fall of the ocean's height at a given location, caused by the gravitational pull of the moon and sun.
**Times Beach** A town in Missouri whose residents were evacuated and whose buildings were demolished after being contaminated by *dioxin* from waste oil sprayed on its roads. The well-publicized event helped spark passage of the *Superfund* legislation. Compare *Love Canal.*
**tissue** In an organism, a collection of *cells* that perform the same function.
**topsoil** That portion of the *soil* that is most nutritive for plants and is thus of the most direct importance to *ecosystems* and to *agriculture.* Also known as the *A horizon.*
**tort law** A system of law addressing harm caused by one entity to another, which operates primarily through lawsuits.
**total fertility rate (TFR)** The average number of children born per female member of a *population* during her lifetime.
**toxic** Poisonous; able to harm health of people or other organisms when a substance is inhaled, ingested, or touched. One criterion for defining *hazardous waste.*
**toxic air pollutant** *Air pollutant* that is known to cause cancer, reproductive defects, or neurological, developmental, immune system, or respiratory problems in humans, and/or to cause substantial *ecological* harm by affecting the health of nonhuman animals and plants. The *Clean Air Act of 1990* identifies 188 toxic air pollutants, ranging from the heavy metal mercury to *volatile organic compounds* such as benzene and methylene chloride.

**toxicant** A substance that acts as a poison to humans or wildlife.
**toxicity** The degree of harm a chemical substance can inflict.
**toxicology** The scientific field that examines the effects of poisonous chemicals and other agents on humans and other organisms.
**toxin** A *toxic* chemical stored or manufactured in the tissues of living organisms. For example, a chemical that plants use to ward off *herbivores* or that insects use to deter *predators.*
**trade winds** Prevailing winds between the equator and 30° latitude that blow from east to west.
**traditional agriculture** Biologically powered *agriculture,* in which human and animal muscle power, along with hand tools and simple machines, perform the work of cultivating, harvesting, storing, and distributing crops. Compare *industrialized agriculture.*
**transboundary** Crossing a political boundary such as a national border.
**transboundary park** A reserve of protected land that overlaps national borders.
**transform plate boundary** Area where two tectonic plates meet and slip and grind alongside one another. For example, the Pacific Plate and the North American Plate rub against each other along California's San Andreas Fault.
**transgene** A *gene* that has been extracted from the *DNA* of one organism and transferred into the DNA of an organism of another *species.*
**transgenic** Term describing an organism that contains *DNA* from another *species.*
**transitional stage** The second stage of the *demographic transition* model, which occurs during the transition from the *pre-industrial stage* to the *industrial stage.* It is characterized by declining death rates but continued high birth rates. See also *post-industrial stage.* Compare *industrial stage; post-industrial stage; pre-industrial stage.*
**transmissible disease** See *infectious disease*
**transit-oriented development** A development approach in which compact communities in the *new urbanist* style are arrayed around stops on a major rail transit line.
**transpiration** The release of water vapor by plants through their leaves.
**trawling** Fishing method that entails dragging immense cone-shaped nets through the water, with weights at the bottom and floats at the top to keep the nets open. Compare *bottom-trawling.*
**treatment** The portion of an *experiment* in which a *variable* has been manipulated in order to test its effect. Compare *control.*
**treaty** See *convention.*
**tributary** A smaller river that flows into a larger one.
**triple bottom line** An approach to sustainability that attempts to meet environmental, economic, and social goals simultaneously.
**trophic cascade** A series of changes in the *population* sizes of organisms at different *trophic levels* in a *food chain,* occurring when *predators* at high trophic levels indirectly promote populations of organisms at low trophic levels by keeping species at intermediate trophic levels in check. Trophic cascades may become apparent when a top predator is eliminated from a system.
**trophic level** Rank in the feeding hierarchy of a *food chain.* Organisms at higher trophic levels consume those at lower trophic levels.
**tropical dry forest** A *biome* that consists of deciduous trees and occurs at tropical and subtropical latitudes where wet and dry seasons each span about half the year. Widespread in India, Africa, South America, and northern Australia.
**tropical rainforest** A *biome* characterized by year-round rain and uniformly warm temperatures. Found in Central America, South America, southeast Asia, west Africa, and other tropical regions. Tropical rainforests have dark, damp interiors; lush vegetation; and highly diverse *biotic communities.*
**tropopause** The boundary between the *troposphere* and the *stratosphere.* Acts like a cap, limiting mixing between these atmospheric layers.
**troposphere** The bottommost layer of the *atmosphere;* it extends to 11 km (7 mi) above sea level. See also *stratosphere.*
**tropospheric ozone** *Ozone* that occurs in the *troposphere,* where it is a *secondary pollutant* created by the interaction of sunlight, heat, nitrogen oxides, and volatile *carbon*-containing chemicals. A major component of *smog,* it can injure living tissues and cause respiratory problems. An EPA *criteria pollutant.*
**tundra** A *biome* that is nearly as dry as *desert* but is located at very high latitudes along the northern edges of Russia, Canada, and Scandinavia. Extremely cold winters with little daylight and moderately cool summers with lengthy days characterize this landscape of lichens and low, scrubby vegetation.
**turbine** A rotary device that converts the *kinetic energy* of a moving substance, such as steam, into mechanical energy. Used widely in commercial power generation from various types of energy sources.
**type I** A *survivorship curve* that shows higher death rates at older ages.
**type II** A *survivorship curve* that shows equal rates of death at all ages.
**type III** A *survivorship curve* that shows highest death rates at young ages.

**umbrella species** A *species* for which meeting its *habitat* needs automatically helps meet those of many other species. Umbrella species generally are species that require large areas of habitat.
**unconfined aquifer** A water-bearing, porous layer of rock, *sand,* or gravel that lies atop a less-permeable substrate. The water in an unconfined aquifer is not under pressure because there is no impermeable upper layer to confine it. Compare *confined aquifer.*
**undernutrition** A condition of insufficient *nutrition* in which people receive less than 90% of their daily caloric needs.
**uneven-aged** Term describing stands of trees in timber plantations that are of different ages. Uneven-aged stands more closely approximate a natural forest than do *even-aged* stands.
**uniform distribution** Distribution pattern in which individuals are evenly spaced (as when individuals hold territories or otherwise compete for space).
**United Nations (U.N.)** Organization founded in 1945 to promote international peace and to cooperate in solving international economic, social, cultural, and humanitarian problems.
**United Nations Environment Programme (UNEP)** Agency within the United Nations that deals with *environmental policy.* Created in 1972.
**United Nations Framework Convention on Climate Change** See *Framework Convention on Climate Change.*
**universalist** An *ethicist* who maintains that there exist objective notions of right and wrong that hold across cultures and situations. Compare *relativist.*
**upwelling** In the ocean, the flow of cold, deep water toward the surface. Upwelling occurs in areas where surface *currents* diverge. Compare *downwelling.*
**urban ecology** A scientific field that views cities explicitly as *ecosystems.* Researchers in this field seek to apply the fundamentals of *ecosystem ecology* and *systems* science to urban areas.
**urban growth boundary (UGB)** In *city planning,* a geographic boundary intended to separate areas desired to be urban from areas desired to remain rural. Development for housing, commerce, and industry are encouraged within urban growth boundaries, but beyond them such development is severely restricted.
**urbanization** The shift from rural to city and *suburban* living.
**utility** An *ethical standard,* elaborated by British philosophers Jeremy Bentham and *John Stuart Mill,* holding that something is right when it produces the greatest practical benefits for the most people.

**variable** In an *experiment,* a condition that can change. See *dependent variable* and *independent variable.*
**variable gas** An atmospheric gas that varies in concentration from time to time or place to place, as a result of natural processes or human activities.
**vector** An organism that transfers a *pathogen* to its *host.* An example is a mosquito that transfers the malaria pathogen to humans.
**vested interest** A strong personal interest in the outcome of a decision that may result in one's private gain or loss.
**virtue** An *ethical standard* that, as the ancient Greek philosopher Aristotle held, involves the personal achievement of moral excellence in character through reasoning and moderation.
**virulence** See *pathogenicity.*
**volatile organic compound (VOC)** One of a large group of potentially harmful organic chemicals used in industrial processes.

**Wallace, Alfred Russell (1823–1913)** English naturalist who proposed, independently of *Charles Darwin*, the concept of *natural selection* as a mechanism for *evolution* and as a way to explain the great variety of living things.
**warm front** The boundary where a mass of warm air displaces a mass of colder air. Compare *cold front.*
**waste** Any unwanted product that results from a human activity or process.
**waste management** Strategic decision making to minimize the amount of *waste* generated and to dispose of waste safely and effectively.
**waste stream** The flow of *waste* as it moves from its sources toward disposal destinations.
**waste-to-energy (WTE) facility** An incinerator that uses heat from its furnace to boil water to create steam that drives electricity generation or that fuels heating systems.
**wastewater** Any water that is used in households, businesses, industries, or public facilities and is drained or flushed down pipes, as well as the polluted *runoff* from streets and storm drains.
**waterlogging** The saturation of *soil* by water, in which the *water table* is raised to the point that water bathes plant roots. Waterlogging deprives roots of access to gases, essentially suffocating them and eventually damaging or killing the plants.
**water mining** The withdrawal of water at a rate faster than it can be replenished.
**watershed** The entire area of land from which water drains into a given river.
**water table** The upper limit of *groundwater* held in an *aquifer.*
**wave energy** *Energy* harnessed from the motion of wind-driven waves at the ocean's surface. Many designs for machinery to harness wave energy have been invented, but few have been adequately tested.
**wax** A type of *lipid* digestible by some but not all organisms, which plays structural roles.
**weather** The local physical properties of the *troposphere,* such as temperature, pressure, humidity, cloudiness, and wind, over relatively short time periods. Compare *climate.*
**weathering** The physical, chemical, and biological processes that break down rocks and minerals, turning large particles into smaller particles.
**weed** A pejorative term for any plant that competes with our crops. The term is subjective and defined by our own economic interest, and is not biologically meaningful. Compare *pest.*
**westerlies** Prevailing winds from 30° to 60° latitude that blow from west to east.
**wetland** A system that combines elements of *fresh water* and dry land. These biologically productive systems include *freshwater marshes, swamps,* and *bogs.*
**Whitman, Walt (1819–1892)** American poet who espoused transcendentalism. See also *Emerson, Ralph Waldo* and *Thoreau, Henry David.*
**wilderness area** Federal land that is designated off-limits to development of any kind but is open to public recreation, such as hiking, nature study, and other activities that have minimal impact on the land.
**windbreak** See *shelterbelt.*
**wind energy** *Energy* from the motion of wind. In this source of renewable energy, the passage of wind through *wind turbines* is used to generate *electricity.*
**wind farm** A development involving a group of *wind turbines.*
**wind turbine** A mechanical assembly that converts the wind's *kinetic energy,* or energy of motion, into electrical energy.
**wise-use movement** A loose confederation of individuals and groups that coalesced in the 1980s and 1990s as a response to the increasing success of environmental advocacy. The movement favors extracting more resources from public lands, obtaining greater local control of lands, and obtaining greater motorized recreational access to public lands.
**World Bank** Institution founded in 1944 that serves as one of the globe's largest sources of funding for *economic* development, including such major projects as *dams, irrigation* infrastructure, and other undertakings.
**world heritage site** A location internationally designated by the *United Nations* for its cultural or natural value. There are over 830 such sites worldwide.
**World Trade Organization (WTO)** Organization based in Geneva, Switzerland, that represents multinational corporations and promotes free trade by reducing obstacles to international commerce and enforcing fairness among nations in trading practices.
**worldview** A way of looking at the world that reflects a person's (or a group's) beliefs about the meaning, purpose, operation, and essence of the world.

**xeriscaping** Landscaping using plants adapted to arid conditions.

**zoning** The practice of classifying areas for different types of development and land use.
**zooplankton** Tiny aquatic animals that feed on *phytoplankton* and generally comprise the second *trophic level* in an aquatic *food chain.* Compare *phytoplankton.*
**zooxanthellae** Symbiotic algae that inhabit the bodies of *corals* and produce food through *photosynthesis.*

# Photo Credits

**Part Opening Photos: Part One** Bill Hatcher/National Geographic Collection **Part Two** AP Photo/Bill Haber

**Chapter 1 Opening Photo** NASA/Johnson Space Center **1.2b** Charles O'Rear/CORBIS **1.3a** Art Resource **1.3b** Bettman/CORBIS **1.6** George Konig/Hulton Archive Photos **1.7** Joel W. Rogers/CORBIS **1.8a** Warren Montague and George Bukenhofer **1.8b** Reuters New Media Inc./CORBIS **1.13** CORBIS **1.14** Reuters **1.15** Paul Gendell/Alamy Images **1.16** AP Photo/Dario Lopez-Mills **The Science behind the Story: Easter Island miniphoto** Richard T. Nowitz/CORBIS **main** Hiroyuki Matsumoto/Getty Images, Inc.

**Chapter 2 Opening Photo** Reuters New Media Inc./CORBIS **Case Study** The Gundjehmi Aboriginal Corporation **2.2a** The Gundjehmi Aboriginal Corporation **2.5** Library of Congress **2.6** CORBIS **2.7** CORBIS **2.8** Bettmann/CORBIS **2.9** Mario Tama/Getty Images-BC **2.10** Loomis Dean/Time Life Pictures/Getty Images **2.13** AFP/CORBIS **2.15** Imagebroker/Alamy Images **2.17a** Konrad Wothe/Minden Pictures **2.17b** Bill Hatcher/National Geographic Image Collection **2.17c** Bruce Forster/The Image Bank **2.17d** CORBIS **2.17e** Charles O'Rear/CORBIS **2.17f, g** Frans Lanting/Minden Pictures **The Science behind the Story: Ethics in Economics** AP Wide World Photos **The Science behind the Story: Economic Value of Ecosystems** Dr. Robert Costanza

**Chapter 3 Opening Photo** CORBIS **Case Study** Jamie Kum/Ocean Imaging, www.oceani.com **3.1** Lori Saldaña **3.3** Annie Griffiths Belt/CORBIS **3.6** Rick Wilking/Reuters/Corbis **3.7a** Bettmann/-CORBIS **3.7b** Museum of History & Industry **3.7c** University of Washington Libraries **3.8** Erich Hartmann/Magnum Photos, Inc. **3.9** Bettmann/CORBIS **3.10a** CWH, Associated Press **3.10b** Binod Joshi/Associated Press **3.11** Bettmann/CORBIS **3.13** Don Boroughs/The Image Works **3.14** Lori Saldana **3.15** Jon Hrusa/AP Photo **3.16** Alex Wong/Getty Images **3.19** Charles Mauzy/CORBIS **3.20** Martin Rugner/age footstock **The Science behind the Story: Transboundary Sewage miniphoto** Joe Kocurek **main** Metropolitan Wastewater Designer **The Science behind the Story: Environmental Regulations** Ted Spiegel/CORBIS

**Chapter 4 Opening Photo** AFP/CORBIS **Case Study** Exxon Corporation **4.1** AP Photo/Jack Smith **4.5a** Digital Vision/Picture Quest **4.13 left** Daniel Zheng/CORBIS **4.13 right** Anne-Marie Weber/Taxi **4.17** Jack Dykinga/Stone **4.18a** Ken MacDonald/SPL/Photo Researchers, Inc. **4.18b** Woods Hole Oceanographic Institute **4.19** Chip Clark **4.20** Dorling Kindersley **The Science behind the Story: Dirty Work** BIOS Van Baelinghem/Peter Arnold, Inc. **The Science behind the Story: Isotopes** Keith Hobson

**Chapter 5 Opening Photo** Michael & Patricia Fogden/CORBIS **Case Study** Michael & Patricia Fogden/CORBIS **5.1a** Nicholas Athanas/Tropical Birding **5.1b** Michael Fogden/DRK Photo **5.1c** Michael & Patricia Fogden/CORBIS **5.1d** Michael Fogden/Photolibrary **5.3** Bishop Museum, Honolulu **5.7** Painting by Marella J. Sibbick/The National History Museum, London **5.8** Chip Clark/National Museum of Natural History, Smithsonian Institution **5.9** Lynda Richardson/CORBIS **5.11a** Wisconsin Historical Society **5.11b** G. I. Bernard/PhotoResearchers, Inc. **5.12a** Keenan Ward/CORBIS **5.12b** Art Wolfe/The Image Bank **5.12c** PhotoDisc Green **5.18a** Mike Danzenbaker **5.18b** A. Witte/C. Mahaney/Getty Images **5.18c** John Waters/Nature Picture Library **5.19** Matthias Clamer/Stone **The Science behind the Story: K-T Mass Extinction miniphoto** Roger Ressmeyer/CORBIS/Bettmann **main** Benjamin Cummings **The Science behind the Story: Climate Change and Monteverde** J. Alan Pounds

**Chapter 6 Opening Photo** Robert Estall/Robert Harding World Imagery **Case Study** Wolfgang Polzer **6.1a** Peter Yates/Photo Researchers, Inc. **6.1b** R. Runk/S. Schoenberger/Grant Heilman Photography **6.4** Michael & Patricia Fogden/CORBIS **6.6a** Peter Johnson/CORBIS **6.6b** Getty Images **6.6c** Michael and Patricia Fogden/CORBIS **6.7** Tom Stack and Associates, Inc. **6.8** Joke Stuurman-Huitema/Foto Nat/Minden Pictures **6.9** Michael & Patricia Fogden/CORBIS **6.14** Charles D. Winters/Photo Researchers, Inc. **6.17** Layne Kennedy/CORBIS **6.20a** Pat O'Hara/CORBIS **6.21a** Philip Gould/CORBIS **6.22a** Charles Mauzy/CORBIS **6.23a** David Samuel Robbins/CORBIS **6.24a** Getty Images/Digital Vision **6.25a** Wolfgang Kaehler/CORBIS **6.26a** Sylvester Adams/Getty Images **6.27a** Darrell Gulin/CORBIS **6.28a** Liz Hymans/CORBIS **6.29a** Charles Mauzy/CORBIS **The Science behind the Story: Zebra Mussels** Heather Malcolm **The Science behind the Story: Otters, Urchins, Kelp** Jim Bodkin

**Chapter 7 Opening Photo** Yann Arthus-Bertrand/CORBIS **Case Study** Owen Franken/CORBIS **7.8** D. W. Schindler **7.9** NASA **7.12a, b** Milos Kalab/CMSP **The Science behind the Story: Landscape-level Planning** Joe McDonald/CORBIS **The Science behind the Story: Hypoxia** Louisiana Universities Marine Consortium

**Chapter 8 Opening Photo** Redlink/CORBIS **Case Study** Louise Gibb/The Image Works **8.1** CORBIS **8.12c** SETBOUN/CORBIS **8.13** David Turnley/CORBIS **8.18 inset** AFP/Getty Images **8.20** Tiziana and Gianni Baldizzone/CORBIS **8.22a** Peter Menzel/Peter Menzel Photography **8.22b** Peter Ginter **8.24** Getty Images **The Science behind the Story: Human Population** Mark L. Imhoff **The Science behind the Story: Fertility Decline** Mark Edwards/Peter Arnold, Inc.

**Chapter 9 Opening Photo** Stephanie Maze/CORBIS **Case Study** Joanna B. Pinneo/AURORA **9.9** Ron Giling/Peter Arnold, Inc. **9.10a** Barry Runk/Stan/Grant Heilman Photography **9.10b** Grant Heilman/Grant Heilman Photography **9.10c, d** U.S. Department of Agriculture **9.12a** Kansas Heritage Center **9.13** Ted Spiegel/CORBIS **9.14a** Sylvan Wittwer/Visuals Unlimited **9.14b** Kevin Horan/Stone **9.14c** Keren Su/Stone **9.14d** Ron Giling/Peter Arnold, Inc. **9.14e** Yann Arthus-Bertrand/CORBIS **9.14f** U.S. Department of Agriculture **9.15** RTH/Grant Heilman Photography 2, Inc. **9.17** China Photos/Getty

Images **9.18a** Photo Disc **9.18b** Getty Images **9.18c** Carol Cohen/CORBIS **9.19a** Peter Dean/Grant Heilman Photography, Inc. **9.19b** CLL/Peter Arnold Photography 2, Inc. **9.23** W. Perry Conway/CORBIS **9.24** Natalie Fobes/CORBIS **The Science behind the Story: Measuring Erosion** Jerry Ritchie **The Science behind the Story: Restoring the Malpai Borderlands** Malpai Borderlands Group

**Chapter 10 Opening Photo** Steve Satushek/Image Bank **Case Study** Macduff Everton/CORBIS **10.3** Alexandra Avakian/CORBIS **10.4** Art Rickerby/Time Life Pictures/Getty Images **10.6** Jack Dykinga/Image Bank **10.6 inset** Barry Runk/Stan/Grant Heilman Photography, Inc. **10.8 a, b** Department of Natural Resources, Queensland, Australia **10.10** Photo Disc **10.11** Bob Rowan, Progressive Image/CORBIS **10.15** John Schmeiser **10.16a** Native Seeds/SEARCH **10.16b** Hal Fritts, Native Seeds/SEARCH **10.18** Arthur C. Smith III/Grant Heilman Photography, Inc. **10.22** Aqua Bounty Farms/AP **10.24** Martin Bourque **10.25** AP Wide World Photos **The Science behind the Story: Maize miniphoto** Peg Skorpinski Photography **main** Josef Polleross **The Science behind the Story: Organic Farming** Mark A. Liebig

**Chapter 11 Opening Photo** Maurice Hornocker **Case Study** Maurice Hornocker **11.11** George Glod, SuperStock, Inc. **11.14** Kathy Ferguson-Johnson/Photo Edit **11.15** Getty Images **11.19** Konrad Wothe/Photolibrary **11.20** Joel Sartore/Getty/National Geographic Society/CORBIS **11.24** Natalie Fobes/CORBIS **11.25** Tom & Pat Leeson/Photo Researchers, Inc. **The Science behind the Story: Amphibian Diversity miniphoto** Asanga Ratnaweera and Madhava Meegaskumbura **main** Craig K. Lorenz/Photo Researchers, Inc. **The Science behind the Story: Island Biogeography miniphoto** Daniel Simberloff **main** Stock Connection, Inc./Alamy

**Chapter 12 Opening Photo** Daniel Dancer/Peter Arnold, Inc. **Case Study** Garth Lenz **12.1** Dennis Hallinan/Alamy **12.7** CORBIS **12.11** Weyerhaeuser Company **12.12** Rob Badger/Getty Photos **12.14** Jonathan Blair/CORBIS **12.15** Reuters/CORBIS **12.16** Jim Brandenburg/Minden Photos **12.17** Authenticated News/Getty Images **12.18** North Carolina Museum of Art/CORBIS **12.19** David Hiser/Stone **12.21a** Gary Braasch/CORBIS **12.21b** James Zipp/Photo Researchers, Inc. **The Science behind the Story: Surveying Earth's Forests** Dan Lamont/Corbis **The Science behind the Story: Fire and Forests** Ross Hamilton/The Oregonian Publishing Co.

**Chapter 13 Opening Photo** Rich Iwaski/age footstock **Case Study** Phil Schermeister/CORBIS **13.3a** Chad Palmer/Shutterstock **13.3b** JupiterImages/Thinkstock/Alamy Images **13.4 a, b** U.S. EPA **13.5a** Lester Lefkowitz/CORBIS **13.5b** Bob Krist/CORBIS **13.5c** David R. Frazier Photolibrary, Inc. **13.5d** Aldo Torelli/Getty Images **13.6a** Picture Desk, Inc./Kobal Collection **13.6b** The Granger Collection **13.6c** Peter Schulz **13.9** Cooper Carry & Associates **13.11a** Steve Semler/Alamy **13.12** Bettmann/CORBIS **13.14** Bohemian Nomad Picture makers/CORBIS **13.16** Kim Karpeles **The Science behind the Story: Sprawl miniphoto** Reid Ewing **main** Editorial Fotos/Alamy **The Science behind the Story: Rail Transit** David M. Grossman/The Image Works

**Chapter 14 Opening Photo** Howard K. Suzuki **Case Study** Howard K. Suzuki **14.1a** Ingram Publishing/Alamy **14.1b** Spencer Grant/Photo Edit **14.1c** Martin Dohrn, Photo Researchers, Inc. **14.1d** SWA-Dann Tardit/CORBIS **14.4** Photofusion Picture Library/Alamy **14.4 inset** SciMAT/Photo Researchers, Inc. **14.5** Joel W. Rogers/CORBIS **14.6** Bettmann/CORBIS **14.7** Bettmann/CORBIS **14.9** Peg Skorpinski **The Science behind the Story: Bisphenol-A miniphoto** Patricia Hunt **main** From Hunt, PA, KE Koehler, M Susiarjo, CA Hodges, A Ilagan, RC Voight, S Thomas, BF Thomas and TJ Hassold. 2003. *Bisphenol A exposure causes meitic aneuploidy in the female mouse.* Current Biology 13: 546–553 **The Science behind the Story: Pesticides and Child Development** Jeff Conant

**Chapter 15 Opening Photo** Lester Lefkowitz/CORBIS **Case Study** Galen Rowell/CORBIS **15.2** William Manning/CORBIS **15.3** Frans Lanting/Minden Pictures **15.8** Laura Rauch/AP Wide World Photos **15.11a** AP Wide World Photos **15.11b** Ian Berry/Magnum Photos **15.12 inset** U.S. Department of the Interior **15.14a** Gilles Saussier/Liaison **15.14b** NASA Visible Earth **15.17** Jim Tuten/Black Star **15.18** Terry Ketron/The Sun/AP Wide World Photos **15.19** U.S. Department of the Interior **15.20a** Robert T. Zappalorti/Photo Researchers, Inc. **15.20b** David Muench/CORBIS **15.22** David Hoffman Photo Library/Alamy Images **15.23** David J. Cross/Peter Arnold, Inc. **The Science behind the Story: Arsenic** Shezad Nooran/Woodfin Camp & Associates **The Science behind the Story: Wastewater miniphoto** Ted Streshinksy/CORBIS **main** Gary Crabbe/Enlightened Images

**Chapter 16 Chapter Opener** Dex Image/CORBIS **Case Study** Jeffrey L. Rotman/CORBIS **16.7** Alcoa Technical Center **16.9** Laguna Design/Science Photo Library/Photo Researchers, Inc. **16.10** Bruce Robison/Minden Pictures **16.11** Ralph A. Clevenger/CORBIS **16.12a** Harvey Lloyd/Taxi **16.12b** Stephen Frink/CORBIS **16.13a** Carol Cohen/CORBIS **16.14** Maresa Pryor/Animals Animals/Earth Scenes **16.15** Mark A. Johnson/CORBIS **16.16** Nik Wheeler/CORBIS **16.17** Momatiuk Eastcott/Animals Animals/Earth Scenes **16.19a** David M. Phillips/Visuals Unlimited **16.19b** Bill Bachman/Photo Researchers, Inc. **16.22a** Norbert Wu/Norbert Wu Productions **16.22b** Norbert Wu Productions **16.25** Profimedia International/Alamy Images **The Science behind the Story: China's Fisheries** Sherman Lai **The Science behind the Story: Marine Reserves** Rachel Graham/Callum Roberts

**Chapter 17 Opening Photo** Bettmann/CORBIS **Case Study** AFP/Getty Images **17.10a** NASA Earth Observatory **17.10b** Bettmann/CORBIS **17.10c** Mark E. Gibson/CORBIS **17.15b** Pittsburgh Post-Gazette **17.16b** Allen Russell/Index Stock Imagery **17.17b** NASA **17.19a** The New York Historical Society **17.19b** Lee Snider/CORBIS **17.21** David Turnley/CORBIS **The Science behind the Story: Ozone Depletion** AP Wide World Photos **The Science behind the Story: Acid Rain miniphoto** HO Photography **main** Will & Deni McIntyre/CORBIS

**Chapter 18 Opening Photo** Rossi, Guido Alberto/Image Bank **Case Study** Adam Woolfitt/CORBIS **18.8a** Ted Spiegel/CORBIS **18.6b** David Whillas, CSIRO Atmospheric Research **18.14 top** BIOS Klein & Hubert/Peter Arnold, Inc. **18.14 middle** Alaska Stock **18.14 bottom** Brian & Cherry Alexander/Photo Researchers, Inc. **18.17 a, b** AP Wide World Photos **18.19** Ben Sklar/AP Photo **18.24** Paramount Classics/Everett Collection **18.26** Bettmann/CORBIS **18.28** Robert Brenner/Photo Edit **The Science behind the Story: World's Longest Ice Core** Pasquale Sorrentino/Photo Researchers, Inc. **The Science behind the Story: Greenland's Glaciers** Rex Stucky/National Geographic Image Collection

**Chapter 19 Opening Photo** Jim Wark/Alaska Stock **Case Study** Steve Armbrust/AlaskaStock.com **19.7** Daniel Esgro Photography/The Image Bank **19.12** Dave Olecko/Bloomberg News/Landov LLC **19.14** Vivian Stockman **19.15** Steven Kazlowski/Alaska Stock **19.23** AGA Infrared/Photo Researchers Inc. **The Science behind the Story: How Electricity Is Generated** Sam Ogden/Photo Researchers, Inc. **The Science behind the Story: How Crude Oil Is Refined miniphoto** John Gress/CORBIS/Reuters America LL **main** CORBIS

**Chapter 20 Opening Photo** HAWKEYE/Alamy **Case Study** Skellefteå Kraft AB **20.4** argus/Peter Arnold, Inc. **20.8** John S. Zeedick/Getty Images **20.9a** Paul Fusco/Magnum Photos **20.9b** Reuters/CORBIS **20.11a** Roger Ressmeyer/CORBIS **20.11b** U.S. Department of Energy/Photo Researchers, Inc. **20.13a** Nuclear Energy Institute **20.13b** U.S. Department of Energy **20.14** Chris Steele-Perkins/ Magnum Photos **20.16a** Larry Lefever/Grant Heilman Photography **20.17** Charles Bensinger/Renewable Energy Partners **20.19** Brent Baker **20.20** Lars-Erik Larsson/Svebio **20.21** Wolfgang Hoffmann/USDA/ARS/ Agricultural Research Service **20.22** GreenWood Resource, Inc/Jake Eaton **20.23a** Earl Roberge/Photo Researchers, Inc. **The Science behind the Story: Assessing Emissions** Liu Liqun/CORBIS **The Science behind the Story: Chernobyl** Caroline Penn/CORBIS

**Chapter 21 Opening Photo** Roger Ressmeyer **Case Study** Islensk NyOrka/Icelandic New Energy Ltd **21.5** Alan Weisman **21.6** CORBIS **21.13** Jorgen Schytte/Peter Arnold, Inc. **21.15b** Pascal Goetgheluck/ Science Photo Library **21.17 inset** Photofusion Picture Library/Alamy Images **The Science behind the Story: Wind Prospectors** Robert Nickelsberg/Getty Images; Photo Researchers, Inc. **The Science behind the Story: Algae miniphoto** Pascal Goetgheluck/Science Photo Library **main** Patrick Holin/CSREES-USDA

**Chapter 22 Opening Photo** Mark Gibson/Index Stock **Case Study** City of New York Department of City Planning **22.4** Nigel Dickinson/Stone **22.7** Lawrence Migdale/Pix **22.13a,b** City of Edmonton **22.17** Joe Sohm/Alamy **22.18** Robert Brook/Photo Researchers, Inc. **The Science behind the Story: Digging Garbage miniphoto** Andrew Holbrooke/CORBIS **main** Louie Psihoyos/©psi-hoyos.com **The Science behind the Story: Toxicity of "E-Waste" miniphoto** Timothy Townsend **main** AP Photo/Keystone, Walter Bieri

**Chapter 23 Opening Photo** Mark Edwards/Peter Arnold, Inc. **Case Study** Ball State Photo Service **23.1** University of North Carolina at Greensboro **23.2a** Oberlin College Archives **23.2b** Timothy Hursley/ The Arkansas Office, Inc. **23.3** Prabjit Virdee **23.4** Stefano Paltera/Solar Decathlon **23.5** Ari Joseph/Middlebury College **23.6** AP Photo/Steve Yeater **23.7** Jon C. Lakey/Salisbury Post **23.8** Galen Rowell/CORBIS **23.11** Costco Wholesale **23.15** AP Wide World Photos **23.16** Bettmann/ CORBIS **23.17** NASA/Langley Research Center **The Science behind the Story: Environmental Performance of Nations** Daniel Esty

# Selected Sources and References for Further Reading


**Chapter 1**

Bahn, Paul, and John Flenley. 1992. *Easter Island, Earth island.* Thames and Hudson, London.

Bowler, Peter J. 1993. *The Norton history of the environmental sciences.* W. W. Norton, New York.

Diamond, Jared. 2005. *Collapse: How societies choose to fail or succeed.* Viking, New York.

Edwards, Andres R. 2005. *The sustainability revolution: Portrait of a paradigm shift.* New Society Publishers, Gabriola Island, British Columbia, Canada.

Ehrlich, Paul. 1968. *The population bomb.* 1997 reprint, Buccaneer Books, Cutchogue, New York.

Flenley, John, and Paul Bahn. 2003. *The enigmas of Easter Island.* Oxford University Press, New York.

Goudie, Andrew. 2005. *The human impact on the natural environment,* 6th ed. Blackwell Publishing, London.

Hardin, Garrett. 1968. The tragedy of the commons. *Science* 162: 1243–1248.

Hunt, Terry L., and Carl P. Lipo. 2006. Late colonization of Easter Island. *Science* 311: 1603–1606.

Katzner, Donald W. 2001. *Unmeasured information and the methodology of social scientific inquiry.* Kluwer, Boston.

Kuhn, Thomas S. 1962. *The structure of scientific revolutions,* 2nd ed., 1970. University of Chicago Press, Chicago.

Lomborg, Bjorn. 2001. *The skeptical environmentalist: Measuring the real state of the world.* Cambridge University Press, Cambridge.

Malthus, Thomas R. *An essay on the principle of population.* 1983 ed. Penguin USA, New York.

Millennium Ecosystem Assessment. 2005. *Ecosystems and human well-being: General synthesis.* Millennium Ecosystem Assessment and World Resources Institute.

Musser, George. 2005. The climax of humanity. *Scientific American* 293(3): 44–47.

Ponting, Clive. 1991. *A green history of the world: The environment and the collapse of great civilizations.* Penguin Books, New York.

Popper, Karl R. 1959. *The logic of scientific discovery.* Hutchinson, London.

Porteous, Andrew. 2000. *Dictionary of environmental science and technology,* 3rd ed. John Wiley & Sons, Hoboken, New Jersey.

Redman, Charles R. 1999. *Human impact on ancient environments.* University of Arizona Press, Tucson.

Sagan, Carl. 1997. *The demon-haunted world: Science as a candle in the dark.* Ballantine Books, New York.

Schneiderman, Jill S., ed. 2003. *The Earth around us: Maintaining a livable planet.* Perseus Books, New York.

Siever, Raymond. 1968. Science: Observational, experimental, historical. *American Scientist* 56: 70–77.

Valiela, Ivan. 2001. *Doing science: Design, analysis, and communication of scientific research.* Oxford University Press, Oxford.

Van Tilburg, Jo Anne. 1994. *Easter Island: Archaeology, ecology, and culture.* Smithsonian Institution Press, Washington, D.C.

Venetoulis, Jason, and John Talberth. 2005. *Ecological footprint of nations: 2005 update.* Redefining Progress, Oakland, California.

Wackernagel, Mathis, and William Rees. 1996. *Our ecological footprint: Reducing human impact on the earth.* New Society Publishers, Gabriola Island, British Columbia, Canada.

Wackernagel, Mathis, et al. 2002. Tracking the ecological overshoot of the human economy. *Proceedings of the National Academy of Sciences of the USA* 99: 9266–9271.

World Bank. 2007. *World development indicators 2007.* World Bank, Washington, D.C.

Worldwatch Institute. 2006. *Vital signs 2006–2007.* Worldwatch Institute and W. W. Norton, Washington, D.C., and New York.

WWF–World Wide Fund for Nature. 2006. *Living planet report 2006.* WWF, Gland, Switzerland.

Young, Emma. 2006. A monumental collapse? *New Scientist,* 29 July 2006: 30–34.

**Chapter 2**

Balmford, Andrew, et al. 2002. Economic reasons for conserving wild nature. *Science* 297: 950–953.

Barbour, Ian G. 1992. *Ethics in an age of technology.* HarperCollins, San Francisco.

Brown, Lester. 2001. *Eco-economy: Building an economy for the Earth.* Earth Policy Institute and W. W. Norton, New York.

Carson, Richard T., Leanne Wilks, and David Imber. 1994. Valuing the preservation of Australia's Kakadu Conservation Zone. *Oxford Economic Papers* 46: 727–749.

Cole, Luke W., and Sheila R. Foster. 2001. *From the ground up: Environmental racism and the rise of the environmental justice movement.* New York University Press, New York.

Costanza, Robert, et al. 1997. The value of the world's ecosystem services and natural capital. *Nature* 387: 253–260.

Costanza, Robert, et al. 1997. *An introduction to ecological economics.* St. Lucie Press, Boca Raton, Florida.

Daily, Gretchen. 1997. *Nature's services: Societal dependence on natural ecosystems.* Island Press, Washington, D.C.

Daly, Herman E. 1996. *Beyond growth.* Beacon Press, Boston.

Daly, Herman E. 2005. Economics in a full world. *Scientific American* 293(3): 100–107.

De Graaf, John, David Wann, and Thomas Naylor. 2002. *Affluenza: The all-consuming epidemic.* Berrett-Koehler Publishers, San Francisco.

Elliot, Robert, and Arran Gare, eds. 1983. *Environmental philosophy: A collection of readings.* Pennsylvania State University Press, University Park.

Esty, Daniel C., and Andrew S. Winston. 2006. *Green to gold: How smart companies use environmental strategy to innovate, create value, and build competitive advantage.* Yale University Press, New Haven, Connecticut.

Field, Barry C., and Martha K. Field. 2005. *Environmental economics,* 4th ed. McGraw-Hill, New York.

Fox, Stephen. 1985. *The American conservation movement: John Muir and his legacy.* University of Wisconsin Press, Madison.

Gardner, Gary, et al. 2004. The state of consumption today. Pp. 3–23 in *State of the world 2004.* Worldwatch Institute and W. W. Norton, Washington, D.C., and New York.

Gardner, Gary, and Erik Assadourian. 2004. Rethinking the good life. Pp. 164–179 in *State of the world 2004.* Worldwatch Institute and W. W. Norton, Washington, D.C., and New York.

Goodstein, Eban. 1999. *The tradeoff myth: Fact and fiction about jobs and the environment.* Island Press, Washington, D.C.

Goodstein, Eban. 2007. *Economics and the environment*, 5th ed. John Wiley & Sons, Hoboken, New Jersey.

Gundjeihmi Aboriginal Corporation. Welcome to the Mirarr site. www.mirarr.net.

Hawken, Paul, Amory Lovins, and L. Hunter Lovins. 1999. *Natural capitalism.* Little, Brown, and Co., Boston.

HM Treasury. 2006. *Stern review on the economics of climate change.* HM Treasury and Cambridge University Press, U.K.

Kolstad, Charles D. 2000. *Environmental economics.* Oxford University Press, Oxford.

Leopold, Aldo. 1949. *A Sand County almanac, and sketches here and there.* Oxford University Press, New York.

Millennium Ecosystem Assessment. 2005. *Ecosystems and human well-being: Opportunities and challenges for business and industry.* Millennium Ecosystem Assessment and World Resources Institute.

Nash, Roderick F. 1989. *The rights of nature.* University of Wisconsin Press, Madison.

Nash, Roderick F. 1990. *American environmentalism: Readings in conservation history*, 3rd ed. McGraw-Hill, New York.

Nordhaus, William. 2007. Critical assumptions in the Stern Review on Climate Change. *Science* 317: 201–202.

O'Neill, John O., R. Kerry Turner, and Ian J. Bateman, eds. 2002. *Environmental ethics and philosophy.* Edward Elgar, Cheltenham, U.K.

Pearson, Charles S. 2000. *Economics and the global environment.* Cambridge University Press, Cambridge.

Ricketts, Taylor, et al. 2004. Economic value of tropical forest to coffee production. *Proceedings of the National Academy of Sciences of the USA* 101: 12579–12582.

Sachs, Jeffrey. 2005. Can extreme poverty be eliminated? *Scientific American* 293(3): 56–65.

Singer, Peter, ed. 1993. *A companion to ethics.* Blackwell Publishers, Oxford.

Smith, Adam. 1776. *An inquiry into the nature and causes of the wealth of nations.* 1993 ed., Oxford University Press, Oxford.

Stone, Christopher D. 1972. Should trees have standing? Towards legal rights for natural objects. *Southern California Law Review* 1972: 450–501.

Tietenberg, Tom. 2006. *Environmental economics and policy*, 5th ed. Addison Wesley, Boston.

Turner, R. Kerry, David Pearce, and Ian Bateman. 1993. *Environmental economics: An elementary introduction.* Johns Hopkins University Press, Baltimore.

Venetoulis, Jason, and Cliff Cobb. 2004. *The genuine progress indicator 1950–2002 (2004 update).* Redefining Progress, Oakland, California.

Wenz, Peter S. 2001. *Environmental ethics today.* Oxford University Press, Oxford.

White, Lynn. 1967. The historic roots of our ecologic crisis. *Science* 155: 1203–1207.

### Chapter 3

Clark, Ray, and Larry Canter. 1997. *Environmental policy and NEPA: Past, present, and future.* St. Lucie Press, Boca Raton, Florida.

Dietz, Thomas, et al. 2003. The struggle to govern the global commons. *Science* 302: 1907–1912.

Fogleman, Valerie M. 1990. *Guide to the National Environmental Policy Act.* Quorum Books, New York.

Fox, Stephen. 1985. *The American conservation movement: John Muir and his legacy.* University of Wisconsin Press, Madison.

French, Hilary. 2000. Environmental treaties gain ground. Pp. 134–135 in *Vital Signs 2000.* Worldwatch Institute and W. W. Norton, Washington D.C., and New York.

Green Scissors. 2004. *Green Scissors 2004: Cutting wasteful and environmentally harmful spending.* Friends of the Earth, Taxpayers for Common Sense, and U.S. Public Interest Research Group.

Herzog, Lawrence A. 1990. *Where north meets south: Cities, space, and politics on the U.S.–Mexico border.* Center for Mexican-American Studies, University of Texas at Austin.

Houck, Oliver. 2003. Tales from a troubled marriage: Science and law in environmental policy. *Science* 302: 1926–1928.

Kraft, Michael E. 2003. *Environmental policy and politics*, 3rd ed. Longman, New York.

Kubasek, Nancy K., and Gary S. Silverman. 2004. *Environmental law*, 5th ed. Prentice Hall, Upper Saddle River, New Jersey.

Myers, Norman, and Jennifer Kent. 2001. *Perverse subsidies: How misused tax dollars harm the environment and the economy.* Island Press, Washington, D.C.

The National Environmental Policy Act of 1969, as amended (Pub. L. 91–190, 42 U.S.C. 4321–4347, January 1, 1970, as amended by Pub. L. 94–52, July 3, 1975, Pub. L. 94–83, August 9, 1975, and Pub. L. 97–258, § 4(b), Sept. 13, 1982). www.nepa.gov/nepa/regs/nepa/nepaeqia.htm.

Office of Management and Budget, Executive Office of the President of the United States, Washington, D.C. 2003. *Informing regulatory decisions: 2003 report to Congress on the costs and benefits of federal and unfunded mandates on state, local, and tribal entities.* Washington, D.C., September 2003.

Percival, Robert V., et al. 2003. *Environmental regulation: Law, science, and policy*, 4th ed. Aspen Publishers, New York.

Shellenberger, Michael, and Ted Nordhaus. 2004. *The death of environmentalism: Global warming politics in a post-environmental world.* Presented at the Environmental Grantmakers Association meeting, October 2004.

Steel, Brent S., Richard L. Clinton, and Nicholas P. Lovrich. 2002. *Environmental politics and policy.* McGraw-Hill, New York.

Tietenberg, Tom. 2006. *Environmental economics and policy*, 5th ed. Addison Wesley, Boston.

Union of Concerned Scientists. 2004. Restoring scientific integrity in policymaking. www.ucsusa.org/scientific_integrity/interference/scientists-signon-statement.html.

United States Congress. House. H.R. 3378. 2000. The Tijuana River Valley Estuary and Beach Sewage Cleanup Act of 2000.

United States Environmental Protection Agency. 2000. *Regulatory impact analysis: Heavy-duty engine and vehicle standards and highway diesel fuel sulfur control requirements.* EPA420-R-00-026. EPA, Washington, D.C.

United States Environmental Protection Agency. 2006. *Acid rain program 2005 progress report.* U.S. EPA, Washington, D.C

Vig, Norman J., and Michael E. Kraft, eds. 2005. *Environmental policy: New directions for the twenty-first century*, 6th ed. CQ Press, Congressional Quarterly, Inc., Washington, D.C.

Wilkinson, Charles F. 1992. *Crossing the next meridian: Land, water, and the future of the West.* Island Press, Washington, D.C.

**Chapter 4**

Alaska Department of Environmental Conservation. 1993. *The* Exxon Valdez *oil spill: Final report, State of Alaska response.* June 1993.

Allen, K. C., and D. E. G. Briggs, eds. 1989. *Evolution and the fossil record.* John Wiley & Sons, Hoboken, New Jersey.

Atlas, Ronald M. 1995. Petroleum biodegradation and oil spill bioremediation. *Marine Pollution Bulletin* 31: 178–182.

Atlas, Ronald M., and Carl E. Cerniglia. 1995. Bioremediation of petroleum pollutants. *BioScience* 45: 332–338.

Azaizeh, Hassan, et al. 2006. Phytoremediation of selenium using subsurface-flow constructed wetland. *International Journal of Phytoremediation* 8: 187–198.

Berry, R. Stephen. 1991. *Understanding energy: Energy, entropy and thermodynamics for every man.* World Scientific Publishing Co.

Bragg, James R., et al. 1994. Effectiveness of bioremediation for the *Exxon Valdez* oil spill. *Nature* 368: 413–418.

Campbell, Neil A., and Jane B. Reece. 2005. *Biology,* 7th ed. Benjamin Cummings, San Francisco.

Fenchel, Tom. 2003. *Origin and early evolution of life.* Oxford University Press, Oxford.

Fortey, Richard. 1998. *Life: A natural history of the first four billion years of life on Earth.* Alfred Knopf, New York.

Gee, Henry. 1999. *In search of deep time: Beyond the fossil record to a new history of life.* Free Press, New York.

Hall, David O., and Krishna Rao. 1999. *Photosynthesis,* 6th ed. Cambridge University Press, Cambridge.

Kertulis-Tartar, Gina M., et al. 2006. Phytoremediation of an arsenic-contaminated site using *Pteris vittata* L.: A two-year study. *International Journal of Phytoremediation* 8: 311–322.

Lancaster, M. 2002. *Green chemistry.* Royal Society of Chemistry, London.

Manahan, Stanley E. 2004. *Environmental chemistry,* 8th ed. Lewis Publishers, CRC Press, Boca Raton, Florida.

McMurry, John E. 2007. *Organic chemistry,* 7th ed. Brooks/Cole, San Francisco.

National Response Team. *NRT fact sheet: Bioremediation in oil spill response.* U.S. EPA. www.epa.gov/oilspill/pdfs/biofact.pdf.

Nealson, Kenneth H. 2003. Harnessing microbial appetites for remediation. *Nature Biotechnology* 21: 243–244.

Ridley, Mark. 2003. *Evolution,* 3rd ed. Blackwell Science, Cambridge, Massachusetts.

Timberlake, Karen C. 2007. *General, organic, and biological chemistry,* 2nd ed. Benjamin Cummings, San Francisco.

United States Environmental Protection Agency. Oil program. www.epa.gov/oilspill.

Van Dover, Cindy Lee. 2000. *The ecology of deep-sea hydrothermal vents.* Princeton University Press, Princeton.

Van Ness, H. C. 1983. *Understanding thermodynamics.* Dover Publications, Mineola, New York.

Ward, Peter D., and Donald Brownlee. 2000. *Rare Earth: Why complex life is uncommon in the universe.* Copernicus, New York.

Wassenaar, Leonard I., and Keith A. Hobson. 1998. Natal origins of migratory monarch butterflies at wintering colonies in Mexico: New isotopic evidence. *Proceedings of the National Academy of Sciences of the USA* 95: 15436–15439.

**Chapter 5**

Alvarez, Luis W., et al. 1980. Extraterrestrial cause for the Cretaceous-Tertiary extinction. *Science* 208: 1095–1108.

Barbour, Michael G., et al. 1998. *Terrestrial plant ecology,* 3rd ed. Benjamin Cummings, Menlo Park, California.

Begon, Michael, Martin Mortimer, and David J. Thompson. 1996. *Population ecology: A unified study of animals and plants,* 3rd ed. Blackwell Scientific, Oxford.

Begon, Michael, Colin R. Townsend, and John L. Harper. 2005. *Ecology: From individuals to ecosystems,* 4th ed. Blackwell Publishing, London.

Campbell, Neil A., and Jane B. Reece. 2005. *Biology,* 7th ed. Benjamin Cummings, San Francisco.

Clark K. L., et al. 1998. Cloud water and precipitation chemistry in a tropical montane forest, Monteverde, Costa Rica. *Atmospheric Environment* 32: 1595–1603.

Crump, L. Martha, et al. 1992. Apparent decline of the golden toad: Underground or extinct? *Copeia* 1992: 413–420.

Darwin, Charles. 1859. *The origin of species by means of natural selection.* John Murray, London.

Endler, John A. 1986. *Natural selection in the wild.* Monographs in Population Biology 21, Princeton University Press, Princeton.

Freeman, Scott, and Jon C. Herron. 2006. *Evolutionary analysis,* 4th ed. Benjamin Cummings, San Francisco.

Futuyma, Douglas J. 2005. *Evolution.* Sinauer Associates, Sunderland, Massachusetts.

Krebs, Charles J. 2001. *Ecology: The experimental analysis of distribution and abundance,* 5th ed. Benjamin Cummings, San Francisco.

Lawton, Robert O., et al. 2001. Climatic impact of tropical lowland deforestation on nearby montane cloud forests. *Science* 294: 584–587.

Molles, Manuel C., Jr. 2005. *Ecology: Concepts and applications,* 3rd ed. McGraw-Hill, Boston.

Nadkarni, Nalini M., and Nathaniel T. Wheelwright, eds. 2000. *Monteverde: Ecology and conservation of a tropical cloud forest.* Oxford University Press, New York.

Pounds, J. Alan. 2001. Climate and amphibian declines. *Nature* 410: 639.

Pounds, J. Alan, et al. 1997. Tests of null models for amphibian declines on a tropical mountain. *Conservation Biology* 11: 1307–1322.

Pounds, J. Alan, et al. 2006. Widespread amphibian extinctions from epidemic disease driven by global warming. *Nature* 439: 161–167.

Pounds, J. Alan, and Martha L. Crump. 1994. Amphibian declines and climate disturbance: The case of the golden toad and the harlequin frog. *Conservation Biology* 8: 72–85.

Pounds, J. Alan, Michael P. L. Fogden, and John H. Campbell. 1999. Biological response to climate change on a tropical mountain. *Nature* 398: 611–615.

Powell, James L. 1998. *Night comes to the Cretaceous: Dinosaur extinction and the transformation of modern geology.* W. H. Freeman, New York.

Raup, David M. 1991. *Extinction: Bad genes or bad luck?* W. W. Norton, New York.

Ricklefs, Robert E., and Gary L. Miller. 2000. *Ecology,* 4th ed. W. H. Freeman, New York.

Ricklefs, Robert E., and Dolph Schluter, eds. 1993. *Species diversity in ecological communities.* University of Chicago Press, Chicago.

Savage, Jay M. 1966. An extraordinary new toad (*Bufo*) from Costa Rica. *Revista de Biologia Tropical* 14: 153–167.

Savage, Jay M. 1998. The "brilliant toad" was telling us something. *Christian Science Monitor,* 14 September 1998: 19.

Smith, Thomas M., and Robert L. Smith. 2006. *Elements of ecology,* 6th ed. Benjamin Cummings, San Francisco.

Ward, Peter. 1994. *The end of evolution.* Bantam Books, New York.

Williams, George C. 1966. *Adaptation and natural selection.* Princeton University Press, Princeton.

Wilson, Edward O. 1992. *The diversity of life.* Harvard University Press, Cambridge, Massachusetts.

**Chapter 6**

Breckle, Siegmar-Walter. 2002. *Walter's vegetation of the Earth: The ecological systems of the geo-biosphere,* 4th ed. Springer-Verlag, Berlin.

Bright, Chris. 1998. *Life out of bounds: Bioinvasion in a borderless world.* Worldwatch Institute and W. W. Norton, Washington, D.C., and New York.

Bronstein, Judith L. 1994. Our current understanding of mutualism. *Quarterly Journal of Biology* 69: 31–51.

Chase, Jonathan M., et al., 2002. The interaction between predation and competition: A review and synthesis. *Ecology Letters* 5: 302.

Connell, Joseph H., and Ralph O. Slatyer, 1977. Mechanisms of succession in natural communities. *American Naturalist* 111: 1119–1144.

Drake, John M., and Jonathan M. Bossenbroek. 2004. The potential distribution of zebra mussels in the United States. *BioScience* 54: 931–941.

Estes, J.A., et al. 1998. Killer whale predation on sea otters linking oceanic and nearshore ecosystems. *Science* 282: 473–476.

Ewald, Paul W., 1987. Transmission modes and evolution of the parasitism-mutualism continuum. *Annals of the New York Academy of Sciences* 503: 295–306.

Gurevitch, Jessica, and Dianna K. Padilla. 2004. Are invasive species a major cause of extinctions? *Trends in Ecology and Evolution* 19: 470–474.

Hobbs, Richard J., et al. 2006. *Foundations of restoration ecology: The science and practice of ecological restoration.* Island Press, Washington, D.C.

Menge, Bruce A., et al. 1994. The keystone species concept: Variation in interaction strength in a rocky intertidal habitat. *Ecological Monographs* 64: 249–286.

Molles, Manuel C. Jr. 2005. *Ecology: Concepts and applications,* 3rd ed. McGraw-Hill, Boston.

Morin, Peter J. 1999. *Community ecology.* Blackwell, London.

Power, Mary E., et al. 1996. Challenges in the quest for keystones. *BioScience* 46: 609–620.

Ricklefs, Robert E., and Gary L. Miller. 2000. *Ecology,* 4th ed. W. H. Freeman and Co., New York.

Schrope, Mark. 2007. Killer in the kelp. *Nature* 445: 703–705.

Shea, Katriona, and Peter Chesson. 2002. Community ecology theory as a framework for biological invasions. *Trends in Ecology and Evolutionary Biology* 17: 170–176.

Sih, Andrew, et al. 1985. Predation, competition, and prey communities: A review of field experiments. *Annual Review of Ecology and Systematics* 16: 269–311.

Smith, Robert L., and Thomas M. Smith. 2001. *Ecology and field biology,* 6th ed. Benjamin Cummings, San Francisco.

Springer, A. M., et al. 2003. Sequential megafaunal collapse in the North Pacific Ocean: An ongoing legacy of industrial whaling? *Proceedings of the National Academy of Sciences of the USA* 100: 12223–12228.

Stokstad, Erik. 2007. Feared quagga mussel turns up in western United States. *Science* 315: 453.

Strayer, David L., et al. 1999. Transformation of freshwater ecosystems by bivalves: A case study of zebra mussels in the Hudson River. *BioScience* 49: 19–27.

Strayer, David L., et al. 2004. Effects of an invasive bivalve (*Dreissena polymorpha*) on fish in the Hudson River estuary. *Canadian Journal of Fisheries and Aquatic Sciences* 61: 924–941.

Thompson, John N. 1999. The evolution of species interactions. *Science* 284: 2116–2118.

Van Andel, Jelte, and James Aronson. 2005. *Restoration ecology: The new frontier.* Blackwell Publishing, London.

Weigel, Marlene, ed. 1999. *Encyclopedia of biomes.* UXL, Farmington Hills, Michigan.

Whittaker, Robert H., and William A. Niering. 1965. Vegetation of the Santa Catalina Mountains, Arizona: A gradient analysis of the south slope. *Ecology* 46: 429–452.

Woodward, Susan L. 2003. *Biomes of Earth: Terrestrial, aquatic, and human-dominated.* Greenwood Publishing, Westport, Connecticut.

**Chapter 7**

Appenzeller, Tim. 2004. The case of the missing carbon. *National Geographic,* February 2004: 88–117.

Capra, Fritjof. 1996. *The web of life: A new scientific understanding of living systems.* Anchor Books Doubleday, New York.

Carpenter, Edward J., and Douglas G. Capone, eds. 1983. *Nitrogen in the marine environment.* Academic Press, New York.

Christopherson, Robert W. 2006. *Geosystems,* 6th ed. Prentice Hall, Upper Saddle River, New Jersey.

Committee on Environment and Natural Resources. 2000. *An integrated assessment: Hypoxia in the northern Gulf of Mexico.* CENR, National Science and Technology Council, Washington, D.C.

Ferber, Dan. 2004. Dead zone fix not a dead issue. *Science* 305: 1557.

Field, Christopher B., et al. 1998. Primary production of the biosphere: Integrating terrestrial and oceanic components. *Science* 281: 237–240.

Jacobson, Michael, et al. 2000. *Earth system science from biogeochemical cycles to global changes.* Academic Press.

Keller, Edward A. 2004. *Introduction to environmental geology,* 3rd ed. Prentice Hall, Upper Saddle River, New Jersey.

Larsen, Janet. 2004. Dead zones increasing in world's coastal waters. *Eco-economy Update #41,* 16 June 2004. Earth Policy Institute. www.earth-policy.org/Updates/Update41.htm.

Mississippi River/Gulf of Mexico Watershed Nutrient Task Force. 2001. *Action plan for reducing, mitigating, and controlling hypoxia in the northern Gulf of Mexico.* Washington, D.C.

Mitsch, William J., et al. 2001. Reducing nitrogen loading to the Gulf of Mexico from the Mississippi River Basin: Strategies to counter a persistent ecological problem. *BioScience* 51: 373–388.

Montgomery, Carla. 2005. *Environmental geology,* 7th ed. McGraw-Hill, New York.

National Oceanic and Atmospheric Administration: National Ocean Service. 2000. Hypoxia in the Gulf of Mexico: Progress toward the completion of an integrated assessment. www.nos.noaa.gov/products/pubs_hypox.html.

National Science and Technology Council, Committee on Environment and Natural Resources. 2003. *An assessment of coastal hypoxia and eutrophication in U.S. waters.* National Science and Technology Council, Washington, D.C.

Pima County, Arizona. Sonoran Desert Conservation Plan. www.pima.gov/cmo/sdcp.

Rabalais, Nancy N., R. E. Turner, and D. Scavia. 2002. Beyond science into policy: Gulf of Mexico hypoxia and the Mississippi River. *BioScience* 52: 129–142.

Rabalais, Nancy N., R. E. Turner, and W. J. Wiseman, Jr. 2002. Hypoxia in the Gulf of Mexico, a.k.a. "The dead zone." *Annual Review of Ecology and Systematics* 33: 235–263.

Raloff, Janet. 2004. Dead waters: Massive oxygen-starved zones are developing along the world's coasts. *Science News* 165: 360–362. June 5, 2004.

Raloff, Janet. 2004. Limiting dead zones: How to curb river pollution and save the Gulf of Mexico. *Science News* 165: 378–380. June 12, 2004.

Ricklefs, Robert E., and Gary L. Miller. 2000. *Ecology*, 4th ed. W. H. Freeman and Co., New York.

Schlesinger, William H. 1997. *Biogeochemistry: An analysis of global change*, 2nd ed. Academic Press, London.

Skinner, Brian J., and Stephen C. Porter. 2003. *The dynamic earth: An introduction to physical geology*, 5th ed. John Wiley and Sons, Hoboken, New Jersey.

Smith, Robert L., and Thomas M. Smith. 2001. *Ecology and field biology*, 6th ed. Benjamin Cummings, San Francisco.

Stiling, Peter. 2002. *Ecology: Theories and applications*, 4th ed. Prentice Hall, Upper Saddle River, New Jersey.

Takahashi, Taro. 2004. The fate of industrial carbon dioxide. *Science* 305: 352–353.

Turner, R. Eugene, and Nancy N. Rabalais. 2003. Linking landscape and water quality in the Mississippi River Basin for 200 years. *BioScience* 53: 563–572.

Vitousek, Peter M., et al. 1997. Human alteration of the global nitrogen cycle: Sources and consequences. *Ecological Applications* 7: 737–750.

Whittaker, Robert H. 1975. *Communities and ecosystems*, 2nd ed. Macmillan, New York.

Wu, Jianguo, and Richard J. Hobbs, eds. 2007. *Key topics in landscape ecology*. Cambridge University Press, Cambridge, U.K.

### Chapter 8

Ausubel, Jesse. H. 1996. Can technology spare the earth? *American Scientist* 84: 166–178.

Balter, Michael. 2006. The baby deficit. *Science* 312: 1894–1897.

Cohen, Joel E. 1995. *How many people can the Earth support?* W. W. Norton, New York.

Cohen, Joel E. 2003. Human population: The next half century. *Science* 302: 1172–1175.

Cohen, Joel E. 2005. Human population grows up. *Scientific American* 293(3): 48–55.

De Souza, Roger-Mark, et al. 2003. Critical links: Population, health, and the environment. *Population Bulletin* 58(3). Population Reference Bureau, Washington, D.C.

Eberstadt, Nicholas. 2000. China's population prospects: Problems ahead. *Problems of Post-Communism* 47: 28.

Ehrlich, Paul R., and John P. Holdren. 1971. Impact of population growth: Complacency concerning this component of man's predicament is unjustified and counterproductive. *Science* 171: 1212–1217.

Ehrlich, Paul R., and Anne H. Ehrlich. 1990. *The population explosion*. Touchstone, New York.

Engelman, Robert, Brian Halweil, and Danielle Nierenberg. 2002. Rethinking population, improving lives. Pp. 127–148 in *State of the world 2002*. Worldwatch Institute and W. W. Norton, Washington, D.C., and New York.

Greenhalgh, Susan. 2001. Fresh winds in Beijing: Chinese feminists speak out on the one-child policy and women's lives. *Signs: Journal of Women in Culture & Society* 26: 847–887.

Harrison, Paul, and Fred Pearce, eds. 2000. *AAAS atlas of population & environment*. University of California Press, Berkeley.

Haub, Carl, and O. P. Sharma. 2006. India's population reality: Reconciling change and tradition. *Population Bulletin* 61(3). Population Reference Bureau, Washington, D.C.

Hesketh, Therese, and Wei Xing Zhu. 1997. Health in China: The one child family policy: The good, the bad, and the ugly. *British Medical Journal* 314: 1685.

Holdren, John P., and Paul R. Ehrlich. 1974. Human population and the global environment. *American Scientist* 62: 282–292.

Imhoff, Marc L., et al. 2004. Global patterns in human consumption of net primary production. *Nature* 429: 870–873.

Kane, Penny. 1987. *The second billion: Population and family planning in China*. Penguin Books, Australia, Ringwood, Victoria.

Kane, Penny, and Ching Y. Choi. 1999. China's one child family policy. *British Medical Journal* 319: 992.

Kent, Mary M., and Carl Haub. 2005. Global demographic divide. *Population Bulletin* 60(4). Population Reference Bureau, Washington, D.C.

Lamptey, Peter R., Jami L. Johnson, and Marya Khan. 2006. The global challenge of HIV and AIDS. *Population Bulletin* 61(1). Population Reference Bureau, Washington, D.C.

Mastny, Lisa, and Richard P. Cincotta. 2005. Examining the connections between population and security. Pp. 22–41 in *State of the world 2005*. Worldwatch Institute and W. W. Norton, Washington, D.C., and New York.

McDonald, Mia, with Danielle Nierenberg. 2003. Linking population, women, and biodiversity. Pp. 38–61 in *State of the world 2003*. Worldwatch Institute and W. W. Norton, Washington D.C., and New York.

Meadows, Donella, Jørgen Randers, and Dennis Meadows. 2004. *Limits to growth: The 30-year update*. Chelsea Green Publishing Co., White River Junction, Vermont.

Notestein, Frank. 1953. Economic problems of population change. Pp. 13–31 in *Proceedings of the Eighth International Conference of Agricultural Economists*. Oxford University Press, London.

Population Reference Bureau. 2007. *2007 World Population Data Sheet*. Population Reference Bureau, Washington, D.C., and John Wiley & Sons, Hoboken, New Jersey.

Riley, Nancy E. 2004. *China's population: New trends and challenges. Population Bulletin* 59(2). Population Reference Bureau, Washington, D.C.

UNAIDS and World Health Organization. 2006. *AIDS epidemic update: December 2006*. UNAIDS and WHO, New York.

United Nations Population Division. 2007. *World population prospects: The 2006 revision*. UNPD, New York.

United Nations Population Fund. UNFPA, the 2005 World Summit and the millennium development goals. UNFPA. www.unfpa.org/icpd.

United Nations Population Fund. 2007. *State of world population 2007*. UNFPA, New York.

United States Census Bureau. www.census.gov.

Wackernagel, Mathis, and William Rees. 1996. *Our ecological footprint: Reducing human impact on the earth*. New Society Publishers, Gabriola Island, British Columbia, Canada.

### Chapter 9

Ashman, Mark R., and Geeta Puri. 2002. *Essential soil science: A clear and concise introduction to soil science*. Blackwell Publishing, Malden, Massachusetts.

Brown, Lester R. 2002. World's rangelands deteriorating under mounting pressure. *Eco-Economy Update #6*, 5 February 2002. Earth Policy Institute. www.earth-policy.org/Updates/Update6.htm.

Brown, Lester R. 2004. *Outgrowing the Earth: The food security challenge in an age of falling water tables and rising temperatures.* Earth Policy Institute, Washington, D.C.

Charman, P. E. V., and Brian W. Murphy. 2000. *Soils: Their properties and management*, 2nd ed. Oxford University Press, South Melbourne, Australia.

Curtin, Charles G. 2002. Integration of science and community-based conservation in the Mexico/U.S. borderlands. *Conservation Biology* 16: 880–886.

Curtin, Charles. 2007. Integrating landscape and ecosystems approaches through science-based collaborative conservation. *Conservation Biology*, in press.

Diamond, Jared. 1999. *Guns, germs, and steel: The fates of human societies.* W. W. Norton, New York.

Diamond, Jared, and Peter Bellwood. 2003. Farmers and their languages: The first expansions. *Science* 300: 597–603.

Food and Agriculture Organization of the United Nations. 2001. Conservation agriculture: Case studies in Latin America and Africa. *FAO Soils Bulletin No. 78.* FAO, Rome.

Fox, Stephen. 1985. *The American conservation movement: John Muir and his legacy.* University of Wisconsin Press, Madison.

Glanz, James. 1995. *Saving our soil: Solutions for sustaining Earth's vital resource.* Johnson Books, Boulder, Colorado.

Goudie, Andrew. 2000. *The human impact on the natural environment*, 5th ed. MIT Press, Cambridge, Massachusetts.

Halweil, Brian. 2002. Farmland quality deteriorating. Pp. 102–103 in *Vital signs 2002.* Worldwatch Institute and W. W. Norton, Washington D.C., and New York.

Harrison, Paul, and Fred Pearce, eds. 2000. *AAAS atlas of population & environment.* University of California Press, Berkeley.

Jenny, Hans. 1941. *Factors of soil formation: A system of quantitative pedology.* McGraw-Hill, New York.

Kaiser, Jocelyn. 2004. Wounding Earth's fragile skin. *Science* 304: 1616–1618.

Larsen, Janet. 2003. Deserts advancing, civilization retreating. *Eco-Economy Update #23*, 27 March 2003. Earth Policy Institute. www.earth-policy.org/Updates/Update23.htm.

Malpai Borderlands Group. Malpai Borderlands Group. www.malpaiborderlandsgroup.org.

Millennium Ecosystem Assessment. 2005. *Ecosystems and human well-being: Desertification synthesis.* Millennium Ecosystem Assessment and World Resources Institute.

Morgan, R. P. C. 2005. *Soil erosion and conservation*, 3rd ed. Blackwell, London.

Natural Resources Conservation Service. 2001. *National resources inventory 2001: Soil erosion.* NRCS, USDA, Washington, D.C.

Natural Resources Conservation Service. Soils. NRCS, USDA. www.soils.usda.gov.

Pieri, Christian, et al. 2002. *No-till farming for sustainable rural development.* Agriculture & Rural Development Working Paper. International Bank for Reconstruction and Development, Washington, D.C.

Pierzynski, Gary M., et al. 2005. *Soils and environmental quality*, 3rd ed. CRC Press, Boca Raton, Florida.

Pretty, Jules, and Rachel Hine. 2001. *Reducing food poverty with sustainable agriculture: A summary of new evidence.* Occasional Paper 2001–2. Center for Environment and Society, University of Essex.

Richter, Daniel D. Jr., and Daniel Markewitz. 2001. *Understanding soil change: Soil sustainability over millennia, centuries, and decades.* Cambridge University Press, Cambridge.

Ritchie, Jerry C. 2000. Combining cesium-137 and topographic surveys for measuring soil erosion/deposition patterns in a rapidly accreting area. TEKTRAN, USDA Division of Agricultural Research, January 14, 2000.

Shaxson, T. F. 1999. The roots of sustainability, concepts and practice: Zero tillage in Brazil. *ABLH Newsletter ENABLE; World Association for Soil and Water Conservation (WASWC) Newsletter.*

Soil Science Society of America. 2001. Internet glossary of soil science terms. www.soils.org/sssagloss.

Stocking, M. A. 2003. Tropical soils and food security: The next 50 years. *Science* 302: 1356–1359.

Trimble, Stanley W., and Pierre Crosson. 2000. U.S. soil erosion rates—myth and reality. *Science* 289: 248–250.

Troeh, Frederick R., J. Arthur Hobbs, and Roy L. Donahue. 2004. *Soil and water conservation for productivity and environmental protection*, 4th ed. Prentice Hall, Upper Saddle River, New Jersey.

Troeh, Frederick R., and Louis M. Thompson. 2004. *Soil and soil fertility*, 6th ed. Blackwell Publishing, London.

United Nations Convention to Combat Desertification. 2001. *Global alarm: Dust and sandstorms from the world's drylands.* UNCCD and others, Bangkok, Thailand.

United Nations Environment Programme. 2002. Land. Pp. 62–89 in *Global environment outlook 3 (GEO-3).* UNEP and Earthscan Publications, Nairobi and London.

Uri, Noel D. 2001. The environmental implications of soil erosion in the United States. *Environmental Monitoring and Assessment* 66: 293–312.

Wilkinson, Bruce H. 2005. Humans as geologic agents: A deep-time perspective. *Geology* 33: 161–164.


**Chapter 10**


Bazzaz, Fakhri A. 2001. Plant biology in the future. *Proceedings of the National Academy of Sciences of the United States of America* 98: 5441–5445.

Brown, Lester R. 2004. *Outgrowing the Earth: The food security challenge in an age of falling water tables and rising temperatures.* Earth Policy Institute, Washington, D.C.

Buchmann, Stephen L., and Gary Paul Nabhan. 1996. *The forgotten pollinators.* Island Press/Shearwater Books, Washington, D.C./Covelo, California.

Center for Food Safety. 2005. *Monsanto vs. U.S. farmers.* Center for Food Safety, Washington, D.C.

Commission for Environmental Cooperation. 2004. *Maize and biodiversity: The effects of transgenic maize in Mexico.* CEC Secretariat.

[Correspondence to *Nature*, various authors]. 2002. *Nature* 416: 600–602, and 417: 897–898.

The Farm Scale Evaluations of spring-sown genetically modified crops. 2003. A themed issue from *Philosophical Transactions of the Royal Society of London B: Biological Sciences* 358(1439), 29 November 2003.

Fedoroff, Nina, and Nancy Marie Brown, 2004. *Mendel in the kitchen: A scientist's view of genetically modified foods.* National Academies Press, Washington, D.C.

Food and Agriculture Organization of the United Nations. 2007. *The state of world fisheries and aquaculture 2006.* FAO Fisheries and Aquaculture Department, Rome.

Gardner, Gary, and Brian Halweil. 2000. *Underfed and overfed: The global epidemic of malnutrition.* Worldwatch Paper #150. Worldwatch Institute, Washington, D.C.

Halweil, Brian. 2004. *Eat here: Reclaiming homegrown pleasures in a global supermarket.* Worldwatch Institute, Washington, D.C.

Halweil, Brian, and Danielle Niereberg. 2004. Watching what we eat. Pp. 68–95 in *State of the world 2004.* Worldwatch Institute and W. W. Norton, Washington, D.C., and New York.

Harrison, Paul, and Fred Pearce, eds. 2000. *AAAS atlas of population & environment.* University of California Press, Berkeley.

International Food Information Council. 2004. Food biotechnology. IFIC, Washington, D.C. www.ific.org/food/biotechnology/index.cfm.

James, Clive. 2006. *Global status of commercialized biotech/GM crops.* International Service for the Acquisition of Agri-biotech Applications.

Kristiansen, P., A. Taji, and J. Reganold, eds. 2006. *Organic agriculture: A global perspective.* CABI Publishing, Oxfordshire, U.K.

Kuiper, Harry A. 2000. Risks of the release of transgenic herbicide-resistant plants with respect to humans, animals, and the environment. *Crop Protection* 19: 773.

Liebig, Mark A., and John W. Doran. 1999. Impact of organic production practices on soil quality indicators. *Journal of Environmental Quality* 28: 1601–1609.

Losey, John E., Linda S. Raynor, and Maureen E. Carter. 1999. Transgenic pollen harms monarch larvae. *Nature* 399: 214.

Maeder, Paul, et al. 2002. Soil fertility and biodiversity in organic farming. *Science* 296: 1694–1697.

Mann, Charles C. 2002. Transgene data deemed unconvincing. *Science* 296: 236–237.

Manning, Richard. 2000. *Food's frontier: The next green revolution.* North Point Press, New York.

Miller, Henry I., and Gregory Conko. 2004. *The frankenfood myth: How protest and politics threaten the biotech revolution.* Praeger Publishers, Westport, Connecticut.

Nestle, Marion. 2002. *Food politics: How the food industry influences nutrition and health.* University of California–Berkeley Press, Berkeley.

Nierenberg, Danielle. 2005. *Happier meals: Rethinking the global meat industry.* Worldwatch Paper #171. Worldwatch Institute, Washington, D.C.

Nierenberg, Danielle. 2006. Meat consumption and output up. Pp. 24–25 in *Vital signs 2006–2007.* Worldwatch Institute and W. W. Norton, Washington, D.C., and New York.

Nierenberg, Danielle, and Brian Halweil. 2005. Cultivating food security. Pp. 62–79 in *State of the world 2005.* Worldwatch Institute and W. W. Norton, Washington, D.C., and New York.

Norris, Robert F., Edward P. Caswell-Chen, and Marcos Kogan. 2002. *Concepts in integrated pest management.* Prentice Hall, Upper Saddle River, New Jersey.

Ortiz-García, Sol, et al. 2005. Absence of detectable transgenes in local landraces of maize in Oaxaca, Mexico (2003–2004). *Proceedings of the National Academy of the United States of America* 102: 12338–12343.

Paoletti, Maurizio G., and David Pimentel. 1996. Genetic engineering in agriculture and the environment: Assessing risks and benefits. *BioScience* 46: 665–673.

Pearce, Fred. 2002. The great Mexican maize scandal. *New Scientist* 174: 14.

Pedigo, Larry P., and Marlin E. Rice. 2006. *Entomology and pest management,* 5th ed. Prentice Hall, Upper Saddle River, New Jersey.

Pimentel, David. 1999. Population growth, environmental resources, and the global availability of food. *Social Research,* Spring 1999.

Pinstrup-Andersen, Per, and Ebbe Schioler. 2001. *Seeds of contention: World hunger and the global controversy over GM (genetically modified) crops.* International Food Policy Research Institute, Washington, D.C.

Polak, Paul. 2005. The big potential of small farms. *Scientific American* 293(3): 84–91.

Pringle, Peter. 2003. *Food, Inc.: Mendel to Monsanto—The promises and perils of the biotech harvest.* Simon and Schuster, New York.

Quist, David, and Ignacio H. Chapela. 2001. Transgenic DNA introgressed into traditional maize landraces in Oaxaca, Mexico. *Nature* 414: 541–543.

Ruse, Michael, and David Castle, eds. 2002. *Genetically modified foods: Debating technology.* Prometheus Books, Amherst, New York.

Schmeiser, Percy. Monsanto vs. Schmeiser. www. percyschmeiser.com.

Shiva, Vandana. 2000. *Stolen harvest: The hijacking of the global food supply.* South End Press, Cambridge, Massachusetts.

Smil, Vaclav. 2001. *Feeding the world: A challenge for the twenty-first century.* MIT Press, Cambridge, Massachusetts.

Stewart, C. Neal. 2004. *Genetically modified planet: Environmental impacts of genetically engineered plants.* Oxford University Press, Oxford.

Teitel, Martin, and Kimberly Wilson. 2001. *Genetically engineered food: Changing the nature of nature.* Park Street Press.

Tuxill, John. 1999. Appreciating the benefits of plant biodiversity. Pp. 96–114 in *State of the world 1999.* Worldwatch Institute and W. W. Norton, Washington D.C., and New York.

Westra, Lauren. 1998. Biotechnology and transgenics in agriculture and aquaculture: The perspective from ecosystem integrity. *Environmental Values* 7: 79.

Wolfenbarger, L. LaReesa. 2000. The ecological risks and benefits of genetically engineered plants. *Science* 290: 2088.


**Chapter 11**


Balmford, Andrew, et al. 2002. Economic reasons for conserving wild nature. *Science* 297: 950–953.

Barnosky, Anthony D., et al. 2004. Assessing the causes of late Pleistocene extinctions on the continents. *Science* 306: 70–75.

Baskin, Yvonne. 1997. *The work of nature: How the diversity of life sustains us.* Island Press, Washington, D.C.

Bright, Chris. 1998. *Life out of bounds: Bioinvasion in a borderless world.* Worldwatch Institute and W. W. Norton, Washington D.C., and New York.

CITES Secretariat. Convention on International Trade in Endangered Species of Wild Fauna and Flora. www.cites.org.

Convention on Biological Diversity. www.biodiv.org.

Daily, Gretchen C., ed. 1997. *Nature's services: Societal dependence on natural ecosystems.* Island Press, Washington, D.C.

Ehrenfeld, David W. 1970. *Biological conservation.* International Thomson Publishing, London.

Gaston, Kevin J., and John I. Spicer. 2004. *Biodiversity: An introduction,* 2nd ed. Blackwell, London.

Groom, Martha J., et al. 2005. *Principles of conservation biology,* 3rd ed. Sinauer Associates, Sunderland, Massachusetts.

Groombridge, Brian, and Martin D. Jenkins. 2002. *Global biodiversity: Earth's living resources in the 21st century.* UNEP, World Conservation Monitoring Centre, and Aventis Foundation; World Conservation Press, Cambridge, U.K.

Groombridge, Brian, and Martin D. Jenkins. 2002. *World atlas of biodiversity: Earth's living resources in the 21st century.* University of California Press, Berkeley.

Hanken, James. 1999. Why are there so many new amphibian species when amphibians are declining? *Trends in Ecology and Evolution* 14: 7–8.

Harris, Larry D. 1984. *The fragmented forest: Island biogeography theory and the preservation of biotic diversity.* University of Chicago Press, Chicago.

Harrison, Paul, and Fred Pearce, eds. 2000. *AAAS atlas of population & environment.* University of California Press, Berkeley.

Jenkins, Martin. 2003. Prospects for biodiversity. *Science* 302: 1175–1177.

Louv, Richard. 2005. *Last child in the woods: Saving our children from nature-deficit disorder.* Algonquin Books, Chapel Hill, North Carolina.

Lovejoy, Thomas E., and Lee Hannah, eds. 2006. *Climate change and biodiversity.* Yale University Press, New Haven, Connecticut.

MacArthur, Robert H., and Edward O. Wilson. 1967. *The theory of island biogeography.* Princeton University Press, Princeton.

Maehr, David S., Reed F. Noss, and Jeffrey Larkin, eds. 2001. *Large mammal restoration: Ecological and sociological challenges in the 21st century.* Island Press, Washington, D.C.

Matthiessen, Peter. 2000. *Tigers in the snow.* North Point Press, New York.

Meegaskumbura, Madhava, et al. 2002. Sri Lanka: An amphibian hot spot. *Science* 298: 379.

Millennium Ecosystem Assessment. 2005. *Ecosystems and human well-being: Biodiversity synthesis.* Millennium Ecosystem Assessment and World Resources Institute.

Miquelle, Dale, Howard Quigley, and Maurice Hornocker. 1999. *A habitat protection plan for Amur Tiger conservation: A proposal outlining habitat protection measures for the Amur Tiger. Hornocker Wildlife Institute.*

Mooney, Harold A., and Richard J. Hobbs, eds. 2000. *Invasive species in a changing world.* Island Press, Washington, D.C.

Newmark, William D. 1987. A land-bridge perspective on mammal extinctions in western North American parks. *Nature* 325: 430.

Pimm, Stuart L., and Clinton Jenkins. 2005. Sustaining the variety of life. *Scientific American* 293(3): 66–73.

Primack, Richard B. 2006. *Essentials of conservation biology*, 4th ed. Sinauer Associates, Sunderland, Massachusetts.

Quammen, David. 1996. *The song of the dodo: Island biogeography in an age of extinction.* Touchstone, New York.

Relyea, Rick, and Nathan Mills. 2001. Predator-induced stress makes the pesticide carbaryl more deadly to gray treefrog tadpoles. *Proceedings of the National Academy of Sciences, USA* 98: 2491–2496.

Rosenzweig, Michael L. 1995. *Species diversity in space and time.* Cambridge University Press, Cambridge.

Sepkoski, John J. 1984. A kinetic model of Phanerozoic taxonomic diversity. *Paleobiology* 10: 246–267.

Simberloff, Daniel S. 1969. Experimental zoogeography of islands: A model for insular colonization. *Ecology* 50: 296–314.

Simberloff, Daniel. 1998. Flagships, umbrellas, and keystones: Is single-species management passé in the landscape era? *Biological Conservation* 83: 247–257.

Simberloff, Daniel S., and Edward O. Wilson. 1969. Experimental zoogeography of islands: The colonization of empty islands. *Ecology* 50: 278–296.

Simberloff, Daniel S., and Edward O. Wilson. 1970. Experimental zoogeography of islands: A two-year record of colonization. *Ecology* 51: 934–937.

Soulé, Michael E. 1986. *Conservation biology: The science of scarcity and diversity.* Sinauer Associates, Sunderland, Massachusetts.

Takacs, David. 1996. *The idea of biodiversity: Philosophies of paradise.* Johns Hopkins University Press, Baltimore.

United Nations Environment Programme. 2002. Biodiversity. Pp. 120–149 in *Global environment outlook 3 (GEO-3).* UNEP and Earthscan Publications, Nairobi and London.

United Nations Environment Programme. 2003. Sustaining life on Earth: How the Convention on Biological Diversity promotes nature and human well-being. www.biodiv.org/doc/publications/guide.asp.

United States Fish and Wildlife Service. The endangered species act of 1973. Accessible online at www.fws.gov/endangered/esa.html.

Wilson, Edward O. 1984. *Biophilia.* Harvard University Press, Cambridge, Massachusetts.

Wilson, Edward O. 1992. *The diversity of life.* Harvard University Press, Cambridge, Massachusetts.

Wilson, Edward O. 1994. *Naturalist.* Island Press, Shearwater Books, Washington, D.C.

Wilson, Edward O. 2002. *The future of life.* Alfred A. Knopf, New York.

Wilson, Edward O., and Daniel S. Simberloff. 1969. Experimental zoogeography of islands: Defaunation and monitoring techniques. *Ecology* 50: 267–278.

World Conservation Union. IUCN Red List. www.iucnredlist.org.

WWF–World Wide Fund for Nature. 2006. *Living planet report 2006.* WWF, Gland, Switzerland.

### Chapter 12

British Columbia Ministry of Forests. Introduction to Silvicultural Systems. www.for.gov.bc.ca/hfd/pubs/SSIntroworkbook/index.htm. British Columbia Ministry of Forests, Victoria, B.C.

Canadian Broadcasting Corporation. 1993. A little place called Clayoquot Sound. CBC broadcast, 13 April 1993. http://archives.cbc.ca/IDC-1-75-679-3918/Science_technology/clearcutting/clip6.

Cascade Resources Advocacy Group Law Center. Scientific integrity and academic freedom. www.crag.org/fire/academicfreedom.php.

Clary, David. 1986. *Timber and the Forest Service.* University Press of Kansas, Lawrence.

Donato, Daniel C., et al. 2006. Post-wildfire logging hinders regeneration and increases fire risk. *Science* 311: 352.

Food and Agriculture Organization of the United Nations. 2005. *Global forest resources assessment.* FAO Forestry Department, Rome.

Foster, Bryan C., and Peggy Foster. 2002. *Wild logging: A guide to environmentally and economically sustainable forestry.* Mountain Press, Missoula, Montana.

Gardner, Gary. 2006. Deforestation continues. Pp. 102–103 in *Vital signs 2006–2007.* Worldwatch Institute and W. W. Norton, Washington, D.C., and New York.

Harrison, Paul, and Fred Pearce, eds. 2000. *AAAS atlas of population & environment.* University of California Press, Berkeley.

Haynes, Richard W., and Gloria E. Perez, tech. eds. 2001. Northwest Forest Plan research synthesis. *Gen. Tech. Rep. PNW-GTR-498.* USDA Forest Service, Pacific Northwest Research Station, Portland, Oregon.

Myers, Norman, and Jennifer Kent. 2001. *Perverse subsidies: How misused tax dollars harm the environment and the economy.* Island Press, Washington, D.C.

National Forest Management Act of 1976. October 22, 1976 (P.O. 94–588, 90 Stat. 2949, as amended; 16 U.S.C.).

Natural Resources Canada. 2006. *The state of Canada's forests, 2005–2006.* Natural Resources Canada, Ottawa.

Runte, Alfred. 1979. *National parks and the American experience.* University of Nebraska Press, Lincoln.

Sedjo, Robert A. 2000. *A vision for the U.S. Forest Service.* Resources for the Future, Washington, D.C.

Shatford, Jeffrey P. A., David E. Hibbs, and Klaus J. Puettman. 2007. Conifer regeneration after forest fire in the Klamath-Siskiyous: How much, how soon? *Journal of Forestry* 105: 139–146.

Singh, Ashbindu, et al. 2001. An assessment of the status of the world's remaining closed forests. United Nations Environmental Program, UNEP/DEWA/TR 01–2l, August 2001.

Smith, David M., et al. 1996. *The practice of silviculture: Applied forest ecology*, 9th ed. Wiley, New York.

Smith, W. Brad, et al. 2004. Forest resources of the United States, 2002. *Gen. Tech. Rep.* NC-241, North Central Research Station, USDA Forest Service, St. Paul, Minnesota.

Soulé, Michael E., and John Terborgh, eds. 1999. *Continental conservation.* Island Press, Washington, D.C.

Stegner, Wallace. 1954. *Beyond the hundredth meridian: John Wesley Powell and the second opening of the West.* Houghton Mifflin, Boston.

Thompson, Jonathan R., Thomas A. Spies, and Lisa M. Ganio. 2007. Re-burn severity in managed and unmanaged vegetation in a large wildfire. *Proceedings of the National Academy of Sciences of the USA* 104: 10743–10748.

United Nations Environment Programme. 2002. Forests. Pp. 90–119 in *Global environment outlook 3 (GEO-3).* UNEP and Earthscan Publications, Nairobi and London.

USDA Forest Service. 2001. *U.S. forest facts and historical trends.* FS-696, March 2001.

U.S. National Park Service. 2005. *National Park Service statistical abstract 2005.* NPS Public Use Statistics Office, U.S. Department of the Interior, Denver, Colorado.

Woodard, Colin. 2006. The sale of the century. *Nature Conservancy*, Autumn 2006: 20–25.

**Chapter 13**

Abbott, Carl. 2001. *Greater Portland: Urban life and landscape in the Pacific Northwest.* University of Pennsylvania Press.

Abbott, Carl. 2002. Planning a sustainable city. Pp. 207–235 in Squires, Gregory D., ed. *Urban sprawl: Causes, consequences, and policy responses.* Urban Institute Press, Washington, D.C.

Beck, Roy, et al. 2003. *Outsmarting smart growth: Population growth, immigration, and the problem of sprawl.* Center for Immigration Studies, Washington, D.C.

Breuste, Jurgen, et al. 1998. *Urban ecology.* Springer-Verlag, Berlin.

Brockerhoff, Martin P. 2000. An urbanizing world. *Population Bulletin* 55(3). Population Reference Bureau, Washington, D.C.

Cronon, William. 1991. *Nature's metropolis: Chicago and the great West.* W. W. Norton, New York.

Duany, Andres, et al. 2001. *Suburban nation: The rise of sprawl and the decline of the American dream.* North Point Press, New York.

Ewing, Reid, et al. 2002. *Measuring sprawl and its impact.* Smart Growth America.

Ewing, Reid, et al. 2003. Measuring sprawl and its transportation impacts. *Transportation Research Record* 1831: 175–183.

Girardet, Herbert. 2004. *Cities people planet: Livable cities for a sustainable world.* Academy Press.

Hall, Kenneth B., and Gerald A. Porterfield. 2001. *Community by design: New urbanism for suburbs and small communities.* McGraw-Hill, New York.

Jacobs, Jane. 1992. *The death and life of great American cities.* Vintage.

Kalnay, Eugenia, and Ming Cai. 2003. Impact of urbanization and land-use change on climate. *Nature* 423: 528–531.

Kirdar, Uner, ed. 1997. *Cities fit for people.* United Nations, New York.

Litman, Todd. 2004. *Rail transit in America: A comprehensive evaluation of benefits.* Victoria Transport Policy Institute and American Public Transportation Association.

Logan, Michael F. 1995. *Fighting sprawl and city hall.* University of Arizona Press, Tucson.

Metro. www.metro-region.org.

New Urbanism. www.newurbanism.org.

Northwest Environment Watch. 2004. *The Portland exception: A comparison of sprawl, smart growth, and rural land loss in 15 U.S. cities.* Northwest Environment Watch, Seattle.

Oppenheimer, Laura. 2006. Measure 37 changed state, but how much? *Oregonian*, 3 Dec. 2006: 1 and A15.

Pearce, Fred. 2005. Cities lead the way to a greener world. *New Scientist*, 4 June 2005: 8–9.

Portney, Kent. E. 2003. *Taking sustainable cities seriously: Economic development, the environment, and quality of life in American cities (American and comparative environmental policy).* MIT Press, Cambridge, Massachusetts.

Pugh, Cedric, ed. 1996. *Sustainability, the environment, and urbanization.* Earthscan Publications, London.

Sheehan, Molly O'Meara. 2001. *City limits: Putting the brakes on sprawl.* Worldwatch Paper #156. Worldwatch Institute, Washington, D.C.

Sheehan, Molly O'Meara. 2002. What will it take to halt sprawl? *WorldWatch*, Jan/Feb 2002: 12–23.

Sprawl City. www.sprawlcity.org.

Stren, R., et al. 1992. *Sustainable cities: Urbanization and the environment in international perspective.* Westview Press, Boulder, Colorado, and San Francisco.

United Nations Environment Programme. 2002. Urban areas. Pp. 240–269 in *Global environment outlook 3 (GEO-3).* UNEP and Earthscan Publications, Nairobi and London.

United Nations Population Division. 2006. *World urbanization prospects: The 2005 revision.* UNPD, New York.

United States Environmental Protection Agency. Smart growth. www.epa.gov/smartgrowth.

Wiewel, Wim, and Jospeph J. Persky, eds. 2002. *Suburban sprawl: Private decisions and public policy.* M. E. Sharpe, Armond, New York.

Worldwatch Institute. 2007. *State of the world 2007: Our urban future.* Worldwatch Institute and W. W. Norton, Washington, D.C., and New York.

**Chapter 14**

Ames, Bruce N., Margie Profet, and Lois Swirsky Gold. 1990. Nature's chemicals and synthetic chemicals: Comparative toxicology. *Proceedings of the National Academy of Sciences of the USA* 87: 7782–7786.

Bloom, Barry. 2005. Public health in transition. *Scientific American* 293(3): 92–99.

Cagen, S. Z., et al. 1999. Normal reproductive organ development in wistar rats exposed to bisphenol A in the drinking water. *Regulatory Toxicology and Pharmacology* 30: 130–139.

Carlsen, Elisabeth, et al. 1992. Evidence for decreasing quality of semen during past 50 years. *British Medical Journal* 305: 609–613.

Carson, Rachel. 1962. *Silent spring.* Houghton Mifflin, Boston.

Colburn, Theo, Dianne Dumanoski, and John P. Myers. 1996. *Our stolen future.* Penguin USA, New York.

Crain, D. Andrew, and Louis J. Guillette Jr. 1998. Reptiles as models of contaminant-induced endocrine disruption. *Animal Reproduction Science* 53: 77–86.

Gross, Liza. 2007. The toxic origins of disease. *PloS Biology* 5(7): e193. doi:10.1371/journal.pbio.0050193.

Guillette, Elizabeth A., et al. 1998. An anthropological approach to the evaluation of preschool children exposed to pesticides in Mexico. *Environmental Health Perspectives* 106: 347–353.

Guillette, Louis J. Jr., et al. 1999. Plasma steroid concentrations and male phallus size in juvenile alligators from seven Florida lakes. *General and Comparative Endocrinology* 116: 356–372.

Guillette, Louis J. Jr., et al. 2000. Alligators and endocrine disrupting contaminants: A current perspective. *American Zoologist* 40: 438–452.

Halweil, Brian. 1999. Sperm counts dropping. Pp. 148–149 in *Vital signs 1999.* Worldwatch Institute and W. W. Norton, Washington, D.C., and New York.

Hayes, Tyrone, et al. 2003. Atrazine-induced hermaphroditism at 0.1 PPB in American leopard frogs (*Rana pipiens*): Laboratory and field evidence. *Environmental Health Perspectives* 111: 568–575.

Hunt, Patricia A., et al. 2003. Bisphenol A exposure causes meiotic aneuploidy in the female mouse. *Current Biology* 13: 546–553.

Kolpin, Dana W., et al. 2002. Pharmaceuticals, hormones, and other organic wastewater contaminants in U.S. streams, 1999–2000: A national reconnaissance. *Environmental Science and Technology* 36: 1202–1211.

Landis, Wayne G., and Ming-Ho Yu. 2004. *Introduction to environmental toxicology,* 3rd ed. Lewis Press, Boca Raton, Florida.

Loewenberg, Samuel. 2003. E.U. starts a chemical reaction. *Science* 300: 405.

Manahan, Stanley E. 2004. *Environmental chemistry,* 8th ed. Lewis Publishers, CRC Press, Boca Raton, Florida.

McGinn, Anne Platt. 2000. *Why poison ourselves? A precautionary approach to synthetic chemicals.* Worldwatch Paper #153. Worldwatch Institute, Washington, D.C.

McGinn, Anne Platt. 2002. Reducing our toxic burden. Pp. 75–100 in *State of the world 2002.* Worldwatch Institute and W. W. Norton, Washington, D.C., and New York.

McGinn, Anne Platt. 2003. Combating malaria. Pp. 62–84 in *State of the world 2003.* Worldwatch Institute and W. W. Norton, Washington, D.C., and New York.

Millennium Ecosystem Assessment. 2005. *Ecosystems and human well-being: Health synthesis.* World Health Organization.

Moeller, Dade. 2004. *Environmental health,* 3rd ed. Harvard University Press, 2004.

National Center for Environmental Health; U.S. Centers for Disease Control and Prevention. 2005. *Third national report on human exposure to environmental chemicals.* NCEH Pub. No. 05-0570, Atlanta.

National Center for Health Statistics. 2004. *Health, United States, 2004, with chartbook on trends in the health of Americans.* Hyattsville, Maryland.

Pirages, Dennis. 2005. Containing infectious disease. Pp. 42–61 in *State of the world 2005.* Worldwatch Institute and W. W. Norton, Washington, D.C., and New York.

Renner, Rebecca. 2002. Conflict brewing over herbicide's link to frog deformities. *Science* 298: 938–939.

Rodricks, Joseph V. 1994. *Calculated risks: Understanding the toxicity of chemicals in our environment.* Cambridge University Press, Cambridge.

Salem, Harry, and Eugene Olajos. 1999. *Toxicology in risk assessment.* CRC Press, Boca Raton, Florida.

Spiteri, I. Daniel, Louis J. Guillette Jr., and D. Andrew Crain. 1999. The functional and structural observations of the neonatal reproductive system of alligators exposed *in ozo* to atrazine, 2,4-D, or estradiol. *Toxicology and Industrial Health* 15: 181–186.

Stancel, George, et al. 2001. Report of the bisphenol A sub-panel. Chapter 1 in *National Toxicology Program's report of the endocrine disruptors low-dose peer review.* U.S. EPA and NIEHS, NIH.

Stephens, Carolyn, and Peter Stair. 2007. Charting a new course for urban public health. Pp. 134–151 in *State of the world 2007: Our urban future.* Worldwatch Institute and W. W. Norton, Washington, D.C., and New York.

Stockholm Convention on Persistent Organic Pollutants. www.pops.int.

United States Environmental Protection Agency. 2003. Pesticide registration program. www.epa.gov/pesticides/factsheets/registration.htm.

United States Environmental Protection Agency. 2003. Toxic Substances Control Act. www.epa.gov/region5/defs/html/tsca.htm.

United States Environmental Protection Agency. 2003. *EPA's draft report on the environment.* EPA 600-R-03-050. EPA, Washington, D.C.

Williams, Phillip L., Robert C. James, and Stephen M. Roberts, eds. 2000. *The principles of toxicology: Environmental and industrial applications,* 2nd ed. Wiley-Interscience, New York.

World Health Organization. 2004. *World health report 2004: Changing history.* WHO, Geneva, Switzerland.

Yu, Ming-Ho. 2004. *Environmental toxicology: Biological and health effects of pollutants,* 2nd ed. CRC Press, Boca Raton, Florida.


**Chapter 15**


American Rivers. 2002. *The ecology of dam removal: A summary of benefits and impacts.* American Rivers, Washington, D.C., February 2002.

British Geographical Society and Bangladesh Department of Public Health Engineering. 2001. *Arsenic contamination of groundwater in Bangladesh.* Technical report WC/00/19, volume 1: Summary.

De Villiers, Marq. 2000. *Water: The fate of our most precious resource.* Mariner Books.

Gleick, Peter H. 2003. Global freshwater resources: Soft-path solutions for the 21st century. *Science* 302: 1524–1527.

Gleick, Peter H., et al. 2006. *The world's water 2006–2007: The biennial report on freshwater resources.* Island Press, Washington, D.C.

Harrison, Paul, and Fred Pearce, eds. 2000. *AAAS atlas of population & environment.* University of California Press, Berkeley.

Harvey, Charles F., et al. 2002. Arsenic mobility and groundwater extraction in Bangladesh. *Science* 298: 1602–1606.

Jenkins, Matt. 2002. California's water binge skids to a halt. *High Country News* 34(17), September 16, 2002.

Jenkins, Matt. 2003. The royal squeeze. *High Country News* 35(1), January 20, 2003.

Jenkins, Matt. 2005. Squeezing water from a stone. *High Country News* 37(17), September 19, 2005.

Jenkins, Matt. 2006. Running on empty in Sin City. *High Country News* 38(17), September 18, 2006.

Liu, Yingling. 2006. Groundwater overdraft problem persists. Pp. 104–105 in *Vital signs 2006–2007.* Worldwatch Institute and W. W. Norton, Washington, D.C., and New York.

Marston, Ed. 2001. Quenching the big thirst. *High Country News* 33(10), May 21, 2001.

Millennium Ecosystem Assessment. 2005. *Ecosystems and human well-being: Wetlands and water synthesis.* Millennium Ecosystem Assessment and World Resources Institute.

Nickson, Ross, et al. 1998. Arsenic poisoning of Bangladesh groundwater. *Nature* 395: 338.

Pala, Christopher. 2006. Once a terminal case, the North Aral Sea shows new signs of life. *Science* 312: 183.

Postel, Sandra. 1999. *Pillar of sand: Can the irrigation miracle last?* W. W. Norton, New York.

Postel, Sandra. 2005. *Liquid assets: The critical need to safeguard freshwater ecosystems.* Worldwatch Paper #170. Worldwatch Institute, Washington, D.C.

Postel, Sandra, and Amy Vickers. 2004. Boosting water productivity. Pp. 46–67 in *State of the world 2004.* Worldwatch Institute and W. W. Norton, Washington, D.C., and New York.

Reisner, Marc. 1986. *Cadillac desert: The American West and its disappearing water.* Viking Penguin, New York.

Sampat, Payal. 2001. Uncovering groundwater pollution. Pp. 21–42 in *State of the world 2001.* Worldwatch Institute and W. W. Norton, Washington, D.C., and New York.

Sibley, George. 1997. A tale of two rivers: The desert empire and the mountain. *High Country News* 29(21), November 10, 1997.

Smith, Lingas Rahman. 2000. Contamination of drinking water by arsenic in Bangladesh: A public health emergency. *Bulletin of the World Health Organization* 78(9).

Stone, Richard. 1999. Coming to grips with the Aral Sea's grim legacy. *Science* 284: 30–33.

United Nations Educational, Scientific and Cultural Organization (UNESCO). 2006. *Water: A shared responsibility.* The United Nations World Water Development Report 2. UNESCO and Berghahn Books, Paris and New York.

United Nations Environment Programme. 2002. Freshwater. Pp. 150–179 in *Global environment outlook 3 (GEO-3).* UNEP and Earthscan Publications, Nairobi and London.

United Nations Environment Programme. 2006. *Challenges to international waters: Regional assessments in a global perspective.* UNEP, Nairobi, Kenya.

United Nations Environment Programme. 2006. *Water quality for ecosystem and human health.* UNEP Global Environment Monitoring System (GEMS)/Water Programme, Burlington, Ontario.

United Nations World Water Assessment Programme. 2003. *U.N. world water development report: Water for people, water for life.* Paris, New York, and Oxford, UNESCO and Berghahn Books.

United States Bureau of Reclamation, Lower Colorado Regional Office. www.usbr.gov/lc/region.

United States Environmental Protection Agency. 1998. *Wastewater primer.* EPA 833-K-98-001. Office of Wastewater Management, May 1998.

United States Environmental Protection Agency. 2003. *Water on tap: What you need to know.* EPA 816-K-03-007. Office of Water, EPA, Washington, D.C.

Wolf, Aaron T., et al. 2005. Managing water conflict and cooperation. Pp. 80–99 in *State of the world 2005.* Worldwatch Institute and W. W. Norton, Washington, D.C., and New York.

World Health Organization. 2000. *Global water supply and sanitation assessment 2000 report.* WHO, Geneva, Switzerland.

Youth, Howard. 2005. Wetlands drying up. Pp. 90–91 in *Vital signs 2005.* Worldwatch Institute and W. W. Norton, Washington, D.C., and New York.

**Chapter 16**

Bellwood, David R., et al. 2004. Confronting the coral reef crisis. *Nature* 429: 827–833.

[Correspondence to *Science*, various authors]. 2001. *Science* 295: 1233–1235.

Embassy of the People's Republic of China. 2001. Fishing statistics "basically correct," ministry says. Press release, 18 December 2001.

Food and Agriculture Organization of the United Nations. 2002. Fishery statistics: Reliability and policy implications. FAO Fisheries Department, Rome. www.fao.org/docrep/field/006/y3354m/y3354m00.htm.

Food and Agriculture Organization of the United Nations. 2007. *The state of world fisheries and aquaculture 2006.* FAO Fisheries and Aquaculture Department, Rome.

Frank, Kenneth T., et al. 2005. Trophic cascades in a formerly cod-dominated ecosystem. *Science* 308: 1621–1623.

Garrison, Tom. 2005. *Oceanography: An invitation to marine science,* 5th ed. Brooks/Cole, San Francisco.

Gell, Fiona R., and Callum M. Roberts. 2003. Benefits beyond boundaries: The fishery effects of marine reserves. *Trends in Ecology and Evolution* 18: 448–455.

Halpern, Benjamin S., and Robert R. Warner. 2002. Marine reserves have rapid and lasting effects. *Ecology Letters* 5: 361–366.

Halpern, Benjamin S., and Robert R. Warner. 2003. Matching marine reserve design to reserve objectives. *Proceedings of the Royal Society of London B*: 270: 1871–1878.

Halweil, Brian. 2006. *Catch of the day: Choosing seafood for healthier oceans.* Worldwatch Paper #172. Worldwatch Institute, Washington, D.C.

Jackson, Jeremy B. C., et al. 2001. Historical overfishing and the recent collapse of coastal ecosystems. *Science* 293: 629–638.

Kurlansky, Mark. 1998. *Cod: A biography of the fish that changed the world.* Penguin Books, New York.

Larsen, Janet. 2005. Wild fish catch hits limits—Oceanic decline offset by increased fish farming eco-economy indicators. Earth Policy Institute. www.earth-policy.org/Indicators/Fish/2005.htm.

Lotze, Heike L., et al. 2006. Depletion, degradation, and recovery potential of estuaries and coastal seas. *Science* 312: 1806–1809.

Mastny, Lisa. 2001. World's coral reefs dying off. Pp. 92–93 in *Vital signs 2001.* Worldwatch Institute and W. W. Norton, Washington D.C., and New York.

Mayo, Ralph K., and Mark Terceiro, eds. 2005. *Assessment of 19 northeast groundfish stocks through 2004.* Northeast Fisheries Science Center Reference Document 05-13 from the 2005 Groundfish Assessment Review Meeting. Northeast Fisheries Science Center (NMFS, NOAA), Woods Hole, Massachusetts.

Myers, Ransom A., and Boris Worm. 2003. Rapid worldwide depletion of predatory fish communities. *Nature* 423: 280–283.

National Academy of Public Administration. 1999. *Protecting our national marine sanctuaries.* Center for the Economy and the Environment, NAPA, Washington, D.C.

National Center for Ecological Analysis and Synthesis (NCEAS) and Communication Partnership for Science and the Sea (COMPASS), sponsors. 2001. *Scientific consensus statement on marine reserves and marine protected areas.* Available online atwww.nceas.ucsb.edu/consensus.

National Oceanic and Atmospheric Administration (NOAA). Florida Keys National Marine Sanctuary.www.fknms.nos.noaa.gov.

National Research Council. 2003. *Oil in the sea III: Inputs, fates, and effects.* National Academies Press, Washington, D.C.

Norse, Elliott, and Larry B. Crowder, eds. 2005. *Marine conservation biology: The science of maintaining the sea's biodiversity.* Island Press, Washington, D.C.

Nybakken, James W., and Mark D. Bertness. 2004. *Marine biology: An ecological approach,* 6th ed. Benjamin Cummings, San Francisco.

O'Brien, L., N. J. Munroe, and L. Col. 2005. A. Georges Bank Atlantic Cod. In Mayo, Ralph K., and Mark Terceiro, eds. *Assessment of 19 northeast groundfish stocks through 2004.* Northeast Fisheries Science Center, Woods Hole, Massachusetts.

Palumbi, Stephen. 2003. *Marine reserves: A tool for ecosystem management and conservation.* Pew Oceans Commission.

Pauly, Daniel, et al. 2002. Towards sustainability in world fisheries. *Nature* 418: 689–695.

Pauly, Daniel, et al. 2003. The future for fisheries. *Science* 302: 1359–1361.

Pew Oceans Commission. 2003. *America's living oceans: Charting a course for sea change.* A report to the nation. May 2003. Pew Oceans Commission, Arlington, Virginia.

Pinet, Paul R. 2003. *Invitation to oceanography,* 3rd ed. Jones & Bartlett, Boston.

Roberts, Callum M., et al. 2001. Effects of marine reserves on adjacent fisheries. *Science* 294: 1920–1923.

Rosenberg, A., et al. 2006. Rebuilding U.S. fisheries: Progress and problems. *Frontiers in Ecology and the Environment* 4(6).

Sumich, James L., and John F. Morrissey. 2004. *Introduction to the biology of marine life,* 8th ed. Jones & Bartlett, Boston.

Thurman, Harold V., and Alan P. Trujillo. 2004. *Introductory oceanography,* 10th ed. Prentice Hall, Upper Saddle River, New Jersey.

United Nations Environment Programme. 2002. Coastal and marine areas. Pp. 180–209 in *Global environment outlook 3 (GEO-3).* UNEP and Earthscan Publications, Nairobi and London.

United States Commission on Ocean Policy. 2004. *An ocean blueprint for the 21st century.* Final report. Washington, D.C.

United States Department of Commerce and United States Department of the Interior. Marine protected areas of the United States. www.mpa.gov.

Watson, Reginald, Lillian Pang, and Daniel Pauly. 2001. The marine fisheries of China: Development and reported catches. *Fisheries Centre Research Reports* 9(2). Fisheries Centre, University of British Columbia, Canada.

Watson, Reginald, and Daniel Pauly. 2001. Systematic distortions in world fisheries catch trends. *Nature* 414: 534–536.

Weber, Michael L. 2001. *From abundance to scarcity: A history of U.S. marine fisheries policy.* Island Press, Washington, D.C.

Weiss, Kenneth R., and Usha Lee McFarling. 2006. Altered oceans (a special five-part series). *Los Angeles Times,* 30 July–3 August, 2006. Available at: www.latimes.com/news/local/oceans/la-oceans-series,0,7842752.special.

Worm, Boris, et al. 2006. Impacts of biodiversity loss on ocean ecosystem services. *Science* 314: 787-790.

**Chapter 17**

Ahrens, C. Donald. 2007. *Meteorology today,* 8th ed. Brooks/Cole, San Francisco.

Akimoto, Hajime. 2003. Global air quality and pollution. *Science* 302: 1716–1719.

Bell, Michelle L., and Devra L. Davis. 2001. Reassessment of the lethal London fog of 1952: Novel indicators of acute and chronic consequences of acute exposure to air pollution. *Environmental Health Perspectives* 109(Suppl 3): 389–394.

Bernard, Susan M., et al. 2001. The potential impacts of climate variability and change on air pollution-related health effects in the United States. *Environmental Health Perspectives* 109(Suppl 2): 199–209.

Biscaye, Pierre E., et al. 2000. Eurasian air pollution reaches eastern North America. *Science* 290: 2258–2259.

Boubel, Richard W., et al., eds. 1994. *Fundamentals of air pollution,* 3rd ed. Academic Press, San Diego, California.

Bruce, Nigel, Rogelio Perez-Padilla, and Rachel Albalak. 2000. Indoor air pollution in developing countries: A major environmental and public health challenge. *Bulletin of the World Health Organization* 78: 1078–1092.

Cooper, C. David, and F. C. Alley. 2002. *Air pollution control,* 3rd ed. Waveland Press.

Davis, Devra. 2002. *When smoke ran like water: Tales of environmental deception and the battle against pollution.* Basic Books, New York.

Davis, Devra L., Michelle L. Bell, and Tony Fletcher. 2002. A look back at the London smog of 1952 and the half century since. *Environmental Health Perspectives* 110: A734.

Driscoll, Charles T., et al. 2001. *Acid rain revisited: Advances in scientific understanding since the passage of the 1970 and 1990 Clean Air Act Amendments.* Hubbard Brook Research Foundation. Science Links™ Publication, vol. 1, no.1.

Ezzati, Majid, and Daniel M. Kammen. 2001. Quantifying the effects of exposure to indoor air pollution from biomass combustion on acute respiratory infections in developing countries. *Environmental Health Perspectives* 109: 481–488.

Godish, Thad. 2003. *Air quality,* 4th ed. CRC Press, Boca Raton, Florida.

Greer, Linda, et al. 2006. Curtailing mercury's global reach. Pp. 96–114 in *State of the world 2006.* Worldwatch Institute and W. W. Norton, Washington, D.C., and New York.

Hoffman, Matthew J. 2005. *Ozone depletion and climate change: Constructing a global response.* SUNY Press, New York.

Hunt, Andrew, et al. 2003. Toxicologic and epidemiologic clues from the characterization of the 1952 London smog fine particulate matter in archival autopsy lung tissues. *Environmental Health Perspectives* 111: 1209–1214.

Jacobson, Mark Z. 2002. *Atmospheric pollution: History, science, and regulation.* Cambridge University Press, New York.

Kunzli, Nino, et al. 2000. Public-health impact of outdoor and traffic-related air pollution: A European assessment. *Lancet* 356: 795–801.

Lelieveld, Jos, et al. 2001. The Indian Ocean experiment: Widespread air pollution from South and Southeast Asia. *Science* 291: 1031–1036.

Likens, Gene E. 2004. Some perspectives on long-term biogeochemical research from the Hubbard Brook ecosystem study. *Ecology* 85: 2355–2362.

Lutgens, Frederick K., and Edward J. Tarbuck. 2007. *The atmosphere.* Pearson Prentice Hall. Upper Saddle River, New Jersey.

London, Stephanie J., and Isabelle Romieu. 2000. Health costs due to outdoor air pollution by traffic. *Lancet* 356: 782–783.

Molina, Mario J., and F. Sherwood Rowland. 1974. Stratospheric sink for chlorofluoromethanes: Chlorine atom catalyzed destruction of ozone. *Nature* 249: 810–812.

Parson, Edward A. 2003. *Protecting the ozone layer: Science and strategy.* Oxford University Press, Oxford.

Seinfeld, John H., and Spyros N. Pandis. 2006. *Atmospheric chemistry and physics,* 2nd ed. Wiley-Interscience, New York.

Transport for London. Congestion charging . . . www.cclondon.com.

United Nations Environment Programme. Montreal Protocol. http://ozone.unep.org/Treaties_and_Ratification/2B_ montreal_ protocol.shtml.

United Nations Environment Programme. 2002. Atmosphere. Pp. 210–239 in *Global environment outlook 3 (GEO-3).* UNEP and Earthscan Publications, Nairobi and London.

United States Environmental Protection Agency. 2003. *Latest findings on national air quality: 2002 status and trends.* EPA 454/K-03-001. Washington, D.C.

United States Environmental Protection Agency, Office of Air and Radiation. www.epa.gov/air.

World Health Organization. Indoor air pollution. WHO, Geneva, Switzerland. www.who.int/indoorair/en/index.html.


**Chapter 18**


Alley, Richard B. 2000. *The two-mile time machine: Ice cores, abrupt climate change, and our future.* Princeton University Press, Princeton, New Jersey.

Appenzeller, Tim. 2007. The big thaw. *National Geographic,* June 2007: 56–71.

Burroughs, William James. 2001. *Climate change: A multidisciplinary approach.* Cambridge University Press, Cambridge.

Caldeira, Kenneth, and Michael E. Wickett. 2003. Anthropogenic carbon and ocean pH. *Nature* 425: 365.

Drake, Frances. 2000. *Global warming: The science of climate change.* Oxford University Press, Oxford.

Dunn, Seth, and Christopher Flavin. 2002. Moving the climate change agenda forward. Pp. 24–50 in *State of the world 2002.* Worldwatch Institute and W. W. Norton, Washington, D.C., and New York.

EPICA community members. 2004. Eight glacial cycles from an Antarctic ice core. *Nature* 429: 623–628.

Flannery, Tim. 2005. *The weather makers: The history and future impact of climate change.* Text Publishing, Melbourne, Australia.

Gelbspan, Ross. 1997. *The heat is on: The climate crisis, the cover-up, the prescription.* Perseus Books, New York.

Gelbspan, Ross. 2004. *Boiling point: How politicians, big oil and coal, journalists, and activists are fueling the climate crisis—and what we can do to avert disaster.* Basic Books, New York.

Gettleman, Jeffrey. 2006. Annan faults "frightening lack of leadership" for global warming. *New York Times,* 16 November 2006.

Gore, Al. 2006. *An inconvenient truth: The planetary emergency of global warming and what we can do about it.* Rodale Press and Melcher Media, New York.

Intergovernmental Panel on Climate Change. 2001. *IPCC third assessment report—Climate change 2001: Synthesis report.* World Meteorological Organization and United Nations Environment Programme.

Intergovernmental Panel on Climate Change. 2007. *Climate change 2007: The physical science basis.* Contribution of Working Group I to the fourth assessment report of the Intergovernmental Panel on Climate Change. World Meteorological Organization and United Nations Environment Programme, Geneva, Switzerland.

Intergovernmental Panel on Climate Change. 2007. *Climate change 2007: Impacts, adaptation, and vulnerability.* Contribution of Working Group II to the fourth assessment report of the Intergovernmental Panel on Climate Change. World Meteorological Organization and United Nations Environment Programme, Geneva, Switzerland.

Intergovernmental Panel on Climate Change. 2007. *Climate change 2007: Mitigation of climate change.* Contribution of Working Group III to the fourth assessment report of the Intergovernmental Panel on Climate Change. World Meteorological Organization and United Nations Environment Programme, Geneva, Switzerland.

Intergovernmental Panel on Climate Change. www.ipcc.ch.

Jonzén, Niclas, et al. 2006. Rapid advance of spring arrival dates in long-distance migratory birds. *Science* 312: 1959–1961.

Karl, Thomas R., and Kevin E. Trenberth. 2003. Modern global climate change. *Science* 302: 1719–1723.

Kerr, Richard A. 2006. A worrying trend of less ice, higher seas. *Science* 311: 1698–1701.

Kerr, Richard A. 2006. A tempestuous birth for hurricane climatology. *Science* 312: 676–678.

Mastny, Lisa. 2005. Global ice melting accelerating. Pp. 88–89 in *Vital signs 2005.* Worldwatch Institute and W. W. Norton, Washington, D.C., and New York.

Mayewski, Paul A., and Frank White. 2002. *The ice chronicles: The quest to understand global climate change.* University Press of New England, Hanover, New Hampshire.

National Assessment Synthesis Team. 2000. *Climate change impacts on the United States: The potential consequences of climate variability and change.* U.S. Global Change Research Program. Cambridge University Press, Cambridge.

National Research Council, Committee on the Science of Climate Change, Division of Earth and Life Studies. 2001. *Climate change science: An analysis of some key questions.* National Academies Press, Washington, D.C.

Pacala, Stephen, and Robert Socolow. 2004. Stabilization wedges: Solving the climate problem for the next 50 years with current technologies. *Science* 305: 968–972.

Parmesan, Camille, and Gary Yohe. 2003. A globally coherent fingerprint of climate change impacts across natural systems. *Nature* 421: 37–42.

Pew Center on Global Climate Change. www.pewclimate.org.

Real Climate. www.realclimate.org.

Rignot, Eric, and Pannir Kanagaratnam. 2006. Changes in the velocity structure of the Greenland Ice Sheet. *Science* 311: 986–990.

Root, Terry L., et al. 2003. Fingerprints of global warming on wild animals and plants. *Nature* 421: 57–60.

The Royal Society. 2005. *Ocean acidification due to increasing atmospheric carbon dioxide.* The Royal Society, London, U.K., June 2005.

Sawin, Janet L. 2005. Climate change indicators on the rise. Pp. 40–41 in *Vital signs 2005.* Worldwatch Institute and W. W. Norton, Washington, D.C., and New York.

Schiermeier, Quirin. 2006. Climate credits. *Nature* 444: 976–977.

Schneider, Stephen H., and Terry L. Root, eds. 2002. *Wildlife responses to climate change: North American case studies.* Island Press, Washington, D.C.

Shapiro, Robert J., Kevin A. Hassett, and Frank S. Arnold. 2002. *Conserving energy and preserving the environment: The role of public transportation.* American Public Transportation Association, July 2002.

Siegenthaler, Urs, et al. 2005. Stable carbon cycle–climate relationship during the Late Pleistocene. *Science* 310: 1313–1317.

Spahni, Renato, et al. 2005. Atmospheric methane and nitrous oxide of the Late Pleistocene from Antarctic ice cores. *Science* 310: 1317–1321.

Speth, James Gustave. 2004. *Red sky at morning: America and the crisis of the global environment.* Yale University Press, New Haven, Connecticut.

Stevens, William K. 1999. *The change in the weather: People, weather and the science of climate.* Delta Trade Paperbacks, New York.

Taylor, David. 2003. Small islands threatened by sea level rise. Pp. 84–85 in *Vital signs 2003.* Worldwatch Institute and W. W. Norton, Washington D.C., and New York.

*Time.* 2006. Special report: Global warming. *Time,* 3 April 2006: 28–62.

Victor, David G. 2004. *Climate change: Debating America's policy options.* U.S. Council on Foreign Relations Press, Washington, D.C.

Victor, David G., Joshua C. House, and Sarah Joy. 2005. A Madisonian approach to climate policy. *Science* 309: 1820–1821.

United Nations. United Nations Framework Convention on Climate Change. http://unfccc.int/2860.php.

United Nations. Kyoto Protocol. http://unfccc.int/kyoto_protocol/items/2830.php.

Zwally, H. Jay, et al. 2002. Surface melt–induced acceleration of Greenland Ice-Sheet flow. *Science* 297: 218–222.


**Chapter 19**


Appenzeller, Tim. The end of cheap oil. *National Geographic,* June 2004: 80–109.

Association for the Study of Peak Oil and Gas. www.peakoil.net.

British Petroleum. 2007. *BP statistical review of world energy 2007.* BP, London.

Campbell, Colin J. 1997. *The coming oil crisis.* Multi-Science Publishing Co., Essex, U.K.

Deffeyes, Kenneth S. 2001. *Hubbert's peak: The impending world oil shortage.* Princeton University Press, Princeton, New Jersey.

Deffeyes, Kenneth S. 2005. *Beyond oil: The view from Hubbert's peak.* Farrar, Straus, and Giroux, New York.

Douglas, D. C., P. E. Reynolds, and E. B. Rhode, eds. 2002. *Arctic Refuge coastal plain terrestrial wildlife research summaries. Biological science report.* USGS/BRD/BSR-2002-0001. United States Geological Survey, Washington, D.C.

Dunn, Seth. 2001. Decarbonizing the energy economy. Pp. 83–102 in *State of the world 2001.* Worldwatch Institute and W. W. Norton, Washington, D.C., and New York.

Energy Information Administration, U.S. Department of Energy. www.eia.doe.gov.

Energy Information Administration, U.S. Department of Energy. 1999. *Petroleum: An energy profile, 1999.* DOE/EIA-0545(99).

Energy Information Administration, U.S. Department of Energy. 2006. *International energy annual 2004.* Washington, D.C.

Energy Information Administration, U.S. Department of Energy. 2007. *Annual energy review 2006.* DOE/EIA, Washington, D.C.

Freese, Barbara. 2003. *Coal: A human history.* Perseus Books, New York.

Goodstein, David. 2004. *Out of gas.* W. W. Norton, New York.

Holmes, Bob, and Nicola Jones. 2003. Brace yourself for the end of cheap oil. *New Scientist,* 2 August 2003: 9–11.

International Energy Agency. 2005. *Key world energy statistics 2005.* IEA Publications, Paris.

International Energy Agency. 2005. *World energy outlook 2005.* IEA Publications, Paris.

International Energy Agency. 2005. *Resources to reserves: Oil and gas technologies for the energy markets of the future.* IEA Publications, Paris.

Kunstler, James H. 2005. *The long emergency.* Atlantic Monthly Press, New York.

Lovins, Amory B. 2005. More profit with less carbon. *Scientific American* 293(3): 74–83.

Lovins, Amory B., et al. 2004. *Winning the oil endgame: Innovation for profits, jobs, and security.* Rocky Mountain Institute, Snowmass, Colorado.

Nellemann, Christian, and Raymond D. Cameron. 1998. Cumulative impacts of an evolving oil-field complex on the distribution of calving caribou. *Canadian Journal of Zoology* 76: 1425–1430.

Pelley, Janet. 2001. Will drilling for oil disrupt the Arctic National Wildlife Refuge? *Environmental Science and Technology* 35: 240–247.

Powell, Stephen G. 1990. Arctic National Wildlife Refuge: How much oil can we expect? *Resources Policy,* Sept. 1990: 225–240.

Prugh, Tom, et al. 2005. Changing the oil economy. Pp. 100–121 in *State of the world 2005.* Worldwatch Institute and W. W. Norton, Washington, D.C., and New York.

Ristinen, Robert A., and Jack J. Kraushaar. 2006. *Energy and the environment,* 2nd ed. John Wiley and Sons, New York.

Roberts, Paul. 2004. *The end of oil: On the edge of a perilous new world.* Houghton Mifflin, Boston.

Rottmann, Katja. 2006. Fossil fuel use continues to grow. Pp. 32–33 in *Vital signs 2006–2007.* Worldwatch Institute and W. W. Norton, Washington, D.C., and New York.

Russell, D. E., and P. McNeil. 2005. *Summer ecology of the Porcupine caribou herd.* Porcupine Caribou Management Board, Whitehorse, Yukon.

Sawin, Janet L. 2004. Making better energy choices. Pp. 24–45 in *State of the world 2004.* Worldwatch Institute and W. W. Norton, Washington, D.C., and New York.

Skinner, Brian J., and Stephen C. Porter. 2003. *The dynamic earth: An introduction to physical geology,* 5th ed. John Wiley and Sons, Hoboken, New Jersey.

United States Environmental Protection Agency. 2006. *Light-duty automotive technology and fuel economy trends: 1975 through 2006.* EPA Office of Transportation and Air Quality, Washington, D.C.

United States Fish and Wildlife Service. 2001. Potential impacts of proposed oil and gas development on the Arctic Refuge's coastal plain: Historical overview and issues of concern. Web page of the Arctic National Wildlife Refuge, Fairbanks, Alaska. http://arctic.fws.gov/issues1.htm.

United States Geological Survey. 2001. *The National Petroleum Reserve–Alaska (NPRA) data archive.* USGS Fact Sheet FS-024-01, March 2001.

United States Geological Survey. 2001. *Arctic National Wildlife Refuge, 1002 Area, petroleum assessment, 1998, including economic analysis.* USGS Fact Sheet FS-028-01, April 2001.

United States Geological Survey. 2002. *Petroleum resource assessment of the National Petroleum Reserve Alaska (NPRA).* USGS, Washington, D.C.

United States Government Accountability Office (GAO). 2007. *Crude oil: Uncertainty about future oil supply makes it important to develop a strategy for addressing a peak and decline in oil production.* Report to Congressional Requesters, February 2007.

Walker, Donald A. 1997. Arctic Alaskan vegetation disturbance and recovery. Pp. 457–479 in R. M. M. Crawford, ed., *Disturbance and recovery in Arctic lands.* Kluwer Academic Publishers, Dordrecht, Netherlands.

**Chapter 20**

British Petroleum. 2007. *BP statistical review of world energy 2007.* BP, London.

Chandler, David. 2003. America steels itself to take the nuclear plunge. *New Scientist*, 9 August 2003: 10–13.

The Chernobyl Forum. 2006. *Chernobyl's legacy: Health, environmental and socio-economic impacts* and *recommendations to the governments of Belarus, the Russian Federation and Ukraine. The Chernobyl Forum: 2003–2005.* Second revised version. World Health Organization and International Atomic Energy Agency, Vienna.

Dunn, Seth. 2001. Decarbonizing the energy economy. Pp. 83–102 in *State of the world 2001.* Worldwatch Institute and W. W. Norton, Washington, D.C., and New York.

Energy Information Administration, U.S. Department of Energy. www.eia.doe.gov.

Energy Information Administration, U.S. Department of Energy. 2006. *International energy annual 2004.* Washington, D.C.

Energy Information Administration, U.S. Department of Energy. 2007. *Annual energy review 2006.* DOE/EIA, Washington, D.C.

Energy Information Administration. 2007. *Annual energy outlook 2007.* Washington, D.C.

European Commission/International Atomic Energy Agency/World Health Organization. 1996. One decade after Chernobyl: Summing up the consequences of the accident. Summary of the conference results. Vienna, Austria, 8–12 April 1996. EC/IAEA/WHO.

Hunt, Suzanne, and Janet L. Sawin, with Peter Stair. 2006. Cultivating renewable alternatives to oil. Pp. 61–77 in *State of the world 2006.* Worldwatch Institute and W. W. Norton, Washington, D.C., and New York.

Hunt, Suzanne, and Peter Stair. 2006. Biofuels hit a gusher. Pp. 40–41 in *Vital signs 2006–2007.* Worldwatch Institute and W. W. Norton, Washington, D.C., and New York.

International Atomic Energy Agency. 2006. *Annual report 2005.* IAEA, Vienna, Austria.

International Atomic Energy Agency. *Nuclear power and sustainable development.* IAEA Information Series 02-01574/FS Series 3/01/E/Rev.1. Vienna, Austria.

International Atomic Energy Agency. 2006. *Environmental consequences of the Chernobyl accident and their remediation: Twenty years of experience.* Report of the U.N. Chernobyl Forum Expert Group "Environment." IAEA, Vienna.

International Energy Agency. 2006. *Key world energy statistics 2006.* IEA Publications, Paris.

International Energy Agency. 2006. *World energy outlook 2006.* IEA Publications, Paris.

International Energy Agency. 2007. *Biomass for power generation and CHP.* IEA Publications, Paris.

Klass, Donald L. 2004. Biomass for renewable energy and fuels. In *The Encyclopedia of Energy.* Elsevier.

Lenssen, Nicholas. 2006. Nuclear power inches up. Pp. 34–35 in *Vital signs 2006–2007.* Worldwatch Institute and W. W. Norton, Washington, D.C., and New York.

Li, Zijun. 2006. Hydropower rebounds slightly. Pp. 46–47 in *Vital signs 2006–2007.* Worldwatch Institute and W. W. Norton, Washington, D.C., and New York.

Lovins, Amory B., et al. 2004. *Winning the oil endgame: Innovation for profits, jobs, and security.* Rocky Mountain Institute, Snowmass, Colorado.

Murray, Danielle. 2005. Ethanol's potential: Looking beyond corn. *Eco-economy Update #49*, 5 June 2005. Earth Policy Institute. www.earth-policy.org/Updates/2005/Update49.htm.

National Renewable Energy Lab, U.S. Department of Energy. www.nrel.gov.

*Nature.* 2006. Special report: Chernobyl and the future. *Nature* 440: 982–989.

Nuclear Energy Agency. 2002. *Chernobyl: Assessment of radiological and health impacts.* 2002 update of *Chernobyl: Ten years on.* OECD, Paris.

Nuclear Energy Agency. 2005. *NEA annual report 2004.* NEA, Organisation for Economic Co-operation and Development. OECD, Paris.

Office of Energy Efficiency and Renewable Energy, U.S. Department of Energy. www.eere.energy.gov.

Organisation for Economic Co-operation and Development. 2000. *Business as usual and nuclear power.* OECD Publications, Paris.

Pearce, Fred. 2006. Fuels gold: Are biofuels really the greenhouse-busting answer to our energy woes? *New Scientist*, 23 Sept. 2006: 36–41.

REN21 Renewable Energy Policy Network. 2005. *Renewables 2005 global status report.* Worldwatch Institute, Washington, D.C.

*Science.* 2005. News Focus: Rethinking nuclear power. *Science* 309: 1168–1179.

Spadaro, Joseph V., Lucille Langlois, and Bruce Hamilton. 2000. Greenhouse gas emissions of electricity generation chains: Assessing the difference. *IAEA Bulletin* 42(2).

Swedish Bioenergy Association (SVEBIO). 2003. *Focus: Bioenergy.* Nos. 1–10. SVEBIO, Stockholm.

Swedish Energy Agency. 2004. *Renewable electricity is the future's electricity.* Swedish Energy Agency, Eskilstuna, Sweden.

Swedish Energy Agency. 2006. *Energy in Sweden: Facts and figures 2005.* Swedish Energy Agency, Eskilstuna, Sweden.

Swedish Energy Agency. 2006. *The Swedish Energy Agency 2005.* Swedish Energy Agency, Eskilstuna, Sweden.

Swedish Energy Agency. 2006. *Energy in Sweden 2006.* Swedish Energy Agency, Eskilstuna, Sweden.

U.N. Food and Agriculture Organization. *Biomass energy in ASEAN member countries.* FAO/ASEAN/EC. FAO Regional Wood Energy Development Programme in Asia, Bangkok, Thailand.

U.S. Environmental Protection Agency. Alternative fuels website. www.epa.gov/otaq/consumer/fuels/altfuels/altfuels.htm.

World Health Organization. 2006. *Health effects of the Chernobyl accident and special health care programmes.* Report of the U.N. Chernobyl Forum Expert Group "Health." WHO, Geneva.

Worldwatch Institute and Center for American Progress. 2006. *American energy: The renewable path to energy security.* Worldwatch Institute and Center for American Progress, Washington, D.C.

**Chapter 21**

American Wind Energy Association. 2005. *Global wind energy market report.* AWEA, Washington, D.C.

Ananthaswamy, Anil. 2003. Reality bites for the dream of a hydrogen economy. *New Scientist,* 15 November 15 2003: 6–7.

Arnason, Bragi, and Thorsteinn I. Sigfusson. 2000. Iceland—a future hydrogen economy. *International Journal of Hydrogen Energy* 25: 389–394.

Ásmundsson, Jón Knútur. 2002. Will fuel cells make Iceland the "Kuwait of the North"? *World Press Review,* 15 February 2002.

Burkett, Elinor. 2003. A mighty wind. *New York Times magazine.* June 15, 2003.

Chow, Jeffrey, et al. 2003. Energy resources and global development. *Science* 302: 1528–1531.

DaimlerChrysler. 2003. *360 DEGREES/DaimlerChrysler Environmental Report 2003.* DaimlerChrysler AG, Stuttgart, Germany.

Dunn, Seth. 2000. The hydrogen experiment. *WorldWatch* 13: 14–25.

Energy Information Administration, U.S. Department of Energy. www.eia.doe.gov.

Energy Information Administration, U.S. Department of Energy. 2006. *International energy annual 2004.* Washington, D.C.

Energy Information Administration, U.S. Department of Energy. 2007. *Annual energy review 2006.* DOE/EIA, Washington, D.C.

Energy Information Administration, U.S. Department of Energy. 2007. *Annual energy outlook 2007.* Washington, D.C.

Flavin, Christopher, and Seth Dunn. 1999. A new energy paradigm for the 21st century. *Journal of International Affairs* 53: 167–190.

Grant, Paul M., Chauncey Starr, and Thomas J. Overbye. 2006. A power grid for the hydrogen economy. *Scientific American,* July 2006: 77–83.

Hirsch, Tim. 2001. Iceland launches energy revolution. *British Broadcasting Corporation News,* 24 December 2001.

Hogan, Jenny, and Philip Cohen. 2004. Is the green dream doomed to fail? *New Scientist,* 17 July 2004: 6–7.

Hydrogen & Fuel Cell Letter. 2003. World's first commercial hydrogen station opens in Iceland. *Hydrogen & Fuel Cell Letter,* May 2003.

Idaho Wind Power Working Group for the Idaho Department of Water Resources Energy Division. 2002. *Idaho wind power development strategic plan.* Boise, Idaho.

Idaho Wind Power Working Group for the Idaho Department of Water Resources Energy Division. 2002. *Wind power potential in Idaho by county.* Boise, Idaho.

International Energy Agency. 2005. *Renewables information 2005.* IEA Publications, Paris.

International Energy Agency. 2006. *Key world energy statistics 2006.* IEA Publications, Paris.

International Energy Agency. 2006. *World energy outlook 2006.* IEA Publications, Paris.

International Energy Agency. 2007. *Renewables in global energy supply: An IEA fact sheet.* IEA Publications, Paris.

International Energy Agency Renewable Energy Working Party. 2002. *Renewable energy . . . into the mainstream.* SITTARD, The Netherlands.

Jacobson, Mark Z., W. G. Colella, and D. N. Golden. 2005. Cleaning the air and improving health with hydrogen fuel-cell vehicles. *Science* 308: 1901–1905.

Knott, Michelle. 2003. Power from the waves. *New Scientist,* 20 Sept. 2003: 33–35.

Lovins, Amory B., et al. 2004. *Winning the oil endgame: Innovation for profits, jobs, and security.* Rocky Mountain Institute, Snowmass, Colorado.

Martinot, Eric, et al. 2002. Renewable energy markets in developing countries. *Annual Review of Energy and the Environment* 27: 309–48.

Martinot, Eric, Ryan Wiser, and Jan Hamrin. 2005. *Renewable energy markets and policies in the United States.* Center for Resource Solutions, San Francisco. www.martinot.info/Martinot_et_al_CRS.pdf.

Melis, Anastasios, et al. 2000. Sustained photobiological hydrogen gas production upon reversible inactivation of oxygen evolution in the green alga *Chlamydomonas reinhardtii. Plant Physiology* 122: 127–135.

National Renewable Energy Lab, U.S. Department of Energy. www.nrel.gov.

Office of Energy Efficiency and Renewable Energy, U.S. Department of Energy. www.eere.energy.gov.

Reeves, Ari, with Fredric Beck. 2003. *Wind energy for electric power: A REPP issue brief.* Renewable Energy Policy Project, Washington, D.C.

REN21 Renewable Energy Policy Network. 2005. *Renewables 2005 global status report.* Worldwatch Institute, Washington, D.C.

Ristinen, Robert A., and Jack J. Kraushaar, 1998. *Energy and the environment.* John Wiley and Sons, New York.

Sawin, Janet. 2004. *Mainstreaming renewable energy in the 21st century.* Worldwatch Paper 169. Worldwatch Institute, Washington, D.C.

Sawin, Janet L. 2006. Wind power blowing strong. Pp. 36–37 in *Vital signs 2006–2007.* Worldwatch Institute and W. W. Norton, Washington, D.C., and New York.

Sawin, Janet L. 2006. Solar industry stays hot. Pp. 38–39 in *Vital signs 2006–2007.* Worldwatch Institute and W. W. Norton, Washington, D.C., and New York.

Wald, Matthew L. 2006. The energy challenge: It's free, plentiful, and fickle. *New York Times,* 28 December 2006.

Weisman, Alan. 1998. *Gaviotas: A village to reinvent the world.* Chelsea Green Publishing Co., White River Junction, Vermont.

World Alliance for Decentralized Energy. 2005. *World survey of decentralized energy 2005.* WADE, Edinburgh, Scotland.

Worldwatch Institute and Center for American Progress. 2006. *American energy: The renewable path to energy security.* Worldwatch Institute and Center for American Progress, Washington, D.C.

**Chapter 22**

Allen, G. H., and R. A. Gearheart, eds. 1988. *Proceedings of a conference on wetlands for wastewater treatment and resource enhancement.* Humboldt State University, Arcata, California.

Ayres, Robert U., and Leslie W. Ayres. 1996. *Industrial ecology: Towards closing the materials cycle.* Edward Elgar Press, Cheltenham, U.K.

Beede, David N., and David E. Bloom. 1995. The economics of municipal solid waste. *World Bank Research Observer* 10: 113–150.

Diesendorf, Mark, and Clive Hamilton. 1997. *Human ecology, human economy*. Allen and Unwin, St. Leonards.

Douglas, Ed. 2007. Better by design. *New Scientist*, 6 Jan. 2007: 31–35.

Edmonton, Alberta, City of. 2003. Waste management. www.edmonton.ca/portal/server.pt/gateway/PTARGS_0_2_104_0_0_35/http%3B/cmsserver/COEWeb/environment+waste+and+recycling/waste.

Gitlitz, Jenny, and Pat Franklin. 2004. *The 10-cent incentive to recycle*, 3rd ed. Container Recycling Institute, Arlington, Virginia.

Graedel, Thomas E., and Braden R. Allenby. 2002. *Industrial ecology*, 2nd ed. Prentice Hall, Upper Saddle River, New Jersey.

Lilienfeld, Robert, and William Rathje. 1998. *Use less stuff: Environmental solutions for who we really are*. Ballantine, New York.

Manahan, Stanley E. 1999. *Industrial ecology: Environmental chemistry and hazardous waste*. Lewis Publishers, CRC Press, Boca Raton, Florida.

McDonough, William, and Michael Braungart. 2002. *Cradle to cradle: Remaking the way we make things*. North Point Press, New York.

McGinn, Anne Platt. 2002. Toxic waste largely unseen. Pp. 112–113 in *Vital signs 2002*. Worldwatch Institute and W. W. Norton, Washington, D.C., and New York.

New York City Department of Parks and Recreation. Fresh Kills Park. www.nycgovparks.org/sub_your_park/fresh_kills_park/html/fresh_kills_park.html.

New York City Department of Planning. Fresh Kills Park Project. www.nyc.gov/html/dcp/html/fkl/fkl3.shtml.

New York City Department of Sanitation. 2000. Closing the Fresh Kills landfill. *The DOS Report*, Feb. 2000.

Rathje, William, and Colleen Murphy. 2001. *Rubbish! The archeology of garbage*. University of Arizona Press.

Simmons, Phil, et al. 2006. The state of garbage in America. *Biocycle* 47(4): 26.

Smith, Ronald S. 1998. *Profit centers in industrial ecology*. Quorum Books, Westport.

Socolow, Robert H., et al., eds. 1994. *Industrial ecology and global change*. Cambridge University Press, Cambridge.

United Nations Environment Programme. 2000. *International source book on environmentally sound technologies (ESTs) for municipal solid waste management (MSWM)*. UNEP IETC, Osaka, Japan.

United States Environmental Protection Agency. 2006. *Municipal solid waste generation, recycling, and disposal in the United States: Facts and figures for 2006*. EPA Office of Solid Waste and Emergency Response.

United States Environmental Protection Agency. Municipal solid waste. www.epa.gov/epaoswer/non-hw/muncpl.

**Chapter 23**

Bartlett, Peggy, and Geoffrey W. Chase, eds. 2004. *Sustainability on campus: Stories and strategies for change*. MIT Press, Cambridge, Massachusetts.

Brower, Michael, and Warren Leon. 1999. *The consumer's guide to effective environmental choices: Practical advice from the Union of Concerned Scientists*. Three Rivers Press, New York.

Brown, Lester. 2001. *Eco-economy: Building an economy for the Earth*. Earth Policy Institute and W. W. Norton, New York.

Brown, Lester. 2006. *Plan B 2.0: Rescuing a planet under stress and a civilization in trouble*. Earth Policy Institute and W. W. Norton, New York.

Carlson, Scott. 2006. In search of the sustainable campus: With eyes on the future, universities try to clean up their acts. *Chronicle of Higher Education*, 20 Oct. 2006: vol 53: A10.

Creighton, Sarah Hammond. 1998. *Greening the ivory tower: Improving the environmental track record of universities, colleges, and other institutions*. MIT Press, Cambridge, Massachusetts.

Daly, Herman E. 1996. *Beyond growth*. Beacon Press, Boston.

Dasgupta, Partha, Simon Levin, and Jane Lubchenco. 2000. Economic pathways to ecological sustainability. *BioScience* 50: 339–345.

Durning, Alan. 1992. *How much is enough? The consumer society and the future of the Earth*. Worldwatch Institute, Washington, D.C.

Erickson, Jon D., and John M. Gowdy. 2002. The strange economics of sustainability. *BioScience* 52: 212.

Esty, Daniel C., et al. 2006. *Pilot 2006 Environmental Performance Index*. Yale Center for Environmental Law & Policy, New Haven, Connecticut.

French, Hilary. 2004. Linking globalization, consumption, and governance. Pp. 144–163 in *State of the world 2004*. Worldwatch Institute and W. W. Norton, Washington, D.C., and New York.

Gardner, Gary. 2001. Accelerating the shift to sustainability. Pp. 189–206 in *State of the world 2001*. Worldwatch Institute and W. W. Norton, Washington, D.C., and New York.

Gardner, Gary, and Erik Assadourian. 2004. Rethinking the good life. Pp. 164–179 in *State of the world 2004*. Worldwatch Institute and W. W. Norton, Washington, D.C., and New York.

Gibbs, W. Wayt. 2005. How should we set priorities? *Scientific American* 293(3): 108–115.

Hawken, Paul. 1994. *The ecology of commerce: A declaration of sustainability*. HarperBusiness, New York.

Kahneman, Daniel, et al. 2006. Would you be happier if you were richer? A focusing illusion. *Science* 312: 1908–1910.

Keniry, Julian. 1995. *Ecodemia: Campus environmental stewardship at the turn of the 21st century*. National Wildlife Federation, Washington, D.C.

McMichael, A. J., et al. 2003. New visions for addressing sustainability. *Science* 302: 1919–1921.

McIntosh, Mary, et al. 2001. *State of the campus environment: A national report card on environmental performance and sustainability in higher education*. National Wildlife Federation Campus Ecology.

Meadows, Donella, Jørgen Randers, and Dennis Meadows. 2004. *Limits to growth: The 30-year update*. Chelsea Green Publishing. Co., White River Junction, Vermont.

Millennium Ecosystem Assessment. 2005. *Ecosystems and human well-being: General synthesis*. Millennium Ecosystem Assessment and World Resources Institute.

National Research Council, Board on Sustainable Development. 1999. *Our common journey: A transition toward sustainability*. National Academies Press, Washington, D.C.

National Wildlife Federation. Campus ecology. www.nwf.org/campusecology.

Sanderson, Eric W., et al. 2002. The human footprint and the last of the wild. *BioScience* 52: 891–904.

Schor, Juliet B., and Betsy Taylor, eds. 2002. *Sustainable planet: Solutions for the twenty-first century*. The Center for a New American Dream. Beacon Press, Boston.

Toor, Will, and Spenser W. Havlick. 2004. *Transportation and sustainable campus communities: Issues, examples, solutions.* Island Press, Washington, D.C.

United Nations. 2002. *Report of the World Summit on Sustainable Development, Johannesburg, South Africa, 26 August–4 September 2002.* United Nations, New York.

United Nations. 2002. *The road from Johannesburg: What was achieved and the way forward.* United Nations, New York.

United Nations Department of Economic and Social Affairs. 2006. *The Millennium Development Goals Report 2006.* U.N. DESA, New York.

United Nations Development Programme. 2002. *Human development report 2002.* Oxford University Press, Oxford.

United Nations Environment Programme. 2002. Outlook: 2002–2032. Pp. 319–400 in *Global environment outlook 3 (GEO-3).* UNEP and Earthscan Publications, Nairobi and London.

University Leaders for a Sustainable Future. www.ulsf.org.

U.S. Green Building Council. www.usgbc.org.

Wackernagel, Mathis, Lillemor Lewan, and Carina Borgström-Hansson. 1999. Evaluating the use of natural capital with the ecological footprint. *Ambio* 28: 604.

Wilson, Edward O. 1998. *Consilience: The unity of knowledge.* Alfred A. Knopf, New York.

World Commission on Environment and Development. 1987. *Our common future.* Oxford University Press, Oxford.

# Index

**Note to reader:** Page numbers in **bold** refer to definitions; page numbers followed by *f* refer to figures; page numbers followed by *t* refer to tables; and page numbers followed by *c* refer to Causes and Consequences exercises.

## B

## C

## D

## J



## K

## N

## Q



## R

## S

## T

## U